NANOELECTROCHEMISTRY

edited by

Michael V. Mirkin
Queens College, New York, USA

Shigeru Amemiya
University of Pittsburgh, Pennsylvania, USA

CRC Press
Taylor & Francis Group
Boca Raton London New York

CRC Press is an imprint of the
Taylor & Francis Group, an **informa** business

CRC Press
Taylor & Francis Group
6000 Broken Sound Parkway NW, Suite 300
Boca Raton, FL 33487-2742

First issued in paperback 2021

© 2015 by Taylor & Francis Group, LLC
CRC Press is an imprint of Taylor & Francis Group, an Informa business

ISBN 13: 978-1-138-89466-2 (pbk)
ISBN 13: 978-1-4665-6119-9 (hbk)
ISBN 13: 978-0-429-09687-7 (ebk)

DOI: 10.1201/b18066

Library of Congress Cataloging-in-Publication Data

Nanoelectrochemistry / editors, Michael V. Mirkin and Shigeru Amemiya.
    pages cm
    Includes bibliographical references and index.
    ISBN 978-1-4665-6119-9 (alk. paper)
    1. Electrochemistry. 2. Nanostructured materials. I. Mirkin, Michael V., 1961- II. Amemiya,
Shigeru, 1971-

QD553.N36 2014
541'.37--dc23   2014025688

**Visit the Taylor & Francis Web site at
http://www.taylorandfrancis.com**

**and the CRC Press Web site at
http://www.crcpress.com**

Publisher's Note
The publisher has gone to great lengths to ensure the quality of this reprint but points out that some imperfections in the original copies may be apparent.

Contents

SECTION I   Theory of Nanoelectrochemistry

SECTION II   Nanoelectrochemical Systems

SECTION III   Nanoelectrochemical Methods

Preface

The number of scientific disciplines whose names begin with the prefix *nano* has grown dramatically since the launch of the National Nanotechnology Initiative in 2000. Some of those specialties were born to be *nano* either because their research subjects are nanometer-sized objects or because of the use of nanometer-scale experimental tools, for example, the scanning tunneling microscope. It took electrochemistry two centuries (counting from Alessandro Volta's creation of the voltaic pile in 1800) to acquire the prefix *nano*. To work on the nanoscale, electrochemists had to produce nanometer-sized probes, learn to accurately measure ultralow currents, and develop a new theory required for data analysis. The evolution of electrochemistry to nanoscience was also accelerated by the development and synthesis of many new nanomaterials and systems. Was the transition to *nano* worth this effort? Our book aims to answer this question with a resounding *Yes!* By the early 1980s, electrochemistry could be seen as a mature and application-driven science. The move to the nanoscale rejuvenated this field and allowed it to generate new electrode systems and begin to gain molecular-level knowledge about heterogeneous processes complementary to that obtained with various spectroscopic and surface science techniques. The nanoscale capabilities also gave electrochemistry a niche in two of the most active areas of current chemical research—biomedical and alternative energy.

This book covers three integral aspects of nanoelectrochemistry. The first two chapters contain theoretical background, which is essential for everyone working in the field, that is, theories of electron transfer (Chapter 1), transport, and double-layer processes (Chapter 2) at nanoscale electrochemical interfaces. The other chapters are dedicated to the electrochemical studies of nanomaterials and nanosystems (Chapters 3 through 14) or the development and applications of nanoelectrochemical techniques (Chapters 15 through 22) and are self-contained. Nanoelectrochemistry has proved useful for a broad range of interdisciplinary research. The applications discussed in this book range from studies of biological systems to nanoparticles and from electrocatalysis to molecular electronics, nanopores, and membranes. Although we did not intend to present even a brief survey of those diverse areas of research, each chapter provides sufficient detail to allow a specialist to evaluate the applicability of nanoelectrochemical approaches to solving a specific problem, and the key ideas are discussed at a level suitable for beginning graduate students.

Our hope is that this book will be useful to all interested in learning about nanoelectrochemical systems. We thank our students, coworkers, and colleagues who have done so much to advance this field.

Editors

Michael V. Mirkin is professor of chemistry at Queens College, City University of New York, New York City, New York. His professional interests are in the application of electrochemical methods to solving problems in physical and analytical chemistry and include charge-transfer reactions at solid–liquid and liquid–liquid interfaces, electrochemical kinetics, and nanoelectrochemistry. He has published more than 110 peer-reviewed papers and book chapters and coedited the first monograph on scanning electrochemical microscopy (second edition, 2012). He earned a PhD in electrochemistry (1987) from Kazakh State University (former USSR) and did postdoctoral research at The University of Texas at Austin from 1990 to 1993.

Shigeru Amemiya is associate professor, Department of Chemistry, University of Pittsburgh, Pittsburgh, Pennsylvania. He is the author or coauthor of more than 60 scholarly papers and book chapters in electroanalytical chemistry. His research interests are electrochemical sensing and imaging for biological and material studies, including the development of nanoscale scanning electrochemical microscopy and ultrasensitive ion-selective electrodes. He earned his BS (1993) and PhD (1998) in chemistry from The University of Tokyo, Japan, and received a postdoctoral fellowship from the Japan Society for the Promotion of Science to work at The University of Tokyo and The University of Texas at Austin.

Contributors

Barak D.B. Aaronson
Department of Chemistry
University of Warwick
Coventry, United Kingdom

Radoslav R. Adzic
Department of Chemistry
Brookhaven National Laboratory
Upton, New York

Christian Amatore
Department of Chemistry
Ecole Normale Supérieure
PSL Research University
Paris, France

Shigeru Amemiya
Department of Chemistry
University of Pittsburgh
Pittsburgh, Pennsylvania

Agnès Anne
Laboratory of Molecular Electrochemistry
Paris Diderot University
Paris, France

Lane A. Baker
Department of Chemistry
Indiana University
Bloomington, Indiana

Allen J. Bard
Center for Electrochemistry
The University of Texas at Austin
Austin, Texas

Gregory W. Bishop
Department of Chemistry
University of Connecticut
Storrs, Connecticut

Aliaksei Boika
Center for Electrochemistry
The University of Texas at Austin
Austin, Texas

Philippe Bühlmann
Department of Chemistry
University of Minnesota
Minneapolis, Minnesota

Tessa M. Carducci
Department of Chemistry
University of North Carolina at Chapel Hill
Chapel Hill, North Carolina

Shengli Chen
Department of Chemistry
Wuhan University
Wuhan, Hubei, People's Republic of China

Christophe Demaille
Laboratory of Molecular Electrochemistry
Paris Diderot University
Paris, France

Jonathon Duay
Department of Chemistry and Biochemistry
University of Maryland
College Park, Maryland

Andrew G. Ewing
Department of Chemical and Biological
    Engineering
Chalmers University of Technology
and
Department of Chemistry and Molecular
    Biology
University of Gothenburg
Göteborg, Sweden

Alicia K. Friedman
Department of Chemistry
Indiana University
Bloomington, Indiana

Kuanping Gong
San Jose Lab
Corporate Research Institute
Samsung Cheil Industries, Inc.
San Jose, California

Jacob M. Goran
Department of Biochemistry and Chemistry
The University of Texas at Austin
Austin, Texas

Aleix G. Güell
Department of Chemistry
University of Warwick
Coventry, United Kingdom

Manon Guille-Collignon
Department of Chemistry
Ecole Normale Supérieure
PSL Research University
Paris, France

Róbert E. Gyurcsányi
Department of Inorganic and Analytical
    Chemistry
Budapest University of Technology and
    Economics
Budapest, Hungary

M.F. Juarez
Institute of Theoretical Chemistry
Ulm University
Ulm, Germany

Shuo Kang
MESA+ Institute for Nanotechnology
University of Twente
Enschede, the Netherlands

Seong Jung Kwon
Center for Electrochemistry
The University of Texas at Austin
Austin, Texas

Wen-Jie Lan
Department of Chemistry
University of Utah
Salt Lake City, Utah

Sang Bok Lee
Department of Chemistry and Biochemistry
University of Maryland
College Park, Maryland

Frédéric Lemaître
Department of Chemistry
Ecole Normale Supérieure
PSL Research University
Paris, France

Serge G. Lemay
MESA+ Institute for Nanotechnology
University of Twente
Enschede, the Netherlands

Stuart Lindsay
Biodesign Institute
Arizona State University
Tempe, Arizona

Stephen Maldonado
Department of Chemistry
University of Michigan
Ann Arbor, Michigan

Richard McCreery
Department of Chemistry
and
National Institute for Nanotechnology
University of Alberta
Edmonton, Alberta, Canada

Kim McKelvey
Department of Chemistry
University of Warwick
Coventry, United Kingdom

Michael V. Mirkin
Department of Chemistry and Biochemistry
Queens College
City University of New York
New York City, New York

Dmitry Momotenko
Department of Chemistry
University of Warwick
Coventry, United Kingdom

Royce W. Murray
Department of Chemistry
University of North Carolina at Chapel Hill
Chapel Hill, North Carolina

Jun Hui Park
Center for Electrochemistry
The University of Texas at Austin
Austin, Texas

Ernö Pretsch
Institute of Biogeochemistry and Pollutant
    Dynamics
ETH Zürich
Zürich, Switzerland

P. Quaino
Programa de Electroquímica Aplicada e
    Ingeniería Electroquímica
Universidad Nacional del Litoral
Santa Fé, Argentina

James F. Rusling
Department of Chemistry
and
Institute of Materials Science
University of Connecticut
Storrs, Connecticut

and

Department of Cell Biology
University of Connecticut Health Center
Farmington, Connecticut

and

School of Chemistry
National University of Ireland at Galway
Galway, Ireland

E. Santos
Institute of Theoretical Chemistry
Ulm University
Ulm, Germany

and

Facultad de Matemáticas
Astronomíay Física
Instituto de Física Enrique Gaviola
Consejo Nacional de Investigaciones Científicas
    y Técnicas
Universidad Nacional de Córdoba
Córdoba, Argentina

W. Schmickler
Institute of Theoretical Chemistry
Ulm University
Ulm, Germany

G.J. Soldano
Institute of Theoretical Chemistry
Ulm University
Ulm, Germany

Keith J. Stevenson
Department of Biochemistry and Chemistry
The University of Texas at Austin
Austin, Texas

Jay A. Switzer
Department of Chemistry and Materials
    Research Center
Missouri University of Science and Technology
Rolla, Missouri

Scott N. Thorgaard
Department of Chemistry
Grand Valley State University
Allendale, Michigan

Patrick R. Unwin
Department of Chemistry
University of Warwick
Coventry, United Kingdom

Jun Wang
Department of Chemical and Biological
    Engineering
Chalmers University of Technology
and
Department of Chemistry and Molecular
    Biology
University of Gothenburg
Göteborg, Sweden

Wen Wen
Department of Chemistry
University of Michigan
Ann Arbor, Michigan

Henry S. White
Department of Chemistry
University of Utah
Salt Lake City, Utah

Section I

Theory of Nanoelectrochemistry

1 Electron Transfer in Nanoelectrochemical Systems

W. Schmickler, E. Santos, P. Quaino,
G.J. Soldano, and M.F. Juarez

CONTENTS

1.1 INTRODUCTION

Ever since the invention of the fuel cell by Grove, electrochemists have looked for better and cheaper catalysts. There have been significant improvements, but mainly in the reduction of the load of precious metals, not in finding a material significantly more active than platinum. All plausible pure metals, many binary, ternary, and even quaternary alloys, have been tried, but without much success. One gets the impression that progress has been slow and incremental and that with bulk materials, we are meeting the law of diminishing returns.

But the development of nanotechnology has given us new hope. Nanostructures have dimensions that lie between those of individual molecules and bulk compounds; they often exhibit special geometries, like nanotubes or nanowires, have different bonding patterns, and exhibit quantum effects. A new world of materials has been discovered and is being intensively investigated. So far, the most fascinating nanostructures that have been discovered are new modifications of carbon, and some of them are finding their way into electrochemical applications. But there are other structures, like core–shell metal particles, metal overlayers or island on foreign metals, and various types of nanotubes or nanopores, which promise to be good electrocatalysts. So there is reason for optimism.

Nanostructures are a nice playground for theoreticians. They are often so small that their electronic properties can be elucidated by quantum chemistry, and it is much easier to design nanomaterials on a computer than to synthesize them in practice. This entails that theoreticians sometimes investigate systems that can never be realized in practice. An example from our own work is the infinitely long unsupported monowires of gold, which should be excellent catalysts for hydrogen evolution. However, here the art of the quantum chemist consists in studying systems from which something general can be learned, like systematic variations of energies or reactivity with size or bond order. Thus, the study of unrealistic systems is not always l'art pour l'art but may show the direction in which one has to search.

An intrinsic problem of nanostructures is that they are often not stable. Graphite is more stable than graphene, bigger metal clusters more stable than smaller ones, etc. Annoyingly, active metal layers on less active metals, like a platinum shell on a copper core or a layer of palladium on gold, are not stable, while the reverse systems are. A little thought shows that this is not malevolence of nature, but a natural consequence of reactivity. Many interesting systems are in metastable states, and how long they will last under practical conditions remains to be seen.

This is the context in which our own theoretical work on electron transfer in nanoelectrochemical systems has to be seen. In contrast to most groups who work in this area, we do not rely on quantum chemical methods like density functional theory (DFT) alone, but have developed our own theory for electrocatalytic reactions. So we shall start this review by giving a brief overview of our method and then give a systematic exposition of some of the nanostructures that we have investigated, focusing on their electronic properties, their stability, and, of course, their catalytic properties, using hydrogen evolution as a test reaction.

1.2   ELECTRON TRANSFER THEORY

1.2.1   LIMITS OF DFT IN MODELING ELECTROCHEMICAL REACTIONS

Electrochemical reactions involve electron transfer and therefore ions, which are stabilized by their solvation shell. For example, in proton discharge on an electrode surface, $H^+ + e^- \rightarrow H_{ad}$, the proton is strongly solvated, with a solvation energy of the order of 11.5 eV. It is this solvation that largely compensates the ionization energy of the hydrogen atom, and without it, the proton would not be locally stable. Therefore, the discharge of the proton requires a large fluctuation of the solvent, a solvent reorganization, for the electron transfer to occur. Such solvent reorganizations during electron transfer are the essence of the Marcus theory [1]. Both the large solvation sphere and the solvent fluctuations are difficult to handle in pure DFT, and so far, there has been no satisfactory solution to this problem.

However, this problem does not affect the *thermodynamics* of electrochemical reactions. The free energy of the solvated proton can be obtained from a thermodynamic argument, that of the adsorbed hydrogen atom from standard DFT, with or without a few water molecules. In this way, the free energy balance for the reaction can be calculated, and from this, the equilibrium potential can be obtained. The same principle can be employed for complicated reactions such as oxygen reduction, which contain many possible intermediate states. Chemical steps not involving charge transfer, such as the recombination reaction $H_{ad} + H_{ad} \rightarrow H_2$, can be treated by pure DFT, and for these, the activation energies can also be obtained by standard methods. Much extremely useful work, based solely on DFT, has been done in this way.

Another principle difficulty is caused by the structure of the electrochemical interface. The distribution of the particles and the electric potential has been investigated by numerous methods, starting from integral equations for hard-sphere ions and dipoles, molecular dynamics and Monte Carlo simulations based on force fields, and to a limited extent (short simulation times, small ensembles) by ab initio molecular dynamics. While the details depend on the system considered and the method employed, they all agree in an important point: for the ionic concentrations usually employed in experiments, 0.1–1 M aqueous solutions, the electrostatic potential and the particle distribution functions vary over a region of at least 10 Å, where they oscillate strongly, creating a small potential drop between the electrode and the bulk of the solution. This feature, which has been observed in many other investigations, is caused by a slight preferential orientation of the water molecules in the first layer due to their interaction with the metal, in this case Ag(111), and by the different interactions of the ions.

DFT calculations can, however, be performed for a given excess charge on the metal or, equivalently, a given electric field. Using experimental data for the interfacial capacity, these can roughly be converted to electrode potentials, with a sizable error. This is not quite satisfactory, since electrochemical reactions are driven by the potential and not by the field.

1.2.2 OUR APPROACH TO ELECTROCATALYSIS

Long before the advent of DFT, there already existed a line of theories for electrochemical electron transfer, which had its origin in the semiclassical theories of Marcus [1] and Hush [2] and was extended to a fully quantum theory by the Soviet school of Levich, Dogonadze, and Kuznetsov [3,4]. These theories explicitly considered the valence orbital of the reactant, the electrons on the metal, their electronic interaction, and the coupling of the electron exchange to the solvent. The latter was modeled as a dielectric continuum [2], a phonon bath [4], or a collection of polarizable dipoles [5], all being mathematically equivalent. The reorganization of the reactant's solvation shell during electron transfer plays a central part. Conceptually, these theories do not suffer from the difficulties that DFT has with electrochemical reactions. The electrode potential simply shifts the energy of the reactant's valence orbital with respect to the metal, and solvation is included from the start.

The older versions of the theory considered the electrode as a reservoir of electrons and hence could not explain catalysis. Substantial progress was achieved when two of us [6,7] connected electron transfer with ideas from the Anderson–Newns theory [8,9] and applied Green's function techniques. This made it possible to consider the electronic structure of the electrode, distinguish between d bands and sp bands, and treat the case of strong electronic interactions, which give rise to catalysis. Since it does not contain many-body effects, it requires input from DFT for quantitative calculations—this will be treated later (see Section 1.2.3). However, the theory by itself already does offer a nice way to understand qualitatively how a catalyst works, which we proceed to present.

As the reactant approaches the electrode surface, its valence orbitals start to interact with the metal electrons. Consequently, the energy of the valence orbital is no longer sharp, but is broadened into a density of states (DOS) $\rho_a(\epsilon)$, where ϵ denotes the electronic energy. Metal sp bands are broad, and they induce a general, unspecific broadening, which resembles a Lorentz function. They behave roughly in the same way on all metals. The d bands are narrower, and they induce a specific structure into the reactant's DOS. From the DOS, one obtains the occupation probability of the valence orbital and its electronic energy:

$$\langle n_a \rangle = \int_{-\infty}^{0} \rho_a(\epsilon)\,d\epsilon \quad E_{elec} = \int_{-\infty}^{0} \epsilon \rho_a(\epsilon)\,d\epsilon \tag{1.1}$$

where a signifies a spin orbital, and the energy of the Fermi level has been taken as the zero.

Solvation is just as important as the electronic energy; both change substantially during the reaction. To be specific, we discuss the hydrogen desorption $H_{ad} \rightarrow H^+ + e^-$, which is illustrated in Figure 1.1 for the cases of Pt(111) and Ag(111) as electrodes. The initial state is the adsorbed hydrogen atom. On both metals, its DOS lies below the d band; it is filled, and the atom carries no charge. Hence, its solvation energy is negligible. The final state is a proton in the solution, which is strongly solvated. Its DOS lies well above the Fermi level, and it is empty. At equilibrium, the loss

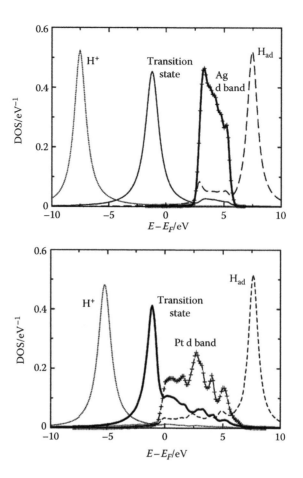

FIGURE 1.1 Changes in the reactant's DOS during hydrogen desorption. The dotted line is the DOS for proton (final state), the dashed line for the adsorbed hydrogen atom, and the full line for the activated state, where the DOS crosses the Fermi level, which is situated at $\epsilon = 0$. The bottom panel is for Pt(111) and the upper panel for Ag(111); the metal d bands are indicated by crossed lines. (From Santos, E. et al., *Phys. Chem. Chem. Phys.*, 14, 11224, 2012.)

in electronic energy of the ion is compensated by the gain in solvation energy. The reaction occurs when a solvent fluctuation lifts the DOS of the hydrogen atom to the Fermi level of the electrode; there, an electron can be transferred to the electrode, the particle becomes charged, and solvation sets in and lifts the DOS to its final equilibrium value. The major part of the energy of activation required for the electron transfer is determined by the electronic energy needed to lift the DOS of the atom to the Fermi level. A good catalyst like platinum (see bottom panel of Figure 1.1) has a d band that spans the Fermi level and interacts strongly with the reactant. It induces a substantial broadening of the DOS as it passes the Fermi level, which lowers the electronic energy of the transition state: the tail of the DOS extending well below the Fermi level reduces the electronic energy according to Equation 1.1. In contrast, on Ag(111), the d band lies well below the Fermi level, and at the transition state, its interaction with the reactant is weak. Therefore, only a very small part of the DOS extends to low energies, and the energy of activation is much higher than on Pt(111). This makes Ag(111) a much worse catalyst than Pt(111).

In summary, a good catalyst for an electron transfer step should have a d band that spans the Fermi level and interacts strongly with the valence level of the reactant. On the basis of these principles, we have explained the trends in the catalysis of hydrogen evolution on pure metals using the position of the metal d bands and their average coupling to the hydrogen 1s orbital [11,12]. A quantitative treatment, which can also consider nanostructured electrodes, requires input from DFT.

1.2.3 Combining Electron Transfer Theory with DFT to Investigate Hydrogen Evolution

The concepts outlined in the previous section can be based on a model Hamiltonian [7,13], which combines ideas from the Marcus–Hush [1,2] theory of electron transfer with the Anderson–Newns model [8,9]. While the resulting theory explains the principles of electrocatalysis well, it suffers from the well-known defects of the Anderson–Newns-type models: it does not account for many-body effects and is therefore not good enough for quantitative calculations. Therefore, we have developed a method to combine our theory with DFT calculations. In the following, we present the main ideas; the mathematical details are given in the appendix.

To explain how our method works, we must first introduce the concept of the solvent coordinate, again using the hydrogen desorption reaction ($H_{ad} \rightarrow H^+ + e^-$) as an example. Initially, the solvent configuration corresponds to that of an adsorbed hydrogen atom with a low or negligible solvation energy. In the final state, the proton is fully solvated. In the course of the reaction, the solvent fluctuates and passes through intermediate configurations between these two limiting cases. By a suitable normalization [14], we can define a solvent coordinate q with the following meaning: When the solvent configuration is characterized by a certain value of q, it would be in equilibrium with a charge of $-q$ on the reactant. Thus, initially the solvent configuration is described by $q=0$, in the final state by $q=-1$, and during the reaction it passes through the whole interval between these values, as the charge on the reactant changes from 0 to 1. Also, during the reaction, the position of the reactant changes: initially, it is adsorbed at a fixed distance, and finally, it is solvated in the solution. Therefore, in order to fully describe the reaction, we require the free energy of the system as a function of the solvent coordinate q and the separation d of the reactant from the surface. DFT gives the electronic energy of the hydrogen atom, corresponding to $q=0$, as a function of the distance. The electronic energy of the proton, at $q=-1$, is zero and thus exactly known. With the aid of our theory, we interpolate the electronic energy between $q=0$ and $q=-1$ and add the solvation energy. The mathematical details of this procedure have been explained in [15].

As an example, we consider in Figure 1.2 the free energy surface for hydrogen adsorption/desorption on an Au(111) surface. The surface is for the case where the overall hydrogen evolution reaction (HER) is in equilibrium (overpotential $\eta=0$), which corresponds to 0 V on the standard hydrogen scale (SHE). The minimum at $q=0$ corresponds to the adsorbed atom at an equilibrium distance of around 0.9 Å, the minimum at $q=-1$ and large separation to the proton in solution. The vertical line at $q=0$ reflects the electronic energy of the neutral adsorbate as a function of d and has been obtained from DFT; it increases with distance d as the interaction with the metal becomes weaker. The interaction of the uncharged atom with the solvent is almost negligible. The vertical line at $q=-1$ corresponds to the proton, whose electronic energy is zero. The increase of the energy with decreasing d reflects the decreasing solvation as the proton approaches the surface. During the course of the reaction, the system passes from one minimum to the other traversing a saddle point with a height of about 0.7 eV. The energy of the adsorbed atom is somewhat higher than that of the proton in solution, indicating that the adsorption is endergonic at the equilibrium potential for the overall reaction.

Before we present a few selected results for hydrogen reaction at other surfaces, let us recapitulate what enters into these calculations. For the electronic energy, we require the DOS and the energy of the hydrogen atom as a function of the distance; from these data, the interaction of the hydrogen 1s orbital with the metal can be derived as well. For the contribution of the solvent, we require the solvation energy of the proton, from which the energy of solvent reorganization can be derived.

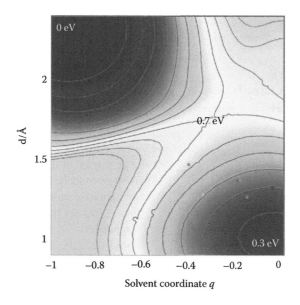

FIGURE 1.2 Potential energy surface for hydrogen adsorption/desorption H $\leftrightarrow$ H$^+$+e$^-$ on an Au(111) surface. All the energies are in eV. (From Santos, E. et al., *Phys. Chem. Chem. Phys.*, 14, 11224, 2012.)

1.2.4 HYDROGEN EVOLUTION REACTION

The total HER in acid media can be written as follows:

$$2H^+ + e^- \rightarrow H_2 \tag{1.2}$$

This process takes place at an electrode supplying the electrons in a two-step mechanism. The first one is always the hydrogen adsorption on the electrode surface:

$$H^+ + e^- \rightarrow H_{ad} \tag{1.3}$$

This is also known as the Volmer reaction. For the second step, there are two alternatives. One of them is the chemical recombination, or the Tafel reaction:

$$H_{ad} + H_{ad} \rightarrow H_2 \tag{1.4}$$

which does not involve charge transfer. The alternative reaction is the electrochemical recombination, or the Heyrovsky reaction:

$$H^+ + H_{ad} + e^- \rightarrow H_2 \tag{1.5}$$

Since this is a two-step reaction, we may expect that near equilibrium Sabatier's principle [16] should apply, which states that for a good catalyst, the first step should involve little change of the reaction free energy, $\Delta G \approx 0$. If the first step is highly endergonic, it will be slow; if it is highly exergonic, then the second step will be uphill and hence slow. Adding this condition to the principles for electron transfer catalysis discussed earlier, we conclude that a good catalyst for hydrogen evolution should fulfill three conditions: $\Delta G \approx 0$ for hydrogen adsorption, a d band that extends across the

Fermi level, and a strong coupling of this band to hydrogen. The latter depends on the extension of the metal d states and thus varies systematically along the periodic table.

To obtain the free energy of adsorption, we must define first the hydrogen adsorption energies at $T = 0$ K:

$$\Delta E_{ad} = \frac{E(surface + n_H) - E(surface) - \frac{1}{2} n_H E(H_2)}{n_H} \tag{1.6}$$

where

ΔE_{ad} is equal to the change in the enthalpy for the reaction

$E(surface + n_H)$, $E(surface)$, and $E(H_2)$ are the energies of the hydrogenated surface, surface, and hydrogen molecule, respectively

The Gibbs free energy now can be defined according to the formula

$$\Delta G_{ad} = \Delta E_{ad} + \Delta E_{ZPE} - T \Delta S_{ad} \tag{1.7}$$

$$\approx \Delta E_{ad} + \frac{1}{2} T S_{H_2}^0 \tag{1.8}$$

$$\approx \Delta E_{ad} + 0.20 \, eV \tag{1.9}$$

where ΔE_{ZPE} and ΔS_{ad} are the vibrational and entropic terms of the reaction, respectively. If we neglected the changes in the zero point energy due to the adsorption of the hydrogen atom, Equation 1.7 could be simplified into expression 1.8. Considering the same procedure as Nørskov et al. [17], we use the fact that the vibrational entropy in the adsorbed state is small meaning that the entropy of adsorption is $\Delta S_{ad} \approx S_{H_2}^0 / 2$, where $S_{H_2}^0 = 0.41 \, eV$ [18] is the entropy of H_2 in the gas phase at standard conditions. This leads to a constant entropic term of 0.20 eV.

1.3 FREE STANDING METALLIC NANOWIRES

1.3.1 THERMODYNAMIC STABILITY OF METALLIC NANOWIRES

The stability of the free standing nanowires can be measured through the calculation of the wire surface energy:

$$\sigma_{wire} = E_{wire} - E_{bulk} \tag{1.10}$$

where E_{wire} and E_{bulk} correspond to the energies of an atom in the wire and in the bulk metal, respectively. According to this equation, a positive value means that external energy is required to ensemble the wire from the bulk system. It is also worth noting that the lower the wire surface energy σ_{wire}, the smaller the force that tends to break the nanowire. Table 1.1 shows energies obtained from DFT calculations of several monometallic nanowires.

Various experimental works have found that nanowires made of Au, Ag, or their combinations are stable for hours [19,20], while wires made of Cu, Pt, or Pd show a mean lifetime of the order of seconds [20–26]. As the results of Table 1.1 show, it is straightforward to observe that the stable nanowires (Ag and Au) have smaller surface energy ($\sigma \leq 1.5$ eV), while the unstable wires (Cu, Ni, Pd, and Pt) have bigger energies ($\sigma \geq 2$ eV). We can conclude that the energy difference between Ag and Ni nanowires is large enough to explain the experimental trends. However, copper and gold wires have similar surface energies, and the previous argument cannot explain their relative stabilities. Fortunately, the energy difference becomes larger if the area A of such

TABLE 1.1

Surface Energy (in eV per Atom for σ and in meV/Å² for γ; See Text) for Ni, Pd, Pt, Cu, Ag, and Au Nanowires

	Wire		Slab	
	σ [eV/at.]	γ [meV/Å²]	σ [eV/at.]	γ [meV/Å²]
Cu	1.93	57	0.46	79
Ag	1.39	32	0.35	48
Au	1.51	35	0.34	45
Ni	2.83	95	0.63	116
Pd	2.57	65	0.53	77
Pt	2.67	74	0.64	92

Note: Surface energies are compared with those of fcc(111) slabs.

structures is considered. A more specific surface energy γ can be defined as σ/A, and the area of wires is obtained by considering them as cylinders with length and radius equal to the wire bond distance (L_x).

The surface energy newly defined as γ shows significant differences for the coinage metal nanowires. It is now clear that the driving force for the chain breaking is the minimization of the metal surface, a force that is larger for Cu wires making them unstable in comparison with the other metals of its group in the periodic table.

For the sake of comparison, the surface energies (σ and γ) for fcc(111) slabs were also calculated using Equation 1.10, replacing the energy of the wire by the surface. As we can see from Table 1.1, the surface energy σ of the wires is at least four times bigger than that of the close-packed metal surfaces. It may look surprising that γ is larger for slab surfaces than for wires, but this is the result of the large area exposed by the nanowires.

The wire surface energy σ, defined according to Equation 1.10, was redefined in order to study the stability of bimetallic nanowires. In a similar way, the surface energy was calculated in reference with the bulk systems:

$$\sigma = \frac{E_{\text{wire}}(\alpha_m\beta_n) - mE_{\text{bulk}}(\alpha) - nE_{\text{bulk}}(\beta)}{m + n} \qquad (1.11)$$

Figure 1.3 shows the variation of the surface energy as a function of the fractional composition χ. We have considered Au β nanowires, where β = Ag or Cu, due to the fact that they were obtained experimentally [20]. For the same fractional composition, we have evaluated different systems by changing the order of the atoms in the unit cell. Compositions with the highest degree of intercalation or mixing are connected with a line in each case. For all the compositions, the bimetallic nanowires with the higher intercalation are the most stable ones.

According to Figure 1.3, most of the bimetallic nanowires studied have a thermodynamic stability in between the corresponding pure wires. Nevertheless, in both cases, there is a local minimum at χ = 0.5 in the surface energy that corresponds to wires with the largest entropy possible for a bimetallic compound. The existence of this local minimum emerges from the increase of the intermetal interactions that stabilize the bimetallic wires.

It is expected that the configuration of the experimental bimetallic wires corresponds to values of χ ≈ 1 since at the working potentials of the experiments, Au does not dissolve [20]. These results agree with the natural assumption that at χ ≈ 1, bimetallic and β pure wires have similar stabilities.

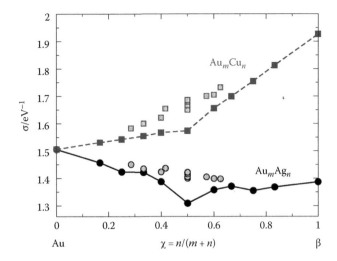

FIGURE 1.3 Surface (σ) energy as a function of the fractional composition (χ) of Au β wires. $\chi=0$ corresponds to a pure Au wire and $\chi=1$ to a pure β wire (Ag or Cu). (From Soldano, G. et al., *J. Chem. Phys.*, 134, 174106:1, 2011.)

1.3.2 HYDROGEN ADSORPTION ON METALLIC NANOWIRES

Table 1.2 shows the hydrogen free adsorption energies of metal nanowires, calculated according to Equations 1.6 and 1.9. All the numbers were obtained for the most stable adsorption site: a threefold fcc site in (111) surfaces and a bridge position between two adjacent metal atoms in the wires. The comparison between the energies in surfaces and wires shows a lowering in the values for all the nanowires, which corresponds to an increase in the interaction between them and the hydrogen atoms. However, the adsorption in Ag nanowires is still not a stable process. The gold nanowire shows the largest increase in the H adsorption energy.

The changes in the interaction between hydrogen and metal wires can be explained by means of the position and the width of the d bands. A brief analysis of Figure 1.4 indicates a narrowing of the d bands in the nanowires, together with a displacement in the position of their center. For Cu and Au, this change is significant and now the d bands end right at the Fermi level, making these nanowires similar to the transition metals! In the case of platinum, the center of the d band is also shifted to higher energies, but the part that lies above the Fermi level remains practically the same, in order to maintain the number of electrons.

While there is no direct evidence for hydrogen adsorption on nanowires, there is good indirect evidence in the form of the conductivity of nanowires, which is known to be affected by adsorbates. In early experiments, Li et al. [29] showed that the electrochemical potential can indirectly affect the conductivity of wires by controlling the adsorption of organic species from the surrounding electrolyte. The adsorbates scatter conduction electrons, so that the conductance

TABLE 1.2

Hydrogen Adsorption Energy (ΔG_{ads}) for Pt, Cu, Ag, and Au Nanowires and fcc(111) Slabs

	Cu	Ag	Au	Pt
ΔG_{ads} [eV] (wire)	−0.31	0.18	−0.51	−0.33
ΔG_{ads} [eV] (111 surface)	0.10	0.34	0.41	−0.25

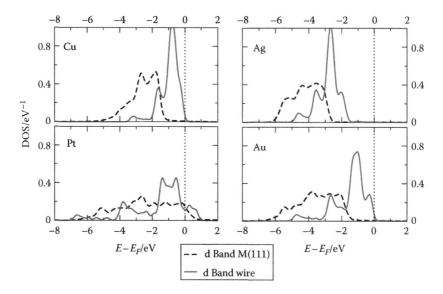

FIGURE 1.4 Projected DOS on the d band of the nanowires and the corresponding metal surfaces. The Fermi level has been taken as the zero energy. (From Santos, E. et al., *Electrochem. Commun.*, 11, 1764, 2009.)

of copper wires, which in the absence of adsorbates is an integral multiple of the quantum G_0 of conductance, was decreased and became fractional. A more interesting case was the observations of fractional conductances on monoatomic gold wires in aqueous solutions of $HClO_4$ and $KClO_4$ by the same group. These electrolytes do not adsorb on bulk gold at negative potentials. Nevertheless, a fractional conductivity occurred only at potentials negative of the reversible hydrogen potential. Recently, this phenomenon has been investigated more extensively by Kiguchi et al. [30,31], who studied Cu, Ag, and Au nanowires in aqueous solutions of $CuSO_4$ and H_2SO_4, which are also nonadsorbing electrolytes. They observed fractional conductances in the hydrogen evolution region on the gold and copper wires, but not on silver. Kiguchi et al. suggested that the observed fractional conductances are caused by hydrogen adsorption, which they propose to take place during hydrogen evolution. They base their interpretation on the fact that in the vacuum, hydrogen has indeed been shown to induce fractional conductances in gold nanowires [32,33].

These arguments are of fundamental importance in order to explain the large decrease of the adsorption energy for copper and gold wires. On both wires, the d band extends right up to the Fermi level. Part of the antibonding DOS of the hydrogen orbital now also extends above the Fermi level and is unfilled, so the d band now contributes to the bonding. The effect is larger for gold than for copper, because gold has the larger orbital radius and interacts more strongly with hydrogen than copper. In contrast, on the silver wire, the d band lies well below the Fermi level, and the antibonding part of the hydrogen DOS is filled, so that the d band still does not contribute to the bonding. This effect can be clearly observed in Figure 1.5. To illustrate this point in detail, we have also calculated the Wannier functions [34] from the occupied orbitals. In the case of silver, we found both a bonding and an antibonding orbital between the wire and the hydrogen atoms; on gold, only the bonding orbital is occupied.

The fact that in an electrochemical environment fractional conductivities are observed only on gold and copper nanowires, but not on silver, can easily be explained by our calculations. On the nanowires, hydrogen is adsorbed on gold and copper, but not on silver (see Table 1.2). A direct investigation of hydrogen evolution on such wires still remains a scientific challenge.

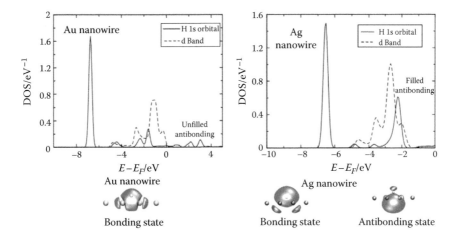

FIGURE 1.5 DOS of the hydrogen 1s orbitals and d metal bands after the adsorption on gold and silver wires. The drawings correspond to the Wannier functions, showing the bonding and antibonding characters of the states. (From Santos, E. et al., *Electrochem. Commun.*, 11, 1764, 2009.)

So far, we have only discussed the reactivity of monometallic wires further. Because of the wide range of possibilities for the adsorption, the study of hydrogen interaction with bimetallic wires is much more complicated. However, it is interesting to examine some aspects that give us a better understanding of the nature of the bond between hydrogen and nanowires. Due to their stability, the most important and illustrative cases are the gold bimetallic wires: $Au_m\beta_2$ and $Au_m\beta_5$, where β correspond to copper or silver atoms.

In order to evaluate the strength of the electronic perturbation that gold atoms produces on copper and silver wires, the number of Au atoms in the unit cell has been varied from 1 to 5. According to the geometry of the wires, the number m corresponds to the number of atoms that separates one pair of β atoms from the next pair in the following cell. Due to the complexity of the systems, several sites have been considered for the adsorption: (1) bridge site between two β metals (bridge(aa)), (2) bridge site between the first Au atom and a β atom (bridge(Aa)), (3) on top of the first gold atom (top(A)), and (4) on top of the β metal atom (top(a)). Figure 1.6 shows a scheme of the wires and the calculated energies.

A detailed examination of Figure 1.6 can lead to several conclusions. In all the cases, the adsorption of hydrogen is most stable on the less coordinated bridge sites. The presence of any amount of gold in the bimetallic wires reduces adsorption and favors the interaction between the hydrogen and the metal. It is interesting to notice that the adsorption energy of hydrogen on the sites top(a) and bridge(aa) remains approximately constant with the molar fraction of gold, but all the values are around 0.3 eV more negative than on pure β wires. We can conclude that the adsorption of hydrogen on copper and silver has been improved due to the presence of gold in the nanowires, but the adsorption is still more favorable in pure gold nanowires.

The observed trends can be explained in terms of the d-band model [35,36]. For the purpose of understanding, these trends (Figure 1.7) show the DOS of the isolated bimetallic wires. The increase of the number of gold atoms has two different consequences. In the gold atoms, it produces a shift to higher energies, increasing the reactivity. Simultaneously, Au induces a subtle upshift of the d band on its β neighbors, making the adsorption of hydrogen stronger than in pure β wires. In this case, the adsorption between H and the metal wires is mainly decided by the position of the d band.

In order to study the extension of the electronic perturbation along the β atoms in the chain, Figure 1.8 shows the adsorption energies of hydrogen in different position on $Au_m\beta_5$ wires.

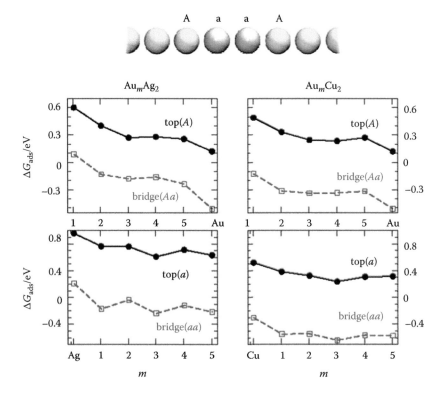

FIGURE 1.6 Free adsorption energy of hydrogen (ΔG_{ads}) on $Au_m\beta_2$ and pure wires for several top and bridge positions. Adsorption on pure wires is denoted with Au, Ag, or Cu correspondingly. (From Soldano, G. et al., *J. Chem. Phys.*, 134, 174106:1, 2011.)

In general, the number of gold atoms does not modify the adsorption energies that depend on the position in the wire. Nevertheless, the number of Au atoms drastically changes the adsorption in the sites close to the gold atoms (bridge(Aa) and top(A)).

According to the results in Figure 1.8, the adsorption of hydrogen in these wires is exergonic for Au_mCu_5 wires and mainly endergonic for the Au_mAg_5 wire, like in the pure surfaces. For the Au_mAg_5 wire, there is one exception; the adsorption in the bridge site close to gold (site(Aa)) has $\Delta G \approx 0$.

The adsorption in the middle of the β chains has roughly the same energy than in pure β wires, showing that the electronic perturbation induced by Au does not go further than to its second β neighbor.

The examination of the thermodynamic stabilities together with the use of the d-band model is always a good starting point in the understanding of the reactivity of any electrode. However, a rigorous and precise study of the kinetics of HER requires the calculation of activation barriers. For this purpose, we have calculated the free energy surface for the adsorption of a hydrogen atom on Au and Cu wires using the method developed in our group [15,37].

The resulting energy surfaces for the hydrogen adsorption are shown in Figure 1.9. They are plotted as a function of the distance d between the H atom and the center of the wire and the solvent coordinate q.

For both nanowires, we have found two minimum on their surfaces: one centered at $q=-1$ and far away from the metal wire and the other one at $q=0$ and close to the nanowire. The first corresponds to a solvated proton (H$^+$) and the last one to an adsorbed atom (H$_{ad}$). The minimum for the adsorbate is deeper, since adsorption is exergonic. The two minimums are separated by a saddle point, which gives the energy of activation.

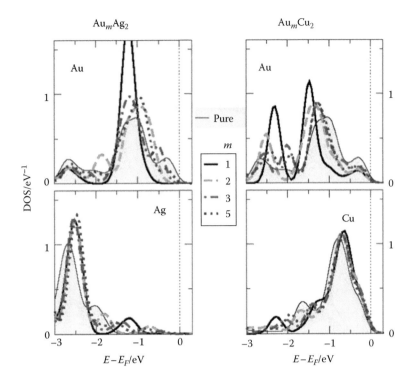

FIGURE 1.7 Projected DOS on the d band of pure (shaded surface) and $Au_m\beta_2$ (lines) wires. The atoms are labeled in the same way as in Figure 1.6. The d band of A and a atoms is plotted. (From Soldano, G. et al., *J. Chem. Phys.*, 134, 174106:1, 2011.)

The activation barrier in the gold nanowire is considerably smaller than in the surface (0.1 vs. 0.7 eV) [15]. This indicates that the reaction is also substantially faster and is in line with the general finding that gold nanostructures show a strongly enhanced reactivity [38–42]. Copper nanowires are also better catalysts than planar surfaces; however, the activation energy is only 0.2 eV lower than for the (111) surface.

1.4 SUPPORTED METALLIC NANOWIRES

1.4.1 THERMODYNAMIC STABILITY OF GRAPHITE-SUPPORTED NANOWIRES (GSW)

In spite of the fact that free standing metal nanowires show a greater reactivity, they cannot be used in practical systems. Nevertheless, nanowires supported on inert substrates are promising systems closer to real devices. In particular, wires adsorbed at the edge of graphite steps show a good compromise between what can be theoretically studied and can be obtained and measured experimentally [43–45].

We have considered two types of graphite edges: zigzag terminations and armchair steps. Figure 1.10 shows the unit cell for both cases and the position of the nanowires. It is important to notice that there are two metal atoms per unit cell, which are completely inequivalent in the armchair edge. The corresponding geometry values for different metals are shown in Table 1.3.

On the zigzag steps, the geometry of adsorption was characterized through the bond distance between the metal and the carbon atoms (d_{GM} (zz)) and the angle θ_{xz} that measures the distortion of the wire in the xz plane. The corrugation of the wires in the xy plane is the result of the mismatch between metal–metal bond distance (d_{MM}) and the distance between the carbon atoms

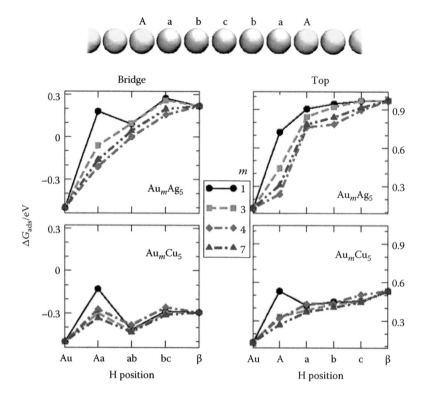

FIGURE 1.8 Hydrogen free energy of adsorption (ΔG_{ads}) on pure and on $Au_m\beta_5$ wires for several bridge (left plots) and top (right plots) positions. The adsorption on pure wires is denoted by Au and β in the abscissas. (From Soldano, G. et al., *J. Chem. Phys.*, 134, 174106:1, 2011.)

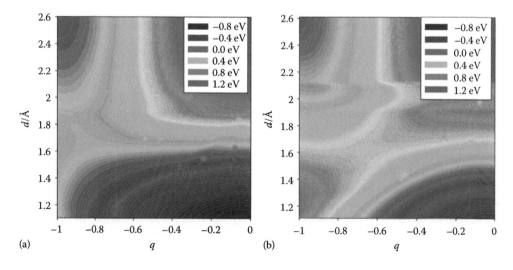

FIGURE 1.9 Free energy surface for the adsorption of a proton on an Au (a) and on Cu (b) wire, as a function of the separation from the center of the wire and of the solvent coordinate. The calculations have been performed for the equilibrium potential at 0 V SHE; all the energies are in eV. (From Santos, E. et al., *Electrochem. Commun.*, 11, 1764, 2009.)

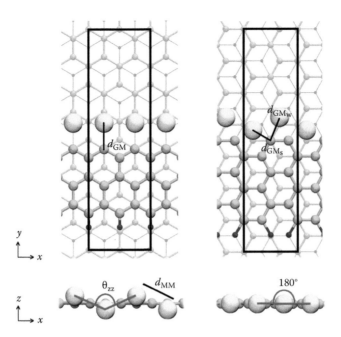

FIGURE 1.10 Lateral unit cell of zigzag and armchair graphite-supported wires. Relevant distances and angles are also included in the figure. The lateral unit cell is sketched with black plain lines. (From Soldano, G.J. et al., *J. Phys. Chem. C*, 117, 19239, 2013.)

TABLE 1.3
Values of Distances and Angles Shown in Figure 1.10

	Ni	Pd	Pt	Cu	Ag	Au
d_{GM} (zz) [Å]	1.80	1.97	1.94	1.92	2.14	2.08
d_{GM_s} (arm) [Å]	1.93	2.06	2.07	1.96	2.13	2.10
d_{GM_w} (arm) [Å]	2.03	3.05	3.05	2.30	3.38	3.29
d_{MM} (NW) [Å]	2.13	2.50	2.39	2.32	2.64	2.61
θ (zz) [°]	180	145	151	180	132	134

Note: The wire bond distance d_{MM} correspond to the bare wires (NWs).

($d_{CC} = 2.46$ Å) in the edge. Thus, it is expected that the nanowires with a larger bond distance d_{MM} will have a greater corrugation and smaller angles θ_{xz}. The values in Table 1.3 show the expected trend.

On the armchair GSWs, the metal atoms are not equivalent anymore. One of them is surrounded by four nearest neighbors and is strongly adsorbed to the step (M_s). The other one is bound only by two carbon atoms and is weakly attached to the step (M_w). Unlike the zigzag GSWs, nanowires adsorbed on the armchair edges do not break the linearity in the xz plane but show a similar corrugation in the xy plane.

The strength of the union between the graphite edges and the metal atoms was measured by the calculation of the binding energy (E_b), according to the following equation:

$$E_b = \frac{E_{GSW} - E_G - E_{NW}}{M} \qquad (1.12)$$

where

E_{GSW}, E_G, and E_{NW} are the energies of the graphite-supported wire, the bare graphite, and the bare wire, respectively

M is the number of metal atoms in the wire

In agreement with this expression, a negative value indicates a strong interaction between the adsorbate (graphite edge) and the substrate (nanowire). Furthermore, the binding energy E_b computes on average the binding energy of all the metal atoms in the wire. Because the atoms in zigzag GSWs are identical, the energy E_b is actually the same for all of them. However, this is not correct for the armchair GSWs, where the metal atoms are bound to the graphite in a rather different manner. In order to distinguish between the metal atoms strongly (M_s) and weakly (M_w) adsorbed in the armchair GSWs, the binding energies of the inequivalent metal atoms must be separately defined in the following equations:

$$E_{b1}(M_s) = E_{G+M_s} - E_G - E_{NW} \qquad (1.13)$$

$$E_{b1}(M_w) = E_{G+M_w} - E_G - E_{NW} \qquad (1.14)$$

where

E_{G+M_s} is the energy of the step with only the strongly adsorbed metal

E_{G+M_w} is the energy of the step with only the weakly adsorbed metal atoms

In this way, the binding energy $E_{b1}(M_s)$ is the energy corresponding only to the bond between the step and the strongly adsorbed metal atoms. All these binding energies are shown in Table 1.4.

In all the metals studied, the adsorption of the whole wire is stronger in the zigzag edges. However, the interaction between the strongly adsorbed metal atoms M_s and the armchair steps is greater than the former interaction. This result also proves that in the armchair steps, the strongly attached metal atoms M_s are adsorbed in the first place.

The nature of the metal atoms is also important to increase the stability of the wire in the support. We have found that the stability of the supported wires decreases in the order Ni $\approx$ Pd $\approx$ Pt > Cu $\approx$ Ag $\approx$ Au.

So far, we have obtained a detailed description of the energetics of the chemical bond between the metal wires and the graphite edges; however, this analysis is not enough to give a complete explanation of the observed trends. We have shown before that the analysis of the bond between

TABLE 1.4

Binding Energy E_b in eV of Zigzag and Armchair GSWs

	Ni	Pd	Pt	Cu	Ag	Au
E_b (zz)	−3.00	−3.29	−3.53	−2.27	−1.72	−1.91
E_b (ac)	−2.20	−2.24	−2.29	−1.70	−1.40	−1.98
$E_{b1}\,M_s$	−3.30	−3.55	−3.81	−2.77	−2.11	−2.33
$E_{b1}\,M_w$	−2.34	−2.56	−2.13	−1.24	−0.68	−0.51

Note: The binding energies of the strong and weak adsorbed species (M_s and M_w, respectively) are also shown.

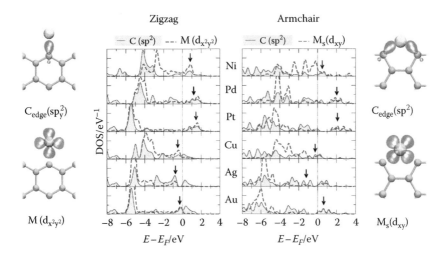

FIGURE 1.11 DOS projected on the atomic orbital of metal atoms in the wire and carbon atoms in the graphite steps. The atomic orbitals are sketched for a better understanding. The positions of the antibonding states are highlighted with arrows. (From Soldano, G.J. et al., *J. Phys. Chem. C*, 117, 19239, 2013.)

substrate and adsorbate can be done efficiently and precisely by means of the calculation of the DOS. The DOS projected on the atomic orbitals of the zigzag and armchair GSWs is shown in Figure 1.11. For this calculation, we have only considered the orbitals that can effectively overlap, based on the symmetry of the newly made bonds.

In all the wires, we can see the formation of a filled bonding state, at very low energies (between −4 and −6 eV below the Fermi level), as the appearance of resonance between the energy levels of the wire and the graphite. The antibonding state is much higher in energy, and in the case of Ni, Pd, and Pt, it is above the Fermi level. The GSWs with lower stability (Au, Ag, and Cu) have their antibonding orbitals partially or completely filled, explaining the trend in the binding energies.

1.4.2 HYDROGEN ADSORPTION VS. NANOWIRE DESORPTION

Hydrogen adsorption on GSWs has a more complicated scenario than the previously examined for freestanding wires. In this case, the adsorption can take place on the metal wire or directly at the edge of the carbon step. The former reaction also involves the desorption of the wire. The reaction energy for these two processes has been distinguishably defined according to Equations 1.15 and 1.16:

$$E_{ads}^{wire} = \frac{E_{GSWH} - (E_{GSW} + E_{H_2})}{2} \tag{1.15}$$

$$E_{ads}^{step} = \frac{E_{GH} + E_{NW} - (E_{GSW} + E_{H_2})}{2} \tag{1.16}$$

where
E_{GSWH} corresponds to the energy of a supported wire covered with adsorbed H atoms
E_{GH} is the energy of the H-terminated graphite step
E_{H_2} is the energy of a hydrogen molecule in the vacuum

TABLE 1.5

Free Energies for the Adsorption of Hydrogen on the Supported Wires: (a) On the Wire and (b) Directly on the Step Edge

		Ni	Pd	Pt	Cu	Ag	Au
Zigzag	(a) Wire	**−0.02**	**0.14**	**−0.09**	0.02	0.20	−0.06
	(b) C_{edge}	0.16	0.45	0.68	**−0.58**	**−1.12**	**−0.93**
Armchair	(a) Wire	**−0.82**	**−0.29**	**−0.58**	−0.53	0.14	0.05
	(b) C_{edge}	−0.07	−0.03	−0.02	**−0.58**	**−0.88**	**−0.29**

Note: The most stable of these two scenarios is highlighted in bold numbers for each metal. All the values are in eV.

These equations correspond to the adsorption energy per hydrogen atom (two per unit cell) for a high coverage (one H per metal atom). Adsorption energy E_{ads}^{step} also includes the desorption energy of the metal wire.

The free energies of adsorption were obtained by using the energies E_{ads} and the equation; the obtained results are summarized in Table 1.5.

The adsorption energy table leads us to an important conclusion. The nature of metal wire is the one that decides the path of the reaction with the hydrogen molecule. On Ni, Pd, and Pt wires, the result of the reaction is the adsorption of H atoms on the metal. On the contrary, when hydrogen reacts with Au, Ag, and Cu support wires, it produces the total desorption of the nanowires. In the latter case, the interaction between the H atoms and the metal wire is extremely weak, and it does not compensate the energy gain for the formation of the covalent C–H bonds. The weak interaction between the graphite steps and the nanowires also reduces the required energy to desorb the metal wires.

The stability of the wires on graphite also plays an important role in the adsorption of hydrogen on the metal atoms. For Ni, Pd, and Pt wires, the H adsorption energy is larger when the wire is supported on the armchair termination proving that a strong interaction between the nanowire and its support becomes the metal inert.

Of all the studied systems, only Ni and Pt wires supported in graphite steps promise to be good catalysts for hydrogen adsorption, showing even better energies than bare gold wires. However, for a definite conclusion, it is crucial to calculate the reaction barriers. The resulting free energy surface for the adsorption of a proton on the armchair graphite-supported Pt wires is shown in Figure 1.12.

The energy surface for the Volmer reaction shows two distinguishable minimums: the solvated proton ($q=−1$), far away from the surface, and the adsorbed H ($q=0$), at a distance $d=1.1$ Å from the wire. The activation barrier for the reaction in the GSW is equal to 0.47 eV, which is a higher value than the barrier of 0.3 eV, obtained for a Pt(111) using the same procedure. The difference is caused by the fact that the interaction between the Pt wire and the hydrogen 1s orbital decays faster with the distance than at the bulk electrode.

Summarizing, we have shown at the beginning that freestanding nanowires of Cu and Au are excellent catalysts for the HER. Later, we have studied the formation and kinetics of nanowires on graphite/graphene edges. The interaction between the wires and the edges was found stronger in the Ni periodic group than in the Cu group. In comparison with freestanding nanowires, hydrogen adsorption is in some cases favored at the supported wire (such as Pt-supported wires on armchair steps), and in other cases, it is favored at bare wires (such as Au freestanding wires). Unfortunately, the reaction is even more exothermic at the graphite step for the three coinage metals. Therefore, the promising catalytic properties of Cu and Au nanowires are not such at graphite-supported media. Platinum armchair-supported nanowires are good

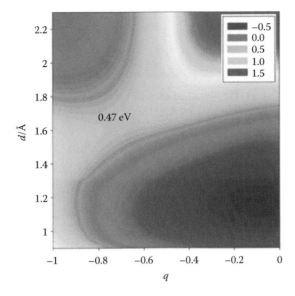

FIGURE 1.12 Free energy surface for the adsorption of a proton on an armchair graphite-supported Pt wire, as a function of the separation from the center of the wire and of the solvent coordinate. (From Soldano, G.J. et al., *J. Phys. Chem. C*, 117, 19239, 2013.)

catalysts for the HER, but not as good as platinum bulk (111) due to the faster decrease of the metal–hydrogen interaction (coupling constant) for the nanowires.

1.5 HYDROGEN ADSORPTION ON EMBEDDED AND ADSORBED NANOSTRUCTURES

In the previous section, we have considered the adsorption of nanostructures on organic and less reactive substrates, as graphite. Nevertheless, another metal can be used as a supporting material for a metallic nanostructure. In this case, the substrate is modifying not only the electronic and catalytic properties of the nanostructure because of the geometric distortion but also the chemical interaction with them.

We will start this section with the study of the hydrogen adsorption on different Pd–Au nanostructures. Figure 1.13 shows the considered systems that include planar surfaces, defects, adatoms, and nanostructures.

It is evident from the analysis of the adsorption energies that palladium atoms are better catalysts than gold for the adsorption of hydrogen in any of the present structures. Actually, when H adsorption occurs on gold atoms, the clean planar (111) surface has the lower reaction energy, and neither a defect nor an adatom reduces the required energy.

On the other side, H adsorption on palladium shows that the nanostructures are more reactive than the planar surface. Further, we have studied the H adsorption reaction on the Pd_3 cluster, increasing the amount of hydrogen.

In order to measure the stability of the newly adsorbed H atom, we have calculated the differential adsorption energy ΔE_{ads} according to equation

$$\Delta E_{ads} = E_{nH} - E_{(n-1)H} - 1/2E_{H_2} \tag{1.17}$$

where E_{nH} is the energy of the nanostructure covered by n hydrogen atoms. The energy defined in this equation corresponds only to the last bond between the hydrogen atom and the nanoparticle,

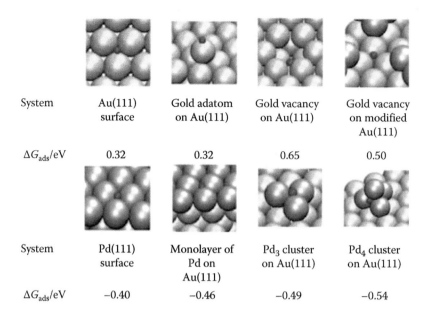

System	Au(111) surface	Gold adatom on Au(111)	Gold vacancy on Au(111)	Gold vacancy on modified Au(111)
ΔG_{ads}/eV	0.32	0.32	0.65	0.50

System	Pd(111) surface	Monolayer of Pd on Au(111)	Pd$_3$ cluster on Au(111)	Pd$_4$ cluster on Au(111)
ΔG_{ads}/eV	−0.40	−0.46	−0.49	−0.54

FIGURE 1.13 Hydrogen adsorption energy (ΔG_{ads}) for Pd–Au nanostructures and fcc(111) slabs.

and it can be used to highlight the interactions (repulsive or attractive) between the adsorbed atoms. The results of the adsorption of hydrogen atoms on the Pd nanostructures are shown in Figure 1.14.

The highly negative adsorption energy of hydrogen atoms on the palladium nanoparticles shows that these nanoparticles have a high capacity to store hydrogen at the studied coverages.

In order to enhance the adsorption, the repulsion between the neighbor atoms is always minimized by moving away the hydrogen atoms. In this sense, the H atoms prefer to adsorb in a Pd–Pd–Au hollow site rather than in a top site on Pd, at any of the present coverages.

It is important to notice that the increase of hydrogen atoms on the nanoparticle produces the formation of hydrogen-like molecules, with a bond length comparable to the free molecule (0.84 Å in comparison with $d_{H_2} = 0.765$ Å). The newly formed molecules have less negative adsorption

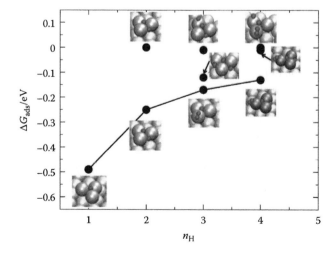

FIGURE 1.14 Adsorption of additional hydrogen atoms at the three-atom cluster of Pd on Au(111) surface.

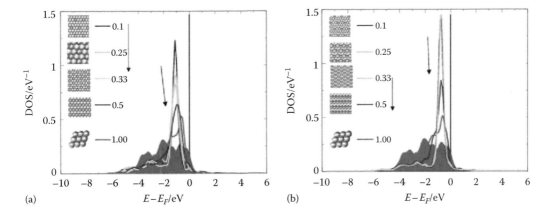

FIGURE 1.15 DOS projected on the d band of palladium homogeneously distributed on Au(111) at different compositions. (a) Embedded and (b) adsorbed palladium atoms in the gold surface. The d band (red shaded area) of Pd(111) is also shown for the sake of comparison. (From Quaino, P. et al., *Catal. Today*, 177, 55, 2011.)

energies and weaker interactions with the surface, mainly because the adsorption occurs at the mixture hollow site. Thus, it is expected that the recombination or the Heyrovsky step in the HER at these nanostructures occurs faster than at pure palladium surfaces.

The morphological changes produced by the modification of the surface can be related with the changes in the DOS. Figure 1.15 shows two sets of results for (a) embedded and (b) adsorbed nanoparticles of the same size and geometry. Besides the similarities, there are two differences that we want to emphasize. The first one is that the corresponding d bands are sharper for the palladium atoms in the adsorbed structures, where the atoms are less coordinated and the localization is larger.

The second effect is due to the interaction with the substrate. On the adsorbed structures, it is easy to observe that there are no more changes on the DOS at compositions lower than 0.33. It indicates that the separation between the atoms is long enough to remove the interactions between them. Nevertheless, the palladium atoms in the embedded structures are always interacting through the gold atoms bonded to them.

In the past, d-band theory has been successfully used to explain the reactivity of a wide range of catalysts. It was taken for granted that the participation of the *structureless* and wide sp band was not relevant. However, we want to emphasize that this approach is an oversimplification that could lead to wrong results. We have included two different pictures of the same system in order to show the behavior of the electron bands during the adsorption process. The electronic properties of a hydrogen atom approaching to the surface of a three palladium atoms cluster on Au(111) are shown in Figure 1.16.

It is clear from Figure 1.16 that the 1s orbital of hydrogen interacts with the sp and d bands in a concerted way. When hydrogen is adsorbed in the palladium cluster, both bands overlap with the 1s orbital at about −6 eV. Another small peak can be observed at about −2 eV, in the same region where the d band appears. All these bonding states are occupied, and consequently, they stabilize the adsorbed species.

At larger distances, the previous interactions show significant changes. There is no overlap of hydrogen and palladium states at energies lower than −4 eV. However, at −1.75 eV, three coincident peaks of the 1s orbital and sp and d bands appear.

Although it is true that the electronic states of the d bands play a key role in the catalyst, when the hydrogen 1s orbital is passing the Fermi level, the interaction with the sp bands is also important in the formation of the bond to the surface at the final state of the reaction.

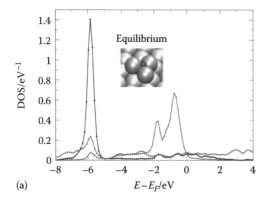
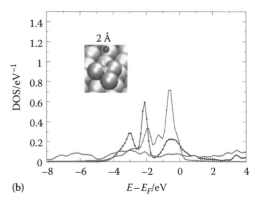

FIGURE 1.16 DOS of hydrogen atom interacting with a three-palladium-atom cluster on Au(111) at (a) the equilibrium distance and (b) 2 Å from the equilibrium position. The states are projected onto the d band (blue line) and sp band (red line) of palladium and the s orbital of hydrogen (black line). (From Quaino, P. et al., *Catal. Today*, 177, 55, 2011.)

1.6 HYDROGEN ADSORPTION ON STEPPED SILVER SURFACES

Although stepped surfaces are mesoscopic structures that are infinite in the direction parallel to the step, they can be considered as nanoscopic objects in the direction perpendicular to it. Stepped surfaces are ideal systems to investigate structural effects on hydrogen adsorption [48]. Cutting single crystals at different angles can systematically create regions of varying local atomic coordination [49]. The use of well-defined stepped surfaces is frequently used as a strategy to approach the behavior of nanoparticles [48]. Sites at steps often seem to be much more active than those at terraces. Therefore, reactivity can be investigated as a function of the density of defects by a systematic variation of the terrace width. Often, a linear trend is obtained, which can be interpreted as a combination of contributions from terraces and step sites.

The catalytic effect at step sites can be explained by local changes of the electronic properties [50]. The redistribution of electronic charges induces a dipole moment pointing outward from the surface as can be observed in the left-hand side of Figure 1.17 for Ag(119). The adsorption of hydrogen on these sites also produces a dipole moment, but it is oriented inwards towards the surface (see Figure 1.17, right).

There are two types of sites for the hydrogen adsorption: terrace and step sites. The terrace sites are the same as in the planar (100) surface: hollow and bridge sites. However, because of the presence of the step, the symmetry in the bridge and hollow sites is broken, and consequently, it gives rise to different dipole moments. We have also to distinguish between the bridge sites parallel (brp) and perpendicular (brs) to the steps. Therefore, the sites in the terrace are different, and they can be sorted according to their distances from the step. Interestingly, a continuous gradient of the dipoles caused by the adsorption of hydrogen between the sites at the terraces near the upper part of one step and the bottom part of the next step has been found. This effect is still noticeable for surfaces with wide terraces, as can be observed from Figure 1.18, left, for the Ag(1 1 17) orientation. This anisotropy indicates a long-range effect.

This effect is correlated with the adsorption energy. In this case also, a continuous energy gradient on the terraces directed from the upper to the bottom of the successive steps, even for surfaces with wide terraces like Ag(1 1 17), is obtained. The step sites show the highest activity for adsorption, even though the central sites of the terraces also show appreciable minima for the adsorption energy. At the upper part of the steps, there are always two favorable sites with low coordination (bridge and hcp), with adsorption energies of about 0.16–0.18 eV independent of the width of the terrace. However, these values are somewhat higher than those at the corresponding sites of a flat

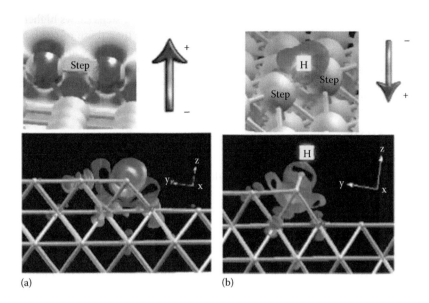

FIGURE 1.17 Electronic charge redistributions at the step. (a) The charge density difference $\Delta\rho$ for the Ag(119) vicinal surface is shown (Ag(100) surface has been taken as reference). (b) Charge density difference $\Delta\rho$ for the Ag(119) covered by hydrogen at bridge sites of the step (the bare surface has been taken as reference). Red and blue contours correspond to the accumulation and depletion of electron density, respectively—0.0015 eÅ^{-3}. (From Juarez, M.F. and Santos, E., *J. Phys. Chem. C*, 117, 4606, 2013.)

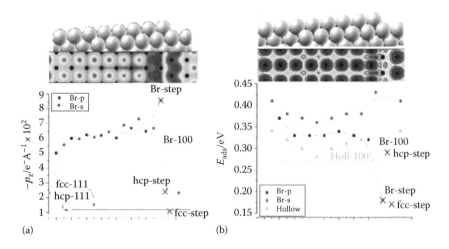

FIGURE 1.18 (a) Dipole moments for hydrogen adsorption at different sites on the Ag(1 1 17) terrace and step. The reference system is the clean surface. The brp are the bridge sites parallel to the step and brs perpendicular to it. (From Juarez, M.F. and Santos, E., *J. Phys. Chem. C*, 117, 4606, 2013.) (b) Adsorption energy for hydrogen at different sites on the Ag(1 1 17) terrace and step. The reference system is the hydrogen molecule. The full lines correspond to the adsorption on equivalent sites of the Ag(100) surface (black (Br-100), bridge site; green (Holl-100), hollow site). (From Juarez, M.F. and Santos, E., *J. Phys. Chem. C*, 117, 4606, 2013.)

(111) surface. The more coordinated fcc step site at the bottom of the steps shows higher values than the equivalent sites of a flat (111) surface and only slightly lower than those of (100) terrace sites. On the terraces, the adsorption energy at a given site depends both on the width of the terrace and on its position on the terrace, sites at the center being more favorable. This quite complicated distribution of the adsorption energies on the surfaces gives account of the nonlinear dependence of the hydrogen evolution rate on the step density [51].

1.7 CONCLUSIONS

As we stated in the introduction, nanostructures are ideal systems for theoretical investigations. They have just the right size for extensive DFT calculations and offer a large variety of geometrical or compositional structures for systematic studies. However, it is not enough to obtain numbers from DFT; in a certain sense, the energies supplied by DFT, which with few exceptions are quite reliable, are like experimental data: the scientific challenge lies in understanding them. Therefore, the most important progress during the last decade has been the development of concepts, models, and theories that explain the behavior of electrochemical systems; these include our own theory of electrocatalysis, which we presented at the beginning of this chapter. We hope that the examples that we treated here, which were taken from our own works, illustrate well the understanding that has been achieved.

The monoatomic freestanding wires, which we treated in Section 1.3, are a good example to demonstrate how greatly the catalytic properties of a metal can change at the nanoscale. Gold and copper nanowires should be excellent catalysts for hydrogen evolution, while bulk copper and gold are rather inert towards hydrogen. Although such wires can be produced experimentally, they are, unfortunately, not practical catalysts. The same systems also demonstrate well the importance of the substrate. When these wires are supported on graphite, they lose much of their activity.

The palladium clusters and islands, which we examined in Section 1.5, show another important property of a good catalyst: They have various adsorption sites for hydrogen with different adsorption energies, so that the reaction can pass through those sites whose energy is optimal in the sense of Sabatier's principle. They are therefore more active than a monolayer of Pd on Ah(111), which offers only two sites, neither of which has the optimum energy.

A theoretician can compose nanostructures as he or she likes—the only limit is the size and complexity. In comparison, the experimental design of nanostructures is much more difficult and is often limited by the lack of stability of the desired structures. In addition, it is often difficult to synthesize structures that are sufficiently small to exhibit nanosize effects, the supported nanowires being a good example. Therefore, in this field, a theory has a particularly important role: it can suggest systems that promise to be good and stable catalysts and can thus act as a guide to experiments.

ACKNOWLEDGMENTS

Financial supports by the Deutsche Forschungsgemeinschaft (Schm 344/34-1,2 and FOR 1376) and by an exchange agreement between the DFG and CONICET are gratefully acknowledged. E. Santos., P.Q., and W.S. thank CONICET for the continued support. E. Santos acknowledges PIP-CONICET 112-2010001-00411 for support. A generous grant of computing time from the Baden-Wurttemberg grid is gratefully acknowledged.

REFERENCES

1. Marcus, R. A. 1956. On the theory of oxidation-reduction reactions involving electron transfer. I. *J. Chem. Phys.* 24: 966–978.
2. Hush, N. S. 1958. Adiabatic rate processes at electrodes. I. Energy-charge relationships. *J. Chem. Phys.* 28: 962–971.

3. Levich, V. G. 1970. *Kinetics of Reactions with Charge Transfer*. New York: Academic Press.
4. Kuznetsov, A. M. 1995. *Charge Transfer in Physics, Chemistry and Biology*. Reading, United Kingdom: Gordon & Breach.
5. Schmickler, W. 1976. A dipole model for the outer solvation sphere and its application to outer sphere electron transfer reactions. *Ber. Bunsenges. Phys. Chem.* 80: 834–838.
6. Schmickler, W. 1986. A theory of adiabatic electron–transfer reactions. *J. Electroanal. Chem.* 204: 31–43.
7. Santos, E. and Schmickler, W. 2007. Fundamental aspects of electrocatalysis. *Chem. Phys.* 332: 39–47.
8. Anderson, P. W. 1961. Localized magnetic states in metals. *Phys. Rev.* 124: 41–53.
9. Newns, D. M. 1969. Self-consistent model of hydrogen chemisorption. *Phys. Rev.* 178: 1123–1135.
10. Santos, E., Quaino, P., and Schmickler, W. 2012. Theory of electrocatalysis: Hydrogen evolution and more. *Phys. Chem. Chem. Phys.* 14: 11224–11233.
11. Santos, E. and Schmickler, W. 2007. Electrocatalysis of hydrogen oxidation-theoretical foundations. *Angew. Chem. Int. Ed.* 46: 8262–8265.
12. Santos, E. and Schmickler, W. 2008. Electronic interactions decreasing the activation barrier for the hydrogen electro-oxidation reaction. *Electrochim. Acta* 53: 6149–6156.
13. Santos, E. and Schmickler, W. 2006. d-Band catalysis in electrochemistry. *Chem. Phys. Chem.* 7: 2282–2285.
14. Schmickler, W. and Santos, E. 2010. *Interfacial Electrochemistry*. Berlin, Germany: Springer Verlag.
15. Santos, E., Lundin, A., Quaino, P., and Schmickler, W. 2009. Model for the electrocatalysis of hydrogen evolution. *Phys. Rev. B* 79: 235436:1–235436:10.
16. Sabatier, F. 1920. *La catalyse en Chimie Organique*. Paris, France: Berauge.
17. Nørskov, J. K., Bligaard, T., Logadottir, A., Kitchin, J. R., Chen, J. G., Pandelov, S., and Stimming, U. 2005. Trends in the exchange current for hydrogen evolution. *J. Electrochem. Soc.* 152: J23–J26.
18. Atkins, P. W. 1998. *Physical Chemistry*. New York: Oxford University Press.
19. Shi, P. and Bohn, P. 2008. Stable atom–scale junctions on silicon fabricated by kinetically controlled electrochemical deposition and dissolution. *ACS Nano* 2: 1581–1588.
20. Shi, P. and Bohn, P. W. 2010. Electrochemical control of stability and restructuring dynamics in Au-Ag-Au and Au-Cu-Au bimetallic atom–scale junctions. *ACS Nano* 4: 2946–2954.
21. Mszros, G., Kronholz, S., Karthuser, S., Mayer, D., and Wandlowski, T. 2001. Electrochemical fabrication and characterization of nanocontacts and nm-sized gaps. *Appl. Phys. A* 87: 569–575.
22. Li, J., Yamada, Y., Murakoshi, K., and Nakato, Y. 2001. Sustainable metal nano-contacts showing quantized conductance prepared at a gap of thin metal wires in solution. *Chem. Commun.* 21: 2170–2171.
23. Jingze, L., Taisuke, K., Kei, M., and Yoshihiro, N. 2002. Metal-dependent conductance quantization of nanocontacts in solution. *Appl. Phys. Lett.* 81: 123–125.
24. Konishi, T., Kiguchi, M., and Murakoshi, K. 2007. Quantized conductance behavior of Pt metal nanoconstrictions under electrochemical potential control. *Surf. Sci.* 601: 4122–4126.
25. Kiguchi, M. and Murakoshi, K. 2006. Fabrication of stable Pd nanowire assisted by hydrogen in solution. *Appl. Phys. Lett.* 88: 253112:1–253112:3.
26. Li, J., Nakato, Y., and Murakoshi, K. 2005. Electrochemical fabrication of Pd-Au heterogeneous nanocontact showing stable conductance quantization under applying high bias voltage. *Chem. Lett.* 34: 374–375.
27. Soldano, G., Santos, E., and Schmickler, W. 2011. Intrinsic stability and hydrogen affinity of pure and bimetallic nanowires. *J. Chem. Phys.* 134: 174106:1–174106:8.
28. Santos, E., Quaino, P., Soldano, G., and Schmickler, W. 2009. Electrochemical reactivity and fractional conductance of nanowires. *Electrochem. Commun.* 11: 1764–1767.
29. Li, C. Z., He, H. X., Bogozi, A., Bunch, J. S., and Tao, N. J. 2000. Molecular detection based on conductance quantization of nanowires. *Appl. Phys. Lett.* 76: 1333–1335.
30. Kiguchi, M., Konishi, T., Hasegawa, K., Shidira, S., and Murakoshi, K. 2008. Three reversible states controlled on a gold monoatomic contact by the electrochemical potential. *Phys. Rev. B* 77: 245421:1–245421:7.
31. Kiguchi, M., Konishi, T., Miura, S., and Murakoshi, K. 2007. The effect of hydrogen evolution reaction on conductance quantization of Au, Ag, Cu nanocontacts. *Nanotechnology* 18: 424011:1–424011:5.
32. Csonka, S., Halbritter, A., Mihalny, G., Jurdik, E., Shklyarevskii, O., Speller, S., and van Kempen, H. 2003. Fractional conductance in hydrogen-embedded gold nanowires. *Phys. Rev. Lett.* 90: 116803:1–116803:4.
33. Csonka, S., Halbritter, A., and Mihalny, G. 2006. Pulling gold nanowires with a hydrogen clamp: Strong interactions of hydrogen molecules with gold nanojunctions. *Phys. Rev. B* 73: 075405:1–075405:6.

34. Thygesen, K. S., Hansen, L. B., and Jacobsen, K. W. 2005. Partly occupied Wannier functions. *Phys. Rev. Lett.* 94: 026405:1–026405:4.

35. Hammer, B. and Nørskov, J. K. 1995. Electronic factors determining the reactivity of metal surfaces. *Surf. Sci.* 343: 211–220.

36. Hammer, B. and Nørskov, J. K. 1995. Why gold is the noblest of all the metals. *Nature* 376: 238–240.

37. Santos, E., Lundin, A., Pötting, K., Quaino, P., and Schmickler, W. 2009. Hydrogen evolution and oxidation—A prototype for an electrocatalytic reaction. *J. Solid State Electrochem.* 13: 1101–1109.

38. Valden, M., Lai, X., and Goodman, D. W. 1998. Onset of catalytic activity of gold clusters on titania with the appearance of nonmetallic properties. *Science* 281: 1647–1650.

39. Zhai, H. J., Kiran, B., and Wang, L. S. 2004. Observation of Au_2 H impurity in pure gold clusters and implications for the anomalous Au-Au distances in gold nanowires. *J. Chem. Phys.* 121: 8231–8236.

40. Bahn, S. R., Lopez, N., Nørskov, J. K., and Jacobsen, K. W. 2002. Adsorption-induced restructuring of gold nanochains. *Phys. Rev. B* 66: 081405:1–081405:4.

41. Barnett, R. N., Hakkinen, H., Scherbakov, A. G., and Landman, U. 2004. Hydrogen welding and hydrogen switches in a monatomic gold nanowire. *Nano Lett.* 4: 1845–1852.

42. Corma, A., Boronat, M., Gonzalez, S., and Illas, F. 2007. On the activation of molecular hydrogen by gold: A theoretical approximation to the nature of potential active sites. *Chem. Commun.* 4: 3371–3373.

43. Walter, E. C., Murray, B. J., Favier, F., Kaltenpoth, G., Grunze, M., and Penner, R. M. 2002. Noble and coinage metal nanowires by electrochemical step edge decoration. *J. Phys. Chem. B* 106: 11407–11411.

44. Cross, C. E., Hemminger, J. C., and Penner, R. M. 2007. Physical vapor deposition of one–dimensional nanoparticle arrays on graphite: Seeding the electrodeposition of gold nanowires. *Langmuir* 23: 10372–10379.

45. Quaino, P. M., de Chialvo, M. R. G., Vela, M. E., and Salvarezza, R. C. 2005. Self-assembly of platinum nanowires on HOPG. *J. Argent. Chem. Soc.* 93 (4/6): 215–224.

46. Soldano, G. J., Quaino, P., Santos, E., and Schmickler, W. 2013. Stability and hydrogen affinity of graphite-supported wires of Cu, Ag, Au, Ni, Pd, and Pt. *J. Phys. Chem. C* 117: 19239–19244.

47. Quaino, P., Santos, E., Wolfschmidt, H., Montero, M., and Stimming, U. 2011. Theory meets experiment: Electrocatalysis of hydrogen oxidation/evolution at Pd-Au nanostructures. *Catal. Today* 177: 55–63.

48. Feliu, J. M., Herrero, E., and Climent, V. 2011. In *Catalysis in Electrochemistry: From Fundamental to Strategies for Fuel Cell Development*, Santos, E. and Schmickler, W. (Eds.), pp. 127–164. Hoboken, NJ: Wiley.

49. Giesen, M. and Beltramo, G. 2011. In *Catalysis in Electrochemistry: From Fundamental to Strategies for Fuel Cell Development*, Santos, E. and Schmickler, W. (Eds.), pp. 67–126. Hoboken, NJ: Wiley.

50. Juarez, M. F. and Santos, E. 2013. Electronic anisotropy at vicinal Ag(11n) surfaces: Work function changes induced by steps and hydrogen adsorption. *J. Phys. Chem. C* 117: 4606–4618.

51. Ruderman, A., Juarez, M. F., Avalle, L. B., Beltramo, G., Giesen, M., and Santos, E. 2013. First insights of the electrocatalytical properties of stepped silver electrodes for the hydrogen evolution reaction. *Electrochem. Commun.* 34: 235–238.

2 Electrical Double-Layer Effects on Electron Transfer and Ion Transport at the Nanoscale

Wen-Jie Lan, Henry S. White, and Shengli Chen

CONTENTS

2.1 INTRODUCTION

Upon bringing an electrode into contact with an electrolyte solution, electron transfer (ET) occurs because of the mismatch of the electronic energy level in the electrode and that in solution, which are controlled and measured, respectively, by the electrode potential (E) and the redox potential of the electroactive species in electrolyte solution.[1,2] As a result, charges are produced at electrode surfaces, and in the meantime, electrolyte ions are accumulated and/or depleted near electrodes due to electrostatic interaction, thus forming electrical double layers (EDLs) (Figure 2.1), which

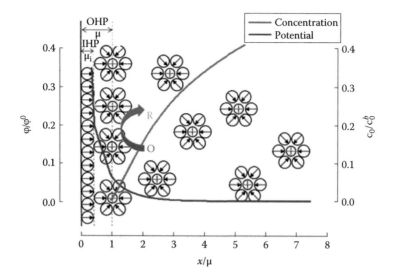

FIGURE 2.1 A schematic illustration of the electrode/electrolyte interface. The blue and red solid lines illustrate the distributions of the electrostatic potential and reactant concentration, respectively.

are *nanoscale* in nature because they are defined by high electric fields screened by mobile charges (ions in solution and electrons in the electrode) and solvent dipoles over distances on the order of 0.1–100 nm in common situations.[1]

If there is no externally imposed electron injection/extraction into/from electrode, ET process will finally cease as E aligns with the redox potential of the electroactive species. The electrode/electrolyte interface thus bears an equilibrium EDL, which at a primitive level can be described by the Gouy–Chapman–Stern–Bockris (GCSB) model.[3] In this primitive level model, the EDL at the electrode/electrolyte interface can be roughly divided into a compact and a diffuse part, with the boundary at the so-called outer Helmholtz plane (OHP), which is considered to the first approximation the plane of closet approach (PCA) of the solvated ions to the solvated electrode surface (Figure 2.1). When no contact adsorption of ions at the electrode surface occurs, the electrostatic potential (φ) should distribute linearly inside the OHP, with the value at the electrode (φ_0) and OHP (φ_1) depending on E, the electronic structure of electrode and the dielectric property of solvent at the interface.[1-3] The charge and electrostatic potential distribution in the diffuse part of EDL can be treated with the Poisson–Boltzmann (PB) theory, as formulated by Equation 2.1, in which c_j and c_j^* are the local and bulk concentration of ion j, respectively, z_j is its charge, ε and ε_0 are the local dielectric constant and the permittivity of vacuum:

$$\nabla^2 \varphi = -\frac{1}{\varepsilon\varepsilon_0} \sum_j z_j c_j = -\frac{1}{\varepsilon\varepsilon_0} \sum_j z_j c_j^* e^{-(z_j F/RT)\varphi} \qquad (2.1)$$

The GCSB models have predicted a variety of interfacial properties, for example, capacitive behavior, charge and potential distributions, and potential dependence of surface tension (the so-called electrocapillary curves), which have been experimentally tested by a variety of electrochemical and physical methods with varying levels of success.[1,3] For instance, much has been learned over the past 70 years about ion adsorption and solvent orientation at Hg and well-defined solid metal electrodes from capacitance measurements.[1,4,5] Similarly, studies in recent decades using *in situ* scanned probe microscopy[6] and surface force microbalance method[7] have been used to map the electrical forces (and thus electric field) extending from electrode surfaces.

In the case when there is an externally imposed electron injection/extraction into/from the electrode, ET processes between the electrode and electroactive species in solution can proceed continuously, which would result in a concentration distribution layer (CDL) of electroactive species at the interface due to their finite mass transport (MT) rates. If assuming that the solution species approximately have the same PCA, one can imagine that the CDL and diffuse EDL would start similarly at the OHP and would merge into each other at the electrode interface when an ET process proceeds continuously (Figure 2.1). To this end, the diffuse EDL will be dynamic in nature, and interfacial MT and ET processes will be impacted by the high electrostatic gradient (on the order of 10^9 V/m) in the EDL.

However, electrochemists rarely report an explicit dependence, much less a quantitative description, of how interfacial fields influence electrochemical reaction rates and currents. This is partly due to the complexity of nonequilibrium models of EDLs that are required when electrostatic forces are coupled to molecular and ion transport. In addition, other physical and chemical phenomena often are rate limiting and very effectively mask the effect of the EDL. At conventional large electrodes, for instance, the CDLs are typically orders of magnitude larger than the width of the EDLs.[1,8] In this case, the electrochemical reactions are limited by transport of the molecule from the bulk solution to the electrode surface, dominating the overall *resistance* of the electrochemical reaction and preventing measurements of the effects of short-range interfacial fields on the electrode reaction. This transport limitation generally occurs in the current–voltage (*i–V*) response of soluble, *kinetically fast* outer-sphere redox couples whose surface concentrations are thermodynamically determined by the applied electrode potential. Similarly, slower inner-sphere reactions involving adsorbed molecules are limited by the kinetics of bond formation and breaking that are poorly understood—it is likely that electric fields have a pronounced effect on the rates of these reactions but are quite difficult to measure due to the complexity of the reaction mechanisms involving adsorption and bond formation.[9]

There are a few situations when the local electric field does manifest itself on the measured *i–V* response and can be quantitatively measured and modeled. One case involves reactions of soluble (nonadsorbing) redox species with very slow ET kinetics, such that the reaction is no longer limited by diffusion of the reactant over large distances in the solution but rather by the kinetic rate of the ET at the interface.[1] In this case, the reaction rate v mol/(cm^2·s) is the product of the ET rate constant (k_{et}, cm/s) and the surface concentrations of the redox species (c^s, mol/cm^3), that is, $v = k_{et}c^s$. Since either or both species of the redox couple $O^z/R^{(z-1)}$ are electrically charged, their concentrations at the interface may be greatly enhanced/suppressed by attraction/repulsion to the EDL fields. This phenomenon is generally referred to as a *Frumkin effect*,[1,8,10] and double-layer correction factors based on equilibrium models (e.g., GCS model) for extracting ET rates from the *i–V* response are available.

A very pronounced double-layer effect is also observed in the voltammetry of adsorbed outer-sphere reactants.[11,12] For instance, the voltammetry of self-assembled monolayers (SAMs) of alkanethiols containing a terminal redox group (e.g., ferrocene) is strongly influenced by the electric field across the SAM, and this is manifested in peak broadening and a shift in the half-wave potential. Because redox-active SAMs are frequently geometrically very well defined, and their dielectric properties can be measured, the electric field across these layers can be readily computed from electrostatics to obtain the electric potential at the redox center. This in turn can be used to compute the influence of the electric potential distribution on voltammetric response, which can then be quantitatively compared to experiment.

As noted earlier, EDL phenomena are frequently masked by slow transport of the redox molecule in solution. Beginning in the 1980s, applications and theoretical treatments of ultramicroelectrode voltammetry were developed, leading to an understanding that very large and steady-state molecular fluxes to the electrode surface could be realized when the electrode dimension was reduced below ~10 μm.[13–16] Such large fluxes are predicted based on Fick's laws and result from the steady-state *radially convergent* flux of molecules to a small disk- or band-shaped electrode. The radially convergent flux of molecules also produces a narrow diffusion layer, δ, that defines the transport resistance and that scales approximately with the smallest dimension (*a*) of the electrode (δ ~ 5*a*). For instance,

$\delta \sim 5$ µm for a 1 µm radius Pt disk, a value independent of the properties of the electrolyte concentration. With the onset of methods to prepare electrodes with hemispherical or quasi disk shapes with a as small as a few nanometers, it became clear that one could reduce diffusion length scales to the dimensions of the EDL and partially remove the masking effect of slow diffusion. As an example, the diffusion length for the oxidation of a positively charged molecule at a 5 nm disk electrode is $\sim 5a = 25$ nm. If this reaction is performed in a 0.01 M KCl solution at room temperature, the Debye screening length, κ^{-1} is on the order of ~ 3 nm.[1,4] Assuming the electric field extends approximately to $5\kappa^{-1}$, the thickness of the EDL is roughly 60% that of the CDL. Thus, the electric field originating from the electrical charge on the metal surface should influence the transport of the redox ions. Qualitatively, the steady-state flux of a cation to a positively charged electrode should be suppressed relative to the diffusion-limited flux, while the flux of an anion at the same electrode potential should be enhanced. In essence, by scaling the size of the electrode, a, down to dimensions on the order of the Debye length, κ^{-1}, the scale of the CDL is reduced to a value that is comparable to that of the EDL.

The following sections provide a quantitative theoretical description of electric field effects on the interfacial MT and ET processes at electrodes of nanometer length scale as well as preliminary experimental validation of the model. The first problem we consider is that of individual electrodes of nanoscale exposed electroactive areas shrouded by an infinitely wide insulating sheath (Figure 2.2a). The influence of the EDL on the voltammetric responses of these electrodes is presented and discussed. We also discuss EDL effects at two related electrochemical systems: a metal electrode recessed at the bottom of a conical nanopore (Figure 2.2b)[17–20] and the nanopore membrane (Figure 2.2c).[21] The recessed nanopore electrode and the nanopore membrane are fabricated by etching the surface of a nanodisk electrode to different depths in the insulating shrouding material,[22] and thus the structures are naturally related to each other in their preparation. More importantly, for reasons analogous to EDL phenomena at nanodisk electrodes, the electric fields associated with the electrically charged surface of the nanopore can greatly influence the rate of ion transport through the nanopore

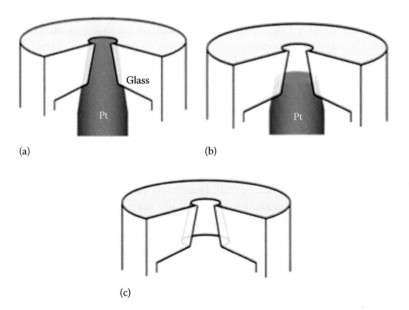

(a)   (b)

(c)

FIGURE 2.2 Schematic illustrations of a nanodisk electrode, a recessed nanopore electrode, and a nanopore membrane. (a) Nanodisk electrode, (b) recessed nanopore electrode, and (c) nanopore membrane. (Reprinted with permission from Zhang, B., Galusha, J., Shiozawa, P.G., Wang, G., Bergren, A.J., Jones, R.M., White, R.J., Ervin, E.N., Cauley, C.C., and White, H.S., A bench-top method of fabricating glass-sealed nanodisk electrodes, glass nanopore electrodes, and glass nanopore membranes of controlled size, *Anal. Chem.*, 2007, 79, 4778–4787. Copyright 2007 American Chemical Society.)

orifice if the orifice dimension, a, is on the order of the Debye screening length. Thus, the same general criterion for observing the influence of EDL on ion transport, $a/\kappa^{-1} \leq 1$, applies in all three cases.

Experimental studies describing the preparation and application of nanoelectrodes, nanopore electrodes, and conical nanopores have been extensively reported over the past few decades, including several reviews. The focus of this chapter is on providing a unified treatment of EDL effects of the behavior of these structures.

2.2 MASS TRANSPORT AND ELECTRON TRANSFER (ET) PROCESSES AT NANOSCALE ELECTRODES

Nanoscale electrodes possess exposed electroactive sizes of nanoscale at least in one dimension.[23,24] The past three decades have seen tremendous growth of electrochemical studies using these electrodes.[23–39] Due to the greatly enhanced radial converging effect, they offer extremely high MT rates of electroactive species by shrinking the CDL into nanometer scales and therefore have become one of the major current approaches to investigate the kinetics of the fast heterogeneous ET reactions that are inaccessible to conventional large electrode systems.[28–31,38,40] Their extremely small electroactive sizes also allow studies of the electrochemistry of single molecules, nanoparticles, biological cells, and enzymatic catalysts.[41] In addition, these electrodes also provide model systems to understand the nanoscopic features in EDL structures and electrochemical processes such as MT and ET at electrode/electrolyte interfaces.[8,41–49]

2.2.1 COUPLING BETWEEN THE ELECTROSTATIC POTENTIAL AND CONCENTRATION DISTRIBUTIONS AT NANOSCALE ELECTRODE/ELECTROLYTE INTERFACES AND THE DYNAMIC ELECTRICAL DOUBLE LAYER (EDL) NATURE OF INTERFACES

As noted earlier, the strong radial convergence of species flux at molecules at nanoscale electrodes produces a very narrow CDL that is comparable in thickness with the diffuse EDL. This means that the diffuse EDL overlaps with a significant part of the CDL. An immediate result of this is the non-deconvolvability of the CDL and EDL at the electrode/electrolyte interface, which makes the conventional electrochemical treatments of MT and ET kinetics problematic.[8,41–49] In most of the current voltammetric theories, the EDL and CDL are physically and therefore mathematically separated, so that the electrostatic potential and concentration distributions can be treated separately at an electrode/electrolyte interface. One thus can assume electroneutrality in the CDL and an equilibrium distribution of electrostatic potential in the diffuse EDL. Accordingly, the transport of electroactive species at the interface is treated either by using Fick's diffuse equations in the presence of excess supporting electrolyte or by using the mixed diffuse/electromigration equation (e.g., the Nernst–Planck equation) in combination with the electroneutrality equation ($\Sigma c_i z_i = 0$) when the solution is weakly supported.[1] In both cases, the EDL structures are totally ignored. In the case when the influence of the EDL on electrode reaction kinetics should be considered, Frumkin correction is employed, in which the electrostatic potential and concentration of electroactive ions obtained from the equilibrium PB theory are used in the ET rate expression, despite the occurrence of ET at the interface.[1,10] To account for the coupling between the EDL and MT effects, the Boltzmann expression for the ion concentration is given as $c_j^s \exp(-z_j RT\varphi_1/F)$, where c_j^s refers to the concentration of species j at the inner boundary of the CDL.[1]

The overlap between CDL and EDL and its effects on the voltammetric responses of nanoscale electrodes were first theoretically treated by White and coworkers[4,44] in the early 1990s using the Poisson–Nernst–Planck equations. The possible failure of the electroneutrality condition at the nanoscale electrode interface has also been studied by Oldham and Bond,[47] Dickinson and Compton,[48] and other researchers.[49] Most of these studies, however, may have overestimated the EDL effect due to the inappropriate treatment of the compact part of the EDL, especially the dielectric screening of solvent in EDL (will be discussed later on). To rationally treat the EDL effect at nanoscopic electrochemical interfaces, He et al.[42] have adopted the concept of dynamic EDL introduced by Levich in the 1940s.[50] The core idea is that the electrochemical interface at a nanoscale electrode as a whole should be considered a dynamic EDL in the course of an ET reaction, rather than a combination of separable EDL and CDL as considered in conventional voltammetric treatment. This treatment emphasizes not only the influence of the electrostatic potential on the concentration distribution of electroactive species but also the alteration of the diffuse EDL structure due to the high transport rates of electroactive ions.[43] The latter makes the EDL dynamic in nature.

The dynamic EDL can be formulated with two closely coupled equations (e.g., given by Equations 2.2 and 2.3 by taking a spherical electrode), which describe, respectively, the interfacial structure (relation between the electrostatic potential and concentration distributions) and the voltammetric behavior (current density–electrode potential dependence):

$$\nabla^2\varphi = -\frac{1}{\varepsilon\varepsilon_0}\left(\sum c_j^* e^{-\frac{z_jF}{RT}\varphi} - i^* c_0^* \sum m_j z_j e^{-\frac{z_jF}{RT}\varphi} \int_r^\infty \frac{r_0+\mu}{r^2} e^{-\frac{z_jF}{RT}\varphi} dr\right) \tag{2.2}$$

$$i^* = \left(\gamma^{-1} e^{\frac{\alpha F(E-E^{0'})}{RT}} e^{(z-\alpha)\frac{F\varphi_1}{RT}} + \int_{r_0+\mu}^\infty \frac{r_0+\mu}{r^2} e^{-\frac{z_0F}{RT}\varphi} dr + e^{\frac{F(E-E^{0'})}{RT}} \int_{r_0+\mu}^\infty \frac{r_0+\mu}{r^2} e^{-\frac{z_RF}{RT}\varphi} dr\right)^{-1} \tag{2.3}$$

In above two equations, m_j is an integer constant which takes values of −1, +1, and 0 for species $R^{(z-1)}$, O^z, and inert electrolyte species respectively if the reduction current is considered positive; μ refers to the thickness of the compact EDL; r_0 refers to the radius of electrode; γ is the ratio between the standard rate constant of ET reaction and the mass transport coefficient of the electroactive species. It can be seen that the current density, which is given in a dimensionless form through normalization with the limiting diffusion current density (i_{dL}), and the electrostatic potential distribution appear simultaneously in the two equations. Equation 2.2 could be approximated to the PB equation at low current density, while Equation 2.3 would reduce to Eq. 2.4, which is the diffusion-corrected Butler-Volmer equation and has been used to perform voltammetric analysis in conventional electrochemistry[1], as $\exp(-z_j RT\varphi/F) = 1$, that is, electrostatic potentials in CDL are close to zero. These conditions are approximately satisfied in large electrode systems, suggesting that the voltammetric behaviour and the EDL structure can be treated separately at large electrode interface:

$$i^* = \frac{1}{\gamma^{-1} e^{\alpha F(E-E^{0'})} + 1 + e^{F(E-E^{0'})/RT}} \tag{2.4}$$

Figure 2.3 compares the profiles of the electrostatic potential and the concentration of redox molecules at interfaces of spherical electrodes with radii of 1 and 100 nm, respectively, obtained by solving Equation 2.2 and 2.3 for one-electron reduction of a −1 valence reactant at an electrode potential of ca. −0.5 V with respect to the potential of zero charge (PZC). Unless specifically stated, we will assume that the formal potential of the considered redox reaction (E^0) is equal to the PZC for the sake of simplicity. The horizontal ordinate represents the distance from the OHP normalized by δ, the distance at which the concentration of reactant reaches 95% of its bulk value. We may consider δ the thickness of the entire interfacial region. One can see that the potential profile (EDL) is nearly squeezed to a vertical line at OHP at an electrode of 100 nm in radius as compared with the concentration profile (CDL), whereas significant penetration of the potential profile into the concentration profile occurs at electrode of 1 nm in radius, which indicates the transition of the interface from a CDL dominated nature to EDL nature as electrode size approaches nanometer scales.

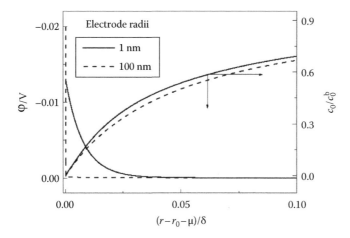

FIGURE 2.3 Calculated electrostatic potential and reactant concentration profiles during the one-electron reduction of 5 mM O^z ($z = -1$) in the presence of 0.5 M 1:1 supporting electrolyte at -0.5 V at spherical electrodes with radii of 1 and 100 nm (r_0), respectively. Other parameters include the following: $k^0 = 1.0$ cm/s, $D = 1 \times 10^{-5}$ cm²/s. The 3-state water dielectric model introduced by Bockris [3] is used in the calculations (see the next section for details). (Reprinted with permission from He, R., Chen, S.L., Yang, F., and Wu, B.L., Dynamic diffuse double-layer model for the electrochemistry of nanometer-sized electrodes, *J. Phys. Chem. B*, 2006, 110, 3262–3270. Copyright 2006 American Chemical Society.)

2.2.2 COUPLING BETWEEN THE ELECTROSTATIC POTENTIAL AND SOLVENT DIELECTRIC DISTRIBUTION AT NANOSCALE ELECTRODE/ELECTROLYTE INTERFACES

As well as the gradient distributions of electrostatic potential and the concentrations of redox and electrolyte species, electrode/electrolyte interfaces are also featured as a gradient distribution of dielectric property of the solvent molecules due to the high electric field within EDL, which, as pointed out earlier by Bockris and Reddy,[3] could significantly impact the EDL structure. The solvent dipole in EDL (Figure 2.1) would be oriented in electric field, with the degree of orientation depending on the electric field strength. The orientation degree of solvent molecule determines its dielectric constant, which determines its screening to the electric field. This coupling between the dielectric property of solvent and the electric field has been seldom considered in classic voltammetric analysis since that in most cases the influence of the EDL can be ignored. In the cases when EDL may have significant impacts on the MT and ET kinetics, for example, at nanoscale electrode interfaces, a relatively accurate description of EDL structure would be crucial for voltammetric analysis. Therefore, it is essential to have a realistic model for the dielectric property of solvent molecules at the interface.

Bockris has introduced a 3-state model for the dielectric distribution of water at the electrode/electrolyte interface.[3] In this simple model, the first layer of water dipoles near the electrode surface (within inner Helmholtz plane [IHP], Figure 2.1) is assumed to be completely oriented due to the strong interactions with the electrode and the high electric field strength, thus forming a *saturated dielectric*. The dielectric constant of this oriented water layer was estimated to be about six. In the region from IHP to OHP, the water molecules are considered partly oriented due to the balanced electrostatic interaction and thermal and hydrogen-bonding interaction. The dielectric constant in this region can be approximately considered a linear change from 6 to that in bulk water (78) or simply an averaged value between 6 and 78 (e.g., 40). Outside OHP, the dielectric constant of water is assumed to be close to 78.

Booth[51] has derived a more delicate relation between the dielectric constant of water and electric field strength ($\bar{E}$), which is formulated by

$$\varepsilon = n_d^2 + \frac{7N_0\mu_v(n_D^2+2)}{3\varepsilon_0\sqrt{73}\bar{E}} L\left[\frac{\sqrt{73}\bar{E}\mu_v(n_D^2+2)}{6k_BT}\right] \tag{2.5}$$

where

n_d and μ_v refer to the optical refractive index and dipole moment of water

k_B is the Boltzmann constant

N_0 is the number of water molecules per unit volume

$L(x)$ is the Langevin function

According to this dielectric model, the dielectric constant of water decreases continuously with increasing $\bar{E}$ in the range of 3×10^7–3×10^{10} V/m. Under lower $\bar{E}$, ε gradually converges to the value of bulk water (78), while it reaches a nearly constant value as $\bar{E}$ is larger than 3×10^{10} V/m due to the dielectric saturation. The prediction of Equation 2.5 has been well verified for a range of $\bar{E}$ by the recent results of molecular dynamics simulations carried out by Yeh and Berkowitz.[52]

Figure 2.4 displays the interfacial distributions of water dielectric constant predicted by the Booth model during the one-electron reduction of a −1 valence reactant at spherical electrodes of different sizes. At large electrodes (>20 nm), the calculated values of ε inside the OHP are close to 20 regardless of electrode sizes and change little with distance to the electrode surface, which means that the electrostatic potential in the compact EDL at large electrodes distributes nearly linearly. This has long been the general belief in electrochemistry for planar electrodes with no specific adsorption of charge species. At electrodes smaller than 10 nm in radii, ε exhibits a significant electrode-size-dependent distribution, as well as a continuous increase with distance from the electrode surface. The varied ε with distance in EDL corresponds a curved distribution of electrostatic potential in the EDL. This should be due to the comparability between the electrode curvature and the EDL thickness at nanoscale. For large spherical electrodes, the curvature of electrode surface is

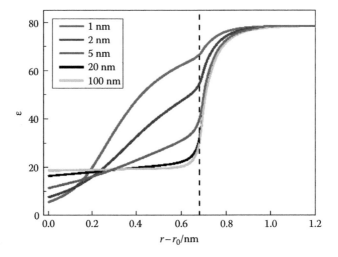

FIGURE 2.4 Dielectric distribution of water molecules at interfaces of spherical electrodes of different radii calculated by combining Equations 2.2, 2.3, and 2.5. The vertical dash line indicates the OHP. The other parameters are the same as that for Figure 2.3. (Reprinted with permission from Chen, S.L. and Liu, Y.W., *Phys. Chem. Chem. Phys.*, 2014, 16, 635–652. Copyright 2014 Royal Chemical Society.)

very small as compared with the thickness of the EDL. Therefore, the EDL property would be more like that at planar electrodes.

The effect of the dielectric properties of water on the EDL structure is seen by the electrostatic potential profiles shown in Figure 2.5, which were obtained by solving Equations 2.2 and 2.3 with a combination of different dielectric models for one-electron reduction of a −1 valence reactant at spherical electrodes of different radii. The uniform dielectric constant of 78 and the 3-state model give electrostatic potential distributions that exhibit much less electrode size dependence than that predicted by the Booth model. This is because the Booth model allows flexible variation of dielectric constant and electric field strength in the calculation, whereas in the uniform and 3-state models, the dielectric constants are fixed. The Booth model gives very similar linear potential distribution in the compact EDL at large electrodes to that predicted by the uniform dielectric model, while its prediction for very small electrodes is close to that of the 3-state model, which produces a very steep potential drop inside IHP and a somewhat dragging

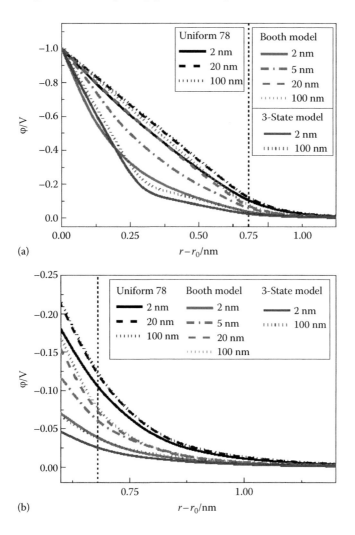

(a)

(b)

FIGURE 2.5 (a) Potential distributions at spherical electrodes of different radii calculated by using the dynamic EDL model (Equations 2.2 and 2.3) together with different dielectric models; (b) Enlarged distribution curves in the diffuse part of EDL. The vertical dash lines indicate the position of OHP. The other parameters are the same as that for Figure 2.3. (Reprinted with permission from Chen, S.L. and Liu, Y.W., *Phys. Chem. Chem. Phys.*, 16, 635, 2014. Copyright 2014 Royal Chemical Society.)

variation outside IHP. As for the electrostatic potentials in the diffuse EDL, the 3-state model predicts the lowest values in magnitude and the narrowest distribution away from OHP, while the uniform model gives the most pronounced diffuse potential distribution.

2.2.3 EDL EFFECTS ON MASS TRANSPORT RATES AT NANOSCALE ELECTRODES

Figure 2.6 compares the steady-state polarization curves for the reduction of $O^z(z=-1)$ at spherical electrodes of different radii given by the dynamic EDL model described earlier with the combination of different dielectric models. The current density is normalized by the limiting diffusion current density (i_{dL}). It is noted that when using a uniform dielectric constant of 78, the polarization curves at electrodes less than 10 nm in radii are peak shaped and no limiting current plateau occurs. A similar phenomenon was also predicted for the oxidation of cations at nanoelectrodes.[44] So far, there has been no experimental evidence to support such a prediction. Instead, sigmoid-shaped polarization curves with limiting current plateaus have been mostly observed for the oxidation of cations (e.g., TMAFc$^+$ and Ru(NH$_3$)$_6^{3+}$)[30-33] or reduction of anions (e.g., Fe(CN)$_6^{3-}$ and IrCl$_6^{2-}$)[26-29,32-34] at various nanoscale electrodes. When the Booth or 3-state dielectric models are used, the calculated polarization curves roughly have sigmoid shapes with limiting current plateaus, regardless of electrode sizes. This indicates that the ignorance of the dielectric saturation in the compact EDL would considerably overestimate the EDL effect on the MT rates of electroactive species at the nanoscale electrode interface. However, this overestimation is not seen at an electrode as large as 100 nm in radius, which suggests that the EDL effect on the MT processes is negligible at large electrodes.

With the Booth dielectric model, the limiting current density approaches to the limiting diffusion value (i_{dL} given by Equation 2.4) as electrode radius becomes larger than 20 nm, while i_{dL} is reached even at electrodes of a few nanometer in radii as the 3-state dielectric model is used. For electrode less than 10 nm in radii, the Booth model predicts limiting current density that are within 20% of magnitude less than the i_{dL} for the reduction of −1 valence anion.

In addition to the dependence on electrode size and the dielectric model, the magnitude of the EDL effect on the MT rate also depends on the charges of the electroactive species. Figure 2.7 gives the ratio of the limiting current density predicated by the dynamic EDL model (i_L) over i_{dL} for

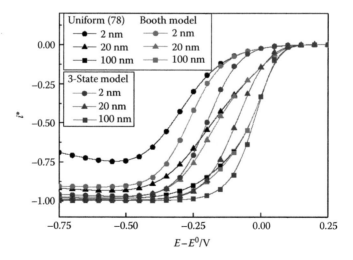

FIGURE 2.6 Calculated steady-state voltammetric curves for the reduction of 5 mM $O^z(z=-1)$ at spherical electrodes of different radii obtained by solving Equations 2.2 and 2.3 with combination of different dielectric models. The other parameters are the same as that for Figure 2.3. (Reprinted with permission from Chen, S.L. and Liu, Y.W., *Phys. Chem. Chem. Phys.*, 2014, 16, 635–652. Copyright 2014 Royal Chemical Society.)

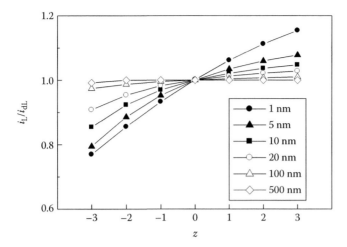

FIGURE 2.7 Ratio of i_L predicated by the dynamic EDL model in combination of the 3-state dielectric model over i_{dL} for the reduction of variously charged reactants at electrodes of different sizes. Other parameters are the same as that for Figure 2.3. (Reprinted with permission from He, R., Chen, S.L., Yang, F., and Wu, B.L., Dynamic diffuse double-layer model for the electrochemistry of nanometer-sized electrodes, *J. Phys. Chem. B*, 2006, 110, 3262–3270. Copyright 2006 American Chemical Society.)

the reduction of variously charged reactants at electrodes of different sizes. The 3-state dielectric model, which predicts the least EDL effect, is used. For −3 valence species, ca. 20% deviation of the limiting current from i_{dL} occurs at an electrode of less than 10 nm in radius. At the 100 nm and larger electrodes, however, nearly no deviation occurs regardless of the reactant charges. This means that the limiting current-based voltammetric analysis can be performed using the conventional theories at electrodes larger than 100 nm. For electrodes with a radius of ca. 20 nm, such analysis would produce inaccuracy of less than 10% magnitude.

The aforementioned prediction for the EDL effects on MT rates of electroactive ions has been tested experimentally by measuring the limiting currents for differently charged redox species at electrodes of nanometer sizes. Figure 2.8 shows the dependence of the measured limiting currents (i_L) as functions of the $Fe(CN)_6^{3-}$ concentrations (c_a) at nanodisk electrodes of different radii (r_0) fabricated by insulating electrochemically etched Pt tips with glass sheath. It can be seen that, although i_L values have a roughly linear relation on c_a at electrodes of various sizes (Figure 2.8a), the i_L/r_0 dependence on c_a for electrodes with a radius of less than 10 nm obviously deviates from that at larger electrodes. According to the diffusion-based voltammetric theory, the slope of the i_L/r_0 versus c_a linear relation is determined by the diffusion coefficient of the reactant (D) and the ET number (n) of the reaction, but independent of r_0. This seems true only for electrodes larger than 10 nm in radii (Figure 2.8b). Considering that D and n for outer-sphere ET reactions would not change with electrode sizes, the altered i_L versus c_a linearity at electrodes smaller than 10 nm should be a result of the EDL effect on the limiting MT rate of $Fe(CN)_6^{3-}$.

According to the Nernst–Planck equation, one can have the following equation for the relation between the limiting current density and the reactant concentration:

$$i_L = nFD'r_0c_a \tag{2.6}$$

where $D' = D[r_0c/c_a + (zFr_0/RT)\nabla\varphi\nabla c/c_a]$, with c referring to the concentration at OHP. Figure 2.9 gives the calculated electrostatic potential and concentration profiles under the limiting transport condition at an electrode with a radius of 5 nm for different c_a of $Fe(CN_6)^{3-}$ in 1.0 M KCl, which shows that the values of c/c_a, $\nabla c/c_a$, and $\nabla\varphi$ are approximately independent of c_a. That is, D' is independent of c_a. Thus, Equation 2.6 predicts a linear relation between i_L and c_a even in the case when

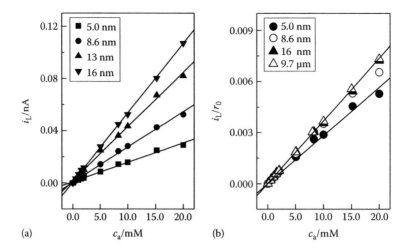

(a)   (b)

FIGURE 2.8 Dependence of (a) the limiting current and (b) the electrode radius normalized limiting current on the concentration of $Fe(CN)_6^{3-}$ at Pt disk electrodes of various sizes in 1.0 M KCl solution. (Reprinted with permission from Sun, Y., Liu, Y.W., Liang, Z.X., Xiong, L., Wang, A.L., and Chen, S.L., On the applicability of conventional voltammetric theory to nanoscale electrochemical interfaces, *J. Phys. Chem. C*, 2009, 113, 9878–9883. Copyright 2009 American Chemical Society.)

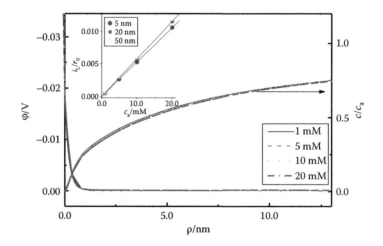

FIGURE 2.9 Electrostatic potential and concentration distributions under limiting transport condition calculated by the dynamic EDL model in combination of the 3-state dielectric model at an electrode of 5 nm in radius for the reduction of different concentrations of $Fe(CN_6)^{3-}$ in 1.0 M KCl. The *x*-axis (ρ) refers to the distance from OHP normalized by r_0 and the concentration axis is normalized by c_a. The insets are the calculated $i_L/r_0 \sim c_a$ dependence for different r_0. (Reprinted with permission from Sun, Y., Liu, Y.W., Liang, Z.X., Xiong, L., Wang, A.L., and Chen, S.L., On the applicability of conventional voltammetric theory to nanoscale electrochemical interfaces, *J. Phys. Chem. C*, 2009, 113, 9878–9883. Copyright 2009 American Chemical Society.)

the EDL effect functions at nanometer-sized electrodes. The slopes of this linearity would give an apparent diffusion coefficient, D', which changes with the electrode sizes. This prediction agrees well with the experimental results shown in Figure 2.8.

According to the $i_L \sim c_a$ data given in Figure 2.8, the apparent diffusion coefficient for $Fe(CN)_6^{3-}$ at an electrode with a radius of 5 nm would be about 80% of that at larger electrode, which meant a

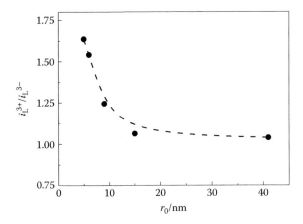

FIGURE 2.10 Ratio between the limiting currents for the reduction of $Ru(NH_6)^{3+}$ and $Fe(CN_6)^{3-}$ in 1.0 M KCl. (Reprinted partly with permission from Sun, Y., Liu, Y.W., Liang, Z.X., Xiong, L., Wang, A.L., and Chen, S.L., On the applicability of conventional voltammetric theory to nanoscale electrochemical interfaces, *J. Phys. Chem. C*, 2009, 113, 9878–9883. Copyright 2009 American Chemical Society.)

ca. 20% depression of limiting current due the EDL effect. This reasonably agrees with the prediction of the dynamic EDL model shown in Figure 2.6. Such a deviation could be easily considered a scattering of experimental data. However, the $i_L/r_0 \sim c_a$ dependence and the corresponding theoretical analysis shown earlier indicated that it is a result of the enhanced EDL effects at nanoscale electrochemical interfaces.

The EDL effect on the MT rate at nanosized electrodes is further confirmed by a comparison of the experimental limiting currents obtained with the reduction of $Ru(NH_6)^{3+}$ and $Fe(CN_6)^{3-}$, a highly charged cation and a highly charged anion, respectively. According to Equation 2.6, opposite EDL effects are expected for anion and cation reactants. The anion reduction will be inhibited while the cation reduction will be enhanced. As shown in Figure 2.10, the ratios between the limiting currents for $Ru(NH_6)^{3+}$ and $Fe(CN_6)^{3-}$ are very close to one for electrodes larger than 20 nm in radius, in accordance with the expectation of the diffusion-based MT theory, while the ratios gradually depart from one as the electrode goes below 10 nm in radius. The smaller the electrode size, the higher the ratio.

Since the EDL thickness can be increased to tens even hundreds of nanometers in dilute solutions of redox probes without supporting electrolyte, the overlap between EDL and CDL can occur at electrodes up to hundreds of nanometers, thus allowing the EDL effect on MT rates more to be readily studied. A number of works in this regard have been reported.[26,32,34–37,39] These studies have clearly showed the deviation of the limiting currents at electrodes of submicro-/nanometer sizes from that predicted by the electroneutrality-based diffusion/electromigration transport theory given in Equation 2.7, in which the sign (±) is positive for $n < z$ and negative for $n > z$. This relation has been derived by Amatore et al.[53] and has been verified experimentally on micrometer-sized electrodes with a range of redox probes:

$$\frac{i_L}{i_{dL}} = 1 \pm z \left\{ 1 + (1 + |z|) \left(1 - \frac{z}{n} \right) \ln \left[1 - \frac{1}{(1 + |z|)(1 - z/n)} \right] \right\} \qquad (2.7)$$

2.2.4 EDL EFFECTS ON HETEROGENEOUS ET KINETICS AT NANOSCALE ELECTRODES

One of the most important applications of nanometer-sized electrodes is to measure the fast heterogeneous ET kinetics.[40] To reliably extract the ET kinetic parameters from the voltammetric responses of nanoscale electrodes, it is crucial to understand various nanoscopic phenomena in ET kinetics at such small electrodes. Since the first report on the ET kinetic study at nanometer-sized electrodes by Penner et al.,[28] there has been a long dispute on whether the ET kinetics at nanoelectrodes is significantly altered by the EDL effect or not. As noted earlier, the EDL effects at nanoscale electrodes are affected by the dielectric properties of the solvent molecule in the compact EDL. As seen in Figure 2.6, the influence of the dielectric models is more pronounced in the potential region where the ET kinetics govern the voltammetric responses. Even at an electrode as large as 100 nm in radius, the polarization curves obtained by using the 3-state and Booth dielectric models differ significantly in the kinetic region, although they predict very similar limiting current. Therefore, a reasonable dielectric model is necessary for addressing the EDL effects on heterogeneous ET kinetics.

Although the Booth model is more delicate in correlating the dielectric constant and the electric field strength, it is somewhat a continuum model without considering the finite sizes of solvent molecules and neglecting the short-range interaction of solvent dipoles with electrode and solute ions, which would become crucially important at the electrode/electrolyte interface. The 3-state model, despite its simplification, seems having captured these features. As shown in Figure 2.11, the steady-state polarization curve given by the dynamic EDL model together with the 3-state dielectric model is nearly identical to that predicted by the conventional diffusion-based voltammetric theory (Equation 2.4) at an electrode of 100 nm in radius, at which the EDL effect should be negligible as suggested in Figure 2.2. This suggests that the 3-state model reasonably accounts for the dielectric distribution of water dipole at the electrode/electrolyte interface.

The half-wave potential ($E_{1/2}$), at which the current density reaches half of the limiting value on the steady-state polarization curve, can be used to estimate the value of k^0 of an ET reaction. For reversible electrode processes, $E_{1/2}$ is nearly identical to E^0, while it varies with MT rates (electrode sizes) in quasi-reversible and irreversible cases. According to Equation 2.4, one can have the following expression for $E_{1/2}$ by assuming that the transfer coefficient $\alpha = 0.5$:

$$E_{1/2} = E^{0'} + 2\frac{RT}{F}\ln\left(\frac{\sqrt{1+4\gamma^2}-1}{2\gamma}\right) \qquad (2.8)$$

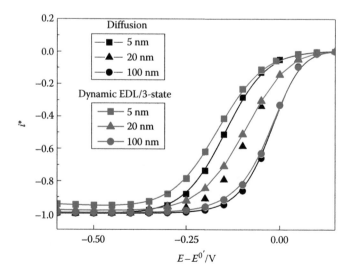

FIGURE 2.11 Comparison of steady-state polarization curves calculated using the dynamic EDL model under 3-state dielectric distribution with that predicted by Equation 2.4 for spherical electrodes of different sizes. The other parameters are the same as that for Figure 2.3. (Reprinted with permission from Chen, S.L. and Liu, Y.W., *Phys. Chem. Chem. Phys.*, 2014, 16, 635–652. Copyright 2014 Royal Chemical Society.)

TABLE 2.1

Values of $E_{1/2}$ (vs. E^0) Predicted by the Dynamic EDL Model for ET Reaction of 1 cm/s k^0 with Reactants of Different Charges at an Electrode of Different Radii and k^{0a} Estimated Using These $E_{1/2}$ Values According to Equation 2.8

	$z = +3$		$z = -1$		$z = -2$		$z = -3$		$z = 0$
r_0/nm	$E_{1/2}$/mV	k^{0a}/cm/s	$E_{1/2}$/mV	k^{0a}/cm/s	$E_{1/2}$/mV	k^{0a}/cm/s	$E_{1/2}$/mV	k^{0a}/cm/s	$E_{1/2}$/mV
5	−132	1.36	−160	0.79	−168	0.67	−176	0.57	−147
10	−102	1.32	−126	0.80	−134	0.69	−142	0.59	−115
20	−73	1.24	−91	0.84	−97	0.75	−103	0.66	−83
100	−23	1.06	−26	0.95	−26	0.92	27	0.90	−25

Source: Chen, S.L. and Liu, Y.W., *Phys. Chem. Chem. Phys.*, 16, 635, 2014. With permission. Copyright 2014 Royal Chemical Society.

in which $\gamma = k^0 r_0/D$, which can be a measure of the reversibility of an electrode process. Usually, γ value of less than 1 would produce quasi-reversible electrode processes, while irreversible electrode processes may be achieved as γ is less than 0.1.[54–56] For irreversible electrode processes, Equation 2.8 can be simplified into[55,56]

$$E_{1/2} = E^{0'} + 2\frac{RT}{F}\ln\gamma \qquad (2.9)$$

To extract the heterogeneous ET kinetic parameters using steady-state voltammetric responses, it is essential that the associated electrode process is irreversible or quasi-reversible. For ET reactions with k^0 larger than 1 cm/s, electrodes smaller than 100 nm in radii should be used for kinetic analysis. As seen from Figure 2.11, the steady-state voltammetric responses of electrodes less than 100 nm in radii may deviate from the prediction of the conventional voltammetric theory in the kinetic region. The magnitude of the deviation depends on the electrode sizes and the reactant charges. Table 2.1 summarizes the $E_{1/2}$ values on polarization curves calculated by combining the dynamic EDL model with the 3-state dielectric model for the reductive ET reactions with k^0 of 1 cm/s and reactants of different charges (z) at spherical electrodes of different radii. The values of apparent rate constants (k^{0a}) estimated using these $E_{1/2}$ values according to Equation 2.8 are also given.

The $E_{1/2}$ values obtained for $z = 0$ are identical to that predicted by Equation 2.8 with 1 cm/s k^0. For reactants with nonzero charges, the $E_{1/2}$ values predicted by the dynamic EDL model deviate from the prediction of Equation 2.8. The deviation becomes more pronounced for higher reactant charges and at smaller electrodes. At an electrode of 100 nm in radius, the deviation of $E_{1/2}$ is only ca. 2 mV. The estimated k^{0a} is very close to the real k^0 value. The most severe deviation occurs for the reduction of highly charged anions. For the reduction of −3 valence reactant at an electrode with radii of 10 and 5 nm, the obtained k^{0a} values are about $0.6k^0$. Intrinsically, this should represent a significant deviation. In practical ET kinetics measurements using micro/nanometer electrodes, however, it might be within the experiment uncertainties. Thus, the heterogeneous ET kinetics at nanometer-sized electrodes would be hardly distinguishable in experiments from that at electrodes of micrometer and larger dimensions, which agrees with the recent results obtained by Mirkin and coworkers[30,40] who found that the rate constants for heterogeneous ET reactions obtained by fitting the steady-state voltammetric responses of nanoelectrodes with conventional voltammetric theories are compared to those previously measured at large electrodes.

2.2.5   HETEROGENEOUS ET KINETICS AT NANOSCALE ELECTRODES: BEYOND THE ELECTRICAL DOUBLE LAYER (EDL) EFFECT

In addition to EDL effects, reduction of electroactive size of an electrode into nanometer scales would raise some other nanoscopic effects on the heterogeneous ET kinetics, two types of which are introduced in the following sections.

2.2.5.1   Failure of the Linear Free Energy and Quasi-Two-State ET Assumptions

There have been considerable recent concerns on the applicability of the widely used BV theory and even the classic Marcus–Hush (MH) theory in treating the voltammetric responses of nanoscale electrodes.[43,46,57–59] Similar to the Brønsted–Evans–Polanyi (BEP) relation in heterogeneous catalysis, the BV theory assumes a linear dependence of the ET activation free energy with the electrode potential (free energy of reaction) without considering the molecular details about the activation process. The linear relation between the activation free energy and reaction free energy is a result of linearization of the free energy curves of reactant and product systems near the crossing point. Considering that any curves can be reasonably linearized in a narrow region around a point by performing the Taylor series expansion, one can imagine that the BV theory should apply only at potentials very close to E^0. This immediately makes it problematic in treating the voltammetric responses of nanoelectrodes, at which the high MT rates would push the ET-kinetics-governed voltammetric responses to potentials far away from E^0. Thus, one may have to use more realistic ET models to perform voltammetric analysis when using electrodes of nanometer sizes.

The MH theory, which was originally developed by Marcus for homogeneous ET kinetics in solutions and was adopted to treat heterogeneous ET kinetics mainly by Marcus and Hush,[60–62] provides more physical insights and a more reasonable description of ET kinetics than the BV theory by relating the activation of the ET process with the reorganization of redox species together with their solvent shells and by employing more realistic quadratic free energy curves. Therefore, it should be valid in wider potential region than the BV theory. In fact, the BV model can be considered a simplified version of the MH model for heterogeneous ET kinetics at electrode potentials close to E^0 and/or for ET reactions with very large reorganization energy (λ), although the BV theory was developed much earlier. This can be seen from the ET rate constant expressions in the two models, which are given in Equations 2.10 and 2.11, in which the subscripts a/c represent the anodic/cathodic direction:

$$k_{a/c}^{MH} = k_{a/c}^{BV} e^{-F^2(E-\varphi_1-E^0)^2/4\lambda RT} \tag{2.10}$$

$$k_{a/c}^{BV} = k^0 e^{\pm F(F-\varphi_1-E^0)/2RT} \tag{2.11}$$

Both of the BV and classic MH models assume that the electrode potential influences the heterogeneous ET rate simply by changing ΔG^0 with FE. This implies that only the electronic states near the Fermi level of electrode participate ET process. Therefore, all the redox molecules have to be brought to the same activation transition state possessing a frontier electronic state aligned with the Fermi level of electrode. This is obviously unrealistic when considering that solid electrodes generally have a continuum of energy band with considerable width and that the frontier electronic states in the redox molecules may fluctuate over a range of energy due to various structural relaxation.[1,2] The classic MH and BV formalisms have simplified the wide multilevel, multistate heterogeneous ET into an equivalent narrow two-state ET around the Fermi level, which should be reasonable only as E is not far away from E^0. To this end, the classic MH theory could also become problematic in treating the voltammetric responses of very small electrodes.

The multilevel, multistate nature of the heterogeneous ET processes at electrode/electrolyte interfaces has been recognized since the early 1970s, mainly Levich[63] and Gerischer.[64] It was brought to attention substantially in the early 1990s when the inapplicability of the quasi-two-state

BV and MH models was demonstrated in the framework of nonadiabatic ETs between electrodes and electroactive agents spaced by SAMs.[65–67] By attaching the redox agents on the end of a SAM, the heterogeneous ET kinetics can be investigated in a wide range of potentials without the MT limitation. For electrode processes involving free moving redox agents, the BV model remains the major theoretical approach for kinetic analysis.[1] As noted earlier, the high MT rates at nanoscale electrodes could magnify the weakness of the BV and even the classic MH formalisms. If assuming that both the electronic couplings between electrode and redox molecules and the density of states (DOS) of an electrode are independent of energy level (ε) and that the ET rate between individual states in electrode and redox agents can be treated with the classic MH model, one can have the following rate constant expression that accounts for the multistate nature of heterogeneous ET[46,57]:

$$k_{a/c}^{MHC} = k_{a/c}^{BV} \frac{\int_{-\infty}^{+\infty} \dfrac{e^{-F^2(\varepsilon-E^0)^2/4\lambda RT}}{2\cosh(F(\varepsilon-E)/2RT)} d\varepsilon}{\int_{-\infty}^{+\infty} \dfrac{e^{-F^2(\varepsilon-E)^2/4\lambda RT}}{2\cosh(F(\varepsilon-E)/2RT)} d\varepsilon} \tag{2.12}$$

The notation of Marcus-Hush-Chidsey (MHC) is used for the rate constant in Equation 2.12 since it is derived by adopting the idea in an earlier report by Chidsey.[65] According to Equation 2.12, the rate constants for a heterogeneous ET will reach a limiting value as E departs from E^0 rather than undergo an inversion as predicted by the classic MH formalism or increase unlimitedly as predicted by the BV formalism. The deviation of the MH and BV prediction from that of the Equation 2.14 depends on λ.[43,46,57–59,65] The larger the λ is, the less deviation of the BV and MH prediction can be seen (Figure 2.12). This is due to that larger λ corresponds to sharper DOS distribution of redox molecule, which makes the contribution of the non-Fermi states in electrode insignificant to the overall ET rate. For typical λ around 100 kJ/mol, the MH formalism predicts very similar potential dependence of rate constants with Equation 2.12 prior to the occurrence of the Marcus inversion, while the BV prediction exhibits deviation as $|E - E^0|$ is larger than 0.2 V.

Voltammetric simulations for ET reactions of various k^0 and λ at electrodes of various sizes have shown that,[43,46] for ET reactions with k^0 near 0.1 cm/s, the BV theory could predict voltammetric responses visibly deviating from that expected by the MHC model as the electrode radii are smaller than 50 nm, while this occurs as r_0 approaches 10 nm for ET reactions with k^0 around 1.0 cm/s. According to the half-wave-potential difference in the polarization curves predicted by the BV and MHC models, the BV-based voltammetric analysis would give standard rate constants of ca. $0.6k^0$ and $0.5k^0$, respectively, for a reaction of 0.1 cm/s k^0 and 100 kJ/mol λ at electrodes with radii of 20 and 10 nm. For the ET reaction with k^0 of 1.0 cm/s, apparent standard rate constants of ~$0.8k^0$ and $0.6k^0$ will be obtained by BV-based analysis at an electrode with radii of 10 and 5 nm. Considering that the EDL effect would result in enhanced apparent ET kinetics for cation reduction or anion oxidation (Table 2.1), the measured polarization curves at nanoelectrodes would be closer to that predicted by the BV formalism without including the EDL effect. For anion reduction or cation oxidation, the EDL effect and the MHC formalism both predict inhibited ET kinetics as compared with the conventional BV model combined with the diffusion-based MT theory. In this case, the measured polarization would significantly deviate from the prediction of conventional voltammetric theory. Therefore, BV-based voltammetric analysis would result in apparent rate constants that are significantly lower than the real k^0.

Figure 2.13 shows the steady-state polarization curves calculated according to different ET models for a relatively facile one-electron reduction reaction (1.0 cm/s k^0) having different λ at spherical electrodes with a radius of 5 nm. In order to clearly compare the prediction of different ET models, the polarization curves have been calculated without considering the possible EDL effects. At such small electrode, the MH model predicts a nearly identical voltammetric response to that given by the MHC model for ET reaction with 100 kJ/mol λ. For reaction of 50 kJ/mol λ, the two models

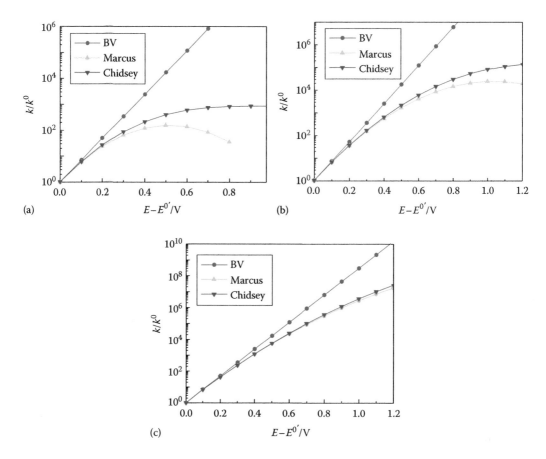

FIGURE 2.12 Dependence of heterogeneous ET rate constants (normalized by k^0) on electrode potential for a one-electron ET process of (a) 50 kJ/mol, (b) 100 kJ/mol, and (c) 200 kJ/mol λ according to different ET kinetics models. (Reprinted with permission from Liu, Y.W. and Chen, S.L., *J. Phys. Chem. C*, 2012, 116, 13594–13602. Copyright 2012 American Chemical Society.)

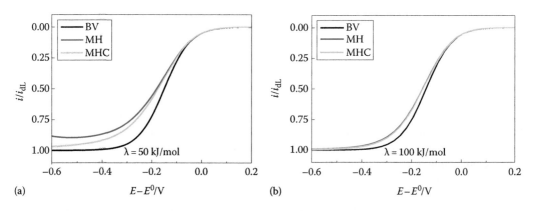

FIGURE 2.13 Steady-state polarization curves predicted by the BV, MH, and MHC models, respectively, at electrodes with a radius of 5 nm for one-electron reduction reactions with $k^0 = 1.0$ cm/s and λ = (a) 50 and (b) 100 kJ/mol. (Reprinted with permission from Chen, S.L. and Liu, Y.W., *Phys. Chem. Chem. Phys.*, 16, 635, 2014. Copyright 2014 Royal Chemical Society.)

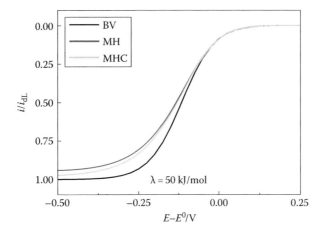

FIGURE 2.14 Steady-state polarization curves predicted by the BV, MH, and MHC models, respectively, at electrodes with a radius of 10 nm for one-electron reduction reaction of 1.0 cm/s k^0 and 50 kJ/mol λ. (Reprinted with permission from Conyers, J.L. and White, H.S., *Anal. Chem.*, 2000, 72, 4441–4446. Copyright 2000 American Chemical Society.)

also give very similar prediction in the kinetic region. In the limiting current region, the MH model significantly underestimates the current.

As shown in Figure 2.14, when electrodes with a radius of 10 nm are considered, the MH and MHC models give nearly identical voltammetric responses in both the kinetic and limiting current regions, regardless of λ. Thus, it seems that the classic MH formalism in most cases can be used to treat the voltammetric responses of nanoelectrodes, unless for the ET reactions with very small λ (e.g., <50 kJ/mol) at electrodes smaller than 5 nm.

2.2.5.2 Importance of Long-Distance Electron Tunneling

Another general voltammetric treatment in conventional electrochemistry that could become questionable at nanoscale electrodes is the use of a single k^0 for a heterogeneous ET reaction at the electrode of certain materials. It has been argued that the comparable sizes of nanoscale electrodes with the effective electron tunneling distance could make the heterogeneous ET rate constants vary with electrode sizes and shapes.[43,45,46] It is known that the heterogeneous ET occurs through electron tunneling between the electrode and the electroactive molecules in the interfacial region. It has also been realized for some time that electron tunneling between an acceptor state and a donor state can occur over a range of distances rather than only at the distance of closest approach. Accordingly, the redox molecules that can transfer electron with electrode should be distributed in an extended region adjacent to HOP rather than only at HOP as considered in most of the current voltammetric treatments. The electron tunneling probability depends on the coupling between the electronic states at the electrode surface and that in the electroactive molecules. The latter varies with the distance through which the electron tunneling takes place.[68] The usually quoted k^0 for an ET reaction in conventional voltammetric treatments should represent an integral of the distance-dependent electron tunneling probability over the entire extended ET region. Feldberg[69–71] has shown that the extended ET may have experimentally measurable implications on the voltammetric responses when the diffusion effect is diminished, which would be satisfied at ultra small electrodes, especially those with nanometer sizes.

Taking a spherical electrode, for example, Figure 2.15 illustrates schematically how the heterogeneous ET rate at a nanoscale electrode may be different from that at a relatively large planar electrode. For a redox molecule approaching a distance of z normal to the electrode surface, the total

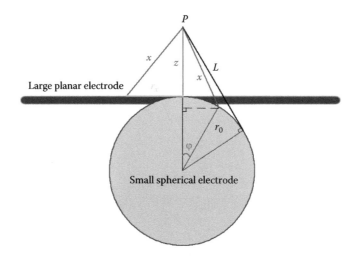

FIGURE 2.15 Schematic illustration of the different ET probability for an electroactive molecule on a large electrode and a spherical nanoelectrode. (Reprinted with permission from Limon-Petersen, J.G., Streeter, I., Rees, N.V., and Compton, R.G., Quantitative voltammetry in weakly supported media: Effects of the applied overpotential and supporting electrolyte concentration on the one electron oxidation of ferrocene in acetonitrile, *J. Phys. Chem. C*, 2009, 113, 333–337. Copyright 2009 American Chemical Society.)

molecule-surface tunneling probability can be expressed by Equations 2.13 and 2.14, respectively, for the large and nanospherical electrodes (see Ref. [46] for derivation details):

$$\Omega^P(z) = \rho 2\pi p_0 \int\limits_z^\infty x e^{-\beta(x-\mu)} dx \tag{2.13}$$

$$\Omega^s(r_0,z)\rho 2\pi p_0 \int\limits_z^L A x e^{-\beta(x-\mu)} dx \tag{2.14}$$

where ρ refers to the site density per surface area, $A = r_0/(r_0+z)$, and

$$L = \sqrt{(r_0 + z)^2 - r_0^2} = \sqrt{2zr_0 + z^2} \tag{2.15}$$

At nanoscale electrodes, L and A vary with the electrode radii r_0. As $r_0 \gg z$, Equation 2.14 will become equivalent to Equation 2.13. Integration of $\Omega^s(r_0, z)$ and $\Omega^P(z)$ over the entire extended ET region would give the total electron tunneling probability, from which an overall ET rate constant can be obtained. At large electrode, this will give the usually quoted k^0. At nanoscale electrodes, however, a radius-dependent rate constant, $k^0(r_0)$, would be obtained (Equation 2.16):

$$\frac{k^0(r_0)}{k^0} = \frac{\int\limits_{r_0+\mu}^\infty \left(r_0(r_0 + z) \int\limits_z^L x e^{-\beta(x-\mu)} dx \right) dz}{(r_0 + \mu)^2 \int\limits_{r_0+\mu}^\infty \int\limits_z^\infty x e^{-\beta(x-\mu)} dx dz} \tag{2.16}$$

Equation 2.16 predicts that $k^0(r_0)$ would start to vary with r_0 as the latter goes smaller than 10 nm. At the same electrode size, the deviation magnitude slightly varies with β. At an electrode with a radius of 5 nm, $k^0(r_0)$ is ca. $0.87k^0$ if $\beta = 1.0$ Å^{-1}.[46] To distinguish such relatively small deviations should be experimentally difficult at current state due to the difficulties in fabrication and accurate characterization of such small electrodes. However, this may be some of significance in dye-sensitized solar cells, in which the ET kinetics between molecular dye and semiconductor nanoparticles sized a few nanometers plays an important role in the energy conversion efficiency.

For nanoscale electrodes with planar geometries, for example, nanodisk electrodes, the extended ET can bring about significant surface heterogeneity in the ET kinetics as well as the electrode size effect.[45] That is, the ET rate constants could vary with the position at an electrode surface. This is mainly due to the increased edge effect of the ET probability at the nanoscale planar electrodes.

2.3 EDL EFFECTS ON ION TRANSPORT AND SOLUTION FLOW IN NANOPORES

Over the last two decades, there has been a great deal of efforts devoted to developing artificial solid-state nanopore membranes that are inexpensive, robust, and tunable in size and surface properties.[72] Nanopores immersed in an electrolyte solution have been widely used in analyte detection, DNA sequencing, diode design, and biosensing.[73–78] In these applications, a potential is applied between the two sides of the membrane containing the nanopore, and the i–V response is recorded. In almost all cases, the interior wall of the nanopore has a fixed charge density due to surface acid/base chemistry, giving rise to an EDL that significantly influences the i–V behavior. Nanopores frequently display a nonlinear response that is influenced by the solution flow within the pore, binding of charged analytes to the pore surface, as well as the concentration of the electrolyte. In the following sections, we present an overview of the EDL effects in conical-shaped nanopores, primarily nanopores prepared in a glass membrane (Figure 2.2c), and recessed metal electrodes positioned at the bottom of a conical nanopore (Figure 2.2b). In these cases, the EDL near the orifice has a significant influence on the transport of electrolyte ions and solution flow through the pore.

The glass nanopore membranes described here comprise a glass or quartz capillary containing an individual conical-shaped nanopore in a membrane (20–75 μm thickness) located at the end of the capillary.[22,79] The glass nanopore membrane and recessed electrode, Figure 2.2, are readily prepared on the bench top and do not require sophisticated nanofabrication or lithographic fabrication methods. A sharpened Pt tip is sealed in a glass capillary and then etched by passing an alternating current between the tip and a Pt counter electrode to produce a conical-shaped pore. The orifice radii of the conical nanopores can be varied from a few nanometers to several tens of micrometers and are readily determined by either measuring the diameter of the Pt disk by voltammetric measurements before etching out the Pt wire or by measuring the ionic conductivity of the nanopore in a concentrated KCl solution.[80]

2.3.1 Ion Current Rectification (ICR) in Conical Nanopores

The unique behavior identified with conical-shaped nanopores is ion current rectification (ICR),[81] which describes the difference in the values of the current passing through the nanopore orifice when potentials of the same magnitude, but of opposite polarity (e.g., ±0.2 V), are applied across the membrane.[82] ICR results from the EDL associated with the charged nanopore surface and generally occurs in geometrically *asymmetric* nanopores or channels when the size of the nanopore or channel orifice is reduced to the dimensions of the ELD (i.e., κ^{-1}).

Wei, Bard, and Feldberg discovered ICR in 1997 when performing i–V measurements using conical quartz nanopipettes, as shown in Figure 2.16. The ICR behavior was found to be highly dependent on the size of the pipette orifice and the ionic concentration of the contacting solution. Nanopore-based ICR is of sustained interest within the large community of analytical chemists, physicists, and materials scientists.[83–85] Siwy et al. reported that ICR relies on the surface characteristics of the inner walls of nanopores.[86] Recently, Baker and coworkers found that ICR can be reversibly controlled by physically bringing a nanopipette in close proximity to a charged substrate.[87]

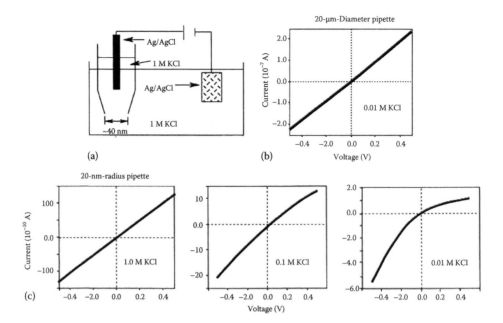

FIGURE 2.16 (a) Typical setup for electrochemical measurements using an Ag/AgCl nanopipette electrode. (b) Typical cyclic voltammogram of 20 μm diameter Ag/AgCl micropipette electrode in 0.01 M KCl. (c) Cyclic voltammograms of Ag/AgCl nanopipette electrodes (20 nm radius) in 1.0, 0.1, and 0.01 M KCl. The potential axis represents that of the Ag/AgCl electrode in the pipette versus another in bulk solution with the same KCl concentration. (Reprinted with permission from Wei, C., Bard, A.J., and Feldberg, S.W., *Anal. Chem.*, 1997, 69, 4627–4633. Copyright 1997 American Chemical Society.)

Surface-charge-dependent nonlinear *i–V* characteristics in ICR allow researchers to build sensors based on the variation of the local surface properties that are triggered by a surface chemical reaction or a binding event. The recent success in ICR-based sensing developments has been extensively discussed in several review papers.[88–90] Applications of conical nanopores have also greatly benefited from the fundamental understanding of the MT behavior within the conical pore and the interaction of ion migration and solution flow with the EDL. In the following sections, we focus our discussion on the ICR mechanism,[91–100] fluid flow effects, and the effect of EDL in nanopore-based particle analysis. Transport gating phenomena observed in electrochemical experiments using the recessed nanopore electrode, which are closely related to ICR, are briefly described in Section 2.3.4.

2.3.1.1 ICR: Theory and Simulation

The surface of a conical glass nanopore is negatively charged at neutral pH due to the dissociation of the surface-bound silanol groups.[101] The EDL associated with the charged surface has a thickness of ~$3\kappa^{-1}$:

$$\kappa^{-1} = \sqrt{\frac{\varepsilon_r \varepsilon_0 RT}{2z^2 F^2 c}} \tag{2.17}$$

where
  κ^{-1} is the Debye screening length
  ε_r is the relative permittivity
  ε_0 is the permittivity in a vacuum
  R is the gas constant

T is the absolute temperature

z is the electrolyte valence

F is Faraday's constant

c is the electrolyte concentration

As a reference point, the Debye length is ~3 nm in a 0.01 M KCl solution.

The transport of ions through the nanopore orifice occurs through a cross-sectional area of πa^2, where a is the orifice radius of the nanopore. Similar to EDL effects becoming important at a nanodisk electrode when the electrode radius approaches the magnitude of κ^{-1}, the EDL influences ion transport through the nanopore when a is of the same order of magnitude as κ^{-1}. The resistance to ion transport and flow and the electric potential drop are all highly localized at the nanopore orifice due to the tapered geometry of the nanopore. Thus, the EDL produced by the nanopore surface charge has a measurable influence on transport when the orifice radius of a conical-shaped pore is smaller than ~100 nm.

Consider, for example, a nanopore filled with and immersed in an aqueous KCl solution. The solution immediately adjacent to the pore orifice becomes enriched in K$^+$ due to the presence of the EDL created by the negative surface charge on the pore surface. When a negative potential is applied inside the pore interior relative to the external solution, K$^+$ are electrophoretically transported into the pore and Cl$^-$ move in the opposite direction. The EDL electrostatically rejects Cl$^-$ and results in a buildup of ions inside the pore orifice.

As noted previously, the resistance of a conical-shaped nanopore is highly localized at the pore orifice.[102] With that in mind, it is no surprise that the buildup of ions in the orifice leads to a higher overall conductivity of the nanopore, as experimentally measured by an increase in the current recording. In contrast, when a positive potential is applied, the flux of Cl$^-$ from the external solution to the pore interior is rejected by the pore orifice, depleting Cl$^-$ within the pore and resulting in a decrease in the nanopore conductivity and measured current. This qualitative description of the effect of EDL is consistent with the nonlinear i–V behavior shown in Figure 2.16.

To quantify the EDL effect, the partial differential equations (PDEs) that govern the MT processes within the nanopore, and the relationship between the electric potential and charges, are simultaneously solved using the finite-element method. The first PDE is the Nernst–Planck equation:

$$\mathbf{J}_i = -D_i \nabla c_i - \frac{z_i F}{RT} D_i c_i \nabla \Phi + c_i \mathbf{u} \tag{2.18}$$

where

$\mathbf{J}_i$, D_i, c_i, and z_i are, respectively, the ion flux vector, diffusion coefficient, concentration, and charge of species i in solution

Φ and $\mathbf{u}$ are the local electric potential and fluid velocity

while F, R, and T are Faraday's constant, the gas constant, and the absolute temperature, respectively

The Nernst–Planck equation describes MT due to diffusion, migration, and convection. Convection in nanopores arises from electroosmotic flow (EOF) or due to a mechanical pressure applied across the membrane containing the nanopore.

The relationship between the local electric potential Φ and ion concentrations c_i is given by the Poisson equation:

$$\nabla^2 \Phi = -\frac{F}{\varepsilon} \sum_i z_i c_i \tag{2.19}$$

where ε is the dielectric constant of the solution.

Finally, the pressure-driven and electroosmosis-driven solution flows are represented by the Navier–Stokes equation:

$$\mathbf{u}\nabla\mathbf{u} = \frac{1}{\rho}\left(-\nabla P + \eta\nabla^2\mathbf{u} - F\left(\sum_i z_i c_i\right)\nabla\Phi\right) \qquad (2.20)$$

where

 ρ and η are the density and viscosity of the fluid
 P is the local pressure

In employing the finite-element method to solve the PDEs, the irregular-shaped conical nanopore is divided into many small components in order to obtain an approximate simultaneous solution to the differential equations.[103] This approach is validated by solving simple electrochemical problems (e.g., EDL structure near a flat metal electrode[21]) and comparing the simulation results with analytical solutions.

Using commercial software (COMSOL Multiphysics), the Nernst–Planck, Poisson, and Navier–Stokes equations are generally solved in a 2D axial symmetric geometry to mimic the 3D structure of conical-shaped nanopores. The more complicated 3D model generates results with similar accuracy but requires additional computational resource. To approximate the semi-infinite solution, the exterior boundary of bulk solution in the model is extended to a distance $r = 20$ μm and $z = 20$ μm away from the pore membrane. The parameters D_i and μ_i for the ionic species are chosen to reflect the electrolyte identity.[21]

Finite-element simulations are performed to obtain solutions for the ionic current, ion concentrations, and electric field distributions at a given potential (and sometimes pressure) applied across the nanopore. As discussed earlier and quantitatively shown in Figure 2.17a, the potential drop across the membrane is focused at the small opening of the pore. The conductivity profiles (in the unit of mS/m) along the central axis of a 50 nm radius pore, Figure 2.17b, indicate that the conductivity inside the pore interior greatly increases at negative potentials as one approaches the pore orifice and that the peak value of the conductivity occurs at the orifice. Conversely, when the potential is reversed to a positive value, a decrease in conductivity is found in the pore interior, with the minimum taking place close to the orifice as well. The potential-dependent conductivity distribution leads to the simulated current rectification curves shown in Figure 2.17c. The results demonstrate that the EDL effect, caused by the surface charge, substantially changes the ion transport behavior within the pore by causing a steady-state redistribution of ions at positive and negative potentials.

2.3.1.2 Electrolyte Flow Coupled with the EDL within a Nanopore

Elimination of ICR and the corresponding EDL effect is possible by applying a pressure-engendered fluid flow through conical glass nanopores.[104] As shown in Figure 2.18, in low ionic strength solutions (0.01 M KCl), ICR decreases with the increasing rate of convective flow through the pore. This pressure dependence also depends on the size of the pore orifice. The equilibrium EDL distributions of cations and anions are disrupted by the pressure-driven flow to a greater extent in larger pores.

Figure 2.18a shows the i–V responses for a nanopore with radius of 185 nm in a 0.01 M KCl solution at applied pressures ranging from ±160 to 0 mmHg. As before, in the absence of flow, the EDL effect manifests itself in producing a significant nonlinear i–V response of the nanopore. These data indicate that a complete overlap of EDL across the nanopore orifice (approximately 20 nm for a 0.01 M KCl solution) in the nanopore is not required to achieve ICR.[105,106]

When a pressure is applied across the nanopore, the resulting convective flow eliminates ICR in the large pore (185 nm), as evidenced by the more ohmic i–V response (Figure 2.18a). We note here that the change in the i–V behavior is fully reversible when the pressure is varied back and

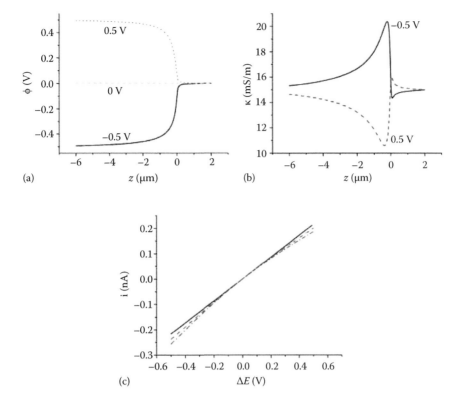

FIGURE 2.17   (a) and (b) Simulated electric potential and conductivity at different potentials (0.5, 0, and −0.5 V) along the centerline axis of a negatively charged 50 nm radius conical nanopore in 1 mM KCl. The pore mouth is located at $z=0$. The bulk value of the conductivity is 15 mS/m, corresponding to a 1 mM KCl solution. (c) Effect of surface charge on the i–V curve of a 50 nm radius nanopore in 1 mM KCl. (–) no surface charge, (- -) −1 mC/m² on the wall, (···) −1 mC/m² around the mouth, and (−··−) −1 mC/m² on the wall and around the mouth. (Reprinted with permission from White, H.S. and Bund, A., *Langmuir*, 2008, 24, 2212–2218. Copyright 2008 American Chemical Society.)

forth between 0 and ±160 mmHg. This reversibility is due to the physical nature of the flow-induced disturbance on the ion distribution within the nanopore.

In contrast to the behavior of the larger 185 nm radius nanopore, an applied pressure had almost no effect on the highly rectified i–V response of the smaller 30 nm radius nanopore, Figure 2.18b. This can be understood by considering the dependence of flow rate on the nanopore orifice radius. The applied pressure generates a pressure-engendered convective flow through the pore, whose volumetric rate, Q, can be estimated by Equation 2.21[107]:

$$Q = \frac{3\pi a^3 \Delta P}{8\eta \cot\theta} \tag{2.21}$$

where
   a is the radius of the pore orifice
   ΔP is the pressure difference across the nanopore
   η is the solution viscosity
   θ is the half-cone angle of the nanopore

Equation 2.21 is analogous to the Hagen–Poiseuille equation that applies to cylindrical channels. The detailed derivation and a more complete description of pressure-driven flow through a conical

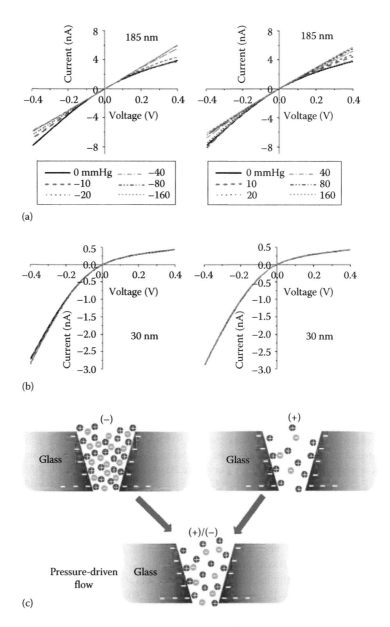

FIGURE 2.18 Pressure-dependent *i–V* responses of conical-shaped nanopores with radii of (a) 185 and (b) 30 nm in a 0.01 M KCl solution. (c) Scheme of ion distributions around the orifice of the negatively charged glass nanopore at positive/negative potentials in the absence and presence of pressure-driven flow. (Reprinted with permission from Lan, W.-J., Holden, D.A., and White, H.S., *J. Am. Chem. Soc.*, 2011, 133, 13300–13303. Copyright 2011 American Chemical Society.)

nanopore can be found in Ref. [107]. According to Equation 2.21, for the same ΔP, the 185 nm radius pore generates a volumetric flow that is ~235 times larger than that for a 30 nm radius nanopore. Finite-element simulation of the Navier–Stokes equation (Equation 2.20) also quantitatively demonstrated that an 80 mmHg pressure produces a significant velocity profile through the 185 nm radius pore but creates very little flow in the 30 nm radius pore. This large difference in volumetric flow rate results in a qualitative difference in the ability to use pressure to bring the external bulk solution containing 0.01 M K$^+$ and Cl$^-$ into the nanopore and disturb the EDL. In the larger nanopore,

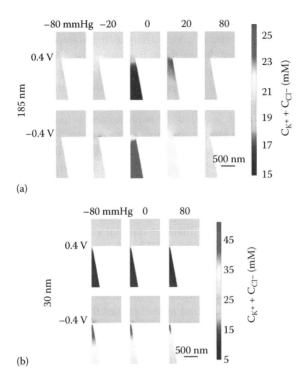

FIGURE 2.19 Simulated distributions of the total ion concentration (K^+ and Cl^-) near the orifice for nanopores with radii of (a) 185 and (b) 30 nm as functions of applied potential and pressure in a 0.01 M KCl solution. (Reprinted with permission from Lan, W.-J., Holden, D.A., and White, H.S., *J. Am. Chem. Soc.*, 2011, 133, 13300–13303. Copyright 2011 American Chemical Society.)

the pressure-driven flow effectively changes the equilibrium ion concentrations in the pore that are established by the EDL (Figure 2.18c) and leads to the experimental observation of pressure-dependent ICR. The greatly reduced flow in the 30 nm radius pore, however, fails to suppress the more pronounced EDL effect associated with the surface charges. Thus, a *pressure-independent* ICR is observed in the smallest nanopore.

To better understand the mechanism underlying the experimental results, finite-element simulations of the coupled Nernst–Planck, Poisson, and Navier–Stokes equations were employed to map the ionic concentration within a conical nanopore in a moving electrolyte. Figure 2.19a shows the distributions of total ion concentrations (K^+ plus Cl^-) at different potentials and pressures for the 185 nm radius nanopore. The middle column shows an EDL-induced depletion/excess of ions at positive/negative potentials in the absence of flow, leading to prominent ICR. When a pressure-induced flow is applied across the nanopore, the EDL effect is largely suppressed because the flow of bulk solution prevents the depletion or buildup of ions responsible for ICR. For a steady-state flow, the ion concentrations at both potentials are close to that in the bulk solution, and a more ohmic behavior is expected. In contrast, the EDL distribution of ions and potential within the 30 nm radius pore is not affected by the applied pressure, Figure 2.19b, in which a stronger EDL effect is not altered by a greatly reduced flow rate. The simulation results are in excellent agreement with the experimental observations.

2.3.2 Negative Differential Resistance in Nanopores Resulting from the EDL

Negative differential resistance (NDR) is used to describe electrical behavior where the current decreases with an increasing applied voltage. A well-known application of NDR is the Esaki or tunnel diode, where electron tunneling between the valence and conduction bands of a heavily doped p–n junction leads to a decrease of conductivity as the voltage is increased.[108] In nanopore systems, electroosmotic and pressure-driven flows have been used to control electrolyte distributions[21,99,104,109] or solvent flux and, thus, alter the nanopore conductance.[75,101,107,110–114] As shown in this section, by carefully balancing these different types of flow within the nanopore, it is possible to establish conditions where electrical NDR is observed based on a complex coupling of the local ion concentration, potential distribution, and flow. This phenomenon has potential applications in the sensing of charged analytes that bind to the nanopore surface and affect EOF velocities.

In the NDR nanopore experiment described by Luo et al.,[115,116] the radius of the small orifice of the nanopore was relatively large (~300 nm) to allow pressure-driven flow through the orifice at reasonable pressures. The internal solution within the nanopore contains a high concentration of electrolyte (e.g., 50 mM KCl) and the external solution contains a lower concentration (e.g., 5–25 mM KCl), as shown schematically in Figure 2.20. The key requirement for NDR is that the external solution has a lower conductivity than the internal solution (alternatively, and rather than using different KCl concentrations, nanopore NDR has been established by using two solvents of different viscosity in the internal and external solutions to establish a difference in electrolyte conductivity, e.g., H_2O and DMSO). After a pressure and a negative voltage are applied across the nanopore, a steady-state EOF (white arrow) drives the lower concentration KCl solution into the nanopore while the pressure-driven flow (red arrow) pushes the higher concentration KCl solution out of the nanopore. At steady-state, the opposing pressure and electroosmotic forces, along with the nanopore surface charge, determine the distributions of cations and anions at the nanopore orifice and, thus, the nanopore conductivity. As shown in Figure 2.20b, by holding the pressure constant while increasing the applied voltage, the balance in flow within the nanopore shifts from an outward pressure-driven dominated flow at low voltages to an inward electroosmotic dominated flow at high

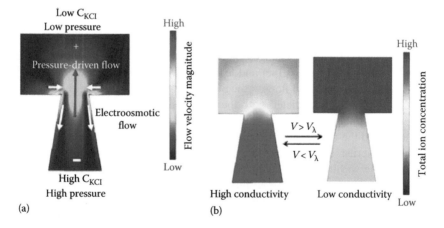

(a)   (b)

FIGURE 2.20 (a) Illustration of pressure-driven and voltage-engendered EOFs that give rise to NDR in the *i*–*V* response of a negatively charged, conical nanopore that separates high and low ionic strength solutions. The color surface indicates the magnitude of the net flow velocity; red and blue denote higher and lower velocities, respectively. Pressure-driven flow *out of the pore* occurs along the central axis of the nanopore (red arrow), while an opposing EOF *into the pore* occurs along the negatively charged nanopore surface (white arrows). NDR observed in the *i*–*V* response of the nanopore results from positive feedback associated with an increase in EOF as the voltage is increased: an increased flux of the external low-conductivity solution into the nanopore orifice leads to a decreased ionic conductivity of solution in the nanopore causing a further increase in EOF and a sudden drop in the nanopore conductivity at a critical voltage, V_λ. (b) Profiles of the total ion concentration (K^+ plus Cl^-) in the nanopore for applied voltages above ($V > V_\lambda$, high conductivity state) and below ($V < V_\lambda$, low-conductivity state) the conductivity switching potential, V_λ. (Reprinted with permission from Luo, L., Holden, D.A., and White, H.S., *ACS Nano*, 2014, 8, 3023–3030. Copyright 2014 American Chemical Society.)

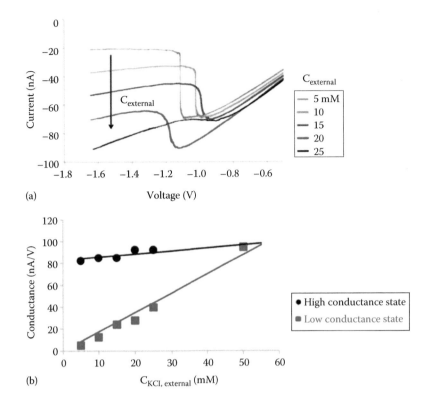

FIGURE 2.21 (a) A series of NDR curves as a function of the external KCl concentration measured using a 260 nm radius nanopore. The KCl concentration of the external solution was varied between 5 and 25 mM KCl, while the internal KCl concentration (50 mM) was held constant; pH = 7.0. A 10 mmHg pressure (internal vs. external) was applied. (b) Conductance values measured from the slopes of $i–V$ responses at voltages positive and negative of the NDR switching potential as a function of the external solution KCl concentration. (Reprinted with permission from Luo, L., Holden, D.A., and White, H.S., *ACS Nano*, 2014, 8, 3023–3030. Copyright 2014 American Chemical Society.)

voltages. The change in flow direction results in a decrease of total ion (K$^+$ and Cl$^-$) concentration near the nanopore orifice, which further enhances the EOF into the pore. The dependence of EOF on ion concentration creates a positive feedback mechanism between the nanopore flow and ion distributions, generating a bistability in the nanopore conductance.

Figure 2.21a shows a series of typical $i–V$ curves exhibiting NDR for a 260 nm radius nanopore containing a 50 mM KCl internal solution obtained while varying the KCl concentration in the external solution between 5 and 25 mM. A constant pressure (10 mmHg) was applied across the nanopore while the voltage was scanned slowly in the negative direction at a rate of 10 mV/s. As shown in Figure 2.21a, a sudden decrease in current behavior occurs at ~ −1.0 and −1.1 V. This $i–V$ response is reversible upon scanning the potential back in the positive direction and corresponds to the first example of solution-phase NDR.

Finite-element simulations are useful to understand the mechanism of NDR and its dependence on the composition in the internal and external solutions, pore geometry, and nanopore surface charge density. Similar to modeling flow effects on nanopore ICR described earlier, the Nernst–Planck equation governing the diffusional, migrational, and convective fluxes of ions (Equation 2.18), the Navier–Stokes equation for low-Reynolds number flow engendered by the external pressure and electroosmosis (Equation 2.20), and Poisson's equation relating the ion distributions to the local electric field (Equation 2.19) were simultaneously solved to obtain local values of the fluid

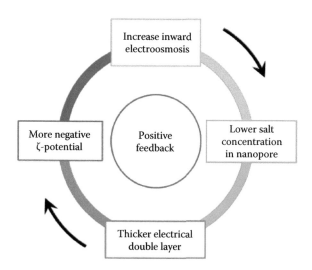

FIGURE 2.22 Positive feedback mechanism associated with the NDR switch. (Reprinted with permission from Luo, L., Holden, D.A., and White, H.S., *ACS Nano*, 2014, 8, 3023–3030. Copyright 2014 American Chemical Society.)

velocity, ion concentrations, and electric potential. The simulation results suggest that NDR represents a sudden transition between high and low conductance states that is associated with a bistability in the electrolyte flow within the nanopore. As schematically illustrated in Figure 2.20, the ion concentration distribution is determined by the combination of the constant outward pressure-driven flow and the voltage-dependent inward EOF.

A discrete jump in flow and current results from the positive feedback mechanism between the ion concentrations and EOF, as qualitatively depicted in Figure 2.22. At potentials positive of the NDR switching potential, V_λ, scanning the applied voltage to more negative potentials results in EOF bringing in the external solution, causing a decrease in the ion concentration within the nanopore orifice. This decrease in ion concentration results in an increased thickness of the EDL, generating a more negative potential of the nanopore surface if the surface charge density σ remains constant, as described by the Grahame equation[117]:

$$\sigma = \sqrt{8c_0 \varepsilon RT} \, \sinh\left(\frac{e\psi_d}{2k_B T}\right) \tag{2.22}$$

where
ψ_d is the diffuse layer potential near the charged surface
c_0 is the bulk concentration of a symmetric monovalent electrolyte

The other parameters in Equation 2.22 have their usual meaning.

The feedback mechanism of NDR is based on the electroosmotic velocity, u, being proportional to the value of zeta potential, ζ, at the velocity slip plane located adjacent to the nanopore surface. The Helmholtz–Smoluchowski equation relates the effective slip electroosmotic velocity to ζ:

$$u = -\frac{\varepsilon \zeta E}{\eta} \tag{2.23}$$

where
E is the electric field parallel to the surface
η is the viscosity of the fluid

The parameters ψ_d and ζ have slightly different physical interpretations,[118] but they have approximately similar values and a similar dependence on electrolyte concentration. Thus, the increase in ζ (and ψ_d) resulting from the decrease in ion concentration at the orifice (resulting from the inward EOF) further enhances the inward EOF of the low-conductivity solution into the nanopore. This dependence of the inward EOF on the ion concentration, *via* the EDL, forms a positive feedback loop between conductance and EOF, Figure 2.22, leading to a sudden increase of flow rate, a drop in the ion concentration, and a concurrent decrease in current over a very narrow potential range, as shown in Figure 2.21.

The NDR effect can be used to sense the binding of charged analytes to the nanopore surface, because the EDL and EOF rates are sensitive to the nanopore surface charge. For instance, Ca^{2+} binds more strongly than K^+ to the dissociated glass silanol groups,[119] and the addition of Ca^{2+} to the KCl solutions reduces the negative surface charge density at the glass nanopore surface, resulting in a shift of the NDR curve to a more negative voltage. When 2 mM $CaCl_2$ was added to the external KCl solution, the NDR curve shifted ~1 V to a more negative potential as shown in Figure 2.23. A positive shift of ~1 V occurs when the solution containing Ca^{2+} is replaced by the original solution containing only KCl.

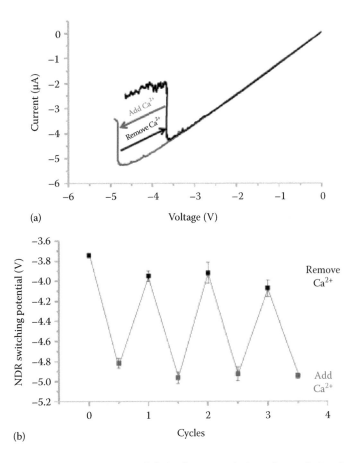

FIGURE 2.23 Reversible NDR response to Ca^{2+} in the external electrolyte solution for a 270-nm-radius nanopore. Experimental conditions: 54 mmHg; 1 M internal and 100 mM external KCl solutions; pH = 7.8; Ca^{2+} concentration (when present in solution) = 2 mM; scan rate: 100 mV/s. (a) i-V curves. (b) Switching potentials measured over 4 voltammetric cycles. (Reprinted with permission from Luo, L., Holden, D.A., and White, H.S., *ACS Nano*, 2014, 8, 3023–3030. Copyright 2014 American Chemical Society.)

2.3.3 EDL Effects in Nanopore-Based Particle Analysis

The EDL also affects the signature electrical signal during resistive-pulse analysis of particles using nanopores. Resistive-pulse analysis, also referred to as Coulter analysis, is a method used to size and count particles suspended in an electrolyte by recording resistive pulses generated by particle translocations through a pore/channel. In a typical resistive-pulse measurement, the steady current flowing through a nanopore between two Ag/AgCl electrodes briefly decreases when a particle translocates through the pore.[120,121] We have recently found that when both the nanopore and the particle are negatively charged, particle translocations can produce *biphasic* pulses (Figure 2.24) in the current–time (i–t) recording, in which both an increase and decrease in current are observed during particle translocation.[122] This biphasic electrical pulse is a consequence of the transient redistribution of ions in the EDLs surrounding both the nanopore and the nanoparticle.

The biphasic pulses are characterized by an initial current increase followed by a sudden current decrease. The behavior is a consequence of the combination of the (1) EDL effect that temporarily increases the ionic conductivity within the pore as the charged particle translocates through the charged orifice and (2) the volume-exclusion effect that results in a decrease of the nanopore current due to simple mechanical blockage of ion fluxes by the particle. Briefly, as a negatively charged nanoparticle translocates through the pore, the EDL associated with the moving particle makes the pore orifice more cation selective. When the nanopore is under a negative potential gradient, as discussed in the previous section, a higher cation selectivity results in an enhanced buildup of ions in the pore interior, thus increasing the nanopore current. If the potential is scanned more negative, the EDL effect becomes stronger and eventually strong enough to override the volume-exclusion effect of the

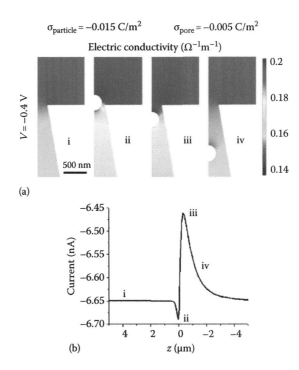

FIGURE 2.24 Simulated (a) distributions of electric conductivity and (b) current–position (i–z) curves as a 160 nm radius nanoparticle translocates through a 215 nm radius nanopore in a 0.01 M KCl solution at an applied voltage of −0.4 V. $z = 0$ corresponds to the location of the pore orifice. The surface charges at the pore wall and particle surface were set equal to −0.005 and −0.015 C/m², respectively. (Reprinted with permission from Lan, W.-J., Kubeil, C., Xiong, J.-W., Bund, A., and White, H.S., *J. Phy. Chem. C*, 2014, 118, 2726–2734. Copyright 2014 American Chemical Society.)

particle translocation that leads to a decrease in the nanopore current. Therefore, a current increase occurs in the *i–t* recording when the particle translocates through the pore orifice. As the particle travels further into the pore, the EDL effect becomes weaker and the pore orifice becomes less cation selective. Thus, the volume-exclusion effect begins to dominate the nanopore current causing a decrease in current that follows the initial increase. When the particle finally leaves the sensing zone of the nanopore, the current returns to its baseline value and a complete biphasic *i–t* pulse is recorded.

In contrast to this behavior at negative applied potentials, the particle-induced EDL effect and corresponding selectivity enhances the ion depletion at positive applied potentials, further decreasing the nanopore conductivity and the observed ionic current. Thus, a single resistive peak is always anticipated when a positive potential is applied across the nanopore.

Analogous to nanopore ICR, a contacting solution with a high ionic strength is also able to effectively screen the EDLs associated with the nanopore and particles and therefore eliminate the biphasic response. Additionally, when a solution flow is superimposed on the system, similar to pressure-dependent ICR, the EDL effect is largely removed and the particle translocation generates a normal decreasing current pulse.

Finite-element simulations based on the Poisson–Nernst–Planck equations have been integrated with a particle trajectory calculation to capture the characteristics of the biphasic response in nanopore *i–V* measurements. The simulation results verify the qualitative explanation presented earlier in showing that the EDL effects, from both the negatively charged particle and nanopore, are responsible for the biphasic pulses and the corresponding surface charge-dependent and voltage-dependent shape evolution. As shown in Figure 2.24a, the electric-field-induced ion accumulation is greatly enhanced as the particle translocates through the pore orifice.

2.3.4 EDL GATING OF REDOX TRANSPORT AT RECESSED CONICAL NANOPORE ELECTRODES

The recessed nanopore electrode shown in Figure 2.2b comprises a Pt or Au microdisk electrode embedded at the bottom of a conical-shaped pore synthesized in a glass membrane. These electrodes are fabricated with pore orifice radii as small as a few nanometers. EDL gating refers to the ability to control the flux of redox-active molecules from the bulk solution to the electrode surface, through the orifice (Figure 2.25), by either chemical (e.g., pH) or external stimuli (e.g., photons) that

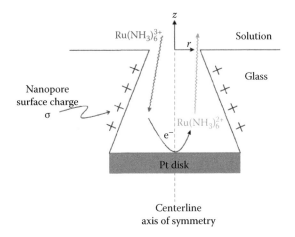

FIGURE 2.25 Schematic sketch of the glass nanopore. As an example, the reduction of $Ru(NH_3)_6^{3+}$ is shown. The interior surface is electrically neutral or positively charged due to functionalization with an aminosilane that may or may not be protonated depending on solution pH. The origin of the underlying coordinate system (r, z) used in the finite-element modeling is at the center of the pore orifice. (Reprinted with permission from White, H.S. and Bund, A., *Langmuir*, 2008, 24, 12062–12067. Copyright 2008 American Chemical Society.)

alter the sign and density of the electrical charge on the interior wall of the nanopore. For instance, attachment of an amine (e.g., 3-aminopropyldimethylethoxysilane) to the nanopore surface provides a means to create a variable surface charge by lowering or raising the pH of the contacting solution.

As with ICR described in the previous sections, the electrical charge on the nanopore surface effectively *gates* the entry (or exit) of charged redox molecules due to strong electrostatic interactions. For instance, as schematically shown in Figure 2.25, the dependence of the transport-limited current of $Ru(NH_3)_6^{3+}$ reduction through a recessed nanopore electrode was investigated as a function of the EDL charge.[123] In a pH-neutral solution containing 5 mM $Ru(NH_3)_6^{3+}$ (without any supporting electrolyte), a nanopore electrode with a 30 nm radius orifice that was modified with 3-aminopropyldimethylethoxysilane displays a diffusion-limited faradaic current of −27 pA for the $1 - e^-$ reduction of $Ru(NH_3)_6^{3+}$. When the bulk pH is decreased to 3.0, the current decreased to ca. −2 pA.[20] As shown by Bund, this behavior is largely due to the protonation of the amine groups inside the pore, reducing the negative charge associated with the glass nanopore surface.[123]

In other experiments, the photon-gated transport of $Fe(bpy)_3^{2+}$ at conical recessed nanopores, functionalized with a spiropyran (SP) moiety, has been described. Upon exposure to UV light, SP is converted in the presence of a weak acid to the protonated merocyanine, MEH^+. MEH^+ is converted back to SP by shining visible light on the nanopore orifice. The effect of the photon-generated charges on the diffusion-limited oxidation of $Fe(bpy)_3^{2+}$ is significant. In the dark (i.e., with the surface attached molecule in the electrically neutral form, SP), i_{lim} for $Fe(bpy)_3^{2+}$ oxidation is ca. 8 pA for a Pt recessed electrode with a 15 nm radius nanopore orifice and a concentration of 5 mM $Fe(bpy)_3^{2+}$ (no supporting electrolyte). Upon exposure to UV, the current dropped below 1 pA (see Figure 2 in Ref. [19]). Exposing the *blocked* electrode to visible light recovered the diffusion-limited voltammetric response. The effect due to the EDL can be demonstrated by adding an excess of supporting electrolyte. In the presence of 0.1 M tetrabutylammonium perchlorate, corresponding to a very short Debye length of ~1 nm, exposure to UV light did not trigger the photon gating. In addition to gating the flux of $Fe(bpy)_3^{2+}$ from the bulk solution to the electrode, it is also possible to electrostatically *trap* $Fe(bpy)_3^{2+}$ inside the nanopore, and subsequently release it, using the same photochemical reactions to control the EDL.[19]

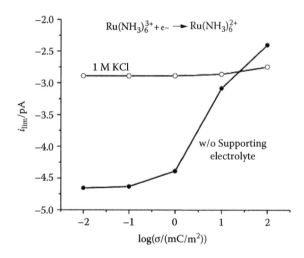

FIGURE 2.26 Dependence of i_{lim} for the reduction of $Ru(NH_3)_6^{3+}$ on the nanopore surface charge (σ) for a 15 nm radius nanopore (depth 300 nm, half-cone angle 10°). The bulk concentration of $Ru(NH_3)_6Cl_3$ is 5 mM. The open symbols are for a solution containing 1 M KCl supporting electrolyte. (Reprinted with permission from White, H.S. and Bund, A., *Langmuir*, 2008, 24, 12062–12067. Copyright 2008 American Chemical Society.)

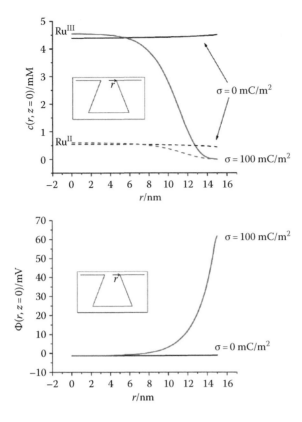

FIGURE 2.27 Effect of the surface charge on (top) the radial dependence of the ruthenium species concentration and (bottom) the electric potential at the orifice ($z=0$) during the transport-limited reduction of $Ru(NH_3)_6^{3+}$ at the bottom of a 15 nm radius nanopore (depth 300 nm, half-cone angle 10°). The bulk concentration of $Ru(NH_3)_6Cl_3$ was 5 mM. The solid lines are for $Ru(NH_3)_6^{3+}$ and the dashed lines are for $Ru(NH_3)_6^{2+}$. The black curves are for $\sigma=0$, and the red curves are for $\sigma=100$ mC/m². The inset shows the line segment (0–15 nm) that corresponds to the x-axis of the plot. (Reprinted with permission from White, H.S. and Bund, A., *Langmuir*, 2008, 24, 12062–12067. Copyright 2008 American Chemical Society.)

Migration of redox species within the EDL near the nanopore surface, in addition to electrostatic rejection of charged molecules at the pore orifice, has been considered in explaining the EDL gating phenomenon. While both components are likely operative, finite-element simulations suggest that electrical migration of redox molecules parallel along the electrical charged walls of the nanopore strongly influences the fluxes of $Ru(NH_3)_6^{3+}$ and $Fe(bpy)_3^{2+}$ (i.e., transport of redox molecules occurs within the width of the EDL). For instance, Figure 2.26 shows the results of computer simulations predicting the transport-limited current for the reduction $Ru(NH_3)_6^{3+}$ as a function of the nanopore surface charge (σ) for a recessed electrode with a 15 nm radius orifice. Figure 2.27 shows the radial distributions of $Ru(NH_3)_6^{3+}$ and the electrogenerated product, $Ru(NH_3)_6^{2+}$, in addition to the electrostatic potential at the orifice of the nanopore. It is clear from these results that varying the charge on the nanopore walls has a tremendous effect on the redox concentrations, and thus their fluxes, when the EDL potential distribution extends across a significant fraction of the nanopore orifice.

2.4 CONCLUSIONS

In this chapter, we have highlighted the effect of the EDL on ET and ion transport when the dimensions of the electrochemical system (i.e., electrode or nanopore radius) are reduced to the nanometer scale, becoming comparable to the width of the EDL. For instance, the greatly enhanced MT rates at nanoscale metal electrodes accentuate kinetic phenomena that are generally inappreciable at conventional micro-/macroelectrode systems. One of these features is the overlapping field nature at the electrode/electrolyte interface: the strongly coupled distributions of chemical potentials (concentrations), electrostatic potential, and solvent dielectric structure at the nanoscale interface preclude independent treatment of the voltammetric response and the EDL structure, as is done in conventional voltammetric theory. The solvent dielectric distribution also significantly impacts the voltammetric response of nanoscale electrodes. Other features evoked by nanoscale electrode sizes are the multistate nature of the heterogeneous ET between electrode and redox molecules and the weakness of the linear free energy relation in the widely used BV kinetic theory.

We have also described how the EDL plays a central role in determining the i–V behavior of electrically charged nanopores. Analogous to EDL phenomena at metal nanodisk electrodes, the EDL associated with the electrically charged surface of the nanopore is coupled to the local distributions of the electrolyte ions and, thus, largely determines the conductivity of the nanopore. When the nanopore size approaches the EDL thickness, the equilibrium ion distributions can be varied by application of an external electric field across the nanopore, resulting in highly i–V behaviors (i.e., ICR) and EOF. When convective flow is additionally coupled to the system, the EDL structure can be disrupted to produce unusual nonlinear i–V behavior not exhibited by macroscale systems, including the phenomena of pressure-dependent ICR and negative differential electrolyte resistance. In addition, EDL effects manifest themselves in producing biphasic pulses in nanopore-based resistive-pulse analysis of particles, and in electrostatic gating of the transport of redox species at recessed nanopore electrodes.

ACKNOWLEDGMENTS

H.S.W. and W.-J.L. acknowledge coworkers who contributed directly to work presented in this chapter (A. Bund, C. Kubeil, G. Wang, B. Zhang, D. Holden, and L. Luo), as well as the funding from the U.S. National Science Foundation and the Office of Naval Research. Similarly, S.C. acknowledges coworkers in Wuhan University who contributed to work presented in this chapter (Y.W. Liu, R. He, Q.F. Zhang), as well as the funding from the National Basic Research Program of China (Grant No. 2012CB932800) and National Natural Science Foundation of China (21173162, 21073137 and 20973131).

REFERENCES

1. Bard, A. J.; Faulkner, L. R. *Electrochemical Methods: Fundamentals and Applications*, 2nd ed.; John Wiley & Sons: New York, 2001.
2. Sato, N. *Electrochemistry at Metal and Semiconductor Electrodes*, Elsevier Science: Amsterdam, the Netherlands, 1998.
3. Bockris, J. O. M.; Reddy, A. K. N. The electrified interface, In *Modern Electrochemistry*, vol. 2, Plenum Press: New York, 1970.
4. Grahame, D. C. Electrode processes and the electrical double layer. *Annu Rev Phys Chem* 1955, *6*, 337–358.
5. Hamelin, A. Double-layer properties at sp and sd metal single-crystal electrodes, In *Modern Aspects of Electrochemistry*, Conway, B. E.; White, R. E.; Bockris, J. O. M. (Eds.), Plenum Press: New York 1985, *16*, 1.
6. Hurth, C.; Li, C.; Bard, A. J. Direct probing of electrical double layers by scanning electrochemical potential microscopy. *J Phys Chem C* 2007, *111*, 4620–4627.
7. Fréchette, J.; Vanderlick, T. K. Double layer forces over large potential ranges as measured in an electrochemical surface forces apparatus. *Langmuir* 2001, *17*, 7620–7627.
8. Norton, J. D.; White, H. S.; Feldberg, S. W. Effect of the electrical double layer on voltammetry at microelectrodes. *J Phys Chem* 1990, *94*, 6772–6780.

9. Bard, A. J. Inner-sphere heterogeneous electrode reactions. Electrocatalysis and photocatalysis: The challenge. *J Am Chem Soc* 2010, *132*, 7559–7567.
10. Frumkin, A. N. Hydrogen overpotential and the double layer structure. *Z Physik Chem* 1933, *164A*, 121.
11. Smith, C. P.; White, H. S. Theory of the voltammetric response of electrodes coated with electroactive molecular films. *Anal Chem* 1992, *64*, 2398–2405.
12. Eggers, P. K.; Darwish, N.; Paddon-Row, M. N.; Gooding, J. J. Surface-bound molecular rulers for probing the electrical double layer. *J Am Chem Soc* 2012, *134*, 7539–7544.
13. Dayton, M. A.; Ewing, A. G.; Wightman, R. M. Response of microvoltammetric electrodes to homogeneous catalytic and slow heterogeneous charge-transfer reactions. *Anal Chem* 1980, *52*, 2392–2396.
14. Dayton, M. A.; Brown, J. C.; Stutts, K. J.; Wightman, R. M. Faradaic electrochemistry at microvoltammetric electrodes. *Anal Chem* 1980, *52*, 946–950.
15. Wightman, R. M. Microvoltammetric electrodes. *Anal Chem* 1981, *53*, 1125A–1134A.
16. Kittlesen, G. P.; White, H. S.; Wrighton, M. S. Chemical derivatization of microelectrode arrays by oxidation of pyrrole and N-methylpyrrole: Fabrication of molecule-based electronic devices. *J Am Chem Soc* 1984, *106*, 7389–7396.
17. Zhang, B.; Zhang, Y.; White, H. S. Steady-state voltammetric response of the nanopore electrode. *Anal Chem* 2006, *78*, 477–483.
18. Zhang, Y.; Zhang, B.; White, H. S. Electrochemistry of nanopore electrodes in low ionic strength solution. *J Phys Chem* 2006, *110*, 1768–1774.
19. Wang, G.; Bohaty, A. K.; Zharov, I.; White, H. S. Photon gated transport at the glass nanopore electrode. *J Am Chem Soc* 2006, *128*, 13553–13558.
20. Wang, G.; Zhang, B.; Wayment, J.; Harris, J. M.; White, H. S. Electrostatic-gated transport in chemically modified glass nanopore electrodes. *J Am Chem Soc* 2006, *128*, 7679–7686.
21. White, H. S.; Bund, A. Ion current rectification at nanopores in glass membranes. *Langmuir* 2008, *14*, 2212–2218.
22. Zhang, B.; Galusha, J.; Shiozawa, P. G.; Wang, G.; Bergren, A. J.; Jones, R. M.; White, R. J.; Ervin, E. N.; Cauley, C. C.; White, H. S. A bench-top method of fabricating glass-sealed nanodisk electrodes, glass nanopore electrodes, and glass nanopore membranes of controlled size. *Anal Chem* 2007, *79*, 4778–4787.
23. Watkins, J. J.; Zhang, B.; White, H. S. Electrochemistry at nanometer-scale electrodes. *J Chem Educ* 2005, *82*, 712–719.
24. Cox, J. T.; Zhang, B. Nanoelectrodes: Recent advances and new directions. *Annu Rev Anal Chem* 2012, *5*, 253–272.
25. Morris, R. B.; Franta, D. J.; White, H. S. Electrochemistry at Pt band electrodes of width approaching molecular dimensions—breakdown of transport equations at very small electrodes. *J Phys Chem* 1987, *91*, 3559–3564.
26. Chen, S. L.; Kucernak, A. The voltammetric response of nanometer-sized carbon electrodes. *J Phys Chem B* 2002, *106*, 9396–9404.
27. Chen, S. L.; Kucernak, A. Fabrication of carbon microelectrodes with an effective radius of 1 nm. *Electrochem Commun* 2002, *4*, 80–85.
28. Penner, R. M.; Heben, M. J.; Longin, T. L.; Lewis, N. S. Fabrication and use of nanometer-sized electrodes in electrochemistry. *Science* 1990, *250*, 1118–1121.
29. Slevin, C. J.; Gray, N. J.; Macpherson, J. V.; Webb, M. A.; Unwin, P. R. Fabrication and characterisation of nanometre-sized platinum electrodes for voltammetric analysis and imaging. *Electrochem Commun* 1999, *1*, 282–288.
30. Sun, P.; Mirkin, M. V. Kinetics of electron-transfer reactions at nanoelectrodes. *Anal Chem* 2006, *78*, 6526–6534.
31. Shao, Y.; Mirkin, M. V.; Fish, G.; Kokotov, S.; Palanker, D.; Lewis, A. Nanometer-sized electrochemical sensors. *Anal Chem* 1997, *69*, 1627–1634.
32. Conyers, J. L.; White, H. S. Electrochemical characterization of electrodes with submicrometer dimensions. *Anal Chem* 2000, *72*, 4441–4446.
33. Sun, Y.; Liu, Y. W.; Liang, Z. X.; Xiong, L.; Wang, A. L.; Chen, S. L. On the applicability of conventional voltammetric theory to nanoscale electrochemical interfaces. *J Phys Chem C* 2009, *113*, 9878–9883.
34. Watkins, J. J.; White, H. S. The role of the electrical double layer and ion pairing on the electrochemical oxidation of hexachloroiridate(III) at Pt electrodes of nanometer dimensions. *Langmuir* 2004, *20*, 5474–5483.

35. Limon-Petersen, J. G.; Streeter, I.; Rees, N. V.; Compton, R. G. Quantitative voltammetry in weakly supported media: Effects of the applied overpotential and supporting electrolyte concentration on the one electron oxidation of ferrocene in acetonitrile. *J Phys Chem C* 2009, *113*, 333–337.

36. Streeter, I.; Compton, R. G. Numerical simulation of potential step chronoamperometry at low concentrations of supporting electrolyte. *J Phys Chem C* 2008, *112*, 13716–13728.

37. Limon-Petersen, J. G.; Streeter, I.; Rees, N. V.; Compton, R. G. Voltammetry in weakly supported media: The stripping of thallium from a hemispherical amalgam drop. Theory and experiment. *J Phys Chem C* 2008, *112*, 17175–17182.

38. Watkins, J. J.; Chen, J. Y.; White, H. S.; Abruña, H. D.; Maisonhaute, E.; Amatore, C. Zeptomole voltammetric detection and electron-transfer rate measurements using platinum electrodes of nanometer dimensions. *Anal Chem* 2003, *75*, 3962–3971.

39. Zhang, B.; Fan, L. X.; Zhong, H. W.; Liu, Y. W.; Chen, S. L. Graphene nanoelectrodes: Fabrication and size-dependent electrochemistry. *J Am Chem Soc* 2013, *135*, 10073–10080.

40. Wang, Y. X.; Velmurugan, J.; Mirkin, M. V. Kinetics of charge-transfer reactions at nanoscopic electrochemical interfaces. *Isr J Chem* 2010, *50*, 291–305.

41. Oja, S. M.; Wood, M.; Zhang, B. Nanoscale electrochemistry. *Anal Chem* 2013, *85*, 473–486.

42. He, R.; Chen, S. L.; Yang, F.; Wu, B. L. Dynamic diffuse double-layer model for the electrochemistry of nanometer-sized electrodes. *J Phys Chem B* 2006, *110*, 3262–3270.

43. Chen, S. L.; Liu, Y. W. Electrochemistry at nanometer-sized electrodes. *Phys Chem Chem Phys* 2014, *16*, 635–652.

44. Smith, C. P.; White, H. S. Theory of the voltammetric response of electrodes of submicron dimensions. violation of electroneutrality in the presence of excess supporting electrolyte. *Anal Chem* 1993, *65*, 3343–3353.

45. Liu, Y. W.; He, R.; Zhang, Q. F.; Chen, S. L. Theory of electrochemistry for nanometer-sized disk electrodes. *J Phys Chem C* 2010, *114*, 10812–10822.

46. Liu, Y. W.; Chen, S. L. Theory of interfacial electron transfer kinetics at nanometer-sized electrodes. *J Phys Chem C* 2012, *116*, 13594–13602.

47. Oldham, K. B.; Bond, A. M. How valid is the electroneutrality approximation in the theory of steady-state voltammetry? *J Electroanal Chem* 2001, *508*, 28–40.

48. Dickinson, E. J. F.; Compton, R. G. Diffuse double layer at nanoelectrodes. *J Phys Chem C* 2009, *113*, 17585–17589.

49. Yang, X. L.; Zhang, G. G. Simulating the structure and effect of the electrical double layer at nanometre electrodes. *Nanotechnology* 2007, *18*, 335201.

50. Levich, B. Theory of the non-equilibrium double layer. *Dokl Akad Nauk SSSR* 1949, *67*, 309–312.

51. Booth, F. The dielectric constant of water and the saturation effect. *J Chem Phys* 1951, *19*, 391–394.

52. Yeh, I.-C.; Berkowitz, M. L. Dielectric constant of water at high electric fields: Molecular dynamics study. *J Chem Phys* 1999, *110*, 7935–7942.

53. Amatore, C.; Fosset, B.; Bartelt, J.; Deakin, M. R.; Wightman, R. M. Electrochemical kinetics at microelectrodes: Part V. Migrational effects on steady or quasi-steady-state voltammograms. *J Electroanal Chem* 1988, *256*, 255–268.

54. Mirkin, M. V.; Bard, A. J. Simple analysis of quasi-reversible steady-state voltammograms. *Anal Chem* 1992, *64*, 2293–2302.

55. Oldham, K. B.; Zoski, C. G. Comparison of voltammetric steady states at hemispherical and disc microelectrodes. *J Electroanal Chem* 1988, *256*, 11–19.

56. Oldham, K. B. Steady-state voltammetry. In *Microelectrode: Theory and Application*; Montenegro, M. I., Queiros, M. A., Daschbach, J. L. (Eds.), Kluwer: Dordrecht, the Netherlands, 1991.

57. Feldberg, S. W. Implications of Marcus–Hush theory for steady-state heterogeneous electron transfer at an inlaid disk electrode. *Anal Chem* 2010, *82*, 5176–5183.

58. Henstridge, M. C.; Ward, K. R.; Compton, R. G. The Marcus–Hush model of electrode kinetics at a single nanoparticle. *J Electroanal Chem* 2014, *712*, 14–18.

59. Suwatchara, D.; Henstridge, M. C.; Rees, N. V.; Compton, R. G. Experimental comparison of the Marcus–Hush and Butler–Volmer descriptions of electrode kinetics. The one-electron oxidation of 9,10-diphenylanthracene and one-electron reduction of 2-nitropropane studied at high-speed channel microband electrodes. *J Phys Chem C* 2011, *115*, 14876–14882.

60. Marcus, R. Chemical and electrochemical electron-transfer theory. *Annu Rev Phys Chem* 1964, *15*, 155–196.

61. Hush, N. S. Adiabatic rate processes at electrodes. I. Energy-charge relationships. *J Chem Phys* 1958, *28*, 962–972.

62. Hush, N. S. Electron transfer in retrospect and prospect 1: Adiabatic electrode processes. *J Electroanal Chem* 1999, *470*, 170–195.

63. Levich, V. G. In *Physical Chemistry: An Advanced Treatise*, vol. 9B; Eyring, H.; Henderson, D.; Jost, W. (Eds.), Academic Press: New York, 1970.

64. Gerischer, H. Electrochemical techniques for the study of photosensitization. *Photochem Photobiol* 1972, *16*, 243–260.

65. Chidsey, C. E. D. Free energy and temperature dependence of electron transfer at the metal-electrolyte interface. *Science* 1991, *251*, 919–922.

66. Finklea, H. O.; Hanshew, D. D. Electron-transfer kinetics in organized thiol monolayers with attached pentaammine(pyridine)ruthenium redox centers. *J Am Chem Soc* 1992, *114*, 3173–3181.

67. Becka, A. M.; Miller, C. J. Electrochemistry at omega-hydroxy thiol coated electrodes. 3. Voltage independence of the electron tunneling barrier and measurements of redox kinetics at large overpotentials. *J Phys Chem* 1992, *96*, 2657–2668.

68. Li, T. T. T.; Weaver, M. J. Intramolecular electron transfer at metal surfaces. 4. Dependence of tunneling probability upon donor-acceptor separation distance. *J Am Chem Soc* 1984, *106*, 6107–6108.

69. Feldberg, S. W.; Sutin, N. Distance dependence of heterogeneous electron transfer through the nonadiabatic and adiabatic regimes. *Chem Phys* 2006, *324*, 216–225.

70. Gavaghan, D. J.; Feldberg, S. W. Extended electron transfer and the frumkin correction. *J Electroanal Chem* 2000, *491*, 103–110.

71. Feldberg, S. W. Implications of extended heterogeneous electron transfer: Part I. Coupling of semiinfinite linear diffusion and heterogeneous electron transfer with a decaying exponential dependence upon distance from the outer helmholtz plane. *J Electroanal Chem* 1986, *198*, 1–18.

72. Dekker, C. Solid-state nanopores. *Nature Nanotech* 2007, *2*, 209–215.

73. Murray, R. W. Nanoelectrochemistry: Metal nanoparticles, nanoelectrodes, and nanopores. *Chem Rev* 2008, *108*, 2688–2720.

74. Bayley, H.; Martin, C. R. Resistive-pulse sensing–from microbes to molecules. *Chem Rev* 2000, *100*, 2575–2594.

75. Lan, W.-J.; White, H. S. Diffusional motion of a particle translocating through a nanopore. *ACS Nano* 2012, *6*, 1757–1765.

76. Holden, D. A.; Hendrickson, G.; Lan, W.-J.; Lyon, L. A.; White, H. S. Electrical signature of the deformation and dehydration of microgels during translocation through nanopores. *Soft Matter* 2011, *7*, 8035–8040.

77. Luo, L.; German, S. R.; Lan, W.-J.; Holden, D. A.; Mega, T. L.; White, H. S. Resistive pulse analysis of nanoparticles. *Annu Rev Anal Chem* 2014, *7*, 513–535.

78. Howorka, S.; Siwy, Z. Nanopore analytics: Sensing of single molecules. *Chem Soc Rev* 2009, *38*, 2360–2384.

79. Schibel, A. E. P.; Edwards, T.; Kawano, R.; Lan, W.; White, H. S. Quartz nanopore membranes for suspended bilayer ion channel recordings. *Anal Chem* 2010, *82*, 7259–7266.

80. White, R. J.; Zhang, B.; Daniel, S.; Tang, J. M.; Ervin, E. N.; Cremer, P. S.; White, H. S. Ionic conductivity of the aqueous layer separating a lipid bilayer membrane and a glass support. *Langmuir* 2006, *22*, 10777–10783.

81. Wei, C.; Bard, A. J.; Feldberg, S. W. Current rectification at quartz nanopipette electrodes. *Anal Chem* 1997, *69*, 4627–4633.

82. Siwy, Z. S. Ion-current rectification in nanopores and nanotubes with broken symmetry. *Adv Funct Mater* 2006, *16*, 735–746.

83. Guo, W.; Tian, Y.; Jiang, L. Asymmetric ion transport through ion-channel-mimetic solid-state nanopores. *Acc Chem Res* 2013, *46*, 2834–2846.

84. Cheng, L.-J.; Guo, L. J. Nanofluidic diodes. *Chem Soc Rev* 2010, *39*, 923–938.

85. Zhou, K. M.; Perry, J. M.; Jacobson, S. C. Transport and sensing in nanofluidic devices. *Annu Rev Anal Chem* 2011, *4*, 321–341.

86. Siwy, Z.; Heins, E.; Harrell, C. C.; Kohli, P.; Martin, C. R. Conical-nanotube ion-current rectifiers: The role of surface charge. *J Am Chem Soc* 2004, *126*, 10850–10851.

87. Sa, N.; Lan, W.-J.; Shi, W.; Baker, L. A. Rectification of ion current in nanopipettes by external substrates. *ACS Nano* 2013, *7*, 11272–11282.

88. Hou, X.; Zhang, H.; Jiang, L. Building bio-inspired artificial functional nanochannels: From symmetric to asymmetric modification. *Angew Chem, Int Ed* 2012, *51*, 5296–5307.

89. Siwy, Z. S.; Howorka, S. Engineered voltage-responsive nanopores. *Chem Soc Rev* 2010, *39*, 1115–1132.

90. Hou, X.; Guo, W.; Jiang, L. Biomimetic smart nanopores and nanochannels. *Chem Soc Rev* 2011, *40*, 2385–2401.

91. Woermann, D. Electrochemical transport properties of a cone-shaped nanopore: High and low electrical conductivity states depending on the sign of an applied electrical potential difference. *Phys Chem Chem Phys* 2003, *5*, 1853–1858.

92. Woermann, D. Electrochemical transport properties of a cone-shaped nanopore: Revisited. *Phys Chem Chem Phys* 2004, *6*, 3130–3132.

93. Woermann, D. Analysis of non-ohmic electrical current–voltage characteristic of membranes carrying a single track-etched conical pore. *Nucl Instrum Methods Phys Res, Sect B* 2002, *194*, 458–462.

94. Cervera, J.; Schiedt, B.; Ramírez, P. A Poisson/Nernst-Planck Model for ionic transport through synthetic conical nanopores. *Europhys Lett* 2005, *71*, 35–41.

95. Cervera, J.; Schiedt, B.; Neumann, R.; Mafa, S.; Ramírez, P. Ionic conduction, rectification, and selectivity in single conical nanopores. *J Chem Phys* 2006, *124*, 104706.

96. Daiguji, H.; Oka, Y.; Shirono, K. Nanofluidic diode and bipolar transistor. *Nano Lett* 2005, *5*, 2274–2280.

97. Vlassiouk, I.; Siwy, Z. S. Nanofluidic diode. *Nano Lett* 2007, *7*, 552–556.

98. Liu, Q.; Wang, Y.; Guo, W.; Ji, H.; Xue, J.; Ouyang, Q. Asymmetric properties of ion transport in a charged conical nanopore. *Phys Rev E* 2007, *75*, 051201.

99. Ai, Y.; Zhang, M.; Joo, S. W.; Cheney, M. A.; Qian, S. Effects of electroosmotic flow on ionic current rectification in conical nanopores. *J Phys Chem C* 2010, *114*, 3883–3890.

100. Kubeil, C.; Bund, A. The role of nanopore geometry for the rectification of ionic currents. *J Phys Chem C* 2011, *115*, 7866–7873.

101. Lan, W.-J.; Holden, D. A.; Zhang, B.; White, H. S. Nanoparticle transport in conical-shaped nanopores. *Anal Chem* 2011, *83*, 3840–3847.

102. Lee, S.; Zhang, Y.; Harrell, C. C.; Martin, C. R.; White, H. S. Electrophoretic capture and detection of nanoparticles at the opening of a membrane pore using scanning electrochemical microscopy. *Anal Chem* 2004, *76*, 6108–6115.

103. Chapra, S. C.; Canale, R. P. *Numerical Methods for Engineers*, 2nd ed.; McGraw-Hill: New York, 1988.

104. Lan, W.-J.; Holden, D. A.; White, H. S. Pressure-dependent ion current rectification in conical-shaped glass nanopores. *J Am Chem Soc* 2011, *133*, 13300–13303.

105. Kovarik, M. L.; Zhou, K. M.; Jacobson, S. C. Effect of conical nanopore diameter on ion current rectification. *J Phys Chem B* 2009, *113*, 15960–15966.

106. Feng, J.; Liu, J.; Wu, B.; Wang, G. Impedance characteristics of amine modified single glass nanopores. *Anal Chem* 2010, *82*, 4520–4528.

107. Lan, W.-J.; Holden, D. A.; Liu, J.; White, H. S. Pressure-driven nanoparticle transport across glass membranes containing a conical-shaped nanopore. *J Phys Chem C* 2011, *115*, 18445–18452.

108. Esaki, L. New phenomenon in narrow germanium P-N junctions. *Phys. Rev.* 1958, *109*, 603–604.

109. Cao, L. X.; Guo, W.; Wang, Y. G.; Jiang, L. Concentration-gradient-dependent ion current rectification in charged conical nanopores. *Langmuir* 2012, *28*, 2194–2199.

110. Firnkes, M.; Pedone, D.; Knezevic, J.; Doblinger, M.; Rant, U. Electrically facilitated translocations of proteins through silicon nitride nanopores: Conjoint and competitive action of diffusion, electrophoresis, and electroosmosis. *Nano Lett* 2010, *10*, 2162–2167.

111. Paik, K.; Liu, Y; Tabard-Cossa, V.; Waugh, M. J.; Huber, D. E.; Provine, J.; Howe, R. T.; Dutton, R. W.; Davis, R. W. Control of DNA capture by nanofluidic transistors. *ACS Nano* 2012, *6*, 6767–6775.

112. Davenport, M.; Healy, K.; Pevarnik, M.; Teslich, N.; Cabrini, S.; Morrison, A. P.; Siwy, Z. S.; Létant, S. E. The role of pore geometry in single nanoparticle detection. *ACS Nano* 2012, *6*, 8366–8380.

113. He, Y. H.; Tsutsui, M.; Fan, C.; Taniguchi, M.; Kawai, T. Controlling DNA translocation through gate modulation of nanopore wall surface charges. *ACS Nano* 2011, *5*, 5509–5518.

114. Ai, Y.; Liu, J.; Zhang, B. K.; Qian, S. Field effect regulation of DNA translocation through a nanopore. *Anal Chem* 2010, *82*, 8217–8225.

115. Luo, L.; Holden, D. A.; White, H. S. Negative differential electrolyte resistance in a solid-state nanopore resulting from electroosmotic flow bistability. *ACS Nano* 2014, *8*, 3023–3030.

116. Luo, L.; Holden, D. A.; Lan, W.-J.; White, H. S. Tunable negative differential electrolyte resistance in a conical nanopore in glass. *ACS Nano* 2012, *6*, 6507–6514.

117. Grahame, D. C. Diffuse double layer theory for electrolytes of unsymmetrical valance types. *J Chem Phys* 1953, *21*, 1054–1060.

118. Probstein, R. F. *Physicochemical Hydrodynamics: An Introduction*, 2nd ed.; John Wiley & Sons, Inc.: New York, 1994.

119. Datta, S.; Conlisk, A. T.; Li, H. F.; Yoda, M. Effect of divalent ions on electroosmotic flow in microchannels. *Mech Research Commun* 2009, *36*, 65–74.
120. Coulter, W. H. Means for counting particles suspended in a fluid. U.S. Patent No. 2656508, 1953.
121. Kozak, D.; Anderson, W.; Vogel, R.; Chen, S.; Antaw, F.; Trau, M. Simultaneous size and ζ-potential measurements of individual nanoparticles in dispersion using size-tunable pore sensors. *ACS Nano* 2012, *6*, 6990–6997.
122. Lan, W.-J.; Kubeil, C.; Xiong, J.-W.; Bund, A.; White, H. S. Effect of surface charge on the resistive pulse waveshape during particle translocation through glass nanopores. *J Phys Chem C* 2014, *118*, 2726–2734.
123. White, H. S.; Bund, A. Mechanism of electrostatic gating at conical glass nanopore electrodes. *Langmuir* 2008, *24*, 12062–12067.

Section II

Nanoelectrochemical Systems

3 Electrochemistry of Monolayer-Protected Clusters

Tessa M. Carducci and Royce W. Murray

CONTENTS

3.1   INTRODUCTION

The concept of monolayer-protected cluster (MPC) was born in the research published by Brust and Schiffrin[1] in 1994. The authors had at that time become interested in preparing metal nanoparticles that, in the fashion of already-known self-assembled monolayers (SAMs),[2,3] could be isolated, handled, and rationally manipulated.[4–7] The Schiffrin contribution was an excellent synthetic pathfinder and was adopted in the author's lab with thanks to these pioneers.

Over the ensuing years, the Brust–Schiffrin synthesis has seen an evolution of variants. The metal most chosen for the nanoparticle core is gold because the nanoparticle's surface(s) can be rather effectively protected from the formation of an oxidized layer (which is problematic for Ag cores) by bonding (i.e., ligating) them with a monolayer shell of organothiolate ligands. The organothiolate ligands can bear a great variety of additional chemical functionalities, singly or as a mixture of different ligands. For smaller nanoparticles—of which those containing 25 gold atoms are important examples—the ligand *monolayer* itself has structural elements in the form of –SR–Au–SR– ring-like *staple* structures like those shown in Figure 3.1B.[8,9] This particular nanoparticle has been structurally defined by single crystal determinations for its reduced *native* $Au_{25}L_{18}{}^{-9}$ and its one-electron oxidized forms.[10] Further important insights have been supplied by a density function theory representation.[11] It is evident from Figure 3.1 that a monolayer-protected Au cluster is not simply a small clump of Au atoms to which organothiolate ligands are attached to prevent their coalescence. The nanoparticle *clump* of atoms has a structure both in arrangement of its internal Au and of its organothiolate ligands.[12]

A coarse analogy to MPCs can be drawn with dendrimer structures.[13] These have similarity to MPC nanoparticles by having multiple peripheral chemical functionalities. The analogy is

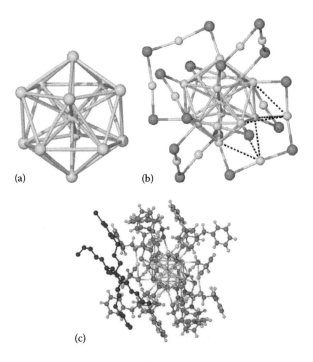

(a)   (b)

(c)

FIGURE 3.1   X-ray crystal structure of [TOA⁺] $\left[Au_{25}(SCH_2CH_2Ph)_{18}{}^{-} \right]$. (a) Arrangement of the Au_{13} core with 12 atoms on the vertices of an icosahedron and one in the center. (b) Depiction of gold and sulfur atoms, showing six orthogonal $-Au_2(SCH_2CH_2Ph)_3{}^{-}$ *staples* surrounding the Au_{13} core (two examples of possible aurophilic bonding shown as dashed lines). (c) [TOA⁺] $\left[Au_{25}(SCH_2CH_2Ph)_{18}{}^{-} \right]$ structure with the ligands and TOA⁺ cation (depicted in blue) (Legend: gold = yellow; sulfur = orange; carbon = gray; hydrogen = off-white; the TOA⁺ counterion is over two positions with one removed for clarity). (From Heaven, M.W. et al., *J. Am. Chem. Soc.*, 130, 3754, 2008.)

structurally imperfect in the sense that dendrimers become less dense near their centers, while the opposite is true of MPCs.

The general idea of preparing a nanometer-scale particle that can be chemically functionalized is quite appealing since so much more can be varied than simply the nanoparticle diameter. Table 3.1 shows some of the different organothiolate ligands and functional groupings that have been used in Au MPC chemistry and references to their syntheses and characterizations. The chemical behavior of the MPC is dominated by the organothiolate ligands and its appended functional groups. The different thiolate ligands also determine (or influence) nanoparticle solubility, since the nanoparticle presents its external ligand shell to the solvating medium. The ligands also contribute to the MPC optical spectrum, but the overall optical spectrum of an Au MPC tends to be dominated by the strong absorbance of the Au core. The ligands on an MPC can be identical (chemically speaking), but MPCs that bear mixtures of different ligands can also be prepared. In the latter case, one might use a mixture of ligands in the MPC synthesis, or the original ligands may have been replaced (in part or fully) by new ones in a process akin to ligand exchange for a metal complex (also termed *place exchange*).[37–39] The author has often chosen the latter synthetic course for a pragmatic reason: preparing MPCs from a limited set of thiols (such as hexanethiol or the thiol $HSCH_2CH_2Ph$) allows establishing experimental parameters that constrain the core size(s) produced. Afterward, plus perhaps some core-size fractionation in the interest of improving monodispersity, ligands may then be changed. The chemical details of how ligand exchanges occur comprise another interesting, and still evolving, aspect of MPC chemistry. For example, consider that for ligands bound in the *staple* structures shown in Figure 3.1 to become replaced with a different organothiolate ligand, multiple bonds are broken for each organothiolate ligand. The details of such reactions remain to be adequately analyzed.

There are also chemical processes that can transport Au moieties between nanoparticles. These operate in so-called annealing steps in MPC nanoparticle preparation.[24] An initially formed mixture of different sizes of MPC nanoparticles can often be reformed into a more uniform set of nanoparticle composition(s), because there exist innate preferences for different sizes (e.g., atom numbers). The entities that are released from nanoparticles during annealing are probably similar to the tetrameric versions $(AuL)_4$ observable by mass spectrometry as seen by Cliffel et al.[40] While there is some research activity in extending MPC chemistry to nanoparticles in which the cores contain other metals, or mixtures thereof, this article will focus on Au MPCs.

Another important variable in regard to electrochemical experiments is the state of the nanoparticle sample. Is it dissolved in solution, or present as a film on an electrode? Figure 3.2 shows cartoons of how the nanoparticle could be presented to an electrode; there is clearly an abundance of variables. However, a variety of other kinds of measurements and ideas have been invoked which aid our understanding of electron transfer (ET) events. These include, for example, transient absorption showing aspects of the quantum confinement effect for Au MPCs,[41] ET catalysis reactions,[42] mass spectrometry of mixed ligand nanoparticles,[43] ETs at liquid–liquid interfaces,[44] and an important theoretical perspective regarding thiolate ligand bonding, termed *divide and protect*.[45] These and selected others are further described in the sections that follow.

A variety of reviews of MPC nanoparticle electrochemistry have appeared,[8,12,46–50] as well as journal themed issues.[51]

3.2 ABOUT MPCs: PREPARATION AND METHODS FOR STUDY AND CHARACTERIZATION

3.2.1 MPC PREPARATION AND CHARACTERIZATION

A simple plan for preparing a thiolated Au nanoparticle is to reduce an Au salt in the presence of the selected thiol. A closer look at this reaction reveals an enormous range of choices of reaction conditions that can, variously, produce nanoparticles with the same Au core size or nanoparticles

TABLE 3.1
Known MPC Nanoparticles and Their Ligands

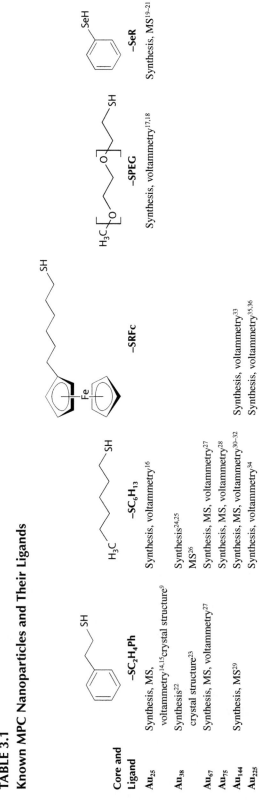

Core and Ligand	–SC₂H₄Ph	–SC₆H₁₃	–SRFc	–SPEG	–SeR
Au_{25}	Synthesis, MS, voltammetry[14,15]crystal structure[9]	Synthesis, voltammetry[16]		Synthesis, voltammetry[17,18]	Synthesis, MS[19-21]
Au_{38}	Synthesis[22] crystal structure[23]	Synthesis[24,25] MS[26]			
Au_{67}	Synthesis, MS, voltammetry[27]	Synthesis, MS, voltammetry[27]			
Au_{75}		Synthesis, MS, voltammetry[28]			
Au_{144}	Synthesis, MS[29]	Synthesis, MS, voltammetry[30-32]	Synthesis, voltammetry[33]		
Au_{225}		Synthesis, voltammetry[34]	Synthesis, voltammetry[35,36]		

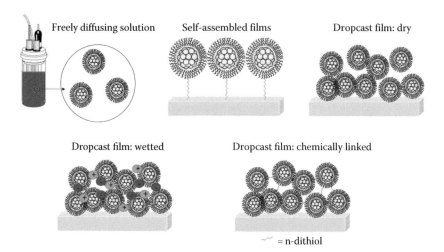

FIGURE 3.2 Different physical states of monolayer-protected Au nanoparticles.

having different sizes. A mixture of produced sizes can be induced to digest toward more uniform nanoparticle size(s) by using a prolonged reaction time, the presence or addition of excess thiol, a change in temperature, isolation of the nanoparticles following a change in solvent medium—there is a long list of options available to the nanoparticle synthesizer. Details of the list change according to whether the desired nanoparticles are aqueous or organic-soluble. The diversity of preparation chemistries is actually rather large.

The first deliberately ligand-stabilized small Au nanoparticles were described by Schmid.[52] The ligands were triphenylphosphine and chloride, and the Au nanoparticles were labeled as $Au_{55}(Ph_3P)_{12}Cl_6$. The claimed Au_{55} composition of these nanoparticles was for some time not generally accepted—it may have been a mixture instead. However, more recent work[53] based on MALDI mass spectrometry has confirmed the existence of alkanethiolate stabilized $Au_{55}(SC_{18}H_{37})_{31}$ and $Au_{54}(SC_{18}H_{37})_{30}$ nanoparticles, as well as a sensitivity of the obtained composition to the synthetic pathway employed.

As noted earlier, the modern era of work on thiolate-protected Au nanoparticles dawned[1] in work by Brust and Schiffrin at Liverpool University (UK) two decades ago (1994). In the spirit of the analytical chemistry of nanoparticles, and with a band of capable collaborators, this author hurled a variety of measurement approaches at the question of their exact chemical description. Our first contribution[4] to this topic described nanoparticles with ca. 1.2 nm (diameter) cores coated with C8, C12, or C16 alkanethiolate ligands. Methodology covered a wide range of measurements, including ^{1}H and ^{13}C NMR, elemental analysis, differential scanning calorimetry (DSC), thermogravimetry (TGA), diffusion-ordered NMR spectroscopy (DOSY), small-angle x-ray scattering (SAXS) data on MPC solutions, and images of films from scanning tunneling (STM), and atomic force microscopy (AFM). This early paper also presented measurements of electron hopping through dry films of these materials, demonstrating distance-dependent electron tunneling through the intervening alkanethiolate chains surrounding the Au nanoparticles. Attention next turned to their solution voltammetry,[5] notably nanoparticles labeled with thiolated Fc groups, producing a recognizable electrochemical signal (Figure 3.3).

Besides showing the expected signature of the Fc wave, the experiment in Figure 3.3 revealed that further valuable information about the nanoparticles lies in currents associated with electrochemical charging of the electrical double layers of the Au core. This charging current, like that of the Fc reaction, is mass transport controlled. Such transport-controlled charging was an already-recognized behavior of colloidal metal nanoparticles.[54] Nanoparticles bearing no electroactive

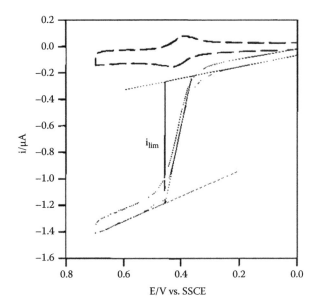

FIGURE 3.3 Cyclic voltammograms at 5 mV s^{-1} for the 1 μM 1:9.5 C8Fc/C8 cluster compound (avg 9 Fc/cluster) in 0.1 Bu$_4$NClO$_4$/CH$_2$Cl$_2$ at 0.15 cm^2 stationary (dashed line) and rotated (solid line, 1600 rpm) glassy carbon disk electrode. (From Hostetler, M.J. et al., *J. Am. Chem. Soc.*, 118, 4212, 1996.)

ligands also produce a mass transport-controlled charging current. Additionally, if their polydispersity can be reduced, they can also produce a distinctive voltammetric pattern[6] based on one-electron charging steps of the nanoparticle's electrical double layer, and over a certain range of sizes, show a pattern termed *quantized double layer* (QDL) *charging*, illustrated in Figure 3.4. This is an electrostatically driven phenomenon; the current peaks arise for increments of charging the nanoparticle's electrical double layer by successive one-electron changes.

Electron microscopy is an obvious and widely available tool for the study of nanoparticles. The information content from standard TEM instrumentation, for nanoparticles with very small (1–3 nm diameter) dimensions, is somewhat constrained by the need to register multiple images to provide a representative histogram of the range of nanoparticle sizes present. Determining the atom arrangements in such small nanoparticles at very high resolution requires care with regard to avoiding *beam damage* caused by inadequate thermal dissipation. When heated sufficiently, MPCs can lose thiolate ligands and experience core-sintering. A special electron microscopy approach called high-angle annular dark field scanning transmission electron microscopy (HAADF-STEM) can produce images interpretable in terms of the numbers of Au atoms per core. Figure 3.5 presents an example[55] where the image intensities could be used to estimate that the average cores in two samples contained only about 13–14 (Panels a, b, d, and e) Au atoms, and in another (Panels c and f), a broader range of 183 ± 116 atoms. Also, while conventional TEM in and of itself does not provide chemical identity of the nanoparticle, many TEM instruments are equipped with x-ray emission detectors, which give valuable elemental composition data.

Clearly the *analytical chemistry of very small nanoparticles* requires multiple kinds of measurements among which nanoparticle diameter is just one of several important parameters. Mass spectrometry can provide important and definitive measurements on MPCs, particularly when ligand loss and/or core fragmentation can be avoided. For the smallest nanoparticles, electrospray[56] and MALDI-TOF[57] mass spectrometry measurements have successfully provided atomically precise mass measurements. In an exploration of ligand exchange chemistry of Au$_{25}$L$_{18}$ nanoparticles, unfragmented nanoparticles of masses >7 kDa could be observed, and the ligand exchange reaction was demonstrated to be a stochastic process. That is, exchange of an initially uniform thiolate ligand

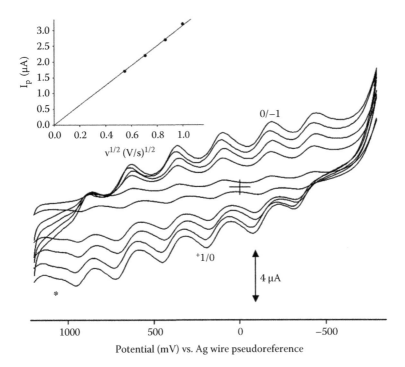

FIGURE 3.4 Cyclic voltammetry at a 0.02 cm² Pt working electrode, of 200 μM annealed EtOH soluble C6 MPCs (annealed with C16SH thiol, as in the experimental section) in 0.1 M Bu$_4$NPF$_6$/CH$_2$Cl$_2$; voltammograms at potential sweep rates (v) of 50, 300, 500, 700, and 1000 mV s^{-1}, potentials vs. Ag wire pseudoreference, Pt flag counterelectrode. Inset shows variation of peak current with v$^{1/2}$ for the MPC$^{0/+1}$ wave. (From Hicks, J.F. et al., *J. Am. Chem. Soc.*, 124, 13322, 2002.)

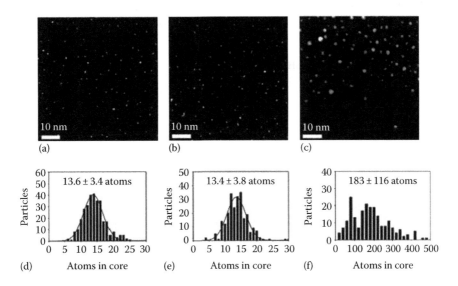

FIGURE 3.5 Representative HAADF-STEM images of (a) Au$_{13}$[PPh$_3$]$_4$[S(CH$_2$)$_{11}$CH$_3$]$_2$Cl$_2$, (b) Au$_{13}$[PPh$_3$]$_4$[S(CH$_2$)$_{11}$CH$_3$]$_4$, and (c) thiolate-protected MPCs. All images were collected at 1 M×magnification. Distribution of core atom counts for gold clusters measured using the quantitative HAADF-STEM technique for (d) 1 (N=313), (e) 2 (N=243), and (f) thiolate-protected MPCs (N=219). Gaussian fits to the histograms of (a) and (b) are provided. (From Menard, L.D. et al., *J. Phys. Chem. B*, 110, 14564, 2006.)

shell on an Au nanoparticle by a different thiolate ligand quickly leads to a polydisperse ligand mixture,[58] as illustrated by Figure 3.6. If there is no preference for bonding one ligand over the other, to the Au core, the equilibrium distribution of ligand populations would follow a binominal distribution. The maximum of mixed ligand mixture population for an Au_{25} nanoparticle shown in Figure 3.6 is not exactly centered, so there is a slight preference for the $-SCH_2CH_2Ph$ ligand over the hexanethiolate ligand. This preference would be difficult to detect from a kinetically based nonequilibrium measurement.

It is generally challenging to obtain mass spectra of larger sized MPCs (such as Au_{144}) without incurring some nanoparticle fragmentation. This is mildly surprising since very heavy and unfragmented ions formed from protonations of multiple base sites on large biological entities are commonly observable in electrospray experiments. The multiple protonations of base sites reduce the m/z (mass/charge) ratios of such ions into m/z ranges commensurate with common mass spectrometer limits. When the same tactic (multiple charging) was attempted, however, to reduce the m/z values for Au_{144} nanoparticles (MW approaching 40 kDa), they could not be charged to sufficiently large m/z values. This likely reflects how the small (relative to a biological entity) dimension of the nanoparticle core leads to severe electrostatic repulsion effects. This problem was addressed[59] for nanoparticles with Au_{144} cores, by place-exchange introduction of multiple intrinsically charged thiolate ligands (i.e., ligands bearing cationic quaternary ammonium groupings). The tactic of using intrinsically charged ligands produced Au_{144} nanoparticles with charge (z) values from 10+ to 15+ on the nanoparticle core, allowing successful electrospray observations. The Au-thiolate bonding was sufficiently strong to resist ligand dissociation.

Another powerful route to observe Au nanoparticle structure is based on crystallography of single crystals of purified nanoparticles. This tactic of course classically requires obtaining a high level of nanoparticle purity, or a period of nanoparticle refining. This was successfully done for the first time by Kornberg et al.,[60,61] producing $Au_{102}(p\text{-}MBA)_{44}$ (p-MBA = p-mercaptobenzoic acid) crystals. These forerunner nanoparticles, like the $Au_{25}L_{18}$ variety,[8,9] showed a proclivity for the formation of semi-ring or *staple* ring structures (Figure 3.1). The central Au atoms of the Au_{102} nanoparticle were packed in a Marks decahedron, with the surrounding Au atoms in somewhat complex and unusual arrangements (Figure 3.7). This paper was accompanied by a useful commentary by Whetten and Price.[62] The structural assessment of the $\left[\left(Oct \right)_4 N^+ \cdot Au_{25}L_{18}^- \right]$

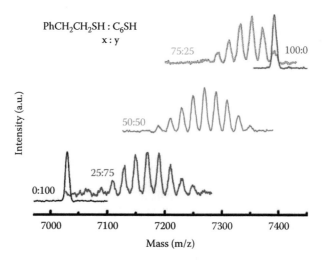

FIGURE 3.6 Monolayer ligand distribution of the mixed Brust reaction product $Au_{25}(SCH_2CH_2Ph)_{18-x}(SC6)_x$ as observed by MALDI-MS spectrum using different starting ligand ratios 25:75, 50:50, and 75:25. (From Dass, A. et al., *J. Phys. Chem. C*, 112, 20276, 2008.)

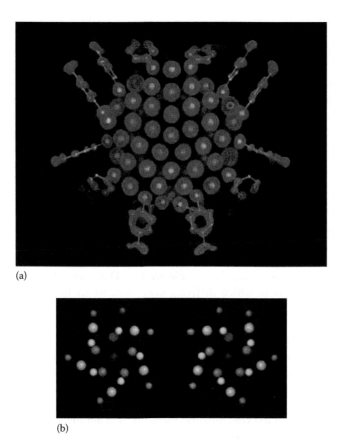

FIGURE 3.7 X-ray crystal structure determination of the $Au_{102}(p\text{-MBA})_{44}$ nanoparticle. (a) Electron density map (red mesh) and atomic structure (gold atoms depicted as yellow spheres, and p-MBA shown as framework and with small spheres [sulfur in cyan, carbon in gray, and oxygen in red]). (b) View down the cluster axis of the two enantiomeric particles. Color scheme as in (A), except only sulfur atoms of p-MBA are shown. (From Jadzinsky, P.D. et al., *Science*, 318, 430, 2007.)

nanoparticle anion (Figure 3.1) also was preceded by a storage of a nanoparticle sample under selected solvents and isolation by solvent-induced precipitation. After several days, needle-shaped black crystals could be isolated.[9,12] This synthetic beginning has been followed by further reports of single crystal structure determinations for $Au_{38}L_{24}$ ($L = SC_2H_4Ph$) and $Au_{36}L_{24}$ ($L = SC_2H_4Ph$) nanoparticles.[23,63]

It is not several-day-long surprising that the Au nanoparticles which have been thus far structurally defined by single crystal measurements are all in the lower size range; this reflects the challenges of attaining sufficiently monodisperse material for single crystal formation. For example, a mass spectral study[29] of the $Au_{144}(SC_2H_4Ph)_{60}$ nanoparticle produced a high quality mass determination, but the materials available apparently did not yield to single crystal isolation.

A different synthetic approach, devised by the Crooks laboratory, invokes dendrimer stabilization of the nanoparticles by encapsulating them. Several different nanoparticle types[64–70] have been sequestered by poly(amidoamine) dendrimers, including Pd, Pt, Au, ferromagnetic Ni, and core-shell Pd–Au and bimetallic nanoparticles. Au and Pd nanoparticles were prepared by sequestering the metal ions into the dendrimers, and then reducing them. Size control was sufficient to observe the distinctive pattern of QDL charging.[34]

3.2.2 Voltammetry and Other Properties as a Function of Nanoparticle Size

We now take up the voltammetry of nanoparticles of known or identifiable sizes. Figure 3.8 provides a general classification of MPC voltammetry according to the number of Au atoms. Nanoparticles with less than about 100 Au atoms develop an energy gap between HOMO and LUMO electronic levels, and are *molecule-like*. Au_{25} is by far the most thoroughly investigated molecule-like nanoparticle. The energy gap decreases with increasing size and closes at about 100 atoms, where the voltammetric population pattern changes to that of Figure 3.4, which is exemplified by Au_{144}. This pattern persists over a range of ca. 100–300 Au atoms per nanoparticle, which is determined by the individual double layer capacitances of the nanoparticles. Thus, Au_{144} nanoparticles behave as quantum capacitors, chargeable in discrete one-electron increments well spaced on the potential scale, as in Figure 3.4. The range of this behavior ends when the nanoparticle and its double layer capacitance are large enough that waves for individual one-electron increments of charge start to overlap and merge. This occurs when the voltage spacing of the current peaks approaches ca. 100 mV. Larger nanoparticles, with larger double layer capacitances, ultimately yield a continuum of charging current—which is the familiar *normal* behavior of electrochemical double layer charging of electrodes. There are also a number of reported Au nanoparticle preparations in which different core dimensions have been obtained, but not explored electrochemically.

Besides the delineation of Figure 3.8, the discussion can be divided according to whether the nanoparticle is organic-soluble or water soluble; this is substantially a function of the chosen ligands. Hydrophobic ligands like hexanethiolate and phenylethanethiolate confer organic solvent solubility on the nanoparticle, whereas ligands like glutathiolate yield water-soluble nanoparticles. These statements refer to nanoparticles in the core-size range of ca. Au_{11} to cores larger than a few hundreds of atoms. Unfortunately, electrochemical observations on MPCs are much more constrained in aqueous media, and most of the voltammetric literature is in aprotic, organic phases. The chosen ligands of course also impact the task of preparing samples, which are monodisperse in core size, or only roughly so.

Nanoparticles that are organic soluble frequently bear ligands that are alkanethiolates or similar hydrophobic thiolates. The largest Au core size for which a distinct QDL charging pattern has been seen[34] is $Au_{225}(S(CH_2)_5CH_3)_{75}$. This nanoparticle sample was admixed with Au_{144} species, but in the context of an HPLC separation, the Au_{225} double layer charging peaks, spaced about 185 mV apart, could be seen. Another core-size property is the optical (UV–vis) spectrum, which can display a

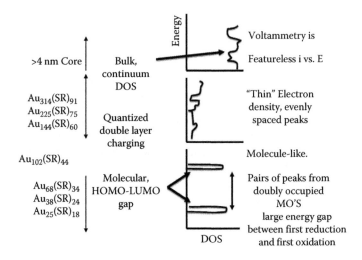

FIGURE 3.8 General classification of MPC voltammetries.

surface plasmon band at about 520 nm. This band appears for the Au_{225} and larger nanoparticles but is largely absent for Au_{144} and smaller nanoparticles.

3.2.2.1 Voltammetry of Au_{25} Nanoparticles

The voltammetry of the 1.1 nm diameter Au_{25}^- nanoparticle, with $-SCH_2CH_2Ph$ ligands, was described in detail in 2004 (Figure 3.9).[71] Unfortunately, for a period of time, owing to the limitations of TEM imaging, it was mislabeled[18,35,71–74] as an Au_{38} nanoparticle. This was corrected in 2007[43,56] when definitive electrospray ionization mass spectrometry of the nanoparticle was obtained. Subsequently, as noted earlier, the crystal structure of this nanoparticle was obtained (Figure 3.1). Figure 3.9 shows that there are two pairs of oxidation waves (e.g., two doubly occupied molecular orbitals) and one pair of reduction steps. The voltammetry was observed at lowered temperature in order to stabilize the products of the most positive and negative ET steps. There is a 1.6 V separation between the first oxidation and the first reduction, which after a 0.29 V charging energy correction shows that the nanoparticle exhibits a HOMO-LUMO energy gap of ca. 1.33 V, definitely a molecule-like characteristic.

The voltammetry for the first oxidation step, $Au_{25}^{-1/0}$, shows that the ET is mildly sluggish, as was reported[75] by Maran et al. The sluggishness opened the door to an NMR investigation[76] of its ET kinetics, which showed that the homogeneous solution self-exchange rate constant was

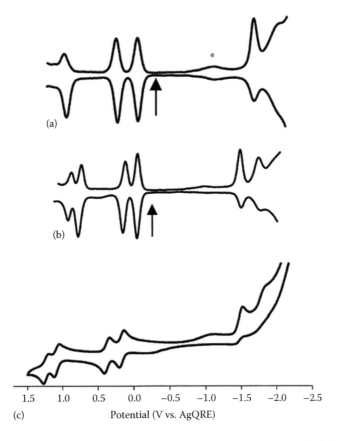

(a)

(b)

| 1.5 | 1.0 | 0.5 | 0.0 | −0.5 | −1.0 | −1.5 | −2.0 | −2.5 |

(c)   Potential (V vs. AgQRE)

FIGURE 3.9 (a) 25°C and (b) −70°C differential pulse voltammograms (DPVs) at 0.02 V s^{-1}, and (c) −70°C cyclic voltammogram (0.1 V s^{-1}) of an Au_{25}^- nanoparticle, with $-SCH_2CH_2Ph$ ligands (previously mislabeled as Au_{38}) in 0.1 M Bu_4NPF_6 in degassed CH_2Cl_2 at 0.4 mm diameter Pt working, Ag wire quasi-reference (AgQRE), and Pt wire counterelectrode. Arrows indicate solution rest potentials, and * indicates wave for incompletely removed O_2. (From *J. Am. Chem. Soc.*, 126, 6193, 2004.)

3×10^7 M^{-1} s^{-1} with a 25 kJ mol^{-1} activation energy barrier (E$_A$). This barrier energy is larger than expected for an outer sphere ET of a species of MPC size but was consistent with the voltammetry reported by Maran et al.[75] The larger barrier is also consistent with a detected difference[76] in Raman Au–S stretch energies for the Au–S bond in the −1 versus zero-valent MPC oxidation states.

The electrochemical potentials for charging of Au MPC cores are affected by the chemical nature of the thiolate ligands, exhibiting substituent effects on the Au$_{25}$ redox potentials.[72] This was illustrated in ligand exchange experiments on the Au$_{25}$ nanoparticle,[77] where serially replacing the ligands –SCH$_2$CH$_2$Ph on Au$_{25}$ cores with different numbers of –SPh(p-X) (where X = –NO$_2$ or –Br) ligands produced shifts in formal potential (Figure 3.10). These shifts are commensurate with general molecular knowledge of how electron induction and changes in electron density for a redox species become reflected in its redox potentials. It is interesting that the changes in Figure 3.10 are linear with the number of ligands. Figure 3.10 also shows DFT theory predictions for the effect of serially replacing –SCH$_3$ ligands with –CH$_2$Cl ligands; the changes are again linear. The different slopes in Figure 3.10 simply reflect differences in electron-withdrawing properties of the ligands.

As shown earlier in Figure 3.9, doublet waves appear in the Au$_{25}$ nanoparticle voltammetry for the successive one-electron oxidations of Au$_{25}$L$_{18}$$^-$ and Au$_{25}$L$_{18}$^{0} and successive one-electron reductions of Au$_{25}$L$_{18}$$^-$. Having an odd number of metal centers, the *native* Au$_{25}$L$_{18}$$^-$ nanoparticle (L = SC$_2$H$_4$Ph$^-$) is a charged, −1 species; it would otherwise have a singly occupied HOMO level. This nanoparticle is readily oxidized when air exposed.[78] The oxidized form (Au$_{25}$L$_{18}$^{0}) contains an unpaired electron and is paramagnetic.[79]

Electrochemically generated luminescence (ECL) refers to experiments in which reduced and oxidized states of a species are alternately generated and allowed to react in a common diffusion layer around the electrode(s).[80] An alternative ECL generating mode is to reduce or oxidize the proposed emitter in the presence of another species whose reduced or oxidized form rapidly decomposes, producing a strongly oxidizing or reducing state. The additional species is referred to as a coreactant. It is known[73,81] that Au$_{25}$ (and Au$_{144}$) nanoparticles exhibit luminescence in the near-IR. The Au$_{25}$ luminescence can also be evoked by ECL, as shown by Li et al.[82] Li also showed that Au$_{25}$ ECL emission could be observed by use of the common coreactant S$_2$O$_8$$^{2-}$ whose reduction leads to the highly electron-deficient SO$_4$$^-$ species. In a more complete report, Ding et al.[83] showed that ECL is produced when Au$_{25}$ reduced to the −2 state reacts with its +2 oxidized state. This reaction

$$Au_{25}^{2-} + Au_{25}^{2+} \rightarrow Au_{25}^{-*} + Au_{25}^{+}$$

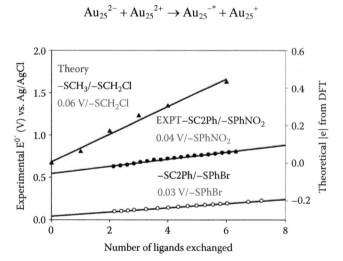

FIGURE 3.10 Comparison of experiment with theory for how Au$_{25}$ E^0 changes with the number of new ligands. (From *J. Phys. Chem. C*, 113, 9440, 2010.)

produces a broad emission at 893 nm. The emission is seen only in the reduction cycle since the Au_{25}^{2-} dianion is only fleetingly stable. The emission intensity could be enhanced by using the common ECL coreactant benzoyl peroxide. The detected 893 nm photon energy is very close to the HOMO-LUMO gap energy of Au_{25}^{+}, as estimated from the Au_{25} voltammetry.

3.2.2.2 Voltammetry of Au_{38} Nanoparticles

The Au_{38} species was isolated for electrochemical study[24,25] by Quinn in 2008. This study was important in another way as it illustrated[25] how sensitive the observed nanoparticle current–potential patterns can be to the nanoparticle's preparation. The nanoparticles were size-focused by prolonged and repeated annealing in solutions containing excess hexanethiol ligand. The Au_{38} study also showed how subtle differences in the preparation protocol can lead to predominance of an Au_{38} nanoparticle product rather than Au_{25}. The voltammetric pattern—doublets of oxidation and reduction waves[25]—that appeared (Figure 3.11) for Au_{38} nanoparticle solutions in DCE qualitatively resembles that of Au_{25} nanoparticles but with a definite difference in the voltage separation (1.2 V) between the first oxidation and first reduction steps, for example, the HOMO-LUMO gap. The peak potential separation for Au_{38}, after a 0.3 V correction for charging energy, corresponds to a 0.9 V HOMO-LUMO energy gap, distinctly different from that (ca. 1.3 V) for Au_{25} nanoparticles.

Another interesting aspect of the Au_{38} nanoparticle voltammetry,[24] evident in Figure 3.11, is that oxidations of the HOMO state electron pair (and reduction into the LUMO) occur as doublet waves. The doublet waves reflect successive removal or addition of pairs of electrons at voltages separated by the ca. 0.3 V nanoparticle charging energy. The clearness of the voltammetry, in regard to this doublet from occupied electron levels, is obviously quite dependent on annealing of the prepared nanoparticle sample.

3.2.2.3 Voltammetry of Au_{67} Nanoparticles

This core size was established in a report by Dass et al. in 2013 that included an exceptionally complete characterization[27] by optical, mass spectrometric, electrochemical, and first-principles theoretical analysis. The nanoparticle was formulated as $Au_{67}(SR)_{35}^{2-}$ where $R = -CH_2CH_2Ph$ or hexyl. Its voltammetry (in THF) is qualitatively reminiscent of the Au_{25} pattern of levels, showing doublets of anodic and cathodic pulse voltammetry peaks. The initial oxidation and reduction peaks are separated by a gap of 0.74 V. Based on a general similarity of its voltammetry and energy gap to that

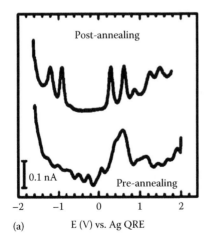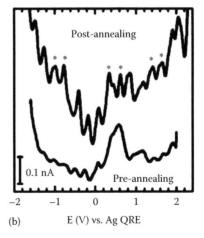

FIGURE 3.11 Microelectrode square wave voltammograms (SWVs) for particles post-annealing (upper traces) in either (a) DCE or (b) CB compared with the response pre-annealing (lower trace in both plots). All SWVs were recorded in DCE solution containing 10 mM base electrolyte. In (B), the peaks ascribable to Au_{38} are marked with asterisks. (From Toikkanen, O. et al., *J. Phys. Chem. Lett.*, 1, 32, 2010.)

reported for a nanoparticle tentatively labeled as Au_{75},[28] it is speculated that the latter's approximate Au atom count could be revised to Au_{67}.

3.2.2.4 Voltammetry of Au_{144} Nanoparticles

The phenomenon of QDL charging[6,12] in differential pulse voltammetry is illustrated in Figure 3.14 (later in the chapter). An idealized view of this phenomenon is that the current peaks correspond to increments of charging the nanoparticle's electrical double layer by successive one-electron changes of the nanoparticle electron population. The clearest examples of this property have been observed[30] for $Au_{144}(SC_6H_{13})_{60}$. A highly simplified model of QDL charging[84] is the nanoparticle, dissolved in an electrolyte solution, as a concentric sphere capacitor. The nanoparticle's ligands (e.g., alkanethiolates or $-SCH_2CH_2Ph$) serve as the capacitor dielectric, and the MPC capacitance C_{CLU} can be thereby stated as

$$C_{CLU} = 4\pi\epsilon\epsilon_0 \frac{r}{d}(r+d), \tag{3.1}$$

where

ϵ_0 is the permittivity of free space
ϵ is the dielectric constant of the monolayer
r is the radius of the core
d is the chain length of the thiolate ligand

This relation shows that the core radius and monolayer chain length are the manipulable MPC variables influencing individual cluster capacitances. The cluster capacitance increases with core radius and decreases with monolayer chain length, and these effects are indeed seen in prepared Au_{144} nanoparticle samples, for both solutions[84] and films.[85]

Although remarkably effective, the concentric sphere view for the capacitance of C_{CLU} is in fact a substantial oversimplification. While it does represent the expected change in capacitance with increasing alkanethiolate chain length, the model does not account for nonidealities like permeation of solvent into the ligand monolayer—which means that the QDL peak spacing can vary with solvent.[85] The model would obviously also be impacted by the presence of multifunctionalized thiolate ligands. Another issue is that the degree of nanoparticle molecularity may become significant. Using density functional theory, a more molecular picture of quantized charging for Au_{144} nanoparticles was put forth by Hakkinen and coworkers.[86] The projected densities of electron states (PDOS) immediately around the HOMO and LUMO levels for the 29 kDa (Au_{145}) nanoparticle show a series of levels that crudely resembles the QDL model, but the level spacing is somewhat different and less regular—as indeed the experiment reveals. The average spacing between states in the HOMO-LUMO is only 0.02 eV. This small spacing means that if one defines molecularity as possession of a HOMO-LUMO gap, the Au_{144} nanoparticle is very modestly so.

Mertens et al.[87] have shown that QDL voltammetry of hexanethiolate-coated MPCs of the Au_{144} size can also be observed in a room temperature ionic liquid. The DPV peak spacing involving the $z=0$ charge state was larger than the potential spacing between other voltammetric peaks.

An optical bandgap, the threshold for photons to be absorbed by a material, differs from the electrochemically detected HOMO-LUMO. The absorption involves no change in the electrical charge state of the material, unlike an ET process. The optical gap for Au_{144} nanoparticles depends somewhat on the ligands but for $-SH$ ligands, it is calculated[88] to be 0.175 eV. The experimental gap for Au_{144} nanoparticles with $-SCH_2CH_2Ph$ ligands is 0.186 eV.

It is important to understand that the double layer charging peaks of MPC solutions—such as those in Figure 3.14—are only *formally* analogous to the current peaks seen in conventional one-electron redox reactions. Thus, the quantized DL charging currents[30,89,90] are controlled by rates of diffusion of the MPCs, and mixtures of MPCs with adjacent states of core charge (z) are mixed-valent solutions that follow the Nernst equation in regard to the *average* core potential. The potentials at which the Figure 3.14 quantized DL charging events appear can be described[84] by

$$E^0_{z,z-1} = E_{PZC} + \frac{(z-1/2)e}{C_{CLU}} \tag{3.2}$$

where

$E^0_{z,z-1}$ is the formal potential of the $z/(z-1)$ charge state *couple* and is given by DPV peak potentials

E_{PZC} is the potential of zero charge (i.e., $z=0$) of the cluster

z is signed such that $z>0$ and $z<0$ correspond to core *oxidation* and *reduction*, respectively

This relation predicts a linear plot of $E^0_{z,z-1}$ versus charge state (termed a *z-plot*, Figure 3.12) and is useful in allowing evaluation of an average value of C_{CLU} from its slope and inspection for irregular or systematic changes in C_{CLU} as a function of potential and charge state. Values of C_{CLU} can also, of course, be obtained from the spacing of any adjacent pair of charging peaks.

We have seen earlier (Figure 3.3) that Au nanoparticles can be labeled with redox moieties (such as Fc) as part of their organothiolate ligand shell. If the nanoparticle's ligand shell is composed *entirely* of somewhat bulky redox labeled organothiolates, their density on the exterior surface of

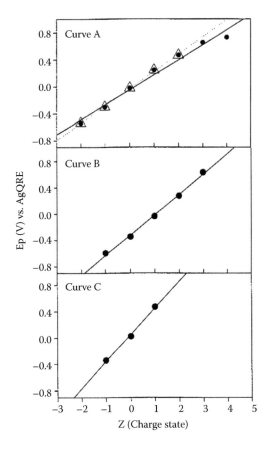

FIGURE 3.12 Formal potentials of charging events versus cluster charge state (Z), taking $E_{PZC} \approx -0.2$ V, for C6 (curve A), C8 (curve B), and C12 (curve C). (•) potentials of all charge states plotted (linear regression,); (△) only charge states −2 to +2 plotted (linear regression). (From Hicks, J.F. et al., *Anal. Chem.*, 71, 3703, 1999.)

the ligand shell can invoke steric constraints on the ligand population.[35] Thus, the population of hexanethiolate ligands on Au_{144} nanoparticles has been estimated[35] to be 53, whereas the estimated population of hexanethiolate ligands with terminal Fc groupings on the same core-size nanoparticle is 39. This apparent steric crowding effect increases for nanoparticles of smaller core size.

3.3 ELECTRON TRANSFERS OF AU MPCs

Quantifying electron transport between Au MPCs or Au MPCs and another redox species is critical for understanding their core-size and ligand-dependent properties and how to tune them for application in electronic devices. As depicted earlier in Figure 3.2, there are three main ways in which ET in Au MPC films can be studied: in solutions, in SAMs, and in films that are in direct contact with the electrode surface. Within the latter category, films can be self-assembled or dropcast, dry or solvent-wetted, or in their native state or chemically linked. Electron transport between Au MPCs for each state differs dramatically, and their ET characteristics and parameters must be distinguished.

3.3.1 ET of Au MPCs Freely Diffusing in Solution

ET in solutions—either in aqueous or organic media—is typically studied as an electron exchange between a freely diffusing Au MPC and an electrode surface. Little is known about the exact details of the interaction between MPC and electrode surface, but in an oxidation reaction, the electron is hypothesized to be transferred from within the metal core of the cluster, tunneling through the insulating ligand to the electrode surface. In a reduction reaction, the reverse process occurs.

ET in solutions of Au MPCs are typically studied by cyclic voltammetry or scanning electrochemical microscopy (SECM). A report from the Cliffel group[91] established a methodology for measuring the forward heterogeneous ET rate through the thiol monolayer of Au MPCs in solution, using the SECM feedback mode. They develop a formula for extracting the ET rate constant from SECM approach curves (Figure 3.13) for monodisperse thiolate-protected Au_{144} MPCs. In freely

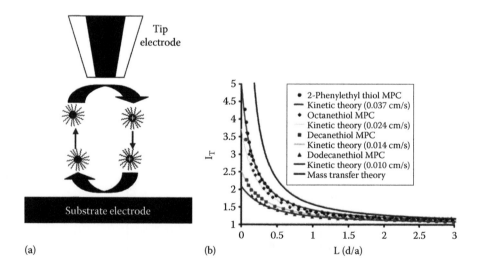

(a)   (b)

FIGURE 3.13 (a) Mass transfer and kinetic transfer limiting processes. The diffusion of the species through the bulk solution to the electrode is the mass transfer limited process while the movement of the electron from the electrode to species is the kinetically limited process. (b) Typical SECM approach curves of octanethiol, decanethiol, dodecanethiol, and 2-phenylethylthiol Au MPCs. 10 μm Pt UME at 1 V, a 2 mm Pt substrate electrode at 0 V, Ag/Ag⁺ nonaqueous reference electrode, and a Pt wire counterelectrode. The samples consisted of 20 mg of sample in 5 mL of 0.1 M $TBAPF_6$ in CH_2Cl_2. (From Peterson, R.R. and Cliffel, D.E., *Langmuir*, 22, 10307, 2006.)

diffusing solution, the MPC ligand length tends to be the more dominant factor in determining the rate constant of ET; a C_6 ligand gave an ET rate constant (0.11 cm s^{-1}) two orders of magnitude larger than C_{12} (0.0048 cm s^{-1}). For phenylethanethiol-protected Au_{144}, the rate of ET was 0.035 cm s^{-1}, on the same order of magnitude as C_8 (0.024 cm s^{-1}) and one order slower than C_6, showing that ligand aromaticity is not a necessarily significant player in ET kinetics in solution. In comparing ET kinetics among different clusters with respect to core size, the ET rate constants for hexanethiolate-protected Au_{25} clusters were determined by cyclic voltammetry and electrochemical impedance to be 0.015–0.022 cm s^{-1},[16] approximately one order of magnitude lower than the ET rate constant for hexanethiolate-protected Au_{144}. Looking at differences in ET kinetics among redox couples of the same cluster, Maran et al.[75] measured the ET rate in hexanethiolate-protected Au_{25} clusters via CV and found that ET rates for the +1/0 charge state transition were slower than those of the 0/−1 redox couple, by about one-half an order of magnitude. The aforementioned results indicate that the effects on ET rate, in order of most to least influential, are ligand length, core size, and redox couple.

ET kinetics have also been characterized in Au MPC molecular melts. Electron transport through the undiluted room temperature melt is proposed to occur by a diffusion-like core-to-core electron-hopping process. In a molecular melt of polyethylene glycol (PEG)-protected Au_{25} MPCs, a first-order electron-hopping rate constant of 2×10^4 s^{-1} and a second-order rate constant of 3.8×10^5 M^{-1} s^{-1} were obtained via potential step chronoamperometry.[17] In a later report, addition of electrolyte, and free PEG or CO_2 as plasticizers, to a molecular melt of PEG-protected Au_{25} MPCs resulted in the formation of a conductive room temperature molten salt. Activation energies of transport and rates of both electron and counterion transport within the melt were measured using voltammetry, chronoamperometry, and electrochemical impedance. Rates and activation energies were found to correlate closely with each other. Plasticization by the addition of small molecule solvent components resulted[18] in a fourfold increase in k_{EX} (from 1.15 to 4.4×10^{-7} M^{-1} s^{-1}) and halving of E_A (from 50 to 22 kJ mol^{-1}).

As an aside, note the homogeneous and heterogeneous ET rate constants have different units. It is, in fact, not straightforward to compare these rate constants obtained through different methods and conversions between units can be approximate. Different theories can model the ET transfer as an electron diffusion coefficient (D_E, cm^2 s^{-1}), a first-order reaction (k_{HOP}, s^{-1}) or a second-order (k_{EX}, M^{-1} s^{-1}) process. This produces the diversity of units found in the literature for electron transport in MPCs. Their equivalences are[17]

$$D_E = \frac{k_{HOP}\delta^2}{6} = \frac{k_{EX}\delta^2 C}{6}. \tag{3.3}$$

Calculating the ET rate constant in different ways is sometimes helpful for comparison of values across the literature. Sometimes, a more general k_{ET} is used.

It was also discovered that ET between Au MPCs in solution can be manipulated, due to electron parity of the cluster, by an applied external magnetic field. This topic is called magnetoelectrochemistry. External magnetic fields can influence ET kinetics by changing the field-induced splitting of originally degenerate energy states. Redox potentials of Au MPCs and subsequent currents become dependent on electrode orientation within the magnetic field and magnetic field strength.[92,93]

ETs in solutions of Au MPCs can also be described in terms of their exchanges with an additional redox species in solution; that is, MPCs can undergo[94] ET reactions with other redox-active species in their solutions. In other experiments by Kontturi et al.,[95] ET reactions between organic-soluble Au MPCs and an aqueous redox species (Ce(IV), $Fe(CN)_6^{3-/4-}$, $Ru(NH_3)_6^{3+}$, and $Ru(CN)_6^{4-}$) were measured with both feedback and potentiometric modes of SECM. Charge transfers occurring heterogeneously at the liquid–liquid interface were slow compared to electron self-exchanges between MPCs. In a report by Murray et al.,[44] the bimolecular rate constant of the reaction between organic-soluble Au_{25} clusters (mislabeled at that time as Au_{38}) and an aqueous redox species was determined through negative-feedback SECM approach curves. In accord with predictions from the Marcus theory, this process involving Au MPCs at the liquid–liquid interface was faster than ET reactions between conventional aqueous and organic redox species.

3.3.2 ET OF AU MPCs IN SELF-ASSEMBLED FILMS

In typical Au MPC ET experiments with SAMs, the clusters are linked to a metal electrode surface through a self-assembled alkyl chain monolayer with specific linking sites. The electron is presumed to be transferred to/from the tethered MPC through the chain linker to/from the electrode surface. Techniques used to measure ET rates in Au MPCs attached to SAMs are potential step voltammetry and electrochemical AC impedance. Rate constants determined through impedance measurements have the virtue of less distortion due to uncompensated resistance. Also, rate constants determined using cyclic voltammetry are explicitly based on the kinetic behavior of the subpopulation of MPCs that produces the most prominent current peaks. Potential step voltammetry resolves more peaks and thus gives a better indication of the diversity of rate constants of the redox species in the monolayer.[96]

Chen[97] first constructed Au MPC SAMs by incorporating some alkanedithiols into the Au MPC ligand shell in order to attach the MPCs to a gold surface. The result was a film of Au MPCs that exhibited distinctive step-like charging features. When the capacitances are in the order $C_{SAM} > C_{EL}$, QDL charging phenomena could be observed in the voltammetry of the Au MPC SAMs. The charging also appeared to be rectified, suggesting that ion-pairing can alter the value of C_{EL}, depending on potential, and can be tuned to give transistor-like behavior.[98] The ET rate constants of C_4 to C_8 chain length MPC SAMs were measured via cyclic voltammetry and electrochemical impedance. In dichloromethane, the values of k_{ET} decreased from 150 to 3 s^{-1} with increasing alkyl chain length of the SAMs and from 15.5 to 7 s^{-1} in water.[99]

A recent study[100] measuring the ET rate constants of various metal MPCs on alkanedithiol SAMs showed that the ET rate depends on the nature of the electrode metal. This can be explained based on the varying densities of electronic states of the different metals; the apparent ET rate constant for decanedithiol SAMs was 1170, 360, and 14 s^{-1} for MPCs of Au, Pt, and Pd, respectively.[100] This is an interesting result, although it is unclear if MPC size and capping ligand were completely uniform among the clusters of different metal cores. SAMs of Au MPCs have also been formed using biferrocene dithiol derivatives.[101] The main claim to fame of Au MPC SAMs and films is that the Coulomb staircase can be observed. For disordered films, such charging is not distinctly observed.

Multilayer films can also be formed upon alkanedithiol SAMs and exhibit charging features. In an experiment looking at the effect of film thickness and chain length effect on ET in dry films, IDA electrodes (the gold fingers part) were modified with dithiol and silane (the glass part), then immersed in dithiol Au MPCs for film growth. k_{ET} was found in a steady-state rotated disk electrode experiment to be dependent on film thickness and charge state.[102]

Another way to assemble films of Au MPC on an electrode surface is through carboxylate–metal ion linkages. For example,[96] monolayers of mercaptoundecanoic acid (MUA) on Au electrodes can be formed by immersion in an ethanol solution of MUA. Au MPCs can be attached by immersing the MUA-functionalized electrodes first into ethanol solutions of Zn and then into solutions of mixed monolayer hexanethiolate/MUA-protected MPCs, subsequently rinsing away MPCs not bound to the electrode surface. This yields a monolayer of MUA–carboxylate–Zn^{+2}—carboxylate–MUA Au MPCs on the electrode surface, as depicted in Figure 3.14. The ET rate constant was measured using cyclic voltammetry (100 s^{-1}), electrochemical impedance (90–160 s^{-1}), and potential step voltammetry (40–150 s^{-1}).

Quantized charging properties were also studied in multilayered films of this type where layers of Au MPCs were linked controllably and reversibly to the underlying monolayer via the carboxylate/(Cu^{2+} or Zn^{2+})/carboxylate linkage.[85] The electron transport within these multilayered, or so-called polymer network films, was characterized in several ways. Murray et al.[103] used cyclic voltammetry

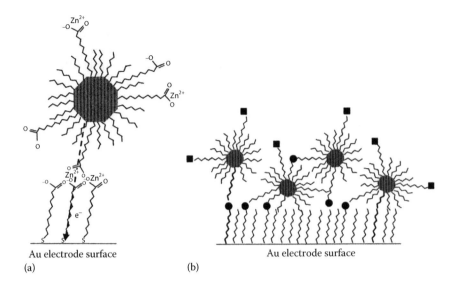

FIGURE 3.14 (a) Cartoon outlining carboxylate–metal ion–carboxylate bridge used to bind MPCs to the electrode surface, and depiction of the ET reaction between a gold MPC and the electrode surface. MPC core and ligand size drawn roughly to scale. (b) Monolayer of $Au_{140}C6_{33}(MUA)_{20}$ MPCs attached to an MUA-functionalized Au electrode. The scheme illustrates differing positions of MPC on MUA monolayer. (■) is an acid-terminated ligand. Bold lines at left and right indicate possible *bonded* and *nonbonded* electron tunneling pathways, respectively. (From Hicks, J.F. et al., *J. Phys. Chem. B*, 106, 7751, 2002.)

and potential step chronoamperometry to determine D_E (~10^{-9} cm^2 s^{-1}), k_{Hop} (2×10^6 s^{-1}), and k_{EX} (2×10^8 M^{-1} s^{-1}). (Note the use of second-order units for k_{EX}; the electron self-exchange rate constant reflects the bimolecularity of the process.) In another experiment, Murray et al.[104] determined the electronic conductivity (σ_{EL}) of such multilayered films on IDA electrodes immersed in hexane (1.8×10^{-4} Ω^{-1} cm^{-1}), acetone (1.3×10^{-4} Ω^{-1} cm^{-1}), ethanol (1.1×10^{-4} Ω^{-1} cm^{-1}), dichloromethane (2×10^{-4} Ω^{-1} cm^{-1}), and as a dry film in air (3.9×10^{-4} Ω^{-1} cm^{-1}). These data show that differences in solvation have minor effects on electron transport within the film. Increasing the length of the ligand from C_4 to C_{12} results, however, in much larger changes (ca. 3 orders of magnitude decrease in σ_{EL}). Also significantly, the electronic conductivity of the dry film was enhanced by the presence of organic vapors, suggesting a possible application in vapor sensing. Later,[102] k_{ET} was measured in the network polymer films by a steady-state rotated disk electrode voltammetry method (1×10^5 s^{-1}), and in the dry state using IDAs (4×10^6 s^{-1}). Cu^{2+}-bridged multilayer films have also been formed with Au MPCs containing a ferrocenated alkanethiolate ligand for enhanced ET.[105]

Dyer et al.[106] reported controlled assembly of films of Au MPCs by Cu^{2+}-pyridine complexation. These films are analogous to Cu^{2+}-carboxylate linked films. A monolayer of pyridine-terminated thiol ligand is attached to the Au electrode surface, and then mixed monolayer Au MPCs containing some pyridine-terminated ligand bind Cu^{2+} along with the pyridine-terminated ligand attached to the electrode surface. Multiple layers of MPCs can be formed by alternating dipping cycles, leading to pyridine–Cu^{2+} linkages between neighboring MPCs. Rectified charging in the presence of some hydrophobic counterions in aqueous solution was observed for these types of films as well, presumed to be caused by ion-pairing.

The Langmuir–Blodgett (LB) method is another important self-assembly strategy and results in a film made up of from one to several monolayers of MPCs on an electrode surface. In an early report by Shiffrin et al.,[107] a self-assembled multilayered thin film was prepared consisting of alternating layers of ~6 nm alkanedithiol-protected Au MPCs. Electron transport throughout the film was determined to occur through an electron-hopping mechanism, suggesting that the Au MPCs in the film retain their individual character (as opposed to fusing together into larger units of bulk-like gold). σ_{EL} was ca. 10^{-4} Ω^{-1} cm^{-1} and thermally activated by 0.02 eV. Electronic conductivity decreased by an order of magnitude with each additional three $-CH_2$ units in the dithiol chain. σ_{EL} of the film also increases with mechanical compression; the pressure likely affects the distance between clusters.[108] A chain length effect in multilayered self-assembled films was also observed by Vossmeyer et al.[109] Additionally, electronic conductivity was found to increase dramatically with the number of deposition layers, the largest takeoff point occurring at around five layers. This is important for understanding the quantitative nature of the current response of an Au MPC film to organic vapors, in relation to possible applications in sensing.

In order to better quantify current dependence on film thickness and the influence of counterion permeation on the onset potential, the Quinn group[110] proposed a theoretical model to describe the cyclic voltammetric responses of Langmuir–Schafer films (which are analogous to LB films but assembled differently). It was observed that cathodic peak potential shifts occur when the negative counterion is made progressively more hydrophobic. The importance of understanding ion penetration comes into play in studying electron transport in Au MPC films at a phase boundary. The Quinn group also measured k_{ET} in an LB monolayer of ~6 nm Au MPCs by using SECM at the nanocluster–electrode interface. The hydrophilicity of the redox couple, $FcCH_2OH$ versus $Fe(CN)_6^{4-}$, influenced the apparent ET rate in the Au MPC film presumably due to differences in penetration.[111] Chen and coworkers[112] observed single electron charging features and measured the electronic conductivity of ~2 nm alkanethiolate-protected Au MPCs in an LB monolayer (see Figure 3.15) at the air–water interface on interdigitated array (IDA) electrodes. Films of Au MPCs with C_4 and C_5 chains exhibited ohmic, linear i–V behavior while those with longer chains underwent rectifying charge transfer attributed to differences in electrolyte penetration between the ligand shells. σ was on the order of 10^{-3} Ω^{-1} cm^{-1} and varied with the interparticle spacing of the monolayer in the LB trough.[113]

Kim and Lee[114] studied electron transport at the air–water interface in LB monolayers of dithiol-protected Au_{25} and the effect of interparticle spacing. Through the use of atomically precise clusters,

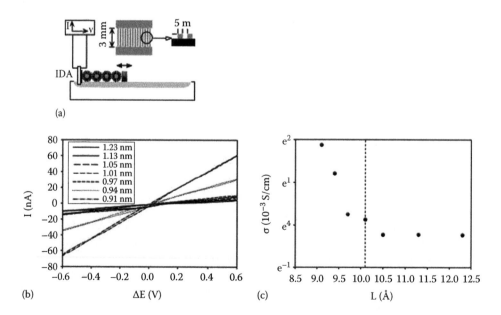

FIGURE 3.15 (a) Schematic setup of the electrical conductivity measurements of nanoparticle Langmuir monolayers. Inset shows the dimensions of the interdigitated array (IDA) electrode. (b) I–V curves of C5Au nanoparticle monolayers at varied edge-to-edge interparticle distances (L), which are shown in the figure insert. Potential scan rate 10 mV s^{-1}. (c) Variation of electronic conductivity with the interparticle spacings. (From Chen, S., *Anal. Chim. Acta*, 496, 29, 2003.)

the effects of ligand and core size can be distinguished. Clusters with dimensions in the quantum confinement region have become increasingly relevant to applications due to their unique electronic properties and highly monodisperse synthesis. Au_{25} nanoparticles show particular promise for utility in sensing devices because of their large surface area to volume ratio. k_{HOP} through the LB monolayer was ~10^6 s^{-1} for C_5 dithiol and ~10^4 s^{-1} for C_9 dithiol.

3.3.3 ET OF Au MPCs IN WETTED FILMS

The discussion now moves from ordered films and monolayers to films that are formed by dropcasting or some other method that yields a *disordered* film structure. Interparticle distance is likely to be less uniform in dropcast films than in self-assembled films. Dropcasting is simpler than self-assembly methods and therefore more attractive for use in device fabrication, but variability of interparticle spacing likely contributes to the fact that quantized charging features are typically not observed. Another important difference to note is that for sufficiently thick dropcast films, currents passed through the film tend to plateau, whereas in films of distinct monolayers, electron transport rate increases with each deposition layer.[110] This has important implications for devices that aim at quantitatively utilizing electronic conductivity of MPCs.

The study of Au MPC films in contact with an aqueous electrolyte has important implications for sensing and electronic devices. k_{HOP} (~10^4 s^{-1}) and k_{EX} (~10^5 M^{-1} s^{-1}) were measured by chronoamperometry for ~1.8 nm hexanethiolate-protected Au MPC dropcast films contacted by an aqueous electrolyte.[115] Deng and Chen[116] investigated, using SECM, rectifying charge transfer for dropcast Au MPC films contacted by an aqueous electrolyte. They observed a mass change at the electrode with de/adsorption of counterions and their de/solvation at more positive potentials than the PZC where charging features are observed that clearly depend upon the hydrophobicity of the counterion. Electrochemical impedance revealed that electron transport within the film was slower than counterion transport by at least an order of magnitude. In another report,[117] they examined rectifying charge transfer in multilayered dip-dried films (i.e., a gold electrode dipped into Au MPC solution then dried), and determined that the presence of hydrophobic anions like nitrate allowed the observation of quantized charging phenomena previously unobservable in dropcast Au MPCs. Further exploring the interactions between Au MPC film, water, and counterion, Wang and Murray[118] observed a decrease in contact angle of a sessile ionic liquid droplet upon oxidation of the Au MPC film from Au_{144}^{0} to Au_{144}^{+1}. This was attributed to the increasing hydrophilicity of the film.

There is some debate in the field about the nature of ion permeation through films of Au MPCs contacted by an aqueous electrolyte and the cause(s) of rectifying charge transfer. Quinn et al.[119] suggest that an ion permeation rate-limited model should replace the ion rectified model of charge transfer for Au MPCs in water. Instead of assuming that an association between Au clusters and the counterion leads to rectification, the ion-limited model suggests that the observed rectification effect is due to the polarizability of the film–water interface. This would also explain rectifying effects in organic solvents and shifts in redox potentials with different counterions.

SECM was also used to determine the conductance of dropcast MPC assemblies at certain potentials using an electrochemical gating mechanism. In this experiment, a redox species in solution is reduced or oxidized at the SECM tip, then re-oxidized or re-reduced at the dropcast MPC film surface. The resulting approach curves can be fitted to calculate the electrical conductance of the film. The potential of the film is set by the relative amounts of reduced and oxidized forms of redox mediator, so different potentials can be effected by switching between different types and different proportions of redox mediators in solution. For a dropcast film of 0.81 nm hexanethiolate-protected Au MPCs, the conductance ranged between 0.05 and 0.13 G Ω^{-1} between 0.1 and 0.5 V vs. Ag/AgCl.[120]

ET reactions of solution species can be inhibited[121] by a film of molecule-like Au MPCs (such as Au_{25}) dropcast on the electrode surface if the formal potential of the redox species falls within the energy gap of the clusters. A noble metal electrode has a continuum of energy levels, so there is always overlap at their surfaces with the formal potential of the redox species. However, molecular MPCs possess distinct energy levels and a bandgap. In 4% CH_3CN/H_2O solution, the redox reaction of methyl viologen is completely inhibited because its formal potential lies within the band gap of Au_{25}. In addition, Li et al.[122] studied superlattices of highly monodisperse 4.3 nm and found that the superlattice structure opened up an artificial forbidden gap where ET is prohibited. Taking advantage of the band gap of Au MPCs in films might lead to future development of selective redox sensors or transistors.

3.3.4 ET OF Au MPCs IN DRY FILMS

ET in dry films of Au MPCs is more directly applicable to real world electronic devices than in solutions. Early studies[4] of ET in dropcast Au MPC films demonstrated the effect of ligand length and interparticle spacing on electronic conductivity. Electron self-exchange, a bimolecular process, is typically used to describe the ET process between Au MPCs in a dry film. The reaction is taken to be

$$MPC^Z + MPC^{Z+1} \leftrightarrow MPC^{Z+1} + MPC^Z \tag{3.4}$$

and is depicted in Figure 3.16. Saveant[123] proposed using a cubic lattice structure as an approximate but convenient model to describe ET in a film with fixed redox sites. White and White[124] later developed a theory to describe electron hopping within a film as a random walk model, which includes variable distance ET probably as a more accurate reflection of the highly disordered film structure.

Electron transport rates within an MPC film become maximized when the concentration of MPC^Z becomes equal to that of MPC^{Z+1}, due to the bimolecularity of the reaction. The notion of a bimolecular mechanism for ET is found in experiments of network polymer films with fixed charge gradient redox sites[125] and mixed-valent osmium bipyridine clusters.[126] Later, it was shown that MPCs with a specific charge state in the solution retain that same charge when dried, allowing selection of specific relative proportions of donors and acceptors within a dry film.[94] Rate constants of ET in mixed-valent Au MPC films can therefore be expressed as[123]

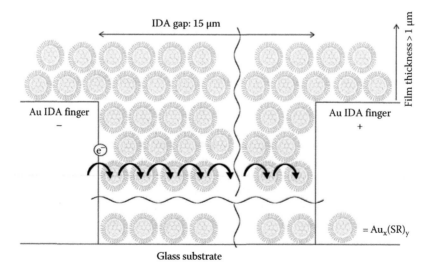

FIGURE 3.16 Cartoon of electron-hopping conductivity in a mixed-valent Au MPC film on an IDA electrode. (From Carducci, T.M. and Murray, R.W., *J. Am. Chem. Soc.*, 135, 11351, 2013.)

$$k_{EX} = \frac{6(10^3)RT\sigma_{EL}}{F^2\delta^2 \left[MPC^{Z+1} \right][MPC^Z]}, \tag{3.5}$$

where δ is the core-to-core edge separation (cm), and relative values of $[MPC^{Z+1}]$ and $[MPC^Z]$ are calculated from the Nernst equation. Once again, rate constants of this type are typically expressed in units of M^{-1} s^{-1}, reflecting the bimolecularity of the reaction. σ_{EL} of solid-state films is for the most part measured using IDA electrodes and calculated by obtaining an i–V curve and applying the geometric cell constant of the IDA film.

In an early study, Murray et al. calculated σ_{EL}, E_A, and k_{EX} for films of mixed-valent alkane-thiolate[127] and arenethiolate[128] protected Au MPC films that were dropcast on IDA electrodes. It was concluded that the rate of electron hopping is controlled by the tunneling transport of charge along ligand chains and by the degree of mixed valency of the clusters. For alkanethiolates, the electronic coupling coefficient (β) was 0.15 nm^{-1} per CH_2 unit of the chain, yielding a rate constant of 4×10^{10} M^{-1} s^{-1} for mixed-valent hexanethiolate-protected $Au_{144}^{3-/4-}$. E_A increased approximately threefold from C_6 to C_{12}. Higher rates of ET were observed for Au MPCs with aromatic ligands presumably due to enhanced electron transport within and between the overlapping, conjugated bonds.

Choi and Murray[74] discovered a significant dependence of Au MPC electron transport rate on nanoparticle core size, finding that k_{EX} for transport in $Au_{25}(SC_2Ph)_{18}^{0/-1}$ films was approximately three orders of magnitude smaller than that in $Au_{144}(SC_6H_{13})_{60}^{+1/0}$ films and E_A approximately threefold larger (20.3 vs. ~6 kJ mol^{-1}), despite Au_{25} having an aromatic ligand. They recalculated k_{EX} for hexanethiolate-protected Au_{144} (3×10^9 M^{-1} s^{-1}) due to their finding that electronic conductivity of films on IDAs become independent of film height after 1 μm. Exploiting improved monodispersity of synthesis and IDA cleaning methods, Carducci and Murray[31] later reported k_{EX} for the hexanethiolate-protected Au_{144} (1×10^8 M^{-1} s^{-1}) and phenylC_2-protected Au_{25} (2×10^6 M^{-1} s^{-1}). They additionally looked at the +1/0 charge couple of Au_{25} and found that ET was slower by about an order of magnitude in comparison to the 0/−1 charge couple, possibly due to changes in electronic structure.

In dropcast MPC films, only linear, featureless, current–potential profiles are observed in electronic conductivity measurements. However, following thermal annealing at temperatures above 300 K, well-defined staircase features of single ET can emerge, indicating lateral ET is occurring through the film.[129] Solid-state quantized charging in annealed films of ~$Au_{314}(SC_6)_{91}$ was observed only over a narrow temperature range (300–320 K), presumably where film structure is optimized to promote interparticle ET. It is unclear what effect annealing has on the chemical structures of the MPCs and films, but lateral ET is an indication that there is some kind of structural change.[130] For quantifying ET parameters of atomically precise Au MPCs, films are typically not annealed to avoid possible thermal damage to the particles, films, and IDA electrodes. However, annealed films could potentially be very useful in novel electronic devices given their staircase charging effects.

3.3.5 ET OF AU MPCs IN DRY, CHEMICALLY LINKED FILMS

There are many reports of electron transport in dry Au MPC films that are chemically linked, such as carboxylate–Cu^{2+}–carboxylate bridged films or Au MPCs connected via dithiol bridges forming a tunnel junction, and covalent Au–S bonds linking the Au MPCs into a network and onto gold electrode surfaces. A comparison of electron transport within solid-state films of linked (dithiol-protected Au MPCs) versus unlinked (thiol) on gold electrodes revealed an increase in current by three orders of magnitude for the linked as compared to the unlinked film.[131] Linked and unlinked films are discussed separately due to this large increase in electronic conductivity that this tunnel junction gives to the film.

Murray et al.[132] examined ET parameters for dry, carboxylate–Cu^{2+}–carboxylate bridged films via AFM. Like unlinked MPC films, σ_{EL} is exponentially dependent on the cluster spacing. β for transport between MPCs was 1.2 Å^{-1}, which is comparable to values obtained for hydrocarbon chains, and yielded values of k_{ET} between 10^5 s^{-1} and 10^7 s^{-1} for C_4/MUA to C_{12}/MUA. The film's electronic conductivity decreases in the presence of organic vapors, possibly due to film swelling that alters the distance between clusters among which ET occurs. The change in film mass due to swelling could be monitored concurrently using QCM.[104] In comparison to unlinked films, the electronic conductivity of the carboxylate–Cu^{2+}–carboxylate bridged films is actually lower. The metal ion/carboxylate linkers hold the MPC network polymer film together, but electronic conductivity of the film is dominated by electron transport through the *non*-linking alkanethiolate ligands. Thus, replacement of alkanethiolate ligands with linking MUA ligands lowers the overall electronic conductivity by providing fewer nonbonded pathways for ET.[133]

The most common type of chemically linked film in the literature is dithiol linked. An early report[134] found that Au MPC dithiol-linked films are photoconductive and that E_A for ET within the films is strongly dependent on linker length. When films are heated at 175°C under a nitrogen atmosphere, the electronic conductivity of the film increased, to nearly that of bulk gold. Film thickness effects on ET, also called percolation phenomena, are often studied in dithiol-linked films. Film thickness seems to have a large effect on electronic conductivity near the metal–insulator transition. In a series of experiments by Dhirani et al.[135] using films of dithiol-linked >15 nm diameter Au MPCs, films with fewer than six MPC layers exhibit electron-hopping conductivity; thicker films transition to metallic conductive-like behavior. Thinner films also exhibit current suppression below a threshold voltage, possibly due to Coulomb blockade effects.[136]

In looking at films of smaller Au MPCs (2 or 4 nm) using IDA electrodes, Zhong et al.[137] observed trends in ET similar to films of unlinked Au MPCs in this size range. In accordance with the electron-hopping model, the results showed that E_A increases with ligand chain length and decreases with particle size. E_A fell into the range of 22–33 meV for 2 nm Au MPCs and 62–88 meV for 4 nm Au MPCs. β was only slightly larger for these dithiol-linked films than unlinked films.

Another interesting and potentially useful difference between dithiol-linked and dithiol-unlinked films is that the electronic conductivity of unlinked Au MPC films that are under CO_2 atmosphere increases with pressure, whereas that of linked films decreases. Plasticization for unlinked films is postulated to increase short-range thermal motion of MPCs, enhancing ET, whereas in linked films, thermal motion is restricted by the dithiol linkages, so that CO_2-induced swelling of the film results in slower ET. This report suggests that the direction of response to gas sorption can be manipulated by linking the MPCs together.[138]

3.3.6 Low Temperature ET of Au MPCs

Some research efforts have focused on deciphering ET behavior of MPC films at low temperatures. Determining the metal–insulator transition[139] is important, for example, in future implementation of Au MPC films in electronic devices.[140] Snow and Wohltjen[141] in early work demonstrated an effect of core size (0.86–3.61 nm) and film thickness on the metal–insulator transition temperature of an Au MPC film. Electronic conductivity increased nearly linearly with film thickness over 0.03–0.7 μm. Between 0°C and 20°C, the film ET transitions from semiconductor type (thermally activated) to metallic type (thermally deactivated) as the core size of the Au MPC increases. This transition, which manifests as a maximum in electronic conductivity, occurs at increasingly lower temperatures for clusters with larger core sizes.

In a series of reports from the Dhirani group,[142] a metal–insulator transition was found to occur for Au MPC dithiol-linked films of n > 5. They investigated[143] the metal–insulator transition in terms of percolation effects. Films below a certain thickness threshold exhibit thermally activated electronic conductivity and conductance suppression near zero bias, a consequence of a possible Coulomb blockade effect. This transistor-like behavior is tunable with film thickness and temperature and could be useful in electronic device fabrication.

Nair and Kimura[144] studied the metal–insulator transition in thin films of water-soluble Au MPCs by four-point probe conductivities. For mercaptopropionyl gycine–protected Au clusters of core sizes ~1.6, 3, and 4 nm, electronic conductivity increased with Au core size. Electronic conductivity was metal-like at very low temperatures but crossed over to semiconductor behavior around 60 K. Mercaptosuccinic acid–protected Au clusters of 2, 4, and 7 nm diameter exhibited a linear increase in resistivity with temperature at low temperatures and a *decrease* in resistivity with temperature at high temperatures, a somewhat surprising result. The transition temperature decreased with increasing cluster size.[145] Wieczorek et al.[146] aptly suggest that these transitions between semiconductor-like and metal-like conductive behavior with temperature may not signal a metal–insulator transition but instead be a result of the change in ET distance between neighboring clusters when the films expand thermally. It is fair to say that low temperature electronic conductivity of Au MPC films is not completely understood and remains a fertile area of research.

The electronic conductivity of a substance exhibiting only thermally activated, or Arrhenius (linear $\ln \sigma_{EL}$ vs. T^{-1}), behavior will decrease with temperature until absolute zero is reached. However, Murray et al.[126] discovered ET behavior indicating the presence of a *minimum* in ETs in mixed-valent osmium bipyridine polymers. An electron tunneling mechanism, that is, a temperature-independent ET rate, can be observed at very low temperatures where thermal energy is depleted. At ambient temperatures, the greater occupancy of the upper vibrational states allows the reaction to proceed over the classical thermal barrier. With decreasing temperature, the reaction is hypothesized to proceed increasingly by tunneling because of the depletion of the upper vibrational states.[147] In films of Au MPCs, this behavior has been observed in large, polydisperse Au MPC dithiol-linked films. For example, in a two-probe configuration, differential conductance of ~5 nm, butanedithiol-linked Au MPC films exhibited temperature independence below 10 K due to a tunneling mechanism.[148] The variable range hopping model (linear $\ln R$ vs. $T^{-1/2}$) was used to describe the ET, with contributions from both tunneling and activated mechanisms.[142]

Carducci and Murray[31] investigated the low temperature ET behavior of films of highly monodisperse, mixed-valent MPCs, namely, $Au(SC_2Ph)_{18}^{0/-1}$, $Au(SC_2Ph)_{18}^{+1/0}$, and $Au_{144}(SC_6H_{13})_{60}$. Non-Arrhenius behavior at low temperatures (see Figure 3.17) was observed for all clusters and redox couples, and electronic conductivity minima were observed for the Au_{25} redox couples. The tunneling currents at 77 K fell in the same order of their ambient-temperature ET rates: $Au_{144}^{+1/0} > Au_{25}^{0/1-} > Au_{25}^{1+/0}$, showing that core size and mixed valency dependence applies to low temperature ET in Au MPCs as well.

3.3.7 OBSERVING SINGLE ET EVENTS OF AU MPCs

A promising research direction in nanoelectrochemistry is measurement of ET events of individual (single) clusters. In an early work,[6] quantized charging of a single Au_{144} MPC adsorbed onto an STM tip was observed—a Coulomb staircase with six charging steps regularly spaced at 0.34 V increments between −1 and 1 V bias. Observing charging events of single clusters not only yields insight into size-dependent properties but may be of use in miniaturization of electronic devices.

In a report from the Zhang group,[149] steady-state electrochemical responses were obtained for single, 10–30 nm citrate-stabilized Au MPCs. A single Au MPC was chemically immobilized by an amine-terminated silane onto an SiO_2-encapsulated Pt disk nanoelectrodes tip. The single Au MPC electrode was found to greatly enhance ET from the bare Pt electrode to a redox species in solution, as depicted in Figure 3.18. In looking at the electrocatalytic activity of the single Au MPC electrode for the oxygen reduction reaction (ORR) in a solution of KOH, the electrode exhibited good electrocatalytic activity that was tunable based on cluster size. The voltammetric response of the electrode depended on the size of the cluster; $E_{1/2}$ shifted to more positive potentials with increasing size of the Au MPC. The limiting current was higher for the Pt electrode bearing a single Au MPC than

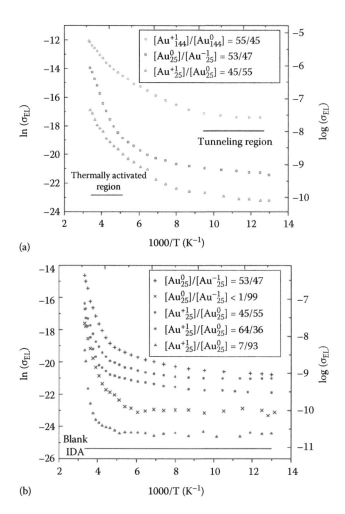

FIGURE 3.17 (a) Arrhenius plot of mixed-valent solid-state Au MPC films containing indicated molar proportions of each MPC charge state. (b) Arrhenius plots of mixed-valent solid-state Au_{25} films containing the indicated proportions of the 0/−1 or +1/0 charge states. The blank IDA line is for an IDA electrode bearing no film. (From Carducci, T.M. and Murray, R.W., *J. Am. Chem. Soc.*, 135, 11351, 2013.)

for the bare Pt electrode and increased with particle size, indicating higher catalytic activity. When considering limiting currents normalized to the radius of the MPC, the 18 nm Au MPC electrode had the best catalytic activity for the ORR. This is an example of ET events not only at a single cluster but of the core size–dependent ET properties of Au MPCs.

Another recent area of study involving single MPC ET events is particle-electrode impact electrochemistry. Bard and coworkers[150] have investigated collisions of metal nanoparticles with an inert electrode surface that result in amplification of an electrocatalytic current. Depicted in Figure 3.19 and accompanying scheme, in a study involving 14 nm citrate-protected Au MPCs, oxidation of BH_4^- was inhibited by forming a layer of Pt oxide on the surface of a Pt ultramicroelectrode. When immersed in a solution of BH_4^- and Au MPCs, current *blips* were observed that were attributed to elastic collisions between individual Au MPCs and the electrode surface. Upon each collision, there was a discrete current pulse corresponding to Au MPC-catalyzed oxidation of BH_4^-. The frequency of collisions increased linearly with concentration of Au MPCs.[151] These single particle experiments represent significant advances toward harnessing the unique ET properties of Au MPCs on the smallest scale possible.

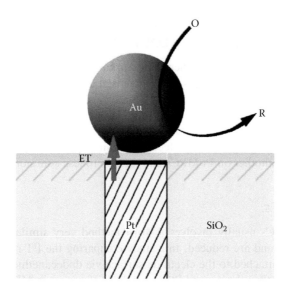

FIGURE 3.18 Schematic drawing of an Au single nanoparticle ET. The single nanoparticle occludes the electrode. (From Li, Y. et al., *J. Am. Chem. Soc.*, 132, 3047, 2010.)

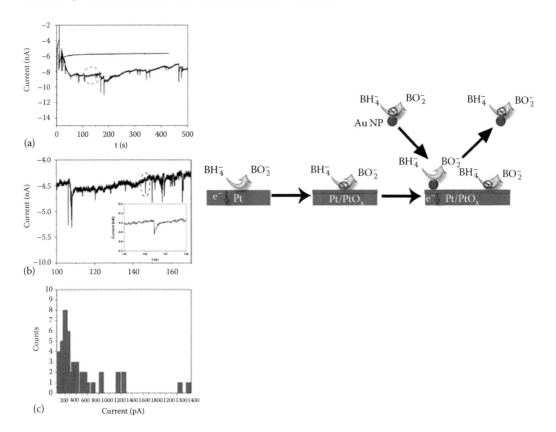

FIGURE 3.19 (a) The i–t curve without (upper curve) and with (bottom curve) injected Au NPs at the preoxidized Pt UME. UME diameter, 10 μm; NP concentration, 24 pM; potential, 0 V. (b) Zoom-in of the marked region in panel (A). The inset is a single-collision peak. (c) Statistical distribution of amplitudes of current peaks. (right) Scheme of a single Au NP collision event on a PtO$_x$-covered Pt UME. (From Zhou, H. et al., *J. Phys. Chem. Lett.*, 1, 2671, 2010.)

3.4 SYNTHESIS, ELECTRONIC, AND OPTICAL PROPERTIES OF PD, PT, RU, AG, AND CU ALLOY MPCs

Syntheses of MPCs with metal cores other than Au have seen many advances over the years and applications involving these particles have increased as a result. Less is known about their structures but the range of core sizes obtained is analogous to that of Au MPCs, that is, a range from molecule-like particles to larger ones whose ET properties resemble that of the bulk metal. Similar to Au MPCs, MPCs of other core metals have highly modifiable surfaces, which can be functionalized with redox ligands or biomolecules to suit a specific application. A discussion of Pd MPCs is followed by another on Pt MPCs and then on to less common core metal MPCs.

3.4.1 PALLADIUM MPCs

The synthesis of Pd MPCs usually involves a Brust method very similar to that for Au MPCs wherein a Pd salt and ligand are reduced. In a study comparing the ET rate constants for MPCs (of various core metals) attached to the electrode surface via dodecanethiol SAMs, the Pd cluster exhibited slower ET rates than the Au and Pt clusters.[100] Differences in MPC electronic conductivity among core metals represent yet another opportunity for the design of clusters to serve a desired purpose. Also, Pd clusters can be coated using either thiol ligands or covalent Pd–C bonds. The latter is unique to Pd and presents a new opportunity for surface functionalization that is distinct from Au clusters.

Chen and Huang[152] were the first to synthesize and perform electrochemical experiments on Pd MPCs. They synthesized water-soluble Pd nanoclusters was coated with a monolayer of N,N,N-trimethyl(8-mercaptooctyl)ammonium chloride (TMMAC) via the reduction of $PdCl_2$ and TMMAC with gaseous H_2. The particles were then further functionalized with viologen moieties incorporated into the particle protecting monolayers using surface place-exchange reactions. Surface functionalization of nanoclusters with viologen holds promise as another ET sensing platform. The two single-ET steps of the viologen moieties were observed in cyclic voltammetry (Figure 3.20) for both freely diffusing viologen-Pd MPCs and as SAMs. Electrochemical impedance gave estimates of k_{ET} for the particle-bound viologens to be 1.2×10^3 s^{-1}, which is comparable to k_{ET} rates measured for viologen SAMs on an Au electrode.

Ghosh and Chen[153] prepared Pd MPCs passivated by metal–carbon covalent linkages by the reduction of Pd salt with diazonium derivatives. Aliphatic radicals generated from the reduction of diazonium ligands was the proposed mechanism for formation of stable Pd–C linkages. The group measured the electronic conductivity of solid-state films of Pd MPCs and found that the Pd–C clusters had much higher electronic conductivity than the alkanethiolate-protected Pd MPCs, exhibiting metal-like conductive properties at temperatures below 180 K, whereas alkanethiolate-Pd MPCs maintained semiconductor-like ET over the entire temperature range studied (Figure 3.21). Metal–ligand linkages are evidently rather important in governing the nature of ET and the metal–insulator transition temperature.

Pd MPCs have also been used in electrocatalytic applications. Chen et al.[154] synthesized butylphenyl diazonium-protected Pd MPCs in the same manner as the earlier study, and more recently, the same group synthesized 1-octyne-stabilized Pd MPCs.[155] Single particle electrocatalysis with Pd MPCs has also been studied.[156,157] Further examples of electrocatalytic applications of MPCs are found in Section 3.5.2.

Pd MPCs have also been synthesized in PAMAM dendrimers. Similar to dendrimer syntheses of Au MPCs, these syntheses result in relatively monodisperse samples due to reduced agglomeration and can be extracted from the dendrimer template or immobilized onto surfaces via SAMs for making nanocluster arrays. The Crooks group[64] synthesized Pd nanoclusters encapsulated in polycationic PAMAM dendrimers that were subsequently linked to an electrode surface via a SAM

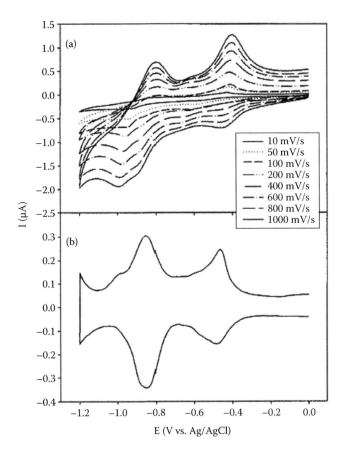

FIGURE 3.20 (a) Cyclic and (b) differential pulse voltammograms of viologen-derivative TMMAC-protected Pd particles, ca. 6 μM in 0.1 M NaCl, at various potential sweet rates. Au electrode area, 0.23 mm². In (A), the sweep rates are shown in the key, while in (b) the scan rate was 10 mV s⁻¹ and the pulse amplitude was 50 mV. (From Chen, S. and Huang, K., *J. Cluster Sci.*, 11, 405, 2000.)

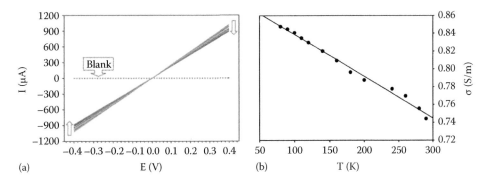

FIGURE 3.21 (a) Current–potential profiles of a biphenyl-stabilized palladium nanoparticle dropcast thick film at various temperatures. Potential scan rate 20 mV s⁻¹. (b) Variation of the ensemble conductivity with temperature. (From Ghosh, D. and Chen, S., *J. Mater. Chem.*, 18, 755, 2008.)

and were robust enough to withstand electrochemical cycling. The same group utilized PAMAM dendrimers as a template for synthesis of relatively monodisperse Pd clusters, and then extracted the MPCs from the dendrimer into the organic phase with hexanethiol replacing the dendrimer as the passivating ligand shell. They were able to observe QDL voltammetry (without further purification steps) corresponding to Pd MPCs with cores of ca. 40, 80, and 140 Pd atoms.[70]

3.4.2 Platinum MPCs

Like Au and Pd MPCs, the syntheses of Pt MPCs usually involve coreduction of a metal salt with the desired ligand, as in the Brust method.[158] An early study by van Kempen et al.[159] measured, via STM, the charging steps and Coulomb blockade of a single particle of phenanthroline-protected Pt_{309} MPCs. A claim to fame of Pt MPCs is their good electrocatalytic activity for ORRs, showing an improvement over traditional Pt black catalysts.[158]

Dendrimer syntheses have also been developed by the Crooks group[64] for Pt MPCs extracted from dendrimer templates, as well as seven different core metal ratios of bimetallic dendrimer-encapsulated Pt/Pd[160] clusters, all with core-size distributions at or lower than ±0.3 nm. This group found that doping or mixing together one metal with another in the MPC core dramatically alters the electrochemical properties of the cluster. The cluster with a Pd to Pt ratio of 5:1 had the best catalytic activity for the ORR, with a mass activity enhancement of 2.4 compared to monometallic dendrimer-encapsulated Pt clusters. However, a 50–100 mV higher overpotential was required.[160] The ligand shells of Pt MPCs can also be functionalized with biomolecules for sensing applications.[161]

3.4.3 Ruthenium MPCs

The majority of synthesized Ru MPCs are protected by ligands through Ru–C bonds. This nanoparticle synthesis involves reaction of an alkyl lithium reagent with a ruthenium chloride salt; the nanoparticles can be further functionalized using a ligand exchange reaction. Chen and coworkers synthesized[162] 2–3 nm octyne-protected Ru MPCs and Ru MPCs with a mixed monolayer of octyne and ethynylferrocene ligands through the formation of Ru–C≡ bonds (Figure 3.22A). The apparent bond order of the ligand C≡C increases upon reduction of the cluster and decreases upon oxidation, suggesting that charge on the cluster is delocalized throughout its ligand shell. In voltammetry of

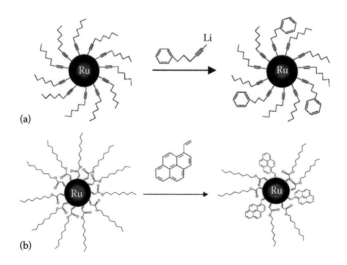

FIGURE 3.22 (a) Synthesis of alkyne-protected Ru MPCs through the formation of Ru–C≡ bonds. (b) Olefin metathesis reactions of carbene-stabilized Ru MPCs with vinylpyrene. (From Chen, W. et al., *J. Phys. Chem. C*, 113, 16988, 2009; Chen, W. et al., *J. Phys. Chem. C*, 114, 18146, 2010.)

the ferrocenated nanoparticles, the group observed two peaks corresponding to ET of the Fc moieties, which suggests that the strong Ru–C≡ bond facilitates intervalence electron transfers between the Fc groups through the metallic core.[163] The strength of the Ru–C≡ interaction also results in a metal-like temperature dependence of electronic conductivity for films of this cluster. β for these clusters (0.3–0.5 Å^{-1}) is much smaller than in alkanethiolate-protected Au MPCs (1.2 Å^{-1}), indicating a small contact resistance for electrons transferring between the metallic core and insulating ligand shell, a possible result of the charge delocalization capability.[164,165]

Chen and coworkers[166] also synthesized pyrene-functionalized Ru MPCs passivated by metal–ligand pi bonds (Figure 3.22B) through olefin metathesis reactions. Similar to the Ru–C≡ clusters, these particles also exhibit intraparticle delocalization, yielding interesting optical fluorescence properties that can be manipulated by chemical linkages. Anthracene-functionalized Ru MPCs stabilized by carbenes had activation energies for ET about one order of magnitude lower than clusters passivated by alkanethiolates due to interparticle delocalization.[167] Ru=N π-bonds in a nitrene-stabilized Ru MPC did not permit interparticle charge delocalization to as great a degree as in the Ru=C and Ru–C≡ clusters; the photoluminescence properties were diminished and only one redox peak due to Fc was observed for the ferrocene–nitrene-functionalized Ru MPC.[168]

3.4.4 Silver MPCs

Ag MPCs represent an interesting target because Ag yields surface plasmon bands with higher extinction coefficients than Au MPCs. These MPCs can also display QDL charging voltammetry, as illustrated in Figure 3.23. Focusing on electronic properties, Wang et al.[169] synthesized ~3.3 nm alkyl ammonium-protected Ag clusters by reducing a metal salt in the presence of the desired ligand, in a Brust-like method. They linked the MPCs to a gold electrode through SAMs, and measured the capacitance (modeling them after the concentric sphere capacitor-like Au MPCs) to be ca. 0.52 aF (per nanoparticle), through differential pulse voltammetry. The metal–insulator transition of ~14 nm thiolate-protected Ag MPCs was less sharp than that for Au MPCs of comparable size, suggesting that the ligand monolayer of Ag MPCs is more disordered.[170] Murray et al.[139] synthesized polydisperse butylbenzylthiolate-protected Ag clusters with an average formula of $Ag_{140}BBT_{53}$ that display a surface plasmon in the UV–visible region and QDL charging voltammetry. The room temperature electronic conductivity of these clusters was determined to be about four orders of magnitude smaller than Au_{140} MPCs. Feature-rich voltammetry is difficult to obtain for Ag clusters because of interference by formation of oxide sites on the surface of the particles. Ag clusters have

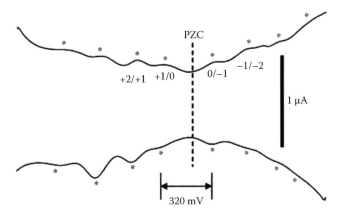

FIGURE 3.23 Square-wave voltammetry of 0.3 mM AgBBT in 0.1 M Bu_4NClO_4/CH_2Cl_2. The upper and lower traces correspond to negative- and positive-going scans of potential. (From Branham, M.R. et al., *Langmuir*, 22, 11376, 2006.)

also been electrodeposited on Au MPC templates with different architectures, offering possible platforms for sensing or catalysis.[171] Like Au, Ag MPCs can be functionalized with biomolecules with an eye toward applications in immunosensing.[172]

With regard to optical properties, water-soluble ~1.6 nm tiopronin-protected Ag MPCs synthesized by Huang and Murray[173] have luminescent properties that vary with solvent polarity and pH. Jin and coworkers reported the highly monodisperse synthesis of dimercaptosuccinic acid–protected Ag_7 MPCs[140] as well as structural studies of a series of silver sulfide cluster ions containing 1–7 Ag atoms.[174] Further insight into the structure and stability of Ag clusters were gained through experiments and computations with the newly synthesized[175] $Ag_{152}(SCH_2CH_2Ph)_{60}$. This cluster is proposed to be comprised of a 92-atom core having icosahedral–dodecahedral symmetry and a shell of 60 Ag atoms and 60 thiolates in a network of six-membered rings. The bandgap was estimated to be ~0.4 eV with absorption at 460 nm and two luminescence peaks, one attributed to the ligated shell and the other to the core.

3.4.5 OTHER METAL MPCS

Cu MPCs have been synthesized but performing voltammetry is difficult because like Ag, Cu is easily oxidized, and the nanoparticles readily gain oxide film surfaces. Chen and Sommers[176] synthesized 1–2 nm Cu MPCs exhibiting QDL charging voltammetry (that disappeared with aging) with capacitances of 0.82 aF. The metal–insulator transition for Cu MPCs occurs at a higher temperature than either Au or Ag, possibly due to a larger Fermi energy of the metal.[170] In a larger size regime, 4 nm tridecylamine-protected Rh MPCs synthesized via a Brust method were found to exhibit double layer charging voltammetry with a peak spacing of 80 mV near the PZC.[177]

3.4.6 AU ALLOY MPCS

Many research efforts have probed the structure-dependent properties of Au MPCs, utilizing core doping as a tactic. As with the Pd/Pt bimetallic clusters, doping of the core can result in dramatically different nanoparticle properties, inviting further study and understanding. Most known examples of alloyed nanoclusters contain Au.

As early examples of doping of Au clusters, Hg was found to weaken and shift the surface plasmon absorption band to shorter wavelengths; a broad absorption band develops at 360 with increasing amalgamation.[178] In a study by Huang and Murray,[173] replacing atoms in the core of Ag_{140} MPCs with Au via a galvanic metal exchange reaction resulted in the loss of the Ag MPC surface plasmon absorbance and emission and the growth in Au MPC emission, with the final product being a bimetallic cluster containing ca. 85 Au and 55 Ag atoms. Cao et al.[179] functionalized Ag/Au core–shell clusters with DNA for biorecognition applications. These core–shell clusters are formed by modifying an Ag MPC with a gold salt in the presence of a reducing agent, typically $NaBH_4$. Bimetallic Ag/Au core–shell clusters such as these hold promise in applications, because they retain their surface plasmon absorption features but are more stable than monometallic Ag clusters.

Considerable recent research has focused on the structural details of doped Au MPCs. Kumara and Dass[148] synthesized $(AuAg)_{144}(SR)_{60}$ alloy MPCs via coreduction of gold and silver salts. These clusters exhibited enhanced surface plasmons relative to monometallic Au_{144} clusters. Malola and Hakkinen[180] deciphered structural details of the alloy cluster using a density functional theory calculation, finding that Ag was likely being incorporated into the 60 atom shell, based on a structural model of the icosahedral $Au_{144}(SR)_{60}$ that features a 114-atom metal core with 60 symmetry-equivalent surface sites and a protecting layer of 30 RS–Au–SR units (Figure 3.24). With increasing Ag content, the optical gap of the alloy cluster was found to decrease from that of the monometallic Au_{144} cluster.[88]

The structural distortion of the Au_{25} nanoparticle upon atom doping is interesting because the crystal structure of Au-only Au_{25} is known. In the $CuAu_{24}(SC_2H_4Ph)_{18}$ cluster, copper occupies

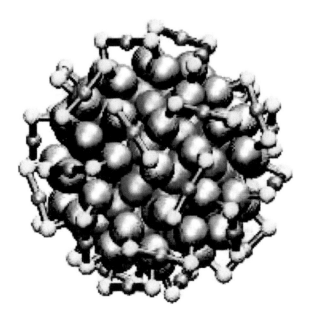

FIGURE 3.24 Optimized structure of cluster of composition $Au_{54}Ag_{60}(RSAuSR)_{30}$. Brown = gold; shiny gray = silver; yellow = sulfur. The hydrogen atoms are not shown for clarity. (From Malola, S. and Häkkinen, H., *J. Phys. Chem. Lett.*, 2, 2316, 2011.)

the center of the metal core, and provokes shrinking of bond distances in the contracted core.[181] Doping with Pd produces a core–shell $Pd_1@Au_{24}(SC_{12}H_{25})_{18}$ structure in which the central Pd atom is surrounded by a frame of $Au_{24}(SC_{12}H_{25})_{18}$.[182] Pd doping also increases stability of the cluster and accelerates ligand exchange rates.[183] The degree of distortion of geometric structure likely depends on the nature of the metal and/or doping site as this degree of distortion is not seen with Ag.[181]

ET properties of doped Au MPCs have also been studied. The Crooks group[68] synthesized dendrimer-encapsulated Au/Pd core–shell (Au@Pd) clusters which could then be extracted by dodecanethiol into organic solvent. The dendrimer-encapsulated Au@Pd clusters were found to not catalyze the conversion of resazurin to resorufin, for which monometallic dendrimer-encapsulated Au clusters are catalysts.[69] The Stevenson group[184] synthesized bimetallic Ni–Au MPCs in dendrimers, followed by extraction with thiol, as catalysts for CO_2 production from CO and O_2. Incorporation of Ni into the clusters results in improved absorption of O_2 onto the surface of the cluster catalyst. These observations are consistent with the importance of structure in catalytic function. Ibanez and Zamborini[185] found that films of Ag-doped Au MPCs would be lower in cost than monometallic Au MPC films, but this weakens their sensitivity for chemiresistive detection of organic compounds by up to two orders of magnitude, likely due to formation of oxides. In $Cu_nAu_{25-n}(SR)_{18}$, Negishi and coworkers[181] found that doping reduces the stability of the MPC in solution and shifts the redox potentials to more negative values. Understanding the structural and electronic properties of doped particles can potentially lead to the manipulation of MPC properties, such as their surface plasmon and electronic conductivity.

3.5 ELECTROCHEMISTRY AND ELECTROCHEMICAL APPLICATIONS OF MPCs

The unique electrochemistry of MPCs that has been the focus of this chapter has potential usefulness in applications. The MPCs might serve as electrocatalysts or be incorporated into electrochemical sensing devices. The advantages of atomically precise MPCs (i.e., distinct energy levels and band gaps) can be envisioned in applications in sensors and as electrocatalysts. The unique electrochemistry of specific types of MPC surface functionalization will be discussed first, followed by applications of these and of unmodified Au MPCs in electrocatalytic and sensing applications.

3.5.1 Unique Electrochemistry of Surface Functionalized MPCs

The surface ligands of the MPC can influence its electronic structure and voltammetric responses. Modifying the surface of an MPC with ligands that suit a particular application is possible, and incorporating a ligand that contains a redox moiety allows the MPC to be electroactive in a chosen potential region.

3.5.1.1 Redox Ligand–Functionalized MPCs

In significant early work,[5] it was shown that functionalization of Au MPCs with a ferrocene (Fc) thiol via a place-exchange reaction leads to surprising multi-ET reactions for both freely diffusing clusters and for a partial monolayer of electrode-adsorbed clusters. The most highly functionalized Au MPCs were determined, via bulk electrolysis, to bear 15 Fc units. They can be viewed as multielectron donor/acceptor reagents and catalysts; previously multi-ET reagents were mostly electroactive polymers. Later syntheses produced fully ferrocenated Au MPCs with core sizes ~55, 140, 225, and 314 Au atoms and average ligand numbers of 37, 39, 43, and 58 ferrocenated ligands, respectively. Their synthesis followed the Brust method where the amount of Fc thiol relative to Au was varied to produce different sized clusters. The core sizes obtained for the ferrocenated forms are similar to those when nonferrocenated thiolate ligands are employed, but the total number of ligands is lower, which was ascribed to the Fc steric bulk. These MPCs have the capacity to transfer many electrons within a small range of potential, up to 60 per Au core for the largest sized MPC. An example cyclic voltammogram is shown in Figure 3.25.

More recently, Schiffrin and coworkers[33] synthesized 1.8 nm Au MPCs containing 4, 7, or 10 Fc thiolate ligands, using a place-exchange reaction. The place-exchange method incorporates a smaller amount of Fc thiol into the ligand shell. The group was able to distinguish between voltammetric peaks from quantized charging events of the MPC core, from ET events of the Fc moieties. Multi-ET was observed and attributed to fast rotational diffusion of the MPC at the electrode surface, as opposed to electron self-exchange between individual Fc moieties. Fc-labeled Au

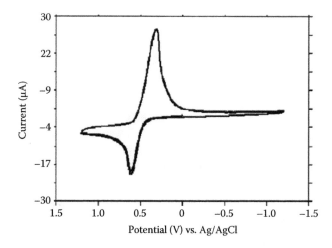

FIGURE 3.25 Cyclic voltammogram of 0.2 mM $Au_{75}(SC6Fc)_{37}$ in 0.1 M Bu_4NClO_4/CH_2Cl_2 under Ar atmosphere, at 284 K, potential scan rate 500 mV s^{-1}, current sampling interval of 1 mV. (From Wolfe, R.L., *Langmuir*, 23, 2247, 2007.)

MPCs can be further functionalized with biomolecules,[186] enhancing current response in biosensor applications.

Films of one to several monolayers of $Au_{225}(SC_6Fc)_{43}$ MPCs on Pt electrodes can be formed by applying a potential to oxidize the Fc sites. Their associated electrolyte anions are a significant part of the film structure, forming bridges with the positively charged Fc moieties (Figure 3.26B). At slow potential scan rates, voltammetric peaks (Figure 3.26A) of the Fc film are significantly narrowed (35 mV FWHM as compared to ideal 90 mV), likely due to *attractive* interactions between neighboring nanoclusters.[36] Multilayered thin films of Fc Au MPCs in the LB style have also been prepared and studied electrochemically.[187] Films become less conductive with increasing thickness, resulting in an anodic shift of the Fc formal potential.

To study ET responses of Fc Au MPCs on more explicitly controlled electrode surfaces, Murray et al.[188] performed voltammetry on $Au_{225}(SC_6Fc)_{43}$ adsorbed onto anionic SAMs (Figure 3.27). From this study, it was concluded that there are three different interactions that can contribute to MPC adsorption: (1) interaction between MPCs and SAM surface (if present), (2) lateral interactions between MPCs due to bridging electrolyte ions, and (3) specific adsorption between electrolyte ions and the bare electrode surface. The use of SAMs for Fc MPC attachment yields the strongest absorption, which is important for stable Fc voltammetry. In another study,[189] some positively charged tetraethyl ammonium-thiolate ligands were incorporated into the shell of the Fc MPC to yield a similarly robust adsorption onto the electrode surface, without the use of SAMs. This supports the hypothesis that multiple ion-pair interactions are central in the irreversibility of adsorption.

Fc Au MPCs have also been shown to greatly enhance the capacitance behavior of highly porous electrodes. When carbon *nanofoams* are loaded with $Au_{225}(SC_6Fc)_{43}$, the Fc redox capacity and the double layer capacity of the intercalated clusters give the nanostructured carbon material supercapacitative properties, making it a potent electrochemical charge storage system.[190] SEM images of the MPC-loaded nanofoams and voltammograms are shown in Figure 3.28. Some other noteworthy examples of Fc-functionalized Au MPCs are as follows. Chang et al.[191] synthesized Au MPCs functionalized with terpyridyl Fc and bi-Fc ligands through an Ru^{2+} bridge that interact with the electrode surface; reaction of all three redox species can be observed in

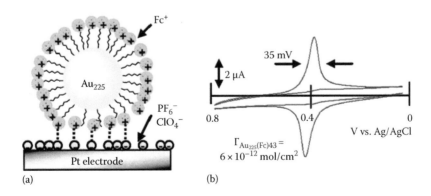

(a)   (b)

FIGURE 3.26 (a) Cartoon of ion-induced adsorption, where ferrocenium cations on the MPC form ion-pair bridges with electrolyte anions specifically adsorbed to the Pt electrode. It can be imagined that the ligand shell may become deformed to form ion-pair bridges of similar dimensions. Formation of ion-pair bridges stabilizes successive ion-pair bridges and can cause a shift in Fc/Fc+ formal potentials. (b) Cyclic voltammetry (blue, 0.20 V s⁻¹) of an adsorbed $Au_{225}(SC6Fc)_{43}$ film formed on a clean Pt electrode from 0.1 mM MPC, 1.0 M Bu_4NPF_6/CH_2Cl_2 solution and transferred to an MPC-free 1.0 M Bu_4NPF_6/CH_2Cl_2 solution. The blue curve is cyclic voltammetry (0.20 V s⁻¹) of an identically treated Pt electrode, except that the Pt electrode was first coated with a dodecanethiolate self-assembled monolayer (SAM).

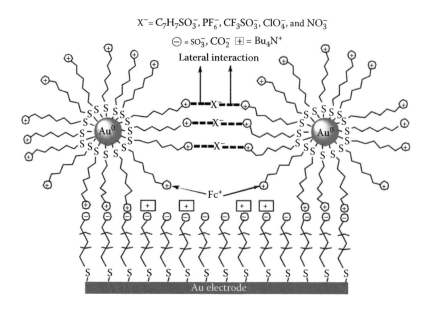

FIGURE 3.27 Cartoon representing adsorption of NPs on a negatively charged SAM surface. The adsorption is promoted by lateral ion bridges between neighbor NPs. (From *Anal. Chem.*, 81, 6980, 2009.)

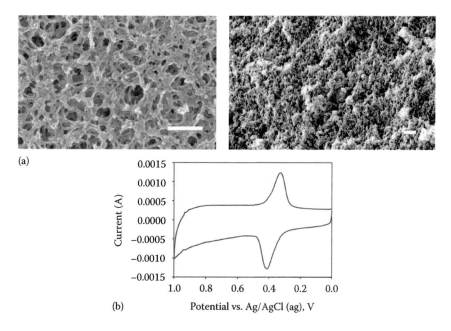

FIGURE 3.28 (a) Microscopic SEM images (scale bar 200 nm for both) of nanofoam materials loaded with the ferrocenylated nanoparticles. (b) Cyclic voltammetry of nanoparticle-loaded nanofoam electrode in 2 M Bu$_4$NPF$_6$/CH$_3$CN at a 20×10^{-5} mV s^{-1} potential scan rate. (From Chow, K.-F. et al., *J. Phys. Chem. C*, 116, 9283, 2012.)

CV. Au MPCs functionalized with bipyridinium were found via STM to exhibit differences in electronic conductivity as a function of the charge state of the redox moiety and could therefore serve as nanoscale electronic switches.[192] Au MPCs have additionally been functionalized with biferrocene,[193–196] anthraquinone,[197–199] Fe-carbonyl,[200] phenothiazine,[201] nitrothiophenolate,[93,202] and viologen[203] ligands.

3.5.1.2 Bioconjugated MPCs

Bioconjugation is another type of surface modification giving rise to unique designed electrode surfaces and responses. Bioconjugation of MPCs presents an exciting opportunity for electrochemical detection of biological molecules, an important part of the future of biosensors and clinical diagnostic tools.

Most bioconjugated MPCs have been synthesized through place-exchange reactions. The Maran group[204] studied the electrochemistry of Au_{25} and Au_{140} nanoparticles modified by ligand place exchanges with thiolated peptides and found that the substitution of electron-withdrawing peptide ligand(s) for phenylethanethiolate caused positive shifts in oxidation potentials by as much as 0.8 V. These shifts were attributed to the orientation of the peptide ligand dipole, which would increase the work function of the metal core if its ligand is electron withdrawing.

Another class of bioconjugated MPCs acts by binding redox-active biological molecules. The Rotello group[205] showed that flavins can be bound to Au MPCs via noncovalent interactions with aminopyridine surface ligands, and the binding be detected voltammetrically due to binding-induced changes in cluster redox potential. Factors such as charge and length of the receptor ligand and redox state of the flavin influence the cluster redox potentials and binding strengths of flavin to the MPCs.[206] The same group also used trimethylammonium-functionalized Au MPCs to provide a high-affinity binding surface for anionic flavin mononucleotide. The surface charges of mixed monolayer Au MPCs effectively modulated the flavin reduction potential, mimicking flavoenzymes. Binding events were measured voltammetrically using these shifts in redox potential.[207] The implications of these studies are that bioconjugation of an MPC surface can be fine-tuned for electrochemical applications such as sensors, and that MPCs could potentially serve as model systems for the study of redox-active enzymes that are otherwise difficult to isolate and manipulate.

Another potential advantage of binding MPCs to redox-active biomolecules is current amplification. The Schiffrin group[208] used MPCs as a linker between the Cu metal center of glucose oxidase and the electrode surface, enabling electrochemical detection and study of the metalloenzyme. They were able to quantify the rate of ET between the electrode and the metal center, as well as the pH dependence of the formal potentials of the two processes involved, which are oxidation/reduction of the tyrosyl radical and the $Cu^{+2/+1}$ redox couple. The metalloenzyme-Au MPC complex showed effective electrocatalysis for O_2 reduction as well. The possibilities for bioconjugated MPCs are extensive and will be discussed further in the sensors section.

3.5.2 CATALYSIS

The electrocatalytic properties of MPCs are tunable based on core size, capping ligand, and even the method of anchoring of the clusters onto a support within a device. There are numerous opportunities for MPCs to be used as catalysts, and recently, the field has pointed toward optimizing MPC catalyst systems by choosing *atomically precise* clusters that best suit the specific application.[209]

3.5.2.1 Redox Catalysts

In an early work, it was shown[197] that electrocatalytic currents can be enhanced by utilizing MPCs. The electrocatalytic reduction of 1,1-dinitrocyclohexane by electrogenerated anthraquinone radical anions incorporated into the ligand shell of a ~2 nm Au MPC was compared to the reactivity of monomeric anthraquinone. ET rate constants of MPC-bound anthraquinone were nearly identical to the monomeric rates, but catalytic currents were higher for the anthraquinone-MPC catalysts, which was attributed to its smaller diffusion coefficient and consequent compressed reaction layer. This represents an example of decorating the surface of the MPC with catalytically active moieties, but the core–shell structure of a nonredox ligand-modified MPC can be exploited for its electrocatalytic properties as well.

There is an abundance of potential redox catalysis applications of large clusters, aided by the ease of synthesis and fast ET kinetics. As an example, Mirkhalaf and Schiffrin[210] studied the electrocatalytic reduction of oxygen, which is important in the design of fuel cells, on ~7 nm 4-diazoniumdecylbenzene fluoroborate-protected Au MPCs, as films on decylphenyl-coated glassy carbon electrodes. Electrons hop from the electrode surface to the nanocluster metal centers to participate in ET reactions. The choice of a hydrophobic ligand provides an apt environment for stabilization of reactive superoxide and peroxide intermediates.

Following the trend of utilizing smaller and core-size specific clusters, the Maran group[42] has reported on the electrocatalytic reduction of benzoyl peroxides by highly monodisperse $Au_{25}(SC2Ph)_{18}$ (Figure 3.29). Recall that the Au_{25} nanoparticle is molecule-like and has a HOMO-LUMO energy gap. The gap improves the charge separation ability of the MPC catalyst, in turn making precise control over the charge state of the MPC very important. Electron donation occurs from the HOMO of the Au_{25} cluster over a distance of approximately the length of the ligand monolayer, suggesting that varying the length of the ligand will also alter the rate of ET. The rate constants of dissociative ET of the peroxides were determined to be on the order of 1 M^{-1} s^{-1} using $Au_{25}{}^{0/-1}$ and $Au_{25}{}^{+1/0}$ redox couples electron donors, with $Au_{25}{}^{-1}$ being a slightly more effective electron donor catalyst than the neutral species. The use of freely diffusing MPCs as electrocatalysts represents an improvement over traditional catalytic surfaces (i.e., redox species attached to a metal electrode through SAMs), because reaction encounters can occur with higher frequency in three-dimensional space in the solution. Atomically precise $Au_{25}L_{18}$ was demonstrated[211] to have good catalytic activity and high selectivity for hydrogenation of α,β-unsaturated ketones and aldehydes to unsaturated alcohols. The electron-deficient Au_{12} shell and electron-rich Au_{13} core structure of Au_{25} is hypothesized to aid the catalysis in two ways. The low coordination number of outer Au atoms favors adsorption and activation of H_2, while the core activates the C=O bond by donating charge.[211] Another possible factor influencing electrocatalytic capabilities of small clusters is co-adsorption. In a theoretical study, co-adsorption of C_2H_4 onto MPCs of 1–9 Au atoms promoted charge transfers from the cluster to O_2 when otherwise they do not occur favorably.[212]

In regard to the effect of core size on electrocatalytic activity, a DFT study dissecting the catalytic mechanism of C=O oxidation by Au_{55} at low temperatures has revealed how tremendously complex the interplay is between energy levels of MPCs and substrate bond activation.[213] Another advantage

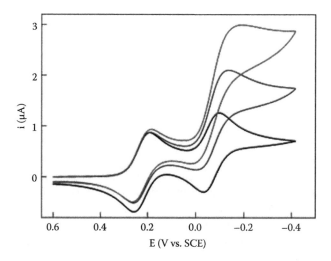

FIGURE 3.29 Cyclic voltammetry for the reduction of 1.12 mM $Au_{25}L_{18}{}^{+}$ in the absence (black line) and in presence of Equation 3.1 (blue line) and Equation 3.2 (red line) of bis(para-cyanobenzoyl) peroxide at 0.05 V s^{-1} in DCM/0.1 M TBAH on a Pt electrode. T = 25°C. (From Antonello, S. et al., *Nanoscale*, 4, 5333, 2012.)

of using small clusters is achieving better catalyst loading capacities. Larger surface area:volume ratio allows a higher total quantity of active sites as compared to conventional methods with less loading, which is important for photocatalytic systems.[214] Therefore, small, atomically precise clusters show potential for utility optimizing rate, selectivity, and efficiency of catalysis.

The possibilities for tailoring an MPC catalyst are substantially enlarged by synthesis of bioconjugated MPCs. Prins et al.,[215] for example, found that peptide-Au MPC complexes facilitate transesterification rates by more than two orders of magnitude. They attribute the catalytic activity to the multivalent scaffold nature of the Au MPC, which mimics an enzyme by bringing the catalyst and substrate into close proximity and creating a favorable local environment. Catalytic applications of MPCs of different core metals have also been studied. Highly monodisperse butylphenyl-protected Pd MPCs were found to be good catalysts for formic acid oxidation, giving fourfold rate increase over the typical Pd black catalyst.[154] Also, alkyne-functionalized Pd MPCs exhibited a twofold increase in catalytic activity for oxidation of ethylene glycol over typical Pt/C catalysts.[155] The oxygen reduction catalytic activity of para-substituted phenol-protected ~2 nm Pt clusters was shown to be highly influenced by the electron-withdrawing ability of the ligand substituent.[158]

There has also been progress in characterizing electrocatalysis by single clusters. Observing reaction kinetics at the single particle level has the potential to yield elusive information about catalytic mechanism and to push increases in the sensitivity of measurement capabilities, the basis of many sensor technologies. Novo et al.[216] measured the rate of oxidation of ascorbic acid on the surface of single ~50 nm Au nanocrystals using surface plasmon spectroscopy. As mentioned earlier, the Bard group[217] has observed electrocatalysis at a single cluster upon its collision with an electrode surface. For Pt MPCs, ET rate, and subsequent catalytic rate, was dependent on the ligand length.[217,218] The observed current transients accompanying MPC collisions with the electrode could be used to estimate the particle size; larger clusters showed higher catalytic currents for oxidation of hydrazine.[219] Electrocatalysis of $NaBH_4$ oxidation by single ~14 nm Au MPCs was also measured by the particle-electrode impact method.[151]

3.5.3 ELECTROCHEMICAL SENSORS

The electroactive properties of MPCs and highly modifiable surfaces present an attractive target for electrochemical sensing platforms. A *sensing platform* has two meanings; it can be the carrier or support of the sensor and/or the mechanism by which sensing occurs. As films, MPCs can provide a physical surface upon which a sensing device can be built. As electroactive species, sensing occurs through detection of analytes via current amplification. The ability to be tailored for selective and sensitive detection of specific analytes, either through core size and/or ligand functionalization, is broad, which can lead to many possibilities for detection of various biomarkers. Three main types account for the majority of analytes targeted through MPC sensors: vapors, small redox-active molecules, and biomolecules. The majority of sensing supports involve a film of MPCs either physically adsorbed or chemically linked to a metal electrode. Other examples of platforms include MPC-decorated microspheres[220] and even kimwipes.[221]

3.5.3.1 Detection of Gas/Vapors and Chemiresistors

We begin discussion of specific sensors that utilize MPCs for the detection of volatile organic compounds (VOCs). The organization is by analyte, noting recent progress for each area and important trends in sensing platforms. Au MPCs represent a good sensing platform for organic vapors because their film conductivities σ_{EL} can be altered reversibly in the presence of adsorbed gases. Flexible carboxylate–Cu^{2+}-linked polymer MPC films swell on exposure to organic vapors, thereby altering interparticle distance and decreasing σ_{EL}.[104] Xu and Chen[222] found that core size influences MPC response to organic vapors. They tested a series of single hexanethiolate-protected Au MPCs of various core sizes for changes in electronic conductivity in the presence of hexane, toluene, THF, etc. using STM and found that 4.9 nm MPCs gave the best sensitivity in measurements at room temperature. Larger and smaller clusters exhibit smaller changes in electronic conductivity, due to reduced or increased Coulomb blockade effects, respectively.

A long-standing trend in sensor technology is miniaturization. An early report from Shao et al.[223] established the potential of extremely small electrodes as sensor platforms. The advantage of micro- and even nanoscale level sensing platforms include precise position resolution and low detection and quantification limits.[223] Merging the research areas of micro- and nanoelectrodes with MPCs has led to some interesting advances in analyte detection, especially of organic vapors. There is a class of organic vapor sensors called chemiresistors that employ MPC films atop microband IDA electrodes. Wohltjen and Snow[224] were early pioneers, first utilizing films of 2 nm octanethiol-protected Au MPCs on IDA electrodes for detection and quantification of VOCs. In another early work in the development of chemiresistors, the Zellers group[225] fabricated devices made from IDAs with 50 pairs of electrodes, gaps of 7–30 μm, and coated with either ~8 nm octanethiolate or ~6 nm phenyl-C_2 thiolate-protected Au MPC films (Figure 3.30A). They exposed the film to vapors, and measured the resistance of the film between the two sets of IDA fingers. The limits of detections achieved were between 0.1 and 10 ppb depending on the specific analyte and gave an improvement over surface acoustic wave (SAW) sensors. They also saw a ligand effect, as C_8 MPC films were slightly more sensitive than the phenyl-C_2 films.[225] More recently, they presented a model to predict the response of chemiresistors taking into account various factors such as core size and ligand of MPCs, vapor-film partition coefficients, and analyte densities.[226] The importance of ET characterization and utilizing clusters of definite core and ligand compositions again comes into play.

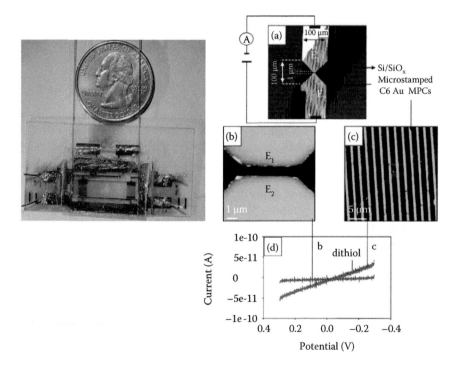

FIGURE 3.30 (Left) Dual-chemiresistor array housed in a 60 μL detector cell. (From Cai, Q.-Y. and Zellers, E.T., *Anal. Chem.*, 74, 3533, 2002.) (Right) (a) Optical image of microscale lines of C6 Au MPCs stamped across two Au electrodes separated by a 1 μm gap at the shortest point. AFM images of the electrodes (b) before and (c) after stamping the C6 Au MPCs across them. (d) I–V curves of the (B, blue) bare electrodes, (C, red) electrodes bridged with stamped C6 Au MPCs, and (*dithiol*, green) electrodes bridged with C6 Au MPCs and exposed to hexanedithiol for 30 min. (From Ibanez, F.J. et al., *Anal. Chem.*, 78, 753, 2006.)

Chemiresistors have also been adapted for sensing of organics dissolved in aqueous solutions, a possible test for safe drinking water. A problem associated with solution measurements is the interference from double layer charging current (large C_{DL}) on the voltammetric response. Wieczorek et al.[227] found that by inkjet printing a thin film of hexanethiolate-protected Au MPCs onto 5 μm IDA electrodes, the charging current was reduced significantly, allowing highly sensitive measurements via electrochemical impedance, ~0.1 ppm for toluene, as an example.

In a departure from IDA electrodes, the Zamborini group[228] designed a method for printing microcontact patterned films of 1.6 nm hexanethiolate-protected Au MPCs across two Au electrodes separated by a 1 μm gap (Figure 3.30B). Upon heating and cross-linking with dithiol vapor, the films become like flexible wires and can be lifted, transported, and reattached to new electrodes for multiple measurements, achieving S/N = 22. While IDA electrodes are small, they are typically not low cost or widely available. These workers found that film flexibility was the key for sensitivity to organic vapors to allow for swelling, which alters the electronic conductivity. They also stressed the importance of high portability, high throughput, and simultaneous detection of multiple analytes in the future of sensor development. The group has also had success with sensing using Au MPC films containing tetraoctylammonium as the surfactant, which are more stable over time and more highly sensitive to VOCs than thiolate-protected MPCs, an improvement from ~100 to 10 ppm. Another interesting finding was that the sensitivity of the MPC sensor also depended on the metal content of the core; doping the Au core with Ag lowers the sensitivity.[185]

3.5.3.2 Detection of Biomolecules

Biomolecules are important target analytes, especially the primary goal of disease diagnosis/prognosis. Sensors utilizing the ET properties of Au MPCs have been designed for small redox-active molecules important in biological systems, both redox-active and redox-inactive biomolecules. These sensors are sometimes used in combination with another redox-active label for current amplification. There is also a large class of sensors for analytes not necessarily limited to biological molecules made using bioconjugated MPCs that will be discussed in the subsequent section. A major challenge in electrochemical sensing is to overcome interference from other analytes in complex matrices such as human serum.

Here are a few examples of detection of small redox-active molecules that are highly relevant in living things. Lee et al.[229] was able to detect ascorbic acid and uric acid down to concentrations of 70 nm by enlisting Au_{25} in a sol–gel framework (Figure 3.31A) as an electrocatalyst. As discussed previously in the electrocatalysis section, the electron-rich core and electron-poor shell of Au_{25} allows it to function as a redox mediator, separating charge and thus lowering the oxidation potential of the analytes. Because Au_{25} is not only a good redox mediator, but a fairly conductive species, there is fast ET between the analyte through the sol–gel film to the electrode surface. Good current response is obtained, with sensitivities around 1.5 μA μm$^{-○}$ (Figure 3.31B). Many MPC sensors utilize their electrocatalytic properties, blurring the line between electrocatalytic applications of Au MPCs and sensing applications.

Another biologically relevant class of small molecule targets is free radicals. Oxidative stress, the imbalance of reactive oxygen species (ROS) including superoxide anions ($\bullet O^{2-}$), hydroxyl radical ($\bullet HO$), singlet oxygen ($^1O_2$), and hydrogen peroxide (H_2O_2)—can occur as a response to injury and in the progression of some diseases, like cancer, making monitoring of the concentrations of these species very important. Tian et al.[230] designed an electrochemical detection method for $\bullet HO$ using a Fc-labeled Au MPC-modified electrode. Typically, electron spin resonance is used to monitor $\bullet HO$, but an electrochemical method of detection such as this could conceivably be adapted to in vivo monitoring of ROS with sufficient miniaturization of the sensor and cost. An LOD of 0.37 nm and linear range of 5–45 nm were achieved for $\bullet HO$ released from living cells, signifying some promise of this sensing platform.

The development of sensors that do not require extensive pretreatment of blood samples is also highly desirable. Concentrations of lysine (an amino acid) in the body can be used to determine

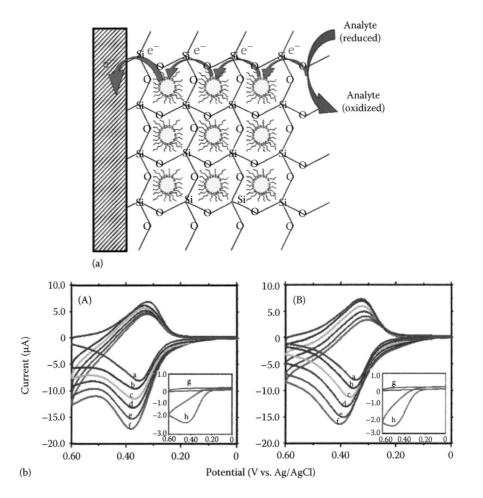

FIGURE 3.31 (a) Mechanism depicting the mediated electrocatalytic oxidation and ensuing electron transport across entrapped Au_{25} in the sol–gel framework ($Au_{25}SGE$). (b) Voltammograms showing the electrocatalytic oxidation of (A) ascorbic acid and (B) uric acid in 0.1 M KCl at 20 mV s^{-1}. (a) CV of $Au_{25}SGE$ in the absence of analyte and the presence of 1, 2, 3, 4, and 5 micromolar analyte. (From Kumar, S.S. et al., *Anal. Chem.* 83, 3244, 2011.)

the nutrition level of an individual, as in a sensor that utilizes citrate-capped Au MPCs on a carbon nanotube–decorated Au electrode surface.[231] A layer of lysine oxidase is physisorbed onto the surface of the Au MPCs and carbon nanotubes. The other purpose of the Au MPCs is to relay the charge transfer from the lysine oxidase to the electrode surface during cyclic voltammetry and electrochemical impedance measurements. These sensors achieved good quantification of ~200 µM lysine in human serum samples (±5 µM) and were washable and reusable.

Larger biomolecules are also important target analytes but present substantial challenges due to their low concentrations. A relatively complex immunosensor for carcinoembryonic antigen (CEA) was designed utilizing the increased HRP-electrocatalyzed reduction of hydrogen peroxide, which was locally generated by the glucose oxidase. The large surface area of a dendrimer is advantageous for immobilizing the CEA, aiding in the low concentration problem. The encapsulation of Au MPCs in the interior of the dendrimer accelerated the ET between the enzymes and the electrode surface, enhancing the immunosensor's response. The limit of quantification in human serum samples was around 10 pg mL^{-L}, a sensitivity higher than the typical enzyme-linked immunosorbent assay (ELISA) method.[232] ELISA methods, which are typically used for detection of antigens, involve many wash steps, and electrochemical sensing platforms could conceivably evolve into the point-of-care diagnostics area with further miniaturization and reduction of cost.

A specific category of protein sensing, termed protein monolayer electrochemistry (PME) by the Leopold group, allows for measuring ET rates of redox-active proteins via current amplification. Steric bulkiness and low biological concentrations of certain redox-active proteins make studying their ET kinetics difficult due to low sensitivity. Using dithiol-linked films of Au_{225} to immobilize cytochrome C^{233} and azurin (Az)[234,235] at the electrode (Figure 3.32), ET rates for each of these proteins could be quantified via cyclic voltammetry. The hydrophilic dithiol-linked MPC film creates a uniform surface for effective binding of the protein and facilitates rapid ET to/from the electrode surface, thus amplifying signal. Also, recall that dithiol linkages between

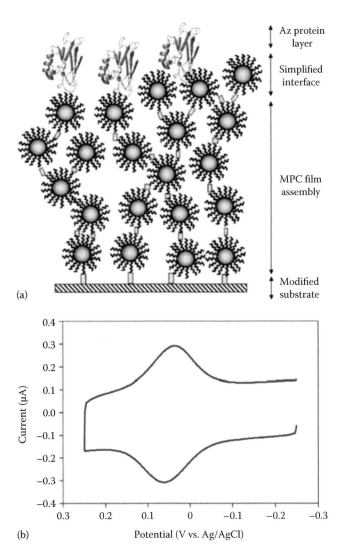

FIGURE 3.32 (a) Schematic representation of azurin adsorbed to a dithiol-linked MPC film assembly. (b) Typical cyclic voltammogram for Az adsorbed to MPC film assembly collected at 100 mV s^{-1} in 4.4 mM potassium phosphate buffer at pH = 7. (From Vargo, M.L. et al., *Langmuir*, 26, 560, 2010.)

clusters increase the electronic conductivity of the film. Multilayer film structure (i.e., dropcast) is preferred over SAMs because it results in more uniform voltammetric peaks and more surface coverage by the protein of interest. In addition, the rate of ET calculated for proteins adhered to film assemblies is not necessarily distance dependent as is with attachment through SAMs.[235] Pulse techniques and electrochemical impedance were also utilized to quantify the ET characteristics of azurin to minimize the influence of charging current in ET rate measurements and better estimate surface coverage.[236]

3.5.3.3 Bioconjugated MPCs as Sensors

Lastly, we consider electrochemical sensors based on bioconjugated MPCs, grouping these applications by their type of bioconjugation. It is important to note that while the measured analyte for a bioconjugated MPC biosensor is not necessarily a biological molecule, that is most often the case, and detection is mostly indirect, measuring the signal using another redox species. The versatility of syntheses of MPCs modified with biomolecules—neurotransmitters, nucleic acids, antibodies, enzymes, etc. gives a wide realm of possibilities for target analytes and sensor designs. This has created an abundance of publications in the recent literature describing advances in sensing technology. A few key examples will be highlighted.

Beginning with Au MPCs modified with small biomolecules, Liu et al.[237] recently synthesized dopamine-capped Au MPCs and incorporated them into a sandwich-type sensor for the capture of glycoproteins with avidin and prostate-specific antigen (PSA) as model analytes. Sandwich-type sensors are based on specific affinity of the sensor to the target (dopamine to glycoprotein in this case) and are thus highly selective. *Dual amplification* was achieved by decorating the surface of the electrode with 4-mercaptophenylboronic acid–capped Au MPCs, which provided multiple binding sites per cluster for the dopamine-capped Au MPCs and more surface area for analyte binding. The sensitivity of this type of sensor was outstanding, with LODs below the pM level.

As an example of an aptasensor, a specific type of sensor where a DNA sequence is the target analyte, Pt MPCs bioconjugated with short DNA sequences complementary to the target analyte sequence allow indirect target detection via the electrocatalytic reduction of H_2O_2. Thrombin bound to its aptamer was also detected by this method using thrombin aptamer–functionalized Pt MPCs. MPCs are advantageous for use in sensors due to fast ET kinetics, electrocatalytic activity, and functionalizable surfaces; this sensor makes use of all three. The limits of detection of thrombin were 100-fold better than traditional aptasensors not utilizing MPCs, demonstrating their significant role in the sensing platform. Some schematics[161] are depicted in Figure 3.33.

MPCs functionalized with large biomolecules including proteins, enzymes, and antibodies (immunosensors) are also abundant in the literature. A TiO_2 nanotube array was decorated with antibody-labeled Au MPCs to allow for binding with a target analyte and amperometric detection. Typically, sensors must achieve detection limits of at least ng mL^{-1} to be a viable sensing platform for low-abundance proteins, and this sensor could lead to detection limits as low was 0.01 ng mL^{-1}. Similar to the vast improvement in sensitivity for the chemiresistor that IDA electrodes made, miniaturization of electrodes into high aspect ratio arrays for this sensor led to enhanced electrochemical detection in comparison to flat electrode surfaces by 10-fold.[238]

Progress toward the specific binding of antibodies to the surface of atomically precise MPCs has been made as well. Kornberg et al.[239] designed a method for controlled addition of MPCs to proteins at sulfhydryl groups. In addition to use of a single core size of MPCs to optimize performance, controlling the reactivity of sulfhydryl groups on target analytes minimizes interference from functional groups on other species in complex matrices. The addition of exactly four glutathione-capped Au_{71} MPCs to the single chain Fv antibody fragment is depicted in Figure 3.34. This was accomplished by oxidation of the MPC using permanganate for addition of Au_{71} followed by reduction by tiopronin for deactivation to stop further addition of Au_{71}.

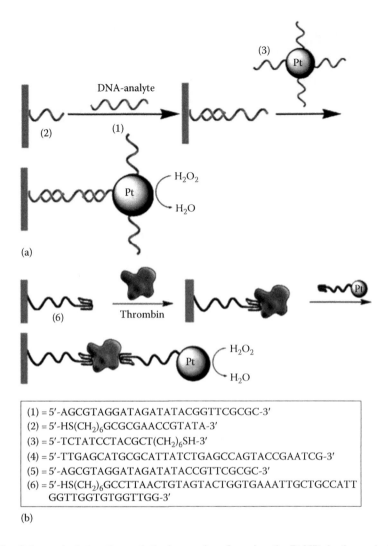

(1) = 5′-AGCGTAGGATAGATATACGGTTCGCGC-3′
(2) = 5′-HS(CH$_2$)$_6$GCGCGAACCGTATA-3′
(3) = 5′-TCTATCCTACGCT(CH$_2$)$_6$SH-3′
(4) = 5′-TTGAGCATGCGCATTATCTGAGCCAGTACCGAATCG-3′
(5) = 5′-AGCGTAGGATAGATATACCGTTCGCGC-3′
(6) = 5′-HS(CH$_2$)$_6$GCCTTAACTGTAGTACTGGTGAAATTGCTGCCATT
    GGTTGGTGTGGTTGG-3′

(b)

FIGURE 3.33 Scheme depicting the analytical procedure for using the Pt-NPs in the analysis of (a) DNA and (b) thrombin. (From Polsky, R. et al., *Anal. Chem.*, 78, 2268, 2006.)

Perhaps practical implementation of highly sensitive MPC biosensors could be facilitated by moving toward a cheaper and more automated system. A study utilizing low-cost and highly automated screen-printed carbon electrodes examined the core-size effect on voltammetric response for sterically hindered antibody-Au MPCs, finding that a large core size yields higher current due to reduced Brownian motion.[240] The Polsky group conducted a study of sensing target DNA with streptavidin-Au MPCs using single-use thick film carbon electrodes.[241] MPC-based electrochemical sensors and biosensors make a strong case as the forefront of analyte detection research due to their synthetic diversity, selectivity, low LODs/LOQs, rapid analysis times, and potential for miniaturization. However, enormous challenges remain. They include issues of stability, which is helped somewhat by the addition of electrolyte to and/or cross-linking of films, and high cost, making them relatively impractical for current use in industry and clinical settings at this point in time. Inexpensive, easy to use electrode materials are going to be paramount if these technologies are translated from bench to bedside.

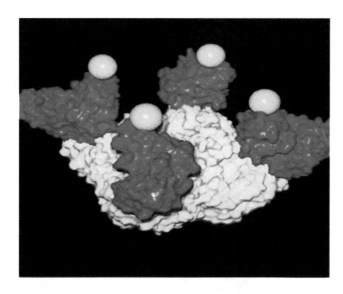

FIGURE 3.34 Rigid, specific labeling of proteins by the conjugation of a glutathione monolayer-protected gold cluster with a single chain Fv antibody fragment (scFv) mutated to present an exposed cysteine residue. (From Ackerson, C.J. et al., *J. Am. Chem. Soc.*, 128, 2635, 2006.)

REFERENCES

1. Brust, M.; Walker, M.; Bethell, D.; Schiffrin, D. J.; Whyman, R. *J. Chem. Soc., Chem. Commun.* 1994, 801.
2. Whitesides, G. M.; Grzybowski, B. *Science* 2002, *295*, 2418.
3. Mrksich, M.; Whitesides, G. M. *Annu. Rev. Biophys. Biomol. Struct.* 1996, *25*, 55.
4. Terrill, R. H.; Postlethwaite, T. A.; Chen, C-h.; Poon, C-D.; Terzis, A.; Chen, A.; Hutchison, J. E.; Clark, M. R.; Wignall, G.; et al. *J. Am. Chem. Soc.* 1995, *117*, 12537.
5. Hostetler, M. J.; Green, S. J.; Stokes, J. J.; Murray, R. W. *J. Am. Chem. Soc.* 1996, *118*, 4212.
6. Ingram, R. S.; Hostetler, M. J.; Murray, R. W.; Schaaff, T. G.; Khoury, J.; Whetten, R. L.; Bigioni, T. P.; Guthrie, D. K.; First, P. N. *J. Am. Chem. Soc.* 1997, *119*, 9279.
7. Green, S. J.; Stokes, J. J.; Hostetler, M. J.; Pietron, J.; Murray, R. W. *J. Phys. Chem. B* 1997, *101*, 2663.
8. Sardar, R.; Funston, A. M.; Mulvaney, P.; Murray, R. W. *Langmuir* 2009, *25*, 13840.
9. Heaven, M. W.; Dass, A.; White, P. S.; Holt, K. M.; Murray, R. W. *J. Am. Chem. Soc.* 2008, *130*, 3754.
10. Zhu, M.; Aikens, C. M.; Hollander, F. J.; Schatz, G. C.; Jin, R. *J. Am. Chem. Soc.* 2008, *130*, 5883.
11. Akola, J.; Walter, M.; Whetten, R. L.; Hakkinen, H.; Gronbeck, H. *J. Am. Chem. Soc.* 2008, *130*, 3756.
12. Murray, R. W. *Chem. Rev.* 2008, *108*, 2688.
13. Astruc, D.; Boisselier, E.; Ornelas, C. *Chem. Rev.* 2010, *110*, 1857.
14. Parker, J. F.; Weaver, J. E. F.; McCallum, F.; Fields-Zinna, C. A.; Murray, R. W. *Langmuir* 2010, *26*, 13650.
15. Parker, J. F.; Fields-Zinna, C. A.; Murray, R. W. *Acc. Chem. Res.* 2010, *43*, 1289.
16. García-Raya, D.; Madueño, R.; Blázquez, M.; Pineda, T. *J. Phys. Chem. C* 2009, *113*, 8756.
17. Lee, D.; Donkers, R. L.; DeSimone, J. M.; Murray, R. W. *J. Am. Chem. Soc.* 2003, *125*, 1182.
18. Wang, W.; Lee, D.; Murray, R. W. *J. Phys. Chem. B* 2006, *110*, 10258.
19. Kurashige, W.; Yamaguchi, M.; Nobusada, K.; Negishi, Y. *J. Phys. Chem. Lett.* 2012, *3*, 2649.
20. Negishi, Y.; Kurashige, W.; Kamimura, U. *Langmuir* 2011, *27*, 12289.
21. Meng, X.; Xu, Q.; Wang, S.; Zhu, M. *Nanoscale* 2012, *4*, 4161.
22. Qian, H.; Zhu, Y.; Jin, R. *ACS Nano* 2009, *3*, 3795.
23. Qian, H.; Eckenhoff, W. T.; Zhu, Y.; Pintauer, T.; Jin, R. *J. Am. Chem. Soc.* 2010, *132*, 8280.
24. Toikkanen, O.; Ruiz, V.; Ronnholm, G.; Kalkkinen, N.; Liljeroth, P.; Quinn, B. M. *J. Am. Chem. Soc.* 2008, *130*, 11049.
25. Toikkanen, O.; Carlsson, S.; Dass, A.; Ronnholm, G.; Kalkkinen, N.; Quinn, B. M. *J. Phys. Chem. Lett.* 2010, *1*, 32.

26. Wang, Z. W.; Toikkanen, O.; Yin, F.; Li, Z. Y.; Quinn, B. M.; Palmer, R. E. *J. Am. Chem. Soc.* 2010, *132*, 2854.
27. Nimmala, P. R.; Yoon, B.; Whetten, R. L.; Landman, U.; Dass, A. *J. Phys. Chem. A* 2013, *117*, 504.
28. Balasubramanian, R.; Guo, R.; Mills, A. J.; Murray, R. W. *J. Am. Chem. Soc.* 2005, *127*, 8126.
29. Qian, H.; Jin, R. *Nano Lett.* 2009, *9*, 4083.
30. Hicks, J. F.; Miles, D. T.; Murray, R. W. *J. Am. Chem. Soc.* 2002, *124*, 13322.
31. Carducci, T. M.; Murray, R. W. *J. Am. Chem. Soc.* 2013, *135*, 11351.
32. Qian, H.; Jin, R. *Chem. Mater.* 2011, *23*, 2209.
33. Cabo-Fernández, L.; Bradley, D. F.; Romani, S.; Higgins, S. J.; Schiffrin, D. J. *ChemPhysChem* 2012, *13*, 2997.
34. Wolfe, R. L.; Murray, R. W. *Anal. Chem.* 2006, *78*, 1167.
35. Wolfe, R. L.; Balasubramanian, R.; Tracy, J. B.; Murray, R. W. *Langmuir* 2007, *23*, 2247.
36. Stiles, R. L.; Balasubramanian, R.; Feldberg, S. W.; Murray, R. W. *J. Am. Chem. Soc.* 2008, *130*, 1856.
37. Hostetler, M. J.; Templeton, A. C.; Murray, R. W. *Langmuir* 1999, *15*, 3782.
38. Templeton, A. C.; Hostetler, M. J.; Kraft, C. T.; Murray, R. W. *J. Am. Chem. Soc.* 1998, *120*, 1906.
39. Ingram, R. S.; Hostetler, M. J.; Murray, R. W. *J. Am. Chem. Soc.* 1997, *119*, 9175.
40. Gies, A. P.; Hercules, D. M.; Gerdon, A. E.; Cliffel, D. E. *J. Am. Chem. Soc.* 2007, *129*, 1095.
41. Varnavski, O.; Ramakrishna, G.; Kim, J.; Lee, D.; Goodson, T. *J. Am. Chem. Soc.* 2009, *132*, 16.
42. Antonello, S.; Hesari, M.; Polo, F.; Maran, F. *Nanoscale* 2012, *4*, 5333.
43. Tracy, J. B.; Crowe, M. C.; Parker, J. F.; Hampe, O.; Fields-Zinna, C. A.; Dass, A.; Murray, R. W. *J. Am. Chem. Soc.* 2007, *129*, 16209.
44. Georganopoulou, D. G.; Mirkin, M. V.; Murray, R. W. *Nano Lett.* 2004, *4*, 1763.
45. Häkkinen, H.; Walter, M.; Grönbeck, H. *J. Phys. Chem. B* 2006, *110*, 9927.
46. Templeton, A. C.; Wuelfing, W. P.; Murray, R. W. *Acc. Chem. Res.* 2000, *33*, 27.
47. Oja, S. M.; Wood, M.; Zhang, B. *Anal. Chem.* 2012, *85*, 473.
48. Devarajan, S.; Sampath, S. Electrochemistry with nanoparticles. In *The Chemistry of Nanomaterials: Synthesis, Properties, and Applications*, C. N. R. Rao, Achim Muller, A. K. Cheetham, (eds.); Wiley-VCH Verlag GmbH & Co. KGaA, Berlin, Germany, published on-line January 2005 (DOI 10.1002/352760247X.ch20) 2004, Chapter 20.
49. Love, J. C.; Estroff, L. A.; Kriebel, J. K.; Nuzzo, R. G.; Whitesides, G. M. *Chem. Rev.* 2005, *105*, 1103.
50. Zamborini, F. P.; Bao, L.; Dasari, R. *Anal. Chem.* 2011, *84*, 541.
51. A themed collection of papers. *Nanoscale* 2012, *4*, 4009–4276;.
52. Schmid, G.; Pfeil, R.; Boese, R.; Brandermann, F.; Meyer, S.; Calis, G. H. M.; Van, d. V. J. W. A. *Chem. Ber.* 1981, *114*, 3634.
53. Tsunoyama, R.; Tsunoyama, H.; Pannopard, P.; Limtrakul, J.; Tsukuda, T. *J. Phys. Chem. C* 2010, *114*, 16004.
54. Mulvaney, P. *Langmuir* 1996, *12*, 788.
55. Menard, L. D.; Xu, H.; Gao, S.-P.; Twesten, R. D.; Harper, A. S.; Song, Y.; Wang, G. et al. *J. Phys. Chem. B* 2006, *110*, 14564.
56. Tracy, J. B.; Kalyuzhny, G.; Crowe, M. C.; Balasubramanian, R.; Choi, J.-P.; Murray, R. W. *J. Am. Chem. Soc.* 2007, *129*, 6706.
57. Dass, A.; Stevenson, A.; Dubay, G. R.; Tracy, J. B.; Murray, R. W. *J. Am. Chem. Soc.* 2008, *130*, 5940.
58. Dass, A.; Holt, K.; Parker, J. F.; Feldberg, S. W.; Murray, R. W. *J. Phys. Chem. C* 2008, *112*, 20276.
59. Fields-Zinna, C. A.; Sardar, R.; Beasley, C. A.; Murray, R. W. *J. Am. Chem. Soc.* 2009, *131*, 16266.
60. Jadzinsky, P. D.; Calero, G.; Ackerson, C. J.; Bushnell, D. A.; Kornberg, R. D. *Science* 2007, *318*, 430.
61. Levi-Kalisman, Y.; Jadzinsky, P. D.; Kalisman, N.; Tsunoyama, H.; Tsukuda, T.; Bushnell, D. A.; Kornberg, R. D. *J. Am. Chem. Soc.* 2011, *133*, 2976.
62. Whetten, R. L.; Price, R. C. *Science* 2007, *318*, 407.
63. Zeng, C.; Qian, H.; Li, T.; Li, G.; Rosi, N. L.; Yoon, B.; Barnett, R. N.; Whetten, R. L.; Landman, U.; Jin, R. *Angew. Chem. Int. Ed.* 2012, *51*, 13114.
64. Oh, S.-K.; Kim, Y.-G.; Ye, H.; Crooks, R. M. *Langmuir* 2003, *19*, 10420.
65. Garcia-Martinez, J. C.; Crooks, R. M. *J. Am. Chem. Soc.* 2004, *126*, 16170.
66. Knecht, M. R.; Garcia-Martinez, J. C.; Crooks, R. M. *Langmuir* 2005, *21*, 11981.
67. Knecht, M. R.; Garcia-Martinez, J. C.; Crooks, R. M. *Chem. Mater.* 2006, *18*, 5039.
68. Knecht, M. R.; Weir, M. G.; Frenkel, A. I.; Crooks, R. M. *Chem. Mater.* 2008, *20*, 1019.
69. Weir, M. G.; Knecht, M. R.; Frenkel, A. I.; Crooks, R. M. *Langmuir* 2010, *26*, 1137.
70. Kim, Y.-G.; Garcia-Martinez, J. C.; Crooks, R. M. *Langmuir* 2005, *21*, 5485.

71. Jimenez, V. L.; Georganopoulou, D. G.; White, R. J.; Harper, A. S.; Mills, A. J.; Lee, D.; Murray, R. W. *Langmuir* 2004, *20*, 6864.
72. Guo, R.; Murray, R. W. *J. Am. Chem. Soc.* 2005, *127*, 12140.
73. Wang, G.; Guo, R.; Kalyuzhny, G.; Choi, J.-P.; Murray, R. W. *J. Phys. Chem. B* 2006, *110*, 20282.
74. Choi, J.-P.; Murray, R. W. *J. Am. Chem. Soc.* 2006, *128*, 10496.
75. Antonello, S.; Holm, A. H.; Instuli, E.; Maran, F. *J. Am. Chem. Soc.* 2007, *129*, 9836.
76. Parker, J. F.; Choi, J.-P.; Wang, W.; Murray, R. W. *J. Phys. Chem. C* 2008, *112*, 13976.
77. Parker, J. F.; Kacprzak, K. A.; Lopez-Acevedo, O.; Häkkinen, H.; Murray, R. W. *J. Phys. Chem. C* 2010, *114*, 8276.
78. Zhu, M.; Eckenhoff, W. T.; Pintauer, T.; Jin, R. *J. Phys. Chem. C* 2008, *112*, 14221.
79. Zhu, M.; Aikens, C. M.; Hendrich, M. P.; Gupta, R.; Qian, H.; Schatz, G. C.; Jin, R. *J. Am. Chem. Soc.* 2009, *131*, 2490.
80. Bard, A. J.; Faulkner, L. R. In *Electrochemical Methods Fundamentals and Applications*; 2 edn.; John Wiley & Sons: Hoboken, NJ, 2001, p. 736.
81. Wang, G.; Huang, T.; Murray, R. W.; Menard, L.; Nuzzo, R. G. *J. Am. Chem. Soc.* 2004, *127*, 812.
82. Li, L.; Liu, H.; Shen, Y.; Zhang, J.; Zhu, J.-J. *Anal. Chem.* 2011, *83*, 661.
83. Swanick, K. N.; Hesari, M.; Workentin, M. S.; Ding, Z. *J. Am. Chem. Soc.* 2012, *134*, 15205.
84. Hicks, J. F.; Templeton, A. C.; Chen, S.; Sheran, K. M.; Jasti, R.; Murray, R. W.; Debord, J.; Schaaff, T. G.; Whetten, R. L. *Anal. Chem.* 1999, *71*, 3703.
85. Zamborini, F. P.; Hicks, J. F.; Murray, R. W. *J. Am. Chem. Soc.* 2000, *122*, 4514.
86. Lopez-Acevedo, O.; Akola, J.; Whetten, R. L.; Grönbeck, H.; Häkkinen, H. *J. Phys. Chem. C* 2009, *113*, 5035.
87. Mertens, S. F. L.; Blech, K.; Sologubenko, A. S.; Mayer, J.; Simon, U.; Wandlowski, T. *Electrochim. Acta* 2009, *54*, 5006.
88. Koivisto, J.; Malola, S.; Kumara, C.; Dass, A.; Häkkinen, H.; Pettersson, M. *J. Phys. Chem. Lett.* 2012, *3*, 3076.
89. Chen, S.; Ingram, R. S.; Hostetler, M. J.; Pietron, J. J.; Murray, R. W.; Schaaff, T. G.; Khoury, J. T.; Alvarez, M. M.; Whetten, R. L. *Science* 1998, *280*, 2098.
90. Chen, S.; Murray, R. W.; Feldberg, S. W. *J. Phys. Chem. B* 1998, *102*, 9898.
91. Peterson, R. R.; Cliffel, D. E. *Langmuir* 2006, *22*, 10307.
92. Chen, S.; Yang, Y. *J. Am. Chem. Soc.* 2002, *124*, 5280.
93. Yang, Y.; Grant, K. M.; White, H. S.; Chen, S. *Langmuir* 2003, *19*, 9446.
94. Pietron, J. J.; Hicks, J. F.; Murray, R. W. *J. Am. Chem. Soc.* 1999, *121*, 5565.
95. Quinn, B. M.; Liljeroth, P.; Kontturi, K. *J. Am. Chem. Soc.* 2002, *124*, 12915.
96. Hicks, J. F.; Zamborini, F. P.; Murray, R. W. *J. Phys. Chem. B* 2002, *106*, 7751.
97. Chen, S. *J. Phys. Chem. B* 2000, *104*, 663.
98. Chen, S. *J. Am. Chem. Soc.* 2000, *122*, 7420.
99. Chen, S.; Pei, R. *J. Am. Chem. Soc.* 2001, *123*, 10607.
100. Liu, F.; Khan, K.; Liang, J.-H.; Yan, J.-W.; Wu, D.-Y.; Mao, B.-W.; Jensen, P. S.; Zhang, J.; Ulstrup, J. *ChemPhysChem*, 2013, *14*, 952–957.
101. Men, Y.; Kubo, K.; Kurihara, M.; Nishihara, H. *Phys. Chem. Chem. Phys.* 2001, *3*, 3427.
102. Brennan, J. L.; Branham, M. R.; Hicks, J. F.; Osisek, A. J.; Donkers, R. L.; Georganopoulou, D. G.; Murray, R. W. *Anal. Chem.* 2004, *76*, 5611.
103. Hicks, J. F.; Zamborini, F. P.; Osisek, A. J.; Murray, R. W. *J. Am. Chem. Soc.* 2001, *123*, 7048.
104. Zamborini, F. P.; Leopold, M. C.; Hicks, J. F.; Kulesza, P. J.; Malik, M. A.; Murray, R. W. *J. Am. Chem. Soc.* 2002, *124*, 8958.
105. Uosaki, K.; Kondo, T.; Okamura, M.; Song, W. *Faraday Discuss.* 2002, *121*, 373.
106. Chen, S.; Pei, R.; Zhao, T.; Dyer, D. J. *J. Phys. Chem. B* 2002, *106*, 1903.
107. Brust, M.; Bethell, D.; Kiely, C. J.; Schiffrin, D. J. *Langmuir* 1998, *14*, 5425.
108. Liljeroth, P.; Vanmaekelbergh, D.; Ruiz, V.; Kontturi, K.; Jiang, H.; Kauppinen, E.; Quinn, B. M. *J. Am. Chem. Soc.* 2004, *126*, 7126.
109. Joseph, Y.; Besnard, I.; Rosenberger, M.; Guse, B.; Nothofer, H.-G.; Wessels, J. M.; Wild, U. et al. *J. Phys. Chem. B* 2003, *107*, 7406.
110. Laaksonen, T.; Ruiz, V.; Liljeroth, P.; Quinn, B. M. *J. Phys. Chem. C* 2008, *112*, 15637.
111. Liljeroth, P.; Quinn, B. M. *J. Am. Chem. Soc.* 2006, *128*, 4922.
112. Yang, Y.; Pradhan, S.; Chen, S. *J. Am. Chem. Soc.* 2004, *126*, 76.
113. Chen, S. *Anal. Chim. Acta* 2003, *496*, 29.
114. Kim, J.; Lee, D. *J. Am. Chem. Soc.* 2006, *128*, 4518.

115. Li, W.; Su, B. *Electrochem. Commun.* 2012, *22*, 8.
116. Deng, F.; Chen, S. *Langmuir* 2007, *23*, 936.
117. Deng, F.; Chen, S. *Phys. Chem. Chem. Phys.* 2005, *7*, 3375.
118. Wang, W.; Murray, R. W. *Anal. Chem.* 2006, *79*, 1213.
119. Laaksonen, T.; Ruiz, V.; Murtomaeki, L.; Quinn, B. M. *J. Am. Chem. Soc.* 2007, *129*, 7732.
120. Ahonen, P.; Ruiz, V.; Kontturi, K.; Liljeroth, P.; Quinn, B. M. *J. Phys. Chem. C* 2008, *112*, 2724.
121. Ranganathan, S.; Guo, R.; Murray, R. W. *Langmuir* 2007, *23*, 7372.
122. Xu, R.; Sun, Y.; Yang, J.-Y.; He, L.; Nie, J.-C.; Li, L.; Li, Y. *Appl. Phys. Lett.* 2010, *97*, 113101/1.
123. Savéant, J. M. *J. Electroanal. Chem. Interfacial Electrochem.* 1988, *242*, 1.
124. White, R. J.; White, H. S. *Anal. Chem.* 2005, *77*, 214A.
125. Terrill, R. H.; Hutchison, J. E.; Murray, R. W. *J. Phys. Chem. B* 1997, *101*, 1535.
126. Jernigan, J. C.; Surridge, N. A.; Zvanut, M. E.; Silver, M.; Murray, R. W. *J. Phys. Chem.* 1989, *93*, 4620.
127. Wuelfing, W. P.; Green, S. J.; Pietron, J. J.; Cliffel, D. E.; Murray, R. W. *J. Am. Chem. Soc.* 2000, *122*, 11465.
128. Wuelfing, W. P.; Murray, R. W. *J. Phys. Chem. B* 2002, *106*, 3139.
129. Pradhan, S.; Kang, X.; Mendoza, E.; Chen, S. *Appl. Phys. Lett.* 2009, *94*, 042113/1.
130. Pradhan, S.; Sun, J.; Deng, F.; Chen, S. *Adv. Mater.* 2006, *18*, 3279.
131. Wallner, A.; Jafri, S. H. M.; Blom, T.; Gogoll, A.; Leifer, K.; Baumgartner, J.; Ottosson, H. *Langmuir* 2011, *27*, 9057.
132. Zamborini, F. P.; Smart, L. E.; Leopold, M. C.; Murray, R. W. *Anal. Chim. Acta* 2003, *496*, 3.
133. Leopold, M. C.; Donkers, R. L.; Georganopoulou, D.; Fisher, M.; Zamborini, F. P.; Murray, R. W. *Faraday Discuss.* 2003, *125*, 63.
134. Fishelson, N.; Shkrob, I.; Lev, O.; Gun, J.; Modestov, A. D. *Langmuir* 2001, *17*, 403.
135. Trudeau, P. E.; Orozco, A.; Kwan, E.; Dhirani, A. A. *J. Chem. Phys.* 2002, *117*, 3978.
136. Trudeau, P. E.; Escorcia, A.; Dhirani, A. A. *J. Chem. Phys.* 2003, *119*, 5267.
137. Wang, G. R.; Wang, L.; Rendeng, Q.; Wang, J.; Luo, J.; Zhong, C.-J. *J. Mater. Chem.* 2007, *17*, 457.
138. Choi, J.-P.; Coble, M. M.; Branham, M. R.; DeSimone, J. M.; Murray, R. W. *J. Phys. Chem. C* 2007, *111*, 3778.
139. Branham, M. R.; Douglas, A. D.; Mills, A. J.; Tracy, J. B.; White, P. S.; Murray, R. W. *Langmuir* 2006, *22*, 11376.
140. Wu, Z.; Lanni, E.; Chen, W.; Bier, M. E.; Ly, D.; Jin, R. *J. Am. Chem. Soc.* 2009, *131*, 16672.
141. Snow, A. W.; Wohltjen, H. *Chem. Mater.* 1998, *10*, 947.
142. Zabet-Khosousi, A.; Trudeau, P.-E.; Suganuma, Y.; Dhirani, A.-A.; Statt, B. *Phys. Rev. Lett.* 2006, *96*, 156403/1.
143. Suganuma, Y.; Dhirani, A.-A. *J. Phys. Chem. B* 2005, *109*, 15391.
144. Nair, A. S.; Kimura, K. *Langmuir* 2009, *25*, 1750.
145. Nair, A. S.; Kimura, K. *J. Chem. Phys.* 2008, *129*, 184117.
146. Muller, K. H.; Herrmann, J.; Wei, G.; Raguse, B.; Wieczorek, L. *J. Phys. Chem. C* 2009, *113*, 18027.
147. McCreery, R. L. *Chem. Mater.* 2004, *16*, 4477.
148. Kumara, C.; Dass, A. *Nanoscale* 2011, *3*, 3064.
149. Li, Y.; Cox, J. T.; Zhang, B. *J. Am. Chem. Soc.* 2010, *132*, 3047.
150. Kwon, S. J.; Zhou, H.; Fan, F.-R. F.; Vorobyev, V.; Zhang, B.; Bard, A. J. *Phys. Chem. Chem. Phys.* 2011, *13*, 5394.
151. Zhou, H.; Fan, F.-R. F.; Bard, A. J. *J. Phys. Chem. Lett.* 2010, *1*, 2671.
152. Chen, S.; Huang, K. *J. Cluster Sci.* 2000, *11*, 405.
153. Ghosh, D.; Chen, S. *J. Mater. Chem.* 2008, *18*, 755.
154. Zhou, Z.-Y.; Kang, X.; Song, Y.; Chen, S. *Chem. Commun.* 2011, *47*, 6075.
155. He, G.; Song, Y.; Kang, X.; Chen, S. *Electrochim. Acta* 2013, *94*, 98.
156. Meier, J.; Friedrich, K. A.; Stimming, U. *Faraday Discuss.* 2002, *121*, 365.
157. Meier, J.; Schiotz, J.; Liu, P.; Norskov, J. K.; Stimming, U. *Chem. Phys. Lett.* 2004, *390*, 440.
158. Zhou, Z.-Y.; Kang, X.; Song, Y.; Chen, S. *J. Phys. Chem. C* 2012, *116*, 10592.
159. Dubois, J. G. A.; Gerritsen, J. W.; Shafranjuk, S. E.; Boon, E. J. G.; Schmid, G.; Kempen, H. v. *EPL (Europhys. Lett.)* 1996, *33*, 279.
160. Ye, H.; Crooks, R. M. *J. Am. Chem. Soc.* 2007, *129*, 3627.
161. Polsky, R.; Gill, R.; Kaganovsky, L.; Willner, I. *Anal. Chem.* 2006, *78*, 2268.
162. Kang, X.; Zuckerman, N. B.; Konopelski, J. P.; Chen, S. *Angew. Chem. Int. Ed.* 2010, *49*, 9496.
163. Chen, W.; Zuckerman, N. B.; Kang, X.; Ghosh, D.; Konopelski, J. P.; Chen, S. *J. Phys. Chem. C* 2010, *114*, 18146.

164. Ghosh, D.; Chen, S. *Chem. Phys. Lett.* 2008, *465*, 115.
165. Kang, X.; Chen, S. *Nanoscale* 2012, *4*, 4183.
166. Chen, W.; Zuckerman, N. B.; Lewis, J. W.; Konopelski, J. P.; Chen, S. *J. Phys. Chem. C* 2009, *113*, 16988.
167. Chen, W.; Pradhan, S.; Chen, S. *Nanoscale* 2011, *3*, 2294.
168. Kang, X.; Song, Y.; Chen, S. *J. Mater. Chem.* 2012, *22*, 19250.
169. Cheng, W.; Dong, S.; Wang, E. *Electrochem. Commun.* 2002, *4*, 412.
170. Aslam, M.; Mulla, I. S.; Vijayamohanan, K. *Appl. Phys. Lett.* 2001, *79*, 689.
171. Taleb, A.; Xue, Y.; Munteanu, S.; Kanoufi, F.; Dubot, P. *Electrochim. Acta*, 2013, *88*, 621–631.
172. Skewis, L. R.; Reinhard, B. R. M. *ACS Appl. Mater. Interfaces* 2009, *2*, 35.
173. Huang, T.; Murray, R. W. *J. Phys. Chem. B* 2003, *107*, 7434.
174. Wu, Z.; Jiang, D.-E.; Lanni, E.; Bier, M. E.; Jin, R. *J. Phys. Chem. Lett.* 2010, *1*, 1423.
175. Chakraborty, I.; Govindarajan, A.; Erusappan, J.; Ghosh, A.; Pradeep, T.; Yoon, B.; Whetten, R. L.; Landman, U. *Nano Lett.* 2012, *12*, 5861.
176. Chen, S.; Sommers, J. M. *J. Phys. Chem. B* 2001, *105*, 8816.
177. Kakade, B. A.; Shintri, S. S.; Sathe, B. R.; Halligudi, S. B.; Pillai, V. K. *Adv. Mater.* 2007, *19*, 272.
178. Henglein, A.; Giersig, M. *J. Phys. Chem. B* 2000, *104*, 5056.
179. Cao, Y.; Jin, R.; Mirkin, C. A. *J. Am. Chem. Soc.* 2001, *123*, 7961.
180. Malola, S.; Häkkinen, H. *J. Phys. Chem. Lett.* 2011, *2*, 2316.
181. Negishi, Y.; Munakata, K.; Ohgake, W.; Nobusada, K. *J. Phys. Chem. Lett.* 2012, *3*, 2209.
182. Negishi, Y.; Kurashige, W.; Niihori, Y.; Iwasa, T.; Nobusada, K. *Phys. Chem. Chem. Phys.* 2010, *12*, 6219.
183. Niihori, Y.; Kurashige, W.; Matsuzaki, M.; Negishi, Y. *Nanoscale* 2013, *5*, 508.
184. Chandler, B. D.; Long, C. G.; Gilbertson, J. D.; Pursell, C. J.; Vijayaraghavan, G.; Stevenson, K. J. *J. Phys. Chem. C* 2010, *114*, 11498.
185. Ibanez, F. J.; Zamborini, F. P. *ACS Nano* 2008, *2*, 1543.
186. Kim, J. M.; Koo, C. M.; Kim, J. *Electroanalysis* 2011, *23*, 2019.
187. Chen, S. *Langmuir* 2001, *17*, 6664.
188. Sardar, R.; Beasley, C. A.; Murray, R. W. *Anal. Chem.* 2009, *81*, 6960.
189. Beasley, C. A.; Sardar, R.; Barnes, N. M.; Murray, R. W. *J. Phys. Chem. C* 2010, *114*, 18384.
190. Chow, K.-F.; Sardar, R.; Sassin, M. B.; Wallace, J. M.; Feldberg, S. W.; Rolison, D. R.; Long, J. W.; Murray, R. W. *J. Phys. Chem. C* 2012, *116*, 9283.
191. Dong, T.-Y.; Shih, H.-W.; Chang, L.-S. *Langmuir* 2004, *20*, 9340.
192. Gittins, D. I.; Bethell, D.; Schiffrin, D. J.; Nichols, R. J. *Nature* 2000, *408*, 67.
193. Horikoshi, T.; Itoh, M.; Kurihara, M.; Kubo, K.; Nishihara, H. *J. Electroanal. Chem.* 1999, *473*, 113.
194. Yamada, M.; Nishihara, H. *Chem. Commun.* 2002, 2578.
195. Yamada, M.; Tadera, T.; Kubo, K.; Nishihara, H. *J. Phys. Chem. B* 2003, *107*, 3703.
196. Yamada, M.; Nishihara, H. *Langmuir* 2003, *19*, 8050.
197. Pietron, J. J.; Murray, R. W. *J. Phys. Chem. B* 1999, *103*, 4440.
198. Ingram, R. S.; Murray, R. W. *Langmuir* 1998, *14*, 4115.
199. Yamada, M.; Tadera, T.; Kubo, K.; Nishihara, H. *Langmuir* 2001, *17*, 2363.
200. Lopez-Acevedo, O.; Rintala, J.; Virtanen, S.; Femoni, C.; Tiozzo, C.; Gronbeck, H.; Pettersson, M.; Hakkinen, H. *J. Am. Chem. Soc.* 2009, *131*, 12573.
201. Miles, D. T.; Murray, R. W. *Anal. Chem.* 2001, *73*, 921.
202. Chen, S.; Huang, K. *Langmuir* 2000, *16*, 2014.
203. Gittins, D. I.; Bethell, D.; Nichols, R. J.; Schiffrin, D. J. *Adv. Mater.* 1999, *11*, 737.
204. Holm, A. H.; Ceccato, M.; Donkers, R. L.; Fabris, L.; Pace, G.; Maran, F. *Langmuir* 2006, *22*, 10584.
205. Boal, A. K.; Rotello, V. M. *J. Am. Chem. Soc.* 1999, *121*, 4914.
206. Boal, A. K.; Rotello, V. M. *J. Am. Chem. Soc.* 2002, *124*, 5019.
207. Bayir, A.; Jordan, B. J.; Verma, A.; Pollier, M. A.; Cooke, G.; Rotello, V. M. *Chem. Commun.* 2006, 4033.
208. Abad, J. M.; Gass, M.; Bleloch, A.; Schiffrin, D. J. *J. Am. Chem. Soc.* 2009, *131*, 10229.
209. Li, G.; Jin, R. *Acc. Chem. Res.* 2013, *46*, 1749–1758.
210. Mirkhalaf, F.; Schiffrin, D. J. *Langmuir* 2010, *26*, 14995.
211. Zhu, Y.; Qian, H.; Drake, B. A.; Jin, R. *Angew. Chem. Int. Ed.* 2010, *49*, 1295.
212. Lyalin, A.; Taketsugu, T. *J. Phys. Chem. Lett.* 2010, *1*, 1752.
213. Tang, D.; Hu, C. *J. Phys. Chem. Lett.* 2011, *2*, 2972.
214. Negishi, Y.; Mizuno, M.; Hirayama, M.; Omatoi, M.; Takayama, T.; Iwase, A.; Kudo, A. *Nanoscale* 2013, *5*, 7188.
215. Zaramella, D.; Scrimin, P.; Prins, L. J. *J. Am. Chem. Soc.* 2012, *134*, 8396.

216. Novo, C.; Funston, A. M.; Mulvaney, P. *Nat. Nanotechnol.* 2008, *3*, 598.
217. Xiao, X.; Bard, A. J. *J. Am. Chem. Soc.* 2007, *129*, 9610.
218. Xiao, X.; Pan, S.; Jang, J. S.; Fan, F.-R. F.; Bard, A. J. *J. Phys. Chem. C* 2009, *113*, 14978.
219. Xiao, X.; Fan, F.-R. F.; Zhou, J.; Bard, A. J. *J. Am. Chem. Soc.* 2008, *130*, 16669.
220. Baron, R.; Wildgoose, G. G.; Compton, R. G. *J. Nanosci. Nanotechnol.* 2009, *9*, 2274.
221. Nakashima, D.; Marken, F.; Oyama, M. *Electroanalysis* 2013, *25*, 975–982.
222. Xu, L.-P.; Chen, S. *Chem. Phys. Lett.* 2009, *468*, 222.
223. Shao, Y.; Mirkin, M. V.; Fish, G.; Kokotov, S.; Palanker, D.; Lewis, A. *Anal. Chem.* 1997, *69*, 1627.
224. Wohltjen, H.; Snow, A. W. *Anal. Chem.* 1998, *70*, 2856.
225. Cai, Q.-Y.; Zellers, E. T. *Anal. Chem.* 2002, *74*, 3533.
226. Steinecker, W. H.; Rowe, M. P.; Zellers, E. T. *Anal. Chem.* 2007, *79*, 4977.
227. Raguse, B.; Chow, E.; Barton, C. S.; Wieczorek, L. *Anal. Chem.* 2007, *79*, 7333.
228. Ibanez, F. J.; Gowrishetty, U.; Crain, M. M.; Walsh, K. M.; Zamborini, F. P. *Anal. Chem.* 2006, *78*, 753.
229. Kumar, S. S.; Kwak, K.; Lee, D. *Anal. Chem.* 2011, *83*, 3244.
230. Li, L.; Zhu, A.; Tian, Y. *Chem. Commun.* 2013, *49*, 1279.
231. Chauhan, N.; Singh, A.; Narang, J.; Dahiya, S.; Pundir, C. S. *Analyst* 2012, *137*, 5113.
232. Jeong, B.; Akter, R.; Han, O. H.; Rhee, C. K.; Rahman, M. A. *Anal. Chem.* 2013, *85*, 1784.
233. Loftus, A. F.; Reighard, K. P.; Kapourales, S. A.; Leopold, M. C. *J. Am. Chem. Soc.* 2008, *130*, 1649.
234. Tran, T. D.; Vargo, M. L.; Gerig, J. K.; Gulka, C. P.; Trawick, M. L.; Dattelbaum, J. D.; Leopold, M. C. *J. Colloid Interface Sci.* 2010, *352*, 50.
235. Vargo, M. L.; Gulka, C. P.; Gerig, J. K.; Manieri, C. M.; Dattelbaum, J. D.; Marks, C. B.; Lawrence, N. T.; Trawick, M. L.; Leopold, M. C. *Langmuir* 2010, *26*, 560.
236. Campbell-Rance, D. S.; Doan, T. T.; Leopold, M. C. *J. Electroanal. Chem.* 2011, *662*, 343.
237. Xia, N.; Deng, D.; Zhang, L.; Yuan, B.; Jing, M.; Du, J.; Liu, L. *Biosens. Bioelectron.* 2013, *43*, 155.
238. Gao, Z.-D.; Guan, F.-F.; Li, C.-Y.; Liu, H.-F.; Song, Y.-Y. *Biosens. Bioelectron.* 2013, *41*, 771.
239. Ackerson, C. J.; Jadzinsky, P. D.; Jensen, G. J.; Kornberg, R. D. *J. Am. Chem. Soc.* 2006, *128*, 2635.
240. Escorura-Muniz, A. de la; Parolo, C.; Maran, F.; Mekoci, A. *Nanoscale* 2011, *3*, 3350.
241. Wang, J.; Xu, D.; Kawde, A.-N.; Polsky, R. *Anal. Chem.* 2001, *73*, 5576.

4 Platinum-Monolayer Oxygen-Reduction Electrocatalysts
Present Status and Future Prospects

Radoslav R. Adzic and Kuanping Gong

CONTENTS

4.1 INTRODUCTION

Fuel cells, the electrochemical direct energy-conversion power sources that generate electricity with high efficiency at mild operating conditions, have been heralded as ideally useful to secure sustainable supply of clean energy.[1] Recently, their performance has been significantly improved by advances in fuel-cell electrocatalysts that brought this technology close to applications. Applying fuel cells in portable electronics and power vehicles is peculiarly desired since they ensure negligible carbon footprint upon energy conversion.[2] At the heart of such applications, there is a long-standing issue with the oxygen-reduction reaction (ORR) at the cathode, whose fairly slow kinetics causes the fuel-cell efficiency to fall down into the range of 50%–60%, much lower than the thermodynamically calculated value of 83% at 298 K. To accelerate the ORR, most fuel cells use platinum particles supported on porous carbon supports (Pt/C), the most active electrocatalysts known to date.[3] However, such superior activity is far too compromised by the high cost and low abundance of Pt to be well appreciated in commercializing fuel cells on a large scale. By 2015, realistic Pt-based electrocatalysts should accomplish the Pt mass activity roughly four times higher that of the standard Pt/C to accord with the latest protocol of the US Department of Energy.[4–7]

In addition to several partly successful approaches that have been studied to resolve high Pt content and insufficient activity and stability (cf. review[1]), in the quest of cost-effective ORR electrocatalysts, a strategy has been explored that rests on the active center of nitrogen-coordinated transition metals, that is, MN_x (M = Fe, Co; x = 2, 4); examples include nonprecious transition-metal macrocycles and nitrogen-doped carbon materials.[8,9] However, quite a few of these nonprecious electrocatalysts show the activity close or equivalent to the typical Pt/C in acid media—even with

125

their components, composition, and structures being well optimized. Another notorious drawback long has prevented their practical applications in fuel cells, largely due to the loss of the metal center that not only degrades the ORR activity but also perhaps causes dysfunction of the polymeric membrane.[10] Ultimately, noble metal-based ORR nanocrystals have been considered as a primary solution to the trade-off between activity, stability, and cost-effectiveness.

The Pt monolayer catalyst concept appears suitable to meet the aforementioned challenges. Solutions to make the best usage of platinum lie in employing a second transition metal to form Pt-shelled nanoparticles that play a dual role in the ORR: (1) fine-tuning the *d-band* center of Pt to balance the kinetics of the elementary chemical and electrochemical reactions involved in the ORR and (2) replacing a significant amount of Pt while achieving comparable and even higher performance.[11,12] These ideal outcomes would be expected to culminate in the most efficient utilization of Pt if an atomic Pt monolayer (Pt_{ML}) with the proper electronic structure can be placed on stable, inexpensive metallic cores. Indeed, Adzic and coworkers have been developing Pt_{ML} electrocatalysts of this kind that exhibited an order-of-magnitude higher mass activity than commercial Pt nanoparticles.[13,14]

For Pt_{ML} electrocatalysts supported on various single crystals, theoretical and experimental results both revealed a volcano-like dependence of the ORR kinetics on the *d-band* center of Pt that was associated with the nature and composition of the second metal.[15] On the top of the *volcano* plot is the Pt monolayer on a Pd core or its alloys, that is, Pt_{ML}/Pd or Pt_{ML}/PdM, the most active electrocatalyst. Not only does the underlying Pd induce a slight contraction of the atomic arrangement of the surface Pt but also the core–shell electronic interaction causes the *d-band* center of Pt to upshift by ~0.12 eV relative to the Fermi level. These favorable geometric and ligand effects ensure a facile electron transfer during the reaction of oxygen with Pt while expediting desorptions of the intermediates and/or the end product, so as to improve the overall kinetics of the ORR. The Pd–Pt_{ML} core–shell electrocatalysts were formed by galvanically displacing with Pt of an underpotentially deposited (UPD) copper monolayer on commercial Pd nanoparticles.[16,17] In membrane electrode assembly (MEA) test under practical operating conditions, these Pt_{ML} electrocatalysts performed much better than did their commercial Pt counterpart in terms of activity, stability, and the utilization of Pt. However, there was a certain loss of the Pd core, diffusing out of the core–shell structure to the PEM, which, if not circumvented, may paralyze a PEMFC stack by causing dysfunction in the polymeric membrane.[17] The Pd^{2+} cations are usually reduced to Pd^0 by H_2 diffusing from the anode. A Pd band may form (Figure 4.15a [later in the chapter]), which does not affect the membrane's conductivity, unlike Co^{2+} or Ni^{2+} cations that are not reduced by H_2.

The recent availability of shape-defined Pd nanocrystals with cleaned surfaces allowed evaluating the facet-specific electrochemical properties in terms of a new support of Pt_{ML} catalysts.[18,19] Such well-defined Pd–Pt_{ML} core–shell electrocatalysts were examined by combining structural analyses and density functional theory (DFT) with electrochemical techniques.[20] The surfaces of the Pd cores are composed of specific facets wherein the Pd atoms are highly coordinated and have low surface energy. Our results revealed that in comparison with sphere Pd-supported Pt_{ML} or pure Pt, these Pd-supported Pt_{ML} catalysts features surface contraction and a downshift of *d-band* relative to the Fermi level. These effects were demonstrated to determine the high activity in the ORR in close conjugation with their surface atomic arrangement and coordination. Such shape-property interdependence promised new approaches to basic and applied research on Pt-based ORR electrocatalysts of pivotal importance to the widespread adoption of fuel cells.

4.2 TUNABLE ORR ACTIVITY OF PT MONOLAYER ELECTROCATALYSTS

A monolayer of a metal deposited on different metals has been shown to fundamentally differ in reactivity from the counterpart pure metals.[21,22] This behavior can mainly be attributed to the geometric (strain) and electronic (ligand) effects, both resulting from the interaction between the host/substrate metal and the deposited metal monolayer.[23–25] This opens up opportunities for modifying

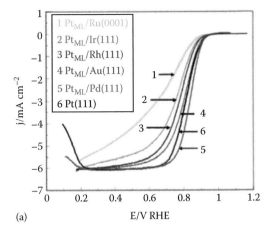
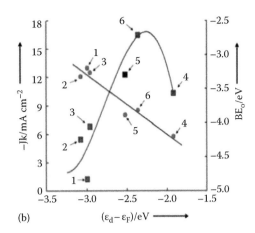

(a)

(b)

FIGURE 4.1 (a) Polarization curves for O_2 reduction on platinum monolayers (Pt_{ML}) on Ru(0001), Ir(111), Rh(111), Au(111), and Pd(111) in a 0.1 M $HClO_4$ solution on a disk electrode. The curve for Pt(111) is taken from the previous publication[26] and included for comparison. The rotation rate is 1600 rpm, and the sweep rate is 20 mV s^{-1} (50 mV s^{-1} for Pt(111)); j, current density; RHE, reversible hydrogen electrode.[15] (b) Kinetic currents (j_K, square symbols) at 0.8 V for O_2 reduction on the platinum monolayers supported on different single-crystal surfaces in a 0.1 M $HClO_4$ solution and calculated binding energies of atomic oxygen (BE_O, filled circles) as functions of calculated d-band center ($\varepsilon_d - \varepsilon_F$), relative to the Fermi level of the respective clean platinum monolayers. The current data for Pt(111) are taken from Ref. [26] and included for comparison. Labels: 1. Pt_{ML}/Ru(0001), 2. Pt_{ML}/Ir(111), 3. Pt_{ML}/Rh(111), 4. Pt_{ML}/Au(111), 5. Pt(111), and 6. Pt_{ML}/Pd(111). (From Zhang, J. et al., *Angew. Chem.*, 44, 2132, 2005. Markovic, N.M. et al., *J. Electroanal. Chem.*, 467, 157, 1999.)

the properties of a metal surface. Rotating disk measurements of the ORR activities of Pt monolayers clearly show that their properties have been modified; Pt_{ML}/Pd(111) and Pt_{ML}/Ru(0001) are the most and least active of these surfaces, respectively (Figure 4.1a).

The kinetic currents on various surfaces, obtained from Koutecký–Levich plots for ORR, are plotted in Figure 4.1 as a function of the calculated *d*-band centers (ε_d), showing a *volcano* shape. At the topmost is Pt_{ML}/Pd(111), that is, the best candidate in term of activity. Importantly, Pt_{ML}/Pd(111) is more reactive than Pt(111), so far best known for its great catalytic activity for the ORR in $HClO_4$. The lattice mismatch between the Pt and the host metals—namely, Au(111), Pd(111), Ir(111), Rh(111), and Ru(0001) (single-crystal surfaces)—induces either compressive or expansive strain in the Pt monolayers. The exact position of the ε_d of the Pt monolayers depends both on the amount of strain and on the electronic coupling between the Pt_{ML} and the substrate. Specifically, the Pt monolayer is compressed on Ir(111), Ru(0001), and Rh(111), whereas it is stretched by almost 4% on Au(111), compared to an all-Pt Pt(111) surface. As shown before, compressive strain lowers the ε_d, which, in turn, leads to weaker oxygen binding to a surface, whereas expansive strain has the opposite effect.[27] As predicted, Figure 4.1b shows a linear correlation between the calculated binding energy of oxygen (BE_O) and ε_d. The BE_O is higher on Pt_{ML}/Au(111) and lower on Pt_{ML}/Ir(111), Pt_{ML}/Ru(0001), and Pt_{ML}/Rh(111). Since the Pt_{ML} on Pd(111) is only mildly compressed because of similar lattice constants of Pt and Pd, Pt(111) and Pt_{ML}/Pd(111) bind oxygen with similar strength (within about 0.1 eV).

Irrespective of its microscopic mechanism, the ORR process must involve both the breaking of an O–O bond and the formation of O–H bonds.[28] On metal surfaces, the rate-limiting step of the ORR is the electrochemical activation of adsorbed O_2 if the metals bind oxygen too weakly, and, if too strongly, it varies to the dissociation of the ORR intermediates to renew the metallic active sites. Consequently, previous studies have shown that surfaces binding an adsorbate strongly tend to enhance the kinetics of bond-breaking steps in which the adsorbate is a product, and vice versa.[27,29,30]

Hence, pure Pt should have an intermediate BE_O since the optimum ORR catalyst needs to facilitate both bond-breaking and bond-making steps without hindering one or the other. The $Pt_{ML}/Au(111)$ surface, which binds atomic O the strongest, should facilitate the bond-breaking step in O–O containing species the most at the cost of slowing down bond-making steps, including the formation of OH, OOH, H_2O, and H_2O_2. Slow O and OH hydrogenation would cause the surface coverage of these species to build up, impeding the adsorption of the main reactant, O_2. $Pt_{ML}/Ru(0001)$, $Pt_{ML}/Ir(111)$, and $Pt_{ML}/Rh(111)$ are at the other end of the BE_O spectrum, binding O more weakly than Pt, and hence should facilitate bond-making steps, including the formation of H_2O_2. The volcano-type dependence of ORR activity on ε_d, as shown in Figure 4.1, confirms these predictions. Furthermore, the experimental data exhibited that bond-making steps are indeed facilitated on the compressed Pt monolayers, through enhanced H_2O_2 formation rates (e.g., on $Pt_{ML}/Ru(0001)$ and $Pt_{ML}/Ir(111)$). Because $Pt_{ML}/Pd(111)$ is calculated to bind O similarly to Pt(111), it raises the intriguing possibility that this bimetallic system might show promising activity for the ORR. Overall, the experimental results are in excellent agreement with the prediction: $Pt_{ML}/Pd(111)$ performs better than all the other Pt monolayers and even shows about 30% increase in current density (Figure 4.1) compared to the Pt(111) surface.

4.3 MIXED PT_{ML} ELECTROCATALYSTS: SURFACE ENSEMBLE EFFECT ON THE ORR ACTIVITY

Compared with Pt(111), another key factor acting in favor of the $Pt_{ML}/Pd(111)$ surface is the reduced affinity for OH on the alloy surface, since high OH coverage is known to inhibit the ORR.[31] DFT calculations suggest that the adsorption of OH is weaker on $Pt_{ML}/Pd(111)$ than on Pt(111), which, in turn, binds OH more weakly than Pd(111). This observation is in line with the enhanced hydrogenation rates of OH on $Pt_{ML}/Pd(111)$ that should follow a trend similar to that of O hydrogenation. Voltammetry and in situ x-ray absorption near-edge structure spectroscopy (XANES) have been used to demonstrate that OH adsorption on Pt sites shifts to more positive potentials on Pt_{ML}–Pd–C, compared to Pt–OH onset formation potentials on Pt/C.[32] Pd–OH, on the other hand, forms at less positive potentials than Pt–OH does, in agreement with the calculations. At low potentials, these extra metal atoms should attract oxygenated groups (e.g., OH) to them and, through enhanced lateral repulsion among the OHs, further decrease the OH coverage on the adjacent Pt sites.

For instance, mixed Pt–Ru monolayers of varying compositions were deposited on Pd(111), and the ORR reactivity of these surfaces was studied by performing rotating disk experiments. The kinetic current densities observed in these studies, when plotted against the Pt–Ru molar ratio, show a similar volcanic dependence. There is a substantial increase in the ORR activity of $PtRu_{ML}/Pd(111)$ surfaces as the Ru mole fraction on the surface is increased up to 0.5 ML, with the maximum being at a Pt–Ru ratio of 4:1. Similar results were obtained for $Pt–Ir_{ML}$–Pd(111) at various Pt–Ir ratios, with the kinetic current density being higher for Pt–Ir at all coverages (Figure 4.2a). The kinetic current density for the 4:1 (Pt–M) composition is significantly enhanced compared to that on the Pt_{ML}–Pd(111) surface, which, as shown earlier, has already a higher ORR activity than Pt(111). The volcano-type behavior observed here can be attributed to the opposing effects of two factors, namely, the decrease in Pt–OH coverage with increasing Ru coverage on the surface and, on the other hand, the decrease in ORR activity with decreasing Pt content of the surface.

Additional rotating disk experiments were carried out on $Pt_{0.8}M_{0.2}/Pd(111)$ with M = Au, Pd, Rh, Re, or Os. The OH–OH repulsion energy was also calculated on these surfaces. Os and Re were found to be significantly more reactive than the other metals (M) considered in this study: the coadsorption of two OHs in the same unit cell, with M = Re or Os, would result in a spontaneous reaction between them to yield an H_2O molecule and an adsorbed O atom. This last finding led to extend the definition of adsorbate-induced destabilization of OH on the $Pt_3Os/Pd(111)$ and $Pt_3Re/Pd(111)$ surfaces by calculating the interaction energy between an OH group and a coadsorbed O atom. Therefore, the DFT calculations also helped identify the type of adsorbate–adsorbate

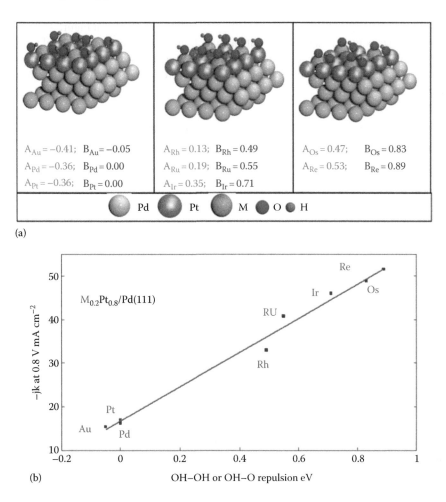

(a)

(b)

FIGURE 4.2 (a) Calculated most stable configurations for OH or OH+O (1/2 ML total coverage) on seven different $(Pt_3M)_{ML}/Pd(111)$ surfaces and on a $Pt_{ML}/Pd(111)$, which is used for reference for the data in Figure 4.2b. Energies are in eV. First two panels: 'A' provides the attractive (when negative) or repulsive (when positive) interaction between two OHs in the unit cell, referenced to the BE of OH at the 1/4 ML coverage. 'B' gives the same quantity, but referenced to the corresponding value on $Pt_{ML}/Pd(111)$. Last panel: 'A' provides the repulsion between adsorbed O and OH, whereas 'B' reflects the same quantity referenced to the interaction between two OHs on the $Pt_{ML}/Pd(111)$ surface. (b) Kinetic current at 0.80 V as a function of the calculated interaction energy between two OHs, or OH and O in a (2×2) unit cell (Figure 4.2a). Positive energies indicate a more repulsive interaction compared to $Pt_{ML}/Pd(111)$.

interaction responsible for reduced OH coverage on these ternary alloy surfaces. The calculated interaction energies and geometries for the most stable coadsorption states of two OHs or one OH and one O on various $Pt_3M/Pd(111)$ surfaces are shown in Figure 4.2a. For metals with comparable or higher oxidation potential to Pt, the two OHs have an attractive interaction and prefer to adsorb both on Pt sites of the Pt–M overlayer, in a bridge and a top configuration. For slabs where M is Rh, Ru, or Ir, all of which having lower oxidation potential compared to Pt, the two OHs have increasingly repulsive interaction energy and adsorb in a configuration involving a M-top and a Pt-top combination of adsorption sites. In the case where M is Re or Os, which yields the largest repulsive interactions among all the surfaces studied here, the O atoms adsorb on top of the M atoms and the OH groups occupy a Pt–Pt bridge site. The kinetic current densities measured on the $(Pt–M)_{ML}/Pd$ alloys are plotted against the effective repulsion energies (calculated relative to the repulsion energy on Pt_{ML}/Pd(111)). The resulting correlation is shown in Figure 4.2b. A defining linear relation exists between these two variables, which magnifies that the increased ORR activity of these Pt–M/Pd(111) catalysts can be attributed to the increase in repulsive interaction between surface-bound OH species. Alternatively, this may be interpreted as an increased destabilization of OH species on the Pt sites of the surface.

4.4 TAILORED SMOOTH SURFACES FOR FAVORABLE CORE–SHELL INTERACTION

Compared with the Pd(111) as a Pt_{ML} support, Pd nanoparticles exhibit much more difficulty in removing OH species because their lowly-coordinated sites (e.g., edge, vertex, and defect atoms) bind oxygen species much stronger.[33,34] Surface smoothing of Pd nanoparticles was realized by bromide treatment to remove significant amounts of such sites on Pd–C and Pd_3Co/C and produce more (111) facets while maintaining the particle size.

Figure 4.3 illustrates the bromide-treatment procedure for removing lowly-coordinated Pd atoms by potential cycling. It starts with chemisorption of a bromine layer in an alkaline solution, followed by the reductive desorption of bromine. Immediately after immersing the electrode in the bromide-containing solution, Br^- is adsorbed on the surface of Pd, preferentially on the low-coordination sites, since the atoms on those sites bind Br more strongly than those on terrace sites. In the cathodic scan, the reductive desorption of the bromine triggers the migration of lowly-coordinated Pd to terrace sites to minimize the global surface energy. Moreover, the reduction process causes the adjacent Pd to be charged positively, and the lowly-coordinated atoms have a relatively loose structure. Hence, the low-coordinated Pd most likely forms a $Pd–Br_2$ pair that is redeposited onto the surface in the subsequent anodic scan, accompanying the oxidative adsorption of bromide. Meanwhile, during the cycling, the adsorbed bromine layer undergoes rearrangement to attain the stable adlayer structure on a given surface, for example, $(\sqrt{3} \times \sqrt{3})R30°$-Br on Pd(111), (2×2)-Br on Pd(100).[35] Therefore, during this Br rearrangement on the surface, the adsorbed Br may draw the dangling atoms to fill the defect sites to form large terrace patches containing an ordering Br adlayer if the original terrace size is large enough to stabilize it.

Figure 4.4 shows the TEM images of palladium nanoparticles before and after bromide treatment. Various shapes of particles with obtruding edges are present in the commercial E-TEK sample. In contrast, the Br-treated Pd–C nanoparticles are rounded, and their size distribution is narrow.

The formation of a well-ordered 2D metal overlayer via UPD is known to depend on the crystallographic orientation and defect density of substrates.[36] The surface with well-arranged atoms is often required for the formation of ordered commensurate 2D phases, provided that the atomic sizes are not too dissimilar. In this case, more Pt(111) 2D patches instead of 3D clusters would form on the bromide-treated surface, which, in turn, should enhance the ORR.

Figure 4.5 shows the polarization curves for the ORR. The sample with Pt_{ML} on Br-treated Pd–C nanoparticles shows significant enhancement in the ORR kinetics, compared with the untreated nanoparticles, especially at potentials more negative than the half-wave potential (marked by a circle), the combinational diffusion-kinetic control region. In this region, the adsorption and

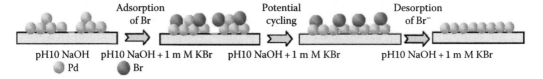

FIGURE 4.3 Illustration of the removal of surface low-coordination sites via the oxidative adsorption and reductive desorption of a bromine layer. (From Cai, Y. et al., *Langmuir*, 27, 8540, 2011.)

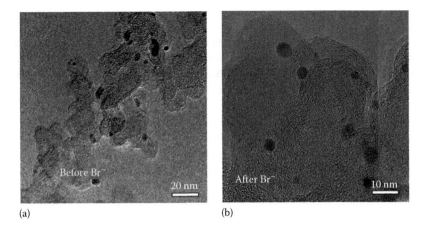

FIGURE 4.4 High-resolution TEM images of Pd–C (a) before and (b) after Br treatment. (From Cai, Y. et al., *Langmuir*, 27, 8540, 2011.)

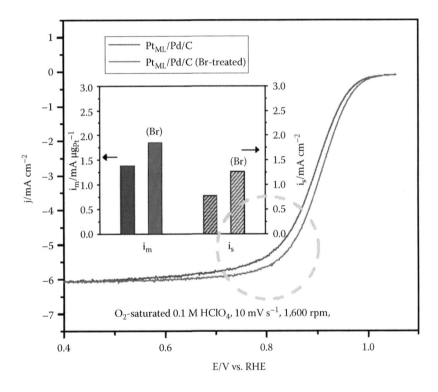

FIGURE 4.5 Polarization curves of oxygen reduction at Pt_{ML} on Br-treated and Br-untreated Pd–C. Inset: comparison of mass activity and specific activity on Pt_{ML}–Pd–C. (From Cai, Y. et al., *Langmuir*, 27, 8540, 2011.)

dissociation of molecular oxygen competes with strongly adsorbed hydroxyl species (OH_{ads}) for the same sites.[26,33,37] Furthermore, OH_{ads} does not only block the active sites on Pt but also change the adsorption energy of intermediates adjacent to it formed during the reaction.[26] Smoothing surface weakens OH binding and thus mitigates the adverse effect of OH on the ORR kinetics. The corresponding mass activity and specific activity for the ORR is inset in Figure 4.5, denoting 0.25-fold and 0.5-fold enhancement, respectively.

4.5   BR TREATMENT OF PD₃CO/C

The ORR on $Pt_{ML}/Pd_3Co/C$ showed the Pt mass activity is 2–3-fold higher that of commercial Pt/C.[38] However, upon the operating conditions, the Co component is subject to diffusion out of the nanoparticles, which engenders more low-coordination sites and so inflicts the catalytic activity. Like Pd–C nanoparticles, Pd_3Co/C can also be surface tailored by bromine treatment to remove the low-coordination sites and improve the ORR.

Figure 4.6 shows the morphology of the original Pd_3Co/C nanoparticles to compare with that of $Pt_{ML}/Pd_3Co/C$ after Br treatment. The particle size of the original Pd_3Co/C spans from 1.1 to 10 nm, averaging 4.25 nm, with many small particles measuring 3 nm in diameter and containing considerable defects and edges (Figure 4.6a). In contrast, the Pt_{ML} supported on Br-treated Pd_3Co/C nanoparticles shows a much more uniform particle size of 5.8 nm. In view of the thickness of a Pt monolayer of 0.25 nm, the average Br-treated Pd_3Co/C nanoparticles could be around 5.5 nm. These Br-treated nanoparticles exhibit a well-defined (111)-crystalline structure over a large area, with and without the Pt_{ML}, while the untreated Pd_3Co/C particles have much smaller patches with poorly defined crystalline structures. Likely, bromide treatment caused most particles of less than 3 nm to be dissolved and redeposited on larger particles, namely, the Ostwald ripening.[39]

Figure 4.7 shows the cyclic voltammogram (CV) and polarization curves for ORR on Pt_{ML} catalysts supported on Br-treated Pd_3Co/C after various potential cycles. The activity and surface area remains unchanged after 25,000 potential cycles between 0.6 and 1.0 V, indicating robust stability, which was not obtained at Johnson Matthey's $Pt_{ML}/Pd_3Co/C$; instead, the latter exhibited a negative shift of the half-wave potential of more than 30 mV in the first 5,000 cycles. These examples of surface modifications of Pd and Pd_3Co core demonstrated that removal of surface low-coordination sites via bromide treatment enables to obtain nanoparticles with smooth surfaces having a high density of (111)-oriented facets and a slightly contracted structure. The treated cores become excellent substrate for Pt monolayer and ensure significant enhancements in the mass and specific activities for the ORR.

4.6   LOW-INDEX FACETS: TETRAHEDRAL PD CORES

Electrocatalytic activity can also be adjusted by using nanocrystals with defining facets as substrates. One recent discovery in the material science of nanocrystals is that their shape and size determine their properties in close conjugation with their surface atomic arrangement and coordination.[3,40,41] These correlations offer new insights into designing Pt_{ML} electrocatalysts; examples of their support include concave tetrahedral Pd (Pd_{TH}) nanoparticles. Pd_{TH} nanoparticles (NPs) have a

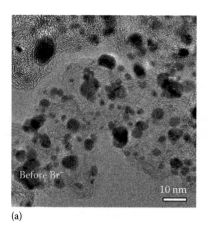
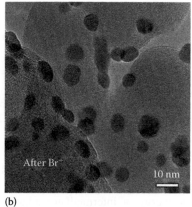

(a)   (b)

FIGURE 4.6   High-resolution TEM images of the original Pd_3Co/C (a) and of Pt_{ML} on Br-treated Pd_3Co/C (b). (From Cai, Y. et al., *Langmuir*, 27, 8540, 2011.)

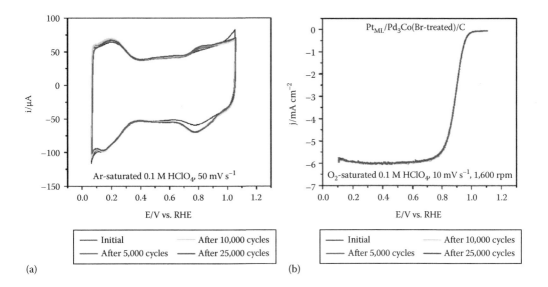

FIGURE 4.7 (a) Voltammetric curves for the Pt_{ML}/Br-treated-Pd_3Co/C and (b) polarization curves for the ORR in a stability test involving 25,000 potential cycles. (From Cai, Y. et al., *Langmuir*, 27, 8540, 2011.)

small fraction of low-coordination sites and defects and high content of the shape-determined high-coordination facets. The recent controlled synthesis of Pd_{TH} with a clean surface allowed studying the facet-specific electrochemistry of Pd in Pd/Pt_{ML} by comparing the overall performance in the ORR of nanocrystals of different shapes and morphologies.[18–20] The Pd_{TH}–Pt_{ML} core–shell electrocatalyst was examined by combining structural analyses and DFT with electrochemical techniques. It was found that this electrocatalyst has remarkable all-component stability and desirable activity.

Pd_{TH} NPs were synthesized via a combination of a hydrothermal route and a CO-adsorption-induced cleaning procedure as reported previously. Afterwards, on the as-prepared Pd_{TH} NPs, a Pt_{ML} was placed by the galvanic displacement with platinum of a UPD copper monolayer. A typical TEM image of Pd_{TH} NPs is shown in Figure 4.8a with the corresponding high-magnification image of an individual particle (Figure 4.8b). The particles are almost monodispersed on carbon black and have a uniform particle size with an edge length averaging 18 nm. Further, they have well-defined edges and an identical crystal lattice in the Pd nanoparticles, denoting the high efficiency of the cleaning procedure in removing nonmetallic adsorbates. Unlike the concave tetrahedral nanocrystals that were earlier prepared with an extended growth period of 5 h, whose surface is composed of both high-energy (110) and relatively low-energy (111) facets,[19,20] these Pd_{TH} NPs are solid and smaller and have flat surfaces consisting of the (111) arrangement only, as illustrated in Figure 4.8c.

The shape-determined ORR kinetics of the Pt_{ML} electrocatalysts on various supports was elucidated by comparing their polarization curves that were acquired in oxygen-saturated 0.1 M $HClO_4$ with the rotating-disk-electrode (RDE) technique. The red and blue curves shown in Figure 4.9 reproduce the typical polarization curves of the Pt_{ML}/Pd_{TH}/C and Pt_{ML}/Pd_{SP}/C, respectively. Close inspection of the two curves therein indicated that Pt_{ML}/Pd_{TH}/C exhibited a half-wave potential ($E_{1/2}$) 9.0 mV higher than that of Pt_{ML}/Pd_{SP}/C, indicating a significant improvement of ORR activity at the former. Furthermore, the Pt_{ML}/Pd_{TH} produced a specific and mass activity of 0.64 mA cm^{-2}_{Pt} and 1.02 A mg^{-1}_{Pt}, respectively; they are much greater than those of the Pt_{ML}/Pd_{SP}/C. For comparison, Figure 4.9 also includes the polarization curve of a Pt(111) single crystal, whose Pt-specific and mass activities are *ca.* 0.8 mA cm^{-2}_{Pt} and *ca.* 1.6 A mg^{-1}_{Pt}, respectively, if its top surface is viewed as a Pt(111) monolayer electrocatalyst.[14,42]

To interrogate the origin of the activities in ORR of the Pt_{ML} electrocatalysts, their mass-averaged electrochemical surface areas (ECSAs) were examined with the hydrogen adsorption peak in their

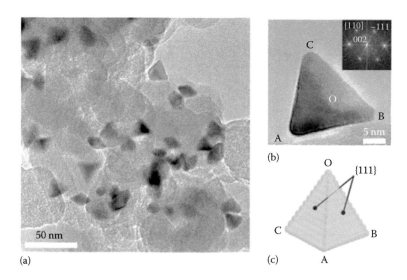

FIGURE 4.8 (a) Low- and (b) high-magnification TEM images of the Pd_{TH}/C. (c) Schematic of the Pd tetrahedron. Inset in Figure 4.8b is the corresponding diffractogram from the individual Pd tetrahedron. (From Gong, K. et al., *Z. Phys. Chem.*, 226, 1025, 2012.)

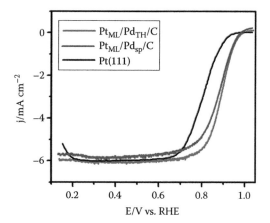

FIGURE 4.9 Polarization curves of the $Pt_{ML}/Pd_{TH}/C$, $Pt_{ML}/Pd_{SP}/C$, and Pt(111) recorded in O_2-saturated 0.1 M $HClO_4$ at a sweep rate of 10 mV s^{-1} and a rotating speed of 1600 rpm.

corresponding voltammograms. The $Pt_{ML}/Pd_{TH}/C$ material yielded values for ECSAs slightly lower than did $Pt_{ML}/Pd_{SP}/C$, which, coupled with the shape-determined surface area, disclosed the relatively large particle size of the Pd tetrahedron, agreeing with the previous TEM visualizations. However, the theoretical calculations for icosahedrons showed that increasing the particle size should lower mass activity.[43] These experimental results, counterintuitive when compared to the theoretical predictions, demonstrated the essential role of the (111) surface of the Pd tetrahedrons, since the ORR activity varied with particles' size in accord with the relative fraction of the surface low-index atoms. The specific DFT calculations for the $Pt_{ML}/Pd_{TH}/C$ supported this correlation, showing a relatively low surface strain at the Pt_{ML}/Pd_{TH}. On the other hand, specific activity is correlated with the fraction of surface high-coordination atoms of platinum that ascends with increasing of particle size. These causal links apparently suggested that, in the case of $Pt_{ML}/Pd_{TH}/C$, the bigger particle size and shape-determined surface (111) arrangement both contributed to the high specific activity in ORR. This attribution of the ORR activities was confirmed further by the fact that the increasing rate of the specific activity was far more significant than that of the mass activity (i.e., 24.6% vs. 4.9%).[20]

These results were attributed to the unique features of the Pt_{ML} on Pt(111) single crystals, including the full utilization of the Pt_{ML} atoms, the minimal fraction of lowly-coordinated atoms, and the low-energy atomic arrangement; the same explanation might be true for the higher ORR activities obtained with the $Pt_{ML}/Pd_{TH}/C$ compared to $Pt_{ML}/Pd_{SP}/C$. Furthermore, having minimal crystal defects and lowly-coordinated atoms, the extended surface of Pt(111) can be considered a nearly ideal Pt_{ML} electrocatalyst that indeed produced the extremely high ECSA of 205 cm^2 mg^{-1}_{Pt}, in addition to the Pt-specific and mass activities. Interestingly, $Pt_{ML}/Pd_{TH}/C$ showed a high performance in the ORR comparable to that of the Pt(111) single crystal although a substantial amount of Pt had been replaced with Pd, a relatively inexpensive metal. Apart from the shape-determined atomic arrangement, the Pd–Pt core–shell interaction (i.e., the ligand effect) was an additional factor playing a decisive role in improving the activity of the $Pt_{ML}/Pd_{TH}/C$ toward the ORR by enhancing the removal of the intermediates of the reaction.

4.7   ULTRATHIN PD NANOWIRE CORES

The Pt_{ML} electrocatalyst supported on Pd nanowires (i.e., Pt_{ML}/Pd_{NW}) is a good example showing the shape-determined ORR activity. Pd nanowires measuring 2 nm in diameter and more than 100 nm in length were prepared through a microemulsion reaction and then placed on multiwalled carbon nanotubes (MWNTs), forming $Pd_{NW}/MWNTs$ (20 wt%), by an electrostatically induced self-assembly method. Insights into the shape-determined performance in ORR of $Pt_{ML}/Pd_{NW}/MWNTs$ were gained by comparatively studying its polarization curve, obtained in oxygen-saturated 0.1M $HClO_4$ with the RDE technique.

Figure 4.10a reproduces the typical voltammogramic response of $Pd_{NW}/MWNTs$ (without Pt_{ML}) to compare with that of the commercial Pd–C (30 wt%). Examination of the underpotential deposition of hydrogen peaks (H_{UPD}, 0.1–0.4 V) gives a comparable ECSA, which, if normalized by mass, clearly unravels that the quasi 1D structure enables much more surface atoms than does the counterpart nanoparticle. This shape-determined physical property can be further confirmed by calculating the geometric surface areas (GSAs) for individual 2 nm Pd NWs with varied lengths and an aggregation of identical 5 nm nanoparticles. Also compared in Figure 4.10a is the oxidation ability of the Pd components in the two materials, which correlates to the binding strength of oxygenated

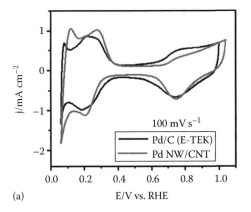
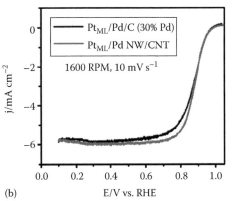

(a)   (b)

FIGURE 4.10   (a) CVs of the commercial Pd–C and $Pd_{NW}/MWNTs$ in argon-blanketed 0.1 M $HClO_4$ at a sweep rate of 100 mV s^{-1}. (b) Corresponding polarization curves of the commercial Pt_{ML}–Pd–C and $Pt_{ML}/Pd_{NW}/$ MWNTs in oxygen-saturated 0.1 M $HClO_4$ at a scan rate of 10 mV s^{-1} and a rotating speed 1600 rpm.

species (e.g., OH) on the metal surface. The oxidation of Pd in Pd_{NW}/MWNTs shows a suppressed coulombic response accompanying a slightly higher onset potential in relation to that in the Pd–C, namely, the former is electrochemically more stable than the latter. Such observations are also true for the 20 nm tetrahedral Pd nanocrystals when comparing them to the Pd–C of 5 nm in diameter; these size- and shape-determined properties are most likely related to the surface roughness and fraction of lowly-coordinated surface atoms.[18] The same attributions might be elicited for the comparison between the Pd_{NW}/MWNTs and Pd–C in view of the features of synthesizing the Pd NWs.

Figure 4.10b shows the typical polarization curve of Pt_{ML}/Pd_{NW}/MWNTs (red curve), along with that of Pt_{ML}–Pd–C for comparison (black curve). Inspection indicates that Pt_{ML}/Pd_{NW}/MWNTs exhibit the overall ORR kinetics slightly higher than Pt_{ML}–Pd–C does, since the RDE voltammograms of the former in the potential range of low overpotential shift somewhat toward the positive direction. Combining this indication with their specific ECSA and loading allows calculation that Pt_{ML}/Pd_{NW}/MWNTs exhibits Pt mass and Pt-specific activities of 1.45 A mg^{-1}_{Pt} and 0.65 mA cm^{-2}_{Pt}, respectively, which both are remarkably higher than those of Pt_{ML}–Pd–C. It is actually surprising that, like in tetrahedral Pd particles, in the present work, the ORR activity of the Pt_{ML} electrocatalyst can be improved so significantly by engineering the shape of the Pd core into ultrathin nanowires. This improvement related to the electrochemical and geometrical properties of the Pd_{NW} that fundamentally differ from those of other shapes (*vide supra*). As Pt binds oxygen a little too strongly from the kinetic standpoint,[11,15] the relatively smooth surface of the Pd_{NW} core and its energy-minimized atomic configuration are most likely to allow the supported Pt_{ML} to be closely contiguous and highly coordinated, so expediting the ORR.

To determine the origin of the ORR activity, the XRD pattern of Pt_{ML}/Pd_{NW}/MWNTs was compared with those of freestanding 2 nm Pd NWs and MWNT-supported 2 nm Pd NPs (i.e., Pd_{NP}/MWNTs) (Figure 4.11). On the basis of the Scherrer equation,[44] their particle sizes are found consistent with the previous TEM visualizations. Furthermore, comparing the four patterns suggests that the 2θ values for each facet relatively shifted; this observation is particularly clear for the (220) and (311) facets. Note that this relative shift is causally linked to the compression or expansion in the crystal lattice. Using the Pd–C as reference, it was found that the peaks of the Pd NWs, both freestanding and supported, shift the most toward the negative direction. Therefore, the influence of the nanowire shape is most noticeable in causing the expansion of the Pd surface, agreeing well with our analysis of shape-determined electrochemistry of Pd. Likewise, changing the nanoparticle size from 5 to 2 nm leads to a subtle negative shift, contrasting with the same peak positions that are obtained within Pd nanoparticles of >5 nm. These observations in the 2 nm Pd NPs are attributed to their extremely small size that causes their surfaces to be tensile to some degree. It is peculiarly noteworthy that expansion in the crystal lattice of Pd plays an adverse role in adjusting the ORR kinetics of supported Pt_{ML}. This adverse effect seemingly contradicts with the actual improvement in the mass and specific activities of Pd_{NW}/MWNTs. Such counterintuitional coexisting clearly reveals that the other exceptional features of the Pd NW—such as high mass-averaged surface area, excellent contiguousness, and low-energy atomic configuration—not only counterweigh the shape-induced surface expansion but also achieve a desirable trade-off between the electron transfer and removal of oxygenated intermediates.

4.8 HOLLOW CORE-INDUCED CONTRACTION OF PT MONOLAYER

Increasing the activity of Pt monolayer can be accomplished by placing it on the surface having a smaller interatomic distance that will cause contraction of a Pt monolayer. Since the slight compressive strain weakens the strength of oxygen binding and thus enhances the ORR activity, the core structure with an appropriate lattice contraction would be a desired substrate for Pt_{ML}. The lattice contraction not only enhances the ORR kinetics but also prevents the instability caused by dissolution of core materials. Hollow Pd was synthesized using Ni nanoparticles as a template that was displaced galvanically by Pd.[45]

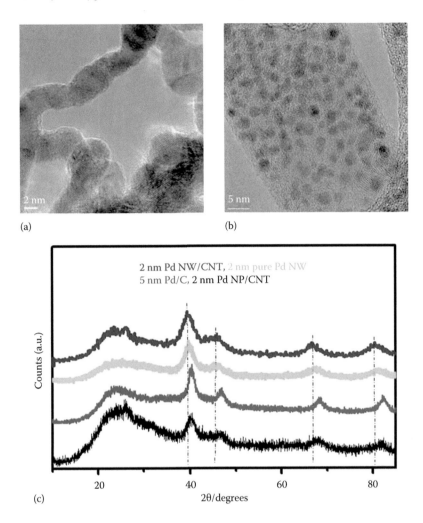

(a)   (b)

(c)

FIGURE 4.11 TEM images of the (a) freestanding Pd NWs and (b) Pd_{NP}/MWNTs. (c) XRD patterns for Pd_{NP}/MWNTs, freestanding Pd NWs, Pd–C, and Pd_{NP}/MWNTs (from top to bottom).

Ni nanoparticles measuring less than 9 nm in diameters were produced by pulse electrodeposition and dispersed on carbon black (Figure 4.12a). In the absence of oxygen, partial galvanic replacement of Ni atoms by mixed Pd and Au ions yielded noble metal shells on Ni particles, which further formed Pd–Au hollow particles upon dissolution of the remaining Ni in acidic solutions at room temperature. Figure 4.12b and c shows TEM images of the Pt monolayer catalysts made with as-prepared hollow Pd–Au cores. The high-resolution image in Figure 4.12c also reveals the catalysts' polycrystalline structure.

Figure 4.13 compares the specific and mass activities determined from the ORR kinetic currents at 0.9 V (vs. reversible hydrogen electrode [RHE]). The $Pt_{ML}/Pd_{20}Au(h)$/C sample yielded a total metal mass activity of 0.57 A mg^{-1}, which was 3.5 times that of solid Pt nanoparticles on carbon support fabricated by pulse electrodeposition (0.16 A mg^{-1}). The enhanced Pt mass and total metal mass activities of Pt monolayer catalysts supported on hollow cores can be ascribed to the smooth surface morphology and hollow-induced lattice contraction. The rounded and smooth surface morphology of the Pt monolayer catalyst supported on hollow Pd-based cores is evident from comparing the TEM images for a Pt monolayer catalyst on hollow $Pd_{10}Au$ particles. In addition, lattice contraction was found for Pt monolayer on hollow cores by x-ray powder diffraction measurements, consistent

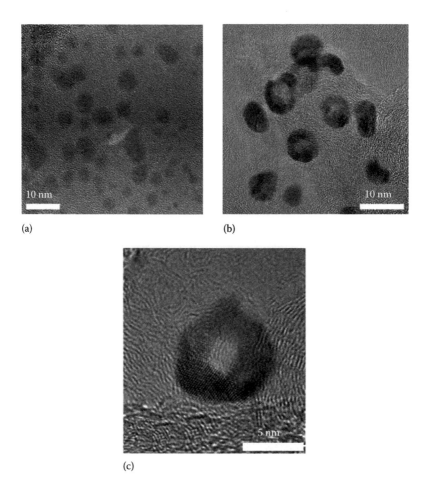

(a)   (b)

(c)

FIGURE 4.12 TEM images of (a) pulse electrodeposited Ni nanoparticles, (b and c) $Pt_{ML}/Pd_{20}Au(h)/C$ nanoparticles fabricated using Ni nanoparticles as templates.

with previously found hollow-induced lattice contraction on hollow Pt particles. As shown in Figure 4.14, diffraction peaks shift to higher angles for the Pt monolayer catalyst with Pd_9Au (hollow) core compared to that with Pd_9Au (solid) and Pd (solid) cores, indicating that a smaller lattice spacing was induced by hollow cores. This synthetic strategy can be applied to other core–shell systems for the formation of metallic or bimetallic, double- or multilayered, hollow or porous nanostructured electrocatalysts.

4.9   STABILITY OF PT ML: SELF-HEALING MECHANISM

One of the major obstacles to commercializing fuel cells comes from Pt dissolution from the cathode catalyst under potential cycling occurring during stop-and-go driving conditions. Figure 4.15 shows the findings from fuel-cell tests of $Pt_{ML}/Pd_9Au/C$ catalysts under potential cycling conditions. Figure 4.15a displays the TEM image of the cross section of MEA after 200,000 potential cycles from 0.6 to 1.0 V and the corresponding distribution of Pt, Au, and Pd in the catalytic nanoparticles after the stability test is displayed in Figure 4.15b. A Pd band forms in the middle due to Pd dissolution and Pd^{2+} reduction by H_2 diffusing from the anode; Pt and Au remains in the cathode. Pd is a slightly more reactive metal than Pt and so dissolves at slightly lower potentials (0.92 [Pd] vs. 1.19 [Pt] V). The small dissolution limits the excursions of potential in an operating fuel cell or at least

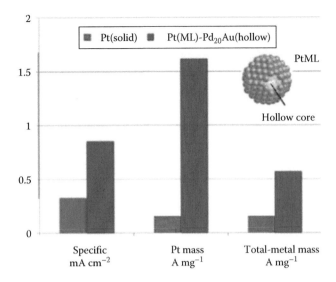

FIGURE 4.13 Comparison of the ORR specific and mass activities derived from the kinetic currents at 0.9 V (vs. RHE) in Figure 4.2 for hollow $Pt_{ML}/Pd_{20}Au(h)/C$ nanocatalysts (red, on the right) made using electrodeposited Ni templates and solid Pt/C nanoparticles (blue, on the left) made by electrodeposition.

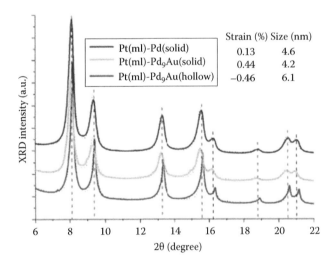

FIGURE 4.14 Profiles of x-ray powder diffraction intensity for three Pt monolayer catalysts on Pd(solid) (blue, top), Pd_9Au(solid) (green, middle), and Pd_9Au(hollow) (red, bottom) cores. The dotted lines are the fits, yielding the average particle diameters and lattice constants. The lattice strains listed are calculated with respect to the lattice constant of bulk Pt, 3.923 Å.

minimizes it. Such partial dissolution of Pd entails a small contraction on the Pt_{ML} shell, giving rise to a more stable structure with increased dissolution resistance and specific activity. This is the self-healing effect observed with this core–shell system as depicted in the model insert in Figure 4.15b: the slow dissolution of Pd causes the decrease in the particle's size, leading to some contraction of the Pt layer. The excess of Pt atoms from a monolayer shell can form a partial bilayered structure, or hollow particles may be formed due to the Kirkendall effect.[17]

The distribution of Au in Pd–Au alloy has a great impact on the stabilization of Pd under potential cycling in the acid media. Figure 4.15c displays the cross section of MEA and the posttest

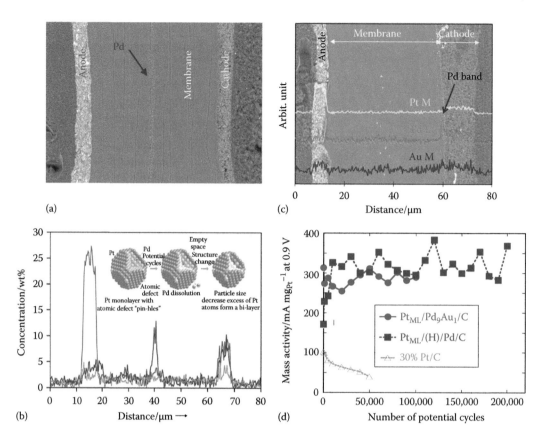

FIGURE 4.15 (a) Cross section of the MEA of $Pt_{ML}/Pd_9Au/C$ after 200,000 potential cycles from 0.6 to 1.0 V. (b) Corresponding distribution of Pt, Au, and Pd, cvs. distance after the test. The insert shows the model for the slow dissolution of Pd, and the decrease in the particle's size, leading to some contraction of the Pt layer. (c) The cross section of the MEA and overlaid posttest distribution of Pt, Au, and Pd in the $Pt_{ML}/Pd_9Au/C$ catalyst based on a highly uniform Pd–Au alloy after 200,000 potential cycles. (d) Comparison of the Pt mass activity for $Pt/Pd_9Au/C$ (open circles), Pt_{ML}/Pd_9Au catalyst based on a highly uniform Pd–Au alloy (red circles), Pt_{ML}–Pd–C electrocatalyst with highly compact Pt-ML (green squares), and a commercial Pt/C catalyst (open triangles). (From Sasaki, K. et al., *Angew. Chem.*, *49*, 8602, 2010.)

distribution of Pt, Au, and Pd in the Pt_{ML}/Pd_9Au catalyst based on a highly uniform Pd–Au alloy. Such a uniform alloy causes a positive shift of Pd oxidation, in accord with its stabilization potential and reduced Pd–OH formation, as evidenced from voltammetry and in situ extended x-ray absorption fine structure studies, in particular confirming the changes in coordination number of Pd–O. Potential cycling did not entail any decrease in Pd, Pt, or Au.

The Pt mass activity of the $Pt/Pd_9Au/C$ electrocatalyst in a test involving 100,000 potential cycles decreased negligibly (Figure 4.15d, red circles). The DOE's target for 30,000 potential cycles under the same protocol is a loss of 40%. For comparison, the mass activity of a commercial Pt/C catalyst shows a terminal loss below 50,000 cycles (Figure 4.15d, open triangles). The preparation of a highly compact Pt–ML, using the combined processes of H absorption and H adsorption on Pd to reduce Pt^{2+}, leads to the production of a new generation of Pt_{ML}–Pd–C catalysts with outstanding, unprecedented stability: there was no loss of activity over 200,000 potential cycles (Figure 4.15d, green squares). Furthermore, under more severe conditions with a potential range of 0.6–1.4 V, there were no significant losses of platinum and gold, although the dissolution of palladium was apparent.

4.10 REMARKS AND OUTLOOK

The last decade witnessed tangible developments in formulating the Pt_{ML} ORR electrocatalysts toward the high activity and stability compared to the state-of-the-art Pt/C electrocatalysts. For most of the Pt_{ML} electrocatalysts constructed so far, their mass activities are one order-of-magnitude higher than that of the conventional all-Pt electrocatalysts while the accelerated stability tests in fuel cells established them as a viable practical concept. These features inherent in Pt_{ML} electrocatalysts, together with their large-scale synthesis, have made them ready for applications including power vehicles. Indeed, producing such electrocatalysts in an industrial way is already a reality with N.E. Chemcat Co. In contrast to the content of platinum in the cathode catalysts currently being tested in fuel-cell vehicles, that is, $400\ \mu g_{Pt}\ cm^{-2}$, Pt_{ML} electrocatalysts require only 40–$80\ \mu g_{Pt}\ cm^{-2}$ and 60–$100\ \mu g_{Pd}\ cm^{-2}$. Thus, the requirement for a 100 kW fuel cell in a medium-sized electric car, with the catalyst's performance of $1\ W\ cm^{-2}$, is 4–8 g of Pt and about 10 g of Pd. Currently, catalytic converters use 5 g of Pt per vehicle. Therefore, there would be no need for considerable increase of the present rate of supply of Pt.

The unique features of Pt_{ML} electrocatalysts include tunable activity and stability, the self-healing effect, and high utilization of noble metals, largely due to the nature of the support, its composition, and its structure. Hence, opportunities remain for furthering performance improvement at reduced cost, a major concern in the development of real-world fuel cells. Future Pt_{ML} electrocatalysts could be designed by adjusting the composition, shape and size of cores to optimize the core–shell interaction. Specifically, the research can be performed from the following aspects.

First, the noble metal (e.g., Pd) content is expected to be further reduced by alloying it with less noble, refractory metals, using its hollow nanoparticle counterparts, or by combining it with metal-free supports. The ultimate reduction in usage of noble metals in catalysts can be achieved by using metal clusters, that is, nanoparticles consisting of less than ~100 atoms. Their properties differ from those of the bulk or bigger nanoparticles since they lose their metallic nature with decreasing size and behave more like a semiconductor. The existence of band gaps in their electronic structure together with geometrical arrangements confers novel properties. Investigation of electrochemical reactions on clusters is a recent, very active, and promising field with numerous possibilities, many of which remain unexplored.

By annealing in an NH_3 gas,[46] nanocrystals can undergo profound structural and chemical changes, forming metal nitride cores and 2–4 monolayer thick metal shells. Catalysts thus prepared exhibited high mass and specific activities with very good stability for the ORR. The experimental data and the DFT calculations indicate that the nitride has the bifunctional effect facilitating formation of the core–shell structures and improving the ORR activity and stability of the Pt shell by inducing both geometric and electronic effects from the cores under high oxidizing conditions. Despite improved activity for the ORR, there is a serious drawback under the fuel-cell operation— that is, dissolution of less noble metals—which often results in degradation and/or dysfunction of separation membranes. Seeking for new nitride-stabilized core–shell nanoparticles may be reasonably expanded to alloy systems, such as AuM, PdM, and RuM (M = transition metal) as substrates for Pt_{ML} catalysts.

Second, crystallographic engineering of the top atomic layer of the cores allows us to employ a variety of core materials and so modification of the properties of supported Pt_{ML}. Electrodeposition in nonaqueous solvents may open a fundamentally distinct area for the design and synthesis of core–shell nanoparticles. Considerable possibilities arise for studying core–shell interactions that are inaccessible in aqueous solutions. Our preliminary data show that yttrium nanoparticles, electrodeposited on carbon black in an organic solution, exhibited a certain unique interaction with codeposited Pt; a 10-fold lower concentration of such Pt does not affect much the cathodic peak while it suppresses the dissolution of yttrium at the anode. On the other hand, we propose that carbon quantum dots, for example, small fragments of graphene oxides or carbon nanotubes, can support a metal adlayer adjacent to their functionalized surfaces. This can produce unexpected properties of significant importance in studying the supported Pt_{ML} for catalyzing the ORR.

Third, core materials with highly-coordinated facets in general, (111) facets in particular, are greatly desirable to be designed and synthesized as a new support of Pt_{ML}; examples include nanowires and nanorods that possess a large fraction of (111) facets and so have high resistance to surface oxidation. Electrodeposition of nanowires is particularly promising since by its nature, this technique is capable of positioning the catalyst optimally in MEA, thus facilitating its highest utilization.

Fourth, incorporation of the synthesis of Pt_{ML} electrocatalysts with the fuel-cell stack is predicted to further reduce the production cost and better accommodation of the catalysts on the electrode. These studies would substantially reduce the technical barriers to produce durable, economical fuel cells. The same stratagems may be true for realizing very high selectivity. Finally, in view of the limited availability of Pt, the concept of Pt_{ML} catalysts will have a broad impact on future catalysis research and technology. Indeed, most recently, we demonstrated that Pt_{ML} under the tensile strain (Pt_{ML}/Au(111)) has high activity for methanol and ethanol oxidation reactions.[47]

ACKNOWLEDGMENTS

This work was performed at Brookhaven National Laboratory under contract DE-AC02-98CH10886 with the US Department of Energy, Office of Basic Energy Science.

REFERENCES

1. Debe, M. K. *Nature* **2012**, *486*, 43.
2. Kim, W. B.; Viotl, T.; Rodriguez-Rivera, G. J.; Dumesic, J. A. *Science* **2004**, *305*, 1280.
3. Guo, S.; Wang, E. *Nano Today* **2011**, *6*, 240.
4. Wagner, F. T.; Lakshmanan, B; Mathisas, M. F. *J. Phys. Chem. Lett.* **2010**, *1*, 2204.
5. Gasteiger, H.; Kocha, S.; Sompalli, B.; Wagner, F. *Appl. Catal. B* **2005**, *56*, 9.
6. Nørskov, J. K.; Bligaard, T.; Rossmeisl, J.; Christensen, C. H. *Nat. Chem.* **2009**, *1*, 37.
7. Markovic, N.; Schmidt, T.; Stamenkovic, V.; Ross, P. *Fuel Cells* **2001**, *1*, 105.
8. Bezerra, C. W. B.; Zhang, L.; Lee, K.; Liu, H.; Marques, A. L. B.; Marques, E. P.; Wang, H.; Zhang, J. *Electrochim. Acta* **2008**, *53*, 4937.
9. Gong, K.; Du, F.; Xia, Z.; Durstock, M.; Dai, L. *Science* **2009**, *323*, 760.
10. Lefèvre, M.; Proietti, E.; Jaouen, F.; Dodelet, J. P. *Science* **2009**, *324*, 71.
11. Stamenkovic, V.; Mun, B. S.; Mayrhofer, K. J. et al. *Angew. Chem.* **2006**, *45*, 2897.
12. Greeley, J.; Stephen, I. E. L.; Bondarenko, A. S. et al. *Nat. Chem.* **2009**, *1*, 552.
13. Adzic, R. R.; Zhang, J.; Sasaki, K. et al. *Top. Catal.* **2007**, *46*, 249.
14. Cai, Y.; Adzic, R. R. *Adv. Phys. Chem.* **2011**, *2011*, 1.
15. Zhang, J.; Vukmirovic, M. B.; Xu, Y.; Mavrikakis, M.; Adzic, R. R. *Angew. Chem.* **2005**, *44*, 2132.
16. Wang, J. X.; Inada, H.; Wu, L. et al. *J. Am. Chem. Soc.* **2009**, *131*, 17298.
17. Sasaki, K.; Naohara, H.; Cai, Y. et al. *Angew. Chem.* **2010**, *49*, 8602.
18. Gong, K.; Vukmirovic, M. B.; Ma, C.; Zhu, Y.; Adzic, R. R. *J. Electroanal. Chem.* **2011**, *662*, 213.
19. Huang, X.; Tang, S.; Zhang, H.; Zhou, Z.; Zheng, N. *J. Am. Chem. Soc.* **2009**, *131*, 13916.
20. Gong, K.; Choi, Y.; Vukmirovic, M. B. et al. *Z. Phys. Chem.* **2012**, *226*, 1025.
21. Greeley, J. *Catal. Today* **2006**, *111*, 52.
22. Greeley, J. *Nat. Mater.* **2004**, *3*, 810.
23. Schlapka, A.; Lischka, M.; Groß, A.; Käsberger, U.; Jakob, P. *Phys. Rev. Lett.* **2003**, *91*, 016101.
24. Kitchin, J. R.; Nørskov, J. K.; Barteau, M. A.; Chen, J. G. *Phys. Rev. Lett.* **2004**, *93*, 156801.
25. Kibler, L. A.; El-Aziz, A. M.; Hoyer, R.; Kolb, D. M. *Angew. Chem. Intl. Ed.* **2005**, *44*, 2080.
26. Markovic, N. M.; Gasteiger, H. A.; Grgur, B. N.; Ross, P. N. *J. Electroanal. Chem.* **1999**, *467*, 157.
27. Mavrikakis, M.; Hammer, B.; Nørskov, J. K. *Phys. Rev. Lett.* **1998**, *81*, 2819.
28. Nilekar, A. U.; Xu, Y.; Zhang, J. et al. *Top. Catal.* **2007**, *46*, 276.
29. Xu, Y.; Ruban, A. V.; Mavrikakis, M. *J. Am. Chem. Soc.* **2004**, *126*, 4717.
30. Wintterlin, J.; Tomaso, Z.; Trost, J.; Greeley, J.; Mavrikakis, M. *Angew. Chem. Intl. Ed.* **2003**, *42*, 2850.
31. Vielstich, W.; Lamm, A.; Gasteiger, H. A. *Handbook of Fuel Cells: Fundamentals, Technology, Applications*; Wiley, West Sussex, U.K., **2003**.

32. Zhang, J.; Mo, Y.; Vukmirovic, M. B.; Klie, R.; Sasaki, K.; Adzic, R. R. *J. Phys. Chem. B* **2004**, *108*, 10955.
33. Mayrhofer, K. J. J.; Blizanac, B. B.; Arenz, M.; Stamenkovic, V. R.; Ross, P. N.; Markovic, N. M. *J. Phys. Chem. B* **2005**, *109*, 14433.
34. Cai, Y.; Ma, C.; Zhu, Y.; Wang, J. X.; Adzic, R. R. *Langmuir* **2011**, *27*, 8540.
35. Carrasquillo, A.; Jeng, J. J.; Barriga, R. J.; Temesghen, W. F.; Soriaga, M. P. *Inorg. Chim. Acta* **1997**, *255*, 249.
36. Budevski, E.; Staikov, G.; Lorenz, W. J. *Electrochemical Phase Formation and Growth—An Introduction to the Initial Stages of Metal Deposition*, VCH Verlagsgesellschaft mbH, Weinheim, Germany, **1996**.
37. Wang, J. X.; Markovic, N. M.; Adzic, R. R. *J. Phys. Chem. B* **2004**, *108*, 4127.
38. Shao, M.; Sasaki, K.; Marinkovic, N. S.; Zhang, L.; Adzic, R. R. *Electrochem. Commun.* **2007**, *9*, 2848.
39. Voorhees, P. W. *J. Stat. Phys.* **1985**, *38*, 231.
40. Cuenya, B. R. *Thin Solid Films* **2010**, *518*, 3127.
41. Habas, S. E.; Lee, H.; Radmilovic, V.; Somorjai, G. A.; Yang, P. *Nat. Mater.* **2007**, *6*, 692.
42. Gong, K.; Su, D.; Adzic, R. R. *J. Am. Chem. Soc.* **2010**, *132*, 14364.
43. Giordano, N.; Passalacqua, E.; Pino, L. et al. *Electrochim. Acta* **1991**, *36*, 1979.
44. Azároff, L. V. *X-Ray Diffraction*, McGraw-Hill, 1974.
45. Zhang, Y.; Ma, C.; Zhu, Y. et al. *Catal. Today* **2013**, *202*, 50.
46. Kuttiyiel, K. A.; Sasaki, K.; Choi, Y. M.; Su, D.; Liu, P.; Adzic, R. R. *Nano Lett.* **2012**, *12*, 6266.
47. Li, M.; Liu, P.; Adzic, R. R. *J. Phys. Chem. Lett.*, **2012**, *3*, 3480.

5 Photoelectrochemistry with Nanostructured Semiconductors

Wen Wen and Stephen Maldonado

CONTENTS

5.1 INTRODUCTION

5.1.1 CONTEXT

Semiconductor photoelectrodes immersed in a solution and under illumination can drive useful electrochemical transformations without any external power supply or bias. This premise has fueled the field of semiconductor photoelectrochemistry since its inception. Early work in this area predominantly involved polished, high-quality single-crystalline semiconductor electrodes serving as the light absorber, charge converter/separator, and electrocatalyst support for heterogeneous electrochemical transformations. The impetus for using such materials was both that planar electrodes were easy to prepare and that these materials were available through the semiconductor industry.

The use of planar, single-crystalline semiconductor electrodes has been vital to the quantitative study and understanding of photoelectrochemical charge-transfer processes. Flat, planar semiconductor electrodes have afforded collection of data that tested heterogeneous charge-transfer models and their relevancy to semiconductor electrochemistry,[1–4] that identified surface modification strategies that work in liquid electrolytes,[5–8] and that described the merits of hybrid constructs that incorporate multiple light-harvesting materials.[9–11] Landmark studies from Tributsch and Gerischer,[12] Morrison and Freund,[13] and Fujishima and Honda[14] separately showed that, when illuminated, crystalline planar semiconductor electrodes immersed in water were a very simple blueprint for a photosynthetic reactor.

Still, the continued reliance of bulk single-crystalline semiconductor materials for semiconductor electrochemistry has also been the Achilles heel of parallel efforts to make photoelectrochemical energy technologies practical and viable. Simply, the costs associated with pure single crystals of common semiconductors, in conjunction with limitations imposed by their sensitivity towards corrosion and light absorption, limit prohibitively their use as photoelectrode materials in a photoelectrochemical cell for energy conversion and storage at a large scale. Recognizing this materials challenge, the photoelectrochemical research community now encompasses a considerable amount of materials science focused on producing alternative crystalline semiconductor electrode platforms. Specifically and in contrast to the primeval days of semiconductor photoelectrochemistry, modern photoelectrochemical research has moved towards semiconductor electrode architectures that are decidedly nonplanar. The push is now to develop semiconductor electrode materials with at least one (or all) spatial dimension(s) at or below 10^{-6} m. Such *nanostructured* semiconductor photoelectrodes are being hotly pursued under the hypothesis that good photoactivity can be realized with nanostructured semiconductor photoelectrodes without the prohibitive expense and limitations of planar single-crystalline semiconductors. The principal thesis is that the lessons learned from planar semiconductor photoelectrodes can be applied to crystalline, nanostructured analogs that exhibit comparable (or better) photoactivity but without sacrificing the promise of low cost and viability.

5.1.2 Scope of the Chapter

The objectives of this chapter are to contextualize and describe the basic hypothesis that nanostructured semiconductor materials are useful for photoelectrochemistry and to relay the current state of the art in the field. Accordingly, a basic description of the general operation of any semiconductor photoelectrode is first provided. The effects that miniaturization has on the transport of species in the electrolyte to a *nanosized* semiconductor electrode interface are identical to those discussed in Chapter 2. One noteworthy study with semiconductor ultramicroelectrodes described a way to exploit fast mass transport to more definitively and rapidly measure heterogeneous charge-transfer rate constants at semiconductor/liquid junctions.[15] In practice, clean kinetic measurements at these interfaces are at best extremely tedious and (more typically) very imprecise. In this capacity, nanoscale semiconductor electrodes may represent a practical testbed for fundamental tests of charge-transfer theories. Still, the emphasis in this chapter is not on mass transport in solution to a semiconductor electrode. Rather, the focus is on the effect that small dimensions have on the transport of photogenerated charge carriers *within* the semiconductor electrode material to the solid/liquid interface. This distinction significantly impacts the electrochemical behaviors of small semiconductor electrodes relative to macroscale analogs in ways unlike similarly sized metal electrodes.

In principle, there are several possible *nanostructured* morphologies. Notably, flat, contiguous planar films with ultrathin (submicron) thicknesses have been exploited in photovoltaic research for many decades. With the exception of a few semiconductor materials with exceedingly poor electrical properties (e.g., Fe_2O_3),[16–18] this morphological tactic has not been widely pursued in photoelectrochemistry, and so this chapter does not exhaustively focus on ultrathin film semiconductor photoelectrodes. Rather, this review focuses principally on tall, thin photoelectrode morphologies with large aspect ratios and small, individual particles. Although a myriad of other semiconductor morphologies are possible, these so-called 1- and 0-D architectures, respectively, represent the primary activity in modern photoelectrochemical research and are the simplest starting points for rationalizing the operation of related but more complex nanostructured semiconductors. Further, although nanostructuring can be applied to all semiconductor materials, the emphasis here is on semiconductors with small-to-medium-sized bandgaps, that is, semiconductors that intrinsically have the capacity to absorb sunlight appreciably. Accordingly, strategies like sensitization with separate chromophores to augment the poor visible light absorption characteristics of large bandgap semiconductors like ZnO and TiO_2 are not discussed here.

The intent of this chapter is to define limiting factors in the operation of nanostructured photoelectrodes that function as both the light-harvesting and charge-separating medium. A working knowledge of semiconductor electronic properties is assumed, with only aspects relevant to the steady-state operation under illumination highlighted here. A more comprehensive background of semiconductor device physics can be found in many semiconductor solid-state device and photovoltaic textbooks.[19,20]

5.1.3 EFFICIENCY DEFINITIONS

The operation of a semiconductor photoelectrode is influenced by both intensive and extensive properties. Fundamentally, sustained photoelectrochemical conversion is a process that works not at equilibrium but rather under (ideally) steady-state conditions. Accordingly, intensity and type of illumination can greatly affect the behavior of a given semiconductor photoelectrode beyond just increasing or decreasing the total number of reaction turnovers of an electrochemical reaction. The principal loss mechanisms can change for the same semiconductor photoelectrode with changes to the steady-state condition.

The net energy conversion efficiency of a working semiconductor photoelectrode, η_{PEC}, is a product of several factors:

$$\eta_{PEC} = \eta_{optical} \times \eta_{separation} \times \eta_{echem} \tag{5.1}$$

where $\eta_{optical}$, $\eta_{separation}$, and η_{echem} are the respective descriptors for light absorption, charge separation, and electrochemistry. The first term, $\eta_{optical}$, encompasses the efficiency of the semiconductor to generate an electron–hole pair with a thermalized energy equal to the bandgap energy for light absorption of incident suprabandgap photons. For perspective, for a perfectly absorbing planar Si photovoltaic without reflection/transmission losses, the maximum value of $\eta_{optical}$ for sunlight at room temperature is approximately 48.95%.[21] The second term, $\eta_{separation}$, describes the internal conversion efficiency of the semiconductor for supplying fluxes of photogenerated electrons and holes at the maximum possible electrochemical potential difference. This term accounts for all loss mechanisms within the semiconductor material. For reference, the upper bounds on the value of $\eta_{separation}$ for an optimized thick (ca. 100 μm) Si photovoltaic under AM 1.5 illumination at room temperature is approximately 60.9%.[22] The third term, η_{echem}, is the efficiency for driving the electrochemical transformation reaction(s) at the driving force supplied by the photogenerated charge carriers.

For any complete system analysis of a photoelectrochemical cell, all three efficiency factors must be rigorously assessed. However, not every component in Equation 5.1 is intractably coupled to each other. For example, analysis of the activity of an electrocatalyst supported on a semiconductor photoelectrode does not principally involve consideration of the semiconductor support.

In this context, the contents of Chapter 4 and the extant literature on electrocatalysis of fuel-forming redox transformations[23–26] speak to η_{echem} in a photoelectrochemical system. Although the details of how to best present/load extrinsic electrocatalysts on nanostructured semiconductors has practical value,[27–29] this chapter will not focus on details pertinent purely to the development or loading of active electrocatalysts. Rather, the effect that nanostructured morphologies have on the *rate* and *driving force* of photogenerated charge carriers supplied to electrocatalysts by semiconductor materials is germane here.

Detailed treatment of the absorption of sunlight by a nanostructured semiconductor electrode to understand $\eta_{optical}$ requires rigorous consideration of the optical properties of both the semiconductor and surrounding media. For feature sizes commensurate with the wavelengths of incident light, light–matter interactions are complex. Consider the simple analysis by Gerischer and Heller on the absorptance efficiency of a small semiconductor nanoparticle suspended in water.[30] Figure 5.1 is a graphical representation of their tabulated results (Table II of Ref. [30]). For a semiconductor nanoparticle with an absorptivity of 10^4 cm^{-1} and illuminated at 10^{15} photons cm^{-2}, Gerischer and Heller showed that the number of photons absorbed relative to the number of photons incident on the nanoparticle is a strong function of the size of the particle, that is, smaller nanoparticles will have a lower apparent photoactivity if the light absorption properties are not properly accounted for.

Conversely, for thin semiconductor nanowires, the absorption at certain wavelengths of light that correspond to leaky mode resonances is substantially enhanced.[31] Accordingly, continued development of optical models for light absorption by nanostructured semiconductors is an active area of research but is outside the scope of this chapter. Interested readers are directed to several existing reviews on light absorption by nanostructured materials.[32–36] Nevertheless, the location within a semiconductor (near/far from the semiconductor/solution interface) where charge carriers are photogenerated is important since it affects the probability for their separation, transport, and ultimate participation in electrochemical reactions. So, while $\eta_{separation}$ cannot be discussed exhaustively without some comment on the details of light absorption, this chapter will principally focus on the quantitative framework necessary to understand how

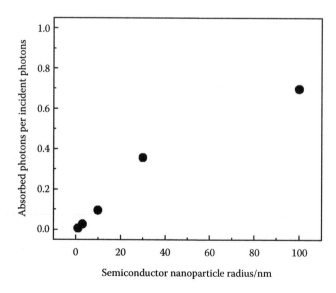

FIGURE 5.1 A plot of the absorbed photons per incident photons as a function of radius for a hypothetical semiconductor particle suspended in water with an absorptivity of 10^4 cm^{-1} and illuminated at 10^{15} photons cm^{-2}. (From Gerischer, H. and Heller, A., *J. Electrochem. Soc.*, 139(1), 113, 1992.)

nanostructured photoelectrode form factors impact the values of $\eta_{separation}$ apart from extensive consideration of optical aspects.

The key observable metrics in assessing $\eta_{separation}$ for a photoelectrochemical cell under illumination are the induced photovoltage and photocurrent. Efficiency losses within a nanostructured semiconductor decrease both parameters. Occasionally, analysis of the internal quantum yield, Φ, for net photocurrent is more useful than absolute photocurrent values to understand losses at steady-state conditions. Φ is the fraction of total charge carriers collected/used relative to the number of photons absorbed by the semiconductor material. The major cause of diminution of Φ from unity comes from either the total carrier recombination in the interior volume of the semiconductor (Φ_R), the total carrier recombination at the interface where charge carriers are collected ($\Phi_{contact}$), or the total carrier recombination at the surface between contacts (Φ_S).[37] The advantage in considering Φ is that these fractional losses in photocurrent flux at steady state are separable and their relative magnitudes are more readily interpretable[22,37]:

$$\Phi = 1 - \Phi_R - \Phi_{contact} - \Phi_S \tag{5.2}$$

5.2 RELEVANT SEMICONDUCTOR PRINCIPLES

5.2.1 CHARGE CARRIERS WITHIN A SEMICONDUCTOR

The flux of charge carriers across the semiconductor/electrolyte interface under illumination is dependent on the potential (E), the electron density (n), and the hole density (p) throughout the semiconductor. The calculation of electron and hole current flow requires the use of the Poisson, steady-state carrier continuity equations and carrier transport equations[38]:

$$\nabla^2 E = -\frac{q}{\varepsilon_s}\left(p - n + N_D - N_A + N_t\right) \tag{5.3}$$

$$\nabla \cdot \vec{J}_n = qG_{ph} + qR_{tot} \tag{5.4a}$$

$$-\nabla \cdot \vec{J}_p = -qG_{ph} + qR_{tot} \tag{5.4b}$$

$$\vec{J}_n = -q\mu_n n\nabla\phi + qD_n\nabla n \tag{5.5a}$$

$$\vec{J}_p = -q\mu_p p\nabla\phi - qD_p\nabla p \tag{5.5b}$$

where
$\vec{J}_n$ and $\vec{J}_p$ are the respective electron and hole current densities
ε_s is the semiconductor permittivity
q is the unsigned charge of an electron
N_D and N_A are the donor and acceptor concentrations
N_t is the trap density
R_{tot} is the sum rate of all carrier recombination processes
G_{ph} is the photogeneration rate at a specific set of illumination conditions

Solution of these coupled equations requires both appropriate boundary conditions and a description of the geometry of the system for the appropriate form of the Laplacian operator.

TABLE 5.1

Laplacian Operators for Planar, Rod, and Particle Electrode Form Factors

Coordinate System	Full Operator on F	Simplified Operator for High Symmetry
Cartesian	$\dfrac{\partial^2 F}{\partial x^2} + \dfrac{\partial^2 F}{\partial y^2} + \dfrac{\partial^2 F}{\partial z^2}$	$\dfrac{\partial^2 F}{\partial x^2}$
Cylindrical	$\dfrac{1}{r}\dfrac{\partial}{\partial r}\left(r\dfrac{\partial F}{\partial r}\right) + \dfrac{1}{r^2}\dfrac{\partial^2 F}{\partial \phi^2} + \dfrac{\partial^2 F}{\partial z^2}$	$\dfrac{1}{r}\dfrac{\partial F}{\partial r} + \dfrac{\partial^2 F}{\partial r^2}$
Spherical	$\dfrac{1}{r^2}\dfrac{\partial}{\partial r}\left(r^2\dfrac{\partial F}{\partial r}\right) + \dfrac{1}{r^2\sin\theta}\dfrac{\partial}{\partial \theta}\left(\sin\theta\dfrac{\partial F}{\partial \theta}\right) + \dfrac{1}{r^2\sin^2\theta}\dfrac{\partial^2 F}{\partial \phi^2}$	$\dfrac{2}{r}\dfrac{\partial F}{\partial r} + \dfrac{\partial^2 F}{\partial r^2}$

Table 5.1 summarizes these operators for the macroscopic planar, particle, and high-aspect-ratio rod electrode shapes, using the standard notations for each coordinate system.[39]

As with the transport of dissolved species to an electrode interface, the solution of Equations 5.3 through 5.5 is greatly simplified in cases with high symmetry. For a thick planar semiconductor electrode, only the dimension normal to the electrode surface plane (x) needs to be considered. That is, transport in the semiconductor can be treated as semi-infinite when the dimensions of length and width are large. For spherical semiconductors, perfectly symmetric particles require only operation in the radial (r) dimension. For rod semiconductors with symmetry around the short axis, operation along both the radial dimension and long axis is required. If the aspect ratio is large, that is, the rod is much longer than it is wide, then the structure can be treated as a semi-infinite cylinder and only differentiation along the radial dimension is needed.

5.2.2 FERMI-LEVEL CONCEPT

To solve the system of equations for the steady-state operation of semiconductor photoelectrodes, a definition of equilibrium must first be introduced. The law of mass action, in the context of semiconductor carrier concentration statistics, dictates that *at equilibrium*, the values of n and p are bound to a constant value (n_i) in the following way:

$$np = n_i^2 \tag{5.6}$$

By Equation 5.6, if the equilibrium concentration of one charge-carrier type is large, then the concentration of the other charge carrier is necessarily small. Accordingly, *n-type* and *p-type* semiconductors represent materials with high electron and hole concentrations, respectively, and low hole and electron concentrations, respectively, at equilibrium. In these materials, electrically active impurities/defects add to the *intrinsic* amount of charge carriers in the semiconductor. Throughout this chapter, the term *dopants* is limited to this description specifically, that is, not the inclusion of any additional element for alloying. In an *n*-type semiconductor that is appropriate for solar energy conversion applications, the added dopant level, N_D, is so large that it effectively can be substituted for n in Equation 5.6. The same applies to N_A substituting for p for a doped *p*-type semiconductor. In either case, the carrier type that has the much larger concentration is denoted as the majority carrier type. Following, the carrier type that has the much smaller concentration is referred to as the minority carrier type.

The Fermi level in electrochemical systems describes the average energy of charge carriers *at equilibrium*. The occupancy of electrons at a particular energy in a semiconductor is given by the Fermi–Dirac relation[19,20]:

$$f(\mathbf{E}) = \frac{1}{1 + e^{\frac{\mathbf{E} - \mathbf{E_F}}{k_B T}}} \approx e^{\frac{\mathbf{E_F} - \mathbf{E}}{k_B T}} \tag{5.7}$$

where

$f(\mathbf{E})$ is the probability of finding an electron at energy, $\mathbf{E}$

$\mathbf{E_F}$ is the energy of the Fermi level

Rigorously, the Fermi level in a given phase is defined as the energy level with a probability of 0.5 for being occupied by an electron. Equation 5.7 also shows the Boltzmann approximation, which is appropriate when the term $(\mathbf{E} - \mathbf{E_F})/k_B T$ is greater than 4.

If the density of available states at a given energy, $g(\mathbf{E})$, is known and the probability of those states holding electrons at that energy is known, then the concentration of electrons at that energy, $n(\mathbf{E})$, is $g(\mathbf{E})f(\mathbf{E})$. The total concentration of transferable (i.e., in the conduction band) electrons is then given by integrating over the entire conduction band:

$$n = \int_{\mathbf{E} = \mathbf{E}_{cb}}^{\mathbf{E} = \infty} g(\mathbf{E})f(\mathbf{E})d\mathbf{E} \tag{5.8}$$

When the Boltzmann approximation holds and assuming $g(\mathbf{E})$ has a parabolic energy dependence at bottom of the conduction band, Equation 5.8 can be simplified:

$$n = N_{cb}e^{\frac{-(\mathbf{E}_{cb} - \mathbf{E_F})}{k_B T}} \tag{5.9}$$

where N_{cb} is the effective density of states (cm^{-3}) at the conduction band edge, $\mathbf{E_{cb}}$. Using the probability of finding an absence of an electron in a state (i.e., $1 - f(\mathbf{E})$) and an analogous approach, an expression relating the Fermi-level energy and the concentration of transferable holes can separately be written as

$$p = N_{vb}e^{\frac{-(\mathbf{E_F} - \mathbf{E}_{vb})}{k_B T}} \tag{5.10}$$

where N_{vb} is the effective density of states (cm^{-3}) at the valence band edge, $\mathbf{E_{vb}}$. From either Equation 5.9 or 5.10, $\mathbf{E_F}$ can be determined. The important point from these equations is that equilibrium *is* the condition where the average energies of both electrons and holes are equivalent and constant throughout the semiconductor. Additionally, from both Equations 5.9 and 5.10, the explicit value of n_i can be calculated as $N_{cb}N_{vb}e^{-(E_{cb} - E_{vb})/k_B T}$.

In a pure semiconductor without any dopants at $T = 0$ K, the Fermi level is located in the middle of the semiconductor bandgap. Adding electrically active impurities/defects (i.e., dopants) to the semiconductor shifts the Fermi energy, with dopants that add to the electron density at equilibrium (i.e., rendering n-type character) raising the Fermi energy closer to the conduction band edge. Conversely, adding dopants that increase the hole density at equilibrium (i.e., effecting p-type character) lowers the Fermi energy closer to the valence band edge. When a sufficiently high density of either n- or p-type dopants is added so that the Fermi level is pushed very close to either the conduction band or valence band edges so that the Boltzmann approximation no longer is valid, then Equations 5.9 and 5.10 no longer hold. Such materials are degeneratively doped and the full Fermi–Dirac statistics must be used to determine their precise Fermi-level energies.

5.2.3   Models for Charge-Carrier Recombination

To solve the system of equations for the steady-state operation of semiconductor electrodes, expressions for R_{tot} are necessary. As indicated earlier, recombination of charge carriers occurs both in the bulk of a semiconductor electrode and at all interfaces.

5.2.3.1   Recombination in the Bulk

The total recombination of charge carriers in a semiconductor electrode is the sum of three recombination rates. Their relative impacts depend on multiple material properties, and even for the same material, the dominant recombination pathway(s) varies with changes in illumination, electrolyte, and target redox process.

Process 1 in Figure 5.2 represents direct band-to-band recombination, that is, re-pairing of conduction band electrons and valence band holes through the emission of a photon with bandgap energy without the involvement of any energy states in the semiconductor bandgap. Process 1 is a *bimolecular* reaction with a rate law that necessarily involves both n and p[19]:

$$R_{rad} = kn_i^2 \left(\frac{np}{n_i^2} - 1 \right) \tag{5.11}$$

where

R_{rad} is the rate of radiative recombination (cm^{-3} s^{-1})

k is a constant describing the thermal generation of electrons in the conduction band and holes in the valence band (cm^{-3} s^{-1})

n_i is the intrinsic carrier density in the semiconductor material

At equilibrium in the dark, Equation 5.6 holds. Application of either a bias or illumination increases n or p (or both). For a direct bandgap semiconductor, transitions between the conduction and valence bands involve only the absorption/emission of photons. Transitions between the conduction and valence bands in indirect bandgap semiconductors require both absorption/emission of photons and phonons (lattice vibrations). Since both energy and momentum must be conserved, direct band-to-band transitions in indirect bandgap semiconductors are much less probable because of the additional requirement of momentum conservation, and so values of k tend to be much smaller than for direct bandgap semiconductors.

Process 2 in Figure 5.2 represents indirect charge-carrier recombination through a single energy (trap) state inside the bandgap. The trap arises from impurities or defects in the semiconductor lattice. Process 2 requires the sequential capture of both electrons and holes at the trap state inside the bandgap. For indirect charge-carrier recombination through a single population of trap states at a

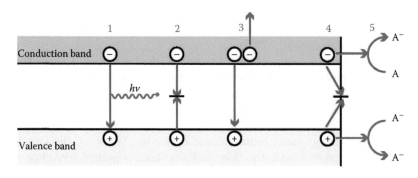

FIGURE 5.2   Schematic depiction of primary bulk and surface recombination processes for a crystalline inorganic semiconductor photoelectrode.

specific energy, the rate law for this type of recombination has been described by several workers and is referred to as the Shockley–Read–Hall (SRH) model.[40,41] The expression for this recombination rate, R_{SRH} (cm^{-3} s^{-1}), is

$$R_{SRH} = \frac{np - n_i^2}{\left(\dfrac{n + n_i e^{\frac{E_t - E_i}{k_B T}}}{\sigma_p N_t v_{th,p}}\right) + \left(\dfrac{p + n_i e^{\frac{E_i - E_t}{k_B T}}}{\sigma_n N_t v_{th,n}}\right)} \tag{5.12}$$

where

E_t is the energy of the trap states
E_i is the energy at the midpoint in the semiconductor bandgap
σ_n and σ_p are the respective capture cross sections (cm^2) for electrons and holes of the trap state
N_t is the trap state density (cm^{-3})
k_B is the Boltzmann constant
T is the temperature
$v_{th,n}$ and $v_{th,p}$ are the respective thermal velocities of electrons and holes inside the semiconductor

For a single trap state, the rate is maximized for a trap energy located at the midpoint of the bandgap. For a distribution of multiple populations of trap states with different energies, the individual recombination rates must be integrated over the entire bandgap energy to determine R_{SRH}. In either case, the expression simplifies when either n or p is much larger than the other type of charge-carrier concentration. The terms $\sigma_n N_t v_{th,n}$ and $\sigma_p N_t v_{th,p}$ are noteworthy since they have units of s^{-1} and are often described as the reciprocal lifetimes for each charge carrier inside the semiconductor.

Process 3 in Figure 5.2 illustrates one type of Auger recombination, a three body processes in which two charge carriers recombine and transfer energy to another carrier. This energy is then partially or wholly dissipated via thermalization. The Auger recombination rate, R_{Auger}, is given by

$$R_{Auger} = \left(C_n n + C_p p\right)\left(np - n_i^2\right) \tag{5.13}$$

where

n and p are the electron and hole carrier densities, respectively
C_n and C_p are the respective temperature-dependent Auger coefficients for electrons and holes[20]

Because Auger recombination involves more than two charge carriers, a distinguishing characteristic of Auger recombination is that at low carrier concentrations, R_{Auger} is negligible and only becomes sizeable at high charge-carrier concentrations.

5.2.3.2 Recombination at Interfaces

SRH-type recombination can also occur through trap states physically located at the surface of a semiconductor electrode (Process 4 in Figure 5.2). For a single population of trap states with an energy $E_{t,s}$, Equation 5.14 is the corresponding expression for the rate of SRH recombination at the surface, $R_{SRH,s}$:

$$R_{SRH,s} = \frac{n_s p_s - n_i^2}{\left(\dfrac{n_s + n_i e^{\frac{E_{t,s} - E_i}{k_B T}}}{\sigma_p N_{t,s} v_{th,p}}\right) + \left(\dfrac{p_s + n_i e^{\frac{E_i - E_{t,s}}{k_B T}}}{\sigma_n N_{t,s} v_{th,n}}\right)} \tag{5.14}$$

where the subscript s denotes these values are the concentrations at the surface plane. For the surface trap state density, $N_{t,s}$, the units are cm^{-2}. Since this rate describes recombination at a surface plane, the units for $R_{SRH,s}$ are explicitly a flux (cm^{-2} s^{-1}). Similarly, the terms $\sigma_n N_{t,s} v_{th,n}$ and $\sigma_p N_{t,s} v_{th,p}$ have units of cm s^{-1} and are often referred to as the electron and hole surface recombination velocities (S_n and S_p), respectively. High-quality interfaces with minimal surface recombination nominally correspond to $\ll 100$ cm s^{-1}. Similarly, heavily defective interfaces typically support surface recombination velocities in excess of 10^5 cm s^{-1}. If there are distributions of traps with different energies, then Equation 5.14 must be integrated over all possible energies.

Process 5 in Figure 5.2 describes charge-carrier recombination through heterogeneous charge transfer with dissolved species in solution. As drawn, heterogeneous charge transfer is deleterious if both electron and hole transfers occur at the surface on the same redox couple. That is, there is no net energy storage since both halves of the same redox couple (i.e., "A" and "A$^-$") are generated at the same time. The total rate of heterogeneous charge transfer is dependent on the concentrations in solution ([A] and [A$^-$], respectively) and is the sum of the electron and hole fluxes[13,42,43]:

$$R_{ct} = k_{et}[\text{A}]\left(n_s - n_{s0}\right) - qk_{ht}\left[\text{A}^-\right]\left(p_s - p_{s0}\right) \tag{5.15}$$

where

k_{et} and k_{ht} are the electron- and hole-transfer constants

n_{s0} and p_{s0} are the surface concentrations of electrons and holes in the dark at equilibrium

5.2.4 Quasi-Fermi-Level Concept

The law of mass action only holds at equilibrium. The Fermi-level concept has no meaning away from equilibrium (e.g., under illumination) when np does not equal n_i^2 because there is no single energy that describes simultaneously the average energies of electrons and holes. Instead, the populations of electrons and holes are uncoupled and the average energy of each is different. Accordingly, the concept of quasi-Fermi levels can be invoked where each carrier type has a descriptive average energy value using modifications of Equations 5.9 and 5.10:

$$\mathbf{E_{F,n}} = \mathbf{E_{cb}} - k_B T \ln \frac{n}{N_{cb}} \tag{5.16}$$

$$\mathbf{E_{F,p}} = \mathbf{E_{vb}} - k_B T \ln \frac{N_{vb}}{p} \tag{5.17}$$

where $\mathbf{E_{F,n}}$ and $\mathbf{E_{F,p}}$ are the quasi-Fermi levels that describe the average energy of electrons and holes, respectively. The corresponding electrochemical potentials for each quasi-Fermi level are $E_{F,n}$ (i.e., $\mathbf{E_{f,n}}/q$) and $E_{F,p}$ (i.e., $\mathbf{E_{f,p}}/q$), respectively. In effect, Equations 5.16 and 5.17 have the same functional forms as the dependence of the electrochemical potential of a liquid solution with the concentrations of dissolved redox species through the Nernst equation.

The electromotive force at which a semiconductor can supply photogenerated charge carriers is equal to the difference between the electrochemical potentials of the electron and hole quasi-Fermi levels, that is, $E_{F,n} - E_{F,p}$. Using Equations 5.16 and 5.17 and the explicit value of n_i, the splitting of the quasi-Fermi levels at any position in the semiconductor by the action of a perturbation away from equilibrium is known:

$$E_{F,n} - E_{F,p} = \frac{k_B T}{q} \ln \frac{np}{n_i^2} \tag{5.18}$$

Equation 5.18 states that conditions that substantially increase np relative to n_i^2 lead to greater electromotive forces from a semiconductor under illumination.

An important consideration in the operation of a semiconductor photoelectrode under illumination is how the quasi-Fermi-level splitting is achieved. Specifically, three separate scenarios are possible that each makes np larger than n_i^2 and thereby would generate a photovoltage that can be used to drive electrochemical reactions. First, under illumination, the value of n could dramatically increase while p remains unchanged. Second, under illumination, the value of p could dramatically increase while n remains unaffected. Third, under illumination, both the values of n and p change significantly. The first scenario occurs with a doped semiconductor photoelectrode that is nominally p-type and illuminated with a light intensity too weak to photogenerate additional holes at a concentration equal to or greater than N_A. The second case occurs with a doped semiconductor photoelectrode that is nominally n-type illuminated by light at an intensity that is insufficient to photogenerate additional electrons at a concentration comparable to N_D. These two possibilities represent low-level injection conditions. The third possibility occurs with an undoped or comparatively lightly doped semiconductor that is illuminated with intense enough light that both n and p change appreciably. This last scenario corresponds to high-level injection conditions. In principle, it is possible to move from low-level injection to high-level injection conditions with the same photoelectrode simply by increasing the illumination intensity.

5.2.5 SEMICONDUCTOR PHOTOELECTRODE AT EQUILIBRIUM IN THE DARK

5.2.5.1 Accumulation, Depletion, and Inversion

When a doped semiconductor is put in contact with a separate conducting phase, heterogeneous charge transfer occurs to equilibrate the Fermi levels of the two respective materials (Figure 5.3). If equilibration involves no net transfer of charge to/from the semiconductor and the concentration of carriers is uniform throughout the semiconductor, then flat band conditions are maintained (Figure 5.3a). Assuming the majority of charges that transfer during the equilibration process originate from the semiconductor, there are three possible resultant scenarios. Figure 5.3 also shows these possibilities for an n-type semiconductor. Analogous cases exist for p-type semiconductors. When equilibrium is attained through the transfer of additional electron density into the semiconductor, n increases at the semiconductor surface relative to n in the semiconductor bulk. The imbalance of electron density manifests a small potential drop inside the semiconductor across a finite distance extending from the surface into the bulk. The associated electric field can be strong and can impact the movement of charge carriers from the bulk to the surface of the semiconductor. In this instance, the n-type semiconductor is in accumulation conditions (Figure 5.3b). For this case, electrons (majority carriers) entering the electric field are directed preferentially towards the interface. Conversely, holes entering the electric field are directed away from the interface. Depending on the extent of accumulation, the n-type semiconductor at the surface may reach degenerate levels with respect to the surface electron concentration, resulting in a sufficiently strong electric field that effectively precludes holes from participating in heterogeneous charge transfer.

If the equilibration process requires the transfer of electrons from the semiconductor into the contacting phase, then the n-type semiconductor is under depletion conditions (Figure 5.3c). If the initial surface concentration of electrons is too low to reach equilibrium, the necessary charge-transfer process requires additional electron density pulled from deeper within the semiconductor. The result is that both the surface and the interior region of the semiconductor near the surface are depleted of electron density relative to n in the bulk of the semiconductor. As before, the concentration imbalance manifests a potential drop across a finite length extending into the interior of the semiconductor. The corresponding electric field impacts the movement of charge carriers from the bulk to the surface in an exactly opposite direction as compared to accumulation conditions. That is, under depletion conditions, the electric field in this n-type semiconductor inhibits electrons (majority carriers) from moving from the bulk to the semiconductor surface and facilitates holes

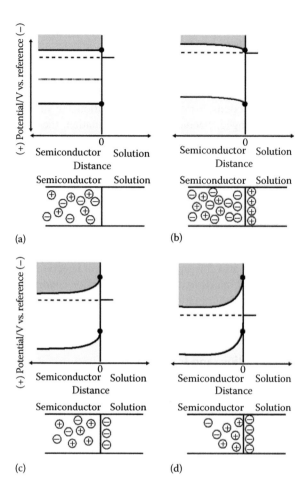

FIGURE 5.3 Schematic depictions of (a) flat band, (b) accumulation, (c) depletion, and (d) inversion conditions for an n-type semiconductor in equilibrium with a solution. (Top) Energy band diagrams of the semiconductor/liquid contact. (Bottom) Charge distribution in the bulk semiconductor and at the semiconductor/liquid interface. (Adapted from Linselbigler, A.L. et al., *Chem. Rev.* 95, 735, 1995.)

moving from the bulk to the interface. Further, at all positions, the law of mass action still holds. Accordingly, p increases in the near-surface region.

When equilibration of an n-type semiconductor in contact with another conductive phase requires considerable electron transfer from the semiconductor, then a special depletion condition arises, inversion (Figure 5.3d). In inversion, the value of n at the surface becomes so small that $p > n$ at the surface, that is, the ratio of surface concentrations of charge carriers, reverses. Since the law of mass action still holds, the Fermi level at the very near-surface region is actually closer to the valence band edge rather than the conduction band edge. Effectively, at the surface, the semiconductor is inverted from n-type to p-type character. For an n-type semiconductor electrode operating under low-level injection, inversion is desirable since a $p–n$ junction is naturally formed between the p-type surface region and the n-type interior. A strongly inverted semiconductor photoelectrode will yield the largest possible quasi-Fermi-level splitting under low-level injection.

5.2.5.2 Potential Drop inside Semiconductor Electrode at Equilibrium

The magnitude of the potential dropped within nanostructured semiconductors under any condition can be determined through Equation 5.3, the appropriate Laplacian operator, and the boundary conditions on either side of the region supporting the potential drop. In the region supporting the potential drop within an n-type semiconductor, a simplification can be made to Equation 5.3. Provided that there are no trapped charges, all the charge in the region where the potential is dropped in an n-type semiconductor comes from the ionized dopant and the free carriers:

$$\nabla^2 E = -\frac{q(N_D - n + p)}{\varepsilon_s} \tag{5.19}$$

Following the approach of Albery,[44] a convenient point of reference for E is setting $E_{cb} = 0$ specifically in the bulk of the semiconductor where there is no electric field (i.e., $n = N_D$). Accordingly, the potential difference between E_{cb} at a point within the electric field and E_{cb} at a point in the bulk semiconductor outside of the electric field is $-k_B T/q \ln n/N_D$. Inserting this definition into Equation 5.19 and recognizing the value of p is small under accumulation and depletion conditions, the following relation is useful for determining the explicit potential at every point within the semiconductor:

$$\nabla^2 E = -\frac{qN_D\left(1 - e^{\frac{-qE}{k_B T}}\right)}{\varepsilon_s} \tag{5.20}$$

Equation 5.20 is valid for all accumulation and depletion conditions and can be solved to calculate the magnitude of the potential drop in a semiconductor electrode at equilibrium. For a semiconductor under strong accumulation, the exponential term dominates and the Poisson relation is further simplified:

$$\nabla^2 E = \frac{qN_D e^{\frac{-qE}{k_B T}}}{\varepsilon_s} \tag{5.21}$$

For a strongly depleted semiconductor, the exponential term is small and drops out:

$$\nabla^2 E = -\frac{qN_D}{\varepsilon_s} \tag{5.22}$$

Under inversion conditions, p in Equation 5.19 cannot be neglected and the Poisson relation cannot be simplified.

In a semiconductor electrode operating under depletion conditions, the total potential dropped within a semiconductor between the surface and the interior at equilibrium is called the built-in potential, V_{bi}. Equation 5.22 yields the following expression for the built-in potential for a large planar semiconductor electrode under depletion conditions:

$$V_{bi} = \frac{qN_D}{2\varepsilon_0 \varepsilon} W^2 \tag{5.23}$$

where W is the distance from the semiconductor surface into the bulk over which the potential is dropped. Typically, W values span 10^{-9} to 10^{-6} m for a planar semiconductor electrode. V_{bi} is maximized when a semiconductor is pushed from depletion to inversion conditions, which yield the same V_{bi} in addition to a potential drop at the semiconductor/liquid interface.

For a given semiconductor/liquid contact with V_{bi} dropped inside the semiconductor, the values of W and N_D are coupled. Operationally, changes in N_D directly impact the physical distance over which the built-in potential is dropped inside a semiconductor electrode. The interplay between this dimension and the morphological feature size of nanomaterial is a primary factor that impacts the attainable photocurrents, photovoltages, and ultimately η_{PEC} in a nanostructured semiconductor photoelectrode.

5.3 PHOTOELECTROCHEMISTRY WITH HIGH-ASPECT-RATIO SEMICONDUCTOR NANOSTRUCTURES

A high-aspect-ratio form factor in a nanostructured semiconductor electrode can naturally decouple the directions of light absorption and photogenerated charge-carrier collection (Figure 5.4). When optimal, the high-aspect-ratio design makes it possible to collect photogenerated charge carriers with high probability and at an appreciable difference in quasi-Fermi levels even with low-grade semiconductor materials. In principle, high-aspect-ratio photoelectrodes can operate under low- or high-level injection conditions. The former has been explored much more fervently than the latter. A discussion of both operational conditions is presented in the following sections.

5.3.1 LOW-LEVEL INJECTION

For $\Phi = 1$, the key condition is to have the liquid electrolyte in contact with at least one dimension of the nanostructured semiconductor that is shorter than the bulk diffusion length, L, of the minority carrier type. In this way, the size of the nanostructured semiconductor becomes the parameter that defines the path length that photogenerated carriers travel to reach the semiconductor/liquid interface for charge transfer. Accordingly, semiconductor materials with L values so short that $\Phi_{Recombination} \sim 1$ are decidedly ineffective for solar energy conversion in a planar macroelectrode morphology.

The first experimental demonstration that nanostructuring with a high aspect ratio was a means for maximizing Φ was reported by a group from the Netherlands.[45,46] They used macroporous GaP photoelectrodes prepared from low-grade GaP wafers (Figure 5.5).

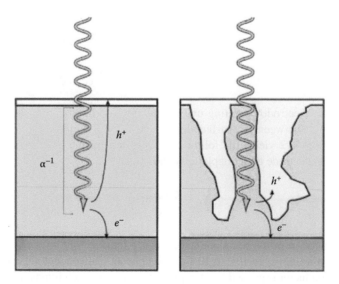

FIGURE 5.4 Cross-sectional view for the comparison of photogenerated charge-carrier collection at planar and high-aspect-ratio semiconductor photoelectrodes. (From Price, M.J. and Maldonado, S., *J. Phys. Chem C*, 113, 11988, 2009.)

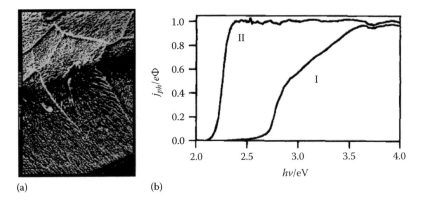

(a)   (b)

FIGURE 5.5 (a) Topside view of a porous n-GaP layer of about 25 μm thickness on a GaP substrate (visible in the bottom right corner). White parts correspond to GaP and dark parts to the pores. The surface of the porous layer can be seen on the upper part of the photograph. (b) Photocurrent quantum yield for a planar (curve I) n-GaP photoelectrode and for a 30 mm thick macroporous (curve II) n-GaP photoelectrode as a function of illumination wavelength. (From Marin, F.I. et al., *J. Electrochem. Soc.*, 143, 1137, 1996.)

Commercially available GaP wafers are prototypical examples of semiconductor materials where the values of L (ca. 10^{-7} m) are inadequate to sustain high Φ with a planar form factor. In planar form, deeply penetrating photons with energies near the bandgap are lost to recombination in the bulk. Their seminal study showed that anodic etching could be used to impart a high-aspect-ratio macroporous morphology, that is, deep pores separated by thin walls. Upon nanostructuring, the *n*-type GaP photoelectrodes supported near unity values of Φ while immersed in water and under 1 sun illumination when an extremely large (2 V) bias was applied. A complicating factor in their study was n-GaP as a photoanode in acidic water since GaP photodegrades quickly. A later report by a separate research team verified and expanded the initial findings using a nonaqueous electrolyte and further showed unambiguously that efficient nanostructuring of GaP allowed both a large quasi-Fermi-level splitting (ca. 1 V) at open circuit (i.e., no net current) and $\Phi \sim 1$ at short circuit (i.e., no bias applied), augmenting η_{PEC} by over an order of magnitude relative to the same material in planar form.[47] Similar nanostructuring-by-etching tactics have been used for other low-grade forms of a variety of semiconductors including Si,[48] GaAs,[49] and CdZnTe.[50]

A disadvantage of the top-down approach for nanostructuring is the intrinsic material loss, limiting the interest in this approach for preparation and study of high-aspect-ratio semiconductor photoelectrodes. Interest in such semiconductor photoelectrode designs did not significantly increase until simultaneous reports from groups at Caltech and Penn State.[51,52] Both publications focused on Si nanowire films acting as nanostructured photoelectrodes, analogous to the macroporous films but prepared from the bottom up. Both works showed driving redox reactions with nanowire films at underpotentials was possible (Figure 5.6).

Although neither report achieved appreciable values of η_{PEC}, they generated considerable excitement in the community since they showed the premise nanostructured photoelectrodes could be synthesized and tailored. Subsequently, numerous research groups have prepared other nanostructured semiconductors to effect more efficient optical-to-chemical energy conversion. A table summarizing such investigations was published previously.[53]

Although syntheses of Si nanowires with large values of L have led researchers towards larger critical feature sizes in Si, most synthetic methods favor the formation of semiconductor nanowires with small radii and most semiconductors (even at the highest purity levels) tend to possess $L < 10^{-6}$ m. Hence, the development specifically of thin semiconductor nanowire photoelectrodes continues to be a vibrant area of research.

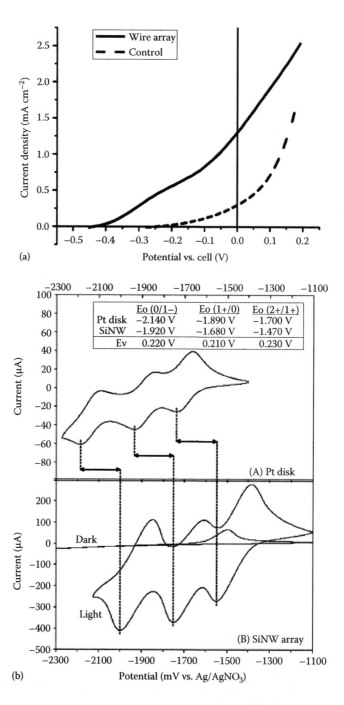

FIGURE 5.6 (a) Current density versus voltage behavior for an Si wire array (solid) and control samples (dashed). The electrode potential was measured versus a Pt reference poised at the Nernst potential of the 0.2 M Me₂Fc/0.5 mM Me₂Fc⁺/1.0 M LiClO₄-CH₃OH cell. (From Maiolo, J. R. et al., *J. Am. Chem. Soc.*, 129, 12346, 2007. (b) Cyclic voltammograms (200 mV s⁻¹) of (A) a Pt disk electrode and (B) a silicon nanowire array photocathode, both light and dark, as indicated. (Inset) Midpoint potentials (E_0) of the Ru(bpy)₃ anodic and cathodic peaks at the metallic and semiconductor electrodes and the calculated photovoltages (E_v). (From Goodey, A.P. et al., *J. Am. Chem. Soc.*, 129, 12344, 2007.)

In many of thin nanowire photoelectrode reports to date, the implied operation is low-level injection with (assumed) strong depletion conditions. Still, the demonstrated values of Φ at short circuit are insufficient to attain high values of η_{PEC} and much less than what has been observed in macroporous nanostructured films with similar feature sizes. Excessive surface recombination amplified by the larger total surface area of nanostructured semiconductors is often invoked to rationalize the low observed η_{PEC} values. This possibility as a universal explanation is somewhat inconsistent with the macroporous photoelectrodes prepared by top-down etching, where good steady-state energy conversion properties have been observed without any effort to mitigate surface defects. An analysis of the operational features of high-aspect-ratio semiconductor photoelectrodes has therefore been sought by several groups to identify design criteria for optimal performance. Of primary importance is an understanding of the potential dropped within a semiconductor at equilibrium and under illumination.

The analytical expression for the position-dependent potential inside a thin semiconductor nanowire in strong depletion through Equation 5.22 has been described and is given here[54–56]:

$$E(r) = E_{r=0} + \frac{qN_D}{2\varepsilon_0\varepsilon}(r_0 - W)^2 \left(\frac{1}{2} - \frac{r^2}{2(r_0 - W)^2} + \ln\left(\frac{r}{(r_0 - W)} \right) \right) \tag{5.24}$$

where
$E_{r=0}$ is the potential at the center of the nanowire and everywhere outside of the depletion region
r is the position inside the nanowire along the radial dimension
r_0 is the radius of the nanowire, and the rest of the terms have their usual meanings

Figure 5.7 shows the potential profile of the conduction band edge within a depleted n-type semiconductor nanowire with a uniform doping profile as a function of dopant concentration, the equilibrium barrier height, and nanowire radius.[54]

A primary finding of these analyses is that the depletion width in a nanowire is not necessarily equivalent to that in a planar photoelectrode. This discrepancy arises because of a difference in where charge is depleted from the semiconductor to reach equilibrium, that is, just the near-surface versus the bulk volume. In a thin nanowire, the charge needed to reach equilibrium comes from both the surface and the volume, perhaps even to the core of the nanowire. That is, the heterogeneous charge transfer necessary to attain equilibrium perturbs the total charge density throughout an entire semiconductor nanowire if it is sufficiently thin. Further, for a given set of nanowire attributes (e.g., dopant density, band edge energetics), the depletion width extends further into nanowires with small radii. In contrast, the charge density in the bulk volume of a macroscale planar semiconductor photoelectrode is not affected at all.

For nanowires that are insufficiently doped and immersed in an electrolyte that should induce a large internal potential drop, r_0 may not be large enough to support the entire potential drop. In this case, thin nanowires are fully depleted, that is, equilibration changes n and p everywhere inside the semiconductor. The maximum built-in potential between the center of the nanowire and the surface of the nanowire is a function of r_0 and is given by[54,56]

$$V_{bi} = \frac{qN_D}{4\varepsilon_0\varepsilon}r_0^2 \tag{5.25}$$

The size-limited built-in potential within a thin nanowire has consequences on the attainable η_{PEC} under low-level injection. For example, an Si nanowire ($\varepsilon_0\varepsilon = 1.04 \times 10^{-12}$ F cm^{-1}) with $r_0 = 50$ nm can support a maximum built-in potential of nearly 1 V when it is doped at 10^{18} cm^{-3} but only 0.01 V when it is doped at 10^{16} cm^{-3}. Accordingly, the electric field inside the nanowire is 100-fold smaller

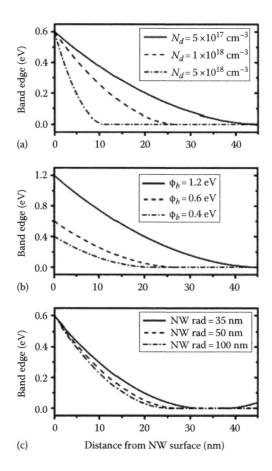

FIGURE 5.7 Calculated potential profiles demonstrating the effect of varying the (a) background dopant concentration, (b) surface barrier height, and (c) NW radius. (From Simpkins, B.S. et al., *J. Appl. Phys.*, 103, 104313(6), 2008.)

in the less doped nanowires, that is, the drift effect to push majority carriers away from the semiconductor/electrolyte interface is diminished. A recent study quantitatively explored the impact of this effect on Φ under 1 sun illumination.[57]

Figure 5.8 shows data for two types of *n*-type Si photoelectrodes immersed in an electrolyte with a fast redox couple that induces a large internal potential drop in n-Si. Here the photoelectrodes are composed of a film of dense Si nanowires on top of a single-crystalline Si substrate. The photoresponses for both the composite films and just the underlying Si substrate are shown.[57]

One of the nanowire films is sufficiently doped to sustain the entire potential drop within the nanowire, while the other is insufficiently doped. In the former, the concentration of majority carriers (electrons) is low at the surface. In the latter case, with the potential drop limited by Equation 5.25, the concentration of majority carriers at the surface is much higher. The difference in the surface concentration of electrons impacts the value of Φ_S predicted from Equation 5.15. Further measurements with monochromatic light showed that the respective photocurrent increase and decrease relative to the planar photoresponse were directly a result of the capacity of the nanowires to efficiently separate photogenerated charges. Effectively, these results showed only highly doped nanowire function as efficient photoelectrode materials with respect to majority carrier recombination at the semiconductor/liquid interface.

Nonuniform doping profiles and shapes can also lower Φ in nanostructured semiconductors. For semiconductor nanowires, spatial variation of dopant concentration occurs often[58-62] as does

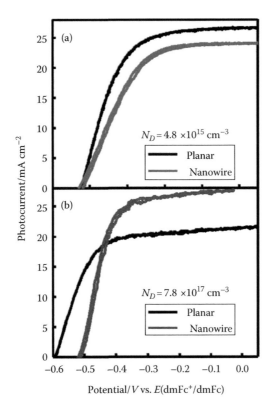

FIGURE 5.8 Steady-state photocurrent–potential responses for Si photoelectrodes immersed in deaerated methanol solutions containing 1 M LiClO$_4$, 195 mM dimethylferrocene, and 5 mM dimethylferrocenium. (a) Representative responses for (black) a planar and (red) a nanowire film photoelectrode with $N_D = 4.8 \times 10^{15}$ cm^{-3}. (b) Representative responses for (black) a planar and (blue) a nanowire film photoelectrode with $N_D = 7.8 \times 10^{17}$ cm^{-3}. (From Hagedorn, K. et al., *J. Phys. Chem. C*, 114, 12010, 2010.)

tapering along the length.[63–65] The influence of such nonuniformities can similarly impact Φ by increasing the concentration of majority carriers at the surface. Figure 5.9 shows modeled results for thin Si nanowires pushed strongly into inversion by equilibration with a solution containing an ideally fast (outer-sphere) redox couple.[53]

In these profiles, the center of the nanowire corresponds to 0 nm. All nanowire photoelectrodes have a uniform dopant density of 2×10^{16} cm^{-3} and equivalent material properties but differed in their shape. The thinnest nanowire yielded the lowest photoresponse, implying exceedingly low values of $\eta_{separation}$. Conversely, a thicker nanowire with the exact same bulk and surface properties yielded a significantly enhanced photoresponse, indicating the nanowire was sufficiently large to support the full potential drop that results in an attenuated concentration of majority carriers at the surface. The tapered cylindrical Si nanowires had some fraction of their total length with a thickness less than necessary to support V_{bi}. Accordingly, the tapered nanowires all suffered majority carrier recombination in those regions. The net impact was that appreciable photocurrent is lost even for moderately tapered nanowires if tapering is enough to limit V_{bi}.

The maximum quasi-Fermi-level splitting in a semiconductor under illumination is also affected by nanostructuring. This potential difference (V_{oc}) will be smaller for the nanowire photoelectrode for several reasons. For example, in a traditional planar photoelectrode geometry under low-level injection, the highest achievable np product in Si is ultimately limited by SRH recombination in the bulk.[66] For a macroscale, planar n-type semiconductor photoelectrode operating under this limitation, the maximum V_{oc} possible is given by[66]

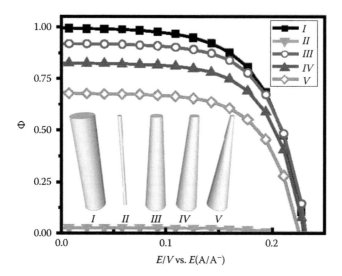

FIGURE 5.9 Comparison of the simulated photoresponses under AM 1.5 (direct–circumsolar) illumination for five different high-aspect-ratio semiconductor photoelectrode morphologies (*I–V*) all modeled with a uniform dopant density of 2×10^{16} cm^{-3} and an interfacial equilibrium barrier height of 1 eV. For *I*, $r_0 = 300$ nm. For *II*, $r_0 = 50$ nm. For *III*, *IV*, and *V*, $r_{0,\text{bottom}} = 300$ nm and $r_{0,\text{top}} = 150$, 115, and 50 nm, respectively. (From Foley, J.M. et al., *Energy Environ. Sci.*, 5, 5203, 2010.)

$$V_{oc,planar} = \frac{k_B T}{q} \ln\left(\frac{q\Phi G_0 L_p N_D}{A_{planar} k_B T \mu_p n_i^2} \right) \tag{5.26}$$

where

G_0 is the total number of photogenerated carriers generated per second (i.e., G_{ph} integrated both over all wavelengths and the whole semiconductor volume)

A_{planar} is the surface area across which current is passed

μ_p is the mobility of the minority carriers (holes)

Collectively, the term $q\Phi G_0/A_{planar}$ is the photocurrent density passed at the planar photoelectrode at short circuit. If a thin nanowire morphology is applied to this exact same semiconductor material and charge-carrier recombination is predominantly in the core of the nanowire (i.e., not in the depletion region), then the photovoltage will follow Equation 5.27 instead[53]:

$$V_{oc,nanowire} = \frac{k_B T}{q} \ln\left(\frac{4q\Phi G_0 L_p^2 N_D}{A_{nanowire} k_B T \mu_p n_i^2 W} \right) \tag{5.27}$$

In this case, the attainable photovoltage by the nanowire differs from that for the planar analog by a factor of $\ln\left(A_{nanowire} W / A_{planar} 4 L_p \right)$.

The diminution in V_{oc} for a nanowire photoelectrode is further complicated when nanostructuring changes which recombination process is dominant. Specifically, for exceedingly thin nanowires where the volume fraction of the nanowire that possesses the internal potential drop is large, most of the charge-carrier photogeneration occurs in the depletion region. Accordingly, the highest rate of charge-carrier recombination occurs inside the depletion region rather than in the (field-free)

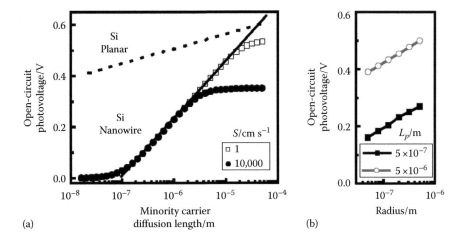

FIGURE 5.10 Simulation results under AM 1.5 (direct + circumsolar) illumination for cylindrical n-Si nanowire photoelectrode with $r_0 = 50$ nm and $l = 100$ μm in contact with an electrolyte resulting in an interfacial equilibrium barrier height of 1 eV. (a) Analysis of the attainable open-circuit photovoltage with an Si nanowire as a function of the minority carrier diffusion length calculated at two different values of the surface recombination velocity. The dashed line indicates the expected dependence of the open-circuit photovoltage with changes in the minority carrier diffusion length for a comparable planar photoelectrode operating under bulk recombination limitations. (b) Analysis of the attainable open-circuit photovoltage with an Si nanowire as a function of nanowire radius at two different values of the minority carrier diffusion length. (From Foley, J.M. et al., *Energy Environ. Sci.*, 5, 5203, 2012.)

core. In this case, the relevant analytical expression for the open-circuit photovoltage is no longer Equation 5.27 and becomes significantly more complex. Finite-element simulations have been used[53] to identify the sensitivity of V_{oc} of a thin nanowire photoelectrode operating under this condition. Figure 5.10 compares the photovoltage dependence with L_p and S_p for two hypothetical Si photoelectrodes (planar and thin nanowire) in contact with an electrolyte that induces an interfacial barrier height of 1 eV. For the exact same material properties, the nanowire will yield much lower photovoltages (by more than a factor of $A_{nanowire}/A_{planar}$) when L_p is small.

Conversely, the gains in V_{oc} are much more significant for a nanowire photoelectrode when L_p is increased. Somewhat counter to conventional wisdom, the modeling results showed that for thin nanowires operating under depletion conditions and low-level injection, the surface recombination velocity was not a significant factor in to the attainable value of V_{oc}. Instead, surface recombination only becomes a significantly limiting process when L_p is much larger than r_0. In other words, for thin, high-aspect-ratio semiconductor photoelectrodes, decreasing surface recombination will not yield the largest or most immediate gains in η_{PEC} if Φ_s is already small. Instead, strategies that improve L_p will have a much greater impact. Such aspects should be considered in efforts to refine nanowire photoelectrode materials for optimal energy conversion efficiencies.

5.3.2 High-Level Injection

To date, macroporous or nanowire photoelectrodes operating under high-level injection conditions have not been extensively explored as a concept or exploited as a possible water-splitting strategy. For high-level injection, illumination has to substantially change n and p from their equilibrium values through extreme illumination intensities and/or through the use of semiconductor materials with n and p values near n_i. The latter condition is true when either the semiconductor has few (or no) intentional dopants or the nanostructured semiconductor is fully depleted. At first glance, both conditions necessary for high-level injection could be unfavorable/detrimental to the prospects for high efficiency. In general, the high illumination intensities can result in an intolerable level of

Auger recombination.[67–69] Further, simultaneously low n and p concentrations are disadvantageous for the reasons enumerated in Section 5.3.1. However, recent data indicate that nanowire photoelectrode operation in high-level injection may be fruitful.

A new experimental and modeling study compared the photoelectrochemical activity of chemically synthesized n-type ZnO nanowires with planar, single-crystalline n-type ZnO photoelectrodes.[70] The nanowires in the study were doped but just slightly undersized in relation to the expected depletion width, implying full depletion. The results showed that in an aqueous cell with the O_2/H_2O redox couple, these nanowires did not exhibit poor photoresponse characteristics but instead demonstrated behaviors comparable to the planar analog (Figure 5.11).

This same group subsequently reported on lightly doped Si microwires immersed in methanol containing the 1–1′-dimethylferrocene/1–1′-dimethylferrocenium redox couple.[71] These high-aspect-ratio Si photoelectrodes were prepared via catalyzed chemical vapor deposition, had radii of slightly in excess of 1000 nm, and had dopant concentrations of only 10^{13} cm^{-3}. These high-aspect-ratio Si photoelectrodes also natively possessed a minority carrier (hole) diffusion length much larger (by a factor of ~70) than the radius. The employed methanolic electrolyte naturally induced both inversion in n-type Si and an effective surface recombination velocity less than 50 cm s^{-1}.[72,73] Again unexpectedly high photocurrents and photovoltages were observed (Figure 5.12).

The cumulative results led the researchers to propose certain conditions where nanowire photoelectrodes could function under high-level injection conditions. Specifically, if (1) the semiconductor/liquid junction is carrier selective, that is, it permits heterogeneous transfer of just one carrier type, (2) the semiconductor surface is rigorous-free of electrical traps, and (3) the charge-carrier lifetimes are long, then moderate to high energy conversion efficiencies can be obtained with high-aspect-ratio semiconductors under high-level injection. Operationally, these studies suggested that surface recombination was tolerable below a threshold value:

$$S = \frac{k_B T \mu r}{l^2} \tag{5.28}$$

where
  μ is the carrier charge mobility
  l is the length of the nanowire

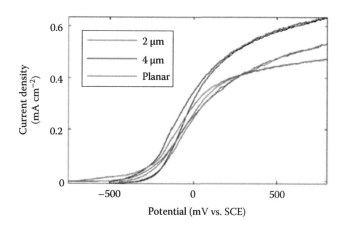

FIGURE 5.11 AM 1.5 response for ZnO single crystal (red) and optimized 4 μm wire arrays (black) and 2 μm wire arrays (blue) measured in 0.5 M K_2SO_4 at pH 6.4. (From Fitch, A. et al., *J. Phys. Chem. C*, 117, 2008, 2013.)

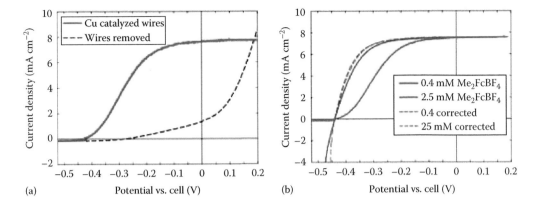

FIGURE 5.12 Current–potential data for Cu-catalyzed Si microwire array photoelectrodes in contact with the $Me_2Fc^{+/0}$–CH_3OH redox system (a) under 100 mW cm^{-2} of AM 1.5 G illumination and (b) with varying amounts of Me_2FcBF_4. (From Santori, E.A. et al., *Energy Environ. Sci.*, 5, 6867, 2012.)

Similarly, the lifetime of charge carriers in the bulk must be above a threshold value according to the following equation:

$$\tau > \frac{ql^2}{k_BT}\mu \tag{5.29}$$

The criterion involving the selectivity of heterogeneous charge transfer is a function of the employed redox couple. For simple outer-sphere redox couples like 1–1′-dimethylferrocene, a key condition is that the standard potential is sufficiently positive/negative to induce strong inversion with the semiconductor. In this capacity, the energetics of the semiconductor/solution junction are paramount. For inner-sphere redox systems like O_2/H_2O or I_3^-/I^-, carrier selectivity is possible by virtue of the chemically complex redox process. Similarly, heterogeneous catalysts that require appreciable overpotential to operate could be selective to one charge-carrier type. So long as the nanowire photoelectrode satisfies these collective criteria, meaningful energy conversion may be performed under high-level injection. If these criteria are not met, then high-level injection affords no advantage over low-level injection (i.e., η_{PEC} will be small).

The encouraging conclusion from the experimental observations is that the bottleneck that prevents large values of $\eta_{separation}$ with insufficiently doped high-aspect-ratio semiconductor photoelectrodes can be overcome in certain instances by switching to the high-level injection regime. The importance of this point cannot be understated as the ability to dope reliably and with precision is not a universal trait to all semiconductor materials. In fact, outside of a few common semiconductors (e.g., Si, Ge, GaAs) where the metallurgical issues relevant to doping have been worked out, many semiconductors cannot even be doped both *n*-type and *p*-type. This issue is significant in nanostructured photoelectrodes under low-level injection where it may not even be possible to dope a semiconductor sufficiently. Accordingly, the use of high-level injection as a separate means to *relax* the dopant-level constraint could enable the use of many more types of semiconductors in the nanostructured motif. However, as described by Equations 5.28 and 5.29, the premium on the surface and bulk properties could still prove prohibitive for many semiconductors with intrinsically low charge-carrier mobilities and carrier lifetimes (e.g., Fe_2O_3).

A separate study was recently published that proposed a strategy for relaxing the tolerances on the values of μ and τ for a thin semiconductor nanowire photoelectrode in high-level injection. The basic premise involved a set of discrete ohmic-selective contacts along the nanowire length instead of a single, conformal contact along the whole surface of the nanowire.

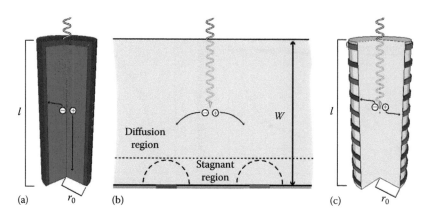

FIGURE 5.13 Cross-sectional view of select semiconductor photovoltaic designs under illumination. (a) A nanowire solar cell featuring a conformal and uniform contact. The terms l and r_0 denote the nanowire height and radius, respectively. (b) A planar point-contact solar cell featuring discrete, ohmic-selective contacts on the backside. The term W denotes the wafer thickness. (c) A nanowire solar cell featuring a series of discrete, ohmic-selective contacts. The terms l and r_0 are as in (a). Blue and red colorings denote regions where electrons and holes are collected, respectively. Gray coloring indicates regions where the semiconductor has no added dopants. Not drawn to scale. (From Chitambar, M.J. et al., *J. Appl. Phys.*, 114, 174503(11), 2014.)

Figure 5.13 illustrates this concept in comparison to a nanowire with a continuous, conformal-contact and a planar point-contact solar cell.

The planar *point-contact* solar cell employs the same tactic with lightly doped planar semiconductor photoelectrodes and small, circular, and ohmic-selective contacts on the backside.[22,74–82] The planar point-contact design is the basis of one of the most commercially successful solar cell products, pushing $\eta_{separation}$ to the theoretical maximum value.[82] A major disadvantage in the planar point-contact scheme is that high efficiencies can only be obtained under high-level injection conditions with semiconductor materials possessing large values of μ and τ. A series of finite-element simulations showed that a nanowire contacted with discrete, ohmic-selective band contacts has much higher tolerances in μ and τ. Figure 5.14 compares directly the dependences of photoresponse characteristics with the bulk charge-carrier lifetime in both planar point-contact and nanowire discrete-contact photoelectrodes. The analyses showed that the discrete-contact geometry supported much higher values of $\eta_{separation}$ with materials with small to extremely small values of the charge-carrier lifetime. Similar results were found for the tolerance towards surface recombination velocities. In effect, the discrete-contact motif allows the principal advantage of low-grade semiconductor nanowires, that is, lower than what is predicted by Equations 5.28 and 5.29, to be realized under high-level injection conditions. To date, a semiconductor nanowire photoelectrode with discrete, ohmic-selective contacts has yet to be realized experimentally.

5.4 PHOTOELECTROCHEMISTRY WITH SEMICONDUCTOR NANOPARTICLES

Historically, nanoparticles were the first *nanostructured* photoelectrochemical motif to be explored and studied. A number of studies have been performed to assess the photocatalytic properties of many types of semiconductor (particularly metal oxide) nanoparticles as catalysts for oxidation reactions and purification/cleaning applications.[83–85] Separately, the photoelectrode behaviors of compact, particulate semiconductor films have been analyzed.[86] Such films serve as the backbone of the dye-sensitized photoelectrodes first popularized by O'Regan and Gratzel.[87] Separately, semiconductor nanoparticles small enough to exhibit quantum-confinement effects have been used to prepare Schottky-type heterojunctions with the capacity for attaining values

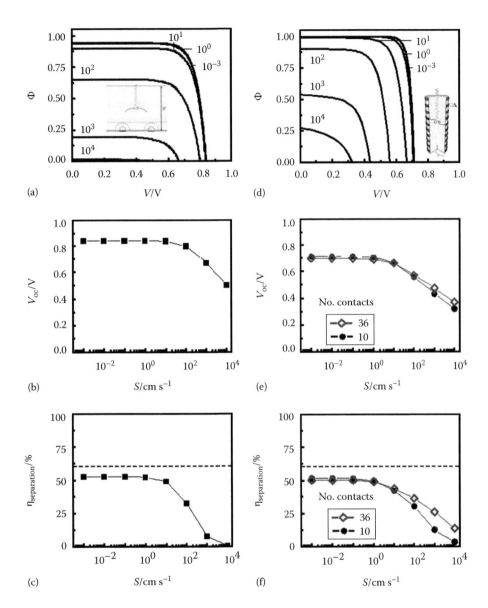

FIGURE 5.14 Comparison of the modeled responses for a (a–c) planar and a (d–f) nanowire solar cell with discrete, doped, and ohmic-selective contacts as a function of the surface recombination velocity, S. (a) Internal quantum yield-potential (Φ–V) response for a planar solar cell with discrete (point) ohmic-selective contacts, $W = 50\ \mu m$, and S ranging between 10^4 and 10^{-1} cm s^{-1}. (b) The photovoltage (V_{oc}) values from (a) as a function of S. (c) The internal energy conversion efficiency ($\eta_{separation}$) as a function of S. The dashed horizontal line indicates the thermodynamic limit for $\eta_{separation}$. (d) Internal quantum yield-potential (Φ–V) response for a nanowire solar cell with discrete, ring ohmic-selective contacts, $r_0 = 50\ \mu m$, and S ranging between 10^4 and 10^{-1} cm s^{-1}. (e) The photovoltage (V_{oc}) values from (d) as a function of S. (f) The internal energy conversion efficiency ($\eta_{separation}$) as a function of S. The dashed horizontal line indicates the thermodynamic limit for $\eta_{separation}$. (From Chitambar, M.J. et al., *J. Appl. Phys.*, 114, 174503(11), 2014.)

of $\Phi > 1$ at short circuit through multiexciton generation.[88,89] Readers interested in the specific details of these uses of semiconductor nanoparticle networks are encouraged to review the listed references.

5.4.1 COMPARISON OF THE OPERATION MODES OF A SUSPENDED SEMICONDUCTOR NANOPARTICLE AND A HIGH-ASPECT-RATIO PHOTOELECTRODE UNDER ILLUMINATION

In the context of solar energy storage, an isolated semiconductor nanoparticle suspended in an aqueous solution is arguably the simplest possible design for a solar-powered water electrolyzer. As in the case with high-aspect-ratio semiconductor photoelectrodes, semiconductor nanoparticles naturally possess short path lengths for collection of photogenerated charge carriers at an interface. In fact, a semiconductor nanoparticle represents the extreme, where the path length for a photo-generated carrier from the interior to the surface is small in all directions. The result is that even the transport of photogenerated carriers to the surface by diffusion alone can be fast and strongly dependent on size. The effective transit time from the core of nanoparticle to the surface by diffusion ($\tau_{diffusion}$) is given by

$$\tau_{diffusion} = \frac{qr_0^2}{k_B T \mu} \tag{5.30}$$

where r_0 is the nanoparticle radius. In even a low-mobility semiconductor (e.g., $\mu = 10^{-2}$ cm^2 V^{-1} s^{-1}) nanoparticle with $r_0 = 50$ nm, both photogenerated carrier types will diffuse to the surface within 100 ns at room temperature. In practical terms, the ready access to the surface weights the surface recombination processes over the bulk recombination processes in a small semiconductor nanoparticle.

Suspended semiconductor nanoparticles also differ from high-aspect-ratio semiconductor photoelectrodes in another aspect. High-aspect-ratio nanowire/macroporous photoelectrodes in low-level injection are connected to current collectors and are thus intended to operate at a specific power point (i.e., at a particular combination of current–potential values that maximizes the product of the quasi-Fermi-level splitting and the *net* photocurrent). In contrast, a suspended semiconductor nanoparticle functions without any external contacts at precisely *open-circuit* conditions. That is, at the operational conditions, photoexcited semiconductor nanoparticles suspended in a solution pass **no** *net* current, that is, $\Phi = 0$, and their quasi-Fermi levels are offset by the maximum value possible under the operative illumination and recombination conditions.

The difference in operational conditions is not subtle. An n-type photoelectrode that operates at a maximum power point must suppress the majority charge-carrier (electron) photocurrent and maximize the minority charge-carrier (hole) photocurrent. The suppression of the majority charge-carrier photocurrent is achieved by sacrificing some of the quasi-Fermi-level splitting through lowering $E_{F,n}$ at the surface. For an n-type semiconductor particle performing solar-to-chemical energy transfer and storage, operating at open circuit means rigorously that all electron-transfer reactions are exactly matched by hole-transfer reactions at the semiconductor nanoparticle surface[90]:

$$k_{et}[\text{A}]\left(n_s - n_{s0}\right) = qk_{ht}[\text{B}]\left(p_s - p_{s0}\right) \tag{5.31}$$

As long as "A" and "B" are chemically distinct species, a net chemical conversion in solution will be effected under illumination since the values of Φ for each respective electron and hole current are themselves not zero. Further, the quasi-Fermi-level splitting induced by illumination is not necessarily applied evenly to the electron and hole currents, that is, n_s and p_s can be different in order to satisfy Equation 5.31. Thus, the equilibrium condition of the nanoparticle in the dark strongly impacts the operation under illumination.

A special comment should be made regarding the charge-transfer rate constants, k_{et} and k_{ht}, at a semiconductor nanoparticle. Recent measurements of charge transfer across solid/liquid interfaces have indicated that ultrasmall semiconductor nanoparticles whose bandgaps are a function of size (i.e., less than the Bohr exciton radius)[91] also show size-dependent charge-transfer rate constants.[92–97] The effect arises because as the bandgap changes with size, the band energetics are a function of size. Correspondingly, the thermodynamic driving force for charge transfer to a dissolved redox molecule changes with the semiconductor nanoparticle size. Measurements have shown a variation between nanoparticle radius and charge-transfer rate constants in accord with basic predictions from macroscopic and microscopic theories for charge transfer (Figure 5.15).[97]

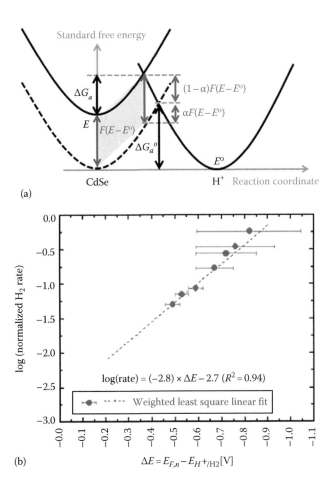

FIGURE 5.15 (a) Butler–Volmer diagram for a charge-transfer process involving a CdSe nanoparticle. As the nanoparticle size diminishes, the Fermi energy of electrons shifts up from $E°$ to E, causing an increase of $F(E - E°)$ in the free energy change ΔG as well as a decrease of $\alpha F(E - E°)$ in the kinetic activation energy ΔG_a. The value of the electron-transfer coefficient α depends on the shape of the energy surface of the reactants and the products. (b) Weighted least square linear fit of the experimental data. Error bars for the four smallest dots are estimated to reach from $E_{F,n}$ for the 2.8 nm nanoparticle to 0.059 V below the measured conduction band edge values and are taken into the weighted least square linear fit. The lowest point is excluded from the fit (dashed line) because the rate is close to the experimental error ($\pm 10^{-3}$) of the H_2 rate measurement. (From Zhao, J. et al., *ACS Nano* 7, 4316, 2013.)

5.4.2 POTENTIAL DROP IN A SUSPENDED SEMICONDUCTOR NANOPARTICLE AT EQUILIBRIUM

Albery,[44] Goossens,[98] and Bisquert[99] have independently solved Equation 5.22 for the potential–distance profile inside a spherical semiconductor particle at equilibrium in the dark. For an n-type semiconductor nanoparticle with r_0 larger than the depletion region with width, W, and with a field-free core (i.e., $E(r) = E_{r=0}$ for $0 \leq r \leq (r_0 - W)$), the potential within the depletion region is

$$E(r) = E_{r=0} + \frac{qN_D}{6\varepsilon_0\varepsilon}\left(r^2 - 3(r_0 - W)^2 + 2\frac{(r_0 - W)^3}{r} \right) \qquad (5.32)$$

Similarly, for a uniformly doped n-type semiconductor nanoparticle that is fully depleted, that is, r_0 is smaller than W induced by charge equilibration with the solution, the distance-dependent potential within the depletion region in the nanoparticle is

$$E(r) = E_{r=0} + \frac{qN_D}{6\varepsilon_0\varepsilon}r^2 \qquad (5.33)$$

As indicated by Equation 5.33 and similar to the case with semiconductor nanowire photoelectrodes, the largest possible potential drop between the surface and core of the semiconductor nanoparticle at equilibrium is a strong function of the nanoparticle radius:

$$V_{bi} = \frac{qN_D}{6\varepsilon_0\varepsilon}r_0^2 \qquad (5.34)$$

Goossens has further developed the related expressions for the potential drop within a fully depleted semiconductor nanoparticle that is *not* uniformly doped.[98]

Irrespective of any quantum mechanical effects, the uncertainties in dopant concentration in an individual semiconductor nanoparticle suspended in a solution have profound impact when predicting and interpreting η_{PEC} under illumination. Consider three pure semiconductor nanoparticles with $r_0 = 5$, 50, and 500 nm, respectively. Adding one dopant atom (or stoichiometric defect) to each nanoparticle effects significantly different dopant densities of 2×10^{18}, 2×10^{15}, and 2×10^{12} cm^{-3}, respectively. This discrepancy has direct implications on the values of n_{s0} and p_{s0} for each individual semiconductor nanoparticle and across an ensemble of particles with a distribution of sizes. Further, if the same illumination intensity is applied to nanoparticles with substantially different dopant concentrations, significantly different modes of steady-state operation can be expected, for example, high-level injection versus low-level injection in depletion. As a result, the predominant recombination process is not always clear, obfuscating predictions for the maximum attainable electromotive force (quasi-Fermi-level splitting) that is available to transform A and B in solution.

In practice, the uncertainty from unknown dopant levels in semiconductor nanoparticles is large in most materials preparation methods.[100] In truth, there is not a clear consensus on how to view dopants in ultrasmall semiconductor nanoparticles.[101,102] Still, even in scenarios where the dopant level is known accurately, there is additional debate in the literature about what the equilibrium condition should be for a suspended semiconductor nanoparticle. Consider the case of a suspended n-type semiconductor nanoparticle in equilibrium with a solution with a defined redox potential. In general, there is no ambiguity in the literature for the equilibrium condition of an n-type nanoparticle if it is moderately doped and large. The convenient definition of *large* is with respect to the Debye length in the semiconductor, L_{SC}. Formally, L_{SC} is the distance that a charge is effectively screened in the semiconductor. For a nondegenerately doped semiconductor, L_{SC} is a function of the dopant concentration and dielectric properties:

$$L_{SC} = \sqrt{\frac{\varepsilon_0 \varepsilon k_B T}{q^2 N_D}} \tag{5.35}$$

Figure 5.16a shows the case of a large, doped n-type nanoparticle that is pushed into inversion at the surface at equilibrium (as indicated by the Fermi level at the surface relative to the Fermi level for semiconductor if it were not doped at all). Here, r_0 is large enough so that the core of the semiconductor nanoparticle is outside of the depletion region. The potential is dropped over a short distance near the surface, resulting in a strong internal electric field akin to that for a planar semiconductor photoelectrode. Figure 5.16b shows the equilibrium state of the same sized semiconductor nanoparticle but now without any extrinsic dopants. The equilibrium condition still results in inversion but not just at the surface. Since the potential drop in the semiconductor nanoparticle extends over a long distance, the inversion condition extends further towards the core. In both of these systems, the semiconductor nanoparticle volume is large enough that the bulk values of n and p near the core are essentially unchanged after reaching equilibrium.

The discrepancies in the literature center on the equilibrium state of very small semiconductor nanoparticles. Figure 5.16c shows the equilibrium state of a nondegenerately doped n-type nanoparticle whose radius is at or below the Debye length. Here again, a large potential is dropped within the semiconductor nanoparticle. However, in this case, the whole semiconductor volume is depleted of charge carriers and the potential drop over the entire nanoparticle is limited by r_0 through Equation 5.34. Still, a finite of charge must transfer across the interface so that a single Fermi level describes the entire system. The divergence in the literature is over where specifically the remaining charge

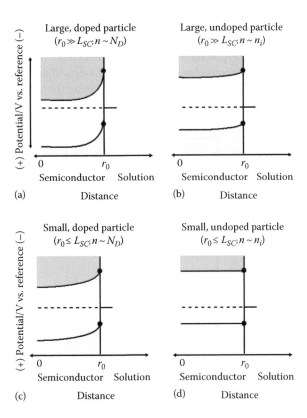

FIGURE 5.16 Schematic depiction of the equilibrium conditions in the absence of illumination for (a and b) a large and (c and d) a small semiconductor particle, either (a and c) doped or (b and d) undoped, suspended in a solution with a defined redox potential. The Fermi level is denoted by the dashed line.

comes from (i.e., how the remaining potential is dropped) to reach equilibrium. One viewpoint[9] is that the balance of the potential not accounted for by Equation 5.34 is dropped entirely across the Helmholtz layer in solution as a double-layer capacitance. Several researchers have raised issue with this point,[99] arguing that dropping the remaining potential across the semiconductor/liquid contact is not physically possible. The contention is that a significant potential dropped across the semi-conductor/liquid interface requires the semiconductor density of donor/acceptor states to be high, comparable to a metal (i.e., the semiconductor is degenerately doped). Such a degenerate condition cannot be the case for a small nanoparticle that has become fully *depleted* of its transferrable charge carriers.

Liver and Nitzan modeled exactly this system using n-type CdS nanoparticles with $r_0 = L_{SC} = 25$ nm (for $N_D = 2 \times 10^{16}$ cm^{-3}) in contact with a solution containing the methylviologen redox couple (MV$^{2+/+}$).[103] Experimentally, this CdS/electrolyte junction requires a potential difference of 0.133 V to reach equilibrium, but the employed CdS nanoparticles could only support a 5 mV drop. Through Equation 5.3 (assuming no surface defects/traps) and the Gouy–Chapman double-layer theory,[104] they calculated the potential–distance profile in this system at equilibrium (Figure 5.17).

Their analysis showed unambiguously that even in the case where the semiconductor nanopar-ticle could only accommodate a vanishingly small potential drop, the remaining 0.128 V was **not** dropped across the semiconductor/liquid interface. In fact, here the potential drop across the Helmholtz layer in solution was small (~1 mV) for the CdS/MV$^{2+/+}$ system, in accord with previous electrochemical observations with nondegenerately doped planar semiconductors.[105]

The manifestation of the remaining potential drop is most readily identified by considering an even more extreme case: a small semiconductor nanoparticle that is entirely undoped (Figure 5.16d). In this scenario, essentially no potential drop occurs within the semiconductor nanoparticle (i.e., the bands are flat throughout). Charge equilibration occurs by changing n and p throughout the *entire* volume of the semiconductor. That is, equilibration through charge transfer from the semiconductor into solution does not stop once the potential drop magnitude dictated by Equation 5.34 is reached. Instead, equilibration via heterogeneous charge transfer continues without inducing further increase to the electric field within the semiconductor—essentially doping the whole nanoparticle through

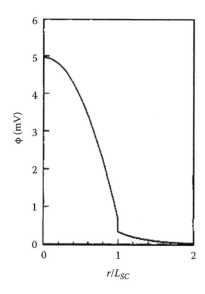

FIGURE 5.17 Potential profile (as a function of the distance from the particle center in units of r/L_{SC}) in a system of n-CdS nanoparticles in methylviologen redox solution. (From Liver, N. and Nitzan, A., *J. Phys. Chem.*, 96, 3366, 1992.)

the act of equilibration with the solution. An alternative viewpoint is that the whole nanoparticle is being oxidized/reduced.[103] A variation of this concept has recently been proposed as a means to dope controllably small semiconductor nanoparticle films, that is, by immersion into a solution with a sufficiently positive/negative redox couple.[102,106]

For optimal energy conversion with a semiconductor nanoparticle suspended in solution, it is not clear which of the four scenarios (large or small? doped or undoped?) in Figure 5.16 represents the optimal scenario. If any steady-state illumination results in some remaining internal potential drop as in Figure 5.15a through c, an electrostatic barrier, however slight, for heterogeneous charge transfer could serve as a limit on a surface redox reaction (in this case preventing electrons from reaching the surface). A recent study suggests that any field-effect diminution of the heterogeneous reactions at a small semiconductor nanoparticle is minimal,[97] but it is not clear if this is universally true in all cases (especially for larger semiconductor nanoparticles). Since the rate of both surface redox reactions is coupled at the open-circuit condition, the observed activity of the semiconductor nanoparticle will be less than optimal. In the case embodied by Figure 5.16d, the nanoparticle should be pushed into high-level injection conditions.

Despite the numerous reports modeling the operation of semiconductor nanoparticles under illumination, the implications of high-level injection in a suspended semiconductor nanoparticle have not been rigorously explored. To date, detailed analyses of a semiconductor nanoparticle acting as a photocatalyst assume certain properties or ideal conditions or even neglect outright pertinent features for the sake of simplicity.[44,98,99,103,107,108] The possibility for misinterpretation of the rate-limiting step in a suspended semiconductor nanoparticle photocatalyst in the absence of knowledge on the heterogeneous charge-transfer rate constants and electron–hole pair generation and recombination has been discussed previously.[90] Thus, a comprehensive analysis using all available kinetic information on the system would be highly informative to show what recombination process(es) is most critical, perhaps aiding subsequent refining chemistries/material syntheses. As a corollary, such studies should also elucidate the maximum quasi-Fermi-level spitting available in an illuminated nanoparticle, thereby addressing a lingering question in the literature about which dissolved species in solution can be photoreduced and photoxidized.[109,110] These issues are paramount for the continued advancement of semiconductor photocatalysts for solar fuel generation.

ACKNOWLEDGMENTS

This work is supported by the National Science Foundation under Grant No. DMR-1054303.

REFERENCES

1. Hamann, T. W.; Gstrein, F.; Brunschwig, B. S.; Lewis, N. S., Measurement of the driving force dependence of interfacial charge-transfer rate constants in response to Ph changes at N-ZnO/H$_2$O interfaces. *Chem. Phys.* **2006**, *326*, 15–23.
2. Koval, C. A.; Howard, J. N., Electron transfer at semiconductor electrode-liquid electrolyte interfaces. *Chem. Rev.* **1992**, *92*, 411–433.
3. Howard, J. N.; Koval, C. A., Kinetics of reduction of dimethylferrocenium ion in acetonitrile at nearly ideal regions of N-tungsten diselenide electrodes. *Anal. Chem.* **1994**, *66*, 4525–4531.
4. Fajardo, A. M.; Lewis, N. S., Rate constants for charge transfer across semiconductor-liquid interfaces. *Science* **1996**, *274*, 969–972.
5. Ma, Z.; Zaera, F., Organic chemistry on solid surfaces. *Surf. Sci. Rep.* **2006**, *61*, 229–281.
6. Grimm, R. L.; Bierman, M. J.; O'Leary, L. E.; Strandwitz, N. C.; Brunschwig, B. S.; Lewis, N. S., Comparison of the photoelectrochemical behavior of H-terminated and methyl-terminated Si(111) surfaces in contact with a series of one-electron, outer-sphere redox couples in Ch$_3$cn. *J. Phys. Chem. C* **2012**, *116*, 23569–23576.
7. Johansson, E.; Boettcher, S. W.; O'Leary, L. E.; Poletayev, A. D.; Maldonado, S.; Brunschwig, B. S.; Lewis, N. S., Control of the Ph-dependence of the band edges of Si(111) surfaces using mixed methyl/allyl monolayers. *J. Phys. Chem. C* **2011**, *115*, 8594–8601.

8. Mukherjee, J.; Peczonczyk, S.; Maldonado, S., Wet chemical functionalization of III–V semiconductor surfaces: Alkylation of gallium phosphide using a grignard reaction sequence. *Langmuir* **2010**, *26*, 10890–10896.

9. Hagfeldt, A.; Gratzel, M., Light-induced redox reactions in nanocrystalline systems. *Chem. Rev.* **1995**, *95*, 49–68.

10. Archer, M. D.; Nozik, A. J., *Nanostructured and Photoelectrochemical Systems for Solar Photon Conversion*. Imperial College Press: London, U.K., 2008; Vol. 3, p. 760.

11. Khaselev, O.; Turner, J. A., A monolithic photovoltaic-photoelectrochemical device for hydrogen production via water splitting. *Science* **1998**, *280*, 425–427.

12. Tributsch, H.; Gerischer, H., The use of semiconductor electrodes in the study of photochemical reactions. *Ber. Bunsen-Ges. Phys. Chem.* **1969**, *73*, 850–854.

13. Morrison, S. R.; Freund, T., Chemical role of holes and electrons in ZnO photocatalysis. *J. Chem. Phys.* **1967**, *47*, 1543–1551.

14. Fujishima, A.; Honda, K., Electrochemical photolysis of water at a semiconductor electrode. *Nature* **1972**, *238*, 37–38.

15. Rodman, S.; Spitler, M. T., Determination of rate constants for dark current reduction at semiconductor electrodes using ZnO single-crystal microelectrodes. *J. Phys. Chem. B* **2000**, *104*, 9438–9443.

16. Riha, S. C.; Devries Vermeer, M. J.; Pellin, M. J.; Hupp, J. T.; Martinson, B. F., Hematite-based photooxidation of water using transparent distributed current collectors. *ACS Appl. Mater. Interf.* **2013**, *5*, 360–367.

17. Klahr, B. M.; Martinson, A. B. F.; Hamann, T. W., Photoelectrochemical investigation of ultrathin film iron oxide solar cells prepared by atomic layer deposition. *Langmuir* **2011**, *27*, 461–468.

18. Lin, Y.; Xu, Y.; Mayer, M. T.; Simpson, Z. I.; McMahon, G.; Zhou, S.; Wang, D., Growth of P-type hematite by atomic layer deposition and its utilization for improved solar water splitting. *J. Am. Chem. Soc.* **2012**, *134*, 5508–5511.

19. Fonash, S., *Solar Cell Device Physics*, 2nd edn. Academic Press: Burlington, MA, 2010.

20. Sze, S. M. N., Kwok, K., *Physics of Semiconductor Devices*. 3rd edn., John Wiley & Sons: Hoboken, NJ, 2007.

21. Tiedje, T.; Yablonovitch, E.; Cody, G. D.; Brooks, B. G., Limiting efficiency of silicon solar cells. *IEEE Trans. Electr. Dev.* **1984**, *31*, 711–716.

22. Swanson, R. M. *Point-Contact Silicon Solar Cells*; EPRI AP-2859; Electric Power Research Institute: Stanford, CA, 1983.

23. Artero, V.; Chavarot-Kerlidou, M.; Fontecave, M., Splitting water with cobalt. *Angew. Chem., Int. Ed.* **2011**, *50*, 7238–7266.

24. Concepcion, J. J.; Jurss, J. W.; Brennaman, M. K.; Hoertz, P. G.; Patrocinio, A. O. T.; Iha, N. Y. M.; Templeton, J. L.; Meyer, T. J., Making oxygen with ruthenium complexes. *Acc. Chem. Res.* **2009**, *42*, 1954–1965.

25. Geletii, Y. V.; Yin, Q. S.; Hou, Y.; Huang, Z. Q.; Ma, H. Y.; Song, J.; Besson, C. et al., Polyoxometalates in the design of effective and tunable water oxidation catalysts. *Isr. J. Chem.* **2011**, *51*, 238–246.

26. Sadakane, M.; Steckhan, E., Electrochemical properties of polyoxometalates as electrocatalysts. *Chem. Rev.* **1998**, *98*, 219–237.

27. Subramanian, V.; Wolf, E.; Kamat, P. V., Semiconductor-metal composite nanostructures. To what extent do metal nanoparticles improve the photocatalytic activity of TiO_2 films? *J. Phys. Chem. B* **2001**, *105*, 11439–11446.

28. Dasgupta, N.; Liu, C.; Andrews, S.; Prinz, F. B.; Yang, P. D., Atomic layer deposition of platinum catalysts on nanowire surfaces for photoelectrochemical water reduction. *J. Am. Chem. Soc.* **2013**, *135*, 12932–12935.

29. Dai, P.; Xie, J.; Mayer, M. T.; Yang, X.; Zhan, J.; Wang, D., Solar hydrogen generation by silicon nanowires modified with platinum nanoparticle catalysts by atomic layer deposition. *Angew. Chem.* **2013**, n/a-n/a.

30. Gerischer, H.; Heller, A., Photocatalytic oxidation of organic-molecules at TiO_2 particles by sunlight in aerated water. *J. Electrochem. Soc.* **1992**, *139*, 113–118.

31. Cao, L. Y.; White, J. S.; Park, J. S.; Schuller, J. A.; Clemens, B. M.; Brongersma, M. L., Engineering light absorption in semiconductor nanowire devices. *Nat. Mater.* **2009**, *8*, 643–647.

32. Hu, L.; Chen, G., Analysis of optical absorption in silicon nanowire arrays for photovoltaic applications. *Nano Lett.* **2007**, *7*, 3249–3252.

33. Kupec, J.; Stoop, R. L.; Witzigmann, B., Light absorption and emission in nanowire array solar cells. *Opt. Expr.* **2010**, *18*, 27589–27605.

34. Lagos, N.; Sigalas, M. M.; Niarchos, D., The optical absorption of nanowire arrays. *Photon. Nanostruct.* **2011**, *9*, 163–167.

35. Muskens, O. L.; Rivas, J. G.; Algra, R. E.; Bakkers, E.; Lagendijk, A., Design of light scattering in nanowire materials for photovoltaic applications. *Nano Lett.* **2008**, *8*, 2638–2642.

36. Yu, E. T.; van de Lagemaat, J., Photon management for photovoltaics. *MRS Bull.* **2011**, *36*, 424–432.

37. Chitambar, M. J.; Wen, W.; Maldonado, S., Discrete-contact nanowire photovoltaics. *J. Appl. Phys.* **2013**, *114*, 174503(11).

38. Van Roosbroeck, W., Theory of the flow of electrons and holes in germanium and other semiconductors. *Bell Labs Tech. J.* **1950**, *29*, 560–607.

39. Crank, J., *The Mathematics of Diffusion*, 2nd edn. Oxford Science Publications: Oxford, U.K., 1976.

40. Shockley, W.; Read, W. T., Statistics of the recombinations of holes and electrons. *Phys. Rev.* **1952**, *87*, 835–842.

41. Hall, R. N., Recombination processes in semiconductors. *Proc. IEE B. Electron. Commun. Eng.* **1959**, *106*, 923–931.

42. Lewis, N. S., An analysis of charge transfer rate constants for semiconductor/liquid interfaces. *Annu. Rev. Phys. Chem.* **1991**, *42*, 543–580.

43. Kumar, A.; Santangelo, P. G.; Lewis, N. S., Electrolysis of water at $SrTiO_3$ photoelectrodes: Distinguishing between the statistical and stochastic formalisms for electron-transfer processes in fuel-forming photo-electrochemical systems. *J. Phys. Chem.* **1992**, *96*, 834–842.

44. Albery, W. J.; Bartlett, P. N., The transport and kinetics of photogenerated carriers in colloidal semiconductor electrode particles. *J. Electrochem. Soc.* **1984**, *131*, 315–325.

45. Vanmaekelbergh, D.; Erne, B. H.; Cheung, C. W.; Tjerkstra, R. W., On the increase of the photocurrent quantum efficiency of gap photoanodes due to (photo)anodic pretreatments. *Electrochim. Acta* **1995**, *40*, 689–698.

46. Marin, F. I.; Hamstra, M. A.; Vanmaekelbergh, D., Greatly enhanced sub-bandgap photocurrent in porous gap photoanodes. *J. Electrochem. Soc.* **1996**, *143*, 1137–1142.

47. Price, M. J.; Maldonado, S., Macroporous N-gap in nonaqueous regenerative photoelectrochemical cells. *J. Phys. Chem. C* **2009**, *113*, 11988–11994.

48. Maiolo, J. R.; Atwater, H. A.; Lewis, N. S., Macroporous silicon as a model for silicon wire array solar cells. *J. Phys. Chem. C* **2008**, *112*, 6194–6201.

49. Ritenour, A. J.; Levinrad, S.; Bradley, C.; Cramer, R. C.; Boettcher, S. W., Electrochemical nanostructuring of N-Gaas photoelectrodes. *ACS Nano* **2013**, *7*, 6840–6849.

50. Erne, B. H.; Mathieu, C.; Vigneron, J.; Million, A.; Etcheberry, A., Porous anodic etching of $P-Cd_{1-x}Zn_xTe$ studied by photocurrent spectroscopy. *J. Electrochem. Soc.* **2000**, *147*, 3759–3767.

51. Maiolo, J. R.; Kayes, B. M.; Filler, M. A.; Putnam, M. C.; Kelzenberg, M. D.; Atwater, H. A.; Lewis, N. S., High aspect ratio silicon wire array photoelectrochemical cells. *J. Am. Chem. Soc.* **2007**, *129*, 12346–12347.

52. Goodey, A. P.; Eichfeld, S. M.; Lew, K.-K.; Redwing, J. M.; Mallouk, T. E., Silicon nanowire array photoelectrochemical cells. *J. Am. Chem. Soc.* **2007**, *129*, 12344–12345.

53. Foley, J. M.; Price, M. J.; Feldblyum, J. I.; Maldonado, S., Analysis of the operation of thin nanowire photoelectrodes for solar energy conversion. *Energy Environ. Sci.* **2012**, *5*, 5203–5220.

54. Simpkins, B. S.; Mastro, M. A.; Eddy, C. R.; Pehrsson, P. E., Surface depletion effects in semiconducting nanowires. *J. Appl. Phys.* **2008**, *103*.

55. Calahorra, Y.; Ritter, D., Surface depletion effects in semiconducting nanowires having a non-uniform radial doping profile. *J. Appl. Phys.* **2013**, *114*, 124310.

56. Chia, A. C. E.; LaPierre, R. R., Analytical model of surface depletion in Gaas nanowires. *J. Appl. Phys.* **2012**, *112*, 063705.

57. Hagedorn, K.; Forgacs, C.; Collins, S.; Maldonado, S., Design considerations for nanowire heterojunctions in solar energy conversion/storage applications. *J. Phys. Chem. C* **2010**, *114*, 12010–12017.

58. Garnett, E. C.; Tseng, Y. C.; Khanal, D. R.; Wu, J. Q.; Bokor, J.; Yang, P. D., Dopant profiling and surface analysis of silicon nanowires using capacitance-voltage measurements. *Nat. Nanotechnol.* **2009**, *4*, 311–314.

59. Putnam, M. C.; Filler, M. A.; Kayes, B. M.; Kelzenberg, M. D.; Guan, Y. B.; Lewis, N. S.; Eiler, J. M.; Atwater, H. A., Secondary ion mass spectrometry of vapor-liquid-solid grown, Au-catalyzed, Si Wires. *Nano Lett.* **2008**, *8*, 3109–3113.

60. Radovanovic, P. V., Nanowires: Keeping track of dopants. *Nat. Nanotechnol.* **2009**, *4*, 282–283.

61. Schubert, E. F., *Doping in III-V Semiconductors*. Cambridge University Press: Cambridge, U.K., **2005**; p. 606.

62. Perea, D. E.; Hemesath, E. R.; Schwalbach, E. J.; Lensch-Falk, J. L.; Voorhees, P. W.; Lauhon, L. J., Direct measurement of dopant distribution in an individual vapour-liquid-solid nanowire. *Nat. Nanotechnol.* **2009**, *4*, 315–319.

63. Krylyuk, S.; Davydov, A. V.; Levin, I., Tapering control of Si nanowires grown from $SiCl_4$ at reduced pressure. *ACS Nano.* **2011**, *5*, 656–664.

64. Fahrenkrug, E.; Gu, J.; Jeon, S.; Veneman, A.; Goldman, R. S.; Maldonado, S., Room-temperature epitaxial electrodeposition of single-crystalline germanium nanowires at the wafer scale from an aqueous solution. *Nano Lett.* **2014**, *14*, 847–852.

65. Nebol'sin, V. A.; Shchetinin, A. A., Role of surface energy in the vapor-liquid-solid growth of silicon. *Inorg. Mater.* **2003**, *39*, 1050–1055.

66. Lewis, N. S., A quantitative investigation of the open-circuit photovoltage at the semiconductor/liquid interface. *J. Electrochem. Soc.* **1984**, *131*, 2496–2503.

67. Green, M. A., Limits on the open-circuit voltage and efficiency of silicon solar cells imposed by intrinsic auger processes. *IEEE Trans. Electron Dev.* **1984**, *31*, 671–678.

68. Kerr, M. J.; Cuevas, A.; Campbell, P., Limiting efficiency of crystalline silicon solar cells due to Coulomb-enhanced auger recombination. *Prog. Photovolt.: Res. Appl.* **2003**, *11*, 97–104.

69. Vossier, A.; Hirsch, B.; Gordon, J. M., Is auger recombination the ultimate performance limiter in concentrator solar cells? *Appl. Phys. Lett.* **2010**, *97*.

70. Fitch, A.; Strandwitz, N. C.; Brunschwig, B. S.; Lewis, N. S., A comparison of the behavior of single crystalline and nanowire array ZnO photoanodes. *J. Phys. Chem. C* **2013**, *117*, 2008–2015.

71. Santori, E. A.; Maiolo, J. R., III; Bierman, M. J.; Strandwitz, N. C.; Kelzenberg, M. D.; Brunschwig, B. S.; Atwater, H. A.; Lewis, N. S., Photoanodic behavior of vapor-liquid-solid-grown, lightly doped, crystalline Si microwire arrays. *Energy Environ. Sci.* **2012**, *5*, 6867–6871.

72. Forbes, M. D. E.; Lewis, N. S., Real-time measurements of interfacial charge transfer rates at silicon/liquid junctions. *J. Am. Chem. Soc.* **1990**, *112*, 3682–3683.

73. Gstrein, F.; Michalak, D. J.; Royea, W. J.; Lewis, N. S., Effects of interfacial energetics on the effective surface recombination velocity of Si/liquid contacts. *J. Phys. Chem. B* **2002**, *106*, 2950–2961.

74. Swanson, R. M.; Beckwith, S. K.; Crane, R. A.; Eades, W. D.; Kwark, Y. H.; Sinton, R. A.; Swirhun, S. E., Point-contact silicon solar cells. *IEEE Trans. Electr. Dev.* **1984**, *31*, 661–664.

75. Sinton, R. A.; Kwark, Y.; Swirhun, S.; Swanson, R. M., Silicon point contact concentrator solar-cells. *IEEE Electr. Dev. Lett.* **1985**, *6*, 405–407.

76. Swanson, R. M., Point-contact solar-cells—Modelling and experiment. *Solar Cells* **1986**, *17*, 85–118.

77. Sinton, R. A.; Swanson, R. M., Design criteria for Si point-contact concentrator solar-cells. *IEEE Trans. Electr. Dev.* **1987**, *34*, 2116–2123.

78. Gruenbaum, P. E.; Sinton, R. A.; Swanson, R. M., Light-induced degradation at the silicon/silicon dioxide interface. *Appl. Phys. Lett.* **1988**, *52*, 1407–1409.

79. King, R. R.; Sinton, R. A.; Swanson, R. M. In *Front and Back Surface Fields for Point-Contact Solar Cells, Photovoltaics Specialists Conference, IEEE*: 1988; pp. 538–544.

80. King, R. R.; Sinton, R. A.; Swanson, R. M., Doped surfaces in one sun, point-contact solar cells. *Appl. Phys. Lett.* **1989**, *54*, 1460–1462.

81. Mulligan, W. P.; Terao, A.; Smith, D. D.; Verlinden, P. J.; Swanson, R. M., Development of chip-size silicon solar cells. In *Conference Record of the Twenty-Eighth IEEE Photovoltaic Specialists Conference—2000*, **2000**; pp. 158–163.

82. Swanson, R. M. In *Approaching the 29% Limit Efficiency of Silicon Solar Cells*, Photovoltaic Specialists Conference, IEEE: **2005**; pp. 889–894.

83. Mrowetz, M.; Balcerski, W.; Colussi, A. J.; Hoffmann, M. R., Oxidative power of nitrogen-doped TiO_2 photocatalysts under visible illumination. *J. Phys. Chem. B* **2004**, *108*, 17269–17273.

84. Linsebigler, A. L.; Lu, G.; Yates, J. T., Photocatalysis on TiO_2 surfaces: Principles, mechanisms, and selected results. *Chem. Rev.* **1995**, *95*, 735–758.

85. Hoffmann, M. R.; Martin, S. T.; Choi, W. Y.; Bahnemann, D. W., Environmental applications of semiconductor photocatalysis. *Chem. Rev.* **1995**, *95*, 69–96.

86. Burr, T. A.; Seraphin, A. A.; Werwa, E.; Kolenbrander, K. D., Carrier transport in thin films of silicon nanoparticles. *Phys. Rev. B* **1997**, *56*, 4818–4824.

87. O'Regan, B.; Gratzel, M., A low-cost, high-efficiency solar cell based on dye-sensitized colloidal TiO_2 films. *Nature* **1991**, *353*, 737–740.

88. Sambur, J. B.; Novet, T.; Parkinson, B. A., Multiple exciton collection in a sensitized photovoltaic system. *Science* **2010**, *220*, 63–66.

89. Beard, M. C.; Luther, J. M.; Semonin, O. E.; Nozik, A. J., Third generation photovoltaics based on multiple exciton generation in quantum confined semiconductors. *Acc. Chem. Res.* **2013**, *46*, 1252–1260.

90. Kesselman, J. M.; Shreve, G. A.; Hoffmann, M. R.; Lewis, N. S., Flux-matching conditions at TiO_2 photoelectrodes—Is interfacial electrontransfer to O_2 rate limiting in the TiO_2 catalyzed photochemical degradation of organics. *J. Phys. Chem.* **1994**, *98*, 13385–13395.

91. Brus, L. E., A simple model for the ionization potential, electron affinity, and aqueous redox potentials of small semiconductor crystallites. *J. Chem. Phys.* **1983**, *79*, 5566–5571.

92. Huang, J.; Stockwell, D.; Huang, Z.; Mohler, D. L.; Lian, T., Photoinduced ultrafast electron transfer from CdSe quantum dots to re-bipyridyl complexes. *J. Am. Chem. Soc.* **2008**, *130*, 5632–5633.

93. Bang, J. H.; Kamat, P. V., CdSe quantum dot–fullerene hybrid nanocomposite for solar energy conversion: Electron transfer and photoelectrochemistry. *ACS Nano* **2011**, *5*, 9421–9427.

94. Scholz, F.; Dworak, L.; Matylitsky, V. V.; Wachtveitl, J., Ultrafast electron transfer from photoexcited CdSe quantum dots to methylviologen. *ChemPhysChem* **2011**, *12*, 2255–2259.

95. Holmes, M. A.; Townsend, T. K.; Osterloh, F. E., Quantum confinement controlled photocatalytic water splitting by suspended CdSe nanocrystals. *Chem. Commun.* **2012**, *48*, 371–373.

96. Sabio, E. M.; Chamousis, R. L.; Browning, N. D.; Osterloh, F. E., Photocatalytic water splitting with suspended calcium niobium oxides: Why nanoscale is better than bulk—A kinetic analysis. *J. Phys. Chem. C* **2012**, *116*, 3161–3170.

97. Zhao, J.; Holmes, M. A.; Osterloh, F. E., Quantum confinement controls photocatalysis: A free energy analysis for photocatalytic proton reduction at CdSe nanocrystals. *ACS Nano* **2013**, *7*, 4316–4325.

98. Goossens, A., Potential distribution in semiconductor particles. *J. Electrochem. Soc.* **1996**, *143*, L131–L133.

99. Bisquert, J.; Garcia-Belmonte, G.; Fabregat-Santiago, F., Modelling the electric potential distribution in the dark in nanoporous semiconductor electrodes. *J. Solid State Electrochem.* **1999**, *3*, 337–347.

100. Mocatta, D.; Cohen, G.; Schattner, J.; Millo, O.; Rabani, E.; Banin, U., Heavily doped semiconductor nanocrystal quantum dots. *Science* **2011**, *332*, 77–81.

101. Choi, W. Y.; Termin, A.; Hoffmann, M. R., Role of metal-ion dopants in quantum-sized TiO_2—Correlation between photoreactivity and charge-carrier recombination dynamics. *J. Phys. Chem.* **1994**, *98*, 13669–13679.

102. Engel, J. H.; Alivisatos, A. P., Postsynthetic doping control of nanocrystal thin films: Balancing space charge to improve photovoltaic efficiency. *Chem. Mater.* **2013**, *26*, 153–162.

103. Liver, N.; Nitzan, A., Redox properties of small semiconductor particles. *J. Phys. Chem.* **1992**, *96*, 3366–3373.

104. Bard, A. J.; Faulkner, L. R., *Electrochemical Methods: Fundamentals and Applications.* 2nd edn.; John Wiley & Sons, Inc.: 2001.

105. Royea, W. J.; Kruger, O.; Lewis, N. S., Frumkin corrections for heterogeneous rate constants at semiconducting electrodes. *J. Electroanal. Chem. Inter. Electrochem.* **1997**, *438*, 191–197.

106. Koh, W.-k.; Koposov, A. Y.; Stewart, J. T.; Pal, B. N.; Robel, I.; Pietryga, J. M.; Klimov, V. I., Heavily doped N-type PbSe and PbS nanocrystals using ground-state charge transfer from cobaltocene. *Sci. Rep.* **2013**, *3*, 2004.

107. Gerischer, H., Photocatalysis in aqueous-solution with small TiO_2 particles and the dependence of the quantum yield on particle-size and light intensity. *Electrochim. Acta* **1995**, *40*, 1277–1281.

108. Curran, J. S.; Lamouche, D., Transport and kinetics in photoelectrolysis by semiconductor particles in suspension. *J. Phys. Chem.* **1983**, *87*, 5405–5411.

109. Leland, J. K.; Bard, A. J., Photochemistry of colloidal semiconducting iron oxide polymorphs. *J. Phys. Chem.* **1987**, *91*, 5076–5083.

110. Leland, J. K.; Bard, A. J., Electrochemical investigation of the electron-transfer kinetics and energetics of illuminated tungsten oxide colloids. *J. Phys. Chem.* **1987**, *91*, 5083–5087.

6 Single-Molecule Nanoelectronics

Stuart Lindsay

CONTENTS

6.1 INTRODUCTION

Molecular electronics has its genesis in the idea that electron transfer in molecules, so central to chemical reactions, might form a basis for building electronic components on a molecular scale. At the time that Aviram and Ratner proposed a mechanism for a molecular diode,[1] this was far from an obvious idea, and one with enormous potential impact, because the scale of electronic devices was measured in microns or even millimeters. In the four decades that have passed since the publication of their paper, the electronics industry has come a long way. At the time of writing, Intel is making silicon devices with dimensions down to 11 nm, achieved with optical lithography! Nowadays, less mileage is to be gained from the relative smallness of single-molecule electronic devices. Molecular films play critical roles in the electronics industry, in displays, flexible electronics, and sensors. But that is not the focus of this chapter. Here, we are concerned with measurements on and applications of *single-molecule* devices. The historic goal of molecular electronics has been to make better, smaller transistors for denser computer circuits. There are three main reasons why molecular electronic circuits by themselves seem unlikely to bring revolutionary new capabilities to the computer industry. The first, alluded to above, is that conventional semiconductor manufacturing is not so far from achieving molecular scale densities. The second is that the promise of 3D fabrication, offered by molecular self-assembly,[2] is also being overtaken by advances in conventional device architectures that now enable remarkably complex stacking and interconnection of circuit elements. And the third is the oft-cited problem of the statistics of small numbers—How small can a semiconductor element be and still contain one doping atom? As we shall see, the same considerations apply to molecules, with polarization heterogeneities and fluctuations replacing doping as the problem.

So what will be the disruptive applications of single-molecule devices? My prejudice is that they will be found in applications that interface electronics directly to chemistry and biochemistry.[3] The primary economic driver of this must surely be the transformation of medicine into a quantitative, data-driven science. Computers have the power to digest and to analyze the enormous volumes of data required to characterize and analyze the detailed molecular composition of individual humans, but we lack the tools to make the needed chemical measurements, except, as of recently, in the field of genomics. Indeed, the first glimpse of this new age comes from the Human Genome Project. Started in 1990 and completed in 2003, the first human genome cost nearly \$3 billions. Genomes are now sequenced in weeks at costs of hundreds of dollars. So why would medicine want anything more? There are several reasons for seeking a vast increase in our ability to quantify a much wider range of analytes and do so at the single-molecule level:

- The genome is only, at its core, a biological *operating system* with many possible implementations. While genomic defects play a role in a number of diseases, the course of a disease, response to external stresses, and the detailed development of an animal lie in the way that gene expression is controlled. Repeated sequencing of a patient's genome is unlikely to map the impact of a particular course of treatment. Mapping of gene expression, and, more importantly, the proteins that are produced, is much more likely to be valuable.
- Genes and their products are only rather indirectly related. One of the surprises of the Human Genome Project was the discovery that humans have only about 20,000 genes (an order of magnitude less than many plants). Yet these genes are used to make millions of different types of proteins, through variations in splicing of exons and/or posttranslational modifications of various amino acid residues.
- The success of genomics owes much to the polymerase chain reaction (PCR), the repeated use of strand separation, followed by synthesis of complementary double helices. Using PCR, tiny amounts of nucleic acid can be amplified and concentrated easily. There is no such universal amplification reaction for proteins, metabolites, sugars, and lipids. For this reason, it is quite likely that we know of only a tiny fraction of the proteins in existence,[4] the concentration of most of them lying below current detection limits.

Thus, the applications I will focus on here are biochemical and analytical in nature.

The capability for making reliable electronic measurements on single molecules is relatively recent. The Weiss group used scanning probe microscopy to examine the transmission of single molecules inserted into an otherwise inert monolayer in 1996.[5] In the same year, Tao demonstrated a direct link between electronic transmission and redox activity, thereby connecting electrochemistry and molecular electronics.[6] Mark Reed's group demonstrated electrical contacts to single molecules in 1997.[7] Statistical analysis was applied to self-assembled junctions to determine accurate values for single-molecule conductance in 2001.[8] (See the paper by Morita and Lindsay for a discussion of contact-size effects.[9]) The same type of statistical analysis was applied to molecular break junction data to yield the first experimental single-molecule conductance values in good agreement with first-principle calculations.[10] Since then, this method has been used by the Columbia group for quantitative studies of structural control of electron transport, by, for example, ring rotations in oligomeric aromatic molecules.[11]

Despite the pioneering work of Tao, the number of studies that utilize electrochemical potential control in single-molecule measurements is limited, probably because it is complicated. However, it is also often essential. For example, voltage switching of oligo(phenylene ethynelene) molecules was shown to be a consequence of redox processes,[12] and negative differential resistance (NDR) in ferrocenylundecanethiol was shown to be a consequence of reactions with oxygen.[13] Indeed, this important element is not yet fully incorporated into the two developments focused on at the end of this chapter, so we will end this introduction with some comments on electrochemistry and those applications.

We will discuss the use of carbon nanotubes (CNTs) as field-effect transistors (FETs) for single-molecule sensing, an application that requires submersion of the CNT in electrolytes.[14] The stabilizing effect of using a reference electrode (RE) was noted recently.[15]

The second application that will be discussed here is recognition tunneling (RT), a technique we believe to have the capability of being a universal single-molecule analytical platform.[16]

It involves the application of significant biases (ca. 0.5 V) across nanometer scale gaps in electrolyte. Electrochemical processes undoubtedly play a significant role, one that is being analyzed at the time of writing.

Finally, plans to build any type of polymer sequencing platform around the methods discussed here will involve ion current manipulation of analytes. Indeed, ion current measurements make an excellent analytical tool by themselves[17] and have already been applied to single-molecule DNA sequencing.[18,19] There is much yet to be done in integrating electronic and ionic techniques, a topic for a future review.

The present chapter begins with a simplified discussion of electron transport in molecular junctions, a topic treated more thoroughly by Schmickler elsewhere in this volume. We then turn to electrostatic and electrochemical methods for gating molecular conductance and then discuss fluctuations in molecular junctions. We end by showing how measurement of these fluctuations yields new insights. Direct readout of enzyme fluctuations in a nanotube FET characterizes kinetics at the single-molecule level, while the fluctuations of analytes trapped in a molecular junction may prove to be as a valuable a diagnostic tool as vibrational spectroscopy. This type of analysis is applicable to single molecules and therefore can work with tiny amounts of sample, offering a way round the roadblock presented by the lack of a PCR-like amplification for analytes other than nucleic acids.

6.2 GETTING FROM ONE ELECTRODE TO ANOTHER VIA A MOLECULE

Vacuum tunneling: If two clean metal surfaces are placed a distance d (generally less than 1 nm) apart in a vacuum, with a bias V applied between them (Figure 6.1a), a current will flow owing to electron tunneling between occupied states in the left electrode and unoccupied states in the right electrode. If the bias energy, electron volt, is small compared to the work function, φ (a large bias is shown in the figure for clarity), the tunneling current decays with d according to

$$i = i_0 e^{-2\sqrt{\frac{2m\phi}{\hbar^2}}d} \tag{6.1}$$

where
m is the mass of an electron
i_0 is the current at zero distance

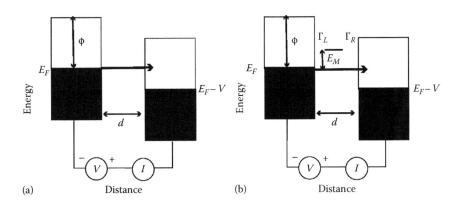

FIGURE 6.1 Direct electron tunneling between left and right electrodes (a) in vacuum and (b) as modified by the presence of an intermediate state.

For a single quantum channel where the contact diameter is comparable to, or less than, the Fermi wavelength, i_0 is given by

$$i_0 = \frac{2e^2}{h}V = \frac{V}{12.9k\Omega} \tag{6.2}$$

where e is the electronic charge. For gold electrodes ($\phi \sim 5$ eV), this current decays as $e^{-22.8d}$ or about a factor 8×10^9 per nanometer. Thus, a vacuum tunnel gap of 1 nm between a pair of electrodes of diameter less than a Fermi wavelength has a resistance on the order of 10^{14} Ω.

How big can a vacuum gap be for tunnel current to be detected? This depends upon the band-width required and the input capacitance of the device recording the current. For example, if this input capacitance was 1 pF (a small value, quite hard to achieve) and the required bandwidth, Δf, was 5 kHz, then the current-to-voltage converter equivalent resistance could not be larger than $\sim 3 \times 10^7$ Ω (calculated from $\omega RC = 1$ at the -3 dB point of 5 kHz). The associated Johnson noise, calculated from

$$\langle i \rangle_{RMS} = \sqrt{\frac{4kT\Delta f}{R}} \tag{6.3}$$

is a little over 1 pA, setting a lower bound on the current that can be measured. This detection limit can be lowered further by reducing the bandwidth and/or the input capacitance of the measuring system. We see from Equation 6.2 that for a bias of 1 V, the maximum possible current at contact is ~ 0.1 mA, falling to a less than a picoampere when the gap is increased by about 0.8 nm (Equation 6.1). In reality, 1 V is already too much for many molecule–electrode systems where molecules may be irreversibly reduced or oxidized at even quite small biases. Thus, it might appear that tunneling could only occur with rather small molecules in the gap.

Tunneling with an allowed state in the gap: Figure 6.1b illustrates a pair of electrodes with a state owing to the presence of a molecule at an energy E_M above the Fermi energy. This state could be a lowest unoccupied molecular orbital (LUMO) (as illustrated) or a highest occupied molecular orbital (HOMO). Whether or not it is occupied is irrelevant because tunneling electrons do not occupy the states that mediate tunneling. Of course, the presence of a molecule will place many states in the gap, but only those closest to the Fermi energy dominate the tunneling. This is with the caveat that the states that mediate tunneling have a nonzero overlap integral (Γ_L or Γ_R) with states in the left (L) or right (R) electrodes. With the molecular potential present in the gap, the effective barrier is reduced to approximately E_M. For example, if E_M is 1 eV, then the decay constant at a gold electrode surface is reduced from 22.8 nm^{-1} to about 10 nm^{-1}. This allows the gap to be increased to nearly 2 nm and still yield picoampere signals.

Tunneling through a double wall containing an intermediate state, an energy E_M away from the Fermi energy (as illustrated in Figure 6.1b) was first investigated by Breit and Wigner, and com-bining their formula for the energy dependence of the transmission coefficient with the quantum conductance of a point contact (the inverse of the quantum resistance in Equation 6.2) leads to the following expression for the conductance of the system[20]:

$$G = \frac{4e^2}{h} \frac{\Gamma_L \Gamma_R}{E_M^2 + \left(\Gamma_L + \Gamma_R\right)^2} \tag{6.4}$$

The appearance of E_M in this formula shows how tunneling rates can be sensitive to chemistry. More realistically, the current is calculated using a sum carried out over all states that connect the left and right electrodes (Green's function propagator). This sum includes an energy-dependent denomina-tor similar to that in Equation 6.4. Zwolak and Di Ventra used propagator methods together with a density functional analysis of the states of a DNA molecule to calculate the tunnel current as each of the four DNA bases was placed between closely spaced tunneling electrodes.[21,22] The calculated currents changed with the nucleobase that sat between the electrodes, suggesting that passing a single-stranded DNA molecule through a tunnel gap would allow the sequence to be read out.

The Breit–Wigner formula points to an interesting effect of having a molecular level lie exactly at the Fermi energy. When $E_M \rightarrow 0$ in Equation 6.4, the formula reduces to

$$G = \frac{4e^2}{h} \frac{\Gamma_L \Gamma_R}{\left(\Gamma_L + \Gamma_R\right)^2} \qquad (6.5)$$

One remarkable consequence appears to be that if the molecule lies in the middle of the barrier so that $\Gamma_L = \Gamma_R$, the conductance remains equal to the quantum of conductance (e^2/h) *independent* of tunneling distance. This is not the case in reality, because Equation 6.4 is only valid when the overlap integrals themselves are close to 1. $\Gamma_L = \Gamma_R \sim 1$ corresponds to metallic bonding where electrons are indeed delocalized. For weaker coupling (normally the case in metal–molecule systems), we have to consider charging of the molecular state, with consequent change in its Coulomb energy.

Tunneling with charging of an intermediate state: Figure 6.2a illustrates a more complex process in which an electron occupies a particle in the tunnel gap. In this case, the particle is taken to change its energy purely as a consequence of charging, a process referred to as the Coulomb blockade. (Charged molecules usually reorganize in response to charging, as does the solvent around them, leading to further changes in the energy of the charged state.) For a rigid particle in a vacuum, the charging energy, $e\Delta V$, is controlled by its (fixed) value of capacitance C according to

$$\Delta V = \frac{e}{2C} \qquad (6.6)$$

where both sides of the equation have been divided by e to give the energy in volts. Inspecting Figure 6.2a, we see that even if the particle has allowed states at the Fermi energy, no current can flow until enough excess energy is available to charge the particle. Thus, for an applied voltage $V < \Delta V$, no current flows. The capacitance of a sphere of radius a is given by $4\pi\varepsilon\varepsilon_0 a$, so for a 2 nm diameter metal particle, this blockade voltage is $\Delta V \sim 0.7$ V. Below 0.7 V, no current flows, leading to the Coulomb blockade. It is repeated as each additional electron is added to the particle, forming steps in the differential conductance. This system of steps is referred to as a Coulomb staircase. In the case of a molecule, as opposed to a rigid particle in vacuum, reorganization plays a major role.

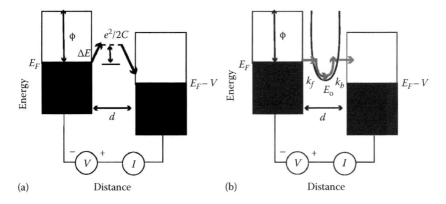

FIGURE 6.2 Tunneling with occupation of an intermediate state. (a) Coulomb blockade. The charging energy, $e^2/2C$, is provided thermally when $kT > \Delta E$. (b) Tunneling via a first reduction step followed by an oxidation with the intermediate state stabilized by solvent and internal reorganization.

Tunneling with sequential reduction and oxidation of a molecule: The reorganization of solvent and the internal structure of a molecule when it is charged leads to a significant change in its electronic energy, referred to as the reorganization energy, λ. This can be on the order of 1 eV. This is the origin of an interesting paradox. Since the energy of the charged and uncharged states is different, Fermi's golden rule states that there can be no tunneling between them. The resolution of this paradox lies in fluctuations, as pointed out by Marcus in his groundbreaking paper.[23] There is a crossing of the free-energy landscapes of the reactants and products at a transition state, $\Delta G^* = \lambda/4$ (for the case of a parabolic electron–phonon coupling). Thus, when thermal excitation takes the appropriate electronic state of the reactants $\lambda/4$ V above their ground state energy, excited states of reactants and products are degenerate, electron transfer can occur by tunneling, and the system can then decay into the reduced products. Inserting this transition state energy into the standard expression for reaction rate, and including a free-energy difference between reactants and products, ΔG^0, leads to the Marcus formula for the probability of an electron transfer reaction as a function of the free-energy difference and the reorganization energy:

$$P(\Delta) = \frac{1}{\pi\sqrt{4k_BT\lambda}} \exp\left[-\frac{(\Delta G^0 + \lambda)^2}{4k_BT\lambda} \right] \tag{6.7}$$

Thus far, our discussion has been confined to electron transfer between molecules. The case of a molecule trapped between a pair of electrodes (Figure 6.2b) is more complicated. Rather than considering the intersection of two sets of states that depend quadratically on a reaction coordinate, one must now consider the interaction of the molecular states with a *continuum* of states in the electrode. This problem leads to a sum over all available states, weighted both by the Fermi–Dirac occupation number and the Marcus probability, as was shown first by Chidsey.[24] Chidsey provided expressions for the backward and forward transfer rates for reduction and subsequent oxidation of a molecule at an electrode as a function of bias (relative to an RE), λ, the reorganization energy, and the formal potential of the molecule. When a second electrode is placed close to the molecule, generation of a net current requires two steps: first, the reduction of the molecule by the left electrode; second, the oxidation (of the charged state) by the right electrode. Ignoring other types of back reaction (e.g., a return to a neutral state by electron transfer back into the left electrode), the probability of these events occurring in the right sequence is proportional to

$$P \sim e^{\left[\frac{\lambda + (V - E_0)}{kT}\right]^2} \times P \sim e^{\left[\frac{\lambda - (V - E_0)}{kT}\right]^2} \tag{6.8}$$

where
   E_0 is the formal potential for the electron transfer reaction (potential where forward and backward rates are equal)
   V is expressed as volts relative to that potential

Equation 6.8 has an important consequence: the tunneling current is a maximum at the formal potential. This was first shown experimentally by Tao.[6]

One of the unresolved questions in *tunneling* experiments on molecules is whether or not the process is electrochemical, mediated by states of the molecule that are occupied by an electron, or true tunneling where molecular states act only as a virtual intermediate in the process (in effect, the electron *sees* the potential of the molecule, which reduces the barrier for tunneling, but no charge accumulates on the molecule).

Single-molecule electrochemical signals, owing to true redox processes, can be observed when a very fast electron transfer molecule can shuttle rapidly between electrodes, alternately being oxidized and reduced. The time required for a molecule to transfer from one electrode to the other is on the order of

$$t \sim \frac{d^2}{2D} \qquad (6.9)$$

where
   d is the gap size
   D is the diffusion constant

Taking $D = 5 \times 10^{-6}$ cm^2/s and $d = 10$ nm leads to $t \sim 10^{-7}$ s or a current of a picoampere. Electron transfer rates will be different from this diffusion-limited rate when the molecule is fixed in place (i.e., in an attached monolayer), but tunneling rates into and out of the molecule can be high if the coupling is significant. Single-molecule electrochemical currents generated by this charge shuffling were demonstrated by Fan and Bard.[25] The key point is that these currents are only observable when both electrodes are poised near the formal potential (cf., Equation 6.8), decaying very rapidly as the electrodes are moved away from the formal potential (Equation 6.8). In contrast, tunnel currents are only indirectly related to the formal potential, in as much as the molecular orbital energies (i.e., the value of E_M) are related to the electrochemical properties.[26] Furthermore, the dependence of tunnel current on E_M is only quadratic (Equation 6.4). Thus, we will call a process *tunneling* if it appears to be insensitive to changes in the redox properties of the molecule.

6.3   GATING MOLECULES

Can molecules be turned *on* and *off* in a manner that enables a single-molecule transistor? A schematic energy diagram of a gated device is shown in Figure 6.3a. This is similar to the schemes shown in Figure 6.2, but now an additional gate electrode is placed near the molecule. In a normal FET, the gate field shifts the conduction or valence band so as to alter the carrier density in a channel. In the case of a molecular transistor, levels that mediate tunnel or hopping transport would be shifted by an amount proportional to the gate voltage, αV_G, where α reflects the geometry that sets up the electric field at the molecule, and α is generally much less than 1. This shift then moves levels that mediate transport out of the energy window between the two Fermi levels, reducing direct transmission or hopping conductance. The problem with this scheme for a molecular transistor is illustrated in Figure 6.3b, which shows the physical layout of a gated molecule (represented by the zigzag line). In general, the drain–source voltage, V_{DS}, cannot be very different from the gate voltage, V_G, because both are limited by dielectric breakdown of the components of the transistor. In order for the gate field to influence molecular levels, the dielectric layer that separates the gate from the molecule must have a thickness t that is much less than the length of the molecule, L. For a molecule of ~1 nm length, this imposes unrealistic constraints on the dielectric layer.

   This limitation does not rule out molecular gating, as pointed out by Ghosh et al.[27] The local electric field may cause a molecule to alter its geometry in the gap, altering the strength of the contacts to the electrode, as illustrated in Figure 6.3c. Another possibility is an internal conformation change caused by the gate field as illustrated in Figure 6.3d. This can lead to a change in transmission through the molecule.

   Gating of single-molecule devices was demonstrated convincingly by Song et al.[28] In this case, electron tunneling spectroscopy showed clear evidence of a shift in molecular orbital energy with gate voltage. At first sight, this result seems to be at odds with the geometrical arguments of Ghosh et al. Song et al. used benzene dithiol, a tiny molecule for which an atomic-scale gate would be required. Nonetheless, Song et al. present convincing evidence that molecular levels move in response to changing gate bias, based on tunneling spectroscopy. One explanation may lie with the possibility of a polarizable layer surrounding the active molecule. This could be composed of the surrounding molecules in the monolayer and/or residue from the solvents used in the deposition process. Such a polarizable film would confine the potential drop to the Debye layer,

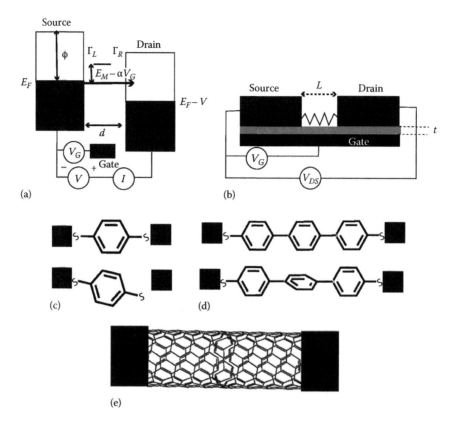

FIGURE 6.3 Gating transmission through molecular junctions: (a) a gate below the molecule applies an electric field that shifts the molecular orbital energy by some fraction, α, of the applied gate voltage, V_G. (b) Geometry of a molecular (zigzag line) FET showing how the source–drain separation, L, needs to be bigger than the gate dielectric thickness, t, for gating to work. Gating mechanisms that can produce large on/off ratios even when the shift in a molecular orbital energy is small: (c) electric field–induced bond breaking at a surface and (d) rotation of aromatic rings in a multiring molecule. (e) A nanowire or CNT can be long enough to satisfy the requirement that $L \gg t$.

resulting in a much larger field close to the gated molecule (see the discussion of electrochemical gating later).

A second solution to the geometry problem is simply to make the molecule longer. A good example of this is the CNT FET (Figure 6.3e). Since CNTs can be many microns (even centimeters) long, gating is not a problem and many types of CNT FET have been fabricated since the early work of the Dekker and coworkers.[29]

Yet a third approach lies in using the electric double layer in an electrolyte as a very thin gate. The thickness of the Debye double layer, in the limit of low potential and salt concentration, is given by

$$\ell_D = \sqrt{\frac{2\pi\varepsilon\varepsilon_0 k_B T}{C_0 Z^2 e^2}} \tag{6.10}$$

where
   C_0 is the concentration of the electrolyte solution
   Z is the ion charge in units of the electronic charge, e

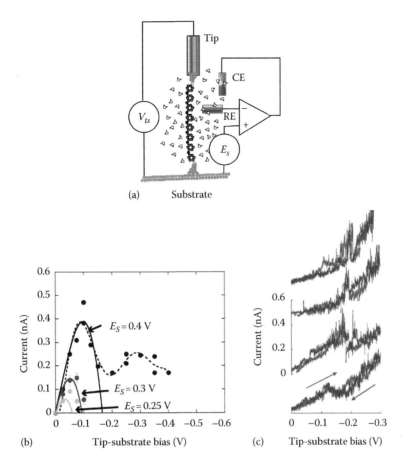

(a) Substrate

(b) Tip-substrate bias (V)

$E_S = 0.4$ V

$E_S = 0.3$ V

$E_S = 0.25$ V

(c) Tip-substrate bias (V)

FIGURE 6.4 Electrochemical gating: (a) A CE is used to polarize the surface of a substrate where a molecule (an aniline octamer is shown) is located inside the intense electric field within the Debye double layer. The molecule is connected to a top electrode (usually an STM tip) and biased by the voltage V_{ts}. (b) Current–voltage characteristics, averaged from thousands of individual I–V_{ts} curves for an oligo aniline molecule under potential control (E_S is the surface potential with respect to an Ag/AgCl reference). Oxidation from the leucoemeraldine (neutral) form of the molecule to the conducting emeraldine salt occurs at $E_S = 0.15$ V. A second wave near 0.6 V reflects a further oxidation to the fully oxidized (and insulating) pernigraniline form. Held at $E_S = 0.4$ V, the conducting molecule oxidizes again to the less-conductive pernigraniline form when the tip bias exceeds about 0.15 V, evidence of a strong local field induced by the tip. This results in a fall in current with increasing voltage, an effect called NDR. The solid lines are fits to the data (dots) obtained at several values of E_S showing how bulk electrochemical data account for these I–V_{ts} curves. (c) Examples of I–V_{ts} curves for single molecules at $E_S = 0.4$ V (red sweeping up, blue sweeping down). There is considerable molecule to molecule variation, but the complex shape of the I–V_{ts} curve is largely reproducible on successive sweeps for a given molecule.

For a monovalent salt, $\ell_D \sim 0.3\,\text{nm}/\sqrt{C_0}$. Thus, in 1 M salt solution, the Debye length is about the thickness of a water molecule. The Debye model does not hold well in this limit where the Debye length approaches the atomic scale, but these numbers make the point that a very high electric field can be generated at the surface of a conductor by placing it in contact with an electrolyte. Tao first demonstrated electrochemical gating of single molecules containing an $Fe^{2/3}$ redox center.[6] Another example of such an experiment is illustrated in Figure 6.4. Here, a molecule was trapped between an STM probe and a conducting surface. The surface was polarized with respect to an RE using a counter electrode (CE) and potentiostat (Figure 6.4a). The figure is not to scale—the reference and CEs reside well outside the double layer, while the molecule at least partially penetrates the double layer. Thus, the field felt by the molecule is a combination of the field, E_{TS}, generated by the tip-to-substrate potential difference and the field owing to polarization of the metal surface, E_S. The molecules used in these experiments were a series of oligoanilines.[30] The leucoemeraldine (neutral) form of the molecule is insulating. Oxidation results in the conducting emeraldine salt. A further oxidation to the fully oxidized pernigraniline results in a material that is again an insulator. Thus, if the surface is poised above the first oxidation potential of the molecule (0.2 V vs. Ag/AgCl), increasing the tip bias at first increases the current in this conducting state, as expected. However, as the local field is increased still further, the onset of the second oxidation process starts, and the current now *falls* as the tip–substrate bias is increased (Figure 6.4b). This phenomenon, of falling current with increasing bias, is known as negative differential resistance (NDR) and it can be exploited to make switches and oscillators. If the surface potential is reduced down toward the first oxidation at 0.2 V, the NDR peak shifts to lower bias, following exactly the predictions of a simple electrochemical model of the process (lines in Figure 6.4b). The data points in this panel were obtained from the average of many single-molecule measurements. What happens at the single-molecule level? Figure 6.4c shows a series of current–voltage traces obtained with a single molecule trapped in the tunnel gap. The peak in conductance is clearly seen in each trace, but it is not reproducible from one molecule to the next. Note, however, that the traces for each molecule are somewhat reproduced on the upward (red) and downward (blue) sweeps. Thus, the molecule to molecule variation is not noise. Rather, it reflects differences in the static polarization surrounding each molecule. This illustrates one of the limitations of single-molecule electronic devices: reproducible characteristics require reproducible polarization at the atomic scale.

Measurements of this sort are useful for probing the role of electrochemical processes in determining the molecular electronic response of devices. For example, switching in oligo(phenylene ethynylene) molecules was shown to coincide with oxidation of the molecues.[12] In another example, NDR in ferrocene molecules was shown to be the result of electrochemical reactions involving dissolved oxygen.[13] In yet another example, electron transport of carotene wires was investigated as a function of potential, showing how enhanced transport in the oxidized state allows carotene to act as a protective fuse in photosynthesis.[31]

6.4   FLUCTUATIONS IN MOLECULAR TUNNEL JUNCTIONS

Thus far, we have described transport processes in terms of a static electron transmission factor for tunneling events. Activated transport (the Coulomb blockade at high temperatures or hopping via oxidation and reduction of molecules—Figure 6.2a and b) is of course driven by thermal energy and thus stochastic in nature. But fluctuations also play an important role in molecular tunnel junctions. In 2001, Donhauser et al. reported that STM images of oligo(phenylene ethynylene) molecules spontaneously switched *on* and *off*, appearing and disappearing from a given location in an STM image, attributed, in this case, to fluctuations in the internal structure of the molecule.[32] However, Ramachandran et al.[33] showed that an alkane molecule, functionalized with a thiol as an upper contact[8] (Figure 6.5a), also showed stochastic switching of the STM image (Figure 6.5b, i–iii). Since this alkane molecule has no obvious internal rearrangements that could turn its transmission on and off, the implication is that the contact to the metal surface is labile. The metal–sulfur bond is strong, but on surfaces like gold, the process of forming a thiol–gold bond results in a very labile gold atom, moving in a liquid-like manner on the gold surface.[33] These fluctuations were exploited by Haiss et al. as a novel way to make measurements of single-molecule conductance.[34] They poised an STM probe in a fixed position over a monolayer of alkane dithiols and just in contact with the top thiol group. They recorded current in the junction as the probe was left in place, observing distinct telegraph noise (on–off switching in the current-time recording, cf., Figure 6.5d). These fluctuations were interpreted as breaks in the contact to the molecule, as shown in Figure 6.5c. This method has been extended to molecular complexes, using an STM probe functionalized with a DNA base contacting a monolayer of the complementary DNA base

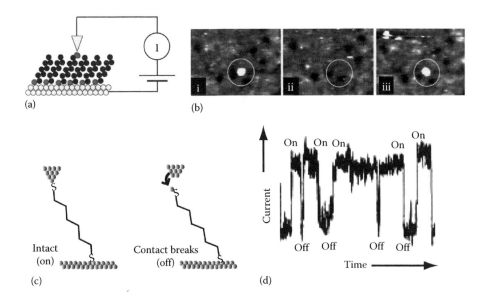

FIGURE 6.5 Contact fluctuations in molecular junctions: (a) An STM probe was scanned over an octane-thiol monolayer containing inserted octanedithiol molecules (red dot at each end). (b) The top thiol contact appears as a bright spot in the STM image (i) but will spontaneously disappear (ii) only to reappear in a later scan at the same point (iii). (c) The same phenomenon can be seen when the probe is stopped over octanedithiol molecules. The contact turns *on* and *off* on a timescale of tens of milliseconds to produce random telegraph signals in the current as shown in panel (d).

on a substrate.[35] An example of a current-time recording from this system is shown in Figure 6.5d. The current switches between two levels, interpreted in this work as stochastic making and breaking of the contacts between the molecule and the metal, or breaking of the hydrogen bonds that hold the complex together.

We will revisit measurements of this type when we discuss RT. But first, we turn to measurements of molecular dynamics based on random telegraph signals generated as molecules interact with CNTs.

6.5 PROTEIN DYNAMICS MEASURED DIRECTLY IN A NANOTUBE CIRCUIT

CNTs can act as single-molecule sensors as a consequence of strong scattering at a defect site,[36] in effect inserting a large series resistance into an otherwise low-resistance circuit. Thus, despite the fact that only a small part of the CNT is modified by a single molecule, large changes in conductance are possible. Chemically modified CNTs have found applications as biosensors,[37] gas sensors,[38] and electrochemical sensors.[39] The sensitivity of CNT FETs is a function of their chirality, metallic tubes being insensitive to gating. For this reason, semiconductor nanowires have an advantage as sensors, because they can be prepared with a controlled band structure.[40]

A single-molecule sensing FET is illustrated in Figure 6.6a. A single-walled CNT is contacted by source and drain electrodes and immersed in an electrolyte solution. The tube is polarized by means of a gate electrode, consisting of a highly doped p^{++} region underneath the device. Application of a bias to the gate with respect to the CNT results in formation of electric double layers on the gate and the CNT. A large (larger than the Debye length) molecule attached to the CNT displaces ions in the double layer and creates a localized charge defect that causes a change in the conductance of the CNT.[41]

Devices like this have been used to detect binding events on CNTs[42] and semiconductor nanowires.[43] More recently, CNT FETs have been used to study enzyme motion at the single-molecule

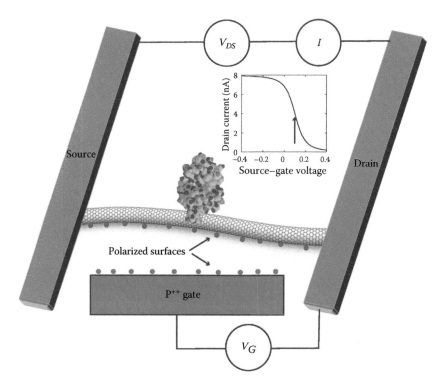

FIGURE 6.6 FET for measuring enzyme fluctuations: A CNT is contacted by source and drain electrodes and functionalized with an enzyme (here lysozyme) modified with a pyrene ring that pi stacks on the CNT. The system is submerged in electrolyte and polarized by a gate electrode under the device with the device poised at the point of maximum gate sensitivity (arrow on inset).

level,[14,15] a remarkable feat that we describe further here. These measurements rely on tethering an enzyme to the CNT using a modified amino acid residue. The modification is made on a part of the protein that undergoes large functional motions and consists of coupling an inserted cysteine residue to a pyrene ring that is absorbed strongly onto the surface of the CNT. Importantly, this region also contains at least one charged residue, so that, as the protein fluctuates, the charged residue moves in and out of the Debye layer where the enzyme is tethered, modulating the local electric field and hence the conductance of the CNT. The measured fluctuation timescales are found to be in agreement with single-molecule FRET experiments. However, the electronic measurements have the advantage of giving additional data points at very short timescales (shorter than the acquisition times for single-molecule optical signals) and out to very long timescales (longer than the photobleaching times in FRET experiments).

Choi et al. used a cysteine-modified lysozyme, with the modification chosen to lie in a part of the enzyme that undergoes substantial motion over the catalytic cycle. By controlling the concentration of the enzyme and the incubation time, it proved relatively straightforward to attach a single molecule to the CNT FET, as illustrated in Figure 6.6. Lysozyme is an enzyme secreted in tears, breast milk, saliva, and mucus, and its function is to destroy bacterial cell walls by hydrolyzing peptidoglycans on their surface. Figure 6.7 shows a current-time trace recorded from a CNT FET functionalized with a single lysozyme molecule. The FET was biased at the point of maximum transconductance (dI_{DS}/dV_G—inset in Figure 6.6), and the current trace was filtered to remove slow $1/f$ noise fluctuations. When the peptidoglycan substrate was added (at $t=0$ in Figure 6.7), rapid fluctuations in current were immediately observed. An expanded trace (inset) shows that these fluctuations are random telegraph noise (RTS), consisting of jumps between two current levels. Two families of RTS were observed: rapid (ms) fluctuations that were shown to be a consequence of

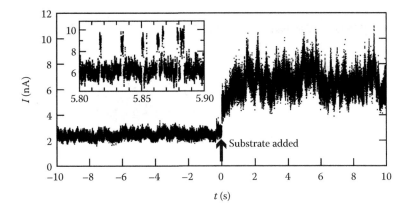

FIGURE 6.7 Showing the onset of current fluctuations ($t=0$) as the peptidoglycan substrate is added to the CNT FET device shown in Figure 6.6, inducing the structural fluctuations associated with enzyme motion. The inset is an expanded trace showing the random telegraph signals that result from the two-state motion of the enzyme between its resting and catalytic states with bound substrate (a $1/f$ noise slow background has been subtracted). The enzyme also displays much more rapid telegraph noise associated with unproductive bindings of the substrate. (Data are courtesy of P.G. Collins.)

abortive substrate-binding events and slower (10 ms) fluctuations that reflect completed hydrolysis of the peptidoglycan (these are the fluctuations illustrated in the inset in Figure 6.7). Each type of event produced an exponential distribution of *on* times for the RTS, a result expected for a Poisson occupation of a simple two-state system. The Poisson statistics were confirmed by showing that the ratio of the variance, σ^2, of the distribution to the sum of the squares of *on* times, $\sum_i t_i^2$, was unity, that is, $\sigma^2 = \sum_i t_i^2$. The current modulation measured in each experiment varied from CNT to CNT, but, using measured values of transconductance, the fluctuations in current, ΔI_{DS}, were translated into fluctuations in gate voltage, ΔV_G. An almost constant value was found for all the devices of $\Delta V_G \sim 0.2\,\text{V}$. This demonstrates the electrostatic origin of the current modulation. The magnitude of the field change is consistent with the known hinge motion of the enzyme and the consequent motion of charged residues near the attachment point to the CNT.

The quality of the recordings, and the long recording times, mean that rare events can be captured. A similar device was used by Sims et al.[15] to characterize a more complex process in which a substrate is modified in an adenosine triphosphate (ATP)-dependent manner, that is, the phosphorylation of a peptide target by cAMP-dependent protein kinase A (PKA). PKA binds ATP and the target residue (serine or threonine) simultaneously. In the presence of Mg^{++}, PKA adds a phosphate to the hydroxyl group of the target residue. This change in charge can result in transport of the modified protein into the cell nucleus where it acts as a signal to modify gene expression. PKA undergoes a transition from an open to a closed form on carrying out this reaction, returning to the open form when the reaction product is released. Sims et al. studied the phosphorylation of the protein Kemptide, observing signals like those shown in Figure 6.8a. The signals consisted of three current levels, one corresponding to the open state, an intermediate level corresponding to the binding of the Kemptide substrate and ATP, and a third level corresponding to the catalytically active (closed) state. Most of the time, catalysis occurred in a relatively short (millisecond) timescale. However, a small population of events (0.2%) occurred on much longer (tens of milliseconds) timescales (Figure 6.8b). These longer events act as a bottleneck in the overall reaction rates and contribute significantly to the overall reaction kinetics, even though they are rare. This study illustrates the new insights that can be obtained from studying single-molecule kinetics over a large range of timescales, a unique capability of these direct electronic measurements.

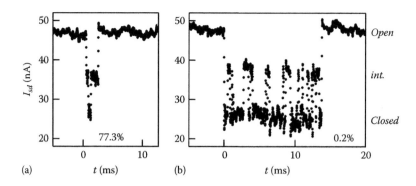

FIGURE 6.8 The FET transducer captures data more rapidly than single-molecule FRET and over longer timescales too. This means that rare events are readily captured. Panel (a) shows the three-level signal associated with the open state of PKA (top) level, the intermediate state where an ATP molecule and the protein substrate are captured, and the low-current (closed) state where catalysis occurs. Normally, the entire catalytic cycle, followed by release of the phosphorylated protein, takes place in about 2 ms. However, a bottleneck is presented by rare states (b, 0.2% of events) where many attempts are made before successful catalysis occurs. (Data are courtesy of P.G. Collins.)

A rather recent paper claims that single-molecule dynamics can also be measured with an enzyme wired directly into a tunnel junction.[44] Changes in current fluctuations were interpreted as the incorporation of different bases into an extending double helix by DNA polymerase. Among the questions raised by this paper are as follows: (1) The device was operated with 9 V applied across a 10 nm gap in aqueous buffer, apparently without electrolysis of water. (2) Superconducting contacts were used, yet the water remained in its liquid form, not a possibility with known superconductors. (3) The polymerase apparently operated efficiently at Mg concentrations many orders of magnitude below those normally required. (4) The polymerase processed its substrates in an absolutely regular manner, in sharp contrast to the stochastic dynamics expected (and described earlier for other enzymes). It will be interesting to see if such measurements can be repeated.

6.6 CHEMICAL ANALYSIS USING TUNNEL CURRENT FLUCTUATIONS: RECOGNITION TUNNELING

Electron tunneling is sensitive to the molecular species in the tunnel gap, as was pointed out in the discussion of Equation 6.4, and the basis of Di Ventra's proposal for DNA sequencing.[21,22] A practical demonstration of single-molecule analysis by tunneling is very challenging, for the following reasons:

- Handling of biomolecular targets (i.e., delivering them to the tunnel junction) requires that they be hydrated. Water molecules have a large HOMO–LUMO gap, in effect placing a large resistance in series with the target to be measured.
- Metal surfaces are contaminated with hydrocarbons outside of a UHV environment.[45] Thus, unless displaced by chemical groups that interact with the metals directly, the tunneling is dominated by the hydrocarbon contamination, which, with its wide HOMO–LUMO gap, further reduces chemical sensitivity.
- Even for analytes that interact chemically with metal surfaces (as the DNA bases do[46]), the tunnel junction must be made small enough so that direct bonding is possible between the electrodes and the target. In the case of the DNA bases, this requires sub–nanometer gaps. Indeed some chemical selectivity has been demonstrated for DNA bases adsorbed in a sub–nanometer tunnel junction.[47] However, the measured current distributions are very broad.

- Interactions with metal surfaces are not particularly specific, because of the near uniform electron density at a metallic surface. The combined effects of random absorption and thermal fluctuations make the signals difficult to interpret.

We have proposed a new approach we call RT to overcome many of these obstacles.[16]

RT readout is illustrated in Figure 6.9a. The metal electrodes are functionalized with recognition molecules (mercaptobenzamide in Figure 6.9a). These are strongly bound to the metals via the thiol groups, displacing contamination and giving the electrodes a chemically well-defined surface. The amide groups provide hydrogen bond donors and acceptors that can form weaker noncovalent bonds with target analytes, here shown as an adenine base in a DNA molecule. The bonding motifs of a next-generation reader molecule (imidazolecarboxamide) with all four DNA bases are shown in Figures 6.9b through e.[48] Binding of the recognition molecules to the target analyte displaces water molecules and ions from the tunneling path and also serves to *clamp* the molecule in place. The resulting signals are much like the *bond-breaking* fluctuation signals described earlier, with millisecond duration spikes separated from each other by several milliseconds.

We have called the process tunneling because the magnitude of the signals from all four DNA bases is very similar, despite the large difference in the oxidation potentials of the bases[49] and also because, in organic solvent (where comparison with simulations is feasible), the relative magnitudes of the signals are in agreement with density functional simulations.[50]

Clamping of the target analyte by the recognition molecules is remarkably effective, leading to a long lifetime for single-molecule complexes in the gap. Figure 6.10 illustrates a force spectroscopy experiment in which a three-body complex (consisting of a recognition molecule on the probe, a recognition molecule on the surface, and a trapped adenosine monophosphate) was pulled apart.[51] Single-molecule events were identified via the characteristic stretching of a PEG tether, as illustrated in the inset in Figure 6.10a and the bond-breaking force measured from the jump at the end of the stretch. Plotting the appropriate function of the loading rate against the bond-breaking force for various pulling rates (Figure 6.10b) gives a series of curves that can be fitted to yield the off rate at zero force, $K_{OFF}^{(0)}$, and the distance to the transition state for unbinding, X_{TS}. The solid lines are fits with $K_{OFF}^{(0)} = 0.3\,s^{-1}$ and $X_{TS} = 0.8\,nm$. The off rate implies that the complexes remain bound for 3 s at zero force. This would be unusual for a free complex in solution, but is expected for a complex confined by surfaces in a nanogap.[52]

This result casts an entirely different light on the origin of the random telegraph signals obtained from complexes like this. If the bonds last for times on the order of seconds, then how are the rapid (millisecond timescale) fluctuations in current to be explained, if they are not a consequence of bond breaking? One explanation is illustrated in Figure 6.11. The bonded complex (6.11a) will undergo thermal fluctuations in position, $X(t)$. However, a sinusoidal fluctuation (blue line, Figure 6.11b) gives rise to a series of spikes in current (red lines) because of the exponential dependence of tunnel current on distance. Simulations[51] of random thermal fluctuations in position (blue lines) and the consequent calculated tunnel currents (red lines) are shown in Figure 6.11c and d. Under a constant thermal driving energy, the more weakly bonded system (6.11d) moves further and therefore exhibits a much larger range of current fluctuations (note the change of scale on the figure). Thus, the spectrum of tunnel current fluctuations reflects the bonding in the gap and is diagnostic of the chemistry of the trapped molecule. Thus, the *spikes* reflect random thermal motion of a trapped molecule that remains bonded in the trap for periods that can be as long as seconds. This gives rise to clusters of signal spikes (cf. 6.13f) each one of which corresponds to a particular single-molecule binding event.

This is shown for DNA bases by experiments in which a functionalized STM probe was passed over DNA oligomers consisting of an alternating base sequence.[53] Figure 6.12b shows data obtained by passing a probe over a polymer consisting of alternating C and 5-methyl-C bases. The signals change in a reproducible way as the probe passes over the C or 5-methyl-C base. The duration of each signal burst was inversely proportional to scanning speed and corresponded to a spatial

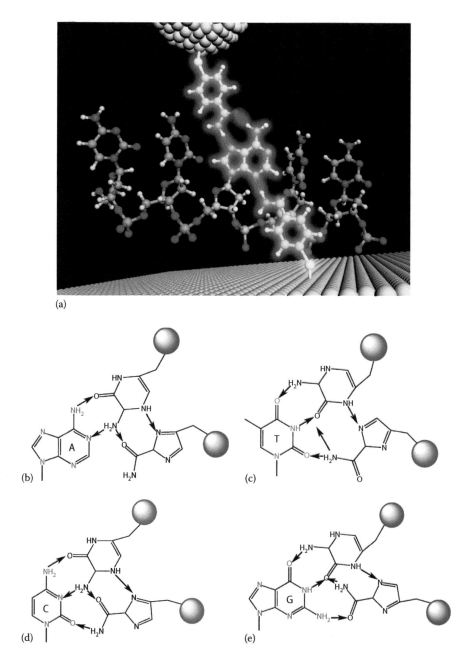

FIGURE 6.9 RT: (a) A probe functionalized with mercaptobenzamide molecules captures an A-base in a DNA oligomer on a flat electrode also functionalized with mercaptobenzamide. Each surface is coated with the mercaptobenzamide recognition molecules, but only the complex forming the shortest path between the electrodes (highlighted) contributes significantly to the signal. Hydrogen bonding for a pair of imidazolecarboxamide molecules capturing (b) adenine, (c) thymine, (d) cytosine, and (e) guanine.

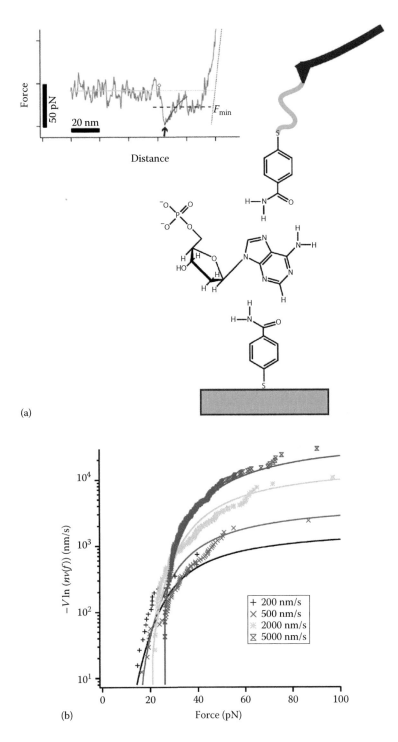

(a)

(b)

FIGURE 6.10 Bonds in an RT junction have a long lifetime: (a) AFM force spectroscopy of adenine trapped between a pair of benzamide reader molecules. The blue line represents a PEG tether that holds the reader molecule on the AFM probe, generating a characteristic force signal (red line in the inset) when a single-molecule pulling event occurs. (b) Data taken over a wide range of pulling velocities can be fitted (lines) using an off rate (at zero force) $K_{OFF}^{(0)} = 0.3\,\mathrm{s}^{-1}$ and a distance from the bound minimum to the top of the transition state of 0.8 nm.

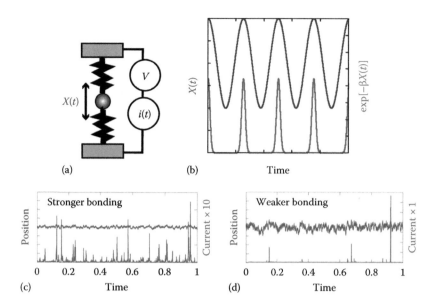

FIGURE 6.11 Generation of RT signals: (a) The trapped molecule is represented here as the sphere held by springs that represent the bonds to the recognition molecules. (b) A normal vibrational mode of the bound molecule (blue trace) generates current spikes because of the exponential dependence of tunnel current on distance. Though driven by random thermal noise, the spectrum of current fluctuations reflects bonding as seen by comparing the signals generated by (c) relatively strong bonds and (d) weaker bonds. Here, the blue line shows a simulated random thermal motion along one direction. The displacements are larger when bonds are weaker, leading to a much larger range of tunnel current spikes (red lines).

distance of about 0.3 nm, that is, about the size of stacked bases in the polymer. In this case, there is a clear correspondence between signal amplitude and chemical composition, the smaller-amplitude spikes coming from the 5-methyl cytosine. This turns out not to be a robust phenomenon, because the signal amplitude varies a lot from tunnel junction to tunnel junction. However, other aspects of the signal are less sensitive to the details of the tunnel junction. These include the pulse frequency, and pulse shape, as characterized by Fourier components. Using multiple features to characterize each pulse and training a machine learning algorithm made it possible to identify DNA bases with quite high accuracy based on just a *single* signal pulse.[53]

The complexity of the RT signals opens up the possibility that the technique may form the basis of a rather general type of single-molecule spectroscopy, and in an unpublished work, we have shown that many types of molecules may be identified. We illustrate this here with the example of identifying the amino acid enantiomers L- and D-asparagine.[54] Figure 6.13 shows RT signals obtained from L- and D-asparagine (6.13a and b, respectively). The insets in the upper right of each panel are expanded to show the complex pulse shapes of the signal spikes. It is often asked how achiral recognition molecules can bind chiral molecules in ways that reflect chirality. The point is that binding at a surface imposes constraints that allow for chiral recognition.[55]

Steps in the process of identifying the analytes based on single spikes are illustrated in Figures 6.13c through e. The particular parameters shown in these panels were identified as significant using a machine learning algorithm. Probability densities for the amplitude of a low-frequency Fourier component (0–2.7 kHz) and a high-frequency component (19–22 kHz) are shown in Figures 6.13c and d, respectively. The red lines give the distributions for L-asparagine, and the green lines give the distributions for D-asparagine. The distributions are quite similar, so that the usefulness of any one of these features is limited. For example, taking the distributions in 6.13d, and assigning all values of amplitude above 0.006 as belonging to L-asparagine spikes, would call them correctly 57% of the time, just a little better than random (50%). However, a remarkable

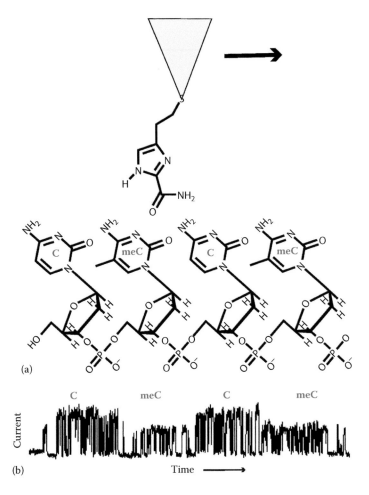

(a)

(b)

FIGURE 6.12 RT signals from DNA: (a) A probe, functionalized with imidazolecarboxamide, was scanned over a DNA oligomer consisting of alternating C and 5-methyl-C bases. The oligomer sat on top of a conducting substrate that was also functionalized with imidazolecarboxamide (not shown). (b) The tunnel current fluctuations changed in a characteristic manner as the probe was moved from base to base.

improvement occurs when the two distributions are used together, as shown in Figure 6.13e. This figure uses exactly the same data as Figures 6.13c and d, but now the two distributions are combined in one plot using a third (colored) dimension to plot the probability of L (red) or D (green). Calling all the points to the red of yellow L-asparagine now results in an 87% calling accuracy. This improved separation in higher dimension is an illustration of Cover's theorem, which states that the separability of pattern recognition problems improves as a function dimension.[56] This is a consequence of the nonlinear correlation between parameters (our analysis excludes linearly correlated signal features[54]).

The probability density plots just described are difficult, if not impossible to construct in yet higher dimensions. A simple way to classify points is to divide the space into two parts. In the case of the two-component plot in Figure 6.13e, this is done using a line (labeled *SVM*), where all the points to the left of the line are assigned as L-asparagine. This is less accurate than using the detailed probability density function, but readily extended to higher dimensions where the accuracy is further improved. The method for partitioning space in higher dimensions is called the support vector machine (SVM).[57] Trained on seven amino acids, it was able to assign each single spike to an accuracy of 95% or better.[54] Figure 6.13f shows an SVM analysis of a signal obtained from a

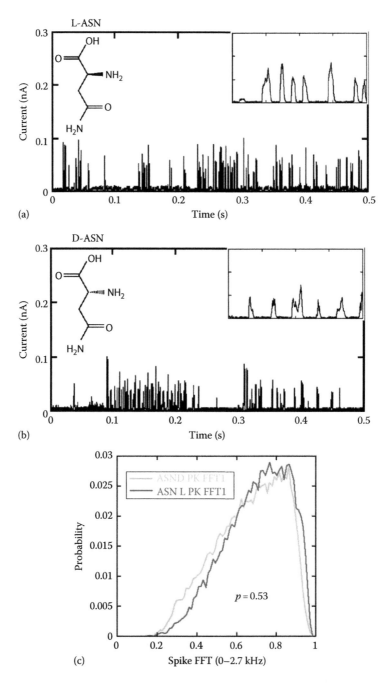

FIGURE 6.13 RT signals from amino acids, illustrated here for a pair of enantiomers, L-asparagine (a) and D-asparagine (b). The insets are expanded traces (20 ms) to show the complex shapes of individual spikes. Many features can be used to characterize these spikes: shown here are (c) distributions of normalized amplitudes of the Fourier components in a low frequency band (0–2.7 kHz). The red curves are distributions for L-asparagine and the green curves are distributions for D-asparagine. Thus D-asparagine (green) is more likely to have a low value for this component than L-asparagine (red), but the distributions are highly overlapped. So using these distributions to assign new data to either L- or D-asparagine would produce a correct answer only 53% of the time (random calling is 50%).   (*Continued*)

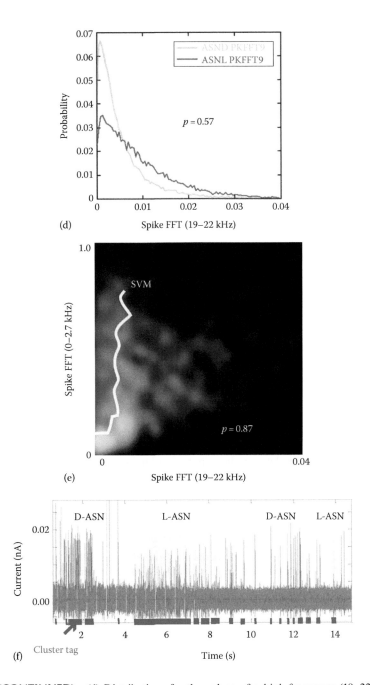

FIGURE 6.13 (CONTINUED) (d) Distributions for the values of a high-frequency (19–22 kHz) band of FFT amplitudes. Once again the distributions are highly overlapped, resulting in only a 57% calling accuracy. However, when these two distributions are plotted together (e) separation improves dramatically. In this 3D plot, the red points are for L-asparagine and the green points are for D-asparagine. This plot is color mixed, so overlap only occurs in the regions that are yellow. Assigning all the "more red" points to L-asparagine and assigning all the "more green" points to D-asparagine results in 87% correct calls. The support vector machine (SVM) makes an assignment by partitioning space to give the best separation (white line). (f) Machine calling of spikes from a mixed sample (1:1 L-asparagine (yellow) and D-asparagine (purple). Note how all the calls in a cluster are the same, consistent with the long trapping time for single molecules in the gap, and the notion that each cluster represents a distinct single-molecule binding event.

solution containing a 1:1 mixture of L- and D-asparagine. The SVM was trained on pure solutions, and here it calls all the spikes in a given cluster as the same analyte, either L-asparagine (yellow) or D-asparagine (purple). The clusters are identified with a Gaussian broadening algorithm[54] and marked by the red bars under the signal trace. The black spikes are signals not recognized by the SVM as trained on pure samples and probably represent new types of signals arising from interactions between the two types of molecule in the mixed solution.

It appears that RT may be a widely applicable analytical technique with the advantages of single-molecule detection and direct electronic readout, enabling its integration into solid-state electronic devices. We are investigating its application to DNA and protein sequencing, as well as methods for fabricating manufacturable tunneling devices. While the electrochemical behavior of these junctions is still under investigation, it is clear that the signals are not very sensitive to the potential of the metal surface, unless driven into a potential range where the layer of recognition molecules is destabilized.

6.7 CONCLUDING REMARKS

Twenty years ago, it was not clear what the conductance of a single molecule meant much less what means might be used to measure it reproducibly. The construction of a single-molecule transistor seemed a remote possibility. Now that these milestones are passed, we are beginning to see the first applications of single-molecule electronics, as opposed to basic experiments designed to elucidate mechanisms. Two of these applications, enzyme dynamics and analytical chemistry, were discussed here. Each may prove revolutionary, but much is to be done before these technologies become robust. In particular, these systems need much more study from an electrochemical standpoint if stable and reproducible devices are to become routine.

ACKNOWLEDGMENTS

This work owes much to many lab members, in particular Jin He, Peiming Zhang, Pei Pang, Brian Ashcroft, and Brett Gyarfas. I am grateful to Phil Collins for supplying original figures. This work was supported in part by a DNA sequencing technology grant from the NHGRI, HG 006323.

REFERENCES

1. Aviram, A. and Ratner, M. A. Molecular rectifiers. *Chem. Phys. Lett.* **29**, 277–283 (1974).
2. Rothemund, P. W. K. Folding DNA to create nanoscale shapes and patterns. *Nature* **440**, 297–302 (2006).
3. Lindsay, S. Biochemistry and semiconductor electronics—The next big hit for silicon? *J. Phys. Condensed Matter.* **24**, 164201–164208 (2012).
4. Archakov, A. I., Ivanov, Y. D., Lisitsa, A. V., and Zgoda, V. G. AFM fishing nanotechnology is the way to reverse the Avogadro number in proteomics. *Proteomics* **7**, 4–9 (2007).
5. Bumm, L. A. et al. Are single molecular wires conducting? *Science* **271**, 1705–1707 (1996).
6. Tao, N. Probing potential-tuned resonant tunneling through redox molecules with scanning tunneling microscopy. *Phys. Rev. Letts.* **76**, 4066–4069 (1996).
7. Reed, M. A., Zhou, C., Muller, C. J., Burgin, T. P., and Tour, J. M. Conductance of a molecular junction. *Science* **278**, 252–254 (1997).
8. Cui, X. D. et al. Reproducible measurement of single-molecule conductivity. *Science* **294**, 571–574 (2001).
9. Morita, T. and Lindsay, S. M. Determination of single molecule conductances of alkanedithiols by conducting-atomic force microscopy with large gold nanoparticles. *J. Am. Chem. Soc.* **129**, 7262–7263 (2007).
10. Xu, B. and Tao, N. J. Measurement of single-molecule resistance by repeated formation of molecular junctions. *Science* **301**, 1221–1223 (2003).
11. Venkataraman, L., Klare, J. E., Nuckolls, C., Hybertsen, M. S., and Steigerwald, M. L. Dependence of single-molecule junction conductance on molecular conformation. *Nature* **442**, 905–907 (2006).

12. He, J., Fu, Q., Lindsay, S. M., Ciszek, J. W., and Tour, J. M. Electrochemical origin of voltage-controlled molecular conductance switching. *J. Am. Chem. Soc.* **128**, 14828–14835 (2006).

13. He, J. and Lindsay, S. On the mechanism of negative differential resistance in ferrocenylundecanethiol self-assembled monolayers. *J. Am. Chem. Soc.* **127**, 11932–11933 (2005).

14. Choi, Y. et al. Single molecule lysozyme dynamics monitored by an electronic circuit. *Science* **335**, 319–323 (2012).

15. Sims, P. C. et al. Electronic measurements of single-molecule catalysis by cAMP-dependent protein kinase A. *J. Am. Chem. Soc.* **135**, 7861–7868 (2013).

16. Lindsay, S. et al. Recognition tunneling. *Nanotechnology* **21**, 262001–262013 (2010).

17. Branton, D. et al. Nanopore sequencing. *Nat. Biotechnol.* **26**, 1146–1153 (2008).

18. Cherf, G. M. et al. Automated forward and reverse ratcheting of DNA in nanopore at 5-Å precision. *Nat. Biotechnol.* **14**, 344–348 (2012).

19. Manrao, E. A. et al. Reading DNA at single-nucleotide resolution with a mutant MspA nanopore and phi29 DNA polymerase. *Nat. Biotechnol.* **30**, 349–353 (2012).

20. Lindsay, S. M. *Introduction to Nanoscience*. Oxford University Press, New York (2009).

21. Zwolak, M. and Di Ventra, M. Electronic signature of DNA nucleotides via transverse transport. *Nano Lett.* **5**, 421–424 (2005).

22. Zwolak, M. and Di Ventra, M. Physical approaches to DNA sequencing and detection. *Rev. Mod. Phys.* **80**, 141–165 (2008).

23. Marcus, R. A. On the theory of electron-transfer reactions. VI: Unified treatment for homogeneous and electrode reactions. *J. Phys. Chem.* **43**, 679–701 (1965).

24. Chidsey, C. E. D. Free energy and temperature dependence of electron transfer at the metal-electrolyte interface. *Science* **251**, 919–922 (1991).

25. Fan, F. R. F. and Bard, A. J. Electrochemical detection of single molecules. *Science* **267**, 871–874 (1995).

26. Mazur, U. and Hipps, K. W. Resonant tunneling bands and electrochemical reduction potentials. *J. Phys. Chem.* **99**, 6684–6688 (1995).

27. Ghosh, A. W., Rakshit, T., and Datta, S. Gating of a molecular transistor: Electrostatic and conformational. *Nano Lett.* **4**, 565–568 (2004).

28. Song, H. et al. Observation of molecular orbital gating. *Nature* **462**, 1039–1043 (2009).

29. Tans, S. J., Verscheuren, A. R. M., and Dekker, C. Room-temperature transistor based on a single carbon nanotube. *Nature* **393**, 49–52 (1998).

30. Chen, F. et al. A molecular switch based on potential-induced changes of oxidation state. *Nano Letts.* **5**, 503–506 (2005).

31. Visoly-Fisher, I. et al. Conductance of a biomolecular wire. *Proc. Nat. Acad. Sci. U. S. A.* **103**, 8686–8690 (2006).

32. Donhauser, Z. J. et al. Conductance switching in single molecules through conformational changes. *Science* **292**, 2303–2307 (2001).

33. Ramachandran, G. K. et al. A bond-fluctuation mechanism for stochastic switching in wired molecules. *Science* **300**, 1413–1415 (2003).

34. Haiss, W. et al. Measurement of single molecule conductivity using the spontaneous formation of molecular wires. *Phys. Chem. Chem. Phys.* **6**, 4330–4337 (2004).

35. Chang, S. et al. Tunnel conductance of Watson-Crick nucleoside-base pairs from telegraph noise. *Nanotechnology* **20**, 185102–185110 (2009).

36. Goldsmith, B. R. et al. Conductance-controlled point functionalization of single-walled carbon nanotubes. *Science* **315**, 77–81 (2007).

37. Allen, B. L., Kichambare, P. D., and Star, A. Carbon nanotube field-effect transistor-based biosensors. *Adv. Mat.* **19**, 1439–1451 (2007).

38. Trojanowicz, M. Analytical applications of carbon nanotubes: A review. *Trends Anal. Chem.* **25**, 480–489 (2006).

39. Gao, C., Guo, Z., Liua, J.-H., and Huang, X.-J. The new age of carbon nanotubes: An updated review of functionalized carbon nanotubes in electrochemical sensors. *Nanoscale* **4**, 1948–1963 (2012).

40. Law, M., Goldberger, J., and Yang, P. Semiconductor nanowires and nanotubes. *Annu. Rev. Mater. Res.* **34**, 83–122 (2004).

41. Besteman, K., Lee, J.-O., Wiertz, F. G. M., Heering, H. A., and Dekker, C. Enzyme-coated carbon nanotubes as single-molecule biosensors. *Nano Letts.* **3**, 727–730 (2003).

42. Star, A., Gabriel, J.-C. P., Bradley, K., and Grüner, G. Electronic detection of specific protein binding using nanotube FET devices. *Nano Letts.* **3**, 459–463 (2003).

43. Cui, Y., Wei, Q., Park, H., and Lieber, C. M. Nanowire nanosensors for highly sensitive and selective detection of biological and chemical species. *Science* **293**, 1289–1292 (2001).
44. Chen, Y. S. et al. DNA sequencing using electrical conductance measurements of a DNA polymerase. *Nat. Nanotechnol.* **8**, 452–458 (2013).
45. Smith, T. The hydrophilic nature of a clean gold surface. *J. Colloid Interface Sci.* **75**, 51–53 (1980).
46. Kimura-Suda, H., Petrovykh, D. Y., Tarlov, M. J., and Whitman, L. J. Base-Dependent competitive adsorption of single-stranded DNA on gold. *J. Am. Chem. Soc.* **125**, 9014–9015 (2003).
47. Tsutsui, M., Taniguchi, M., Yokota, K., and Kawai, T. Identification of single nucleotide via tunnelling current. *Nat. Nanotechnol.* **5**, 286–290 (2010).
48. Liang, F., Li, S., Lindsay, S., and Zhang, P. Synthesis, physicochemical properties, and hydrogen bonding of 4(5)-substituted-1H-imidazole-2-carboxamide, a potential universal reader for DNA sequencing by recognition tunneling. *Chem. Eur. J.* **18**, 5998–6007 (2012).
49. Palecek, E., Lukasova, E., Jelen, F., and Vojtiskova, M. Electrochemical analysis of polynucleotides. *Bioelectrochem. Bioenerg.* **8**, 497–506 (1981).
50. Chang, S. et al. Electronic signature of all four DNA nucleosides in a tunneling gap. *Nano Letts.* **10**, 1070–1075 (2010).
51. Huang, S. et al. Identifying single bases in a DNA oligomer with electron tunneling. *Nat. Nanotechnol.* **5**, 868–873 (2010).
52. Friddle, R. W., Noy, A., and James J. De Yoreoa. Interpreting the widespread nonlinear force spectra of intermolecular bonds. *Proc. Natl. Acad. Sci. U.S.A.* **109**, 13573–13578 (2012).
53. Chang, S. et al. Chemical recognition and binding kinetics in a functionalized tunnel junction. *Nanotechnology* **23**, 235101–235115 (2012).
54. Zhao, Y., Ashcroft, B., Zhang, P., Liu, H., Sen, S., Song, W., Im, J., Gyarfas, B., Manna, S., Biswas, S., Borges, C., and Lindsay S. Single molecule spectroscopy of amino acids and peptides by recognition tunneling. *Nature Nanotechnoloy* **9**, 466–473 (2014).
55. Lorenzo, M. O., Baddeley, C. J., Muryn, C., and Raval, R. Extended surface chirality from supramolecular assemblies of adsorbed chiral molecules. *Nature* **404**, 376–379 (2000).
56. Cover, T. M. Geometrical and statistical properties of systems of linear inequalities with applications in pattern recognition. *IEEE Trans. Elect. Comp.* **EC-14**, 326–334 (1965).
57. Chang, C.-C. and Lin, C.-J. LIBSVM: A library for support vector machines. *ACM Trans. Intell. Syst. Technol.* **2**, 27–52 (2011).

7 Electron Transport and Redox Reactions in Solid-State Molecular Electronic Devices

Richard McCreery

CONTENTS

7.1 INTRODUCTION AND SCOPE

A core concept in electrochemistry is activated electron transfer (ET) between an electrode, usually a conducting solid, and a redox system in the nearby solution. The vast literature on ET kinetics describes the importance of ET to chemical and biological processes, and the underlying phenomenon of coupling a chemical reaction to the flow of current is the basis of >$300 billion of annual gross national product. Chapter 1 in this volume describes ET in nanoscale systems, mainly at an interface between an electrode and an electrolyte solution. A widely studied example of ET kinetics of relevance to the current chapter deals with ET occurring through a self-assembled monolayer (SAM) to a redox molecule (e.g., ferrocene) bonded to the SAM at the solution interface,[1–3] as shown in Figure 7.1a. Such experiments stimulated a large research effort to understand the relationship between ET from a solid to a redox system through a nonredox active SAM and the thoroughly investigated dependence of ET within molecules, such as occur in biological metabolism and photosynthesis. An important conclusion about ET at electrodes as well as between two molecules in solution is the fact that the electrode and redox center (or the two redox centers in solution) need not be in direct contact to transfer electrons. It is possible, and quite common, for electrons to transfer through a SAM or intervening spacer (even a vacuum) by quantum mechanical tunneling, as described in Chapters 1 and 6 and in Section 7.3. For ET at both electrodes in solution and between redox centers within molecules, the ET rate depends on the driving force in terms of free energy and on the composition of the intervening solution or molecular structure. In addition to tunneling, ET in such systems may occur by other mechanisms, such as redox exchange, superexchange, and a sequence of ETs between distinct redox centers.[4]

The current chapter deals with ET in solid-state devices, which lack a solution but may involve electron transport, activated redox reactions, and/or ion motion. Extensive past investigations of transport in organic semiconductors, conducting polymers, and redox polymers provide important

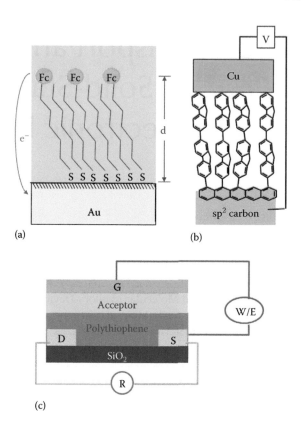

FIGURE 7.1 (a) Schematic of ET between ferrocene (Fc) bonded to Au and a SAM. (b) Two-terminal MJ with a conducting carbon substrate and Cu top contact. (c) Three-terminal molecular memory device with *read* (R) and *W/E* circuits shown.

precedents, in that they all involve electron transport across organic (usually) materials over distances of a few nanometer up to several micrometer. *Organic electronics* is focused mainly on organic films capable of electron transport, with commercial applications in organic light-emitting diodes (OLEDs) and organic field-effect transistors (OFETs). An active area of electrochemistry of the 1980s and 1990s considered ET through polymers such as poly(vinylferrocene) by a series of redox exchange reactions between ferrocene centers in the polymer.[5–8] Such experiments involved variants of the *molecular junction (MJ)* shown in Figure 7.1b, consisting of an organic film between two conducting contacts, with little or no solvent present. ET through a 100–1000 nm thick molecular layer was measured as a current between the two contacts and in some cases was accompanied by ion motion. Starting in the late 1990s, ET in MJs with thicknesses of <10 nm was investigated, and such studies comprise the rapidly growing area of molecular electronics (ME).[9–15] In many cases, the molecular *layer* is a single molecule suspended between two contacts, with one contact being an STM or AFM probe.[16–18] The important distinction between *organic* and *molecular* electronics is one of scale, with the latter involving transport distances of a few nanometer. As summarized in Table 7.1, the >100 nm transport distance common to organic electronics dictates transport consisting of a series of steps, usually activated redox exchange.

The carrier mobility is governed in part by the reorganization energy of the radical ions required for hopping, and transport generally has a positive temperature dependence. The low mobility common to organic semiconductors compared to silicon has been a serious drawback of organic electronics, as has low stability resulting from carriers consisting of radical ions. Reduction of the transport distance to <10 nm enables other ET mechanisms such as tunneling and may avoid

TABLE 7.1

Comparison of Organic and Molecular Electronics Highlighting Differences in Distance Scale, Mechanism, and Electric Fields across the Molecules

	Transport Distance	Temperature Dependence (σ = Conductivity)	Transport Mechanism	Electric Fields (V cm^{-1})
Organic electronics	0.1–100 μm	$d\sigma/dt > 0$	Redox exchange variable range hopping	10^3–10^4
Molecular electronics	1–10 nm	??	Tunneling, *off resonance injection*, ??	10^6–10^7

formation of reactive intermediates. Furthermore, the electric fields in MJs may be much higher than those in OLEDs and OTFTs, often exceeding 10^6 V cm^{-1}.[13] As noted in Section 7.3, these high fields can have major consequences to the ET mechanism.

In addition to ET across short distances in MJs, this chapter also considers memory devices in which a redox reaction results in a large change in conductance in a device with the geometry shown in Figure 7.1c. Although its arrangement is similar to that of an OFET, its operation is fundamentally different, as described in Section 7.4. Solid-state nonvolatile memory (NVM) has enabled portable electronics, with broad applications in cell phones and portable music players, and replaced disk drives in laptop computers.[19,20] The large demand for NVM has driven an extensive research effort on alternatives to existing dynamic random access memory (DRAM) and *flash* memory based on the silicon floating gate structure. *Flash* memory is a successful example of NVM based on the silicon floating gate field-effect transistor (FET), widely used in portable consumer electronics such as smartphones and USB drives.[21] Flash memory has retention times exceeding 10 years; can be very dense, for example, >10 GB cm^{-2}; and is inexpensive enough for widespread use. Flash uses a relatively high-voltage pulse (~10 V) to inject electrons into a floating gate of silicon on silicon oxide insulator. The resulting field from the trapped charge then modulates the conductance of a FET, and the low and high conductance states can be read nondestructively to provide a *bit* of information storage. However, the high-voltage requirement leads to relatively high energy requirements for write/erase (W/E) operations, and fatigue of the SiOx *gate oxide* leads to limited cycle life.[21] In addition, flash is subject to cross talk between adjacent cells, which ultimately limits its information density. Furthermore, flash is much slower than DRAM based on charge storage on a capacitor, and its limited cycle life prevents many applications where long retention is desirable.

Many *alternative* NVM devices involve redox processes of both organic[20,22,23] and inorganic[24–27] materials, although none of these have penetrated the commercial market. An example of redox-based memory where commercialization was attempted is a dynamic memory element in which charge is stored in a porphyrin redox center instead of a capacitor. The porphyrin device had potentially higher density, lower cost, and longer retention time than currently very common DRAM.[23,28] A single bit of the porphyrin-based memory is essentially a very small battery having two redox systems and mobile ions, with significantly higher charge density than possible with a capacitor of similar size. An important but fairly obvious aspect of redox-based memory devices is that redox reactions are difficult or impossible in silicon, and the active memory element nearly always involves addition of a metal, metal oxide, or organic redox element.

An additional important feature of more recently developed memory devices, including *flash*, is their readout based on resistance[29–33] rather than electronic charge or magnetism. The charge-storage mechanism used in DRAM limits retention time due to leakage from a very small capacitor, limits density to a minimum readable charge on the capacitor, and also increases cost due to the need for a high aspect ratio *trench* capacitor.[23] The long retention, nondestructive readout, and high bit density of *flash* are the main reasons for its widespread use, but the cycle life of ~10^4 W/E cycles constrains its use to applications with relatively low duty cycle. Some of the newer NVM alternatives based on redox processes, including that shown in Figure 7.1c, also involve resistance readout to exploit its advantages, but seek to increase lifetime and reduce energy consumption. As an indication of the breadth of the research effort into alternative NVM, a SciFinder search yielded >14,000 citations for either of the terms *resistance memory* and *conductance switching*. There is a wide range of mechanisms for achieving a persistent change in conductance, which can then be read by an electronic circuit. However, there is also a significant uncertainty about the mechanism operative in particular devices, with the result that improving performance is not straightforward.[22,24] For the case of redox-based memory devices, electrochemical concepts and techniques should be invaluable for understanding and improving performance.

In the context of nanoelectrochemistry, this chapter describes the relationship between some familiar electrochemical concepts and solid-state devices of potential interest in microelectronics. Specifically, what charge transport mechanisms are operative when the transport distance decreases below the ~100 nm typical of organic electronic devices and redox polymers? Do the electrified interfaces or high electric fields in molecular layers with 1–20 nm thicknesses result in redox events? Can redox-based resistance memory exhibit improved performance over commonly used commercial devices? What electrochemical processes control such memory devices and what are the fundamental limits to performance? Section 7.2 describes fabrication of MJs of both the two-terminal (Figure 7.1b) and three-terminal (Figure 7.1c) configurations. Section 7.3 describes electron transport across two-terminal devices with thickness in the range of 4–22 nm, with emphasis on transport mechanism and electronic behavior. Section 7.4 describes resistance memory devices based on redox reactions in conducting polymers and their potential applications as alternative NVM. Finally, Section 7.5 considers future prospects for nanometric molecular electronic devices, both in fundamental science and commercial applications.

7.2 FABRICATION OF MOLECULAR ELECTRONIC DEVICES

As already noted in Figure 7.1, an MJ is derived from the modified electrodes in common use in solution phase electrochemistry. Some additional comments are appropriate when considering modified surfaces in the context of possible microelectronic applications. Obviously, many existing components of today's microelectronics are also two- or three-terminal, and their functions depend strongly on which configuration is employed. The two-terminal geometry shown in Figure 7.1b is equivalent to a modified electrode with the solution replaced by a conductor and is intended to lack ions and solvent and usually mass transport. As will be discussed in Section 7.4, the three-terminal geometry of Figure 7.1c provides a *control* electrode (denoted as "G" in the figure) and can also have some of the properties of an electrochemical cell. There have been many variations on fabrication of both two- and three-terminal molecular devices, usually classified by the type of surface bonding between the molecular component and the conducting substrate.[13–15,34–36] These many variations will not be reviewed here, but some special concerns about the structures shown in Figure 7.1 will be considered, in order to illustrate the transition from modified electrodes (Figure 7.1a) to molecular electronic devices (Figure 7.1b and c), namely, substrate patterning and flatness, molecular layer bonding and formation, and top contact deposition.

7.2.1 FABRICATION OF TWO-TERMINAL MOLECULAR JUNCTIONS

Most applications of ME will likely involve the wafer-scale, massively parallel processing currently employed in the microelectronics industry. Therefore, MJ fabrication must be compatible with photolithography and reasonable temperature excursions during processing and operation. In addition, the substrate must be very flat, so that its root-mean-square (RMS) roughness is significantly less than the thickness of the molecular layer. Our approach to addressing these issues is the use of the unique properties of carbon–carbon bonding possible by diazonium reduction on sp^2 carbon

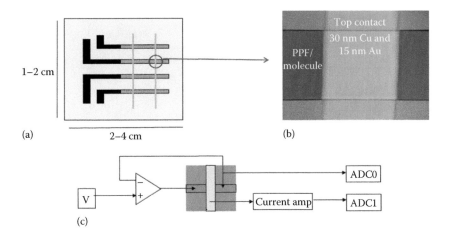

FIGURE 7.2 (a) Substrate pattern for two-terminal MJs, starting with four *stripes* of PPF on an insulating Si/SiO$_2$ chip and then a magnified image of the finished MJ (b). (c) The three-electrode schematic that corrects for ohmic potential error in the PPF.

surfaces.[14,37,38] Pyrolyzed photoresist film (PPF) is a form of glassy carbon made by pyrolysis of commercial photoresist resins (mainly *novolac*), which can be patterned by conventional photolithography.[39–42] Heat treatment in 5% H$_2$ in N$_2$ (i.e., *forming gas*) provides a reducing atmosphere that produces a very flat (<0.5 nm RMS by AFM) sp^2 hybridized carbon surface in the same pattern as the original photoresist.[40] PPF has electronic, electrochemical, and structural properties similar to glassy carbon, and its surface presumably consists of a random array of basal and edge regions of the sp^2 carbon. Figure 7.2a shows the pattern used for many *laboratory* devices, in which a given sample has 4–12 MJs, consisting of a 200–500 μm wide PPF strip with a 100–250 μm wide *top contact*, in this case Cu and Au.

An image of an MJ is shown in Figure 7.2b, and contact is made with probes (usually tungsten) positioned on the MJ as shown in Figure 7.2c. Notice that the three-wire measurement geometry shown in Figure 7.2c is analogous to that used in electrochemistry and provides compensation of the ohmic potential drop in the PPF lead.[43] Microfabrication of MJs at the level of a 100 mm diameter wafer is shown in Figure 7.3, starting with a mask (panel a) followed by pyrolysis to generate a wafer of PPF patterns (panel b), which are then diced into individual samples and modified further (panels c and d).[44]

Diazonium reduction to form C–C bonds on sp^2 surfaces has been studied extensively[45–50] and is well suited to MJ fabrication for several reasons. First, the C–C bond is thermally stable, at least up to >500°C, and is formed mainly at the edges of graphitic sheets on the PPF surface.[51,52] Second, surface modification is mediated by a phenyl radical, which aggressively *patches* pinholes on the PPF surface, leading to high coverage. Although the radical mechanism can also produce multilayers, the layer is covalent and conjugated, and its thickness can be verified by AFM *scratching*.[53] Third, the conjugated phenyl–phenyl bond between the molecular layer and the PPF may have special electronic consequences, as it represents an example of strong electronic coupling, as discussed in Section 7.3. The molecular layer is usually deposited electrochemically, by immersing the PPF *stripes* shown in Figure 7.2a or the microfabricated chip of 7.3 in an electrolyte solution containing the appropriate diazonium reagent. Repeated negative voltammetric scans are then used to build up a molecular layer with thickness in the range of 2–25 nm. The RMS roughness of the modified layer matches that of the substrate (<0.5 nm) for molecular layers up to ~6 nm and increases slightly to <1 nm for 22 nm films.[54] An example of the statistical analysis currently used to determine layer thickness is shown in Figure 7.4, with the 10.3±0.9 nm result reflecting the combined (quadrature) standard deviation of both the PPF substrate and the molecular layer,

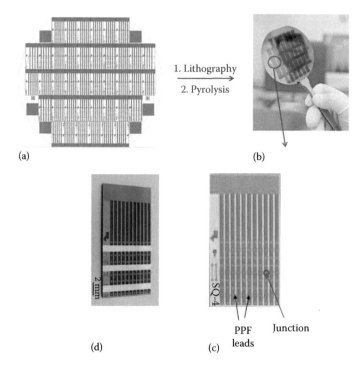

FIGURE 7.3 Photolithography mask (a) and photograph (b) for a PPF pattern on a 100 mm diameter wafer. Pattern includes 40 sample substrates (c) with 32 junctions each, and top contact is applied through a shadow mask to yield finished devices (d).

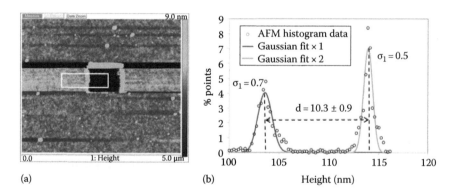

FIGURE 7.4 (a) A modified PPF surface showing a *scratch* made with contact mode AFM with sufficient force to remove the molecular layer. (b) Statistical analysis of the heights determined with tapping mode in the area of the white rectangle. (Used from supporting information of Yan, H. et al., *Proc. Nat. Acad. Sci. U. S. A.*, 110, 5326, 2013. With permission.)

determined at many positions inside and outside an AFM scratch made with the contact mode but analyzed with the *tapping* mode.[54]

In most cases, the thickness determination is performed directly on the sample studied electronically, on the exposed PPF/molecule region adjacent to one of the junctions.

Application of a top contact to complete the MJ has been a very active and often controversial topic, with no consensus emerging yet regarding the best method. The main concern is that common methods based on metal sputtering or vapor deposition can damage and/or penetrate the molecular

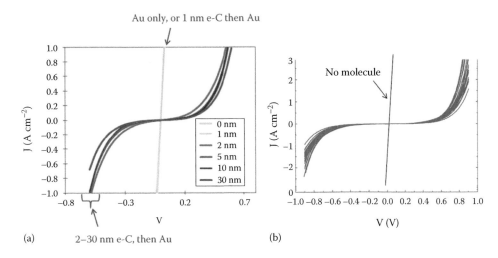

FIGURE 7.5 (a) JV response for MJs made from a 4.1 nm thick layer of NAB on PPF after deposition of various top contacts. The vertical line results from direct deposition of 15 nm of Au on the molecular layer and from deposition of 1 nm of e-C followed by 15 nm of Au. The remaining curves result from deposition of 15 nm of Au on top of the indicated thicknesses of e-C. (b) Overlay of JV curves from 32 NAB junctions on four separate *chips* (red curves) compared to one with the molecule absent. (Used from Yan, H. et al., *J. Am. Chem. Soc.*, 133, 19168, 2011. With permission.)

layer, often with the formation of *shorts*, that is, direct electrical contact between the substrate and top electrodes.[55-57] We reviewed these methods[14] and their associated pitfalls, including the importance of structural characterization of completed MJs with optical spectroscopy.[43,58-63] The high coverage and thermal stability of diazonium-derived PPF/molecule layers enable them to tolerate electron beam deposition of Cu, Ag, and Au with minimal effects on the Raman spectrum of the molecular layer. However, deposition of Ti or Pt on the same substrates results in significant changes to the Raman spectrum as well as *shorts*, indicating serious alteration of molecular layer structure.[58] Although Au does not change the Raman spectrum of the molecular layer, it does penetrate the film and results in a short circuit, as shown in Figure 7.5a.

Similar problems with Au deposition on SAMs and Langmuir–Blodgett films have been reported extensively, as has layer damage by Ti.[55-57,65] An alternative *cold* deposition method based on diffusion of metal atoms to contact the molecular layer yields very similar current-voltage response for PPF/molecule/Au MJs, indicating that direct metal deposition does not alter electronic behavior.[66,67] Microscopy images of the MJs depicted in Figure 7.4d at progressively higher magnification are shown in Figure 7.6, starting with a close-up of the junction region in panel a.

A recent alternative to direct metal deposition uses electron beam–deposited carbon (e-C) as the initial top contact, followed by a layer of 15 nm of Au to permit electrical contact.[64] Figure 7.5a shows that a 2 nm thick layer of e-C prevents penetration of subsequently deposited Au, and the final MJ shows response comparable to a similar junction with a Cu/Au top contact.

Although e-C has a higher resistivity than PPF, its thickness may be increased to 30 nm without significant effects on the observed current density. Raman spectra before and after deposition of e-C and Au showed no change in peak positions or relative intensities of the bands of nitroazobenzene (NAB), implying minimal changes to the molecular layer structure. Reproducibility of devices made with e-C (10 nm) and Au (15 nm) top contacts is shown in Figure 7.5b for 32 MJs from four separate samples with eight junctions on each. The 100% yield and rsd of 10%–20% for the current density indicate reliable fabrication and good consistency for finished devices. The e-C/Au top contact also permits a higher range of applied bias and fabrication of thicker MJs, up to at least 22 nm.[54,64]

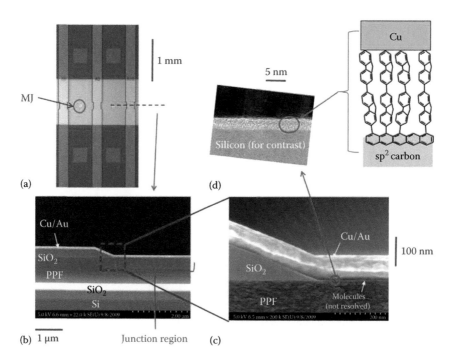

FIGURE 7.6 Micrographs of microfabricated MJs at progressively higher magnifications: (a) Photomicrograph of two junctions and their contact pads. (b) SEM cross section of a junction cleaved at the dotted line in panel A. (c) Magnified SEM of one edge of the MJ showing the metal top contact layer. (d) TEM of molecular layer between silicon and copper, with silicon replacing PPF to provide contrast. (Images a, b, and c used from Ru, J. et al., *ACS Appl. Mater. Inter.*, 2, 3693, 2010. With permission.)

7.2.2 FABRICATION OF THREE-TERMINAL MOLECULAR MEMORY DEVICES

Since the early experimental work on ME in the late 1990s, memory devices have been prominent, due to the large commercial incentive to make solid-state memory that is more dense and faster or requires less power than current technology. Many of these devices were two-terminal, often a *crossbar* arrangement, and were in principle scalable to very small device size, with possibly even a single molecule representing one *bit*.[9,60] As noted in Section 7.1, a wide range of devices has been reported for both organic and inorganic memory devices with both two- and three-terminal geometry having mechanisms involving distinct physical phenomena. We investigated a two-terminal device based on conductance changes in a layer of TiO_2 adjacent to a molecular layer of NAB or polypyrrole (PPy). Although these devices showed initial promise, they proved to be unreliable due in part to the requirement that the two contacts must perform both W/E and read (R) operations. As described in Section 7.4, we then adopted the configuration of Figure 7.1c, consisting of a conducting polymer (poly(3,3′-didodecylquaterthiophene) [PQT] in the case shown) and a third electrode (labeled "G" by analogy to the gate of a FET). Note that the W/E and R circuits are now separated, so these operations can be controlled independently. In addition, the *read* process is nondestructive and in principle very fast. We use the "S," "D," "G" labeling from FET technology due to their familiarity, but the operation of the memory device differs fundamentally from that of a FET.

Fabrication of three-terminal memory devices is derived in part from the lithography and spin coating methods common to the very active OFET literature, but with some special requirements dictated by the need for mobile ions and redox-based conductance changes. A research test substrate for a one-bit memory device is shown in Figure 7.7a, with a 20 μm gap between the S and D electrodes. The pattern on the substrate includes S and D electrodes, onto which the polymer layers

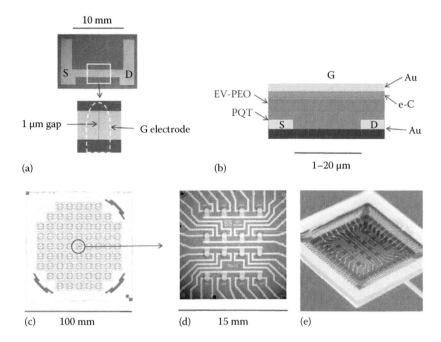

FIGURE 7.7   (a) Top view of three-terminal substrate showing S and D electrodes (upper) and the location of the G electrode (lower). (b) Side view schematic of redox-gated memory device. (c) Mask for microfabricated devices shown enlarged in panel (d) and wire bonded in a package in E.

are spin coated. Regioregular PQT is a polythiophene derivative made by Xerox Research Centre of Canada for application in OFETs and printable electronics,[68–70] and it is first applied to the SD substrate by spin coating. The EV–PEO layer is then either drop cast or spin coated on top of the PQT layer, to build up the structure shown from a side view in Figure 7.7b.

Finally, an e-C (15 nm) and Au (15 nm) are applied by e-beam deposition through a shadow mask to complete the device. Most of our research on redox-gated memory devices to date was done with the *laboratory* substrate of Figure 7.7a,[34,71–73] but the entire process can be adapted to parallel fabrication using the substrate shown in Figure 7.7c through 7.7e. Each unit in a pattern on a 100 mm wafer (8C) has 16 test cells (8D) onto which polymers can be applied. The test chip can then be mounted in commercial packaging by wire bonding (Figure 7.7e) to permit environmental and lifetime testing.

7.3   ELECTRON TRANSPORT ACROSS NANOMETER-SCALE MOLECULAR LAYERS

As noted in Section 7.1, electron transport in MJs has important similarities to ET through monolayers on modified electrodes. Transport by tunneling in both cases is shown in Figure 7.8 for a conducting carbon electrode. The density of electronic states in a conductor is high, and the distribution of electrons in these states is dictated by the Fermi function, which defines the Fermi level as the energy level that is half filled with electrons. The wave nature of electrons dictates that there is finite electron density outside the carbon electrode that extends into space or into a solution with an exponentially decreasing *tail*, as shown in Figure 7.8a.

Stated differently, an electron has finite probability of being located *outside* the electrode itself, with the probability dictated by both distance and the nature of the material adjacent to the electrode. A reducible compound such as anthracene (AN) in solution approaching the electrode surface will

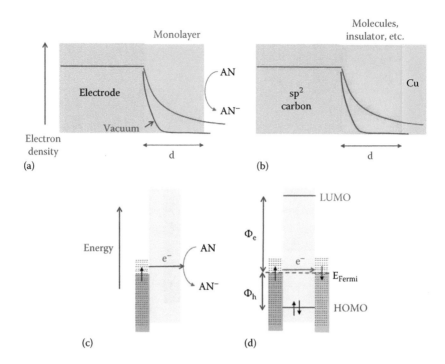

FIGURE 7.8 (a) Schematic of electron density vs. distance for a modified electrode reducing a redox system in solution. (b) Same schematic for a MJ with Cu replacing the solution. (c) Energy levels for the process of panel (a), with shaded area representing filled states in the electrode and an electron tunneling through the molecular layer. (d) Energy levels for an MJ, with HOMO and LUMO representing the highest occupied and lowest unoccupied molecular orbitals of the molecular layer. Φ_e and Φ_h indicate the electron and hole tunneling barriers, respectively.

first encounter the *tail* of electrons, and an electron may transfer to the electrode even though the AN is not actually in direct contact with the electrode surface. The rate of ET through the monolayer is strongly dependent on the nature of the monolayer, as well as the usual parameters related to reorganization energy, transfer coefficient, etc.[1–4,74] A wealth of experimental and theoretical results supports an exponential dependence of ET on the thickness of the monolayer (d), according to Equation 7.1:

$$k°(obs) = k°(d=0) \exp(-\beta d) \tag{7.1}$$

where

    $k°(obs)$ is the observed heterogeneous ET rate constant
    $k°(d=0)$ is the rate constant with the monolayer absent
    β is the attenuation parameter, with units of Å^{-1} or nm^{-1} (nm^{-1} will be used here for consistency
    with the figures)

Therefore, a β equal to 1.0 nm^{-1} indicates that the ET rate decreases by 1/e (or 63%) for every nanometer of monolayer thickness. Equation 7.1 has been applied to a variety of ET reactions in electrochemistry as well as intramolecular and intermolecular ET occurring in homogeneous solutions. This literature is extensive, but relevant benchmark β values include 8–9 nm^{-1} for ET through alkane chains; 3–5 nm^{-1} for conjugated, aromatic molecules; and as low as 1–2 nm^{-1} for the conjugated backbone of polyolefins.[14] β for a vacuum depends on the electrode work function, but is high, typically ~25 nm^{-1}. So although an electron can tunnel through a vacuum, the range of distances is very short, with the probability of ET decreasing to less than 1% with a vacuum gap less than

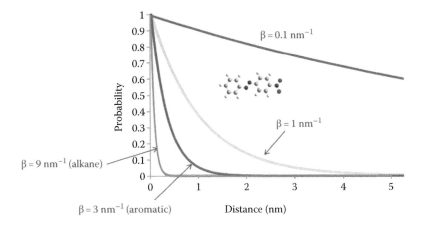

FIGURE 7.9   Prediction of probability of finding an electron at some distance from an electrode as a simple function of exp(−βd) for the indicated values of β. A 1.3 nm NAB molecule is shown for scale.

2 Å thick. Figure 7.9 shows the probability distribution for electrons with distance from a surface for various β values, along with a typical molecule (NAB) to indicate the scale.

An MJ can be considered a modified electrode with the solution replaced by a conductor, as shown in Figure 7.8b. This case is the classical metal/insulator/metal (MIM) device that was first treated theoretically in the 1960s[75–77] to describe tunneling in metal or silicon oxides. A redox reaction need not be involved, and the observed current follows the empirical relationship of Equation 7.2. The absence of a redox reaction

$$J \text{ (A cm}^{-2}) = B(\text{A cm}^{-2}) \exp (-\beta d) \tag{7.2}$$

simplifies the process significantly, since there is no mass transport, reorganization energy, or reaction kinetics. The common depiction of energy levels in the MIM system is shown in Figure 7.8d for a vacuum gap and an MJ. The levels in the conductors are shown as a continuum, with a constant density of electronic states usually assumed for simplicity. The shaded areas represent filled electronic states, up to a Fermi level that may differ for various conductors. For a vacuum gap, the tunneling barrier equals the work function of the contact material, since removal of an electron from the contact produces a free electron in vacuum. Adding a molecular layer between the two contacts to make the MJ shown in Figure 7.10d introduces HOMO and LUMO (highest occupied and lowest unoccupied molecular orbitals) levels into the *gap*, and these are generally assumed to straddle the contact Fermi levels, as shown. If the HOMO occurred at a higher energy than the Fermi level (or the LUMO at lower energy), there would be spontaneous ET between the molecule and the contact and the energies of both phases would change, leading to an entirely different situation. Note that the simple picture of Figure 7.10d ignores charge transfer and energy level realignment when the molecule and contacts are brought together, an assumption often referred to as the *Mott–Schottky limit* in the semiconductor literature.[78–80] As described later, this assumption can often be violated, with significant consequences to device behavior.

The tunneling barriers predicted in the Mott–Schottky limit are shown in Figure 7.8d, for both electron and hole tunneling. For the case of equal Fermi levels of the two contacts, the electron tunneling barrier Φ_e equals $E_{LUMO} - E_F$, with both referenced to the vacuum level. This barrier may be significantly lower than the contact work function; hence, tunneling through a molecular layer may be much more efficient than through a vacuum. *Hole tunneling* is less intuitive but occurs when $\Phi_h = E_F - E_{HOMO}$ is smaller than Φ_e. The only particle actually moving is an electron, of course, but *hole tunneling* may be viewed as starting with transport of an electron in the HOMO level rather than the contact and may become quite efficient if the HOMO level is close to E_F.

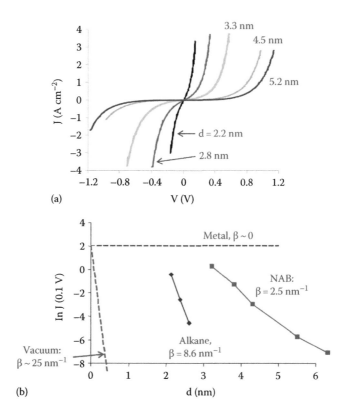

FIGURE 7.10 (a) JV responses of PPF/NAB/Cu MJs with varying NAB layer thickness between 2.2 and 5.2 nm. (b) Plot of ln J(0.1 V) vs. layer thickness for experimental results for alkane and NAB devices and predicted lines for a metal and vacuum. (Alkane data from Bonifas, A.P. and McCreery, R.L., *Nat. Nanotechnol.*, 5, 612, 2010.)

Extensive theoretical treatments of tunneling in MIM devices are based either on the *Simmons* model, which assumes a continuous *insulator* layer with an average barrier height,[75–77,81] or the *Landauer* approach, which is based on predicting the probability of tunneling through a single molecule between the contacts.[12,82,83] These models will not be discussed here in any detail, but they both are consistent with the observed thickness dependence, that is, Equation 7.2. Although both approaches lead to similar predictions of the tunneling rate based on molecule and contact energy levels, the Simmons approach is more readily applied to junctions containing a large number of molecules, while the Landauer treatment is often used for single-molecule devices.[12]

Many types of MJs have been fabricated in order to determine electron transport mechanisms, and many of those have been analyzed in the context of Equation 7.2, to determine if tunneling is the major contributor to the current. We have reviewed the various paradigms for studying conduction in single molecules and *large area* MJs, including some advantages and problems with each approach.[13–15,38] Chapter 6 in this volume describes many experiments on single molecules, including factors that determine electron transport in single-molecule junctions. The results discussed in the current section are based on carbon/molecule/metal devices described in Section 7.2, which illustrate several important factors affecting device conduction. Figure 7.5a shows the current density vs. voltage (JV) response for PPF/NAB(4.5)/e-C(10)/Au(15) MJs, compared to the same device with the molecular layer absent.[64] As indicated in Figure 7.5b, such MJs are quite reproducible, with JV curves that differ dramatically from that of a direct PPF/e-C/Au *short circuit*. A ln(J) vs. V plot is linear above about 0.3 V, indicating that the current increases exponentially with applied bias[37,64,78] and can exceed 400 A cm^{-2} without degradation.[44] The response is weakly dependent on temperature, with an apparent Arrhenius barrier of <0.2 meV for the range of 5–250 K and <100 meV for the 260–450 K range.[37] The <100 meV *activation* can be attributed to broadening of the Fermi function in the contact, implying that ET is not dependent on molecular reorganization or nuclear motion over the entire 5–450 K temperature range.[37] The JV response was independent of scan rate between 0.1 and 1000 V s^{-1}, and scans to ±1 A cm^{-2} could be repeated for at least 10^9 cycles without observable changes in response. Similar PPF/NAB/Cu devices were tolerant of processing temperatures up to 350°C, as well as a complete photolithography cycle (spin coating, UV exposure, and developing) performed directly on a finished MJ.[58]

Figure 7.10a shows the dependence of JV response on molecular layer thickness (d) for NAB multilayers in the range of 2.2–5.2 nm, with the thicknesses verified with AFM *scratching*.[53] The strong variation of current density with thickness is not expected if the molecules were behaving like resistors, which should yield a 1/d dependence. Figure 7.10b shows a plot of ln J (0.1 V) vs. d for the curves of Figure 7.10a, demonstrating an exponential dependence of J on thickness, as predicted from Equation 7.2. The attenuation coefficient β is 2.5 nm^{-1} for NAB, corresponding to a decrease of J by a factor of 1/e$^{2.5}$ or 0.082 for each nm of molecular layer thickness. Also shown in Figure 7.10b are results for alkane junctions made by bonding alkyl amines to PPF by oxidation, which exhibited a β=8.8 nm^{-1}.[66] β for the alkane MJs compares favorably with that noted earlier for ET through molecules in solution, and ET across SAMs on electrodes, implying that similar effects control transport rate. Furthermore, the weak temperature dependence over a 5–450 K range is consistent with a tunneling mechanism, presumably controlled by barriers defined by either the molecular HOMO or LUMO, as depicted in Figure 7.8d. The β observed for NAB is similar to that reported for ET through similar aromatic layers on glassy carbon electrodes in solution[74] and also with the 2–5 nm^{-1} reported for aromatic SAMs[3] and single aromatic molecules in break junctions.[12]

The consistently lower β observed for aromatic compared to aliphatic molecules in MJs implies either a HOMO or LUMO that is closer to the contact Fermi level and also indicates an important variable for using molecular structure to *tune* the electronic properties of the MJ. Analysis of JV curves such as Figure 7.10a with the Simmons equations incorporating effective mass and image charge effects yielded tunneling barriers of 1.07–1.25 eV for NAB in the range of d=2.8–5.2 nm, implying that the main transport mode is hole tunneling mediated by the HOMO level.[37] An independent determination of energy levels using photocurrents yielded a $E_{HOMO} - E_F$ offset of 1.1 eV for an aromatic MJ and 1.7 eV for an aliphatic MJ,[84] completely consistent with the attenuation plots of Figure 7.10b.

The prospect of controlling junction conductance by changes in molecular structure is quite attractive, since there are many molecules available, and *rational design* of MJs with useful electronic behaviors may be possible. The approach of Figure 7.10 was extended to a series of seven aromatic molecules, chosen to have a range of 2.3 eV for both the LUMO energies and HOMO energies of the free molecules.[78] PPF/molecule/Cu MJs were constructed from the seven molecules with various molecular layer thicknesses, for a total of >400 junctions, and then JV curves were obtained at room temperature. The range of molecular energy levels should have resulted in a similar range of tunneling barriers for either electrons or holes if the free molecule energies did not change significantly upon formation of the MJ. As shown in Figure 7.11, the aromatic junctions differed little from NAB, with an average β value of 2.7±0.6 nm^{-1}. While the difference between the aromatic and aliphatic devices was clearly statistically significant, those for the aromatic junctions were not, counter to the expectation from the free molecule HOMO or LUMO energies.[78]

Not only did the aromatic molecules have similar slopes for the lnJ vs. d plots, their current magnitudes were also similar and statistically indistinguishable. The reason for this similarity was revealed by independent measurements of the work functions and HOMO levels using ultraviolet photoelectron spectroscopy (UPS)[78] and photocurrent measurements.[84] Strong electronic coupling between the PPF substrate and the bonded, aromatic molecule resulted in charge exchange between the PPF and the molecule, leading to a *leveling effect* that resulted in similar tunneling barriers of

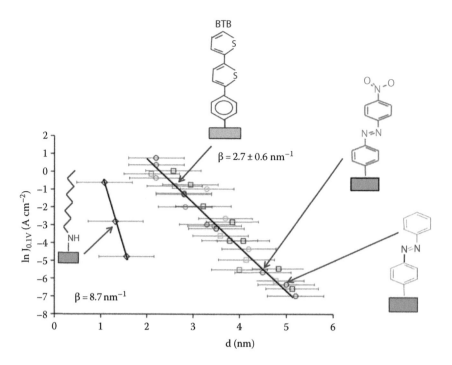

FIGURE 7.11 Attenuation plots from >400 MJs with the indicated molecules plus bromophenyl, ethynyl benzene, anthraquinone, and nitrophenyl molecular layers. (Adapted from Sayed, S.Y. et al., *Proc. Natl. Acad. Sci. U. S. A.*, 109, 11498, 2012. With permission.)

1.3 ± 0.2 eV for all seven aromatic molecules. In effect, the addition of electron donating or withdrawing groups to the aromatic molecule caused shifts in both the apparent PPF work function and the molecular HOMO level, which *compressed* the tunneling barriers expected for the free molecules to a common value of ~1.3 eV, as shown schematically in Figure 7.12. In the semiconductor literature, charge transfer between two materials placed into contact causes similar shifts in energy levels and is known as a violation of the Mott–Schottky assumption that the two phases maintain their *free* energy levels when in contact.

The phenomenon has been discussed extensively for interfaces between metals and organic semiconductors and often has major consequences to the device behavior.[34,85–87] In the case of molecular tunnel junctions such as PPF/molecule/Cu, it results in significantly diminished ability to vary the tunneling barrier by varying substituents on a strongly coupled aromatic molecule. As apparent from the clear difference between aromatic and aliphatic MJs shown in Figure 7.11, the electronic properties of the molecule can still have a major effect on transport, but the electronic properties of the *system*, consisting of both contacts and molecules, must be considered.[78]

The MJs considered thus far were limited to thicknesses below 6 nm, due mainly to experimental limitations on molecular layer growth and top contact deposition. A different MJ structure using e-C as a top contact[64] permitted a wider range of thicknesses (4.5–22 nm) in order to investigate transport mechanisms beyond the tunneling limit of ~5 nm. Figure 7.13a shows JV curves for bis-thienyl benzene (BTB) of the type PPF/BTB/e-C/Au, where e-C represents a 10 nm thick layer of e-C. For d < 5 nm, such junctions are quite similar electronically to the analogous PPF/molecule/Cu devices, but are more stable at higher bias.[64] Note that significant current densities (J >1 A cm⁻²) are achieved for d = 22 nm, where tunneling should be negligible. Figure 7.14 shows the attenuation plot for V = +1 V and a range of temperatures from 6 to 300 K. For d < 8 nm, the slope of 3.0 nm⁻¹ is similar to that observed previously (Figure 7.11) and is independent of temperature. For d >15 nm, β ≈ 0 at 300 K, but the current is temperature dependent with an Arrhenius

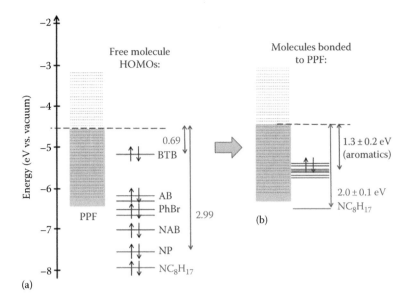

FIGURE 7.12 (a) Energy level diagrams for free molecules calculated with density functional theory [B3LYP 6-31G(d)] on the same scale as the observed work function for PPF (dashed line), predicting tunneling barriers ranging from ~0.7 to ~3 eV. (b) Observed energy levels based on UPS and transport results demonstrating *compression* of the aromatic barriers to values near 1.3 eV. (From Sayed, S.Y. et al., *Proc. Natl. Acad. Sci. U. S. A.*, 109, 11498, 2012.)

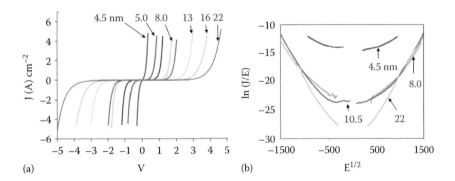

FIGURE 7.13 (a) JV curves for PPF/BTB/e-C/Au MJs with molecular layer thickness from 4.5 to 22 nm. (b) $\ln(J/E)$ vs. $E^{1/2}$ plots for selected thickness. Temperature was 300 K in all cases shown. E is the electric field across the molecular layer, assumed to be linear across the device.

slope of 160 meV. For $d = 10–16$ nm, $\beta = 1$ nm^{-1} and is weakly temperature dependent. Note also that the low-temperature behavior for $d = 22$ nm is close to the $\beta = 1$ nm^{-1} line observed for all temperatures and $d < 16$ nm.

A detailed analysis of JV curve shape and temperature dependence resulted in the proposal that three mechanisms were active over the 4.5–22 nm thickness range.[54] For $d < 8$ nm, transport is dominated by tunneling, and extrapolation of the $\beta = 3$ nm^{-1} line to greater distances indicates that tunneling will decrease below the detection limit at ~12 nm. For $d = 22$ nm, transport is consistent with thermally activated hopping with an activation energy of 160 meV, with a decrease at low temperature to the same $\beta = 1$ nm^{-1} line observed for $d = 12–16$ nm. Reported activation energies for transport in bulk polythiophene are in the range of 130–280 meV, so transport in structurally

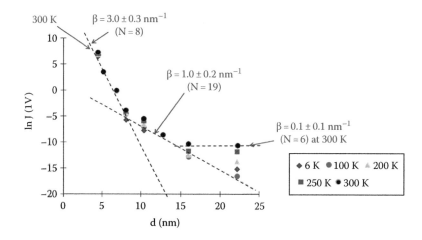

FIGURE 7.14 Attenuation plots for BTB junctions at five temperatures, all obtained with a 1 V bias. Statistics indicate β values for fixed thickness ranges, across several samples and temperatures, as described in Ref. [54].

similar BTB films with $d=22$ nm is consistent with redox exchange commonly encountered in organic semiconductors. It is quite likely that the 4.5–22 nm distances shown in Figures 7.14 and 7.15 bridge the *gap* between tunneling transport at short distances and redox exchange operative for *bulk* organic semiconductors.[54]

The nature of transport in the intermediate region with $\beta=1$ nm^{-1} is at least partially revealed by a plot of $\ln(J/E)$ vs. $E^{1/2}$ where E is the electric field, assumed to be V/d for a given MJ. Such plots are expected to be linear for Schottky and Poole–Frenkel transport, with the unusual $E^{1/2}$ dependence resulting from the effect of electric field on transport barriers.[88] Figure 7.13b shows this plot for four thicknesses of BTB at room temperature, which leads to two important observations. First, the curves coalesce at high E for $d=8$, 10.5, and 22 nm, indicating that transport is *field* dependent over a wide range of thickness, completely inconsistent with tunneling. Second, $\ln(J/E)$ vs. $E^{1/2}$ is nearly linear for these thicknesses and high field and is very linear for $d=22$ nm and $T=6$ K ($R^2=0.9899$ over eight orders of magnitude of J/E).[54] These results are consistent with Schottky and Poole–Frenkel transport in one respect; that is, there is a transport barrier that varies with $E^{1/2}$. However, both these mechanisms are based on thermal activation of a carrier over such a barrier, and the $\beta=1$ nm^{-1} mechanism clearly is effective even at 6 K. Therefore, whatever mechanism is proposed for this intermediate region should be *activationless* and effective over a wide range of electric field and temperature. We recently proposed that the intermediate mechanism is based on field ionization of molecular HOMO levels in the molecular layer,[54] conceptually similar to the ionization of electrons in *traps* that is the basis of the Poole–Frenkel mechanism. For $d=10$ nm, and a bias of 1 V, the average electric field across the molecule is 10^6 V cm^{-1} and much higher than that present in typical organic electronic devices with thicknesses in the 100 nm–10 μm range. Given that the HOMO level of BTB is predicted to be 0.7–1.2 eV below the system Fermi level, Poole–Frenkel theory predicts that ionization of an electron *trapped* in the HOMO may occur when E exceeds 10^6 V cm^{-1}, which is the point where the curves coalesce in Figure 7.13b.

As outlined in Section 7.5, there is much remaining to be learned about electron transport through nanometric molecular layers, but the current results in addition to those from the literature clearly demonstrate that molecules can act as circuit components and have unusual transport properties.[13] In all of the cases described thus far, transport occurs without conventional redox reactions and is generally weakly activated. We now turn to an approach to molecular memory in which redox reactions are both intentional and essential to the intended application of the molecular electronic device.

7.4 REDOX-MEDIATED SOLID-STATE MOLECULAR MEMORY DEVICES

As noted in Section 7.1, redox reactions have been associated with memory functions in a variety of microelectronic devices, based on both inorganic materials such as TiO_2 and metal filaments[24] and organic materials, notably organic semiconductors.[22] An example of a two-terminal memory device based on resistance switching is shown in Figure 7.15a, consisting of thin layers of PPy and TiO_2 between conducting contacts.

The conductivity of PPy increases by a factor of ~10^8 when it is oxidized by one electron, while that of TiO_2 increases a comparable amount when it is reduced from Ti^{IV} to Ti^{III} oxide.[90] When the PPF electrode shown in Figure 7.15a is biased with the PPF positive, an electron can move in the external circuit from the PPy to the TiO_2, resulting in a major decrease in the resistance between the PPF and Au electrodes. As shown in Figure 7.15c, the resistance may be *read* by standard voltammetry, and the device may be switched repeatedly between the low and high conductance states. Although the electronic characteristics of the PPy/TiO_2 device such as speed, retention, and cycle life were attractive, the mechanism proved complex. Solid-state spectroelectrochemistry of a partially transparent device is shown in Figure 7.15d, demonstrating reversible oxidation of the PPy.[89] However, it was also determined that H_2O was a critical reagent in the process, possibly due to its involvement in redox reactions at one or both electrodes. A detailed examination of TiO_2 revealed a water-mediated reduction and conductance change in TiO_2 alone, without PPy present.[26] These redox events may be involved in any nominally solid-state device containing TiO_2, particularly if they are studied in ambient air. TiO_2 and its bias-induced conductance changes are the basis of significant recent activity on inorganic memory devices dubbed *memristors*, but have not yet been realized commercially.[25,27,91–93]

In addition to the complexity associated with TiO_2, the two-terminal geometry of Figure 7.15a is subject to more fundamental problems with retention and readout. While it has some properties of

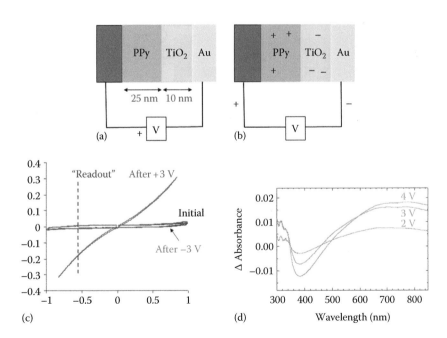

FIGURE 7.15 (a) Two-terminal memory device containing PPy and TiO_2, in its initial low conductance state. (b) The same device after an external voltage causes the conductivity of both layers to increase. (c) IV response of the device following *write* and *erase* pulses. (d) Solid-state spectroelectrochemistry of a PPy/Al_2O_3 device, plotted as the change in absorbance in response to the indicated voltages. (From Bonifas, A.P. and McCreery, R.L., *Anal. Chem.*, 84, 2459, 2012.)

a conventional battery with two redox systems, the redox systems are not physically separated and are subject to recombination reactions at their common interface. Batteries require a separator to prevent such reactions, but introducing a separator into the two-terminal geometry would produce high dc resistance and mask the conductance changes required for memory operation. These recombination reactions will *discharge* the conducting states of the PPy and TiO_2 until the conductance of the device reverts to its initial low value. As long as the two redox systems are in physical contact, retention time will be limited by recombination, and attempts to introduce a separator will unavoidably introduce unacceptably high dc resistance.

In order to avoid both the recombination problem and the complex redox chemistry of TiO_2, we adopted the three-terminal geometry of Figure 7.1c. Initially, the polymer was poly(3-hexyl thiophene) (P3HT), which has a conductivity of $<10^{-7}$ S cm^{-1} in its native form and >1 S cm^{-1} when oxidized by one electron. The configuration of Figure 7.16 was studied with Raman spectroscopy to monitor the oxidation of PQT and correlate it with conductance.[72,73] The conductivity of P3HT is $<10^{-7}$ S cm^{-1} in its native form, which increases to >1 S cm^{-1} when oxidized to its conducting *polaron*.

Although the SiO_2/Pt "G" contact does not include an intentional redox system, it was possible to demonstrate oxidation of the P3HT by changes in its Raman spectrum. Figure 7.16b shows Raman spectra of P3HT in the three-terminal P3HT/SiO_2 device during progressive application of a +4 V bias to the S electrode relative to G. The shift in the prominent Raman band at ~1460 cm^{-1} to lower energy and the appearance of a new band at ~1400 cm^{-1} are the same changes that accompany P3HT oxidation in solution, providing direct evidence for the formation of P3HT polaron.[72] The change in conductance between the S and D electrodes accompanying the spectroscopy experiments is

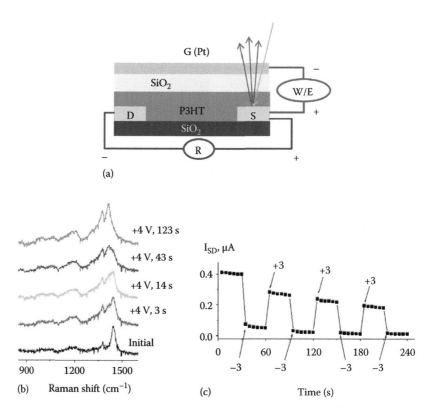

FIGURE 7.16 (a) Schematic of a three-terminal P3HT/SiO_2 memory device showing the incident and Raman scattered light from the source region. (b) Raman spectra of the source region when a +4 V *write* voltage was applied for the times indicated. (c) Repeated changes in SD current in response to *write* (+3 V) and *erase* (−3 V) pulses. (Used from Shoute, L.C.T. et al., *Electrochim. Acta*, 110, 437, 2013. With permission.)

indicated in Figure 7.16c, which is the SD current at a constant applied bias of 0.1 V. Therefore, the P3HT/SiO$_2$ memory device provides a direct correlation between spectroscopic changes indicating polaron formation and the resulting change in polymer conductivity.

Although the P3HT/SiO$_2$ device is not suitable for practical application due to its slow W/E times, it does illustrate some merits of the three-terminal geometry and also requirements for useful redox-based memory operation. First, the resistance readout is nondestructive and does not significantly perturb the redox state of the polythiophene. Second, direct spectroscopic monitoring of device operation is possible, which was extended to the spatially resolved experiments described below. Third, the complementary redox reaction accompanying the oxidation of PQT was identified as the reduction of H$_2$O at the platinum counter electrode (G), and memory operation ceased in a vacuum. While dependence on trace water is undesirable for several reasons, it does illustrate that the redox counterreaction can be spatially separated from the polymer redox events. At least in principle, this situation is required to reduce recombination and extend the retention time. Fourth, the polythiophene active layer avoids the proton transfer reactions of PPy,[89] as well as the complex chemistry of TiO$_2$. Fifth, the dependence on the presence of water may result from either water's redox reactions or its ability to *solvate* ions and increase ionic mobility, even in a nominally solid-state device. These considerations derived in part from the P3HT/SiO$_2$ three-terminal example resulted in important design changes intended to improve performance.

It is informative to consider the design introduced in Figure 7.16a as a simple battery, as it illustrates some of the requirements on redox-gated memory devices. Suppose that the two redox systems are a polythiophene on top of the S and D electrodes, and an acceptor layer that contains a viologen derivative, as shown in Figure 7.17a. If the S and D electrodes were one continuous conducting layer, the device is a battery lacking a separator between the anodic and cathodic reactions. It would be possible to charge or discharge the battery, but there would also be a parasitic

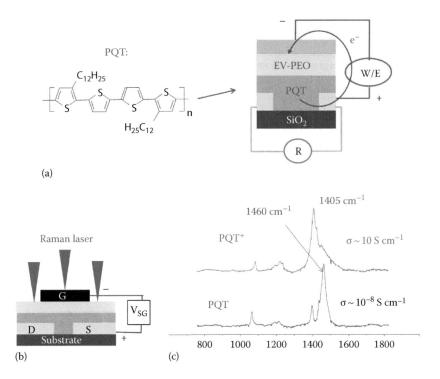

(a)

(b)

(c)

FIGURE 7.17   (a) Three-terminal PQT/viologen memory device, showing EV acceptor in PEO and the PQT structure. (b) Position of incident laser for sampling different regions of the devices. (c) Raman spectra in CHCl$_3$ solution. PEO = polyethylene oxide, EV = ethyl viologen diperchlorate.

recombination reaction at the polythiophene–viologen interface. Separation of the bottom electrode into S and D provides a means to measure the conductance of the polythiophene layer during the charge/discharge process. If a separator were placed between the polythiophene and viologen layers, charging and discharging could still occur, provided the separator permitted ion transport but not electronic conduction. As is well known to electrochemists, a good separator is required in commercial batteries to prevent *chemical* short circuits and is essential to providing long shelf life and long duration of the charged state. By analogy, the three-terminal redox-gated memory configuration is conceptually similar to a conventional battery, with the addition of a second electrode at one contact to permit measurement of conductivity of one of the redox phases. In principle, the speed of such memory devices should depend on the rate of the redox reactions and their associated ion motion, and the retention should depend directly on the properties of the separator.

A design that incorporates all of these design considerations except the separator is shown in Figure 7.17a.[34,71] The polythiophene is regioregular PQT, which was developed for OFETs[68–70] and is chemically and electronically similar to P3HT. Its attractive spin coating properties and higher stability to air oxidation make it amenable to possible commercial applications. The second redox system was ethyl viologen (EV) diperchlorate, an electron acceptor and also a source of mobile ClO_4^- ions. As noted in Section 7.2.2, PQT was spin coated onto a substrate containing the S and D electrodes, and then a mixture of polyethylene oxide (PEO) and $EV(ClO_4)_2$ was drop cast onto the PQT layer to cover the gap between the S and D electrodes. The *write* operation of the intended memory device is indicated by the arrow in Figure 7.18a, consisting of a positive pulse that oxidizes PQT to its conducting polaron and reduces EV^{+2} to EV^{+1}, thus producing a large increase in conductivity in the PQT layer. As is well known, the resulting space charge generated in the PQT and EV layers by the redox reaction must be compensated by ion motion, in this case ClO_4^- ion moving from the EV/PEO layer into the PQT. Formation of a PQT^+ ClO_4^- ion pair in the PQT layer stabilizes the polaron charge to permit a persistent conducting state. As was the case with the P3HT/SiO$_2$ devices, the top contact is thin enough to permit Raman spectroscopy directly in the *gap* region between S and D, as well as other locations on the device, as shown in Figure 7.17b. The Raman spectra of PQT and PQT$^+$ in chloroform are shown in Figure 7.17d, with the polaron formed by chemical oxidation with $FeCl_3$.

Figure 7.18a shows the solid-state voltammetry of the PQT/EV–PEO memory cell when the voltage between the S and G electrodes is scanned, with the S electrode considered positive by convention. Note that unlike conventional voltammetry, there is no reference electrode, and one

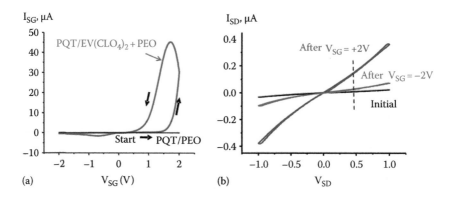

FIGURE 7.18 (a) Voltammetry for three-terminal memory device when V_{SG} is scanned at 50 mV s^{-1} in the presence and absence of EV in the PEO layer, as indicated. (b) SD current when the SD voltage was scanned before and after the indicated *write* (blue) and *erase* (red) voltages were applied. (Used from Kumar, R. et al., *J. Am. Chem. Soc.*, 134, 14869, 2012. With permission.)

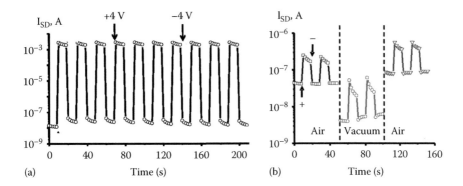

FIGURE 7.19 (a) Ten R/W/E/R cycles for a PQT/EV–PEO memory device in vacuum. Arrows indicate the timing of a typical *write* voltage (+4 V) and *erase* voltage (−4 V) for two cycles. (b) Effect of vacuum treatment and return to air on memory cycles. (Used from Kumar, R. et al., *J. Am. Chem. Soc.*, 134, 14869, 2012. With permission.)

redox system (PQT) is immobile, while the other (EV) is expected to diffuse very slowly in PEO. With EV absent and just the PQT and PEO layers present, the voltammetric current between S and G is very small, as shown by the horizontal black line in Figure 7.18a, indicating that no observable Faradaic reactions occur.

With EV present, I_{SG} increases when the voltage applied between the S and G electrodes exceeds ~1.5 V, with the polarity indicating ET from the S electrode to G in the external SG circuit. The standard potentials (E°) values for PQT/PQT⁺ and EV²⁺/¹⁺ are 0.76 and −0.45 V vs. NHE in solution, so we expect that 1.5 V is sufficient to *charge* the PQT/EV *cell*, thereby generating PQT⁺ and reducing EV²⁺ to EV¹⁺. Since the conductivity of PQT⁺ is much higher than that of the initial PQT, we expect an increase in the SD conductance to accompany the generation of PQT⁺ indicated by the positive I_{SG} current. Figure 7.19b shows the voltammetric response of the SD circuit obtained before and after successive V_{SG} pulses lasting 1 s each. Initially (black curve), the SD current is very small, since the PQT is in its low conductivity state. After $V_{SG} = +2$ V for 1 s, the SD current increases by several orders of magnitude and is approximately linear with voltage, as expected for a partially doped conducting polymer.[71] Reversing the polarity of the V_{SG} pulse to −2 V reduces most of the PQT⁺ back to PQT, and the low conductance state is nearly completely restored.

The voltammetric results of Figure 7.18 illustrate the memory *cycle* depicted in Figure 7.17a, with the SG circuit used for W/E operations and the SD circuit for readout. Consider a constant V_{SD} *readout* voltage of 0.5 V, imposed during various W/E operations and indicated by "R" in Figure 7.17a. The I_{SD} current in response to this fixed V_{SD} would then indicate a "1" or ON state when it exceeds 100 nA in this case or a "0" or OFF state when it is less than 20 nA. The readout process is nondestructive and potentially very fast, since it only requires a transient current measurement. The ON/OFF ratio is often quoted for memory devices and indicates the magnitude of the conductance change between the two states. The $V_{SG} = +2$ V pulse is considered the *write* process, while $V_{SG} = -2$ V is *erase*. In addition to the ON/OFF ratio, important performance indicators for memory devices are the write and erase speed, indicated by the length of the W and E pulses required to yield the desired ON and OFF currents, and *retention* indicated by the stability of the ON and OFF states when the device is left at rest. *Endurance* is a measure of cycle life and indicates the number of W/R/E/R cycles possible before the device fails, with some measure of error rate included in the specification. To provide some context for these parameters, today's DRAM has W/E speeds of 10's of ns and endurance of >10¹⁵ cycles, but short retention (<100 ms) and destructive readout. *Flash* memory has W speeds of a few μs, ~ms E speeds, and >10 year retention, but limited cycle life (10³–10⁴ W/E cycles). These properties determine the applications of each memory type in finished products, since *flash* is excellent for long-term data storage, but would rapidly fail if used in processors, while DRAM has excellent speed, but needs to refresh frequently due to short retention.

Repeated W/R/E/R cycles for the PQT/PEO-EV memory cell are shown in Figure 7.19a, for $V_{SD} = 0.5$ V and W/E pulses of $+4/-4$ V for 2 s each. Note the logarithmic scale of the I_{SD} readout current and the observed ON/OFF ratio of $>10^4$.

As noted earlier, related P3HT/SiO$_2$ memory devices ceased operation completely in a vacuum, due mainly to the lack of a redox counterreaction to accompany polythiophene oxidation. When EV is provided to accept electrons as well as provide mobile ClO$_4^-$ ions, memory operation continues after 12 h exposure to vacuum. As shown in Figure 7.19b, the magnitudes of the currents are diminished in vacuum, but the ON/OFF ratio is not significantly affected.[71] Figure 7.20 shows 200 and 2000 W/R/E/R memory cycles performed in vacuum using the same parameters.[34]

Although the magnitudes of the ON and OFF currents decrease with repeated operations, the ON/OFF ratio remains above 50 after 2000 complete cycles. Examination of repeated readout currents in the ON state shows a decrease with time, indicating limited retention, at least for the ON state. This decay is partly due to the recombination reactions of EV^{1+} with PQT$^+$, resulting in reduction of the conducting polaron state to the initial low conductance neutral form. A separator between the EV and PQT layers that conducts ions but not electrons should greatly reduce or eliminate recombination and extend retention.

Raman spectroscopy provides a means to confirm the memory mechanism, by monitoring the spectrum before and after W/E pulses. The solution spectra of Figure 7.17d show the shift in the prominent band for polythiophene from 1460 to 1405 cm^{-1} when PQT is oxidized to the polaron form, and PQT scattering is sufficiently intense to monitor using a 780 nm laser and commercial spectrometer. As shown in Figure 7.17b, laser illumination could be localized to the S or D

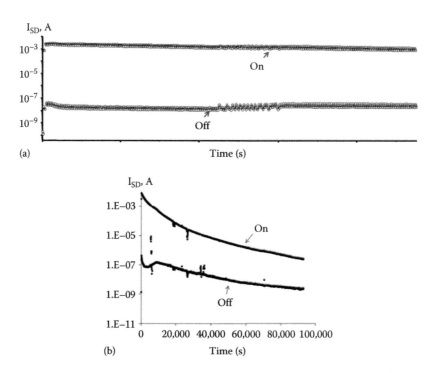

FIGURE 7.20 High (*ON*) and low (*OFF*) currents for $V_{SD} = 0.5$ V from a PQT/EV–PEO memory device during repetitive R/W/R/E cycles in vacuum for a total of 200 (a) and 2000 (b) cycles. (Used from Das, B.C. et al., *ACS Appl. Mater. Inter.* 5, 11052, 2013. With permission.)

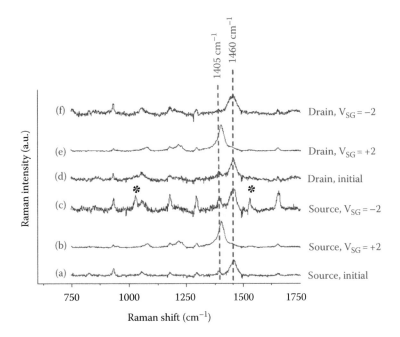

FIGURE 7.21 Raman spectra of a PQT/PEO-EV memory device during operation, over the source and drain regions, as described in the text. (Used from Kumar, R. et al., *J. Am. Chem. Soc.*, 134, 14869, 2012. With permission.)

regions adjacent to the G electrode, or sampling could occur through a G electrode that was partially transparent.

The spectra of Figure 7.21 were obtained at the S and D positions, before and after W and E pulses, with *initial* designating a fresh device not exposed to any bias voltages.[71] Spectra (a) and (d) were obtained at the source and drain electrodes, respectively, for the initial device and show the prominent 1460 cm^{-1} band of the neutral PQT in addition to some small bands related to EV. Spectrum (b) was obtained over the S electrode with $V_{SD} = +2$ V and demonstrates a shift of the 1460 band to 1405 cm^{-1}, indicating formation of the PQT$^+$ polaron at the source. The Raman spectra independently confirm the expected formation of PQT$^+$ by a W pulse, consistent with the proposed memory mechanism and associated voltammetry. A similar shift was also observed over the D electrode (spectrum e), which was not expected since the D electrode was not in the SG circuit and could not itself generate PQT$^+$.

Since Raman sampling may occur at various positions on the memory cell structure, it is possible to determine the spatial distribution of structural changes during operation. The main question in the case of the PQT/EV device is the distribution of the conducting PQT$^+$ polaron, particularly over the channel. Since the gap region between the S and D electrodes determines the conductance for the *read* process, it is important to know how fast the channel is oxidized. For this purpose, the Raman laser passed through a microscope objective to generate a 3 µm diameter spot, which could be incrementally moved along the path indicated by the dashed white line in Figure 7.22a. A Raman spectrum was obtained at 15 points along the white line, covering S, D, and gap regions, with the results displayed as false color images in Figure 7.22.[71]

Panel B shows the initial image, with Raman shift along the abscissa, position on the ordinate and intensity as false color (with red most intense and dark blue weakest). The red spots show the main PQT band at 1460 cm^{-1}, where it is not attenuated by the gate electrode. Since the gate is e-carbon and Au, it attenuates both the incident laser and the Raman scattering, significantly reducing the apparent Raman intensity. However, it is clear that the PQT is distributed vertically over the S–G–D

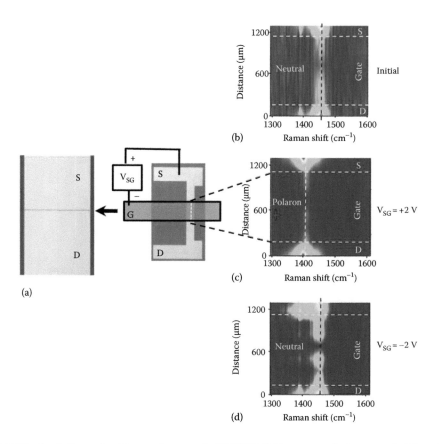

FIGURE 7.22 Spatially resolved Raman images of a PQT/EV–PEO memory device compiled from 15 individual spectra obtained along the line between the S and D electrodes, as shown by the dashed white line in panel (a). Each image shows Raman shift along the abscissa and position on the ordinate, with Raman intensity in false color (red highest). (b) is the initial image, (c) is with $V_{SG} = +2$ V, and (d) is with $V_{SG} = -2$ V. (Used from Kumar, R. et al., *J. Am. Chem. Soc.*, 134, 14869, 2012. With permission.)

regions in its neutral, initial form, as expected for an unperturbed device. Figure 7.22c shows an identical scan performed after imposing $V_{SG} = +2$ V. Within the sampling time of the Raman system (a few seconds), the PQT band shifts from 1460 to 1405 cm^{-1} over the S and D electrodes, both the exposed regions and those under the gate. Recall that the D electrode is not part of the SG circuit and has no bias applied during the entire experiment. Figure 7.22d shows a third scan after V_{SG} was set to -2 V, showing nearly complete return to the initial PQT spectrum. Although the 1 μm wide gap between S and D is not well resolved by the 3 μm Raman sampling region, it is clear that the entire region under the gate and on the D electrode is undergoing oxidation to the conducting PQT$^+$ polaron, even though the gap and D electrode are not electrically connected to the external circuit.

A mechanism that accounts for the unexpected oxidation of PQT over the D electrode is shown in Figure 7.23 and depends on the fact that PQT becomes a conductor when it is oxidized to PQT$^+$. Figure 7.23a shows the device shortly after initiation of the *write* pulse, with "N" designating neutral PQT, EV^{2+} the initial oxidized form of EV, and A$^-$ the perchlorate anion. Application of a sufficiently positive V_{SG} bias generates PQT$^+$ polaron (P$^+$) at the S electrode, which is accompanied by migration of the perchlorate anion into the PQT layer to compensate the polaron charge. Since PQT is both a redox polymer and a conducting polymer (in the PQT$^+$ state), it is possible for ET to occur from a nearby neutral molecule to one of the polarons generated at the S electrode, as shown in Figure 7.23b.

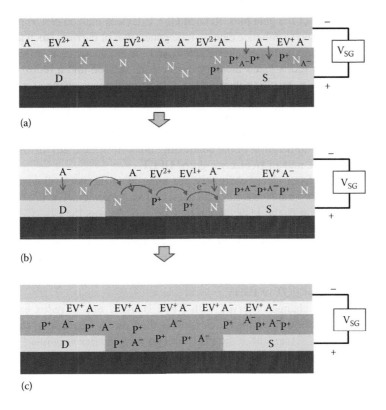

FIGURE 7.23   Schematic of the *write* process, with N = neutral PQT, P$^+$ = PQT$^+$ polaron, EV = ethyl viologen, and A$^-$ = perchlorate ion. (a) Immediately after initiation of the write pulse, (b) during polaron propagation across the SD gap, and (c) after complete oxidation of PQT. See text for details.

Such redox exchange reactions occurring in bulk redox polymers have been investigated intensely, and the redox process can extend through micrometer-thick films between two electrodes.[5,6] For the PQT/EV memory device, propagation of the polaron extends into the gap between S and D, and oxidation of the neutral polymer continues until the entire channel and Drain regions are fully oxidized to a conducting film. In effect, the S electrode is increasing in size, as it generates a conducting *front* in the polymer that continues to oxidize PQT to form polarons. Perchlorate ions continue to enter the PQT film from PEO as PQT is oxidized, and such transport may occur anywhere along the axis between the S and D electrodes. As shown in Figure 7.23c, PQT oxidation is accompanied by EV^{2+} reduction, with the end result being a *charged battery* with laterally homogeneous layers and an electronically conducting region between the S and D electrodes. Although the *erase* process with S polarized negatively follows the reverse process to regenerate the initial PQT and EV^{2+} layers, the reduction is less complete due to the loss of conductivity of the PQT$^+$ film as it is reduced.

We refer to the memory mechanism just described as *redox-gated* organic memory, to highlight the control of SD conductance by an electrochemical reaction. It is clearly an example of solid-state electrochemistry, with the added feature of a second circuit to measure changes in the conductivity of one of the phases. The similarity in geometry to an OFET can be misleading, since the mechanism of operation is fundamentally different in the two devices. The gate bias in an OFET generates charge carriers electrostatically by charging the capacitance between the gate and the S–D electrodes. There is no intentional motion of ions, and the conductance change vanishes as soon as the gate bias is removed. This process is often called *electrostatic doping*, and it may generate the same conducting polarons as the electrochemical analog. But electrostatic doping is not useful for memory devices, since the conductance change induced electrostatically reverts to its initial state immediately upon removal of the gate bias. Related OFET devices include *electrolyte-gated* FETs, in which mobile ions are used to enhance electrostatic generation of carriers (such as polarons).[95–98] Such devices do not include a redox counterreaction, and the conductance is modified only when the gate bias is present. In redox-gated memory devices, ion motion and a redox counterreaction stabilize the conducting form, resulting in retention of the conductance change for ~30 min after a 1 s *write* pulse for the devices shown in Figure 7.23. For redox gating, the W/E speed depends on the rate of the redox reactions and the motion of counterions, and retention is determined by recombination reactions between the oxidized and reduced products of the *write* reactions. The motion of ions and the associated stabilization of the conducting polaron is the main distinction between a *redox-gated* memory device and an *electrolyte-gated* OFET.

Several basic concepts from electrochemistry are relevant to the performance of redox-gated memory devices and provide guidance for improving speed and retention. The EV^{2+} electron acceptor is essential, since without it, the PQT cannot be oxidized. As was the case with SiO_2 devices, adventitious H_2O can act as a receptor, but replacement of EV–PEO with PEO containing $LiClO_4$ and residual H_2O causes memory operation to cease completely in a vacuum. As shown in Figure 7.24, the active polymer must be both redox active and have high and low conductivity states, since replacement of PQT with poly(vinyl-ferrocene) (PVFc) results in an ON/OFF ratio four orders of magnitude lower than that of PQT or P3HT.

The small conductance change observed with PVFc is likely due to redox exchange current between the S and D electrodes[5,6] and is much smaller than the dc conduction through the PQT polaron. Not surprisingly, replacement of the PQT with polystyrene that is neither redox active nor conducting results in a very small observable conductance change.

The dynamics of the *write* operation were monitored with the circuit shown in Figure 7.25a, which amounts to a bipotentiostat familiar to electrochemists. The S electrode is biased by the *write* voltage, and the S and D currents are monitored independently and simultaneously. Control of the D voltage permits different bias values for V_{SD} and V_{SG}, and the entire experiment is conducted by a data acquisition system with LabVIEW software. The SG current for PQT and PVFc devices, both containing PEO-EV, is shown in Figure 7.25b for a $V_{SG} = +3$ V *write* pulse lasting 2 s.[34]

The current magnitudes are comparable for PQT and PVFc, since in both cases, the SG current is oxidizing the polymers and there is little or no dc conduction. The increase in the PQT response with time is likely due to the propagation of polaron away from the S electrode. The SD current response accompanying the PQT *write* pulse for a constant V_{SD} of 0.5 V is shown in Figure 7.25c, overlaid

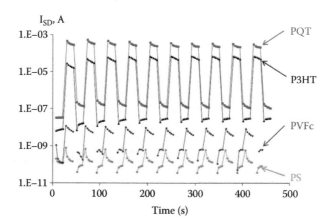

FIGURE 7.24 Ten R/W/R/E memory cycles for polymer/EV–PEO devices that are identical except for the active polymer layer, as indicated. W/E pulses were ±4 V and 1 s long in all cases. PS = polystyrene and PVFc = poly(vinyl ferrocene). (Used from Das, B.C. et al., *ACS Appl. Mater. Inter.*, 5, 11052, 2013. With permission.)

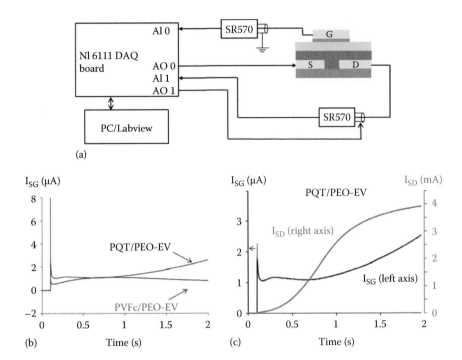

FIGURE 7.25 (a) Schematic for dual pulse experiment permitting simultaneous monitoring of SG and SD circuits. SR570 = Stanford Research current amplifier, NI 6111 = National Instruments data acquisition board. (b) SG current in response to $V_{SG} = 3$ V *write* pulse for polymer/PEO-EV devices containing different polymers, as indicated. (c) SD and SG currents for PQT/PEO-EV device obtained simultaneously during a 3 V write pulse. Not large difference in current scales. (b and c used from Das, B.C. et al., *ACS Appl. Mater. Inter.*, 5, 11052, 2013. With permission.)

with the SG current recorded simultaneously. Note that the current scales are very different in panel (c), with the SD current increasing to >3 mA, while the SG current remains below 3 μA. There is an effectively large *gain* in the devices, since a small SG current can produce a large SD response due to formation of the conducting PQT$^+$ layer between S and D. This effect is similar to that observed in *electrochemical transistors* made from conducting polymers in electrolyte solution,[8] except that the present devices are all solid state. The dual pulse experiment of Figure 7.25 provides direct information about the *write* speed of the memory device, as judged by the time required to reach some threshold I_{SD} value. Obviously, a write speed of hundreds of milliseconds is not competitive with commercial memory devices, so the factors controlling write speed need to be determined.

An important clue is provided by the magnitude of the I_{SG} response of Figure 7.25b, which directly indicates the rate of polaron generation at the S electrode and then propagation into the SD gap. A ~ 1 μA SG current is predicted to completely oxidize the PQT between the S and G electrodes in a few 100 ms, approximately the time required to produce a significant I_{SD} response. This inference implies that the rate of polaron formation is simply too slow to rapidly fill the SG or SD regions and must be increased to improve *write* speed. Figure 7.26 shows the strong effect of local atmosphere on device speed, for a single device that was first run in air, then after 12 h in vacuum (<1 × 10^{-5} torr), and then after admitting acetonitrile vapor into the vacuum chamber. Note that the vacuum removes most of the water from the device, which should not only prevent its possible redox reactions but also reduce ionic conductivity.

An acetonitrile atmosphere with no added water vapor shows much faster *write* speed, and much higher I_{SG}, indicating more rapid formation of polarons. These results are consistent with device speed controlled by ion transport, which must accompany PQT$^+$ formation. Both water

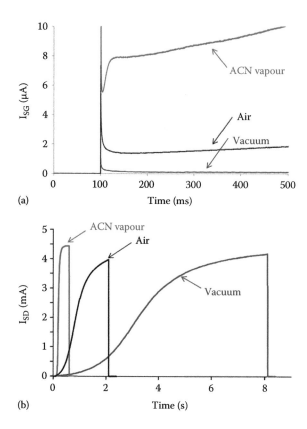

FIGURE 7.26 Effect of atmosphere on (a) SG and (b) SD current responses for a PQT/PEO-EV memory device during a $V_{SG} = +3$ V pulse, initially in air, then after 12 h in vacuum, and then again after 10 min in acetonitrile vapor in the same vacuum chamber. Duration of the V_{SG} in panel B due to the varying device response speeds. (Used from Das, B.C. et al., *ACS Appl. Mater. Inter.*, 5, 11052, 2013. With permission.)

and acetonitrile can stabilize mobile ions in the PEO (and presumably in PQT), thus reducing the ohmic losses in the cell during *write* operations.[34] These losses can be partially overcome with an increased V_{SG} and can be used to advantage. Figure 7.27 shows not only that increased V_{SG} causes higher conductance for a given *write* time, but also that different conductance states are possible and repeatable.[34] The prospect of *multistate memory* increases the effective data density and has been considered for many solid-state devices but never achieved commercially. Control of conductance by *write* voltage is at least a conceivable mechanism for increasing storage capacity in redox-gated memory devices.

The PQT/EV memory devices are currently deficient in both W/E speed and retention, but appreciation of the mechanism provides a *road map* for improvement, as well as quantitative indications of what performance is ultimately possible. The conductivity of ClO_4^- is approximately 10^{-7} S cm^{-1} in PEO, determined from the PQT/EV devices. Solid-state electrolytes vary greatly in ionic conductivity, but many examples have conductivities above 0.01 S cm^{-1}. Implementation of one of these electrolytes to conduct anions could decrease *write* time by up to five orders of magnitude. Furthermore, incorporation of acetonitrile in solid-state electronic devices is not impractical, given the wide use of carbonate solvents in lithium batteries. As already noted, a second parameter of direct interest to practical applications is retention time, which must be much greater than that of DRAM (<100 ms) and approach today's *flash* (>10 years). The redox-gated devices described here lack separators, so recombination is expected and responsible for the short retention time of several minutes. But adding an ionic conductor that is electronically insulating between the PQT and EV layers should in

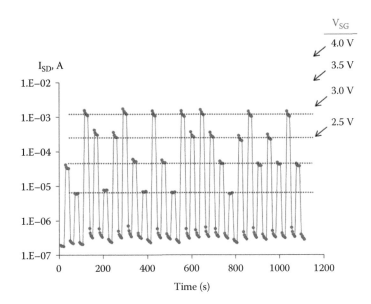

FIGURE 7.27 Demonstration of five distinct conductance states for a single PQT/PEO-EV memory in response to random application of four different V_{SG} pulses. V_{SD} was 0.5 V in all cases. (Used from Das, B.C. et al., *ACS Appl. Mater. Inter.*, 5, 11052, 2013. With permission.)

principle extend retention indefinitely by preventing recombination. Mechanistically, retention is analogous to *shelf life* in common batteries, and those can easily exceed 10 years. The potentially low W/E energy requirement of redox gating combined with better cycle life than current *flash* memory provides the incentive for further improvements in performance of redox-gated devices.

7.5 CONCLUSIONS AND FUTURE PROSPECTS

This chapter explored relationships between electrochemical phenomena and the behavior of nanoscopic layers of molecules in microelectronic devices. Some useful generalizations about ME are available, which will likely continue to be significant as the field develops. First, a modified electrode or a MJ must be considered as one electronic system, with properties quite different from those of the isolated electrode material and unbound molecules. The strong electronic coupling discussed in Section 7.3 significantly affects electronic properties such as transport and tunneling barriers and must be considered when designing molecular electronic devices. A visual demonstration of electronic coupling is provided in Figure 7.28, which shows the molecular orbitals for a NAB molecule bonded to a nine-ring graphene molecule (G9). G9 is used to model the sp^2 lattice in the PPF electrode and the entire system is optimized, yielding a low-energy dihedral angle between the G9 and NAB of 37°.

The HOMO is localized on G9, while the next highest occupied orbital (HOMO-1) has electron density on both the G9 and NAB. One consequence of the sharing of electrons between the G9 and NAB is the *leveling effect* noted in Figure 7.12, and strong coupling is clearly an important factor in choosing molecules and surface bonding schemes for MJs.[13,78] It is likely that theoretical methods such as density functional theory will be critical for the design and understanding of practical ME, once the structural factors affecting transport are understood.

A second important observation is that tunneling is not the only transport mechanism mediating electron transport in MJs. As noted in Section 7.1, electron transport across >100 nm distances in organic semiconductors has been studied extensively and is generally considered a hopping mechanism by redox exchange, variable range hopping, etc. Figure 7.29 shows tunneling and hopping at the short and long extremes of the scale of transport distances relevant to solid-state electronics.

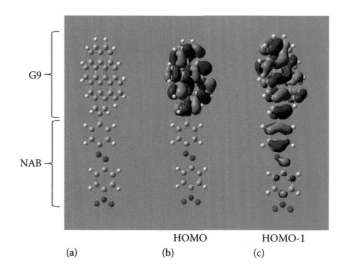

FIGURE 7.28   (a) Structure of graphene–NAB model compound used to calculate orbitals for NAB bonded to PPF. Optimized structure has a dihedral angle between G9 and NAB of 37°. (b) HOMO orbital for the G9–NAB system, calculated with Gaussian 09[99] B3LYP 6-31G(d). (c) HOMO-1 orbital, showing distribution of electron density over both G9 and NAB.

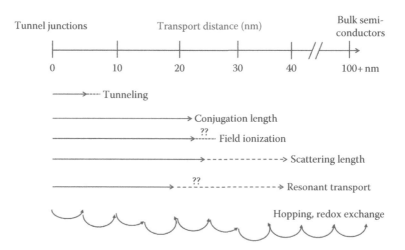

FIGURE 7.29   Transport phenomena in molecular and organic devices ranging from tunnel junctions to bulk semiconductors (upper scale). Arrows indicate the length scales of various effects, with the dashed lines indicating uncertainty. (Adapted from McCreery, R. et al., *Phys. Chem. Chem. Phys.*, 15, 1065, 2013. With permission.)

Hopping across >100 nm is temperature dependent due to reorganization associated with redox exchange and often leads to low carrier mobility. As discussed in Section 7.3, transport by tunneling can extend for 5–8 nm for aromatic molecules, and the range of possible tunneling barriers is compressed by the leveling effect of strong electronic coupling. However, Figure 7.14 shows at least one additional mechanism between tunneling and redox exchange that is effective at least over the range of 8–22 nm, which is consistent with field ionization. As indicated in Figure 7.29, conjugation length can exceed 10 nm and raises the possibility that a single molecular orbital can bridge between the two contacts. *Resonant* transport should occur when a molecular orbital of the contact/molecule/contact electronic system has an energy close to the contact Fermi level. The possible range of distances for resonant transport is not known, but is likely equal to or greater than the conjugation length and may resemble band transport in metals. Note that field ionization and resonant transport do not necessarily involve redox exchange and may be much more efficient than the *hopping* in bulk organic semiconductors. Note also that transport through >20 nm is likely possible without scattering, hence could be *ballistic*, that is, very fast, with lower power and heat dissipation demands than current microelectronic devices.

As described in Section 7.4, solid-state redox reactions can affect device behavior and are useful for creating two or more metastable states in molecular memory devices. Such reactions are controlled by many of the same phenomena as solution electrochemistry, notably activated ET, and charge compensation by ion motion. It is likely that less conventional redox events can occur in MJs, such as reorganization of a conjugated molecule during electron or hole transport through it.[82,100,101] The term *nanoionics* has been used to describe ion motion across nanoscopic dimensions, notably in memory devices.[102–104] Although the mobility of ions in solid-state devices can be low, the transport distances involved are often very short, and the ion transport time can easily be in the nanosecond to microsecond range. Ion transport and redox reactions are unlikely (and usually avoided) in conventional microelectronic devices and provide avenues to novel electronic behavior not currently possible.

After an initial burst of excitement over the prospect of molecules acting as circuit elements, we might consider the prospects of ME for practical integration in widely used microelectronic products. A prerequisite to such adoption is the requirement for compatibility with processing techniques and temperature excursions common to microelectronic manufacturing, as it appears likely that ME will be used initially in hybrid devices with conventional semiconductors. The existing industry is sophisticated and well established, so augmentation by integration of molecular devices is much more likely than replacement of silicon. The main driving force for the adoption of ME in the real world will be novel functions and properties not currently possible with silicon. Resonant transport, field ionization, redox reactions, ion motion, and chemical recognition are phenomena that are possible in molecular electronic systems that are not readily achieved with silicon. Assuming that these or other functions become possible with ME and yield new types of electronic devices, the additional processing concerns associated with integration of molecules in microelectronic manufacturing will likely be addressed readily.

ACKNOWLEDGMENTS

The author's work described in this chapter was supported by the National Research Council (Canada), the Natural Science and Engineering Research Council, the University of Alberta, and Alberta Innovates Technology Futures. The author thanks Bryan Szeto and Nikola Pekas for fabrication expertise and Adam Bergren and Haijun Yan for many useful discussions.

REFERENCES

1. Smalley, J. F., Finklea, H. O., Chidsey, C. E. D. et al. 2003 Heterogeneous electron-transfer kinetics for ruthenium and ferrocene redox moieties through alkanethiol monolayers on gold. *J. Am. Chem. Soc.* 125:2004–2013.
2. Sikes, H. D., Smalley, J. F., Dudek, S. P. et al. 2001 Rapid electron tunneling through oligophenylenevinylene bridges. *Science* (Washington, DC) 291:1519–1523.
3. Creager, S., Yu, C. J., Bamdad, C. et al. 1999 Electron transfer at electrodes through conjugated "Molecular Wire" bridges. *J. Am. Chem. Soc.* 121:1059–1064.
4. Finklea, H. O. 1996 Electrochemistry of organized monolayers of thiols and related molecules on electrodes. *Electroanalytical Chemistry*, ed Bard AJ (Dekker, New York), Vol. 19, pp. 109–335.

5. Sullivan, M. G. and Murray, R. W. 1994 Solid state electron self-exchange dynamics in mixed valent poly(vinylferrocene) films. *J. Phys. Chem.* 98:4343–4351.

6. Daum, P., Lenhard, J. R., Rolison, D., and Murray, R. W. 1980 Diffusional charge transport through ultrathin films of radiofrequency plasma polymerized vinylferrocene at low temperature. *J. Am. Chem. Soc.* 102:4649–4653.

7. Buttry, D. A. and Anson, F. C. 1983 Effects of electron exchange and single-file diffusion on charge propagation in nafion films containing redox couples. *J. Am. Chem. Soc.* 105:685–689.

8. Ofer, D., Crooks, R. M., and Wrighton, M. S. 1990 Potential dependence of the conductivity of highly oxidized polythiophenes, polypyrroles, and polyaniline: Finite windows of high conductivity. *J. Am. Chem. Soc.* 112:7869–7879.

9. Heath, J. R. and Ratner, M. A. 2003 Molecular electronics. *Phys. Today* 56:43–49.

10. Jortner, J. and Ratner, M. 1997 *Molecular Electronics* (Blackwell Science Ltd., Maiden, MA), p. 485.

11. Mirkin, C. A. and Ratner, M. A. 1992 Molecular electronics. *Annu. Rev. Phys. Chem.* 43:719–754.

12. Lindsay, S. M. and Ratner, M. A. 2007 Molecular transport junctions: Clearing mists. *Adv. Mater.* 19:23–31.

13. McCreery, R., Yan, H., and Bergren, A. J. 2013 A critical perspective on molecular electronic junctions: There is plenty of room in the middle. *Phys. Chem. Chem. Phys.* 15:1065–1081.

14. McCreery, R. L. and Bergren, A. J. 2009 Progress with molecular electronic junctions: Meeting experimental challenges in design and fabrication. *Adv. Mater.* 21:4303–4322.

15. McCreery, R. 2004 Molecular electronic junctions. *Chem. Mat.* 16:4477–4496.

16. Chen, F. and Tao, N. J. 2009 Electron transport in single molecules: From benzene to graphene. *Acc. Chem. Res.* 42:429–438.

17. Huang, Z. F., Chen, F., Bennett, P. A., and Tao, N. J. 2007 Single molecule junctions formed via Au-Thiol contact: Stability and breakdown mechanism. *J. Am. Chem. Soc.* 129:13225–13231.

18. Tao, N. J. 2006 Electron transport in molecular junctions. *Nat. Nanotechnol.* 1:173.

19. Caldwell, M. A., Jeyasingh, R. G. D., Wong, H. S. P., and Milliron, D. J. 2012 Nanoscale phase change memory materials. *Nanoscale* 4:4382–4392.

20. Heremans, P., Gelinck, G. H., Muller, R. et al. 2011 Polymer and organic nonvolatile memory devices. *Chem. Mater.* 23:341–358.

21. Chih-Yuan, L. and Kuan, H. 2009 Nonvolatile semiconductor memory revolutionizing information storage. *IEEE Nanotechnol. Mag.* 3:4–9.

22. Scott, J. C. and Bozano, L. D. 2007 Nonvolatile memory elements based on organic materials. *Adv. Mater.* 19:1452–1463.

23. Kuhr, W. G. 2004 Integration of molecular components into silicon memory devices. *Interface (The Electrochemical Society)* 13:34–38.

24. Waser, R., Dittmann, R., Staikov, G., and Szot, K. 2009 Redox-based resistive switching memories—Nanoionic mechanisms, prospects, and challenges. *Adv. Mater.* 21:2632–2663.

25. Strukov, D. B., Snider, G. S., Stewart, D. R., and Williams, R. S. 2008 The missing memristor found. *Nature* 453:80–83.

26. Wu, J. and McCreery, R. L. 2009 Solid-state electrochemistry in molecule/TiO$_2$ molecular heterojunctions as the basis of the TiO$_2$ "Memristor". *J. Electrochem. Soc.* 156:P29–P37.

27. Lee, M.-J., Lee, C. B., Lee, D. et al. 2011 A fast, high-endurance and scalable non-volatile memory device made from asymmetric Ta$_2$O$_{5-x}$/TaO$_{2-x}$ bilayer structures. *Nat. Mater.* 10:625–630.

28. Roth, K. M., Yasseri, A. A., Liu, Z. et al. 2003 Measurements of electron-transfer rates of charge-storage molecular monolayers on Si(100). Toward hybrid molecular/semiconductor information storage devices. *J. Am. Chem. Soc.* 125:505–517.

29. Wei, D., Baral, J. K., Osterbacka, R., and Ivaska, A. 2008 Electrochemical fabrication of a nonvolatile memory device based on polyaniline and gold particles. *J. Mater. Chem.* 18:1853–1857.

30. Chauhan, A. K., Aswal, D. K., Koiry, S. P. et al. 2008 Resistive memory effect in self-assembled 3-aminopropyltrimethoxysilane molecular multilayers. *Phys. Status Solid A* 205:373–377.

31. Frank, V., Stefan, C. J. M., Rene, A. J. J. et al. 2007 Reproducible resistive switching in nonvolatile organic memories. *Appl. Phys. Lett.* 91:192103.

32. Kim, D. C., Seo, S., Ahn, S. E. et al. 2006 Electrical observations of filamentary conductions for the resistive memory switching in NiO films. *Appl. Phys. Lett.* 88:202102.

33. Kozicki, M. N., Park, M., and Mitkova, M. 2005 Nanoscale memory elements based on solid-state electrolytes. *IEEE Trans. Nanotechnol.* 4:331–338.

34. Haick, H., Niitsoo, O., Ghabboun, J., and Cahen, D. 2007 Electrical contacts to organic molecular films by metal evaporation: Effect of contacting details. *J. Phys. Chem. C* 111:2318–2329.

35. Kim, B., Choi, S. H., Zhu, X. Y., and Frisbie, C. D. 2011 Molecular tunnel junctions based on pi-conjugated oligoacene thiols and dithiols between Ag, Au, and Pt contacts: Effect of surface linking group and metal work function. *J. Am. Chem. Soc.* 133:19864–19877.
36. Kim, B., Beebe, J. M., Jun, Y., Zhu, X. Y., and Frisbie, C. D. 2006 Correlation between HOMO alignment and contact resistance in molecular junctions: Aromatic thiols versus aromatic isocyanides. *J. Am. Chem. Soc.* 128:4970–4971.
37. Bergren, A. J., McCreery, R. L., Stoyanov, S. R., Gusarov, S., and Kovalenko, A. 2010 Electronic characteristics and charge transport mechanisms for large area aromatic molecular junctions. *J. Phys. Chem. C* 114:15806–15815.
38. McCreery, R., Wu, J., and Kalakodimi, R. J. 2006 Electron transport and redox reactions in carbon based molecular electronic junctions. *Phys. Chem. Chem. Phys.* 8:2572–2590.
39. Ranganathan, S., McCreery, R. L., Majji, S. M., and Madou, M. 2000 Photoresist-derived carbon for microelectrochemical applications. *J. Electrochem. Soc.* 147:277–282.
40. Ranganathan, S. and McCreery, R. L. 2001 Electroanalytical performance of carbon films with near-atomic flatness. *Anal. Chem.* 73:893–900.
41. Kostecki, R., Song, X., and Kinoshita, K. 1999 Electrochemical analysis of carbon interdigitated microelectrodes. *Electrochem. Solid State Lett.* 2:465.
42. Fairman, C., Yu, S., Liu, G. et al. 2008 Exploration of variables in the fabrication of pyrolysed photoresist. *J. Solid State Electrochem.* 12:1357–1365.
43. Bergren, A. J. and McCreery, R. L. 2011 Analytical chemistry in molecular electronics. *Annu. Rev. Anal. Chem.* 4:173–195.
44. Ru, J., Szeto, B., Bonifas, A., and McCreery, R. L. 2010 Microfabrication and integration of diazonium-based aromatic molecular junctions. *ACS Appl. Mater. Inter.* 2:3693–3701.
45. Belanger, D. and Pinson, J. 2011 Electrografting: A powerful method for surface modification. *Chem. Soc. Rev.* 40:3995–4048.
46. Lehr, J., Garrett, D. J., Paulik, M. G. et al. 2010 Patterning of metal, carbon, and semiconductor substrates with thin organic films by microcontact printing with aryldiazonium salt inks. *Anal. Chem.* 82:7027–7034.
47. Garrett, D. J., Lehr, J., Miskelly, G. M., and Downard, A. J. 2007 Microcontact printing using the spontaneous reduction of aryldiazonium salts. *J. Am. Chem. Soc.* 129:15456–15457.
48. Pinson, J. and Podvorica, F. 2005 Attachment of organic layers to conductive or semiconductive surfaces by reduction of diazonium salts. *Chem. Soc. Rev.* 34:429–439.
49. Brooksby, P. A., Downard, A. J., and Yu, S. S. C. 2005 Effect of applied potential on arylmethyl films oxidatively grafted to carbon surfaces. *Langmuir* 21:11304–11311.
50. Delamar, M., Hitmi, R., Pinson, J., and Saveant, J. M. 1992 Covalent modification of carbon surfaces by grafting of functionalized aryl radicals produced from electrochemical reduction of diazonium salts. *J. Am. Chem. Soc.* 114:5883–5884.
51. Kariuki, J. K. and McDermott, M. T. 2001 Formation of multilayers on glassy carbon electrodes via the reduction of diazonium salts. *Langmuir* 17:5947–5951.
52. Kariuki, J. K. and McDermott, M. T. 1999 Nucleation and growth of functionalized aryl films on graphite electrodes. *Langmuir* 15:6534–6540.
53. Anariba, F., DuVall, S. H., and McCreery, R. L. 2003 Mono- and multilayer formation by diazonium reduction on carbon surfaces monitored with atomic force microscopy "Scratching". *Anal. Chem.* 75:3837–3844.
54. Yan, H., Bergren, A. J., McCreery, R. et al. 2013 Activationless charge transport across 4.5 to 22 nm in molecular electronic junctions. *Proc. Nat. Acad. Sci. U. S. A.* 110:5326–5330.
55. Walker, A. V., Tighe, T. B., Haynie, B. C. et al. 2005 Chemical pathways in the interactions of reactive metal atoms with organic surfaces: Vapor deposition of Ca and Ti on a methoxy-terminated alkanethiolate monolayer on Au. *J. Phys. Chem. B* 109:11263–11272.
56. Walker, A. V., Tighe, T. B., Cabarcos, O. M. et al. 2004 The dynamics of noble metal atom penetration through methoxy-terminated alkanethiolate monolayers. *J. Am. Chem. Soc.* 126:3954–3963.
57. Fisher, G. L., Walker, A. V., Hooper, A. E. et al. 2002 Bond insertion, complexation, and penetration pathways of vapor-deposited aluminum atoms HO- and CH_3O-terminated organic monolayers. *J. Am. Chem. Soc.* 124:5528–5541.
58. Mahmoud, A. M., Bergren, A. J., Pekas, N., and McCreery, R. L. 2011 Towards integrated molecular electronic devices: Characterization of molecular layer integrity during fabrication processes. *Adv. Funct. Mater.* 21:2273–2281.

59. Nowak, A. M. and McCreery, R. L. 2004 In situ Raman spectroscopy of bias-induced structural changes in nitroazobenzene molecular electronic junctions. *J. Am. Chem. Soc.* 126:16621–16631.
60. DeIonno, E., Tseng, H. R., Harvey, D. D., Stoddart, J. F., and Heath, J. R. 2006 Infrared spectroscopic characterization of [2]rotaxane molecular switch tunnel junction devices. *J. Phys. Chem. B* 110:7609–7612.
61. Wang, W., Scott, A., Gergel-Hackett, N. et al. 2008 Probing molecules in integrated silicon-molecule-metal junctions by inelastic tunneling spectroscopy. *Nano Lett.* 8:478–484.
62. Richter, C. A., Hacker, C. A., and Richter, L. J. 2005 Electrical and spectroscopic characterization of metal/monolayer/Si devices. *J. Phys. Chem. B* 109:21836–21841.
63. Hacker, C., Batteas, J. D., Garno, J. C. et al. 2004 Structural and chemical characterization of mono-fluoro substituted oligo(phenylene-ethynylene) thiolate self-assembled monolayers on gold. *Langmuir* 20:6195–6205.
64. Yan, H., Bergren, A. J., and McCreery, R. L. 2011 All-carbon molecular tunnel junctions. *J. Am. Chem. Soc.* 133:19168–19177.
65. Haynie, B. C., Walker, A. V., Tighe, T. B., Allara, D. L., and Winograd, N. 2003 Adventures in molecular electronics: How to attach wires to molecules. *Appl. Surf. Sci.* 203–204:433–436.
66. Bonifas, A. P. and McCreery, R. L. 2010 "Soft" Au, Pt and Cu contacts for molecular junctions through surface-diffusion-mediated deposition. *Nat. Nanotechnol.* 5:612–617.
67. Bonifas, A. P. and McCreery, R. L. 2011 Assembling molecular electronic junctions one molecule at a time. *Nano Lett.* 11:4725–4729.
68. Pan, H., Li, Y., Wu, Y. et al. 2007 Low-temperature, solution-processed, high-mobility polymer semiconductors for thin-film transistors. *J. Am. Chem. Soc.* 129:4112–4113.
69. Wu, Y., Liu, P., Ong, B. S. et al. 2005 Controlled orientation of liquid-crystalline polythiophene semiconductors for high-performance organic thin-film transistors. *Appl. Phys. Lett.* 86:142102–142103.
70. Ong, B. S., Wu, Y., Liu, P., and Gardner, S. 2004 High-performance semiconducting polythiophenes for organic thin-film transistors. *J. Am. Chem. Soc.* 126:3378–3379.
71. Kumar, R., Pillai, R. G., Pekas, N., Wu, Y., and McCreery, R. L. 2012 Spatially resolved Raman spectroelectrochemistry of solid-state polythiophene/viologen memory devices. *J. Am. Chem. Soc.* 134:14869–14876.
72. Shoute, L. C. T., Wu, Y., and McCreery, R. L. 2013 Direct spectroscopic monitoring of conductance switching in polythiophene memory devices. *Electrochim. Acta* 110:437–445.
73. Shoute, L., Pekas, N., Wu, Y., and McCreery, R. 2011 Redox driven conductance changes for resistive memory. *Appl. Phys A: Mater Sci. Process.* 102:841–850.
74. Yang, H.-H. and McCreery, R. L. 1999 Effects of surface monolayers on the electron transfer kinetics and adsorption of methyl viologen and phenothiazine derivatives on glassy carbon electrodes. *Anal. Chem.* 71:4081–4087.
75. Simmons, J. G. 1964 Generalized thermal J-V characteristic for the electric tunnel effect. *J. Appl. Phys.* 35:2655–2658.
76. Simmons, J. G. 1963 Generalized formula for the electric tunnel effect between similar electrodes separated by a thin insulating film. *J. Appl. Phys.* 34:1793–1803.
77. Simmons, J. G. 1963 Electric tunnel effect between dissimilar electrodes separated by a thin insulating film. *J. Appl. Phys.* 34:2581–2590.
78. Sayed, S. Y., Fereiro, J. A., Yan, H., McCreery, R. L., and Bergren, A. J. 2012 Charge transport in molecular electronic junctions: Compression of the molecular tunnel barrier in the strong coupling regime. *Proc. Natl. Acad. Sci. U. S. A.* 109:11498–11503.
79. Har-Lavan, R., Yaffe, O., Joshi, P. et al. 2012 Ambient organic molecular passivation of Si yields near-ideal, Schottky-Mott limited, junctions. *AIP Adv.* 2:012164.
80. Koch, N. and Vollmer, A. 2006 Electrode-molecular semiconductor contacts: Work-function-dependent hole injection barriers versus Fermi-level pinning. *Appl. Phys. Lett.* 89:162107–162113.
81. Simmons, J. G. 1964 Potential barriers and emission-limited current flow between closely spaced parallel metal electrodes. *J. Appl. Phys.* 35:2472–2481.
82. Burin, A. L., Berlin, Y. A., and Ratner, M. A. 2001 Semiclassical theory for tunneling of electrons interacting with media. *J. Phys. Chem. A* 105:2652–2659.
83. Buttiker, M. and Landauer, R. 1982 Traversal time for tunneling. *Phys. Rev. Lett.* 49:1739.
84. Fereiro, J. A., McCreery, R. L., and Bergren, A. J. 2013 Direct optical determination of interfacial transport barriers in molecular tunnel junctions. *J. Am. Chem. Soc.* 135:9584–9587.
85. Hwang, J., Wan, A., and Kahn, A. 2009 Energetics of metal–organic interfaces: New experiments and assessment of the field. *Mater. Sci. Eng.: R: Reports* 64:1–31.

86. Salomon, A., Shpaisman, H., Seitz, O., Boecking, T., and Cahen, D. 2008 Temperature-dependent electronic transport through alkyl chain monolayers: Evidence for a molecular signature. *J. Phys. Chem. C* 112:3969–3974.

87. Haick, H. and Cahen, D. 2008 Making contact: Connecting molecules electrically to the macroscopic world. *Prog. Surf. Sci.* 83:217–261.

88. Sze, S. M. 1981 *Physics of Semiconductor Devices* (Wiley, New York) 2nd edn.

89. Bonifas, A. P. and McCreery, R. L. 2012 Solid state spectroelectrochemistry of redox reactions in polypyrrole/oxide molecular heterojunctions. *Anal. Chem.* 84:2459–2465.

90. Barman, S., Deng, F., and McCreery, R. 2008 Conducting polymer memory devices based on dynamic doping. *J. Am. Chem. Soc.* 130:11073–11081.

91. Strachan, J. P., Pickett, M. D., Yang, J. J. et al. 2010 Direct identification of the conducting channels in a functioning memristive device. *Adv. Mater.* 22:3573–3577.

92. Strukov, D. B. and Williams, R. S. 2009 Exponential ionic drift: Fast switching and low volatility of thin-film memristors. *Appl. Phys. A: Mater. Sci. Process.* 94:515–519.

93. Gergel-Hackett, N., Hamadani, B., Dunlap, B. et al. 2009 A flexible solution-processed memristor. *IEEE Electr. Device Lett.* 30:706–708.

94. Das, B. C., Pillai, R. G., Wu, Y., and McCreery, R. L. 2013 Redox-gated three-terminal organic memory devices: Effect of composition and environment on performance. *ACS Appl. Mater. Inter.* 5:11052–11058.

95. Herlogsson, L., Noh, Y.-Y., Zhao, N. et al. 2008 Downscaling of organic field-effect transistors with a polyelectrolyte gate insulator. *Adv. Mater.* 20:4708–4713.

96. Kaake, L. G., Zou, Y., Panzer, M. J., Frisbie, C. D., and Zhu, X. Y. 2007 Vibrational spectroscopy reveals electrostatic and electrochemical doping in organic thin film transistors gated with a polymer electrolyte dielectric. *J. Am. Chem. Soc.* 129:7824–7430.

97. Panzer, M. J. and Frisbie, C. D. 2005 Polymer electrolyte gate dielectric reveals finite windows of high conductivity in organic thin film transistors at high charge carrier densities. *J. Am. Chem. Soc.* 127:6960–6961.

98. Xia, Y., Cho, J., Paulsen, B., Frisbie, C. D., and Renn, M. J. 2009 Correlation of on-state conductance with referenced electrochemical potential in ion gel gated polymer transistors. *Appl. Phys. Lett.* 94:013304.

99. Frisch, M. J., Trucks, G. W., Schlegel, H. B. et al. 2009 Gaussian 09 (Gaussian, Inc., Pittsburgh, PA), A.7.

100. McCreery, R. L. 2009 Electron transport and redox reactions in molecular electronic junctions. *Chem. Phys. Chem.* 10:2387–2391.

101. Yeganeh, S., Galperin, M., and Ratner, M. A. 2007 Switching in molecular transport junctions: Polarization response. *J. Am. Chem. Soc.* 129:13313–13320.

102. Tsuruoka, T., Terabe, K., Hasegawa, T. et al. 2012 Effects of moisture on the switching characteristics of oxide-based, gapless-type atomic switches. *Adv. Funct. Mater.* 22:70–77.

103. Chan, W. K., Haverkate, L. A., Borghols, W. J. H. et al. 2011 Direct view on nanoionic proton mobility. *Adv. Funct. Mater.* 21:1364–1374.

104. Waser, R. and Aono, M. 2007 Nanoionics-based resistive switching memories. *Nat. Mater.* 6:833–840.

8 Stochastic Events in Nanoelectrochemical Systems

Allen J. Bard, Aliaksei Boika, Seong Jung Kwon,
Jun Hui Park, and Scott N. Thorgaard

CONTENTS

8.1 INTRODUCTION

In the realm of nanoscience, studies of single molecules, nanostructures, and nanoparticles (NPs) are an active research area that often involves comparison of results obtained from single-molecule studies with those of the more familiar ensemble investigations of the same material. Most of the work in this area has employed immobilized single entities studied by spectroscopic methods (e.g., fluorescence and surface-enhanced Raman spectroscopy [SERS]). While these powerful methods have provided very useful results, they are limited in the particular molecules that can be studied and their stability under intense laser radiation. The basic spatial resolution optical imaging is limited by diffraction, and very special configurations are often needed to carry out such studies. Electrochemical methods can provide a complementary approach, and this chapter deals with such studies.

8.1.1 ENSEMBLES VERSUS SINGLE MOLECULE/NPs

In most electrochemical studies, one employs solutions where the concentration of the electroactive species, i, is ~1 mM. With these concentrations, the diffusion flux of electroactive species to the electrode, f_i, is of the order of $m_i C_i^*$, where m_i is the mass transfer coefficient (cm/s) and C_i^* is the bulk concentration (mol/cm³). With $m_i \sim 10^{-3}$ cm/s, this produces fluxes of the order of 10^{-9} mol/s/cm² or 6×10^{14} molecules/s/cm², producing currents of 10^{-4} A/cm². Under these conditions, even with very small electrodes, one measures the behavior of large ensembles of molecules. However, if the concentration of electroactive species is dropped to ~1 pM, these fluxes drop to $f_i = 10^{-18}$ mol/s/cm² or 6×10^5 molecules/s/cm² with a current density of 10^{-13} A/cm². Thus, with an ultramicroelectrode (UME) with about a 10 μm size or area of about 10^{-6} cm², the number of molecules arriving by diffusion to the electrode is about 1/s. In our previous work, we showed that by using very small (~μm) nanoelectrocatalytic C or Au electrodes with relatively small background currents, that nanometer-size electrocatalytic NPs, for example, of Pt, amplify the current of an appropriate inner-sphere (IS) reaction (e.g., hydrazine oxidation or proton reduction) to the pA level, and the frequency, size, and shape of collision events could be investigated.[1] More recent work with different approaches has shown that interactions of the NPs with the electrode can be detected, even for outer-sphere (OS) reactions, as described in later sections of this chapter.

This chapter reviews approaches to observing and analyzing single-particle collision events for different types of NPs. NPs can be classified as *hard* or *soft*; hard particles include conductive NPs, like metals (Pt, Au, and Ag) and carbon. Insulating (dielectric) particles include polymer (Teflon, Latex) and inorganic ones (SiO_2 and Al_2O_3). Also included in hard NPs are the widely investigated semiconductor NPs (CdSe and TiO_2), sometimes also called *quantum dots* or *nanocrystals*. Soft NPs include liquid droplets in emulsions, vesicles, micelles, and various biological structures, like viruses and proteins. The characterization of such NPs includes finding the shape (geometry) (e.g., sphere, rod and tube), size concentration, charge, and properties (energy levels, band gap, diffusion coefficient, etc.). The most widely used methods are probably electron microscopy, especially transmission electron microscopy (TEM) for observing the size and shapes, and dynamic light scattering (DLS) that investigates the response of an ensemble and uses modeling to find a particle size distribution.

This chapter on electrochemical approaches to NP characterization is organized considering both the different types of NPs that are appropriate for different techniques. Thus, particles that can carry out electrocatalysis can show amperometric responses over the appropriate potential range. They can also be studied by observing changes in the open circuit potential (OCP). Insulating particles can be observed by how they block an electrochemical reaction of a solution redox species by landing on an electrode and decreasing its effective area. Noncatalytic conducting particles can be observed by an analogous increase in the electrode area of the electrode on sticking. In these latter two approaches, the size of the electrode compared to that of the NPs is clearly important. The analysis of the behavior is based on measuring the frequency of the collisions, the magnitude, and the shape of the response.

It is useful to describe a given system with the shorthand notation NP/UME/electrochemistry.[2] For example, the system where Pt NP collisions on a Au electrode involve hydrazine oxidation can be represented Pt NP/Au UME/hydrazine oxidation. This can be amplified by adding the nature of the electrode reaction (catalytic amplification, area blocking, etc.) and the measurement approach (amperometry, OCP change, voltammetry, etc.), for example, Pt NP/Au UME/hydrazine oxidation/catalytic amplification/amperometry.

There are actually a number of ways that single NP collisions can be observed, as long as the background in the absence of NPs is small and clear signals from a particle interaction with the electrode can be recorded above the noise level. For example, as described in the following sections, one can observe a faradaic reaction involving oxidative decomposition of the particle. One can also note on how a collision affects the UME behavior of a reaction of a solution electrode reaction by either blocking a portion of the electrode area (blocking) or increasing the area (area amplification). Collisions can also affect the OCP or another property (e.g., double-layer capacitance) of the UME.

8.2 COLLISION EXPERIMENTS: ELECTROCATALYTIC AMPLIFICATION

8.2.1 EXAMPLES

As explained earlier, when NPs in solution occasionally collide with an inert (less active electrochemically) electrode and make electrical contact (i.e., the NP resides within tunneling distance), the NP behaves as a nanoelectrode. To detect a single electrically contacted NP at a UME, an IS electron transfer reaction is used, wherein the reactant strongly interacts with the electrode material. Therefore, the electrocatalytic current response varies with the electrode material. An inert electrode is obtained by choosing an electrode material at which the catalytic reaction is sluggish within a certain potential region and where the NP material shows good electrocatalytic current at the same potential. With an OS electron transfer reaction, there is little interaction between the electrode and the reactant, so there is less difference in response between different electrode materials.

For amperometric collision detection using electrocatalytic amplification (EA), one finds an appropriate electrode potential where the particle materials give its maximum current while the inert electrode displays a minimal current response. As shown in Figure 8.1, by comparing the electrochemical behavior of the electrocatalytic reaction of interest on two different electrode materials, the useful applied potential is easily determined. In other words, the catalytic difference between the two materials is used to detect catalytic NPs at a less catalytic electrode. NPs can be detected

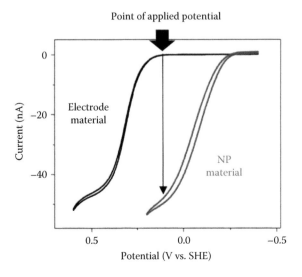

FIGURE 8.1 Schematic illustration of single NP detection using electrocatalytic differences of two materials.

using various IS reactions. For example, Pt NPs have been detected using the proton reduction reaction,[1] the hydrazine oxidation reaction,[3] and the hydrogen peroxide reduction reaction.[1] Au NPs were detected using the sodium borohydride oxidation reaction[4]; iridium oxide (IrOx) NPs were also detected using the water oxidation reaction.[5] Other reactions like oxygen reduction, formic acid oxidation, and methanol oxidation are also conceivable. However, the amount of current from oxygen reduction is limited by its poor solubility in water, and also the oxidation of small organic molecules, such as methanol and formic acid, can lead to poisoning of the surface by adsorbed intermediates (e.g., CO).[3] Therefore, each system can have a unique response depending on the intrinsic property of the metal, reactant, reaction intermediate, and reaction environment. Types of responses used to detect NPs are described in following sections.

8.2.2 Types of Response

The nature of the current response upon a single NP collision event depends on how strongly the NP is adsorbed on the electrode surface. An adsorbed NP behaves like a nanosphere electrode on an inert plane and produces a steplike current increase. The step height can be estimated from Equation 8.1 when the rate of charge transfer from the NP to the solution species is fast (i.e., diffusion-limited conditions):

$$i_{ss} = 4\pi(ln2)n_0FD_0C_0r_0 \qquad (8.1)$$

where
  n_0 is the number of electrons transferred
  F is the Faraday
  D_0 is the diffusion coefficient of reactant
  C_0 is the concentration of reactant
  r_0 is the radius of the NP[6]

A nonadsorbing NP or one that decomposes or whose surface is deactivated can show spike or blip responses, and the shape is affected from the residence time, if the adsorbed NP is ideally stable on the UME surface. The step current response may be attenuated by impurity adsorption, entrapment of a bubble, NP decomposition, or NP detachment. Thus, the current response for single NP collisions is different for each system.

Staircase current response: A staircase current is observed when the particles collide with the electrode surface and stick (or adsorb). A steady-state current response is obtained with the following electrocatalytic systems: proton reduction,[1] hydrazine oxidation,[3] and hydrogen peroxide reduction.[1] In many cases, the staircase current is distorted by a current decay after collision events. This decay is caused by a deactivation process, such as the adsorption of impurities, and the rate of decay depends on the experimental conditions.

Proton reduction is sluggish at carbon electrodes in 50 mM sodium dihydrogen citrate at an applied potential of −0.5 V (vs. NHE). However, a Pt NP-covered carbon electrode showed a steady-state limiting current at −0.5 V as shown in Figure 8.2b. According to Equation 8.1, a steady-state limiting current is estimated to be ~60 pA when we assume an attached Pt NP is an ideal sphere 4 nm in diameter. The Pt NPs were made by $NaBH_4$ reduction of H_2PtCl_6 in the presence of sodium citrate.[7] Synthesized citrate-capped Pt NPs having diameters of 4 ± 0.8 nm were characterized by TEM. Figure 8.2c shows the current transients in the absence and presence of Pt NPs and carbon NPs, respectively. When Pt NPs are added, the expected magnitude of current steps (i.e., 40–80 pA) is observed. However, the shape of the current was not steplike, and the current decayed over tens of seconds. This decay has been attributed to deactivation of the NP surface, either by the adsorption of reaction intermediates, adventitious impurities, or hydrogen gas bubbles. When the potential of the electrode was shifted positive, the amplitude of the current step decreased, which is in

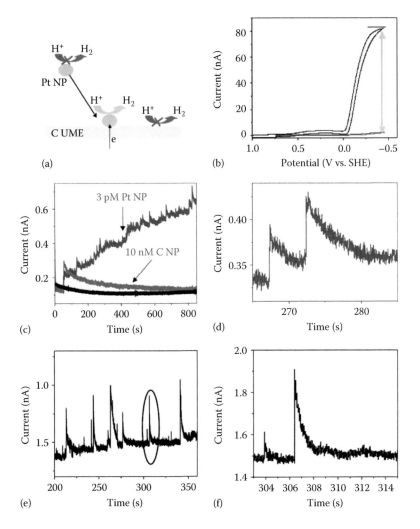

FIGURE 8.2 (a) Schematic of a single platinum NP collision event for the Pt NP/C UME/proton reduction system. (b) Electrochemical reduction of proton at carbon fiber electrode without and with Pt NP on the surface in air-saturated 50 mM sodium dihydrogencitrate solution (fiber diameter, 8 μm; scan rate, 100 mV/s). (c) Current–time curve recorded before (black) and after injection of carbon (red) and Pt (blue) NP solutions in 50 mM sodium dihydrogen citrate (electrode potential, −0.5 V). (d) Zoom in of panel (c). (e) Current–time curve in 10 mM perchloric acid and 20 mM sodium perchlorate in the presence of Pt NP(12.5 pM). (f) Zoom in of panel (e).

agreement with Figure 8.2b. As expected from Equation 8.1, the proton concentration affects the steady-state limiting current. In this case, large current spikes were seen in the presence of Pt NPs as expected; however, the spikes eventually disappeared after ~600 s, which can be explained by the unstable status of citrated Pt NPs at low pH where carboxylic acid groups are protonated and the NPs may aggregate and settle out.

Hydrazine oxidation shows obvious catalytic differences with different electrode materials (Pt and Au) as shown in Figure 8.3b. A Pt electrode has a more negative onset potential for hydrazine oxidation compared to a Au electrode. Therefore, Pt NP detection at a Au UME is possible. Figure 8.3c shows a current transient at a Au UME in the presence and absence of Pt NPs.

A steplike current increase was found only in the presence of Pt NPs. In each current step, the current increased rapidly and then remained at a steady level, which is a clear evidence for Pt NP

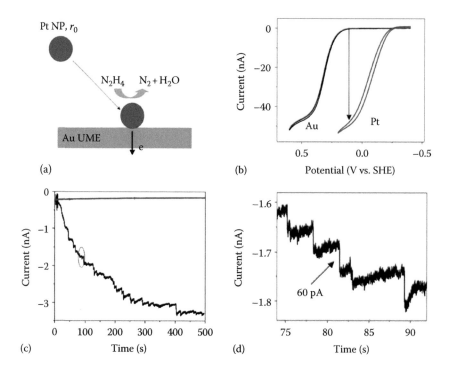

FIGURE 8.3 (a) Schematic illustration of detecting a single Pt NP at the Au UME using the hydrazine oxidation reaction. (b) Electrochemical oxidation of hydrazine at Pt and Au UMEs. (c) Current–time curve recorded before and after injection of Pt NP solution in 15 mM hydrazine and 50 mM phosphate buffer, pH ~ 7.5. (d) Zoom in of panel (c).

attachment at the Au UME. Most of current steps were in the range of 40–65 pA, which correspond to a 2.6–4.3 nm diameter spherical NP from Equation 8.1, the equation of steady-state diffusion-limited current. The Pt NP size as estimated by the current agreed with measurements obtained by TEM. The occasional large step currents are probably due to the adsorption of aggregated NPs. Collision experiments were also performed under different experimental conditions to confirm the proposed collision model (i.e., that an attached NP behaves as an individual nanoelectrode giving a steady-state current). When the hydrazine concentration was increased, the amplitude of the current step increased proportionally. When the NP concentration was increased, the step frequency increased accordingly while the magnitude of the step currents was unchanged. Use of larger diameter Pt NPs increased the current step height.

The adhesion of NPs on the UME surfaces was investigated by pretreatment of the UME. A piranha solution–treated C UME showed about five times higher collision events as compared to an untreated C UME under the same conditions as shown in Figure 8.4. This result implies that only some of the NPs that collide with the UME actually stick to the electrode surface as the NP flux to the electrode is unchanged in these cases. The effect of electrode modification is discussed in the following section.

Hydrogen peroxide reduction has also been used to detect Pt NPs at a Au electrode. Au is less catalytic than Pt for the hydrogen peroxide reduction reaction, and moreover, Au electrodes are easily modified using thiolate self-assembled monolayers (SAMs). Here, benzenedimethanethiol, which allows electron tunneling to solution species, was used to bind the collided particle to the electrode by its terminal thiolate group. A steplike current increase is observed in the presence of the Pt NPs as shown in Figure 8.5. Besides clear current steps, current fluctuations with a high frequency were also observed. This may be attributed to either smaller Pt NPs in the preparation or

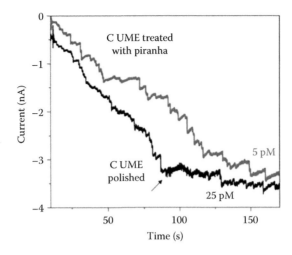

FIGURE 8.4 Current transients recorded before and after injection of Pt NPs at a C UME polished (black) and further treated with piranha solution (red). Electrode potential, 0.5 V; Pt NP size, ~3.6 nm; test electrolyte, 15 mM hydrazine and 50 mM phosphate buffer.

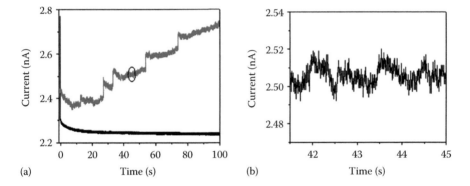

FIGURE 8.5 (a) Current transients at a benzenedimethanethiol modified 25 μm diameter Au electrode at 0.1 V versus SHE in 0.1 M phosphate buffer (pH 7.4) solution containing 50 mM hydrogen peroxide in the absence (black) and presence of 25 pM Pt NP (red). The background current in black curve was increased by 1.5 nA for clarity. Panel (b) is a zoom in of panel (a).

from heterogeneous catalytic decomposition of hydrogen peroxide. Gas bubbles are observed when Pt NPs are injected into a fresh hydrogen peroxide solution. Though the presence of Pt NPs can be established using the hydrogen peroxide reaction at a benzenedimethanethiol-modified Au UME, the hydrogen peroxide reduction is not adequate for the discrimination of single NP collisions.

Blip (or spike) response: A *blip*-type response can occasionally be observed when a particle makes electrical contact with the electrode surface only for a short time, such as when the NP contacts the surface but cannot stick well. The amperometric response for an elastic (i.e., nonsticking) collision event should have a spiky shape with its width determined by the particle residence time. When the NP does not irreversibly stick to the UME surface, the collision frequency cannot be estimated from the diffusion-controlled flux of NPs because a concentration gradient is not generated. For the case of a blip response, it is not possible to estimate the number of attached NPs on the UME from the final current, in contrast to the current step response where the number of attached particles can be estimated using the final current, assuming there is no deactivation process for the particles. However, systems exhibiting a blip-type response still retain a stable background current,

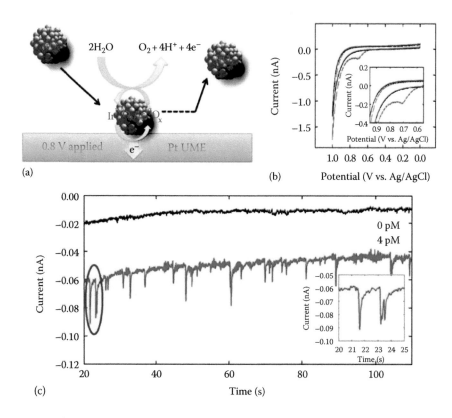

FIGURE 8.6 (a) Schematic illustration of single IrOx NP collision event. (b) Cyclic voltammograms of water oxidation at Pt UME (radius 5 μm) in pH 13 solution (0.1 M NaOH) containing 0 (black solid) or 12 pM IrOx NPs (red dashed). Scan rate is 50 mV/s. (c) Chronoamperometric curves for single IrOx NP (radius ~14 nm) collisions at the NaBH$_4$-treated Pt UME (radius 5 μm) in pH 13 solution (0.1 M NaOH) without (black) and with (red) 4 pM IrOx NPs. Applied potential is 0.8 V (vs. Ag/AgCl). Data acquisition time is 50 ms. The small change in the background current may be caused by a small number of IrOx NPs adhering to the electrode.

which makes it easy to distinguish the collision events and hinders the signal interference of consecutive NP collision. Thus, NP detection based on blip-type responses may be advantageous in analytical applications (e.g., investigating size distributions of NPs).

IrOx is a good electrocatalyst for water oxidation. As shown in Figure 8.6b, the water oxidation reaction is sluggish at Au UME in 0.1 M NaOH solution at +0.8 V versus Ag/AgCl. However, the reaction is accelerated in the presence of citrated IrOx NPs in solution. Electrochemically generated hexavalent IrOx NPs have been proposed to act as a catalyst for water oxidation.[8] After injecting IrOx NPs, blip responses were observed as shown in Figure 8.6c. The blip responses of IrOx NP are attributed to poisoning of the NPs by reaction intermediate oxygen.[9] Thereby, the addition of an oxygen scavenger (e.g., sodium sulfite) changed the collision response from spiky current transient to stair-like decaying current (data not shown). Current transients at potentials above +0.7 V versus Ag/AgCl showed that the amplitude of the current spikes increased with the applied potential, while with a potential of +0.7 V or below, spiky current transients were not observed. This is in agreement with that shown in Figure 8.6b. As in the staircase current response, the blip-type current response is also affected by the size of NPs, their concentration, the applied potential, and the surface state of the electrode. The current spikes were also sensitive to the electrode surface conditions, similar to the experiments where a piranha solution–treated C UME was used for detection of Pt NPs by the hydrazine oxidation reaction. A Pt UME pretreated in 10 mM sodium

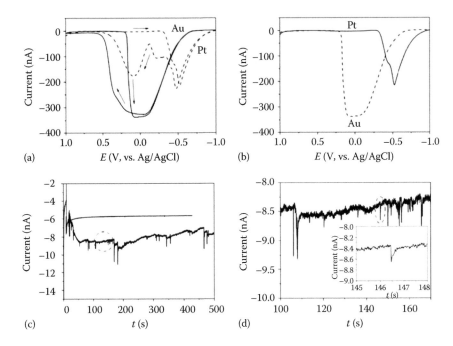

FIGURE 8.7 (a) Cyclic voltammetry of Au and Pt UME in 10 mM NaBH4 and 0.1 M NaOH solution. Scan rate, 100 mV/s. (b) Negative scan in panel (a). (c) Current–time curve without (upper curve) and with (bottom curve) injected Au NPs at the preoxidized Pt UME. (d) Zoom in of marked region in panel (c). UME diameter, 10 µm; NP concentration, 24 pM; potential, 0 V.

borohydride for 10 min showed a higher collision frequency than an untreated Pt UME under the same conditions.

To demonstrate the detection of Au NPs at a Pt UME, the $NaBH_4$ oxidation reaction was used. In most IS electron transfer reactions, Pt showed better catalytic activity than Au. Thereby, detecting a Au NP at a Pt UME seemed challenging given the procedures used in earlier experiments wherein amperometric detection of NP collisions is accomplished by EA under conditions where the NP has better catalytic activity than the underlying UME material. Here, this condition was met by first forming a layer of Pt oxide on the UME surface, which is capable of blocking surface sites active for the catalytic reaction. Gold oxide is reduced at more positive potentials than Pt oxide in a solution containing 0.1 M $NaBH_4$ and 0.1 M NaOH, as shown in Figure 8.7b. To detect a Au NP, a two-step approach was used: (step 1) A potential of +0.9 V versus Ag/AgCl was applied to grow an oxide layer on the Pt surface, and (step 2) the potential was then changed to 0 V versus Ag/AgCl where the Pt oxide layer is still preserved and Au NP can oxidize BH_4^-. The current–time curve of preoxidized Pt UME in the presence of citrated Au NPs (diameter ~14 nm) is shown in Figure 8.7c. For each collision, the spiky current transient is observed, which is similar to the one observed in IrOx NP detection experiments. The collision transients lasted only about 1 s, and the current then returned to the background level, which indicates the NPs leave or are deactivated after the collision. The amplitude of each current spike depends on the particle size, residence time, electrode reaction rate constants, and the strength of the interaction between the NPs and the electrode. Most of the peaks were in the range of 100–400 pA, which are smaller than the estimated value (~1 nA) from Equation 8.1. Integrating these peaks finds the total charge transferred to be about 2×10^{-11} C. This amount of charge is much higher than that required to oxidize the whole Au NP (around $1.4–4 \times 10^{-14}$ C, assuming one to three electrons transferred per Au atom). Therefore, these current transients originate from the catalyzed $NaBH_4$ oxidation reaction rather than reactions of the Au NP itself.

8.2.3 ELECTRODE AND PARTICLE MODIFICATION

The collision responses are affected by the nature of interactions between the collided NPs and the electrode. The adhesion of NPs to the electrode is a complex process that is still poorly understood. The adhesion may depend on many factors, such as the electrode material, particle material, particle capping agent, point of zero charge, electrode surface treatments, the electrolyte solution composition, and the electrode reaction. Therefore, chemical modification of the NPs or the UME can change the particles' adhesive as well as nature of the electron transfer during the experiment.

To study the viability of using linker molecules to modulate the NP–UME interactions, the effects (e.g., electronic and electrochemical) of small molecules adsorbed on the UME have been investigated,[10] especially electrode modification using alkanethiol molecules.[11]Alkanethiol molecules form SAMs on many metal surfaces, which can hinder surface reactions (i.e., IS electron transfer reactions) and suppress OS reaction rates through length-dependent electron transport (e.g., tunneling) through the SAM. Electrocatalytically amplified collision detection methods use only IS reactions (e.g., proton reduction and hydrazine oxidation). For these, even a short-chain thiol, for example, C3, will block the electrode reaction. The NP surface should thus be partly exposed for catalytic amplification. However, the electron transfer from UME to bare NP surface is probably an OS reaction, which is only affected by longer-chain thiols. Moreover, electron transfer rates to a metal NP should be less affected than rates to individual ions or molecules. The detection of NP collisions at a SAM-modified UME has been investigated using the hydrazine oxidation reaction. The electron tunneling distance from the UME to the NP was varied by using alkanethiol SAMs of different chain length. The instance of the particle collision current events depended on the length of SAM as shown in Figure 8.8. Therefore, if electrode modification involving adhesion using a SAM is desired, shorter molecules are recommended in order to maintain the amplitude of current response.

The influence of the NP surface characteristics during collision experiments also needs to be considered. In general, NPs are made by colloidal synthetic methods that require the use of stabilizing molecules (capping agents) to control the growth processes and to prevent aggregation. The stabilizing molecules are usually alkyl chains with thiol, amine, or carboxylic acid end groups or polymers. These capping agents can affect the NP stability in solution and the catalytic properties of the NP surface. Therefore, the use of partially modified NPs in collision experiments at clean UMEs was investigated. When Pt NPs are stabilized by citrate ions, the citrate ions are strongly

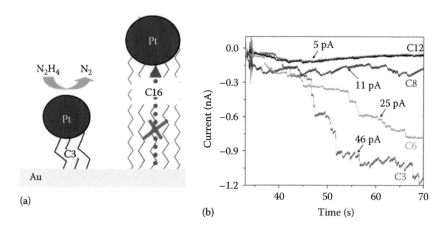

(a)   (b)

FIGURE 8.8   (a) Schematic illustration of Pt NP collision detection using the hydrazine oxidation reaction at the SAM-modified electrode. (b) Current–time curve recorded at a Au UME modified with different lengths of alkanethiolate SAMs. Au UME diameter, 10 µm; electrode potential, 0.1 V (vs. SHE); electrolyte, 12 mM hydrazine and 50 mM phosphate buffer (pH ~ 7.5).

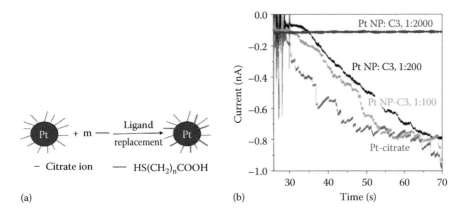

FIGURE 8.9 (a) Schematic illustration of the surfactant replacement. (b) Current–time curve recorded at Au UMEs before and after injection of Pt NPs capped by a mixture of citrate ions and C3 (propanethiol) SAMs. Au UME diameter, 10 µm; electrode potential, 0.1 V (vs. SHE); electrolyte, 12 mM hydrazine and 50 mM phosphate buffer (pH ~ 7.5).

solvated with water molecules, and the electrocatalytic activity of the Pt NPs for hydrazine oxidation is mostly maintained. When thiol molecules are incorporated into an NP surface layer of mixed composition, the weak electrostatic interaction of citrate is replaced by the stronger Pt–S bond. The partial replacement of citrate ions by the thiol led to a decrease in NP activity because the Pt–S bonds reduce the number of active sites for the catalytic reaction. Therefore, the surface coverage of thiols on the NP also affects the particle collision current as shown in Figure 8.9. Well-known surfactant molecules such as polyvinylpyridine (PVP) and cetyltrimethylammonium bromide (CTAB) are also not favored for stabilizing NPs when they are prepared for a collision detection experiment. Similarly, if modification of the particles with DNA is desired, the amount of DNA applied to the NPs as a detection probe must be controlled. In work described by Kwon and coworkers, NPs were produced that were modified with an average of 10 DNA molecules per NP (~4 nm in diameter). As a result, the DNA-modified NPs (i.e., detection probes) were successfully detected at a target DNA-tethered Au UME using a sandwich-type DNA hybridization.[12] Results from this DNA-NP collision experiment are shown in Section 8.9.

8.2.4 MIGRATION EFFECTS IN ELECTROCATALYTIC AMPLIFICATION SYSTEMS

In understanding NP collision experiments at UMEs, it is crucial to know the mode of the mass transport of the NPs to the detecting electrode. The earliest reports on EA systems were based on the assumption that the particles arrive at the UME from the bulk solution exclusively by diffusion.[1,3] However, advances in the understanding of the particle mass transport based on knowledge gained from works on blocking collisions[13,14] (Section 8.4) have led to the finding of migration effects also in EA systems.[15]

Migration of metal NPs is possible because they are charged (usually negatively, due to citrate or other ions used as capping agents) and due to the electric field created in solution by the faradaic process occurring at the detector electrode. The theory of migration in electrochemical systems has been described in fundamental work by Oldham et al.[16–19] Migration effects specific to the NP collision experiments are explained in detail in Section 8.7.3. Here, a general overview of migration effects in electrocatalytic systems is given. The importance of migration to NP collisions lies in the enhancement of the mass transfer rate for particles to the electrode. Migration of the particles both improves their detection limit and shortens the time needed for the analysis, leading to the possible detection of sub-pM to fM NP concentrations.

The total flux of NPs in solution can be written as the sum of diffusion and migration components according to the following equation:

$$J_{total} = J_{diff} + J_{migr} \tag{8.2}$$

Depending on the parameters of the system such as the concentration of the supporting electrolyte and the faradaic current due to redox processes at the working electrode (which are determined by the applied potential, the concentration of the redox species, and the size of the electrode), one can expect that either the diffusion or migration of the NPs will be the dominant factor in their mass transport. This is illustrated by the following examples.[15]

In Figure 8.10, one can see Pt NP collision events observed at a gold electrode where hydrazine oxidation takes place using different applied potentials and, therefore, different background faradaic currents. As determined from these experiments, at −100 mV (faradaic current ~1 nA) (Figure 8.10b), the frequency of collisions was about 17 times larger than that at −150 mV (current ~90 pA) (Figure 8.10a). Such an increase in the collision frequency can be attributed to the presence of migrational flux of the particles, since their diffusion is not affected by the change in the applied potential or the

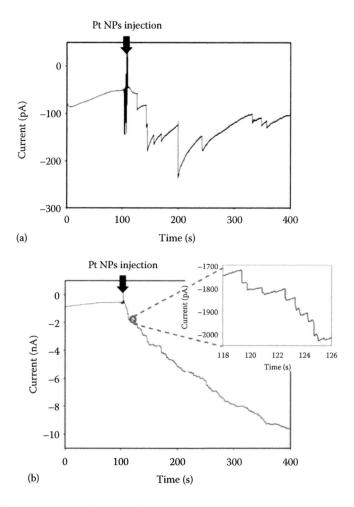

(a)

(b)

FIGURE 8.10 Chronoamperometric curves for single Pt NP (radius ~16 nm) collisions at the Au UME (radius 5 μm) in the presence of 7.5 pM Pt NPs in 5 mM phosphate buffer (pH 7.0) and 15 mM hydrazine. Applied potentials are (a) −0.15 V and (b) −0.1 V (vs. Ag/AgCl). Data acquisition time was 50 ms.

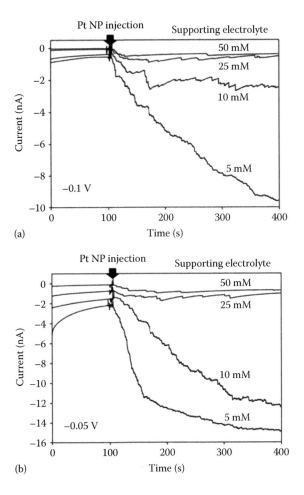

FIGURE 8.11 Chronoamperometric curves for single Pt NP (radius ~16 nm) collisions at the Au UME (radius 5 μm) with various phosphate buffer concentrations (50, 25, 10, and 5 mM; pH 7) in the presence of 7.5 pM Pt NPs and 15 mM hydrazine at −0.1 V (a) and −0.05 V (b) (vs. Ag/AgCl). Data acquisition time was 50 ms.

faradaic current due to the hydrazine oxidation, and the bulk concentration of the particles is the same in both cases. Another confirmation of the presence of the migration effect comes from the observed effect of the supporting electrolyte concentration on the frequency of collisions; this is illustrated in Figure 8.11. At a high concentration of the supporting electrolyte (50 mM PBS), the frequency of Pt NP collisions was ~0.017 pM/s for the first 100 s (Figure 8.11a). However, once the concentration of the buffer was decreased to 5 mM, the frequency of collisions increased to ~0.14 pM/s.

The observed results in Figures 8.10 and 8.11 arise from both the electric field that develops in solution to cause migration of NPs and the effect of the faradaic current and the supporting electrolyte ion concentration on its magnitude. By oxidizing hydrazine on a detector electrode, a positively charged species is produced in solution (H_3O^+). As a result of this anodic faradaic reaction, an electric field is set up in solution with its strength vector directed away from the working electrode surface. This field is responsible for the migration of all charged species in solution, including the ions of the supporting electrolyte and the NPs. The magnitude of this field is determined by the rate of production of the positively charged species (hence the effect of the current magnitude in Figure 8.10) and the concentration of the supporting electrolyte ions (Figure 8.11). The higher the background faradaic current, the more positively charged species are produced, the stronger the field, and the higher the frequency of collisions of Pt NPs with the Au electrode surface. On the other hand, the higher the concentration of the supporting electrolyte in solution, the more effective the field is *neutralized* by ions present, thus making the NP migration smaller. As suggested by Park et al.,[15] one can also use the transference number formalism to explain the observed experimental results in Figures 8.10 and 8.11.

The transference number (*t*) serves as a measure of the fraction of the current transferred by the considered species in the bulk solution. Thus, its magnitude also provides information about the relative flux of the species compared to all other components of the system:

$$t_k = \frac{i_k}{i_{total}}$$ (8.3)

and

$$t_k = \frac{|z_k| u_k C_k}{\sum_n |z_n| u_n C_n}$$ (8.4)

where
  i is the current
  z is the charge
  u is the mobility
  C is the concentration of the species

From Equation 8.4, one can relate the transference number of the NPs to their flux (J_{migr}) according to the following relation:

$$J_{migr} = \frac{t_{NP} i_{avg}}{|z_{NP}| F}$$ (8.5)

where
  i_{avg} is the average total current flowing through the working electrode
  F is the Faraday constant

Thus, from Equation 8.5, one can see that by increasing the faradaic current through the working electrode, one can increase the flux of the NPs and, thereby, the frequency of collisions (Figure 8.10). However, as the concentration of the supporting electrolyte is increased, according to Equations 8.3 and 8.4, the transference number of the NPs can be reduced effectively to zero and no NP migration is observed. The particles then arrive at the electrode exclusively by diffusion as in Equation 8.2, and the frequency of collisions becomes much smaller (Figure 8.11). The validity of Equation 8.5 has been confirmed by experiments, as shown in Figure 8.12. Here, the frequency of collisions depends linearly on the magnitude of the faradaic current and also on the concentration of the supporting electrolyte.

Using the semiquantitative considerations presented in this section, one can propose guidelines for observing/controlling the frequency of NP collisions at the detector electrode. By maximizing the transference number of NPs (i.e., high charge and small radius) and the faradaic current (i.e., the concentration of the redox species and the potential) and minimizing the concentration of the supporting electrolyte, one can fine-tune the frequency of collisions to achieve very low limits of detection (sub-fM).

8.2.5 OCP MEASUREMENTS

To this point, the single NP collision detection techniques described have been based on amperometric methods exploiting the catalytic difference of two materials (i.e., NP and UME). Using the

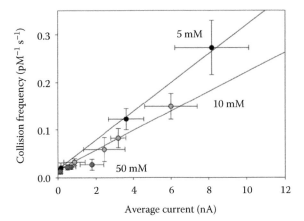

FIGURE 8.12 Collision frequency versus average current at three different phosphate buffer concentrations (5, 10, and 50 mM) at various potentials (−150, −100, −50, and 0 mV), which set the background current. Collision frequency was counted for the first 100 s after particle injection. The average current was obtained from integrating the charge and dividing by time. Data obtained by 3–5 replicate measurements.

same system (e.g., the hydrazine oxidation reaction), single NP collision events can be detected by measuring the OCP change upon particle adsorption.[20] The OCP of a working electrode is its potential measured versus a reference electrode when no external current flows.* The OCP seen with a reversible redox couple produces a poised solution, where the two half reactions (anodic and cathodic) of the same redox couple establish the thermodynamically expected potential where the external current is zero. The type of OCP of interest here is not of a reversible or poised type seen with a redox couple, but one that is kinetically controlled by two or more half reactions—a so-called mixed potential. When the system has $E_{ox}^0 - E_{red}^0 > 0$ and with very slow kinetics, this potentiometric system is sensitive to small current changes as shown in Figure 8.13a. For example, the hydrazine oxidation reaction and proton reduction reaction satisfy the aforementioned conditions. If an electrocatalytic NP collides and sticks to an inert electrode, it will change the anodic or cathodic half-reaction current. Depending on the reaction catalyzed, this will cause a shift in the potential as schematically represent in Figure 8.13b. Indeed, as long as the collision results in some charge transfer with the electrode at an open circuit, some potential change will occur.

Pt is a better electrocatalyst than Au for hydrazine oxidation as shown in Figure 8.3b. The mixed OCP, which is determined by hydrazine oxidation, proton reduction, and trace amount of oxygen reduction, is shifted by approximately −0.4 V between Au and Pt as shown in Figure 8.14a. If a Pt NP contacts a Au UME, because of the Pt electrocatalysis of hydrazine oxidation at the OCP of Au, the overall oxidation current on the Pt NP together with that on the Au UME becomes larger than the overall reduction current. To maintain the OCP condition, the OCP shifts negatively to produce a zero net current as shown in Figure 8.14b, and the OCP exhibits stepwise changes at every NP collision event. Similar to the amperometric detection methods, the OCP change is related to the sizes of the UME and NPs and also the concentration of hydrazine. Compared with the amperometric technique, this OCP-based approach has the advantages of simpler apparatus, higher sensitivity, and fewer problems associated with NP deactivation.

The principle of OCP change upon NP collision events is confirmed by a mimic experiment, which is performed by monitoring the OCP change when two different electrodes (Pt and Au) are connected and brought into contact with the same solution as shown in Figure 8.15a.[21] From this experiment, the factors affecting the magnitude of the OCP change were studied. A micrometer-sized Au UME, when

---

* In reality, during the measurement of the OCP with an electrometer, a very small current flows, depending on the input impedance of the measuring device, which can be as large as 200 TΩ.

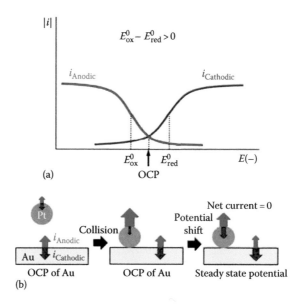

FIGURE 8.13 (a) Schematic representation of the half reaction i–E curve when $E_{ox}^0 - E_{red}^0 > 0$. (b) Pictorial representation of the relative changes in the anodic (red arrow) and cathodic (blue arrow) currents on a single Pt NP and on the Au UME before and during particle collision event.

connected to a Pt nanoelectrode, showed clearly measurable OCP changes as shown in Figure 8.15b. The magnitude of the OCP changes matches well with that predicted by a simplified mixed potential theory, which contains Butler–Volmer formalism and neglects mechanistic complications.

E_{OCP} for the coupled electrodes, the condition is

$$\left|i_{C,j=1}\right| + \left|i_{C,j=2}\right| = \left|i_{A,j=1}\right| + \left|i_{A,j=2}\right| \tag{8.6}$$

$$\left|i_C\right| = \frac{i_{dC}}{\left\{1 + \dfrac{m_0}{k_C^0}\exp\left[\alpha_C f\left(E - E_C^0\right)\right]\right\}} \tag{8.7}$$

$$\left|i_A\right| = \frac{i_{dA}}{\left\{1 + \dfrac{m_R}{k_A^0}\exp\left[-(1-\alpha_A)f\left(E - E_A^0\right)\right]\right\}} \tag{8.8}$$

where

$i_{C,j}$ and $i_{A,j}$ are half-reaction currents ($j = 1$(Au) or 2(Pt))
k_C^0, k_A^0 are the apparent rate constants in the heterogeneous rate expression
m_0, m_R are the mass transfer coefficients
i_{dC} and i_{dA} are the diffusion-limited currents
E_A^0, E_C^0 are the assumed standard potentials of the overall half reactions
subscripts C and A represent cathodic and anodic half reactions

The summation of currents $\left|i_{C,j=1}\right| + \left|i_{C,j=2}\right|$ and $\left|i_{A,j=1}\right| + \left|i_{A,j=2}\right|$ can then be plotted and where Equation 8.6 is satisfied, $E = E_{OCP}$. Figure 8.16 shows a representative example for the case of a disk electrode and a spherical NP. The intersection point of the $\left|i_j\right|$ – E plot indicates the OCP where the net current equals zero.

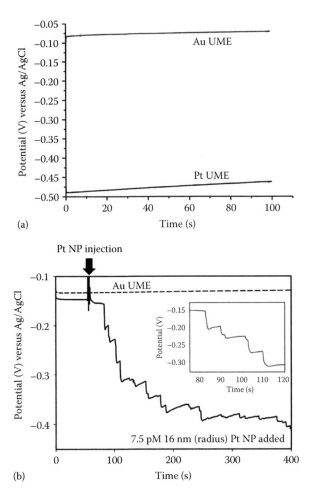

FIGURE 8.14 (a) OCP of Au and Pt UMEs in a solution of 50 mM phosphate buffer (pH 7.0) and 15 mM hydrazine; radius of Pt and Au UMEs, 5 μm; sweep rate, 100 mV/s. (b) OCP versus time plot for single Pt NP (radius ~16 nm) collisions at the Au UME (radius 5 μm) in the presence of 5 mM phosphate buffer (pH 7.0) and 15 mM hydrazine without (dashed line) and with (solid line) 7.5 pM Pt NPs. Inset shows magnified image of staircase potential response. Data acquisition time is 100 ms.

From the results of theoretical approaches, the OCP change after connection with a different electrode material reflects the kinetic (rate constant) differences for the reaction at the different materials. The essence of particle detection for this system then relies on the heterogeneous catalytic effects for both the anodic and cathodic reactions.

8.3 DIRECT PARTICLE ELECTROLYSIS

NP collision electrochemical responses mostly measure the amplified current of electrocatalytic reactions occurring on the surface of collided NPs. Large amplification is essential to distinguish the collision signal current from the background noise as described in the preceding section. However, beyond fast electrocatalytic reactions, other types of reactions, for example, direct metal NP oxidation,[22,23] deposition,[24,25] or OS electron transfer reactions,[26] can be used for NP detection. The Compton group has described NP collision experiments by the direct reactions of NPs.

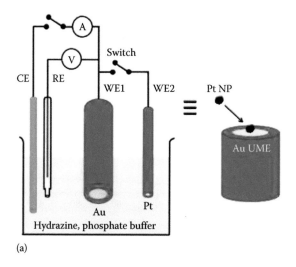

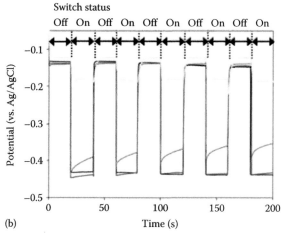

(b)

FIGURE 8.15 (a) Pictorial representation of the experimental setup of parallel-connected different electrode materials to mimic the single Pt NP collision event on Au electrode. (b) OCP versus time plots for 12.5 μm radius Au UME connected with 12.5 μm (black line), 5 μm (red line), 1 μm (green line), 50 nm (blue line), and 20 nm (yellow line) radius Pt electrode in the presence of 5 mM phosphate buffer (pH 7) and 15 mM hydrazine. The connection was controlled by a switch, and the switch status changes every 20 s. Data acquisition time is 100 ms.

The direct characterization of Ag NPs by instantaneous oxidation during collisions was investigated.[22] The process occurring during the impact is

$$Ag(NP) - e^- \rightarrow Ag^+(aq) \tag{8.9}$$

A Ag NP-modified GC electrode was scanned to increasingly positive potentials in citrate solutions to find an appropriate potential for Ag NP oxidation. Under potentiostatic conditions where the Ag is oxidized (+0.5 V vs. Ag/AgCl in this experiment), oxidation current spikes were observed as shown in Figure 8.17. The onset of these spikes was found to vary with the potential, and the collision frequency was also investigated. Assuming that the NPs are spherical (radius r_{np}), the maximum charge passed as a result of complete oxidation of the Ag NP is given by

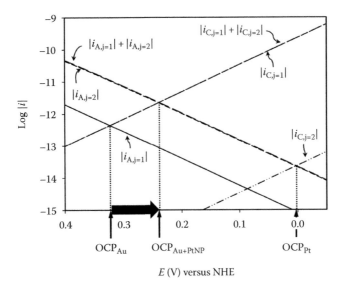

FIGURE 8.16 Simulated series of anodic current and cathodic current versus potential plot of Au disk electrode ($j = 1$) and Pt spherical NP ($j = 2$). OCP is decided by definition; net current equals zero ($|i_{A,j}| = |i_{C,j}|$). OCP of two electrode materials ($OCP_{Au+PtNP}$) is decided when Equation 8.6 is satisfied. The large arrow represents the OCP shift upon contact with Pt NP to a Au UME. (Au disk radius, 12.5 μm; Pt NP radius, 20 nm).

$$Q_{max} = \frac{4F\pi\rho r_{np}^3}{3A_r} \tag{8.10}$$

where

ρ is the bulk density

A_r is the relative atomic mass

The size of NPs used in these experiments must be of sufficient size to give a distinguishable current spike. The sizes of Ag NPs used in this experiment ranged from 20 to 50 nm.

The authors named this methodology *anodic particle coulometry* (APC). APC has also been applied to analysis of Au NPs.[23] The chloride-assisted dissolution of Au at oxidative potentials was due to the complexation equilibria[27,28]:

$$Au + 4Cl^- \rightleftharpoons AuCl_4^- + 3e^- \tag{8.11}$$

$$AuCl_4^- + 2Au + 2Cl^- \rightleftharpoons 3AuCl_2^- \tag{8.12}$$

This effect enables the identification of Au NPs. Au NPs (10 nm radius) in 0.1 M HCl solution were used for the experiment. The electrode potential was held at a series of potentials between +0.8 and +1.3 V with a sampling time of 0.5 ms. Collision transients were not observed at potentials less than +1.0 V but were recorded above this threshold potential and showed a current spike shape. The current spikes were similar to those observed for Ag NP oxidation.

In addition to these direct NP oxidation detection experiments, the underpotential deposition (UPD) of metal ions from solution onto metal NPs during collisions between the NPs and an inert electrode was also reported (see Figure 8.18). Reactions for UPD of thallium and bulk electrodeposition of cadmium onto Ag NPs were used for detection, which formed bimetallic core–shell NPs (denoted Ag@Tl and Ag@Cd, respectively).[24,25] For the case of thallium, it was shown that up to a

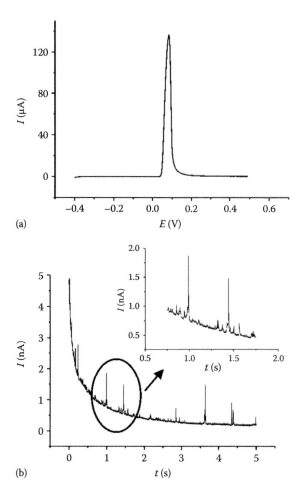

(a)

(b)

FIGURE 8.17 (a) Anodic stripping voltammogram for a Ag NP-modified GC electrode and (b) typical APC transients showing oxidation spikes due to collisions of Ag NPs at the GC electrode (radius 11 μm, $E = +0.5$ V vs. Ag/AgCl). (From Zhou, Y.-G., Rees, N. V., and Compton, R.G., The electrochemical detection and characterization of silver NPs in aqueous solution, *Angew. Chem. Int. Ed.*, 50, 4219–4221, 2011. Reprinted by permission of The Royal Society of Chemistry.)

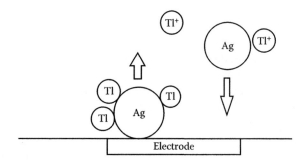

FIGURE 8.18 UPD of metal ions from solution onto metal NPs during collisions between the NPs and an inert electrode. (Reprinted with permission from Wiley.)

monolayer, it could be deposited onto Ag NPs, and for the electroplating of cadmium onto Ag NPs, the average number of layers deposited was about 19[29]:

$$Ag(NP) + mTl^+(aq) + me^- \rightarrow Ag @ Tl_m \tag{8.13}$$

$$Ag(NP) + mCd^{2+}(aq) + 2me^- \rightarrow Ag @ Cd_m \tag{8.14}$$

In both cases, deposition occurs only during NP–electrode contact, which lasts approximately 1–10 ms. A random walk model was adapted to investigate the contact time required to electro deposit a monolayer of metal onto an NP during collision and found this to be on the order of 10^{-4} s for a 45 nm radius NP in 10 mM solution of analyte.[30] This provided confirmation that the thermally driven NP collision contacts were of a sufficient duration to allow an experimentally observable faradaic process to occur. A theoretical model was also developed for the oxidation of an NP contacting an electrode held at a suitably oxidizing potential. The model incorporates Brownian motion to account for the experimentally observed timescale of such reactions.[31]

8.4 DETECTION OF SINGLE NP COLLISIONS BY UME BLOCKING

8.4.1 AREA BLOCKING BY INSULATING PARTICLES AT UMEs

Another alternative approach to detect NP collisions involves detection by a particle's ability to block the diffusion of a redox species to the surface of an UME. The time-resolved detection of single particles by this approach was first demonstrated in a 2004 paper by Quinn, van't Hof, and Lemay (QHL),[13] although earlier work by Gorschlüter et al. first proposed single-particle detection using diffusion blocking.[32] The experiment has also been revisited more recently by our group[14] and later by Crooks and coworkers.[33] The approach described here detects particles based on the surface area blocked on the underlying UME, so it detects insulating particles.

The essential features of the approach are shown in Figure 8.19. A potential is applied to a UME sufficient to produce diffusion-limited oxidation or reduction of a dissolved redox species. When an insulating particle arrives at the electrode, its adsorption is accompanied by a decreasing step feature

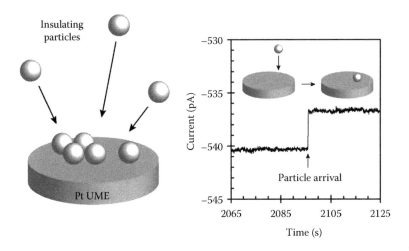

FIGURE 8.19 Schematic of single-particle detection using area blocking. Right panel shows a portion of a chronoamperogram recorded for the diffusion-limited oxidation of FcMeOH at a 1 μm Pt UME. Adsorption of a single 310 nm diameter silica sphere is indicated by a decreasing step feature to the steady-state current, as marked by an arrow in the figure. (Reprinted with permission from ACS.)

to the diffusion-limited current at the UME due to the decrease in redox species flux imposed by the blocking particle. The size of the step feature is related not only to the size of the blocking particle but also to its location because of the difference in diffusive flux of redox species at different locations on the disk-shaped electrode. This section describes general and practical aspects of these experiments, while theoretical treatments of the blocking step heights are described in Section 8.7.2.

8.4.2 Description of Particle Detection Experiments Using Diffusion Blocking

In the initial work by QHL, adsorption events of individual carboxylated latex spheres were detected by their blocking of ferrocenemethanol (FcMeOH) oxidation at a Pt disk UME.[13] Chronoamperograms from this work for a 5 μm radius UME in 0.31 mM FcMeOH, varying concentrations of the supporting electrolyte, and with 4×10^8 spheres/cm^3 (sphere radius = 0.5 μm) suspended in the solution are shown in Figure 8.20. The spheres are electrically insulating, and each decreasing step feature in the chronoamperograms corresponds to blocking of FcMeOH diffusion by one adsorbed sphere. Adsorption of the spheres was later verified by ex situ optical microscopy, which found a nearly close-packed layer of spheres covering the UME surface (as shown in Figure 8.20c).

Using the Stokes–Einstein equation[34] and an expression for diffusive flux at a disk electrode under hemispherical diffusion,[35] one can estimate the expected frequency of collisions for spherical particles at a UME:

$$f = 4D_s C_s r \tag{8.15}$$

where
  f is the collision frequency in s^{-1}
  D_s is the diffusion coefficient of the particles in cm^2/s
  C_s is the concentration of the particles in cm^{-3}
  r is the radius of the electrode in cm

For the experiments in the work by QHL where 0.5 μm radius spherical beads arrived at a 5 μm radius UME, this expression yields an expected collision frequency of only 0.003 s^{-1}. The larger

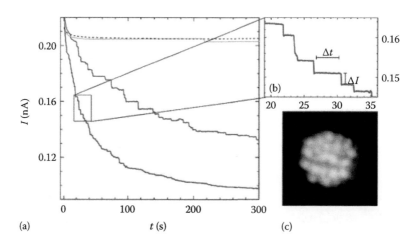

(a)   (b)   (c)

FIGURE 8.20 (a) Current–time (I–t) transients for the diffusion-limited oxidation of 0.31 mM FcMeOH (black line) at a disk Au microelectrode ($r_e = 2.5$ mm) showing that a discrete current step decreases in the presence of dispersed beads ($r_b = 0.5$ mm, $c_{bead} = 4 \times 10^8$ cm^{-3}) at 0.5 (red line), 5 (blue line), and 50 mM (green line) KCl supporting electrolyte concentrations. (b) Detail of panel (a). (c) Ex situ optical micrograph taken after the amperometric measurements showing the electrode surface covered by beads. (Reprinted by permission of ACS.)

than expected number of step features ($f=0.43$ s^{-1} for the plot in 0.5 mM KCl) found over the 300 s observation time in Figure 8.20 is due to electrophoretic migration. The spheres were negatively charged by surface carboxylate groups and migrated from the bulk solution to the UME in the electric field resulting from the oxidation of FcMeOH. The arrival rate of the spheres was found by QHL to be proportional to the FcMeOH oxidation current, which is consistent with the interpretation that the spheres arrive at the UME primarily by migration rather than diffusion. The rate of sphere arrivals at the UME by migration could be estimated using the following expression[13]:

$$J_{mig} = \frac{1}{\Delta t} \approx \frac{I c_{bead}}{e c_{KCl}} \left(\frac{u_{bead}}{u_{K^+} + u_{Cl^-}} \right) \tag{8.16}$$

where
  J_{mig} is the arrival rate of the spheres by migration
  I is the FcMeOH oxidation current
  e is the elementary charge
  c and u are the concentrations and mobilities of the spheres (beads) and supporting electrolyte ions, respectively

As shown in Figure 8.21, a plot of experimentally measured bead arrival rates by QHL versus the combined current and concentration term ($I c_{bead}^{*} / c_{KCl}$) was in agreement with the earlier expression. The sphere arrivals are also shown in Figure 8.20 to decrease with the concentration of supporting electrolyte in the cell, because the particle transference number decreases. When the concentration of supporting electrolyte was increased to 50 mM (green curve), the measured sphere arrival frequency was only 0.005 s^{-1}, indicating a greatly reduced effect of migration on sphere transport to the electrode.

In the reported particle collision experiments with negatively charged latex, polystyrene, and silica spheres, most of the step features in the recorded chronoamperograms showed a current decrease due to the blocking of redox mediator diffusion by adsorbed particles. However, a small number of steps showed a current increase, suggesting removal of a particle from the surface. However, in

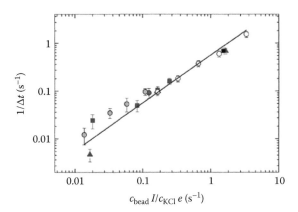

FIGURE 8.21 Experimental values for the bead arrival rate plotted versus the grouped variable $c_{bead} I / c_{KCl}$ (log–log scale). Each point is from a separate measurement. Individual variables were varied systematically. (Blue triangles) $c_{KCl}=0.5$, 5, and 50 mM ($I=0.2$ nA, $c_{bead}=4 \times 10^8$ cm^{-3}), (red circle) $I=0.2$ nA, $c_{bead}=3.2 \times 10^9$ cm^{-3}, $c_{KCl}=50$ mM, (magenta squares) $I=0.02$, 0.1, 0.2, and 0.3 nA ($c_{bead}=4 \times 10^8$ cm^{-3} $c_{KCl}=5$ mM), (black square) $I=1.89$ nA, $c_{bead}=0.4 \times 10^9$ cm^{-3}, $c_{KCl}=5$ mM, (open circles) $c_{bead}=0.4$, 0.8, 1.6, 3.2, and 8×10^9 cm^{-3} ($I=0.2$ nA, $c_{KCl}=5$ mM), (green circles) $I=0.06$, 0.16, 0.22, and 0.53 nA ($c_{bead}=1 \times 10^8$ cm^{-3} $c_{KCl}=5$ mM). Error bars were evaluated from counting statistics. (Reprinted by permission of ACS.)

general, in these experiments, stepping the electrode potential after particle accumulation to cause a cathodic current at the UME did not cause particle removal. These observations suggest that the sphere adsorption is irreversible, with the spheres experiencing significant attractive interactions with the electrode surface.

8.4.3 Interpretation of Current Step Features

For collision experiments involving insulating particles, the heights of current steps depend on the sizes of the insulating particles and UME, the concentration of the redox mediator, the location on the UME where the NP lands, and also the degree of surface coverage by previously adsorbed particles. A plot of observed current step heights for the adsorption events of 310 nm diameter silica spheres at a 2 μm Pt UME recorded over 5000 s observation times is shown in Figure 8.22.[14] For the plot recorded in 1 mM KCl, the heights of the individual current steps decreased over the observation time, eventually becoming unresolvable from the noise. Complete blocking of the surface (a decrease to negligible current at the UME) is not observed in these experiments because of mediator flux through voids between the adsorbed spheres, even as the degree of coverage reaches and exceeds one monolayer. Therefore, the observed decrease in the average step height with time reflects increasing surface coverage of the blocking particles: as additional particles adsorb to the surface, the amount of relative blocking decreases because the total flux to the electrode becomes dominated by voids in the sphere layer rather than pristine electrode area. In contrast to the silica sphere collision experiment in 1 mM KCl, the average heights of current steps for an analogous experiment recorded in 5 mM KCl do not decrease significantly, because the rate of sphere arrival over the 5000 s observation time with migration greatly decreased is insufficient to reach monolayer (or near-monolayer) coverage.

Simulations show that particles migrating to a UME in these experiments appreciably block diffusion of the redox mediator, even when the particle is located several bead radii away from the electrode edge. However, the stability of most of the step features observed in experiments (lack of additional current variation with time after the initial step) suggests that single-particle arrivals from the bulk solution to the UME occur quickly relative to the experiment time, making it unlikely that longer distance *flybys* of particles are recorded. QHL as well as our own group has performed simulations to relate the observed current step height to the location of the particle on the UME surface, which varies due to differences in redox mediator flux at different regions of the electrode.[13,14] A more in-depth description of these simulation results is given in Section 8.7.2. In general, the current step height (and likelihood of adsorption) was greatest at the electrode edge due to maximal diffusion from the bulk at the UME disk edge versus the center. Likewise, particles were also found to influence the electrode current even when adsorbed slightly outside the UME disk (on the electrode insulation), again because of blocking of diffusion to the electrode edge.

It is also possible to draw conclusions about particle behavior by further inspection of the chronoamperograms recorded during collision experiments with insulating particles. In experiments with silica and polystyrene spheres, unusual features were infrequently (but reproducibly) observed that cannot be explained simply as the arrival and irreversible sticking of single particles. Examples of some of these features are shown in Figure 8.23. In Figure 8.23a, one decreasing current step feature is preceded immediately by an increasing current step feature of equal magnitude. We ascribe this reversal of the step to loss of the same particle after adsorption, possibly due to a surface defect at the particle adsorption site. Figure 8.23b shows a single step punctuated by two smaller step features, with no additional steps occurring over several seconds before and after this group of features. Such tightly spaced step *bunches* were repeatedly observed, even when the sphere concentration in solution was reduced (or the supporting electrolyte concentration was increased) such that the overall step frequency (sphere arrival rate) was small. We have suggested that such step bunches are the result not of the adsorption of multiple particles but rather the adsorption of a single particle and then subsequent movement of the particle to different locations on the surface, eventually stopping when a stable adsorption site is found (such a site may be a microscale defect on the metal

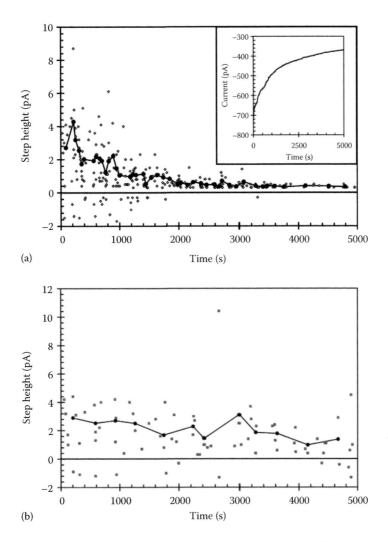

FIGURE 8.22   Plot of the height of individual current step events during 5000 s chronoamperograms recorded using a 2 μm diameter Pt UME in aqueous solutions containing 2.5 mM FcMeOH, 50 fM silica spheres (310 nm diameter), and either (a) 1 mM KCl or (b) 5 mM KCl. Each data point reflects the height of one current step event with positive currents here indicating a decrease of the UME current magnitude. The connected black dots in each panel show a moving average of every five positive steps. Positively signed steps indicate a decrease in FcMeOH oxidation current (more blocking of FcMeOH diffusion), while negatively signed steps indicate an increase in FcMeOH oxidation current (less blocking of FcMeOH diffusion). Instrumental noise prevented discrimination of steps below 300 fA. Inset in part (a): Raw chronoamperogram for the 1 mM KCl case, showing eventual near-maximal blocking due to the adsorbed spheres. (Reprinted by permission of ACS.)

surface or an aggregate of previously adsorbed particles). Moreover, applying a Poisson distribution to the observed step frequencies found that the coincidence of several individual adsorption events within such short time windows (as opposed to movements of one particle) should be very unlikely as compared to the frequency with which the step bunches were observed. Such features may reflect a small degree of aggregation of the particles in solution or on the surface, with the step bunches corresponding to the adsorption and rearrangement of one cluster of particles on the surface.

Finally, Figure 8.23c shows a feature where the current oscillates rapidly between two values. These oscillations occur at a much higher frequency than the more common single step decreases

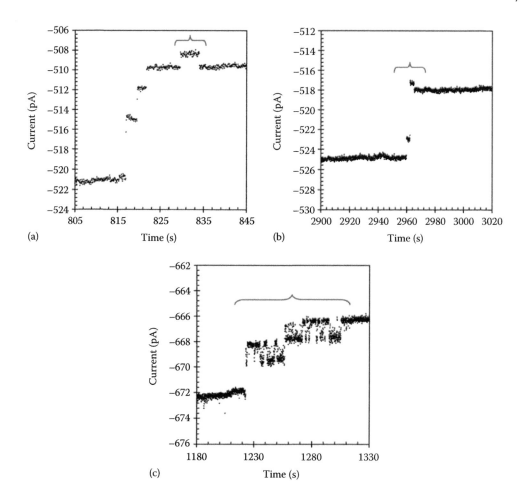

FIGURE 8.23 Highlighted features found in chronoamperograms for adsorption of 310 nm diameter silica spheres at a 2 μm Pt UME in 2.5 mM FcMeOH: (a) desorption of a particle, with 50 fM silica spheres and 1 mM KCl; (b) blocking events suggesting movement of one particle, with 5 fM silica spheres and 1 mM KCl; (c) apparent instability of one or more adsorbed particles on the electrode, with 50 fM silica spheres and 5 mM KCl. (Reprinted by permission of ACS.)

for successive particle adsorptions. With silica and polystyrene spheres, such oscillations were occasionally observed (regularly 1 or 2 oscillating periods per 1000–5000 s observation period) in experiments with 310 nm diameter silica spheres, but never with larger sizes. QHL also observed such oscillation features in collision experiments with 50 nm and 300 nm diameter latex spheres, but not for 1 μm diameter spheres. Such oscillations perhaps reflect instability of a single particle as it moves between two closely spaced adsorption sites (e.g., surface defects or collections of particles) on the electrode surface. QHL attributed such current oscillations only with smaller (<500 nm diameter) spheres to competition between electrostatic forces at the surface and Brownian motion, where with larger NPs, Brownian motion on the surface is overwhelmed by electrostatics, and only stable current steps are seen.[13] Features such as those previously attributed to particle removal and rearrangement may also show particle size dependence, but detailed studies have not yet been reported.

Recently, this diffusion blocking approach to monitoring the adsorption events was used with large, 1 μm diameter, fluorescent carboxylated polystyrene beads while simultaneously using fluorescence microscopy to track the movements and adsorption of individual beads on the UME.[33] In this work, the electrochemical cell for the collision experiments was on the stage of a fluorescence

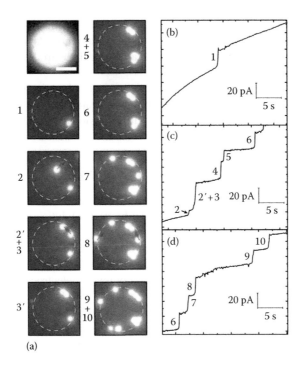

(a)

FIGURE 8.24 (a) A series of micrographs showing bead collisions at a Pt UME. (Top left) Optical micrograph of the 10 μm Pt UME (the white scale bar indicates 5 μm). The remaining frames show fluorescence micrographs. The numbers to the left of each frame indicate which bead collision is represented in that frame. Arrows have been added to show the motion of the beads on the electrode surface following collisions. (b–d) Chronoamperometric (i–t) traces obtained. The time-correlated collision from the movie frames (part a) and the i–t traces (parts b–d) are indicated by the numbers in each frame. The current convention used here shows increasing anodic current down, so that collisions cause a decrease in current magnitude from FcDM oxidation (values of i_{step} are positive). Notice that after beads 7 and 8 collide with the surface, they are observed to move on the surface in the direction of the red arrows. This leads to an increase in the apparent noise between the landing time of beads 8 and 9 in the i–t trace shown in part (d). (Reprinted with permission from Fosdick, S.E., Anderson, M.J., Nettleton, E.G., and Crooks, R.M., Correlated electrochemical and optical tracking of discrete collision events, *J. Am. Chem. Soc.*, 135, 5994–5997, 2013. Copyright 2013, American Chemical Society.)

microscope, allowing optical particle tracking. The general results agreed with the conclusions from the electrochemical experiments (typical data in Figure 8.24).

8.4.4 ADDITIONAL EXPERIMENTAL CONSIDERATIONS AND FUTURE DIRECTIONS

A key characteristic that controls the usefulness of this diffusion blocking approach is the interplay between the sizes of the particles under investigation and the size of the UME used for detection. If the UME diameter is much larger than the NP diameter (roughly >10 times), individual particle arrivals cannot be resolved because the effective area decrease is too small. For example, in attempting to monitor the adsorption of 390 nm carboxylated polystyrenes using a 100 μm diameter Pt electrode, individual particle collisions could not be resolved from the instrument noise, and only a gradual current decrease was observed.

Smaller UMEs, for example, where the size is comparable to the adsorbing NPs, show large current decrease transients but of course are limited to observing only one or two NP collisions. Generally, for observing multiple collision events with good resolution, an optimal size range for the UME is a diameter roughly 3–20 times the diameter of the NPs. A longer-range goal of these experiments is to bring down the size of the UME to a level where objects of molecular size can be detected. This requires nm-size UMEs. The ambiguity of location on a disk electrode affecting the current step size could, in principle, be solved with a spherical electrode that is uniformly accessible; this has not yet been demonstrated however.

A second key experimental consideration is the role of electrophoretic migration in bringing particles to the UME. These experiments, as well as those with catalytic NPs, are further described in Section 8.7.3. Finally, analogous blocking is found with emulsion droplets (emulsion droplet blocking [EDB]) with systems described in Section 8.3.3.

8.5   COLLISION EXPERIMENTS: AREA AMPLIFICATION

With the exceptions of the earlier approach using OCP changes and those using area blocking, our single NP detection methods described have mainly used catalytic amplification.[1–5] This catalytic current amplification method was essential for detecting NPs at a micrometer-scale electrode. However, deactivation of the NP for the electrocatalytic reaction (i.e., IS electron transfer reaction) can cause a decay of the current, and this effect makes it more difficult to model the NP current after the collision. Studies with electrode reactions that do not involve electrocatalysis (e.g., OS electron transfer reactions) suffer less from these deactivation effects. With OS reactions, a stepwise current increase because of the electroactive area increase can be obtained after an NP collision.[36,37] To detect NP attachment by an increase in electrode area, the size of the particle and the size of the electrode should be comparable, so the observed current change is readily detectable. Therefore, a nanometer-scale UME is necessary for NP detection using an OS reaction. It should be recognized that as the size of the electrode decreases, the collision frequency decreases proportionally:

$$f_{\text{diff}} = 4D_p C_p r \tag{8.17}$$

where

    D_p is the diffusion coefficient of the NP
    C_p is the NP concentration
    r is the radius of the nanoelectrode

An additional driving force is required to increase the collision frequency and enable the detection of single NPs at nanoelectrode at smaller C_p. As for the catalytic approaches described earlier, the electric field at the electrode is used to attract the NPs by migration.[15] However, at a nano-sized electrode, only a small electric field is generated because of the small faradaic current flow, so it was necessary to enhance the mobility of the NPs by increasing the charge on each NP. This was accomplished by modifying the NPs with single-wall carbon nanotubes (SWCNTs) that had been previously dispersed in water. SWCNTs can migrate easily in an electric field because of their high charge and low density. SWNTs modified with Au NPs (designated as Au-SWCNTs) were obtained by SWCNT-templated growth of Au NPs.[38]

As suggested in Figure 8.25a, Au-SWCNTs migrate to a Pt nanoelectrode because the oxidation of FcMeOH to positively charged ferrocenium methanol creates a charge imbalance (electric field) near the nanoelectrode that attracts negatively charged particles. Using a high concentration of FcMeOH and a low concentration of supporting electrolyte (e.g., 100 µM KNO_3), the electric field at the nanoelectrode could be enhanced, thus increasing the mass transfer rate for the particles.

The higher electric field also plays a role in holding the NPs on the electrode. Figure 8.25b showed instantaneous current increase to a steady-state current that remained constant for over 250 s. This was difficult to accomplish in electrocatalytic-based studies and is consistent with the fact that OS reactions are less affected by impurity adsorption than IS reactions. The sizes of the attached Au-SWCNT can be estimated from the current step size. Figure 8.26 shows a dimensionless plot of the simulated current change for Au-SWCNT collisions based on the assumption that the

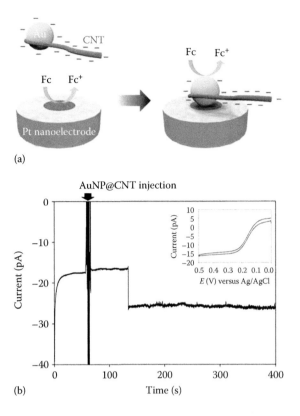

FIGURE 8.25 (a) Pictorial representation of the attachment of a CNT-modified Au NP at a Pt nanoelectrode. (b) Chronoamperometric curve for attachment of a single Au-SWCNT at the Pt nanoelectrode (~30 nm diameter) in the presence of 4 mM FcMeOH, 100 µM KNO_3, and 85 ng/mL Au-SWCNTs. The inset shows the cyclic voltammogram for the Pt nanoelectrode. The data acquisition time was 50 ms, and the applied potential was +0.4 V versus Ag/AgCl. The large noise upon injection of the NPs was caused by opening of the Faraday cage.

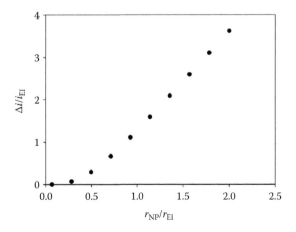

FIGURE 8.26 Simulated results for the relative magnitude of the electrode current increase ($\Delta i/i_{EI}$) after NP attachment as a function of the relative size of the NP (r_{NP}/r_{EI}), where r_{NP} and r_{EI} are the radii of a spherical NP and a disk electrode, respectively. The NPs are assumed to be ideally spherical.

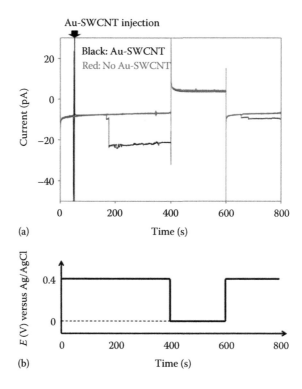

Au-SWCNT injection

(a)

(b)

FIGURE 8.27 (a) Current–time curves for single Au-SWCNT attachment and detachment at the Pt nano-electrode by controlling the electrode potential in the presence of 4 mM FcMeOH and 100 μM KNO$_3$ with Au-SWCNT (85 ng/mL, black line) and without Au-SWCNT (red line). (b) Applied electrode potential is plotted versus time. Data acquisition time is 50 ms. Pt nanoelectrode (diameter ~15 nm) was used.

Au-SWCNT lands at the center of the disk electrode. Removing the electric field by changing the electrode potential resulted in Au-SWCNTs detaching from the nanoelectrode as shown in Figure 8.27. After Au-SWCNT detachment, the FcMeOH current decreased to the background current level. These experiments suggest the possibility of single NP separations that can be used in single NP studies. NPs have been widely studied for catalytic applications based on ensemble-averaged data. However, the catalytic properties of well-characterized single NPs have yet to be reported.[39] In view of the recent interest in electrochemical studies at the single NP level, there is value in techniques that allow the capture of single NPs from solution. This method might also be applicable to the study of the behavior of single biomolecules (e.g., enzymes) using sufficiently small electrodes.[37]

8.6 COLLISION EXPERIMENTS—WITH OTHER DETECTION RESPONSES—ECL, PEC

Reports describing NP collisions that were detected by methods other than electrochemical ones are still scarce. Here, we review such alternative NP collision generation/detection schemes involving electrogenerated chemiluminescence (ECL) and photoelectrochemistry (PEC).

8.6.1 ELECTROGENERATED CHEMILUMINESCENCE

The possibility of observing NP collisions using ECL has been demonstrated by Fan and Bard.[40] In their approach (Scheme 8.1), the authors use the well-known ECL reaction of electrochemical oxidation of Ru(bpy)$_3$$^{2+}$ and tri-n-propylamine (TPrA) coreactant.[41]

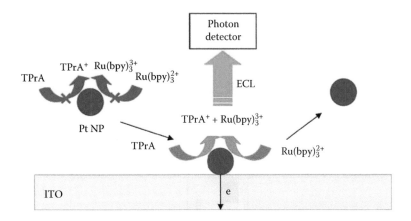

SCHEME 8.1

The reaction mechanism is complex; however, the oxidation results in $Ru(bpy)_3^{3+}$ and a radical TPrA, which upon reaction with one another produce $Ru(bpy)_3^{2+*}$. Observing ECL with higher intensity at the NP depends upon a rather small difference in the rate of TPrA electrooxidation at Pt versus indium tin oxide (ITO). The second problem is the relatively low ECL intensity at the small NP compared to the usual macroelectrodes. The experimental conditions are chosen so that the reaction between the species and the coreactant at relatively high concentrations in solution does not result in the production of a high ECL signal at the conductive measuring electrode (ITO) at a given potential. However, when Pt NPs are added into the system, the particles are able to electrocatalyze the oxidation of TPrA upon their collision with the ITO surface. Thus, a significant enhancement in the ECL intensity is observed as shown by the red curve in Figure 8.28b. This enhancement is observed at a somewhat lower bias potential, although no significant change in the faradaic current can be noticed (Figure 8.28a). The enhancement in the ECL signal depends on the concentration of the Pt particles, thus indicating that it is associated with an electrode reaction on the surface of the Pt NPs upon their collision with ITO.

The current transients before and after the addition of the Pt NPs show smooth decay in both cases due to the double-layer charging and the oxidation of TPrA and only very small contribution due to the electron transfer associated with the Pt NPs. On the contrary, the ECL signal observed

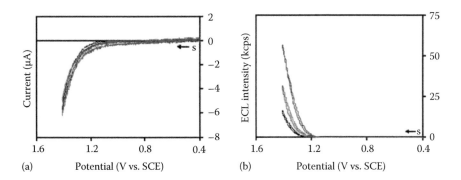

FIGURE 8.28 Cyclic voltammograms (frame a) and ECL intensity (kilocounts per second, kcps) versus potential curves (frame b) at an ITO electrode in a solution before (black curves) and after injecting 1 nM (red curves) and 2 nM (blue curves) Pt NPs. The solution contains 0.1 M NaClO$_4$, phosphate buffer (pH 7.0), 10 µM Ru(bpy)$_3$(ClO4)$_2$, and 50 mM TPrA. Potential scan rate is 20 mV/s from points. (Reprinted with permission from Fan, F.-R.F. and Bard, A.J., Observing single nanoparticle collisions by electrogenerated chemiluminescence Amplification, *Nano Lett.*, 8(6), 1746–1749, 2008. Copyright 2008, American Chemical Society.)

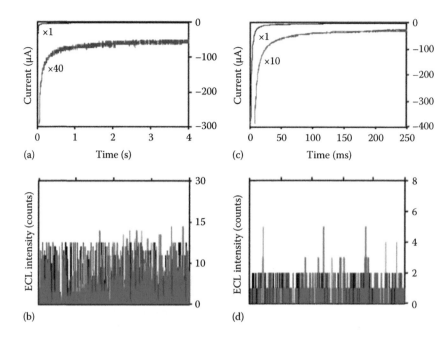

FIGURE 8.29 Current transients (frames a and c) and ECL intensity versus time records (frames c and d) at the ITO electrode before (black curves) and after (red curves) the injection of 2 nM Pt NPs colloidal solution. Current axis expansions (blue and violet curves): 40 times in frame a and 10 times in frame c. The solution contains 0.1 M NaClO$_4$, phosphate buffer (pH 7.0), 1.3 μM Ru(bpy)$_3$(ClO4)$_2$, and 5 mM TPrA. Potential is stepped from 0 to 1.29 V versus SCE for 4 s (channel dwell [or binning] time, $\tau_{ch} = 15.6$ ms) in frame b and for 250 ms ($\tau_{ch} = 975$ μs) in frame d.

upon addition of the NPs is transient and stochastic in nature, which is indicative of the random motion of the particles in solution and their collision with the electrode (Figure 8.29).

ECL intensity profile (intensity vs. time) includes the contributions from the two processes: one due to ECL generated on the ITO electrode itself and the other due to Pt NP collisions; the latter contribution is attributed to large photon spikes above the steady background. It has been established that the frequency of these spikes depends strongly on the bias potential: at more positive potentials, the frequency increases until the potential value (>1.7 V vs. SCE) where ITO surface is believed to deactivate. Analysis of the ECL intensity profiles in terms of the calculation of the probability density functions allowed discriminating the ECL signal from the reaction occurring directly on the ITO and the signal from the NP collisions with the electrode. Similarly, the power spectral density function (PSDF) analysis of the intensity profiles also confirmed the presence of the ECL fluctuations associated with individual particle collisions with the electrode surface.

The ECL intensity associated with single NP collisions with the electrode is believed to be affected by the size of the NPs, the nature of the interaction between a particle and the electrode surface (e.g., the particle residence time), the concentrations and the electron transfer kinetics of indicator species and a coreactant on NPs, and the lifetimes of active intermediate precursors and excited states. Thus, the developments of these research directions are yet to be demonstrated; however, the research on ECL detection is ongoing. Because of the high amplification, ECL collisions should be useful as a very sensitive analytical tool, perhaps eventually down to the single-molecule level. Improvements could be based on the NP itself being the source of the ECL reaction or fabrication of the UME where the electrochemical reactions of emitter precursor or coreactant are greatly diminished.

8.6.2 Photoelectrochemistry

The basis of PEC detection of semiconductor NPs is the well-studied phenomenon where irradiation of a semiconductor with radiation of energy larger than the semiconductor band gap causes generation of electrons and holes capable of carrying out redox reduction. Capture by a solution reactant of one of these thus charges the NP and that charge can be collected at an electrode.[42,43] As with ECL, a problem is that the observed photocurrent response for a single NP is very small. In a recent report by Alpuche-Aviles and coworkers stochastic photoelectrochemical currents were observed upon collision of TiO$_2$ NPs with a Pt UME surface under UV irradiation.[44] This is the first example of individual collision events of semiconductor NPs undergoing photochemical process.

Current steps, mostly in anodic direction, were observed upon irradiation (Xe arc lamp) of the electrode with neat methanol as the hole scavenger and anatase TiO$_2$ NPs (18–70 nm) present in the system. Since methanol is a well-known scavenger of holes (h$^+$, see Figure 8.30a), it is believed that the steps in the current response are due to the transfer of electrons from the NPs that are irreversibly interacting with the electrode, as seen in ensemble measurements.[43] Only a small fraction of collisions (ca. 1 in 10^5) were found to result in observable current steps. Nevertheless, the authors determined that the magnitude of the current steps (mostly 1–10 pA) depends on the size of the NPs suggesting that the steps are due to the NPs interacting with the electrode (Figure 8.30b). Stepwise changes in current in the cathodic direction were also sometimes observed, as well as the current transients in the dark (after the illumination was turned off). In general, the background currents were not constant and depended on the history of the electrode. All these findings suggest that there could be possible contribution to the current response from photogenerated intermediates or by-products of methanol oxidation. Responses were also not observed with lower concentrations of methanol.

8.7 THEORY

8.7.1 Signal Amplification of Single NP Collision

When a single metal NP contacts an electrode, the NP will become charged, but only a few electrons will be transferred, and the current will be too small to be distinguished from the background level or noise. However, if the NP can electrocatalyze some reaction that the contacting electrode cannot, the current can be much higher because of the continued electron flow involved with this reaction. This provides the needed amplification for single-particle detection. If the NP sticks to the electrode surface, the steady-state diffusion-controlled current at the NP (assuming spherical diffusion and a very fast electrocatalytic reaction) is

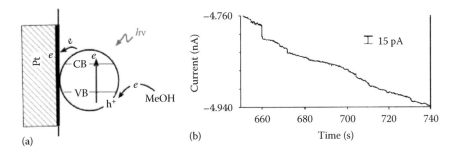

FIGURE 8.30 (a) Schematics of photooxidation of MeOH by a NP attached to a Pt UME and (b) plot of current versus time for 25 nM anatase NPs suspended in MeOH and a 10 μm diameter Pt UME with $E_{app}=0.68$ V versus NHE under illumination with a 150 W Xe lamp. (Reprinted with permission from Fernando, A., Parajuli, S., and Alpuche-Aviles, M.A., Observation of individual semiconducting nanoparticle collisions by stochastic photoelectrochemical currents, *J. Am. Chem. Soc.*, 135(30), 10894–10897, 2013. Copyright 2013, American Chemical Society.)

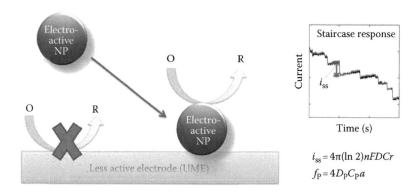

FIGURE 8.31 Schematic illustration of sticky NP collision and equations for staircase current increase and collision frequency.

$$i_{ss} = 4\pi\left(\ln 2\right)nFDCr \tag{8.18}$$

where

  n is the number of electrons transferred
  F is the Faraday constant
  D is diffusion coefficient of reactant
  C is the concentration of reactant
  r is the radius of the NP[6]

This equation differs from that for a spherical UME by the *ln2* term, which accounts for blocking of the diffusion path to the NP by the supporting planar surface.[6,35] In this situation, the current from each new NP contact adds to the previous one and a *staircase* response is observed as shown in Figure 8.31.

If the NP does not stick to the electrode, the total charge transferred is determined by the NP size, its residence time, the rate of the electrocatalytic reaction, and the rate of charge transfer from the electrode. In this situation, each NP contact may result in a transient current or *blip*, superimposed on a constant background current. However, the nature of the blip response observed in the IrO_x NP system was interpreted as a *sticky* collision (in which the colliding particle does not desorb from the surface) with an accompanying deactivation of the surface by-products of the electrode reaction.[9]

Another blip-type response observed was for the direct electrolysis of the collided NPs. This also occurs with a sticky collision as described in Section 8.3. Considering the short residence time (~ns) of nonsticky (elastic) collisions, the small amount of charge transfer in this time, and the relatively long data sampling time (~ms) of potentiostat, it is believed that the current transients for nonsticky collisions of NPs are extremely difficult to detect. Therefore, we can consider that most observable current transients are due to only the irreversible (sticky) collisions of NPs.

Particles that irreversibly stick thus generate a concentration gradient, so the collision frequency can be calculated exactly from the steady-state diffusion-controlled flux of particles to the UME surface:

$$J_p = \frac{4D_p C_p^{bulk}}{\pi a} \tag{8.19}$$

where

  D_p is the particle diffusion coefficient
  C_p^{bulk} is the concentration of particle
  a is the radius of the UME disk electrode

Therefore, at a sufficiently extreme potential, a diffusion-limited flux of particles can be assumed, and with irreversible adsorption, the frequency of arrival at the electrode surface is given by

$$f_p = J_p A = 4D_p C_p^{bulk} a \qquad (8.20)$$

However, the estimated diffusion arrival frequency is about 10 times higher than the experimental value, suggesting that all arrivals at the electrode surface do not result in measureable transients. Therefore, if the particle adsorption is not sufficiently fast, a more general expression can be used to describe mixed kinetic and diffusion control:

$$f_p = 4D_p a \left(C_p^{bulk} - C_p^{pca} \right) \qquad (8.21)$$

and

$$f_s = \pi a^2 k_{ads} C_p^{pca} \qquad (8.22)$$

where
C_p^{pca} is the time-averaged concentration of NPs at the electrode surface
k_{ads} (cm/s) is the rate constant for adsorption

The stochastic character of k_{ads} will be a function of k_{coll} (rate constant for collision of particles on the electrode, in cm/s) and p_{ads}, the probability that any given collision leads to adsorption. Thus, on average,

$$k_{ads} = p_{ads} k_{coll} \qquad (8.23)$$

With this conceptual model, adsorption cannot occur unless there has been a collision; however, not every collision will lead to adsorption or an observable event. A slightly different way of thinking about this is to assume that adsorption occurs only at particular surface sites, for example, particular structural defects on the electrode surface or places with surface adsorbates that promote adsorption of the NP. A collision is still required, but the probability of adsorption, p_{ads}, will also be a function of the location on the surface. Combining Equation 8.20 with Equation 8.23 gives

$$f_p = 4D_p a \left(C_p^{bulk} - C_p^{pca} \right) = 4D_p a \left(C_p^{bulk} - \frac{f_p}{\pi a^2 k_{ads}} \right) \qquad (8.24)$$

Then

$$f_p = \frac{4D_p a C_p^{bulk}}{1 + 4D_p / \pi a k_{ads}} \qquad (8.25)$$

Therefore, when the k_{ads} is large, the maximum collision frequency, Equation 8.20 is obtained, and when the sticking frequency, k_{ads}, is small ($k_{ads} \ll 4D_p/\pi a$), Equation 8.25 reduces to Equation 8.22 with C_p^{pca} replaced by C_p^{bulk}. The same collision frequency formula, which is proportional to $D_p C_p^{bulk} a$, was derived by a random walk simulation and supported by a C-language program simulation.[45]

In experimental studies, the observed step frequency depends on the electrode materials and surface modification. Applying this equation for Pt NP/Au, C electrode system, we could estimate a value for k_{ads} from the experiment. This approach for the NP collision frequency could be applied not only for the staircase-type response but also the blip-type response, wherein the particle sticks and becomes deactivated.

Frequency analysis of NP collisions is not simple because the shape and frequency of the current transients are affected not only by the NPs but also by the material and nature of the surface of the measuring electrode. For example, the current transient frequency of citrate-stabilized IrO_x NPs differed by the electrode material: current transients for IrO_x NP collisions were frequent on bare Au, rare on bare Pt, and not observed at all on carbon fiber UMEs. The electrocatalytic redox recycling behavior also depends strongly on the electrode material. The current spikes are sensitive to the electrode surface, and we find that the current transient behavior can be modified with different surface treatments, for example, by immersing the Pt UME in a 10 mM aqueous $NaBH_4$ solution for 5 min.[5] The influence of the electrode surface properties on NP behavior is still not well understood, but the single NP collision detection techniques described here can be useful tools to study such phenomena.

8.7.2 SIMULATION OF THE MAGNITUDE OF CURRENT STEPS DUE TO THE BLOCKING OF THE ELECTRODE ACTIVE AREA BY INSULATING PARTICLES

When an insulating particle sticks to the surface of an electrode where an electrochemical reaction takes place, it impedes the flow of the reactant to it, and as a result, a decrease in the measured faradaic current is observed. Thus, the approach that was used to estimate the magnitude of the current steps due to blocking of the active electrode area consisted of two parts: in the first part, the steady-state current was calculated and compared to an analytical solution for a disk electrode,[36] and in the second part, the current was calculated assuming the presence of an insulating body on the disk surface. Thereby, it was possible to calculate the change in the faradaic current due to the adsorption of a single insulating particle at the electrode surface.

The simulation discussed here was performed using the convection and diffusion application mode of Multiphysics v3.5a (COMSOL, Stockholm, Sweden). The Nernst–Planck equation, which is solved numerically, can be written assuming diffusion and migration mass transfer modes as

$$J_i(r,z) = -D_i \nabla C_i(r,z) - \frac{z_i F}{RT} D_i C_i(r,z) \nabla \varphi(r,z) \qquad (8.26)$$

where
  J_i is the flux of the species i
  D_i is the diffusion coefficient
  C_i is the concentration
  z_i is the charge on the species in signed units of electronic charge
  F is the Faraday constant
  φ is the electric potential responsible for the migration effect, and other symbols have their usual
    meaning

The expressions for the boundary conditions used in the simulations are given by Equations 8.27 through 8.29.

For the electrode surface, a fully reversible redox process is assumed (such as FcMeOH oxidation):

$$C(r,z) = 0 \qquad (8.27)$$

where C is the concentration of the redox species (FcMeOH).

For the bulk solution boundary,

$$\lim_{r,z \to \infty} C(r,z) = C_b \qquad (8.28)$$

where C_b is the bulk concentration of the species.

For the axial symmetry axis, insulating ring, and a glass sheath,

$$\bar{n} \cdot \bar{J} = 0 \tag{8.29}$$

where
$\bar{n}$ is the normal vector
J is the flux of the redox species given by the Nernst–Planck equation (8.26)
The initial condition is given by an equation similar to Equation 8.28:

$$C(r, z) = C_b \tag{8.30}$$

Finally, the current through the disk is obtained by a simple integration:

$$I = nFD \int_0^{r_0} 2\pi r \frac{\partial C(r, z)}{\partial z} \, dr \tag{8.31}$$

where
n is the number of electrons transferred in the electrode reaction
F is the Faraday constant
D is the diffusion coefficient of the redox species ($1 \cdot 10^{-5}$ cm^2/s)
r_0 is the radius of the electrode
$\partial C/\partial z$ is the species concentration gradient at the electrode surface

The geometry of the model used in the simulations is shown in Figure 8.32.

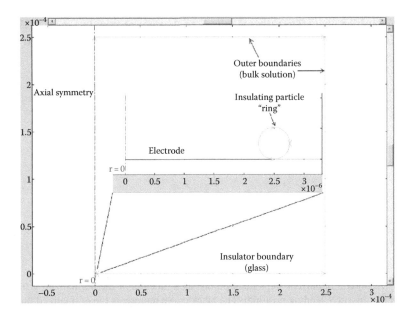

FIGURE 8.32 Geometry of the model used in simulations of the magnitude of current steps due to adsorption of insulating particles. An inset shows an electrode disk (2.5 µm radius) embedded in a glass sheath and a spherical ring representing an insulating feature positioned right on top of the disk edge.

Note that if the meshing of the geometry in COMSOL is not sufficiently fine, then the numerical solution for the limiting current through the disk differs from the analytical by more than 0.4% (ca. 8 pA for a 2.5 μm radius electrode and 2 mM concentration of the redox species). This creates a problem, since the simulated change in the faradaic current due to blocking by a single particle (the magnitude of the current step) is then comparable to the simulation error. Therefore, to improve the accuracy of the calculations within the multiphysics environment, we simulate the change in the current due to the presence of an insulating *ring* rather than a particle. The number of spherical particles (N) that *fit* into the ring is given by the following equation:

$$N = \frac{3\pi}{2}\left(\frac{r}{a}\right)$$ (8.32)

where
  a is the particle radius
  r is the distance from the disk center to the particle position on the electrode surface

Consequently, the magnitude of a step due to a single NP (I_s) is equal to the value for the ring (I_r) divided by the number of particles in the ring:

$$I_s = \frac{I_r}{N}$$ (8.33)

The results of the simulations indicate that the magnitude of the observed current steps depends on a number of factors, such as the concentration of the redox species, that is, FcMeOH, the size of both the particle and the electrode, and the particle landing position on the electrode surface. The flux of the redox species toward the electrode is directly proportional to its concentration. Therefore, the magnitude of a current step due to the flux blockage by a particle will also be directly proportional to the concentration of the species. However, a complication arises here from the fact that the particle landing position also affects the magnitude of the current step. In Figure 8.33, the relative change in the faradaic current due to the adsorption of an insulating particle is plotted as a function

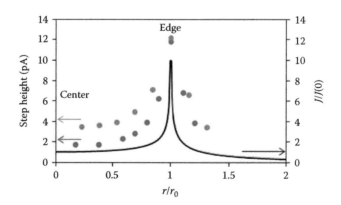

FIGURE 8.33 Magnitude of current steps (data points) as a function of the normalized distance from the electrode disk center (particle landing position). The black curve represents the normalized flux of the redox species through the disk. Simulations were done for (red points) an electrode disk radius r_0 of 2.5 μm, a particle radius of 260 nm, and a FcMeOH concentration of 2 mM and (blue points) an electrode disk radius r_0 of 1 μm, a particle radius of 155 nm, and a FcMeOH concentration of 2.5 mM.

of the distance from the disk center; the magnitude of the current step increases when the particle lands closer to the electrode edge. This is because the flux of the redox species is much higher at the edge of the electrode than at its center, and therefore, the particle blocks its flux to a larger extent when it adsorbs closer to the edge.

The solid curve in Figure 8.33 corresponds to the normalized flux of the redox species (such as FcMeOH) at the electrode disk, and the data points correspond to the magnitude of current steps. Thus, the change in the current step magnitude follows the change in the flux of the redox species. In addition, for the same ratio of the radius of an electrode to the radius of a particle, it is possible to correlate the position where a particle lands on the electrode surface to the magnitude of the current step observed in the collision experiment.

Note that $r/r_0 > 1$ represents particles landing on the insulator sheath surrounding the electrode and that even particles landing in this region cause a perturbation of the current (because mass transfer involves spherical coordinates).

8.7.3 SIMULATION OF THE MIGRATION EFFECTS IN NP COLLISION SYSTEMS

The simulations of the flux of charged particles (both insulating and catalytic) can be carried out using the Nernst–Planck equation with electroneutrality (NPE) application mode of the Chemical Engineering Module in Multiphysics v3.5a.

The two general equations solved are the Nernst–Planck equation (Equation 8.26) and the electroneutrality condition (Equation 8.34):

$$\sum_i z_i C_i = 0 \tag{8.34}$$

where
z is the charge
C is the concentration of all species in the system

These two equations apply both to the oxidized and reduced forms of the redox species, the supporting electrolyte ions, as well as the charged NPs. Equations 8.26 and 8.34 are valid for all cases, that is, reduction or oxidation electrochemical processes. However, it is easier to consider the migration effects on a single example, such as the oxidation of FcMeOH. A similar approach has been proposed for the migration of catalytic particles in a system Pt NP/Au electrode/hydrazine oxidation.[15]

Oxidation of FcMeOH (A), which occurs on the working electrode, leads to the formation of a positively charged species (hydroxymethylferrocenium ion, A$^+$). Uncharged FcMeOH only arrives at the electrode by diffusion; however, the electric field is responsible for the migration of the charged NPs, ions of the supporting electrolyte, and the charged product of the reaction. If the concentration of the supporting electrolyte is low, then this field will substantially affect the concentration of all charged species in the system, particularly near the working electrode. Therefore, in addition to the diffusive mode of mass transport (first term in Equation 8.26), we need to consider the migration of ions and particles (second term in Equation 8.26).

The electroneutrality condition (Equation 8.34) requires that in any volume element of solution, the number of positive and negative species is equal. The use of this condition is needed so that the number of unknowns (concentrations of the species and the electric potential) and solved equations are equal.

For the boundary conditions, one has to specify the relations that describe concentrations of all the species and the electric (migration) field at all system boundaries. In the bulk solution, concentrations of the species are given by their bulk values (same as Equation 8.28) and the electric (migration) potential is zero:

$$\phi = 0 \tag{8.35}$$

At the electrode surface, the flux boundary conditions are used to describe the flux of the redox species (Equations 8.36 and 8.37), and a no-flux condition is used for the cations of the supporting electrolyte (Equation 8.28):

$$\bar{n} \cdot \bar{J} = k_f C_A - k_b C_{A^+} \tag{8.36}$$

$$\bar{n} \cdot \bar{J} = -k_f C_A + k_b C_{A^+} \tag{8.37}$$

$$\bar{n} \cdot \bar{J} = 0 \tag{8.38}$$

It should be noted that the concentration of anions of the supporting electrolyte in our model, assuming an oxidation reaction, is determined using the electroneutrality condition, so no condition needs to be specified specifically for these ions. The electrode surface boundary condition for the NP concentration is identical to Equation 8.27, meaning that all particles reaching the surface of the electrode *stick* to it, thus leaving the solution phase.

The rate constants k_f and k_b in Equations 8.36 and 8.37 are given assuming Butler–Volmer formalism:

$$k_f = k_0 \exp\left[\frac{(1-\alpha)F}{RT} \left(E - E^0 \right) \right] \tag{8.39}$$

$$k_b = k_0 \exp\left[\frac{-\alpha F}{RT} \left(E - E^0 \right) \right] \tag{8.40}$$

where
 k_0 is the standard rate constant
 α is the electron transfer coefficient
 E is the electrode potential
 E^0 is the standard potential of FcMeOH⁺/FcMeOH couple

For the migration electric field, a current inflow boundary condition is used:

$$-\bar{n} \cdot F \sum_i z_i J_i = z_{A^+} F \left(-D_{A^+} \frac{\partial C_{A^+}}{\partial z} - z_{A^+} u_{A^+} F C_{A^+} \frac{\partial \phi}{\partial z} \right) \tag{8.41}$$

where u is the mobility of the ionic species indicated by the subscript and all other symbols have their usual meaning. One can see that this current density is proportional to the flux of the hydroxy-methylferricinium ions.

All other boundary conditions are given similarly to Equation 8.29 (insulating boundaries).

For initial conditions, it is assumed that the potential ϕ is zero everywhere in the system and the concentrations of all species are equal to their values in the bulk solution.

Besides the boundary and initial conditions, one also has to specify the mobility and charge of all ionic species, as well as the NPs. For ions in solution, the mobility and diffusion coefficients are directly related:

$$u_i = D_i \frac{|z_i| F}{RT} \tag{8.42}$$

where the diffusion coefficient is given by the Einstein–Smoluchowski equation (Equation 8.43):

$$D_i = \frac{kT}{6\pi\eta a} \tag{8.43}$$

where
  k is the Boltzmann constant
  η is the viscosity of solution
  a is the radius of a species

However, the behavior of large colloidal particles (compared to ions) is more complex, and their mobility is expected to depend, among other parameters in Equations 8.42 and 8.43, on the concentration of the supporting electrolyte ions in solution. Therefore, the mobility of NPs can be calculated using an approximate analytical relation reported by Ohshima et al.[46] Also, the average charge on NPs, z_p, can be determined from their electrokinetic charge density σ^{ek}, which in turn is related to the zeta potential of the particles, ζ, in aqueous solutions at 25°C[47]:

$$\sigma^{ek} = -11.73\sqrt{C} \sinh\left(0.0195 z_p \zeta\right) \tag{8.44}$$

where
  C is the molar concentration of the supporting electrolyte ions
  σ^{ek} has units μC/cm and ζ is in mV

For further details regarding the NPE approach in simulations of the migration effect, the interested reader is referred to the works by Oldham and Feldberg.[16,17]

8.8 NP SYNTHESIS AND CHARACTERIZATION

It is beyond the scope of this chapter to discuss the large literature on NP synthesis and characterization. General references are given; however, the properties of NPs, especially in the EA involving IS reactions, are very dependent on the surface of the NPs and especially the capping agents that help maintain stability.

8.8.1 Synthetic Methods and Stabilizing Agents for NP Collision Experiments

Collectively, the generation of particles having nanometer dimensions has been the subject of an enormous number of investigations. For a more expansive description of synthetic strategies for producing specific types of NPs, numerous chapters and review articles are available.[48–57] Here, we present a brief overview of methods related specifically to types of NPs used in the electrochemical experiments described in this chapter.

For experiments involving NP detection by EA, metal NPs have been synthesized by the chemical reduction of dissolved metal salts in the presence of adsorbing anions.[1,4] In these works, gold and platinum NPs were generated by reduction of their chloride salts in the presence of citrate, a method first applied for gold NPs by Turkevich and later advanced by Frens and others.[58–60] The catalytic amplification experiments described here present a restriction on the choice of stabilizing (or capping) agents in that the surface of the NP must remain catalytic for IS electron transfers as required for detection of the NP. Therefore, electrostatic stabilization by the adsorption of small anions is preferred as diffusing redox species may still reach catalytic NP surface sites. Steric stabilization methods, for example, by the inclusion of polymer surfactants or chemical modification of the NP surface using even short-chain alkyl ligands, are not used because the adsorbed species layer will limit electron transfer at the NP surface. Even for experiments in which catalytic amplification is not being applied, such as those by Zhou et al.,[22] conductive NPs must still be able to make a reliable electrical OS connection to the working electrode to be detected, similarly limiting the choice of stabilizing agents to those that do not form a thick layer on the NP surface.

As compared to metallic NPs, semiconductor NPs used in NP collision experiments have the same requirement that any stabilization method must not prevent electron transfer at the NP surface, although their synthesis differs. IrOx NPs used by Kwon et al. were prepared by the hydrolysis of $IrCl_6^{2-}$ by NaOH in the presence of citrate 5.[61] The semiconductor NP collision experiments now using titanium (II) oxide NPs were prepared by a hydrothermal method originally reported by Zaban and coworkers.[44,62]

For the experiments described in Sections 8.4 and 8.7 involving NP detection by diffusion blocking, polymer and silica spheres were used with diameters between 0.3 and 1 µm.[13,14,33] For these, commercially available NPs were preferred for their monodispersity, although the literature contains numerous synthetic routes to polymer and silica NPs.[63,64] One strength here of the diffusion blocking approach is that the choice of stabilizing agents is less restricted because no electron transfer is occurring at the NP, although with larger NPs, charged surface moieties may be required to provide for electrophoretic migration of the NPs.

8.8.2 CHARACTERIZATION OF NPs FOR ELECTROCHEMICAL NP COLLISION EXPERIMENTS

The stochastic electrochemical experiments described in this chapter focus on the detection of single NPs and the elucidation of their electrochemistry as observed through single collision events at electrodes under potential control. To this end, supplemental characterization of the NPs is primarily concerned with obtaining NP size, morphology, dispersity, and concentration information, as well as assessing the state of the NPs in situ, with reduced emphasis on the assessment of the material composition of the NPs.

TEM and scanning electron microscopy (SEM) have been the primary techniques used for obtaining NP size and morphology.[1] TEM and SEM are effective for determining individual NP size and shape characteristics due to their ability to resolve single NPs, a key requirement for characterization of NPs used for single NP electrochemistry. However, electron microscopy must be performed in vacuum and is therefore unable to gauge the characteristics of the NP inside an electrolyte, which may depend on the solution composition and choice of stabilizing agent. Moreover, sample preparation problems, such as aggregation of the NPs during drying and also adverse effects of the electron beam, make these methods poorly suited to measuring NP size dispersity.[65]

Methods using DLS may be applied to NP samples dispersed in electrolyte solutions allowing for in situ characterization of the NPs. In contrast to NP sizing techniques based on diffraction, DLS measurements effectively record Doppler broadening of Rayleigh scattered light due to diffusion of NPs in a sample.[66] Therefore, DLS may provide size, dispersity, and diffusion coefficients for NPs with diameters of a few nm to around 5 µm. Furthermore, coupled with electrophoresis apparatus, DLS instrumentation may also provide the NP zeta potential, which allows estimation of the surface charge, a key parameter effecting NP stability.[67]

One challenge in the application of DLS to NP samples is the interrelatedness of NP size, size distribution, and the effects of NP aggregation in the measurement. NP aggregation effects cause widening of size distribution estimates due to their relatively larger contribution to Doppler broadening for the NP sample. Moreover, larger NPs may screen Doppler broadening from smaller NPs in the sample, resulting in skewed estimates of both the size distribution and diffusion coefficient for samples where the real size distribution is large.[68] Further compounding these challenges is the inclusion of electrolyte to the solution, which can profoundly affect aggregation/settling behavior by reducing the electrical double-layer thickness at the NP surface.[69] Thus, coupled zeta potential measurements can prove especially valuable in electrochemical experiments, as a low zeta potential (roughly smaller than ±30 mV)[65] may indicate instability of the NP and thus reason to suspect aggregation effects in the DLS data.

Importantly DLS involves an ensemble measurement, with the light scattered by all the particles in the irradiated area. Thus, the results are model dependent and, as indicated earlier, may be biased toward the larger particles. It also does not provide information about NP concentration. An alternative is single-particle tracking methodology. In this method, a small slab of solution, about 10 μm high, 80 μm wide, and 100 μm long, is irradiated with a ribbon of laser light, and individual particle locations are imaged by the light they scatter or by their fluorescence. A movie is made for a given time as the individual particles are tracked, and from a diffusion coefficient calculated from the Brownian motion, one can find the size. The number of particles in the volume yields the concentration. As in DLS, an electric field can be applied across the cell, and the change in particle velocity from migration allows determination of the zeta potential. Of course a solution must be sufficiently dilute that there is a suitable number of particles in the test volume that can be individually tracked.

While electron microscopy and DLS have been the most frequently applied characterization techniques for NPs used in electrochemical collision experiments, an array of other characterization methods are used in the larger field of NP chemistry, which may prove useful in applications related to these experiments.[50,53,65] In particular, x-ray diffraction methods as well as UV-visible light spectroscopy have been used extensively for the characterization of NPs since the earliest experiments in the field, although x-ray diffraction is generally only applicable ex situ. On the other hand, electrochemical scanning probe methods including electrochemical scanning tunneling microscopy (EC-STM),[70–72] electrochemical atomic force microscopy (EC-AFM),[73,74] and scanning electrochemical microscopy (SECM)[75,76] can offer characterization of even single NPs in situ. However, for these techniques, NPs must usually be affixed to a surface for characterization, and thus, the techniques may be less useful for experiments concerned primarily with diffusing or migrating NPs.

8.9 APPLICATIONS AND CLOSING COMMENTS

Many applications have been suggested for NP collisions both in analysis and fundamental studies. Here, several examples are reviewed.

Parameters of collision chronoamperograms such as the magnitude and frequency of current events provide information about the properties of NPs in solution, that is, their size and concentration. From the earliest reports, it was noticed that the shape of the current step and particle size distributions are very similar (Figure 8.34) and correlated well with the analytical relation between the faradaic current and radius of a particle (Section 8.2).[3]

Similarly, in the direct particle electrolysis scheme (Section 8.3), there is a direct relationship between the charge passing during the particle electrolysis and its size. Thus, from the magnitude of collision events, one can determine the size of the colliding NPs. The examples include sorting of the NPs by size for silver,[77–79] gold,[23] nickel,[80] and magnetite[81] particles. One can also tag or label NPs with electroactive molecules such as 1,4-nitrothiophenol[26] and monitor the size of the NPs

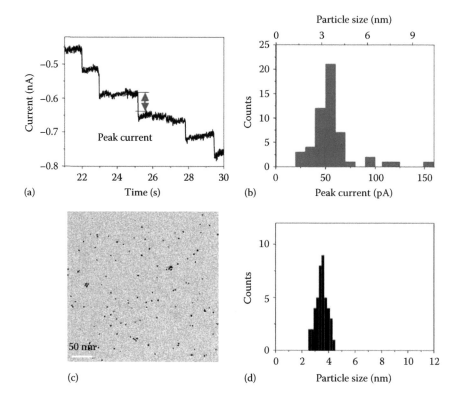

FIGURE 8.34 (a) Representative current steps from a collision experiment of ~36 pM Pt NP solution with 10 μm Au UME in 15 mM hydrazine and 50 mM PBS buffer; pH ~ 7.5; particle size, ~3.6 nm. (b) Statistical peak current versus peak frequency analyzed for a 200 s interval. (c, d) TEM image and size distribution of the corresponding Pt NPs.

from the charge passing during the electrochemical transformation of the tag; the method had been termed as tag-redox coulometry (TRC).[82] The size of droplets in emulsions can similarly be determined in the EDR method.

The frequency of collisions is, in principle, the most sensitive measure of particle concentration in solution. One, however, needs to know the dominant mode of mass transfer of particles to the electrode (i.e., diffusion, migration, and convection), since that determines the exact form of the Nernst–Plank equation (Section 8.7). For example, with a low supporting electrolyte concentration, the electrophoretic migration can be the dominant mode of NP mass transfer in the detection of both insulating and catalytic particles in the low pM–fM concentration range.[14,15] When the concentration of the supporting electrolyte is relatively high and particles move by diffusion, the concentration of the NPs that can be detected in a reasonable time from the frequency of collisions is usually in the high pM and even nM concentration range. The examples of quantification of NP concentrations assuming only the diffusive mode of mass transfer include the detection of IrO_x particles[5,9] and direct electrolysis of silver and nickel NPs.[80] Besides the size and concentration studies, one can also investigate the agglomeration state of the NPs in solution using the collision approach; the electrochemical method is believed to be a reasonable alternative to the NP tracking analysis systems.[83]

The ability to resolve individual collision events using the EA scheme (Section 8.2) makes this approach also very attractive for analytical detection of many analytes where the electrocatalytic metal NPs act as electrochemical labels. In a recent report, Kwon and Bard demonstrated a highly

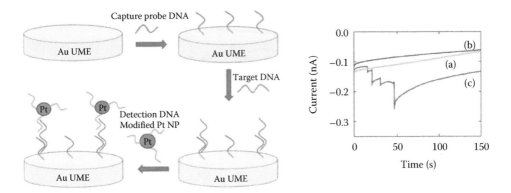

FIGURE 8.35 Illustration of sandwich-type DNA sensor on the Au UME (5 μm radius) and representative chronoamperometric (CA) curves for single Pt NP collisions recorded with it. The UME was first incubated in a capture probe DNA and then incubated in (a) 10 nM (green) noncomplementary DNA, (b) 0 M (black), and (c) 10 pM (red) complementary target DNA solution. The UMEs were then incubated in detection DNA-modified Pt NP solution. After the hybridization was finished, the electrode was rinsed and tested by CA in 50 mM pH 7 PB containing 10 mM hydrazine at 0 V versus Ag/AgCl.

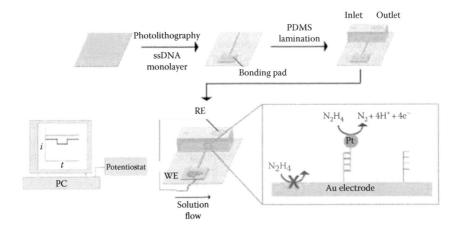

SCHEME 8.2

sensitive approach to the detection of DNA, in principle down to the single-molecule level.[12] In the developed detection scheme, the surface of a gold UME was modified with a 16-base oligonucle-otide with a C6 spacer thiol (capture probe), and the Pt NP surface was modified with a 20-base oligonucleotide with a C6 spacer thiol (detection probe). As a result, in the presence of a target oli-gonucleotide (31-base) that hybridized with both capture and detection probes, a Pt NP was brought close to the detector electrode surface where a flow of faradaic current was observed due to the electrochemical oxidation of hydrazine on Pt (Figure 8.35). As a result, target DNA molecules were detected at the 10 pM concentration level. A similar approach has been adopted by Alligrant et al. for the EA detection of DNA in a microfluidic channel.[84] A gold microband electrode was used as a detector placed in a PDMS channel (6 mm long, 25 μm wide, and 20 μm high, Scheme 8.2). The gold surface was modified with a 25 bp ssDNA (probe), while the Pt NP was modified with a 25 bp cDNA (target). Therefore, upon introduction of the target-modified Pt NPs into the channel, elec-trocatalytic collision events were observed due to the oxidation of hydrazine at the NP surface. The observed shape of the current response varied between sharp transients and more *steplike* features (Figure 8.36).

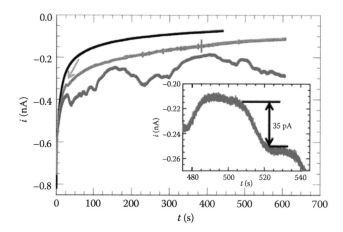

FIGURE 8.36 Curves of *i–t* measured at 0.15 V versus Ag/AgCl using the prepared device (Scheme 8.2). The Au electrode surface was modified with ssDNA, and in addition to 5.0 mM N_2H_4, the solution contained (black) no target DNA, (green) 25 pM PtNP-ncDNA, and (red) 25 pM PtNP-cDNA. The inset shows an expanded region of the red *i–t* curve. Conditions: 50.0 mM phosphate buffer, pH 7.0, flow rate = 50 nL/min.

Such variations in the shape of current transients are not yet understood. However, as pointed out by the authors, it may be due to a combination of new and old hybridization events (i.e., a hybridization event gives rise to a current transient upon initial binding, but later, the same Pt NP may yield subsequent current pulses).

Observation of EA collisions is dependent on the effectiveness of the electron transfer between the electrode and the particles. Therefore, by using the EA approach, one is able to study electron transfer on the level of single NPs. Such studies were reported (Figures 8.37 and 8.38).[10] In this work, the authors used the modification of the surface of either the Pt NPs or the detector electrode (Au) with alkyl thiols of variable length to investigate its effect on the electrocatalytic activity of the particles toward the electrochemical oxidation of hydrazine.

If NPs were only capped with citrate ions and the detector electrode surface was covered with alkyl thiols of various lengths, the electron transfer decreased exponentially with an increase in the alkyl chain length for both OS and IS reactions. Thus, the decrease of the overall particle activity (as measured by the magnitude of current transients) was mainly limited by the exponential decay of electron transport from Pt NPs through SAMs to the gold electrode surface (Figure 8.37).

However, if the Pt NPs were capped by alkyl thiols or a mixture of alkyl thiols and citrate ions, while the working electrode remained clean, the electrocatalytic activity of the NPs toward hydrazine oxidation gradually decreased with an increase in the concentration of the alkyl thiol of the same carbon chain length (Figure 8.38).

Thus, the decrease of particle activity was mainly due to the blockage of the catalytic surface sites for IS reactions such as hydrazine electrooxidation. The examples of applications of NP collision measurements listed earlier suggest the possibility of a wider range of applications. In terms of analytical applications, stochastic electrochemical detection based on observation of NP label individual collision events provides the single-molecule sensitivity, that is, individual analyte species are detected one at a time. As with other similar analytical schemes, the lowest concentrations that can actually be detected are limited by factors such as strength of interactions, selectivity, and the time required to detect the particle. In the field of fundamental studies, the study of individual NP collisions should allow extraction of information that is buried or averaged out in ensemble measurements. An example is an examination of factors that contribute to adsorption or sticking

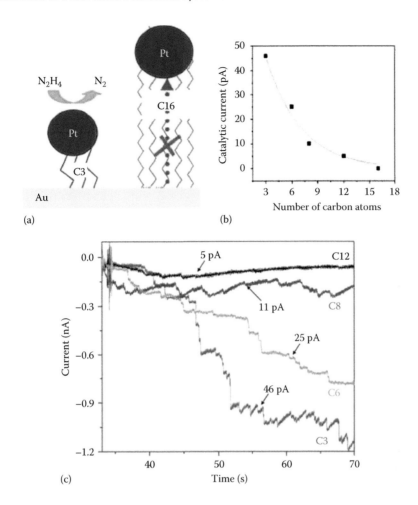

FIGURE 8.37 Collisions of Pt NPs at SAM-covered Au electrodes. (a) Schematic illustration of particle collisions at the SAMs: catalytic hydrazine oxidation and suppression of electron transport through long molecular spacers. (b) Plot of representative catalytic current versus the number of carbon atoms in SAMs. (c) Current transients recorded at different SAMs. Au electrode 10 μm in diameter; electrode potential, 0.1 V; particle size, ~3.6 nm; particle concentration, 50 pM; electrolyte, 12 mM hydrazine + 50 mM PBS buffer; pH ~ 7.5.

of an NP to a surface or the shape changes that occur when a liquid droplet interacts with a surface. The examples given earlier certainly show great promise in helping to unveil the unknown aspects of electron transfer that are important for the advancement of basic science and potential applications.[85,86]

It seems clear that interest in stochastic (vs. ensemble) electrochemistry will continue and, with time, progress to similar experiments with large biomolecules and eventually to single molecules of ordinary dimensions. Such studies result in deeper insights into the interactions of molecules with electrode surfaces and the nature of electron transfer processes at interfaces. Such advances will rely on smaller, nm dimension, UMEs and surfaces that show much lower background currents (or other responses). Such studies will also place great demands on solvent and electrolyte purity and instrument sensitivity. The challenges are considerable, but with continued advances as have been seen over the last decades, success is assured.

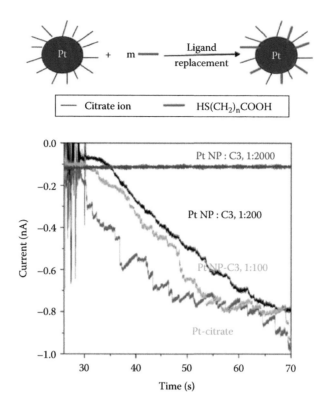

FIGURE 8.38 Current transients recorded at Au UMEs before and after injection of Pt NPs capped by a mixture of citrate ions and C3 SAMs. Replacement of citrate by SAMs is illustrated at the top of the figure. Au UMEs 10 μm in diameter; electrode potential, 0.1 V; particle size, ~3.6 nm; particle concentration, ~50 pM; electrolyte, 12 mM hydrazine + 50 mM PBS buffer; pH ~ 7.5.

ACKNOWLEDGMENTS

We acknowledge the support of this work from the National Science Foundation (CHE-1111518), Defense Threat Reduction Agency (HDTRA1-11-1-0005), and the Robert A. Welch Foundation (F-0032).

REFERENCES

1. Xiao, X.; Bard, A. J. 2007. Observing single nanoparticle collisions at an ultramicroelectrode by electro-catalytic amplification. *J. Am. Chem. Soc.* 129: 9610–9612.
2. Bard, A. J.; Zhou, H.; Kwon, S. J. 2010. Electrochemistry of single nanoparticles via electrocatalytic amplification. *Isr. J. Chem.* 50: 267–276.
3. Xiao, X.; Fan, F.-R. F.; Zhou, J.; Bard, A. J. 2008. Current transients in single nanoparticle collision events. *J. Am. Chem. Soc.* 130(49): 16669–16677.
4. Zhou, H.; Fan, F.-R. F.; Bard, A. J. 2010. Observation of discrete Au nanoparticle collisions by electro-catalytic amplification using PT ultramicroelectrode surface modification. *J. Phys. Chem. Lett.* 1(18): 2671–2674.
5. Kwon, S. J.; Fan, F.-R. F.; Bard, A. J. 2010. Observing iridium oxide (IrO$_x$) single nanoparticle collisions at ultramicroelectrodes. *J. Am. Chem. Soc.* 132(38): 13165–13167.
6. Bobbert, P. A.; Wind, M. M.; Vlieger, J. 1987. Diffusion to a slowly growing truncated sphere on a substrate. *Physica A* 141(1): 58–72.
7. Yang, J.; Lee, J. Y.; Too, H. P. 2006. Size effect in thiol and amine binding to small Pt nanoparticles. *Anal. Chim. Acta* 571: 206–210.

8. Nakagawa, T.; Bjorge, N. S.; Murray, R. W. 2009. Electrogenerated IrO_x nanoparticles as dissolved redox catalysts for water oxidation. *J. Am. Chem. Soc.* 131(43): 15578–15579.

9. Kwon, S. J.; Bard, A. J. 2012. Analysis of diffusion-controlled stochastic events of iridium oxide single nanoparticle collisions by scanning electrochemical microscopy. *J. Am. Chem. Soc.* 134(16): 7102–7108.

10. Xiao, X.; Pan, S.; Jang, J. S.; Fan, F.-R. F.; Bard, A. J. 2009. Single nanoparticle electrocatalysis: Effect of monolayers on particle and electrode on electron transfer. *J. Phys. Chem. C* 113: 14978–14982.

11. Chidsey, C. E. D.; Loiacono, N. D. 1990. Chemical functionality in self-assembled monolayers: Structural and electrochemical properties. *Langmuir* 6(3): 682–691.

12. Kwon, S. J.; Bard, A. J. 2012. DNA analysis by application of pt nanoparticle electrochemical amplification with single label response. *J. Am. Chem. Soc.* 134(26): 10777–107779.

13. Quinn, B. M.; van't Hof, P. G.; Lemay, S. G. 2004. Time-resolved electrochemical detection of discrete adsorption events. *J. Am. Chem. Soc.* 126(27): 8360–8361.

14. Boika, A.; Thorgaard, S. N.; Bard, A. J. 2013. Monitoring the electrophoretic migration and adsorption of single insulating nanoparticles at ultramicroelectrodes. *J. Phys. Chem. B* 117(16): 4371–4380.

15. Park, J. H.; Boika, A.; Park, H. S.; Lee, H. C.; Bard, A. J. 2013. Single collision events of conductive nanoparticles driven by migration. *J. Phys. Chem. C* 117(13): 6651–6657.

16. Oldham, K. B. 1988. Theory of microelectrode voltammetry with little electrolyte. *J. Electroanal. Chem.* 250(1): 1–21.

17. Oldham, K. B.; Feldberg, S. W. 1999. Principle of unchanging total concentration and its implications for modeling unsupported transient voltammetry. *J. Phys. Chem. B* 103(10): 1699–1704.

18. Amatore, C.; Deakin, M. R.; Wightman, R. M. 1987. Electrochemical kinetics at microelectrodes: Part IV. Electrochemistry in media of low ionic strength. *J. Electroanal. Chem.* 225(1–2): 49–63.

19. Streeter, I.; Compton, R. G. 2009. Numerical simulation of potential step chronoamperometry at low concentrations of supporting electrolyte. *J. Phys. Chem. C* 112(35): 13716–13728.

20. Zhou, H.; Park, J. H.; Fan, F.-R. F.; Bard, A. J. 2012. Observation of single metal nanoparticle collisions by open circuit (mixed) potential changes at an ultramicroelectrode. *J. Am. Chem. Soc.* 134(32): 13212–13215.

21. Park, J. H.; Zhou, H.; Percival, S. J.; Zhang, B.; Fan, F.-R. F.; Bard, A. J. 2013. Open circuit (mixed) potential changes upon contact between different inert electrodes-size and kinetic effects. *Anal. Chem.* 85(2): 964–970.

22. Zhou, Y.-G.; Rees, N. V.; Compton, R. G. 2011. The electrochemical detection and characterization of silver NPs in aqueous solution. *Angew. Chem. Int. Ed.* 50: 4219–4221.

23. Zhou, Y.-G.; Rees, N. V.; Pillay, J.; Tshikhudo, R.; Vilakazi, S.; Compton, R. G. 2012. Gold NPs show electroactivity: Counting and sorting NPs upon impact with electrode. *Chem. Commun.* 48: 224–226.

24. Zhou, Y.-G.; Rees, N. V.; Compton, R. G. 2011. NP–Electrode collision processes: The underpotential deposition of thallium on silver NPs in aqueous solution. *ChemPhysChem* 12: 2085–2087.

25. Zhou, Y.-G.; Rees, N. V.; Compton, R. G. 2011. NP–electrode collision processes: The electroplating of bulk cadmium on impacting silver NPs. *Chem. Phys. Lett.* 511: 183–186.

26. Zhou, Y.-G.; Rees, N. V.; Compton, R. G. 2012. The electrochemical detection of tagged NPs via particle-electrode collisions: Nanoelectroanalysis beyond immobilisation. *Chem. Commun.* 48: 2510–2512.

27. Kolics, A.; Thomas, A. E.; Wieekowski, A. 1996. Cl labelling and electrochemical study of chloride adsorption on a gold electrode from perchloric acid media. *J. Chem. Soc. Faraday Trans.* 92: 3727–3736.

28. Dias, M. A.; Kelsall, G. H.; Welham, N. J. 1993. Electrowinning coupled to gold leaching by electrogenerated chlorine: I. Au(III) Au(I)/Au kinetics in aqueous Cl2/Cl—electrolytes. *J. Electroanal. Chem.* 361: 25–38.

29. Rees, N. V.; Zhou, Y.-G.; Compton, R. G. 2012. Making contact: Charge transfer during particle-electrode collisions. *RSC Adv.* 2: 379–384.

30. Cutress, I. J.; Rees, N. V.; Zhou, Y.-G.; Compton, R. G. 2011. NP–electrode collision processes: Investigating the contact time required for the diffusion-controlled monolayer underpotential deposition on impacting NPs. *Chem. Phys. Lett.* 514: 58–61.

31. Dickinson, E. J. F.; Rees, N. V.; Compton, R. G. 2012. NP–electrode collision studies: Brownian motion and the timescale of NP oxidation. *Chem. Phys. Lett.* 528: 44–48.

32. Gorschlüter, A.; Sundermeier, C.; Roß, B.; Knoll, M. 2002. Microparticle detector for biosensor application. *Sens. Actuators B* 85: 158–165.

33. Fosdick, S. E.; Anderson, M. J.; Nettleton, E. G.; Crooks, R. M.; 2013. Correlated electrochemical and optical tracking of discrete collision events. *J. Am. Chem. Soc.* 135: 5994–5997.

34. Cussler, E. L. *Diffusion: Mass Transfer in Fluid Systems*, 3rd ed. Cambridge, U.K.: Cambridge University Press, 2009.

35. Bard, A. J.; Faulkner, L. R. *Electrochemical Methods: Fundamentals and Applications*, 2nd ed. New York: John Wiley & Sons.

36. Park, J. H.; Thorgaard, S. N.; Zhang, B.; Bard, A. J. 2013. Single particle detection by area amplification: Single wall carbon nanotube attachment to a nanoelectrode. *J. Am. Chem. Soc.* 135(14): 5258–5261.

37. Li, Y.; Cox, J. T.; Zhang, B. 2010. Electrochemical responses and electrocatalysis at single Au nanoparticles. *J. Am. Chem. Soc.* 132(9): 3047–3054.

38. Choi, H. C.; Shim, M.; Bangsaruntip, S.; Dai, H. 2002. Spontaneous reduction of metal ions on the sidewalls of carbon nanotubes. *J. Am. Chem. Soc.* 124(31): 9058–9059.

39. Bard, A. J. 2008. Toward single enzyme molecule electrochemistry. *ACS Nano* 2(12): 2437–2440.

40. Fan, F.-R. F.; Bard, A. J. 2008. Observing single nanoparticle collisions by electrogenerated chemiluminescence amplification. *Nano Lett.* 8(6): 1746–1749.

41. Bard, A. J. 2004. *Electrogenerated Chemiluminescence*. New York: Marcel Dekker, Inc.

42. Dunn, W.; Aikawa, Y.; Bard, A. J. 1981. Semiconductor electrodes. XXXV. Slurry electrodes based on semiconductor powder suspensions. *J. Electrochem. Soc.*, 128: 222–224.

43. Dunn, W. W.; Aikawa, Y.; Bard, A. J. 1981. Characterization of particulate titanium dioxide photocatalysts by photoelectrophoretic and electrochemical measurements. *J. Am. Chem. Soc.* 103: 3456–3459.

44. Fernando, A.; Parajuli, S.; Alpuche-Aviles, M. A. 2013. Observation of individual semiconducting nanoparticle collisions by stochastic photoelectrochemical currents. *J. Am. Chem. Soc.* 135(30): 10894–10897.

45. Kwon, S. J.; Zhou, H.; Fan, F.-R. F.; Vorobyev, V.; Zhang, B.; Bard, A. J. 2011. Stochastic electrochemistry with electrocatalytic NPs at inert ultramicroelectrodes-theory and experiments. *Phys. Chem. Chem. Phys.* 13: 5394–5402.

46. Ohshima, H.; Healy, T. W.; White, L. R. 1983. Approximate analytic expressions for the electrophoretic mobility of spherical colloidal particles and the conductivity of their dilute suspensions. *Chem. Soc., Faraday Trans.* 2(79): 1613–1628.

47. Lyklema, J. 1995. *Fundamentals of Interface and Colloid Science*, vol. II. London, U.K.: Academic Press, p. 3.21.

48. Burda, C.; Chen, X.; Narayanan, R.; El-Sayed, M. A. 2005. Chemistry and properties of nanocrystals of different shapes. *Chem. Rev.* 105: 1025–1102.

49. Trindade, T.; O'Brien, P.; Picket, N. L. 2001. Nanocrystalline semiconductors: Synthesis, properties, and perspectives. *Chem. Mater.* 13: 3843–3858.

50. Eychmüller, A.; Banin, U.; Dehnen, S. et al. 2004. Syntheses and characterizations. In *NPNPs*, Schmid, G. (ed.), pp. 50–238. Weinheim, Germany: Wiley-VCH.

51. Daniel, M.; Astruc. D. 2004. Gold NPNPs: Assembly, supramolecular chemistry, quantum-size-related properties, and applications toward biology, catalysis, and nanotechnology. *Chem. Rev.* 104: 293–346.

52. Cao, G. 2004. Zero-dimensional nanostructures: NPNPs. In *Nanostructures & Nanomaterials: Synthesis, Properties, and Applications*, pp. 51–109. London, U.K.: Imperial College Press.

53. Capek, I. 2006. *Nanocomposite Structures and Dispersions: Science and Nanotechnology—Fundamental Principles and Colloidal NPs*. Amsterdam, the Netherlands: Elsevier.

54. Schmid, G. 1992. Large clusters and colloids. Metals in the embryonic state. *Chem. Rev.* 92: 1709–1727.

55. Zabet-Khosousi, A.; Dhirani, A. 2008. Charge transport in NPNP assemblies. *Chem. Rev.* 108: 4072–4124.

56. Masala, O.; Seshadri, R. 2004. Synthesis routes for large volumes of NPNPs. *Annu. Rev. Mater. Res.* 34: 41–81.

57. Templeton, A. C.; Wuelfing, P. W.; Murray, R. W. 2000. Monolayer-protected cluster molecules. *Acc. Chem. Res.* 33: 27–36.

58. Turkevich, J.; Stevenson, P. C.; Hillier, J. 1951. A study of the nucleation and growth processes in the synthesis of colloidal gold. *Discuss. Faraday Soc.* 11: 55–75.

59. Frenz, G. 1973. Controlled nucleation for the regulation of the NP size in monodisperse gold suspensions. *Nat. Phys. Sci.* 241: 20–22.

60. Kimling, J.; Maier, M.; Okenve, B.; Kotaidis, V.; Ballot, H.; Plech, A. 2006. Turkevich method for gold NPNP synthesis revisited. *J. Phys. Chem. B* 110: 15700–15707.

61. Hara, M.; Waraksa, C. C.; Lean, J. T.; Lewis, B. A.; Mallouk, T. E. 2000. Photocatalytic water oxidation in a buffered tris(2,2'-bipyridyl)ruthenium complex-colloidal IrO_2 system. *J. Phys. Chem. A* 104: 5275–5280.

62. Zaban, A.; Ferrerre, S.; Sprague, J.; Gregg, B. A. 1997. pH-dependent redox potential induced in a sensitizing dye by adsorption onto TiO_2. *J. Phys. Chem. B* 101: 55–57.

63. Rao, J. P.; Geckeler, K. E. 2011. Polymer NPNPs: Preparation techniques and size-control parameters. *Prog. Polym. Sci.* 36: 887–913.

64. Rao, K. S.; El-Hami, K.; Kodaki, T.; Matsushige, K.; Makino, K. 2005. A novel method for synthesis of silica NPNPs. *J. Coll. Int. Sci.* 289: 125–131.
65. Cho, E. J.; Holback, H.; Liu, K. C.; Abouelmagd, S. A.; Park, J.; Yeo, Y. 2013. Nanoparticle characterization: State of the art, challenges, and emerging technologies. *Mol. Pharm.* 10: 2093–2110.
66. Skoog, D. A.; Holler, J. F.; Crouch, S. R.; 2007. NP size determination. In *Principles of Instrumental Analysis*, pp. 950–963. Belmont, New South Wales, Australia: Thomson Brooks/Cole.
67. Lu, X.; Wu, D.; Li, Z.; Chen, G. 2011. Polymer NPNPs. *Prog. Mol. Biol. Transl. Sci.* 104: 299–323.
68. Murdock, R. C.; Braydich-Stolle, L.; Schrand, A. M.; Schlager, J. J.; Hussain, S. M. 2008. Characterization of nanomaterial dispersion in solution prior to in vitro exposure using dynamic light scattering technique. *Toxicol. Sci.* 101: 239–253.
69. Kun, R.; Fendler, J. H. 2004. Use of attenuated total internal reflection-Fourier transform infrared spectroscopy to investigate the adsorption of and interactions between charged latex NPs. *J. Phys. Chem. B* 108: 3462–3468.
70. Ingram, R. S.; Hostetler, M. J.; Murray, R. W.; Schaaff, T. G.; Khoury, J. T.; Whetten, R. L.; Bigioni, T. P.; Guthrie, D. K.; First, P. N. 1997. 28 kDa Alkanethiolate-protected Au clusters give analogous solution electrochemistry and STM Coulomb staircases. *J. Am. Chem. Soc.* 119: 9279–9280.
71. Gilbert, S. E.; Cavalleri, O.; Kern, K. 1996. Electrodeposition of Cu NPNPs on Decanethiol-covered Au(111) surfaces: An in situ STM investigation. *J. Phys. Chem.* 100: 12123–12130.
72. Tang, L.; Han, B.; Persson, K.; Friesen, C.; He, T.; Sieradzki, K.; Ceder, G. 2010. Electrochemical stability of nanometer-scale Pt NPs in acidic environments. *J. Am. Chem. Soc.* 132: 596–600.
73. Dai, X.; Nekrassova, O.; Hyde, M. E.; Compton, R. G.; 2004. Anodic stripping voltammetry of arsenic(III) using gold NPNP-modified electrodes. *Anal. Chem.* 76: 5924–5929.
74. Huang, K.; Anne, A.; Bahri, M. A.; Demaille, C. 2013. Probing individual redox PEGylated gold NPNPs by electrochemical-atomic force microscopy. *ACS Nano* 7: 4151–4163.
75. Noel, J.; Zigah, D.; Simonet, J.; Hapiot, P. 2010. Synthesis and immobilization of Ag(0) NPNPs on diazonium modified electrodes: SECM and cyclic voltammetry studies of the modified interfaces. *Langmuir* 26: 7638–7643.
76. Tel-Vered, R.; Bard, A. J.; 2006. Generation and detection of single metal NPNPs using scanning electrochemical microscopy techniques. *J. Phys. Chem. B* 110: 25279–25287.
77. Tschulik, K.; Palgrave, R. G.; Batchelor-McAuley, C.; Compton, R. G.; 2013. 'Sticky electrodes' for the detection of silver nanoparticles. *Nanotechnology* 24(29): 295502–295508.
78. Stuart, E. J. E.; Rees, N. V.; Cullen, J. T.; Compton, R. G.; 2013. Direct electrochemical detection and sizing of silver nanoparticles in seawater media. *Nanoscale* 5(1): 174–177.
79. Stuart, E. J. E.; Rees, N. V.; Compton, R. G.; 2012. Particle-impact voltammetry: The reduction of hydrogen peroxide at silver nanoparticles impacting a carbon electrode. *Chem. Phys. Lett.* 531: 94–97.
80. Stuart, E. J. E.; Zhou, Y.-G.; Rees, N. V.; Compton, R. G.; 2012. Determining unknown concentrations of nanoparticles: The particle-impact electrochemistry of nickel and silver. *RSC Adv.* 2(17): 6879–6884.
81. Tschulik, K.; Haddou, B.; Omanović, D.; Rees, N. V.; Compton, R. G.; 2013. Coulometric sizing of nanoparticles: Cathodic and anodic impact experiments open two independent routes to electrochemical sizing of Fe3O4 nanoparticles. *Nano Res.* 6(11): 836–841.
82. Rees, N. V.; Zhou, Y.-G.; Compton, R. G.; 2012. The non-destructive sizing of nanoparticles via particle–electrode collisions: Tag-redox coulometry (TRC). *Chem. Phys. Lett.* 525–526: 69–71.
83. Ellison, J.; Tschulik, K.; Stuart, E. J. E. et al. 2013. Get more out of your data: A new approach to agglomeration and aggregation studies using nanoparticle impact experiments. *Chem. Open* 2(2): 69–75.
84. Alligrant, T. M.; Nettleton, E. G.; Crooks, R. M. 2013. Electrochemical detection of individual DNA hybridization events. *Lab Chip* 13(3): 349–354.
85. Adams, D. M.; Brus, L.; Chidsey, C. E. D. et al. 2003. Charge transfer on the nanoscale: Current status. *J. Phys. Chem. B* 107(28): 6668–6697.
86. Kissling, G. P.; Miles, D. O.; Fermín, D. J. 2011. Electrochemical charge transfer mediated by metal nanoparticles and quantum dots. *Phys. Chem. Chem. Phys.* 13(48): 21175–21185.

9 Nanoelectrochemistry of Carbon

Jacob M. Goran and Keith J. Stevenson

CONTENTS

9.1 CARBON ELECTRODES

9.1.1 INTRODUCTION

Carbon materials have a rich history in electrochemistry, going back over 150 years.[1] No other electrode material has such an expansive array of allotropes, structural polymorphisms, variations of synthetic procedures, or breadth of applications. The liberal use of carbon is due to its relative abundance (low cost), ease of functionalization, wide potential window, and biocompatibility compared to traditional noble metal electrode materials. Furthermore, the various allotropes and structural polymorphisms of carbon display vastly different physicochemical properties, with similar types being differentiated by pretreatment, source material, postsynthetic treatment, and even small deviations in material processing. Thus, tailored carbon electrodes can comprise various forms and properties for individualized applications. Advances in the synthesis, characterization, and separation of carbon materials have caused a surge of interest, largely due to the fact that many of these carbon forms are on the nanoscale.[2] Carbon nanotubes (CNTs) and the 2D graphene are the two most important forms of nanoscale carbon materials, but material processing techniques are allowing more traditional forms such as graphite, carbon fibers (CFs), and pyrolyzed precursors to be configured to increasingly smaller profiles. Carbon thin films, such as pyrolyzed photoresist films (PPFs), are creating opportunities in spectroelectrochemistry, since the nanoscale dimensions of the film allow for optically transparent (OT) electrodes. The small size of nanocarbons and matching biocompatibility opens the door to a vast array of electroanalytical measurements in biological systems, including exocytosis of individual cells and subcellular systems. Herein, we highlight the recent uses of carbon in electrochemistry on the nanoscale. This review includes both nanoelectrodes that are derived from carbon materials and the electrochemical interactions of electroactive molecules with carbon surfaces at the nanoscale. In order to focus the enormous volume of research performed on carbon in electrochemistry, we have narrowed our scope to carbons in electroanalytical applications and nanocarbon electrodes.[3] This excludes nanocarbons used in energy applications[4–6] such as electrochemical capacitors,[7] Li-ion batteries,[8] catalysis,[9,10] microbial fuel cells,[11] and enzymatic fuel cells.[12,13]

9.1.2 GRAPHENE TO GRAPHITE

We must start with the main building block of the vast majority of carbon electrode materials discussed here, graphene. Graphene contains sp^2-hybridized carbons with a C–C bond length of 1.42 Å. The sp^2 hybridization means that the four valence electrons in carbon are arranged so that three form σ bonds with adjacent carbons, while the fourth is delocalized in overlapping π states across the honeycomb lattice. Graphene is the 2D structure and the fundamental building block of the 0D fullerenes, 1D CNTs, and 3D graphite but was the last to be isolated with credit given to Novoselov and Geim in 2004.[14]

Figure 9.1 presents the family of graphene materials and how each can be formed from graphene. Fullerenes were first discovered in 1985[15] and incorporate alternating five- and six-membered carbon rings that form closed structures. Fullerenes exhibit molecular-like electrochemistry, with discrete electron transfer (ET) reactions at electrode surfaces.[16] The most common fullerenes contain 60 or 70 carbons, but many others exist. Fullerenes display limited conductivity due to their reduced π bond overlap and are generally not used as electrode materials but are good electron acceptors. Graphene, however, displays distinctly different electronic properties, exhibiting both relativistic and quantum features. These include a high charge carrier mobility, called ballistic transport,[17,18] or the observation of the quantum Hall effect at room temperature.[19] Graphene is a zero bandgap semiconductor, in its pristine state. Its electronic structure is such that the valence and conduction bands touch each other at a point, the Dirac point, which is the Fermi level (E_F). This *pristine* electronic structure is only theoretical since thermodynamics require some degree of disorder in crystalline materials. The electronic structure is subsequently altered by the introduction of disorder such as defects and dopants or simply adding another layer of graphene as in bilayer graphene.[20,21] Intrinsic defects such as Stone–Wales defects and lattice vacancies or extrinsic defects such as the presence of foreign atoms can significantly alter the electronic structure, chemical reactivity, and hence electrochemical reactivity.[21,22] Figure 9.2 displays the local density of states (LDOS) before and after a single C vacancy was introduced into graphite by Ar$^+$ ion irradiation, exhibiting a resonance peak near E_F.

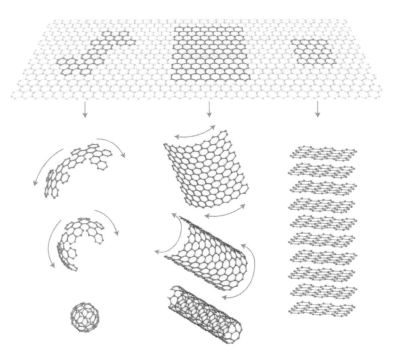

FIGURE 9.1 The derivation of 0D fullerenes, 1D CNTs, and 3D graphite from the 2D building block graphene. (Reprinted by permission from Macmillan Publishers Ltd. *Nature Mater.*, Geim, A. and Novoselov, K., The rise of graphene, 183–192, 2007. Copyright 2007.)

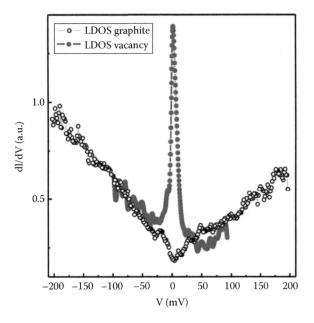

FIGURE 9.2 LDOS for graphite with a single C vacancy. (Reprinted with permission from Ugeda, M.M., Brihuega, I., Guinea, F. et al., Missing atom as a source of carbon magnetism, *Phys. Rev. Lett.*, 104, 096804, 2010. Copyright 2010 by the American Physical Society.)

Although graphene can be thought of as a pure 2D material, in practice and in theoretical calculations, it contains ripples or corrugations in a suspended state due to thermal fluctuations.[23] Since monolayers of graphene are usually synthesized and studied on supporting substrates, electronic effects due to the underlying substrate must be considered.[24] Indeed, a graphene sheet on SiO_2 displays electron-rich and hole-rich puddles and thus fluctuations in its LDOS and surface potential.[25] This effect can also be seen in the difference in carbon vacancies if graphene is supported on a metal substrate[26] or on the surface of graphite.[27] The electronic structure of graphene evolves with each additional graphene layer until about 10 layers, where graphene mimics the behavior of graphite, a semimetal with approximately a 41 meV overlap between the valence and conduction bands.[17,28,29] The graphene layers of graphite are stacked in a Bernal ABAB orientation. This means that a carbon atom will be located in the center of each hexagon in the plane above and below. The crystallographic d_{002} spacing, which is the interplanar distance between graphene sheets, is 0.335 nm. The sheets are held together mainly by van der Waals forces, which permits easy delamination between graphene sheets, the reason why graphite is used as a solid lubricant. Other crystallographic characteristics include L_a, the in-plane crystallite size (basal plane crystallites), and L_c, the crystallite size perpendicular to the graphene sheets (edge plane crystallites). The electrical conductivity of graphite is anisotropic, being much higher parallel to the hexagonal lattice (a axis) rather than perpendicular to the lattice (c axis).

Figure 9.3 presents a chart of L_a and L_c values for some common carbon materials. L_a and L_c are determined mostly by x-ray diffraction, but Raman has been shown to be very useful for characterizing carbon structures.[30] There are two main Raman bands: the G-band (typically around 1350 cm^{-1}) and the D-band (typically around 1580 cm^{-1}). The inverse of L_a is directly proportional to the ratio of the intensity of the D-band/G-band ($1/L_a \propto I_D/I_G$), which is known as the Tuinstra–Koenig relation.[31] The G-band is observed in graphite and graphite-like materials and is sharp and strong for highly crystalline, defect-free graphite. The G-band is expected from the E_{2g} mode, while the D-band is observed in small graphitic crystallites, the edge plane of highly ordered pyrolytic graphite (HOPG), or by introducing defects in HOPG, a band assigned to the A_{1g} mode.[32] Thus, the D-band or I_D/I_G ratio is also a general indicator for degree of disorder or edge plane character in graphitic structures.[33,34] An additional peak at 1620 cm^{-1}, called the D′-band, is indicative of increased disorder in the d_{002} spacing.[35] The G-band and the overtone of the D-band at 2700 cm^{-1} (2D-band) can be used to monitor the number of graphene layers as they merge into the band structure indicative of graphite.[36]

Glassy carbon (GC), made from the heat treatment of polymers such as polyacrylonitrile, displays small values for L_a and L_c. Also known as vitreous carbon, GC is impermeable to liquids or gases and is one of the most commonly used carbon electrodes. The C–C bonds in the polymeric resin do not break during heat treatment and retain their initial structure, preventing enlargement of the graphitic crystallites. Thus, GC is considered a highly *defective* material, with a higher capacitance, coordinating density of states (DOS), and more intense Raman D-band than basal plane HOPG.[37] The structure is often described as interwoven graphitic ribbons with a typical interplanar spacing slightly larger than polycrystalline graphite.[33,34] CFs have similar L_c values to GC but slightly larger values of L_a. CFs are made by chemical vapor deposition (CVD) or by heating petroleum pitch or polymer resin, such as polyacrylonitrile, which is extruded or spun to form fibers in the molten state. CFs are extensively used in biological applications, such as in the detection of neurotranmitters.[38] CFs can also be fabricated to a very fine point, as will be shown in the carbon nanoelectrode section.

The edge of graphitic carbons plays an important role in their electrochemistry, with vastly different ET rates reported between the edge plane and basal plane of HOPG. Figure 9.4 depicts a graphene sheet but highlights the two types of edge structures: *zigzag* and *armchair*. The edges, however, are not stagnant and can reconstruct or possess structures other than zigzag and armchair configurations.[39] Transmission electron microscopy (TEM) studies suggest that the zigzag edge has better stability, based on reconstruction after an edge atom is ejected.[40] Carbon atoms at the edge are

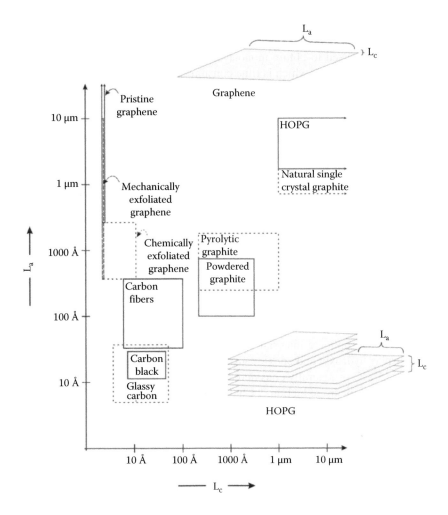

FIGURE 9.3 Some common values of L_a and L_c for selected carbon materials. (Brownson, D.A.C., Kampouris, D.K., and Banks, C.E., Graphene electrochemistry: Fundamental concepts through to prominent applications, *Chem. Soc. Rev.*, 41, 6944–6976, 2012. Reproduced by permission of The Royal Society of Chemistry.)

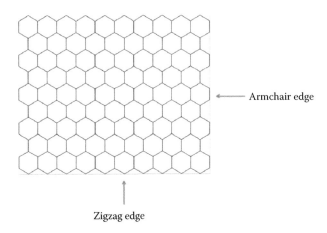

FIGURE 9.4 Graphene edge structure: zigzag edge and armchair edge.

coordinated differently than in the basal plane and thus can have significantly different properties as observed by energy-loss near-edge fine structure analysis.[41]

The original method used to obtain monolayer graphene involved micromechanical exfoliation, or the *scotch tape* method, which is time consuming and only produces a small quantity of material.[14,42] Epitaxial growth of graphene on SiC produces high-quality graphene but requires high-temperature, high-vacuum, and single-crystal substrates.[43] CVD methods show promise as a viable route to produce high-quality graphene at larger quantities, but the growth process is difficult to control, subsequently creating polycrystalline and usually multilayer graphene. Furthermore, the resulting product is difficult to manipulate postsynthesis (such as removing the graphene sheet from the original substrate and successfully placing it at another location).[44] For electroanalytical purposes, the need to produce larger amounts of material and easily manipulate the product postsynthesis has led most researchers to obtain graphene via wet chemical methods.[45–47] Typically, these methods chemically oxidize graphite, which, after full delamination into individual graphene oxide (GO) layers, can be thermally, chemically, or electrochemically reduced to graphene, albeit highly defective. The most common techniques involve oxidizing graphite through the use of strong mineral acids and oxidizers such as HNO_3, H_2SO_4, $KClO_3$, and $KMnO_4$ by methods developed by Brodie, Staudenmaier, or Hummers.[42,48,49] The harsh treatments cause intercalation of ions into the graphitic lattice, which will subsequently exfoliate the graphite oxide into graphite oxide lamella or even GO. Exfoliation can be further ensured through sonication or stirring in an appropriate solvent such as water, since incorporated oxygen functionalities allow the normally hydrophobic material to disperse in aqueous solutions. Graphite oxide and GO generally display hydroxyl and epoxide groups on the basal plane and carbonyl and carboxyl groups at the edge.[42] GO is generally considered an insulator but can be conductive depending on the degree of oxidation or by reduction.[50] Challenges remain to practically use graphene, since it is both difficult to physically handle and requires modification or functionalization in order to open the bandgap for use in field-effect transistors (FETs).[51] Furthermore, wet chemical methods only allow limited control over the size and composition of the GO and the coordinating reduced GO (rGO).[52]

9.1.3 CARBON NANOTUBES

How a single graphene sheet is rolled into the cylindrical form of a single-walled carbon nanotube (SWCNT) will determine many of the physicochemical properties, such as the conductivity (metallic or semiconducting) and the DOS.[53,54] This structure-dependent physicochemical behavior is one of the reasons why CNTs and other nanocarbons are so unique. The curvature or diameter of CNTs will define the electronic properties of the material, allowing for controlled design of electronic properties such as the size of the bandgap in semiconducting CNTs.[55] More specific details on the symmetry of a rolled graphene sheet are supplied elegantly elsewhere.[53] Briefly, the spiral symmetry of a CNT is classified as either achiral or chiral, with only two achiral types, zigzag and armchair (reminiscent of the graphene edge types). The chiral vector C_h defines the chirality of the nanotube and can be expressed as $C_h = a_1 n + a_2 m = (n, m)$, where a_1 and a_2 are unit vectors of the hexagonal lattice and n and m are associated integers such that $0 \leq |m| \leq n$. C_h expresses the vector from any starting point A to the ending point (A') shown in Figure 9.5. There is also a chiral angle θ, associated with the C_h, defined with respect to the zigzag edge along the rolling direction from the starting point (A). Figure 9.5 displays an example of a graphene sheet being rolled into an SWCNT with C_h (5, 3). The special cases where n = m or where m = 0 give the achiral types of armchair and zigzag, respectively.

A simple rule of (2n + m)/3 defines the type of CNT, with integers being metallic, while anything else is semiconducting. Multiwalled CNTs (MWCNTs) will display different morphologies such as *hollow tube*, *bamboo*, and *herringbone* depending on the synthesis procedure. CNTs are generally

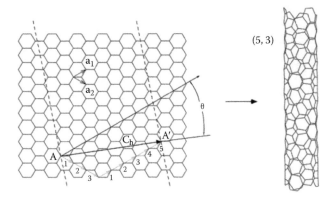

FIGURE 9.5 Schematic of rolling a graphene sheet into an SWCNT. (Reprinted with permission from Charlier, J.-C., Defects in carbon nanotubes, *Acc. Chem. Res.*, 35, 1063–1069. Copyright 2002, American Chemical Society.)

prepared by either an arc-discharge method, laser ablation, or CVD on thin metal films or metal particles of Fe, Ni, or Co[56]; however, there are now many different methods to synthesize CNTs without metal catalysts.[57] CNT rediscovery has been attributed to a paper by Iijima in 1991.[58] The word rediscovery is used here to emphasize that the discovery of a material can often be difficult to pinpoint, as a number of articles in CARBON have highlighted with regard to the discovery of CNTs.[59,60] Furthermore, the different types of CNTs, such as SWCNTs, double-walled CNTs (DWCNTs), and MWCNTs, can be further differentiated from one another, with SWCNTs being characterized in 1993.[61] In general, SWCNTs will have a diameter on the order of 1–2 nm, while MWCNTs will vary (depending on the number of walls) from 2 to 100 nm, with lengths on the order of microns for both types. The end caps of the tubes can either be opened or closed (Figure 9.6), electrochemically being viewed as an edge plane or a basal plane, respectively. The C–C bond in CNTs is slightly longer than that for graphene at 1.44 Å, presumably due to the structural curvature.[53] The 1D structure of CNTs makes them extremely sensitive to sp^3 defects, with just one in 10^6 carbon atoms showing an observable decrease in the conductance.[62] The DOS for CNTs varies considerably in both structure and magnitude, keeping in mind that graphene is expected to have a low DOS near E_F, zero for a *theoretical* defect-free graphene sheet. Figure 9.7 presents two types of SWCNTs and their respective DOS. The structure-dependent DOS and their subsequent effects on ET is especially apparent in the large variety of SWCNTs, which should yield different rate constants due to atomic structure.[63]

The DOS for carbon is lower than metals, leading to lower conductivity.[33] According to Marcus theory, an ET will occur only if the energy states of the donor and acceptor are degenerate. If there is a low DOS near the E_F, then ET is less probable. To increase the probability of ET, one can increase the DOS near E_F so that the probability of degenerate states between the electrode and redox species increases. Disorder and defects in the graphene plane create more energy states near E_F.[54] Basal plane HOPG has been shown to exhibit unusually low capacitance values compared to the edge plane,[64] which is also partially due to the higher density of surface functionalities at the edge.[65] The low capacitance values are indicative of a low DOS around the E_F,[66,67] and hence, low ET rates compared to highly *defective* carbon materials such as GC.[68] Capacitance values are around 1.5–3 $\mu F/cm^2$ for basal plane, while edge plane displays significantly higher capacitance values of >50 $\mu F/cm^2$, greater than that of metals.[69] Induced defects can be described as an increase in the edge plane density, which subsequently corresponds to an increase in the capacitance, coordinating DOS near E_F, and the overall ET rate.[32,68,70]

Hollow tube MWCNT

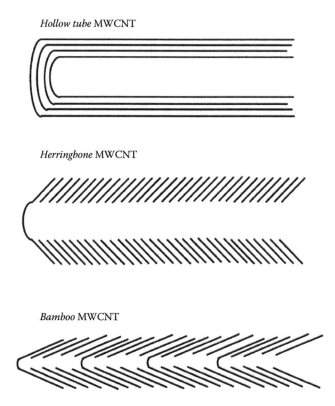

Herringbone MWCNT

Bamboo MWCNT

FIGURE 9.6 Schematic of the alignment of the graphene sheets in *hollow tube*, *herringbone*, and *bamboo* MWCNTs. Right side displays an open end; left side displays a closed end. (Banks, C.E., Davies, T.J., Wildgoose, G.G. et al., Electrocatalysis at graphite and carbon nanotube modified electrodes: Edge-plane sites and tube ends are the reactive sites, *Chem. Commun.*, 829–841, 2005. Reproduced by permission of The Royal Society of Chemistry.)

9.1.4 DOPED NANOCARBONS AND DIAMOND

CNTs and graphene can also be doped with heteroatoms as a substitute for carbon in the hexagonal lattice, known as substitutional doping. The most common heteroatoms are boron, a p-type dopant in carbon, and nitrogen, a n-type dopant in carbon.[71–73] Dopants significantly effect the electronic properties like the DOS near E_F and the band structure.[54,74] Nitrogen-doped CNTs (N-CNTs) exhibit strong electron donor states on the conduction band near E_F.[75,76] Likewise, boron-doped CNTs (B-CNTs) exhibit strong acceptor states on the valence band near E_F.[77] N-doped carbons are more commonly synthesized and used in applications.[78] The inclusion of a heteroatom breaks up the homogeneous surface of the carbon lattice, creating acidic or basic sites on the surface that can be beneficial for many electrochemical reactions, often forming sites favorable for ET. The acidic or basic sites also change the pH of zero charge, such as those seen in N-CNTs where the pH of zero charge is shifted to higher values due to the incorporation of basic nitrogen functionalities.[79]

The types of functionalities incorporated into the carbon surface are determined by the source material, the synthesis process, and any postsynthesis treatments, but an obvious parameter is temperature. In particular, nitrogen is incorporated substitutionally at higher temperatures, with the level of pyridinic-N increasing concurrently with temperature.[80] As the temperature goes even higher, quaternary-N is favored over pyridinic-N.[81] Figure 9.8 displays examples of oxygen and nitrogen functional groups incorporated into a graphitic carbon matrix. The disorder caused by the inclusion of a heteroatom can also be observed in the D-band through Raman spectroscopy,[82] often described as an increase in the edge plane sites or edge plane density.[79] Figure 9.9 presents TEM images of a

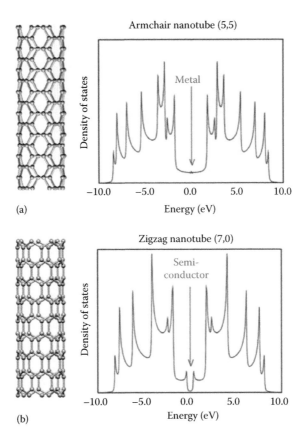

FIGURE 9.7 The DOS for a (5,5) armchair nanotube (a) and a (7,0) zigzag nanotube (b). (Reprinted with permission from Charlier, J.-C., Defects in carbon nanotubes, *Acc. Chem. Res.*, 35, 1063–1069. Copyright 2002, American Chemical Society.)

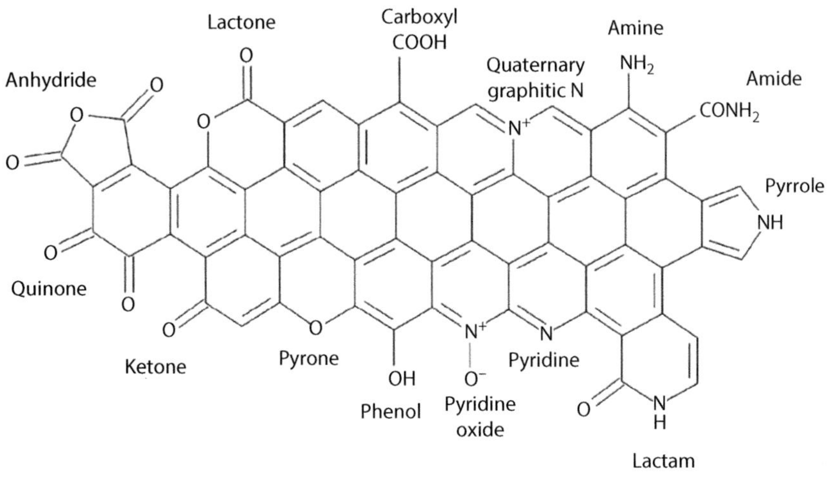

FIGURE 9.8 Examples of possible oxygen and nitrogen functionalities in graphitic carbon. (Reprinted with permission from Arrigo, R., Hävecker, M., Wrabetz, S. et al., Tuning the acid/base properties of nanocarbons by functionalization via amination, *J. Am. Chem. Soc.*, 132, 9616–9630. Copyright 2010, American Chemical Society.)

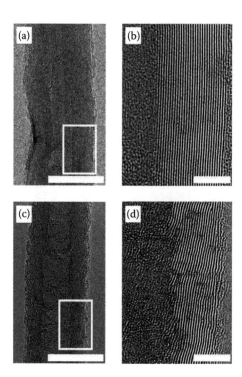

FIGURE 9.9 TEM images of a single CNT (a and b) or N-CNT (c and d) showing the increased disorder caused by N-doping (scale bar 20 nm a and c; 5 nm for b and d). (Reprinted with permission from Maldonado, S. and Stevenson, K.J., Influence of nitrogen doping on oxygen reduction electrocatalysis at carbon nanofiber electrodes, *J. Phys. Chem. B*, 109, 4707–4716. Copyright 2010, American Chemical Society.)

single CNT and a single N-CNT, clearly showing the increased disorder from nitrogen inclusion. Density functional theory (DFT) has also shown that B- and N-doped graphene nanoribbon edges are more energetically favorable for charge carrier transfer, such as proton exchange.[83]

Diamond is sp^3 hybridized with a C–C bond of 1.54 Å and has very low intrinsic conductivity, considered a wide-bandgap semiconductor (5.5 eV). Doping is required to increase the conductivity to a level sufficient to support electrochemical measurements. Boron, a p-type dopant, is the most common element used, but nitrogen is also used as an n-type dopant. Doped diamond materials, such as boron-doped diamond (BDD), are chemically inert and can often display a wider potential window than other carbon electrodes. Nanocrystalline diamond (NCD) contains significant sp^2 character due to defects, which increase the material's conductivity without additional doping. The Raman band at 1332 cm^{-1} is associated with the sp^3 diamond lattice. The intensities of the sp^3 band and the D-band for sp^2 carbons at 1360 cm^{-1} can be used to determine the relative amounts of sp^2 impurities in diamond, but the cross section for the D-band is approximately 50 times larger than that for diamond.[33]

9.2 CARBON SURFACE CHEMISTRY AND REDOX REACTIONS

9.2.1 Inner-Sphere and Outer-Sphere Electron Transfers

We must first highlight the important differences that redox molecules will exhibit during an ET with an appropriate electrode material. It is well known that ET reactions involving intimate interaction with the electrode surface, such as adsorption, are denoted as inner sphere, while ET reactions that do not involve a strong interaction with the electrode surface are denoted as outer sphere.[84] The *sphere* was originally referring to the coordination of ligands around a solution-based metal complex during ET with another metal complex as described by Taube.[85] ET reactions that did not perturb the coordination sphere of metal complexes were denoted as outer sphere, while ET reactions that occurred through a bridging ligand were called inner sphere. This terminology has since transferred over to the mechanism of ET between an electroactive molecule and an electrode surface. ET by an outer-sphere mechanism occurs through a tunneling process, since an outer-sphere ET will inherently have a barrier, the solvation shell and inner Helmholtz plane between the electroactive molecule and the electrode. $Ru(NH_3)_6^{3+/2+}$ is a common example of a nearly ideal outer-sphere redox couple. Although most outer-sphere redox reactions do not show appreciable changes in ET rates due to varying surface chemistry, outer-sphere ET is observed to be dependent on the electronic structure, or the DOS around E_F, a property unique to both the electrode material and its coordinating structure with respect to carbon. Inner-sphere ET reactions will show significant deviations based on surface chemistry, surface structure, and the electronic structure of the electrode, since it requires intimate contact with the electrode surface, or through a bridging ligand, for ET to occur. Of particular note is the ferri-/ferrocyanide ($Fe(CN)_6^{3-/4-}$) couple, which has been used as a benchmark for carbon electrodes for many years. This redox couple was originally considered a outer-sphere ET (referencing the original use of the term as described earlier by Taube)[86,87] but is far from an ideal outer-sphere redox probe.[88] The $Fe(CN)_6^{3-/4-}$ couple has been shown to proceed with additional side reactions,[89] which can cause products to adsorb onto the surface of the working electrode.[90] It is surface sensitive and shows voltammetric deviations if the electrode surface has been exposed to organic solvent, such as dimethyl formamide, commonly used to drop cast CNT suspensions on supporting electrodes.[91] Additionally, conflicting reports are given about the ET characteristics of $Fe(CN)_6^{3-/4-}$ at carbon electrodes. For instance, oxidation seems to increase the ET kinetics of $Fe(CN)_6^{3-/4-}$ at the edge plane sites of SWCNTs, supposedly due to carboxylic acid functionalities.[92] However, others have not observed a difference in the ET kinetics upon oxidation of MWCNTs[93] or shown that the ET kinetics decrease upon oxidation, as observed at both the basal and edge planes of HOPG.[94] Tsierkezos and Ritter have recently performed a thorough study on the ET behavior of $Fe(CN)_6^{3-/4-}$ at vertically aligned MWCNTs as a function of temperature.[95] The ET properties of inner-sphere redox couples at carbon electrodes must be addressed on an individual basis, but general patterns have already been noted, as shown in Figure 9.10.[33] It must also be mentioned that unless the ET rate of a redox couple is in a regime that can be measured, assignments cannot be made (i.e., ET may be too fast to observe any surface-sensitive effects).

9.2.2 Surface State of Carbon and Electron Transfer Reactions

The long history of carbon's use as an electrode material has been confounded by the difficulty in preparing a well-characterized electrode surface. This is partially due to the wide variety of surface functional groups potentially present on the electrode surface and their electrochemical interactions (which either enhance or hinder ET) with redox-active molecules. Furthermore, separating/identifying surface functionalities is not a trivial task. Previous reports have shown that electroactive molecules $Fe(CN)_6^{3-/4-}$, ascorbic acid (AA), and dopamine (DA) do not show a strong dependence on the rate of ET based on the presence of surface oxides, while aquated ions $Fe^{3+/2+}$, $V^{3+/2+}$, or $Eu^{3+/2+}$ show significant variation.[33,34,96–99] This does not mean, however, that they are not surface sensitive, but rather that the oxygen-containing surface functional groups do not have an appreciable effect on the rate of ET at the electrode surface. The redox couples $Ru(NH_3)^{3+/2+}$, ferrocenium/ferrocene (Fc^+/Fc), and $IrCl_6^{2-/3-}$ are neither surface sensitive nor dependent on surface oxides and are considered outer-sphere redox reactions. Redox couples that are surface sensitive to oxygen functionalities can be further categorized if they show a dependence on a specific type of oxygen-containing functional group, such as $Fe^{3+/2+}$, $V^{3+/2+}$, or $Eu^{3+/2+}$, which show a dependence on the presence of surface carbonyl (C=O) groups.[97,100] The surface density of carbonyl groups, created by electrochemical oxidation, is also a function of the anodization potential.[101]

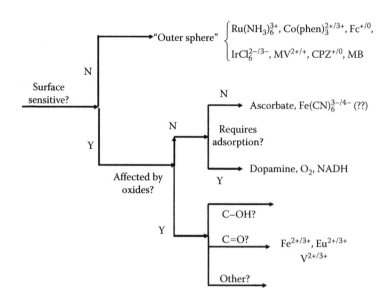

FIGURE 9.10 A flowchart of the electrochemical behavior of some redox couples based on the observed ET kinetics at carbon electrodes. (Reprinted with permission from McCreery, R.L., Advanced carbon electrode materials for molecular electrochemistry, *Chem. Rev.*, 108, 2646–2687. Copyright 2008, American Chemical Society.)

The inherent functionalities of a carbon surface are a result of many factors, a few being source material, synthesis process, heat treatment, and electrode preparation methods such as polishing with an abrasive (alumina or diamond) and/or sonication in various solvents. Before use, carbon electrodes are often electrochemically pretreated or put through an *activation* process which generally includes poising the electrode at a high positive potential (anodization), further increasing the amount of oxygen functional groups.[102,103] Additionally, anodization roughens the surface as visualized through scanning tunneling microscopy (STM) images of HOPG during oxidation.[104] Oxygen functionalities such as carboxylic acids can also impart a pH dependence in the electrode, not observed at traditional metallic electrodes, which can influence electrochemical behavior even if outer-sphere redox couples are employed.[105] Figure 9.10 displays a flowchart created by McCreery and coworkers classifying a number of redox molecules based on observed changes in the ET kinetics at characterized carbon electrodes. The flowchart gives some general classifications, but as will be seen herein, they are only general. Firstly, the flowchart is mainly based on HOPG and GC electrodes, not nanocarbons such as CNTs and graphene. Secondly, as already mentioned, the carbon surface, especially for nanocarbons, is rarely characterized before performing electrochemical experiments. These *as-prepared* or *as-obtained* nanocarbons may possess unintentional impurities and/or functionalities, which can influence or even dominate the observed electrochemical behavior. Thirdly, the structure of nanocarbons can vary significantly, and the carbon structure influences reactivity. Fourthly, the interplay between structure/defect density and the introduction of surface functionality are hard to isolate or separate. For instance, oxidation can both increase oxygen functionalities and increase defect density. Determining which process produced the greatest effect on ET requires careful attention to detail and well-planned controls.

9.2.3 Surface State of Nanocarbons and Electron Transfer Reactions

CNTs are commonly purified by refluxing or sonication in strong oxidizing acids such as HNO_3 or H_2SO_4.[106–109] Oxidation has long been known to purify CNTs through gaseous reactions.[110]

Acid oxidation removes impurities such as amorphous carbon from the synthesis procedure, opens up closed end caps of CNTs, removes metallic residuals, introduces oxygen functional groups, and can *cut* tubes to shorter sizes.[111–114] The introduction of oxygen functionalities also significantly improves the solubility of CNTs.[115,116] The formation of carboxylic acid groups, which can be electrochemically observed as a surface wave,[117] facilitates further functionalization, often through carbodiimide coupling.[118] Electrochemical *activation* of CNTs by anodization has also been performed but seems to indicate that the CVD-produced CNTs are unaffected by the treatment, while the electric arc-discharge-produced CNTs display increased electroactivity to AA and small increases towards NADH and H_2O_2.[119] The ideal outer-sphere ET of $Ru(NH_3)_6^{3+/2+}$ was found to be unaffected by acid oxidation or plasma oxidation of SWCNTs, while DA oxidation was enhanced after either treatment.[114] Besides adding functionalities such as carboxylic acid groups to CNT tube ends, acid purification can also introduce defects at CNT sidewalls.[120] Sidewall functionalization induces defect states near E_F and alters electronic transport along the tube, especially when the sp^2-hybridized structure is altered to a sp^3, creating scattering centers.[121] SWCNTs were fabricated into single nanotube devices to monitor the conductance in real time as individual oxidation events created oxygen functionalities on the SWCNT surface.[62] The oxidation of SWCNTs in strong acids and subsequent electrochemical reduction, which regained most of the initial conductance, suggested that the remaining functionalities from these redox-cycled CNTs were mainly ethers, which are sp^2 hybridized and only weakly scatter free carriers.

Oxidation of CNTs with strong acids generally increases background capacitances despite decreasing the BET surface area, indicating the large role surface functionalities play in the electrochemical capacitance.[122,123] The type of oxygen functionalities is dependent on the oxidants, but HNO_3 seems to produce the greatest increase in defect density.[124] The harsh chemical methods used to create GO from graphite also introduce oxygen functionalities. Oxidation decreases conductivity, but electrochemically reduced GO displays greatly increased conductivity (eight orders of magnitude), with a coordinating decrease in the O/C ratio from about 69% to about 4%.[125] Electrochemical reduction of GO has been shown to be a viable method to gain control over the O/C ratio and reactivity of rGO materials.[126] Figure 9.11 displays x-ray photoelectron spectroscopy (XPS) survey spectra and resulting C/O ratio as a function of the applied reduction potential for GO. Likewise, electrochemical oxidation can be used to increase the amount of oxygen functionalities and subsequent capacitance on the graphene surface.[127] As mentioned earlier, oxidation, either chemically or electrochemically, causes an increase in the disorder or number of defect sites. Colina et al. used Raman to observe the increase in the D-band after anodization of an SWCNT film.[128]

9.3   ELECTROACTIVE SITES ON CARBON ELECTRODES

9.3.1   SENSING AND BIOSENSING

Nanocarbons have been used heavily in sensing and biosensing applications. The first reported use of CNTs as an electrode material for electroanalytical purposes was in 1996, where Britto et al. used bromoform as a binder to form a CNT paste electrode. The CNT electrode displayed an enhanced DA oxidation current compared to other carbon electrodes.[129] Since then, nanocarbon electrodes such as graphene and CNTs have been shown to be electrocatalytic towards many more biologically active molecules such as NADH and H_2O_2.[47,130–137] Figure 9.12 presents cyclic voltammograms (CVs) of graphene-modified basal plane pyrolytic graphite (BPPG) and edge plane pyrolytic graphite (EPPG) electrodes in the presence of H_2O_2. In addition to their electrocatalytic activity, graphene and CNT electrodes have also displayed increased resistance to fouling associated with the electrochemical detection of biological molecules such as NADH or serotonin, a property which is particularly useful for *in vivo* applications.[138,139] Thus, there is an immense

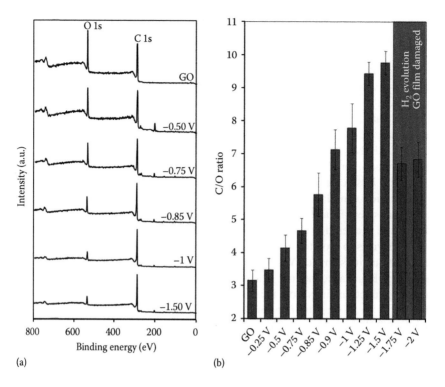

FIGURE 9.11 (a) XPS survey spectra and (b) the C/O ratio content as a function of the reduction potential for GO. (Ambrosi, A. and Pumera, M.: *Chemistry*. 4748–4753. 2013. Copyright Wiley-VCH Verlag GmbH & Co. KGaA. Reproduced with permission.)

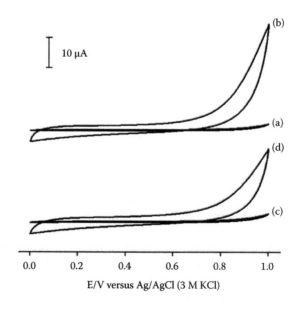

FIGURE 9.12 CVs of (a) bare BPPG, (b) graphene-modified BPPG, (c) bare EPPG, and (d) graphene-modified EPPG in the presence of 0.5 mM H_2O_2. (Reprinted from *Electrochem. Commun.*, 11, Lin, W.-J., Liao, C.-S., Jhang, J.-H. et al., Graphene modified basal and edge plane pyrolytic graphite electrodes for electrocatalytic oxidation of hydrogen peroxide and β-nicotinamide adenine dinucleotide, 2153–2156, Copyright 2009, with permission from Elsevier.)

number of publications either elucidating the electrocatalytic properties for electrochemical sensing[140–142] or putting them to use in various applications, especially with regard to electrochemical biosensing.[13,143–148] Since H_2O_2 and NADH are the byproducts of oxidase and dehydrogenase enzymes, respectively, many biosensing applications have applied enzymes for the selective detection of analytes.[147–151] Coupling nanocarbons with proteins has a rich history going back to Davis et al. who observed well-behaved voltammograms of the redox proteins cytochrome c and azurin using CNT electrodes.[152] These studies have since been repeated with cytochrome c[153,154] and followed up with microperoxidase MP-11[155] and other heme enzymes such as horseradish peroxidase[156] and myoglobin[157] where direct electron transfer (DET) has been observed since the redox-active center is located close to the protein surface. By far the most commonly used redox protein is glucose oxidase (GOx),[158] a model enzyme due to its prevalence in the management of diabetes and detection of glucose.[159] Glucose biosensors dominate the electrochemical biosensing field,[160] and many reported glucose biosensors already incorporate nanocarbon materials.[151,161] DET has also been observed between CNTs and large proteins such as glucose oxidase, where the redox-active center is deeply embedded in the protein/glycoprotein shell.[162–164] A hypothesis has been proposed that the small-diameter CNTs are able to pierce the protein shell and get within tunneling distance of the cofactor flavin adenine dinucleotide (FAD) while still allowing the protein to retain enzymatic activity.[162] DET from GOx has also been observed for graphene materials as well.[165–169] The original hypothesis, then, must also apply to the edges of graphene, but this concept also raises some questions. Although there are many reports of DET between nanocarbons and GOx, there are also conflicting reports on the reason FAD is observed electrochemically or the demonstration of DET without the use of an appropriate mediator.[170,171] FAD, however, is a useful redox probe for nanocarbon materials since the surface-adsorbed FAD displays Langmuir adsorption behavior, which can subsequently be extrapolated to determine the electroactive surface area of the nanocarbon electrode.[172] Additionally, surface-adsorbed FAD displays a pK_a, which reflects the relative hydrophobicity of the nanocarbon surface. The best evidence for DET with GOx using CNTs was produced by Patolsky et al. who attached FAD to CNTs aligned normal to a Au surface by covalently attaching thiols on the opposite end of the CNT.[173] The apoenzyme of GOx (GOx with FAD removed) was then reconstituted around the covalently attached FAD, subsequently displaying bioelectrocatalytic behavior. However, DET was not attributed to a unique property of CNT, but rather the well-developed enzyme reconstitution method.[174,175] CNTs were simply acting as an electrical connector or wire, which displayed a length-dependent ET rate constant (slower rate constant = longer CNT). Vertically aligned CNTs (VA-CNTs) display better ET characteristics than traditional drop cast CNT mesh electrodes. Ferrocenemethylamine was attached to the ends of CNTs that were either vertically aligned on a Au electrode or randomly dispersed on the surface.[176] The study pointed out that ET is faster with VA-CNTs than randomly dispersed CNTs due to the differing electron pathways. Randomly dispersed CNTs are inhibited by either counter-ion diffusion through the CNT network or CNT–CNT junctions, which must be traversed by an electron making its way to the underlying Au surface. Presumably, VA-CNTs allow electrons to travel directly to the Au electrode through a single CNT, while the ferrocenemethylamine is easily accessible to the electrolyte.

Chemically reduced graphene oxide (CR-GO), reduced with hydrazine, has been shown to significantly enhance the detection of many biomolecules such as DNA bases, H_2O_2, NADH, DA, AA, uric acid (UA), and acetaminophen compared to GC and graphite/GC electrodes.[177] Figure 9.13 displays differential pulse voltammograms (DPVs) of the four DNA bases guanine (G), adenine (A), thymine (T), and cytosine (C) clearly separated at CR-GO, as opposed to graphite or GC electrodes. Tominaga et al. investigated NADH oxidation on SWCNTs and observed both a diffusion-controlled peak and adsorption-controlled peak in CVs, where the ratio of the adsorption peak to the diffusive peak increased concurrently with increasing defect density created by anodization.[178] Figure 9.14 presents the Raman spectra and coordinating CVs of NADH oxidation at anodized SWCNTs. Their results correlate well with Pumera et al. who investigated the adsorption of NAD^+

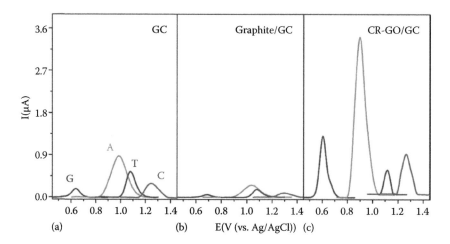

FIGURE 9.13 DPVs of (a) GC graphite/(b) GC or (c) CR-GO/GC electrodes in the presence of DNA bases. (Reprinted with permission from Zhou, M., Zhai, Y., and Dong, S., Electrochemical sensing and biosensing platform based on chemically reduced graphene oxide, *Anal. Chem.*, 81, 5603–5613. Copyright 2009, American Chemical Society.)

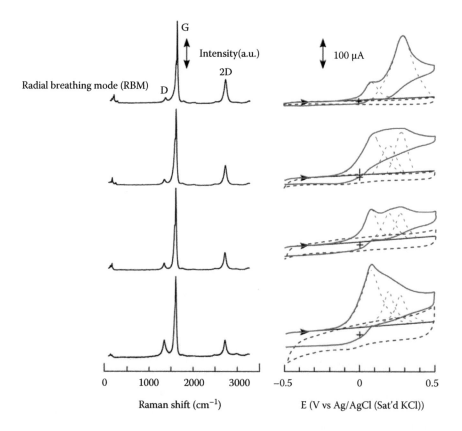

FIGURE 9.14 Raman spectra and coordinating CVs of electrochemically oxidized SWCNTs and their subsequent reaction with NADH. (Reprinted from *Electrochem. Commun.*, 31, Tominaga, M., Iwaoka, A., Kawai, D. et al., Correlation between carbon oxygenated species of SWCNTs and the electrochemical oxidation reaction of NADH, 76–79, Copyright 2013, with permission from Elsevier.)

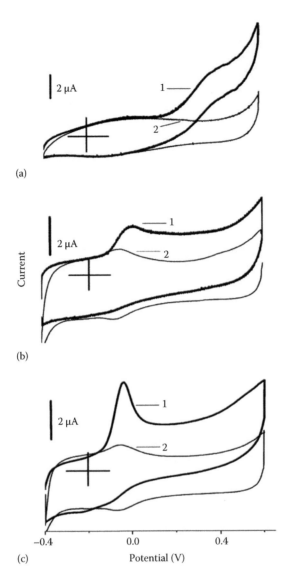

FIGURE 9.15 CVs of (a) CNTs, (b) water-boiled CNTs, and (c) acid-microwaved CNT in the presence (curve 1) or absence (curve 2) of 1 mM NADH. (Reprinted with permission from Wooten, M. and Gorski, W., Facilitation of NADH electro-oxidation at treated carbon nanotubes, *Anal. Chem.*, 82, 1299–1304. Copyright 2010, American Chemical Society.)

at graphene electrodes and determined that adsorption preferentially occurs at the edge and edge-like defects containing oxygen functionalities.[179] The edges were also observed to be responsible for increased activity from NADH and H_2O_2 by Lawrence et al., who looked at arc-discharge and CVD-produced CNTs dispersed by three different methods (Nafion, DMF, and HNO_3).[180] The CVD CNTs showed higher electroactivity than the arc-produced CNTs, due to increased edge plane sites from the open ends. Wooten and Gorski lowered the overpotential for NADH oxidation by boiling CNTs in water or microwaving CNTs in nitric acid.[181] Figure 9.15 presents CVs of the CNT treatments and their activity towards NADH. Both boiling in water and microwaving in acid increased the Raman D-band, with the microwave treatment creating the most defects. The authors suggest that surface quinones created by either pretreatment mediate the reaction, with an increase in the peak current in the microwave-treated CNTs due to increased adsorption on defect sites. NADH was oxidized on ionic liquid-functionalized graphene at 440 mV lower potential than GC. The graphene electrode also showed an increased resistance to NADH fouling and was coupled with alcohol dehydrogenase to create an alcohol biosensor.[139] Graphene quantum dots of about 30 nm were positioned on a gold electrode through a cysteamine monolayer and carbodiimide coupling to the graphene. The graphene electrode showed electrocatalytic reduction of H_2O_2 at -0.4 V versus SCE and was used to monitor H_2O_2 release from human breast adenocarcinoma cells upon the addition of phorbol myristate acetate.[182]

Graphene nanosheets were directly compared to SWCNTs using DA and serotonin. The results suggest that the graphene nanosheets had better sensitivity, conductivity, stability, and signal-to-noise ratio using voltammetric detection.[183] Another comparative study used graphene oxide nanosheets (GONs), chemically reduced GON, electrochemically reduced GONs, and SWCNTs with $Fe(CN)_6^{3-/4-}$, NADH, and AA. For all three redox probes, the ET kinetics were more facile for the electrochemically reduced GONs and SWCNTs compared to the pristine GONs and chemically reduced GONs.[184] Tang et al. used rGO, chemically reduced with hydrazine, to study redox couples $Ru(NH_3)_6^{3+/2+}$, $Fe(CN)_6^{3-/4-}$, $Fe^{3+/2+}$, DA, NADH, and the oxygen reduction reaction (ORR) versus their electrochemical behavior on GC.[185] In all cases, faster ET was observed at the graphene films. The number of graphene layers seems to influence the electroanalytical performance for certain analytes. In the oxidation of DNA bases adenine and guanine, few-layer graphene exhibited a better response than a single monolayer.[186] For detection of AA, monolayer graphene provides the highest sensitivity, while UA seems to be indifferent to the number of graphene layers, with graphite nanoparticles providing better sensitivity.[187]

9.3.2   EDGE PLANE SITES VERSUS BASAL PLANE SITES AND HOPG AS A MODEL

The specific site or reason for the observed electrocatalytic behavior of nanocarbons has been focused on the surface microstructure of graphene and graphene-derived materials, which contain both edge plane sites and basal plane sites. For CNTs, the open ends of the tube are considered the edge plane sites, while the cylindrical tube wall is the basal plane. For MWCNTs, an outer tube end can reside on the surface of an inner tube, since the concentric tubes may terminate at different points, allowing for more edge plane sites. The vast majority of electrochemical experiments performed with graphene and CNTs are in the form of networks or jumbled dispersions that are drop cast forming a 3D mesh on the supporting electrode surface. Thus, differentiating between the edge plane and basal plane becomes difficult due to the random orientation of the network. Additionally, random networks of drop cast CNT films often suffer from poor reproducibility compared to their well-defined GC or HOPG counterparts.[188] Nonetheless, CNT mesh electrodes can offer advantages such as increased loading of enzymes and transport of substrates/products.[189] Furthermore, the spontaneously adsorbed enzyme at CNT mesh electrodes retains their enzymatic activity.[190] Transition metals (such as those used for CNT growth) and N-doping significantly increase the network conductivity.[191]

The edge plane versus basal plane reactivity has spurred a renewed interest in the use of graphite as an electrode material or, more specifically, HOPG, as a model reference for the electrochemical behavior of graphene and CNTs. HOPG can be easily manipulated (by cleaving to obtain a pristine surface) or oriented (so that only the edge planes or the basal plane may be allowed to interact with the electrochemical probe molecule) than graphene or CNTs. Studies have aligned CNTs and graphene in a single direction to clarify electrocatalytic behavior, but defects and step-edge sites still make electrocatalytic assignments difficult. This is especially true with *herringbone* and *bamboo*-type MWCNTs, which have significantly more edge plane sites on the tube surface compared to the *hollow tube* version.[192] Furthermore, edge plane density versus the effects of oxidation also makes assignments difficult, since oxidation increases edge plane–like defects and shortens tube lengths.[93] A number of studies suggest that the edge plane/edge plane–like defect sites at CNTs are solely responsible for the electrocatalytic behavior of CNTs,[192–194] such as their electrocatalytic behavior towards electroactive molecules NADH,[195] epinephrine, and norepinephrine.[196] In contrast, other studies have clearly shown sidewall activity at CNTs.[88] Nonetheless, the conversation has spurred a renewed interest in electrochemistry at carbon electrodes, especially using the well-studied model, HOPG.[37]

9.3.3 EDGE PLANE REACTIVITY

The edge plane/basal plane reactivity was initially investigated by McCreery and coworkers who, through careful electrochemical analysis and corroborating methods such as STM and Raman, correlated the structure/microstructure of carbon electrodes with observed ET kinetics.[33,34] Their studies indicated that edge plane/defect sites significantly improved ET kinetics on either HOPG or GC electrodes.[32,69,70,197–200] It must be noted that in the case of HOPG, edge plane sites were created, whereas at GC, inherent edge plane sites were effectively cleaned or exposed, by the removal of impurities. The *cleaning* of GC electrodes was clearly demonstrated by observing an identical ET rate for both a freshly fractured GC surface and a laser-activated surface.[199] Edge plane sites were created in HOPG by two methods: laser irradiation and electrochemical pretreatment. Both methods increased the Raman D-band (indicative of disorder and edge plane defects), while simultaneously increasing the observed ET kinetics.[32] For electrochemical pretreatment, it was necessary to hold the electrode at a sufficiently high potential to observe an increase in the Raman D-band and simultaneous increase in ET kinetics.[201] As mentioned earlier, electronic factors of a carbon electrode can have a considerable effect on the observed ET rates, irrespective of the redox molecule employed.[33,202] Thus, for outer-sphere redox couples that display an increase in the ET rate with the introduction of defects, the change in electronic structure rather than the introduction of an active site on the electrode surface can cause the increased kinetics.[68]

For nanocarbons, the edge plane/defects are also observed to enhance ET compared to the basal plane. Yuan et al. elegantly compared the edge and basal plane electroactivity of monolayer graphene by creating electrodes with only the edge or basal plane exposed to the electrolyte solution.[203] The edge plane electrodes exhibited significantly higher current density and specific capacitance (two and four orders of magnitude, respectively) than basal plane electrodes. Additionally, NADH and AA were oxidized at edge plane electrodes at 110 and 200 mV lower overpotential, respectively. The ORR also exhibited a 230 mV lower peak potential at edge planes, although the onset potential for ORR was identical to the basal plane. Figure 9.16 displays CVs of AA, NADH, and the ORR at the monolayer graphene edge plane or basal plane electrodes.

A unique study measured the ET characteristics at folded or open edges of graphene layers using $Fe(CN)_6^{3-/4-}$, AA, DA, and NADH. In all cases, the open edges displayed better ET kinetics than their folded counterparts, created by thermally annealing the multilayer graphene.[204] Superlong VA-CNTs were formed into electrodes so that only their sidewalls or tips were exposed to the electrolyte solution containing NADH, H_2O_2, AA, cysteine, $Fe(CN)_6^{3-/4-}$, or dioxygen as electroactive molecules.[205] The oriented electrodes were also oxidized electrochemically at a high positive potential and cycled in H_2SO_4. The results indicated that electroactivity was based on redox couple, with CVs of NADH, H_2O_2, and AA presented in Figure 9.17. NADH and cysteine displayed enhanced electroactivity upon oxidation but did not show a clear differentiation between the sidewall and tip. H_2O_2 displayed sensitivity to the orientation of CNTs but was not influenced by oxidation, while AA and the ORR showed a preference for both the CNT tip and the presence of surface oxides. Another study used long and short CNTs to electrochemically detect H_2O_2 before and after CNTs were oxidized by various acid treatments.[123] Although no clear difference was observed in the reduction of H_2O_2, the oxidation exhibited an increased current with the short CNTs compared to the long CNTs, while oxidation only slightly increased the activity.

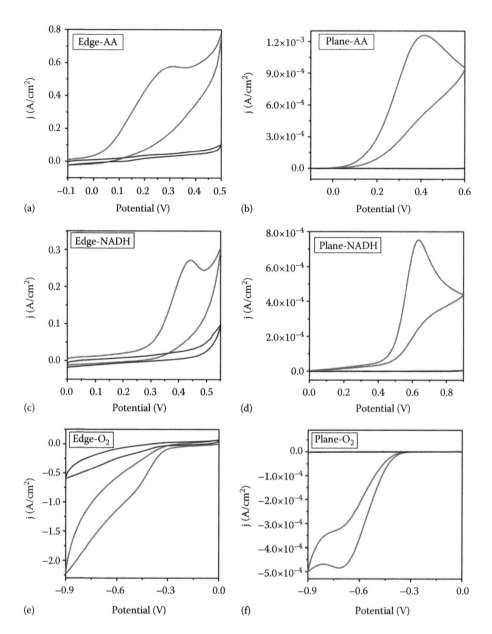

FIGURE 9.16 CVs of monolayer graphene edge plane electrodes in the absence and presence of (a) AA, (c) NADH, and (e) O_2; and monolayer graphene basal plane electrodes in the absence and presence of (b) AA, (d) NADH, and (f) O_2. (Reprinted by permission from Macmillan Publishing Ltd. *Sci. Rep.*, Yuan, W., Zhou, Y., Li, Y. et al., The edge-and basal-plane-specific electrochemistry of a single-layer graphene sheet, 3, 1–7, 2013. Copyright 2013.)

Scanning electrochemical microscopy (SECM) was used to study the electrochemical reactivity of graphene imperfections with $Fe(CN)_6^{3-/4-}$. Exposed edges and mechanically created defects displayed an order of magnitude faster ET than the pristine surface.[206]

DA (2-(3,4-dihydroxyphenyl)ethylamine) is an important neurotransmitter, but its electrochemical detection by oxidation is often hindered by interferents such as UA and AA. Anodized epitaxial graphene (grown on SiC) displayed resolved peaks in DPVs of DA, AA, and UA by creating edge plane defects through anodization.[207] The anodized epitaxial graphene was also applied to

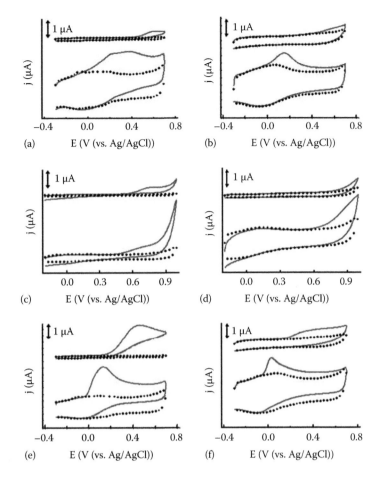

FIGURE 9.17 Superlong VA-CNT electrodes oriented so that only the (a, c, e) sidewall or (b, d, f) tip is exposed to (a, b) NADH, (c, d) H_2O_2, (e, f) AA. The dotted lines are in the absence of an analyte and solid lines in the presence of analyte. The upper trace in each (a through f) is from unoxidized CNTs, while the lower trace is from oxidized CNTs. (Gong, K., Chakrabarti, S., and Dai, L.: *Angew. Chem. Int. Ed.* 5446–5450. 2008. Copyright Wiley-VCH Verlag GmbH & Co. KGaA. Reproduced with permission.)

the detection of nucleic acids, where all four DNA bases (ATCG) exhibited well-separated peaks. Catalyst-free graphene nanoflakes, fabricated by a microwave-enhanced CVD, also exhibited well-separated peaks for the voltammetric detection of DA, UA, and AA.[208] Figure 9.18 displays CVs of the graphene nanoflake or GC electrodes in the presence of DA, AA, and UA.

Valota et al. used mechanically exfoliated graphene to fabricate electrodes of *defect-free* monolayer graphene, defective monolayer graphene (where edges and holes were visible), bilayer graphene, and multilayer graphene/graphite to study ferricyanide reduction. They found significantly improved ET kinetics at monolayer and bilayer graphene compared to multilayer but surprisingly reported that defects present on monolayer graphene make almost no difference in the ET kinetics.[209] Chemical reactions involving ET have also observed significant increases in the reactivity of the edges as compared to the basal plane for monolayer graphene and increased reactivity of single layers compared to bi- or multilayers.[210] SECM has been employed to determine the electroactive sites on a single strand of CF using the feedback mode with α-methyl ferrocene methanol as a redox mediator.[211] The results correlated increased D/G-band ratio from micro-Raman with increased electrochemical activity.

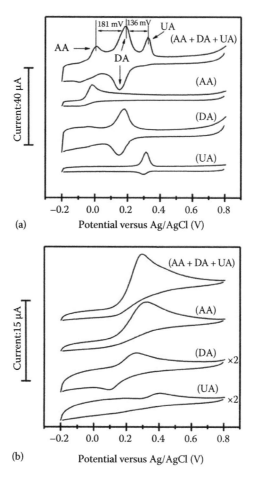

FIGURE 9.18 CVs of (a) graphene nanoflake or (b) GC electrodes in the presence of DA, AA, and UA. (Shang, N.G., Papakonstantinou, P., McMullan, M. et al.: *Adv. Funct. Mater.* 3506–3514. 2008. Copyright Wiley-VCH Verlag GmbH & Co. KGaA. Reproduced with permission.)

9.3.4 REACTIVITY OF THE BASAL PLANE

A number of studies have clearly demonstrated that the basal plane of graphitic carbons supports fast ET kinetics. Güell et al. used a nanoscale scanning electrochemical cell microscopy (SECCM) probe with ferrocenylmethyl trimethylammonium ($FcTMA^{+/2+}$) and $Ru(NH_3)_6^{3+/2+}$ to investigate the ET behavior of sparse dispersions of SWCNTs on an insulating support.[212] Figure 9.19 presents a schematic of the SECCM probe alongside an atomic force microscopy (AFM) image of the 2D SWCNT network. The small size of the mobile electrochemical cell permitted individual regions of SWCNTs to be investigated, so that tube sidewalls, ends, and adjacent tube responses could be separated. The results indicated that ET at the CNT sidewalls was nearly identical to the tube ends, tube intercrossings, and nearly homogeneous across individual tubes in the 2D dispersion. Kim et al. observed a similar behavior using SECM to study the ET at unbiased individual SWCNTs.[213] Snowden et al. used $FcTMA^{+/2+}$ in Nafion to investigate 2D SWCNT dispersions. Nafion slows the diffusion of $FcTMA^+$, which spatially decouples active and inactive areas on the electrode. The current transients were compared to models with varying defect density, demonstrating that SWCNTs were significantly more active than the typical defect densities, indicating that the tube sidewalls must be participating in the observed current.[214] Similarly, Dumitrescu et al. used SECM

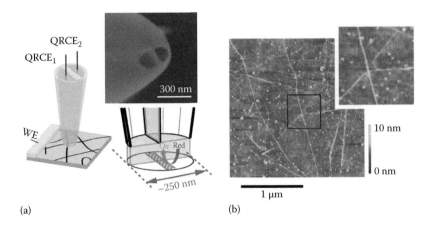

(a)   (b)

FIGURE 9.19 (a) Schematic of the SECCM probe and (b) an AFM image of the 2D SWCNT network. (Güell, A.G., Ebejer, N., Snowden, M.E. et al., Quantitative nanoscale visualization of heterogeneous electron transfer rates in 2D carbon nanotube networks, *Proc. Natl. Acad. Sci. USA*, 109, 11487–11492. Copyright 2012, National Academy of Sciences, U.S.A.)

in a dual-electrode thin-layer cell arrangement to investigate the ET kinetics of SWCNTs using FcTMA$^{+/2+}$. As before, results demonstrated that defects are not solely responsible for electrochemical activity of SWCNTs and that the sidewalls must be electrochemically active.[215] Miller et al. demonstrated fast ET using both $Ru(NH_3)_6^{3+/2+}$ and FcTMA$^{+/2+}$ at the sidewalls and closed ends of SWCNT forests.[216] Figure 9.20 presents field-emission scanning electron microscopy (FE-SEM) and TEM images of the VA-SWCNT forests and the current transients at various points along the sidewalls or closed ends using $Ru(NH_3)_6^{3+/2+}$. SECCM was also used to study the surface of

(a)

FIGURE 9.20 (a, top) FE-SEM image of the aligned SWCNT forest; (a, bottom) TEM image of an SWCNT; (Miller, T.S., Ebejer, N., Güell, A.G. et al., Electrochemistry at carbon nanotube forests: Sidewalls and closed ends allow fast electron transfer, *Chem. Commun.*, 48, 7435–7437, 2012. Reproduced by permission of The Royal Society of Chemistry.)   *(Continued)*

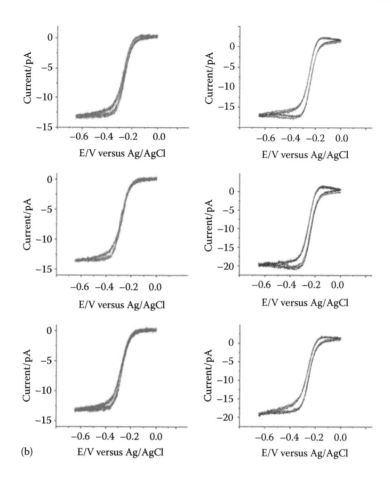

FIGURE 9.20 (CONTINUED)   (b) nanopipette electrochemical cell transients at VA-SWCNT sidewalls (red) or closed ends (blue) using $Ru(NH_3)_6^{3+/2+}$ at three different regions. (Miller, T.S., Ebejer, N., Güell, A.G. et al., Electrochemistry at carbon nanotube forests: Sidewalls and closed ends allow fast electron transfer, *Chem. Commun.*, 48, 7435–7437, 2012. Reproduced by permission of The Royal Society of Chemistry.)

graphene using FcTMA$^{+/2+}$. The ET at step edges or boundaries between graphene flakes was similar to that of the basal plane; however, increased ET rates were observed as the number of graphene layers increased.[217]

9.3.5   HOPG

The argument for HOPG as a reference material for nanocarbons has incited a renewed interest in the fundamental electrochemical behavior of this material. Patel et al. performed a thorough investigation of five different grades of HOPG using the two most common redox probes $Ru(NH_3)_6^{3+/2+}$ and $Fe(CN)_6^{4-/3-}$.[218] The step-edge height and density were fully characterized by AFM, in order to address subsequent ET kinetics at edge planes/defects versus the basal plane. Fast heterogeneous ET was observed at both basal and edge planes for both redox probes. More importantly, this work highlighted the time-dependent properties of voltammetry conducted on freshly cleaved HOPG surfaces. In particular, $Fe(CN)_6^{4-/3-}$ initially displays fast ET with reversible behavior but was quickly diminished by continued cycling, exposure to the $Fe(CN)_6^{4-}$ solution without cycling, or simple exposure to air before CV measurements (HOPG also showed increased resistance via conducting AFM after being left in air). In contrast, $Ru(NH_3)_6^{3+/2+}$ showed little to no change

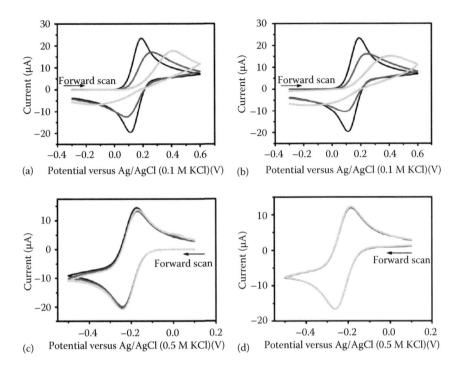

FIGURE 9.21 CVs of freshly cleaved HOPG after it was left in solution (a and c) or air (b and d) for 0 min (black line), 1 h (red line), and 3 h (green line) before the oxidation of $Fe(CN)_6^{4-}$ (a and b) or reduction of $Ru(NH_3)_6^{3+}$ (c and d). (Reprinted with permission from Patel, A.N., Collignon, M.G., O'Connell, M.A. et al., A new view of electrochemistry at highly oriented pyrolytic graphite, *J. Am. Chem. Soc.*, 134, 20117–20130. Copyright 2012, American Chemical Society.)

based on the same experimental conditions. Figure 9.21 presents CVs of freshly cleaved HOPG left in solution or in the air for specific time frames before the oxidation of $Fe(CN)_6^{4-}$ or reduction of $Ru(NH_3)_6^{3+}$. The results suggest that surface oxides hinder ET for $Fe(CN)_6^{4-/3-}$, which was also observed by Ji et al.[94] AFM analysis also indentified adsorbed side products after the oxidation of $Fe(CN)_6^{4-}$. Higher concentrations of $Fe(CN)_6^{4-}$ caused the ET kinetics to depreciate more rapidly. Patel et al. followed this up with a closer look at DA oxidation on HOPG. They used SECCM to carefully deliver DA on the basal plane and across step edges to quantify electrochemical behavior with HOPG surface structure through coordinating AFM and scanning electron microscopy (SEM) measurements. Fast ET kinetics were observed at both sites and could be readily observed due to the adsorbed oxidation products.[219] Well-characterized basal plane ZYA grade HOPG samples were investigated using SECCM with redox probes $Fe(CN)_6^{4-/3-}$ and $Ru(NH_3)_6^{3+/2+}$.[220] As before, highly intrinsic ET was observed on the basal plane with almost no change as the cell traversed a step edge. If a freshly cleaved HOPG was left out in air, the ET of $Fe(CN)_6^{4-/3-}$ decreased. Nafion films embedded with redox molecules ($Ru(NH_3)_6^{3+/2+}$ and $Ru(bpy)_3^{2+/3+}$) were applied to HOPG to diffusionally decoupled areas of fast and slow ET kinetics on the time scale of a CV measurement.[221] The results indicated that the basal plane was sufficiently active and dominated the current response, as opposed to only the step edges being active.

9.3.6 Metallic Impurities and Carbonaceous Debris

In addition to the edge plane/basal plane controversy, many nanocarbons also contain residual metal nanoparticles. Figure 9.23 displays an Fe nanoparticle enclosed inside an MWCNT.

Some consider the metal/metal oxides to be the source of electrocatalytic activity, such as in H_2O_2 reduction[222] or hydrazine oxidation.[223] Figure 9.22 presents the reduction of H_2O_2 at CNT or Fe(III) oxide–modified BPPG electrodes, displaying nearly identical current responses. The presence and type of transition metal impurity influences the electronic properties of nanocarbons, such as the DOS and work function in CNTs.[224] Common techniques like washing or *super washing* the CNTs in acid do not remove all of the residual metal.[225] Metallic impurities also remain from the common processes used to create graphene.[226] Although some of the residual metal can be removed by acid washing, the remainder are *sheathed* in graphene layers and thus inaccessible to the acid, as shown in Figure 9.23.[227] Metallic impurities may also be electrochemically accessible. Lyon and Stevenson showed that Fe impurities in CNTs or N-CNTs display both potential-dependent and supporting electrolyte-dependent redox behaviors.[228] Figure 9.24 presents CVs of the CNTs and N-CNTs in four different electrolytes, displaying redox peaks associated with the electrochemically accessible Fe. In addition, they also showed that the redox-active Fe could be passivated by potential cycling.

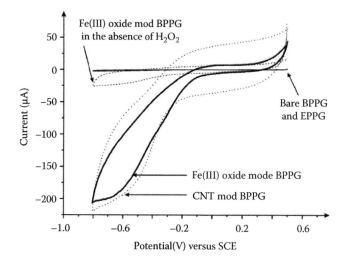

FIGURE 9.22   The reduction of H_2O_2 at Fe(III) oxide–modified BPPG or CNT-modified BPPG. (Reprinted with permission from Sljukić, B., Banks, C.E., and Compton, R.G., Iron oxide particles are the active sites for hydrogen peroxide sensing at multiwalled carbon nanotube modified electrodes, *Nano Lett.*, 6, 1556–1558. Copyright 2006, American Chemical Society.)

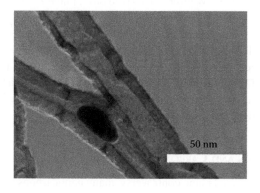

FIGURE 9.23   An Fe nanoparticle enclosed in an MWCNT.

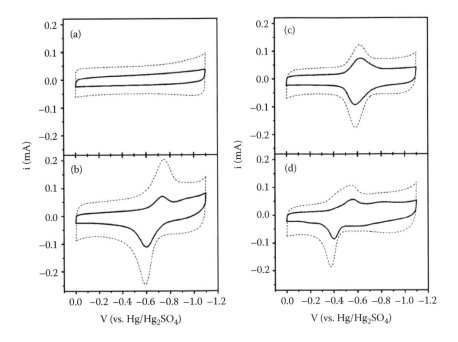

FIGURE 9.24 CVs of CNTs (solid line) and N-CNTs (dotted line) in 1 M (pH = 6.40) solutions of (a) potassium nitrate, (b) sodium phosphate, (c) sodium citrate, (d) and sodium acetate. (Reprinted with permission from Lyon, J.L. and Stevenson, K.J., Anomalous electrochemical dissolution and passivation of iron growth catalysts in carbon nanotubes, *Langmuir*, 23, 11311–11318. Copyright 2007, American Chemical Society.)

Wong et al. showed that oxidatively opened CNTs, which create GO nanoribbons, contain metallic impurities as well, significantly impacting the electroactivity of HS⁻ and hydrazine.[229] Sonication of CNTs, often used to create nanotube dispersions, helps to *bring out* more of the metallic impurities.[230] Metal-free CNTs have been developed,[57] which should help identify the electrocatalytic behavior inherent to the CNT, as has been shown with hydrazine oxidation.[231,232]

Besides metallic impurities, other impurities such as carbonaceous debris (e.g., amorphous carbon or nanographite) can also significantly affect the electrochemical behavior of redox molecules, such as acetaminophen.[233,234] Nanographite impurities have been reported to be responsible for the electrocatalytic behavior of NADH, tyrosine, tryptophan, and AA.[235,236] Onion-like graphite particles, present in electric arc-discharge-produced CNTs, could be responsible for the electrocatalytic behavior towards NADH, epinephrine, norepinephrine, cysteine, and glutathione due to increased edge plane defects present in the particles.[237] Nano-onions have also been shown to detect DA in the presence of UA and AA.[238] GO and rGO exhibit different electrochemical behaviors after the removal of carbonaceous debris by NaOH treatment. GO displayed significantly improved ET kinetics with AA, NADH, aminophenol, and DA, while rGO displayed inhibited ET kinetics after NaOH treatment using the same analytes.[239]

9.3.7 DIAMOND ELECTRODES

Diamond is sp³ hybridized and nonconductive, but doping diamond with boron significantly alters the electronic properties, allowing diamond to support ET. Patten et al. used SECCM as a way to correlate the heterogeneous surface of polycrystalline BDD (pBDD) electrodes with the electroactivity of various redox probes (an outer sphere [FcTMA⁺], an inner sphere [Fe²⁺], and a complex redox process [serotonin]).[240]

I(pA)

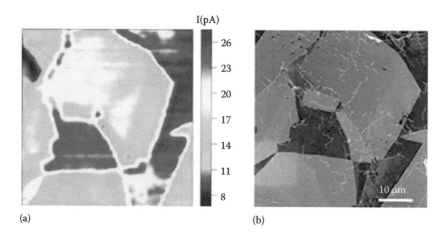

26

23

20

17

14

11

8

(a)   (b)

FIGURE 9.25 (a) SECCM image using FcTMA$^{+/2+}$ and (b) an FE-SEM image of the same area of a pBDD electrode surface. (Reprinted with permission from Patten, H.V., Lai, S.C.S., Macpherson, J.V. et al., Active sites for outer-sphere, inner-sphere, and complex multistage electrochemical reactions at polycrystalline boron-doped diamond electrodes (pBDD) revealed with scanning electrochemical cell microscopy (SECCM), *Anal. Chem.*, 84, 5427–5432. Copyright 2012, American Chemical Society.)

Oxidation of the redox probes displayed similar patterns of electroactivity across the pBDD surface. Areas of high activity were correlated with crystal facets containing higher dopant concentrations, as visualized by the FE-SEM (dark areas). Figure 9.25 presents a SECCM image and an FE-SEM image of the same area on a pBDD surface. pBDD is known to be heterogeneous,[241] with different doping levels dependent on the crystal facet.[242,243] The complex redox process of serotonin, which typically fouls the electrode surface, was mitigated in the SECCM arrangement, since the scanning pipette continually passed over a fresh electrode surface during electrochemical measurements. The ET characteristics of oxygen-terminated pBDD electrodes were directly linked to the local dopant levels and LDOS. Electroactivity was acquired via SECM with outer-sphere redox couples $Ru(NH_3)_6^{3+/2+}$ and FcTMA$^{2+/+}$, Raman mapping (by integration of the diamond band at 1332 cm^{-1}) and FE-SEM allowed evaluation of the local dopant level, and SECCM measured the local capacitance (and subsequently the LDOS at E_F).[244] Areas of high ET correlated well with higher dopant levels and higher capacitance values. Undoped nanodiamond in the form of a film or nanoparticle powder was investigated with CV and SECM.[245] Although undoped diamond is nonconductive, the nanodiamond form displays limited conductivity due to the increased amount of defect sites and sp^2 graphitic character. The redox behavior of $Ru(NH_3)_6^{2+/3+}$, $Fe(CN)_6^{3-/4-}$, and FcMeOH$^{0/+}$ displayed sluggish ET kinetics as expected, but at high overpotentials, the reduction of $Fe(CN)_6^{3-}$ was found to be excessively sluggish compared to the oxidation of $Fe(CN)_6^{4-}$. It was postulated that the electroactive surface area (about 2% at −0.4 to 0.5 V versus Ag/AgCl) increased with potential above 0.5 V up to 10% at 0.8 V, due to the oxidation of defect sites that widen the impurity bands in the bandgap and hence increased the DOS for ET.

9.3.8 DOPED NANOCARBONS

For heteroatom-doped carbon electrodes, the active site may now include an additional atom. In such instances, it is often easier to identify the site of electroactivity, since it can be clearly differentiated from the carbon background. This is observed in STM images of N-doped graphene, where a single nitrogen dopant can be clearly identified in the carbon lattice due to changes in the local electronic structure.[246] The dopant may also be incorporated into a number of different environments/coordinations,[247,248] identified by different binding energies via XPS.[35,249]

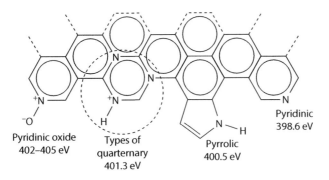

FIGURE 9.26 Nitrogen coordinations and their respective binding energies for N 1 s. (With kind permission from Springer Science+Business Media: *Top. Catal.*, Nitrogen-containing carbon nanostructures as oxygen-reduction catalysts, 52, 2009, 1566–1574, Biddinger, E.J., Deak, D., and Ozkan, U.S., Copyright 2009.)

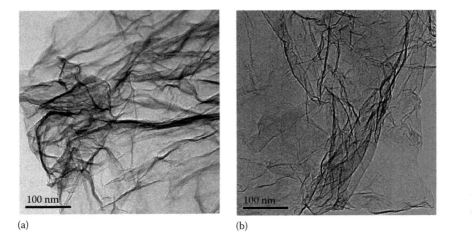

(a)   (b)

FIGURE 9.27 TEM image of (a) graphene and (b) N-doped graphene. (Reprinted with permission from Wang, Y., Shao, Y., Matson, D. et al., Nitrogen-doped graphene and its application in electrochemical biosensing, *ACS Nano*, 4, 1790–1798. Copyright 2010, American Chemical Society.)

Figure 9.26 displays a number of different ways nitrogen can be incorporated into a carbon lattice and their respective N 1s binding energies. The type of functionality depends on the source material and synthesis process, which can vary considerably. N-doped graphene can be formed by thermally annealing GO in NH_3.[250] In this case, nitrogen was incorporated through reactions with oxygen functional groups, and higher temperatures facilitated increased nitrogen incorporation into the carbon lattice such as quaternary-N, while lower temperatures displayed amide, amine, and pyrrolic functionalities. Due to the decomposition temperature of certain oxygen functional groups, the most likely functionalities to facilitate nitrogen incorporation are carbonyl, carboxylic, lactone, and quinone groups.

Similar to their undoped counterpart, electroanalytical interest in doped nanocarbons is partially due to their enhanced sensing/biosensing properties, especially with respect to NADH and to H_2O_2.[251–257] Figure 9.27 presents TEM images of graphene and N-doped graphene. Also like their nondoped counterparts, enzyme-coupled nitrogen and boron-doped CNTs have reported DET with GOx,[258–260] which also remains contentious.[171] N-CNTs have been shown to be more biocompatible than their nondoped counterparts[261] and allow increased loading of active proteins,[262] presumably due to the increased surface area and hydrophilicity of N-CNTs.[172] B-CNTs exhibited a decreased

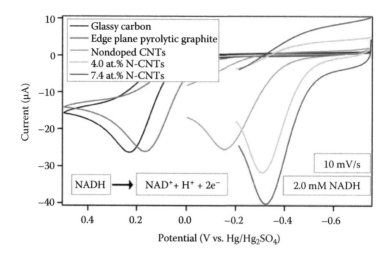

FIGURE 9.28 CVs of GC, EPPG, nondoped CNT, 4.0 atom% N-CNT, and 7.4 N-CNT electrodes in the presence of 2 mM NADH. (Reprinted with permission from Goran, J.M., Favela, C.A., and Stevenson, K.J., Electrochemical oxidation of dihydronicotinamide adenine dinucleotide at nitrogen-doped carbon nanotube electrodes, *Anal. Chem.*, 85, 9135–9141. Copyright 2013, American Chemical Society.)

overpotential for NADH oxidation compared to nondoped CNTs.[256] Likewise, N-CNTs have also been shown to be electrocatalytic towards NADH oxidation compared to GC, EPPG, and nondoped CNTs.[254] Undoped CNTs lowered the overpotential by 370 mV compared to GC, while 7.4 atom% N-CNTs lowered the overpotential by 170 mV compared to undoped CNTs. Figure 9.28 presents CVs of GC, EPPG, nondoped CNT, 4.0 atom% N-CNT, and 7.4 atom% N-CNT electrodes in the presence of 2 mM NADH. Additionally, CNTs and N-CNTs displayed higher resistance to NADH fouling than EPPG or GC, based on the sensitivity to NADH before and after extensive cycling in NADH. The electrocatalytic behavior of N-CNTs compared to undoped CNTs can be attributed to both a change in the electronic properties (such as an increased carrier density and DOS near E_F) and morphology such as an increased density of edge plane sites.[263] Undoped graphene was observed to have a slightly lower overpotential than B-doped graphene for NADH oxidation, but both nanocarbons were significantly better than GC or BDD electrodes.[264] Other nanocarbon structures have also been doped, such as N-doped porous carbon polyhedra, which were shown to simultaneously detect AA, DA, and UA using DPV.[265] N-doped carbon hollow spheres also displayed similar behavior, allowing DA or UA to be quantified in the presence of AA.[266] Chemical reduction of GO may also incorporate nitrogen into the lattice, depending on the reducing agent. N-doped graphene, prepared from GO chemically reduced with hydrazine, was shown to contain nitrogen after chemical reduction and be electrocatalytic towards the reduction of H_2O_2.[255,267] The reduced material was used to monitor H_2O_2 release from neutrophil, RAW 264.7 macrophage, and MCF-7 cells by chronoamperometry after stimulation from phorbol 12-myristate-13-acetate (PMA), N-formylmethionylleucylphenylalanine (fMLP), ADP (adenosine 5′-diphosphate), and AA.

9.3.9 Oxygen Reduction Reaction at Doped Nanocarbons

Doped nanocarbons have shown promise as catalysts for the ORR.[268–276] The reason is primarily due to a disruption in the all carbon lattice, from any dopant, which produces polar or charged sites favorable to O_2 adsorption.[274,277] Although the dopant is known to be involved, the exact catalytic site has been difficult to identify.[273] Edge plane sites/defects were already known to enhance the ORR at carbon electrodes.[99] Increased disorder also accounts for higher ORR activity at N-CNTs with similar elemental compositions.[278] For N-doped nanocarbons, gross nitrogen content does not necessarily improve the ORR and must be incorporated into the graphitic matrix to increase activity. Edge plane sites provide appropriate locations for nitrogen to incorporate into the graphitic matrix.[279] Thus, probable active sites could be quaternary-N located at the edge $(N - Q_{Valley})$[280] or pyridinic-N also located at the edge.[249,281–283] Zhang et al. used DFT to show the effect of potential on the catalytic site. At low potentials, quaternary-N and pyridinic-N are similar, while at high potentials, pyridinic-N is more energetically favorable.[284] In N-doped graphene, both sites are suggested to work in tandem, with pyridinic-N controlling the onset potential, while quaternary-N determines the limiting current.[285] In N-doped graphene nanoribbons, DFT calculations suggest that the graphitic-N site opens and becomes a pyridinic-N during the catalytic cycle.[286] The benefits of the ORR at N-doped nanocarbons have also been observed from N-doped carbon nanocapsules,[287] carbon onions,[288] and holy graphene.[289] N-doped graphene quantum dots exhibit a size-dependent ORR activity, with the larger quantum dots having higher electrocatalytic activity.[290] Boron and nitrogen codoped graphenes, synthesized by thermally annealing GO in the presence of boric acid under an ammonia atmosphere, have also shown promise as an efficient ORR catalyst, displaying a higher current density than the standard 20% Pt/C ORR catalyst.[291] Vertically aligned boron and nitrogen codoped CNTs (VA-BCNs) exhibited better ORR activity than their individually doped counterparts.[292] Figure 9.29 presents CVs of VA-CNT, B-CNT (VA-BCNT), N-CNT (VA-NCNT), and boron–nitrogen codoped CNT (VA-BCN) electrodes in nitrogen- or oxygen-saturated solutions.

In order to clarify the active site for N-doped nanocarbons, N-doped graphene was fabricated with only pyridinic-N, but results indicated that pyridinic-N alone may not be effective for the ORR.[293] Other studies have also indicated that one isolated parameter may not be sufficient to account for the high ORR activity at N-doped nanocarbons, especially with regard to the complete catalytic cycle.[294,295] As with any ET, electronic considerations must also be taken into account,

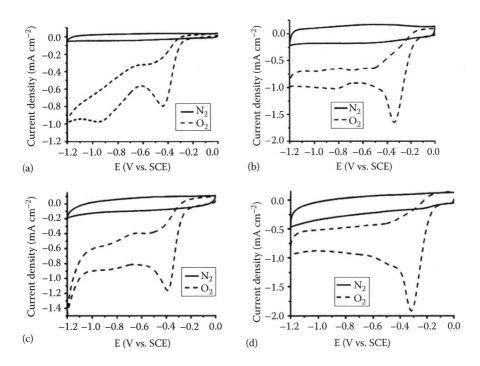

FIGURE 9.29 CVs of (a) VA-CNT, (b) VA-BCNT, (c) VA-NCNT, and (d) VA-BCN in nitrogen-saturated (solid line) or oxygen-saturated (dashed line) 0.1 M KOH solutions. (Wang, S., Iyyamperumal, E., Roy, A. et al.: *Angew. Chem. Int. Ed.* 11756–11760. 2011. Copyright Wiley-VCH Verlag GmbH & Co. KGaA. Reproduced with permission.)

such as the charge carrier density, the capacitance and DOS near E_F, or the work function, all of which are increased due to N-doping.[263,296] Metallic impurities may also play a major role. Parvez et al. were able to produce N-doped graphene with or without the incorporation of Fe nanoparticles. The Fe-incorporated N-doped graphene exhibited significantly higher activity for the ORR than N-doped graphene without Fe or the Pt/C standard in either acidic or alkaline media.[297] A loading of 5 wt% Fe in N-doped graphene produced the highest activity, while more Fe did not further enhance the ORR activity, an effect observed by others.[298] It was postulated that pyridinic-N coordinated with Fe to form active sites for the ORR. Increasing Fe content beyond the amount of pyridinic-N available would not result in increased ORR activity. Furthermore, acid treatment, which should remove residual Fe, did not significantly influence the ORR in either media. For N-CNTs, Fe precursors increase the edge plane character compared to Ni and Co nanoparticle growth, due to an increased *bamboo* N-CNT structure.[81]

Accounting for the multiple aspects at play, a mechanism was purposed by Wiggins-Camacho et al., which involves both a surface Fe site and an N-coordinated site (dual-site mechanism). The first step of the ORR at nondoped and N-doped CNTs involves the reduction of O_2 to the intermediate HO_2^- via a two-electron reduction. At CNTs, the formation of HO_2^- is followed by another two-electron reduction to OH^-, while at N-CNTs, HO_2^- can disproportionate to regenerate O_2 at a Fe_xO_y/Fe surface site stabilized by the presence of pyridinic-N (shown in Figure 9.30). The disproportion rates, which were measured both gasometrically and electrochemically, are comparable to the best known peroxide decomposition catalysts, outrunning the traditional second two-electron reduction step at nondoped CNTs (shown in the CVs of Figure 9.30 as two peaks for nondoped CNTs and one peak for N-CNTs).[299–301]

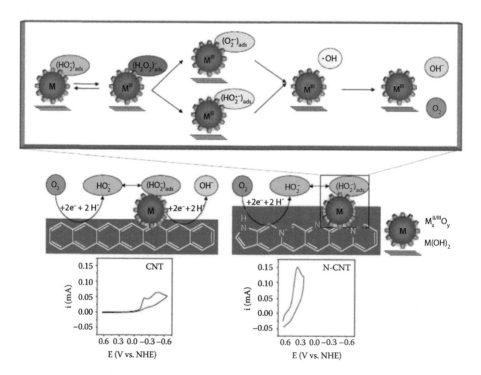

FIGURE 9.30 Schematic of the purposed mechanism of ORR at CNTs and N-CNTs. CVs display the respective steps of the ORR at nondoped CNTs, which presents two electrochemical reductions, while only one is observed at N-CNTs. (Reprinted with permission from Wiggins-Camacho, J.D. and Stevenson, K.J., Mechanistic discussion of the oxygen reduction reaction at nitrogen-doped carbon nanotubes, *J. Phys. Chem. C*, 115, 20002–20010. Copyright 2011, American Chemical Society.)

Ab initio studies indicate that nitrogen stabilizes Fe on CNTs as identified by a higher binding energy compared to Fe on pristine CNTs. The study also indicated that pyridinic configurations are favored over pyrrolic, and that the incorporation of Fe-4N sites are favored in larger-diameter CNTs and graphene rather than small-diameter CNTs.[302] The catalytic disproportionation of HO_2^- (or H_2O_2) mentioned earlier, however, is not at a heme-like porphyrin site, since neither CN^- [303] nor CO[299] inhibits ORR activity. Rather, the *pseudo* four-electron pathway permits a unique biosensing technique, where increased ORR (from regenerated O_2) can be used to detect enzyme turnover from peroxide-producing enzymes.[257,304]

9.4   CARBON NANOELECTRODES

9.4.1   ELECTRIC DOUBLE-LAYER STRUCTURE

At the electrode/solution interface, there is a layer of solvent molecules with dipoles oriented in accordance with the charge on the electrode surface and/or ions that are specifically adsorbed from the supporting electrolyte. This layer is called the compact, inner Helmholtz plane, or Stern layer.[84] Beyond this layer, ions in solution will be arranged in order to screen any charge remaining from the electrode surface, becoming increasingly disordered as the remaining charge or electric field is diminished.[305] The region from the inner Helmholtz plane to the point where the electric field is essentially zero is called the diffuse layer and is dependent on the concentration and type of ions in solution. Typically, at concentrations greater than 10^{-2} M, this layer is less than 10 nm.[84] The inner Helmholtz plane and the diffuse layer together make up the electric double layer (EDL). Solvated redox-active molecules that undergo outer-sphere ET can only approach the electrode surface as close as their solvated shell and the inner Helmholtz plane will allow. This is called the outer Helmholtz plane and is effectively the plane of closest approach for an outer-sphere ET. Since the electric field emanating from the electrode is screened or counterbalanced by the electric double layer structure, the structure will directly influence the potential drop and hence the driving force for ET.[306] If the electric field is not effectively screened or if the supporting electrolyte does not efficiently counterbalance the electric field, the potential drop will be significantly smaller, ET with a redox molecule will be influenced, and the diffuse layer will be extended in order to collect enough opposite charge to neutralize the electric field. At the nanoscale, these factors play an important role in the observed electrochemical behavior since their effect is at or larger than the dimensions of the electrode.

9.4.2   ELECTROCHEMICAL BEHAVIOR OF ELECTRODES AT THE NANOSCALE

There has been a great amount of work done to minimize the size of electrodes all the way down to the nanoscale.[307–309] This is reflected in the title of this section, the title of this book, and fittingly, highlighted by some of the luminaries in electrochemistry as one of the current trends of electrochemical research.[310] The enhanced spatial resolution of nanoelectrodes will find important applications in SECM[311] and measuring biological reactions such as exocytosis[312] and intracellular events.[313] Like microelectrodes, nanoelectrodes display radial diffusion where enhanced flux renders the steady-state current almost independent of the scan rate for CVs with Nernstian reactions (sufficiently fast ET kinetics). Initial studies used nanoelectrode bands that would amplify the small currents obtained at nanoelectrodes by increasing the length.[314] Unlike microelectrodes, the enhanced mass transport and smaller size of nanoelectrodes increases the current density and extends the upper limit of fast ET rate constants that can be experimentally measured.[315] One must be careful to correctly assess the geometry of the electrode, which can have significant impacts on the voltammetric behavior and subsequent kinetic analyses.[316–318] At the nanoscale, the size of the electrode begins to reach the size of the EDL, the interionic spacing of the supporting electrolyte, the Debye length, the size of the reactant molecule, or even the electron tunneling distance which

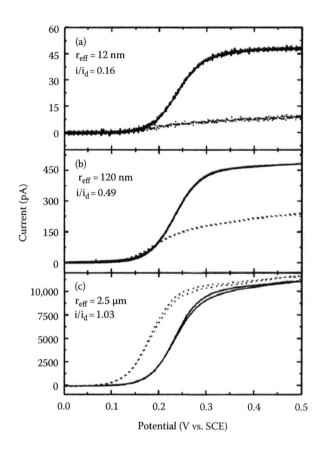

FIGURE 9.31 Steady-state CVs of $K_4Fe(CN)_6$ being oxidized in the presence (solid line) and absence (dotted line) of 0.5 M KCl, as the effect radius (r_{eff}) of the carbon electrode is decreased to (a) 2.5 μm, (b) 120 nm, and (c) 12 nm (limiting current in the presence (i_d) or absence (i) of 0.5 M KCl). (Reprinted with permission from Chen, S. and Kucernak, A., The voltammetric response of nanometer-sized carbon electrodes, *J. Phys. Chem. B*, 106, 9396–9404. Copyright 2002, American Chemical Society.)

inevitably causes significant effects on the mass transport, the charge distribution, and the observed current-potential response. For instance, the diffuse layer (defined by the Debye length) at nano-electrodes will often be comparable to the length scale of the depletion layer (defined by the electrode size, geometry, and mode of mass transport).[306,319,320] This overlap can be achieved by either decreasing the electrode size, which will cause the depletion layer to shrink while maintaining a relatively constant diffuse layer, or lowering the ionic strength of the solution, thereby increasing the Debye length and coordinating diffuse layer (the Debye length is inversely proportional to the square root of the ionic concentration). The diffuse layer and depletion layer overlap is especially apparent when both techniques are applied simultaneously, often contrasting the current-potential response in the presence or absence of supporting electrolyte.[306,321–324]

Figure 9.31 presents the steady-state current as $K_4Fe(CN)_6$ is oxidized in the presence and absence of supporting electrolyte, displaying significant deviations as the size of the carbon electrode decreases. As a consequence of the diffuse layer/depletion layer overlap, the electric field from the charged electrode interface can cause significant deviations from traditional transport theory, usually observed at electrodes with radii lower than 10 nm.[306,321,323,325,326] Excess inert supporting electrolyte is used to create a *compact* EDL by screening the charge on the electrode surface within a short distance, buffering local charge density variations due to faradaic redox processes, maintaining electroneutrality, and minimizing migration as a mode of mass transport. In low ionic strength solutions, migration will increasingly contribute to the observed current response as discussed with microelectrodes.[327] In nanoelectrodes, the electric field overlap with the depletion layer will also cause migration to influence ionic transport, where rapid ionic fluxes will further cause migration to either enhance or hinder limiting currents as a function of charge on the reactant and product, the concentration of reactant, the concentration and composition of the inert supporting electrolyte, and the potential of zero charge on the electrode surface.[319,324,328] Since decreasing the ionic strength extends the Debye length and reach of the diffuse layer, the potential drop associated with the electric field is also extended. This means that the driving force for ET (φ), which is dependent on the magnitude of the potential drop between the electrified surface and the plane of ET (PET) (i.e., the tunneling distance for outer-sphere ET), is significantly less, which subsequently influences observed ET kinetics.[306,323,324] Figure 9.32 presents a diagram of the potential drop (φ) at a Pt electrode as function of distance (x) and electrolyte and an example of its effect on $IrCl_6^{3-/2-}$.

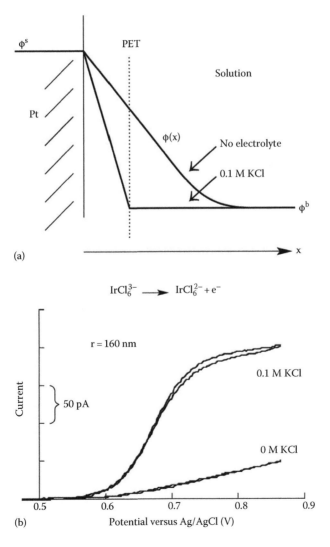

FIGURE 9.32 (a) A schematic of the driving force for ET (φ) on a Pt electrode as a function of distance (x) with or without supporting electrolyte (PET). (b) An example of the decreased driving force in ET with $IrCl_6^{3-/2-}$. (Reprinted with permission from Watkins, J., Zhang, B., and White, H., Electrochemistry at nanometer-scaled electrodes, *J. Chem. Ed.*, 82, 712–719. Copyright 2005, American Chemical Society.)

The rapid ionic fluxes at nanoelectrodes also call into question the traditional assumptions of electroneutrality,[329] even in the presence of excess electrolyte.[328] Significant deviations are also observed in the voltammetric wave shape, which is a function of the formal potential, the charge on the reactant and product, the concentration of supporting electrolyte, and the potential of zero charge on the electrode. Deviations are even observed for neutral species, which are unaffected by the electric field.[328] Thus, normal formalisms of the electric double layer and double layer effects from Gouy–Chapman–Stern and Frumkin, the distribution of ions based on the Poisson–Boltzmann equation, and the transport of ions from Nernst–Planck–Poisson, which all rely on maintained chemical equilibria and electroneutrality, break down due to the high mass transport rates, large faradaic currents, and diffuse double layer effects at nanoelectrodes.[323,326,328,330] Additionally, formalisms of ET kinetics may also be inappropriate, as is the case for Butler–Volmer kinetics at large overpotentials.[331,332] Modified treatments of traditional formalisms must be used to accurately model the dynamic electrochemistry at nanoelectrodes.[330,333,334]

9.4.3 CNT and Graphene Nanoelectrodes

The first individual CNT electrode was created by Campbell et al., who attached 80–200 nm diameter CNTs to a Pt tip and electrodeposited a polyphenol layer on the CNT. The electrode surface was exposed by applying a bias to the tip in solution (which exposes the CNT) or at a specific site along the CNT in air, which cuts the CNT/polyphenol at the desired point.[335] The fabrication of another individual CNT nanoelectrode was done by Kawano et al., where CNTs made through Joule heating on a silicon support structure were subsequently coated with parylene in a core–shell-type arrangement, shown in Figure 9.33.[336]

The electrochemistry of individual SWCNTs was observed to be identical for both metallic and semiconducting types, exhibiting fast ET rates at the sidewalls during the oxidation of FcTMA+.

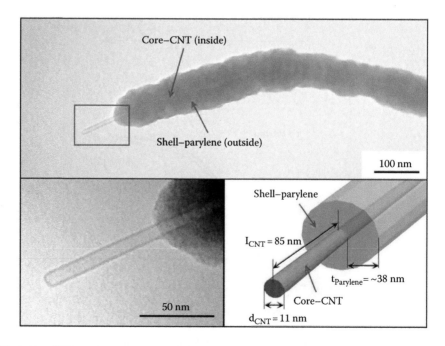

FIGURE 9.33 CNT–parylene core–shell nanoelectrode. (Reprinted from *Sens. Actuators, A*, 187, Kawano, T., Cho, C. Y., and Lin, L., An overhanging carbon nanotube/parylene core–shell nanoprobe electrode, 79–83, Copyright 2012, with permission from Elsevier.)

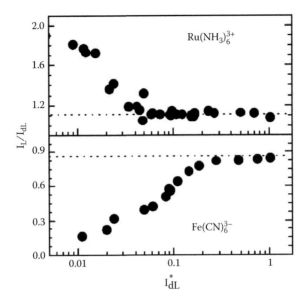

FIGURE 9.34 Ratios of the limiting current in the absence (I_L) and presence (I_{dL}) of 1 M KCl as a function of the nanographene electrode size (based on I_{dL}). (Reprinted with permission from Zhang, B., Fan, L., Zhong, H. et al., Graphene nanoelectrodes: Fabrication and size-dependent electrochemistry, *J. Am. Chem. Soc.*, 135, 10073–10080. Copyright 2013, American Chemical Society.)

The similar electrochemical responses, even though the SWCNTs have significantly different electronic structures, are explained by considering water as a gate (which can influence the E_F) and the contribution of electronic states close to E_F (which can overlap with the width of the reactant states).[337] Using *n*-dodecanethiol-modified Au ultramicroelectrodes (UMEs), graphene nanoelectrodes were formed by simply dipping the modified Au UME in a dilute suspension of hydrazine-rGO flakes, which were annealed and separated by high-speed centrifugation and size-selected ultrafiltration.[338] The subsequent study of ET at these graphene nanoelectrodes exhibited an enhanced or inhibited limiting transport current from the reduction of $Ru(NH_3)_6^{3+}$ and $Fe(CN)_6^{3-}$, respectively, when supporting electrolyte was removed from the solution. Figure 9.34 presents deviations in the limiting current (with or without 1 M KCl) as a function of the graphene nanoelectrode size. The effect was more pronounced as the graphene nanoelectrode decreased in size, pointing to the effects in the nanoscopic size regime. More specifically, the diffuse layer caused deviations in both the electroneutrality-based mass transport and the Frumkin treatment of the EDL towards ET kinetics. For the reduction of $Fe(CN)_6^{3-}$, this deviation was manifested in both the ET kinetics and the mass transport dynamics, while only the mass transport was observed to be affected for $Ru(NH_3)_6^{3+}$, due to fast ET kinetics of this redox molecule. Another single CNT electrode was fabricated and covered with an atomic layer deposition (ALD)–grown HfO_2 layer as an insulating cover. The 20–30 nm diameter CNT electrodes were manipulated in an SEM to mount the electrode and expose the CNT by a flash process after ALD.[339] FETs are increasingly being made from CNT materials since the conductance is highly sensitive to surface perturbations.[340] One such nanodevice used a single SWCNT as a FET, which was applied to single bovine chromaffin cells to detect exocytosis of chromogranin A (CgA) after stimulation from histamine.[341] The SWCNT-FET was functionalized with the CgA antibodies and subsequently covered with a blocking agent (TWEEN-20) to ensure measured binding events correlated directly with the release of CgA. The SWCNT-FET device was able to discriminate relative strengths of CgA release from a single cell, which correlated well with the strength of stimulation.

9.4.4 Carbon Fiber Nanoelectrodes and Carbon Nanoprobes

The formation of CF nanoelectrodes with effective radii as low as 46 nm were formed using shear force–based SECM positioning to place an anodically etched CF (via a 3 V bias inside a Pt cathodic ring in 0.1 M NaOH solution) within a soft silicon rubber while the rest of the fiber was covered with electrodeposited polymer Canguard.[342] Double-barrel carbon nanoprobes (DBCNPs) were developed for use with scanning ion conductance microscopy (SICM) and SECM, simultaneously, with radii ranging from 10 nm to 1 µm.[343] In this arrangement, the ion current was used to control the tip–substrate distance (which will decrease as the nanoprobe gets close to the surface due to occlusion of the ion flux), while a redox probe was used to obtain electrochemical information. Figure 9.35 presents a schematic of the fabrication and operation process of the DBCNP along with some images. The authors demonstrate the acquisition of neuron images using ferrocenylmethanol, which can cross the cell membrane. They further demonstrate the utility of using the DBCNPs by stimulating the release of neurotransmitters by K^+ using another pipette for whole-cell stimulation, or the DBCNP itself for local stimulation, by voltage-driven K^+ release.

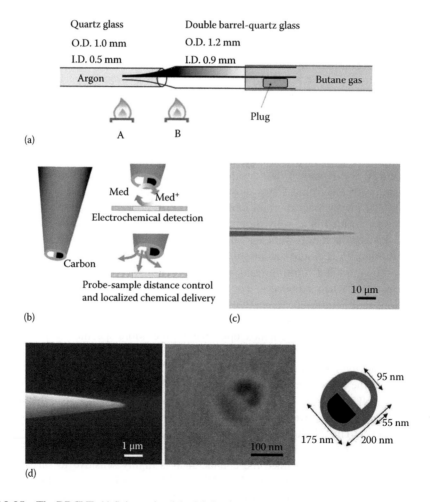

FIGURE 9.35 The DBCNP. (a) Schematic of the fabrication process, (b) schematic of operation using the ion current to detect position and the carbon electrode to obtain electrochemical information, (c) optical image of the DBCNP, and (d) SEM images of the DBCNP and the dimensions at the tip. (Takahashi, Y., Shevchuk, A.I., Novak, P. et al.: *Angew. Chem. Int. Ed.* 9638–9642. 2011. Copyright Wiley-VCH Verlag GmbH & Co. KGaA. Reproduced with permission.)

This idea was further developed to form dual-carbon nanoelectrodes by pyrolyzing butane gas in both barrels of a quartz theta nanopipette.[344] The nanoelectrodes were used in SECM in a generation–collection mode, which displayed enhanced efficiencies at insulating substrates and depressed efficiencies at conducting substrates. The dual-carbon nanoelectrode was used to image thylakoid membranes with an artificial electron acceptor FcTMA^{2+} during photosynthesis. The activity of alkaline phosphatase (ALP) in HeLa cells confined to a single droplet was also investigated with the use of a double-barrel carbon nanoelectrode. The ALP-catalyzed hydrolysis of p-aminophenyl phosphate to the electroactive p-aminophenol exhibited an increased current concomitant with the number of cells in the droplet.[345] Pyrolytic carbon nanoelectrodes (radii 6.5–100 nm) were developed for constant-current SECM (diffusion-limited reduction of $Ru(NH_3)_6^{3+}$ was used for distance control) where high-resolution topographical images of cells with convoluted structures such as differentiated PC12 cells and A431 cells were effectively obtained.[346] In the same report, a voltage-switching mode for SECM (VSM-SECM) was designed for the simultaneous acquisition of topographical and electrochemical information. Examples included the electrochemical imaging of the epidermal growth factor receptor on a cell surface (via $Ru(NH_3)_6^{3+}$ reduction for topography and p-aminophenol oxidation for electrochemical information) or the neurotransmitter release from single PC12 cells (oxygen reduction provided distance control). Figure 9.36 presents a schematic of the VSM-SECM and examples of the images obtained from A431 cells.

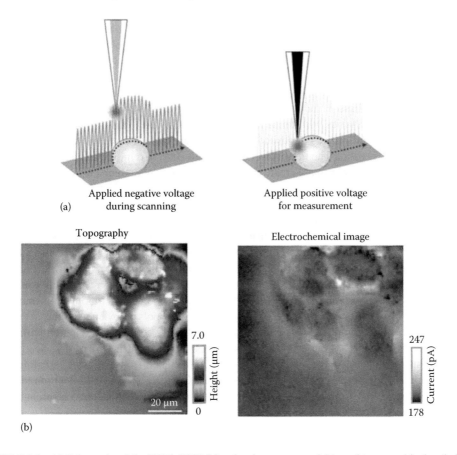

FIGURE 9.36 (a) Schematic of the VSM-SECM for simultaneous acquisition of topographical and electrochemical information. (b) Example of the VSM-SECM mode displaying topographical and electrochemical images of A431 cells. (Takahashi, Y., Shevchuk, A.I., Novak, P. et al., Topographical and electrochemical nanoscale imaging of living cells using voltage-switching mode scanning electrochemical microscopy, *Proc. Natl. Acad. Sci. U. S. A.*, 109, 11540–11545, Copyright 2012, National Academy of Sciences, U.S.A.)

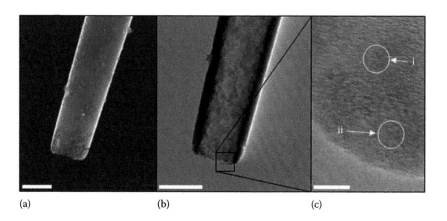

(a)   (b)   (c)

FIGURE 9.37   Carbon nanopipette. (a) SEM micrograph, scale bar 500 nm. (b) TEM micrograph, scale bar 500 nm. (c) TEM micrograph, scale bar 10 nm showing (i) amorphous or (ii) graphitic carbon. (Schrlau, M.G., Falls, E.M., Ziober, B.L. et al., *Nanotechnology*, 2008, Copyright IOP Publishing.)

A glass-encased CF nanoelectrode was modified with SWCNTs by immersion in an SDS suspension of SWCNTs.[347] The resulting SWCNT-modified nanoelectrode (diameter 100–300 nm) showed a linear increase with peak current for DA, epinephrine, and norepinephrine using CV at 0.1–100 μM, 0.3–100 μM, and 0.5–100 μM, respectively. A CF nanoelectrode was coupled with PC12 cells to determine exocytosis with high spatial resolution.[348] The 100 nm diameter nanoelectrodes were placed at various sites on the cell surface and used to monitor DA release after stimulation with a high K⁺ solution. The results indicated that the majority of DA release happens from multiple vesicles that release at the same site. Carbon nanopipettes were fabricated by filling the inner bore of quartz capillaries with ferric nitrate in IPA, allowed to air dry, and pulled into fine-tipped micropipettes. Carbon was deposited on the catalyst surface by CVD at 900°C (thickness was controlled by CVD time). The tip was wet etched with BHF to remove the quartz at the end and further etched with oxygen plasma to create carbon nanopipettes with diameters of tens to hundreds of nanometers. The nanopipettes have the advantage of being able to perform intracellular injections and concurrent electrical measurements.[349] Figure 9.37 presents SEM and TEM images of the carbon nanopipettes. The nanopipettes were then applied to measure the membrane potential of mouse hippocampal cell line HT-22 during ionic enrichment or the application of extracellular agents such as γ-aminobutyric acid (GABA). Extracellular addition or perfusion of K⁺ caused the membrane potential to change in accordance with the Nernst equation, while the extracellular introduction of GABA caused a hyperpolarization. Perfusion of the cells with bicuculline, a competitive antagonist for the GABA$_A$ receptor, could block the GABA response in the cell membrane.[350] Nanodisk CF electrodes were electrochemically etched and insulated with polypropylene for the detection of transmitters of *Caenorhabditis elegans* NSM neurons and PC12 cells upon excitation with K⁺. The nanodisk electrodes exhibited better spatiotemporal resolution than the microdisk electrodes for kinetic analysis of exocytosis.[351] CF nanoelectrodes, flame etched from CF, were integrated into a microchip capillary electrophoresis to detect DA from PC12 cells.[352] The 100–300 nm electrodes displayed a low detection limit of 59 nM and better resolution between DA and isoprenaline than 7 μm CF electrodes.

9.5   CARBON NEAs AND NEEs

9.5.1   Electrochemical Properties of NEAs and NEEs

Nanoelectrodes have been widely used in the form of nanoelectrode arrays (NEAs) or nanoelectrode ensembles (NEEs) where individual nanoelectrodes are isolated from one another by various insulating materials.[353,354] Nanoelectrode arrangements are classified by their degree of control during the fabrication process. If there is precise control over the interelectrode distance, the arrangement is considered a NEA. Arrangements with a lower degree of precision or a random dispersion are considered NEEs.[353] The benefits of micro- and nanoelectrodes, such as their ability to operate in solutions of high resistance due to a small iR drop, increased mass transport via radial diffusion, and a small RC time constant (fast background charging due to a small electroactive area) remains intact for NEAs/NEEs, while the very small currents created at single nanoelectrodes are amplified from the parallel orientation. During measurements, the interelectrode distance becomes important, as well as the time frame of the measurement, the geometry of the nanoelectrode, and the diffusion constant of the analyte, since overlapping diffusion zones in NEAs/NEEs will render many of the benefits of the individual nanoelectrodes obsolete. Dense populations of nanoelectrodes will essentially display macroelectrode behavior, mimicking the linear diffusion of a macroelectrode with an identical geometric area, rather than nanoelectrodes working in parallel from radial diffusion. It is important, then, to ensure the electrochemical behavior of the NEA/NEE is representative of small electrodes, such that a sigmoidal peak is observed when a CV is performed with an appropriate redox molecule. The background charging current is proportional to the total area of the nanoelectrode elements in a NEA/NEE. This means that the signal-to-noise ratio (faradaic current to the background charging current) will be higher, even if the resulting diffusion layers between nanoelectrode elements overlap, since the total electroactive area of NEEs/NEAs is much smaller than an equivalently sized macroelectrode. In some applications, diffusion layer overlap is desired to capitalize on the lower detection limit while maintaining a relatively high signal current.[355]

9.5.2 Carbon NEAs and NEEs

VA-CNT NEAs and NEEs have been fabricated on a Si wafer via a process compatible with current Si microfabrication techniques, developed by Meyyappan and coworkers.[354] The basic procedure consists of depositing a layer of metal on a Si wafer (covered with an insulating barrier of SiO_2 or Si_3N_4) as an electrical contact. A thin Ni film (usually 10–30 nm) is then deposited on top of the metal, which can be patterned by e-beam or UV lithography to control the density of VA-CNTs. A plasma-enhanced CVD (PE-CVD) process forms CNTs normal to the Ni film, which are subsequently encapsulated or insulated from one another by a low-pressure CVD process using tetraethoxysilane (TEOS). Chemical–mechanical polishing exposes the CNT tips but needs annealing and/or anodic pretreatment in 1.0 M NaOH in order to observe acceptable electrochemical behavior. One study applied this process to form a DNA sensor using BRCA1 DNA probes.[356] The 30–160 nm diameter CNT electrodes were functionalized with DNA through carbodiimide coupling causing the carboxylic groups on the carbon surface (from anodic treatment in NaOH) to form amide bonds with a primary amine group on the probe DNA. The difference in oxidation currents from AC voltammetry, coupled with $Ru(bpy)_3^{2+}$ as a redox mediator for the oxidation of guanine, could effectively detect the hybridization of target DNA. This idea was advanced by multiplexing the Si chip with nine arrays, set up as a 3×3 matrix for a positive control, a negative control, and the specific hybridization of DNA. The multiplexed VA-CNT NEAs were about 70 nm in diameter and spaced 1 μm apart, covalently attached to oligonucleotide probes, which are hybridized with DNA targets from *Escherichia coli* O157:H7.[357]

Figure 9.38 displays the as-grown VA-CNTs on the Ni film, producing an NEE, or a patterned Ni film (100 nm dots spaced 1 μm apart), producing a NEA, before and after being insulated and exposing the CNT tips. The voltammograms display the unique electrochemical behavior of dense NEEs, where the diffusion layers overlap, or NEAs, when the nanoelectrodes are well spaced, diffusional boundaries do not overlap, and voltammograms display sigmoidal behavior. Another 3×3 array of VA-CNT NEAs (diameters ~102 nm) was integrated with the wireless instantaneous neurotransmitter concentration sensor system (WINCS) to determine DA levels by fast-scan CV. The NEA matrix shows promise for the detection of neurochemical release with high spatial

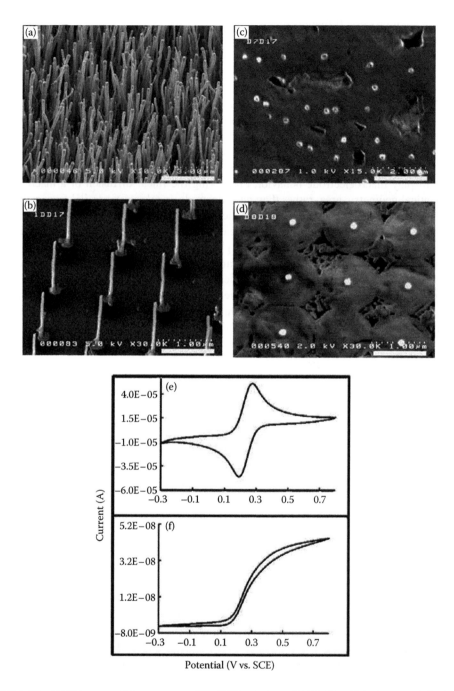

FIGURE 9.38 SEM images of the as-grown VA-CNTs on the solid Ni film (a) before and (c) after SiO_2 encapsulation and tip exposure; patterned Ni film (100 nm dots spaced 1 µm apart) (b) before and (d) after SiO_2 encapsulation and tip exposure; (e) CVs of the Ni film–grown VA-CNT NEE and (f) Ni-patterned-grown VA-CNT NEA during oxidation of $K_4Fe(CN)_6$ at 20 mV/s. (Reprinted from *Biosens. Bioelectron.*, 24, Arumugam, P.U., Chen, H., Siddiqui, S. et al., Wafer-scale fabrication of patterned carbon nanofiber nanoelectrode arrays: A route for development of multiplexed, ultrasensitive disposable biosensors, 2818–2824, 2009, with permission from Elsevier.)

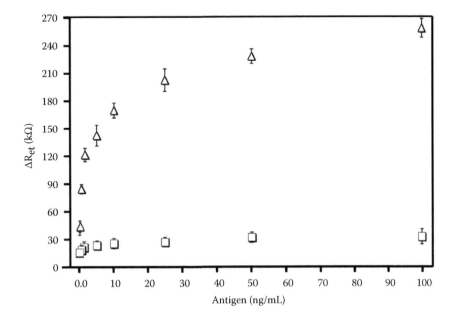

FIGURE 9.39 The change in ET resistance as a function of the concentration of cTnI (Δ) or a nonspecific control myoglobin (♦). (Reprinted with permissions from Periyakaruppan, A., Gandhiraman, R.P., Meyyappan, M. et al., Label-free detection of cardiac troponin-I using carbon nanofiber based nanoelectrode arrays, *Anal. Chem.*, 85, 3858–3863. Copyright 2013, American Chemical Society.)

resolution compared to conventional CF microelectrodes.[358] Siddiqui et al. used electrochemical impedance spectroscopy (EIS) to characterize the VA-CNT NEAs (tip diameters ~80 nm) and found that the interelectrode spacing controls the spectrum. Randomly grown NEEs on the Ni film gave a straight line, but patterned NEAs with an interelectrode spacing of 1 μm displayed a near-perfect semicircle with low noise and linearity at larger amplitudes.[359] Randomly grown VA-CNT NEEs (average interelectrode distance of about 700 nm with diameters ~80 nm) were covalently attached to either a ricin-A antibody or an antiricin RNA aptamer for the detection of the toxic glycoprotein ricin by EIS via an increase in ET resistance.[360] The aptamer-based detection scheme displayed better sensitivity than the antibody scheme and also exhibited reusability through denaturation and release of the ricin protein. Similarly, an identical scheme was used with antibody cardiac troponin-I (cTnI) for the selective detection of cTnI, an early biomarker for the detection of acute myocardial infarction. Figure 9.39 presents the change in ET resistance as a function of cTnI concentration or myoglobin (as a nonspecific control). The antibody–antigen interaction for cTnI exhibited a very low detection limit of 0.2 ng/mL.[361] VA-CNT NEAs (200–300 nm diameter with interelectrode distance of 1 μm) were used to detect phosphorylation and dephosphorylation of peptides covalently attached to the CNT tip ends. The real-time EIS detection displayed reversible behavior of the phosphorylation or dephosphorylation using c-Src tyrosine kinase and tyrosine phosphatase 1B, respectively. Additionally, the dephosphorylation showed well-defined kinetics, which were quantitatively analyzed by the Michaelis–Menten heterogeneous enzymatic model.[362] Swisher et al. were able to measure enzyme kinetics of cancer-mediated proteases at a VA-CNT NEA using high-frequency AC voltammetry.[363] Two types of tetrapeptides specific to the cancer-mediated proteases, legumain and cathepsin B, were covalently attached to the exposed VA-CNTs, with a ferrocene (Fc) molecule attached to the distal end of the tetrapeptides. The change in redox signal from Fc as the protease cleaved the tetrapeptide proved to be both an effective detection method for the cancer proteases and a unique way to measure their enzyme kinetics, subsequently validated by a fluorescence assay.

VA-CNT NEAs have also been fabricated via electrodeposition of Ni nanoparticles as nucleation sites for CNT growth by Ren and coworkers.[364] VA-CNT NEAs were insulated with a thin coat of SiO$_2$ by magnetron sputtering and M-Bond 610 (an epoxy-phenolic adhesive) before exposing the CNT tips by polishing (interelectrode spacing of 10 μm). Similarly, VA-CNT NEAs were fabricated by insulating the CNTs with a spin-coated epoxy resin Epon 828, which produced 50–80 nm diameter electrodes spaced 5 μm apart. The NEAs could be used for the voltammetric detection of Pb^{2+} at the ppb level[365] or be covalently attached to GOx by carbodiimide coupling.[366] The resulting glucose biosensor could efficiently detect enzymatically generated H$_2$O$_2$ by reduction at a low potential of −0.2 versus Ag/AgCl, which eliminates deleterious effects from easily oxidized interferents such as UA, AA, and acetaminophen. A glutamate biosensor was developed from VA-CNT NEAs using covalently attached glutamate dehydrogenase. The biosensor exhibited two linear regimes, 0.01–20 μM and 20–300 μM, with a detection limit of 10 nM. The VA-CNT NEAs were created from a photolithography/reactive ion etching technique, which grew VA-CNTs spaced 3 μm apart with an average effective radius of 35 nm based on steady-state sigmoidal currents.[367] Figure 9.40 displays a schematic of the fabrication process for the glutamate biosensor.

Hees et al. fabricated recessed boron-doped NCD (B-NCD) NEAs and NEEs with radii in the range of 150–250 nm. The NEAs were created by e-beam lithography/photolithography of a SiO$_2$ layer on B-NCD, which was filled in with a layer of insulating NCD (i-NCD). The SiO$_2$ was removed by HF, leaving recessed B-NCD nanoelectrodes spaced 10 μm apart. The NEEs were created by allowing SiO$_2$ spheres to adsorb onto the surface of B-NCD and filling the remaining space with i-NCD before removing the spheres with HF. The density of the NEE was directly related to the concentration of spheres in the adsorbing solution. Figure 9.41 displays the B-NCD

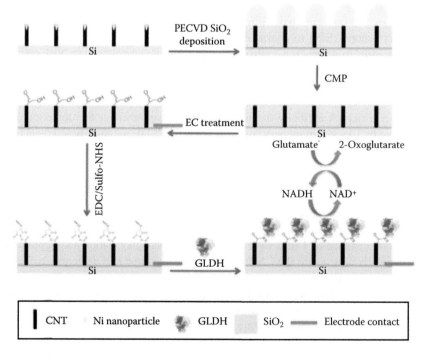

FIGURE 9.40 Schematic of the fabrication process for a VA-CNT NEA glutamate biosensor using glutamate dehydrogenase. (Reprinted with permission from Gholizadeh, A., Shahrokhian, S., Iraji zad, A. et al., Fabrication of sensitive glutamate biosensor based on vertically aligned CNT nanoelectrode array and investigating the effect of CNTs density on the electrode performance, *Anal. Chem.*, 84, 5932–5938. Copyright 2012, American Chemical Society.)

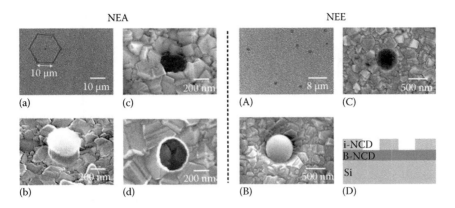

FIGURE 9.41 SEM images of the NEA, (a) distribution of nanoelectrodes, (b) growth of SiO_2 islands on B-NCD layer, (c) growth of i-NCD around the SiO_2 island for the B-NCD, and (d) final recessed nanoelectrode after removal of the SiO_2 islands; NEE, (A) distribution of nanoelectrodes, (B) SiO_2 spheres after growth of i-NCD, (C) final recessed nanoelectrode after removal of SiO_2 spheres, and (D) schematic of the cross section for the B-NCD NEAs and NEEs. (Reprinted with permission from Hees, J., Hoffmann, R., Kriele, A. et al., Nanocrystalline diamond nanoelectrode arrays and ensembles, *ACS Nano*, 5, 3339–3346. Copyright 2011, American Chemical Society.)

recessed NEAs and NEEs at various stages of the fabrication process, as well as a schematic of the cross section of the B-NCD nanoelectrodes. The B-NCD NEAs and NEEs showed that unlike BDD macroelectrodes, nanoelectrodes display a difference in the ET kinetics of common redox probes based on treatment of the diamond surface. Although it has already been shown that the ET rate of $Fe(CN)_6^{3-/4-}$ decreases if BDD films are terminated with oxygen rather than hydrogen, no difference was observed in the ET rate of $Ru(NH_3)_6^{3+/2+}$ and $IrCl_6^{2-/3-}$ for either surface termination.[368] Using the B-NCD NEAs and NEEs, a faster ET rate was observed for an oxygen-terminated surface for $Ru(NH_3)_6^{3+/2+}$, while $IrCl_6^{2-/3-}$ displayed a faster ET rate at a hydrogen-terminated surface.[369]

This study was followed up with another that looked at the voltammetric behavior of methyl viologen (MV) at B-NCD NEAs with an average electrode radius of 170 nm and interelectrode distance of 10 μm.[370] The hydrogen-terminated electrodes displayed a sigmoidal transient, consistent with radial diffusion, but included a surface wave above 0.4 mM MV^{2+}. This surface wave was due to adsorption of MV^0 on the electrode surface via hydrophobic interactions, which are dependent on the type and concentration of supporting electrolyte according to the Hofmeister series. The oxygen-terminated B-NCD surface did not display a surface wave due to its more hydrophilic nature. NEEs of carbon nanopipettes were created from a simple dip-coating step of conical graphite sheets (10–15 nm at the tips) formed from PE-CVD and later dipped in a UV light curable acrylate coating.[371] The amount of polymer could be applied to ensure only the conical tips remained accessible to electrochemical reactions and verified by observing a sigmoidal steady-state voltammogram using $Fe(CN)_6^{3-/4-}$. Another VA-CNT NEE was developed by Sun et al. (average diameter of 10.5 nm, insulated with epoxy resin), which could effectively detect $Fe(CN)_6^{3-}$ and DA at 5 nM and 10 nM, respectively.[372]

9.6 CARBON THIN FILMS AND SPECTROELECTROCHEMISTRY

Carbon optically transparent electrodes (COTEs), which usually have one dimension in the nanoscale, are easily coupled to optical techniques for spectroelectrochemistry. COTEs are attractive due to their wide potential window, chemical stability under acidic or alkaline conditions, and ease of surface modification. COTEs are often made from PFFs, resistive sputtering, CVD and

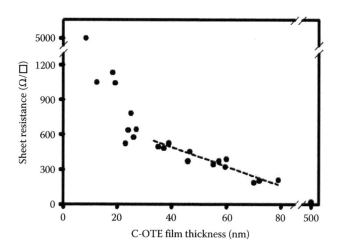

FIGURE 9.42 PPF sheet resistance as a function of the film thickness. (Reprinted with permission from Donner, S., Li, H.-W., Yeung, E.S. et al., Fabrication of optically transparent carbon electrodes by the pyrolysis of photoresist films: Approach to single-molecule spectroelectrochemistry, *Anal. Chem.*, 78, 2816–2822. Copyright 2006, American Chemical Society.)

microwave plasma CVD, or vapor deposition of carbon or carbon nanomaterials such as CNTs or graphene.[373] PPFs are incredible flat with lower intrinsic capacitance and O/C ratios than GC.[374] PPFs are created by spin coating commercially available photoresist on a suitable substrate, followed by pyrolysis. Although the resistivity and electrochemical and properties of PPFs depend on the pyrolysis temperature and atmosphere, they are similar to GC.[375] OT PPF electrodes are made by diluting the photoresist before spin coating, lowering the viscosity, and creating thinner films. The film thickness of PPFs dictates both the transparency of the film (thinner films have higher transparency) and the sheet resistance (thinner films have higher sheet resistance), requiring a balance of properties depending on the application.[376] Figure 9.42 displays the measured sheet resistance as a function of the PPF thickness.

McCreery and coworkers used 6 nm thick OT PPF electrodes to study chemisorbed molecular layers of nitroazobenzene, nitrobiphenyl, and azobenzene by UV–vis spectroelectrochemistry. They observed red shifts in the chemisorbed molecular layers with respect to the molecules in solution, attributed to electronic interactions from the π system of the substrate or adjacent molecules. Additionally, upon reduction, new spectral features appeared, indicative of a *methide* species possessing more extensive electron delocalization than the parent molecules.[377] PFFs 11 nm thick were developed for electrogenerated chemiluminescence (ECL) of C8S3 J-aggregates with 2-(dibutylamino)ethanol as a coreactant. The films exhibited excellent ECL yield in both oxidative–reductive ECL with the J-aggregates and reductive–oxidative ECL using Ru(bpy)$_3$$^{2+}$. Both ECL modes outperformed the traditionally used transparent ITO electrodes.[378] The same PPFs were also used to optically observe the formation of an electrogenerated graphitic oxide (EGO) solid electrolyte interphase (SEI) upon the application of anodic potentials to the PPF electrodes in alkali chloride supporting electrolyte.[379] UV–vis spectra revealed an increase in adsorption from the π–π* aromatic carbon transition of graphitic oxide at 230 nm (or reduced graphitic oxide at 270 nm) at anodic potentials. Figure 9.43 displays the EGO SEI in the absorbance spectra during a CV in 1 M KCl. The EGO SEI at PPFs is formed when cations are drawn to the surface of the oxidized carbon due to negatively charged oxygen functionalities. Upon cycling at negative potentials, the cations are electrostatically driven into the PPFs, forming a SEI layer, which scales with the size of the cation (Li$^+$, Na$^+$, and K$^+$). The Cl$^-$ appears to act as a passivating agent, protecting the carbon from dissolution.

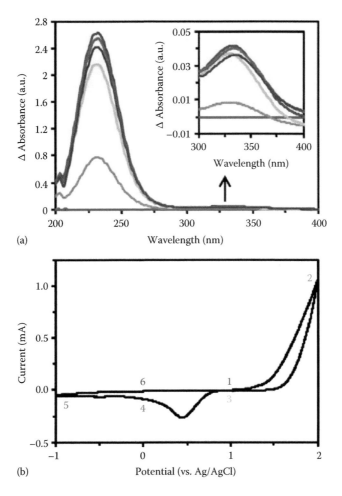

FIGURE 9.43 (a) Absorbance spectra at specific points (color coordinated) during a (b) CV displaying the appearance of the EGO in 1 M KCl. (Reprinted with permission from Walker, E.K., Vanden Bout, D.A., and Stevenson, K.J., Spectroelectrochemical investigation of an electrogenerated graphitic oxide solid-electrolyte interphase, *Anal. Chem.*, 84, 8190–8197. Copyright 2012, American Chemical Society.)

A unique aspect of PPFs has been explored by Xiao et al. who through lithography defined a 3D porous network, which was conformally sputtered with nickel and annealed to convert the porous PPF architecture into graphene, before removing the nickel in acid. Figure 9.44 presents SEM images of the 3D graphene structure at various points in the fabrication process. The 3D graphene network displayed sigmoidal voltammograms from 5 to 100 mV/s in 2 mM $K_3Fe(CN)_6$.[380] Another interesting COTE was made by the pyrolysis of spontaneously adsorbed protein layers of BSA. The number of layers was characterized up to 12 iterations of adsorption/pyrolysis with the number of layers corresponding to an increase in conductivity, decrease in transparency, and increase in thickness. The electrodes were electrochemically characterized using $Fe(CN)_6^{3-/4-}$ and $Ru(NH_3)_6^{3+/2+}$ and coupled with GOx to form a glucose biosensor.[381] Thin pyrolytic carbon films of about 150 nm, although not coupled with optical techniques, were shown to electrochemically detect DA and paracetamol simultaneously by CV or DPV techniques.[382] The noncatalytic CVD-grown films were treated with an oxygen plasma, which significantly increased the ET kinetics and peak separation between DA and paracetamol (225 mV) due to the introduction of edge plane sites/defects and oxygen functionalities (verified electrochemically through the reaction with $Fe^{3+/2+}$).[383]

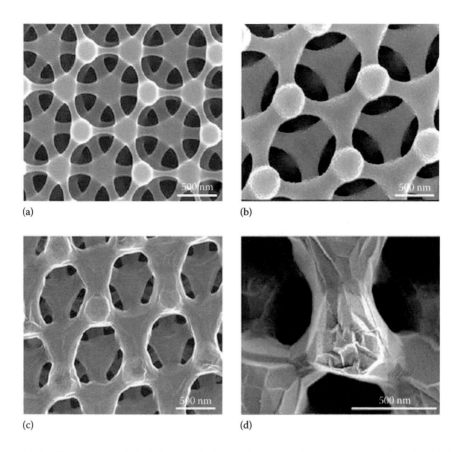

FIGURE 9.44 SEM images of the lithographic 3D PPF structure (a) after coating with nickel (b) and as graphene 3D structures (c and d). (Reprinted with permission from Xiao, X., Beechem, T.E., Brumbach, M.T. et al., Lithographically defined three-dimensional graphene structures, *ACS Nano*, 6, 3573–3579. Copyright 2012, American Chemical Society.)

Although CNT[384] and graphene[385] COTEs have been produced, they are rarely applied to electrochemical applications.[386] Garoz-Ruiz fabricated an COTE with SWCNTs by using pressure to transfer the SWCNT dispersed in an organic solvent onto a polymer substrates.[387] The subsequent electrochemistry of ferrocenemethanol was relatively independent of the amount of SWCNTs applied, as long as the mesh was well interconnected. The SWCNT OT electrode (OTE) was also used to electropolymerize aniline while monitoring the UV–vis spectral changes. Flexible COTEs made from SWCNTs were also fabricated by pressing aerosol CVD-grown SWCNTs onto a polyethylene terephthalate support.[388] The COTEs exhibited 92% transmittance at 550 nm and were used to monitor the UV–vis spectroelectrochemistry of $Ru(bpy)_3^{2+}$. Spectroelectrochemistry was also performed with graphene COTEs, where exfoliated graphite oxide was spin coated onto quartz substrates and thermally reduced to graphene.[389] The graphene OTEs were between 6 and 25 nm thick and were applied to the oxidation of C8S3 J-aggregates while monitoring the changes in the UV–vis spectrum. Rutkowska et al. used CNTs to create a disk-shaped electrode on quartz with diameters of about 50 μm.[390] The UME CNT OTE was used to simultaneously monitor the oxidation of $Ru(bpy)_3^{2+}$ by CV and the optical signal via fluorescence confocal laser scanning microscopy. A BDD thin film was grown on undoped silicon for use in the mid and far IR regions for spectroelectrochemistry.[391] The BDD thin film electrode showed excellent ET behavior with aqueous redox couples $Fe(CN)_6^{3-/4-}$, $IrCl_6^{2-/3-}$, methyl viologen, and $Ru(NH_3)_6^{3+/2+}$ and the nonaqueous $Fc^{0/+}$. The C–H bending of the cyclopentadienyl ring of ferrocene was observed to shift from 823 to 857 cm^{-1} upon oxidation, while the C–N stretching mode of ferrocyanide shifted from 2039 to 2116 cm^{-1} upon oxidation. A nitrogen-doped diamond-like carbon (N-DLC) thin film electrode was fabricated by pulsed laser deposition from the ablation of a graphite target in the presence of nitrogen gas.[392] The N-DLC films were used for IR-ATR spectroelectrochemistry during the *in situ* oxidation of the N-DLC films at high anodic potentials and monitoring the electropolymerization of polyaniline.

REFERENCES

1. Kinoshita, K. 1988. *Carbon: Electrochemical and Physicochemical Properties*. New York: John Wiley & Sons.
2. Hirsch, A. 2010. The era of carbon allotropes. *Nat. Mater.* 9:868–871.
3. Wallace, G. G., Chen, J., Li, D. et al. 2010. Nanostructured carbon electrodes. *J. Mater. Chem.* 20:3553–3562.
4. Su, D. S. and Schlögl, R. 2010. Nanostructured carbon and carbon nanocomposites for electrochemical energy storage applications. *ChemSusChem* 3:136–168.
5. Pumera, M. 2011. Graphene-based nanomaterials for energy storage. *Energy Environ. Sci.* 4:668–674.
6. Xu, C., Xu, B., Gu, Y. et al. 2013. Graphene-based electrodes for electrochemical energy storage. *Energy Environ. Sci.* 6:1388–1414.
7. Simon, P. and Gogotsi, Y. 2012. Capacitive energy storage in nanostructured carbon-electrolyte systems. *Acc. Chem. Res.* 46:1094–1103.
8. Lahiri, I. and Choi, W. 2013. Carbon nanostructures in lithium ion batteries: Past, present, and future. *Crit. Rev. Solid State Mater. Sci.* 38:128–166.
9. Huang, C., Li, C., and Shi, G. 2012. Graphene based catalysts. *Energy Environ. Sci.* 5:8848–8868.
10. Su, D. S., Perathoner, S., and Centi, G. 2013. Nanocarbons for the development of advanced catalysts. *Chem. Rev.* 113:5782–5816.
11. Ghasemi, M., Daud, W. R. W., Hassan, S. H. A. et al. 2013. Nano-structured carbon as electrode material in microbial fuel cells: A comprehensive review. *J. Alloys Compd.* 580:245–255.
12. Willner, I., Yan, Y.-M., Willner, B. et al. 2009. Integrated enzyme-based biofuel cells—A review. *Fuel Cells* 9:7–24.
13. Walcarius, A., Minteer, S. D., Wang, J. et al. 2013. Nanomaterials for bio-functionalized electrodes: Recent trends. *J. Mater. Chem. B* 1:4878–4908.
14. Novoselov, K. S., Geim, A. K., Morozov, S. V. et al. 2004. Electric field effect in atomically thin carbon films. *Science* 306:666–669.
15. Kroto, H., Heath, J., O'Brien, S. et al. 1985. C60: Buckminsterfullerene. *Nature* 318:162–163.
16. Echegoyen, L. and Echegoyen, L. 1998. Electrochemistry of fullerenes and their derivatives. *Acc. Chem. Res.* 4842:593–601.
17. Geim, A. and Novoselov, K. 2007. The rise of graphene. *Nat. Mater.* 183–192.
18. Castro Neto, A. H., Peres, N. M. R., Novoselov, K. S. et al. 2009. The electronic properties of graphene. *Rev. Mod. Phys.* 81:109–162.
19. Novoselov, K., Jiang, Z., and Zhang, Y. 2007. Room-temperature quantum Hall effect in graphene. *Science* 315:1379.
20. Ohta, T., Bostwick, A., Seyller, T. et al. 2006. Controlling the electronic structure of bilayer graphene. *Science* 313:951–954.
21. Banhart, F., Kotakoski, J., and Krasheninnikov, A. V. 2011. Structural defects in graphene. *ACS Nano* 5:26–41.
22. Boukhvalov, D. W. and Katsnelson, M. I. 2008. Chemical functionalization of graphene with defects. *Nano Lett.* 8:4373–4379.
23. Meyer, J. C., Geim, A. K., Katsnelson, M. I. et al. 2007. The structure of suspended graphene sheets. *Nature* 446:60–63.
24. Deshpande, A. and LeRoy, B. J. 2012. Scanning probe microscopy of graphene. *Physica E* 44:743–759.
25. Martin, J., Akerman, N., Ulbricht, G. et al. 2007. Observation of electron—Hole puddles in graphene using a scanning single-electron transistor. *Nat. Phys.* 4:144–148.
26. Ugeda, M. M., Fernández-Torre, D., Brihuega, I. et al. 2011. Point defects on graphene on metals. *Phys. Rev. Lett.* 107:116803.

27. Ugeda, M. M., Brihuega, I., Guinea, F. et al. 2010. Missing atom as a source of carbon magnetism. *Phys. Rev. Lett.* 104:096804.
28. Partoens, B. and Peeters, F. 2006. From graphene to graphite: Electronic structure around the K point. *Phys. Rev. B* 74:075404.
29. Pumera, M. 2010. Graphene-based nanomaterials and their electrochemistry. *Chem. Soc. Rev.* 39:4146–4157.
30. Dresselhaus, M. S., Dresselhaus, G., Saito, R. et al. 2005. Raman spectroscopy of carbon nanotubes. *Phys. Rep.* 409:47–99.
31. Tuinstra, F. and Koenig, J. L. 1970. Raman spectrum of graphite. *J. Chem. Phys.* 53:1126–1130.
32. Bowling, R., Packard, R., and McCreery, R. 1989. Activation of highly ordered pyrolytic graphite for heterogeneous electron transfer: Relationship between electrochemical performance and carbon microstructure. *J. Am. Chem. Soc.* 111:1217–1223.
33. McCreery, R. L. 2008. Advanced carbon electrode materials for molecular electrochemistry. *Chem. Rev.* 108:2646–2687.
34. McCreery, R. 1991. Carbon electrodes: Structural effects on electron transfer kinetics. In *Electroanalytical Chemistry*, Bard, A. J. (ed.), pp. 221–374. New York: Marcel Dekker.
35. Maldonado, S., Morin, S., and Stevenson, K. J. 2006. Structure, composition, and chemical reactivity of carbon nanotubes by selective nitrogen doping. *Carbon* 44:1429–1437.
36. Ferrari, A. C. 2007. Raman spectroscopy of graphene and graphite: Disorder, electron–phonon coupling, doping and nonadiabatic effects. *Solid State Commun.* 143:47–57.
37. McCreery, R. L. and McDermott, M. T. 2012. Comment on electrochemical kinetics at ordered graphite electrodes. *Anal. Chem.* 84:2602–2605.
38. Robinson, D. L., Hermans, A., Seipel, A. T. et al. 2008. Monitoring rapid chemical communication in the brain. *Chem. Rev.* 108:2554–2584.
39. Koskinen, P., Malola, S., and Häkkinen, H. 2009. Evidence for graphene edges beyond zigzag and armchair. *Phys. Rev. B* 80:073401.
40. Girit, Ç., Meyer, J., Erni, R. et al. 2009. Graphene at the edge: Stability and dynamics. *Science* 666:1705–1708.
41. Suenaga, K. and Koshino, M. 2010. Atom-by-atom spectroscopy at graphene edge. *Nature* 468:1088–1090.
42. Park, S. and Ruoff, R. S. 2009. Chemical methods for the production of graphenes. *Nat. Nanotechnol.* 4:217–224.
43. Chen, X., Wu, G., Jiang, Y. et al. 2011. Graphene and graphene-based nanomaterials: The promising materials for bright future of electroanalytical chemistry. *Analyst* 136:4631–4640.
44. Brownson, D. A. C. and Banks, C. E. 2012. The electrochemistry of CVD graphene: Progress and prospects. *Phys. Chem. Chem. Phys.* 14:8264–8281.
45. Li, D., Müller, M. B., Gilje, S. et al. 2008. Processable aqueous dispersions of graphene nanosheets. *Nat. Nanotechnol.* 3:101–105.
46. Brownson, D. A. C., Kampouris, D. K., and Banks, C. E. 2012. Graphene electrochemistry: Fundamental concepts through to prominent applications. *Chem. Soc. Rev.* 41:6944–6976.
47. Ratinac, K. R., Yang, W., Gooding, J. J. et al. 2011. Graphene and related materials in electrochemical sensing. *Electroanalysis* 23:803–826.
48. Dreyer, D. R., Park, S., Bielawski, C. W. et al. 2010. The chemistry of graphene oxide. *Chem. Soc. Rev.* 39:228–240.
49. Rümmeli, M. H., Rocha, C. G., Ortmann, F. et al. 2011. Graphene: Piecing it together. *Adv. Mater.* 23:4471–4490.
50. Zhu, Y., James, D. K., and Tour, J. M. 2012. New routes to graphene, graphene oxide and their related applications. *Adv. Mater.* 24:4924–4955.
51. Chua, C. K. and Pumera, M. 2013. Covalent chemistry on graphene. *Chem. Soc. Rev.* 42:3222–3233.
52. Dreyer, D. R., Ruoff, R. S., and Bielawski, C. W. 2010. From conception to realization: An historial account of graphene and some perspectives for its future. *Angew. Chem. Int. Ed.* 49:9336–9344.
53. Saito, R., Dresselhaus, G., and Dresselhaus, M. 1998. *Physical Properties of Carbon Nanotubes.* London, U.K.: Imperial College Press.
54. Charlier, J.-C. 2002. Defects in carbon nanotubes. *Acc. Chem. Res.* 35:1063–1069.
55. Weisman, R. and Bachilo, S. 2003. Dependence of optical transition energies on structure for single-walled carbon nanotubes in aqueous suspension: An empirical Kataura plot. *Nano Lett.* 3:1235–1238.
56. Dai, H. 2002. Carbon nanotubes: Opportunities and challenges. *Surf. Sci.* 500:218–241.
57. Hirsch, A. 2009. Growth of single-walled carbon nanotubes without a metal catalyst—A surprising discovery. *Angew. Chem. Int. Ed.* 48:5403–5404.

58. Iijima, S. 1991. Helical microtubules of graphitic carbon. *Nature* 354:56–58.
59. Boehm, H. 1997. The first observation of carbon nanotubes. *Carbon* 35:581–584.
60. Monthioux, M. and Kuznetsov, V. L. 2006. Who should be given the credit for the discovery of carbon nanotubes? *Carbon* 44:1621–1623.
61. Iijima, S. and Ichihashi, T. 1993. Single-shell carbon nanotubes of 1-nm diameter. *Nature* 363: 603–605.
62. Goldsmith, B. R., Coroneus, J. G., Khalap, V. R. et al. 2007. Conductance-controlled point functionalization of single-walled carbon nanotubes. *Science* 315:77–81.
63. Heller, I., Kong, J., Williams, K. A. et al. 2006. Electrochemistry at single-walled carbon nanotubes: The role of band structure and quantum capacitance. *J. Am. Chem. Soc.* 128:7353–7359.
64. Randin, J.-P. and Yeager, E. 1971. Differential capacitance study of stress-annealed pyrolytic graphite electrodes. *J. Electrochem. Soc.* 118:711–714.
65. Randin, J.-P. and Yeager, E. 1975. Differential capacitance study on the edge orientation of pyrolytic graphite and glassy carbon electrodes. *J. Electroanal. Chem. Interfacial Electrochem.* 58:313–322.
66. Gerischer, H. 1985. An interpretation of the double layer capacity of graphite electrodes in relation to the density of states at the Fermi level. *J. Phys. Chem.* 89:4249–4251.
67. Gerischer, H. and McIntyre, R. 1987. Density of the electronic states of graphite: Derivation from differential capacitance measurements. *J. Phys. Chem.* 91:1930–1935.
68. Cline, K., McDermott, M., and McCreery, R. 1994. Anomalously slow electron transfer at ordered graphite electrodes: Influence of electronic factors and reactive sites. *J. Phys. Chem.* 98:5314–5319.
69. Robinson, R., Sternitzke, K., McDermott, M. et al. 1991. Morphology and electrochemical effects of defects on highly oriented pyrolytic graphite. *J. Electrochem. Soc.* 138:2412–2418.
70. Rice, R. J. and McCreery, R. L. 1989. Quantitative relationship between electron transfer rate and surface microstructure of laser-modified graphite electrodes. *Anal. Chem.* 61:1637–1641.
71. Panchakarla, L. S., Govindaraj, A., and Rao, C. N. R. 2010. Boron- and nitrogen-doped carbon nanotubes and graphene. *Inorg. Chim. Acta* 363:4163–4174.
72. Ayala, P., Arenal, R., Loiseau, A. et al. 2010. The physical and chemical properties of heteronanotubes. *Rev. Mod. Phys.* 82:1843–1885.
73. Panchakarla, L. S., Subrahmanyam, K. S., Saha, S. K. et al. 2009. Synthesis, structure, and properties of boron- and nitrogen-doped graphene. *Adv. Mater.* 560012:4726–4730.
74. Espejel-Morales, R. A., López-Moreno, S., Calles, A. G. et al. 2013. Structural, electronic, vibrational, and elastic properties of SWCNTs doped with B and N: An ab initio study. *Eur. Phys. J. D* 67:164.
75. Czerw, R., Terrones, M., Charlier, J. et al. 2001. Identification of electron donor states in N-doped carbon nanotubes. *Nano Lett.* 1:457–460.
76. Villalpando-Paez, F., Zamudio, A., Elias, A. L. et al. 2006. Synthesis and characterization of long strands of nitrogen-doped single-walled carbon nanotubes. *Chem. Phys. Lett.* 424:345–352.
77. Carroll, D., Redlich, P., Blase, X. et al. 1998. Effects of nanodomain formation on the electronic structure of doped carbon nanotubes. *Phys. Rev. Lett.* 81:2332–2335.
78. Ayala, P., Arenal, R., Rümmeli, M. et al. 2010. The doping of carbon nanotubes with nitrogen and their potential applications. *Carbon N.Y.* 48:575–586.
79. Maldonado, S., Morin, S., and Stevenson, K. J. 2006. Electrochemical oxidation of catecholamines and catechols at carbon nanotube electrodes. *Analyst* 131:262–267.
80. Arrigo, R., Hävecker, M., Wrabetz, S. et al. 2010. Tuning the acid/base properties of nanocarbons by functionalization via amination. *J. Am. Chem. Soc.* 132:9616–9630.
81. Van Dommele, S., Romero-Izquirdo, A., Brydson, R. et al. 2008. Tuning nitrogen functionalities in catalytically grown nitrogen-containing carbon nanotubes. *Carbon* 46:138–148.
82. Wang, Y., Alsmeyer, D., and McCreery, R. 1990. Raman spectroscopy of carbon materials: Structural basis of observed spectra. *Chem. Mater.* 2:557–563.
83. Liao, T., Sun, C., Du, A. et al. 2012. Charge carrier exchange at chemically modified graphene edges: A density functional theory study. *J. Mater. Chem.* 22:8321–8326.
84. Bard, A. J. and Faulkner, L. R. 2001. *Electrochemical Methods: Fundamentals and Applications.* New York: John Wiley & Sons.
85. Taube, H. 1970. *Electron Transfer Reactions of Complex Ions in Solution.* Orlando, FL: Academic Press.
86. Curtis, J. C. and Meyer, T. J. 1978. Outer-sphere intervalence transfer. *J. Am. Chem. Soc.* 100:6284–6286.
87. Curtis, J. C. and Meyer, T. J. 1982. Outer-sphere charge transfer in mixed-metal ion pairs. *Inorg. Chem.* 21:1562–1571.
88. Dumitrescu, I., Unwin, P. R., and Macpherson, J. V. 2009. Electrochemistry at carbon nanotubes: Perspective and issues. *Chem. Commun.* 7345:6886–6901.

89. Heras, A., Colina, A., Ruiz, V. et al. 2003. UV-visible spectroelectrochemical detection of side-reactions in the hexacyanoferrate(III)/(II) electrode process. *Electroanalysis* 15:702–708.

90. Pharr, C. M. and Griffiths, P. R. 1997. Infrared spectroelectrochemical analysis of adsorbed hexacyanoferrate species formed during potential cycling in the ferrocyanide/ferricyanide redox couple. *Anal. Chem.* 69:4673–4679.

91. Xiong, L., Batchelor-McAuley, C., Ward, K. R. et al. 2011. Voltammetry at graphite electrodes: The oxidation of hexacyanoferrate (II) (ferrocyanide) does not exhibit pure outer-sphere electron transfer kinetics and is sensitive to pre-exposure of the electrode to organic solvents. *J. Electroanal. Chem.* 661:144–149.

92. Chou, A., Bocking, T., Singh, N. K. et al. 2005. Demonstration of the importance of oxygenated species at the ends of carbon nanotubes for their favourable electrochemical properties. *Chem. Commun.* 842–844. DOI:10.1039/b415051a.

93. Banks, C. E., Ji, X., Crossley, A. et al. 2006. Understanding the electrochemical reactivity of bamboo multiwalled carbon nanotubes: The presence of oxygenated species at tube ends may not increase electron transfer kinetics. *Electroanalysis* 18:2137–2140.

94. Ji, X., Banks, C. E., Crossley, A. et al. 2006. Oxygenated edge plane sites slow the electron transfer of the ferro-/ferricyanide redox couple at graphite electrodes. *ChemPhysChem* 7:1337–1344.

95. Tsierkezos, N. G. and Ritter, U. 2012. Electrochemical and thermodynamic properties of hexacyanoferrate(II)/(III) redox system on multi-walled carbon nanotubes. *J. Chem. Thermodyn.* 54:35–40.

96. McDermott, C. A., Kneten, K. R., and McCreery, R. L. 1993. Electron transfer kinetics of aquated $Fe^{3+/2+}$, $Eu^{3+/2+}$, and $V^{3+/2+}$ at carbon electrodes: Inner sphere catalysis by surface oxides. *J. Electrochem. Soc.* 140:2593–2599.

97. Chen, P. and McCreery, R. L. 1996. Control of electron transfer kinetics at glassy carbon electrodes by specific surface modification. *Anal. Chem.* 68:3958–3965.

98. DuVall, S. H. and McCreery, R. L. 1999. Control of catechol and hydroquinone electron-transfer kinetics on native and modified glassy carbon electrodes. *Anal. Chem.* 71:4594–4602.

99. Chu, X. and Kinoshita, K. 1997. Surface modification of carbons for enhanced electrochemical activity. *Mater. Sci. Eng. B* 49:53–60.

100. Chen, P., Fryling, M., and McCreery, R. 1995. Electron transfer kinetics at modified carbon electrode surfaces: The role of specific surface sites. *Anal. Chem.* 67:3115–3122.

101. Vettorazzi, N. R., Sereno, L., Katoh, M. et al. 2008. Correlation between the distribution of oxide functional groups and electrocatalytic activity of glassy carbon surface. *J. Electrochem. Soc.* 155:F110–F115.

102. Engstrom, R. 1982. Electrochemical pretreatment of glassy carbon electrodes. *Anal. Chem.* 2314:2310–2314.

103. Engstrom, R. C. and Strasser, V. A. 1984. Characterization of electrochemically pretreated glassy carbon electrodes. *Anal. Chem.* 56:136–141.

104. Gewirth, A. and Bard, A. 1988. In situ scanning tunneling microscopy of the anodic oxidation of highly oriented pyrolytic graphite surfaces. *J. Phys. Chem.* 1:5563–5566.

105. Deakin, M., Stutts, K., and Wightman, M. 1985. The effect of pH on some outer-sphere electrode reactions at carbon electrodes. *J. Electroanal. Chem. Interfacial Electrochem.* 182:113–122.

106. Rao, C. N. R., Govindaraj, A., and Satishkumar, B. C. 1996. Functionalised carbon nanotubes from solutions. *Chem. Commun.* 3:1525–1526.

107. Kuznetsova, A., Popova, I., Yates, J. T. et al. 2001. Oxygen-containing functional groups on single-wall carbon nanotubes: NEXAFS and vibrational spectroscopic studies. *J. Am. Chem. Soc.* 123:10699–10704.

108. Liu, J., Rinzler, A., Dai, H. et al. 1998. Fullerene pipes. *Science* 280:1253–1256.

109. Dujardin, E., Ebbesen, T. W., Krishnan, A. et al. 1998. Purification of single-shell nanotubes. *Adv. Mater.* 10:611–613.

110. Ebbesen, T., Ajayan, P., Hiura, H. et al. 1994. Purification of nanotubes. *Nature* 367:519.

111. Strong, K., Anderson, D., Lafdi, K. et al. 2003. Purification process for single-wall carbon nanotubes. *Carbon* 41:1477–1488.

112. Li, Y., Zhang, X., Luo, J. et al. 2004. Purification of CVD synthesized single-wall carbon nanotubes by different acid oxidation treatments. *Nanotechnology* 15:1645–1649.

113. Chiang, I. W., Brinson, B. E., Huang, A. Y. et al. 2001. Purification and characterization of single-wall carbon nanotubes (SWNTs) obtained from the gas-phase decomposition of CO (HiPco Process). *J. Phys. Chem. B* 105:8297–8301.

114. Dumitrescu, I., Wilson, N. R., and Macpherson, J. V. 2007. Functionalizing single-walled carbon nanotube networks: Effect on electrical and electrochemical properties. *J. Phys. Chem. C* 111: 12944–12953.

115. Smith, B., Wepasnick, K., Schrote, K. E. et al. 2009. Influence of surface oxides on the colloidal stability of multi-walled carbon nanotubes: A structure-property relationship. *Langmuir* 25:9767–9776.

116. Hu, H., Bhowmik, P., Zhao, B. et al. 2001. Determination of the acidic sites of purified single-walled carbon nanotubes by acid–base titration. *Chem. Phys. Lett.* 345:6–9.

117. Luo, H., Shi, Z., Li, N. et al. 2001. Investigation of the electrochemical and electrocatalytic behavior of single-wall carbon nanotube film on a glassy carbon electrode. *Anal. Chem.* 73:915–920.

118. Karousis, N., Tagmatarchis, N., and Tasis, D. 2010. Current progress on the chemical modification of carbon nanotubes. *Chem. Rev.* 110:5366–5397.

119. Musameh, M., Lawrence, N. S., and Wang, J. 2005. Electrochemical activation of carbon nanotubes. *Electrochem. Commun.* 7:14–18.

120. Mawhinney, D. and Naumenko, V. 2000. Surface defect site density on single walled carbon nanotubes by titration. *Chem. Phys. Lett.* 324:213–216.

121. Zhao, J., Park, H., Han, J. et al. 2004. Electronic properties of carbon nanotubes with covalent sidewall functionalization. *J. Phys. Chem. B* 108:4227–4230.

122. Jang, I. Y., Muramatsu, H., Park, K. C. et al. 2009. Capacitance response of double-walled carbon nanotubes depending on surface modification. *Electrochem. Commun.* 11:719–723.

123. Cañete-Rosales, P., Ortega, V., Álvarez-Lueje, A. et al. 2012. Influence of size and oxidative treatments of multi-walled carbon nanotubes on their electrocatalytic properties. *Electrochim. Acta* 62:163–171.

124. Datsyuk, V., Kalyva, M., Papagelis, K. et al. 2008. Chemical oxidation of multiwalled carbon nanotubes. *Carbon* 46:833–840.

125. Zhou, M., Wang, Y., Zhai, Y. et al. 2009. Controlled synthesis of large-area and patterned electrochemically reduced graphene oxide films. *Chemistry* 15:6116–6120.

126. Ambrosi, A. and Pumera, M. 2013. Precise tuning of surface composition and electron-transfer properties of graphene oxide films through electroreduction. *Chemistry* 19:4748–4753.

127. Liu, F. and Xue, D. 2013. An electrochemical route to quantitative oxidation of graphene frameworks with controllable C/O ratios and added pseudocapacitances. *Chemistry* 19:10716–10722.

128. Colina, A., Ruiz, V., Heras, A. et al. 2011. Low resolution Raman spectroelectrochemistry of single walled carbon nanotube electrodes. *Electrochim. Acta* 56:1294–1299.

129. Britto, P. J., Santhanam, K. S. V., and Ajayan, P. M. 1996. Carbon nanotube electrode for oxidation of dopamine. *Bioelectrochem. Bioenerg.* 41:121–125.

130. Musameh, M., Wang, J., Merkoci, A. et al. 2002. Low-potential stable NADH detection at carbon-nanotube-modified glassy carbon electrodes. *Electrochem. Commun.* 4:743–746.

131. Wang, J., Musameh, M., and Lin, Y. 2003. Solubilization of carbon nanotubes by Nafion toward the preparation of amperometric biosensors. *J. Am. Chem. Soc.* 125:2408–2409.

132. Wang, J., Deo, R. P., Poulin, P. et al. 2003. Carbon nanotube fiber microelectrodes. *J. Am. Chem. Soc.* 125:14706–14707.

133. Wang, J. and Musameh, M. 2003. Carbon nanotube/teflon composite electrochemical sensors and biosensors. *Anal. Chem.* 75:2075–2079.

134. Chen, J., Bao, J., Cai, C. et al. 2004. Electrocatalytic oxidation of NADH at an ordered carbon nanotubes modified glassy carbon electrode. *Anal. Chim. Acta* 516:29–34.

135. Pumera, M., Ambrosi, A., Bonanni, A. et al. 2010. Graphene for electrochemical sensing and biosensing. *Trend. Anal. Chem.* 29:954–965.

136. Shao, Y., Wang, J., Wu, H. et al. 2010. Graphene based electrochemical sensors and biosensors: A review. *Electroanalysis* 22:1027–1036.

137. Lin, W.-J., Liao, C.-S., Jhang, J.-H. et al. 2009. Graphene modified basal and edge plane pyrolytic graphite electrodes for electrocatalytic oxidation of hydrogen peroxide and β-nicotinamide adenine dinucleotide. *Electrochem. Commun.* 11:2153–2156.

138. Swamy, B. E. K. and Venton, B. J. 2007. Carbon nanotube-modified microelectrodes for simultaneous detection of dopamine and serotonin in vivo. *Analyst* 132:876–884.

139. Shan, C., Yang, H., Han, D. et al. 2010. Electrochemical determination of NADH and ethanol based on ionic liquid-functionalized graphene. *Biosens. Bioelectron.* 25:1504–1508.

140. Gooding, J. J. 2005. Nanostructuring electrodes with carbon nanotubes: A review on electrochemistry and applications for sensing. *Electrochim. Acta* 50:3049–3060.

141. Ji, X., Kadara, R. O., Krussma, J. et al. 2010. Understanding the physicoelectrochemical properties of carbon nanotubes: Current state of the art. *Electroanalysis* 22:7–19.

142. Yáñez-Sedeño, P., Pingarrón, J. M., Riu, J. et al. 2010. Electrochemical sensing based on carbon nanotubes. *Trends Anal. Chem.* 29:939–953.

143. Katz, E. and Willner, I. 2004. Biomolecule-functionalized carbon nanotubes: Applications in nanobio-electronics. *ChemPhysChem* 5:1084–1104.
144. Wang, J. 2005. Carbon-nanotube based electrochemical biosensors: A review. *Electroanalysis* 17:7–14.
145. Lin, Y., Yantasee, W., and Wang, J. 2005. Carbon nanotubes (CNTs) for the development of electrochemical biosensors. *Front. Biosci.* 2:492–505.
146. Yang, W., Ratinac, K. R., Ringer, S. P. et al. 2010. Carbon nanomaterials in biosensors: Should you use nanotubes or graphene? *Angew. Chem. Int. Ed.* 49:2114–2138.
147. Kuila, T., Bose, S., Khanra, P. et al. 2011. Recent advances in graphene-based biosensors. *Biosens. Bioelectron.* 26:4637–4648.
148. Jacobs, C. B., Peairs, M. J., and Venton, B. J. 2010. Review: Carbon nanotube based electrochemical sensors for biomolecules. *Anal. Chim. Acta* 662:105–127.
149. Balasubramanian, K. and Burghard, M. 2006. Biosensors based on carbon nanotubes. *Anal. Bioanal. Chem.* 385:452–468.
150. Vashist, S. K., Zheng, D., and Al-Rubeaan, K. 2011. Advances in carbon nanotube based electrochemical sensors for bioanalytical applications. *Biotechnol. Adv.* 29:169–188.
151. Zhu, Z., Garcia-Gancedo, L., Flewitt, A. J. et al. 2012. A critical review of glucose biosensors based on carbon nanomaterials: Carbon nanotubes and graphene. *Sensors* 12:5996–6022.
152. Davis, J., Coles, R., Allen, H. et al. 1997. Protein electrochemistry at carbon nanotube electrodes. *J. Electroanal. Chem.* 0728:22–25.
153. Wang, J., Li, M., Shi, Z. et al. 2002. Direct electrochemistry of cytochrome c at a glassy carbon electrode modified with single-wall carbon nanotubes. *Anal. Chem.* 74:1993–1997.
154. Wang, G., Xu, J.-J., and Chen, H.-Y. 2002. Interfacing cytochrome c to electrodes with a DNA–carbon nanotube composite film. *Electrochem. Commun.* 4:506–509.
155. Gooding, J. J., Wibowo, R., Liu, J. et al. 2003. Protein electrochemistry using aligned carbon nanotube arrays. *J. Am. Chem. Soc.* 125:9006–9007.
156. Zhao, Y.-D., Zhang, W.-D., Chen, H. et al. 2002. Direct electrochemistry of horseradish peroxidase at carbon nanotube powder microelectrode. *Sens. Actuators, B* 87:168–172.
157. Yu, X., Chattopadhyay, D., Galeska, I. et al. 2003. Peroxidase activity of enzymes bound to the ends of single-wall carbon nanotube forest electrodes. *Electrochem. Commun.* 5:408–411.
158. Wilson, R. and Turner, A. P. F. 1992. Review article glucose oxidase: An ideal enzyme. *Biosens. Bioelectron.* 7:165–185.
159. Heller, A. and Feldman, B. 2008. Electrochemical glucose sensors and their applications in diabetes management. *Chem. Rev.* 108:2482–2505.
160. Wang, J. 2008. Electrochemical glucose biosensors. *Chem. Rev.* 108:814–825.
161. Alwarappan, S., Liu, C., Kumar, A. et al. 2010. Enzyme-doped graphene nanosheets for enhanced glucose biosensing. *J. Phys. Chem. C* 114:12920–12924.
162. Guiseppi-Elie, A., Lei, C., and Baughman, R. H. 2002. Direct electron transfer of glucose oxidase on carbon nanotubes. *Nanotechnology* 13:559–564.
163. Cai, C. and Chen, J. 2004. Direct electron transfer of glucose oxidase promoted by carbon nanotubes. *Anal. Biochem.* 332:75–83.
164. Liu, J., Chou, A., Rahmat, W. et al. 2005. Achieving direct electrical connection to glucose oxidase using aligned single walled carbon nanotube arrays. *Electroanalysis* 17:38–46.
165. Shan, C., Yang, H., Song, J. et al. 2009. Direct electrochemistry of glucose oxidase and biosensing for glucose based on graphene. *Anal. Chem.* 81:2378–2382.
166. Kang, X., Wang, J., Wu, H. et al. 2009. Glucose oxidase-graphene-chitosan modified electrode for direct electrochemistry and glucose sensing. *Biosens. Bioelectron.* 25:901–905.
167. Wang, Z., Zhou, X., Zhang, J. et al. 2009. Direct electrochemical reduction of single-layer graphene oxide and subsequent functionalization with glucose oxidase. *J. Phys. Chem. C* 113:14071–14075.
168. Fu, C., Yang, W., Chen, X. et al. 2009. Direct electrochemistry of glucose oxidase on a graphite nanosheet–Nafion composite film modified electrode. *Electrochem. Commun.* 11:997–1000.
169. Wu, P., Shao, Q., Hu, Y. et al. 2010. Direct electrochemistry of glucose oxidase assembled on graphene and application to glucose detection. *Electrochim. Acta* 55:8606–8614.
170. Wang, Y. and Yao, Y. 2012. Direct electron transfer of glucose oxidase promoted by carbon nanotubes is without value in certain mediator-free applications. *Microchim. Acta* 176:271–277.
171. Goran, J. M., Mantilla, S. M., and Stevenson, K. J. 2013. Influence of surface adsorption on the interfacial electron transfer of flavin adenine dinucleotide and glucose oxidase at carbon nanotube and nitrogen-doped carbon nanotube electrodes. *Anal. Chem.* 85:1571–1581.

172. Goran, J. M. and Stevenson, K. J. 2013. Electrochemical behavior of flavin adenine dinucleotide adsorbed onto carbon nanotube and nitrogen-doped carbon nanotube electrodes. *Langmuir* 29:13605–13613.

173. Patolsky, F., Weizmann, Y., and Willner, I. 2004. Long-range electrical contacting of redox enzymes by SWCNT connectors. *Angew. Chem. Int. Ed.* 43:2113–2117.

174. Xiao, Y., Patolsky, F., Katz, E. et al. 2003. Plugging into enzymes: Nanowiring of redox enzymes by a gold nanoparticle. *Science* 299:1877–1881.

175. Zayats, M., Willner, B., and Willner, I. 2008. Design of amperometric biosensors and biofuel cells by the reconstitution of electrically contacted enzyme electrodes. *Electroanalysis* 20:583–601.

176. Gooding, J. J., Chou, A., Liu, J. et al. 2007. The effects of the lengths and orientations of single-walled carbon nanotubes on the electrochemistry of nanotube-modified electrodes. *Electrochem. Commun.* 9:1677–1683.

177. Zhou, M., Zhai, Y., and Dong, S. 2009. Electrochemical sensing and biosensing platform based on chemically reduced graphene oxide. *Anal. Chem.* 81:5603–5613.

178. Tominaga, M., Iwaoka, A., Kawai, D. et al. 2013. Correlation between carbon oxygenated species of SWCNTs and the electrochemical oxidation reaction of NADH. *Electrochem. Commun.* 31:76–79.

179. Pumera, M., Scipioni, R., Iwai, H. et al. 2009. A mechanism of adsorption of beta-nicotinamide adenine dinucleotide on graphene sheets: Experiment and theory. *Chemistry* 15:10851–10856.

180. Lawrence, N., Deo, R., and Wang, J. 2005. Comparison of the electrochemical reactivity of electrodes modified with carbon nanotubes from different sources. *Electroanalysis* 17:65–72.

181. Wooten, M. and Gorski, W. 2010. Facilitation of NADH electro-oxidation at treated carbon nanotubes. *Anal. Chem.* 82:1299–1304.

182. Zhang, Y., Wu, C., Zhou, X. et al. 2013. Graphene quantum dots/gold electrode and its application in living cell H2O2 detection. *Nanoscale* 5:1816–1819.

183. Alwarappan, S., Erdem, A., Liu, C. et al. 2009. Probing the electrochemical properties of graphene nanosheets for biosensing applications. *J. Phys. Chem. C* 113:8853–8857.

184. Wang, J., Yang, S., Guo, D. et al. 2009. Comparative studies on electrochemical activity of graphene nanosheets and carbon nanotubes. *Electrochem. Commun.* 11:1892–1895.

185. Tang, L., Wang, Y., Li, Y. et al. 2009. Preparation, structure, and electrochemical properties of reduced graphene sheet films. *Adv. Funct. Mater.* 19:2782–2789.

186. Goh, M. S. and Pumera, M. 2012. Number of graphene layers exhibiting an influence on oxidation of DNA bases: Analytical parameters. *Anal. Chim. Acta* 711:29–31.

187. Goh, M. and Pumera, M. 2010. Single-, few-, and multilayer graphene not exhibiting significant advantages over graphite microparticles in electroanalysis. *Anal. Chem.* 82:8367–8370.

188. Scott, C. L. and Pumera, M. 2011. Electroanalytical parameters of carbon nanotubes are inferior with respect to well defined surfaces of glassy carbon and EPPG. *Electrochem. Commun.* 13:213–216.

189. Nejadnik, M. R., Deepak, F. L., and Garcia, C. D. 2011. Adsorption of glucose oxidase to 3-D scaffolds of carbon nanotubes: Analytical applications. *Electroanalysis* 23:1462–1469.

190. Felhofer, J. L., Caranto, J. D., and Garcia, C. D. 2010. Adsorption kinetics of catalase to thin films of carbon nanotubes. *Langmuir* 26:17178–17183.

191. Li, E. Y. and Marzari, N. 2011. Improving the electrical conductivity of carbon nanotube networks: A first-principles study. *ACS Nano* 5:9726–9736.

192. Banks, C. E., Davies, T. J., Wildgoose, G. G. et al. 2005. Electrocatalysis at graphite and carbon nanotube modified electrodes: Edge-plane sites and tube ends are the reactive sites. *Chem. Commun.* 829–841.

193. Banks, C. E., Moore, R. R., Davies, T. J. et al. 2004. Investigation of modified basal plane pyrolytic graphite electrodes: Definitive evidence for the electrocatalytic properties of the ends of carbon nanotubes. *Chem. Commun.* 1804–1805.

194. Banks, C. E. and Compton, R. G. 2006. New electrodes from old: From carbon nanotubes to edge plane pyrolytic graphite. *Analyst* 131:15–21.

195. Banks, C. E. and Compton, R. G. 2005. Exploring the electrocatalytic sites of carbon nanotubes for NADH detection: An edge plane pyrolytic graphite electrode study. *Analyst* 130:1232–1239.

196. Moore, R. R., Banks, C. E., and Compton, R. G. 2004. Basal plane pyrolytic graphite modified electrodes: Comparison of carbon nanotubes and graphite powder as electrocatalysts. *Anal. Chem.* 76:2677–2682.

197. Poon, M. and McCreery, R. 1986. In situ laser activation of glassy carbon electrodes. *Anal. Chem.* 58:2745–2750.

198. Poon, M., McCreery, R. L., and Engstrom, R. 1988. Laser activation of carbon electrodes. Relationship between laser-induced surface effects and electron transfer activation. *Anal. Chem.* 60:1725–1730.

199. Rice, R., Pontikos, N., and McCreery, R. 1990. Quantitative correlations of heterogeneous electron-transfer kinetics with surface properties of glassy carbon. *J. Am. Chem. Soc.* 112:4617–4622.

200. McDermott, M. T., McDermott, C. A., and McCreery, R. L. 1993. Scanning tunneling microscopy of carbon surfaces: Relationships between electrode kinetics, capacitance, and morphology for glassy carbon electrodes. *Anal. Chem.* 937–944.
201. Bowling, R., Packard, R., and McCreery, R. 1988. Raman spectroscopy of carbon electrodes correlation between defect density and heterogeneous electron transfer rate. *J. Electrochem. Soc.* 135:1605–1606.
202. Kneten, K. and McCreery, R. 1992. Effects of redox system structure on electron-transfer kinetics at ordered graphite and glassy carbon electrodes. *Anal. Chem.* 2518–2524.
203. Yuan, W., Zhou, Y., Li, Y. et al. 2013. The edge- and basal-plane-specific electrochemistry of a single-layer graphene sheet. *Sci. Rep.* 3:1–7.
204. Ambrosi, A., Bonanni, A., and Pumera, M. 2011. Electrochemistry of folded graphene edges. *Nanoscale* 3:2256–2260.
205. Gong, K., Chakrabarti, S., and Dai, L. 2008. Electrochemistry at carbon nanotube electrodes: Is the nanotube tip more active than the sidewall? *Angew. Chem. Int. Ed.* 47:5446–5450.
206. Tan, C., Rodríguez-López, J., Parks, J. J. et al. 2012. Reactivity of monolayer chemical vapor deposited graphene imperfections studied using scanning electrochemical microscopy. *ACS Nano* 6:3070–3079.
207. Lim, C. X., Hoh, H. Y., Ang, P. K. et al. 2010. Direct voltammetric detection of DNA and pH sensing on epitaxial graphene: An insight into the role of oxygenated defects. *Anal. Chem.* 82:7387–7393.
208. Shang, N. G., Papakonstantinou, P., McMullan, M. et al. 2008. Catalyst-free efficient growth, orientation and biosensing properties of multilayer graphene nanoflake films with sharp edge planes. *Adv. Funct. Mater.* 18:3506–3514.
209. Valota, A. T., Kinloch, I. A., Novoselov, K. S. et al. 2011. Electrochemical behavior of monolayer and bilayer graphene. *ACS Nano* 5:8809–8815.
210. Sharma, R., Baik, J. H., Perera, C. J. et al. 2010. Anomalously large reactivity of single graphene layers and edges toward electron transfer chemistries. *Nano Lett.* 10:398–405.
211. Joshi, V. S., Haram, S. K., Dasgupta, A. et al. 2012. Mapping of electrocatalytic sites on a single strand of carbon fiber using scanning electrochemical microscopy (SECM). *J. Phys. Chem. C* 116:9703–9708.
212. Güell, A. G., Ebejer, N., Snowden, M. E. et al. 2012. Quantitative nanoscale visualization of heterogeneous electron transfer rates in 2D carbon nanotube networks. *Proc. Natl. Acad. Sci. U. S. A.* 109:11487–11492.
213. Kim, J., Xiong, H., and Hofmann, M. 2010. Scanning electrochemical microscopy of individual single-walled carbon nanotubes. *Anal. Chem.* 82:1605–1607.
214. Snowden, M. E., Edwards, M. A., Rudd, N. C. et al. 2013. Intrinsic electrochemical activity of single walled carbon nanotube-Nafion assemblies. *Phys. Chem. Chem. Phys.* 15:5030–5038.
215. Dumitrescu, I., Dudin, P. V., Edgeworth, J. P. et al. 2010. Electron transfer kinetics at single-walled carbon nanotube electrodes using scanning electrochemical microscopy. *J. Phys. Chem. C* 114:2633–2639.
216. Miller, T. S., Ebejer, N., Güell, A. G. et al. 2012. Electrochemistry at carbon nanotube forests: Sidewalls and closed ends allow fast electron transfer. *Chem. Commun.* 48:7435–7437.
217. Güell, A. G., Ebejer, N., Snowden, M. E. et al. 2012. Structural correlations in heterogeneous electron transfer at monolayer and multilayer graphene electrodes. *J. Am. Chem. Soc.* 134:7258–7261.
218. Patel, A. N., Collignon, M. G., O'Connell, M. A. et al. 2012. A new view of electrochemistry at highly oriented pyrolytic graphite. *J. Am. Chem. Soc.* 134:20117–20130.
219. Patel, A. N., McKelvey, K., and Unwin, P. R. 2012. Nanoscale electrochemical patterning reveals the active sites for catechol oxidation at graphite surfaces. *J. Am. Chem. Soc.* 134:20246–20249.
220. Lai, S. C. S., Patel, A. N., McKelvey, K. et al. 2012. Definitive evidence for fast electron transfer at pristine basal plane graphite from high-resolution electrochemical imaging. *Angew. Chem. Int. Ed.* 51:5405–5408.
221. Edwards, M. A., Bertoncello, P., and Unwin, P. R. 2009. Slow diffusion reveals the intrinsic electrochemical activity of basal plane highly oriented pyrolytic graphite electrodes. *J. Phys. Chem. C* 9218–9223.
222. Sljukić, B., Banks, C. E., and Compton, R. G. 2006. Iron oxide particles are the active sites for hydrogen peroxide sensing at multiwalled carbon nanotube modified electrodes. *Nano Lett.* 6:1556–1558.
223. Banks, C. E., Crossley, A., Salter, C. et al. 2006. Carbon nanotubes contain metal impurities which are responsible for the "electrocatalysis" seen at some nanotube-modified electrodes. *Angew. Chem. Int. Ed.* 45:2533–2537.
224. Azevedo, S., Chesman, C., and Kaschny, J. R. 2010. Stability and electronic properties of carbon nanotubes doped with transition metal impurities. *Eur. Phys. J. B* 74:123–128.
225. Jurkschat, K., Ji, X., Crossley, A. et al. 2007. Super-washing does not leave single walled carbon nanotubes iron-free. *Analyst* 132:21–23.
226. Ambrosi, A., Chua, C. K., Khezri, B. et al. 2012. Chemically reduced graphene contains inherent metallic impurities present in parent natural and synthetic graphite. *Proc. Natl. Acad. Sci. U. S. A.* 109:12899–12904.

227. Pumera, M. 2007. Carbon nanotubes contain residual metal catalyst nanoparticles even after washing with nitric acid at elevated temperature because these metal nanoparticles are sheathed by several graphene sheets. *Langmuir* 23:6453–6458.

228. Lyon, J. L. and Stevenson, K. J. 2007. Anomalous electrochemical dissolution and passivation of iron growth catalysts in carbon nanotubes. *Langmuir* 23:11311–11318.

229. Wong, C. H. A., Chua, C. K., Khezri, B. et al. 2013. Graphene oxide nanoribbons from the oxidative opening of carbon nanotubes retain electrochemically active metallic impurities. *Angew. Chem.* 125:8847–8850.

230. Toh, R. J., Ambrosi, A., and Pumera, M. 2012. Bioavailability of metallic impurities in carbon nanotubes is greatly enhanced by ultrasonication. *Chemistry* 18:11593–11596.

231. Jones, C. P., Jurkschat, K., Crossley, A. et al. 2007. Use of high-purity metal-catalyst-free multiwalled carbon nanotubes to avoid potential experimental misinterpretations. *Langmuir* 23:9501–9504.

232. Pumera, M. and Iwai, H. 2009. Multicomponent metallic impurities and their influence upon the electrochemistry of carbon nanotubes. *J. Phys. Chem. C* 113:4401–4405.

233. Wang, L., Ambrosi, A., and Pumera, M. 2013. Carbonaceous impurities in carbon nanotubes are responsible for accelerated electrochemistry of acetaminophen. *Electrochem. Commun.* 26:71–73.

234. Pumera, M., Ambrosi, A., and Chng, E. L. K. 2012. Impurities in graphenes and carbon nanotubes and their influence on the redox properties. *Chem. Sci.* 3:3347–3355.

235. Stuart, E. J. E. and Pumera, M. 2011. Signal transducers and enzyme cofactors are susceptible to oxidation by nanographite impurities in carbon nanotube materials. *Chemistry.* 17:5544–5548.

236. Scott, C. L. and Pumera, M. 2011. Carbon nanotubes can exhibit negative effects in electroanalysis due to presence of nanographite impurities. *Electrochem. Commun.* 13:426–428.

237. Henstridge, M. C., Shao, L., Wildgoose, G. G. et al. 2008. The electrocatalytic properties of arc-MWCNTs and associated "carbon onions." *Electroanalysis* 20:498–506.

238. Breczko, J., Plonska-Brzezinska, M. E., and Echegoyen, L. 2012. Electrochemical oxidation and determination of dopamine in the presence of uric and ascorbic acids using a carbon nano-onion and poly(diallyldimethylammonium chloride) composite. *Electrochim. Acta* 72:61–67.

239. Li, X., Yang, X., Jia, L. et al. 2012. Carbonaceous debris that resided in graphene oxide/reduced graphene oxide profoundly affect their electrochemical behaviors. *Electrochem. Commun.* 23:94–97.

240. Patten, H. V., Lai, S. C. S., Macpherson, J. V. et al. 2012. Active sites for outer-sphere, inner-sphere, and complex multistage electrochemical reactions at polycrystalline boron-doped diamond electrodes (pBDD) revealed with scanning electrochemical cell microscopy (SECCM). *Anal. Chem.* 84:5427–5432.

241. Holt, K. B., Bard, A. J., Show, Y. et al. 2004. Scanning electrochemical microscopy and conductive probe atomic force microscopy studies of hydrogen-terminated boron-doped diamond electrodes with different doping levels. *J. Phys. Chem. B* 108:15117–15127.

242. Szunerits, S., Mermoux, M., Crisci, A. et al. 2006. Raman imaging and Kelvin probe microscopy for the examination of the heterogeneity of doping in polycrystalline boron-doped diamond electrodes. *J. Phys. Chem. B* 110:23888–23897.

243. Wilson, N. R., Clewes, S. L., Newton, M. E. et al. 2006. Impact of grain-dependent boron uptake on the electrochemical and electrical properties of polycrystalline boron doped diamond electrodes. *J. Phys. Chem. B* 110:5639–5646.

244. Patten, H. V, Meadows, K. E., Hutton, L. A. et al. 2012. Electrochemical mapping reveals direct correlation between heterogeneous electron-transfer kinetics and local density of states in diamond electrodes. *Angew. Chem. Int. Ed.* 51:7002–7006.

245. Holt, K. B., Ziegler, C., Zang, J. et al. 2009. Scanning electrochemical microscopy studies of redox processes at undoped nanodiamond surfaces. *J. Phys. Chem. C* 113:2761–2770.

246. Zhao, L., He, R., Rim, K. T. et al. 2011. Visualizing individual nitrogen dopants in monolayer graphene. *Science* 333:999–1003.

247. Zhou, J., Wang, J., Liu, H. et al. 2010. Imaging nitrogen in individual carbon nanotubes. *J. Phys. Chem. Lett.* 1:1709–1713.

248. Florea, I., Ersen, O., Arenal, R. et al. 2012. 3D analysis of the morphology and spatial distribution of nitrogen in nitrogen-doped carbon nanotubes by energy-filtered transmission electron microscopy tomography. *J. Am. Chem. Soc.* 134:9672–9680.

249. Biddinger, E. J., Deak, D., and Ozkan, U. S. 2009. Nitrogen-containing carbon nanostructures as oxygen-reduction catalysts. *Top. Catal.* 52:1566–1574.

250. Li, X., Wang, H., Robinson, J. T. et al. 2009. Simultaneous nitrogen doping and reduction of graphene oxide. *J. Am. Chem. Soc.* 131:15939–15944.

251. Tang, Y., Allen, B. L., Kauffman, D. R. et al. 2009. Electrocatalytic activity of nitrogen-doped carbon nanotube cups. *J. Am. Chem. Soc.* 131:13200–13201.
252. Xu, X., Jiang, S., Hu, Z. et al. 2010. Nitrogen-doped carbon nanotubes: High electrocatalytic activity toward the oxidation of hydrogen peroxide and its application for biosensing. *ACS Nano* 4:4292–4298.
253. Wang, Y., Shao, Y., Matson, D. et al. 2010. Nitrogen-doped graphene and its application in electrochemical biosensing. *ACS Nano* 4:1790–1798.
254. Goran, J. M., Favela, C. A., and Stevenson, K. J. 2013. Electrochemical oxidation of dihydronicotinamide adenine dinucleotide at nitrogen-doped carbon nanotube electrodes. *Anal. Chem.* 85:9135–9141.
255. Wu, P., Qian, Y., Du, P. et al. 2012. Facile synthesis of nitrogen-doped graphene for measuring the releasing process of hydrogen peroxide from living cells. *J. Mater. Chem.* 22:6402–6412.
256. Deng, C., Chen, J., Chen, X. et al. 2008. Boron-doped carbon nanotubes modified electrode for electroanalysis of NADH. *Electrochem. Commun.* 10:907–909.
257. Goran, J. M., Lyon, J. L., and Stevenson, K. J. 2011. Amperometric detection of L-lactate using nitrogen-doped carbon nanotubes modified with lactate oxidase. *Anal. Chem.* 83:8123–8129.
258. Jia, N., Liu, L., Zhou, Q. et al. 2005. Bioelectrochemistry and enzymatic activity of glucose oxidase immobilized onto the bamboo-shaped CNx nanotubes. *Electrochim. Acta* 51:611–618.
259. Deng, S., Jian, G., Lei, J. et al. 2009. A glucose biosensor based on direct electrochemistry of glucose oxidase immobilized on nitrogen-doped carbon nanotubes. *Biosens. Bioelectron.* 25:373–377.
260. Deng, C., Chen, J., Chen, X. et al. 2008. Direct electrochemistry of glucose oxidase and biosensing for glucose based on boron-doped carbon nanotubes modified electrode. *Biosens. Bioelectron.* 23:1272–1277.
261. Carrero-Sanchez, J. C., Elías, A. L., Mancilla, R. et al. 2006. Biocompatibility and toxicological studies of carbon nanotubes doped with nitrogen. *Nano Lett.* 6:1609–1616.
262. Burch, H. J., Contera, S. A., de Planque, M. R. R. et al. 2008. Doping of carbon nanotubes with nitrogen improves protein coverage whilst retaining correct conformation. *Nanotechnology* 19:384001.
263. Wiggins-Camacho, J. D. and Stevenson, K. J. 2009. Effect of nitrogen concentration on capacitance, density of states, electronic conductivity, and morphology of N-doped carbon nanotube electrodes. *J. Phys. Chem. C* 113:19082–19090.
264. Tan, S. M., Poh, H. L., Sofer, Z. et al. 2013. Boron-doped graphene and boron-doped diamond electrodes: Detection of biomarkers and resistance to fouling. *Analyst* 138:4885–4891.
265. Gai, P., Zhang, H., Zhang, Y. et al. 2013. Simultaneous electrochemical detection of ascorbic acid, dopamine and uric acid based on nitrogen doped porous carbon nanopolyhedra. *J. Mater. Chem. B* 1:2742–2749.
266. Li, Y., Yao, M., Li, T.-T. et al. 2013. Simultaneous electrochemical determination of uric acid and dopamine in the presence of ascorbic acid using nitrogen-doped carbon hollow spheres. *Anal. Methods* 5:3635–3638.
267. Wu, P., Cai, Z., Gao, Y. et al. 2011. Enhancing the electrochemical reduction of hydrogen peroxide based on nitrogen-doped graphene for measurement of its releasing process from living cells. *Chem. Commun.* 47:11327–11329.
268. Gong, K., Du, F., Xia, Z. et al. 2009. Nitrogen-doped carbon nanotube arrays with high electrocatalytic activity for oxygen reduction. *Science* 323:760–764.
269. Qu, L., Liu, Y., Baek, J.-B. et al. 2010. Nitrogen-doped graphene as efficient metal-free electrocatalyst for oxygen reduction in fuel cells. *ACS Nano* 4:1321–1326.
270. Chen, Z., Higgins, D., and Chen, Z. 2010. Nitrogen doped carbon nanotubes and their impact on the oxygen reduction reaction in fuel cells. *Carbon N.Y.* 48:3057–3065.
271. Alexeyeva, N., Shulga, E., Kisand, V. et al. 2010. Electroreduction of oxygen on nitrogen-doped carbon nanotube modified glassy carbon electrodes in acid and alkaline solutions. *J. Electroanal. Chem.* 648:169–175.
272. Byon, H. R., Suntivich, J., and Shao-Horn, Y. 2011. Graphene-based non-noble-metal catalysts for oxygen reduction reaction in acid. *Chem. Mater.* 23:3421–3428.
273. Wong, W. Y., Daud, W. R. W., Mohamad, A. B. et al. 2012. Nitrogen-containing carbon nanotubes as cathodic catalysts for proton exchange membrane fuel cells. *Diam. Relat. Mater.* 22:12–22.
274. Yang, L., Jiang, S., Zhao, Y. et al. 2011. Boron-doped carbon nanotubes as metal-free electrocatalysts for the oxygen reduction reaction. *Angew. Chem. Int. Ed.* 50:7132–7135.
275. Wu, G. and Zelenay, P. 2013. Nanostructured nonprecious metal catalysts for oxygen reduction reaction. *Acc. Chem. Res.* 46:1878–1889.
276. Yang, Z., Nie, H., Chen, X. et al. 2013. Recent progress in doped carbon nanomaterials as effective cathode catalysts for fuel cell oxygen reduction reaction. *J. Power Sources* 236:238–249.

277. Hu, X., Wu, Y., Li, H. et al. 2010. Adsorption and activation of O_2 on nitrogen-doped carbon nanotubes. *J. Phys. Chem. C* 114:9603–9607.
278. Chen, Z., Higgins, D., and Chen, Z. 2010. Electrocatalytic activity of nitrogen doped carbon nanotubes with different morphologies for oxygen reduction reaction. *Electrochim. Acta* 55:4799–4804.
279. Biddinger, E. J. and Ozkan, U. S. 2010. Role of graphitic edge plane exposure in carbon nanostructures for oxygen reduction reaction. *J. Phys. Chem. C* 114:15306–15314.
280. Sharifi, T., Hu, G., Jia, X. et al. 2012. Formation of active sites for oxygen reduction reactions by transformation of nitrogen functionalities in nitrogen-doped carbon nanotubes. *ACS Nano* 6:8904–8912.
281. Rao, C. V., Cabrera, C. R., and Ishikawa, Y. 2010. In search of the active site in nitrogen-doped carbon nanotube electrodes for the oxygen reduction reaction. *J. Phys. Chem. Lett.* 1:2622–2627.
282. Matter, P., Zhang, L., and Ozkan, U. 2006. The role of nanostructure in nitrogen-containing carbon catalysts for the oxygen reduction reaction. *J. Catal.* 239:83–96.
283. Lee, K. R., Lee, K. U., Lee, J. W. et al. 2010. Electrochemical oxygen reduction on nitrogen doped graphene sheets in acid media. *Electrochem. Commun.* 12:1052–1055.
284. Zhang, P., Lian, J. S., and Jiang, Q. 2012. Potential dependent and structural selectivity of the oxygen reduction reaction on nitrogen-doped carbon nanotubes: A density functional theory study. *Phys. Chem. Chem. Phys.* 14:11715–11723.
285. Lai, L., Potts, J. R., Zhan, D. et al. 2012. Exploration of the active center structure of nitrogen-doped graphene-based catalysts for oxygen reduction reaction. *Energy Environ. Sci.* 5:7936–7942.
286. Kim, H., Lee, K., Woo, S. I. et al. 2011. On the mechanism of enhanced oxygen reduction reaction in nitrogen-doped graphene nanoribbons. *Phys. Chem. Chem. Phys.* 13:17505–17510.
287. Shanmugam, S. and Osaka, T. 2011. Efficient electrocatalytic oxygen reduction over metal free-nitrogen doped carbon nanocapsules. *Chem. Commun.* 47:4463–4465.
288. Wu, G., Nelson, M., Ma, S. et al. 2011. Synthesis of nitrogen-doped onion-like carbon and its use in carbon-based CoFe binary non-precious-metal catalysts for oxygen-reduction. *Carbon* 49:3972–3982.
289. Yu, D., Wei, L., Jiang, W. et al. 2013. Nitrogen doped holey graphene as an efficient metal-free multifunctional electrochemical catalyst for hydrazine oxidation and oxygen reduction. *Nanoscale* 5:3457–3464.
290. Li, Q., Zhang, S., Dai, L. et al. 2012. Nitrogen-doped colloidal graphene quantum dots and their size-dependent electrocatalytic activity for the oxygen reduction reaction. *J. Am. Chem. Soc.* 134:18932–18935.
291. Wang, S., Zhang, L., Xia, Z. et al. 2012. BCN graphene as efficient metal-free electrocatalyst for the oxygen reduction reaction. *Angew. Chem. Int. Ed.* 51:4209–4212.
292. Wang, S., Iyyamperumal, E., Roy, A. et al. 2011. Vertically aligned BCN nanotubes as efficient metal-free electrocatalysts for the oxygen reduction reaction: A synergetic effect by co-doping with boron and nitrogen. *Angew. Chem. Int. Ed.* 50:11756–11760.
293. Luo, Z., Lim, S., Tian, Z. et al. 2011. Pyridinic N doped graphene: Synthesis, electronic structure, and electrocatalytic property. *J. Mater. Chem.* 21:8038–8044.
294. Kurak, K. A. and Anderson, A. B. 2009. Nitrogen-treated graphite and oxygen electroreduction on pyridinic edge sites. *J. Phys. Chem. C* 113:6730–6734.
295. Bao, X., Nie, X., Deak, D. et al. 2013. A first-principles study of the role of quaternary-N doping on the oxygen reduction reaction activity and selectivity of graphene edge sites. *Top. Catal.* 56:1623–1633.
296. Wang, P., Wang, Z., Jia, L. et al. 2009. Origin of the catalytic activity of graphite nitride for the electrochemical reduction of oxygen: Geometric factors vs. electronic factors. *Phys. Chem. Chem. Phys.* 11:2730–2740.
297. Parvez, K., Yang, S., Hernandez, Y. et al. 2012. Nitrogen-doped graphene and its iron-based composite as efficient electrocatalysts for oxygen reduction reaction. *ACS Nano* 6:9541–9550.
298. Wu, G., Johnston, C. M., Mack, N. H. et al. 2011. Synthesis–structure–performance correlation for polyaniline–Me–C non-precious metal cathode catalysts for oxygen reduction in fuel cells. *J. Mater. Chem.* 21:11392–11405.
299. Maldonado, S. and Stevenson, K. J. 2005. Influence of nitrogen doping on oxygen reduction electrocatalysis at carbon nanofiber electrodes. *J. Phys. Chem. B* 109:4707–4716.
300. Maldonado, S. and Stevenson, K. J. 2004. Direct preparation of carbon nanofiber electrodes via pyrolysis of iron(II) phthalocyanine: Electrocatalytic aspects for oxygen reduction. *J. Phys. Chem. B* 108:11375–11383.
301. Wiggins-Camacho, J. D. and Stevenson, K. J. 2011. Mechanistic discussion of the oxygen reduction reaction at nitrogen-doped carbon nanotubes. *J. Phys. Chem. C* 115:20002–20010.
302. Titov, A., Zapol, P., Kra, P. et al. 2009. Catalytic Fe-xN sites in carbon nanotubes. *J. Phys. Chem. C* 113:21629–21634.

303. Wiggins-Camacho, J. D. and Stevenson, K. J. 2011. Indirect electrocatalytic degradation of cyanide at nitrogen-doped carbon nanotube electrodes. *Environ. Sci. Technol.* 45:3650–3656.
304. Lyon, J. L. and Stevenson, K. J. 2009. Peroxidase mimetic activity at tailored nanocarbon electrodes. *ECS Trans.* 16:1–12.
305. Wang, J. 2006. *Analytical Electrochemistry*. Hoboken, NJ: John Wiley & Sons.
306. Watkins, J., Zhang, B., and White, H. 2005. Electrochemistry at nanometer-scaled electrodes. *J. Chem. Ed.* 82:712–719.
307. Cox, J. T. and Zhang, B. 2012. Nanoelectrodes: Recent advances and new directions. *Annu. Rev. Anal. Chem.* 5:253–272.
308. Oja, S. M., Wood, M., and Zhang, B. 2013. Nanoscale electrochemistry. *Anal. Chem.* 85:473–486.
309. Murray, R. W. 2008. Nanoelectrochemistry: Metal nanoparticles, nanoelectrodes, and nanopores. *Chem. Rev.* 108:2688–2720.
310. Bard, A. J. and Murray, R. W. 2012. Electrochemistry. *Proc. Natl. Acad. Sci. U. S. A.* 109:11484–11486.
311. Amemiya, S., Bard, A. J., Fan, F.-R. F. et al. 2008. Scanning electrochemical microscopy. *Annu. Rev. Anal. Chem.* 1:95–131.
312. Huang, Y., Cai, D., and Chen, P. 2011. Micro- and nanotechnologies for study of cell secretion. *Anal. Chem.* 83:4393–4406.
313. Sun, P., Laforge, F. O., Abeyweera, T. P. et al. 2008. Nanoelectrochemistry of mammalian cells. *Proc. Natl. Acad. Sci. U. S. A.* 105:443–448.
314. Wehmeyer, K. R., Deakin, M. R., and Wightman, R. M. 1985. Electroanalytical properties of band electrodes of submicrometer width. *Anal. Chem.* 57:1913–1916.
315. Penner, R. M., Heben, M. J., Longin, T. L. et al. 1990. Fabrication and use of nanometer-sized electrodes in electrochemistry. *Science* 250:1118–1121.
316. Oldham, K. 1992. A hole can serve as a microelectrode. *Anal. Chem.* 64:646–651.
317. Mirkin, M. V., Fan, F.-R. F., and Bard, A. J. 1992. Scanning electrochemical microscopy part 13. Evaluation of the tip shapes of nanometer size microelectrodes. *J. Electroanal. Chem.* 328:47–62.
318. Sun, P. and Mirkin, M. V 2006. Kinetics of electron-transfer reactions at nanoelectrodes. *Anal. Chem.* 78:6526–6534.
319. Norton, J., White, H., and Feldberg, S. 1990. Effect of the electrical double layer on voltammetry at microelectrodes. *J. Phys. Chem.* 281:6772–6780.
320. Dickinson, E. J. F. and Compton, R. G. 2011. Influence of the diffuse double layer on steady-state voltammetry. *J. Electroanal. Chem.* 661:198–212.
321. Conyers, J. and White, H. 2000. Electrochemical characterization of electrodes with submicrometer dimensions. *Anal. Chem.* 72:4441–4446.
322. Chen, S. and Kucernak, A. 2002. Fabrication of carbon microelectrodes with an effective radius of 1 nm. *Electrochem. Commun.* 4:80–85.
323. Chen, S. and Kucernak, A. 2002. The voltammetric response of nanometer-sized carbon electrodes. *J. Phys. Chem. B* 106:9396–9404.
324. Watkins, J. J. and White, H. S. 2004. The role of the electrical double layer and ion pairing on the electrochemical oxidation of hexachloroiridate(III) at Pt electrodes of nanometer dimensions. *Langmuir* 20:5474–5483.
325. Morris, R., Franta, D. and White, H. 1987. Electrochemistry at platinum bane electrodes of width approaching molecular dimensions: Breakdown of transport equations at very small electrodes. *J. Phys. Chem.* 3559–3564.
326. Krapf, D., Quinn, B. M., Wu, M.-Y. et al. 2006. Experimental observation of nonlinear ionic transport at the nanometer scale. *Nano Lett.* 6:2531–2535.
327. Amatore, C., Fosset, B., Bartelt, J. et al. 1988. Electrochemical kinetics at microelectrodes: Part V. Migrational effects on steady or quasi-steady-state voltammograms. *J. Electroanal. Chem. Interfacial Electrochem.* 256:255–268.
328. Smith, C. and White, H. 1993. Theory of the voltammetric response of electrodes of submicron dimensions. Violation of electroneutrality in the presence of excess supporting electrolyte. *Anal. Chem.* 65:3343–3353.
329. Dickinson, E. J. F., Limon-Petersen, J. G., and Compton, R. G. 2011. The electroneutrality approximation in electrochemistry. *J. Solid State Electrochem.* 15:1335–1345.
330. He, R., Chen, S., Yang, F. et al. 2006. Dynamic diffuse double-layer model for the electrochemistry of nanometer-sized electrodes. *J. Phys. Chem. B* 110:3262–3270.
331. Feldberg, S. W. 2010. Implications of Marcus-Hush theory for steady-state heterogeneous electron transfer at an inlaid disk electrode. *Anal. Chem.* 82:5176–5183.

332. Liu, Y. and Chen, S. 2012. Theory of interfacial electron transfer kinetics at nanometer-sized electrodes. *J. Phys. Chem. C* 116:13594–13602.

333. Liu, Y., He, R., Zhang, Q. et al. 2010. Theory of electrochemistry for nanometer-sized disk electrodes. *J. Phys. Chem. C* 114:10812–10822.

334. Liu, Y., Zhang, Q., and Chen, S. 2010. The voltammetric responses of nanometer-sized electrodes in weakly supported electrolyte: A theoretical study. *Electrochim. Acta* 55:8280–8286.

335. Campbell, J., Sun, L., and Crooks, R. 1999. Electrochemistry using single carbon nanotubes. *J. Am. Chem. Soc.* 3779–3780.

336. Kawano, T., Cho, C. Y., and Lin, L. 2012. An overhanging carbon nanotube/parylene core–shell nanoprobe electrode. *Sens. Actuators, A* 187:79–83.

337. Heller, I., Kong, J., Heering, H. A. et al. 2005. Individual single-walled carbon nanotubes as nanoelectrodes for electrochemistry. *Nano Lett.* 5:137–142.

338. Zhang, B., Fan, L., Zhong, H. et al. 2013. Graphene nanoelectrodes: Fabrication and size-dependent electrochemistry. *J. Am. Chem. Soc.* 135:10073–10080.

339. Shen, J., Wang, W., Chen, Q. et al. 2009. The fabrication of nanoelectrodes based on a single carbon nanotube. *Nanotechnology* 20:245307.

340. Huang, Y. and Chen, P. 2010. Nanoelectronic biosensing of dynamic cellular activities based on nanostructured materials. *Adv. Mater.* 22:2818–2823.

341. Tsai, C.-C., Yang, C.-C., Shih, P.-Y. et al. 2008. Exocytosis of a single bovine adrenal chromaffin cell: The electrical and morphological studies. *J. Phys. Chem. B* 112:9165–9173.

342. Hussien, E. M., Schuhmann, W., and Schulte, A. 2010. Shearforce-based constant-distance scanning electrochemical microscopy as fabrication tool for needle-type carbon-fiber nanoelectrodes. *Anal. Chem.* 82:5900–5905.

343. Takahashi, Y., Shevchuk, A. I., Novak, P. et al. 2011. Multifunctional nanoprobes for nanoscale chemical imaging and localized chemical delivery at surfaces and interfaces. *Angew. Chem. Int. Ed.* 50:9638–9642.

344. McKelvey, K., Nadappuram, B. P., Actis, P. et al. 2013. Fabrication, characterization, and functionalization of dual carbon electrodes as probes for scanning electrochemical microscopy (SECM). *Anal. Chem.* 85:7519–7526.

345. Ino, K., Ono, K., Arai, T. et al. 2013. Carbon-Ag/AgCl probes for detection of cell activity in droplets. *Anal. Chem.* 85:3832–3835.

346. Takahashi, Y., Shevchuk, A. I., Novak, P. et al. 2012. Topographical and electrochemical nanoscale imaging of living cells using voltage-switching mode scanning electrochemical microscopy. *Proc. Natl. Acad. Sci. U. S. A.* 109:11540–11545.

347. Chen, R., Huang, W., and Tong, H. 2003. Carbon fiber nanoelectrodes modified by single-walled carbon nanotubes. *Anal. Chem.* 75:6341–6345.

348. Wu, W.-Z., Huang, W.-H., Wang, W. et al. 2005. Monitoring dopamine release from single living vesicles with nanoelectrodes. *J. Am. Chem. Soc.* 127:8914–8915.

349. Schrlau, M. G., Falls, E. M., Ziober, B. L. et al. 2008. Carbon nanopipettes for cell probes and intracellular injection. *Nanotechnology* 19:015101.

350. Schrlau, M. G., Dun, N. J., and Bau, H. H. 2009. Cell electrophysiology with carbon nanopipettes. *ACS Nano* 3:563–568.

351. Li, Z.-Y., Zhou, W., Wu, Z.-X. et al. 2009. Fabrication of size-controllable ultrasmall-disk electrode: Monitoring single vesicle release kinetics at tiny structures with high spatio-temporal resolution. *Biosens. Bioelectron.* 24:1358–1364.

352. Cheng, H., Huang, W.-H., Chen, R.-S. et al. 2007. Carbon fiber nanoelectrodes applied to microchip electrophoresis amperometric detection of neurotransmitter dopamine in rat pheochromocytoma (PC12) cells. *Electrophoresis* 28:1579–1586.

353. Arrigan, D. 2004. Nanoelectrodes, nanoelectrode arrays and their applications. *Analyst* 129:1157–1165.

354. Li, J., Koehne, J., Cassell, A. M. et al. 2005. Inlaid multi-walled carbon nanotube nanoelectrode arrays for electroanalysis. *Electroanalysis* 17:15–27.

355. Ugo, P., Moretto, L. M. and Vezzà, F. 2002. Ionomer-coated electrodes and nanoelectrode ensembles as electrochemical environmental sensors: Recent advances and prospects. *ChemPhysChem* 3:917–925.

356. Koehne, J., Li, J., Cassell, A. M. et al. 2004. The fabrication and electrochemical characterization of carbon nanotube nanoelectrode arrays. *J. Mater. Chem.* 14:676–684.

357. Arumugam, P. U., Chen, H., Siddiqui, S. et al. 2009. Wafer-scale fabrication of patterned carbon nanofiber nanoelectrode arrays: A route for development of multiplexed, ultrasensitive disposable biosensors. *Biosens. Bioelectron.* 24:2818–2824.

358. Koehne, J. E., Marsh, M., Boakye, A. et al. 2011. Carbon nanofiber electrode array for electrochemical detection of dopamine using fast scan cyclic voltammetry. *Analyst* 136:1802–1805.

359. Siddiqui, S., Arumugam, P. U., Chen, H. et al. 2010. Characterization of carbon nanofiber electrode arrays using electrochemical impedance spectroscopy: Effect of scaling down electrode size. *ACS Nano* 4:955–961.

360. Periyakaruppan, A., Arumugam, P. U., Meyyappan, M. et al. 2011. Detection of ricin using a carbon nanofiber based biosensor. *Biosens. Bioelectron.* 28:428–433.

361. Periyakaruppan, A., Gandhiraman, R. P., Meyyappan, M. et al. 2013. Label-free detection of cardiac troponin-I using carbon nanofiber based nanoelectrode arrays. *Anal. Chem.* 85:3858–3863.

362. Li, Y., Syed, L., Liu, J. et al. 2012. Label-free electrochemical impedance detection of kinase and phosphatase activities using carbon nanofiber nanoelectrode arrays. *Anal. Chim. Acta* 744:45–53.

363. Swisher, L. Z., Syed, L. U., Prior, A. M. et al. 2013. Electrochemical protease biosensor based on enhanced AC voltammetry using carbon nanofiber nanoelectrode arrays. *J. Phys. Chem. C* 117:4268–4277.

364. Tu, Y., Lin, Y., and Ren, Z. F. 2003. Nanoelectrode arrays based on low site density aligned carbon nanotubes. *Nano Lett.* 3:107–109.

365. Tu, Y., Lin, Y., Yantasee, W. et al. 2005. Carbon nanotubes based nanoelectrode arrays: Fabrication, evaluation, and application in voltammetric analysis. *Electroanalysis* 17:79–84.

366. Lin, Y., Lu, F., Tu, Y. et al. 2004. Glucose biosensors based on carbon nanotube nanoelectrode ensembles. *Nano Lett.* 4:191–195.

367. Gholizadeh, A., Shahrokhian, S., Iraji zad, A. et al. 2012. Fabrication of sensitive glutamate biosensor based on vertically aligned CNT nanoelectrode array and investigating the effect of CNTs density on the electrode performance. *Anal. Chem.* 84:5932–5938.

368. Granger, M. and Swain, G. 1999. The influence of surface interactions on the reversibility of ferri/ferrocyanide at boron-doped diamond thin-film electrodes. *J. Electrochem. Soc.* 146:4551–4558.

369. Hees, J., Hoffmann, R., Kriele, A. et al. 2011. Nanocrystalline diamond nanoelectrode arrays and ensembles. *ACS Nano* 5:3339–3346.

370. Hees, J., Hoffmann, R., Yang, N. et al. 2013. Diamond nanoelectrode arrays for the detection of surface sensitive adsorption. *Chemistry* 19:11287–11292.

371. Lowe, R. D., Mani, R. C., Baldwin, R. P. et al. 2006. Nanoelectrode ensembles using carbon nanopipettes. *Electrochem. Solid State Lett.* 9:H43–H47.

372. Sun, G., Huang, Y., Zheng, L. et al. 2011. Ultra-sensitive and wide-dynamic-range sensors based on dense arrays of carbon nanotube tips. *Nanoscale* 3:4854–4858.

373. Dai, Y., Swain, G., Porter, M. et al. 2008. New horizons in spectroelectrochemical measurements: Optically transparent carbon electrodes. *Anal. Chem.* 80:14–22.

374. Ranganathan, S. and McCreery, R. L. 2001. Electroanalytical performance of carbon films with near-atomic flatness. *Anal. Chem.* 73:893–900.

375. Ranganathan, S., McCreery, R., Majji, S. M. et al. 2000. Photoresist-derived carbon for microelectromechanical systems and electrochemical applications. *J. Electrochem. Soc.* 147:277–282.

376. Donner, S., Li, H.-W., Yeung, E. S. et al. 2006. Fabrication of optically transparent carbon electrodes by the pyrolysis of photoresist films: Approach to single-molecule spectroelectrochemistry. *Anal. Chem.* 78:2816–2822.

377. Tian, H., Bergren, A. J., and McCreery, R. L. 2007. Ultraviolet-visible spectroelectrochemistry of chemisorbed molecular layers on optically transparent carbon electrodes. *Appl. Spectrosc.* 61:1246–1253.

378. Walker, E. K., Vanden Bout, D. A., and Stevenson, K. J. 2012. Carbon optically transparent electrodes for electrogenerated chemiluminescence. *Langmuir* 28:1604–1610.

379. Walker, E. K., Vanden Bout, D. A., and Stevenson, K. J. 2012. Spectroelectrochemical investigation of an electrogenerated graphitic oxide solid-electrolyte interphase. *Anal. Chem.* 84:8190–8197.

380. Xiao, X., Beechem, T. E., Brumbach, M. T. et al. 2012. Lithographically defined three-dimensional graphene structures. *ACS Nano* 6:3573–3579.

381. Alharthi, S. A., Benavidez, T. E., and Garcia, C. D. 2013. Ultrathin optically transparent carbon electrodes produced from layers of adsorbed proteins. *Langmuir* 29:3320–3327.

382. Keeley, G. P., McEvoy, N., Nolan, H. et al. 2012. Simultaneous electrochemical determination of dopamine and paracetamol based on thin pyrolytic carbon films. *Anal. Methods* 4:2048.

383. Keeley, G. P., McEvoy, N., Kumar, S. et al. 2010. Thin film pyrolytic carbon electrodes: A new class of carbon electrode for electroanalytical sensing applications. *Electrochem. Commun.* 12:1034–1036.

384. Doherty, E. M., De, S., Lyons, P. E. et al. 2009. The spatial uniformity and electromechanical stability of transparent, conductive films of single walled nanotubes. *Carbon* 47:2466–2473.

385. Bae, S., Kim, H., Lee, Y. et al. 2010. Roll-to-roll production of 30-inch graphene films for transparent electrodes. *Nat. Nanotechnol.* 5:574–578.
386. Hecht, D. S., Hu, L., and Irvin, G. 2011. Emerging transparent electrodes based on thin films of carbon nanotubes, graphene, and metallic nanostructures. *Adv. Mater.* 23:1482–1513.
387. Garoz-Ruiz, J., Palmero, S., Ibañez, D. et al. 2012. Press-transfer optically transparent electrodes fabricated from commercial single-walled carbon nanotubes. *Electrochem. Commun.* 25:1–4.
388. Heras, A., Colina, A., López-Palacios, J. et al. 2009. Flexible optically transparent single-walled carbon nanotube electrodes for UV–vis absorption spectroelectrochemistry. *Electrochem. Commun.* 11:442–445.
389. Weber, C. M., Eisele, D. M., Rabe, J. P. et al. 2010. Graphene-based optically transparent electrodes for spectroelectrochemistry in the UV-vis region. *Small* 6:184–189.
390. Rutkowska, A., Bawazeer, T. M., Macpherson, J. V. et al. 2011. Visualisation of electrochemical processes at optically transparent carbon nanotube ultramicroelectrodes (OT-CNT-UMEs). *Phys. Chem. Chem. Phys.* 13:5223–5226.
391. Dai, Y., Proshlyakov, D. A., Zak, J. K. et al. 2007. Optically transparent diamond electrode for use in IR transmission spectroelectrochemical measurements. *Anal. Chem.* 79:7526–7533.
392. Menegazzo, N., Kahn, M., Berghauser, R. et al. 2011. Nitrogen-doped diamond-like carbon as optically transparent electrode for infrared attenuated total reflection spectroelectrochemistry. *Analyst* 136:1831–1839.

10 Template-Directed Controlled Electrodeposition of Nanostructure and Composition

Jonathon Duay and Sang Bok Lee

CONTENTS

10.1 INTRODUCTION

Nanostructured materials are a heavily researched topic in today's scientific community as many new and exciting properties have been attributed to these materials with confined dimensions.[1–10] However, there is still a lot unknown about what aspects of these structures initiate and/or enhance these new found phenomena. Therefore, methods are needed to tune their morphology and composition in order to make model nanomaterials that can be precisely controlled so accurate knowledge of their performance versus structure can be achieved.

An excellent example of this was the discovery of Raman signal enhancement when an analyte is placed on a roughened silver substrate.[11,12] This discovery led to a new analytical technique termed surface-enhanced Raman spectroscopy (SERS). However, a lot is unknown about the mechanism of this anomalous signal enhancement. In the literature, the enhancement has been attributed to a combination of electromagnetic and chemical mechanisms[13,14]; however, until a great control over the substrates atomistic composition and surface morphology can be achieved, the full extent of each mechanism will not be understood. Of course, the production of controllable nanostructured substrates through SERS is currently a heavily studied research area.[14–17]

Accordingly, a great method that can provide this precise control over structure, composition, and size is electrodeposition, utilizing a versatile template. Whereas the template provides an adjustable nanostructured, electrochemically inactive, mold for synthesis, the electrodeposition process can provide accurate control over composition and even further control over the nanostructures' dimensions. Therefore, this chapter is designed to discuss the current science of the controllable formation of nanostructures through electrochemical synthesis of materials utilizing templates with feature sizes in the nanoscale.

The fundamental aspect for creating these templated nanomaterials is relatively straightforward. First, the template, a variety of which will be discussed in the next section, is positioned on a current collector, or one is applied to it through a thin film deposition technique such as thermal or e-beam evaporation. The template is then placed in the electrochemical plating bath. Electrodeposition of the material occurs through the template pores or along their walls. This is followed by the removal of the template through mostly chemical processes such as dissolution; however, other processes such as thermal decomposition can be used.

After the removal of the template, sometimes called a membrane and in some instances a mask, an ordered film is produced. Depending on the template used, this film can consist of either a well-ordered porous film or countless nanostructures (more than 10^{12} structures per cm^2 electrode area are possible) with the same morphology and material composition. For the latter, the material can be analyzed as an array of nanostructures or can be dispersed into high concentration nanoparticulate suspensions. Consequently, these nanoparticles can be investigated individually or as ensembles for applications such as electrochemical energy storage,[18–21] energy conversion,[3,22,23] biomedical,[24–26] photoluminescence,[27,28] magnetism,[29–33] and nanomotors,[34,35] among others.[5,36–38]

As electrochemically synthesized nanomaterials using templates have been accomplished innumerable times and even reviewed a small number of times including by us, this chapter will deeply focus on the mechanism of formation for these structures. It is believed that detailed knowledge of this mechanism can help produce heterogeneous and hierarchical materials that have been shown to have better functionality than single component nanomaterials.[4,39–44] This includes nanomaterials with more than one component and/or more than a single architecture as well as materials that maintain the porous nature of the template.

Accordingly, the main thrust of this chapter is the scientific work studying the underlying synthetic mechanism of how electrodeposited materials may or may not fill the pores of the templates completely or homogeneously. This information can be used for the production of heterogeneous and/or hierarchical nanostructures through one- or multistep electrochemical or chemical methods. Finally, this chapter will include additional electrochemical processes, whereas the nanostructures themselves act as templates for novel synthesis of nanostructures. This will include electrochemical phase transitions, electrochemical dealloying, and galvanic replacement methods.

10.2 ELECTROCHEMICAL DEPOSITION USING TEMPLATES

Electrochemical deposition (ECD) has been a technique for material synthesis since Luigi Brugnatelli first electroplated gold in 1805.[45] After this historic achievement, ECD has advanced to include the deposition of other metals and alloys through the use of earth-abundant metal salts. In more recent times, metal oxides,[46–48] metal sulfides,[49,50] semiconductors,[51–53] and conductive polymers[54–56] among other materials have been added to the list of possible electrochemically deposited materials.

The fundamentals of this deposition process rely on the diffusion of charged species in the solution toward the working electrode. At this electrode, the ECD process is governed by the Nernst equation:

$$E = E_0 + \frac{RT}{nF}\ln(a)$$

where

E is the equilibrium electrode potential
E_0 is the standard electrode potential when all of the variables are at standard values
R is the gas constant
T is the temperature
n is the number of electrons exchanged
F is the Faraday's constant
a is the activity of the species in solution

When the potential applied is greater than the equilibrium electrode potential, oxidation of the ion species will occur and *vice versa* with oxidation supplanted with reduction. Depending on the electrochemical nature of the species, oxidation or reduction will cause the deposition of the solvated ion into a bound species in the solid state.

As the equilibrium electrode potential can sometimes be outside of the electrochemical stability window of most common solvents. Recent advances have included the use of ionic liquids[49,52,57–59] that have wider electrochemical potential windows to allow the ECD of additional materials that were previously not obtainable. For example, Edstrom et al. were able to electrodeposit aluminum into the pores of an alumina template (see later for description of this template) by pulsed potential electrodeposition utilizing $Al_2Cl_7^-$ in a 1:2 1-ethyl-3-methylimidazolium chloride/aluminum chloride ionic liquid.[60] Before the advent of ionic liquids, this achievement could not have been possible.

As mentioned previously, there is a growing interest in making nanomaterials with controllable sizes that have well-defined size distributions. Template synthesis, sometimes called membrane synthesis depending on the template, can provide us with these exciting materials. For even improved control, ECD can be employed to better fine-tune sizes, masses, and morphologies, as will be discussed later. As the list of materials capable of being electrodeposited and the ingenuity that governs what is achievable expands, more and more research continues toward this exciting area.

A wide variety of template synthesis methods have been developed for preparing nanomaterials by ECD. At a minimum, the ECD template process requires a conductive substrate that acts as a current collector and electrolyte wet pores. The growth of the material then proceeds from the conductive substrate through the template either along the surface of the template pores or filling them completely. In some cases, the material continues as overgrowth after the template has been filled. The variety of templates currently utilized can be found in the following text. They have been divided between hard and soft templates with a new, innovative example for each one.

10.2.1 HARD TEMPLATES

10.2.1.1 Anodic Aluminum Oxide Template Synthesis

When aluminum is anodized at high overpotentials from 2 to 500 V in acidic electrolytes, a porous aluminum oxide structure forms on the aluminum surface often called anodic aluminum oxide (AAO) (although other acronyms include porous anodic aluminum [PAA], anodic alumina membrane [AAM]). The hexagonal nature of this porous oxide structure was first observed in 1953 by Keller.[61]

Control over this oxide coating can be achieved through modulation of the voltage used for pore synthesis with larger potentials resulting in larger pore diameters and longer anodization times resulting in longer pore lengths. This allows great control over the aspect ratio of the pores along with materials synthesized utilizing this AAO structure as a template. For a detailed review of the aluminum oxide structure formation, please see Sulka reference.[62]

Electrochemical synthesis of materials into this pore structure was first accomplished by Asada in 1963 although only as a way of coloring the oxide coating.[63] These materials were produced utilizing the underlying aluminum as a current collector; however, at the base of the pores, there is an oxide barrier layer that required large overpotentials to electrochemically produce these materials in the pores. Therefore, for easier material synthesis, the aluminum along with the barrier oxide is usually removed followed by the addition of a new current collector through mostly thin film deposition techniques. Electrodeposition can then proceed from the bottom of the pore and continue up to fill the pore to produce nanowires or continue up just along the pore walls to produce nanotubes, the mechanisms of which will be discussed later in this chapter. Many diverse nanowire and nanotube structures have been prepared using this method.[4,20,21,41,64–66]

A recent example of the versatility of this template is the work done by Meng et al.[67] As mentioned, synthesized materials utilizing AAO normally results in nanowire or nanotube morphologies; however, additional morphologies can be formed. Accordingly, Meng et al. were able to produce tuneable nanodots and nanorings utilizing this templating process.[67] The nanodots are desirable in applications such as optoelectronics, data storage, and gas sensors, whereas the nanorings are desirable morphologies for magnetic memory and biosensors. The nanodot or nanoring formation is controlled by the morphology of the base current collector, as depicted in Figure 10.1, with nanodot formation being completed utilizing a base electrode that completely covers the pore while the nanoring formation utilizes a base electrode that only covers the edge of the pore. This base electrode morphology is an important matter in nanowire/nanotube morphology as well, a topic that will be discussed in detail in the nanotube formation mechanism section of this chapter.

Some other recent materials electrochemically synthesized utilizing this template have included copper azide nanowires for microdetonators,[68] polyaniline–Ag nanocables for supercapacitors,[69] segmented silver nanowire/nanotubes and nanospheroid/nanotubes for surface plasmon resonance sensing,[70] segmented Ag–Cu_2O nanowires for optics,[71] mesoporous silica nanowires,[72] segmented Fe–Ga/Cu nanowires for magnetics,[73] and segmented Cu/Pt nanowires for nanomotors.[34]

10.2.1.2 Anodic Titanium Oxide Nanotubes

In addition to AAO, other anodized porous metal oxides have been attempted. The most studied of these is titanium dioxide, anodized from titanium, due to its large use in solar cell and photocatalysis applications.[36] The first report of the ordered porous nature of anodized titanium was presented in 1999 by Zwilling, who used an aqueous hydrofluoric acid electrolyte to anodize the titanium.[74] More practical, high aspect ratio TiO_2 nanotubes were first accomplished in 2005 by Schmuki using an organic electrolyte.[75] These high aspect ratio TiO_2 nanotubes are better served for creating controllable nanostructures. The TiO_2 nanotubes can then be used as a scaffold for the electrochemical production of nanowires or nanotubes.[3,36,76]

An interesting recent utilization of this template was accomplished by Cui and coworkers.[77] Whereas in the literature, TiO_2 as a template is normally utilized for its bandgap in addition to its high surface area, these authors were able to convert it to a conductive substrate. The TiO_2 nanotube arrays were calcined under ammonia to produce a TiN nanotube array. They then used this conductive nitride to electrodeposit materials directly on the surface of the nanotube walls. A schematic of this process along with SEM and TEM images of the resulting material can be

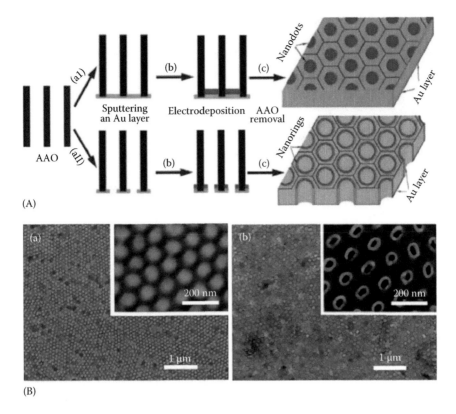

FIGURE 10.1   (A) Synthesis process for the controlled production of nanodots and nanorings. (B) SEM images of an array of (a) nanodots and (b) nanorings. (Yang, D., Meng, G., Zhu, X., and Zhu, C., *Chem. Commun.*, 7110–7112, 2009. Reproduced by permission of The Royal Society of Chemistry.)

found in Figure 10.2. The structure was then tested as a supercapacitor electrode exhibiting promising performance metrics.

Other recent uses have included Bi_2O_3-decorated TiO_2 nanotubes for supercapacitors,[78] CdTe/Au–TiO_2 nanotube array for photoelectrochemical sensors,[79] PbO_2-coated TiO_2 nanotubes for the electrocatalytic degradation of phenol,[80] ZnO/TiO_2 nanotube array for dye-sensitized solar cells,[81] and PbS nanoparticle-decorated TiO_2 nanotubes for photocatalysis.[82]

10.2.1.3   Track-Etched Polymer Template

Track-etched membranes were first developed by Fleischer at GE in the 1960s.[83] These membranes are made by creating radiation damage tracks using heavy ion bombardment in thin films of polymer (polycarbonate, polyethylene terephthalate, etc.) or mica and then etching these tracks into pores. The development of these membranes was short with commercial polycarbonate membranes available in the early 1970s.[84] However, it was not until the mid-1980s when Penner and Martin used these membranes as templates to produce conductive polymer fibrils.[85] Since then, many groups have used this template to produce 1D electrodeposited nanomaterials.[52,56,86]

One advantage of using these membranes over AAO is that they can be removed by organic solvents such as methylene chloride, dichloromethane, or chloroform that can provide protection for amphoteric materials or metals that can, respectively, dissolve or react in the high or low pH solvent that is needed for AAO removal.

Recently, Rauber et al.[87] was able to create 3D nanowire networks through a modification of this template synthesis method. This was done by irradiating the polymer foil at different angles,

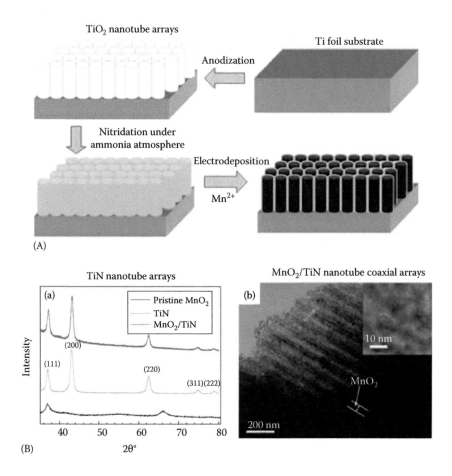

FIGURE 10.2 (A) Schematic representing the synthesis of manganese oxide–coated TiN nanotubes. (B) (a) XRD results supporting the MnO$_2$/TiN synthesis mechanism and (b) TEM images of the resulting MnO$_2$/TiN nanotube coaxial arrays. (Dong, S.; Chen, X.; Gu, L. et al., *Energy Environ. Sci.*, 4, 3502–3508, 2011. Reproduced by permission of The Royal Society of Chemistry.)

creating networked pores. Subsequent electrodeposition of platinum followed by template removal produced interconnected networks of platinum wires. The schematic for the synthesis process as well as SEM images of the nanonetwork can be found in Figure 10.3. The resulting freestanding 3D platinum network has potential to be used in catalysis, optoelectronics, as well as energy storage and energy conversion applications.

Other recent materials synthesized utilizing these templates include single-crystal Au, Ag, and Cu as well as polycrystalline Ni, Co, and Rh nanowires;[88] aluminum nanowires for microbatteries;[89] Bi$_2$Te$_3$ nanowires for thermoelectrics;[90] Fe$_{70}$Pd$_{30}$ nanotubes for drug delivery;[91] patterned chitosan hydrogel for biotechnology;[92] gallium nanowires for compound semiconductors or battery anodes;[93] and zinc nanowire arrays for batteries.[94]

10.2.1.4 Colloid Crystal Template

Colloid crystal templates originate from studies done by Luck in 1963 where they observed that the color of dried latexes related to the Bragg diffraction of the incident light indicating crystal planes that they attributed to face-centered cubic (FCC) orientations of stacked latex particles.[95,96] The small size distribution of the latex particles allowed for a close-packed ordered structure forming the FCC orientation. Under controlled drying, three- as well as two- dimensional films

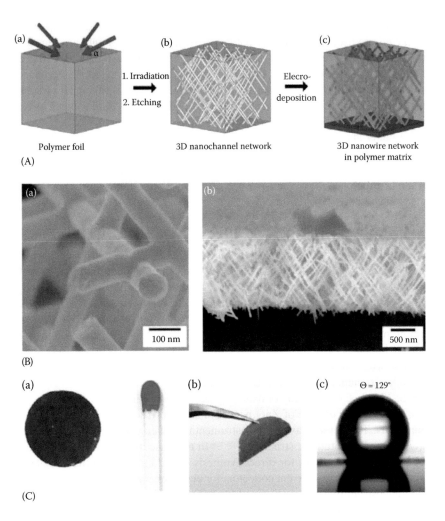

FIGURE 10.3 (A) Schematic representing the synthesis of a 3D nanowire network by irradiating a polymer film at a variety of angles. (B) SEM images of the 3D freestanding nanowire network and (C) optical images of flexible platinum wire network film and water contact angle measurements. (Reprinted with permission from Rauber, M., Alber, I., Muller, S. et al., *Nano Lett.*, 11, 2304–2310, 2011. Copyright 2011, American Chemical Society.)

can be produced that can then be used as templates. In addition to latex, SiO_2 colloidal crystal templates originate from the synthesis of monodispersed (narrow-size distribution) SiO_2 microspheres first reported by Stober et al.[97]

The close-packed films of these spheres were not used for colloid crystal templating until Velev et al. first developed it by infiltrating silica precursors into a latex colloid crystal film in 1997.[98] In 1999, Braun developed what is now called electrochemical colloid crystal templating by electrodepositing CdSe and CdS through a polystyrene and a silica colloidal close-pack assembly, respectively, on a conductive substrate.[99] In addition to electrochemical growth of the material through the colloid crystal lattice, these templates can be used as a mask to photolithographically pattern electrodes to produce ordered arrays of controllable nanostructures.[100] Many groups have utilized these methods to produce porous thin films as well as ordered arrays of electrodeposited nanomaterials.[101–106]

A new and interesting method along these lines was recently developed by Cai et al. in which polystyrene spheres are used to chemically pattern a 1-hexadecanethiol onto an electrode followed by removal of the spheres resulting in what is termed an invisible template.[107] Basically, after removal of the spheres, an electrode with disk-shaped electroactive regions is produced. In this case, the authors produced gold nanoflowers on the electroactive area producing ordered arrays of these nanostructures. Depiction of this synthesis method along with SEM images of the resulting nanoflowers can be found in Figure 10.4.

Other recent materials created using these ordered arrays include porous nickel phosphide film for lithium ion batteries,[108] porous aluminum films for lithium electrodeposition,[109] macroporous silicon films for optics,[110] cobalt metal nanoarrays of various morphologies for biomedicine,[111] macro-/nanoporous gold films for electrocatalysis,[112,113] silver bowls as SERS substrates,[114] and complex macroporous gold and polypyrrole films.[115]

10.2.1.5 Other Hard Templates

Besides the aforementioned templates, many other research groups have developed their own template-based electrochemical synthesis techniques. For example, photolithography methods such as the lithographically patterned nanoscale electrodeposition (LPNE) method developed by Penner are excellent ways to produce nanostructures with controllable dimensions.[116–118] With this process, the Penner group has been able to produce nanowires with length scales in the macroscale while still maintaining precise control over the nanowires' height and width with resolutions in the tens of nanometers.[118]

Halpern and Corn[100] were actually able to combine this method along with the colloid crystal template to synthesize nanoring arrays with controllable dimensions. A schematic of this process along with an SEM image of the array of nanorings can be found in Figure 10.5.

Other templates include the use of ZnO nanorods as nanostructured current collectors to electrodeposit materials. Tong et al. have been instrumental in using these structures as templates to produce core/shell and nanotubular structures.[119–123] In this case, the template has a relatively good electronic conductivity that allows for direct synthesis of the material on the template. This also allows the ZnO nanorods act as current collectors for electrochemical testing.

These ZnO nanostructures are synthesized by a variety of methods including chemical vapor deposition,[124] sputtering, thermal evaporation, pulsed laser deposition,[125,126] vapor-phase techniques,[127] as well as ECD.[128–130]

10.2.2 Soft Template

10.2.2.1 Block Copolymer Thin Film Template

Block copolymers are polymers that consist of two or more polymer chains attached at their ends. They can self-assemble into thin films having segregated *blocks* with domain sizes from 10 to 200 nm. This domain size and shape can be controlled by the molecular weight, segment size, and block interactions.[131] By selective etching of one of the blocks, an ordered thin film that can be used as a template is produced with the remaining block. This type of template synthesis is a fairly new and emerging method with a variety of copolymers as well as applications.[131–134]

An innovative example of electrodeposition utilizing this template was done by Steiner and coworkers.[134] Utilizing a poly(4-fluorostyrene-r-styrene)-b-poly(D,L-lactide) diblock copolymer containing 39.9% poly(D,L-lactide) that is selectively removed, vanadium oxide is electrodeposited into the bicontinuous double gyroid structure of the remaining polymer block. Subsequent removal of the remaining polymer template results in an interconnected mesoporous V_2O_5 film with a gyroid structure. A schematic of this synthesis process can be found in Figure 10.6. This material was tested for its electrochromic properties, which were found to be vastly superior when compared to its planar thin film counterpart.

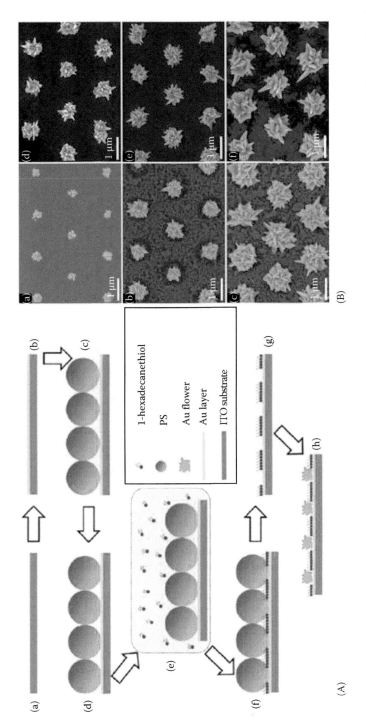

FIGURE 10.4 (A) Schematic of the synthesis of ordered arrays of gold nanoflowers utilizing an invisible template. (a) ITO coated glass. (b) Gold was deposited on the ITO via sputtering. (c) A polystyrene (PS) sphere monolayer was transferred onto the Au/ITO substrate. (d) The PS/Au/ITO was then heat treated. (e) The PS/Au/ITO substrate was soaked in a 2 mM 1-hexadecanthiol ethanol solution. (f) The 1-hexadecanthiol molecules formed a SAM on the bare Au/ITO substrate between the PS spheres. (g) The invisible template was then synthesized by electrodeposition. (h) The gold nanoflower array was then synthesized by electrodeposition. (B) SEM images of arrays of gold nanoflowers. (a,d) The electrodeposition was carried out with a current density of 0.05 mA/cm^2 with total deposition times of 1 min, (b,e) 2 min, and (c,f) 5 min. (a–c) The thickness of the gold layer is 10 nm and (d–f) 30 nm. (Reprinted with permission from Wang, J., Duan, G., Li, Y., Liu, G., Dai, Z., Zhang, H., and Cai, W., *Langmuir*, 29, 3512–3517, 2013. Copyright 2013, American Chemical Society.)

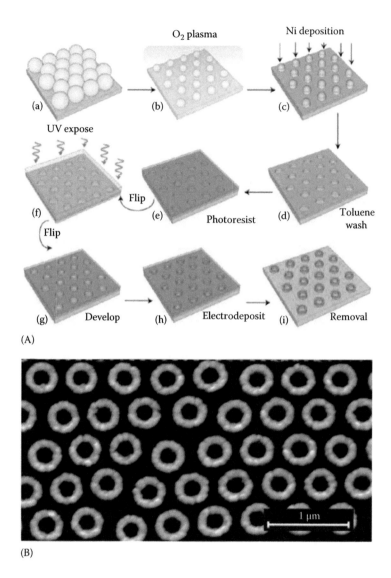

FIGURE 10.5 (A) Schematic for the combination of colloidal lithography with lithographically patterned nanoscale electrodeposition. (a) A colloidal monolayer was formed and then (b) etched under an O_2 plasma. (c) Ni (or Ag) was then evaporated on top followed by (d) dissolution of the colloidal monolayer revealing a nanohole array. (e) A photoresist layer was spin coated on top, and (f) the backside was exposed through the nanohole array. (g) The photoresist was then developed and (h) a metal was electrodeposited inside each nanohole. (i) Finally, the sacrificial Ni or Ag was selectively removed. (B) Gold rings resulting from the combination of the two processes. (Reprinted with permission from Halpern, A.R. and Corn, R.M., *ACS Nano*, 7, 1755–1762, 2013. Copyright 2013, American Chemical Society.)

Other materials synthesized using this template include nickel metal nanodots or nanowires,[135] Au–TiO_2 mesoporous nanocomposites,[136] double-gyroid-patterned platinum films,[137] CdSe nanorods for solar cells,[138] Ni nanopillars on GaAs for optoelectronics,[139] and gold nanoparticles for biosensors.[140]

10.2.2.2 Liquid Crystal or Micellar Template

The utilization of surfactants in electrochemical baths can result in micelle or even liquid crystal formation at high enough surfactant concentrations (>30 wt%). Subsequent electrodeposition results in materials with pores created by these micelles or liquid crystals.

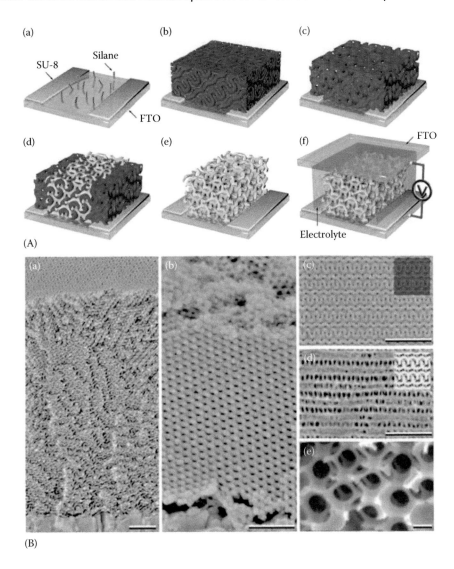

FIGURE 10.6 (A) Schematic for the synthesis of an interconnected mesoporous V_2O_5 film with a gyroid morphology utilizing a block copolymer template synthesis method. (B) SEM images of the resulting mesoporous V_2O_5 film. (Lee, J.I., Cho, S.H., and Park, S.-M. et al.: Highly aligned ultrahigh density arrays of conducting polymer nanorods using block copolymer templates. *Nano Lett.* 2008. 8. 2315–2320; Scherer, M.R.J., Li, L., Cunha, P.M.S., Scherman, O.A., and Steiner, U.: Enhanced electrochromism in gyroid-structured vanadium pentoxide. *Adv. Mater.* 2012. 24. 1217–1221. Copyright Wiley-VCH Verlag GmbH & Co. KGaA. Reproduced with permission.)

In 1997, Wang et al. were the first to show that ordered porous materials can be synthesized through electrodeposition using a lyotropic liquid crystal phase to electrodeposit porous platinum.[141] Since then, many other scientists have used this approach to make porous materials through electrodeposition.[141–147]

A current example investigating the mechanism of this method was employed by Borguet and coworkers.[148] Utilizing small concentrations of silver ions and an electrochemical scanning tunnelling microscope, they were able to visualize the electrodeposition of Ag utilizing a sodium dodecyl sulfate (SDS) surfactant. In this case, the SDS adsorbs in parallel lines on the Au(111) current collector surface. The lines appear wavy due to the adsorption of silver ions. After the application of

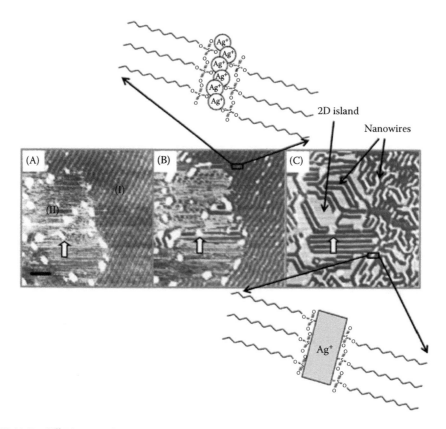

FIGURE 10.7 STM images of silver deposition on Au(111) utilizing an SDS surfactant. (A) Substrate at positive 0.2 V potential where no reaction takes place. (B) Substrate at −0.1 V where only adsorption of Ag+ takes place. (C) Substrate at −0.2 V where electroplating of Ag occurs. (Reprinted with permission from Seo, S., Ye, T., and Borguet, E., *Langmuir*, 28, 17537–17544, 2012. Copyright 2012, American Chemical Society.)

the deposition voltage, the Ag ions are reduced and the SDS adlayer rearranges due to the potential change. Removing the voltage reveals silver metal deposition where the SDS adlayer was absent. The authors suggest adequate control over the Ag deposition, and SDS adlayer rearrangement can result in lateral deposition of metal nanostructures with controlled sizes and shapes. Figure 10.7 shows a representation of this deposition process.

Other recent materials produced using this method include hexagonal nanoporous nickel hydroxide film for supercapacitors,[149] silver particles and gold nanoparticles,[150] ZnO nanobelt arrays for dye-sensitized solar cells,[151] heteroaromatic conjugated polymer films for optical activity measurements,[152] hexagonal mesoporous platinum films for electronic and optical devices,[153] and mesoporous Ni/Co alloys for energy storage applications.[154]

10.2.2.3 Other Soft Templates

A newly developed method by Valles and coworkers[155] involves utilizing microemulsions as templates during the electrochemical synthesis of metals and metal alloys. In their latest work, the authors utilized a microemulsion consisting of an aqueous electrolyte solution, Triton X-100, and diisopropyl adipate.[156] A phase diagram for these three components can be found in Figure 10.8A. They found that under certain composition, these components formed a bicontinuous microemulsion. When electrodeposition took place using this bicontinuous microemulsion, porous films were formed. An SEM image of a NiCo film utilizing this method can be found in Figure 10.8B.

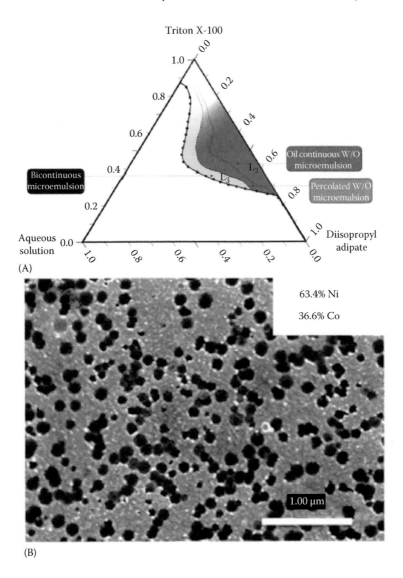

(A)

(B)

FIGURE 10.8 (A) Phase diagram for microemulsions containing an aqueous solution, Triton X-100, and diisopropyl adipate. (B) Porous NiCo film electrodeposited from the bicontinuous region of the phase diagram. (Reproduced from Calderó, G. et al., *Phys. Chem. Chem. Phys.*, 15, 14653, 2013. With permission from the PCCP Owner Societies.)

10.3 PORE-FILLING ELECTROCHEMICAL DEPOSITION MECHANISM

As template pores can range from the macro- (>50 nm diameter) to the meso- (2–50 nm diameter) and even to the micropore (<2 nm diameter) range, there is still a lot to be learned about migration through these pores of the active ions during electrodeposition. A detailed knowledge of the dynamics of this diffusion, so called ionics in these pores, can extend the ability to create multifunctional nanostructures in these pores through electrodeposition.

There have been various theoretical papers in the literature that discuss the current response during the ECD of materials in pores.[157,158] Most of these papers discuss only deposition in 1D pores. The conclusions of which will be reviewed here.

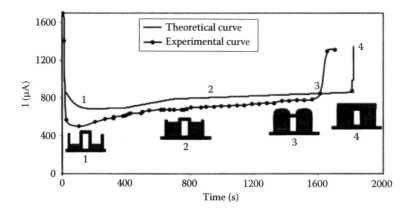

FIGURE 10.9 Experimental and theoretical chronoamperometric electrodeposition curves of Co nanowires in the 1D pores of polycarbonate template. (Reprinted from *Electrochim. Acta*, 47, Valizadeh, S., George, J.M., Leisner, P., and Hultman, L., Electrochemical deposition of Co nanowire arrays; quantitative consideration of concentration profiles, 865–874, 2001. Copyright 2013, with permission from Elsevier.)

Accordingly, the current response during electrodeposition in 1D pores is usually divided into four main regions. The first region involves the immediate nucleation of the material at the surface of the electrode resulting in a current spike in the chronoamperometric graph that quickly depletes once the diffusion layer is formed. The second region is noted as the growth of the material in the pore, and the current is close to stagnant or slightly decreasing or increasing due to the interplay between the imperfect conductivity of the material and the diminishing diffusion length of the ions in the filling pores. The third region involves an increase in the current that happens as the material grows out of the pore as hemispherical caps resulting in a larger active surface area for deposition; however, it has been theoretically shown that this current increase occurs even before the material emerges from the pore due to the increasingly small diffusion layer thickness in the pores.[159] A fourth region that consists of the overfilling of the pores, which can be found in the experimental and theoretical curves in Figure 10.9, eventually results in essentially a linear thin film growth with a steady-state current.[31,158]

Martin et al. were able to divide the second region where the material is growing in the pores into three subdivisions by plotting the current versus $t^{-1/2}$ as demonstrated in Figure 10.10 represented by sections 3–5.[86] They associated these different regions with the differences in the geometries of the Nernst layer. This was explained by noting that there are essentially three different diffusion zones as illustrated in Figure 10.10. There is the linear diffusion from the bulk of the electrolyte toward the pore, radial diffusion at the pore opening, and linear diffusion again in the pore toward the growing surface. At short times, linear diffusion in the pore is the rate limiting fragment; however, as the Nernst layer becomes thicker than the pore length, radial diffusion at the pore openings becomes the rate limiting fragment. At even longer times, the radial diffusion layers at each pore opening begin to overlap, and the linear diffusion from the bulk of the solution toward the pores becomes the limiting fragment.

A qualitative and quantitative theoretical kinetic model of the first three regions was done by Philippe et al.[157] The work was based on previous work by Szabo et al. on recessed microelectrodes done in 1987 where the global diffusion coefficient for the steady-state current and the geometry of the nanoelectrode are taken into account. Conclusions from this theoretical work, utilizing cobalt electrodeposition into track-etched polymer membranes for an experimental analogue, suggest the three important parameters that influence the current during electrodeposition in the pores are the geometry of the nanoelectrode defined by the size of the pores, the concentration of the active ion in solution, and the potential applied.

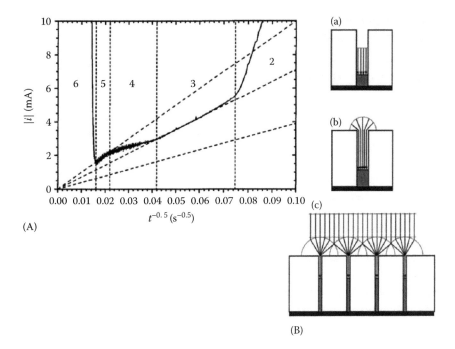

FIGURE 10.10   (A) Current density vs. $t^{-1/2}$ plot of copper electrodeposition in a track-etched polycarbonate template. (B) Limiting diffusion regimes during electrodeposition (a) representing linear diffusion in the pores, (b) representing radial diffusion at the pore openings, and (c) representing linear diffusion from the bulk solution to the membrane. (Reproduced from Dobrev, D. et al., *J. Electrochem. Soc.*, 150, C189, 2003. With permission from The Electrochemical Society.)

Just recently, Dolati et al. revisited this model based on Szabo's recessed microelectrode work this time for gold electrodeposition.[126] They found that the radial diffusion coefficient of the active $AuCl_4^-$ species was about eight times greater than the electrostatic component of its diffusion coefficient. This means that the limiting diffusion mechanism during electrodeposition utilizing small electrolyte concentrations is through either linear diffusion or electrostatic diffusion control. However, it was shown that at higher concentrations only the electrostatic contributions to the diffusion coefficient control the rate of mass transfer.

Although, still not entirely understood, the pore-filling electrodeposition is much more complex than previously thought. It is believed that through understanding of the diffusion regimes presented here, one could achieve any desirable composition through advanced scientific techniques such as microfluidics and/or localized electromagnetic fields among others. Therefore, it is hoped that experimental and theoretical work in this area will continue.

10.4   NANOTUBE GROWTH MECHANISM

In any 1D template such as AAO or track-etched polymers, there is the ability to electrochemically grow either nanowires or nanotubes. This has been achieved by both multistep and single-step methods. The growth mechanism for nanowires and multistep growth mechanism of nanotubes is mostly straightforward; however, the growth mechanism for single-step nanotubes is debated in the literature.

There has been much discussion whether the growth is caused by the shape of the electrode at the bottom of the pore, by surface affinity at the propagation front of the growing material, by chemical additives such as boric acid, or by gaseous bubble formation that excludes the center of the pore to material growth as well as other proposed mechanisms. The mechanism of growth for both polymer and metal nanotubes is presented in detail below.

10.4.1 POLYMER NANOTUBE SYNTHESIS

Polymer nanotubes have advantages over their nanowire counterparts by way of their increased surface area that allows for higher gravimetric chemical activities, in addition to their inner void that can function as a container as well as their thin shell that can promote rapid conversion between their oxidized and reduced states.[18,37,56,160] These properties allow them to be used in applications such as electrochromic windows,[56,161,162] supercapacitors,[18,162,163] chemical sensors,[42] molecular encapsulation and release,[27,164,165] controlled drug delivery,[39] organic photonic devices,[166] neural electrode interfaces[24–26] along with many other usages. As will be discussed, electrochemical template synthesis of these nanotubes offers a great way to control the tube lengths as well as both the outer and inner diameter, thus tuning the wall thickness.

Electrochemical template–synthesized polymer nanotubes were first prepared by Martin et al.[167] They were able to galvanostatically create polypyrrole tubules in track-etched polycarbonate. They hypothesized that the formation mechanism was due to electrostatic interactions between the anionic walls of the template and the cationic polymer causing adsorption and growth of the polymer along the pore walls starting at the base of the pore although Szklarczyk later postulated that the monomer polymerized simultaneously at the base and along the pore walls due to the deposition of the current collector during the thin film metal deposition.[168]

Demoustier-Champagne et al. further investigated the formation of these polypyrrole tubules in polycarbonate membranes.[54] They used a pulsed chronoamperometric technique and found the tubular morphology was preserved with similar wall thickness as compared to nonpulsed techniques. They also found that monomer and dopant concentration did not affect the nanotubular morphology; however, they did find that by using a NaPSS instead of a $LiClO_4$ supporting electrolyte, the polypyrrole wall thickness was increased.

These authors did similar studies with polyaniline in different polycarbonate membranes (pore diameters between 20 and 1000 nm) and concluded again that the polymerization has an affinity for the pore wall along with additional insight that the diffusion of the monomer in the pore contributes to the tubular morphology.[55] However, additional studies by these authors for poly(3,4-ethylenedioxythiophene) revealed only a small affinity for growth along the pore walls with only the tip exhibiting any nanotubular morphology.[169]

Conversely, in 2005, Lee et al. were indeed able to create poly(3,4-ethylenedioxythiophene) nanotubes.[161] These were potentiostatically deposited in an unmodified AAO template. They were able to modulate the morphology by controlling the monomer concentration and electrodeposition potential with higher concentrations and lower polymerization potentials resulting in thicker nanotubes. In addition, they were able to show this same control while utilizing a polycarbonate template.[56] However, they noticed that below a critical potential, the polymer obtained nanotubular morphology without regard to the monomer concentration. They associated this with the nanoelectrode morphology at the base of the pores, which results from the sputtering technique of the current collector.

Further studies along these lines from this author indeed demonstrate different mechanisms at different polymerization potentials.[66] For example, poly(3,4-ethylenedioxythiophene) (PEDOT's) morphology in the template at low (<1.2 V vs. Ag/AgCl) potentials depends on the morphology of the base electrode only with nanotubes being formed on annular-shaped electrodes and nanowires being formed on flat-topped electrodes. While at higher potentials (>1.4 V vs. Ag/AgCl), the morphology was heavily dependent on the monomer concentration and applied potential with low monomer concentration and high potentials favoring nanotube formation and the inverse favoring nanowire formation. TEM images of these nanostructures along with the data indicating the two mechanism electrosynthesis processes can be found in Figure 10.11.

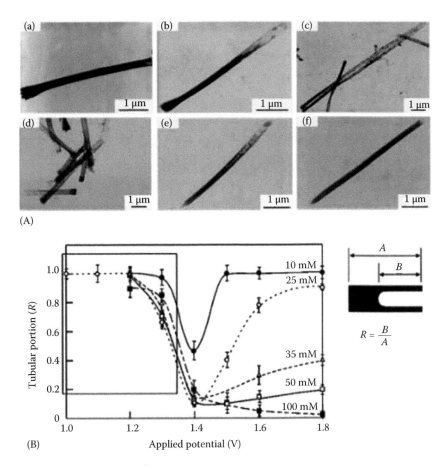

FIGURE 10.11 (A) The electropolymerization time was fixed at 100 s at the following potential values and precursor concentrations: (a) 1.4 V, (b) 1.5 V, and (c) 1.8 V in 25 mM EDOT; (d) 1.2 V, (e) 1.3 V, and (f) 1.4 V in 50 mM EDOT. (B) Graph demonstrating the two different mechanisms with the low potentials representing base electrode morphology mediated growth and the high potentials representing kinetic mediated growth. (Reprinted with permission from Xiao, R., Cho, S. I., Liu, R., and Lee, S.B., *J. Am. Chem. Soc.*, 129, 4483–4489, 2007. Copyright 2007, American Chemical Society.)

These authors were able to exploit this mechanism to produce coaxial nanowires consisting of MnO_2 cores and PEDOT shells.[170] They hypothesized that the PEDOT shells preferentially deposited along the walls due to the annular base electrode morphology. This formation mechanism for this material was supported by additional work with flat-topped electrodes that only produced segmented MnO_2/PEDOT nanowires.[65] A schematic for the formation mechanism and cyclic voltammetry curves for both the annular and flat-topped electrode can be found in Figure 10.12. The cyclic voltammetry (CV) curves for PEDOT indicate a lower polymerization potential for PEDOT at the annular-shaped electrodes.

It was shown here that polymer nanotube growth can proceed through one of three ways: through pore-wall affinity of the monomer, nanoelectrode directed growth, and interplay between the concentration of the monomer and the potential or current density applied.

10.4.2 METAL NANOTUBE SYNTHESIS

For metal synthesis, the polarization potential is usually negative as positive metal ions are reduced to zerovalent metals. This switch from the previous positive oxidation synthesis results

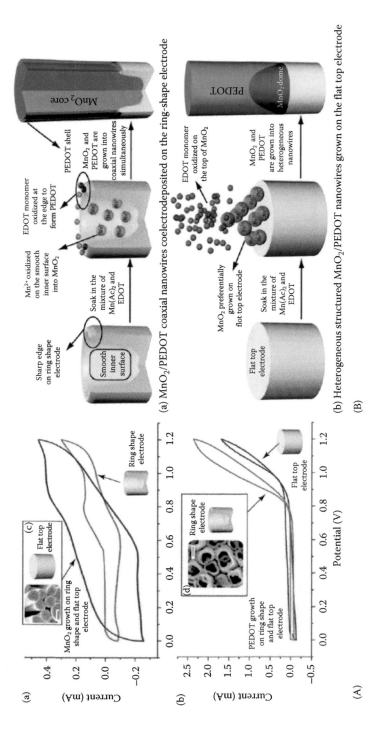

FIGURE 10.12 (A) Cyclic voltammetry curves for (a) MnO_2 and (b) PEDOT electrodeposition on (c, d) ring-shaped and flat-topped electrodes denoted in the inset SEM images. (B) Schematic for the electrodeposition mechanism of MnO_2/PEDOT nanowires on (a) annular-shaped and (b) flat-topped base electrodes. (Reprinted with permission from Liu, R., Duay, J., and Lee, S.B., *ACS Nano*, 5, 5608–5619, 2011. Copyright 2011, American Chemical Society.)

in a potentially different mechanism for nanotube growth. Although multistep along with rotating electric field processes can be used for metal nanotube synthesis,[171–173] only the mechanism of formation of metal nanotubes through single-step and linear electromagnetic field methods will be discussed.

In 1991, Martin et al. were the first to utilize a template to produce metal nanotubes (200 nm diameter and 2 μm long).[174] They used AAO templates to electrodeposit gold *microtubules*. The synthesis mechanism was thought to be affiliated with molecular anchoring of cyanosilanes on the AAO pore walls as the template was first sonicated in a solution of (2-cyanoethyl) triethoxysilane prior to gold electrodeposition. The gold was then galvanostatically formed along the pore walls due to the interaction between the Au ions in solution and the cyano group along the pore walls. Xu et al. were also able to create metal nanotubes through the functionalization of AAO pore walls.[33] In this case, they galvanostatically synthesized Ni nanotubes (160 nm diameter and 35 μm length) utilizing an organoamine to functionalize the AAO pores.

However, Tourillon et al. were able to create Co and Fe nanotubes (30 nm diameter and 6 μm long) without the use of a molecular anchor by utilizing a track-etched polycarbonate template.[175] They hypothesized that the metal cations in the solution complexed with the $-CO_3^{2-}$ groups along the template wall. The metal cations were then reduced by pulsed ECD for short total deposition times leading to the formation of 1–2 nm thick nanotubes. If the deposition time was increased, then nanoclusters would result and eventually the clusters would coalescence into nanowires.

In 2004, Lee et al. observed a different mechanism for nanotube formation.[64] While preparing platinum nanowires by galvanostatic electrodeposition in an unmodified AAO template, they noticed that at high current densities, the deposition process occurs quickly along the pore walls resulting in the current collector of hollow metal nanotubes. They hypothesized that this was the result of a high electric field created at the tips of the initial tube-shaped formation that results from the plasma-deposited base electrode. The rest of the deposition was then kinetically driven up the pore walls. This was the first demonstration of a kinetics-related mechanism.

Another possible mechanism was proposed by Davis and Podlaha[176] who reasoned that metal nanotube formation was caused by the hydrogen evolution simultaneously occurring during the metal electrodeposition. They came to this conclusion by electrodepositing Cu from two different Cu^{2+} ion concentration baths in an unmodified polycarbonate template. The more concentrated bath produced nanowires, while the less concentrated bath produced nanotubes. They associated the lower concentrations with lower efficiencies resulting from an increase in the hydrogen evolution side reaction resulting in bubble formation creating a void in the center of the pore thus producing nanotubes, while high concentrations produced high efficiencies with little to no hydrogen evolution producing materials that completely fill the pores, that is, nanowires.

Other evidence of hydrogen formation playing a role includes Fukunaka et al.,[177] who were able to observe an increase in Ni nanotube wall thickness when the electrodeposition bath's pH was increased. These nanotubes were produced in an unmodified polycarbonate template. They associate the nanotube formation with the depression of the H_2 bubble formation, a schematic of which is found in Figure 10.13. It can be inferred that H_2 generation is less at higher pHs resulting in smaller diameter bubble formation and thus thicker-walled nanotubes. Thus, H_2 gas bubble formation and/or suppression must be considered as a possible aspect of the metal nanotube formation mechanism.

In 2006, Xu et al. were able to synthesize Ni nanotubes in AAO utilizing P123 as a surfactant in the electrodeposition bath.[178] The surfactant is thought to have a strong affinity for both the AAO and the Ni^{2+} ions, thus encouraging deposition along the pore walls leading to nickel nanotube synthesis. An increase in nanotube wall thickness was observed with increased electrodeposition time; however, in contrast to the previous kinetics mechanism, they found that increasing the current density also leads to an increase in nanotube wall thickness. Certainly, there may be different mechanism between the unmodified pores and these surfactant modified pores.

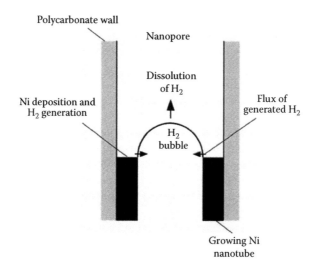

FIGURE 10.13 Proposed metal nanotube electrochemical synthesis method by hydrogen bubble suppression. (Reproduced from Fukunaka, Y. et al., *Electrochem. Solid State Lett.*, 9, C62, 2006. With permission from The Electrochemical Society.)

Cao et al. proposed their own general method for metal nanotube growth for unmodified AAO pores that they termed current directed tubular growth (CDTG).[179] In this mechanism, they attribute the growth to competing deposition directions. One is parallel to the current direction while the other one is perpendicular. At low current densities, the parallel and perpendicular growths have the same probability of proceeding; therefore, the pore will completely fill resulting in nanowire formation. At high current densities, parallel growth is preferential, and since nucleation normally occurs at the bottom edge of the template pore, the deposition will proceed from there along the pore wall forming nanotubes. This mechanism, a schematic of which can be observed in Figure 10.14, is an extension of the previous kinetics mechanism.

Although the CDTG mechanism has merit, it neglects to provide a good explanation for the nucleation of the metal at bottom pore edge. In other words, they neglect the influence of the morphology of the base current collector. As shown before for polymer nanotube synthesis, the base electrode can direct growth along the pore walls if it is annular or it can promote growth along the whole cross section of the pore if it is flat.[66] This was also demonstrated for reductive deposition by Li et al. for ZnO nanotube formation in an unmodified AAO template.[180] This is not to say that the CDTG mechanism is incorrect, but that other aspects of the deposition process should be included in this mechanism. For example, Liu et al. were able to exploit both the base electrode morphology and the current density mechanism to make Co–Cu alloy nanotubes in AAO.[181] Figure 10.15 shows the formation mechanism of these ferromagnetic nanotubes along with their SEM images.

The base electrode morphology mechanism was further exploited by Han et al. who varied sputtering times of the base electrode to produce different nanotube wall thicknesses of electrodeposited ferromagnetic alloys.[30] It is inferred that the increased sputtering of the gold current collector resulted in a larger coverage of the pore resulting in nucleation of the alloys at greater distance from the pore edge thus creating thicker-walled nanotubes.

Li and coworkers[182] have proposed a new mechanism that assumes different nanotube to nanowire transitions for different melting point metals. As mentioned earlier, the morphology of sputtered base electrodes are annular, and the authors assume that all metals synthesized with this base electrode morphology initially form tubes, and depending on the conditions, some materials

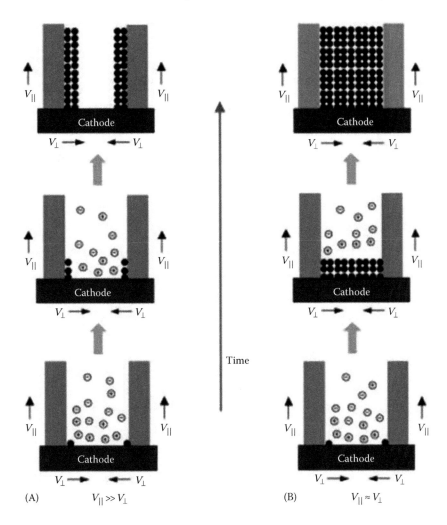

FIGURE 10.14 Proposed formation mechanism termed current directed tubular growth (CDTG) (A) representing nanotube formation and (B) representing nanowire formation. (Cao, H., Wang, L., Qiu, Y., Wu, Q., Wang, G., Zhang, L., and Liu, X.: Generation and growth mechanism of metal (Fe, Co, Ni) nanotube arrays. *Chem Phys Chem.* 2006. 7. 1500–1504. Copyright Wiley-VCH Verlag GmbH & Co. KGaA. Reproduced with permission.)

eventually begin to deposit as nanowires. This transition from nanotube to nanowire growth has indeed been demonstrated for Ni metal electrodeposition with shorter times producing nanotubes and longer times producing nanowires.[29,183] Thus, the authors propose that the quickness of this nanotube to nanowire transition is dependent on the metal. They suggest that high melting point metals, such as Ni, and metal alloys have adatoms that tend to form new nuclei. As the electrodeposition proceeds, the concentration gradient in the pore promotes a higher and higher nucleation rate that eventually leads to the gradual closing of the pore that completes the nanotube to nanowire transition as depicted in Figure 10.16A. However, for lower melting point metals and metal alloys, such as BiSb, adatoms tend to diffuse relatively quickly forming large grain sizes. Thus, grain growth is preferential to nucleation, meaning perpendicular growth will occur preferentially to parallel growth resulting in an abrupt transition from nanotube to nanowire formation as depicted in Figure 10.16B.

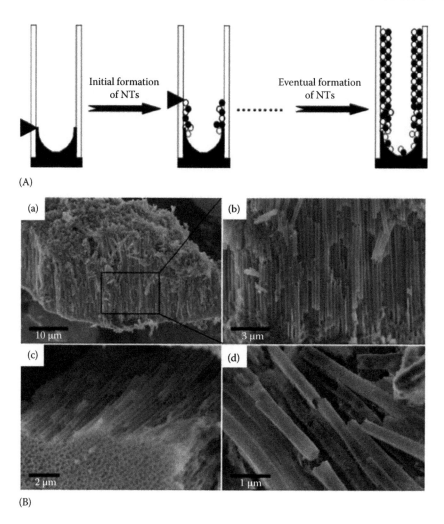

(A)

(B)

FIGURE 10.15 (A) Proposed electrochemical formation mechanism for ferromagnetic nanotubes. (B) SEM images of ferromagnetic nanotubes. (a) Overview SEM image of a Co-Cu alloy nanotube array after the removal of the AAO template. (b) Side-view SEM image of the rectangle area marked in panel a. (c) Top-view SEM image of the Co-Cu nanotube array. (d) SEM image of the dispersed Co-Cu alloy NTs. (Reprinted with permission from Liu, L., Zhou, W., Xie, S. et al., *J. Phys. Chem. C*, 112, 2256–2261, 2008. Copyright 2008, American Chemical Society.)

Recently, Zhang et al.[184] proposed the idea of a critical potential for nanotubular growth while observing the electrodeposition of Co into unmodified AAO templates. In this mechanism, nanowires are formed when the applied potential is below a critical value, while nanotubes are formed when the applied potential is above this critical value. A depiction of this for Co nanostructures formed at different applied potentials can be found in Figure 10.17. This work was extended for other metals and metal alloys including Zn, Cu, Ni, ZnNi, and CoNi, all of which had a critical potential for nanotube formation.[185]

Work by Lee et al.[205] proposes a surface-directed growth mechanism for nickel nanotubes, primarily controlled by the adsorption of borate onto the AAO pore wall. In this work, it is found that nanotube growth only occurs when the borate ion is present, with boric acid being a common electrolyte component in the prevalent Watts nickel electroplating bath. Modulation of pH, applied voltage, and base electrode morphology had no effect as nanotubes were always

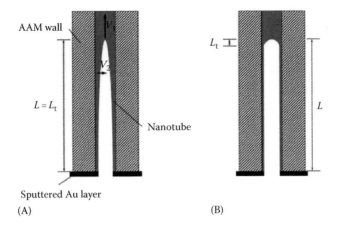

FIGURE 10.16 Nanotube to nanowire growth transition of (A) a high melting point metal and (B) a low melting point metal. (Reprinted with permission from Dou, X., Li, G., Huang, X., and Li, L., *J. Phys. Chem. C*, 112, 8167–8171, 2008. Copyright 2008, American Chemical Society.)

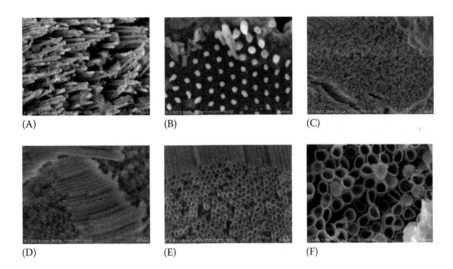

FIGURE 10.17 SEM images of cobalt nanostructures deposited into an unmodified AAO template at applied potentials of (A) −1.0 V, (B) −1.2 V, (C) −1.25 V, (D) −1.5 V, (E) −2.0 V, and (F) −3.0 V. (With kind permission from Springer Science+Business Media: *J. Mater. Sci.*, Template-based synthesis and discontinuous hysteresis loops of cobalt nanotube arrays, 48, 2013, 7392–7398, Zhang, H., Zhang, X., Wu, T., Zhang, Z., Zheng, J., and Sun, H. Copyright 2013.)

produced as long as the borate ion was existent. However, when boric acid was replaced with citric acid, nanowires were formed. Also, interesting was when ethylene glycol was added to the deposition bath, nanowires were formed. This was thought to be caused by the complexation of the diol compound with boric acid, eliminating boric acid's efficacy. A schematic of the complexation of the borate ion with the pore wall with and without ethylene glycol can be found in Figure 10.18.

Work done by Dolati et al., mentioned earlier in this chapter, shows that the cyclic voltammetry response is different at different scan rates for gold nanotube growth electrodeposition as seen in Figure 10.19.[126] At slower scan rates, the CV curves were sigmoidally shaped representing

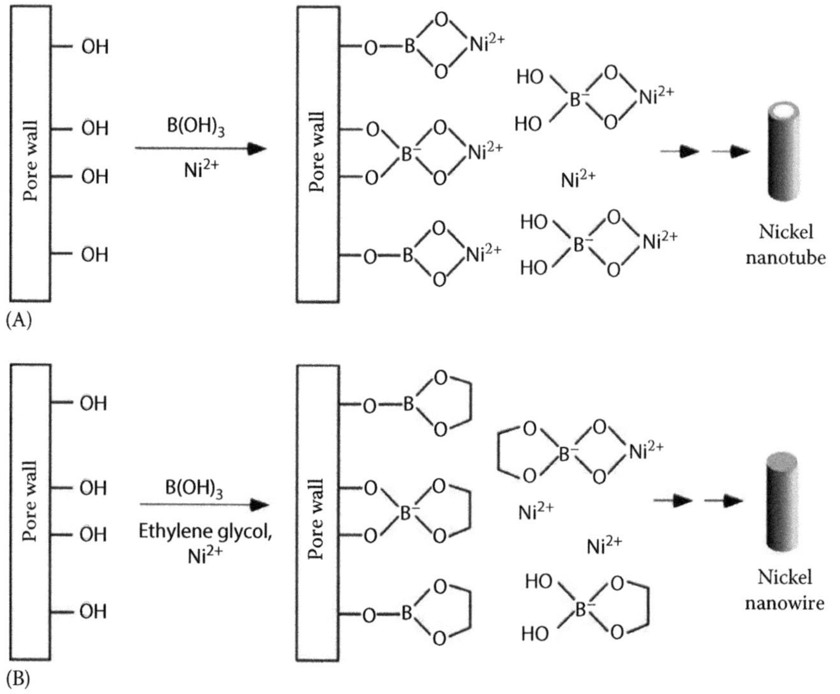

FIGURE 10.18 (A) Schematic representing the adsorption of borate ions on the AAO pore wall facilitating Ni nanotube growth. (B) Schematic representing an ethylene glycol blocking mechanism where ethylene glycol forms covalent bonds with the adsorbed borate ion thus preventing nanotube growth. (From Graham, L.M. et al., *Chem. Commun.*, 50, 527, 2014.)

radial diffusion at the pore openings, while, at faster scan rates, the CV curves were peak shaped, representing linear diffusion in the pores. This resulting gold deposition was in the morphology of nanotubes. Recently, the CV method was also shown to create gold nanotubes.[186] It is thought that the CV deposition represents a type of pulsed electrodeposition, resulting in alternating deposition rates causing concentration polarizations that result in nanotube rather than nanowire formation.

A great example of what structures can be produced when knowledge of the electrodeposition mechanism is exploited is palladium nanosprings electrochemically produced by Park et al.[187] Previously, the authors were able to create thin-walled palladium nanotubes in AAO by exploiting the hydroxyl-terminated pore walls of the template.[188] They accomplished this by utilizing a strongly acidic electrodeposition bath that influences hydrogen ions to adsorb on the alumina surface resulting in a compact double layer enriched with hydrogen at the alumina surface. Upon administration of the deposition voltage, hydrogen will be preferentially generated at the pore walls. Hydrogen then reduces the palladium ions in the solution forming palladium metal on the pore walls resulting in palladium nanotubes.

Taking this mechanism a step further by adding Cu^{2+} to the electrodeposition bath, the deposition becomes more interesting due to the copper ion's lower affinity for the adsorbed hydrogen. The resulting structure was a double-spiral alloy of copper and palladium with the copper spiral being indented due to its lower affinity for the pore wall. It is thought that the double spiral structure is due to screw dislocations forming during the metal electrodeposition. The energy required for growth of these screw dislocations is a thousand times smaller than any new crystal nucleation resulting in continuous spiral formation. Selective etching of the copper segments results in palladium

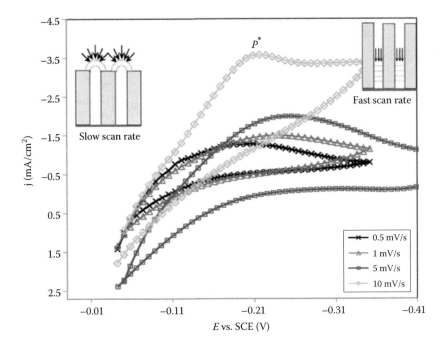

FIGURE 10.19 Cyclic voltammetry curves at different scan rates for a track-etched polymer template in an $AuCl_4^-$ showing sigmoidal shapes at slow scan rates and peak shapes at fast scan rates. (Reproduced from Hariri, M.B. et al., *J. Electrochem. Soc.*, 160, D279, 2013. With permission from the Electrochemical Society.)

nanospring formation. A schematic of the synthesis along with TEM and SEM images of these nanosprings can be found in Figure 10.20.

It is shown that metal nanotube growth can proceed through a variety of ways including pore-wall modification, nanoelectrode morphology at the base of the nanopore, hydrogen bubble–mediated formation, interplay between kinetics and thermodynamics, intrinsic aspects of specific metals such as favorable crystal structure, and metal ion concentration and electrochemical potential effects. Unless there is a systematic study developed that can take all of these proposed mechanisms into account, the metal nanotube formation mechanism cannot be generalized. Therefore, it must be concluded that all proposed mechanisms here must be thought to contribute to the nanotube formation.

10.5 ELECTROCHEMICAL TRANSFORMATIONS OF NANOSTRUCTURES

Although template synthesis is indeed the focus of this chapter, it is noteworthy to include other controlled electrochemical methods where the nanostructures themselves act as sacrificial templates for production of novel nanostructures. The resulting products of these transformations can result in even more improved performance metrics than the previously untransformed nanomaterial.

10.5.1 Electrochemical Phase Transformation

One way to modify the properties of materials is by changing their phase. This has been accomplished electrochemically through oxidation and reduction of the material. For example, Zhang et al. were able to electrochemically transform NiS nanoparticles into $Ni(OH)_2$.[189] They accomplished this by cyclic voltammetry of NiS in a KOH electrolyte.

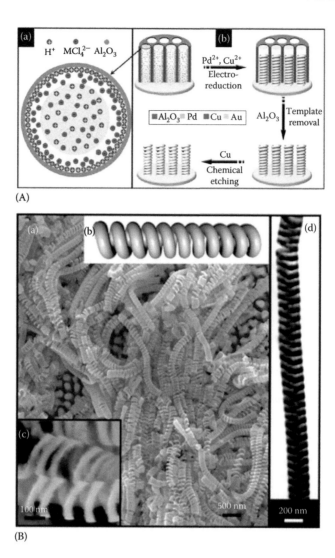

FIGURE 10.20 (A) Synthesis mechanism of the palladium nanosprings. (B) SEM and TEM of the resulting Pd nanosprings. (Reprinted with permission from Liu, L., Yoo, S.-H., Lee, S.A., and Park, S., *Nano Lett.*, 11, 3979–3982, 2011. Copyright 2011, American Chemical Society.)

Other phase transformations can actually change the morphology of the material. For example, Lee et al. were able to produce hierarchical MnO_2 nanowire/nanofibrils by reducing bare MnO_2 nanowires in a neutral Mn^{2+} electrolyte.[19] In this process, MnO_2 is reduced to Mn_2O_3 simultaneously producing hydroxide along the surface of the nanowire as a by-product. This local increase in pH causes precipitation of the Mn^{2+} in solution as $Mn(OH)_2$ nanofibrils. Subsequent electrochemical oxidation converts the nanowires and nanofibrils to MnO_2. A depiction of this synthesis mechanism along with TEM images of the product at different charge densities can be found in Figure 10.21.

An additional electrochemical phase transformation is a transition from pure crystalline Si nanowires to lithiated amorphous Si nanowires that was accomplished through an electrochemical short-circuit method where lithium metal was directly contacted with the nanowires in a lithium ion electrolyte.[190] This electrochemical short-circuit method, which is essentially a battery without a

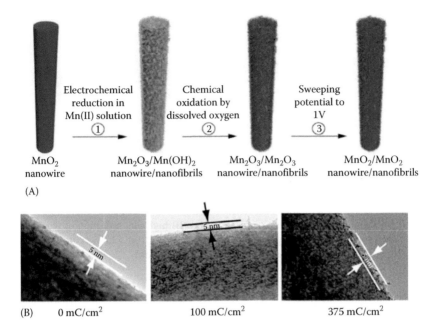

FIGURE 10.21   (A) Schematic representing the mechanism of formation for MnO_2 nanowire/nanofibril arrays. (B) TEM images of the MnO_2 nanofibril growth at different charge densities. (Reprinted with permission from Duay, J., Sherrill, S.A., Gui, Z., Gillette, E., and Lee, S.B., *ACS Nano*, 7, 1200–1214, 2013. Copyright 2013, American Chemical Society.)

separator, was first developed by Delmas et al. in the 1980s to study lithiated oxides with intermediate compositions.[191] This method allows one to prelithiate compounds below 1.0 V vs. Li, which is unobtainable through chemical means such as n-butyllithium.[192]

10.5.2   ELECTROCHEMICAL DEALLOYING

One way of creating nanostructured materials is first to create bulk heterogeneous materials and then selectively etching one of the components of the bulk material. A good method for accomplishing this is through first synthesizing a bulk alloy and then selectively etching one of the components through a method called electrochemical dealloying.

This method has been shown to create a variety of materials, including porous ruthenium oxide with 2–3 nm pores through the anodic removal of copper in a Ru–Cu bulk alloy by Chung and coworkers,[193] porous gold film created by electrodeposition of porous copper that is partially galvanically exchanged with gold creating a porous Au/Cu alloy followed by electrochemical dealloying of the copper,[194] nanoporous gold created from electrochemical dealloying of a Ag–Au alloy,[195] nanoporous Ag ribbons by adding creating Ce–Ag–Cu ribbons followed by electrochemical dealloying,[196] and nanoporous Pd from a $Cu_{75}Pd_{25}$ alloy that can be seen in Figure 10.22.[197]

10.5.3   GALVANIC REPLACEMENT

Hollow noble metal nanostructures are proposed as containers for drug delivery as well as low density inert packing materials for composite materials. In addition, they have unique plasmonic and catalysis properties.[198] Many methods have been used to produce these materials including hard template synthesis of nanotubes mentioned earlier in this chapter.

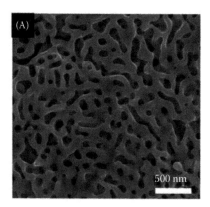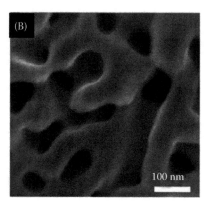

FIGURE 10.22 SEM images at (A) low magnification and (B) high magnification of nanoporous Pd synthesized by the electrochemical dealloying of $Cu_{75}Pd_{25}$. (Reproduced from Li, W. et al., *Phys. Chem. Chem. Phys.*, 13, 5565, 2011. With permission from the PCCP Owner Societies.)

A relatively new research direction that produces hollow metal nanostructures is galvanic replacement.[198–200] This method originates from the activity series of metals. When high activity metals are placed in solutions containing the ions of metals with low activity, a spontaneous replacement reaction ensues where the more active metal species becomes ionic, and the ions of the less active metal become reduced to the zerovalent metal.[201]

This process was essentially adapted for nanoapplications by Xia et al. in 2002 by using silver nanostructures as the sacrificial relatively active metal.[202] These were placed in a gold, platinum, and palladium noble metal ion solutions to produce hollow noble metal nanostructures. The synthesis mechanism along with TEM and SEM images of the products of this reaction can be found in Figure 10.23.

As shown, great control over morphology of these hollow materials can be obtained by choosing the appropriate metal nanoparticle structure. Recent studies have shown that halogen ions can promote the replacement reaction between Pd nanocrystals and $PtCl_6^-$ ions.[203] Also, many different morphologies such as spherical, cylindrical, and cubic topologies were shown to be created by simple room temperature interplay between the galvanic replacement mechanism and the Kirkendall growth effect.[204]

10.6 SUMMARY

This chapter has demonstrated a variety of ways to produce controlled nanostructures through the use of ECD utilizing templates both hard and soft as well as electrochemical techniques employing nanostructures as templates themselves.

In addition, the nanopore filling and nanotube formation mechanism of both polymers and metals is analyzed in detail. The results indicate that many factors affect the deposition morphology including electrochemical bath pH, pore-anchoring agents, nanopore geometry, nanoelectrode morphology, pore-wall functionalization, active ion concentration, as well as electrodeposition potential and/or current density.

Accordingly, this chapter is meant to show the current state of the art and novel techniques for preparing controllable nanostructures as well as detailed mechanisms for electrochemical growth. Utilizing this information, it is hoped that it can serve as a template in its own right for scientists to create future nanostructured materials through electrochemical template synthesis.

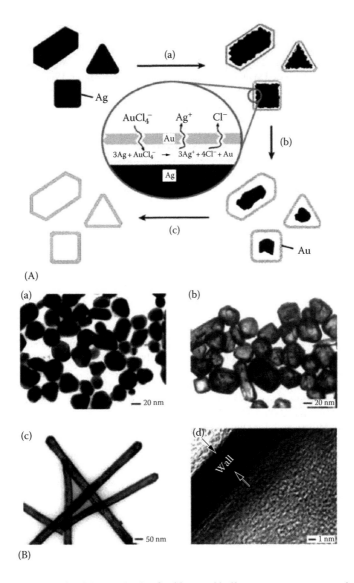

FIGURE 10.23 (A) Schematic of the synthesis of noble metal hollow nanostructures synthesized by galvanic replacement of silver nanostructures. (B) SEM and TEM images of the resulting hollow metal nanostructured products. (Reprinted with permission from Sun, Y., Mayers, B.T., and Xia, Y., *Nano Lett.*, 2, 481–485, 2002. Copyright 2002, American Chemical Society.)

These controllable nanostructures can then be used as model systems to scientifically investigate the enhanced properties of materials with confined dimensions.

ACKNOWLEDGMENT

The work was supported by the Nanostructures for Electrical Energy Storage, an Energy Frontier Research Center funded by the US Department of Energy, Office of Science, Office of Basic Energy Sciences under Award Number DESC0001160.

REFERENCES

1. Bruce, P. G.; Scrosati, B.; Tarascon, J.-M. Nanomaterials for rechargeable lithium batteries. *Angew. Chem. Int. Ed.* 2008, 47, 2930–2946.
2. Martin, C. R. Membrane-based synthesis of nanomaterials. *Chem. Mater.* 1996, 8, 1739–1746.
3. Mor, G. K.; Varghese, O. K.; Paulose, M.; Shankar, K.; Grimes, C. A. A review on highly ordered, vertically oriented TiO₂ nanotube arrays: Fabrication, material properties, and solar energy applications. *Sol. Energy Mater. Sol. Cells* 2006, 90, 2011–2075.
4. Liu, R.; Duay, J.; Lee, S. B. Heterogeneous nanostructured electrode materials for electrochemical energy storage. *Chem. Commun.* 2011, 47, 1384–1404.
5. Barth, S.; Hernandez-Ramirez, F.; Holmes, J. D.; Romano-Rodriguez, A. Synthesis and applications of one-dimensional semiconductors. *Prog. Mater. Sci.* 2010, 55, 563–627.
6. Simon, P.; Gogotsi, Y. Materials for electrochemical capacitors. *Nat. Mater.* 2008, 7, 845–854.
7. Xia, Y.; Gates, B.; Yin, Y.; Lu, Y. Monodispersed colloidal spheres: Old materials with new applications. *Adv. Mater.* 2000, 12, 693–713.
8. Liu, C.; Li, F.; Ma, L.-P.; Cheng, H.-M. Advanced materials for energy storage. *Adv. Mater.* 2010, 22, E28–E62.
9. Guo, Y.-G.; Hu, J.-S.; Wan, L.-J. Nanostructured materials for electrochemical energy conversion and storage devices. *Adv. Mater.* 2008, 20. 2878–2887.
10. Szabo, A.; Cope, D.: Tallman, D.; Kovach, D.; Wightman, P. Chronoamperometric current at hemicylinder and band microelectrodes—Theory and experiment. *J. Electroanal. Chem.* 1987, 217, 417–423.
11. Albrecht, M. G.; Creighton, J. A. Anomalously intense Raman spectra of pyridine at a silver electrode. *J. Am. Chem. Soc.* 1977, 99, 5215–5217.
12. Jeanmaire, D. L.; Van Duyne, R. P. Surface Raman spectroelectrochemistry: Part I. Heterocyclic, aromatic, and aliphatic amines adsorbed on the anodized silver electrode. *J. Electroanal. Chem. Interfacial Electrochem.* 1977, 84, 1–20.
13. Dieringer, J.; McFarland, A.; Shah, N.; Stuart, D.; Whitney, A.; Yonzon, C.; Young, M.; Zhang, X.; Van, D. R. Surface enhanced Raman spectroscopy: New materials, concepts, characterization tools, and applications. *Faraday Discuss.* 2006, 132, 9–26.
14. Banholzer, M. J.; Millstone, J. E.; Qin, L.; Mirkin, C. A. Rationally designed nanostructures for surface-enhanced Raman spectroscopy. *Chem. Soc. Rev.* 2008, 37, 885–897.
15. Jensen, T. R.; Malinsky, M. D.; Haynes, C. L.; Van Duyne, R. P. Nanosphere lithography: Tunable localized surface plasmon resonance spectra of silver nanoparticles. *J. Phys. Chem. B* 2000, 104, 10549–10556.
16. Li, X.; Zhang, Y.; Shen, Z. X.; Fan, H. J. Highly ordered arrays of particle-in-bowl plasmonic nanostructures for surface-enhanced Raman scattering. *Small* 2012, 8, 2548–2554.
17. Li, J. F.; Huang, Y. F.; Ding, Y. et al. Shell-isolated nanoparticle-enhanced Raman spectroscopy. *Nature* 2010, 464, 392–395.
18. Liu, R.; Duay, J.; Lane, T.; Lee, S. B. Synthesis and characterization of RuO₂/poly(3,4-ethylenedioxythiophene) composite nanotubes for supercapacitors. *Phys. Chem. Chem. Phys.* 2010, 12, 4309–4316.
19. Duay, J.; Sherrill, S. A.; Gui, Z.; Gillette, E.; Lee, S. B. Self-limiting electrodeposition of hierarchical MnO₂ and M(OH)₂/MnO₂ nanofibril/nanowires: Mechanism and supercapacitor properties. *ACS Nano* 2013, 7, 1200–1214.
20. Liu, R.; Duay, J.; Lee, S. B. Redox exchange induced MnO₂ nanoparticle enrichment in poly (3,4-ethylenedioxythiophene) nanowires for electrochemical energy storage. *ACS Nano* 2010, 4, 4299–4307.
21. Sherrill, S. A.; Duay, J.; Gui, Z.; Banerjee, P.; Rubloff, G. W.; Lee, S. B. MnO₂/TiN heterogeneous nanostructure design for electrochemical energy storage. *Phys. Chem. Chem. Phys.* 2011, 13, 15221–15226.
22. Syed, M.; Singh, N.; Lee, J.; Stucky, G. D.; Moskovits, M.; McFarland, E. W. Synthesis of chemicals using solar energy with stable photoelectrochemically active heterostructures. *Nano Lett.* 2013, 13, 2110–2115.
23. Hossain, M. A.; Yang, G.; Parameswaran, M.; Jennings, J. R.; Wang, Q. Mesoporous SnO₂ spheres synthesized by electrochemical anodization and their application in CdSe-sensitized solar cells. *J. Phys. Chem. C* 2010, 114, 21878–21884.
24. Sealy, C. Polymer nanotubes improve connection to brain implants. *Nano Today* 2009, 4, 453–453.
25. Abidian, M.; Corey, J.; Kipke, D.; Martin, D. Conducting-polymer nanotubes improve electrical properties, mechanical adhesion, neural attachment, and neurite outgrowth of neural electrodes. *Small* 2010, 6, 421–429.

26. Abidian, M.; Ludwig, K.; Marzullo, T.; Martin, D.; Kipke, D. Interfacing conducting polymer nanotubes with the central nervous system: Chronic neural recording using poly(3,4-ethylenedioxythiophene) nanotubes. *Adv. Mater.* 2009, 21, 3764–3770.

27. Iacopino, D.; Redmond, G. Reversible modulation of photoluminescence from conjugated polymer nanotubes by incorporation of photochromic spirooxazine molecules. *Chem. Commun.* 2011, 47, 9170–9172.

28. Fujihara, H.; Nakahodo, T.; Sato, K. Redox-active π-conjugated polymer nanotubes with viologen for encapsulation and release of fluorescent dye in the nanospace. *Chem. Commun.* 2011, 47, 10067–10069.

29. Li, X.; Wang, Y.; Song, G. et al. Fabrication and magnetic properties of Ni/Cu shell/core nanocable arrays. *J. Phys. Chem. C* 2010, 114, 6914–6916.

30. Chen, J.; Liu, D.; Shamaila, S.; Liu, H.; Sharif, R.; Han, X. Structural and magnetic properties of various ferromagnetic nanotubes. *Adv. Mater.* 2009, 21, 4619–4624.

31. Whitney, T. M.; Searson, P. C.; Jiang, J. S.; Chien, C. L. Fabrication and magnetic properties of arrays of metallic nanowires. *Science* 1993, 261, 1316–1319.

32. Xu, L.; Tung, L. D.; Spinu, L.; Zakhidov, A. A.; Baughman, R. H.; Wiley, J. B. Synthesis and magnetic behavior of periodic nickel sphere arrays. *Adv. Mater.* 2003, 15, 1562–1564.

33. Bao, J.; Tie, C.; Xu, Z.; Zhou, Q.; Shen, D.; Ma, Q. Template synthesis of an array of nickel nanotubules and its magnetic behavior. *Adv. Mater.* 2001, 13, 1631–1633.

34. Liu, R.; Sen, A. Autonomous nanomotor based on copper–platinum segmented nanobattery. *J. Am. Chem. Soc.* 2011, 133, 20064–20067.

35. Garcia-Gradilla, V.; Orozco, J.; Sattayasamitsathit, S. et al. Functionalized ultrasound-propelled magnetically guided nanomotors: Toward practical biomedical applications. *ACS Nano* 2013, 7(10), 9232–9240. doi:10.1021/nn403851v.

36. Kowalski, D.; Kim, D.; Schmuki, P. TiO_2 nanotubes, nanochannels and mesosponge: Self-organized formation and applications. *Nano Today* 2013, 8, 235–264.

37. Long, Y.-Z.; Li, M.-M.; Gu, C.; Wan, M.; Duvail, J.-L.; Liu, Z.; Fan, Z. Recent advances in synthesis, physical properties and applications of conducting polymer nanotubes and nanofibers. *Prog. Polym. Sci.* 2011, 36, 1415–1442.

38. Zhang, D.; Wang, Y. Synthesis and applications of one-dimensional nano-structured polyaniline: An overview. *Mater. Sci. Eng. B* 2006, 134, 9–19.

39. Han, J.; Wang, L.; Guo, R. Facile synthesis of hierarchical conducting polymer nanotubes derived from nanofibers and their application for controlled drug release. *Macromol. Rapid Commun.* 2011, 32, 729–735.

40. Yuan, Y. F.; Xia, X. H.; Wu, J. B. et al. Hierarchically porous Co_3O_4 film with mesoporous walls prepared via liquid crystalline template for supercapacitor application. *Electrochem. Commun.* 2011, 13, 1123–1126.

41. Wang, D.; Zhang, L.; Lee, W.; Knez, M.; Liu, L. Novel three-dimensional nanoporous alumina as a template for hierarchical TiO_2 nanotube arrays. *Small* 2013, 9, 1025–1029.

42. Kwon, O. S.; Park, S. J.; Lee, J. S. et al. Multidimensional conducting polymer nanotubes for ultrasensitive chemical nerve agent sensing. *Nano Lett.* 2012, 12, 2797–2802.

43. Rolison, D. R.; Long, J. W.; Lytle, J. C. et al. Multifunctional 3D nanoarchitectures for energy storage and conversion. *Chem. Soc. Rev.* 2009, 38, 226–252.

44. Reddy, A. L. M.; Shaijumon, M. M.; Gowda, S. R.; Ajayan, P. M. Multisegmented Au-MnO_2/carbon nanotube hybrid coaxial arrays for high-power supercapacitor applications. *J. Phys. Chem. C* 2010, 114, 658–663.

45. Venable, S. L. *Gold: A Cultural Encyclopedia*; ABC-CLIO, Santa Barbara, CA, 2011.

46. Kim, K.-H.; Kim, K. S.; Kim, G.-P.; Baeck, S.-H. Electrodeposition of mesoporous ruthenium oxide using an aqueous mixture of CTAB and SDS as a templating agent. *Curr. Appl. Phys.* 2012, 12, 36–39.

47. Benedetti, T. M.; Gonçales, V. R.; Córdoba de Torresi, S. I.; Torresi, R. M. In search of an appropriate ionic liquid as electrolyte for macroporous manganese oxide film electrochemistry. *J. Power Sources* 2013, 239, 1–8.

48. Chen, Q.-P.; Xue, M.-Z.; Sheng, Q.-R.; Liu, Y.-G.; Ma, Z.-F. Electrochemical growth of nanopillar zinc oxide films by applying a low concentration of zinc nitrate precursor. *Electrochem. Solid State Lett.* 2006, 9, C58–C61.

49. Murugesan, S.; Akkineni, A.; Chou, B. P.; Glaz, M. S.; Vanden Bout, D. A.; Stevenson, K. J. Room temperature electrodeposition of molybdenum sulfide for catalytic and photoluminescence applications. *ACS Nano* 2013, 7, 8199–8205.

50. Zhang, F.; Wong, S. S. Controlled synthesis of semiconducting metal sulfide nanowires. *Chem. Mater.* 2009, 21, 4541–4554.

51. Gu, J.; Fahrenkrug, E.; Maldonado, S. Direct electrodeposition of crystalline silicon at low temperatures. *J. Am. Chem. Soc.* 2013, 135, 1684–1687.

52. Al-Salman, R.; Mallet, J.; Molinari, M.; Fricoteaux, P.; Martineau, F.; Troyon, M.; El, A., S. Zein; Endres, F. Template assisted electrodeposition of germanium and silicon nanowires in an ionic liquid. *Phys. Chem. Chem. Phys.* 2008, 10, 6233–6237.

53. Mallet, J.; Martineau, F.; Molinari, M.; Namur, K. Electrodeposition of silicon nanotubes at room temperature using ionic liquid. *Phys. Chem. Chem. Phys.* 2013, 15, 16446–16449.

54. Demoustier-Champagne, S.; Ferain, E.; Jérôme, C.; Jérôme, R.; Legras, R. Electrochemically synthesized polypyrrole nanotubules: Effects of different experimental conditions. *Eur. Polym. J.* 1998, 34, 1767–1774.

55. Delvaux, M.; Duchet, J.; Stavaux, P.-Y.; Legras, R.; Demoustier-Champagne, S. Chemical and electrochemical synthesis of polyaniline micro- and nano-tubules. *Synth. Met.* 2000, 113, 275–280.

56. Cho, S. I.; Choi, D. H.; Kim, S.-H.; Lee, S. B. Electrochemical synthesis and fast electrochromics of poly(3,4-ethylenedioxythiophene) nanotubes in flexible substrate. *Chem. Mater.* 2005, 17, 4564–4566.

57. Yan, J.; Fu, Y.; Mao, B.; Wei, Y.; Su, Y. The electrode/ionic liquid interface: Electric double layer and metal electrodeposition. *ChemPhysChem* 2010, 11, 2764–2778.

58. McKenzie, K. J.; Abbott, A. P. Application of ionic liquids to the electrodeposition of metals. *Phys. Chem. Chem. Phys.* 2006, 8, 4265–4279.

59. Pomfret, M. B.; Brown, D. J.; Epshteyn, A.; Purdy, A. P.; Owrutsky, J. C. Electrochemical template deposition of aluminum nanorods using ionic liquids. *Chem. Mater.* 2008, 20, 5945–5947.

60. Perre, E.; Nyholm, L.; Gustafsson, T.; Taberna, P.-L.; Simon, P.; Edstroem, K. Direct electrodeposition of aluminum nano-rods. *Electrochem. Commun.* 2008, 10, 1467–1470.

61. Keller, F.; Robinson, D. L.; Hunter, M. S. Structural features of oxide coatings on aluminum. *J. Electrochem. Soc.* 1953, 100, 411–419.

62. Sulka, G. D. Highly ordered anodic porous alumina formation by self-organized anodizing. In *Nanostructured Materials in Electrochemistry*. Weinheim, Germany: Wiley-VCH Verlag GmbH & Co. KGaA, 2008; pp. 1–116.

63. Asada, T. Japanese Patent No. 310,401 1963.

64. Yoo, W.; Lee, J. Field-dependent growth patterns of metals electroplated in nanoporous alumina membranes. *Adv. Mater.* 2004, 16, 1097–1101.

65. Liu, R.; Duay, J.; Lee, S. B. Electrochemical formation mechanism for the controlled synthesis of heterogeneous MnO₂/poly(3,4-ethylenedioxythiophene) nanowires. *ACS Nano* 2011, 5, 5608–5619.

66. Xiao, R.; Cho, S. I.; Liu, R.; Lee, S. B. Controlled electrochemical synthesis of conductive polymer nanotube structures. *J. Am. Chem. Soc.* 2007, 129, 4483–4489.

67. Yang, D.; Meng, G.; Zhu, X.; Zhu, C. Size-tunable nano-dots and nano-rings from nanochannel-confined electrodeposition. *Chem. Commun.* 2009, 7110–7112.

68. Zhang, F.; Wang, Y.; Bai, Y.; Zhang, R. Preparation and characterization of copper azide nanowire array. *Mater. Lett.* 2012, 89, 176–179.

69. Xie, Y.; Song, Z.; Yao, S.; Wang, H.; Zhang, W.; Yao, Y.; Ye, B.; Song, C.; Chen, J.; Wang, Y. High capacitance properties of electrodeposited PANI-Ag nanocable arrays. *Mater. Lett.* 2012, 86, 77–79.

70. Giallongo, G.; Durante, C.; Pilot, R.; Garoli, D.; Bozio, R.; Romanato, F.; Gennaro, A.; Rizzi, G. A.; Granozzi, G. Growth and optical properties of silver nanostructures obtained on connected anodic aluminum oxide templates. *Nanotechnology* 2012, 23, 325604.

71. Hou, J.-W.; Yang, X.-C.; Cui, M.-M.; Huang, M.; Wang, Q.-Y. Synthesis and optical property of one-dimensional Ag–Cu₂O heterojunctions. *Mater. Lett.* 2012, 74, 159–162.

72. Ren, X.; Lun, Z. Mesoporous silica nanowires synthesized by electrodeposition in AAO. *Mater. Lett.* 2012, 68, 228–229.

73. Reddy, S. M.; Park, J. J.; Na, S.-M.; Maqableh, M. M.; Flatau, A. B.; Stadler, B. J. H. Electrochemical synthesis of magnetostrictive Fe–Ga/Cu multilayered nanowire arrays with tailored magnetic response. *Adv. Funct. Mater.* 2011, 21, 4677–4683.

74. Zwilling, V.; Aucouturier, M.; Darque-Ceretti, E. Anodic oxidation of titanium and TA6V alloy in chromic media. An electrochemical approach. *Electrochim. Acta* 1999, 45, 921–929.

75. Macak, J. M.; Tsuchiya, H.; Schmuki, P. High-aspect-ratio TiO₂ nanotubes by anodization of titanium. *Angew. Chem. Int. Ed.* 2005, 44, 2100–2102.

76. Yang, Y.; Kim, D.-H.; Yang, M.; Schmuki, P. Vertically aligned mixed V₂O₅-TiO₂ nanotube arrays for supercapacitor applications. *Chem. Commun.* 2011, 47, 7746–7748.

77. Dong, S.; Chen, X.; Gu, L. et al. One dimensional MnO₂/titanium nitride nanotube coaxial arrays for high performance electrochemical capacitive energy storage. *Energy Environ. Sci.* 2011, 4, 3502–3508.

78. Sarma, B.; Jurovitzki, A. L.; Smith, Y. R.; Mohanty, S. K.; Misra, M. Redox-induced enhancement in interfacial capacitance of the titania nanotube/bismuth oxide composite electrode. *ACS Appl. Mater. Interfaces* 2013, 5, 1688–1697.

79. Feng, H.; Zhou, L.; Li, J.; Yuan, L.; Wang, N.; Yan, Z.; Cai, Q. A photoelectrochemical immunosensor for tris(2,3-dibromopropyl) isocyanurate detection with a multiple hybrid CdTe/Au-TiO$_2$ nanotube arrays. *Analyst* 2013, 138, 5726–5733.

80. Li, S.; Wang, F.; Xu, M.; Wang, Y.; Fang, W.; Hu, Y. Fabrication and characteristics of a nanostructure PbO$_2$ anode and its application for degradation of phenol. *J. Electrochem. Soc.* 2013, 160, E44–E48.

81. Liu, R.; Yang, W.-D.; Qiang, L.-S.; Liu, H.-Y. Conveniently fabricated heterojunction ZnO/TiO$_2$ electrodes using TiO$_2$ nanotube arrays for dye-sensitized solar cells. *J. Power Sources* 2012, 220, 153–159.

82. Kang, Q.; Liu, S.; Yang, L.; Cai, Q.; Grimes, C. A. Fabrication of PbS nanoparticle-sensitized TiO$_2$ nanotube arrays and their photoelectrochemical properties. *ACS Appl. Mater. Interfaces* 2011, 3, 746–749.

83. Fleischer, R. L.; Price, P. B.; Symes, E. M. Novel filter for biological materials. *Science* 1964, 143, 249–250.

84. Apel, P. Track etching technique in membrane technology. *Radiat. Meas.* 2001, 34, 559–566.

85. Penner, R. M.; Martin, C. R. Controlling the morphology of electronically conductive polymers. *J. Electrochem. Soc.* 1986, 133, 2206–2207.

86. Dobrev, D.; Martin, M.; Neumann, R.; Vetter, J.; Schuchert, I. U.; Molares, M. E. T. Electrochemical copper deposition in etched ion track membranes experimental results and a qualitative kinetic model. *J. Electrochem. Soc.* 2003, 150, C189–C194.

87. Rauber, M.; Alber, I.; Muller, S. et al. Highly-ordered supportless three-dimensional nanowire networks with tunable complexity and interwire connectivity for device integration. *Nano Lett.* 2011, 11, 2304–2310.

88. Tian, M.; Wang, J.; Kurtz, J.; Mallouk, T. E.; Chan, M. H. W. Electrochemical growth of single-crystal metal nanowires via a two-dimensional nucleation and growth mechanism. *Nano Lett.* 2003, 3, 919–923.

89. Zein El Abedin, S.; Endres, F. Free-standing aluminium nanowire architectures made in an ionic liquid. *ChemPhysChem* 2012, 13, 250–255.

90. Picht, O.; Müller, S.; Alber, I. et al. Tuning the geometrical and crystallographic characteristics of Bi$_2$Te$_3$ nanowires by electrodeposition in ion-track membranes. *J. Phys. Chem. C* 2012, 116, 5367–5375.

91. Žužek Rožman, K.; Pečko, D.; Šturm, S. et al. Electrochemical synthesis and characterization of Fe$_{70}$Pd$_{30}$ nanotubes for drug-delivery applications. *Mater. Chem. Phys.* 2012, 133, 218–224.

92. Wei, X.-Q.; Payne, G. F.; Shi, X.-W.; Du, Y. Electrodeposition of a biopolymeric hydrogel in track-etched micropores. *Soft Matter* 2013, 9, 2131–2135.

93. Al Zoubi, M.; Al-Salman, R.; El Abedin, S. Z.; Li, Y.; Endres, F. Electrochemical synthesis of gallium nanowires and macroporous structures in an ionic liquid. *ChemPhysChem* 2011, 12, 2751–2754.

94. Endres, F.; Liu, Z.; Shapouri, M.; El Abedin, S. Z. Electrochemical synthesis of vertically aligned zinc nanowires using track etched polycarbonate membranes as templates. *Phys. Chem. Chem. Phys.* 2013, 15, 11362–11367.

95. Hiltner, P. A.; Krieger, I. M. Diffraction of light by ordered suspensions. *J. Phys. Chem.* 1969, 73, 2386–2389.

96. Luck, W.; Klier, M.; Wesslau, H. Uber Bragg-Reflexe mit sichtbarem licht an monodispersen kunststofflatices. 1. Berichte der Bunsen-gesellschaft fur. *Phys. Chem.* 1963, 67, 75–83.

97. Stober, W.; Fink, A.; Bohn, E. Controlled growth of monodisperse silica spheres in micron size range. *J. Colloid Interface Sci.* 1968, 26, 62–69.

98. Lobo, R. F.; Jede, T. A.; Velev, O. D.; Lenhoff, A. M. Porous silica via colloidal crystallization. *Nature* 1997, 389, 447–448.

99. Braun, P. V.; Wiltzius, P. Microporous materials: Electrochemically grown photonic crystals. *Nature* 1999, 402, 603–604.

100. Halpern, A. R.; Corn, R. M. Lithographically patterned electrodeposition of gold, silver, and nickel nanoring arrays with widely tunable near-infrared plasmonic resonances. *ACS Nano* 2013, 7, 1755–1762.

101. Xia, X.; Tu, J.; Wang, X.; Gu, C.; Zhao, X. Hierarchically porous NiO film grown by chemical bath deposition via a colloidal crystal template as an electrochemical pseudocapacitor material. *J. Mater. Chem.* 2010, 21, 671–679.

102. Cai, J.; Li, S.; Li, Z.; Wang, J.; Ren, Y.; Qin, G. Electrodeposition of Sn-doped hollow α-Fe$_2$O$_3$ nanostructures for photoelectrochemical water splitting. *J. Alloys Compd.* 2013, 574, 421–426.

103. Park, J. Y.; Hendricks, N. R.; Carter, K. R. Hierarchically structured porous cadmium selenide polycrystals using polystyrene bilayer templates. *Langmuir* 2012, 28, 13149–13156.

104. Ho, C.-L.; Wu, M.-S. Manganese oxide nanowires grown on ordered macroporous conductive nickel scaffold for high-performance supercapacitors. *J. Phys. Chem. C* 2011, 115, 22068–22074.

105. Stein, A.; Schroden, R. C. Colloidal crystal templating of three-dimensionally ordered macroporous solids: materials for photonics and beyond. *Curr. Opin. Solid State Mater. Sci.* 2001, 5, 553–564.

106. Zhang, J.; Li, Y.; Zhang, X.; Yang, B. Colloidal self-assembly meets nanofabrication: From two-dimensional colloidal crystals to nanostructure arrays. *Adv. Mater.* 2010, 22, 4249–4269.

107. Wang, J.; Duan, G.; Li, Y.; Liu, G.; Dai, Z.; Zhang, H.; Cai, W. An invisible template method toward gold regular arrays of nanoflowers by electrodeposition. *Langmuir* 2013, 29, 3512–3517.

108. Xiang, J. Y.; Wang, X. L.; Zhong, J.; Zhang, D.; Tu, J. P. Enhanced rate capability of multi-layered ordered porous nickel phosphide film as anode for lithium ion batteries. *J. Power Sources* 2011, 196, 379–385.

109. Gasparotto, L. H. S.; Prowald, A.; Borisenko, N.; El Abedin, S. Z.; Garsuch, A.; Endres, F. Electrochemical synthesis of macroporous aluminium films and their behavior towards lithium deposition/stripping. *J. Power Sources* 2011, 196, 2879–2883.

110. Liu, X.; Zhang, Y.; Ge, D.; Zhao, J.; Li, Y.; Endres, F. Three-dimensionally ordered macroporous silicon films made by electrodeposition from an ionic liquid. *Phys. Chem. Chem. Phys.* 2012, 14, 5100–5105.

111. Li, Z.; Liu, Y.; Liu, P.; Chen, W.; Feng, S.; Zhong, W.; Yu, C. Fabrication and morphology dependent magnetic properties of cobalt nanoarrays via template-assisted electrodeposition. *RSC Adv.* 2012, 2, 3447–3450.

112. Wang, J.; Sattayasamitsathit, S.; Gu, Y.; Kaufmann, K.; Polsky, R.; Minteer, S. D. Tunable hierarchical macro/mesoporous gold microwires fabricated by dual-templating and dealloying processes. *Nanoscale* 2013, 5, 7849–7854.

113. Reculusa, S.; Heim, M.; Gao, F.; Mano, N.; Ravaine, S.; Kuhn, A. Design of catalytically active cylindrical and macroporous gold microelectrodes. *Adv. Funct. Mater.* 2011, 21, 691–698.

114. Tian, S.; Gu, X.; Zheng, J.; Zhou, Q.; Gu, Z. Fabrication of a bowl-shaped silver cavity substrate for SERS-based immunoassay. *Analyst* 2013, 138, 2604–2612.

115. Heim, M.; Reculusa, S.; Ravaine, S.; Kuhn, A. Engineering of complex macroporous materials through controlled electrodeposition in colloidal superstructures. *Adv. Funct. Mater.* 2012, 22, 538–545.

116. Xiang, C.; Yang, Y.; Penner, R. M. Cheating the diffraction limit: Electrodeposited nanowires patterned by photolithography. *Chem. Commun.* 2009, (8), 859–873.

117. Yan, W.; Kim, J. Y.; Xing, W.; Donavan, K. C.; Ayvazian, T.; Penner, R. M. Lithographically patterned gold/manganese dioxide core/shell nanowires for high capacity, high rate, and high cyclability hybrid electrical energy storage. *Chem. Mater.* 2012, 24, 2382–2390.

118. Yan, W.; Ayvazian, T.; Kim, J. et al. M. Mesoporous manganese oxide nanowires for high-capacity, high-rate, hybrid electrical energy storage. *ACS Nano* 2011, 5, 8275–8287.

119. Li, G.-R.; Wang, Z.-L.; Zheng, F.-L.; Ou, Y.-N.; Tong, Y.-X. ZnO@MoO₃ core/shell nanocables: Facile electrochemical synthesis and enhanced supercapacitor performances. *J. Mater. Chem.* 2011, 21, 4217–4221.

120. Lu, X.; Zheng, D.; Zhai, T.; Liu, Z.; Huang, Y.; Xie, S.; Tong, Y. Facile synthesis of large-area manganese oxide nanorod arrays as a high-performance electrochemical supercapacitor. *Energy Environ. Sci.* 2011, 4, 2915–2921.

121. He, Y.-B.; Li, G.-R.; Wang, Z.-L.; Su, C.-Y.; Tong, Y.-X. Single-crystal ZnO nanorod/amorphous and nanoporous metal oxide shell composites: Controllable electrochemical synthesis and enhanced supercapacitor performances. *Energy Environ. Sci.* 2011, 4, 1288–1292.

122. Li, Q.; Wang, Z.-L.; Li, G.-R.; Guo, R.; Ding, L.-X.; Tong, Y.-X. Design and synthesis of MnO₂/Mn/MnO₂ sandwich-structured nanotube arrays with high supercapacitive performance for electrochemical energy storage. *Nano Lett.* 2012, 12, 3803–3807.

123. Wang, Z.-L.; Guo, R.; Li, G.-R. et al. Polyaniline nanotube arrays as high-performance flexible electrodes for electrochemical energy storage devices. *J. Mater. Chem.* 2012, 22, 2401–2404.

124. Hirate, T.; Sasaki, S.; Li, W.; Miyashita, H.; Kimpara, T.; Satoh, T. Effects of laser-ablated impurity on aligned ZnO nanorods grown by chemical vapor deposition. *Thin Solid Films* 2005, 487, 35–39.

125. Sun, Y.; Fuge, G. M.; Ashfold, M. N. R. Growth of aligned ZnO nanorod arrays by catalyst-free pulsed laser deposition methods. *Chem. Phys. Lett.* 2004, 396, 21–26.

126. Hariri, M. B.; Dolati, A.; Moakhar, R. S. The potentiostatic electrodeposition of gold nanowire/nanotube in HAuCl₄ solutions based on the model of recessed cylindrical ultramicroelectrode array. *J. Electrochem. Soc.* 2013, 160, D279–D288.

127. Xu, C. X.; Sun, X. W.; Dong, Z. L.; Yu, M. B. Self-organized nanocomb of ZnO fabricated by Au-catalyzed vapor-phase transport. *J. Cryst. Growth* 2004, 270, 498–504.

128. Elias, J.; Levy-Clement, C.; Bechelany, M.; Michler, J.; Wang, G.-Y.; Wang, Z.; Philippe, L. Hollow urchin-like ZnO thin films by electrochemical deposition. *Adv. Mater.* 2010, 22, 1607–1612.

129. Xu, L.; Guo, Y.; Liao, Q.; Zhang, J.; Xu, D. Morphological control of ZnO nanostructures by electrodeposition. *J. Phys. Chem. B* 2005, 109, 13519–13522.

130. Illy, B.; Shollock, B. A.; MacManus-Driscoll, J. L.; Ryan, M. P. Electrochemical growth of ZnO nanoplates. *Nanotechnology* 2005, 16, 320–324.

131. Hamley, I. W. Nanostructure fabrication using block copolymers. *Nanotechnology* 2003, 14, R39.

132. Lee, J. I.; Cho, S. H.; Park, S.-M. et al. Highly aligned ultrahigh density arrays of conducting polymer nanorods using block copolymer templates. *Nano Lett.* 2008, 8, 2315–2320.

133. Bang, J.; Jeong, U.; Ryu, D. Y.; Russell, T. P.; Hawker, C. J. Block copolymer nanolithography: Translation of molecular level control to nanoscale patterns. *Adv. Mater.* 2009, 21, 4769–4792.

134. Scherer, M. R. J.; Li, L.; Cunha, P. M. S.; Scherman, O. A.; Steiner, U. Enhanced electrochromism in gyroid-structured vanadium pentoxide. *Adv. Mater.* 2012, 24, 1217–1221.

135. Sidorenko, A.; Tokarev, I.; Minko, S.; Stamm, M. Ordered reactive nanomembranes/nanotemplates from thin films of block copolymer supramolecular assembly. *J. Am. Chem. Soc.* 2003, 125, 12211–12216.

136. Pérez, M. D.; Otal, E.; Bilmes, S. A.; et al. Growth of gold nanoparticle arrays in TiO$_2$ mesoporous matrixes. *Langmuir* 2004, 20, 6879–6886.

137. Urade, V. N.; Wei, T.-C.; Tate, M. P.; Kowalski, J. D.; Hillhouse, H. W. Nanofabrication of double-gyroid thin films. *Chem. Mater.* 2007, 19, 768–777.

138. Kwon, S.; Shim, M.; Lee, J. I.; Lee, T.-W.; Cho, K.; Kim, J. K. Ultrahigh density array of CdSe nanorods for CdSe/polymer hybrid solar cells: enhancement in short-circuit current density. *J. Mater. Chem.* 2011, 21, 12449–12453.

139. Chang, C.-C.; Botez, D.; Wan, L.; Nealey, P. F.; Ruder, S.; Kuech, T. F. Fabrication of large-area, high-density Ni nanopillar arrays on GaAs substrates using diblock copolymer lithography and electrodeposition. *J. Vac. Sci. Technol. B* 2013, 31, 031801–031805.

140. Del Río, R.; Armijo, F.; Schrebler, R.; Gutierrez, C.; Amaro, A.; Biaggio, S. R. Modification of composites of block copolymers–gold nanoparticles with enzymes and their characterization by electrochemical techniques. *J. Solid State Electrochem.* 2011, 15, 697–702.

141. Attard, G. S.; Bartlett, P. N.; Coleman, N. R.; Elliott, J. M.; Owen, J. R.; Wang, J. H. Mesoporous platinum films from lyotropic liquid crystalline phases. *Science* 1997, 278, 838–840.

142. Palmqvist, A. E. C. Synthesis of ordered mesoporous materials using surfactant liquid crystals or micellar solutions. *Curr. Opin. Colloid Interface Sci.* 2003, 8, 145–155.

143. Wang, H.; Wang, L.; Sato, T.; Sakamoto, Y.; Tominaka, S.; Miyasaka, K.; Miyamoto, N.; Nemoto, Y.; Terasaki, O.; Yamauchi, Y. Synthesis of mesoporous Pt films with tunable pore sizes from aqueous surfactant solutions. *Chem. Mater.* 2012, 24, 1591–1598.

144. Nelson, P. A.; Owen, J. R. A high-performance supercapacitor/battery hybrid incorporating templated mesoporous electrodes. *J. Electrochem. Soc.* 2003, 150, A1313–A1317.

145. Zhao, D.-D.; Xu, M. W.; Zhou, W.-J.; Zhang, J.; Li, H. L. Preparation of ordered mesoporous nickel oxide film electrodes via lyotropic liquid crystal templated electrodeposition route. *Electrochim. Acta* 2008, 53, 2699–2705.

146. Dong, B.; Xue, T.; Xu, C.-L.; Li, H.-L. Electrodeposition of mesoporous manganese dioxide films from lyotropic liquid crystalline phases. *Microporous Mesoporous Mater.* 2008, 112, 627–631.

147. Zhou, W.; Zhang, J.; Xue, T.; Zhao, D.; Li, H. Electrodeposition of ordered mesoporous cobalt hydroxide film from lyotropic liquid crystal media for electrochemical capacitors. *J. Mater. Chem.* 2008, 18, 905–910.

148. Seo, S.; Ye, T.; Borguet, E. Electrochemical nanoscale templating: Laterally self-aligned growth of organic/metal nanostructures. *Langmuir* 2012, 28, 17537–17544.

149. Zhao, D.-D.; Bao, S.-J.; Zhou, W.-J.; Li, H.-L. Preparation of hexagonal nanoporous nickel hydroxide film and its application for electrochemical capacitor. *Electrochem. Commun.* 2007, 9, 869–874.

150. Dobbs, W.; Suisse, J.-M.; Douce, L.; Welter, R. Electrodeposition of silver particles and gold nanoparticles from ionic liquid-crystal precursors. *Angew. Chem.* 2006, 118, 4285–4288.

151. Lin, C.; Lin, H.; Li, J.; Li, X. Electrodeposition preparation of ZnO nanobelt array films and application to dye-sensitized solar cells. *J. Alloys Compd.* 2008, 462, 175–180.

152. Kawabata, K.; Takeguchi, M.; Goto, H. Optical activity of heteroaromatic conjugated polymer films prepared by asymmetric electrochemical polymerization in cholesteric liquid crystals: Structural function for chiral induction. *Macromolecules* 2013, 46, 2078–2091.

153. Asghar, K. A.; Elliott, J. M.; Squires, A. M. 2D hexagonal mesoporous platinum films exhibiting biaxial, in-plane pore alignment. *J. Mater. Chem.* 2012, 22, 13311–13317.

154. Bartlett, P. N.; Pletcher, D.; Esterle, T. F.; Low, C. T. J. The deposition of mesoporous Ni/Co alloy using cetyltrimethylammonium bromide as the surfactant in the lyotropic liquid crystalline phase bath. *J. Electroanal. Chem.* 2013, 688, 232–236.

155. Serrà, A.; Gómez, E.; Calderó, G.; Esquena, J.; Solans, C.; Vallés, E. Microemulsions for obtaining nanostructures by means of electrodeposition method. *Electrochem. Commun.* 2013, 27, 14–18.

156. Calderó, G.; Vallés, E.; Gómez, E.; Solans, C.; Serrà, A.; Esquena, J. Conductive microemulsions for template CoNi electrodeposition. *Phys. Chem. Chem. Phys.* 2013, 15, 14653–14659.

157. Philippe, L.; Kacem, N.; Michler, J. Electrochemical deposition of metals inside high aspect ratio nanoelectrode array: Analytical current expression and multidimensional kinetic model for cobalt nanostructure synthesis. *J. Phys. Chem. C* 2007, 111, 5229–5235.

158. Valizadeh, S.; George, J. M.; Leisner, P.; Hultman, L. Electrochemical deposition of Co nanowire arrays; quantitative consideration of concentration profiles. *Electrochim. Acta* 2001, 47, 865–874.

159. Bograchev, D. A.; Volgin, V. M.; Davydov, A. D. Simple model of mass transfer in template synthesis of metal ordered nanowire arrays. *Electrochim. Acta* 2013, 96, 1–7.

160. Greiner, A.; Wendorff, J. H.; Yarin, A. L.; Zussman, E. Biohybrid nanosystems with polymer nanofibers and nanotubes. *Appl. Microbiol. Biotechnol.* 2006, 71, 387–393.

161. Cho, S.; Kwon, W.; Choi, S. et al. Nanotube-based ultrafast electrochromic display. *Adv. Mater.* 2005, 17, 171–175.

162. Cho, S. I.; Lee, S. B. Fast electrochemistry of conductive polymer nanotubes: Synthesis, mechanism, and application. *Acc. Chem. Res.* 2008, 41, 699–707.

163. Liu, R.; Cho, S. I.; Lee, S. B. Poly(3,4-ethylenedioxythiophene) nanotubes as electrode materials for a high-powered supercapacitor. *Nanotechnology* 2008, 19, 215710.

164. Sato, K.; Nakahodo, T.; Fujihara, H. Redox-active π-conjugated polymer nanotubes with viologen for encapsulation and release of fluorescent dye in the nanospace. *Chem. Commun.* 2011, 47, 10067–10069.

165. Lee, K. J.; Min, S. H.; Oh, H.; Jang, J. Fabrication of polymer nanotubes containing nanoparticles and inside functionalization. *Chem. Commun.* 2011, 47, 9447–9449.

166. Huby, N.; Luc Duvail, J.; Duval, D.; Pluchon, D.; Bêche, B. Light propagation in single mode polymer nanotubes integrated on photonic circuits. *Appl. Phys. Lett.* 2011, 99, 113302.

167. Martin, C. R.; Van Dyke, L. S.; Cai, Z.; Liang, W. Template synthesis of organic microtubules. *J. Am. Chem. Soc.* 1990, 112, 8976–8977.

168. Szklarczyk, M.; Strawski, M.; Donten, M. L.; Donten, M. A study of tubular nanostructures formation in the pores of membrane electrode. *Electrochem. Commun.* 2004, 6, 880–886.

169. Duvail, J. L.; Rétho, P.; Garreau, S.; Louarn, G.; Godon, C.; Demoustier-Champagne, S. Transport and vibrational properties of poly(3,4-ethylenedioxythiophene) nanofibers. *Synth. Met.* 2002, 131, 123–128.

170. Liu, R.; Lee, S. B. MnO$_2$/poly(3,4-ethylenedioxythiophene) coaxial nanowires by one-step coelectrodeposition for electrochemical energy storage. *J. Am. Chem. Soc.* 2008, 130, 2942–2943.

171. Hoyer, P. Formation of a titanium dioxide nanotube array. *Langmuir* 1996, 12, 1411–1413.

172. Mu, C.; Yu, Y.-X.; Wang, R. M.; Wu, K.; Xu, D. S.; Guo, G.-L. Uniform metal nanotube arrays by multi-step template replication and electrodeposition. *Adv. Mater.* 2004, 16, 1550–1553.

173. Venkata Kamalakar, M.; Raychaudhuri, A. K. A novel method of synthesis of dense arrays of aligned single crystalline copper nanotubes using electrodeposition in the presence of a rotating electric field. *Adv. Mater.* 2008, 20, 149–154.

174. Brumlik, C. J.; Martin, C. R. Template synthesis of metal microtubules. *J. Am. Chem. Soc.* 1991, 113, 3174–3175.

175. Tourillon, G.; Pontonnier, L.; Levy, J. P.; Langlais, V. Electrochemically synthesized Co and Fe nanowires and nanotubes. *Electrochem. Solid State Lett.* 2000, 3, 20–23.

176. Davis, D. M.; Podlaha, E. J. CoNiCu and Cu nanotube electrodeposition. *Electrochem. Solid State Lett.* 2005, 8, D1–D4.

177. Fukunaka, Y.; Motoyama, M.; Konishi, Y.; Ishii, R. Producing shape-controlled metal nanowires and nanotubes by an electrochemical method. *Electrochem. Solid State Lett.* 2006, 9, C62–C64.

178. Tao, F.; Guan, M.; Xue, Z.; Jiang, Y.; Xu, Z.; Zhu, J. An easy way to construct an ordered array of nickel nanotubes: The triblock-copolymer-assisted hard-template method. *Adv. Mater.* 2006, 18, 2161–2164.

179. Cao, H.; Wang, L.; Qiu, Y.; Wu, Q.; Wang, G.; Zhang, L.; Liu, X. Generation and growth mechanism of metal (Fe, Co, Ni) nanotube arrays. *ChemPhysChem* 2006, 7, 1500–1504.

180. Li, L.; Pan, S.; Dou, X. et al. Direct electrodeposition of ZnO nanotube arrays in anodic alumina membranes. *J. Phys. Chem. C* 2007, 111, 7288–7291.

181. Liu, L.; Zhou, W.; Xie, S. et al. Highly efficient direct electrodeposition of Co–Cu alloy nanotubes in an anodic alumina template. *J. Phys. Chem. C* 2008, 112, 2256–2261.

182. Dou, X.; Li, G.; Huang, X.; Li, L. Abnormal growth of electrodeposited BiSb alloy nanotubes. *J. Phys. Chem. C* 2008, 112, 8167–8171.

183. Li, X.; Wang, Y.; Song, G.; Peng, Z.; Yu, Y.; She, X.; Li, J. Synthesis and growth mechanism of Ni nanotubes and nanowires. *Nanoscale Res. Lett.* 2009, 4, 1015–1020.

184. Zhang, H.; Zhang, X.; Wu, T.; Zhang, Z.; Zheng, J.; Sun, H. Template-based synthesis and discontinuous hysteresis loops of cobalt nanotube arrays. *J. Mater. Sci.* 2013, 48, 7392–7398.

185. Sun, H.; Zhang, H.; Zhang, X.; Zhang, J.; Li, Z. Template-based electrodeposition growth mechanism of metal nanotubes. *J. Electrochem. Soc.* 2013, 160, D41–D45.

186. Yang, G.; Li, L.; Jiang, J.; Yang, Y. Direct electrodeposition of gold nanotube arrays of rough and porous wall by cyclic voltammetry and its applications of simultaneous determination of ascorbic acid and uric acid. *Mater. Sci. Eng. C* 2012, 32, 1323–1330.

187. Liu, L.; Yoo, S.-H.; Lee, S. A.; Park, S. Wet-chemical synthesis of palladium nanosprings. *Nano Lett.* 2011, 11, 3979–3982.

188. Liu, L.; Park, S. Direct formation of thin-walled palladium nanotubes in nanochannels under an electrical potential. *Chem. Mater.* 2011, 23, 1456–1460.

189. Hou, L.; Yuan, C.; Li, D. et al. Electrochemically induced transformation of NiS nanoparticles into $Ni(OH)_2$ in KOH aqueous solution toward electrochemical capacitors. *Electrochim. Acta* 2011, 56, 7454–7459.

190. Liu, N.; Hu, L.; McDowell, M. T.; Jackson, A.; Cui, Y. Prelithiated silicon nanowires as an anode for lithium ion batteries. *ACS Nano* 2011, 5, 6487–6493.

191. Delmas, C.; Nadiri, A. The chemical short circuit method. An improvement in the intercalation-deintercalation techniques. *Mater. Res. Bull.* 1988, 23, 65–72.

192. Whittingham, M. S.; Dines, M. B. n-Butyllithium—An effective, general cathode screening agent. *J. Electrochem. Soc.* 1977, 124, 1387–1388.

193. Jeong, M.-G.; Zhuo, K.; Cherevko, S.; Kim, W.-J.; Chung, C.-H. Facile preparation of three-dimensional porous hydrous ruthenium oxide electrode for supercapacitors. *J. Power Sources* 2013, 244, 806–811.

194. Li, Y.; Song, Y.-Y.; Yang, C.; Xia, X.-H. Hydrogen bubble dynamic template synthesis of porous gold for nonenzymatic electrochemical detection of glucose. *Electrochem. Commun.* 2007, 9, 981–988.

195. Parida, S.; Kramer, D.; Volkert, C. A.; Rösner, H.; Erlebacher, J.; Weissmüller, J. Volume change during the formation of nanoporous gold by dealloying. *Phys. Rev. Lett.* 2006, 97, 035504.

196. Li, G.; Song, X.; Lu, F.; Sun, Z.; Yang, Z.; Yang, S.; Ding, B. Formation and control of nanoporous Ag through electrochemical dealloying of the melt-spun Cu-Ag-Ce alloys. *J. Mater. Res.* 2012, 27, 1612–1620.

197. Li, W.; Ma, H.; Huang, L.; Ding, Y. Well-defined nanoporous palladium for electrochemical reductive dechlorination. *Phys. Chem. Chem. Phys.* 2011, 13, 5565–5568.

198. Cobley, C. M.; Xia, Y. Engineering the properties of metal nanostructures via galvanic replacement reactions. *Mater. Sci. Eng. R: Rep.* 2010, 70, 44–62.

199. Kim, K.-H.; Kim, J.-Y.; Kim, K.-B. Synthesis of manganese dioxide/poly(3,4-ethylenedioxythiophene) core/sheath nanowires by galvanic displacement reaction. *J. Electroceramics* 2012, 29, 149–154.

200. Hu, J.-Q.; Lai, S.-Q.; Liu, J.-Y. et al. One-pot synthesis and enhanced catalytic performance of Pd and Pt nanocages via galvanic replacement reactions. *RSC Adv.* 2013, 3, 12577–12580.

201. Skrabalak, S. E.; Chen, J.; Sun, Y. et al. Gold nanocages: Synthesis, properties, and applications. *Accounts Chem. Res.* 2008, 41, 1587–1595.

202. Sun, Y.; Mayers, B. T.; Xia, Y. Template-engaged replacement reaction: A one-step approach to the large-scale synthesis of metal nanostructures with hollow interiors. *Nano Lett.* 2002, 2, 481–485.

203. Zhang, H.; Jin, M.; Wang, J. et al. Synthesis of Pd–Pt bimetallic nanocrystals with a concave structure through a bromide-induced galvanic replacement reaction. *J. Am. Chem. Soc.* 2011, 133, 6078–6089.

204. González, E.; Arbiol, J.; Puntes, V. F. Carving at the nanoscale: Sequential galvanic exchange and Kirkendall growth at room temperature. *Science* 2011, 334, 1377–1380.

205. Graham, L. M.; Cho, S.; Kim, S. K.; Noked, M.; Lee, S. B. Role of boric acid in nickel nanotube electrodeposition: a surface-directed growth mechanism. *Chem. Commun.* 2014, 50, 527–529.

11 Nanopores and Nanoporous Membranes

Alicia K. Friedman and Lane A. Baker

CONTENTS

11.1 INTRODUCTION

Efforts in the realm of nanoscience have been made to miniaturize existing technology and to explore the unique phenomena that occur at small length scales. Studies of the nanoscale often blur the interfaces between traditionally defined disciplines such that knowledge of engineering, chemistry, physics, materials science, and biological sciences must be considered in a comprehensive perspective. Investigations and applications of nanopores, found in a wide variety of scientific fields and for a number of diverse applications, exemplify this point. A nanopore is defined as a structure with a characteristic feature (e.g., aperture size) between 1 and 100 nm in diameter; to put this in perspective, this dimension is nominally between the size of a glucose molecule and a typical virus. This scale—essentially the molecular and macromolecular regime—demonstrates a number of size-dependent phenomena and interesting applications for single-molecule detection and analysis. For instance, electrically driven transport, which depends on electrochemical phenomena such as double layer thickness and transference numbers, is altered dramatically at these small scales. This review will highlight platforms commonly used in the development of nanopores, typical measurements accomplished with these systems, unique electrochemical properties of nanopores, and the application of nanopores in a range of venues.

11.2 NANOPOROUS PLATFORMS

Inherent features of different nanoporous systems in use today, such as pore density, size, and material, can vary widely. Selection of the appropriate nanoporous platform on the basis of such characteristics is important to ensure that the measurement or application of interest is best served. Many of these nanoporous systems have been commercialized and find wide use. Nanoporous platforms that will be discussed in this review include solid-state pores and native/engineered biological systems.

11.2.1 SOLID-STATE NANOPORES

Solid-state nanopores, or nanopores comprising primarily synthetic materials, have become ubiquitous because of the ease with which material composition and pore size can be controlled. Background on the most common solid-state systems, which include nanoporous membranes, nanopores within silicon supports, nanopipettes, nanofluidic platforms, and nanoporous graphene, is considered. A direct comparison of sizes and characteristics of nanopores within these systems can be found in Table 11.1.

11.2.1.1 Nanoporous Membranes

Porous membranes have been used for centuries in a number of applications and are especially important for use as filters in the size-dependent sieving of material from solution. The movement to miniaturize the pores contained within such membranes has been quite successful and has generated a number of different types of nanoporous membranes. These membranes find great diversity in application, and membranes can be prepared with a single nanopore or with a high density of nanopores. Additionally, membranes are amenable to production of pores of specific size,

TABLE 11.1
Characteristics and Sizes of Various Solid-State Nanoporous Platforms

Material	Pore Sizes (diam.)	Pore Densities	Comments
Track-etch membranes	2 nm–µms	$1–10^8$ per cm^2	High chemical and thermal resistance, pores randomly distributed
Alumina membranes	5–300 nm with 10–500 nm spacing	$10^9–10^{12}$ per cm^2	Highly uniform pore structure, high thermal resistance
Block copolymer membranes	10–100 nm	High density dependent on the polymer composition	Pore size controlled by the molecular weight of the polymer
Silicon wafers	<10 nm–µms	Dependent on number desired (controlled via nanofabrication)	Well-developed materials due to electronics industry
Nanopipettes	~10 nm–µms	1–7	Easy/cheap to fabricate
Nanofluidics	Single nanochannel: 10 nm–µms with 3-D membrane: 2 nm–µms	Dependent on number desired (controlled via nanofabrication)	Systems typically reusable
Graphene	<10 nm–µms	Dependent on number desired (controlled via nanofabrication)	Conductive material, strong mechanical properties, atomically thin

shape, and surface charge. Membranes can be prepared from a variety of materials, and subsequent chemical modification can be used to impart additional properties of interest.

Various polymers are commonly utilized in the fabrication of nanoporous membranes; some of the most widely utilized materials include poly(carbonate) (PC), poly(ethylene terephthalate) (PET), and poly(imide) (PI). Such polymer membranes possess a number of beneficial attributes. They are compatible with aqueous systems typically involved with biological or biomimetic systems, and the pore sizes and shapes can be controlled through chemical and/or plasma etch processes. Typically, to fabricate nanopores within polymer materials, thin-film membranes are exposed to high-energy fission fragments from a cyclotron/synchrotron source. When the high-energy particles collide with a thin polymer film, damage tracks are produced, which span the thickness of the membrane. Once subjected to an appropriate chemical etchant, these damage tracks are etched preferentially relative to the undamaged bulk polymer material. This *track-etch* process produces pores at locations within the polymer film that were initially hit with the fission fragments and results in a membrane with randomly distributed pores. Membranes with cylindrical pores prepared in this manner can be purchased commercially from a variety of manufacturers. Recently, polymer films with damaged tracks, but not etched, also have become available through select vendors. Nanoporous polymer membranes prepared in this fashion can have a range of pore densities, as determined by the density of initial tracks.[1] The dimensions and shape of nanopores can be controlled through the modification of the postbombardment etching procedure (Figure 11.1).[2] Equal application of chemical etchant to both sides of the membrane results in symmetric cylindrical pores. Introduction of a neutralizing solution to one face of the membrane or plasma etching one face of the membrane can generate asymmetric pores, with a conical or bullet shape and a wide range pore dimensions.[2,3] In addition to polymers, membranes have been prepared via the track-etch method in a variety of materials, which include glass,[4,5] mica,[6] and silicon nitride.[7]

Like the polymer membranes discussed above, block copolymers also are used to produce nanoporous materials.[8,9] Block copolymer-derived nanoporous membranes are fabricated in a series of steps. First, two or more distinct, immiscible monomers self-assemble into a monolithic polymer; heat is then applied to reach a temperature greater than the glass transition temperature of the material to form cylindrical domains within the polymer matrix. Finally, the cylindrical domains are converted into nanopores that traverse the membrane through chemical etchant processes. Polymers have been developed that are diblock and triblock in composition, and surface functionalization of these nanoporous membranes can be accomplished in an effort to better tune the system to an experiment or analyte of interest.[10,11] Nanopores within block copolymers have been utilized in a variety of applications, including as lithographic masks for the preparation of gold nanoparticle[12] and silicon nanorod[13] arrays and as sensors for the separation and detection of biomolecules.[14,15]

Another common material for fabrication of nanoporous membranes is alumina. Nanoporous features within the oxide coating of aluminum were observed in the 1950s,[16] but the popularity of alumina as a nanoporous platform has surged in recent years as a result of the self-organized growth method developed by Fukuda and coworkers in 1995.[17] In this method, under oxidizing and acidic conditions, a porous structure of aluminum oxide grows perpendicularly from the face of an aluminum surface. With temperature stability as high as 1000°C and tunable pore sizes from 5 to 300 nm in diameter, alumina is an advantageous material for membrane preparation.[18] Nanoporous alumina membranes typically have a greater pore density than the polymer membranes described previously, and the pores are arranged regularly in a near-hexagonal/honeycomb array,[1] which can be visualized in the scanning electron microscopy (SEM) images in Figure 11.2.[19]

Properties of alumina membranes and polymer membranes, like chemical/thermal stabilities and pore density, are often quite different, and selection of the proper material for desired applications is important. Chemical modification has proven useful to impart functionality to nanoporous membranes.[3] For alumina membranes in particular, growth of thin sol-gel films on the walls of the membrane provides an opportunity to functionalize membranes with organosilanes (e.g., triethoxy- or trichlorosilanes).[20–22] Deposition of gold films also provides a method for chemical modification

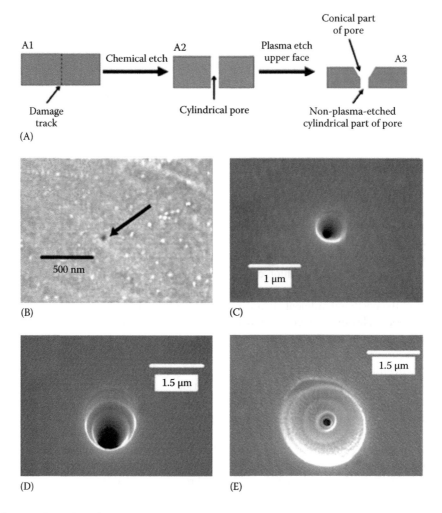

FIGURE 11.1 Overview of nanoporous polymer membranes. (A) Schematic of the postbombardment etch process utilized to attain cylindrical/conical nanopores within a polycarbonate membrane. The initial damage tracks within the polymer material are produced when the membrane is exposed to high-energy nuclear fission fragments from a cyclotron source. SEM images of the membrane surface (B) after chemical etching, (C) after 5 min plasma etching, (D) after 10 min plasma etching, and (E) after 20 min of plasma etching. The plasma-etch process produces pores with a conical, or asymmetric, shape. (Reprinted with permission from Li, N.C., Yu, S.F., Harrell, C.C., and Martin, C.R., Conical nanopore membranes. Preparation and transport properties, *Anal. Chem.*, 76, 2025–2030, 2004. Copyright 2004 American Chemical Society.)

via gold-thiol chemistry.[23–25] Additional strategies, which include layer-by-layer deposition[26–28] and amide-carboxyl group surface reactions,[29–31] have also been employed to impart specific chemical functionality to nanoporous membranes.

A prevalent use for the nanoporous membranes detailed previously, particularly the highly ordered structure of alumina, is to fabricate reproducible nanoscale materials through a method called template synthesis.[1,32,33] Briefly, the composition material of desired nanostructures is grown or deposited into a nanoporous membrane; the new nanostructures are then separated from the original membrane via mechanical or chemical means. Dependent on the synthetic approach used, this process can produce nanoscale arrays[34,35] or individual nanostructures like wires,[36] nanotubes[37–39] or nanoparticles. Template synthesis also has been used to produce biological nanostructures through the deposition of biomolecules into nanoporous membranes.[26,27] For a more detailed explanation on

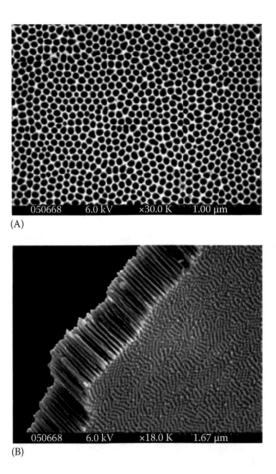

FIGURE 11.2   SEM images of nanoporous alumina membranes. (A) A top-view image highlights the hexagonal/honeycomb packing of the nanopores, and (B) a side-view image shows the overall thickness of the membrane. (Kang, M.C., Yu, S.F., Li, N.C., and Martin, C.R.: Nanowell-array surfaces. *Small.* 2005. 1. 69–72. Copyright Wiley-VCH Verlag GmbH & Co. KGaA. Reproduced with permission.)

template synthesis and the nanomaterials produced with this technique, please refer to Chapter 10 within this book for a review by Lee and Duay.

11.2.1.2   Nanopores within Silicon Supports

The development of nanopores within silicon supports, particularly silicon nitride and silicon oxide, makes use of knowledge derived from electronics and semiconductor fabrication processes. Silicon wafers with thin oxide/nitride films can be readily attained and provide a useful platform for experimentation. Though these materials are not fundamentally nanoporous, techniques have been developed to incorporate nanopores.

Golovchenko and coworkers developed a method in 2001 to fabricate nanopores from silicon materials through the utilization of an ion beam.[40] This nanopore sculpting strategy is outlined in Figure 11.3A, in which a free-standing silicon nitride membrane with a cavity is exposed to sputtering from an ion beam to remove material until a pore is milled through the membrane. A feedback-controlled sputtering system was developed that could monitor the flux of ions through the nanopore opening for precise nanopore size control (Figure 11.3B). Temperature dependence of the milling process was studied, and the pore was found to shrink upon ion beam milling at room temperature as a result of the lateral flow of surface material to fill the pore. The ability of focused

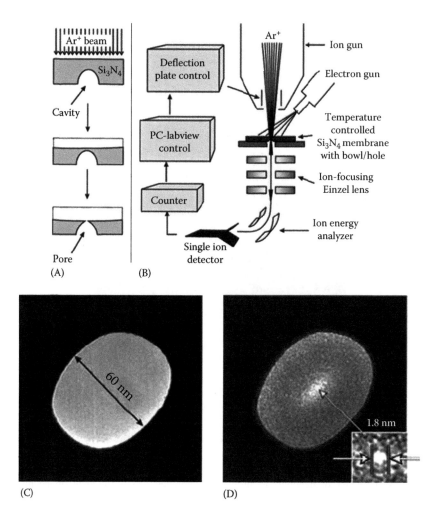

FIGURE 11.3 Overview of the ion beam sculpting strategy to fabricate nanopores within a silicon support. (A) Sputtering with an argon ion beam removes the material from a free-standing silicon nitride membrane with a cavity. (B) A focused ion beam (FIB) is utilized to *sculpt* or tailor the exact dimensions of the nanopore as desired. (C) Transmission electron microscopy (TEM) image of an initial 61 nm pore before and (D) after exposure to argon ion beam sputtering. The final pore area was determined to be shrunk to 1.8 nm by energy dispersive x-ray analysis. (Reprinted by permission from MacMillan Publishers Ltd. *Nature*, Li, J., Stein, D., McMullan, C., Branton, D., Aziz, and M.J., Golovchenko, J.A., Ion-beam sculpting at nanometre length scales, 412, 166–169. Copyright 2001.)

ion beam (FIB) milling to both remove the material to form pores and cause pore shrinkage makes this a powerful method for control of pore size within silicon supports. Ion beam sculpting of a 60 nm diameter pore resulted in a 1.8 nm pore upon sputtering at 28°C (Figure 11.3C, D). Another commonly utilized technique to attain precise size control of the nanopores within silicon supports is similar and employs the electron beam of a transmission electron microscope (TEM) to shrink pores and fine-tune their dimensions as desired.[41,42]

A newer nanoporous material composed of silicon is porous nanocrystalline silicon (pnc-Si), which comprises an extremely thin membrane with thickness typically on the order of tens of nm and average nanopore sizes of 5–55 nm.[43,44] Fabrication of pnc-Si is accomplished through a combination of silicon deposition and etchant processes, and the pores are formed from random nucleation sites on the surface rather than direct patterning. The pnc-Si is robust enough to handle for experimental use but presents pores with lengths on the size scale of molecular analytes.[45] These ultrathin membranes have demonstrated utility as synthetic models of the nuclear envelope[46] and cell culture membranes.[47]

11.2.1.3   Nanopipettes

Glass vessels and pipettes have been used for centuries as tools to collect, store, or deliver desired materials. Shrinking the dimensions of such macroscale devices to the nanoregime has realized so-called micro- and nanopipettes which have been employed in a number of unique applications.[48–50] For instance, pipettes with small diameter openings (typically microns) have been used to investigate cellular ion channels in the field of electrophysiology for decades in patch clamp experiments.[51] Nanopipettes can also be utilized to control and deliver ultrasmall volumes of desired materials in a function similar to their macroscale counterparts. Nanoscale pipettes are also commonly utilized as sensors to optically or electrochemically detect analytes of interest. Finally, nanopipettes and modified nanopipettes have been shown to serve well as the probe in many scanning probe microscopy (SPM) techniques and hybrid platforms, including scanning electrochemical microscopy (SECM),[52] scanning ion conductance microscopy (SICM),[53] atomic force microscopy (AFM),[54,55] near-field scanning optical microscopy (NSOM),[56,57] SECM-SICM,[58,59] and SICM-NSOM.[60–62]

Though there are a number of protocols utilized to fabricate nanopipettes, the most common technique employs a pipette puller. In this system, a glass capillary is heated while a mechanical force is applied to physically separate the capillary into two symmetric sister pipettes. In this case, each pipette will have a large end with the dimensions of the initial capillary (the stem) that tapers to a much smaller end, typically on the order of tens to hundreds of nanometers (the tip). An overview of nanopipettes is depicted in Figure 11.4 with a schematic to illustrate particular regions of a nanopipette (Figure 11.4A) and optical (Figure 11.4B) and electron micrographs (Figure 11.4C) of pulled pipettes. The size of the resulting tip and the angle of taper can be controlled by adjusting various parameters within the pull mechanism such as the force applied, velocity of separation, and heating time. Various glass materials can be used to create nanopipettes, with borosilicate and quartz being the most common. Due to the high temperature necessary to melt quartz, however, a laser is typically required as the heating source within the puller to achieve good separation. Capillaries with two or more barrels are commercially available and can be used similarly to produce multibarrel pipettes.[63]

The smallest capillary-type nanopores prepared to date have been fabricated using a seal and polish approach to produce the *glass nanopore*.[64] Nanopores with tip dimensions as small as a few nanometers have been prepared in this fashion. These nanopores find additional application as supports for lipid bilayers and subsequent biophysical studies.[65–75] Detailed studies of the electrochemical properties of glass nanopores and additional applications that make use of resistive-pulse sensing (method described in the following text) have been reported as well.[76–87]

11.2.1.4   Nanofluidics

As described here, nanofluidic channels consist of a nanoscale pore or channel within a larger microfluidic device. These nanodevices offer several advantages over larger flow systems such as a high surface-to-volume ratios, size scales comparable to the size of important biomolecules, and the ability to interrogate smaller sample volumes.[88]

As barriers to nanofabrication have lowered, the popularity of nanofluidic systems has risen and complex systems have been realized. Nanofluidic channels can be constructed within a wide variety of materials, which include silicon,[89] glass,[90] poly(dimethylsiloxane) (PDMS)[91,92] and poly(methylmethacrylate) (PMMA).[93,94] The method typically employed to fabricate nanochannel platforms is that of nanolithography, a technique that utilizes patterns and reactive etching to create 3-D designs. Various nanolithographic methods have been developed, including electron beam lithography (EBL), FIB milling, nanoimprint lithography (NIL), interoferometric lithography (IL), and sphere lithography (SL).[88] Two common types of nanofluidic designs are highlighted

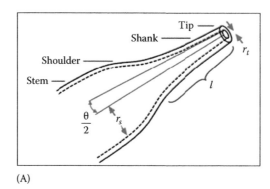

(A)

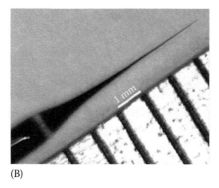

(B)

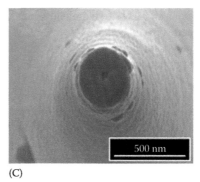

(C)

FIGURE 11.4 Overview of nanopipettes. (A) Illustration that denotes the differently named regions of a nanopipette. The radius of the stem (r_s), length of the pipette stem (l), cone angle of the nanopipette (θ), and radius of the tip (r_t) are labeled. (B) Optical micrograph of a nanopipette filled with dye for easier visualization. (C) End-on SEM image of a nanopipette in which the opening is visualized in the center of the tip. (Morris, C.A., Friedman, A.K., and Baker, L.A., Applications of nanopipettes in the analytical sciences, *Analyst*, 135, 2190–2202, 2010. Reproduced by permission of The Royal Society of Chemistry.)

in Figure 11.5, one that consists of a nanochannel constriction between two larger microchannels[95] (Figure 11.5A) and another that uses an interfacial nanoporous membrane within a 3-D microfluidic device (Figure 11.5B).[96]

11.2.1.5 Nanoporous Graphene

Graphene is the newest nanoporous material of significance to date, and research of the application of this material has exploded recently. This ordered, 2-D carbon with single-atom thickness was isolated experimentally in 2004.[97] The properties of graphene, including conductivity, thermal and mechanical characteristics, and the possibility to form single-monolayer membranes, lead many to consider graphene the ultimate nanoporous platform.[98–100] Many proposed applications of graphene have not been realized experimentally because of the relatively short time the material has been studied. However, computational studies have been performed that demonstrate the possible utility of graphene in areas such as separations.[101]

Native graphene is not inherently nanoporous, but a number of different methods have been developed to produce porous forms of graphene. Hydrothermal steam has been applied to graphene oxide to produce a conductive nanoporous network within the graphene oxide matrix.[102] Nanoporous graphene oxide also has been fabricated with controllable pore sizes of 30–120 nm from silica-particle templates that can be etched from the nanoporous foam with hydrofluoric acid.[103] Pristine graphene can be altered physically to contain nanopores by drilling into the 2-D carbon sheet with the focused electron beam in a TEM.[104] Graphene also can be grown on a porous template via chemical vapor

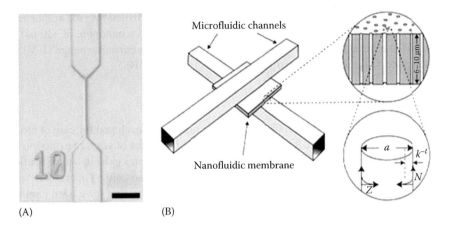

(A)   (B)

FIGURE 11.5 Examples of nanofluidic systems. (A) An optical micrograph of a nanofluidic channel imaged with differential interface contrast microscopy. The narrow region, which is 500 nm wide and 250 nm deep, is used in conjunction with fluorescence microscopy analysis to detect DNA and histones in individual chromatin fragments (scale bar = 10 μm). (B) Schematic diagram of a 3-D microfluidic device with an interfacial nanofluidic membrane (shown schematically in the cross section in the upper-right inset). The lower-right inset shows the relationship between a, the nanochannel diameter, and κ^{-1}, the Debye length under typically encountered nanopore sizes and electrolyte concentrations. Z is the distance from the nanochannel wall, and N is the ion number density. (Reprinted with permission from Cipriany, B.R., Zhao, R.Q., Murphy, P.J., Levy, S.L., Tan, C.P., Craighead, H.G., and Soloway, P.D., Single molecule epigenetic analysis in a nanofluidic channel, *Anal. Chem.*, 82, 2480–2487, 2010; Kuo, T.C., Cannon, D.M., Chen, Y.N., Tulock, J.J., Shannon, M.A., Sweedler, J.V., and Bohn, P.W., Gateable nanofluidic interconnects for multilayered microfluidic separation systems, *Anal. Chem.*, 75, 1861–1867, 2010. Copyright 2010 and 2003 American Chemical Society.)

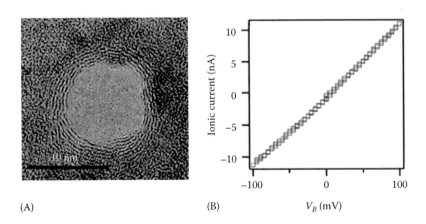

(A)   (B) V_B (mV)

FIGURE 11.6 Overview of nanoporous graphene. (A) TEM image of a nanopore within a graphene membrane (scale bar = 10 nm), and (B) the corresponding I–V curve measured in 1 M KCl, pH 9, which demonstrates no observable rectification. (Reprinted with permission from Merchant, C.A., Healy, K., Wanunu, M., Ray, V., Peterman, N., Bartel, J., and Fischbein, M.D. et al., DNA translocation through graphene nanopores, *Nano Lett.*, 10, 2915–2921, 2010. Copyright 2010 American Chemical Society.)

deposition (CVD). In this method, carbon only deposits on the membrane where the material is present (avoiding the nanopores) to produce a nanoporous graphene that can be separated from the template with an acid wash.[105,106] A brief overview of nanoporous graphene is highlighted in Figure 11.6 in work by Drndic and coworkers.[107] Graphene of a few atomic layers was grown via CVD on copper foils, chemically etched from the support to produce free-standing graphene, and placed on a silicon nitride support for easier handling. Nanopores then were drilled into the graphene material with an electron beam. A transmission electron micrograph shows a nanopore of ~10 nm diameter in the center of graphene in Figure 11.6A, and the corresponding current–voltage (I–V) measurements for this pore demonstrate no rectification effect (Figure 11.6B).

11.2.2 PORES FROM NATURE

The solid-state nanopores described in the previous section were developed for ease of modification and manipulation. However, nature has had the advantage of millions of years of evolution to perfect the size and function of proteinaceous nanochannels. Many nanopores exist in nature to allow selective passage of species across membranes found within eukaryotic cells. This review will briefly highlight both native nanopores from nature and protein nanopores that have been engineered for better application to sensor systems. Many of the unmodified protein nanopores that are utilized in nanotechnology today are transmembrane proteins that can be inserted easily into the support of a lipid bilayer or polymeric membrane. Some of the most commonly used native protein channels include α-hemolysin (αHL),[108,109] *Mycobacterium smegmatis* porin A (MspA),[110,111] outer membrane protein F of *Escherichia coli* (OmpF),[112] and outer membrane protein G of *E. coli* (OmpG).[113,114] A comparative analysis of the specific properties exhibited by these four biological pores can be found in Table 11.2.

The most ubiquitous protein nanopore applied to nanotechnology systems is formed from αHL of *Staphylococcus aureus*. In nature, this protein is secreted by the *S. aureus* toxin and binds to various cells within a host to induce cell death. This cytotoxicity is caused by the assembly of the monomeric protein into a heptameric pore within the cellular membrane, which affects the permeability of the cell to ions, small molecules, and water.[108] Insertion of the αHL pore complex into an *ex-vivo* support, such as a lipid bilayer, can utilize the same porous characteristic of the protein that causes cytotoxicity as a tool for measurement and sensing.[115,116] Ribbon representations of the structure of native αHL, initially presented in a landmark paper by Song and coworkers,[117] are displayed in Figure 11.7A and B. The nanoporous channel remains open under conditions of neutral pH and high ionic strength such that steady ionic current measurements can be undertaken and used as a baseline for resistive pulse-sensing measurements.[118] Efforts to engineer this pore to induce interactions with species of interest will be explained in the following text as αHL does not have an intrinsic attraction to any molecules after formation of the nanopore structure.

Though native biological pores have small pore diameters and very specific selectivity, some sort of manipulation or engineering is often necessary to tune the pores for application within an experimental system.[119,120] Schematic depictions of three modification approaches are shown in Figure 11.7C, in which selective binding sites are introduced into αHL pores within a lipid bilayer

TABLE 11.2
Comparison of Pores from Nature

Native Biological Pore	Pore Origin	Role in Nature	Channel Motif	Pore Diameter (nm)
αHL	*Staphylococcus aureus*	Pathogenic lysis of cells (typically RBCs)	β-barrel in heptameric complex	1–2
MspA	*Mycobacterium smegmatis*	Mediation of hydrophilic nutrients	β-barrel in octameric complex	1–2
OmpF	*Escherichia coli*	Uptake of hydrophilic nutrients	β-barrel in trimeric complex	1–2
OmpG	*Escherichia coli*	Uptake of hydrophilic nutrients	Monomeric p-barrel	~1

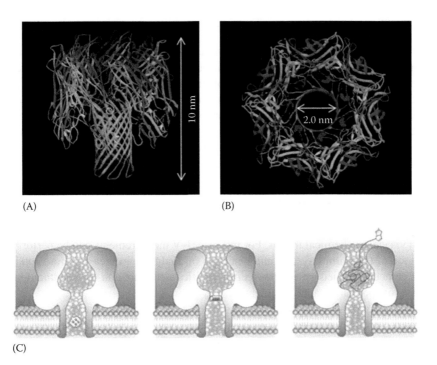

FIGURE 11.7 Overview of native and engineered αHL. (A) Ribbon representation of native αHL with a view perpendicular to the sevenfold symmetry axis and (B) with a top-down view. The diameter of the stem domain, and thus the nanopore itself, is approximately 26 Å or 2.6 nm. (C) Schematic representations of αHL that has been engineered with selective binding sites for analytes of interest: (left) insertion of histidine residues within the pore to bind divalent metal cations; (center) insertion of a β-cyclodextrin ring adaptor capable of performing host–guest chemistry within the pore; and (right) covalent attachment of a target-ligand terminated polyethylene glycol (PEG) chain to a cysteine residue within the pore. Molecular graphics images were produced using the UCSF Chimera package from the Resource for Biocomputing, Visualization, and Informatics at the University of California, San Francisco (supported by NIH P41 RR001081).[287] (C: Reprinted by permission from Macmillan Publishers Ltd. *Nature*, Bayley, H. and Cremer, P.S., Stochastic sensors inspired by biology, 413, 226–230. Copyright 2001.)

for improved interaction with an analyte of interest. Molecular biology techniques can be employed to mutate the protein sequence of the protein directly; often, amino acid residues chains are substituted or inserted such that reactive side chains protrude into the pore and induce interactions with an analyte. Such biological manipulation of the protein sequence has been used to insert histidine residues to bind metal cations,[121,122] hydrophobic residues for hydrophobic interactions,[123,124] and basic or acidic amino acids to induce an overall charge within the pore for electrostatic interactions with charged species like DNA or proteins.[125,126] The αHL pore also can be noncovalently engineered to contain various adaptors to impart selectivity to the pore.[127–129] Finally, the inner surface of the pore can be surface-modified such that functional groups that interact with specific analytes are accessible to target species of interest.[130] These three strategies (genetic mutation, adaptor molecules, and surface pore modification) have been utilized to engineer different biological pores for improved interactions with species of interest.

There are a number of natural pores whose full utility within nanotechnology have not yet been exploited. These native nanoporous systems include the nuclear pore complex (NPC), a megadalton protein complex within the nuclear membrane that selectively permits the translocation of molecules between the cytoplasm and nucleoplasm,[131] and the cavity of the ribosome, through which RNA passes as nucleic acids are translated into a corresponding chain of amino acids.[132]

11.3 MEASUREMENTS AND PROPERTIES OF NANOPORES

Pores in nature and solid-state adaptations serve as platforms to accomplish measurement and analysis. Employment of nanopore devices in various applications such as sensing and efforts to sequence has made nanopores intriguing tools of interdisciplinary interest. Here, we discuss the unique conductance and current rectification properties of nanopores and how they are applied within experimental setups to make measurements of interest. Nanoporous systems are utilized to measure analytes indirectly via resistive-pulse sensing in which they function as channels whose conductivity changes as a species translocates the pore; direct analysis of species also can be accomplished with the advent of nanoporous electrodes.

11.3.1 IONIC CONDUCTANCE AND RECTIFICATION

Molecules and ions that translocate across a nanopore that is filled with an electrolyte solution can be measured via electrochemical means. For instance, I–V measurements can be performed for the characterization of the porous system and detection of analytes. To measure pore conductance, a potential difference is applied across the electrolyte-filled nanopore and the resultant ion current is measured as a function of the applied potential.

Current–voltage curves that demonstrate rectification, or the preferential passage of ions under specific applied potentials, usually indicate an inherent asymmetry within the system. Asymmetry typically arises from factors such as nanopore shape, disproportionate surface charge, or unequal electrolyte solution conditions in the surrounding environments of either side of the pore (Figure 11.8).[133] Conical nanopores, with one side of the pore a larger diameter than the other, and nanopipettes rectify ion current as a result of their asymmetric geometry.[134–136] Likewise, nanopore systems with asymmetric distribution of surface charge have also resulted in rectified ion current measurements. Rectification is also observed when the nanopore is in an environment of uneven ionic strength, as this gradient affects the transport of ions through the nanoscale opening.[137] Improved understanding of rectification-related phenomena has led to the development of various nanoporous transistors[138,139] and diodes.[29,138,140–142]

Often, the extent of rectification is expressed through the ion current rectification ratio (ICR ratio):

$$ICR = \frac{I_{-V}}{I_{+V}} \tag{11.1}$$

This ratio represents the current value at a negative voltage (I_{-V}) divided by the current obtained at the corresponding positive voltage (I_{+V}) and can range in value from zero to infinity. A system that follows Ohm's law and displays no rectification will have an ICR ratio of one; deviations from this value indicate the degree of rectification in either a forward or reverse potential bias mode.

11.3.2 ANALYTE DETECTION VIA RESISTIVE-PULSE SENSING

Many applications with nanopores and nanoporous platforms use measurements collected via a technique called resistive-pulse sensing. Developed on the basis of the Coulter counter of the 1950s,[143,144] the fundamental principle of resistive sensing relies on the altered conductivity of nanopores as material passes through the pore lumen. In typical resistive-pulse measurements, a constant transmembrane potential difference is applied to induce the continuous migration of ions from the surrounding electrolyte solution through the pore, and the ionic conductance of the pore is measured as a function of current over a period of time. Transient fluctuations in current, called resistive pulses, can be observed when a species enters and passes through the pore to momentarily alter the conductivity of the volume of electrolyte held in the pore. Measurable changes in current not only

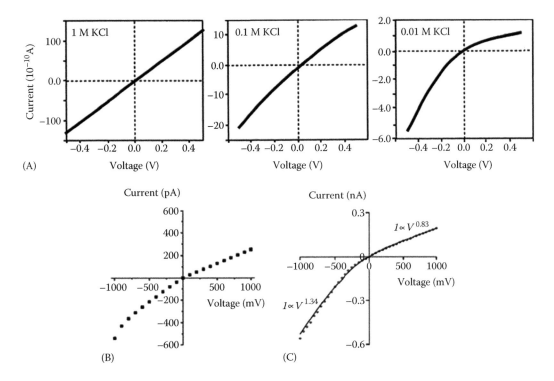

FIGURE 11.8 Overview of ion conductance through nanopores and the rectification effects that often are observed. (A) Current–voltage measurements of a nanopipette (~40 nm diameter) in various concentrations of electrolyte (scan rate = 20 mV/s). Rectification only is observed under low electrolyte-concentration conditions as the negative charge of the quartz surface is not fully screened. (B) Current–voltage measurements of a single cylindrical nanopore (10 nm) in PET with 0.1 M KCl on both sides of the membrane and (C) of a single conical nanopore in PET with 0.1 M KCl on both sides of the membrane. The asymmetric shape of the conical pore leads to increased rectification than what is observed for a similarly sized cylindrical pore. (Reprinted with permission from Wei, C., Bard, A.J., and Feldberg, S.W., Current rectification at quartz nano-pipet electrodes, *Anal. Chem.*, 69, 4627–4633, 1997. Copyright 1997 American Chemical Society; Siwy, Z.S.: Ion-current rectification in nanopores and nanotubes with broken symmetry. *Adv. Funct. Mater.* 2006. 16. 735–746. Copyright Wiley-VCH Verlag GmbH & Co. KGaA. Reproduced with permission.)

detect when species pass through the pore, but also elucidate important structural and chemical information about the analyte. For example, the amplitude of the current spike can be related to the size of the material contained within the nanopore. The width of the current peak is a measure of the time required for an analyte to translocate through the pore and thus provides insight into pore–analyte interactions that may have occurred. This measurement is useful especially for studies that utilize pores modified with a receptor known to interact with a target analyte such that interaction kinetics can be measured. An overview of resistive pulse-sensing experiments and a representative current trace collected as a function of time are illustrated in Figure 11.9.[145]

A major advantage of resistive-pulse sensing is that any analyte able to translocate a nanopore can be probed as the measurement is not contingent upon any interactions or chemical labels. The only requirement for detection with resistive-pulse sensing is that the nanopore must be similarly sized as the species of interest such that the change in current upon translocation is significantly different than the baseline open pore current. The resistive-pulse method has been used to study various analytes,[3] including DNA/RNA,[146–148] antibody complexes,[149,150] peptides/proteins,[151,152] viral components,[153,154] particles,[155,156] droplets within emulsions,[157] and small molecules.[121,123,127,158] The current signal measured during translocation events can be used not just to attain analyte detection

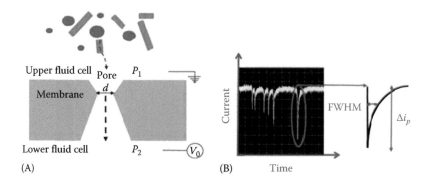

FIGURE 11.9 Overview of resistive pulse-sensing measurements. (A) Schematic of the typical experimental setup in which analytes are placed in one chamber of a U-tube apparatus and driven through a nanopore into the opposite chamber through the application of a transmembrane potential difference. (B) An example of the data readout collected during the course of a resistive sensing experiment. Current is measured as a function of time, and drops in current are observed when material passes through the nanopore from one chamber into the other. The full width half maximum (FWHM) of each event gives information about the time required for translocation, and thus provides information about interactions the analyte undergoes with the pore. The magnitude of current change (Δi_p) gives information about the size of the analyte as larger particles effectively block the pore more completely and cause a larger drop in pore conductance. (Platt, M., Willmott, G.R., and Lee, G.U.: Resistive pulse sensing of analyte—Induced multicomponent rod aggregation using tunable pores. *Small*. 2012. 8. 2436–2444. Copyright Wiley-VCH Verlag GmbH & Co. KGaA. Reproduced with permission.)

but also to provide details of the specific geometry of the individual nanopore.[159,160] Additionally, resistive sensing is an integral part to many applications such as sensing and sequencing that will be discussed further in subsequent sections.

11.3.3 ANALYTE DETECTION VIA NANOPOROUS ELECTRODES

Conductive nanoporous materials have demonstrated interesting signal phenomena as a result of the nanoscale features present within their electrode geometry.[161] Electrochemical detection systems have been designed with nanoporous electrodes fabricated from gold,[162–164] copper,[165] titanium dioxide,[166,167] platinum,[168–170] and graphene.[171] Electrodes fabricated with pores intercalated throughout the electroactive surface often demonstrate enhanced electrochemical signal in comparison to planar electrodes due to both the increased electrode surface area and the increased time of interaction of the electroactive species at the electrode surface.[168]

These nanoporous electrodes can be utilized for the same type of electrochemical studies as planar electrodes but have some distinct differences in electrochemical response due to their nanoscale geometry. The kinetics at the electrode surface can be significantly altered as the pore sizes are often on the same scale as the electrical double layer thickness and thus overlapping double layers can be achieved.[172,173] Voltammetric responses of nanoporous electrodes are distinctly different from their nonporous counterparts as the surface features can provide acceleration of proton or electron transfer steps of electrochemical reactions.[174–176] Despite these differences in electrochemical response, nanoporous electrodes are often utilized to perform chronoamperometry and voltammetry to achieve the detection of various electroactive analytes.

Nanoporous electrodes also have found utility in the study of photoelectrochemical systems.[162,166,167] One such study is highlighted in Figure 11.10, in which nanoporous gold leaf (NPGL) was surface-modified with a protein complex involved with photosynthesis and photocurrent responses directly measured upon exposure to light.[162] NPGL is a free-standing porous thin film fabricated from Au/Ag leaf that is dealloyed in concentrated nitric acid. SEM images show that most

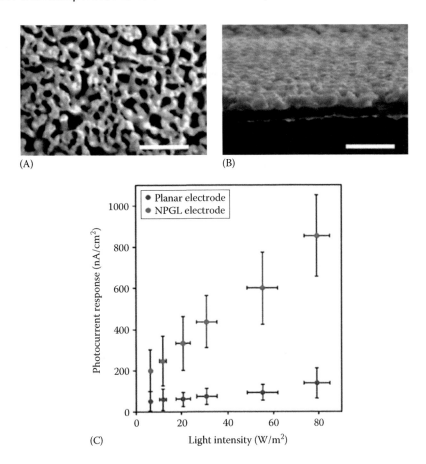

FIGURE 11.10 An example of nanoporous electrodes utilized for measurements. (A) Top-down SEM image and (B) cross-sectional SEM image of a nanoporous gold leaf (NPGL) electrode (scale bar = 300 nm). (C) Photosystem I (PSI), a protein complex involved with photosynthesis, was immobilized to the surface of NPGL and planar Au electrodes and subsequent photocurrent responses measured at various intensities of white light. Vertical and horizontal error bars represent the standard deviations of the photocurrent responses and light intensities, respectively. The larger interfacial surface area of the NPGL electrodes meant more PSI was immobilized on the porous electrodes; thus, an increase in the amount of light absorbed led to a dramatic increase in measured photocurrent in comparison to the planar electrodes. (Reprinted with permission from Ciesielski, P.N., Scott, A.M., Faulkner, C.J., Berron, B.J., Cliffel, D.E., and Jennings, G.K., Functionalized nanoporous gold leaf electrode films for the immobilization of photosystem I, *ACS Nano*, 2, 2465–2472, 2008. Copyright 2008 American Chemical Society.)

pores in NPGL have diameters of 50–100 nm (Figure 11.10A, B). The porous electrode was surface modified with a self-assembled monolayer of a thiol conjugated with terminal aldehyde groups, and the protein complex photosystem I (PSI) was tethered to the system through covalent binding of the amine side chains of lysine residues. Electrodes were then exposed to various intensities of polychromatic light, and the chlorophylls within PSI converted energy from photons into excited electrons through the photosynthetic process. Photocurrent, or the difference between the current measured by the system in the dark and upon exposure to light, was measured through photochronoamperometry to compare NPGL electrodes with planar Au electrodes. The NPGL electrodes displayed at least a threefold enhancement in measured photocurrent over the planar electrodes; this outcome is likely attributed to the increased interfacial surface area of the nanoporous electrodes, which allows for a greater concentration of PSI to be immobilized and subsequently harness the incident photons.

11.4 APPLICATIONS

The nanopore platforms described earlier have become critical tools for advances in analyte detection, separation, delivery, and imaging. Recent applications in which nanoporous platforms are utilized are discussed in the following text.

11.4.1 Sensors

Individual nanopores or nanopore arrays can be developed into a sensor to detect a species of interest.[3,177] The transduction methods described earlier (e.g., I–V response or current–time (I–t) response) can be applied to sense/detect an analyte. For instance, changes in the observed rectification response of a nanoporous system can be used to indicate and identify the presence of an analyte.[178–181] Often, this change in ICR ratio can be attributed to a disruption in the surface charge of the nanopore as the analyte of interest binds to the pore. Direct sensing of electroactive molecules can be accomplished electrochemically with nanoporous electrodes; chronoamperometry and voltammetry measurements are most often applied with these systems for detection.[163,164,182,183] However, the most typical measurement utilized in the application of nanopores as sensors is that of resistive-pulse sensing in which current blockades occur as a molecule passes through a pore and alters the conductivity.[146,158,184] For a detailed list of analytes that have been detected via different nanoporous platforms and measurement techniques, please refer to Table 11.3.

In an elegant example of nanopore resistive-pulse measurements, Bayley and coworkers have developed an engineered αHL pore that is capable of the detection of 2,4,6-trinitrotoluene (TNT) (Figure 11.11).[123] The protein pore was genetically mutated such that the aromatic side chains of phenylalanine (Phe), tyrosine (Tyr), and tryptophan (Trp) were positioned within the pore to initiate aromatic–aromatic interactions with the TNT molecule. In the absence of TNT, the pore remained open at all times and no drop in current was observed, whereas individual current blockades were observed in the presence of TNT when the analyte interacted with the engineered pore. The facile detection of TNT is important for the detection of explosive materials, and nanoporous platforms are intriguing systems for the detection of other compounds of significance with regard to bioterrorism and security.[185–187]

The utilization of nanopores to sense species of interest has resulted in additional applications discussed in the following text. Without the capability of detecting analytes, nanoporous platforms could not be used to accomplish feats like sequencing and microscopy on the nanoscale.

11.4.2 Sequencing

The push in recent years toward individual genome sequencing and personalized medicine has helped to motivate the development of next-generation sequencing technologies. The Human Genome Project and most DNA sequencing over the past several decades has relied on the Sanger method, which utilizes chain termination of DNA fragments with fluorescently labeled dideoxy-nucleotides that are separated via capillary electrophoresis.[188,189]

The proposed utilization of nanopores as a tool to accomplish sequencing dates back to initial experiments by Deamer and coworkers in 1996 in which αHL was utilized to detect the passage of individual polynucleotide molecules.[118] In the nearly 20 years since this landmark paper was published, numerous efforts have been undertaken to employ nanopores in a highly sensitive sequencing method.[190,191] In this next-generation sequencing technique, resistive-pulse sensing is performed to measure the change in current as a biopolymer translocates the pore. The premise that nanopores could move beyond simple sensors and attain the polynucleotide sequences of DNA and RNA chains requires a measurable difference in current blockage signal as the different monomers interact with the pore. DNA and RNA are made up of a combination of purines with fused five- and

TABLE 11.3
Examples of Analytes Detected with Various Nanoporous Platforms

Class of Compounds	Specific Analyte	Nanoporous Platform	Reference(s)
Bioanalytes	Amino acid enantiomers	Engineered protein pore	[288]
	Antibody complexes	Nanoporous membrane	[149]
		Protein pore	[289]
		Silicon nanopores	[150]
		Nanopipettes	[290,291]
	DNA	Nanoporous membrane	[146,292]
		Protein pore	[118,147,293]
		Nanoporous electrode	[294]
		Silicon nanopores	[40,295,296]
		Nanopipette	[180]
	DNA–metal interactions	Protein pore	[297]
		Engineered protein pore	[298]
	DNA repair activity	Protein pore	[299]
	Hepatitis B virus	Nanofluidics	[153,154]
	Hormones	Nanoporous electrode	[300]
	Peptides	Protein pore	[301,302]
		Engineered protein pore	[184,303]
	Proteins	Silicon nanopore	[152,304–306]
		Protein pore	[307]
		Engineered protein pore	[130,308–311]
	RNA–protein complexes	Protein pore	[312]
Drug molecules	Antibiotics	Protein pore	[313]
	Hoescht 33258	Nanoporous membrane	[181]
	Insulin	Nanoporous electrode	[314]
	Tricyclic antidepressant	Engineered protein pore	[127]
Ions	Alkali/alkaline earth metal ions	Nanoporous electrode	[169]
	Au^{3+}	Nanoporous membrane	[315]
	Divalent metal ions	Engineered protein pore	[121,122]
		Nanopipette	[179]
	H^+ (pH)	Nanopipette	[178]
	Hg^{2+}	Nanoporous electrode	[183]
	Nitrite	Nanoporous electrode	[163]
	Redox probes	Nanofluidics	[316]
Organic molecules	Coronene/perylene	Protein pore	[317]
	Cyclofructans/cyclodextrans	Engineered protein pore	[318]
	p-Nitrophenol	Nanoporous electrode	[164]
	Organic molecules	Engineered protein pore	[319]
	Porphyrins	Nanoporous membrane	[158]
	Propanediol	Nanoporous electrode	[170]
	THF	Engineered protein pore	[320]
Particles	Nanoparticles	Silicon nanopores	[235,321]
		Engineered protein pore	[322]
		Nanopipettes	[290]
	Oil droplets within emulsion	Nanoporous membrane	[157]
	Polystyrene beads	Nanoporous membrane	[160]
	Superparamagnetic beads	Nanoporous membrane	[155]

(Continued)

TABLE 11.3 (*CONTINUED*)

Examples of Analytes Detected with Various Nanoporous Platforms

Class of Compounds	Specific Analyte	Nanoporous Platform	Reference(s)
Small molecules	ADP	Engineered protein pore	[114]
	ATP	Nanoporous electrode	[323]
	Glucose	Nanoporous membrane	[324]
		Nanoporous electrode	[165]
	H_2O_2	Nanoporous electrode	[325]
	Humidity	Nanoporous membrane	[326]
	IP_3	Protein pore	[327]
		Engineered protein pore	[328,329]
Chemical/biological threats	Anthrax toxin	Protein pore	[330]
	Botulinum neurotoxin A	Nanoporous membrane	[331]
	Liquid explosives	Protein pore	[332]
	Nitrogen mustards	Engineered protein pore	[186]
	Organoarsenic compounds	Engineered protein pore	[333]
	Ricin	Nanoporous membrane	[185]
	TNT	Engineered protein pore	[123]

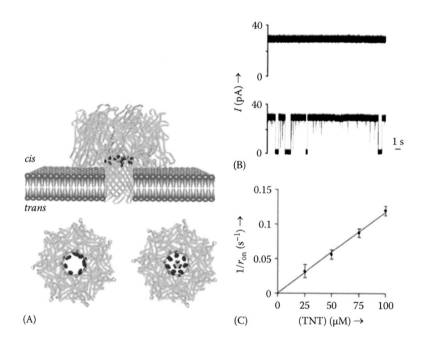

(A)

(B)

(C)

FIGURE 11.11 Nanopores utilized to detect TNT. (A) Molecular graphics representation of the engineered αHL pore utilized in this study. Side view (top), view from the *cis* side of the lipid bilayer (bottom left), and view from *trans* side of the lipid bilayer (bottom right). Seven amino acids with aromatic side chains (Phe, Tyr, and Trp) were mutated into the pore center to enable aromatic–aromatic interactions with the nitroaromatic moiety of TNT. (B) Current measured in the absence of TNT (upper) and 50 μM TNT (lower). Individual current spikes correspond to translocation events when TNT passed through the nanopore. (C) Mean translocation time plotted as a function of TNT concentration produces a linear response, which is the equivalence of a calibration curve. (Guan, X.Y., Gu, L.Q., Cheley, S., Braha, O., and Bayley, H.: Stochastic sensing of TNT with a genetically engineered pore. *Chembiochem.* 2005. 6. 1875–1881. Copyright Wiley-VCH Verlag GmbH & Co. KGaA. Reproduced with permission.)

six-membered rings (adenine and guanine) and pyrimidines with a six-membered ring (thymine/uridine and cytosine); thus, single-stranded DNA feasibly could produce current signals dependent on the size and positioning of an individual nucleotide located at the most confined width of the nanopore.[147]

A number of studies have been performed to attain specific information about DNA strands as they pass through the pore and develop rapid, cost-effective analyses with these next-generation sequencing platforms. Single-stranded DNA can be distinguished from double-stranded polynucleotides on the basis of the duration and magnitude of current blockage,[192] and nanopores have been developed to accomplish the translocation of DNA under extreme conditions that denature the double-stranded biopolymer.[193] DNA duplexes have been unzipped within solid-state nanopores, so information about the kinetics and required energy of the dissociation process can be collected from the changes in pore conductance.[194] Though both solid-state and proteinaceous nanopore systems have been used in these studies, nanoporous graphene is predicted computationally to be an optimistic substrate with which to accomplish sequencing.[195]

Often, DNA passes through a nanopore too quickly for individual base pairs to interact with the pore for any extended period of time, and information that could determine the actual sequence of unknown polynucleotide chains is unable to be attained.[196,197] A DNA sequencing platform that achieves single-nucleotide resolution through the utilization of an engineered nanopore and polymerase enzyme is presented in Figure 11.12.[126] MspA was mutated to replace negatively charged amino acids within the pore with neutral or positively charged residues to promote the translocation of the negatively charged sugar-phosphate backbone of DNA.[126,198] DNA is electrically driven through the pore by the application of a transmembrane potential difference, but the rate of translocation is limited by the action of the polymerase to move the DNA strand. This slower rate of passage of the polynucleotide chain through the pore means that changes in blockage current can be measured when a different nucleotide is transferred to the nanopore constriction to achieve single-nucleotide resolution.

Nanopores also are being explored to collect sequence information about other biopolymers. As discussed previously, RNA has been examined with the aid of nanopores.[118,148] Studies with nanopores are also in progress to determine the location of specific epigenetic modifications, such as methylation sites of DNA necessary to regulate proper organism development.[95,199]

11.4.3 Separations

The ability to separate species from one another is an important area of study within analytical chemistry and is often necessary to accomplish adequate detection of an analyte of interest over the surrounding background matrix. The common lab practices of chromatography, distillation, and centrifugation are examples of separation performed on the macroscale. Nanoporous membranes can be utilized to function as filters and partition analytes on a much smaller size scale. This separation can be accomplished in a size exclusion fashion that divides molecules as a function of their size or as the result of different interaction dynamics between the species and the nanopores themselves.

Separation of biomolecules on the basis of size has been accomplished with ultrathin silicon membranes and is described in Figure 11.13. Fauchet and coworkers fabricated a pnc-Si membrane of ~10 nm thickness with average pore sizes between 5 and 25 nm.[43] The pnc-Si membrane was sandwiched between two reservoirs of solution in a setup similar to a U-tube cell. Fluorescently tagged analytes were present initially in the solution on one side of the membrane, and fluorescent imaging was performed at a position 50 μm away from the membrane edge in the opposite chamber so that the passage of material through the pores could be analyzed over the progression of time. The pnc-Si membrane was demonstrated to separate small molecules from proteins as well as differently sized proteins under physiological conditions. The separation of free Alexa 546 dye (1 kDa) and bovine serum albumin (BSA) (67 kDa) is quantitatively

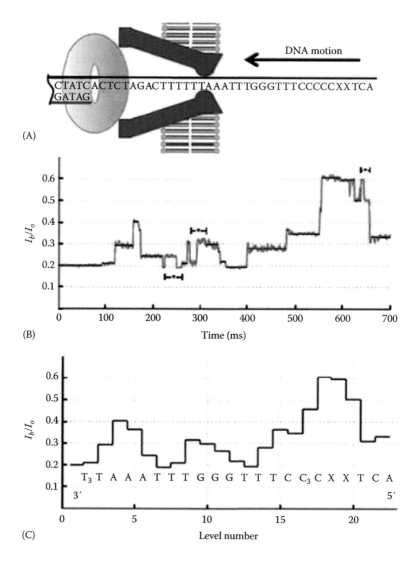

FIGURE 11.12 DNA sequencing through an engineered MspA pore at single-nucleotide resolution. (A) Schematic of the sequencing experiment. DNA is driven through the pore due to a potential difference applied across the membrane. The rate of transcription accomplished by DNA polymerase (green oval) controls the translocation speed of the sDNA. (B) A typical current trace measured over time as DNA passes through the pore. The y-axis plots the measured blockage current (I_b) as a fraction of the current when the pore is empty (I_o). The initial readings are of multiple thymines, which produce a $I_b/I_o \sim 0.2$. At the end of the measurements, two abasic residues (XX) produced a high current ($I_b/I_o \sim 0.61$). (C) The mean currents of levels extracted from (B) are plotted with the known associated DNA sequence. (Reprinted by permission from Macmillan Publishers Ltd. *Nat. Biotechnol.*, Manrao, E.A., Derrington, I.M., Laszlo, A.H., Langford, K.W., Hopper, M.K., Gillgren, N., Pavlenok, M., Niederweis, M., and Gundlach, J.H., Reading DNA at single-nucleotide resolution with a mutant MspA nanopore and phi29 DNA polymerase, 30, 349–353. Copyright 2012.)

displayed in the top graph of Figure 11.13C: translocation of the free dye through the nanopores is observed by the increase in corresponding fluorescence signal over time. The lack of signal from BSA is likely due to the protein's larger molecular weight and hydrodynamic radius that hinder passage through the pnc-Si. Similarly, BSA is passed through the nanoporous membrane over four times faster than the larger IgG (150 kDa) (bottom graph Figure 11.13C). This difference in transit rate across the membrane, despite the bioanalytes' similar diffusion coefficients,

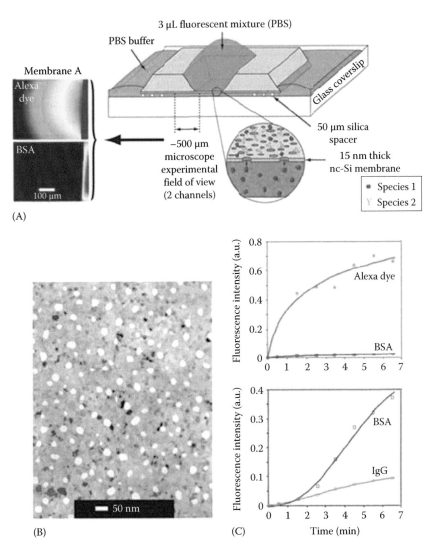

FIGURE 11.13 Separation accomplished with a nanoporous membrane. (A) Schematic of the experimental setup in which fluorescence is utilized to monitor the passage of analytes through an ultrathin porous nanocrystalline silicon (pnc-Si) membrane. Experimental false-color fluorescence images clearly show separation of Alexa dye and fluorescently labeled BSA. (B) A TEM image of the porous membrane; pores appear as bright spots, whereas the pnc-Si is in gray or black contrast. (C) Highly efficient separation of BSA protein (67 kDa) and free dye (1 kDa) is observed upon passage through the pnc-Si membrane (top). A greater than fourfold separation of BSA and IgG (MW = 150 kDa) proteins is observed through the membrane (bottom). (Reprinted by permission from Macmillan Publishers Ltd. *Nature*, Striemer, C.C., Gaborski, T.R., McGrath, J.L., and Fauchet, P.M., Charge- and size-based separation of macromolecules using ultrathin silicon membranes, 445, 749–753. Copyright 2007.)

is because the nanoporous membrane hinders the passage of larger proteins and accomplishes separations on the basis of size. Though not shown here, the pnc-Si membrane also demonstrated the ability to separate analytes on the basis of charge through the modification of membrane surface charges.[43]

Separation of analytes on the basis of size has been accomplished in a varied selection of nanoporous platforms, as discussed previously. Nanopores often are used as filters to separate differently sized species within a mixture by effectively blocking material that is incapable of passing through the dimensions of the pore due to size; this design has been applied to the separation of biomolecule mixtures[43,200] and particles/dextrans.[201,202] The larger proteins of blood serum can be separated from the rest of the matrix with nanoporous silicon to yield a sample with a distinctly different mass spectrometric profile for better analysis of lower molecular weight molecules.[203] Nanoporous alumina is able to accomplish the equivalence of size exclusion chromatography because smaller molecules are able to enter the pores and spend more loss time within them than their larger molecule counterparts.[204] Differently sized DNA molecules have been able to be separated within a nanochannel lined with anodic porous alumina by this size exclusion method.[204]

Though separation on the basis of size is a very common method, some separations with nanopores occur on the basis of electrostatic interactions. Bare alumina membranes,[205] Pt-coated alumina membranes,[206] and Au-coated polycarbonate nanopores[207] have all been used to partition similarly sized biomolecules on the basis of the charge of the analyte. Silica nanoporous membranes are also able to partition positively charged thioflavin T from free solution due to the negative charge imparted from deprotonated silanol groups.[208] One could imagine the fabrication of nanoporous systems that combine the size separation described earlier with charge separation to produce the equivalent of a macroscale 2-D gel electrophoresis experiment.

Sometimes, the nanoporous platform must be surface-modified to create or enhance analyte–nanopore interactions that aid in separation of species. Recently, nanoporous membranes were modified via the adsorption of polyelectrolytes to various solid-state nanoporous membranes of different materials to realize the separation of ions.[209] Ion mobility was altered in a hybrid biological/solid-state nanoporous polymer membrane.[210] The surfaces of nanofluidic channels can be chemically altered to mimic the packing material often found in liquid chromatography columns to separate on the basis of hydrophobicity.[211] Likewise, surface-modified polycarbonate membranes have been developed, which utilize hydrophobic interactions to achieve separation of similarly sized dye molecules.[212]

In recent years, nanoporous graphene has been studied extensively and may prove to be a powerful and exciting platform with respect to separations.[101,213–220] Graphene is an attractive nanoporous membrane to accomplish separations because of the single- or few-atom thickness; the inherent thinness of graphene results in a higher membrane permeance value and improved separation efficiency.[100,101] Though separation with graphene nanopores has not yet been realized experimentally, various simulations conclude that graphene would serve as superior technology for the desalination of water[216] and separation of gases.[101,214,215,217,219,220] Further, bioinspired graphene nanopores have been designed with carboxylate groups from the side chains of glutamate protruding into the pore to discriminate between Na^+ and K^+ ions.[213]

11.4.4 Localized Sample Manipulations and Measurements

One of the tremendous advantages of working with nanoporous platforms, particularly nanopipettes, is the ease with which they can be manipulated spatially through the use of piezoelectric motors. Piezos offer subnanometer control and can be used to move a nanoprobe in all three spatial dimensions such that the nanoporous probe is positioned extremely close to a sample or area of interest; thus, extremely localized functions and measurements can be performed. This ease of 3-D positioning renders nanopores excellent tools to deliver analytes to or sample from localized regions of a sample, pattern a surface, and produce topographical and chemical maps at a sample-solution interface.

11.4.4.1 Sample Delivery and Trapping

Nanopipettes serve as miniaturized versions of sample delivery devices, and precise volumes of material can be both collected and delivered at localized positions. Controlled deposition of material often is accomplished electrically with a potential difference applied between the surrounding

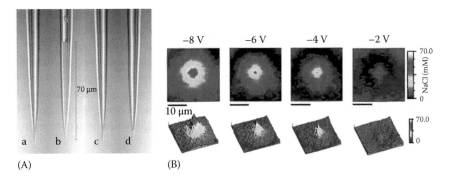

FIGURE 11.14 Delivery of analytes from nanopipettes. (A) Sequential ingress and egress of an aqueous electrolyte solution into a nanopipette filled with dichloroethane and electrolyte. The sampling is controlled by the polarity and magnitude of a potential difference between the nanopipette and bulk. (a) demonstrates initial immersion of the pipette into the aqueous surroundings; (b) shows uptake of water after the potential applied to the nanopipette was stepped to −100 mV; (c) demonstrates complete egress of the sampled aqueous solution via applied potential of +600 mV to the nanopipette; (d) shows ingress of water at the same potential as (a), but for a shorter time period to uptake a smaller volume. (B) Dosing profiles of Na^+ ions delivered by varied potential differences between a 320 nm inner diameter nanopipette with 100 mM NaCl at a constant probe–surface distance of 160 nm and bulk solution. The concentration profiles were measured via total internal reflectance fluorescence (TIRF). (Laforge, F.O., Carpino, J., Rotenberg, S.A., and Mirkin, M.V., Electrochemical attosyringe, *Proc. Natl. Acad. Sci. USA*, 104, 11895–11900, 2007. Copyright 2007 National Academy of Sciences, U.S.A.; Piper, J.D., Li, C., Lo, C.J., Berry, R., Korchev, Y., Ying, L.M., and Klenerman, D., Characterization and application of controllable local chemical changes produced by reagent delivery from a nanopipette, *J. Am. Chem. Soc.*, 130, 10386–10393, 2008. Copyright 2008 American Chemical Society.)

bulk solution and the inner-nanopipette solution. The combination of spatial control of the nanopipette from a piezoelectric positioner and delivery control through the pipette lends this system to the patterning of surfaces, which will be discussed in the following section.

Delivery of analyte solution or sampling of localized regions of interest is accomplished electrophoretically via a potential difference applied between the bulk solution and that solution within the nanopipette. The ingress of different volumes of aqueous solution into a nanopipette that is filled with an immiscible organic solvent and electrolyte is shown in Figure 11.14A. By altering the polarity of the applied potential and the time length potential is applied to the system, the volume loaded into the nanopipette can be tuned as desired; volumes as small as 3 ± 2 aL have been drawn into the nanopipette with this method.[221] This two-phase system has been applied as a delivery system for the direct injection of analytes into single cells. The sharp tip of the nanopipette was used to puncture human breast epithelial cells, and potential-controlled delivery of a fluorescent dye was accomplished into the target cell's cytoplasm.[221]

Klenerman and coworkers have utilized potential-controlled delivery of water-soluble analytes from nanopipettes in a system which is entirely aqueous.[222] Pulsed voltage-driven delivery of ions was performed from nanopipettes at close proximity to a surface, and the localized concentration profiles were measured immediately with total internal reflectance fluorescence (TIRF) microscopy (Figure 11.14B). Their studies found that the total concentration of analyte delivered was dependent on the size of the nanopipette, applied potential difference, height above the surface during delivery, and initial analyte concentration within the pipette. This localized delivery system was then applied to cellular studies in which the sodium-sensitive *E. coli* flagellar motor was dosed with Na^+ ions.[222] The cellular response of the speed of motor movement was observed and matched well with previously determined dosing profiles under defined delivery conditions.

In addition to nanopipettes, other nanoporous platforms have demonstrated tremendous sampling and delivery capabilities. Nanofluidic systems can be fabricated with a T-junction such that pinched injection produces pulsed samples of controlled volume.[223–225] Through incorporation of a T-junction, small quantities of sample may be diverted from the bulk sample flow into a perpendicular cross channel in which they can be concentrated or analyzed individually.[225,226] A unique application of nanofluidics for sample delivery was developed by Zambelli and coworkers who fabricated hollow cantilevers that are able to dispense liquid while concurrently performing AFM.[227,228] Dubbed fluidFM, a nanofluidic channel within the cantilever probe delivers solution through a 100 nm aperture in the AFM tip to localized regions of the sample under investigation;[227] this powerful tool has been used to explore viral infection[229] and will certainly find employment in the study of other biological systems.

Nanopores and nanopipettes also have been utilized in the trapping of samples for subsequent analysis or delivery. Nanopipettes are excellent tools for dielectrophoretic trapping of charged molecules as almost all of the electric field gradient is focused at the tip of the pipette,[230,231] and the manipulation of these forces gives rise to controllable delivery of samples from the tip. Dielectrophoretic entrapment can be used to measure surface conductivity of biomolecules and for the concentration enhancement/delivery of bioanalytes.[233,234] Recently, single nanoparticles were electrically trapped to the surface of silicon nitride nanopores with smaller dimensions in an extremely controllable manner.[235]

Optical tweezers also have been used in combination with nanopores to provide a means of sample manipulation. Optical tweezers use a highly focused laser to provide a force that can physically move (or balanced forces to hold steady) micron- or nanometer-sized beads; analytes of interest are attached to particles for spatial manipulation by the laser. Optical tweezers can be used to measure the electrophoretic forces on biomolecules that are within a nanopore and have measured species that include DNA,[236–238] RNA,[239] protein-coated DNA,[240] and ligand-bound DNA.[241] Though optical tweezers are utilized most often with nanoporous membranes, they have been demonstrated in nanopipette systems as well.[242]

11.4.4.2 Surface Patterning

Due to the ease of local sample delivery, nanopipettes have also been applied to produce arrays or surface patterns. With the ability of piezos to finely control the positioning of a nanopipette, the material can be electrophoretically transferred from a nanopipette onto a surface in any pattern of interest. Surfaces have been patterned from a single-barrel nanopipette with a number of biological analytes; DNA/proteins were the first reported bioanalytes to be surface patterned.[243] DNA and antibodies have also been deposited to create multicomponent and functional submicron features.[244] An antibody nanoarray with ~300 nm diameter features has been patterned through the deposition of antibodies to a nanoporous surface.[245] Nanopipettes also have been used to create surface patterns with inorganic and organic analytes; dot arrays and line patterns of gold colloids were patterned onto silicon surfaces,[246] and double-barrel nanopipettes have been used to produce arrayed water droplets with volumes as small as a few attoliters under a layer of organic solvent.[247] Free-standing Cu nanowire arrays have been fabricated through a method known as electrochemical fountain pen nanofabrication (ec-FPN), which utilizes nanopipettes as templates from which nanowires are electrochemically deposited on a surface.[248,249]

The ability of nanopores to deliver multiple analytes to pattern a surface is demonstrated in Figure 11.15. In this study, small-scale reproductions of famous pictures were recreated with the localized delivery of fluorescently labeled DNA strands from a double-barrel nanopipette.[250] The two individually addressable barrels of a theta pipette allow independent delivery of different analytes from each barrel directly to a surface. This study was accomplished in air, and this environment leads to smaller feature sizes from the lack of lateral diffusion found in bath solutions.[250] The control of deposition necessary to reproduce such detailed images comes from the potential-controlled delivery described previously. In the future, nanopipettes with more than two barrels may be used to deliver an even greater number of different analytes from each pore to produce surface patterns that are more complicated and layered.

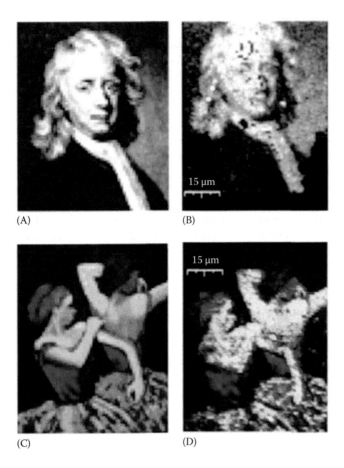

(A)   (B)

(C)   (D)

FIGURE 11.15 Nanopipettes utilized to locally deposit biomolecules to surfaces. (A) An image of Sir Isaac Newton sized with Adobe Photoshop to 75 × 62 pixels and (B) a reproduction of the same image with Alexa 647-labeled DNA deposited on a surface from a double-barrel pipette with 1 μm pixels. The printed area of the reproduced image is 75 × 62 μm². (C) An image of Candelori's painting *Degas Dancers* sized with Adobe Photoshop to 75 × 61 pixels and (D) a reproduction of the same image patterned on a surface with a double-barrel nanopipette. Green and red channels were written consecutively with rhodamine green and Alexa 647-labeled DNA with 1 μm pixels; the reproduction area is 75 × 61 μm². (Rodolfa, K.T., Bruckbauer, A., Zhou, D.J., Korchev, Y.E., and Klenerman, D.: Two-component graded deposition of biomolecules with a double-barreled nanopipette. *Angew. Chem. Int. Ed.* 2005. 44. 6854–6859. Copyright Wiley-VCH Verlag GmbH & Co. KGaA. Reproduced with permission.)

Recently, the fluidFM technique of a nanofluidic channel within a hollow AFM cantilever has been used as a lithography tool to create surface patterns and arrays. Fluorescent polystyrene particles were dispensed with tremendous control to produce lines and arrays on a glass surface.[251] This technique is similar to the surface deposition accomplished with nanopipettes, but offers a new tool with which to expand the field of localized patterning of surfaces.

11.4.4.3 Microscopy

Nanoporous systems have been applied to numerous microscopy techniques. Nanopipettes often are used as probes in SPMs due to the ease of fabrication, small nanopore dimensions at the tip, and ease of spatial positioning via piezoelectric motors. However, this review will focus briefly on SPM techniques which involve electrochemical measurements, namely, SECM, SICM, and the hybrid of the two (SECM-SICM).

11.4.4.3.1 Scanning Electrochemical Microscopy

SECM[52] is a technique utilized to measure localized electroactive analytes and the topography of a surface by applying the measured current to the feedback system. A typical SECM setup utilizes an ultramicroelectrode (UME) to measure the faradaic current produced from the reduction/oxidation of an electroactive mediator within the solution surrounding the sample of interest. For a more in-depth discussion on SECM details and studies, please refer to Amemiya and coworkers' review in Chapter 18; this review will focus only on the application of nanopores to SECM studies.

Nanopipettes have been applied to contact-mode SECM to produce the scanning micropipette contact method (SMCM).[252] In SMCM experiments, a nanopipette filled with a redox-active species is brought into contact with a sample surface via a piezoelectric positioner, and the faradaic current is measured between the surface and the nanopipette solution (Figure 11.16A). With this technique, localized electrochemical measurements of the sample surface can be attained, and a 2-D map of the measured faradaic current is generated as the nanopipette is moved over the plane of the surface. Unwin and coworkers also have utilized theta nanopipettes with two barrels to add high-resolution topography measurements to the electrochemical data already attainable with single-barrel pipettes.[253] This SMCM technique is especially useful in the examination of heterogeneous substrates as localized *hotspots* of electrochemical activity can be observed along the surface of the sample. Additionally, SMCM alleviates the issue of probe–sample distance, a parameter which can play an important role in SECM measurements. For more discussion and details related to this technique, please refer to Unwin and coworkers' review on scanning electrochemical cell microscopy and related techniques in Chapter 19.

Nanopipettes also are employed within SECM experiments to study the kinetics of ion transfer across an interface between two immiscible electrolyte solutions (ITIES).[254–258] In these experiments, the nanopipette is filled with an electrolyte solution that is immiscible with the surrounding bulk environment, and the current is measured as ions pass through the nanopore at the tip (Figure 11.16B).

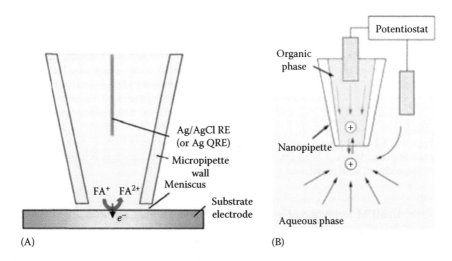

(A)   (B)

FIGURE 11.16 Nanopipettes utilized as probes in SECM experiments. (A) Schematic of SMCM experiments in which trimethylammoniummethylferrocene (FA$^+$) within the nanopipette is oxidized to FA^{2+} as the meniscus is in contact with the surface of the substrate electrode. (B) Schematic of ion flux across the nanopipette/bulk solution ITIES interface due to the applied potential difference between the two solutions. (Reprinted with permission from Keller, F., Hunter, M.S., and Robinson, D.L., Structural features of oxide coatings on aluminum, *J. Electrochem. Soc.*, 100, 411–419, 1953. Copyright 2009 American Chemical Society; Reprinted from *Electrochim. Acta*, 110, Amemiya, S., Kim, J., Izadyar, A., Kabagambe, B., Shen, M., and Ishimatsu, R., Electrochemical sensing and imaging based on ion transfer at liquid/liquid interfaces, 836–845, Copyright 2013, with permission from Elsevier.)

Solutions that contain a specific ionophore are often backfilled within the nanopipette, which is then inserted into a bulk solution of immiscible solvent. With this setup, the nanopipette becomes analogous to an ion-selective electrode, and the current signal measured over the course of the study can be attributed to the flux of the ions of interest across the liquid–liquid interface into the pipette. In these experiments, biologically relevant but redox-inactive ions, such as lithium,[259] sodium,[259,260] potassium,[260–262] and chloride,[259] are often monitored. By utilizing the feedback of SECM to control the position of the nanopipette, spatially resolved ion-selective measurements can be made across various membranes.

11.4.4.3.2 Scanning Ion-Conductance Microscopy

SICM[53] is a contact-free method used to measure the topography of sample surfaces by coupling ion current measurements made within a nanopipette to feedback control of the pipette's positioning. In this microscopy setup, a nanopipette filled with electrolyte solution is placed in a bath solution containing the sample of interest. A potential difference is applied between the solution within the nanopipette and the bulk environment, and an ion current is generated as ions flow through the nanopore at the tip of the pipette. The magnitude of this ion current is determined by the ease through which ions are able to migrate into the nanopipette, a term which is called the access resistance. Access resistance is extremely dependent upon the distance between the nanopipette and sample surface. As the tip moves closer to the interface of the sample, ions are hindered from entering the nanopipette and a smaller ion current is measured. This probe–surface distance dependence means that a desired ion current can be selected as a set-point for feedback, and the piezoelectric system will compensate the positioning of the nanopipette to maintain a constant height above the surface.

SICM is often employed in the study of biological samples because cell media and biological buffers are rich in electrolytes that dissociate into ions in solution, and such a sample environment is extremely compatible with SICM for *in vivo* and *in vitro* measurements. The high spatial resolution of SICM has been demonstrated as the microvilli protruding from cells are often able to be resolved.[263] The highest resolution study that has been achieved to date experimentally was accomplished by Korchev and coworkers in 2006.[264] In this study, subunits ~13 nm apart were able to image a monolayer of periodic S-layer proteins from *Bacillus sphaericus* on mica, and protein subunits ~13 nm apart were able to be resolved.[264] More recently, probes were fabricated that combined the excellent spatial resolution of SICM with ion-selective measurements of the sample in terms of localized pH.[46,47]

One such biological study utilized SICM to study the topography and localized surface changes of porcine aortic endothelial cells (Figure 11.17).[265] Here, a nanopipette filled with electrolyte solution and an electrode is placed in close proximity to the cell surface through the help of a piezo; the cell medium surrounding the pipette contains Cl⁻ ions such that an ion current can be measured between the pipette electrode (PE) and a reference electrode (RE) in the bulk solution. The ion current is utilized as feedback for z-positioning as the nanopipette is scanned over the cells such that the probe–surface distance remains constant, and the adjustments in probe height can produce a topographic image like those shown in Figure 11.17B. This study demonstrated that SICM is an excellent method with which to measure changes in structure and shape upon cellular exposure to a stressor.

More recently, potentiometric measurements have been applied to SICM in a technique aptly named potentiometric-SICM (P-SICM).[266,267] P-SICM utilizes a five-electrode setup to measure localized conductance over sample surfaces—a PE in one barrel of a theta nanopipette to control the pipette positioning and record topography; a potentiometric electrode (UE) in the other barrel to record localized potential at the tip of the pipette; a working electrode (WE) to apply a potential across the sample and induce ion migration; and a counter electrode (CE) and RE to which all other electrodes are referenced (Figure 11.18A). With this technique, topographic images and localized potentiometric images can be collected simultaneously (Figure 11.18B and C, respectively).

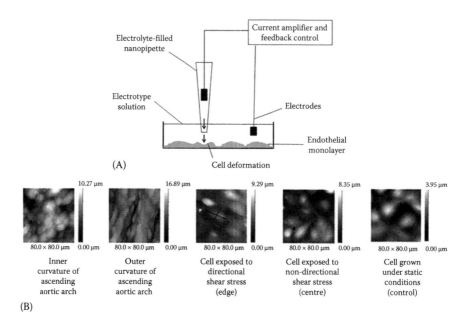

FIGURE 11.17 Nanopipettes utilized as noncontact probes in SICM experiments. (A) Schematic of the experimental setup in which the topography and mechanical properties of porcine aortic endothelial cells are measured via SICM. (B) Representative topographic images collected with SICM of various regions of the aorta. Topographical differences in cell height can be observed through comparison of the control and stressed samples. (From Potter, C.M.F. et al., *PLoS ONE*, 7, e31228, 2012.)

The advantage of this technique over typical SICM experiments is the dramatic increase in measurement sensitivity when differences in potential are measured rather than current. This increased sensitivity and high spatial resolution of P-SICM means that the conductance of the transcellular and paracellular transport pathways can be differentiated in a monolayer of epithelial cells (Figure 11.18D, E).[267]

11.4.4.3.3 SECM-SICM

The hybrid technique of SECM-SICM, first reported concurrently by two separate research groups,[58,59] combines the capability of SICM to attain topographical images of samples within biologically relevant media with the ability of SECM to measure localized electrochemistry. Like SICM, the ion current flow between bulk solution and a nanopipette is utilized in this system as a feedback control for the probe position so that a topographical image of the sample can be obtained. Simultaneously, an independent electrode on the probe measures the faradaic current to provide localized chemical information of redox-active species. This hybrid SPM technique has been utilized to measure the transport of redox probes through polymer nanoporous membranes,[268] immobilized enzymes,[58] and live cells.[58]

A variety of probe types have been developed for SECM-SICM. For one type of probe, gold was deposited onto one side of a nanopipette, which was then insulated entirely with aluminum oxide; the nanopipette opening and a UME of the deposited gold were subsequently exposed via an FIB mill.[59] A similar but simpler technique was also developed and used in SECM-SICM experiments—a nanopipette with a thin gold layer thermally evaporated on one side.[268] Double-barrel carbon nanoprobes have also been fabricated, in which one barrel of the nanopipette is filled with a pyrolized carbon electrode for the measurement of the faradaic current of SECM analysis and the other barrel filled with electrolyte solution to simultaneously measure the ion current for feedback and SICM analysis.[269]

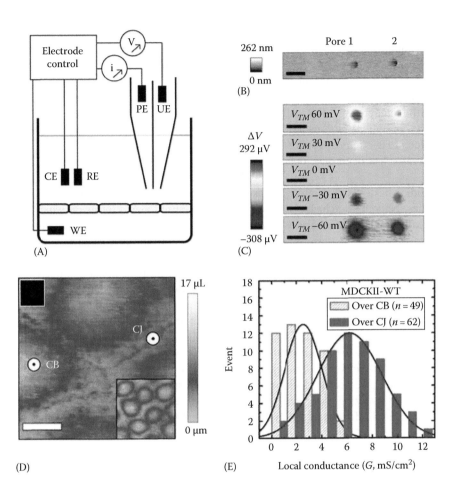

FIGURE 11.18 Nanopipettes utilized as probes in P-SICM studies. (A) Illustration of P-SICM and electrode setup: the pipette electrode (PE) controls the pipette position and records topography; the potentiometric electrode (UE) measures localized potential at the tip; the counter electrode (CE) and reference electrode (RE) are in the bulk solution near the nanopipette, and the RE serves as a common reference to all other electrodes; and a potential is applied between the working electrode (WE) and the RE to induce migration of ions across the sample. (B) Topography map of two nanopores within a polyimide membrane measured with the PE, and (C) simultaneous potential maps measured with the UE for these two pores at a progression of different transmembrane potentials applied between the WE and the RE (scale bars = 1 μm). (D) Topographical image of an epithelial cell monolayer in which a cell body (CB) and cell junction (CJ) can be identified. The black dot in the white marker approximates the size of the pipette tip used in scale with the sample (scale bar = 5 μm). (E) Histogram of conductances obtained with potentiometric measurements over cell bodies and cell junctions; a measurable difference in the conductance over the two locations is observed with the conductances attributed to the transcellular and paracellular pathways, respectively. (Reprinted with permission from Chen, C.C., Zhou, Y., Morris, C.A., Hou, J.H., and Baker, L.A., Scanning ion conductance microscopy measurement of paracellular channel conductance in tight junctions, *Anal. Chem.*, 85, 3621–3628, 2013. Copyright 2013 American Chemical Society.)

11.4.5 ENABLING TOOLS

Nanopores have been applied to various existing techniques and analyses in an effort to expand technological capabilities; here, we present only a small subset of such systems.

Nanoporous platforms recently have found utility in the fields of plasmonics and optical detection. Nanoporous gold films[270] and metallic-coated nanopores[271,272] have been applied to techniques like surface-plasmon resonance (SPR) and surface-enhanced Raman spectroscopy (SERS). Additionally, nanoporous metal has been demonstrated to enhance single-molecule fluorescence intensity of immobilized fluorophores due to the enhanced localized plasmon field present within the nanopores.[273] Optically transparent alumina membranes have been developed and found utility as optical biosensors.[274] Additionally, nanoporous gold has been demonstrated to optically detect Hg^{2+} ions at concentrations smaller than parts per trillion.[183] A fiber-optic ultrasound generator has been developed from the excitation of gold nanopores with a nanosecond laser.[275]

Nanofluidics has extended into systems that previously utilized microfluidic channels for different studies. Nanofluidic channels are now commonly used as platforms on which to grow and immobilize different cells for various analyses. These nanochannels also have been used to preconcentrate analytes of interest prior to analysis.[276–278] The small dimensions of nanofluidic channels have been used to explore electrolysis of solvents.[279]

Nanopores of various types have also found application within the domain of mass spectrometry (MS). Nanoporous membranes have been utilized to improve enrichment[280–283] and proteolysis[284] prior to MS analysis. Nanopipettes recently have found application as emitters for electrospray ionization MS (ESI-MS) (Figure 11.19A).[285] With sub-100 nm sized nanopipettes, dramatic improvement was observed in the signal-to-noise ratio, along with a narrowing of the distribution of observed charge states for a multicharged biologically relevant sample. Additionally, nanopores have been utilized to accomplish single-molecule MS in which analytes pass through individual pores of αHL and are measured via resistive-pulse sensing prior to online MS analysis (Figure 11.19B).[286] By hybridizing the electrochemical nanopore measurements with the MS platform, structure/interaction information and mass analysis/identification can be obtained in real time.

11.5 CONCLUSIONS AND OUTLOOK

Nanoporous systems come in a variety of materials, sizes, and porosities that can be selected or tuned for the best outcome in a specific measurement or application of interest. Electrochemical measurements of analytes with nanopores can be performed directly, as with nanoporous electrodes, or indirectly with resistive-pulse sensing. The applications in which nanopores have been employed are extremely extensive and include fabrication, analyte detection, separations, microscopy, and sample delivery.

As fabrication and analysis techniques improve to better accommodate small systems, porous materials may be pushed past present limits, where smaller pores could lead to resolution beyond the single-molecule resolution that exists today. For this miniaturization to happen, fabrication methods must be evolved and improved electronics developed such that lower electrochemical signals can be confidently measured.

The wide range of applications and technology encompassed by nanopore research is expected to diversify further in the future. Special attention to hybrid applications that make use of biochemical functionality (e.g., enzymes) appended to synthetic or natural nanopores and the integration of spectroscopic/spectrometric tools (e.g., surface-plasmon measurements, optical-trapping, and MS) provide especially fruitful ground for the near future. Ultimately, nanopores provide a versatile technology platform that can span fundamental studies and practical applications, and exciting avenues of inquiry remain.

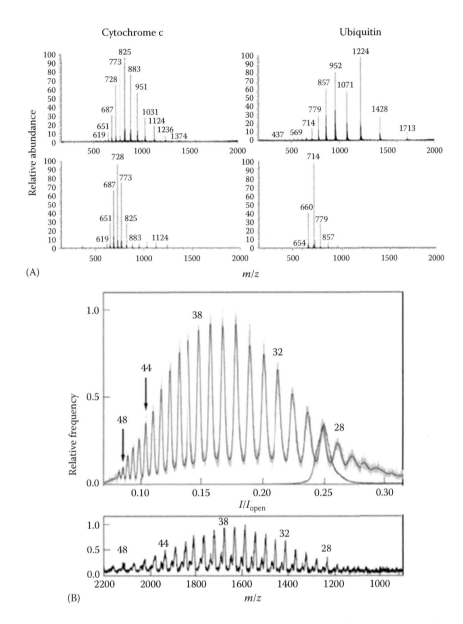

FIGURE 11.19 Nanopores utilized as enabling tools to extend the capabilities of MS. (A) Mass spectra obtained via ESI with commercially available PicoTips (upper) and nanopipette emitters <100 nm in diameter (lower) for cytochrome c and ubiquitin protein standards. Spectra obtained with nanopipettes show a narrower distribution of charges and a shift toward higher charge states than spectra obtained with the larger 1–2 μm emitters. (B) A comparison of mass distributions of polydisperse PEG (M_r = 1500 g/mol) obtained with a single αHL nanopore (upper) and conventional matrix-assisted laser desorption ionization-time-of-flight (MALDI-TOF) mass spectrometry (lower). The conductance of the αHL nanopore can be utilized to discriminate between individual PEG polymers that have different masses in a distribution similar to those achieved via typical mass spectrometric techniques. (Reprinted with permission from Yuill, E.M., Sa, N., Ray, S.J., Hieftje, G.M., and Baker, L.A., Electrospray ionization from nanopipette emitters with tip diameters of less than 100 nm, *Anal. Chem.*, 85, 8498–8502, 2013. Copyright 2013 American Chemical Society; Robertson, J.W.F., Rodrigues, C.G., Stanford, V.M., Rubinson, K.A., Krasilnikov, O.V., and Kasianowicz, J.J., Single-molecule mass spectrometry in solution using a solitary nanopore, *Proc. Natl. Acad. Sci. USA*, 104, 8207–8211, 2007. Copyright 2007 National Academy of Sciences, U.S.A.)

REFERENCES

1. Martin, C. R. Nanomaterials—A membrane-based synthetic approach. *Science* 1994, *266*, 1961–1966.
2. Li, N. C.; Yu, S. F.; Harrell, C. C.; Martin, C. R. Conical nanopore membranes. Preparation and transport properties. *Anal Chem* 2004, *76*, 2025–2030.
3. Howorka, S.; Siwy, Z. Nanopore analytics: Sensing of single molecules. *Chem Soc Rev* 2009, *38*, 2360–2384.
4. Dallanora, A.; Marcondes, T. L.; Bermudez, G. G.; Fichtner, P. F. P.; Trautmann, C.; Toulemonde, M.; Papaleo, R. M. Nanoporous SiO$_2$/Si thin layers produced by ion track etching: Dependence on the ion energy and criterion for etchability. *J Appl Phys* 2008, *104*, 024307.
5. Rajasekaran, P. R.; Wolff, J.; Zhou, C.; Kinsel, M.; Trautmann, C.; Aouadi, S.; Kohli, P. Two dimensional anisotropic etching in tracked glass. *J Mater Chem* 2009, *19*, 8142–8149.
6. Jin, P.; Mukaibo, H.; Horne, L. P.; Bishop, G. W.; Martin, C. R. Electroosmotic flow rectification in pyramidal-pore mica membranes. *J Am Chem Soc* 2010, *132*, 2118–2119.
7. Vlassiouk, I.; Apel, P. Y.; Dmitriev, S. N.; Healy, K.; Siwy, Z. S. Versatile ultrathin nanoporous silicon nitride membranes. *Proc Natl Acad Sci USA* 2009, *106*, 21039–21044.
8. Hillmyer, M. A. Nanoporous materials from block copolymer precursors. *Adv Polym Sci* 2005, *190*, 137–181.
9. Ito, T.; Perera, D. M. N. T. Analytical applications of block copolymer-derived nanoporous membranes. In *Trace Analysis with Nanomaterials*. Weinheim, Germany: Wiley-VCH Verlag GmbH & Co. KGaA, 2010; pp. 341–358.
10. Rzayev, J.; Hillmyer, M. A. Nanochannel array plastics with tailored surface chemistry. *J Am Chem Soc* 2005, *127*, 13373–13379.
11. Li, Y. X.; Ito, T. Surface chemical functionalization of cylindrical nanopores derived from a polystyrene—Poly(methylmethacrylate) diblock copolymer via amidation. *Langmuir* 2008, *24*, 8959–8963.
12. Zschech, D.; Kim, D. H.; Milenin, A. P.; Hopfe, S.; Scholz, R.; Goring, P.; Hillebrand, R. et al. High-temperature resistant, ordered gold nanoparticle arrays. *Nanotechnology* 2006, *17*, 2122–2126.
13. Zschech, D.; Kim, D. H.; Milenin, A. P.; Scholz, R.; Hillebrand, R.; Hawker, C. J.; Russell, T. P.; Steinhart, M.; Gosele, U. Ordered arrays of <100>—Oriented silicon nanorods by CMOS-compatible block copolymer lithography. *Nano Lett* 2007, *7*, 1516–1520.
14. Yang, S. Y.; Ryu, I.; Kim, H. Y.; Kim, J. K.; Jang, S. K.; Russell, T. P. Nanoporous membranes with ultra-high selectivity and flux for the filtration of viruses. *Adv Mater* 2006, *18*, 709–712.
15. Yang, S. Y.; Park, J.; Yoon, J.; Ree, M.; Jang, S. K.; Kim, J. K. Virus filtration membranes prepared from nanoporous block copolymers with good dimensional stability under high pressures and excellent solvent resistance. *Adv Funct Mater* 2008, *18*, 1371–1377.
16. Keller, F.; Hunter, M. S.; Robinson, D. L. Structural features of oxide coatings on aluminum. *J Electrochem Soc* 1953, *100*, 411–419.
17. Masuda, H.; Fukuda, K. Ordered metal nanohole arrays made by a 2-step replication of honeycomb structures of anodic alumina. *Science* 1995, *268*, 1466–1468.
18. Schmid, G. Materials in nanoporous alumina. *J Mater Chem* 2002, *12*, 1231–1238.
19. Kang, M. C.; Yu, S. F.; Li, N. C.; Martin, C. R. Nanowell-array surfaces. *Small* 2005, *1*, 69–72.
20. Alami-Younssi, S.; Kiefer, C.; Larbot, A.; Persin, M.; Sarrazin, J. Grafting gamma alumina microporous membranes by organosilanes: Characterisation by pervaporation. *J Membr Sci* 1998, *143*, 27–36.
21. Wanunu, M.; Meller, A. Chemically modified solid-state nanopores. *Nano Lett* 2007, *7*, 1580–1585.
22. Caro, J.; Noack, M.; Kolsch, P. Chemically modified ceramic membranes. *Microporous Mesoporous Mater* 1998, *22*, 321–332.
23. Menon, V. P.; Martin, C. R. Fabrication and evaluation of nanoelectrode ensembles. *Anal Chem* 1995, *67*, 1920–1928.
24. Nishizawa, M.; Menon, V. P.; Martin, C. R. Metal nanotubule membranes with electrochemically switch-able ion-transport selectivity. *Science* 1995, *268*, 700–702.
25. Kohli, P.; Harrell, C. C.; Cao, Z. H.; Gasparac, R.; Tan, W. H.; Martin, C. R. DNA-functionalized nano-tube membranes with single-base mismatch selectivity. *Science* 2004, *305*, 984–986.
26. Hou, S. F.; Wang, J. H.; Martin, C. R. Template-synthesized DNA nanotubes. *J Am Chem Soc* 2005, *127*, 8586–8587.
27. Hou, S. F.; Wang, J. H.; Martin, C. R. Template-synthesized protein nanotubes. *Nano Lett* 2005, *5*, 231–234.
28. Hou, S. F.; Harrell, C. C.; Trofin, L.; Kohli, P.; Martin, C. R. Layer-by-layer nanotube template synthesis. *J Am Chem Soc* 2004, *126*, 5674–5675.

29. Vlassiouk, I.; Siwy, Z. S. Nanofluidic diode. *Nano Lett* 2007, *7*, 552–556.
30. Ali, M.; Schiedt, B.; Healy, K.; Neumann, R.; Ensinger, A. Modifying the surface charge of single track-etched conical nanopores in polyimide. *Nanotechnology* 2008, *19*, 085713.
31. Ali, M.; Ramirez, P.; Tahir, M. N.; Mafe, S.; Siwy, Z.; Neumann, R.; Tremel, W.; Ensinger, W. Biomolecular conjugation inside synthetic polymer nanopores via glycoprotein-lectin interactions. *Nanoscale* 2011, *3*, 1894–1903.
32. Martin, C. R. Membrane-based synthesis of nanomaterials. *Chem Mater* 1996, *8*, 1739–1746.
33. Krasilnikova, O. K.; Pogosian, A. C.; Serebryakova, N. V.; Grankina, T. Y. Synthesis of carbon materials with high porous alumina as template. *Adv Chem Res* 2012, *12*, 177–204.
34. Losic, D.; Velleman, L.; Kant, K.; Kumeria, T.; Gulati, K.; Shapter, J. G.; Beattie, D. A.; Simovic, S. Self-ordering electrochemistry: A simple approach for engineering nanopore and nanotube arrays for emerging applications. *Aust J Chem* 2011, *64*, 294–301.
35. Valizadeh, S.; Strömberg, M.; Strømme, M. Template synthesis of magnetic nanowire arrays. *Nanostruct Mater Electrochem* 2008, 211–241.
36. Callegari, V.; Demoustier-Champagne, S. Using the hard templating method for the synthesis of metal-conducting polymer multi-segmented nanowires. *Macromol Rapid Commun* 2011, *32*, 25–34.
37. Martin, J.; Maiz, J.; Sacristan, J.; Mijangos, C. Tailored polymer-based nanorods and nanotubes by "template synthesis": From preparation to applications. *Polymer* 2012, *53*, 1149–1166.
38. Perry, J. L.; Martin, C. R.; Stewart, J. D. Drug-delivery strategies by using template-synthesized nanotubes. *Chem Eur J* 2011, *17*, 6296–6302.
39. Bae, C.; Yoo, H.; Kim, S.; Lee, K.; Kim, J.; Sung, M. A.; Shin, H. Template-directed synthesis of oxide nanotubes: Fabrication, characterization, and applications. *Chem Mater* 2008, *20*, 756–767.
40. Li, J.; Stein, D.; McMullan, C.; Branton, D.; Aziz, M. J.; Golovchenko, J. A. Ion-beam sculpting at nanometre length scales. *Nature* 2001, *412*, 166–169.
41. Storm, A. J.; Chen, J. H.; Ling, X. S.; Zandbergen, H. W.; Dekker, C. Fabrication of solid-state nanopores with single-nanometre precision. *Nat Mater* 2003, *2*, 537–540.
42. Storm, A. J.; Chen, J. H.; Ling, X. S.; Zandbergen, H. W.; Dekker, C. Electron-beam-induced deformations of SiO_2 nanostructures. *J Appl Phys* 2005, *98*, 014307.
43. Striemer, C. C.; Gaborski, T. R.; McGrath, J. L.; Fauchet, P. M. Charge- and size-based separation of macromolecules using ultrathin silicon membranes. *Nature* 2007, *445*, 749–753.
44. Fang, D. Z.; Striemer, C. C.; Gaborski, T. R.; McGrath, J. L.; Fauchet, P. M. Methods for controlling the pore properties of ultra-thin nanocrystalline silicon membranes. *J Phys Condens Matter* 2010, *22*, 454134.
45. Shen, M.; Ishimatsu, R.; Kim, J.; Amemiya, S. Quantitative imaging of ion transport through single nanopores by high-resolution scanning electrochemical microscopy. *J Am Chem Soc* 2012, *134*, 9856–9859.
46. Kim, E.; Xiong, H.; Striemer, C. C.; Fang, D. Z.; Fauchet, P. M.; McGrath, J. L.; Amemiya, S. A structure-permeability relationship of ultrathin nanoporous silicon membrane: A comparison with the nuclear envelope. *J Am Chem Soc* 2008, *130*, 4230–4231.
47. Agrawal, A. A.; Nehilla, B. J.; Reisig, K. V.; Gaborski, T. R.; Fang, D. Z.; Striemer, C. C.; Fauchet, P. M.; McGrath, J. L. Porous nanocrystalline silicon membranes as highly permeable and molecularly thin substrates for cell culture. *Biomaterials* 2010, *31*, 5408–5417.
48. Morris, C. A.; Friedman, A. K.; Baker, L. A. Applications of nanopipettes in the analytical sciences. *Analyst* 2010, *135*, 2190–2202.
49. Ying, L. M.; Bruckbauer, A.; Zhou, D. J.; Gorelik, J.; Shevehuk, A.; Lab, M.; Korchev, Y.; Klenerman, D. The scanned nanopipette: A new tool for high resolution bioimaging and controlled deposition of biomolecules. *Phys Chem Chem Phys* 2005, *7*, 2859–2866.
50. Ying, L. M. Applications of nanopipettes in bionanotechnology. *Biochem Soc Trans* 2009, *37*, 702–706.
51. Neher, E.; Sakmann, B. Single-channel currents recorded from membrane of denervated frog muscle-fibers. *Nature* 1976, *260*, 799–802.
52. Bard, A. J.; Fan, F. R. F.; Pierce, D. T.; Unwin, P. R.; Wipf, D. O.; Zhou, F. M. Chemical imaging of surfaces with the scanning electrochemical microscope. *Science* 1991, *254*, 68–74.
53. Hansma, P. K.; Drake, B.; Marti, O.; Gould, S. A. C.; Prater, C. B. The scanning ion-conductance microscope. *Science* 1989, *243*, 641–643.
54. Binnig, G.; Quate, C. F.; Gerber, C. Atomic force microscope. *Phys Rev Lett* 1986, *56*, 930–933.
55. Proksch, R.; Lal, R.; Hansma, P. K.; Morse, D.; Stucky, G. Imaging the internal and external pore structure of membranes in fluid: Tapping mode scanning ion conductance microscopy. *Biophys J* 1996, *71*, 2155–2157.

56. Lewis, A.; Isaacson, M.; Harootunian, A.; Muray, A. Development of a 500-a spatial-resolution light-microscope. 1. Light is efficiently transmitted through gamma-16 diameter apertures. *Ultramicroscopy* 1984, *13*, 227–231.

57. Pohl, D. W.; Denk, W.; Lanz, M. Optical stethoscopy—Image recording with resolution lambda/20. *Appl Phys Lett* 1984, *44*, 651–653.

58. Takahashi, Y.; Shevchuk, A. I.; Novak, P.; Murakami, Y.; Shiku, H.; Korchev, Y. E.; Matsue, T. Simultaneous noncontact topography and electrochemical imaging by SECM/SICM featuring ion current feedback regulation. *J Am Chem Soc* 2010, *132*, 10118–10126.

59. Comstock, D. J.; Elam, J. W.; Pellin, M. J.; Hersam, M. C. Integrated ultramicroelectrode—Nanopipette probe for concurrent scanning electrochemical microscopy and scanning ion conductance microscopy. *Anal Chem* 2010, *82*, 1270–1276.

60. Shevchuk, A. I.; Gorelik, J.; Harding, S. E.; Lab, M. J.; Klenerman, D.; Korchev, Y. E. Simultaneous measurement of Ca^{2+} and cellular dynamics: Combined scanning ion conductance and optical microscopy to study contracting cardiac myocytes. *Biophys J* 2001, *81*, 1759–1764.

61. Rothery, A. M.; Gorelik, J.; Bruckbauer, A.; Yu, W.; Korchev, Y. E.; Klenerman, D. A novel light source for SICM-SNOM of living cells. *J Microsc* 2003, *209*, 94–101.

62. Hong, M. H.; Kim, K. H.; Bae, J.; Jhe, W. Scanning nanolithography using a material-filled nanopipette. *Appl Phys Lett* 2000, *77*, 2604–2606.

63. Brown, K. T.; Flaming, D. G. New micro-electrode techniques for intracellular work in small cells. *Neuroscience* 1977, *2*, 813–827.

64. Zhang, B.; Galusha, J.; Shiozawa, P. G.; Wang, G.; Bergren, A. J.; Jones, R. M.; White, R. J.; Ervin, E. N.; Cauley, C. C.; White, H. S. Bench-top method for fabricating glass-sealed nanodisk electrodes, glass nanopore electrodes, and glass nanopore membranes of controlled size. *Anal Chem* 2007, *79*, 4778–4787.

65. Wolna, A. H.; Fleming, A. M.; An, N.; He, L.; White, H. S.; Burrows, C. J. Electrical current signatures of DNA base modifications in single molecules immobilized in the—Hemolysin ion channel. *Isr J Chem* 2013, *53*, 417–430.

66. White, R. J.; Zhang, B.; Daniel, S.; Tang, J. M.; Ervin, E. N.; Cremer, P. S.; White, H. S. Ionic conductivity of the aqueous layer separating a lipid bilayer membrane and a glass support. *Langmuir* 2006, *22*, 10777–10783.

67. Schibel, A. E. P.; Heider, E. C.; Harris, J. M.; White, H. S. Fluorescence microscopy of the pressure-dependent structure of lipid bilayers suspended across conical nanopores. *J Am Chem Soc* 2011, *133*, 7810–7815.

68. Schibel, A. E. P.; Fleming, A. M.; Jin, Q.; An, N.; Liu, J.; Blakemore, C. P.; White, H. S.; Burrows, C. J. Sequence-specific single-molecule analysis of 8-oxo-7,8-dihydroguanine lesions in DNA based on unzipping kinetics of complementary probes in ion channel recordings. *J Am Chem Soc* 2011, *133*, 14778–14784.

69. Schibel, A. E. P.; Edwards, T.; Kawano, R.; Lan, W.; White, H. S. Quartz nanopore membranes for suspended bilayer ion channel recordings. *Anal Chem* 2010, *82*, 7259–7266.

70. Schibel, A. E. P.; An, N.; Jin, Q.; Fleming, A. M.; Burrows, C. J.; White, H. S. Nanopore detection of 8-oxo-7,8-dihydro-2′-deoxyguanosine in immobilized single-stranded DNA via adduct formation to the DNA damage site. *J Am Chem Soc* 2010, *132*, 17992–17995.

71. Kawano, R.; Schibel, A. E. P.; Cauley, C.; White, H. S. Controlling the translocation of single-stranded DNA through alpha-hemolysin ion channels using viscosity. *Langmuir* 2009, *25*, 1233–1237.

72. Jin, Q.; Fleming, A. M.; Burrows, C. J.; White, H. S. Unzipping kinetics of duplex DNA containing oxidized lesions in an alpha-hemolysin nanopore. *J Am Chem Soc* 2012, *134*, 11006–11011.

73. Ervin, E. N.; White, R. J.; White, H. S. Sensitivity and signal complexity as a function of the number of ion channels in a stochastic sensor. *Anal Chem* 2009, *81*, 533–537.

74. Ervin, E. N.; Kawano, R.; White, R. J.; White, H. S. Simultaneous alternating and direct current readout of protein ion channel blocking events using glass nanopore membranes. *Anal Chem* 2008, *80*, 2069–2076.

75. An, N.; White, H. S.; Burrows, C. J. Modulation of the current signatures of DNA abasic site adducts in the alpha-hemolysin ion channel. *Chem Commun* 2012, *48*, 11410–11412.

76. White, H. S.; Bund, A. Ion current rectification at nanopores in glass membranes. *Langmuir* 2008, *24*, 2212–2218.

77. White, H. S.; Bund, A. Mechanism of electrostatic gating at conical glass nanopore electrodes. *Langmuir* 2008, *24*, 12062–12067.

78. Wang, G. L.; Zhang, B.; Wayment, J. R.; Harris, J. M.; White, H. S. Electrostatic-gated transport in chemically modified glass nanopore electrodes. *J Am Chem Soc* 2006, *128*, 7679–7686.

79. Wang, G.; Bohaty, A. K.; Zharov, I.; White, H. S. Photon gated transport at the glass nanopore electrode. *J Am Chem Soc* 2006, *128*, 13553–13558.

80. Shim, J. H.; Kim, J.; Cha, G. S.; Nam, H.; White, R. J.; White, H. S.; Brown, R. B. Glass nanopore-based ion-selective electrodes. *Anal Chem* 2007, *79*, 3568–3574.

81. Luo, L.; Holden, D. A.; Lan, W.-J.; White, H. S. Tunable negative differential electrolyte resistance in a conical nanopore in glass. *ACS Nano* 2012, *6*, 6507–6514.

82. Lan, W.-J.; White, H. S. Diffusional motion of a particle translocating through a nanopore. *ACS Nano* 2012, *6*, 1757–1765.

83. Lan, W.-J.; Holden, D. A.; Zhang, B.; White, H. S. Nanoparticle transport in conical-shaped nanopores. *Anal Chem* 2011, *83*, 3840–3847.

84. Holden, D. A.; Watkins, J. J.; White, H. S. Resistive-pulse detection of multilamellar liposomes. *Langmuir* 2012, *28*, 7572–7577.

85. Holden, D. A.; Hendrickson, G. R.; Lan, W.-J.; Lyon, L. A.; White, H. S. Electrical signature of the deformation and dehydration of microgels during translocation through nanopores. *Soft Matter* 2011, *7*, 8035–8040.

86. Holden, D. A.; Hendrickson, G.; Lyon, L. A.; White, H. S. Resistive pulse analysis of microgel deformation during nanopore translocation. *J Phys Chem C* 2011, *115*, 2999–3004.

87. German, S. R.; Luo, L.; White, H. S.; Mega, T. L. Controlling nanoparticle dynamics in conical nanopores. *J Phys Chem C* 2013, *117*, 703–711.

88. Duan, C. H.; Wang, W.; Xie, Q. Review article: Fabrication of nanofluidic devices. *Biomicrofluidics* 2013, *7*, 26501.

89. Wu, C. J.; Jin, Z. H.; Wang, H. Q.; Ma, H. L.; Wang, Y. L. Design and fabrication of a nanofluidic channel by selective thermal oxidation and etching back of silicon dioxide made on a silicon substrate. *J Micromech Microeng* 2007, *17*, 2393–2397.

90. Mao, P.; Han, J. Y. Fabrication and characterization of 20 nm planar nanofluidic channels by glass-glass and glass-silicon bonding. *Lab Chip* 2005, *5*, 837–844.

91. Huh, D.; Mills, K. L.; Zhu, X. Y.; Burns, M. A.; Thouless, M. D.; Takayama, S. Tuneable elastomeric nanochannels for nanofluidic manipulation. *Nat Mater* 2007, *6*, 424–428.

92. Park, S. M.; Huh, Y. S.; Craighead, H. G.; Erickson, D. A method for nanofluidic device prototyping using elastomeric collapse. *Proc Natl Acad Sci USA* 2009, *106*, 15549–15554.

93. Guo, L. J.; Cheng, X.; Chou, C. F. Fabrication of size-controllable nanofluidic channels by nanoimprinting and its application for DNA stretching. *Nano Lett* 2004, *4*, 69–73.

94. Mahabadi, K. A.; Rodriguez, I.; Haur, S. C.; van Kan, J. A.; Bettiol, A. A.; Watt, F. Fabrication of PMMA micro- and nanofluidic channels by proton beam writing: Electrokinetic and morphological characterization. *J Micromech Microeng* 2006, *16*, 1170–1180.

95. Cipriany, B. R.; Zhao, R. Q.; Murphy, P. J.; Levy, S. L.; Tan, C. P.; Craighead, H. G.; Soloway, P. D. Single molecule epigenetic analysis in a nanofluidic channel. *Anal Chem* 2010, *82*, 2480–2487.

96. Kuo, T. C.; Cannon, D. M.; Chen, Y. N.; Tulock, J. J.; Shannon, M. A.; Sweedler, J. V.; Bohn, P. W. Gateable nanofluidic interconnects for multilayered microfluidic separation systems. *Anal Chem* 2003, *75*, 1861–1867.

97. Novoselov, K. S.; Geim, A. K.; Morozov, S. V.; Jiang, D.; Zhang, Y.; Dubonos, S. V.; Grigorieva, I. V.; Firsov, A. A. Electric field effect in atomically thin carbon films. *Science* 2004, *306*, 666–669.

98. Castro Neto, A. H.; Guinea, F.; Peres, N. M. R.; Novoselov, K. S.; Geim, A. K. The electronic properties of graphene. *Rev Mod Phys* 2009, *81*, 109–162.

99. Geim, A. K.; Novoselov, K. S. The rise of graphene. *Nat Mater* 2007, *6*, 183–191.

100. Oyama, S. T.; Lee, D.; Hacarlioglu, P.; Saraf, R. F. Theory of hydrogen permeability in nonporous silica membranes. *J Membr Sci* 2004, *244*, 45–53.

101. Jiang, D. E.; Cooper, V. R.; Dai, S. Porous graphene as the ultimate membrane for gas separation. *Nano Lett* 2009, *9*, 4019–4024.

102. Han, T. H.; Huang, Y. K.; Tan, A. T. L.; Dravid, V. P.; Huang, J. X. Steam etched porous graphene oxide network for chemical sensing. *J Am Chem Soc* 2011, *133*, 15264–15267.

103. Huang, X. D.; Qian, K.; Yang, J.; Zhang, J.; Li, L.; Yu, C. Z.; Zhao, D. Y. Functional nanoporous graphene foams with controlled pore sizes. *Adv Mater* 2012, *24*, 4419–4423.

104. Schneider, G. F.; Kowalczyk, S. W.; Calado, V. E.; Pandraud, G.; Zandbergen, H. W.; Vandersypen, L. M. K.; Dekker, C. DNA translocation through graphene nanopores. *Nano Lett* 2010, *10*, 3163–3167.

105. Bai, J. W.; Zhong, X.; Jiang, S.; Huang, Y.; Duan, X. F. Graphene nanomesh. *Nat Nanotechnol* 2010, *5*, 190–194.

106. Ning, G. Q.; Fan, Z. J.; Wang, G.; Gao, J. S.; Qian, W. Z.; Wei, F. Gram-scale synthesis of nanomesh graphene with high surface area and its application in supercapacitor electrodes. *Chem Commun* 2011, *47*, 5976–5978.

107. Merchant, C. A.; Healy, K.; Wanunu, M.; Ray, V.; Peterman, N.; Bartel, J.; Fischbein, M. D. et al. DNA translocation through graphene nanopores. *Nano Lett* 2010, *10*, 2915–2921.

108. Bhakdi, S.; Tranumjensen, J. Alpha-toxin of *Staphylococcus aureus*. *Microbiol Rev* 1991, *55*, 733–751.

109. Gouaux, E. Alpha-hemolysin from *Staphylococcus aureus*: An archetype of beta-barrel, channel-forming toxins. *J Struct Biol* 1998, *121*, 110–122.

110. Faller, M.; Niederweis, M.; Schulz, G. E. The structure of a mycobacterial outer-membrane channel. *Science* 2004, *303*, 1189–1192.

111. Niederweis, M. Mycobacterial porins—New channel proteins in unique outer membranes. *Mol Microbiol* 2003, *49*, 1167–1177.

112. Cowan, S. W.; Schirmer, T.; Rummel, G.; Steiert, M.; Ghosh, R.; Pauptit, R. A.; Jansonius, J. N.; Rosenbusch, J. P. Crystal-structures explain functional-properties of 2 *Escherichia coli* porins. *Nature* 1992, *358*, 727–733.

113. Yildiz, O.; Vinothkumar, K. R.; Goswami, P.; Kuhlbrandt, W. Structure of the monomeric outer-membrane porin OmpG in the open and closed conformation. *Embo J* 2006, *25*, 3702–3713.

114. Chen, M.; Khalid, S.; Sansom, M. S. P.; Bayley, H. Outer membrane protein G: Engineering a quiet pore for biosensing. *Proc Natl Acad Sci USA* 2008, *105*, 6272–6277.

115. Bezrukov, S. M.; Kasianowicz, J. J. Current noise reveals protonation kinetics and number of ionizable sites in an open protein ion channel. *Phys Rev Lett* 1993, *70*, 2352–2355.

116. Bezrukov, S. M.; Vodyanoy, I.; Brutyan, R. A.; Kasianowicz, J. J. Dynamics and free energy of polymers partitioning into a nanoscale pore. *Macromolecules* 1996, *29*, 8517–8522.

117. Song, L. Z.; Hobaugh, M. R.; Shustak, C.; Cheley, S.; Bayley, H.; Gouaux, J. E. Structure of staphylococcal alpha-hemolysin, a heptameric transmembrane pore. *Science* 1996, *274*, 1859–1866.

118. Kasianowicz, J. J.; Brandin, E.; Branton, D.; Deamer, D. W. Characterization of individual polynucleotide molecules using a membrane channel. *Proc Natl Acad Sci USA* 1996, *93*, 13770–13773.

119. Bayley, H.; Cremer, P. S. Stochastic sensors inspired by biology. *Nature* 2001, *413*, 226–230.

120. Grosse, W.; Essen, L. O.; Koert, U. Strategies and perspectives in ion-channel engineering. *Chembiochem* 2011, *12*, 830–839.

121. Braha, O.; Gu, L. Q.; Zhou, L.; Lu, X. F.; Cheley, S.; Bayley, H. Simultaneous stochastic sensing of divalent metal ions. *Nat Biotechnol* 2000, *18*, 1005–1007.

122. Braha, O.; Walker, B.; Cheley, S.; Kasianowicz, J. J.; Song, L. Z.; Gouaux, J. E.; Bayley, H. Designed protein pores as components for biosensors. *Chem Biol* 1997, *4*, 497–505.

123. Guan, X. Y.; Gu, L. Q.; Cheley, S.; Braha, O.; Bayley, H. Stochastic sensing of TNT with a genetically engineered pore. *Chembiochem* 2005, *6*, 1875–1881.

124. Liu, A. H.; Zhao, Q. T.; Krishantha, D. M. M.; Guan, X. Y. Unzipping of double-stranded DNA in engineered alpha-hemolysin pores. *J Phys Chem Lett* 2011, *2*, 1372–1376.

125. Mohammad, M. M.; Prakash, S.; Matouschek, A.; Movileanu, L. Controlling a single protein in a nanopore through electrostatic traps. *J Am Chem Soc* 2008, *130*, 4081–4088.

126. Manrao, E. A.; Derrington, I. M.; Laszlo, A. H.; Langford, K. W.; Hopper, M. K.; Gillgren, N.; Pavlenok, M.; Niederweis, M.; Gundlach, J. H. Reading DNA at single-nucleotide resolution with a mutant MspA nanopore and phi29 DNA polymerase. *Nat Biotechnol* 2012, *30*, 349–353.

127. Gu, L. Q.; Braha, O.; Conlan, S.; Cheley, S.; Bayley, H. Stochastic sensing of organic analytes by a pore-forming protein containing a molecular adapter. *Nature* 1999, *398*, 686–690.

128. Gu, L. Q.; Cheley, S.; Bayley, H. Capture of a single molecule in a nanocavity. *Science* 2001, *291*, 636–640.

129. Sanchez-Quesada, J.; Ghadiri, M. R.; Bayley, H.; Braha, O. Cyclic peptides as molecular adapters for a pore-forming protein. *J Am Chem Soc* 2000, *122*, 11757–11766.

130. Movileanu, L.; Howorka, S.; Braha, O.; Bayley, H. Detecting protein analytes that modulate transmembrane movement of a polymer chain within a single protein pore. *Nat Biotechnol* 2000, *18*, 1091–1095.

131. Rout, M. P.; Aitchison, J. D.; Suprapto, A.; Hjertaas, K.; Zhao, Y. M.; Chait, B. T. The yeast nuclear pore complex: Composition, architecture, and transport mechanism. *J Cell Biol* 2000, *148*, 635–651.

132. Rivas, M.; Tran, Q.; Fox, G. E. Nanometer scale pores similar in size to the entrance of the ribosomal exit cavity are a common feature of large RNAs. *RNA* 2013, *19*, 1349–1354.

133. Cheng, L. J.; Guo, L. J. Nanofluidic diodes. *Chem Soc Rev* 2010, *39*, 923–938.

134. Siwy, Z.; Apel, P.; Baur, D.; Dobrev, D. D.; Korchev, Y. E.; Neumann, R.; Spohr, R.; Trautmann, C.; Voss, K. O. Preparation of synthetic nanopores with transport properties analogous to biological channels. *Surf Sci* 2003, *532*, 1061–1066.

135. Wei, C.; Bard, A. J.; Feldberg, S. W. Current rectification at quartz nanopipet electrodes. *Anal Chem* 1997, *69*, 4627–4633.

136. Siwy, Z. S. Ion-current rectification in nanopores and nanotubes with broken symmetry. *Adv Funct Mater* 2006, *16*, 735–746.

137. Cheng, L. J.; Guo, L. J. Rectified ion transport through concentration gradient in homogeneous silica nanochannels. *Nano Lett* 2007, *7*, 3165–3171.

138. Daiguji, H.; Oka, Y.; Shirono, K. Nanofluidic diode and bipolar transistor. *Nano Lett* 2005, *5*, 2274–2280.

139. Kalman, E. B.; Vlassiouk, I.; Siwy, Z. S. Nanofluidic bipolar transistors. *Adv Mater* 2008, *20*, 293–297.

140. Karnik, R.; Duan, C. H.; Castelino, K.; Daiguji, H.; Majumdar, A. Rectification of ionic current in a nanofluidic diode. *Nano Lett* 2007, *7*, 547–551.

141. Constantin, D.; Siwy, Z. S. Poisson–Nernst–Planck model of ion current rectification through a nanofluidic diode. *Phys Rev E* 2007, *76*, 041202.

142. Miedema, H.; Vrouenraets, M.; Wierenga, J.; Meijberg, W.; Robillard, G.; Eisenberg, B. A biological porin engineered into a molecular, nanofluidic diode. *Nano Lett* 2007, *7*, 2886–2891.

143. Coulter, W. H. US Patent 2656508, October 20, 1953; vol. 2, 656, 508.

144. Graham, M. D. The coulter principle: Foundation of an industry. *J Assoc Lab Automat* 2003, *8*, 72–81.

145. Platt, M.; Willmott, G. R.; Lee, G. U. Resistive pulse sensing of analyte—Induced multicomponent rod aggregation using tunable pores. *Small* 2012, *8*, 2436–2444.

146. Harrell, C. C.; Choi, Y.; Horne, L. P.; Baker, L. A.; Siwy, Z. S.; Martin, C. R. Resistive-pulse DNA detection with a conical nanopore sensor. *Langmuir* 2006, *22*, 10837–10843.

147. Akeson, M.; Branton, D.; Kasianowicz, J. J.; Brandin, E.; Deamer, D. W. Microsecond time-scale discrimination among polycytidylic acid, polyadenylic acid, and polyuridylic acid as homopolymers or as segments within single RNA molecules. *Biophys J* 1999, *77*, 3227–3233.

148. Butler, T. Z.; Gundlach, J. H.; Troll, M. A. Determination of RNA orientation during translocation through a biological nanopore. *Biophys J* 2006, *90*, 190–199.

149. Sexton, L. T.; Horne, L. P.; Sherrill, S. A.; Bishop, G. W.; Baker, L. A.; Martin, C. R. Resistive-pulse studies of proteins and protein/antibody complexes using a conical nanotube sensor. *J Am Chem Soc* 2007, *129*, 13144–13152.

150. Uram, J. D.; Ke, K.; Hunt, A. J.; Mayer, M. Label-free affinity assays by rapid detection of immune complexes in submicrometer pores. *Angew Chem Int Ed* 2006, *45*, 2281–2285.

151. Yusko, E. C.; Prangkio, P.; Sept, D.; Rollings, R. C.; Li, J. L.; Mayer, M. Single-particle characterization of a beta oligomers in solution. *ACS Nano* 2012, *6*, 5909–5919.

152. Han, A. P.; Schurmann, G.; Mondin, G.; Bitterli, R. A.; Hegelbach, N. G.; de Rooij, N. F.; Staufer, U. Sensing protein molecules using nanofabricated pores. *Appl Phys Lett* 2006, *88*, 093901.

153. Harms, Z. D.; Mogensen, K. B.; Nunes, P. S.; Zhou, K. M.; Hildenbrand, B. W.; Mitra, I.; Tan, Z. N.; Zlotnick, A.; Kutter, J. P.; Jacobson, S. C. Nanofluidic devices with two pores in series for resistive-pulse sensing of single virus capsids. *Anal Chem* 2011, *83*, 9573–9578.

154. Zhou, K. M.; Li, L. C.; Tan, Z. N.; Zlotnick, A.; Jacobson, S. C. Characterization of hepatitis B virus capsids by resistive-pulse sensing. *J Am Chem Soc* 2011, *133*, 1618–1621.

155. Willmott, G. R.; Platt, M.; Lee, G. U. Resistive pulse sensing of magnetic beads and supraparticle structures using tunable pores. *Biomicrofluidics* 2012, *6*, 14103–14115.

156. Pevarnik, M.; Healy, K.; Siwy, Z. Polystyrene beads as a model system for virus particles reveal pore substructure as they translocate. *Biophys J* 2012, *6*, 7295–7302.

157. Somerville, J. A.; Willmott, G. R.; Eldridge, J.; Griffiths, M.; McGrath, K. M. Size and charge characterisation of a submicrometre oil-in-water emulsion using resistive pulse sensing with tunable pores. *J Colloid Interf Sci* 2013, *394*, 243–251.

158. Heins, E. A.; Siwy, Z. S.; Baker, L. A.; Martin, C. R. Detecting single porphyrin molecules in a conically shaped synthetic nanopore. *Nano Lett* 2005, *5*, 1824–1829.

159. Davenport, M.; Healy, K.; Pevarnik, M.; Teslich, N.; Cabrini, S.; Morrison, A. P.; Siwy, Z. S.; Letant, S. E. The role of pore geometry in single nanoparticle detection. *ACS Nano* 2012, *6*, 8366–8380.

160. Pevarnik, M.; Healy, K.; Toimil-Molares, M. E.; Morrison, A.; Letant, S. E.; Siwy, Z. S. Polystyrene particles reveal pore substructure as they translocate. *ACS Nano* 2012, *6*, 7295–7302.

161. Bae, J. H.; Han, J. H.; Chung, T. D. Electrochemistry at nanoporous interfaces: New opportunity for electrocatalysis. *Phys Chem Chem Phys* 2012, *14*, 448–463.

162. Ciesielski, P. N.; Scott, A. M.; Faulkner, C. J.; Berron, B. J.; Cliffel, D. E.; Jennings, G. K. Functionalized nanoporous gold leaf electrode films for the immobilization of photosystem I. *ACS Nano* 2008, *2*, 2465–2472.

163. Ge, X. B.; Wang, L. Q.; Liu, Z. N.; Ding, Y. Nanoporous gold leaf for amperometric determination of nitrite. *Electroanalysis* 2011, *23*, 381–386.

164. Xu, J. L.; Kou, T. Y.; Zhang, Z. H. Anodization strategy to fabricate nanoporous gold for high-sensitivity detection of p-nitrophenol. *CrystEngComm* 2013, *15*, 7856–7862.

165. Sattayasamitsathit, S.; Thavarungkul, P.; Thammakhet, C.; Limbut, W.; Numnuam, A.; Buranachai, C.; Kanatharana, P. Fabrication of nanoporous copper film for electrochemical detection of glucose. *Electroanalysis* 2009, *21*, 2371–2377.

166. Monllor-Satoca, D.; Gomez, R. A photoelectrochemical and spectroscopic study of phenol and catechol oxidation on titanium dioxide nanoporous electrodes. *Electrochim Acta* 2010, *55*, 4661–4668.

167. Zhao, H. L.; Jiang, D. L.; Zhang, S. L.; Wen, W. Photoelectrocatalytic oxidation of organic compounds at nanoporous TiO_2 electrodes in a thin-layer photoelectrochemical cell. *J Catal* 2007, *250*, 102–109.

168. Han, J. H.; Lee, E.; Park, S.; Chang, R.; Chung, T. D. Effect of nanoporous structure on enhanced electrochemical reaction. *J Phys Chem C* 2010, *114*, 9546–9553.

169. Bae, J. H.; Chung, T. D. Conductometric discrimination of electro-inactive metal ions using nanoporous electrodes. *Electrochim Acta* 2011, *56*, 1947–1954.

170. Dimos, M. M.; Blanchard, G. J. Electro-catalytic oxidation of 1,2-propanediol at nanoporous and planar solid Pt electrodes. *J Electroanal Chem* 2011, *654*, 13–19.

171. Paek, S. M.; Yoo, E.; Honma, I. Enhanced cyclic performance and lithium storage capacity of SnO_2/graphene nanoporous electrodes with three-dimensionally delaminated flexible structure. *Nano Lett* 2009, *9*, 72–75.

172. Li, F.; Ito, T. Complexation-induced control of electron propagation based on bounded diffusion through nanopore-tethered ferrocenes. *J Am Chem Soc* 2013, *135*, 16260–16263.

173. Sprague, I. B.; Dutta, P. Improved kinetics from ion advection through overlapping electric double layers in nano-porous electrodes. *Electrochim Acta* 2013, *91*, 20–29.

174. Pandey, B.; Khanh, H. T. B.; Li, Y. X.; Diaz, R.; Ito, T. Electrochemical study of the diffusion of cytochrome c within nanoscale pores derived from cylinder-forming polystyrene-poly(methylmethacrylate) diblock copolymers. *Electrochim Acta* 2011, *56*, 10185–10190.

175. Bae, J. H.; Kim, Y. R.; Kim, R. S.; Chung, T. D. Enhanced electrochemical reactions of 1,4-benzoquinone at nanoporous electrodes. *Phys Chem Chem Phys* 2013, *15*, 10645–10653.

176. Li, Y. X.; Ito, T. Size-exclusion properties of nanoporous films derived from polystyrene-poly(methylmethacrylate) diblock copolymers assessed using direct electrochemistry of ferritin. *Anal Chem* 2009, *81*, 851–855.

177. Gu, L. Q.; Shim, J. W. Single molecule sensing by nanopores and nanopore devices. *Analyst* 2010, *135*, 441–451.

178. Umehara, S.; Pourmand, N.; Webb, C. D.; Davis, R. W.; Yasuda, K.; Karhanek, M. Current rectification with poly-L-lysine-coated quartz nanopipettes. *Nano Lett* 2006, *6*, 2486–2492.

179. Sa, N. Y.; Fu, Y. Q.; Baker, L. A. Reversible cobalt ion binding to imidazole-modified nanopipettes. *Anal Chem* 2010, *82*, 9963–9966.

180. Fu, Y. Q.; Tokuhisa, H.; Baker, L. A. Nanopore DNA sensors based on dendrimer-modified nanopipettes. *Chem Commun* 2009, 4877–4879.

181. Wang, J.; Martin, C. R. A new drug-sensing paradigm based on ion-current rectification in a conically shaped nanopore. *Nanomedicine (London, England)* 2008, *3*, 13–20.

182. Neugebauer, S.; Stoica, L.; Guschin, D.; Schuhmann, W. Redox-amplified biosensors based on selective modification of nanopore electrode structures with enzymes entrapped within electrodeposition paints. *Microchim Acta* 2008, *163*, 33–40.

183. Zhang, L.; Chang, H. X.; Hirata, A.; Wu, H. K.; Xue, Q. K.; Chen, M. W. Nanoporous gold based optical sensor for sub-ppt detection of mercury ions. *ACS Nano* 2013, *7*, 4595–4600.

184. Zhao, Q. T.; Wang, D.; Jayawardhana, D. A.; Guan, X. Y. Stochastic sensing of biomolecules in a nanopore sensor array. *Nanotechnology* 2008, *19*, 505504.

185. Siwy, Z.; Trofin, L.; Kohli, P.; Baker, L. A.; Trautmann, C.; Martin, C. R. Protein biosensors based on biofunctionalized conical gold nanotubes. *J Am Chem Soc* 2005, *127*, 5000–5001.

186. Wu, H. C.; Bayley, H. Single-molecule detection of nitrogen mustards by covalent reaction within a protein nanopore. *J Am Chem Soc* 2008, *130*, 6813–6819.

187. Liu, A. H.; Zhao, Q. T.; Guan, X. Y. Stochastic nanopore sensors for the detection of terrorist agents: Current status and challenges. *Anal Chim Acta* 2010, *675*, 106–115.

188. Sanger, F.; Coulson, A. R. Rapid method for determining sequences in DNA by primed synthesis with DNA-polymerase. *J Mol Biol* 1975, *94*, 441–448.

189. Swerdlow, H.; Zhang, J. Z.; Chen, D. Y.; Harke, H. R.; Grey, R.; Wu, S. L.; Dovichi, N. J.; Fuller, C. 3 DNA sequencing methods using capillary gel-electrophoresis and laser-induced fluorescence. *Anal Chem* 1991, *63*, 2835–2841.

190. Maitra, R. D.; Kim, J.; Dunbar, W. B. Recent advances in nanopore sequencing. *Electrophoresis* 2012, *33*, 3418–3428.

191. Yang, Y. Q.; Liu, R. Y.; Xie, H. Q.; Hui, Y. T.; Jiao, R. G.; Gong, Y.; Zhang, Y. Y. Advances in nanopore sequencing technology. *J Nanosci Nanotechnol* 2013, *13*, 4521–4538.

192. Heng, J. B.; Ho, C.; Kim, T.; Timp, R.; Aksimentiev, A.; Grinkova, Y. V.; Sligar, S.; Schulten, K.; Timp, G. Sizing DNA using a nanometer-diameter pore. *Biophys J* 2004, *87*, 2905–2911.

193. Maglia, G.; Henricus, M.; Wyss, R.; Li, Q. H.; Cheley, S.; Bayley, H. DNA strands from denatured duplexes are translocated through engineered protein nanopores at alkaline pH. *Nano Lett* 2009, *9*, 3831–3836.

194. McNally, B.; Wanunu, M.; Meller, A. Electromechanical unzipping of individual DNA molecules using synthetic sub-2 nm pores. *Nano Lett* 2008, *8*, 3418–3422.

195. Wells, D. B.; Belkin, M.; Comer, J.; Aksimentiev, A. Assessing graphene nanopores for sequencing DNA. *Nano Lett* 2012, *12*, 4117–4123.

196. Branton, D.; Deamer, D. W.; Marziali, A.; Bayley, H.; Benner, S. A.; Butler, T.; Di Ventra, M. et al. The potential and challenges of nanopore sequencing. *Nat Biotechnol* 2008, *26*, 1146–1153.

197. Luan, B. Q.; Stolovitzky, G.; Martyna, G. Slowing and controlling the translocation of DNA in a solid-state nanopore. *Nanoscale* 2012, *4*, 1068–1077.

198. Butler, T. Z.; Pavlenok, M.; Derrington, I. M.; Niederweis, M.; Gundlach, J. H. Single-molecule DNA detection with an engineered MspA protein nanopore. *Proc Natl Acad Sci USA* 2008, *105*, 20647–20652.

199. Wallace, E. V. B.; Stoddart, D.; Heron, A. J.; Mikhailova, E.; Maglia, G.; Donohoe, T. J.; Bayley, H. Identification of epigenetic DNA modifications with a protein nanopore. *Chem Commun* 2010, *46*, 8195–8197.

200. Long, Z. C.; Liu, D. Y.; Ye, N. N.; Qin, J. H.; Lin, B. C. Integration of nanoporous membranes for sample filtration/preconcentration in microchip electrophoresis. *Electrophoresis* 2006, *27*, 4927–4934.

201. Gu, Y.; Miki, N. A microfilter utilizing a polyethersulfone porous membrane with nanopores. *J Micromech Microeng* 2007, *17*, 2308–2315.

202. Prabhu, A. S.; Jubery, T. Z. N.; Freedman, K. J.; Mulero, R.; Dutta, P.; Kim, M. J. Chemically modified solid state nanopores for high throughput nanoparticle separation. *J Phys Condens Matter* 2010, *22*, 454107.

203. Geho, D.; Cheng, M. M. C.; Killian, K.; Lowenthal, M.; Ross, S.; Frogale, K.; Nijdam, J. et al. Fractionation of serum components using nanoporous substrates. *Bioconjug Chem* 2006, *17*, 654–661.

204. Sano, T.; Iguchi, N.; Iida, K.; Sakamoto, T.; Baba, M.; Kawaura, H. Size-exclusion chromatography using self-organized nanopores in anodic porous alumina. *Appl Phys Lett* 2003, *83*, 4438–4440.

205. Osmanbeyoglu, H. U.; Hur, T. B.; Kim, H. K. Thin alumina nanoporous membranes for similar size biomolecule separation. *J Membr Sci* 2009, *343*, 1–6.

206. Cheow, P. S.; Zhi, E.; Ting, C.; Tan, M. Q.; Toh, C. S. Transport and separation of proteins across platinum-coated nanoporous alumina membranes. *Electrochim Acta* 2008, *53*, 4669–4673.

207. Chun, K. Y.; Mafe, S.; Ramirez, P.; Stroeve, P. Protein transport through gold-coated, charged nanopores: Effects of applied voltage. *Chem Phys Lett* 2006, *418*, 561–564.

208. D'Amico, M.; Schiro, G.; Cupane, A.; D'Alfonso, L.; Leone, M.; Militello, V.; Vetri, V. High fluorescence of thioflavin T confined in mesoporous silica xerogels. *Langmuir* 2013, *29*, 10238–10246.

209. Armstrong, J. A.; Bernal, E. E. L.; Yaroshchuk, A.; Bruening, M. L. Separation of ions using polyelectrolyte-modified nanoporous track-etched membranes. *Langmuir* 2013, *29*, 10287–10296.

210. Balme, S.; Picaud, F.; Kraszewski, S.; Dejardin, P.; Janot, J. M.; Lepoitevin, M.; Capomanes, J.; Ramseyer, C.; Henn, F. Controlling potassium selectivity and proton blocking in a hybrid biological/solid-state polymer nanoporous membrane. *Nanoscale* 2013, *5*, 3961–3968.

211. Vankrunkelsven, S.; Clicq, D.; Cabooter, D.; De Malsche, W.; Gardeniers, J. G. E.; Desmet, G. Ultra-rapid separation of an angiotensin mixture in nanochannels using shear-driven chromatography. *J Chromatogr A* 2006, *1102*, 96–103.

212. Asatekin, A.; Gleason, K. K. Polymeric nanopore membranes for hydrophobicity-based separations by conformal initiated chemical vapor deposition. *Nano Lett* 2011, *11*, 677–686.

213. He, Z.; Zhou, J.; Lu, X.; Corry, B. Bioinspired graphene nanopores with voltage-tunable ion selectivity for Na and K. *ACS Nano* 2013, *7*, 10148–10157.

214. Hauser, A. W.; Schrier, J.; Schwerdtfeger, P. Helium tunneling through nitrogen-functionalized graphene pores: Pressure- and temperature-driven approaches to isotope separation. *J Phys Chem C* 2012, *116*, 10819–10827.

215. Hauser, A. W.; Schwerdtfeger, P. Methane-selective nanoporous graphene membranes for gas purification. *Phys Chem Chem Phys* 2012, *14*, 13292–13298.

216. Cohen-Tanugi, D.; Grossman, J. C. Water desalination across nanoporous graphene. *Nano Lett* 2012, *12*, 3602–3608.

217. Du, H. L.; Li, J. Y.; Zhang, J.; Su, G.; Li, X. Y.; Zhao, Y. L. Separation of hydrogen and nitrogen gases with porous graphene membrane. *J Phys Chem C* 2011, *115*, 23261–23266.

218. Sint, K.; Wang, B.; Kral, P. Selective ion passage through functionalized graphene nanopores. *J Am Chem Soc* 2008, *130*, 16448–16449.

219. Hauser, A. W.; Schwerdtfeger, P. Nanoporous graphene membranes for efficient ^{3}He/^{4}He separation. *J Phys Chem Lett* 2012, *3*, 209–213.

220. Schrier, J. Fluorinated and nanoporous graphene materials as sorbents for gas separations. *ACS Appl Mater Interfaces* 2011, *3*, 4451–4458.

221. Laforge, F. O.; Carpino, J.; Rotenberg, S. A.; Mirkin, M. V. Electrochemical attosyringe. *Proc Natl Acad Sci USA* 2007, *104*, 11895–11900.

222. Piper, J. D.; Li, C.; Lo, C. J.; Berry, R.; Korchev, Y.; Ying, L. M.; Klenerman, D. Characterization and application of controllable local chemical changes produced by reagent delivery from a nanopipette. *J Am Chem Soc* 2008, *130*, 10386–10393.

223. Jacobson, S. C.; Hergenroder, R.; Koutny, L. B.; Warmack, R. J.; Ramsey, J. M. Effects of injection schemes and column geometry on the performance of microchip electrophoresis devices. *Anal Chem* 1994, *66*, 1107–1113.

224. Han, A. P.; de Rooij, N. F.; Staufer, U. Design and fabrication of nanofluidic devices by surface micromachining. *Nanotechnology* 2006, *17*, 2498–2503.

225. Kovarik, M. L.; Jacobson, S. C. Attoliter-scale dispensing in nanofluidic channels. *Anal Chem* 2007, *79*, 1655–1660.

226. Wang, Y. C.; Stevens, A. L.; Han, J. Y. Million-fold preconcentration of proteins and peptides by nanofluidic filter. *Anal Chem* 2005, *77*, 4293–4299.

227. Meister, A.; Gabi, M.; Behr, P.; Studer, P.; Voros, J.; Niedermann, P.; Bitterli, J. et al. FluidFM: Combining atomic force microscopy and nanofluidics in a universal liquid delivery system for single cell applications and beyond. *Nano Lett* 2009, *9*, 2501–2507.

228. Dorig, P.; Stiefel, P.; Behr, P.; Sarajlic, E.; Bijl, D.; Gabi, M.; Voros, J.; Vorholt, J. A.; Zambelli, T. Force-controlled spatial manipulation of viable mammalian cells and micro-organisms by means of FluidFM technology. *Appl Phys Lett* 2010, *97*, 023701.

229. Stiefel, P.; Schmidt, F. I.; Dorig, P.; Behr, P.; Zambelli, T.; Vorholt, J. A.; Mercer, J. Cooperative vaccinia infection demonstrated at the single-cell level using FluidFM. *Nano Lett* 2012, *12*, 4219–4227.

230. Ying, L. M.; White, S. S.; Bruckbauer, A.; Meadows, L.; Korchev, Y. E.; Klenerman, D. Frequency and voltage dependence of the dielectrophoretic trapping of short lengths of DNA and dCTP in a nanopipette. *Biophys J* 2004, *86*, 1018–1027.

231. White, S. S.; Balasubramanian, S.; Klenerman, D.; Ying, L. M. A simple nanomixer for single-molecule kinetics measurements. *Angew Chem Int Ed* 2006, *45*, 7540–7543.

232. Clarke, R. W.; Piper, J. D.; Ying, L. M.; Klenerman, D. Surface conductivity of biological macromolecules measured by nanopipette dielectrophoresis. *Phys Rev Lett* 2007, *98*, 198102.

233. Clarke, R. W.; White, S. S.; Zhou, D. J.; Ying, L. M.; Klenerman, D. Trapping of proteins under physiological conditions in a nanopipette. *Angew Chem Int Ed* 2005, *44*, 3747–3750.

234. Ying, L. M.; Bruckbauer, A.; Rothery, A. M.; Korchev, Y. E.; Klenerman, D. Programmable delivery of DNA through a nanopipette. *Anal Chem* 2002, *74*, 1380–1385.

235. Tsutsui, M.; Maeda, Y.; He, Y. H.; Hongo, S.; Ryuzaki, S.; Kawano, S.; Kawai, T.; Taniguchi, M. Trapping and identifying single-nanoparticles using a low-aspect-ratio nanopore. *Appl Phys Lett* 2013, *103*, 013108.

236. Dekker, C. Solid-state nanopores. *Nat Nanotechnol* 2007, *2*, 209–215.

237. Keyser, U. F.; Koeleman, B. N.; Van Dorp, S.; Krapf, D.; Smeets, R. M. M.; Lemay, S. G.; Dekker, N. H.; Dekker, C. Direct force measurements on DNA in a solid-state nanopore. *Nat Phys* 2006, *2*, 473–477.

238. Keyser, U. F.; van der Does, J.; Dekker, C.; Dekker, N. H. Optical tweezers for force measurements on DNA in nanopores. *Rev Sci Instrum* 2006, *77*, 105105.

239. van den Hout, M.; Krapf, D.; Keyser, U.; Lemay, S.; Dekker, N.; Dekker, C. Studying RNA folding using solid state nanopores and optical tweezers. *Biophys J* 2007, 227a.

240. Hall, A. R.; van Dorp, S.; Lemay, S. G.; Dekker, C. Electrophoretic force on a protein-coated DNA molecule in a solid-state nanopore. *Nano Lett* 2009, *9*, 4441–4445.

241. Sischka, A.; Spiering, A.; Khaksar, M.; Laxa, M.; König, J.; Dietz, K. J.; Anselmetti, D. Dynamic translocation of ligand-complexed DNA through solid-state nanopores with optical tweezers. *J Phys Condens Matter* 2010, *22*, 454121.

242. Steinbock, L. J.; Otto, O.; Skarstam, D. R.; Jahn, S.; Chimerel, C.; Gornall, J. L.; Keyser, U. F. Probing DNA with micro- and nanocapillaries and optical tweezers. *J Phys Condens Matter* 2010, *22*, 454113.

243. Bruckbauer, A.; Ying, L. M.; Rothery, A. M.; Zhou, D. J.; Shevchuk, A. I.; Abell, C.; Korchev, Y. E.; Klenerman, D. Writing with DNA and protein using a nanopipet for controlled delivery. *J Am Chem Soc* 2002, *124*, 8810–8811.

244. Bruckbauer, A.; Zhou, D. J.; Ying, L. M.; Korchev, Y. E.; Abell, C.; Klenerman, D. Multicomponent submicron features of biomolecules created by voltage controlled deposition from a nanopipet. *J Am Chem Soc* 2003, *125*, 9834–9839.

245. Bruckbauer, A.; Zhou, D. J.; Kang, D. J.; Korchev, Y. E.; Abell, C.; Klenerman, D. An addressable antibody nanoarray produced on a nanostructured surface. *J Am Chem Soc* 2004, *126*, 6508–6509.

246. Iwata, F.; Nagami, S.; Sumiya, Y.; Sasaki, A. Nanometre-scale deposition of colloidal Au particles using electrophoresis in a nanopipette probe. *Nanotechnology* 2007, *18*, 20070314.

247. Rodolfa, K. T.; Bruckbauer, A.; Zhou, D. J.; Schevchuk, A. I.; Korchev, Y. E.; Klenerman, D. Nanoscale pipetting for controlled chemistry in small arrayed water droplets using a double-barrel pipet. *Nano Lett* 2006, *6*, 252–257.

248. Suryavanshi, A. P.; Yu, M. F. Probe-based electrochemical fabrication of freestanding Cu nanowire array. *Appl Phys Lett* 2006, *88*, 083103.

249. Suryavanshi, A. P.; Yu, M. F. Electrochemical fountain pen nanofabrication of vertically grown platinum nanowires. *Nanotechnology* 2007, *18*, 105305.

250. Rodolfa, K. T.; Bruckbauer, A.; Zhou, D. J.; Korchev, Y. E.; Klenerman, D. Two-component graded deposition of biomolecules with a double-barreled nanopipette. *Angew Chem Int Ed* 2005, *44*, 6854–6859.

251. Gruter, R. R.; Voros, J.; Zambelli, T. FluidFM as a lithography tool in liquid: Spatially controlled deposition of fluorescent nanoparticles. *Nanoscale* 2013, *5*, 1097–1104.

252. Williams, C. G.; Edwards, M. A.; Colley, A. L.; Macpherson, J. V.; Unwin, P. R. Scanning micropipette contact method for high-resolution imaging of electrode surface redox activity. *Anal Chem* 2009, *81*, 2486–2495.

253. Ebejer, N.; Schnippering, M.; Colburn, A. W.; Edwards, M. A.; Unwin, P. R. Localized high resolution electrochemistry and multifunctional imaging: Scanning electrochemical cell microscopy. *Anal Chem* 2010, *82*, 9141–9145.

254. Shao, Y. H.; Mirkin, M. V. Scanning electrochemical microscopy (SECM) of facilitated ion transfer at the liquid/liquid interface. *J Electroanal Chem* 1997, *439*, 137–143.

255. Amemiya, S.; Bard, A. J. Scanning electrochemical microscopy. 40. Voltammetric ion-selective micropipet electrodes for probing ion transfer at bilayer lipid membranes. *Anal Chem* 2000, *72*, 4940–4948.

256. Shao, Y. H.; Mirkin, M. V. Probing ion transfer at the liquid/liquid interface by scanning electrochemical microscopy (SECM). *J Phys Chem B* 1998, *102*, 9915–9921.

257. Rodgers, P. J.; Amemiya, S.; Wang, Y. X.; Mirkin, M. V. Nanopipette voltammetry of common ions across the liquid-liquid interface. Theory and limitations in kinetic analysis of nanoelectrode voltammograms. *Anal Chem* 2010, *82*, 84–90.

258. Amemiya, S.; Kim, J.; Izadyar, A.; Kabagambe, B.; Shen, M.; Ishimatsu, R. Electrochemical sensing and imaging based on ion transfer at liquid/liquid interfaces. *Electrochim Acta* 2013, *110*, 836–845.

259. Sun, P.; Laforge, F. O.; Mirkin, M. V. Role of trace amounts of water in transfers of hydrophilic and hydrophobic ions to low-polarity organic solvents. *J Am Chem Soc* 2007, *129*, 12410–12411.

260. Lu, X. Q.; Wang, T. X.; Zhou, X. B.; Li, Y.; Wu, B. W.; Liu, X. H. Investigation of ion transport traversing the ion channels by scanning electrochemical microscopy (SECM). *J Phys Chem C* 2011, *115*, 4800–4805.

261. Evans, N. J.; Gonsalves, M.; Gray, N. J.; Barker, A. L.; Macpherson, J. V.; Unwin, P. R. Local amperometric detection of K$^+$ in aqueous solution using scanning electrochemical microscopy ion-transfer voltammetry. *Electrochem Commun* 2000, *2*, 201–206.

262. Cai, C. X.; Tong, Y. H.; Mirkin, M. V. Probing rapid ion transfer across a nanoscopic liquid-liquid interface. *J Phys Chem B* 2004, *108*, 17872–17878.

263. Hegde, V.; Mason, A.; Saliev, T.; Smith, F. J. D.; McLean, W. H. I.; Campbell, P. A. Scanning ion conductance microscopy of live keratinocytes. *J Phys Conf Ser* 2012, *371*, 012023.

264. Shevchuk, A. I.; Frolenkov, G. I.; Sanchez, D.; James, P. S.; Freedman, N.; Lab, M. J.; Jones, R.; Klenerman, D.; Korchev, Y. E. Imaging proteins in membranes of living cells by high-resolution scanning ion conductance microscopy. *Angew Chem Int Ed Engl* 2006, *45*, 2212–2216.

265. Potter, C. M. F.; Schobesberger, S.; Lundberg, M. H.; Weinberg, P. D.; Mitchell, J. A.; Gorelik, J. Shape and compliance of endothelial cells after shear stress in vitro or from different aortic regions: Scanning ion conductance microscopy study. *PLoS ONE* 2012, *7*, e31228.

266. Chen, C. C.; Zhou, Y.; Morris, C. A.; Hou, J. H.; Baker, L. A. Scanning ion conductance microscopy measurement of paracellular channel conductance in tight junctions. *Anal Chem* 2013, *85*, 3621–3628.

267. Zhou, Y.; Chen, C. C.; Weber, A.; Baker, L. A.; Hou, J. Potentiometric-scanning ion conductance microscopy for measurement at tight junctions. *Tissue Barriers* 2013, *1*, e2558s.

268. Morris, C. A.; Chen, C. C.; Baker, L. A. Transport of redox probes through single pores measured by scanning electrochemical-scanning ion conductance microscopy (SECM-SICM). *Analyst* 2012, *137*, 2933–2938.

269. Takahashi, Y.; Shevchuk, A. I.; Novak, P.; Zhang, Y. J.; Ebejer, N.; Macpherson, J. V.; Unwin, P. R. et al. Multifunctional nanoprobes for nanoscale chemical imaging and localized chemical delivery at surfaces and interfaces. *Angew Chem Int Ed* 2011, *50*, 9638–9642.

270. Dixon, M. C.; Daniel, T. A.; Hieda, M.; Smilgies, D. M.; Chan, M. H. W.; Allara, D. L. Preparation, structure, and optical properties of nanoporous gold thin films. *Langmuir* 2007, *23*, 2414–2422.

271. Cecchini, M. P.; Wiener, A.; Turek, V. A.; Chon, H.; Lee, S.; Ivanov, A. P.; McComb, D. W. et al. Rapid ultrasensitive single particle surface-enhanced Raman spectroscopy using metallic nanopores. *Nano Lett* 2013, *13*, 4602–4609.

272. Kodiyath, R.; Wang, J.; Combs, Z. A.; Chang, S.; Gupta, M. K.; Anderson, K. D.; Brown, R. J. C.; Tsukruk, V. V. SERS effects in silver-decorated cylindrical nanopores. *Small* 2011, *7*, 3452–3457.

273. Fu, Y.; Zhang, J.; Lakowicz, J. R. Largely enhanced single-molecule fluorescence in plasmonic nanogaps formed by hybrid silver nanostructures. *Langmuir* 2013, *29*, 2731–2738.

274. Santos, A.; Balderrama, V. S.; Alba, M.; Formentin, P.; Ferre-Borrull, J.; Pallares, J.; Marsal, L. F. Nanoporous anodic alumina barcodes: Toward smart optical biosensors. *Adv Mater* 2012, *24*, 1050–1054.

275. Tian, Y.; Wu, N.; Zou, X. T.; Felemban, H.; Cao, C. Y.; Wang, X. W. Fiber-optic ultrasound generator using periodic gold nanopores fabricated by a focused ion beam. *Opt Eng* 2013, *52*, 065005.

276. Lee, J. H.; Chung, S.; Kim, S. J.; Han, J. Y. Poly(dimethylsiloxane)-based protein preconcentration using a nanogap generated by junction gap breakdown. *Anal Chem* 2007, *79*, 6868–6873.

277. Chung, S.; Lee, J. H.; Moon, M. W.; Han, J.; Kamm, R. D. Non-lithographic wrinkle nanochannels for protein preconcentration. *Adv Mater* 2008, *20*, 3011–3016.

278. Stein, D.; Deurvorst, Z.; van der Heyden, F. H. J.; Koopmans, W. J. A.; Gabel, A.; Dekker, C. Electrokinetic concentration of DNA polymers in nanofluidic channels. *Nano Lett* 2010, *10*, 765–772.

279. Contento, N. M.; Branagan, S. P.; Bohn, P. W. Electrolysis in nanochannels for in situ reagent generation in confined geometries. *Lab Chip* 2011, *11*, 3634–3641.

280. Dunn, J. D.; Igrisan, E. A.; Palumbo, A. M.; Reid, G. E.; Bruening, M. L. Phosphopeptide enrichment using MALDI plates modified with high-capacity polymer brushes. *Anal Chem* 2008, *80*, 5727–5735.

281. Wang, W. H.; Palumbo, A. M.; Tan, Y. J.; Reid, G. E.; Tepe, J. J.; Bruening, M. L. Identification of p65-associated phosphoproteins by mass spectrometry after on-plate phosphopeptide enrichment using polymer-oxotitanium films. *J Proteome Res* 2010, *9*, 3005–3015.

282. Tan, Y. J.; Sui, D. X.; Wang, W. H.; Kuo, M. H.; Reid, G. E.; Bruening, M. L. Phosphopeptide enrichment with TiO_2-modified membranes and investigation of tau protein phosphorylation. *Anal Chem* 2013, *85*, 5699–5706.

283. Wang, W. H.; Dong, J. L.; Baker, G. L.; Bruening, M. L. Bifunctional polymer brushes for low-bias enrichment of mono- and multi-phosphorylated peptides prior to mass spectrometry analysis. *Analyst* 2011, *136*, 3595–3598.

284. Tan, Y. J.; Wang, W. H.; Zheng, Y.; Dong, J. L.; Stefano, G.; Brandizzi, F.; Garavito, R. M.; Reid, G. E.; Bruening, M. L. Limited proteolysis via millisecond digestions in protease-modified membranes. *Anal Chem* 2012, *84*, 8357–8363.

285. Yuill, E. M.; Sa, N.; Ray, S. J.; Hieftje, G. M.; Baker, L. A. Electrospray ionization from nanopipette emitters with tip diameters of less than 100 nm. *Anal Chem* 2013, *85*, 8498–8502.

286. Robertson, J. W. F.; Rodrigues, C. G.; Stanford, V. M.; Rubinson, K. A.; Krasilnikov, O. V.; Kasianowicz, J. J. Single-molecule mass spectrometry in solution using a solitary nanopore. *Proc Natl Acad Sci USA* 2007, *104*, 8207–8211.

287. http://www.cgl.ucsf.edu/chimera.

288. Boersma, A. J.; Bayley, H. Continuous stochastic detection of amino acid enantiomers with a protein nanopore. *Angew Chem Int Ed* 2012, *51*, 9606–9609.

289. Madampage, C. A.; Andrievskaia, O.; Lee, J. S. Nanopore detection of antibody prion interactions. *Anal Biochem* 2010, *396*, 36–41.

290. Wang, Y. X.; Kececi, K.; Mirkin, M. V.; Mani, V.; Sardesai, N.; Rusling, J. F. Resistive-pulse measurements with nanopipettes: Detection of Au nanoparticles and nanoparticle-bound anti-peanut IgY. *Chem Sci* 2013, *4*, 655–663.

291. Umehara, S.; Karhanek, M.; Davis, R. W.; Pourmand, N. Label-free biosensing with functionalized nanopipette probes. *Proc Natl Acad Sci USA* 2009, *106*, 4611–4616.

292. Deng, J.; Toh, C. S. Impedimetric DNA biosensor based on a nanoporous alumina membrane for the detection of the specific oligonucleotide sequence of dengue virus. *Sensors (Basel)* 2013, *13*, 7774–7785.

293. Meller, A.; Nivon, L.; Brandin, E.; Golovchenko, J.; Branton, D. Rapid nanopore discrimination between single polynucleotide molecules. *Proc Natl Acad Sci USA* 2000, *97*, 1079–1084.

294. Liu, J.; Huo, Q. S. Self-assembled nanoporous electrodes for sensitive and labeling-free biomolecular recognition. *Appl Phys Lett* 2005, *87*, 133902.

295. Heng, J. B.; Aksimentiev, A.; Ho, C.; Marks, P.; Grinkova, Y. V.; Sligar, S.; Schulten, K.; Timp, G. The electromechanics of DNA in a synthetic nanopore. *Biophys J* 2006, *90*, 1098–1106.

296. Wanunu, M.; Sutin, J.; McNally, B.; Chow, A.; Meller, A. DNA translocation governed by interactions with solid-state nanopores. *Biophys J* 2008, *95*, 4716–4725.

297. Yang, C.; Liu, L.; Zeng, T.; Yang, D.; Yao, Z.; Zhao, Y.; Wu, H. C. Highly sensitive simultaneous detection of lead(II) and barium(II) with G-quadruplex DNA in alpha-hemolysin nanopore. *Anal Chem* 2013, *85*, 7302–7307.

298. Wang, G. H.; Zhao, Q. T.; Kang, X. F.; Guan, X. Y. Probing mercury(II)-DNA interactions by nanopore stochastic sensing. *J Phys Chem B* 2013, *117*, 4763–4769.

299. Jin, Q.; Fleming, A. M.; Johnson, R. P.; Ding, Y.; Burrows, C. J.; White, H. S. Base-excision repair activity of uracil-DNA glycosylase monitored using the latch zone of alpha-hemolysin. *J Am Chem Soc* 2013, *135*, 19347–19353.

300. Li, R.; Wu, D.; Li, H.; Xu, C. X.; Wang, H.; Zhao, Y. F.; Cai, Y. Y.; Wei, Q.; Du, B. Label-free amperometric immunosensor for the detection of human serum chorionic gonadotropin based on nanoporous gold and graphene. *Anal Biochem* 2011, *414*, 196–201.

301. Movileanu, L.; Schmittschmitt, J. P.; Scholtz, J. M.; Bayley, H. Interactions of peptides with a protein pore. *Biophys J* 2005, *89*, 1030–1045.

302. Stefureac, R.; Long, Y. T.; Kraatz, H. B.; Howard, P.; Lee, J. S. Transport of alpha-helical peptides through alpha-hemolysin and aerolysin pores. *Biochemistry* 2006, *45*, 9172–9179.

303. Zhao, Q. T.; de Zoysa, R. S. S.; Wang, D. Q.; Jayawardhana, D. A.; Guan, X. Y. Real-time monitoring of peptide cleavage using a nanopore probe. *J Am Chem Soc* 2009, *131*, 6324–6325.

304. Wei, R. S.; Gatterdam, V.; Wieneke, R.; Tampe, R.; Rant, U. Stochastic sensing of proteins with receptor-modified solid-state nanopores. *Nat Nanotechnol* 2012, *7*, 257–263.

305. Fologea, D.; Ledden, B.; McNabb, D. S.; Li, J. L. Electrical characterization of protein molecules by a solid-state nanopore. *Appl Phys Lett* 2007, *91*, 539011–539013.

306. Yusko, E. C.; Johnson, J. M.; Majd, S.; Prangkio, P.; Rollings, R. C.; Li, J. L.; Yang, J.; Mayer, M. Controlling protein translocation through nanopores with bio-inspired fluid walls. *Nat Nanotechnol* 2011, *6*, 253–260.

307. Kasianowicz, J. J.; Henrickson, S. E.; Weetall, H. H.; Robertson, B. Simultaneous multianalyte detection with a nanometer-scale pore. *Anal Chem* 2001, *73*, 2268–2272.

308. Rotem, D.; Jayasinghe, L.; Salichou, M.; Bayley, H. Protein detection by nanopores equipped with aptamers. *J Am Chem Soc* 2012, *134*, 2781–2787.

309. Cheley, S.; Xie, H. Z.; Bayley, H. A genetically encoded pore for the stochastic detection of a protein kinase. *Chembiochem* 2006, *7*, 1923–1927.

310. Xie, H. Z.; Braha, O.; Gu, L. Q.; Cheley, S.; Bayley, H. Single-molecule observation of the catalytic subunit of cAMP-dependent protein kinase binding to an inhibitor peptide. *Chem Biol* 2005, *12*, 109–120.

311. Howorka, S.; Nam, J.; Bayley, H.; Kahne, D. Stochastic detection of monovalent and bivalent protein-ligand interactions. *Angew Chem Int Ed* 2004, *43*, 842–846.

312. Astier, Y.; Kainov, D. E.; Bayley, H.; Tuma, R.; Howorka, S. Stochastic detection of motor protein-RNA complexes by single-channel current recording. *Chemphyschem* 2007, *8*, 2189–2194.

313. Nestorovich, E. M.; Danelon, C.; Winterhalter, M.; Bezrukov, S. M. Designed to penetrate: Time-resolved interaction of single antibiotic molecules with bacterial pores. *Proc Natl Acad Sci USA* 2002, *99*, 9789–9794.

314. Feng, J. D.; Wu, J. M. Nanoporous gold channel with attached DNA nanolock for drug screening. *Small* 2012, *8*, 3786–3790.
315. Kumeria, T.; Santos, A.; Losic, D. Ultrasensitive nanoporous interferometric sensor for label-free detection of gold(III) Ions. *ACS Appl Mater Interfaces* 2013, *5*, 11783–11790.
316. Zevenbergen, M. A. G.; Singh, P. S.; Goluch, E. D.; Wolfrum, B. L.; Lemay, S. G. Stochastic sensing of single molecules in a nanofluidic electrochemical device. *Nano Lett* 2011, *11*, 2881–2886.
317. Shivanna, R.; Pramanik, D.; Kumar, H.; Rao, K. V.; George, S. J.; Maiti, P. K.; Narayan, K. S. Confinement induced stochastic sensing of charged coronene and perylene aggregates in alpha-hemolysin nanochannels. *Soft Matter* 2013, *9*, 10196–10202.
318. Krishantha, D. M. M.; Breitbach, Z. S.; Padivitage, N. L. T.; Armstrong, D. W.; Guan, X. Y. Rapid determination of sample purity and composition by nanopore stochastic sensing. *Nanoscale* 2011, *3*, 4593–4596.
319. Wu, H. C.; Astier, Y.; Maglia, G.; Mikhailova, E.; Bayley, H. Protein nanopores with covalently attached molecular adapters. *J Am Chem Soc* 2007, *129*, 16142–16148.
320. Braha, O.; Webb, J.; Gu, L. Q.; Kim, K.; Bayley, H. Carriers versus adapters in stochastic sensing. *Chemphyschem* 2005, *6*, 889–892.
321. Goyal, G.; Freedman, K. J.; Kim, M. J. Gold nanoparticle translocation dynamics and electrical detection of single particle diffusion using solid-state nanopores. *Anal Chem* 2013, *85*, 8180–8187.
322. Campos, E.; McVey, C. E.; Carney, R. P.; Stellacci, F.; Astier, Y.; Yates, J. Sensing single mixed-monolayer protected gold nanoparticles by the alpha-hemolysin nanopore. *Anal Chem* 2013, *85*, 10149–10158.
323. Kashefi-Kheyrabadi, L.; Mehrgardi, M. A. Aptamer-based electrochemical biosensor for detection of adenosine triphosphate using a nanoporous gold platform. *Bioelectrochemistry* 2013, *94*, 47–52.
324. Li, S.-J.; Xing, Y.; Tang, M.-Y.; Wang, L.-H.; Liu, L. A novel nanomachined flow channel glucose sensor based on an alumina membrane. *Anal Methods* 2013, *5*, 7022–7029.
325. Yin, G.; Xing, L.; Ma, X.-J.; Wan, J. Non-enzymatic hydrogen peroxide sensor based on a nanoporous gold electrode modified with platinum nanoparticles. *Chem Paper* 2013, *68*, 1–7.
326. Fei, T.; Jiang, K.; Liu, S.; Zhang, T. Humidity sensors based on Li-loaded nanoporous polymers. *Sensors Actuat B Chem* 2014, *190*, 523–528.
327. Shim, J. W.; Gu, L. Q. Stochastic sensing on a modular chip containing a single-ion channel. *Anal Chem* 2007, *79*, 2207–2213.
328. Kang, X. F.; Cheley, S.; Rice-Ficht, A. C.; Bayley, H. A storable encapsulated bilayer chip containing a single protein nanopore. *J Am Chem Soc* 2007, *129*, 4701–4705.
329. Cheley, S.; Gu, L. Q.; Bayley, H. Stochastic sensing of nanomolar inositol 1,4,5-trisphosphate with an engineered pore. *Chem Biol* 2002, *9*, 829–838.
330. Halverson, K. M.; Panchal, R. G.; Nguyen, T. L.; Gussio, R.; Little, S. F.; Misakian, M.; Bavari, S.; Kasianowicz, J. J. Anthrax biosensor, protective antigen ion channel asymmetric blockade. *J Biol Chem* 2005, *280*, 34056–34062.
331. Ye, W.; Guo, J.; Cheng, S.; Yang, M. Nanoporous membrane based impedance sensors to detect the enzymatic activity of botulinum neurotoxin A. *J Mater Chem B* 2013, *1*, 6544–6550.
332. Jayawardhana, D. A.; Crank, J. A.; Zhao, Q.; Armstrong, D. W.; Guan, X. Y. Nanopore stochastic detection of a liquid explosive component and sensitizers using boromycin and an ionic liquid supporting electrolyte. *Anal Chem* 2009, *81*, 460–464.
333. Shin, S. H.; Luchian, T.; Cheley, S.; Braha, O.; Bayley, H. Kinetics of a reversible covalent-bond-forming reaction observed at the single-molecule level. *Angew Chem Int Ed* 2002, *41*, 3707–3709.

12 Recent Investigations of Single Living Cells with Ultramicroelectrodes

Christian Amatore, Manon Guille-Collignon, and Frédéric Lemaître

CONTENTS

12.1 INTRODUCTION: MICRO VERSUS ULTRAMICRO NANOELECTRODE

12.1.1 ELECTROCHEMICAL PROPERTIES OF UMEs AND ARTIFICIAL SYNAPSE CONFIGURATION

Communication between cellular organisms mostly occurs through the release of specific biochemical or chemical messengers from an emitting cell to a receiving cell. In a practical view, analytical detection of such a release in real time can be particularly difficult since an infinitely minute number of molecules are generally released during a brief fraction of time (milliseconds to seconds). By mimicking the advantages of the chemical synapse involved during neural communication, electrochemical techniques using ultramicroelectrodes (UMEs) represent a convenient way of bypassing this problem through the *artificial synapse* configuration (Figure 12.1). In practice, the UME is positioned in the close vicinity (a few hundreds of nanometers) of the investigated single-cell organism (Figure 12.1). Provided that a fraction of the content released by the cell is electroactive, intrinsic properties of UMEs as well as the experimental configuration itself allow one to electrochemically monitor (chronoamperometry, fast-scan cyclic voltammetry [FSCV]) the cell release with an excellent signal-to-noise (S/N) ratio and temporal resolution (see Section 12.1.2). Indeed, UME properties perfectly match the analytical requirements for an accurate electrochemical detection

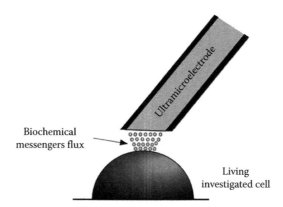

Biochemical
messengers flux

Living
investigated cell

FIGURE 12.1 Scheme of the artificial synapse configuration for which a UME is positioned near an emitting cell. Considering that the species released can be oxidized or reduced at the electrode surface, electrochemical techniques (amperometry at constant potential or FSCV) allow one to monitor the release in real time.

of cell release *at the single cell level in real time*, that is, recording zepto- to attomoles of chemical messengers released during a millisecond timescale, as explained in the following.

The first quite obvious reason for using UMEs for biological release investigations is that the UME's micrometric size directly corresponds to the usual cell dimensions, thus making UMEs perfectly adapted for investigations at the single-cell level.

Moreover, using a UME ensures a better S/N ratio compared to conventional millimetric electrodes. Indeed, the expected faradic current i_F can be distorted by a competitive capacitive current i_C that results from capacitive phenomena related to the double layer formed at the electrode/solution interface. The exponential decay behavior of the capacitive current is governed by a time constant $\tau = RC$ (where R is the resistance of the solution and C the electrical capacitance). At disk electrodes, $\tau \propto r_0$, thus decreasing with r_0. Using UMEs thus allows one to limit the capacitive distortion for short timescales by comparison with millimetric electrodes. Furthermore, such an electrochemical capacitance also induces intrinsic noise depending on the electrode, that is, to $(r_0)^2$. As discussed in the next section, the signal, that is, the faradic current, is proportional to the electrode radius for a UME under spherical diffusion regime. When stray capacitances are negligible, the S/N ratio is thus proportional to $1/r_0$, thus making UMEs much appropriated for highly sensitive electroanalytical measurements.

The last interest to consider UMEs is related to the control of the working electrode potential. In practice, an ohmic drop (Ri, resulting to the current passing through the solution resistance R) can possibly alter the potential value of the working electrode versus that of the reference electrode and is usually more or less compensated through the use of a counter electrode. Under a linear diffusion transient regime at a millimetric electrode (see Section 12.1.2), the ohmic drop directly scales with the electrode radius and decreases when the electrode size is reduced. For UMEs, once diffusion has switched to the spherical diffusion regime, ohmic drop value reaches a minimal and constant value, which is generally negligible in electrolytes suitable with biological experiments. The counter electrode is no longer required: a more basic two-electrode configuration may be used (as soon as the reference electrode has a millimetric size at least), thus facilitating the micromanipulations during the experiments.

Beyond the UME properties, as introduced earlier, the artificial synapse configuration itself corresponds to appropriate experimental conditions for accurate measurements. As an example, the faradic current recorded by a UME is directly proportional to the concentration of electroactive species and not to their amount. On the other hand, positioning a UME at a submicrometric distance from the investigated cell allows one to restrict the volume in which the messengers are released, hence results in large concentrations even for a few thousand released molecules as occurs in real biological synapses. Typically, one attomole delivered in a volume of a few femtoliters creates a

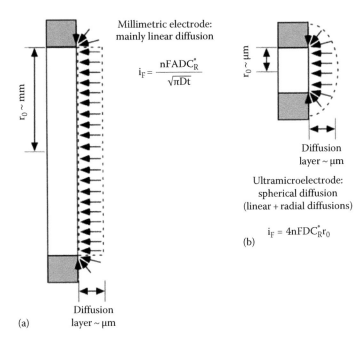

FIGURE 12.2 Schemes of diffusion layers (arrows display the electroactive species diffusive flux at the electrode surface) for a millimeter-sized electrode (a) and a UME (b). The faradic current (i_F) equation is given for both cases and corresponds to a disk electrode of radius r_0.

millimolar concentration rise that can be easily detected with electrochemical techniques. Moreover, minimizing the electrode–cell distance favors the total or partial detection of moderately unstable species, particularly in the case of the oxidative stress investigations (see Section 12.2.1.2). At last, because the UME diffusion layer is comparable to the UME dimension, that is, a few micrometers, the artificial synapse configuration guarantees a total electrochemical detection of molecules present in the artificial synaptic cleft. The collection of electroactive molecules released at the surface of a single living cell is thus quantitative.[1]

In summary, UMEs are particularly adequate for monitoring minute amounts of species released by a single living cell. They rapidly reach a spherical diffusion regime (~ms) leading to a steady state, that is, a constant diffusion layer (a few r_0) and a recorded faradic current only depending on r_0 and the electroactive species concentration (Figure 12.2) and weakly distorted by capacitive phenomena. Therefore, due to the other intrinsic properties of UMEs described earlier and to the artificial synapse configuration itself, *UMEs are prone to detect concentration changes in the millisecond range* with a maximized S/N ratio and collection efficiency and thus to resolve *fast kinetics of cell secretion*. Regarding the definitions introduced in the next section, it has to be emphasized that a micrometer-sized electrode used within the artificial synapse configuration acts as a UME but not as a microelectrode.

12.1.2 What Is a UME?

Simply defining the term *ultramicroelectrode* is far from clear and is furthermore complicated by the vague usage of another term *microelectrode*. Indeed, considering a typical electrochemical experiment (three-electrode setup in which the voltage between the working and reference electrodes is settled), two families of experimental approaches can be considered according to the electrode area A and the volume of the solution V. Thus, for rather small A and large V values, the amount of electroactive species consumed due to the passage of current will not be drastically

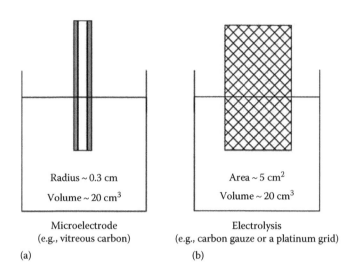

Radius ~ 0.3 cm   Area ~ 5 cm^2
Volume ~ 20 cm^3 Volume ~ 20 cm^3

Microelectrode   Electrolysis
(e.g., vitreous carbon)   (e.g., carbon gauze or a platinum grid)
(a)   (b)

FIGURE 12.3   Schematic conditions defining what a microelectrode is (a) and what a microelectrode is not (b).

changed for experimental time lengths of a few minutes. For this small A/V conditions (typically a millimeter-sized working electrode in a few milliliters solution), the sample is not altered by the electrochemical experiment (Figure 12.3). Acting in an *electroanalytical* scale, the electrode is named here a *microelectrode*. Conversely, for larger A/V conditions corresponding to an electrolysis experiment, the bulk concentrations of electroactive species will be drastically affected, thus corresponding to a *preparative* scale. In that way, the microelectrode term no longer applies, and the electrode is commonly referred to as *the electrode*.

However, for a given temporal scale and given diffusivities, the analytical properties of a given microelectrode also depend on the geometry of the microelectrode. As an example, let us consider a solution containing a species R (bulk concentration C_R^*) and possibly oxidized at an electrode surface ($R = O + ne^-$) in a diffusion-controlled way, the microelectrode surface area being small in comparison with the volume of solution so that the passage of current will not alter the bulk concentrations of electroactive species. For conventional millimetric electrodes (for instance, a disk of radius r_0), diffusion essentially proceeds along the direction orthogonal to the electrode surface due to the micrometric thickness of the diffusion layer (Figure 12.2). If the working electrode potential is maintained constant (versus a reference electrode) at the appropriate value to perform R oxidation, the corresponding current depends on the time and follows Cottrell's law, while the diffusion layer remains contained within the convection-free domain. As a consequence, the diffusion layer thickness ($\sim (Dt)^{1/2}$ where D is the diffusion coefficient of the electroactive species) increases, thus leading to the time faradic current i_F decrease ($\propto t^{-1/2}$). Furthermore, the faradic current is proportional to the electrode surface, that is, $(r_0)^2$. Conversely, edge effects cannot be neglected for *electrodes for which at least one dimension compares to—or are smaller than—the diffusion layer dimensions.* In that case, the radial diffusion must be taken into account, and after a transient period, the previous 1D linear diffusion is ($t > (r_0)^2/D \sim$ ms, where r_0 is the radius of the disk electrode) replaced by a 2D spherical diffusion in which the radial diffusion mostly contributes. At disk or spherical electrodes, this convergent diffusional regime leads to a steady state, that is, to a constant diffusion layer (a few r_0) and a constant faradic current recorded at the electrode, the latter henceforth depending on r_0 (Figure 12.2). To differentiate this behavior among microelectrodes, such electrodes for which the faradic current is controlled by a steady-state spherical diffusion are called a UME. In other words, a UME can be defined as an electrode whose diffusion layer can reach a comparable or even larger size than the dimension of the electrode at infinite timescale for a given electrochemical experiment.[2,3] Consequently, UME term is expected to be not directly related to the UME size but also to diffusivities. However, UME dimensions are restricted by the intrinsic properties of diffusion layers and mainly that of the convection-free domain (a few hundred of micrometers), which prevents a millimeter-sized electrode to respect the definition introduced earlier (a hydrodynamic steady state is reached rather than a pure diffusional one). As a consequence, a UME indirectly becomes an electrode whose at least one dimension (radius for a disk, width for a band, etc.) is below a couple of tens of micrometers. Therefore, while UME properties are not directly related to their dimensions, the UME critical dimension is considered as smaller than 25 μm in classical electrochemical solvents, including aqueous media, from a practical point of view.

Regarding the two definitions given earlier, microelectrodes and UMEs correspond to completely different situations (A/V values and diffusion layer thickness, respectively). As previously stated by Amatore et al., "the electrochemical behaviors of the electrodes not only are related to their dimensions but also depend on the time scale of the experiment and thickness of the convection-free layer. It stresses once more the futility of trying to propose an absolute definition of UMEs based on the objects themselves. Indeed, the same electrode may behave as a microelectrode or a UME, depending on these parameters."[4] Nevertheless, in most cases, UMEs provide small A/V conditions. In that way, UMEs are also microelectrodes within the definition introduced earlier. Therefore, distinction is not always done between UMEs and microelectrodes. This is why the term *microelectrode* is increasingly used for micrometric electrodes, thus being a potential source of confusion.[3] Finally, the recent application of nanoelectrodes also causes confusion since the definition of nanoelectrodes directly refers to their nanometric dimensions, while the validity of the usual steady-state current equation for UMEs is not ensured for dimensions smaller than 10 nm.[5] Indeed, in that case, diffusion layer and double-layer thicknesses (usually a few microns and a few nanometers, respectively, for large-sized electrodes) will therefore be in the same order of magnitude.

12.2 ELECTROCHEMICAL DETECTION AT THE SINGLE-CELL LEVEL

The main applications of electrochemical investigations involving the artificial synapse configuration deal with two important and ubiquitous biological mechanisms: vesicular exocytosis and oxidative stress. Nevertheless, due to the high number of works reported during the past decades, particularly for exocytosis, a selection of representative examples will be offered in this chapter. The reader is thus encouraged to browse the reviews quoted therein for a more comprehensive overview.[6–10]

12.2.1 USING A BASIC UME

12.2.1.1 Detection of Exocytosis

Vesicular exocytosis is an important secretory pathway mainly used during intercellular communication in multicellular organisms, the most famous example being the chemical synapse. Thus, biochemical or chemical messengers are locally and selectively delivered by an emitting cell toward a definite target.[11] In that way, the messengers are initially stored into secretion vesicles. In regulated secretory pathways, such vesicles are prone to dock and then fuse with the cell membrane after an appropriate stimulation (that somehow allows a Ca^{2+} entry/increase into the intracellular medium). The subsequent fusion between cell and vesicular membranes allows the formation of a nanometric fusion pore through which the release of the vesicular content toward the extracellular medium can start. After that, depending on the cell model, the expansion of the pore could take place, leading to a more massive release of the chemical messengers toward the extracellular medium (Figure 12.4).

Because they mimic the chemical synapse by replacing the receptor cell by a UME, electrochemical techniques involving the artificial synapse configuration are suitable for the real-time monitoring of exocytosis at the single-cell level (see Section 12.1.1) and most particularly because the common neurotransmitters (serotonin [5-hydroxytryptamine] and catecholamines like dopamine, noradrenaline, and adrenaline) can be easily oxidized at carbon surface (E = 650 mV versus Ag/AgCl),

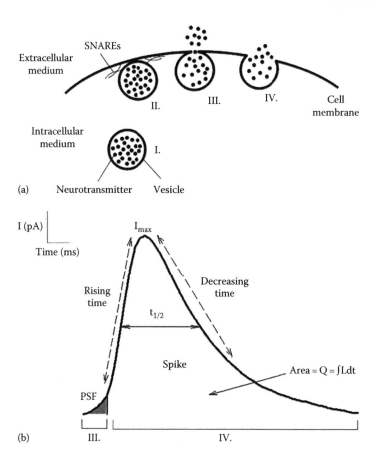

FIGURE 12.4 (a) Main steps of vesicular dense-core exocytosis (see text). (b) The described exocytotic event is correlated to its current spike recorded by amperometry at a carbon fiber UME in the *artificial synapse* configuration.

Contrary to the oxidative stress (see Section 12.2.1.2), the rather short duration of release for a given vesicular event (from a few milliseconds to several hundred milliseconds) requires using UMEs with a fast time of response in order to record precise kinetics. Therefore, modified electrodes are scarcely used,[12–14] and the most commonly used UME is a micrometric (5–10 μm diameter) carbon fiber electrode, which, beyond its ability to oxidize neurotransmitters, gives rise to low capacitive background current. Typically, the fiber is sealed into a glass capillary, and insulation of the outer cylindrical surface can be performed by different methods.[6] Coating the UME exposed surface with a thin film of Nafion (perfluorinated cation-exchange material) enhances selectivity toward cations like catecholamines that are protonated in physiological conditions. However, the reduced diffusivity of detected molecules through the film can alter the temporal resolution (diffusion filtering) and even decrease the collection efficiency.[15] It has to be emphasized that the electrochemical measurements are performed under conditions as close as possible to physiological ones. Detection in reduction is thus precluded due to the presence of dioxygen ($\sim$0.24 mmol·L^{-1}) required in order of not submitting aerobic cells to a stress.

To date, two electroanalytical techniques have been mainly involved for real-time analysis of exocytosis at the single-cell level: amperometry and FSCV. In FSCV, as in all cyclic voltammetry techniques, the current is measured as a function of the potential applied between the UME and the reference electrode (the potential is usually applied in triangular voltage ramps).[8] Because the position and shape of the voltammogram peaks depend on the species analyzed, FSCV is a powerful method that allows identifying and quantifying the neurotransmitter released or differentiating different species within a mixture. Due to the fast release involved during a vesicular release, FSCV requires performing a series of repetitive scans at high scan rates (a few hundreds of Volt per second). However, even at UMEs, capacitive currents at high scan rates can alter recorded measurements and have to be sufficiently stable to be subtracted. In amperometry, the UME potential is maintained at a constant value that corresponds to the oxidization potential of the species analyzed; hence, the capacitive charging currents are negligible. The corresponding current is then purely faradic and is recorded as a function of time. Contrary to FSCV, amperometry is able neither to discriminate different oxidable species nor to give access to their relative concentration. Conversely, the temporal resolution of amperometry is better (submillisecond) and gives access to the dynamics of each vesicular event and also to the exact number of species detected provided the electron stoichiometry is known (see below). In practice, the exocytotic release of a single cell is displayed in amperometry as a succession of amperometric spikes whose frequency (number of detected events as a function of the time) features the whole activity of the cell. Each spike corresponds to one vesicular release event. Each detected exocytotic event can be analyzed through its maximum spike oxidation current (I_{max} [pA] related to the maximal flux of neurotransmitters released during an event) or time parameters such as the half-width time $t_{1/2}$ (ms), the rising or decreasing times (Figure 12.4). Moreover, the area (charge Q [fC or pC]) directly corresponds to the amount of species detected due to Faraday's law ($Q = nNF$, where n is the number of exchanged electrons in the oxidation process, F is Faraday's constant, and N is the species amount) as another quantitative parameter. Finally, the initial fusion pore signature is displayed as a small step of current preceding amperometric spikes as apparent for a fraction of spikes. Such *amperometric foot* or prespike feature (PSF) can be analyzed (duration, maximum current, and charge) to investigate dynamics and size of fusion pore (Figure 12.4).[16,17]

Due to the difficulties to monitor real neuron synapses with a usual micrometric carbon UME (low vesicular size, i.e., ~50 nm diameter, low amount released ~5000 molecules per vesicles, and the short time of release ~1 ms, neuron culture, low synaptic cleft size ~nms) nonsynaptic cell model systems are mainly investigated. The UME detection sensitivity (~1000 molecules per millisecond) when using the artificial synapse configuration is adapted to their quite large-sized vesicles (0.25–1 μm diameter) and their high neurotransmitter concentration (>0.1 mol·L^{-1}). The cell model systems investigated are numerous and well documented in quite recent reviews.[6–9] Despite the fact that electrochemical detections on the cell body of neuronal systems have been also reported on invertebrate systems (leech *Hirudo medicinalis*, pond snail *Planorbis corneus*, superior cervical ganglion neurons from neonatal rats, rat cultured ventral midbrain neurons, dissociated neurons of the sea pansy *Renilla koellikeri*, etc.), adrenal (bovine, calf, rat, etc.) chromaffin cells (which release a mixture of catecholamines [adrenaline, noradrenaline, and dopamine]), and pheochromocytoma (PC12) cells (which release dopamine and are derived from cancerous rat adrenal gland) are the most used cell models, even if mast cells from normal or beige mice (serotonin/histamine) or pancreatic β-cells (insulin/serotonin) are also investigated by some groups. Exocytotic release is mainly evoked by using a glass micropipette positioned close to the emitting cell and a pressure-driven injection of a large variety of stimulating agents (nicotine, digitonin/Ca^{2+}, Ba^{2+}, K^+/Ca^{2+}, etc.). Finally, the electrochemical monitoring of neurotransmitters can be also achieved in brain slices or in rat living brain,[18,19] and even in fruit fly brain,[20] though the single-cell level is less spatially controlled in that case. *Indeed, one should note that although the UME is micrometer sized, all the exocytotic events electrochemically recorded correspond to the detection at a vesicular scale of a hundred nanometers. The advantage of the UME comparatively larger sized is that it may detect all events occurring at different loci on the cell membrane considered by the UME.*

In the next section, we will restrict the electrochemical measurements of exocytosis to a few examples that evidence the important contributions of amperometry to different questions/debates related to the exocytotic mechanism and more particularly to the controlling factors/parameters.

First of all, the role of biological parameters has been confirmed, particularly the assistance of soluble *N*-ethylmaleimide-sensitive fusion protein attachment receptors (*SNAREs*) assemblies.

Indeed, the docking of a given vesicle to the cell membrane needs to be assisted through the formation of a SNARE complex by multiple protein–protein interactions (vesicle membrane proteins *v-SNAREs* linked to cell membrane proteins *t-SNAREs*) in order to overcome the electrostatic repulsions between cell and vesicular membranes. Investigations performed by amperometry evidenced the prominent role of SNAREs.[21–24] For instance, modifying their amino acid sequence or cleaving them by neurotoxins leads to an important decrease in the exocytotic frequency,[25–29] while such an effect could also depend on the stimuli magnitude.[30] Moreover, regulatory-specific proteins (α-SNAP, protein kinase C, Munc-18, dynamin, etc.) involved in the SNARE complex formation are expected to play a role on the exocytotic release itself and the stability of the fusion pore.[31–38] The importance of such specific biological machineries should not lead to undervalue the role of physicochemical parameters. As an example, amperometric investigations evidenced the *pH dependence* of the exocytotic release, particularly for *dense-core* vesicles. In that case, the cationic messengers are compacted into a polyelectrolytic matrix. The ionic exchanges between the matrix and the extracellular medium could induce the matrix destructuration/swelling and the expansion of the pore. While results obtained on different cell types are not still easy to rationalize, amperometry evidenced that modifying the pH gradient (usually 7.4 in the extracellular medium versus 5.5 in the vesicle) induces changes that are observed by amperometry on spike size and shape and on exocytosis frequency, according to the pH dependence of the swelling of the vesicular matrix.[39–42] Similar trends can be obtained by decreasing the intravesicular pH (by using drugs that block vesicle proton pumps) upon which a deceleration of the dynamics of release and the decrease in the neurotransmitters released amount are observed.[43] One has to emphasize that the role of the destructuration of the intravesicular matrix on the dynamics of the exocytotic release has been also corroborated through the changes in the extracellular osmolarity.[44–48] The crucial role of chromogranins A/B (the main soluble proteins in dense-core vesicles) has also been evidenced.[49–54] The role of the local *cell membrane curvature* has been also proved by inserting exogenous compounds into the cell membrane and investigating its effects on the exocytotic release amperometrically.[55,56] Such results contributed to the debate dealing with the nature of the fusion pore and are globally consistent with a lipidic fusion pore formation. Additionally, alterations of the membrane curvature through modifications of lipid cone angle seem to play a role in exocytosis. As already described earlier, amperometry allows detecting the formation of the *fusion pore* that precedes the release itself. Detailed analysis of PSFs suggests that the amperometric foot corresponds to a catecholamine flux whose observation does depend not only on the stability of the pore but also on the secretion granule composition.[57] Particularly, modifications of the intravesicular composition (L-DOPA and reserpine treatments) raised the question whether the amperometric foot could be related to vesicles equipped of a liquid film named *halo* (in which the neurotransmitter's diffusion coefficient is higher than in the matrix) between the vesicle membrane and the vesicular matrix.[58–60] Assuming that the neurotransmitter's intravesicular concentration has a constant value and applying Faraday's law to the spherical volume of the vesicle,[61] the cube root of charge Q for a given amperometric spike results as proportional to the vesicle radius. Statistical distributions using $Q^{1/3}$ evidenced that at least *two vesicular populations* with different properties of secretion are involved in chromaffin cells,[62] which was confirmed by optical microscopy and by others through analyses of statistical distributions of logQ.[45,63] Furthermore, a theoretical analysis of a given amperometric spike (vesicular opening/species diffusion) is expected to give access to important parameters about the corresponding exocytotic event, particularly the maximum opening angle of the vesicle that reveals the state of fusion (partial versus full). While the kinetics characteristics and amplitude of an amperometric spike are not self-sufficient for solving the invert problem, internal calibration involving the PSF and patch-clamp reported data coupled to a semianalytical procedure related to the deconvolution of the released molecule flux has been developed and allows an estimation of the final opening angle less than a few tens of degrees, far from the expected value for a full fusion event (180°).[64–66] Finally, exocytosis can be also investigated by *scanning electrochemical microscopy* (SECM). Indeed, while involving UME probing of a surface, SECM is not intrinsically adapted to investigations of exocytosis in real time because the temporal resolution of SECM (UME speed of a few micrometer per second) cannot be compared with that of exocytotic release (tens of milliseconds). Nevertheless, Schuhmann and coworkers designed a Bio-SECM to analyze the topography of a single cell (chromaffin cell, PC12) through a shear-force-based constant-distance control on the one hand and to achieve the amperometric detection at the UME surface on the other hand.[67,68]

12.2.1.2 Detection of Oxidative Stress

Oxidative stress is a hazardous metabolic situation encountered by aerobic organisms when usual protection pathways (enzymes, antioxidants, etc.) against damages of oxidant species (partially resulting from the respiratory chain) are overwhelmed. The primary species of oxidative stress (superoxide anion [O_2^-] and radical nitric oxide [NO]) as well as the ensuing reactive oxygen species (ROS) and reactive nitrogen species (RNS) like nitrite (NO_2^-), peroxynitrite ($ONOO^-$), and hydrogen peroxide (H_2O_2) can be detected electrochemically on various UME surfaces depending on selectivity, sensitivity, and time response expected. NO has many biological roles as a ubiquitous mediator involved in vascular physiology, neuronal mediation, immune response, and oxidative stress. NO detection on single living cells is challenging because of the low levels and short lifetime of this radical under aerobic conditions. Highly sensitive spectroscopic techniques associated with fluorescent probes as well as Griess tests or electron paramagnetic resonance (EPR) spectrometry are used for in situ detection of NO. But electrochemical methods are more routinely used for NO bioanalytical monitoring due to intrinsic advantages of electrochemical sensors: easily manufacturable and miniaturized; cheap; modifiable by different sensing chemistries for playing on response time, sensitivity, and selectivity depending on the application considered; fast acquisition time by chronoamperometry (up to millisecond range); and size between micrometer and tens of micrometers adapted to cell dimensions. NO is oxidizable at platinum or platinized surface electrodes and to a lesser extent at carbon surfaces. At platinized electrodes, its detection potential with a maximum sensitivity is around 0.7 V versus Ag/AgCl. The detection of nitric oxide by porphyrin-modified UME was first described in 1992 by Malinski et al. based on the electrocatalytic oxidation of NO in aqueous solutions. This work was followed by numerous examples of metalloporphyrin and metallophthalocyanine electrodes, which showed a negative shift of the voltammetric oxidation wave for NO associated with an increase in the current intensity. Filtering of interferences in NO detection by charged species like nitrites, dopamine, and ascorbic acid may be avoided using Nafion or hydrophobic membranes. Another electrochemical strategy based on direct oxidation of NO on platinum electrode was first reported by Shibuki who used a miniature Clark electrode containing a membrane (in chloroprene, which is permeable to gases but not contaminants and/or ions of the sample).[69–72] This last decade, excellent analytical reviews have reported comprehensive and exhaustive work performed for electrochemical detection and quantification of nitric oxide and other ROS and RNS.[6,73–75] Here, we will focus on works reported over the last 5–10 years with a special highlight on innovative technological developments upon selecting specific and illustrative examples among the most widely relevant literature. Selected significant examples are thus described in the following in order to illustrate the different approaches at the micro-, nano-, and array-electrode levels to bring new insights in the physiological roles of oxidative stress species and mainly on NO and its related in vivo biochemistry.

A strategy developed at the beginning of the 2000s consisting of positioning a sensor above a cell for NO detection by SECM was reported by the group of W. Schuhmann. This method allows precise positioning of the electrode and obtention of the spatial distribution of NO release over the sample. For example, Isik et al. detected NO release with a dual-electrode sensor above endothelial cells with a distance sensor (10 μm diameter) aimed at measuring height between cells and electrode tip by reducing oxygen used as redox mediator and a NO sensor (50 μm diameter) modified with porphyrin (NiTmPyP) for NO detection.[76] The accurate positioning of the tip controlled by the feedback signal of oxygen was achieved within a 5 μm distance from the surface of a single HUVEC cell, and the dependencies of the current obtained (featuring the NO fluxes released) on the distance have been evidenced. However, the strong dependence of signal on height and the fact that no topographic information is obtained due to the sensor diameter limited the method scope. These drawbacks were overcome in the following study when another NO UME also proposed by Isik and Schuhmann was used, consisting of a carbon fiber (diameter 7–9 μm), first coated with platinum, then with a layer of Ni tetrasulfonated phthalocyanine (NiTSPc) tetrasodium salt polymerized on the tip, and finally covered with a Nafion film. Thanks to a shear-force-dependent constant-distance mode of SECM, a precise positioning of the NO sensor above adherent endothelial cells was performed. Reproducible detection of NO was consecutively obtained after stimulation with 1 μmol·L^{-1} bradykinin. This analytical performance could be achieved, thanks to the accurate positioning at an extremely small distance of below 300 nm between the NO UME and the cell membrane. The authors reported that single HUVEC cell releases from around 2 to 10 fmol of NO during few hundreds of seconds upon bradykinin stimulation.[77] Borgmann et al. used a double-barrel electrode arrangement made of a Ni(II) tetrakis(p-nitrophenylporphyrin) electrochemically deposited film coated over a 50 μm diameter platinum disk electrode integrated side by side with a 25 μm diameter platinum disk. The first sensor coated with a metalloporphyrin monitored NO, while the second one helped automatic tip positioning by negative feedback over a layer of adherent growing T-HUVEC cells. Plates of 96 wells loaded with cells could then be scanned with automatized z-approach curves and allowed selective and sensitive detection of NO over the array. This system is interesting in the future because it allows screening different drugs and effectors supposed to play a role at different steps of NO signaling pathway (Figure 12.5).[78] In 2006, Patel et al. detected the neuronal release of nitric oxide with a selective film-coated 30 μm carbon fiber disk electrode made of Nafion and electropolymerized polyeugenol or o-phenylenediamine.[79] These UMEs have been used to evaluate the differences in NO production between two different identified neurons in pond snails in intact neuronal preparations using a high Ca^{2+}/K$^+$ stimulus. The different types of neurons investigated were shown to release different quantities of NO that were both suppressed by equal concentrations of L-nitro-arginine methyl ester (L-NAME), an inhibitor of the nitric oxide synthase (NOS). Detection of oxygen and hydrogen peroxide released by cultured living RAW macrophages was performed by Zhao et al. by means of SECM using constant-height and constant-distance mode without requiring any external redox molecules. This was demonstrated by scanning the z-position with an alternating current kept constant and by recording concomitantly the amperometric current at the same SECM probe on cells under normal physiological conditions, in culture medium, and without subsequent stimulation. The method reported here seems to be convenient and efficient in simultaneously recording the topographical and biochemical information of the cells and then deconvoluting both kinds of information. Finally, it provides by amperometric recording the released hydrogen peroxide and oxygen spatial profiles and shows that the nucleus region of the cell releases more ROS than other zones of the cells (Figure 12.5).[80]

A large series of works were proposed by Amatore and his group to study the fundamental mechanisms of oxidant release at the level of a single cell, in real time and quantitatively. To do so, the tip of a platinized carbon fiber UME was positioned at a submicrometric distance in an artificial synapse configuration above an isolated cell cultured in a Petri dish to measure oxidative stress release by means of different stimulations: physical or biochemical ones. The first example was performed on macrophage single cells, responsible for the first immune response of organisms, when it was submitted to a physical stimulation performed by a micrometric microcapillary. The following release of electroactive ROS and RNS was detected by chronoamperometry (Figure 12.6), and the accurate chemical nature of the compounds was obtained through comparison with in vitro electrochemical oxidation of H$_2$O$_2$, ONOO$^-$, NO, and NO$_2^-$ solutions (Figure 12.6). These data enabled the calculation of time variations of emission flux for each species and the reconstruction of the original flux of production of primary species, O$_2^-$ and NO, by the single macrophage.[81] After establishing that the lesion created by the stimulating microcapillary in the membrane of the single immune cell was able to induce localized membrane depolarization and consecutive oxidative bursts, spatial characteristics of these bursts from large cells like fibroblasts were explored by a triangulation mapping approach (Figure 12.7). This geometrical approach was achieved by comparison of amperometric

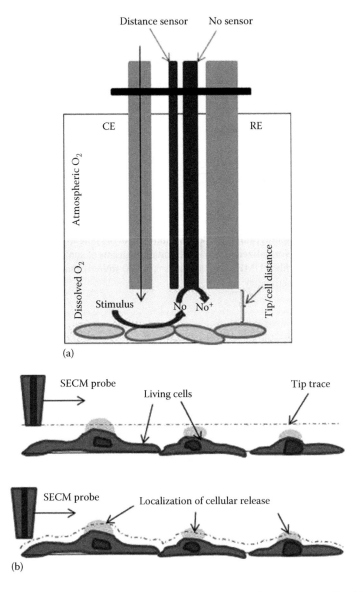

FIGURE 12.5 (a) Scheme of the general assembly used in the work of Borgmann et al., with the integrated distance and NO sensors above the cells. (Adapted from Borgmann, S. et al., *ChemBioChem*, 7, 662, 2006.), (b) with schematic of SECM imaging above living cells in constant-height mode (top) and constant-distance mode (bottom) with the path of the SECM probe in dotted lines and the gray clouds indicating the zones of cellular activity. (Image adapted from Zhao, X. et al., *Anal. Chem.*, 82, 8371, 2010.)

responses detected at four different positions and referenced to the spot where the 1 μm diameter glass microcapillary's end was used to poke a hole in the membrane. Thanks to a simple physico-chemical model, characterization of the ROS and RNS diffusion pattern was obtained like spatial extent of the cell release and extent of propagation of the activation. RNS and ROS release was limited to a 15 μm radius region around the puncture site where the microcapillary was applied.[82] Finally, the oxidative stress responses of another cell type, single MG63 osteosarcoma cells, submitted to the same brief mechanical stress by a microcapillary, have been investigated by Hu et al. by amperometry at platinized carbon fiber electrodes for monitoring and characterizing the nature and the amounts of the various ROS and RNS released. The important *nitric oxide/hydrogen peroxide*

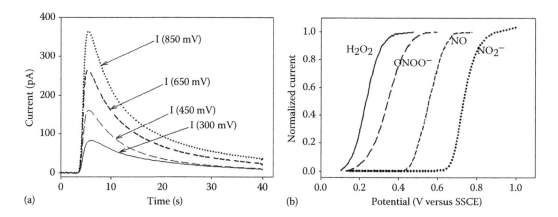

FIGURE 12.6 (a) Typical amperograms recorded as a function of the potential at a single macrophage stimulated by a microcapillary. (b) Normalized steady-state voltammograms of some electroactive species (working electrode = platinized carbon fiber UME, $C = 1$ mmol·L^{-1}, $v = 20$ mV·s^{-1}) that help to select the appropriate potential values in chronoamperometry.

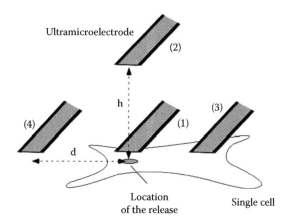

FIGURE 12.7 Schematic view of the triangulation mapping approach reported in Ref. [82]. The ROS/RNS release has been detected at four different positions (1, 2, 3, and 4) from the location of the release (h, d), thus evidencing the propagation of activation in the considered single cell.

and *nitric oxide/superoxide anion* ratios obtained were totally consubstantial with bibliographic data that conjectured that the malignant bone formation ability of osteosarcoma cells is associated with a special small production of O_2^- and a high production of NO.[83]

Another kind of stimulation with biochemical activators was used by Amatore's group to detect oxidative stress from RAW 264.7 macrophages, that is, immunologically activated by interferon-γ and lipopolysaccharide (INFγ- and LPS). Once again, the reactive mixture released by single cells was analyzed, in real time, by amperometry at platinized carbon UMEs. The amounts of nitric oxide, nitrite, and peroxynitrite, which were the main reactive species, were detected. Interestingly, in this work, a direct evidence for the formation of ONOO$^-$ was brought. Incubation with 1400 W, a selective inducible nitric oxide synthase (iNOS) inhibitor, led to no amperometric response from the same activated RAW 264.7 cells. The cocktail of ROS and RNS produced by macrophages is thus likely to originate from NO and superoxide (O_2^-) coproduced by iNOS.[84]

As a methodological extension to that work, Amatore et al. proposed a new electroanalytical method to concomitantly monitor the release of these ROS/RNS released by single

INF-γ/LPS/PMA-stimulated RAW 264.7 macrophages. This method, called triple-potential-step chronoamperometry (at platinized carbon UMEs), provided satisfactory S/N ratio and sensitivity. Basically, meaningful transitory variations in the release of H_2O_2, $ONOO^-$, and NO were evidenced using this methodology, and this would not have been conceivable through constant potential amperometry described earlier or chemical detection through reagents like Griess one. However, the method can be applied only to slow and long-lasting oxidative stress signals (tens of seconds).[85] The simplicity and power of chronoamperometry at platinized carbon fiber UME were also employed at more applied examples in several pharmaceutical investigations unraveling oxidative stress mechanisms. This was performed in artificial synapse configuration above single cell. First, Tapsoba et al. validated the ancient Egyptian customs of applying black eye makeup containing nonnatural lead chlorides (laurionite Pb(OH)Cl and phosgenite $Pb_2Cl_2CO_3$) through the investigation of Pb^{2+}-treated single keratinocytes and showed that the cells then produced high levels of NO. Owing to the bactericidal properties of NO, this confirmed that application of such makeup prevented eye infection, bacterial proliferation, and illness. Submicromolar concentrations of Pb^{2+} ions released by spontaneous dissolution of the lead chloride material were shown perfectly adequate to elicit specific oxidative stress responses of the cells without affecting otherwise the cell status.[86] Different pro- or antioxidant and azido-based drugs were also tested. Ferreira et al. tested a drug used in cancer treatment, β-lapachone, (sought to present only prooxidant properties) on RAW 264.7 macrophages. The antioxidant ability of the molecule was observed after short-time incubation of less than 1 h, leading to a fall in the release of NO, while larger exposition times led to oxidatively induced cell death. These opposite anti- and prooxidant properties as a function of dose and incubation time were rationalized by two conflicting effects: (1) a calcium chelation mediation decreasing ROS and RNS production versus (2) an electron transfer–mediated redox cycling to form radical species.[87] Another example of pharmaceutical investigation appeared in the work of Filipovic et al. and Bernard et al. where mimics of superoxide dismutase (SOD) enzymes, able to convert superoxide radical into hydrogen peroxide, were found, after incubations, to scavenge peroxynitrite and sometimes NO to nitrites.[88,89] This was evidenced by investigations on single macrophages and afforded explanation on the process by which artificial SODs may act as potential candidates for curing inflammation diseases by converting toxic and reactive species into less harmful molecules. At last, the role of the azido moiety in azidothymidine (AZT)-induced oxidative stress was investigated. AZT is a drug used in HIV treatments, so its effect was investigated at single macrophages.[90] Cells preincubated with AZT were found to release significant amounts of ROS and RNS like NO_2^-, $ONOO^-$, H_2O_2, and NO. Surprisingly, these amounts were the highest when macrophages were incubated with azido-containing AZT analogs. Interestingly, these effects decreased when the free azide terminal group was modified. This study performed electrochemically at UMEs implies that the azido moiety plays a crucial role in AZT-induced oxidative stress.

12.2.2　Investigations Involving Microelectrode Arrays or Microdevices

12.2.2.1　Exocytosis

As documented earlier, carbon fiber UMEs are now providing one of the most successful routine analytical techniques for investigating exocytosis at the single-cell level. Nonetheless, beyond the numerous advantages provided by UMEs and the artificial synapse configuration, two main drawbacks remain, for which modifications or adaptations of the electrochemical tool to microdevices have been recently considered, as reported in excellent reviews.[9,91–94]

The first drawback is related to the lack of spatial resolution of the electrochemical technique. Unless small-sized electrodes are considered (see next paragraph), the conventional UMEs (5–10 μm diameter) probe the entire cell surface apex. In other words, a given vesicular release event featured by an amperometric spike is ensured to be located on the investigated cell apex, but its exact location at the cell surface remains unknown. Conversely, optical techniques based on fluorescence microscopy gave access to a good spatial resolution (but still restricted to a few hundred nanometers). Indeed, secretion vesicles can be tagged with fluorescent probes (fluorescent styryl compounds, acridine orange, GFP, quantum dots, etc.) on their surfaces or on their inside, thus allowing direct visualization of individual exocytotic events (vesicle motion, location including docking and fusion) to be visualized.[95,96] Among the different fluorescence microscopies (epifluorescence, confocal microscopy, etc.) that could be considered, total internal reflection fluorescence microscopy (TIRFM) can be viewed as the best analytical compromise due to a low penetration depth (50–300 nm) from the cell surface investigated.[97] Therefore, only the vesicles located near the cell membrane may be observed, the corresponding minimization of the out-of-focus intracellular fluorescence, thus increasing the S/N ratio. Additionally, TIRFM also provides reduced photobleaching and cell damages. Finally, the temporal resolution (10–100 ms) is compatible with real-time measurements of exocytosis though it lacks the excellent kinetic resolution of amperometry. Hence, the two techniques are complementary. This is why microfabricated devices have been elaborated to combine simultaneous electrochemical and fluorescence measurements. In that way, indium tin oxide (ITO) is the most prevalent material for a combined detection at the bottom of a cell owing to its excellent optical (transparency) and electrochemical (conductivity) properties.

As a first example, microwells were implemented onto ITO electrodes (at which one or a few cells are adhered) through lithography by delimiting a 40–100 μm well in a photoresist film covering an underlying ITO electrode.[98–100] While amperometric detection of the exocytosis at the single-cell level or fluorescence observation has been evidenced as feasible,[98] the presence of a few isolated cells into such micrometric wells is not routinely ensured (due to a low probability for a cell to drop at the electrode surface from a cell suspension). This is why more sophisticated ITO devices have been designed with different configurations (four 200 μm width/750 μm length ITO bands whose active surfaces could be delimited by a circular or a spiral well) as a compromise between a limited electrode surface (minimization of capacitive phenomena) and a significant probability of finding observable cells on the electrode.[101,102] In that case, simultaneous electrochemical and TIRFM detections of single exocytotic events were successfully demonstrated using enterochromaffin human BON cells. Please note that the combined fluorescence–electrochemistry measurements are not necessarily precluded when using the well microelectrode configuration. As an example, the fluorescence–electrochemistry coupling has been also reported with SH-SY5Y neuroblastoma cells at platinum micrometric recessed electrodes in wells.[103] In that case, the electrode material (platinum) was not transparent, thus requiring a precise positioning of the cell and making the microelectrode contact with one side only of the cell in order to monitor exocytotic events from the upper lateral vesicles of the cell. Moreover, TIRFM–amperometry coupling has been also achieved at single chromaffin cells with a micrometric well electrode (8 μm diameter) or a micrometric squared zone (~15 μm side) with four electrodes delimited by a photoresist.[104] This latter work examined the question of the most suitable electrode material and considered ITO as well as other conducting/transparent materials such as thin gold or nitrogen-doped diamond-like carbon. Note also that the question of working with carbon nanotubes still remains an interesting issue in this respect.[99,105]

Eventually, another solution for positioning single cells at an ITO micrometric surface is brought by resorting to microchannels, like microchip devices that use one or a few transparent ITO electrodes crossing a microfluidic channel. Such configurations result as 20×20 μm or 20×100 μm cell-sized ITO electrodes, enabling the measurement of exocytosis from individual or population of chromaffin cells with concomitant fluorescent measurements of intracellular Ca^{2+}.[106,107] It has to be emphasized that the analytical strategies for improving the spatial resolution of amperometry at UMEs are not limited to the combination of electrochemistry with another complementary technique. Hence, some works have reported the fabrication of microelectrode arrays (MEAs) whose entire dimension is compared to a single cell surface. Because the size of one UME within the MEA is less than the cell dimension, a mapping of the exocytotic activity is thus expected to be achieved. The first example reproduces the usual detection with the artificial synapse configuration at the cell apex with carbon MEAs embedded. As such, two kinds of MEAs were reported, more specifically seven individually 5 μm diameter carbon fibers inserted into a seven-barrel capillary or a multibarreled silica capillary containing up to 15 tips in which carbon is deposited through pyrolysis of acetylene. Corresponding to a 20–50 μm diameter array, they were shown to perform simultaneous amperometric measurements at different zones of the same single PC12 cell.[108–110] Due to recent advances in microfabrication, MEAs have been also elaborated for which exocytosis is electrochemically investigated at the bottom of the cell. As an example, MEAs based on three or four platinum UMEs implemented on glass can be used to map an isolated chromaffin cell release activity since three or four cell membrane zones of the cell-adhered surface can be simultaneously monitored.[111,112] Finally, similar arrays have been recently reported with four boron-doped nano-crystalline diamond electrodes[113] or four ITO (thin gold and nitrogen-doped diamond-like carbon were also considered)[104] delimited by a photoresist (15 μm diameter well or a 15 μm side square) and used for the detection of exocytosis at a single chromaffin cell.

The second drawback of carbon fiber amperometry using the artificial synapse configuration is directly related to one of its advantages, which is the single-cell level analysis. Indeed, overcoming the cellular variability requires a large and time-consuming number of experiments for extract-ing significant statistical data and rationalizing the results. In that way, numerous microsystems, mainly MEAs, have thus been elaborated to perform a combined electroanalysis of a cell popula-tion. Furthermore, using several micrometric electrodes instead of a unique millimetric one ensures optimizing the cell coverage by the electroactive surface and thus increases the S/N ratio. In a prac-tical view, the electrodes are maintained at an identical potential value. They can be connected to a unique output, thus leading to a single electrochemical measurement of an entire cell population or separately addressed, thus providing multiple parallel recordings of groups of cells within the population. In a noncomprehensive view, numerous MEAs can be mentioned to illustrate the utility of this strategy. First of all, arrays of disk UMEs containing four nanocrystalline diamond wells (20 μm diameter) have been used with chromaffin cells and were prone to simultaneously monitor the release of four cells.[113,114] A MEA compatible with rigorous cell culture conditions has been also reported. Such an array of independent platinum UMEs (6 × 6; 25 μm diameter) is combined to a cell culture chamber/perfusion system. This allowed long time cell cultures to be performed, so that the cell's behavior could be electrochemically investigated along time (as a function of the cell growth, for instance).[115] Moreover, another device with 12 sets of MEAs (24 doped polypyrrole [PPy] film-gold electrodes) has been successfully applied to the population of differentiated PC12 cells.[116] Last, MEAs localized into a microfluidic channel have also been considered, in which 16 nitrogen-doped diamond-like carbon UMEs (prepared by a magnetron sputtering process with nitrogen doping) ensure investigating a few chromaffin cells.[117]

To impose the artificial synapse configuration in MEAs is rather difficult. This would require enforcing single cells to be precisely located at the UME surface instead of randomly distributing them over the MEA from a cell suspension. Though this can be achieved through different ways, notably by using a microchip device containing microfluidic traps able to target individual or small groups of cells (chromaffin, PC12) to different kinds of UMEs (platinum, iridium oxide–platinum, mercaptopropionic acid–gold, or ITO).[118–121] Contrary to the previous mechanical approach, cells can be brought to electrode surfaces by using a surface chemistry method. Then, MEA devices with *cytophilic* UMEs separated by *cytophobic* insulting materials have been recently reported.[122,123] Such MEAs have been incorporating 40 wells (10–20 μm diameter) containing nitrogen-doped dia-mond-like carbon (deposited on ITO and possibly patterned with poly-L-Lysine) UMEs whose cell affinity favors adhesion. Separate or simultaneous electrochemical recordings of the cell releases can be potentially envisioned, while the question remains whether the mechanical stress or the cytophobic environment applied here to the cells may alter their general behavior and, in particular, their exocytotic release properties.

Finally, one can mention that microdevices devoted to the electrochemical investigation of exo-cytosis are not limited to the applications described earlier. As such, microdevices allowing both electrical stimulation and electrochemical detection of exocytosis at the single-cell level have been reported. First of all, a microchannel crossed by eight independent gold working electrodes has been reported. In that case, a single PC12 cell can be positioned onto one of the working electrodes, thus allowing to monitor its exocytotic release evoked through the application of an extracellular voltage pulse or even through the electric field induced by the potentiostat circuit used for the electrochemical recording.[124] Second, a 36 parallel gold UME MEA has been reported where a single chromaffin cell was positioned at a given UME surface, which was used to electroporate the cell and record the resulting electrochemical signal.[125] Another microdevice can be also mentioned in which a screen-printed carbon paste millimetric electrode is initially covered by a thin PDMS film. A UME surface is then delimited through poking a hole with a needle through the PDMS layer and exposing the underlying carbon working electrode surface.[126] Eventually, a microdevice has been fabricated for performing electrochemical cytometry, that is, the separation/lysis/electrochemical detection of individual vesicles. Compared to the experiments usually achieved with carbon fiber UMEs, results obtained with such a device have contributed to the debate about the full/partial fusion.[127]

While nanoelectrode array (NEA) fabrication is now reasonably controlled,[128] no investigation of the electrochemical mapping of a single cell covering an NEA has been reported to date, to the best of our knowledge. Nevertheless, recent works involving MEAs with subcellular micrometric electrodes suggest that this is a growing and important future issue.[129,130]

As a conclusion, the microdevices devoted to the electrochemical detection of exocytosis can be classified into two general categories (Figure 12.8): (1) devices with one or several wells delimiting the UME surface on which one or a few cells may adhere and (2) series of bands, disks, or square UME arrays. It has to be emphasized that both strategies of fabrication can be adapted to the sought analytical purpose, that is, improving the spatial resolution of the electrochemical technique or allowing one to perform a combinatory analysis.

12.2.2.2 Oxidative Stress

As discussed earlier, and despite all the contributions brought by UMEs applications for the detection of ROS and RNS at single-cell level, no evidence of the fine localization of the emitted compounds within the cell can be carried by a micrometric-dimension tip due to the cell's dimensions (ten of micrometers or more). The use of nanoelectrodes can thus start to lift the veil on this special needed characterization and will be developed in Section 12.2.3. But first, another aspect of nanoelectrochemistry on single cells concerns the fabrication and use of UME arrays for oxidative stress detection and is developed in the following. Future directions and perspectives in the simultaneous detection of various oxidative stress species like nitric oxide, hydrogen peroxide, peroxynitrite, and nitrite imply fabrication of MEAs with numerous individual electrodes separately designed and functionalized with the appropriate chemistry to allow multianalyte detection. Such versatile and multitarget platforms start to offer applications on investigating the physiology of individual cells in wells, cultured cells in uniform layer, or cell tissues like slices and biological tissues.

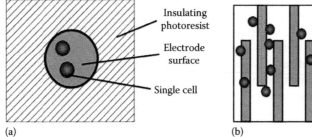
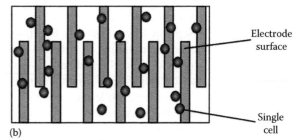

(a)   (b)

FIGURE 12.8 Schematic representation of the two main strategies of microdevices for the electrochemical monitoring of exocytosis. (a) A micrometric well delimiting the electrode surface (ITO, Pt, etc.; see text) in which one or a few cells are adhered. (b) MEAs at which a few cells or a cell population can be cultured.

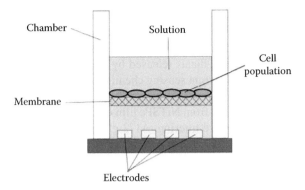

FIGURE 12.9   Scheme of the array used in Refs. [132,133].

Information collected concerns, for example, cartography of responses to local and biochemical or mechanical stimuli by gathering dynamic data about desired specific analytes and requires simultaneous, selective, and sensitive detection.[74,131]

In the last decade, one of the first examples of nitric oxide detection by means of a MEA was brought by the two groups of Schuhmann and Bedioui in 2004. In their work, they described an NO-MEA to detect nitric oxide released from a population of HUVEC cells (Figure 12.9). The NO sensor array was composed of three individually addressable electrodes, each of them being modified by different NO-sensitive catalysts: an NiTSPc tetrasodium salt–based sensor, a pyrrole-functionalized Mn trimethoxyphenylcarboxyphenylporphyrin–modified electrode, and a nickel tetra(4-*N*-methylpyridyl) porphyrin free base/electrodeposition paint mixture–based electrode. This procedure shows the possibility of modifying individual electrodes of an array with different chemistries and paves the way for multiple-analyte detection from cell populations by appropriately designing composition, sensitivity, and selectivity of each electrode in the array for each detectable molecule.[132] Castillo et al. proposed a simultaneous detection of NO and glutamate from adherently growing C6 glioma cells using a MEA coated with a thick layer of biocompatible gel after bradykinin or potassium stimulation.[133] In this configuration, NO was sensed with a positively charged Ni porphyrin entrapped into a negatively charged electrodeposition paint gold electrode followed by the modification of a second working gold UME with a horseradish peroxidase (HRP)/glutamate oxidase hydrogel. The current responses recorded at each electrode validated the ability of the UME within the MEA to detect individually its analyte target without interferences from other UME or from interferences in the solution. We can also cite, in a similar manner, Pekarova et al. who used a porphyrinic NO sensor to monitor nitric oxide release induced by LPS in RAW 264.7 macrophages.[134] Kamei et al. similarly constructed a cell-based assessing system for unraveling nitric oxide pathways in immune systems using an electrochemical NO sensor, based on a gold electrode coated with a polyion complex layer. The immune cell model chosen was the RAW 264.7 cell line. Detection limit achieved was 8.4 nmol·L^{-1} of NO when the NO release was triggered by LPS and INF-γ-stimulated cells cultured on a polyion-coated gold MEA.[135] Pereira-Rodrigues et al. proposed optically and amperometrically simultaneous NO detection produced by glioblastoma multiform cell line (A172). Intracellular production of NO was followed via fluorescence image analysis with a probe named 4,5-diaminofluorescein, while extracellular NO release was monitored via a carbon working electrode in an array, made of electrodeposited NiTSPc and Nafion.[136] In 2008, Patel et al. fabricated six individual addressable gold working electrode microarray to assess responses of oxygen and nitric oxide from fibroblasts.[137] They used a polyion complex film coating to allow convenient cell culture and to prevent biofouling. This surface modification presents a good stability and does not modify significantly electron transfer rates. However, these advantages were counterbalanced by a slight decrease of 20% in the electrochemical collections of oxygen and nitric oxide.

Another amperometric array sensor platform was assembled by Chang et al. from a 24-well culture plate in which different working electrodes were embedded: one by depositing sputtered gold and the other by using screen-printed carbon. This was aimed at simultaneous in vitro detection of superoxide and nitric oxide radicals released by stimulated cells. The selectivity between these two species was offered by different sensing chemistries and electrode surface modifications: the superoxide sensor used covalent immobilization of cytochrome c via a binder onto gold electrodes, while the NO sensor was built from NiTSPc film electrodeposited onto the carbon electrode with a layer of Nafion for interference filtering. A172 glioblastoma cells seeded in the MEA and stimulated by biochemical means (phorbol myristate acetate [PMA]) showed a release of the two aforementioned radicals simultaneously monitored by the MEA. Selective inhibition controls or scavenging of both analytes reported demonstrating that different electrodes were operated without interfering.[138]

The release of four ROS and RNS species by macrophages has been studied by electrochemistry within a microfluidic device by Amatore et al. Cells were seeded, cultured, and then stimulated by a calcium ionophore in the microchip containing the three-electrode system, allowing amperometric detection of the NO, NO_2^-, H_2O_2, and $ONOO^-$ cocktail at the platinized working microelectrode. Kinetics of the responses and reproducibility of the measurement system were offered despite the fact that no separation between the different contributions from each molecule was proposed at that point.[139] In a similar way, Cha et al. described amperometric detection of solely NO in a microfluidic device with a gold/ITO-patterned electrode onto a porous polymer membrane. NO generation from macrophage population cultured within the microdevice was evidenced following LPS stimulation of the cells.[140]

In 2011, Quinton et al. proposed an on-chip device for the simultaneous detection by amperometry of peroxynitrite and nitric oxide in a MEA including individually addressable batch of gold UMEs of 50 µm diameter for the working electrodes. The UMEs were divided into two sets, each having its own reference and counter electrodes; in total, 110 UMEs were specifically dedicated to a specific analyte. Each UME in the network devoted to NO oxidation detection was specifically modified by electrodepositing poly(eugenol) and poly(phenol), while the set for $ONOO^-$ detection was used in reduction without subsequent modifications. Simultaneous detection of NO and $ONOO^-$ in physiological buffer chemical solution was achieved to validate the analytical concept, but further applications are still needed on stimulated macrophages.[131,141] Another work confirmed the analytical interest of using microfabricated platinum-black coated platinum electrodes optimized to achieve optimal detection performances of oxidative stress by-products like NO_2^- and H_2O_2 in microchannels. Oxidation mechanisms of hydrogen peroxide and nitrite at these microelectrodes were investigated, and electrode responses were compared to theoretical predictions based on convective mass transport at microchannel electrodes. In both cases, the active surface area of Pt/Pt-black electrodes allowed to avoid inhibition effect leading to long-term stability in contrast to bare Pt electrodes. Such highly sensitive Pt/Pt-black electrodes allowed measurements over almost five decades of concentration and detection limits down to 10 nmol·L^{-1}, a range suitable for the detection of ROS and RNS released by a few cells.[142] A recent paper of Yan et al. described a micropatterned sensing/cell culture surface integrated into a PDMS-based microfluidic device containing flow-through reference and counter electrodes. Working electrodes were made of layers of HRP/hydrogel/gold with the result that macrophages seeded in the microchip adhered on the glass regions around the HRP-coated electrodes. This microsystem allowed real-time monitoring of extracellular hydrogen peroxide released by macrophages activated with PMA and could be used in future studies as a biosensor platform for elucidating inflammatory responses in micropatterned cell cultures.[143] Recently, a biocompatible array of NO gold sensors coated with J774 mouse macrophages have been proposed by Trouillon et al. to detect a recombinant protective antigen (PA) from *Bacillus anthracis* (anthrax) through its activation of iNOS. Extracellular concentration of NO and nitrites increased after exposure of macrophages to this PA, which probably binds to cell surface receptors.[144]

Despite their creativity and numerous progresses, all these studies about detection of oxidative stress species in MEA envisioned only microarrays, and none involved nanoarrays at that point. In the future, NEAs will probably be proposed more and more often. Multiarrays of nanoelectrodes should gather the following advantages: compatible with applications where good spatial resolution is required (measurements on the level of the single cell) and when simultaneous recording is profitable.

12.2.3 APPLICATIONS BASED ON NANOELECTRODES

From an analytical point of view, decreasing the electrode size to a submicrometer scale should improve its ability to perform electrochemical measurements, particularly in the case of nanometer-sized electrodes, that is, nanoelectrodes. So, for example, considering the UME's behavior described in Section 12.1, the S/N ratio ($\propto 1/r_0$) is expected to be increased as well as the noise level, and the time constant ($\tau \propto r_0$) should be decreased. Additionally, using a nanoelectrode obviously brings the spatial resolution to the nanometric scale. However, beyond key points about nanoelectrodes (detection of low magnitude currents since the faradic current $\propto r_0$, validity of UME equations for dimensions <10 nm),[5] some of the expected advantages outlined previously are not necessarily provided. Indeed, stray capacitances have also to be considered at the nanometric scale. As such, the time constant could be larger than expected for small radius values, while the corresponding S/N ratio may be less than expected.[145]

12.2.3.1 Exocytosis

In comparison to studies involving basic UMEs or MEAs, investigations of exocytosis with nanoelectrodes remain rather scarce.[146] This is probably because a single nanoelectrode gives no supplementary benefits by comparison with conventional UMEs for general detection purposes related to cell exocytosis. For instance, amperometric detection of exocytosis was reported at PC12 cells using 100 nm diameter carbon fiber electrodes. Because the nanoelectrode size was much smaller than that of the cell (10 μm diameter in average) and compared here to a typical vesicle diameter range (200–300 nm), the detection was performed at the single-cell level and also at the single-vesicle level. However, at a given moment, the nanoelectrode may probe only a nanometric area of the cell surface, while exocytosis occurs randomly at different locations. While it indirectly confirms the existence of active/inactive release zones at the cell surface, exocytosis is therefore rarely detected since no amperometric events is recorded in 70% of the recordings (in the remaining 30%, only a few spikes per cell can be detected).[147] Other amperometric investigations at PC12 cells have been achieved with carbon fiber disk nanoelectrodes (500 ± 100 nm diameter). According to the work described earlier, exocytosis was not routinely detected (only 60% of the cells were investigated).[148] More particularly, the spike features (charge, maximum intensity, and time lengths) were slightly different from those obtained with conventional 7 μm diameter UMEs. Exact reasons for these differences are still not clear. Because narrower average spikes were detected with nanoelectrodes, it can be argued that the diffusion broadening effect possibly distorting some spikes with usual UMEs is avoided with nanoelectrodes,[148] but such an explanation cannot be invoked when UMEs are used in the artificial synapse configuration. Furthermore, it cannot be excluded that the scarcity of events recorded per cell reduces the statistical significance of the data. Eventually, because the electrode dimensions compared to those of a vesicle and the nanoelectrode are randomly positioned, the total collection of a single vesicular release cannot be warranted and could lead to only partial detection of the vesicular content. Finally, it has to be emphasized that the demonstration of active/inactive release zones does not require nanoelectrodes since this was already achieved using 1 μm diameter UMEs.[149–152]

Nonetheless, reducing the electrode size should offer possible advantages related to the data treatment. Indeed, electrodes whose dimensions are less than that of the cell will not collect all the cell release, thus decreasing the probability of overlapping spikes, which are usually excluded from the data treatment.[153,154] Investigations of exocytosis with nanoelectrodes are consistent with this statement since no overlapped spikes were observed.[147] However, it has been demonstrated at chromaffin cells by using different electrode areas (from 90 to 25 µm²) that the censorship induced by spike's superimposition did not alter the information (PSF, spike's magnitude or charge, etc.) extracted from the analytical treatment.[154] Hence, with the exception of possible future applications for cell surface mapping, using a single nanoelectrode for amperometric monitoring exocytosis at isolated cells in the artificial synapse configuration does not seem to offer a clear benefit. However, nanoelectrodes could represent an electrochemical tool able to investigate real synapses (nanometer sized) without destructuring them, that is, through allowing their direct insertion between the two nerve terminals.

Conversely, due to their small size compared to micrometric cell dimensions, nanoelectrodes are particularly adapted to be inserted into a single-cell organism without any damages. As a first example, ZnO nanorods (80 nm diameter; 700 nm length) grown on the tip of a borosilicate glass capillary (0.7 µm diameter) have been reported and used as a pH sensor for monitoring biological process within single cells.[155] More precisely, intracellular pH measurements were performed in a single human fat cell using such nanoelectrodes inserted into the investigated cell.

12.2.3.2 Oxidative Stress

Examples of nanoelectrodes used on single cell for electrochemical detection of NO or derivatives are really scarce in the literature probably due to technical limitations and difficulties to produce a stable and robust nanosensor with all the specifications needed for NO detection (selectivity, sensitivity, etc.). First, Malinski and coauthors have evidenced that peroxynitrite was a major mediator of cardiac injuries occurring during ischemia/reperfusion process by simultaneously monitoring O_2^-, NO, and $ONOO^-$ over a layer of endothelial cells with a final 3–4 µm diameter modified carbon microelectrode combining three *nanosensors*. Each electrode in the module of $NO/O_2^-/ONOO^-$ sensors was poised at different potentials relevant for each species and positioned with the help of a computer-controlled micromanipulator at 5 ± 2 µm above a single endothelial cell. Selectivity was offered by different surface modifications: a conductive film of polymeric nickel(II) tetrakis(3-methoxy-4-hydroxyphenyl) porphyrinic covered with Nafion for NO, a polymeric film of Mn(III)-[2,2]paracyclophenylporphyrin for $ONOO^-$, and an immobilized PPy/HRP and SOD enzyme for the O_2^- sensor. Bursts of oxidative stress species were monitored after stimulation by a calcium ionophore A23187 and allow evaluating the ratio between nitric oxide and peroxynitrite concentrations detected in different conditions. This work demonstrated that this ratio dramatically decreased in endothelium originating from hypertensive rats and following reperfusion of vessels. Such electrochemical technique is powerful to study local dysfunction of endothelial isolated cells in vivo conditions. However, beyond its elegancy, this work can hardly be categorized in the nanoelectrode field, despite the fact that the authors claimed that the three sensors are nanosized, because no information was given on the individual size of the three electrodes (apart from the fact that they are made of carbon fiber with a length of 4 µm and a tip diameter of 200–250 µm, which is far to nanometric size).[156]

Second, Heller et al. probe macrophage activity with carbon nanotube sensors as versatile electrode of nanometer dimension (Figure 12.10).[157] This material has the properties of electrostatic field-effect transistor (FET) and nanoscale electrochemical sensors. The principle consisted in using single-walled carbon nanotubes (SWNTs) coated with antibodies (murine IgG) for macrophage recognition, phagocytosis, and then subsequent stimulation and digestion. The sensors were used in an electrolyte-gated transistor configuration and were supposed to follow phagocytosis phenomenon in real time by simultaneous monitoring of changes both in FET signal and in electrochemical current (EC) signal. FET and EC signals were observed after cell adhesion on the nanosensor, but quantification seemed hazardous (signals unstable or suppressed). Electrochemical signals were enhanced when the SWNTs were coated with platinum nanoparticles before the antibody coating, a fact consistent with the advantages of platinized carbon fiber UMEs for ROS and RNS detection

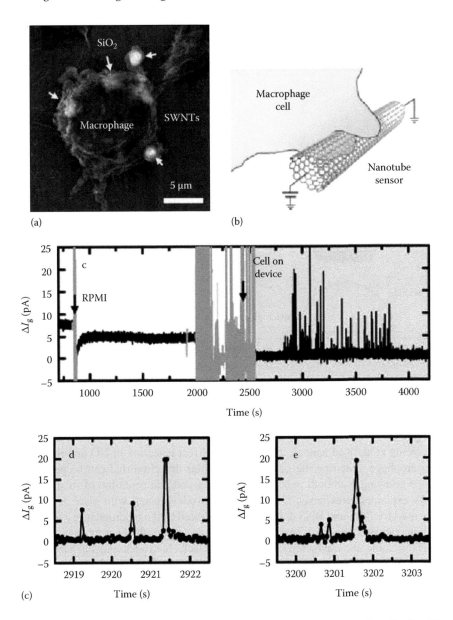

FIGURE 12.10 (a) SEM image of a THP-1 cell deposited on SWNTs coated with IgG antibodies. The arrows indicate locations of polystyrene beads ingested by the cell as a test of phagocytic activity. (b) Schematic view of the interactions between a macrophage and SWNTs. (c) Electrochemical measurement with an SWNT transistor at −700 mV versus Ag/AgCl in the presence of a THP-1 macrophage. (Reproduced from Heller, I. et al., *Small*, 5, 2528, 2009. With permission.)

discussed earlier. A potential of 700 mV versus Ag/AgCl, known to allow electrochemical detection of H_2O_2, NO, and $ONOO^-$, was applied, and then a series of spikes were observed after few hundreds of seconds with a duration of tens of ms and height between 5 and 20 pA, representing 10 attomoles of elementary charge. Hypothesis on the nature of these sharp bursts like vesicular release of ROS and RNS was proposed. Because macrophages can digest the SWNTs, authors claimed that their approach will allow monitoring ROS and RNS release during phagocytosis of pathogens *from within the cell* with minimal perturbation and maximal spatial resolution, thanks to SWNT dimensions. To achieve this goal, a third work was proposed in 2012.

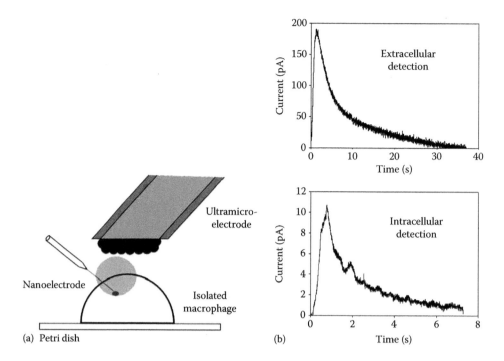

FIGURE 12.11   (a) Scheme of the nanoelectrode insertion into a single macrophage. Extracellular detection can be performed with a carbon platinized UME, or intracellular detection can be achieved with the nanoelectrode. (b) Comparison between the two measurements.

Finally, Wang et al. used nanoelectrodes for the direct detection of NO and related compounds inside a macrophage, reaching the goal of intracellular detection that can be provided only with a nanosensor allowing a smooth penetration within a cell and resealing of its membrane (Figure 12.11).[158] As explained previously, UME-sized electrodes coated with black platinum had been extensively used for ROS and RNS detection on single cells (like macrophages and other cell types) and other biological systems. To transpose this methodology inside macrophages, authors needed to extend the fabrication and characterization of Pt/Pt-black electrodes. Indeed, nanoelectrodes made of Pt wire in a glass capillary and then pulled to reach an end diameter of few hundreds of nanometers were nanofabricated by etching electrodepositing Pt black inside the etched nanocavity (tens of nanometers diameter). Pt-black deposition was followed under the control of AFM images. In vitro stable and reproducible ROS and RNS responses were obtained on these nanoprobes despite their nanometric radius. Here, the platinized nanoelectrode was inserted inside the macrophage with the purpose of eliciting its response, taking advantage of the previous observation that the penetration of the cell membrane by a submicrometer tip activates it and induces fast oxidative burst release.[81] Short-time and weak leakages of ROS and RNS were detected inside the cytoplasm by the nanoelectrodes (few picoamperes height and few seconds time length), while larger oxidative burst signals were monitored outside of the cell by a classical UME in artificial synapse configuration (few hundreds of pA height) persisting on the timescale of tens of seconds. Beyond this first challenging intracellular oxidative stress detection, an interesting aspect of this work is the fact that the sensor plays a dual role of stimulation and detection in the same time by achieving the mechanical depolarization of the cell membrane itself when puncturing the macrophage surface. Another interesting aspect is that these results established that though macrophages produce NO and peroxynitrite, two species prone to cross over lipidic membranes, in their phagolysosomes, none of them leak appreciably inside the macrophage during phagocytosis.

12.3 CONCLUSION

The present review, though certainly not comprehensive, was aimed to provide the reader with an educated feeling of the wide development of miniaturized methods for the investigation of two important biological issues, vesicular release of neurotransmitters and oxidative stress.

The first topic was recently publicized by the awarding of the Nobel Prize 2013 in Medicine to J. Rothman, R. Schekman, and T. Südhof for their detailed studies of the formation and transport of vesicles in cells and their crucial role in signaling and communication in organisms. Yet these seminal works left out the final outcome of the chain, namely, the final acts that condition the ultimate release of neurotransmitters. Frankly speaking, this is not a wonder since this final stage is generally rapid and extremely localized and involves the release of minute amounts of molecules. Electrochemical methods, first with UMEs and now with nanoelectrodes and microsystems, proved to be perfectly adequate to meet these challenges. This review documented the advances thus obtained by analytical electrochemists over the recent years, while former ones (J. Rothman, R. Schekman, T. Südhof) presented the initial steps of this adventure into a crucial area of life. The works documented earlier demonstrate that the adventure is far from being finished. New methods and new systems blossom continuously owing to the increasing power of microfabrications.

The second topic documented in this review is comparatively a younger field, which is not a tantamount to a lesser importance for aerobic organisms. Aerobic cells find their energy in the catalyzed oxidation of organic substrates, a process releasing electrons that are ultimately eliminated through their capture by dioxygen. However, this not always ends into water but into oxygen species whose reactivity leads to a large variety of organic radicals and alteration of organic essential biomolecules. When not controlled, such processes initiate many deleterious biological cascades, giving rise to chronic and deadly illnesses. On the other hand, modes of such ROS and RNS are necessary to the homeostasis of tissues and organisms as well as during fetal development. This brief summary of bad and good aspects of oxidative stress easily explains why this domain was considered fuzzy, confusing, and not a fashionable research area. Changes in the scientific appreciation of this field occurred with the development of molecular biology as applied to cancerogenesis and autoimmune diseases. This led to many important views in medicine supported by the characterization of crucial biochemical cascades. Despite their importance for medical and health issue, it remains that the initial molecular steps of oxidative stress were essentially speculative and mostly focused on the production of superoxide and its follow-up product, hydrogen peroxide. Electrochemical approaches first at UMEs, then at microsystems and nanoelectrodes, have helped enriching the *zoo* of initial ROS and RNS by detecting them and quantifying their production by single cells. In our view, one of the most breakthrough outcomes of these electroanalytical methodologies has been to install peroxynitrite anion as a crucial player in oxidative stress–supported central biological mechanisms as well as in the development of many health-related issues.

We hope that the great successes achieved with miniaturized electroanalytical systems in two important fields of biology with important consequences for a better understanding of their crucial mechanisms will encourage readers to examine other issues related to small molecular effector generation, transport, and traffic in the living world.

REFERENCES

1. Schroeder, T. J., Jankowski, J. A., Kawagoe, K. T. et al. 1992. Analysis of diffusional broadening of vesicular packets of catecholamines released from biological cells during exocytosis. *Anal. Chem.* 64:3077–3083.
2. Amatore, C., 1995. Electrochemistry at ultramicroelectrodes. In *Physical Electrochemistry: Principles, Methods and Applications*, ed. I. Rubinstein, pp. 131–208. M. Dekker, New York.
3. Bard, A. J., Faulkner, L. R.: *Electrochemical Methods: Fundamentals and Applications*. John Wiley & Sons, New York, 2001.

4. Amatore, C., Pebay, C., Thouin, L., Wang, A. F., Warkocz, J. S. 2010. Difference between ultramicroelectrodes and microelectrodes: Influence of natural convection. *Anal. Chem.* 82:6933–6939.

5. Murray, R. W. 2008. Nanoelectrochemistry: Metal nanoparticles, nanoelectrodes, and nanopores. *Chem. Rev.* 108:2688–2720.

6. Amatore, C., Arbault, S., Guille, M., Lemaître, F. 2008. Electrochemical monitoring of single cell secretion: Vesicular exocytosis and oxidative stress. *Chem. Rev.* 108:2585–2621.

7. Wang, W., Zhang, S.-H., Li, L.-M. et al. 2009. Monitoring of vesicular exocytosis from single cells using micrometer and nanometer-sized electrochemical sensors. *Anal. Bioanal. Chem.* 394:17–32.

8. Kim, D., Koseoglu, S., Manning, B. M., Meyer, A. F., Haynes, C. L. 2011. Electroanalytical eavesdropping on single cell communication. *Anal. Chem.* 83:7242–7249.

9. Lin, Y. Q., Trouillon, R., Safina, G., Ewing, A. G. 2011. Chemical analysis of single cells. *Anal. Chem.* 83:4369–4392.

10. Adams, K. L., Puchades, M., Ewing, A. G., 2008. In vitro electrochemistry of biological systems. *Annu. Rev. Anal. Chem, Ed.* 329–355.

11. Burgoyne, R. D., Morgan, A. 2003. Secretory granule exocytosis. *Physiol. Rev.* 83:581–632.

12. Kennedy, R. T., Huang, L., Atkinson, M. A., Dush, P. 1993. Amperometric monitoring of chemical secretions from individual pancreatic beta-cells. *Anal. Chem.* 65:1882–1887.

13. Huang, L., Shen, H., Atkinson, M. A., Kennedy, R. T. 1995. Detection of exocytosis at individual pancreatic beta-cells by amperometry at a chemically-modified microelectrode. *Proc. Natl. Acad. Sci. U.S.A.* 92:9608–9612.

14. Yang, S. Y., Kim, B. N., Zakhidov, A. A. et al. 2011. Detection of transmitter release from single living cells using conducting polymer microelectrodes. *Adv. Mater.* 23:H184–H188.

15. Leszczyszyn, D. J., Jankowski, J. A., Viveros, O. H. et al. 1991. Secretion of catecholamines from individual adrenal-medullary chromaffin cells. *J. Neurochem.* 56:1855–1863.

16. Chow, R. H., Vonruden, L., Neher, E. 1992. Delay in vesicle fusion revealed by electrochemical monitoring of single secretory events in adrenal chromaffin cells. *Nature* 356:60–63.

17. DeToledo, G. A., Fernandezchacon, R., Fernandez, J. M. 1993. Release of secretory products during transient vesicle fusion. *Nature* 363:554–558.

18. Kita, J. M., Wightman, R. M. 2008. Microelectrodes for studying neurobiology. *Curr. Opin. Chem. Biol.* 12:491–496.

19. Robinson, D. L., Hermans, A., Seipel, A. T., Wightman, R. M. 2008. Monitoring rapid chemical communication in the brain. *Chem. Rev.* 108:2554–2584.

20. Trouillon, R., Svensson, M. I., Berglund, E. C., Cans, A. S., Ewing, A. G. 2012. Highlights of selected recent electrochemical measurements in living systems. *Electrochim. Acta* 84:84–95.

21. Jackson, M. B., Chapman, E. R. 2008. The fusion pores of Ca^{2+}-triggered exocytosis. *Nat. Struct. Mol. Biol.* 15:684–689.

22. Fang, Q. H., Berberian, K., Gong, L. W. et al. 2008. The role of the C terminus of the SNARE protein SNAP-25 in fusion pore opening and a model for fusion pore mechanics. *Proc. Natl. Acad. Sci. U.S.A.* 105:15388–15392.

23. Ngatchou, A. N., Kisler, K., Fang, Q. H. et al. 2010. Role of the synaptobrevin C terminus in fusion pore formation. *Proc. Natl. Acad. Sci. U.S.A.* 107:18463–18468.

24. Zhang, Z., Hui, E. F., Chapman, E. R., Jackson, M. B. 2010. Regulation of exocytosis and fusion pores by synaptotagmin-effector interactions. *Mol. Biol. Cell* 21:2821–2831.

25. Criado, M., Gil, A., Viniegra, S., Gutierrez, L. M. 1999. A single amino acid near the C terminus of the synaptosome-associated protein of 25 kDa (SNAP-25) is essential for exocytosis in chromaffin cells. *Proc. Natl. Acad. Sci. U.S.A.* 96:7256–7261.

26. Fisher, R. J., Burgoyne, R. D. 1999. The effect of transfection with botulinum neurotoxin C1 light chain on exocytosis measured in cell populations and by single-cell amperometry in PC12 cells. *Pflugers Arch.* 437:754–762.

27. Gil, A., Viniegra, S., Gutierrez, L. M. 1998. Dual effects of botulinum neurotoxin A on the secretory stages of chromaffin cells. *Eur. J. Neurosci.* 10:3369–3378.

28. Graham, M. E., Fisher, R. J., Burgoyne, R. D. 2000. Measurement of exocytosis by amperometry in adrenal chromaffin cells: Effects of clostridial neurotoxins and activation of protein kinase C on fusion pore kinetics. *Biochimie* 82:469–479.

29. Quetglas, S., Iborra, C., Sasakawa, N. et al. 2002. Calmodulin and lipid binding to synaptobrevin regulates calcium-dependent exocytosis. *EMBO J.* 21:3970–3679.

30. Bretou, M., Anne, C., Darchen, F. 2008. A fast mode of membrane fusion dependent on tight SNARE zippering. *J. Neurosci.* 28:8470–8476.

31. Burgoyne, R. D., Fisher, R. J., Graham, M. E., Haynes, L. P., Morgan, A. 2001. Control of membrane fusion dynamics during regulated exocytosis. *Biochem. Soc. Trans.* 29:467–472.

32. Fisher, R. J., Pevsner, J., Burgoyne, R. D. 2001. Control of fusion pore dynamics during exocytosis by Munc18. *Science* 291:875–878.

33. Graham, M. E., Barclay, J. W., Burgoyne, R. D. 2004. Syntaxin/Munc18 interactions in the late events during vesicle fusion and release in exocytosis. *J. Biol. Chem.* 279:32751–32760.

34. Graham, M. E., Burgoyne, R. D. 2000. Comparison of cysteine string protein (Csp) and mutant alpha-SNAP overexpression reveals a role for Csp in late steps of membrane fusion in dense-core granule exocytosis in adrenal chromaffin cells. *J. Neurosci.* 20:1281–1289.

35. Schutz, D., Zilly, F., Lang, T., Jahn, R., Bruns, D. 2005. A dual function for Munc-18 in exocytosis of PC12 cells. *Eur. J. Neurosci.* 21:2419–2432.

36. Fulop, T., Doreian, B., Smith, C. 2008. Dynamin I plays dual roles in the activity-dependent shift in exocytic mode in mouse adrenal chromaffin cells. *Arch. Biochem. Biophys.* 477:146–154.

37. Anantharam, A., Bittner, M. A., Aikman, R. L. et al. 2011. A new role for the dynamin GTPase in the regulation of fusion pore expansion. *Mol. Biol. Cell* 22:1907–1918.

38. Trouillon, R., Ewing, A. G. 2013. Amperometric measurements at cells support a role for dynamin in the dilation of the fusion pore during exocytosis. *ChemPhysChem* 14:2295–2301.

39. Jankowski, J. A., Schroeder, T. J., Ciolkowski, E. L., Wightman, R. M. 1993. Temporal characteristics of quantal secretion of catecholamines from adrenal-medullary cells. *J. Biol. Chem.* 268:14694–14700.

40. Aspinwall, C. A., Brooks, S. A., Kennedy, R. T., Lakey, J. R. T. 1997. Effects of intravesicular H^+ and extracellular H^+ and Zn^{2+} on insulin secretion in pancreatic beta cells. *J. Biol. Chem.* 272: 31308–31314.

41. Jankowski, J. A., Finnegan, J. M., Wightman, R. M. 1994. Extracellular ionic composition alters kinetics of vesicular release of catecholamines and quantal size during exocytosis at adrenal-medullary cells. *J. Neurochem.* 63:1739–1747.

42. Kennedy, R. T., Lan, H. A., Aspinwall, C. A. 1996. Extracellular pH is required for rapid release of insulin from Zn-insulin precipitates in beta-cell secretory vesicles during exocytosis. *J. Am. Chem. Soc.* 118:1795–1796.

43. Camacho, M., Machado, J. D., Montesinos, M. S., Criado, M., Borges, R. 2006. Intragranular pH rapidly modulates exocytosis in adrenal chromaffin cells. *J. Neurochem.* 96:324–334.

44. Amatore, C., Arbault, S., Bonifas, I. et al. 2003. Dynamics of full fusion during vesicular exocytotic events: Release of adrenaline by chromaffin cells *ChemPhysChem* 4:147–154.

45. Amatore, C., Arbault, S., Bonifas, I., Lemaître, F., Verchier, Y. 2007. Vesicular exocytosis under hypotonic conditions evidences two distinct populations of dense core vesicles in bovine chromaffin cells. *ChemPhysChem* 8:578–585.

46. Schroeder, T. J., Borges, R., Finnegan, J. M. et al. 1996. Temporally resolved, independent stages of individual exocytotic secretion events. *Biophys. J.* 70:1061–1068.

47. Borges, R., Travis, E. R., Hochstetler, S. E., Wightman, R. M. 1997. Effects of external osmotic pressure on vesicular secretion from bovine adrenal medullary cells. *J. Biol. Chem.* 272:8325–8331.

48. Troyer, K. P., Wightman, R. M. 2002. Temporal separation of vesicle release from vesicle fusion during exocytosis. *J. Biol. Chem.* 277:29101–29107.

49. Borges, R., Diaz-Vera, J., Dominguez, N., Arnau, M. R., Machado, J. D. 2010. Chromogranins as regulators of exocytosis. *J. Neurochem.* 114:335–343.

50. Borges, R., Pereda, D., Beltran, B. et al. 2010. Intravesicular factors controlling exocytosis in chromaffin cells. *Cell. Mol. Neurobiol.* 30:1359–1364.

51. Diaz-Vera, J., Morales, Y. G., Hernandez-Fernaud, J. R. et al. 2010. Chromogranin B gene ablation reduces the catecholamine cargo and decelerates exocytosis in chromaffin secretory vesicles. *J. Neurosci.* 30:950–957.

52. Dominguez, N., Estevez-Herrera, J., Pardo, M. R. et al. 2012. The functional role of chromogranins in exocytosis. *J. Mol. Neurosci.* 48:317–322.

53. Machado, J. D., Diaz-Vera, J., Dominguez, N. et al. 2010. Chromogranins A and B as regulators of vesicle cargo and exocytosis. *Cell. Mol. Neurobiol.* 30:1181–1187.

54. Montesinos, M. S., Machado, D., Camacho, M. et al. 2008. The crucial role of chromogranins in storage and exocytosis revealed using chromaffin cells from chromogranin a null mouse. *J. Neurosci.* 28:3350–3358.

55. Amatore, C., Arbault, S., Guille, M., Lemaître, F., Verchier, Y. 2006. Regulation of exocytosis in chromaffin cells by trans-insertion of lyso-phosphatidylcholine and arachidonic acid into the outer leaflet of the cell membrane. *ChemBioChem* 7:1998–2003.

56. Uchiyama, Y., Maxson, M. M., Sawada, T., Nakano, A., Ewing, A. G. 2007. Phospholipid mediated plasticity in exocytosis observed in PC12 cells. *Brain Res.* 1151:46–54.

57. Amatore, C., Arbault, S., Bonifas, I. et al. 2007. Relationship between amperometric pre-spike feet and secretion granule composition in Chromaffin cells: An overview. *Biophys. Chem.* 129:181–189.

58. Sombers, L. A., Hanchar, H. J., Colliver, T. L. et al. 2004. The effects of vesicular volume on secretion through the fusion pore in exocytotic release from PC12 cells. *J. Neurosci.* 24:303–309.

59. Amatore, C., Arbault, S., Bonifas, I. et al. 2005. Correlation between vesicle quantal size and fusion pore release in chromaffin cell exocytosis. *Biophys. J.* 88:4411–4420.

60. Sombers, L. A., Maxson, M. M., Ewing, A. G. 2005. Loaded dopamine is preferentially stored in the halo portion of PC12 cell dense core vesicles. *J. Neurochem.* 93:1122–1131.

61. Wightman, R. M., Jankowski, J. A., Kennedy, R. T. et al. 1991. Temporally resolved catecholamine spikes correspond to single vesicle release from individual chromaffin cells. *Proc. Natl. Acad. Sci. U.S.A.* 88:10754–10758.

62. Grabner, C. P., Price, S. D., Lysakowski, A., Fox, A. P. 2005. Mouse chromaffin cells have two populations of dense core vesicles. *J. Neurophysiol.* 94:2093–2104.

63. van Kempen, G. T. H., vanderLeest, H. T., van den Berg, R. J., Eilers, P., Westerink, R. H. S. 2011. Three distinct modes of exocytosis revealed by amperometry in neuroendocrine cells. *Biophys. J.* 100:968–977.

64. Amatore, C., Oleinick, A. I., Svir, I. 2010. Diffusion from within a spherical body with partially blocked surface: Diffusion through a constant surface area. *ChemPhysChem* 11:149–158.

65. Amatore, C., Oleinick, A. I., Svir, I. 2010. Reconstruction of aperture functions during full fusion in vesicular exocytosis of neurotransmitters. *ChemPhysChem* 11:159–174.

66. Oleinick, A., Lemaître, F., Guille-Collignon, M., Svir, I., Amatore, C. 2013. Vesicular release of neurotransmitters: Converting amperometric measurements into size, dynamics and energetics of initial fusion pores. *Faraday Discuss.* 164:33–55.

67. Hengstenberg, A., Blochl, A., Dietzel, I. D., Schuhmann, W. 2001. Spatially resolved detection of neurotransmitter secretion from individual cells by means of scanning electrochemical microscopy. *Angew. Chem. Int. Ed.* 40:905–908.

68. Bauermann, L. P., Schuhmann, W., Schulte, A. 2004. An advanced biological scanning electrochemical microscope (Bio-SECM) for studying individual living cells. *PCCP* 6:4003–4008.

69. Malinski, T., Taha, Z. 1992. Nitric-oxide release from a single cell measured in situ by a porphyrinic-based microsensor. *Nature* 358:676–678.

70. Malinski, T., Taha, Z., Grunfeld, S. et al. 1993. Diffusion of nitric oxide in the aorta wall monitored in situ by porphyrinic microsensors. *Biochem. Biophys. Res. Commun.* 193:1076–1082.

71. Shibuki, K. 1990. An electrochemical microprobe for detecting nitric-oxide release in brain-tissue. *Neurosci. Res.* 9:69–76.

72. Shibuki, K., Okada, D. 1991. Endogenous nitric-oxide release required for long-term synaptic depression in the cerebellum. *Nature* 349:326–328.

73. Borgmann, S. 2009. Electrochemical quantification of reactive oxygen and nitrogen: Challenges and opportunities. *Anal. Bioanal. Chem.* 394:95–105.

74. Bedioui, F., Quinton, D., Griveau, S., Nyokong, T. 2010. Designing molecular materials and strategies for the electrochemical detection of nitric oxide, superoxide and peroxynitrite in biological systems. *PCCP* 12:9976–9988.

75. Trouillon, R. 2013. Biological applications of the electrochemical sensing of nitric oxide: Fundamentals and recent developments. *Biol. Chem.* 394:17–33.

76. Isik, S., Etienne, M., Oni, J. et al. 2004. Dual microelectrodes for distance control and detection of nitric oxide from endothelial cells by means of scanning electrochemical microscope. *Anal. Chem.* 76:6389–6394.

77. Isik, S., Schuhmann, W. 2006. Detection of nitric oxide release from single cells by using constant-distance-mode scanning electrochemical microscopy. *Angew. Chem. Int. Ed.* 45:7451–7454.

78. Borgmann, S., Radtke, I., Erichsen, T. et al. 2006. Electrochemical high-content screening of nitric oxide release from endothelial cells. *ChemBioChem* 7:662–668.

79. Patel, B. A., Arundell, M., Parker, K. H., Yeoman, M. S., O'Hare, D. 2006. Detection of nitric oxide release from single neurons in the pond snail, *Lymnaea stagnalis*. *Anal. Chem.* 78:7643–7648.

80. Zhao, X., Diakowski, P. M., Ding, Z. 2010. Deconvoluting topography and spatial physiological activity of live macrophage cells by scanning electrochemical microscopy in constant-distance mode. *Anal. Chem.* 82:8371–8373.

81. Amatore, C., Arbault, S., Bouton, C. et al. 2006. Monitoring in real time with a microelectrode the release of reactive oxygen and nitrogen species by a single macrophage stimulated by its membrane mechanical depolarization. *ChemBioChem* 7:653–661.

82. Amatore, C., Arbault, S., Erard, M. 2008. Triangulation mapping of oxidative bursts released by single fibroblasts by amperometry at microelectrodes. *Anal. Chem.* 80:9635–9641.

83. Hu, R., Guille, M., Arbault, S., Lin, C. J., Amatore, C. 2010. In situ electrochemical monitoring of reactive oxygen and nitrogen species released by single MG63 osteosarcoma cell submitted to a mechanical stress. *PCCP* 12:10048–10054.

84. Amatore, C., Arbault, S., Bouton, C. et al. 2008. Real-time amperometric analysis of reactive oxygen and nitrogen species released by single immunostimulated macrophages. *ChemBioChem* 9:1472–1480.

85. Amatore, C., Arbault, S., Koh, A. C. W. 2010. Simultaneous detection of reactive oxygen and nitrogen species released by a single macrophage by triple potential-step chronoamperometry. *Anal. Chem.* 82:1411–1419.

86. Tapsoba, I., Arbault, S., Walter, P., Amatore, C. 2010. Finding out Egyptian Gods' secret using analytical chemistry: Biomedical properties of Egyptian black makeup revealed by amperometry at single cells. *Anal. Chem.* 82:457–460.

87. Ferreira, D. C. M., Tapsoba, I., Arbault, S. et al. 2009. Ex vivo activities of beta-Lapachone and alpha-Lapachone on macrophages: A quantitative pharmacological analysis based on amperometric monitoring of oxidative bursts by single cells. *ChemBioChem* 10:528–538.

88. Filipovic, M. R., Koh, A. C. W., Arbault, S. et al. 2010. Striking inflammation from both sides: Manganese(II) pentaazamacrocyclic SOD mimics act also as nitric oxide dismutases: A single-cell study. *Angew. Chem. Int. Ed.* 49:4228–4232.

89. Bernard, A.-S., Giroud, C., Ching, H. Y. V. et al. 2012. Evaluation of the anti-oxidant properties of a SOD-mimic Mn-complex in activated macrophages. *Dalton Trans.* 41:6399–6403.

90. Amatore, C., Arbault, S., Jaouen, G. et al. 2010. Pro-oxidant properties of AZT and other thymidine analogues in macrophages: Implication of the azido moiety in oxidative stress. *ChemMedChem* 5:296–301.

91. Spegel, C., Heiskanen, A., Skjolding, L. H. D., Emneus, J. 2008. Chip based electroanalytical systems for cell analysis. *Electroanalysis* 20:680–702.

92. Huang, Y. X., Cai, D., Chen, P. 2011. Micro- and nanotechnologies for study of cell secretion. *Anal. Chem.* 83:4393–4406.

93. Johnson, A. S., Selimovic, A., Martin, R. S. 2013. Microchip-based electrochemical detection for monitoring cellular systems. *Anal. Bioanal. Chem.* 405:3013–3020.

94. Mellander, L., Cans, A. S., Ewing, A. G. 2010. Electrochemical probes for detection and analysis of exocytosis and vesicles. *ChemPhysChem* 11:2756–2763.

95. Omiatek, D. M., Cans, A. S., Heien, M. L., Ewing, A. G. 2010. Analytical approaches to investigate transmitter content and release from single secretory vesicles. *Anal. Bioanal. Chem.* 397:3269–3279.

96. Ge, S., Koseoglu, S., Haynes, C. L. 2010. Bioanalytical tools for single-cell study of exocytosis. *Anal. Bioanal. Chem.* 397:3281–3304.

97. Keighron, J. D., Ewing, A. G., Cans, A. S. 2012. Analytical tools to monitor exocytosis: A focus on new fluorescent probes and methods. *Analyst* 137:1755–1763.

98. Amatore, C., Arbault, S., Chen, Y. et al. 2006. Coupling of electrochemistry and fluorescence microscopy at indium tin oxide microelectrodes for the analysis of single exocytotic events. *Angew. Chem. Int. Ed.* 45:4000–4003.

99. Shi, B. X., Wang, Y., Zhang, K., Lam, T. L., Chan, H. L. W. 2011. Monitoring of dopamine release in single cell using ultrasensitive ITO microsensors modified with carbon nanotubes. *Biosens. Bioelectron.* 26:2917–2921.

100. Zhao, H., Li, L., Fan, H. J. et al. 2012. Exocytosis of SH-SY5Y single cell with different shapes cultured on ITO micro-pore electrode. *Mol. Cell. Biochem.* 363:309–313.

101. Meunier, A., Fulcrand, R., Darchen, F. et al. 2012. Indium tin oxide devices for amperometric detection of vesicular release by single cells. *Biophys. Chem.* 162:14–21.

102. Meunier, A., Jouannot, O., Fulcrand, R. et al. 2011. Coupling amperometry and total internal reflection fluorescence microscopy at ITO surfaces for monitoring exocytosis of single vesicles. *Angew. Chem. Int. Ed.* 50:5081–5084.

103. Shi, B. X., Wang, Y., Lam, T. L. et al. 2010. Release monitoring of single cells on a microfluidic device coupled with fluorescence microscopy and electrochemistry. *Biomicrofluidics* 4:1–9.

104. Kisler, K., Kim, B. N., Liu, X. et al. 2012. Transparent electrode materials for simultaneous amperometric detection of exocytosis and fluorescence microscopy. *J. Biomater. Nanobiotechnol.* 3:243–253.

105. Sudibya, H. G., Ma, J. M., Dong, X. C. et al. 2009. Interfacing glycosylated carbon-nanotube-network devices with living cells to detect dynamic secretion of biomolecules. *Angew. Chem. Int. Ed.* 48:2723–2726.

106. Chen, X. H., Gao, Y. F., Hossain, M., Gangopadhyay, S., Gillis, K. D. 2008. Controlled on-chip stimulation of quantal catecholamine release from chromaffin cells using photolysis of caged Ca^{2+} on transparent indium-tin-oxide microchip electrodes. *Lab Chip* 8:161–169.

107. Sun, X. H., Gillis, K. D. 2006. On-chip amperometric measurement of quantal catecholamine release using transparent indium tin oxide electrodes. *Anal. Chem.* 78:2521–2525.

108. Zhang, B., Adams, K. L., Luber, S. J. et al. 2008. Spatially and temporally resolved single-cell exocytosis utilizing individually addressable carbon microelectrode arrays. *Anal. Chem.* 80:1394–1400.

109. Zhang, B., Heien, M., Santillo, M. F., Mellander, L., Ewing, A. G. 2011. Temporal resolution in electrochemical imaging on single PC12 cells using amperometry and voltammetry at microelectrode arrays. *Anal. Chem.* 83:571–577.

110. Lin, Y. Q., Trouillon, R., Svensson, M. I. et al. 2012. Carbon-ring microelectrode arrays for electrochemical imaging of single cell exocytosis: Fabrication and characterization. *Anal. Chem.* 84:2949–2954.

111. Berberian, K., Kisler, K., Fang, Q. H., Lindau, M. 2009. Improved surface-patterned platinum microelectrodes for the study of exocytotic events. *Anal. Chem.* 81:8734–8740.

112. Hafez, I., Kisler, K., Berberian, K. et al. 2005. Electrochemical imaging of fusion pore openings by electrochemical detector arrays. *Proc. Natl. Acad. Sci. U.S.A.* 102:13879–13884.

113. Pasquarelli, A., Carabelli, V., Xu, Y. L. et al. 2011. Diamond microelectrodes arrays for the detection of secretory cell activity. *Int. J. Environ. Anal. Chem.* 91:150–160.

114. Carabelli, V., Gosso, S., Marcantoni, A. et al. 2010. Nanocrystalline diamond microelectrode arrays fabricated on sapphire technology for high-time resolution of quantal catecholamine secretion from chromaffin cells. *Biosens. Bioelectron.* 26:92–98.

115. Li, L.-M., Wang, W., Zhang, S.-H. et al. 2011. Integrated microdevice for long-term automated perfusion culture without shear stress and real-time electrochemical monitoring of cells. *Anal. Chem.* 83:9524–9530.

116. Sasso, L., Heiskanen, A., Diazzi, F. et al. 2013. Doped overoxidized polypyrrole microelectrodes as sensors for the detection of dopamine released from cell populations. *Analyst* 138:3651–3659.

117. Gao, Y. F., Chen, X. H., Gupta, S., Gillis, K. D., Gangopadhyay, S. 2008. Magnetron sputtered diamond-like carbon microelectrodes for on-chip measurement of quantal catecholamine release from cells. *Biomed. Microdev.* 10:623–629.

118. Gao, Y. F., Bhattacharya, S., Chen, X. H. et al. 2009. A microfluidic cell trap device for automated measurement of quantal catecholamine release from cells. *Lab Chip* 9:3442–3446.

119. Spegel, C., Heiskanen, A., Pedersen, S. et al. 2008. Fully automated microchip system for the detection of quantal exocytosis from single and small ensembles of cells. *Lab Chip* 8:323–329.

120. Gao, C. L., Sun, X. H., Gillis, K. D. 2013. Fabrication of two-layer poly(dimethyl siloxane) devices for hydrodynamic cell trapping and exocytosis measurement with integrated indium tin oxide microelectrodes arrays. *Biomed. Microdev.* 15:445–451.

121. Ges, I. A., Currie, K. P. M., Baudenbacher, F. 2012. Electrochemical detection of catecholamine release using planar iridium oxide electrodes in nanoliter microfluidic cell culture volumes. *Biosens. Bioelectron.* 34:30–36.

122. Barizuddin, S., Liu, X., Mathai, J. C. et al. 2010. Automated targeting of cells to electrochemical electrodes using a surface chemistry approach for the measurement of quantal exocytosis. *ACS Chem. Neurosci.* 1:590–597.

123. Liu, X., Barizuddin, S., Shin, W. et al. 2011. Microwell device for targeting single cells to electrochemical microelectrodes for high-throughput amperometric detection of quantal exocytosis. *Anal. Chem.* 83:2445–2451.

124. Dittami, G. M., Rabbitt, R. D. 2010. Electrically evoking and electrochemically resolving quantal release on a microchip. *Lab Chip* 10:30–35.

125. Ghosh, J., Liu, X., Gillis, K. D. 2013. Electroporation followed by electrochemical measurement of quantal transmitter release from single cells using a patterned microelectrode. *Lab Chip* 13:2083–2090.

126. Yakushenko, A., Schnitker, J., Wolfrum, B. 2012. Printed carbon microelectrodes for electrochemical detection of single vesicle release from PC12 cells. *Anal. Chem.* 84:4613–4617.

127. Omiatek, D. M., Dong, Y., Heien, M. L., Ewing, A. G. 2010. Only a fraction of quantal content is released during exocytosis as revealed by electrochemical cytometry of secretory vesicles. *ACS Chem. Neurosci.* 1:234–245.

128. Yeh, J. I., Shi, H. B. 2010. Nanoelectrodes for biological measurements. *Wiley Interdiscip. Rev. Nanomed. Nanobiotechnol.* 2:176–188.

129. Wang, J., Trouillon, R., Lin, Y. Q., Svensson, M. I., Ewing, A. G. 2013. Individually addressable thin-film ultramicroelectrode array for spatial measurements of single vesicle release. *Anal. Chem.* 85:5600–5608.

130. Yakushenko, A., Katelhon, E., Wolfrum, B. 2013. Parallel on-chip analysis of single vesicle neurotransmitter release. *Anal. Chem.* 85:5483–5490.

131. Griveau, S., Bedioui, F. 2013. Overview of significant examples of electrochemical sensor arrays designed for detection of nitric oxide and relevant species in a biological environment. *Anal. Bioanal. Chem.* 405:3475–3488.

132. Oni, J., Pailleret, A., Isik, S. et al. 2004. Functionalised electrode array for the detection of nitric oxide released by endothelial cells using different NO-sensing chemistries. *Anal. Bioanal. Chem.* 378:1594–1600.

133. Castillo, J., Isik, S., Blochl, A. et al. 2005. Simultaneous detection of the release of glutamate and nitric oxide from adherently growing cells using an array of glutamate and nitric oxide selective electrodes. *Biosens. Bioelectron.* 20:1559–1565.

134. Pekarova, M., Kralova, J., Kubala, L. et al. 2009. Continuous electrochemical monitoring of nitric oxide production in murine macrophage cell line RAW 264.7. *Anal. Bioanal. Chem.* 394:1497–1504.

135. Kamei, K., Mie, M., Yanagida, Y., Aizawa, M., Kobatake, E. 2004. Construction and use of an electrochemical NO sensor in a cell-based assessing system. *Sens. Actuators B-Chem.* 99:106–112.

136. Pereira-Rodrigues, N., Zurgil, N., Chang, S. C. et al. 2005. Combined system for the simultaneous optical and electrochemical monitoring of intra- and extracellular NO produced by glioblastoma cells. *Anal. Chem.* 77:2733–2738.

137. Patel, B. A., Arundell, M., Quek, R. G. W. et al. 2008. Individually addressable microelectrode array for monitoring oxygen and nitric oxide release. *Anal. Bioanal. Chem.* 390:1379–1387.

138. Chang, S. C., Pereira-Rodrigues, N., Henderson, J. R. et al. 2005. An electrochemical sensor array system for the direct, simultaneous in vitro monitoring of nitric oxide and superoxide production by cultured cells. *Biosens. Bioelectron.* 21:917–922.

139. Amatore, C., Arbault, S., Chen, Y., Crozatier, C., Tapsoba, I. 2007. Electrochemical detection in a microfluidic device of oxidative stress generated by macrophage cells. *Lab Chip* 7:233–238.

140. Cha, W., Tung, Y.-C., Meyerhoff, M. E., Takayama, S. 2010. Patterned electrode-based amperometric gas sensor for direct nitric oxide detection within microfluidic devices. *Anal. Chem.* 82:3300–3305.

141. Quinton, D., Girard, A., Kim, L. T. T. et al. 2011. On-chip multi-electrochemical sensor array platform for simultaneous screening of nitric oxide and peroxynitrite. *Lab Chip* 11:1342–1350.

142. Li, Y., Sella, C., Lemaitre, F. et al. 2013. Highly sensitive platinum-black coated platinum electrodes for electrochemical detection of hydrogen peroxide and nitrite in microchannel. *Electroanalysis* 25:895–902.

143. Yan, J., Pedrosa, V. A., Enomoto, J., Simonian, A. L., Revzin, A. 2011. Electrochemical biosensors for on-chip detection of oxidative stress from immune cells. *Biomicrofluidics* 5:32008–3200811.

144. Trouillon, R., Williamson, E. D., Saint, R. J., O'Hare, D. 2012. Electrochemical detection of the binding of *Bacillus anthracis* protective antigen (PA) to the membrane receptor on macrophages through release of nitric oxide. *Biosens. Bioelectron.* 38:138–144.

145. Amatore, C., Maisonhaute, E., Schollhorn, B. 2008. Molecular electrochemistry pushed to its limits: From nanosecond kinetics to the dynamic study of nanometric objects. *Actualité Chimique*:69–74.

146. Cox, J. T., Zhang, B., 2012. Nanoelectrodes: Recent advances and new directions. *Annu. Rev. Anal. Chem. Ed.* 5:253–272.

147. Wu, W. Z., Huang, W. H., Wang, W. et al. 2005. Monitoring dopamine release from single living vesicles with nanoelectrodes. *J. Am. Chem. Soc.* 127:8914–8915.

148. Li, Z. Y., Zhou, W., Wu, Z. X., Zhang, R. Y., Xu, T. 2009. Fabrication of size-controllable ultrasmall-disk electrode: Monitoring single vesicle release kinetics at tiny structures with high spatio-temporal resolution. *Biosens. Bioelectron.* 24:1358–1364.

149. Gutierrez, L. M., Gil, A., Viniegra, S. 1998. Preferential localization of exocytotic active zones in the terminals of neurite-emitting chromaffin cells. *Eur. J. Cell Biol.* 76:274–278.

150. Paras, C. D., Qian, W. J., Lakey, J. R., Tan, W. H., Kennedy, R. T. 2000. Localized exocytosis detected by spatially resolved amperometry in single pancreatic beta-cells. *Cell Biochem. Biophys.* 33:227–240.

151. Qian, W. J., Aspinwall, C. A., Battiste, M. A., Kennedy, R. T. 2000. Detection of secretion from single pancreatic beta-cells using extracellular fluorogenic reactions and confocal fluorescence microscopy. *Anal. Chem.* 72:711–717.

152. Robinson, I. M., Finnegan, J. M., Monck, J. R., Wightman, R. M., Fernandez, J. M. 1995. Colocalization of calcium-entry and exocytotic release sites in adrenal chromaffin cells. *Proc. Natl. Acad. Sci. U.S.A.* 92:2474–2478.

153. Mosharov, E. V., Sulzer, D. 2005. Analysis of exocytotic events recorded by amperometry. *Nat. Methods* 2:651–658.

154. Amatore, C., Arbault, S., Bouret, Y. et al. 2009. Invariance of exocytotic events detected by amperometry as a function of the carbon fiber microelectrode diameter. *Anal. Chem.* 81:3087–3093.

155. Al-Hilli, S. M., Willander, M., Ost, A., Stralfors, P. 2007. ZnO nanorods as an intracellular sensor for pH measurements. *J. Appl. Phys.* 102:084304.

156. Kubant, R., Malinski, C., Burewicz, A., Malinski, T. 2006. Peroxynitrite/nitric oxide balance in ischemia/reperfusion injury-nanomedical approach. *Electroanalysis* 18:410–416.

157. Heller, I., Smaal, W. T. T., Lemay, S. G., Dekker, C. 2009. Probing macrophage activity with carbon-nanotube sensors. *Small* 5:2528–2532.

158. Wang, Y., Noel, J.-M., Velmurugan, J. et al. 2012. Nanoelectrodes for determination of reactive oxygen and nitrogen species inside murine macrophages. *Proc. Natl. Acad. Sci. U.S.A.* 109:11534–11539.

13 Nanobioelectrochemistry
Proteins, Enzymes, and Biosensors

Gregory W. Bishop and James F. Rusling

CONTENTS

13.1 INTRODUCTION

Electrochemical investigations of biomolecules have long captured the attention of researchers, not only because of their significance in the fundamental understanding of electron transfer in life processes but also due to their applications in developing sensors capable of measuring important biochemical indicators of health. Due to their fast electron-transfer properties and high specific surface areas compared to bulk materials, nanomaterials have played an important role in advancing the use of electrochemical techniques for studies of biomolecules such as nucleic acids and proteins.

13.1.1 BIOELECTROCHEMISTRY: A BRIEF SUMMARY OF CHALLENGES AND PROGRESS

While direct electrochemistry of DNA was reported over 50 years ago,[1–4] direct voltammetry of proteins often proved difficult prior to the 1980s. These difficulties can be attributed mainly to a poor understanding of how proteins interact with metal surfaces as well as a lack of control over the electrode surface structure. As a result, proteins often denature, adopting unfavorable configurations for direct electron transfer when adsorbed directly onto pure metal or carbon electrodes. Challenges in observing protein redox processes by electrochemical techniques may also stem from the fact that electroactive moieties of proteins may be difficult to access due to their positions within insulating protein shells.

Prior to the wide availability of nanomaterials, direct voltammetry of proteins was usually achieved by careful chemical modification of the electrode surface or of the insulating protein shell. Studies in the late 1970s and 1980s showed that direct voltammetry of proteins in solution can be observed by using a carefully cleaned Ag or ITO electrode surface in a highly purified protein solution,[5,6] by adsorbing a monolayer of small organic electron-transfer promoter molecules onto the electrode surface,[7–9] or by modifying the insulating protein shell with conductive species (i.e., *electron relays*) to improve electron transfer with the electrode.[10,11] In an early demonstration of promoter monolayers, modification of a gold electrode surface with a monolayer of electron-transfer promoter 4,4′-bipyridyl enabled observation of stable, quasireversible cyclic voltammetry of cytochrome (cyt) *c* and other proteins, giving cyclic voltammograms (CVs) with peak separations ~60 mV characteristic of reversible electron transfer.[9] For 4,4′-bipyridyl-modified gold electrodes, hydrogen bonding between cyt *c* lysines and pyridyl nitrogens is thought to prevent denaturation and orient the protein for rapid electron transfer between the cyt *c* heme group and the electrode.

Studies in the 1990s showed that direct voltammetry of proteins can be achieved by coating electrodes with films or layers of surfactants, polyions, redox polymers, or small chemisorbed molecules to which proteins can be electrostatically or covalently attached.[5,11–17] These strategies using film-incorporated proteins for electrochemical studies have been shown to exhibit important advantages over solution-based methods for protein electrochemistry. While solution-based protein voltammetry is often thwarted by processes that inhibit electron transfer such as protein denaturation on the electrode surface, slow diffusion of large proteins, and adsorption of other blocking macromolecules on the electrode, protein-film electrodes can provide a stable electrode–protein electron-transfer interface. Such films are also often easy to prepare, are durable, and require very little protein.[5,16] In addition to eliminating protein denaturation and preventing interference caused by the adsorption of other macromolecules, protein-film electrodes also often exhibit enhanced electrochemical reversibility in comparison to solution-based protein voltammetry.

Examples of protein-film electrodes include polyion–protein assemblies and mono- or multilayered biomembrane-like lipid or surfactant films with entrapped proteins.[5,12–17] Redox-active proteins such as myoglobin and cyt P450$_{cam}$ exhibited stable, reversible voltammetry when trapped in surfactant films on electrodes.[5,15] Prior to the advent of these protein-film electrodes, reversible voltammetry of myoglobin and cyt P450$_{cam}$ had only been possible with carefully purified protein solutions and specially cleaned electrodes.[5] Direct electron transfer between myoglobin or cyt P450$_{cam}$ and electrodes was also obtained by electrostatically adsorbing successive layers of alternately charged macromolecules (e.g., polyion and protein) onto the electrode.[16] When a smooth bulk gold electrode was used as the platform for constructing multilayered, layer-by-layer-assembled polyion–protein films, the protein layer nearest the electrode surface was found to be entirely electroactive, while second-nearest protein layers exhibited only 20%–40% activity, and additional layers were completely inactive.[16,17] However, multilayered myoglobin/poly(styrenesulfonate) (Mb/PSS) films built on rough pyrolytic graphite surfaces displayed electroactivity for Mb in layers as far as seven Mb/PSS layer thicknesses from the electrode surface.[17,18] Such long-range electroactivity is enabled by the morphology of the rough pyrolytic graphite template, which is predicted to introduce disorder in the films, facilitate intermixing of neighboring layers, and promote charge transport by electron hopping.[17]

While studies employing surface-modified electrodes to promote voltammetry of proteins in solution and films mostly focused on relatively small and simple electron-transfer proteins (e.g., cyt *c* MW = 12,400 Da,[9] myoglobin MW ≈ 17,000 Da[15]), achieving direct electron transfer with more complex redox-active proteins (e.g., glucose oxidase [GOx] MW ≈ 160,000 Da[11]) proved to be more elusive. Hill and coworkers commented that enzymes such as GOx lack a *natural* pathway through which electrons may be transferred over long distances because the redox-active center is buried deep within the insulating protein shell and the catalyzed reaction occurs within the active site.[9] Thus, they suggested some strategies that may be successful in direct electron transport with such *intrinsic* redox enzymes could include the employment of electrodes with surfaces that project into enzyme and the preparation of long-range electron-transfer pathways through modification.[9]

Degani and Heller used an idea inspired by nature to achieve direct electron transfer with GOx by *electrically wiring* the enzyme to the electrode through the modification of the insulating protein shell with *electron relays*.[11] Naturally occurring electron transfer between redox centers of proteins and enzymes (e.g., cyt *c* and cyt *c* peroxidase or cyt *c* oxidase) proceeds upon the formation of protein–enzyme complexes (often stabilized through electrostatic interactions), which lessen the distance between redox centers.[11] Degani and Heller showed that direct electron transfer between GOx and an electrode could similarly be attained by covalently coupling *electron relays*, such as ferrocene carboxylate, to the protein backbone.[10,11] They also found that redox polymers electrostatically bound to an electrode could be used to covalently modify GOx for direct electron transfer.[12]

Like film-modified electrodes, *electrical wiring* of proteins to electrodes, and other successful strategies for protein electrochemistry, nanostructured electrodes can also facilitate direct electron transfer between proteins or other biomolecules and the electrical interface. Due to their large surface areas, small size, increasing availability, and, in some cases, electrocatalytic properties, electrodes with nanometer-sized surface structures have been found to serve as excellent platforms for bioelectrochemical studies. The properties and bioelectrochemical applications of nanostructured electrodes will be described in subsequent sections.

13.1.2 BIOELECTROCHEMISTRY AND ELECTROCHEMICAL BIOSENSORS

Electrochemical investigations of biomolecules are often motivated by the desire to develop biosensors for medical diagnostics. Biosensors are devices that transform biochemical interactions into analytical signals such that a specific sample component or components can be measured (Figure 13.1).[19] Clark and Lyons invented the field of biosensors in 1962 with their development of the glucose sensor.[20–22] The glucose sensor was an extension of Clark's work on the oxygen electrode[21,22] and featured the enzyme GOx, trapped in a layer between two membranes that were used to separate the electrode from the sample. As described in the following sections, incorporation of nanomaterials into biosensor design has resulted in great advances in sensitivity, detection limits (DLs), and overall performance of these bioanalytical devices.

Among electrochemical biosensors, electrochemical immunoassays have especially benefited from the assimilation of nanomaterials in sensor design. Immunoassays rely on a capture antibody to isolate the target analyte from the sample (Figure 13.1). Though the formation of the capture antibody–analyte complex may be measured directly through surface plasmon resonance, impedance spectroscopy, or other label-free methods (Figure 13.1d), to increase sensitivity and specificity, a species labeled with a transducing agent is often used to generate a signal that can be measured with the aid of an appropriate analytical technique (Figure 13.1e and f). In competitive assays, the labeled species is of a form similar to the analyte, such that it competitively binds to the antibody (Figure 13.1e). In sandwich-type assays, the labeled species is a second recognition agent (e.g., antibody) that binds to a different epitope of the analyte than that involved in the formation of the capture antibody–analyte complex (Figure 13.1f).

Techniques based on enzyme-linked immunosorbent assays (ELISAs) have emerged as standard biosensing methods. In classic ELISAs, the labels are enzymes that, upon reaction with a

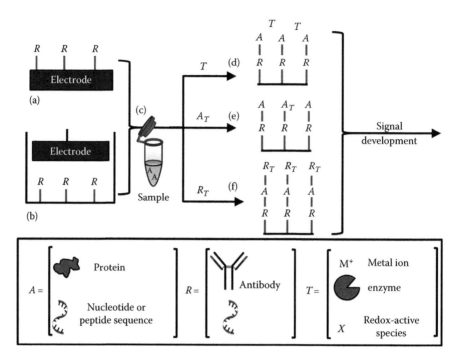

FIGURE 13.1 Some basic strategies for electrochemical biosensing depicting various types of analytes (A), recognition agents (R), and signal-transducing agents (T). Recognition agents (R) at (a) or near (b) the electrode surface are employed to isolate the analyte (A) from the sample (c). Electrochemical signal can be generated through (d) monitoring accumulation, depletion, or changes in transport of an electroactive species or signal-transducing agent (T); through (e) the use of an analyte analog A_T (labeled with a signal-transducing agent), which competes with analyte for binding sites; or through (f) the use of a second recognition agent R_T, which is labeled with a signal-transducing agent.

substrate, generate chromogenic products that can be measured by spectroscopic techniques. In electrochemical immunoassays, labels can be metal ions, redox molecules, or enzymes that generate redox-active products upon reaction with a substrate.

Electrochemical immunoassays were pioneered in the 1980s by Heineman, Halsall, and coworkers.[23–29] Early in their studies, they employed voltammetric and amperometric techniques to measure proteins and small biologically important molecules, such as hormones, via competitive and sandwich-type immunoassays. Heineman and Halsall and other researchers explored many different measurement techniques and incorporated microfluidics,[30–32] reagent delivery systems,[33] interdigitated electrode arrays,[31,34–36] and magnetic beads[30,31,33,35–37] into various designs to improve sensitivity and reduce sample volume requirements. The recent incorporation of nanomaterials into electrochemical immunosensors and protocols has resulted in further improvements in these techniques.

Electrochemical immunoassays for the simultaneous detection of multiple analytes (multiplexed detection) have also been achieved through the use of distinguishable metal labels,[38] spatial separation of antibodies,[39] or electrode arrays designed with individual electrodes that are decorated with specific capture antibodies for different analytes.[40,41] Due to biochemical diversity in living systems, measurements of single biochemical analytes from biological samples are often insufficient in providing accurate and timely information about patient health status.[42–45] Thus, multiplexed measurements of biochemical analytes are essential in advanced diagnostic tests for diseases and health conditions. Multiplexed measurements of disease biomarkers such as RNAs, proteins, and metabolites should enable more accurate future diagnoses and lead to better informed therapy decisions.

Electrochemical immunoassays may offer some advantages over other multiplex sensing technologies, which often require expensive reagents and complex instrumentation. As described in the following sections, the application of nanomaterials to electrochemical immunoassays and other types of electrochemical biosensing devices has resulted in extremely sensitive analytical techniques for multiplexed analyses.

13.2 APPLICATION OF NANOMATERIAL-BASED ELECTRODES IN BIOANALYTICAL ELECTROCHEMISTRY

As mentioned earlier, observations of direct voltammetry of biomolecules, especially proteins, can be limited at flat, bulk-material electrodes due to problems with electrode fouling, biomolecule denaturation, and accessibility of redox centers.[46–52] Various approaches (e.g., protein-film electrodes and *electrical wiring* of proteins to electrodes) have been employed to address these issues. However, strategies that employ nanomaterials to conduct bioelectrochemical studies have become increasingly popular for bioelectrochemical applications as nanomaterials have become readily available and have been shown to possess properties (e.g., high specific areas and excellent electron-transfer characteristics) that are favorable to bioelectrochemistry.

A number of different types of nanostructured electrodes, mostly used in conjunction with biomolecule films, have emerged for bioanalytical applications. These include metal nanoparticle films prepared by casting or electrodeposition, films of randomly oriented carbon nanotubes (CNTs), vertically aligned CNT *forests*, and graphene films (Figure 13.2). Such nanostructured electrodes have been found to serve as excellent platforms for electrochemical studies involving redox-active biomolecules.[46–60] The following section describes how nanostructured electrodes, electrodes with

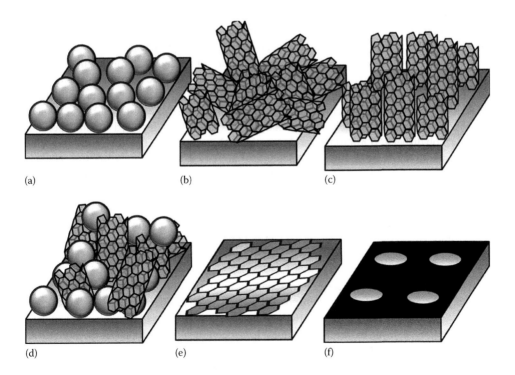

FIGURE 13.2 Examples of nanostructured electrodes. Representative illustrations of (a) metal nanoparticle film, (b) randomly oriented CNT film, (c) vertically aligned CNT film or CNT forest, (d) nanocomposite film, (e) graphene film, and (f) nanodisk electrode array.

nanometer-sized critical dimensions (nanoelectrodes), and nanoelectrode arrays have been applied in electrochemical studies of biomolecules and the development of biosensors.

13.2.1 Bioelectrochemistry with Gold Nanoparticle Film Electrodes

Among the first examples of nanostructured electrodes developed in bioelectrochemical studies were colloidal gold-film electrodes prepared by adsorption, electrodeposition, or covalent binding of gold nanoparticles onto conductive bulk-material surfaces. Crumbliss et al. deposited 30 and 50 nm diameter enzyme-modified gold colloids onto platinum gauze, glassy carbon (GC), and vapor-deposited gold on glass by electrodeposition or evaporation.[53,61,62] Direct electron transfer for a horseradish peroxidase (HRP)–gold sol electrode was observed for the reduction of H_2O_2.[53] The performance of the HRP–Au sol was optimized when ~12 layers of 30 nm HRP–Au were deposited on a 3 mm diameter GC surface. These early nanostructured electrodes were also implemented as biosensors. For example, a cholesterol biosensor was developed using an electrode that featured an HRP–Au sol and a cholesterol oxidase hydrogel.[62]

The importance of Au colloid size used in the preparation of nanostructured electrodes for bioelectrochemical studies was explored by Brown et al. (Figure 13.3).[54] They found that the electrochemical response of cytochrome c (cyt c) with electrodes that consisted of submonolayer Au colloid–modified In- or Sb-doped tin oxide (SnO_2) was determined by the colloid-size-dependent electrode surface morphology.[54] Au colloids were deposited at submonolayer levels on In- or Sb-doped SnO_2 surfaces using (3-aminopropyl)trimethoxysilane. Electrochemical behavior of the electrodes resembled that of a collection of isolated but closely spaced microelectrodes, and voltammetric response for cyt c in solution was best for electrodes that were prepared with 12 nm diameter Au colloids.

Electrodes that featured a submonolayer (~15% of a monolayer) of 12 nm diameter Au colloids exhibited reversible cyt c voltammetry at scan rates of 50, 100, 200, and 500 mV s^{-1} (Figure 13.3). Au particles in the submonolayer films can be "thought of as 'electron antennae', efficiently funneling electrons between the electrode and the electrolyte."[54] However, the authors also reported a wide variation in voltammetric responses, with peak-to-peak separations ranging from 60 to 115 mV, among several similarly prepared electrodes.[54] The heterogeneous rate constant calculated from the dependence of the peak-to-peak separation on scan rate was 7×10^{-3} cm s^{-1}, which is in agreement with previously reported measurements obtained with bulk material electrodes.[54]

Cyt c exhibited quasireversible voltammetry at electrodes that featured monodisperse 36 nm diameter or polydisperse 6 nm diameter Au particles (Figure 13.3b), indicating the importance of the colloid size in determining the electrochemical activity toward cyt c. Moreover, at electrodes composed of 12 nm diameter or 22 nm diameter Au particle aggregates, no voltammetric response was observed for cyt c (Figure 13.3b). Thus, Brown et al. surmised that Au aggregates may provide sites for irreversible protein adsorption on the electrode surface, which had been previously cited as an obstacle to obtaining good electrochemistry of cyt c.[54] In contrast, the inherent negative charge of (citrate-stabilized) colloidal Au particles and morphology of submonolayer Au colloid films may favor the close approach of cyt c, which possesses a positive dipole moment near its redox-active center, to the colloid surface.[54]

Au nanoparticle (AuNP) electrodes prepared through various other deposition strategies have been extensively used for protein voltammetry and in electrochemical biosensing applications.[50,60,63–75] Electrodes with layers of HRP-modified AuNPs (prepared through covalent modification, self-assembly, or layer-by-layer techniques) have been shown to exhibit electrocatalytic activity toward the reduction of H_2O_2 through direct electron transfer between electrode and enzyme redox center.[64,66,67,70,72,74] Direct electron transfer for other proteins such as hemoglobin,[65] myoglobin,[76] and GOx (for applications in glucose sensing[60,69]) has also been observed with these platforms.

Chai et al. demonstrated that electron transfer with myoglobin could be obtained over distances of more than 100 nm when layer-by-layer films assembled from successive layers of positively charged

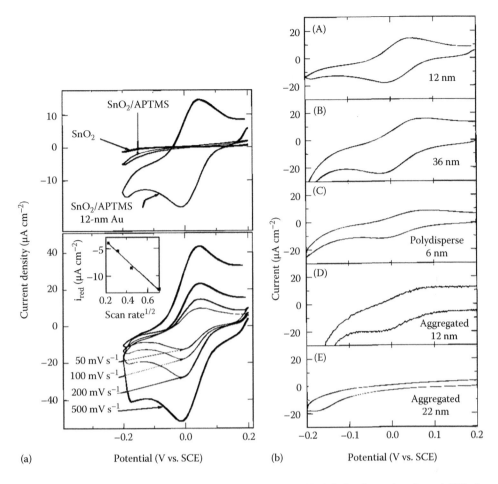

FIGURE 13.3 CV response of cyt c at various Au colloid–modified SnO_2 electrodes. (a, top) CV of cyt c at bare SnO_2, (3-aminopropyl)trimethoxysilane (APTMS)-modified SnO_2 and 12 nm diameter Au colloid submonolayer-modified APTMS/SnO_2 electrodes. (a, bottom) Scan rate–dependent CV response of cyt c at a 12 nm diameter Au colloid–modified APTMS/SnO_2 electrode. (b) Dependence of cyt c CV response on electrode surface morphology presenting the variation in CV due to the size of colloids or aggregates used to prepare submonolayer Au-modified electrodes. All CVs were obtained with 0.5 mM cyt c in 0.1 M $NaClO_4$/18 mM, pH 7 Na_2HPO_4/NaH_2PO_4 at scan rates of 100 mV s^{-1} except where otherwise noted. (Adapted with permission from Brown, K.R., Fox, A.P., and Natan, M.J., Morphology-dependent electrochemistry of cytochrome c at Au colloid-modified SnO_2 electrodes, *J. Am. Chem. Soc.*, 118, 1154–1157, 1996. Copyright 1996 American Chemical Society.)

polyion poly(allyl amine) and 5 nm glutathione-decorated AuNPs are deposited on pyrolytic graphite substrates.[76] While voltammetric peak currents for myoglobin attached to the outermost layer of the films decreased with increasing film thickness, peak separation changed very little for films of 4–11 bilayers.[76] Thus, the apparent heterogeneous electron-transfer rate constant (k_s) was found to reside in the narrow range of 3.6–5.3 s^{-1} for these studies.[76] The insensitivity of k_s toward film thickness is indicative of AuNP-mediated electron hopping, which enables electron transfer though the film between the underlying pyrolytic graphite substrate and myoglobin on the outmost film layer.[76] Myoglobin also exhibited electrocatalytic activity toward H_2O_2 in these AuNP/polyion-film electrodes.

Electrochemical sensors based on AuNP-modified electrodes have also been reported for small biologically important molecules such as dopamine.[68,73] Raj et al. found that well-separated voltammetric peaks for dopamine and ascorbate could be obtained at AuNP-modified electrodes because the high catalytic activity of the AuNP toward ascorbate shifts the oxidation of ascorbate to a less positive potential.[68]

13.2.2 Gold Nanoelectrode Arrays and Ensembles for Bioelectrochemical Applications

Au nanoelectrodes have also been used for bioelectrochemical applications. For example, Kelley and coworkers developed 2D and 3D gold nanoelectrode ensembles (NEEs) (i.e., collections of irregularly spaced individual nanoelectrodes) for the electrocatalytic detection of DNA.[77] NEEs were prepared through a template synthesis method[78] that consisted of electroless plating of gold inside the pores of porous polymer membranes. The membranes contain randomly spaced pores, which become filled with gold nanowires, resulting in 2D Au disk NEEs (Figure 13.4) with disk diameter defined by polymer membrane pore size and NEE density defined by polymer membrane pore density. Oxygen plasma etching was used to remove a thin surface layer of the polymer, thus exposing a desired length of the Au nanowire to yield 3D NEEs. Au NEEs were functionalized with thiolated probe DNA, and hybridization with the target strand increased the amount of negatively charged phosphate groups at the NEE surface. Signal was generated through the use of the electrocatalytic reaction between $Ru(NH_3)_6^{3+}$ and $Fe(CN)_6^{3-}$. The increase in negative surface charge afforded by DNA hybridization leads to an increase in the local $Ru(NH_3)_6^{3+}$ concentration.

Kelley et al. extended their work on array-based electrochemical biosensors by preparing arrays of nanostructured microelectrodes (Figure 13.5).[79–88] Analytical performance characteristics including

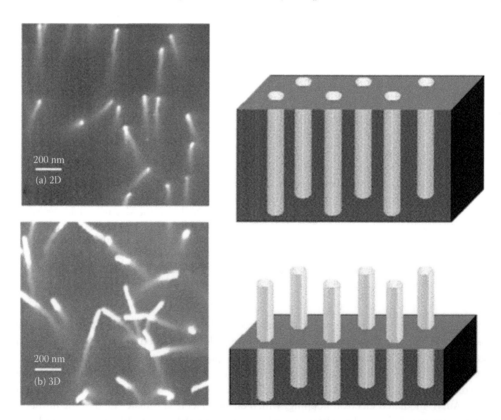

FIGURE 13.4 Scanning electron micrographs and representative illustrations of (a) 2D and (b) 3D NEEs grown by electroless deposition of gold within the pores of a nanoporous polymer membrane. (Reprinted with permission from Gasparac, R., Taft, B.J., Lapierre-Devlin, M.A. et al. Ultrasensitive electrocatalytic DNA detection at two- and three-dimensional nanoelectrodes, *J. Am. Chem. Soc.*, 126, 12270–12271, 2004. Copyright 2004 American Chemical Society.)

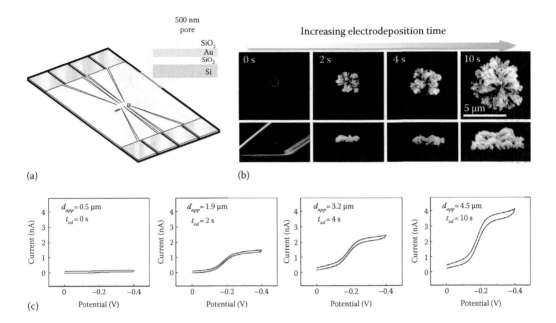

FIGURE 13.5 Fabrication of nanostructured microelectrode arrays. (a) Illustrated representation of a nano-structured microelectrode chip. A photolithographically deposited gold pattern is covered with SiO₂; selective etching of the top SiO₂ layer exposes 500 nm diameter portions of the gold surface. (b) Scanning electron microscope (SEM) images of palladium nanostructures deposited on gold microelectrodes using various deposition times. (c) Cyclic voltammograms, with more negative (reducing) potentials plotted to the right, for nanostructured microelectrodes in $Ru(NH_3)_6^{3+}$ verify that electrode area increases with palladium deposition time. (Adapted with permission from Fang, Z., Soleymani, L., Pampalakis, G. et al., Direct profiling of cancer biomarkers in tumor tissue using a multiplexed nanostructured microelectrode integrated circuit, *ACS Nano*, 3, 3207–3213, 2009. Copyright 2009 American Chemical Society.)

limit of detection[79] and efficiency of biomolecular capture[83] of these biosensors are determined by surface morphology, which is controlled by electrodeposition of gold or palladium. Attomolar DLs for small oligonucleotides and micro-RNA were achieved using the same $Ru(NH_3)_6^{3+}/Fe(CN)_6^{3-}$ mediator system described previously.[81,82,84,85] Nanostructured microelectrode arrays were also used in electrochemical sensing and enzyme-linked electrochemical immunoassays of proteins from serum and whole blood.[85,87]

13.2.3 Carbon Nanotube–Based Bioelectrochemical Strategies

A CNT is a cylindrical arrangement of sp²-hybridized carbon atoms, essentially a graphene sheet rolled up in the form of a tube (Figure 13.6).[52,89] Nanostructures that consist of single tubular graphene sheets are called single-walled CNTs (SWCNTs), which have diameters of 0.4–3 nm.[52,89] Assemblies of multiple concentric tubular graphene sheets (with each sheet separated by 0.34 nm) are known as multiwalled CNTs (MWCNTs). MWCNTs have outer diameters of 2–100 nm and inner diameters typically of 2–10 nm.[52,89–91] CNTs possess specific surfaces areas of 300 m² g⁻¹ with values as great as 1600 m² g⁻¹ estimated for SWCNTs.[52,90] These large specific surface areas facilitate immobilization of substantial quantities of biomolecules and promote electron transfer by presenting many potentially conductive pathways.[52,90] These characteristics have made CNTs very attractive for applications in bioelectrochemistry.

Electron transport properties of SWCNTs are largely determined by the extent of alignment for the sp²-carbon p-orbitals, as governed by the angle at which the graphene sheet is rolled up to form

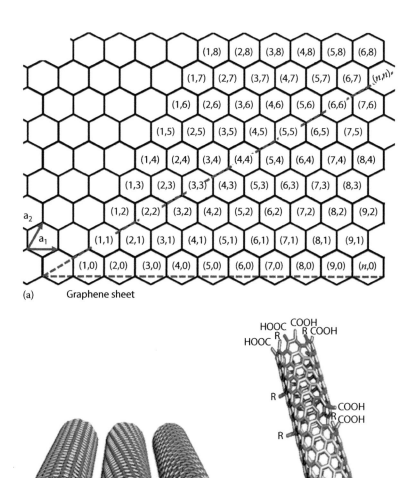

FIGURE 13.6 Illustrated representations of graphene and CNT structures. (a) Illustrated representation of a graphene sheet with lattice unit vectors a_1 and a_2 and a superimposed (n,m) map, which indicates the chirality of the SWCNT that results from rolling up the sheet along a certain vector. Dotted lines depict the roll-up vectors for $m = 0$ and $n = m$, which result in the formation of zigzag and armchair SWCNTs, respectively. (b) Illustrated representations of defect-free (n,m) open-ended SWCNTs. From left, a metallic conducting armchair (10,10), a chiral semiconducting (12,7), and a conducting zigzag (15,0) SWCNT are depicted. (c) Illustrated representation of common defects found in an SWCNT. (Parts b and c from Hirsch, A. Functionalization of single-walled carbon nanotubes, *Angew. Chem. Int. Ed.*, 2002, 41, 1853–1859. Copyright Wiley-VCH Verlag GmbH & Co. KGaA. Weinheim, Fed. Rep. of Germany. Reprinted with permission.)

the nanotube (Figure 13.6).[52,89,91–94] SWCNT structure is described in terms of chirality and a pair of indices (n,m). An SWCNT is essentially formed when a graphene sheet is rolled up such that one of the carbon hexagons in the lattice defined as the origin point and another carbon hexagon in the lattice defined as the lattice point become superimposed. Therefore, (n,m) corresponds to the indices of the carbon hexagon (lattice point) in the graphene sheet lattice that overlaps with the origin carbon hexagon (origin point) of the lattice to produce the tubular structure. If $n = m$, the SWCNT is in the armchair structural configuration and is metallic in terms of conductivity.[52,89] If $n \neq m$ and $n \neq 0$, the SWCNT is chiral and semiconducting. If $n = 0$, the SWCNT is in the zigzag configuration and is semiconducting.[52,89] Zigzag SWCNTs can further be distinguished as semimetallic, that is, small bandgap semiconductors, when the quantity $(n - m)/3$ is an integer or large bandgap semiconductors when the quantity $(n - m)/3$ is not an integer.[52,89] Semimetallic SWCNTs possess such small bandgaps that they are often considered metallic.[52] MWCNTs are also deemed metallic as a single metallic layer in the structure of concentric CNTs results in the entire nanotube exhibiting metallic conductivity.[59] Electronic properties can also be affected by production and processing procedures that can introduce catalytic particles, ion doping, and sidewall functionalization in the CNT structure.[89]

Both SWCNTs and MWCNTs have found widespread application in bioelectrochemistry and biosensors.[46,52,55–57,59,89] CNTs may be covalently or noncovalently functionalized[52,93–99] with biomolecules or small organic molecules and thus serve as sensitive transducers in biosensing applications. Covalent coupling of molecules to CNTs often involves amide bond formation between the primary amine groups of a molecule and the carboxylate groups found at CNT ends and sidewall defects introduced by acid-catalyzed oxidation. Covalent modification may result in some loss of favorable electronic characteristics of CNTs. However, noncovalent coupling strategies, such as wrapping CNTs with polymers or DNA, can be used to prepare CNTs with a desired surface modification while helping preserve the structural integrity and properties of CNTs.[94,98,99]

As with AuNP electrodes, electrodes that consist of collections of CNTs prepared by adsorption of SWCNTs or MWCNTs on a conductive substrate or CNT electrodes prepared by screen printing of CNT-based conductive inks[100] have been used as platforms for observing direct electron transfer in proteins, such as cyt c,[55,101,102] HRP,[103,104] GOx,[105,106] and catalase.[107] CNT electrodes have also been used in the direct oxidation of adenine and guanine nucleobases in DNA[108–110] and for the detection of small biologically important molecules such as dopamine,[111] uric acid,[112] and β-nicotinamide adenine dinucleotide (NADH).[113]

CNT-modified carbon paste electrodes and GC electrodes (GCEs) gave improved oxidation signals for DNA nucleobases compared to unmodified electrodes.[46,108–110] Wang et al. found that MWCNTs or SWCNTs, deposited onto GCEs through drop casting from a nitric acid or N,N-dimethylformamide suspension, facilitated the accumulation of guanine nucleotides at the electrode surface.[46,108,109] Adsorptive accumulation of guanine on MWCNTs led to oxidation signals 11-fold larger than those observed at bare GCEs. This phenomenon was exploited to detect DNA segments related to the breast cancer gene BRCA1 with DLs of 40 ng mL^{-1} (~2 nM).[108] SWCNT electrodes exhibited similar electrochemical responses to intact and denatured DNA.[109] In accordance with spectroscopic studies of SWCNT/DNA interactions, it is suspected that the SWCNT electrode disrupts hydrogen bonding in DNA, thus unwrapping the DNA double helix and exposing the primary redox sites of adenine and guanine residues.[109]

CNT electrodes also offer accelerated electron transfer to NADH, lowering the large overvoltage commonly associated with NADH oxidation at ordinary bulk-material electrodes.[46,113] Thus, amperometric biosensors based on the coimmobilization of dehydrogenase enzymes and cofactor nicotinamide adenine dinucleotide (NAD$^+$) on CNT-modified electrodes can be used at lower potential, giving larger current signals than similar sensors based on bulk-material electrodes. These CNT-electrode biosensors can be used to detect important substrates such as lactate, alcohol, and glucose via oxidation of NADH, which is the product of the dehydrogenase/NAD$^+$ enzymatic reaction.[46,113] For example, Wang and Musameh incorporated alcohol dehydrogenase and NAD$^+$

in an MWCNT/Teflon matrix for low-potential detection of ethanol based on the electrocatalytic oxidation of NADH at MWCNTs.[114]

By comparing the results of electrooxidation of NADH at MWCNTs, basal-plane highly ordered pyrolytic graphite (HOPG), and edge-plane HOPG, Banks and Compton concluded that edge-plane sites and defects, which occur at the open ends of the MWCNTs, were primarily responsible for electrocatalytic activity toward NADH.[115,116] Their results suggested that CNTs may be no better than edge-plane HOPG in terms of electrochemical properties.[91,115,116] In another study, Gong et al. reported CVs of electrodes featuring 5 mm *superlong* CNTs with various redox species.[117] They found that the electrochemical response was not always enhanced at the nanotube ends or oxygen-containing surface defects compared to nanotube sidewalls.[91,117] For example, potassium ferricyanide had enhanced electrochemistry at CNT ends, especially ends presenting oxygenated moieties.[91,117] However, oxidation of hydrogen peroxide was favored at CNT sidewalls rather than ends and was not influenced by oxygenated sites on the nanotube.[91,117] Due to production and processing protocols, CNTs may also contain metal impurities that can give rise to electrocatalytic activity.[118] Overall, structural and compositional heterogeneity of CNTs, variations in CNT preparation and purification, and differences in the construction of CNT electrodes have resulted in widely varied reports of performance for biosensors based on CNT electrodes.[91]

In addition to the previously described dehydrogenase-based CNT electrodes, electrochemical biosensors that employ other types of enzyme-modified CNTs have also been reported. Kowalewska and Kulesza applied CNTs with adsorbed redox mediator tetrathiafulvalene (TTF) for electrochemical detection of glucose.[119] TTF-modified CNTs were found to facilitate electron transfer between GOx and the electrode surface for glucose detection. Jia et al. reported a similar strategy for the detection of lactate using MWCNTs modified with TTF and lactate oxidase.[120] Since TTF does not cause skin irritation and the CNT/TTF platform also enables low-potential sensing of lactate, CNT/TTF/lactate oxidase-based electrochemical biosensors could be used to detect lactate in perspiration directly on human skin. This was accomplished by preparing temporary tattoos from CNT/TTF/lactate oxidase-conductive carbon ink that was transferred onto a human subject's skin.[120]

Sandwich-type electrochemical and electrochemiluminescent[121–125] immunoassays have been developed by functionalizing CNT electrodes with antibodies.[89,126] For example, Wohlstadter et al. covalently coupled streptavidin to CNTs through amide bond formation between CNT carboxyl groups and primary amine groups of streptavidin.[126] Biotinylated antibodies for α-fetoprotein (AFP) were then immobilized on the streptavidin/CNT electrode to capture the target protein analyte. Detection was based on the electrochemiluminescence (ECL) signal generated by a second antibody labeled with ruthenium(II) tris(2,2′-bipyridine) $\left(Ru \left(bpy \right)_3^{2+} \right)$.

13.2.4 VERTICALLY ALIGNED CARBON NANOTUBE ELECTRODES, ARRAYS, AND ENSEMBLES

In terms of electron transfer, CNT ends have been likened to the edge planes of HOPG, at which fast electron-transfer rates are typically observed. CNT sidewalls are thought to behave like basal planes of HOPG, where electron transfer occurs at slower rates.[46,47,57,59,89,115,116,127] Thus, enhanced electron transfer at CNTs compared to other carbon materials has largely been attributed to edge-plane defects at nanotube ends,[115,116,127] though trapped metal impurities in CNTs may also contribute.[118,127] Therefore, special care has been taken to prepare vertically aligned assemblies of CNTs for use as electrodes in contrast to electrodes that feature randomly ordered assemblies of CNTs.

Vertically aligned MWCNTs and SWCNTs can be produced through template synthesis methods by chemical vapor deposition (CVD) of a gaseous hydrocarbon (e.g., acetylene or ethylene) precursor on metal-coated surfaces[128–133] or within porous membranes[134–136] or through covalent, electrostatic, or metal-assisted assembly on surfaces.[56,57,137–142] Electrodes composed of vertically aligned CNTs offer signal enhancement and improved electron-transfer kinetics compared to many types of electrodes modified with randomly oriented CNTs.[59,132,133,142] For example, carbon-fiber microdisks modified with vertically aligned SWCNTs (often called SWCNT forests[139]) exhibited over 30 times larger oxidation currents for dopamine compared to similar electrodes prepared with randomly oriented SWCNTs.[142]

Yu et al.[56] showed that SWCNT forest electrodes could be used to observe direct reversible electrochemistry of proteins HRP and myoglobin, and Gooding et al.[57] obtained reversible voltammetry for the heme-containing peptide microperoxidase MP-11 derived from cyt c. Yu et al. assembled oxidized SWCNTs 20–30 nm in length onto a composite bed of Nafion ionomer and precipitated Fe^{3+} hydroxides on ordinary pyrolytic graphite (Figure 13.7).[56] Myoglobin or HRP was covalently attached to the ends

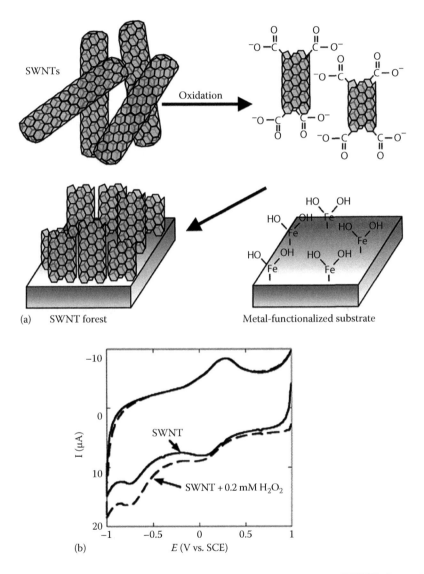

FIGURE 13.7 Preparation and characterization of single-walled nanotube (SWNT) forest formed by metal-assisted deposition of oxidized, shortened SWNTs. (a) Schematic representation of SWNT forest preparation. Shortened, open-ended SWNTs with carboxyl-functionalized ends are produced by oxidation of SWNTs. A suspension of SWNTs is introduced to a metal surface functionalized with iron hydroxides. An SWNT forest results as SWNTs vertically align via self-assembly. (Part a adapted with permission from Chattopadhyay, D., Galeska, I., and Papadimitrakopoulos, F., Metal-assisted organization of shortened carbon nanotubes in monolayer and multilayer forest assemblies, *J. Am. Chem. Soc.*, 123, 9451–9452, 2001. Copyright 2001 American Chemical Society.) Cyclic voltammograms (scan rate 300 mV s^{-1}) of (b) SWNT forest electrodes in pH 5.5 buffer with and without 0.2 mM H_2O_2.   (*Continued*)

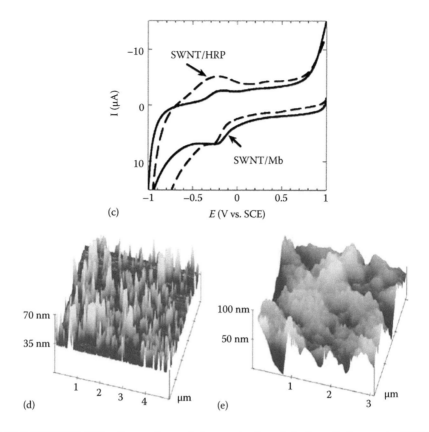

(c)

(d)

(e)

FIGURE 13.7 (CONTINUED)   Preparation and characterization of single-walled nanotube (SWNT) forest formed by metal-assisted deposition of oxidized, shortened SWNTs. (c) SWNT forest electrodes functionalized with myoglobin (Mb) in pH 5.5 buffer and HRP in pH 6.0 buffer (reduction is downward, oxidation is upward on y-axis). Atomic force microscope (AFM) images of an SWNT forest (d) and SWNT forest modified with Mb (e). (Parts b–e reprinted from *Electrochem. Commun.*, 5, Yu, X., Chattopadhyay, D., Galeska, I., Papadimitrakopoulos, F., and Rusling, J.F., Peroxidase activity of enzymes bound to the ends of single-wall carbon nanotube forest electrodes, 408–411. Copyright 2003 with permission from Elsevier.)

of the vertically aligned SWCNTs via amide bonds formed between carboxylate groups on SWCNT ends and primary amine groups on the protein. Catalytic reduction of H_2O_2 was observed using these protein-modified SWCNT forest electrodes, with results indicating that the aligned SWCNTs are efficient conductors of electrons from the external circuit to enzymatic redox sites.

SWCNT forest electrodes have been employed in electrochemical and electrochemiluminescent (ECL) immunosensors. Antibody-modified SWCNT forest electrodes and HRP-labeled detection antibodies were first employed to measure the small molecule biotin[143] and model protein human serum albumin[144] and later for cancer biomarker proteins in serum. These nanostructured immunosensors offer 3–10-fold gains in sensitivity compared to similar sensors based on flat bulk electrodes.[144,145] Analysis of the SWCNT forest electrode-based sensor revealed that dense packing of the carboxylated ends of SWCNTs enables attachment of a large surface concentration of capture antibodies, which, along with other favorable properties of the nanotubes, contributes to the observed sensitivity improvements of SWCNT-modified sensors over sensors based on bulk electrode materials.[145] ECL-based SWCNT forest immunosensors for cancer biomarker proteins relied on $Ru(bpy)_3^{2+}$ encapsulated in silica particles (Section 13.3.6) that were decorated with antibodies.[146,147] For the prostate cancer biomarker prostate-specific antigen, SWCNT forest ECL immunosensors exhibited 34 times higher sensitivities and 10 times lower DLs than analogous

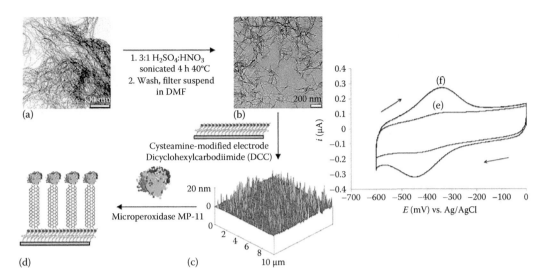

FIGURE 13.8 Preparation and characterization of SWNT forest formed by covalent attachment of oxidized, shortened SWNTs onto a cysteamine monolayer on a gold substrate. Scanning electron micrographs of SWNTs before (a) and after oxidative shortening (b). AFM image of SWNT forest (c) prepared by vertical self-assembly and covalent attachment of SWNTs on a cysteamine-modified gold substrate. Illustrated representation of attachment of MP-11 on SWNT forest electrode (d). CVs of cysteamine-modified gold electrode after immersion in MP-11/DMF solution (e) and SWNT forest electrode modified with MP-11 (f). CVs were obtained in 0.05 M phosphate buffer (pH 7.0) containing 0.05 M KCl under argon at a scan rate of 100 mV s^{-1}. (Adapted with permission from Gooding, J.J., Wibowo, R., Liu, J. et al., Protein electrochemistry using aligned carbon nanotube arrays, *J. Am. Chem. Soc.*, 125, 9006–9007, 2003. Copyright 2003 American Chemical Society.)

ECL immunosensors based on bulk pyrolytic graphite.[146] Venkatanarayanan et al. showed that such SWCNT forest electrodes can also be inkjet printed onto indium tin oxide to produce ECL-based immunosensors.[148]

Gooding et al. prepared vertically aligned SWCNT electrodes through covalent attachment of SWCNTs onto cysteamine-functionalized gold surfaces (Figure 13.8).[57] Electron transfer between SWCNTs and heme-containing peptide MP-11, covalently bound to SWCNT ends through amide bonds, occurred at a rate similar to that observed for MP-11 attached directly to the underlying electrode surface, indicating that SWCNTs exhibit very little resistance to electron transfer. Electron-transfer rates were not affected by the length of the nanotubes for average lengths of ~70, 120, or 490 nm.[57]

Patolsky et al., employing similar vertically aligned SWCNT electrodes, showed that electrons can be transported along distances greater than 150 nm to the enzyme GOx on SWCNT ends.[141] However, the rate of electron transfer in these experiments depended on nanotube length. Electron transfer between the electrode and the redox center of GOx was slower when longer SWCNTs were used. Gooding suspected that the relationship between electron-transfer rate and nanotube length may depend on CNT preparation methods.[89] Particularly, dialysis, which was used by Patolsky et al. to purify SWCNTs after oxidative acid treatment for tube shortening, probably removes acid from the nanotube center.[89] The concentrated acids commonly used to prepare SWCNTs may lead to doping as ions enter the nanotube center and inject electrons or holes into the conduction bands.[89] Thus, removal of ion doping by dialysis may restore semiconducting behavior of SWCNTs, which could explain why electron-transfer rates depend on nanotube length for some SWCNT forest electrodes.[89]

Li et al. prepared arrays of vertically aligned individual MWCNTs and MWCNT bundles from plasma-enhanced chemical vapor deposition (PECVD) on 10–30 nm thick nickel films

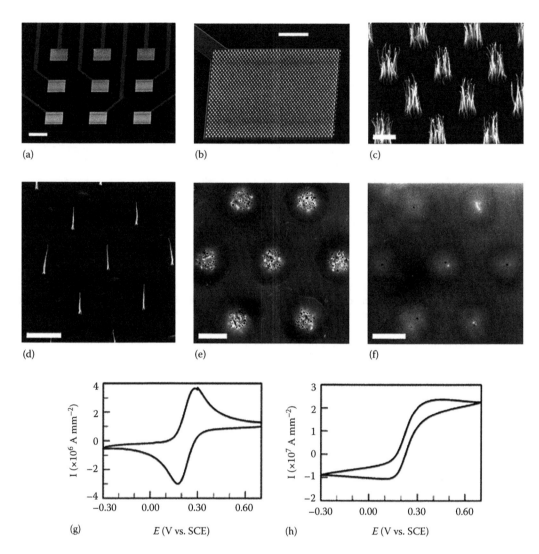

FIGURE 13.9   Vertically aligned MWCNTs prepared by PECVD. SEM images of (a) 3×3 electrode array of MWCNTs, (b) MWCNT bundles on a single electrode pad, (c) MWCNT array prepared by UV lithography, (d) MWCNT array prepared by e-beam lithography, (e) polished MWCNT electrodes prepared by UV lithography, and (f) polished MWCNT electrodes prepared by e-beam lithography. Scale bars for SEM images are 200 (a), 50 (b), 5 (d), and 2 (c, e, f) μm. Cyclic voltammograms of high (g)- and low (h)-density MWCNT arrays in 1 mM potassium ferrocyanide and 0.1 M potassium chloride. (Adapted with permission from Li, J., Ng, H.T., Cassell, A. et al., Carbon nanotube nanoelectrode array for ultrasensitive DNA detection, *Nano Lett.*, 3, 597–602, 2003. Copyright 2003 American Chemical Society.)

(Figure 13.9).[129,130] Array density was controlled by using lithography to define Ni-catalyst spot size. Nanotubes were encapsulated and insulated using a tetraethoxysilane CVD process, and chemical mechanical polishing was done to planarize the array and expose nanotube ends. Probe DNA complementary to the wild-type allele of the BRCA1 gene was covalently attached to the nanotube ends. Electroactive guanine bases were replaced with electrochemically inactive inosines in the probe strand, and the target strand was labeled with a 10 bp polyG sequence. The target was electrochemically detected through the oxidation of guanines mediated by $\left(Ru\left(bpy\right)_3^{2+}\right)$.

Li et al. found that low-density MWCNT arrays ($\sim 7 \times 10^7$ electrodes cm^{-2}) provided ultrahigh sensitivity in electrochemical DNA sensing applications.[129,130] In general, maximum current density is achieved with ultramicro- and nanoelectrode arrays when each electrode is able to act as an individual ultramicro- or nanoelectrode.[149] If the center-to-center separation distance between each electrode is too small, transport of electroactive species, such as a small diffusible mediator like $\left(Ru(bpy)_3^{2+} \right)$, to individual electrodes suffers from interaction and overlapping of diffusion fields for neighboring electrodes.[149] In terms of analytical performance, separation distance and array element density are key considerations in array design.[149,150] Separation distance should be large enough to avoid diffusion layer overlap, but small enough to garner maximum signal from the electrodes and avoid inefficient use of space.[149,150]

Arrays with ultramicro- or nanoelectrodes separated from nearest neighbors by more than 12 times the average radius ($12R_{av}$) are considered sparse enough such that electrodes are sufficiently isolated to exhibit an independent electrochemical response.[130,149–151] This supposition is rooted in theoretical work by Saito, who showed that the concentration of an electroactive species undergoing a redox reaction at a macroscopic circular electrode attains 90% of its bulk value at a distance of ~ 6 times the radius (R) of the electrode.[152] Thus, $6R$ has been considered the diffusion layer thickness for a macroscopic electrode. However, several nanoelectrode arrays and ensembles (NEEs; collections of irregularly spaced electrodes) with spacing $>12R_{av}$ still exhibit diffusion layer overlap as evidenced by nonsigmoidal peak-shaped CV responses.[129,130,151] This discrepancy has often been ascribed to the random placement of individual electrodes in NEEs.[129,130,151] However, recent theoretical studies suggest that a separation distance of $12R_{av}$ is not sufficient for collections of ultramicro- and nanoelectrodes to exhibit ideal, independent sigmoidal responses.[149,150]

Amatore et al. developed a theoretical framework to describe the electrochemical responses of ultramicroelectrode ensemble and NEEs by considering mass transport for assemblies of microdisk and microband electrodes.[149,153,154] Lee et al. used finite element simulation to solve 3D diffusion equations and found that a collection of 10 μm diameter microdisk electrodes required a separation distance of more than $40R_{av}$ to exhibit a sigmoidal simulated CV response typical for radial diffusion.[154] CV response typical of reversible linear diffusion at macroelectrodes was observed when the separation distance was less than $6R_{av}$. Assemblies of microelectrodes for which the separation distances were between $6R_{av}$ and $40R_{av}$ exhibited peak-shaped simulated CVs indicative of a mixture of radial and linear diffusion behavior. Thus, $12R_{av}$ seems to be too small a separation distance for the design of ideal microelectrode arrays.

Davies and Compton completed 2D simulations based on the diffusion domain approximation introduced by Amatore et al. and also found that $12R_{av}$ was not a sufficient guideline for separation distance in collections of regularly and irregularly spaced electrodes with radii <10 μm.[149] Davies and Compton suggested that minimum distance between neighboring electrodes to avoid diffusion layer overlap is not linearly related to electrode size. Rather, ideal separation distance is a function of R_{av}, diffusion coefficient of the electroactive species, and the scan rate of the CV. Simulations can thus be used to correctly predict ideal separation distance. For example, Davies and Compton estimated this distance for 100 nm radius electrodes to be ~ 12 μm at 2 V s^{-1} and 70 μm at 0.005 V s^{-1} and for 10 nm radius electrodes to be ~ 5 μm at 2 V s^{-1} and 20 μm at 0.005 V s^{-1} when the diffusion coefficient of the electroactive species is 10^{-5} cm^2 s^{-1}.[149]

Li et al. showed that high-density MWCNT arrays ($\sim 2 \times 10^9$ electrodes cm^{-2}; ~ 240 nm nearest-neighbor separation distance) gave typical peak-shaped, linear diffusion–limited CVs.[129,130] Low-density MWCNT arrays ($\sim 7 \times 10^7$ electrodes cm^{-2}; ~ 1.3 μm nearest-neighbor separation distance) produced sigmoidal CV characteristic of typical radial-diffusion responses of single ultramicro- or nanoelectrodes (Figure 13.9g and h). The small size of the MWCNTs (~ 17 nm radius in these studies) and large separation distance (~ 1.3 μm) between neighboring MWCNTs in low-density arrays enabled each MWCNT to behave like an independent nanoelectrode with high signal-to-noise ratio and small time constant.[130]

Other biosensors that feature vertically aligned CNT arrays with sufficiently spaced CNTs to minimize diffusion layer overlap have also been reported. Lin et al. prepared vertically aligned, irregularly spaced CNTs, or CNT NEEs for glucose sensing.[155] CNTs were embedded in epoxy, and GOx was covalently bound to exposed CNT tips. Though amperometric detection of glucose at +0.4 V was complicated by the influences of interferents (i.e., ascorbic acid, uric acid, and acetaminophen), the CNT NEEs enabled highly selective, mediator-free detection of glucose at −0.2 V.

Yun et al. developed a label-free immunosensor based on aligned MWCNT arrays.[156] Antibodies were attached to MWCNT ends, and binding of model antigen mouse IgG to the antibody-modified surface resulted in an increase in the electron-transfer resistance. Formation of the antibody–antigen complex was monitored by CV and electrochemical impedance spectroscopy (EIS), resulting in a DL of 200 ng mL^{-1} with a dynamic range up to 100 μg mL^{-1} for IgG.

13.2.5 GRAPHENE-BASED ELECTRODES FOR BIOELECTROCHEMISTRY

CNTs in electrochemical biosensors can be complicated by metal impurities and difficulties in isolating, purifying, and arranging heterogeneous CNTs onto a conductive platform.[91,157,158] Graphene, with its planar sp^2 carbon sheet structure (Figure 13.6a) has recently emerged as another attractive material for preparing electrochemical biosensors.[51,157–159] Graphene can be prepared without using a metal catalyst, thereby decreasing the possibility of metal particle impurities.[91,157–159] Graphene also possesses a larger surface area (2630 m^2 g^{-1}) than CNTs,[157] and the conductivity of graphene, at 64 mS cm^{-1}, is about 60 times greater than that of CNTs.[160]

Mohanty and Berry were the first to demonstrate electrical biosensors based on graphene.[91,161] Chemically modified graphene nanosheets (i.e., graphene oxide [GO] or graphene amine [GA]) were electrostatically deposited between gold electrodes on a silica layer supported by heavily doped *n*-type silicon (Figure 13.10a). The thickness of GO sheets was 1–4 nm, indicating that the graphene derivatives consisted of a few layers. The chemically modified graphene nanosheets exhibited *p*-type semiconducting character with high resistances (10s to 100s of MΩ) and low charge carrier mobilities (0.002–5.9 cm^2 V^{-1} s^{-1}). However, covalent attachment of single-stranded DNA (ss-DNA) to GO resulted in conductivity that is twofold larger than the conductivity of unmodified GO (Figure 13.10b). The attachment of negatively charged DNA was predicted to increase the hole density in *p*-type GO, thus resulting in an increase in conductivity. Hybridization of the GO-anchored ss-DNA to its complementary strand resulted in further 71% increase in conductivity, while exposure to noncomplementary strand did not affect the measured conductivity. Again the increase afforded by hybridization was attributed to the generation of holes in the *p*-type GO brought about by the phosphate ions of the complementary DNA. Dehybridization and rehybridization experiments showed that GO-based DNA sensors are reusable and reproducible. Electrostatic adsorption of bacteria to GA was similarly found to increase conductivity by 42% compared to unmodified GA. Similar to the DNA studies, the attachment of negatively charged bacteria leads to the generation of more holes in the *p*-type GA.

Zhou et al. showed good separation of oxidation peaks for the four free bases of DNA via differential pulse voltammetry using chemically reduced GO on a GC substrate.[162] While guanine (G), adenine (A), thymine (T), and cytosine (C) oxidation peaks are not adequately separated in voltammetry that employs unmodified GCEs or graphite/GCEs, chemically reduced GO/GCEs allow simultaneous detection of G, A, T, and C in both ss-DNA and double-stranded DNA (ds-DNA). Chemically reduced GO/GCEs also gave enhanced peak currents compared to unmodified GC and graphite/GC, thus indicating higher electrocatalytic activity of chemically reduced GO toward G, A, T, and C oxidation (Figure 13.11). Single-nucleotide polymorphisms for short oligomers related to the p53 gene were distinguished, without hybridization or labeling, by employing differential pulse voltammetry at chemically reduced GO/GCEs.[162]

Graphene-based electrodes have also been used as platforms for observing direct electrochemistry of GOx and the development of glucose sensors.[51,163,164] Shan et al. prepared electrodes from poly(vinylpyrrolidone) (PVP)-protected graphene that was immobilized on a GC substrate with the

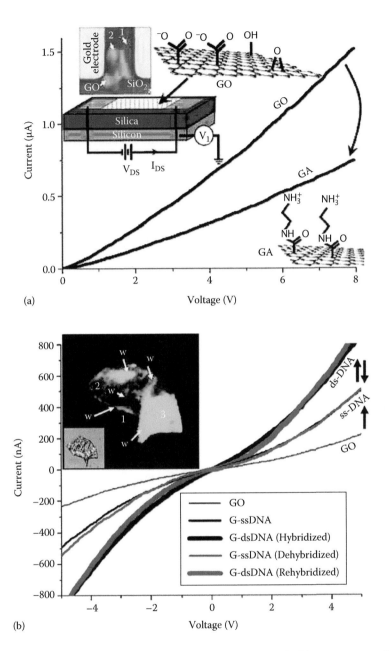

FIGURE 13.10 Electrical biosensors based on graphene sheets deposited between two gold electrodes. (a) Current–voltage behavior of GO and GA devices. Insets show GO between two gold electrodes, an illustrated representation of the device, and illustrated representations of GO and GA structures. (b) Current–voltage responses of GO device, GO device modified with ss-DNA, and ss-DNA-modified GO device exposed to complimentary strand to form ds-DNA. Inset shows wrinkles (W) and folds in a ds-DNA-functionalized graphene sheet. (Adapted with permission from Mohanty, N. and Berry, V., Graphene-based single-bacterium resolution biodevice and DNA transistor: Interfacing graphene derivatives with nanoscale and microscale biocomponents, *Nano Lett.*, 8, 4469–4476, 2008. Copyright 2008 American Chemical Society.)

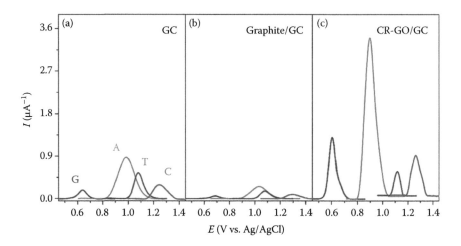

FIGURE 13.11 Differential pulse voltammograms of free single nucleobases guanine (G, blue), adenine (A, orange), thymine (T, violet), and cytosine (C, pink) at (a) GC, (b) graphite/GC, and (c) chemically reduced GO (CR-GO)/GC electrodes. Concentration of each nucleobase was 10 μg mL^{-1} and supporting electrolyte was 0.1 M phosphate-buffered saline (PBS), pH 7.0. (Adapted with permission from Zhou, M., Zhai, Y., and Dong, S., Electrochemical sensing and biosensing platform based on chemically reduced graphene oxide, *Anal. Chem.*, 81, 5603–5613, 2009. Copyright 2009 American Chemical Society.)

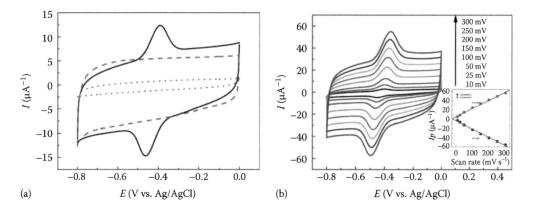

FIGURE 13.12 CV response of GOx at a graphene-modified GCE in 0.05 M PBS, pH 7.4. (a) Response for a GCE modified with graphene through the use of a PFIL (red dashed), with GOx–graphite–PFIL (blue dotted), and with GOx–graphene–PFIL (solid black). (b) Scan rate dependence of voltammetric response for GOx–graphite–PFIL-modified electrode. (Adapted with permission from Shan, C., Yang, H., Song, J., Han, D., Ivaska, A., and Niu, L, Direct electrochemistry of glucose oxidase and biosensing of glucose based on graphene, *Anal. Chem.*, 81, 2378–2382, 2009. Copyright 2009 American Chemical Society.)

aid of a poly(ethyleneimine)-functionalized ionic liquid (PFIL).[163] In CV studies, GOx adsorbed onto the graphene composite electrode presented a pair of peaks centered around −0.43 V versus Ag/AgCl with peak-to-peak separation of ~69 mV and a cathodic-to-anodic peak current ratio of ~1 (Figure 13.12). The resulting glucose sensor based on the reduction of O_2 and H_2O_2 at the enzyme-modified graphene composite electrode exhibited a linear range from 2 to 14 mM and was stable for at least 1 week.[163]

In another study, Kang et al. prepared GOx-modified graphene-based electrodes through the use of a chitosan–graphene film on GC.[164] EIS showed that the graphene–chitosan film electrode

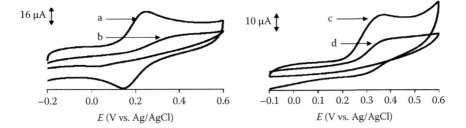

FIGURE 13.13 Cyclic voltammograms for graphene (a, c)- and SWCNT (b, d)-modified electrodes in pH 7.4 PBS with 2.5 mM dopamine (a, b) and 2.5 mM serotonin (c, d). (Adapted with permission from Alwarappan, S., Erdem, A., Liu, C., and Li, C.-Z., Probing the electrochemical properties of graphene nanosheets for biosensing applications, *J. Phys. Chem. C*, 113, 8853–8857, 2009. Copyright 2009 American Chemical Society.)

exhibited a lower electron-transfer resistance than unmodified and chitosan-modified GCEs. The GOx-modified graphene–chitosan film electrode produced a pair of CV peaks centered at −0.477 V versus Ag/AgCl and separated by ~80 mV. Kang et al. reported a linear range of 0.08–12 mM and DL of 0.02 mM for glucose sensing with their graphene composite electrodes. Enzyme- and antibody-modified graphene-based electrodes have also been reported for other sensing applications based on voltammetry, amperometry, and impedance measurements.[165–167]

Electrochemical sensors for small biologically important molecules such as dopamine[160,168,169] and enzyme-reaction products and cofactors such as H_2O_2 and NADH have also been made from graphene.[51] Alwarappan et al. reported that graphene-based electrodes provided better sensitivity, signal-to-noise ratio, and stability than randomly oriented SWCNT electrodes for the electrochemical detection of dopamine (Figure 13.13).[160] Graphene-based electrodes exhibited three distinct oxidation peaks when a solution that contained dopamine along with common interferents serotonin and ascorbic acid was analyzed by differential pulse voltammetry, while SWCNT electrodes displayed a single broad peak. Similarly, Wang et al. showed that graphene electrodes presented superior electroanalytical performance over MWCNT electrodes for the detection of dopamine.[169]

Though the term graphene refers to a single sheet of sp^2 carbon atoms, graphene employed in many biosensors often consists of a few layers due to the difficulty in controllably isolating the single layers.[91] Mohanty and Berry showed that electrostatic immobilization of chemically modified graphene on flat substrates can introduce wrinkles at which biomolecules are more likely to bind.[161] Varying degrees of success have been documented with CNT biosensors due to the diversity of CNTs employed and different methods used to fabricate CNT electrodes. From studies done so far, it can similarly be concluded that the analytical performance of any graphene biosensor is likely to depend on the methods to prepare, purify, and arrange the graphene layer(s).[91]

13.2.6 Bioelectrochemistry with Other Nanomaterials and Nanocomposite Electrodes

Electrodes made from metal or semiconducting nanoparticles paired with carbon-based nanomaterials have been used as platforms for electrochemical and electrochemiluminescent (ECL) biosensing.[170–176] Pairing platinum or gold nanoparticles with CNTs or graphene can result in a nanocomposite electrode that exhibits improved electrocatalytic activity over electrodes composed of either single nanomaterial.[170,176] For example, platinum nanoparticle (PtNP)/MWCNT nanocomposite electrodes exhibited four times better sensitivity than MWCNT electrodes for glucose sensing.[176] In this case, the improved sensitivity was attributed to the combination of electrocatalytic activity of the PtNPs and MWCNTs toward hydrogen peroxide.[176]

ECL-based nanocomposite biosensors have resulted from the combination of semiconducting nanocrystals or quantum dots with other nanomaterials. For example, Jie et al. prepared nanocomposite electrodes consisting of AuNPs or CNTs and cadmium sulfide (CdS) or cadmium selenide (CdSe) quantum dots for ECL-based low-density lipoprotein and human IgG biosensors.[171–173] In ECL of quantum dots, persulfate ($S_2O_3{}^{2-}$), dissolved oxygen (O_2), and hydrogen peroxide (H_2O_2) can serve as coreactants to help facilitate the formation of and stabilize electrogenerated species derived from nanocrystals.[171,172,177,178] In general, reduced quantum dot species (CdX^-, where X may be S, Te, or Se) enter the excited state (CdX^*) upon interaction with strong oxidants that result from the reduction of the coreactant (e.g., $SO_4{}^{-}$ from the reduction of $S_2O_8{}^{2-}$ or OOH^- from the reduction of O_2)[171–173,177,178]:

$$S_2O_8{}^{2-} + e^- \rightarrow SO_4{}^{2-} + SO_4{}^{-\bullet} \tag{13.1}$$

$$O_2 + H_2O + 2e^- \rightarrow OOH^- + OH^- \tag{13.2}$$

$$CdX + e^- \rightarrow CdX^{-\bullet} \tag{13.3}$$

$$CdX^{-\bullet} + SO_4{}^{-\bullet} \rightarrow CdX^* + SO_4{}^{2-} \tag{13.4}$$

$$2CdX^{-\bullet} + OOH^- + H_2O \rightarrow 3OH^- + 2CdX^* \tag{13.5}$$

$$CdX^* \rightarrow CdX + h\nu \tag{13.6}$$

In the work of Jie et al., quantum dot nanocomposite electrodes were functionalized with an antibody or a ligand for a receptor, and a decrease in ECL signal derived from the CdS or CdSe nanocrystals was measured.[171–173] Quantum dot nanocomposite ECL-based biosensors had a linear range of 0.002–500 ng mL^{-1} with a DL of 0.6 pg mL^{-1} using human IgG as a model analyte.[173]

13.2.7 ELECTROCHEMICAL STUDIES OF BIOMOLECULES AT SINGLE NANOELECTRODES

Single particle nanoelectrodes have also been fabricated for electrochemical studies of biomolecules. Crooks and coworkers reported fundamental electrochemical studies using single CNT electrodes.[179] Similar to preparations of single CNT probes for scanning probe microscopy,[180,181] a single MWCNT (80–200 nm in diameter) was attached to a sharpened Pt wire and cut to a length of 15–50 μm.[179] Boo et al. similarly mounted a single MWCNT (~30 nm in diameter) onto a tungsten tip and used bare and enzyme-modified MWCNT electrodes to measure dopamine and glutamate.[182]

Hoeben et al. prepared Au electrodes with dimensions of ~70×70 nm^2 by lithography and used them to study the enzyme [NiFe] hydrogenase via protein-film voltammetry.[183] A submonolayer of [NiFe] hydrogenase was immobilized on the polymyxin-pretreated Au nanoelectrode. Electrochemical response from fewer than 50 enzyme molecules was reported.

13.3 NANOMATERIALS AS LABELS IN ELECTROCHEMICAL BIOSENSING STRATEGIES

Since many biomolecules are not sufficiently electroactive to elicit a measurable electrochemical signal, direct electrochemical detection is often not feasible. Thus, nanomaterials have found widespread use as labels in electrochemical biosensing strategies. Metal and semiconductor nanoparticles can serve as signal-transducing agents in voltammetric and ECL-based biosensors. Other types of nano- and microparticles can be used as vehicles to deliver large numbers of nanoparticle, redox, ECL, or enzyme labels to the electrode. As outlined in the following sections, nanomaterial labeling strategies can be used for highly sensitive detection of biomolecules. Nanomaterial labeling can be combined with nanostructured electrode platforms to produce biosensors with excellent analytical performance and ultrasensitive response.[184]

13.3.1 METAL NANOPARTICLES AS LABELS IN BIOMOLECULE DETECTION STRATEGIES

Many early examples of bioanalyses facilitated by nanoparticle labels were centered on anodic stripping voltammetry (ASV) of metal nanoparticles that were functionalized with biomolecules.[185] For example, González-García and Costa-García showed that differential pulse adsorptive stripping voltammetry on a carbon paste electrode can be used to determine IgG in the sub-nM to nM concentration range when it is labeled with colloidal gold.[186] Colloidal gold labels adsorb onto the carbon paste electrode, where they are oxidized in HCl at 1.25 V vs. Ag/AgCl to yield $AuCl_4^-$. The reduction of adsorbed $AuCl_4^-$ at +0.43 V gives rise to the analytical signal. Alternatively, silver can be deposited on oxidized AuNP labels, and sensing can be conducted by ASV of silver.[187]

González-García et al. also employed carbon paste electrodes with adsorbed biotinylated albumin to monitor biotin–streptavidin binding using 10 nm gold colloid labels with adsorbed streptavidin.[188] Voltammetry of the bound colloidal gold label enabled measurement of streptavidin at concentrations as low as 2.5 nM in phosphate buffer. Ozsoz et al. later demonstrated the use of pencil graphite electrodes modified with target DNA and colloidal gold-labeled probe sequences for voltammetric detection by oxidation of AuNP labels.[189] One drawback to these electrochemical methods of label detection is that the metal nanoparticle label must be in direct contact with the electrode surface.[190,191] Thus, large portions of analyte on nanoparticles not in sufficient contact with the electrode surface are excluded from contributing to the signal. The result is decreased sensitivity in comparison to methods where all metal nanoparticles with analyte can be detected.[190,191]

Alternative strategies for electrochemical detection of metal nanoparticle labels include first dissolving the nanoparticle to form metal ions that can be detected from solution by ASV and improving nanoparticle label-electrode surface contact through the use of a magnetic field and magnetic particles. For example, Wang et al. employed magnetic beads labeled with probe DNA to collect biotinylated target DNA segments related to the BRCA1 breast cancer gene.[192] Biotinylated target DNA was labeled with streptavidin-coated AuNPs, which were subjected to hydroquinone (HQ)-mediated silver deposition. A magnet was placed under a screen-printed carbon electrode to facilitate direct contact between the silver-enhanced AuNPs and the electrode surface. Chronopotentiometric stripping analysis for silver was observed with target DNA at 200 ng mL^{-1} or lower only when the magnet was used to improve contact between nanoparticle labels and electrode. DL for BRCA1 DNA was estimated to be 150 pg mL^{-1}.

Limoges and coworkers first employed the particle dissolution strategy for an electrochemical metalloimmunoassay for IgG based on the detection of dissolved 18 nm colloidal gold labels through ASV (Figure 13.14).[193] IgG bound to a capture antibody on the microwell platform was recognized by a second antibody labeled with the colloidal gold particle. Subsequent oxidative dissolution of the colloidal gold particles in a hydrobromic acid/bromine mixture released ~200,000 Au^{3+} ions per particle. ASV at screen-printed carbon electrodes was used for measurement. With this approach, they detected IgG from 35 µL samples down to 3 pM in buffer, comparable to standard ELISA. They later anchored a single-stranded target DNA to the microwells and used an oligonucleotide-modified gold colloid label to detect a 406-base DNA sequence of herpes virus human cytomegalovirus down to 5 pM.[194]

Wang et al. used a similar strategy but employed magnetic microspheres in place of the microwell platform to capture and detect nucleic acid sequences related to the breast cancer gene BRCA1 (Figure 13.15).[195] Streptavidin-coated magnetic beads with bound biotinylated capture probe were used to separate biotinylated target DNA from microliter sample volumes. Streptavidin-coated gold nanoparticles (AuNP; 5 nm dia.) were then introduced to the captured biotinylated target DNA. The AuNP labels were either immediately dissolved in oxidative acid solution and detected by

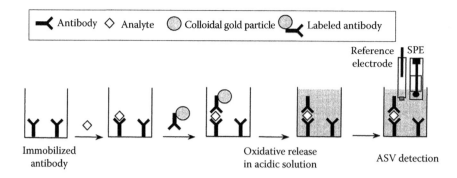

FIGURE 13.14 Schematic representation of a sandwich-type electrochemical metalloimmunoassay that uses a polystyrene microwell platform, a gold nanoparticle label, and anodic stripping voltammetric (ASV) detection at a screen-printed electrode (SPE). (Reprinted with permission from Dequaire, M., C. Degrande, and B. Limoges. An electrochemical metalloimmunoassay based on a colloidal gold label, *Anal. Chem.*, 72, 5521–5528, 2000. Copyright 2000 American Chemical Society.)

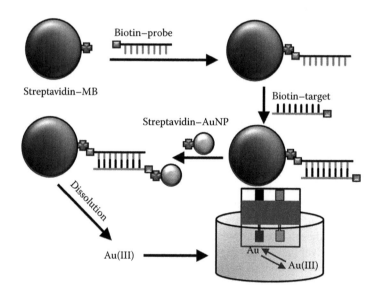

FIGURE 13.15 Schematic representation of metal nanoparticle-based electrochemical detection of DNA hybridization that uses streptavidin-coated magnetic beads to serve as the platform for the biotinylated capture probe DNA sequence. After the biotinylated oligomer target hybridizes with the capture probe, streptavidin-coated gold particles are used to label and detect hybridization via oxidative dissolution of gold and subsequent electrochemical stripping analysis at screen-printed electrodes. (Adapted with permission from Wang, J., Xu, D., Kawde, A.-N., and Polsky, R., Metal nanoparticle-based electrochemical stripping potentiometric detection of DNA hybridization, *Anal. Chem.*, 73, 5576–5581, 2001. Copyright 2001 American Chemical Society.)

chronopotentiometric stripping or first subjected to an amplification step during which gold[195] or silver[196] was catalytically precipitated onto the AuNP labels, followed by subsequent dissolution and stripping analysis. Such amplification strategies had been previously used in electron microscopy to enlarge AuNP-labeled structures for better visualization[197,198] and were exploited by Mirkin et al. in sensitive scanometric DNA hybridization assays[199] and immunoassays.[200] Velev and Kaler also used silver enhancement of AuNP labels to develop an immunoassay based on the measurement of resistance across a micrometer-sized gap between microelectrodes.[201]

Wang et al. have used chronopotentiometric stripping analysis of silver-enhanced AuNPs employing nitric acid solution to oxidize and dissolve deposited silver,[196] whereas the AuNP and gold-enhanced AuNP-based assays require the use of a hydrobromic acid/bromine mixture for oxidative dissolution of gold.[195,202] Catalytic precipitation of gold onto AuNP labels resulted in a 1.5 nM stripping analysis DL for target DNA,[195] while catalytic precipitation of silver resulted in pM DLs.[196] Using gold enhancement of AuNP labels and ASV in a microwell platform, Rochelet-Dequaire et al. demonstrated DLs as low as 600 aM for a 35 base-pair human cytomegalovirus DNA target.[202] A poly(ethylene glycol)–sodium chloride mixture, an aggregating agent during AuNP label enlargement, was necessary to retain enlarged labels on the bottom of the microwell during rinsing steps of the assay.[202]

Metalloimmunoassays for proteins based on stripping analysis of copper-[203] and silver-enhanced[204–206] AuNPs as well as silver nanoparticle[190] labels have also been reported. One problem with all dissolution-based strategies for electrochemical detection of nanoparticle or enhanced nanoparticle labels is that dissolution often involves employing toxic or corrosive reagents, for example, hydrobromic acid/bromine mixture for AuNPs and nitric acid for silver nanoparticles and copper- or silver-enhanced AuNPs. In one study, electrooxidation and complexation with thiocyanate was used as an alternative to dissolve silver nanoparticles in place of nitric acid for anodic stripping detection of myoglobin.[190]

13.3.2 Quantum Dots as Electrochemical and Electrochemiluminescent Labels for Biomolecule Detection

Quantum dots have been used as labels in electrochemical stripping analysis and ECL biosensing.[49,185,207,208] For example, chronopotentiometric stripping of cadmium from quantum dot cadmium sulfide (CdS) labels was explored for the detection of DNA and proteins.[185] Signal amplification can be accomplished by catalytic precipitation of cadmium onto the CdS labels and subsequent dissolution of enlarged labels by exposure to 1 M nitric acid.[209] Enlarged CdS labels produced a response >12 times larger than normal CdS labels, resulting in a DNA DL of 20 ng mL^{-1} (100 fmol in a 50 μL sample). An aptamer-based immunoassay for proteins that involved dissolution of CdS labels in hydrogen peroxide and subsequent detection of Cd^{2+} using an ion-selective microelectrode was also reported.[210] A DL of 0.14 nM (28 fmol in 200 μL) was demonstrated for thrombin with this strategy.

Quantum dots have also been used as labels for ECL-based protein and nucleic acid assays.[124,177,178] Hu et al. reported CdTe quantum dots as electrochemiluminescent labels for DNA hybridization sensors.[177] As described in Section 13.2.6, persulfate ($S_2O_8^{2-}$), dissolved oxygen (O_2), and hydrogen peroxide (H_2O_2) can serve as coreactants to facilitate the formation of and stabilize electrogenerated species derived from quantum dots. Hu et al. demonstrated a linear range of 5 fM to 10 pM for target DNA when CdTe quantum dot labels were used for ECL-based detection.[177]

13.3.3 Distinguishable Nanoparticle Labels and Multiplexed Biomolecule Detection

Multiplexed metalloimmunoassays for proteins have been developed by employing antibodies labeled with different colloidal metal-containing nanoparticles (CdS, zinc sulfide, lead sulfide, and copper sulfide) that give rise to well-resolved peaks in square-wave ASV from stripping of the different metal ions (Figure 13.16).[211] Stripping analysis coding strategies were expanded using nanowires and other nanoparticles fabricated to contain controlled amounts of different metal ions. These distinctly patterned labels are identifiable through stripping analysis due to their specific metal compositions and thus allow simultaneous detection of several analytes from a single sample.[185,212,213]

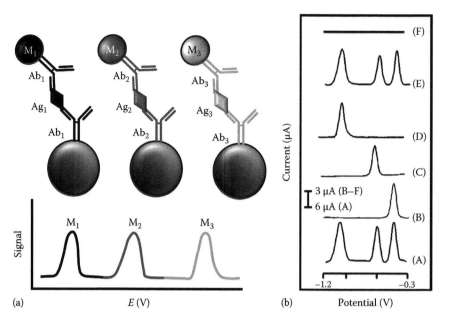

FIGURE 13.16 Metal nanoparticle-based multiplexed electrochemical detection of proteins. (a) Illustrated schematic of basic sensing strategy. Magnetic beads modified with three different antibodies (Ab_1, Ab_2, Ab_3) are used to capture target protein analytes (Ag_1, Ag_2, Ag_3) from the sample. Antibodies labeled with different metal colloids (M_1, M_2, M_3) bind to their respective captured target antigen. Dissolution of metal colloids followed by stripping analysis results in observable signal measurement and identification of targets. (Part a adapted from Wang, J., Nanoparticle-based electrochemical bioassays of proteins, *Electroanalysis*, 2007, 19, 769–776. Copyright Wiley-VCH verlag GmBH & Co. KGaA, Weinheim. Reproduced with permission.) Stripping analysis results (b) for (A) a mixture of three different M–Ab bioconjugates, (from left to right) ZnS–anti-β_2-microglobulin, CdS–anti-IgG, and PbS–anti-BSA; for (B–E) immunoassays of M–Ab/Ag/Ab-magnetic bead bioconjugates, resulting from the mixture of M–Ab bioconjugates with 100 ng mL^{-1} (B) BSA, (C) IgG, (D) β_2 microglobulin, and (E) all three protein analytes; and for (F) the immunoassay control. (Part b adapted with permission from Liu, G., Wang, J.. Kim, J., Jan, M.R., and Collins, G.E., Electrochemical coding for multiplexed immunoassays of proteins, *Anal. Chem.*, 76, 7126–7130, 2004. Copyright 2004 American Chemical Society.)

13.3.4 ELECTROCATALYTIC NANOPARTICLE LABELS AND BIOMOLECULE DETECTION

Nanoparticles can also act as labels by serving as catalysts for electrochemical reactions. For example, PtNPs have been employed in electrochemical biosensors due to their ability to catalyze the reduction of hydrogen peroxide.[214] Picomolar DLs for DNA and nanomolar DLs for thrombin were demonstrated using electrochemical hybridization and sandwich-type assays that employed PtNP labels for this purpose.[214]

Chow et al. used the catalytic properties of PtNP labels to prepare DNA sensors based on bipolar electrodes (BPEs)[215–217] (Figure 13.17).[218] In a BPE, electrochemical reactions take place at each extremity (e.g., each end or pole of a conductive wire or rod) when a sufficient potential difference exists between the electrode surface and electrolyte solution.[217] A pair of driving electrodes that are not in direct contact with the BPE supplies the potential difference necessary to drive the oxidation and reduction reactions at each pole of the BPE.

In DNA sensors based on PtNP labels, capture of the PtNP-labeled target sequence at the cathode end of a BPE resulted in the catalytic reduction of dissolved oxygen and led to the oxidation of the ECL reporter Ru(bpy)$_3^{2+}$ and coreactant tri-*n*-propylamine at the anode end.[218] Since BPEs do not require direct contact with a voltage source, many BPEs can be controlled by the electric field

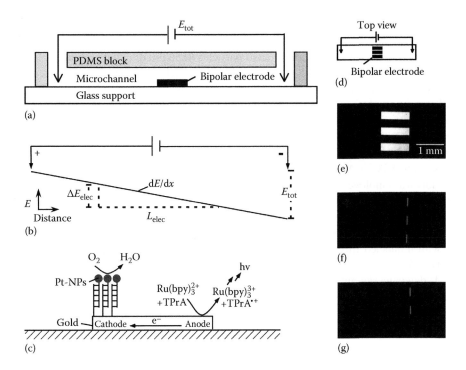

FIGURE 13.17 BPE-based sensor for DNA. (a) Illustrated schematic for microchannel BPE-based sensing device. (b) Schematic representation of electric field (dE/dx) and potential difference (ΔE_{elec}) that develops between the electrode poles and electrolyte solution for a BPE of length L_{elec}. (c) Illustrated schematic for DNA sensing using a BPE. Target DNA labeled with PtNPs binds to probe DNA at the cathodic end of a gold BPE. Catalytic reduction of O_2 at the PtNPs leads to oxidation of $Ru(bpy)_3^{2+}$ and tri-n-propylamine (TPrA) and subsequent ECL at the anodic pole. (d) Illustrated representation of BPE position in optical and luminescence micrographs. (e) Optical micrograph of three parallel BPEs in a single microchannel. (f) Luminescence micrograph for three BPEs after exposure to labeled target DNA. (g) Luminescence micrograph for three BPEs after exposure of only the top two BPEs to labeled target DNA. (Adapted with permission from Chow, K.-F., Mavré, F., and Crooks, R.M., Wireless electrochemical DNA microarray sensor, *J. Am. Chem. Soc.*, 130, 7544–7545, 2008. Copyright 2008 American Chemical Society.)

generated between the two driving electrodes. Bipolar microelectrode arrays with ~2000 sensing elements per square centimeter have been reported.[219] However, since the potential difference across the BPE is equivalent to the product of the electric field strength and the length of the electrode in conventional bipolar electrochemical setups, high voltages are often required to drive the necessary redox reactions. BPEs equivalent to two series-coupled electrochemical cells (i.e., *closed* BPEs) lessen power consumption requirements because the majority of the voltage drop between the two driving electrodes occurs at the BPE/solution interface rather than over the entire separation distance of the driving electrodes.[220]

13.3.5 BIOSENSOR SIGNAL ENHANCEMENT USING NANO- AND MICROPARTICLES AS CARRIERS FOR LARGE NUMBERS OF SIGNAL-TRANSDUCING NANOPARTICLE LABELS

Nano- and microparticles have also been employed as carriers for large numbers of other nanoparticles, enzymes, or redox probes that act as multiple labels for electrochemical or ECL-based biosensing. Using these multilabel carriers, a single analyte capture event becomes amplified by many signal-transducing labels.[185] As with metal nanoparticle label enlargement and dissolution and strategies that employ nanoparticles for electrocatalysis, multilabel carrier sensing can result in impressive signal enhancements and remarkably low limits of detection compared to methods that rely on single nanoparticle labels.

For example, polymeric streptavidin-coated particles (0.56 μm dia.) with bound biotinylated target DNA and ~80 biotinylated AuNPs per polymeric particle have been used to measure breast cancer gene E908X-WT via chronopotentiometric stripping analysis.[221] Combination of multiple AuNP labels with catalytic deposition of gold onto the AuNPs led to a low pM limit of detection for the target DNA. Similarly, multiple CdS labels have been loaded onto SWCNTs (~500 CdS nanocrystals per SWCNT[222]) and graphene sheets[223] for the detection of DNA and proteins using ASV. Polymeric microbeads with up to 10,000 CdS particles led to sub-fM DLs for DNA in hybridization assays based on ASV.[224] An immunosensor using silica nanoparticles coated with multiple CdTe quantum dots was employed for stripping analysis and ECL detection of IgG.[178]

Multinanoparticle carriers have also been employed in biosensors for simultaneous multiplexed detection. In one example, thrombin and lysozyme tagged with multiple quantum dot (CdS or PbS) labels were bound to aptamers on a gold electrode surface.[225] Single-step displacement of labeled proteins with unlabeled thrombin and lysozyme and subsequent ASV was found to result subpicomolar DLs for thrombin and lysozyme.[225] In another study, distinguishable poly(amido amine) (PAMAM) dendrimer/quantum dot nanocomposite labels with large numbers of different identifiable quantum dots (CdS, ZnS, or PbS) were employed for the simultaneous detection of cancer antigen 125 (CA 125), CA 15–3, and CA 19–9 in serum through ASV.[226]

13.3.6 Nano- and Microparticles as Carriers for Multiple Redox and ECL Labels

Nano- and microparticles have also been used as carriers for other types of redox and ECL labels in assays for proteins and nucleic acids. Wang et al. encapsulated ferrocenecarboxaldehyde (FCA) within polystyrene microbeads (~5×10^{11} FCA per microbead) that were labeled with a target DNA sequence.[227] After capturing the target by probe-sequence-labeled magnetic beads, the microbeads were dissolved in acetonitrile. Released FCA was measured by chronopotentiometry at sub-fM DLs for target DNA. Nanoparticles labeled with[228–230] or composed of[231] other redox-active species have also been employed in sandwich immunoassays and hybridization assays that do not rely on organic solvents for label liberation. Since nucleobases guanine and adenine are redox active at different potentials, nanoparticles labeled with sequences that have distinct guanine to adenine ratios can be used as distinguishable electrochemical labels for simultaneous detection of different proteins.[229] Similar strategies that use nanoparticles labeled with other distinguishable redox tags have also been reported for multiplexed detection of proteins.[232–234]

Miao and Bard encapsulated tris(2,2′-bipyridyl)ruthenium(II) tetrakis(pentafluorophenyl)borate or $Ru(bpy)_3[B(C_6F_5)_4]_2$, inside polymeric microspheres (~10^9 $Ru(bpy)_3^{2+}$ species per microsphere) for ECL-based DNA hybridization assays[235] and immunoassays.[236] DLs were 1 fM for DNA[235] and 0.010 μg mL^{-1} (sub-nM) for C-reactive protein.[236] Signal development involved dissolution of the polymeric microspheres in acetonitrile to liberate the ECL probe. To circumvent this issue, Zhan and Bard later used liposomes from which $Ru(bpy)_3^{2+}$ could be released simply by using a nonionic surfactant solution.[237]

Other nanoparticle carriers or attachment of the ECL label on outsides of nanoparticles resulted in strategies that avoid the need to liberate the label from the carrier particle.[124,146–148,238–242] For example, Sardesai et al. encapsulated $Ru(bpy)_3^{2+}$ ions in ~100 nm diameter, mesoporous, antibody-labeled silica particles (~2.5×10^5 $Ru(bpy)_3^{2+}$ per particle) and employed them to develop an ECL-based biosensor for prostate cancer biomarkers prostate cancer biomarkers prostate specific antigen (PSA) and interleukin 6 (IL-6) using SWCNT forest platforms (Figure 13.18).[146,147,243] Since ECL signal is provided in the form of light, ECL-based systems for multiplexed detection can be easily developed by preparing arrays on a conducting substrate such as a carbon block with electrode spots defined by patterning an insulating barrier on the block surface. Using this arrangement, array elements need not be individually electronically addressable in ECL-based biosensors. A charge-coupled

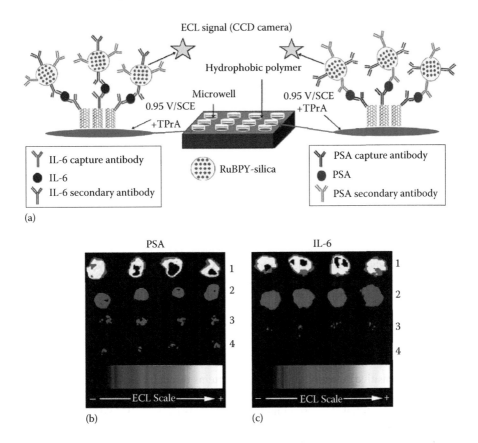

FIGURE 13.18 SWCNT forest/multilabeled particle-based biosensor for ECL detection of two cancer biomarker proteins. (a) Schematic illustration of sensing strategy. Microwells are defined on a pyrolytic graphite block via hydrophobic polymer ink. At the bottom of each microwell is an SWCNT forest electrode onto which antibodies for either IL-6 or prostate-specific antigen (PSA) are attached. ECL signal is generated by $Ru(bpy)_3^{2+}$-containing antibody-coated silica nanoparticles (RuBPY–silica) that are used to capture the target antigen from the sample. (b) ECL image resulting from detection of PSA at concentrations of 10 ng mL^{-1} (row 1), 0.4 ng mL^{-1} (row 2), 1 pg mL^{-1} (row 3), and 0 pg mL^{-1} (row 4). (c) ECL image resulting from detection of IL-6 at concentrations of 2 ng mL^{-1} (row 1), 0.2 ng mL^{-1} (row 2), 0.1 pg mL^{-1} (row 3), and 0 pg mL^{-1} (row 4). (Adapted with permission from Sardesai, N., Barron, J.C., and Rusling, J.F., Carbon nanotube microwell array for sensitive electrochemiluminescent detection of cancer biomarker proteins, *Anal. Chem.*, 83, 6698–6703, 2011. Copyright 2011 American Chemical Society.)

device (CCD) camera can be used to simultaneously measure ECL signals generated at each array element. For example, Sardesai et al. prepared ECL arrays for proteins using a pyrolytic graphite block platform on which the borders of microwell array elements (spots) were defined by patterning hydrophobic ink onto the block surface (Figure 13.18).[146,147,243] SWCNT forests were grown on each array spot and functionalized with a specific antibody for PSA or IL-6 so that the two biomarkers could be simultaneously detected. The combination of silica nanoparticle $Ru(bpy)_3^{2+}$ label carriers and SWCNT forests resulted in DLs as low as 100 fg mL^{-1} for PSA and 10 fg mL^{-1} IL-6.[243]

13.3.7 NANO- AND MICROPARTICLES AS CARRIERS FOR MULTIPLE ENZYME LABELS

Enzyme labels such as alkaline phosphatase (ALP), HRP, and GOx have been loaded onto CNTs, AuNPs, and magnetic, polymeric, and carbon particles for electrochemical detection of proteins and nucleic acids. Wang et al. reported the use of CNTs, each covalently modified with ~9600 ALP

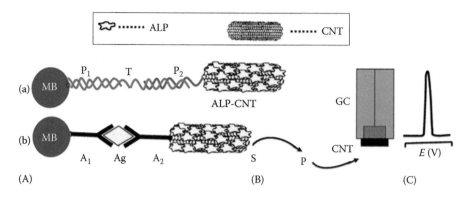

FIGURE 13.19 Schematic representation of sandwich-type electrochemical detection of nucleic acids (a) and proteins (b) through the use of CNTs modified with multiple ALP enzyme labels. (A) The target nucleic acid (T) or antigen (Ag) is captured by magnetic beads (MB) modified with either capture probe sequence (P_1) or antibody (A_1) and labeled with a multi-ALP-CNT probe sequence (P_2) or antibody (A_2) bioconjugate. (B) Enzyme substrate (S) is introduced and converted to a product (P) that is voltammetrically detected (C) at a CNT-modified GCE. (Reprinted with permission from Wang, J., Liu, G., and Jan, M.R., Ultrasensitive electrical biosensing of proteins and DNA: carbon-nanotube derived amplification of the recognition and transduction events, *J. Am. Chem. Soc.*, 126, 3010–3011, 2004. Copyright 2004 American Chemical Society.)

molecules for ultrasensitive detection of DNA and proteins through hybridization assays and immunoassays (Figure 13.19).[244] Reaction of ALP with α-naphthyl phosphate resulted in electrochemically detectable product α-naphthol. The use of multilabeled CNTs led to a 10^4-fold improvement in sensitivity over methods that used recognition elements such as oligonucleotide probe sequences or antibodies modified with single ALP labels. An additional ~30-fold signal enhancement was obtained by employing an MWCNT-modified electrode in comparison to a GCE. This additional enhancement was attributed to strong adsorptive accumulation of α-naphthol on the MWCNTs. In subsequent studies, stepwise layer-by-layer assembly of multilayer poly(diallyldimethylammonium) (PDDA) polyion/enzyme films on CNT label carriers was employed to further increase enzyme loading for ultrasensitive biomolecule assays.[245] The resulting CNT–(PDDA/ALP)$_4$ bioconjugates featured ~196,000 ALP molecules per CNT, which contributed to DLs of 5.4 aM for DNA and 67 aM for protein analytes.

HRP labels have been attached to many types of particles, including CNTs (~100 HRP labels per 100 nm in CNT length),[246,247] polymeric microbeads (~4200 HRP labels per 500 nm bead),[248] paramagnetic microspheres (up to 500,000 HRP labels per 1 μm bead),[71,249–251] metal nanoparticles[252,253] and metal nanoparticle composites,[254–256] silica nanoparticles,[257–259] GO sheets[260,261] and graphene nanocomposites,[262] and other carbon particles.[263] These multienzyme label carriers have been employed for single[71,246–249,252,254,255,257,258,260–263] or multiplexed[250,251,253,256,259] electrochemical detection of oral,[247–249,251] prostate,[71,246,250,260] liver,[252,253,254,256,257,259,263] colorectal,[255,259] and ovarian[258,259] cancer biomarker proteins in serum. Hydrogen peroxide is used to oxidize the iron heme group of the HRP label, which in turn can oxidize an appropriate redox mediator such as HQ,[71,246–251] *o*-phenylenediamine,[252,259,262,263] thionine,[253,254,256,257,258,260,261] or ferrocene[253,256] leading to an electrochemical signal that is observable through amperometric[71,246–251,260] or voltammetric[252–259,261–263] detection. Prussian blue[255] has also been used as a mediator in HRP-based enzyme-labeled strategies, though some care must be taken in selecting an appropriate HRP label carrier since Prussian white (the reduced form of Prussian blue) can also catalytically reduce hydrogen peroxide.[264–266] The mediator can be injected into the sensing solution,[71,246–248,249–252,259,261,263] immobilized on the electrode surface,[254,255] or incorporated in the enzyme-labeled particles.[256–258,260]

Multiplexed detection with these sandwich-type immunoassays can be accomplished through the use of electrode arrays[250,251,259] or through the preparation of distinguishable antibody-modified bioconjugates that contain HRP paired with a mediator.[253,256] HRP-based multienzyme-labeled strategies using massively labeled magnetic particles (up to 500,000 enzyme labels per magnetic particle) in combination with AuNP-modified screen-printed electrode arrays in modular microfluidic devices have enabled DLs in the low fg mL^{-1} (attomolar) range for the simultaneous detection of four biomarkers for head and neck squamous cell carcinoma (HNSCC), that is, interleukin 6 (IL-6), interleukin 8 (IL-8), vascular endothelial growth factor (VEGF), and VEGF-C, in serum (Figure 13.20).[251]

Tang et al. used biofunctionalized labels that consisted of HRP and either thionine or ferrocene mediators encapsulated within nanogold hollow microspheres, prepared by a reverse micelle process, to simultaneously detect carcinoembryonic antigen (CEA) and AFP (Figure 13.21).[256] Antigen bound to the bioconjugate labels was separated from the serum matrix using biofunctionalized GO

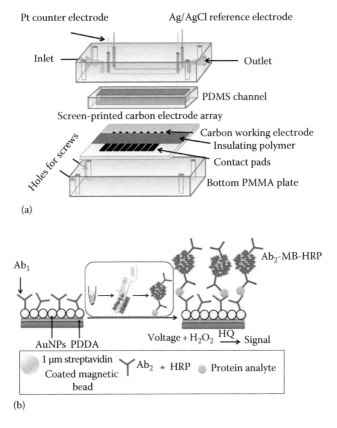

(a)

(b)

FIGURE 13.20 Nanostructured electrode array with enzyme-labeled particles for simultaneous electrochemical detection of four cancer biomarker proteins. (a) Illustrated representation of an electrochemical cell for multiplexed detection of cancer biomarker proteins (PMMA, poly(methylmethacrylate), PDMS, poly(dimethylsiloxane)). The cell features a PDMS channel (volume ~60 μL) that houses an array of 8 screen-printed carbon working electrodes. (b) Illustrated scheme depicting the biosensing strategy for a single electrode of an 8-electrode array. Nanostructured electrode built by layer-by-layer deposition of polyion poly(dimethyl diallyl ammonium) (PDDA) and gold nanoparticles (AuNPs) on a conductive screen-printed carbon substrate is modified with antibody (Ab$_1$). In a microcentrifuge tube, cancer biomarker protein analyte is captured from the sample through the use of magnetic beads that are coated with many copies of a second specific antibody (Ab$_2$) and enzyme label HRP. Labeled analyte is introduced to the nanostructured electrode array via a syringe pump (not shown), and signal is generated by the addition of a solution (via the syringe pump) containing hydrogen peroxide to activate the HRP labels and HQ to serve as an electron mediator.

(Continued)

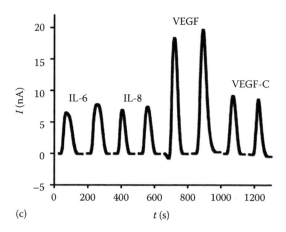

(c)

FIGURE 13.20 (CONTINUED)   Nanostructured electrode array with enzyme-labeled particles for simultaneous electrochemical detection of four cancer biomarker proteins. (c) Peak-shaped amperometric responses for the simultaneous detection of four oral cancer biomarker proteins, IL-6, IL-8, VEGF, and VEGF-C, using an array of eight nanostructured electrodes. Peak-shaped response is generated by the injection of the sensing solution (H_2O_2/HQ mixture) at a flow rate of 100 μL min^{-1}. The four biomarker singular mixture contained 10 fg mL^{-1} IL-6, 15 fg mL^{-1} IL-8, 25 fg mL^{-1} VEGF, and 60 fg mL^{-1} VEGF-C. (Adapted with permission from Malhotra, R., Patel, V., Chikkaveeraiah, B. et al., Ultrasensitive detection of cancer biomarkers in the clinic by use of a nanostructured microfluidic array, *Anal. Chem.*, 84, 6249–6255, 2012. Copyright 2012 American Chemical Society.)

nanosheet/magnetic iron oxide nanoparticle composites. Since mediators thionine and ferrocene are reduced at different potentials, simultaneous detection was achieved through the use of voltammetry, and DLs of 1 pg mL^{-1} (low fM) were reported.

Similarly, Lai et al. reported a GOx multilabel strategy for the simultaneous detection of CEA and AFP based on CNT multilabel carriers.[267] The reduced form of GOx generated upon reaction with glucose can reduce dissolved oxygen to yield H_2O_2. Screen-printed electrodes modified with Prussian blue catalyze the reduction of H_2O_2, leading to the measurable electrochemical signal. Low pg mL^{-1} (low fM) DLs were obtained for CEA and AFP in serum.

13.3.8   LARGE NUMBERS OF ANTIBODIES ON NANO- AND MICROPARTICLE LABEL CARRIERS IMPROVE ANTIGEN CAPTURE

Particle-based sandwich-type assays may not only lead to improvements in sensitivity by enabling one analyte capture event to correspond with many signal-transduction events. Researchers have also shown that incorporating many antibodies on the label-carrying detection particle helps result in ultralow DLs.[268,269] In addition to signal-transducing labels, nano- and microparticles in many of the strategies previously outlined are modified with as many as 100,000 antibodies or more. Although good-quality antibodies typically exhibit dissociation constants (K_D) in the 10 pM to 10 nM range,[268,270] the presence of multiple copies of an antibody on the detection particle favors stable formation of antigen–antibody–particle complexes at concentrations orders of magnitude lower than K_D. This helps to explain the subfemtomolar DLs accessible with multilabel particle-based strategies.[268,269]

13.4   SUMMARY AND OUTLOOK

Nanomaterials have had great impact in bioanalytical electrochemistry. Due to their high specific surface areas and superior electron-transfer properties in comparison to bulk materials,

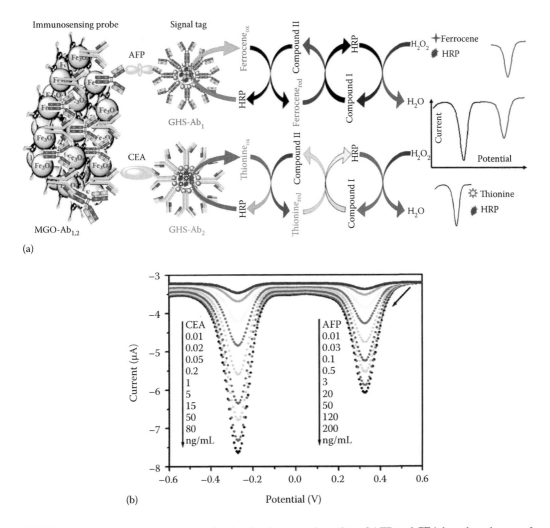

FIGURE 13.21 Immunoassay strategy for the simultaneous detection of AFP and CEA based on the use of HRP enzyme labels and electrochemical mediators (ferrocene or thionine) encapsulated within nanogold hollow microspheres (GHS). (a) Schematic illustration of magnetic nanoparticle-modified GO nanosheet (MGO)/GHS-based electrochemical detection strategy. AFP and CEA are captured by immunosensing probes that consist of antibody-modified MGO and distinguished by the unique voltammetric signature provided by the encapsulated mediators present within GHS signal tags. (b) Differential pulse voltammetric responses for AFP and CEA standards in pH 6.5 buffer containing 3.5 mM H_2O_2. (Adapted with permission from Tang, J., Tang, D., Niessner, R., Chen, G., and Knopp, D., Magneto-controlled graphene immunosensing platform for simultaneous multiplexed electrochemical immunoassay using distinguishable signal tags, *Anal. Chem.*, 83, 5407–5414, 2011. Copyright 2011 American Chemical Society.)

nanomaterials have been found to impart excellent sensitivity and selectivity to electrochemical biosensors. Nanomaterial-based electrodes provide superior electron-transfer environments for biomolecules compared to relatively flat, bulk-material-based electrodes and can also be used to provide advantages in fundamental studies of protein electrochemistry and electrochemical biocatalysis. Like *electrical wiring* strategies and film-modified electrodes, nanomaterial-based electrodes enable direct protein electrochemistry. Nanofabrication strategies allow for biosensor miniaturization and the production of multisensor arrays.

Electrochemical biosensors have also benefited from the employment of nanoparticles as labels and label carriers. Using particle-based labels or label carriers enables each analyte recognition event to produce many signal-transducing events, which results in impressive signal amplification. Particle-based label carriers not only deliver many signal-transducing agents to the electrode surface but also allow for the attachment of many recognition agents. This pushes DLs well beyond limits of K_D for the best single antibodies and into subfemtomolar territory.

While nanomaterial-based electrochemical biosensors are promising candidates for future medical diagnostics, some issues still need to be addressed. Performance is largely determined by the quality of nanomaterials employed, methods used to ensure that the favorable properties of nanomaterials are preserved through biofunctionalization, and strategies to arrange or orient them in biosensing platforms. Therefore, complete characterization of the surface of nanomaterial-based biosensing platforms and of the biofunctionalization and signal-transduction events is desirable to identify best practices and possible failure modes in developing new nanomaterial-based bioelectrochemical sensors.

Nanostructured electrode arrays and nanomaterial-based labeling strategies have resulted in ultrasensitive devices for measuring clinically relevant biomolecules. Due to their small size, electrocatalytic properties, compatibility with microfluidics, and ability to be functionalized with biomolecules, nanomaterial-derived electrochemical biosensor arrays have shown promise in applications that require the simultaneous detection of multiple biomolecules. As personalized medicine and future medical diagnostics require the development of small, sensitive, versatile, low-cost, and energy-efficient devices for multiplexed detection of biomolecules from complex samples, nanomaterial-based bioelectrochemistry is poised to address the rigorous demands of biosensor development.

ACKNOWLEDGMENTS

This work was supported by NIH grants EB014586 and EB016707 from NIBIB. The authors thank collaborators and research students named in joint papers without which their progress in this research would not have been possible.

REFERENCES

1. Paleček, E. 1960. Oscillographic polarography of highly polymerized deoxyribonucleic acid. *Nature* 188: 656–657.
2. Paleček, E. 1996. From polarography of DNA to microanalysis with nucleic acid-modified electrodes. *Electroanalysis* 8: 7–14.
3. Paleček, E. 2002. Past, present and future of nucleic acids electrochemistry. *Talanta* 56: 809–819.
4. Paleček, E. and M. Bartošík. 2012. Electrochemistry of nucleic acids. *Chem. Rev.* 112: 3427–3481.
5. Rusling, J.F. 1998. Enzyme bioelectrochemistry in cast biomembrane-like films. *Acc. Chem. Res.* 31: 363–369.
6. Reed, D.E. and F.M. Hawkridge. 1987. Direct electron transfer reactions of cytochrome *c* at silver electrodes. *Anal. Chem.* 59: 2334–2339.
7. Eddowes, M.J. and H.A.O. Hill. 1977. Novel method for the investigation of the electrochemistry of metalloproteins: Cytochrome c. *J. Chem. Soc., Chem. Commun.* 771–772.
8. Eddowes, M.J. and H.A.O. Hill. 1979. Electrochemistry of horse heart cytochrome *c*. *J. Am. Chem. Soc.* 101: 4461–4464.
9. Armstrong, F.A., H.A.O. Hill, and N.J. Walton. 1988. Direct electrochemistry of redox proteins. *Acc. Chem. Res.* 21: 407–413.
10. Degani, Y. and A. Heller. 1987. Direct electrical communication between chemically modified enzymes and metal electrodes. 1. Electron transfer from glucose oxidase to metal electrodes via electron relays bound covalently to the enzyme. *J. Phys. Chem.* 91: 1285–1289.
11. Heller, A. 1990. Electrical wiring of redox enzymes. *Acc. Chem. Res.* 23: 128–134.
12. Degani, Y. and A. Heller. 1989. Electrical communication between redox centers of glucose oxidase and electrodes via electrostatically and covalently bound redox polymers. *J. Am. Chem. Soc.* 111: 2358–2361.
13. Hamachi, I., T. Honda, S. Noda, and T. Kunitake. 1991. Oriented intercalation of myoglobin into multilayered films of synthetic bilayer membranes. *Chem. Lett.* 1121–1124.

14. Hamachi, I., S. Noda, and T. Kunitake. 1991. Functional conversion of myoglobin bound to synthetic bilayer membranes: From dioxygen storage protein to redox enzyme. *J. Am. Chem. Soc.* 113: 9625–9630.
15. Rusling, J.F. and A.-E.F. Nassar. 1993. Enhanced electron transfer for myoglobin in surfactant films on electrodes. *J. Am. Chem. Soc.* 115: 11891–11897.
16. Lvov, Y.M., Z. Lu, J.B. Shenkman, X. Zu, and J.F. Rusling. 1998. Direct electrochemistry of myoglobin and cytochrome P450cm in alternate layer-by-layer films with DNA and other polyions. *J. Am. Chem. Soc.* 120: 4073–4080.
17. Rusling, J.F. and R.J. Forster. 2003. Electrochemical catalysis with redox polymer and polyion–protein films. *J. Colloid Interf. Sci.* 262: 1–15.
18. Ma, H., N. Hu, and J.F. Rusling. 2000. Electroactive myoglobin films grown layer-by-layer with poly(styrenesulfonate) on pyrolytic graphite electrodes. *Langmuir* 16: 4969–4975.
19. Thevenot, D.R., K. Toth, R.A. Durst, and G.S. Wilson. 2001. Electrochemical biosensors: Recommended definitions and classification. *Biosens. Bioelectron.* 16: 121–131.
20. Clark, L.C. and C. Lyons. 1962. Electrode systems for continuous monitoring in cardiovascular surgery. *Ann. NY Acad. Sci.* 102: 29–45.
21. Wang, J. 2008. Electrochemical glucose biosensors. *Chem. Rev.* 108: 814–825.
22. Grieshaber, D., R. MacKenzie, J. Vörös, and E. Reimhult. 2008. Electrochemical biosensors—Sensor principles and architectures. *Sensors* 8: 1400–1458.
23. Heineman, W.R., C.W. Anderson, and H.B. Halsall. 1979. Immunoassay by differential pulse polarography. *Science* 204: 865–866.
24. Eggers, H.M., H.B. Halsall, and W.R. Heineman. 1982. Enzyme immunoassay with flow-amperometric detection of NADH. *Clin. Chem.* 28: 1848–1851.
25. Doyle, M.J., H.B. Halsall, and W.R. Heineman. 1984. Enzyme-linked immunoadsorbent assay with electrochemical detection for α_1-acid glycoprotein. *Anal. Chem.* 56: 2355–2360.
26. Wehmeyer, K.R., H.B. Halsall, and W.R. Heineman. 1985. Heterogeneous enzyme immunoassay with electrochemical detection: Competitive and "sandwich"-type immunoassays. *Clin. Chem.* 31: 1546–1549.
27. Wehmeyer, K.R., H.B. Halsall, W.R. Heineman, C.P. Volle, and I.-W. Chen. 1986. Competitive heterogeneous enzyme immunoassay for digoxin with electrochemical detection. *Anal. Chem.* 58: 135–139.
28. Ronkainen-Matsuno, N.J., J.H. Tomas, H.B. Halsall, and W.R. Heineman. 2002. Electrochemical immunoassay moving into the fast lane. *Trends Anal. Chem.* 21: 213–225.
29. Ronkainen, N.J., H.B. Halsall, and W.R. Heineman. 2010. Electrochemical biosensors. *Chem. Soc. Rev.* 39: 1747–1763.
30. Choi, J.-W., K.W. Oh, A. Han et al. 2001. Development and characterization of microfluidic devices and systems for magnetic bead-based biochemical detection. *Biomed. Microdevices* 3: 191–200.
31. Choi, J.-W., K.W. Oh, J.H. Thomas et al. 2002. An integrated microfluidic biochemical detection system for protein analysis with magnetic bead-based sampling capabilities. *Lab Chip* 2: 27–30.
32. Bange, A., H.B. Halsall, and W.R. Heineman. 2005. Microfluidic immunosensor systems. *Biosens. Bioelectron.* 20: 2488–2503.
33. Kuramitz, H., M. Dziewatkoski, B. Barnett, H.B. Halsall, and W.R. Heineman. 2006. Application of an automated fluidic system using electrochemical bead-based immunoassay to detect bacteriophage MS2 and ovalbumin. *Anal. Chim. Acta* 561: 69–77.
34. Niwa, O., Y. Xu, H.B. Halsall, and W.R. Heineman. 1993. Small-volume voltammetric detection of 4-aminophenol with interdigitated array electrodes and its application to electrochemical enzyme immunoassay. *Anal. Chem.* 65: 1559–1563.
35. Thomas, J.H., S.K. Kim, P.J. Hesketh, H.B. Halsall, and W.R. Heineman. 2004. Microbead-based electrochemical immunoassay with interdigitated array electrodes. *Anal. Biochem.* 328: 113–122.
36. Thomas, J.H., S.K. Kim, P.J. Hesketh, H.B. Halsall, and W.R. Heineman. 2004. Bead-based electrochemical immunoassay for bacteriophage MS2. *Anal. Chem.* 76: 2700–2707.
37. Wijayawardhana, C.A., H.B. Halsall, and W.R. Heineman. 1999. Micro volume rotating disk electrode (RDE) amperometric detection for a bead-based immunoassay. *Anal. Chim. Acta* 399: 3–11.
38. Hayes, F.J., H.B. Halsall, and W.R. Heineman. 1994. Simultaneous immunoassay using electrochemical detection of metal ion labels. *Anal. Chem.* 66: 1860–1865.
39. Ding, Y., L. Zhou, H.B. Halsall, and W.R. Heineman. 1998. Feasibility studies of simultaneous multianalyte amperometric immunoassay based on spatial resolution. *J. Pharm. Biomed.* 19: 153–161.
40. Wilson, M.S. 2005. Electrochemical immunosensors for the simultaneous detection of two tumor markers. *Anal. Chem.* 77: 1496–1502.
41. Wilson, M.S. and W. Nie. 2006. Electrochemical multianalyte immunoassays using an array-based sensor. *Anal. Chem.* 78: 2507–2513.

42. Kingsmore, S.F. 2006. Multiplexed protein measurement: Technologies and applications of protein and antibody arrays. *Nat. Rev. Drug Discov.* 5: 310–320.

43. Rusling, J.F., C.V. Kumar, J.S. Gutkind, and V. Patel. 2010. Measurement of biomarker proteins for point-of-care early detection and monitoring of cancer. *Analyst* 135: 2496–2511.

44. Rusling, J.F. 2012. Nanomaterials-based electrochemical immunosensors for proteins. *Chem. Rec.* 12: 164–176.

45. Rusling, J.F. 2013. Multiplexed electrochemical protein detection and translation to personalized cancer diagnostics. *Anal. Chem.* 85: 5304–5310.

46. Wang, J. 2005. Carbon-nanotube based electrochemical biosensors: A review. *Electroanalysis* 17: 7–14.

47. Murphy, L. 2006. Biosensors and bioelectrochemistry. *Curr. Opin. Chem. Biol.* 10: 177–184.

48. Guo, S. and E. Wang. 2007. Synthesis and electrochemical applications of gold nanoparticles. *Anal. Chim. Acta* 598: 181–192.

49. Luo, X., A. Morrin, A.J. Killard, and M.R. Smyth. 2006. Application of nanoparticles in electrochemical sensors and biosensors. *Electroanalysis* 18: 319–326.

50. Liu, S., D. Leech, and H. Ju. 2007. Application of colloidal gold in protein immobilization, electron transfer, and biosensing. *Anal. Lett.* 36: 1–19.

51. Shao, Y., J. Wang, H. Wu, J. Liu, I.A. Aksay, and Y. Lin. 2010. Graphene based electrochemical sensors and biosensors: A review. *Electroanalysis* 22: 1027–1036.

52. Kim, S.N., J.F. Rusling, and F. Papadimitrakopoulos. 2007. Carbon nanotubes for electronic and electrochemical detection of biomolecules. *Adv. Mater.* 19: 3214–3228.

53. Zhao, J., R.W. Henkens, J. Stonehuerner, J.P. O'Daly, and A.L. Crumbliss. 1992. Direct electron transfer at horseradish peroxidase–colloidal gold modified electrodes. *J. Electroanal. Chem.* 327: 109–119.

54. Brown, K.R., A.P. Fox, and M.J. Natan. 1996. Morphology-dependent electrochemistry of cytochrome *c* at Au colloid-modified SnO_2 electrodes. *J. Am. Chem. Soc.* 118: 1154–1157.

55. Davis, J.J., R.J. Coles, and H.A.O Hill. 1997. Protein electrochemistry at carbon nanotube electrodes. *J. Electroanal. Chem.* 440: 279–282.

56. Yu, X., D. Chattopadhyay, I. Galeska, F. Papadimitrakopoulos, and J.F. Rusling. 2003. Peroxidase activity of enzymes bound to the ends of single-wall carbon nanotube forest electrodes. *Electrochem. Commun.* 5: 408–411.

57. Gooding, J.J., R. Wibowo, J. Liu et al. 2003. Protein electrochemistry using aligned carbon nanotube arrays. *J. Am. Chem. Soc.* 125: 9006–9007.

58. Katz, E., I. Willner, and J. Wang. 2004. Electroanalytical and bioelectroanalytical systems based on metal and semiconductor nanoparticles. *Electroanalysis* 16: 19–42.

59. Jacobs, C.B., M.J. Peairs, and B.J. Venton. 2010. Review: Carbon nanotube based electrochemical sensors for biomolecules. *Anal. Chim. Acta* 662: 105–127.

60. Xiao, Y., F. Patolsky, E. Katz, J.F. Hainfeld, and I. Willner. 2003. "Plugging into enzymes": Nanowiring of redox enzymes by a gold nanoparticle. *Science* 299: 1877–1881.

61. Crumbliss, A.L., S.C. Perine, J. Stonhuerner et al. 1992. Colloidal gold as a biocompatible immobilization matrix suitable for the fabrication of enzyme electrodes by electrodeposition. *Biotechnol. Bioeng.* 40: 483–490.

62. Crumbliss, A.L., J.G. Stonehuerner, R.W. Henkens, J. Zhao, and J.P. O'Daly. 1993. A carrageenan hydrogel stabilized colloidal gold multi-enzyme biosensor electrode utilizing immobilized horseradish peroxidase and cholesterol oxidase/cholesterol esterase to detect cholesterol in serum and whole blood. *Biosens. Bioelectron.* 8: 331–337.

63. Xiao, Y., H.-X. Ju, and H.-Y. Chen. 1999. Hydrogen peroxide sensor based on horseradish peroxidase-labeled Au colloids immobilized on gold electrode surface by cysteamine monolayer. *Anal. Chim. Acta* 391: 73–82.

64. Xiao, Y., H.-X. Ju, and H.-Y. Chen. 2000. Direct electrochemistry of horseradish peroxidase immobilized on colloid/cysteamine-modified gold electrode. *Anal. Biochem.* 278: 22–28.

65. Gu, H.-Y., A.-M. Yu, and H.-Y. Chen. 2001. Direct electron transfer and characterization of hemoglobin immobilized on a Au colloid—Cysteamine-modified gold electrode. *J. Electroanal. Chem.* 516: 119–126.

66. Liu, S.-Q. and H.-X. Ju. 2002. Renewable reagentless hydrogen peroxide sensor based on direct electron transfer of horseradish peroxidase immobilized on colloidal gold-modified electrode. *Anal. Biochem.* 307: 110–116.

67. Jia, J., B. Wang, A. W, G. Cheng, Z. Li, and S. Dong. 2002. A method to construct a third-generation horseradish peroxidase biosensor: Self-assembling gold nanoparticles to three-dimensional sol–gel network. *Anal. Chem.* 74: 2217–2223.

68. Raj, C.R., T. Okajima, and T. Ohsaka. 2003. Gold nanoparticle arrays for the voltammetric sensing of dopamine. *J. Electroanal. Chem.* 543: 127–133.

69. Liu, S. and H. Ju. 2003. Reagentless glucose biosensor based on direct electron transfer of glucose oxidase immobilized on colloidal gold modified carbon paste electrode. *Biosens. Bioelectron.* 19: 177–183.

70. Wang, L. and E. Wang. 2004. A novel hydrogen peroxide sensor based on horseradish peroxidase immobilized on colloidal Au modified ITO electrode. *Electrochem. Commun.* 6: 225–229.

71. Mani, V., B.V. Chikkaveeraiah, V. Patel, J.S. Gutkind, and J.F. Rusling. 2009. Ultrasensitive immunosensor for cancer biomarker proteins using gold nanoparticles film electrodes and multienzyme-particle amplification. *ACS Nano* 3: 585–594.

72. Chikkaveeraiah, B.V., H. Liu, V. Mani, F. Papadimitrakopoulos, and J.F. Rusling. 2009. A microfluidic electrochemical device for high sensitivity biosensing: Detection of nanomolar hydrogen peroxide. *Electrochem. Commun.* 11: 819–822.

73. Adams, K.L., B.K. Jena, S.J. Percival, and B. Zhang. 2011. Highly sensitive detection of exocytotic dopamine release using a gold-nanoparticle-network microelectrode. *Anal. Chem.* 83: 920–927.

74. Tangkuaram, T., C. Ponchio, T. Kagnkasomboon, P. Katikawong, and W. Veerasai. 2007. Design and development of a highly stable hydrogen peroxide biosensor on screen printed carbon electrode based on horseradish peroxidase bound with gold nanoparticles in the matrix of chitosan. *Biosens. Bioelectron.* 22: 2071–2078.

75. Hu, C., D.-P. Yang, Z. Wang, L. Yu, J. Zhang, and N. Jia. 2013. Improved EIS performance of an electrochemical cytosensor using three-dimensional architecture Au@BSA as sensing layer. *Anal. Chem.* 85: 5200–5206.

76. Chai, H., H. Lu, X. Guo et al. 2012. Long distance electron transfer across >100 nm thick Au nanoparticle/polyion films to a surface redox protein. *Electroanalysis* 24: 1129–1140.

77. Gasparac, R., B.J. Taft, M.A. Lapierre-Devlin, A.D. Lazareck, J.M. Xu, and S.O. Kelley. 2004. Ultrasensitive electrocatalytic DNA detection at two- and three-dimensional nanoelectrodes. *J. Am. Chem. Soc.* 126: 12270–12271.

78. Menon, V. and C.R. Martin. 1995. Fabrication and evaluation of nanoelectrode ensembles. *Anal. Chem.* 67: 1920–1928.

79. Soleymani, L., Z. Fang, E.H. Sargent, and S.O. Kelley. 2009. Programming the detection limits of biosensors through controlled nanostructuring. *Nat. Nanotechnol.* 4: 844–848.

80. Soleymani, L., Z. Fang, X. Sun et al. 2009. Nanostructuring of patterned microelectrodes to enhance the sensitivity of electrochemical nucleic acids detection. *Angew. Chem. Int. Ed.* 48: 8457–8460.

81. Yang, H., A. Hui, G. Pampalakis et al. 2009. Direct, electronic microRNA detection for the rapid determination of differential expression profiles. *Angew. Chem. Int. Ed.* 48: 8561–8564.

82. Fang, Z., L. Soleymani, G. Pampalakis et al. 2009. Direct profiling of cancer biomarkers in tumor tissue using a multiplexed nanostructured microelectrode integrated circuit. *ACS Nano* 3: 3207–3213.

83. Bin, X., E.H. Sargent, and S.O. Kelley. 2010. Nanostructuring determines the efficiency of biomolecular capture. *Anal. Chem.* 82: 5928–5931.

84. Vasilyeva, E., B. Lam, Z. Fang, M.D. Minden, E.H. Sargent, and S.O. Kelley. 2011. Direct genetic analysis of ten cancer cells: Tuning sensor structure and molecular probe design for efficient mRNA capture. *Angew. Chem. Int. Ed.* 50: 4137–4141.

85. Das, J. and S.O. Kelley. 2011. Protein detection using arrayed microsensor chips: Tuning sensor footprint to achieve ultrasensitive readout of CA-125 in serum and whole blood. *Anal. Chem.* 83: 1167–1172.

86. Lam, B., R.D. Holmes, J. Das et al. 2013. Optimized templates for bottom-up growth of high-performance integrated biomolecular detectors. *Lab Chip* 13: 2569–2575.

87. Bhimji, A., A.A. Zaragoza, L.S. Live, and S.O. Kelley. 2013. Electrochemical enzyme-linked immunosorbent assay featuring proximal reagent generation: Detection of human immunodeficiency virus antibodies in clinical samples. *Anal. Chem.* 85: 6813–6819.

88. Lam, B., J. Das, R.D. Holmes et al. 2013. Solution-based circuits enable rapid and multiplexed pathogen detection. *Nat. Commun.* 4: 2001.

89. Gooding, J.J. 2005. Nanostructuring electrodes with carbon nanotubes: A review on electrochemistry and applications for sensing. *Electrochim. Acta* 50: 3049–3060.

90. Veetil, J.V. and K. Ye. 2007. Development of immunosensors using carbon nanotubes. *Biotechnol. Prog.* 23: 517–531.

91. Yang, W., K.R. Ratinac, S.P. Ringer, P. Thordarson, J.J. Gooding, and F. Braet. 2010. Carbon nanomaterials in biosensors: Should you use nanotubes or graphene? *Angew Chem. Int. Ed.* 49: 2114–2138.

92. Hamada, N., S. Sawada, and A. Oshiyama. 1992. New one-dimensional conductors: Graphitic microtubules. *Phys. Rev. Lett.* 68: 1579–1581.

93. Hirsch, A. 2002. Functionalization of single-walled carbon nanotubes. *Angew. Chem. Int. Ed.* 41: 1853–1859.
94. Karousis, N. and N. Tagmatarchis. 2010. Current progress on the chemical modification of carbon nanotubes. *Chem. Rev.* 110: 5366–5397.
95. Chen, R.J., Y. Zhang, D. Wang, and H. Dai. 2001. Noncovalent sidewall functionalization of single-walled carbon nanotubes for protein immobilization. *J. Am. Chem. Soc.* 123: 3838–3839.
96. Star, A., J.F. Stoddart, D. Steuerman et al. 2001. Preparation and properties of polymer-wrapped single-walled carbon nanotubes. *Angew. Chem. Int. Ed.* 40: 1721–1725.
97. Peng, X. and S.S. Wong. 2009. Functional covalent chemistry of carbon nanotube surfaces. *Adv. Mater.* 21: 625–642.
98. Zhao, Y.-L. and J.F. Stoddart. 2009. Noncovalent functionalization of single-walled carbon nanotubes. *Acc. Chem. Res.* 42: 1161–1171.
99. Liu, Z., S. Tabakman, K. Welsher, and H. Dai. 2009. Carbon nanotubes in biology and medicine: In vitro and in vivo detection, imaging and drug delivery. *Nano Res.* 2: 85–120.
100. Wang, J. and M. Musameh. 2004. Carbon nanotube screen-printed electrochemical sensors. *Analyst* 129: 1–2.
101. Wang, J., M. Li, Z. Shi, N. Li, and Z. Gu. 2002. Direct electrochemistry of cytochrome *c* at a glassy carbon electrode modified with single-wall carbon nanotubes. *Anal. Chem.* 74: 1993–1997.
102. Wang, G., J.-J. Xu, and H.-Y. Chen. 2002. Interfacing cytochrome c to electrodes with a DNA—Carbon nanotube composite film. *Electrochem. Commun.* 4: 506–509.
103. Zhao, Y.-D., W.-D. Zhang, H. Chen, Q.-M. Luo, and S.F.Y. Li. 2002. Direct electrochemistry of horseradish peroxidase at carbon nanotube powder microelectrode. *Sens. Actuators, B* 87: 168–172.
104. Yamamoto, K., G. Shi, T. Zhou et al. 2003. Study of carbon nanotubes—HRP modified electrode and its application for novel on-line biosensors. *Analyst* 128: 249–254.
105. Zhao, Y.-D., W.-D. Zhang, H. Chen, and Q.-M. Luo. 2002. Direct electron transfer of glucose oxidase molecules adsorbed onto carbon nanotube powder microelectrode. *Anal. Sci.* 18: 939–941.
106. Guiseppi-Elie, A., C. Lei, and R.H. Baughman. 2002. Direct electron transfer of glucose oxidase on carbon nanotubes. *Nanotechnology* 13: 559–564.
107. Wang, L., J. Wang, and F. Zhou. 2004. Direct electrochemistry of catalase at a gold electrode modified with single-wall carbon nanotubes. *Electroanalysis* 16: 627–632.
108. Wang, J., A.-N. Kawde, and M. Musameh. 2003. Carbon-nanotube-modified glassy carbon electrodes for amplified label-free electrochemical detection of DNA hybridization. *Analyst* 128: 912–916.
109. Wang, J., M. Li, Z. Shi, N. Li, and Z. Gu. 2004. Electrochemistry of DNA at single-wall carbon nanotubes. *Electroanalysis* 16: 140–144.
110. Pedano, M.L. and G.A. Rivas. 2004. Adsorption and electrooxidation of nucleic acids at carbon nanotubes paste electrodes. *Electrochem. Commun.* 6: 10–16.
111. Britto, P.J., K.S.V. Santhanam, and P.M. Ajayan. 1996. Carbon nanotube electrode for oxidation of dopamine. *Bioelectrochem. Bioenerg.* 41: 121–125.
112. Wang, Z., Y. Wang, and G. Luo. 2002. A selective voltammetric method for uric acid detection at b-cyclodextrin modified electrode incorporating carbon nanotubes. *Analyst* 127: 1353–1358.
113. Musameh, M., J. Wang, A. Merkoci, and Y. Lin. 2002. Low-potential stable NADH detection at carbon-nanotube-modified glassy carbon electrodes. *Electrochem. Commun.* 4: 743–746.
114. Wang, J. and M. Musameh. 2003. A reagentless amperometric alcohol biosensor based on carbon-nanotube/Teflon composite electrodes. *Anal. Lett.* 36: 2041–2048.
115. Banks, C.E., T.J. Davies, G.G. Wildgoose, and R.G. Compton. 2004. Electrocatalysis at graphite and carbon nanotube modified electrodes: Edge-plane sites and tube ends are the reactive sites. *Chem. Commun.* 829–841.
116. Banks, C.E. and R.G. Compton. 2005. Exploring the electrocatalytic sites of carbon nanotubes for NADH detection: An edge plane pyrolytic graphite electrode study. *Analyst* 130: 1232–1239.
117. Gong, K., S. Chakrabarti, and L. Dai. 2008. Electrochemistry at carbon nanotube electrodes: Is the nanotube tip more active than the sidewall? *Angew. Chem. Int. Ed.* 47: 5446–5450.
118. Banks, C.E., A. Crossley, C. Salter, S.J. Wilkins, and R.G. Compton. 2006. Carbon nanotubes contain metal impurities which are responsible for the "electrocatalysis" seen at some nanotube-modified electrodes. *Angew. Chem. Int. Ed.* 45: 2533–2537.
119. Kowalewska, B. and P.J. Kulesza. 2009. Application of tetrathiafulvalene-modified carbon nanotubes to preparation of integrated mediating system for bioelectrocatalytic oxidation of glucose. *Electroanalysis* 21: 351–359.

120. Jia, W., A.J. Bandodkar, G. Valdés-Ramírez et al. 2013. Electrochemical tattoo biosensors for real-time noninvasive lactate monitoring in human perspiration. *Anal. Chem.* 85: 6553–6560.

121. Ding, Z., B.M. Quinn, S.K. Haram, L.E. Pell, B.A. Korgel, and A.J. Bard. 2002. Electrochemistry and electrogenerated chemiluminescence from silicon nanocrystal quantum dots. *Science* 296: 1293–1297.

122. Myung, N., Z. Ding, and A.J. Bard. 2002. Electrogenerated chemiluminescence of CdSe nanocrystals. *Nano Lett.* 2: 1315–1319.

123. Bard, A.J. and L.R. Faulkner. 2001. *Electrochemical Methods*, 2nd edn. New York: John Wiley & Sons. pp. 736–768.

124. Bertoncello, P. and R.J. Forster. 2009. Nanostructured materials for electrochemiluminescence (ECL)-based detection methods: Recent advances and future perspectives. *Biosens. Bioelectron.* 24: 3191–3200.

125. Miao, W. and A.J. Bard. 2003. Electrogenerated chemiluminescence. 72. Determination of immobilized DNA and C-reactive protein on Au(111) electrodes using tris(2,2′-bipyridyl)ruthenium(II) labels. *Anal. Chem.* 75: 5825–5834.

126. Wohlstadter, J.N., J.L. Wilbur, G.B. Sigal et al. 2003. Carbon nanotube-based biosensor. *Adv. Mater.* 15: 1184–1187.

127. Ji, X., R.O. Kadara, J. Krussma, Q. Chen, and C.E. Banks. 2010. Understanding the physicoelectrochemical properties of carbon nanotubes: Current state of the art. *Electroanalysis* 22: 7–19.

128. Ren, Z.F., Z.P. Huang, J.W. Xu et al. 1998. Synthesis of large arrays of well-aligned carbon nanotubes on glass. *Science* 282: 1105–1107.

129. Li, J., H.T. Ng, A. Cassell et al. 2003. Carbon nanotube nanoelectrode array for ultrasensitive DNA detection. *Nano Lett.* 3: 597–602.

130. Koehne, J., J. Li, A.M. Cassell et al. 2004. The fabrication and electrochemical characterization of carbon nanoelectrode arrays. *J. Mater. Chem.* 14: 676–684.

131. Cao, A., X. Zhang, C. Xu et al. 2001. Grapevine-like growth of single walled carbon nanotubes among vertically aligned multiwalled nanotube arrays. *Appl. Phys. Lett.* 79: 1252–1254.

132. Jiang, L.-C. and W.-D. Zhang. 2009. Electroanalysis of dopamine at RuO_2 modified vertically aligned carbon nanotube electrode. *Electroanalysis* 21: 1811–1815.

133. Esplandiu, M.J., M. Pacios, L. Cyganek, J. Bartroli, and M. del Valle. 2009. Enhancing the electrochemical response of myoglobin with carbon nanotube electrodes. *Nanotechnology* 20: 355502.

134. Che, G., B.B. Lakshmi, E.R. Fisher, and C.R. Martin. 1998. Carbon nanotubule membranes for electrochemical energy storage and production. *Nature* 393: 346–349.

135. Che, G., B.B. Lakshmi, C.R. Martin, E.R. Fisher, and R.S. Ruoff. 1998. Chemical vapor deposition based synthesis of carbon nanotubes and nanofibers using a template method. *Chem. Mater.* 10: 260–267.

136. Li, J., C. Papadopoulos, J.M. Xu, and M. Moskovits. 1999. Highly-ordered carbon nanotube arrays for electronics aplications. *Appl. Phys. Lett.* 75: 367–369.

137. Liu, Z., Z. Shen, T. Zhu et al. 2000. Organizing single-walled carbon nanotubes on gold using a wet chemical self-assembling technique. *Langmuir* 16: 3569–3573.

138. Wu, B., J. Zhang, Z. Wei, S. Cai, and Z. Liu. 2001. Chemical alignment of oxidatively shortened single-walled carbon nanotubes on silver surface. *J. Phys. Chem. B* 105: 5075–5078.

139. Chattopadhyay, D., I. Galeska, and F. Papadimitrakopoulos. 2001. Metal-assisted organization of shortened carbon nanotubes in monolayer and multilayer forest assemblies. *J. Am. Chem. Soc.* 123: 9451–9452.

140. Diao, P., Z. Liu, B. Wu, X. Nan, J. Zhang, and Z. Wei. 2002. Chemically assembled single-wall carbon nanotubes and their electrochemistry. *ChemPhysChem* 10: 898–901.

141. Patolsky, F., Y. Weizmann, and I. Willner. 2004. Long-range electrical contacting of redox enzymes by SWCNT connectors. *Angew. Chem. Int. Ed.* 43: 2113–2117.

142. Xiao, N. and B.J. Venton. 2012. Rapid, sensitive detection of neurotransmitters at microelectrodes modified with self-assembled SWCNT forests. *Anal. Chem.* 84: 7816–7822.

143. O'Connor, M., S.N. Kim, A.J. Killiard et al. 2004. Mediated amperometric immunosensing using single walled carbon nanotube forests. *Analyst* 129: 1176–1180.

144. Yu, X., S.N. Kim, F. Papadimitrakopoulos, and J.F. Rusling. 2005. Protein immunosensor using single-wall carbon nanotube forests with electrochemical detection of enzyme labels. *Mol. BioSyst.* 1: 70–78.

145. Malhotra, R., F. Papadimitrakopoulos, and J.F. Rusling. 2010. Sequential layer analysis of protein immunosensors based on single wall carbon nanotube forests. *Langmuir* 26: 15050–15056.

146. Sardesai, N., S. Pan, and J. Rusling. 2009. Electrochemiluminescent immunosensor for detection of protein cancer biomarkers using carbon nanotube forests and [Ru-(bpy)3]2+-doped silica nanoparticles. *Chem. Commun.* 4968–4970.

147. Sardesai, N.P., J.C. Barron, and J.F. Rusling. 2011. Carbon nanotube microwell array for sensitive electrochemiluminescent detection of cancer biomarker proteins. *Anal. Chem.* 83: 6698–6703.

148. Venkatanarayanan, A., K. Crowley, E. Lestini, T.E. Keyes, J.F. Rusling, and R.J. Forster. 2012. High sensitivity carbon nanotube based electrochemiluminescence sensor array. *Biosens. Bioelectron.* 31: 233–239.

149. Davies, T.J. and R.G. Compton. 2005. The cyclic and linear sweep voltammetry of regular and random arrays of microdisc electrodes: Theory. *J. Electroanal. Chem.* 585: 63–82.

150. LaFratta, C.N. and D.R. Walt. 2008. Very high density sensing arrays. *Chem. Rev.* 108: 614–637.

151. Baker, W.S. and R.M. Crooks. 1998. Independent geometrical and electrochemical characterization of arrays of nanometer-scale electrodes. *J. Phys. Chem. B* 102: 10041–10046.

152. Saito, Y. 1968. A theoretical study of the diffusion current at the stationary electrodes of circular and narrow band types. *Rev. Polarogr.* 15: 177–187.

153. Amatore, C., J.M. Savéant, and D. Tessier. 1983. Charge transfer at partially blocked surfaces: A model for the case of microscopic active and inactive sites. *J. Electroanal. Chem.* 147: 39–51.

154. Lee, H.J., C. Beriet, R. Ferrigno, and H.H. Girault. 2001. Cyclic voltammetry at a regular microdisc electrode array. *J. Electroanal. Chem.* 502: 138–145.

155. Lin, Y., F. Lu, Y. Tu, and Z. Ren. 2004. Glucose biosensors based on carbon nanotube nanoelectrode ensembles. *ACS Nano* 4: 191–195.

156. Yun, Y., A. Bange, W.R. Heineman et al. 2007. A nanotube immunosensor for direct electrochemical detection of antigen–antibody binding. *Sens. Actuators B* 123: 177–182.

157. Pumera, M. 2009. Electrochemistry of graphene: New horizons for sensing and energy storage. *Chem. Rec.* 9: 211–233.

158. Kuila, T., S. Bose, P. Khanra, A.K. Mishra, N.H. Kim, and J.H. Lee. 2011. Recent advances in graphene-based biosensors. *Biosens. Bioelectron.* 26: 4637–4648.

159. Pumera, M. 2010. Graphene-based nanomaterials and their electrochemistry. *Chem. Soc. Rev.* 39: 4146–4157.

160. Alwarappan, S., A. Erdem, C. Liu, and C.-Z. Li. 2009. Probing the electrochemical properties of graphene nanosheets for biosensing applications. *J. Phys. Chem. C* 113: 8853–8857.

161. Mohanty, N. and V. Berry. 2008. Graphene-based single-bacterium resolution biodevice and DNA transistor: Interfacing graphene derivatives with nanoscale and microscale biocomponents. *Nano Lett.* 8: 4469–4476.

162. Zhou, M., Y. Zhai, and S. Dong. 2009. Electrochemical sensing and biosensing platform based on chemically reduced graphene oxide. *Anal. Chem.* 81: 5603–5613.

163. Shan, C., H. Yang, J. Song, D. Han, A. Ivaska, and L. Niu. 2009. Direct electrochemistry of glucose oxidase and biosensing for glucose based on graphene. *Anal. Chem.* 81: 2378–2382.

164. Kang, X., J. Wang, H. Wu, I.A. Aksay, J. Liu, and Y. Lin. 2009. Glucose oxidase–graphene–chitosan modified electrode for direct electrochemistry and glucose sensing. *Biosens. Bioelectron.* 25: 901–905.

165. Wu, J.-F., M.-Q. Xu, and G.-C. Zhao. 2010. Graphene-based modified electrode for the direct electron transfer of cytochrome c and biosensing. *Electrochem. Commun.* 12: 175–177.

166. Wan, Y., Z. Lin, D. Zhang, Y. Wang, and B. Hou. 2011. Impedimetric immunosensor doped with reduced graphene sheets fabricated by controllable electrodeposition for the non-labelled detection of bacteria. *Biosens. Bioelectron.* 26: 1959–1964.

167. Zeng, Q., J. Cheng, L. Tang et al. 2010. Self-assembled graphene—Enzyme hierarchical nanostructures for electrochemical biosensing. *Adv. Funct. Mater.* 20: 3366–3372.

168. Shang, N.G., P. Papakonstantinou, M. McMullan et al. 2008. Catalyst-free efficient growth, orientation and biosensing properties of multilayer graphene nanoflake films with sharp edge planes. *Adv. Funct. Mater.* 18: 3506–3514.

169. Wang, Y., Y. Li, L. Tang, J. Lu, and J. Li. 2009. Application of graphene-modified electrode for selective detection of dopamine. *Electrochem. Commun.* 11: 889–892.

170. Hrapovic, S., Y. Liu, K.B. Male, and J.H.T. Luong. 2004. Electrochemical biosensing platforms using platinum nanoparticles and carbon nanotubes. *Anal. Chem.* 76: 1083–1088.

171. Jie, G., B. Liu, H. Pan, J.-J. Zhu, and H.-Y. Chen. 2007. CdS nanocrystal-based electrochemiluminescence biosensor for the detection of low-density lipoprotein by increasing sensitivity with gold nanoparticle amplification. *Anal. Chem.* 79: 5574–5581.

172. Jie, G., J. Zhang, D. Wang, C. Cheng, H.-Y. Chen, and J.-J. Zhu. 2008. Electrochemiluminescence immunosensor based on CdSe nanocomposites. *Anal. Chem.* 80: 4033–4039.

173. Jie, G., L. Li, C. Chen, J. Xuan, and J.-J. Zhu. 2009. Enhanced electrochemiluminescence of CdSe quantum dots composited with CNTs and PDDA for sensitive immunoassay. *Biosens. Bioelectron.* 24: 3352–3358.

174. Ho, J.A., Y.-C. Lin, L.-S. Wang, K.-C. Hwang, and P.-T. Chou. 2009. Carbon nanoparticle-enhanced immunoelectrochemical detection for protein tumor marker with cadmium sulfide biotracers. *Anal. Chem.* 81: 1340–1346.

175. Shan, C., H. Yang, D. Han, Q. Zhang, A. Ivaska, and L. Niu. 2010. Graphene/AuNPs/chitosan nanocomposites film for glucose biosensing. *Biosens. Bioelectron.* 25: 1070–1074.

176. Yang, M., Y. Yang, Y. Liu, G. Shen, and R. Yu. 2006. Platinum nanoparticles-doped sol-gel/carbon nanotubes composite electrochemical sensors and biosensors. *Biosens. Bioelectron.* 21: 1125–1131.

177. Hu, X., R. Wang, Y. Ding, X. Zhang, and W. Jin. 2010. Electrochemiluminescence of CdTe quantum dots as labels at nanoporous gold leaf electrodes for ultrasensitive DNA analysis. *Talanta* 80: 1737–1743.

178. Qian, J., C. Zhang, X. Cao, and S. Liu. 2010. Versatile immunosensor using a quantum dot coated silica nanosphere as a label for signal amplification. *Anal. Chem.* 82: 6422–6429.

179. Campbell, J.K., L. Sun, and R.M. Crooks. 1999. Electrochemistry using single carbon nanotubes. *J. Am. Chem. Soc.* 121: 3779–3780.

180. Dai, H., J.H. Hafner, A.G. Rinzler, D.T. Colbert, and R.E. Smalley. 1996. Nanotubes as nanoprobes in scanning probe microscopy. *Nature* 384: 147–150.

181. Wong, S., E. Joselevich, A.T. Woolley, C.L. Cheung, and C.M. Lieber. 1998. Covalently functionalized nanotubes as nanometre-sized probes in chemistry and biology. *Nature* 392: 52–55.

182. Boo, H., R.-A. Jeong, S. Park et al. 2006. Electrochemical nanoneedle biosensor based on multiwall carbon nanotube. *Anal. Chem.* 78: 617–620.

183. Hoeben, F.J.M., F.S. Meijer, C. Dekker, S.P.J. Albracht, H.A. Heering, and S.G. Lemay. 2008. Toward single-enzyme molecule electrochemistry: [NiFe]-hydrogenase protein film voltammetry at nanoelectrodes. *ACS Nano* 2: 2497–2504.

184. Rusling, J.F., G.W. Bishop, N.M. Doan, and F. Papadimitrakopoulos. 2014. Nanomaterials and biomaterials for protein detection. *J. Mater. Chem. B* 2: 12–30.

185. Wang, J. 2007. Nanoparticle-based electrochemical bioassays of proteins. *Electroanalysis* 19: 769–776.

186. González-García, M.B. and A. Costa-García. 1995. Adsorptive stripping voltammetric behavior of colloidal gold and immunogold on carbon paste electrode. *Bioelectroch. Bioenerg.* 38: 389–395.

187. Hernández-Santos, D., M.B. González-García, and A. Costa-García. 2000. Electrochemical determination of gold nanoparticles in colloidal solutions. *Electrochim. Acta* 46: 607–615.

188. González-García, M.B., C. Fernández-Sánchez, and A. Costa-García. 2000. Colloidal gold as an electrochemical label of streptavidin–biotin interaction. *Biosens. Bioelectron.* 15: 315–321.

189. Ozsoz, M., A. Erdem, K. Kerman et al. 2003. Electrochemical genosensor based on colloidal gold nanoparticles for the detection of Factor V Leiden mutation using disposable pencil graphite electrodes. *Anal. Chem.* 75: 2181–2187.

190. Szymanski, M., A.P.F Turner, and R. Porter. 2010. Electrochemical dissolution of silver nanoparticles and its application in metalloimmunoassay. *Electroanalysis* 22: 191–198.

191. de la Escosura-Muñiz, A., A. Ambrosi, and A. Merkoçi. 2008. Electrochemical analysis with nanoparticle-based biosystems. *Trends Anal. Chem.* 27: 568–584.

192. Wang, J., D. Xu, and R. Polsky. 2002. Magnetically-induced solid-state electrochemical detection of DNA hybridization. *J. Am. Chem. Soc.* 124: 4208–4209.

193. Dequaire, M., C. Degrand, and B. Limoges. 2000. An electrochemical metalloimmunoassay based on a colloidal gold label. *Anal. Chem.* 72: 5521–5528.

194. Authier, L., C. Grossiord, P. Brossier, and B. Limoges. 2001. Electrochemical detection of amplified human cytomegalovirus DNA using disposable microband electrodes. *Anal. Chem.* 73: 4450–4456.

195. Wang, J., D. Xu, A.-N. Kawde, and R. Polsky. 2001. Metal nanoparticle-based electrochemical stripping potentiometric detection of DNA hybridization. *Anal. Chem.* 73: 5576–5581.

196. Wang, J., R. Polsky, and D. Xu. 2001. Silver-enhanced colloidal gold electrochemical stripping detection of DNA hybridization. *Langmuir* 17: 5739–5741.

197. Hainfeld, J.F. and R.D. Powell. 2000. New frontiers in gold labeling. *J. Histochem. Cytochem.* 48: 471–480.

198. Liu, R., Y. Zhang, S. Zhang, W. Qiu, and Y. Gao. 2014. Silver enhancement of gold nanoparticles for biosensing: From qualitative to quantitative. *Appl. Spectrosc. Rev.* 49: 121–138.

199. Taton, T.A., C.A. Mirkin, and R.L. Letsinger. 2000. Scanometric DNA array detection with nanoparticle probes. *Science* 289: 1757–1760.

200. Kim, D., W.L. Daniel, and C.A. Mirkin. 2009. Microarray-based multiplexed scanometric immunoassay for protein cancer markers using gold nanoparticle probes. *Anal. Chem.* 81: 9183–9187.

201. Velev, O.D. and E.W. Kaler. 1999. In situ assembly of colloidal particles into miniaturized biosensors. *Langmuir* 15: 3693–3698.

202. Rochelet-Dequaire, M., B. Limoges, and P. Brossier. 2006. Subfemtomolar electrochemical detection of target DNA by catalytic enlargement of the hybridized gold nanoparticle labels. *Analyst* 131: 923–929.

203. Mao, X., J. Jiang, Y. Luo, G. Shen, and R. Yu. 2007. Copper-enhanced gold nanoparticles for electrochemical stripping detection of human IgG. *Talanta* 73: 420–424.

204. Chu, X., X. Fu, K. Chen, G.-L. Shen, and R.-Q. Yu. 2005. An electrochemical stripping metalloimmunoassay based on silver-enhanced gold nanoparticle label. *Biosens. Bioelectron.* 20: 1805–1812.

205. Guo, H., N. He, S. Ge, D. Yang, and J. Zhang. 2005. MCM-41 mesoporous material modified carbon paste electrode for the determination of cardiac troponin I by anodic stripping voltammetry. *Talanta* 68: 61–66.

206. Lai, G., F. Yan, J. Wu, C. Leng, and H. Ju. 2011. Ultrasensitive multiplexed immunoassay with electrochemical stripping analysis of silver nanoparticles catalytically deposited by gold nanoparticles and enzymatic reaction. *Anal. Chem.* 83: 2726–2732.

207. Willner, I., R. Baron, and B. Willner. 2007. Integrated nanoparticle–biomolecule systems for biosensing and bioelectronics. *Biosens. Bioelectron.* 22: 1841–1852.

208. Guo, S. and S. Dong. 2009. Biomolecule-nanoparticle hybrids for electrochemical biosensors. *Trends Anal. Chem.* 28: 96–109.

209. Wang, J., G. Liu, R. Polsky, and A. Merkoçi. 2002. Electrochemical stripping detection of DNA hybridization based on cadmium sulfide nanoparticle tags. *Electrochem. Commun.* 4: 722–726.

210. Numnuam, A., K.Y. Chumbimuni-Torres, Y. Xiang et al. 2008. Aptamer-based potentiometric measurements of proteins using ion-selective microelectrodes. *Anal. Chem.* 80: 707–712.

211. Liu, G., J. Wang, J. Kim, M.R. Jan, and G.E. Collins. 2004. Electrochemical coding for multiplexed immunoassays of proteins. *Anal. Chem.* 76: 7126–7130.

212. Wang, J. 2008. Barcoded metal nanowires. *J. Mater. Chem.* 18: 4017–4020.

213. Xiang, Y., Y. Zhang, Y. Chang, Y. Chai, J. Wang, and R. Yuan. 2010. Reverse-micelle synthesis of electrochemically encoded quantum dot barcodes: Application to electronic coding of a cancer marker. *Anal. Chem.* 82: 1138–1141.

214. Polsky, R., R. Gill, L. Kaganovsky, and I. Willner. 2006. Nucleic acid-functionalized Pt nanoparticles: Catalytic labels for the amplified electrochemical detection of biomolecules. *Anal. Chem.* 78: 2268–2271.

215. Arora, A., J.C.T. Eijkel, W.E. Morf, and A. Manz. 2001. A wireless electrochemiluminescence detector applied to direct and indirect detection for electrophoresis on a microfabricated glass device. *Anal. Chem.* 73: 3282–3288.

216. Mavré, F., R.K. Anand, D.R. Laws et al. 2010. Bipolar electrodes: A useful tool for concentration, separation, and detection of analytes in microelectrochemical systems. *Anal. Chem.* 82: 8766–8774.

217. Fosdick, S.E., K.N. Knust, K. Scida, and R.M. Crooks. 2013. Bipolar electrochemistry. *Angew. Chem. Int. Ed.* 52: 10438–10456.

218. Chow, K.-F., F. Mavré, and R.M. Crooks. 2008. Wireless electrochemical DNA microarray sensor. *J. Am. Chem. Soc.* 130: 7544–7545.

219. Chow, K.-F., F. Mavré, J.A. Crooks, B.-Y. Chang, and R.M. Crooks. 2009. A large-scale, wireless electrochemical bipolar electrode microarray. *J. Am. Chem. Soc.* 131: 8364–8365.

220. Guerrette, J.P., S.M. Oja, and B. Zhang. 2012. Coupled electrochemical reactions at bipolar microelectrodes and nanoelectrodes. *Anal. Chem.* 84: 1609–1616.

221. Kawde, A.-N. and J. Wang. 2004. Amplified electrical transduction of DNA hybridization based on polymeric beads loaded with multiple gold nanoparticle tags. *Electroanalysis* 16: 101–107.

222. Wang, J., G. Liu, M.R. Jan, and Q. Zhu. 2003. Electrochemical detection of DNA hybridization based on carbon-nanotubes loaded with CdS tags. *Electrochem. Commun.* 5: 1000–1004.

223. Yang, M., A. Javadi, and S. Gong. 2011. Sensitive electrochemical immunosensor for the detection of cancer biomarker using quantum dot functionalized graphene sheets as labels. *Sens. Actuators, B* 155: 357–360.

224. Dong, H., F. Yan, H. Ji, D.K.Y. Wong, and H. Ju. 2010. Quantum-dot-functionalized poly(styrene-*co*-acrylic acid) microbeads: Step-wise self-assembly, characterization, and applications for sub-femtomolar electrochemical detection of DNA hybridization. *Adv. Funct. Mater.* 20: 1173–1179.

225. Hansen, J.A., J. Wang, A.-N. Kawde, Y. Xiang, K.V. Gothelf, and G. Collins. 2006. Quantum-dot/aptamer-based ultrasensitive multi-analyte electrochemical biosensor. *J. Am. Chem. Soc.* 128: 2228–2229.

226. Tang, D., L. Hou, R. Niessner, M. Xu, Z. Gao, and D. Knopp. 2013. Multiplexed electrochemical immunoassay of biomarkers using metal sulfide quantum dot nanolabels and trifunctionalized magnetic beads. *Biosens. Bioelectron.* 46: 37–43.

227. Wang, J., R. Polsky, A. Merkoci, and K.L. Turner. 2003. "Electroactive beads" for ultrasensitive DNA detection. *Langmuir* 19: 989–991.

228. Wang, J., J. Li, A.J. Baca et al. 2003. Amplified voltammetric detection of DNA hybridization via oxidation of ferrocene caps on gold nanoparticle/streptavidin conjugates. *Anal. Chem.* 75: 3941–3945.

229. Wang, J., G. Liu, B. Munge, L. Lin, and Q. Zhu. 2004. DNA-based amplified bioelectronics detection and coding of proteins. *Angew. Chem. Int. Ed.* 43: 2158–2161.

230. Wang, J., G. Liu, M.H. Engelhard, and Y. Lin. 2006. Sensitive immunoassay of a biomarker tumor necrosis factor-a based on poly(guanine)-functionalized silica nanoparticle label. *Anal. Chem.* 78: 6974–6979.

231. Mak, W.C., K.Y. Cheung, D. Trau, A. Warsinke, F. Scheller, and R. Renneberg. 2005. Electrochemical bioassay utilizing encapsulated electrochemical active microcrystal biolabels. *Anal. Chem.* 77: 2835–2841.

232. Xiang, Y., X. Qian, B. Jiang, Y. Chai, and R. Yuan. 2011. An aptamer-based signal-on and multiplexed sensing platform for one-spot simultaneous electronic detection of proteins and small molecules. *Chem. Commun.* 4733–4735.

233. Li, Y., Z. Zhong, Y. Chai et al. 2012. Simultaneous electrochemical immunoassay of three liver cancer biomarkers using distinguishable redox probes as signal tags and gold nanoparticles coated carbon nanotubes as signal enhancers. *Chem. Commun.* 537–539.

234. Zhu, Z., Y. Chai, R. Yuan et al. 2013. Amperometric immunosensor for simultaneous detection of three analytes in one interface using dual functionalized graphene sheets integrated with redox-probes as tracer matrixes. *Biosens. Bioelectron.* 43: 440–445.

235. Miao, W. and A.J. Bard. 2004. Electrogenerated chemiluminescence. 77. DNA hybridization detection at high amplification with [Ru(bpy)$_3$]$^{2+}$-containing microspheres. *Anal. Chem.* 76: 5379–5386.

236. Miao, W. and A.J. Bard. 2004. Electrogenerated chemiluminescence. 80. C-reactive protein determination at high amplification with [Ru(bpy)$_3$]$^{2+}$-containing microspheres. *Anal. Chem.* 76: 7109–7113.

237. Zhan, W. and A.J. Bard. 2007. Electrogenerated chemiluminescence. 83. Immunoassay of human C-reactive protein by using Ru(bpy)$_3$$^{2+}$-encapsulated liposomes as labels. *Anal. Chem.* 79:459–463.

238. Wang, H., C. Zhang, Y. Li, and H. Qi. 2006. Electrogenerated chemiluminescence detection for deoxyribonucleic acid hybridization based on gold nanoparticles carrying multiple probes. *Anal. Chim. Acta* 575: 205–211.

239. Li, Y., H. Qi, F. Fang, and C. Zhang. 2007. Ultrasensitive electrogenerated chemiluminescence detection of DNA hybridization using carbon-nanotubes loaded with tris(2,2′-bipyridyl) ruthenium derivative tags. *Talanta* 72: 1704–1709.

240. Zhu, D., Y. Tang, D. Xing, and W.R. Chen. 2008. PCR-free quantitative detection of genetically modified organism from raw materials. An electrochemiluminescence-based bio bar code method. *Anal. Chem.* 80: 3566–3571.

241. Deiss, F., C.N. LaFratta, M. Symer, T.M. Bicharz, N. Sojic, and D.R. Walt. 2009. Multiplexed sandwich immunoassays using electrochemiluminescence imaging resolved at the single bead level. *J. Am. Chem. Soc.* 131: 6088–6089.

242. Hu, L. and G. Xu. 2010. Applications and trends in electrochemiluminescence. *Chem. Soc. Rev.* 39: 3275–3304.

243. Sardesai, N.P., K. Kadimisetty, R. Faria, and J.F. Rusling. 2013. A microfluidic electrochemiluminescent device for detecting cancer biomarker proteins. *Anal. Bioanal. Chem.* 405: 3831–3838.

244. Wang, J., G. Liu, and M.R. Jan. 2004. Ultrasensitive electrical biosensing of proteins and DNA: Carbon-nanotube derived amplification of the recognition and transduction events. *J. Am. Chem. Soc.* 126: 3010–3011.

245. Munge, B., G. Liu, G. Collins, and J. Wang. 2005. Multiple enzyme layers on carbon nanotubes for electrochemical detection down to 80 DNA copies. *Anal. Chem.* 77: 4662–4666.

246. Yu, X., B. Munge, V. Patel et al. 2006. Carbon nanotube amplification strategies for highly sensitive immunodetection of cancer biomarkers. *J. Am. Chem. Soc.* 128: 11199–11205.

247. Malhotra, R., V. Patel, J.P. Vaqué, J.S. Gutkind, and J.F. Rusling. 2010. Ultrasensitive electrochemical immunosensor for oral cancer biomarker IL-6 using carbon nanotube forest electrodes and multilabel amplification. *Anal. Chem.* 82: 3118–3123.

248. Munge, B.S., J. Fisher, L.N. Millord, C.E. Krause, R.S. Dowd, and J.F. Rusling. 2010. Sensitive electrochemical immunosensor for matrix metalloproteinase-3 based on single-wall carbon nanotubes. *Analyst* 135: 1345–1350.

249. Munge, B.S., A.L. Coffey, J.M. Doucette et al. 2011. Nanostructured immunosensor for attomolar detection of cancer biomarker interleukin-8 using massively labeled superparamagnetic particles. *Angew. Chem. Int. Ed.* 50: 7915–7918.

250. Chikkaveeraiah, B.V., V. Mani, V. Patel, J.S. Gutkind, and J.F. Rusling. 2011. Microfluidic electrochemical immunoarray for ultrasensitive detection of two cancer biomarker proteins in serum. *Biosens. Bioelectron.* 26: 4477–4483.

251. Malhotra, R., V. Patel, B.V. Chikkaveeraiah et al. 2012. Ultrasensitive detection of cancer biomarkers in the clinic by use of a nanostructured microfluidic array. *Anal. Chem.* 84: 6249–6255.

252. Tang, J., B. Su, D. Tang, and G. Chen. 2010. Conductive carbon nanoparticles-based electrochemical immunosensor with enhanced sensitivity for α-fetoprotein using irregular-shaped gold nanoparticles-labeled enzyme-linked antibodies as signal improvement. *Biosens. Bioelectron.* 25: 2657–2662.

253. Song, Z., R. Yuan, Y. Chai et al. 2010. Horseradish peroxidase-functionalized Pt hollow nanospheres and multiple redox probes as trace labels for a sensitive simultaneous multianalyte electrochemical immunoassay. *Chem. Commun.* 6750–2675.

254. Su, B., D. Tang, Q. Li, J. Tang, and G. Chen. 2011. Gold-silver-graphene hybrid nanosheets-based sensors for sensitive amperometric immunoassay of alpha-fetoprotein using nanogold-enclosed titania nanoparticles as labels. *Anal. Chim. Acta* 692: 116–124.

255. Zhong, Z., W. Wu, D. Wang et al. 2010. Nanogold-enwrapped graphene nanocomposites as trace labels for sensitivity enhancement of electrochemical immunosensors in clinical immunoassays: Carcinoembryonic antigen as a model. *Biosens. Bioelectron.* 25: 2379–2383.

256. Tang, J., D. Tang, R. Niessner, G. Chen, and D. Knopp. 2011. Magneto-controlled graphene immunosensing platform for simultaneous multiplexed electrochemical immunoassay using distinguishable signal tags. *Anal. Chem.* 83: 5407–5414.

257. Wu, Y., C. Chen, and S. Liu. 2009. Enzyme-functionalized silica nanoparticles as sensitive labels in biosensing. *Anal. Chem.* 81: 1600–1607.

258. Tang, D., B. Su, J. Tang, J. Ren, and G. Chen. 2010. Nanoparticle-based sandwich electrochemical immunoassay for carbohydrate antigen 125 with signal enhancement using enzyme-coated nanometer-sized enzyme-doped silica beads. *Anal. Chem.* 82: 1527–1534.

259. Wu, Y., P. Xue, Y. Kang, and K.M. Hui. 2013. Paper-based microfluidic electrochemical immunodevice integrated with nanobioprobes onto graphene film for ultrasensitive multiplexed detection of cancer biomarkers. *Anal. Chem.* 85: 8661–8668.

260. Yang, M., A. Javadi, and S. Gong. 2010. Ultrasensitive immunosensor for the detection of cancer biomarker based on graphene sheet. *Biosens. Bioelectron.* 26: 560–565.

261. Du, D., L. Wang, Y. Shao, J. Wang, M.H. Engelhard, and Y. Lin. 2011. Functionalized graphene oxide as nanocarrier in a multienzyme labeling amplification strategy for ultrasensitive electrochemical immunoassay of phosphorylated p53 (S392). *Anal. Chem.* 83: 746–752.

262. Liu, K., J.-J. Zhang, C. Wang, and J.-J. Zhu. 2011. Graphene-assisted dual amplification strategy for the fabrication of sensitive amperometric immunosensor. *Biosens. Bioelectron.* 26: 3627–3632.

263. Du, D., Z. Zou, Y. Shin et al. 2010. Sensitive immunosensor for cancer biomarker based on dual signal amplification strategy of graphene sheets and multienzyme functionalized carbon nanospheres. *Anal. Chem.* 82: 2989–2995.

264. Karyakin, A.A., E.E. Karyakina, and L. Gorton. 2000. Amperometric biosensor using Prussian blue-based "artificial peroxidase" as transducer for hydrogen peroxide. *Anal. Chem.* 72: 1720–1723.

265. Karyakin, A.A. 2001. Prussian blue and its analogues: Electrochemistry and analytical applications. *Electroanalysis* 13: 813–819.

266. Karyakin, A.A., E.A. Puganova, I.A. Budashov et al. 2004. Prussian blue based nanoelectrode arrays for H_2O_2 detection. *Anal. Chem.* 76: 474–478.

267. Lai, G., F. Yan, and H. Ju. 2009. Dual signal amplification of glucose oxidase-functionalized nanocomposites as a trace label for ultrasensitive simultaneous multiplexed electrochemical detection of tumor markers. *Anal. Chem.* 81: 9730–9736.

268. Chang, L., D.M. Rissin, D.R. Fournier et al. 2012. Single molecule enzyme-linked immunosorbent assays: Theoretical considerations. *J. Immunol. Methods* 378: 102–115.

269. Mani, V., D.P. Wasalathanthri, A.A. Joshi, C.V. Kumar, and J.F. Rusling. 2012. Highly efficient binding of paramagnetic beads bioconjugated with 100,000 or more antibodies to protein-coated surfaces. *Anal. Chem.* 84: 10485–10491.

270. Karlsson, R., A. Michaelsson, and L. Mattsson. 1991. Kinetic analysis of monoclonal antibody-antigen interactions with a new biosensor based analytical system. *J. Immunol. Methods* 145: 229–240.

14 Electrode Array Probes of Exocytosis at Single-Cell Membranes and Exocytosis Measurements at Cell Biomimetic Systems

Jun Wang and Andrew G. Ewing

CONTENTS

14.1 INTRODUCTION

Membrane fusion involves the merging of two phospholipid bilayers in an aqueous environment and is involved in many cellular processes, such as cell exocytosis. Synaptic vesicles are nanometer-sized organelles, which are packaged with chemical messengers (e.g., neurotransmitters, neuro-hormones, and neuropeptides).[1-3] Each presynaptic nerve terminal contains hundreds of synaptic vesicles. When an action potential depolarizes the presynaptic plasma membrane, Ca^{2+} channels

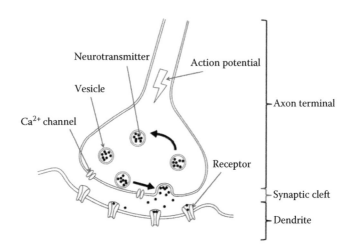

FIGURE 14.1 Schematic representation of the structure of the chemical synapse.

open, and Ca^{2+} flows into the nerve terminal to trigger the exocytosis of synaptic vesicles, thereby vesicles migrate to the plasma membrane of a cell, fuse, and release their contents into the extracellular space. These messengers can then bind to receptors on a target cell, thus inducing a cascade of signaling events in a complex network[3,4] (Figure 14.1). Until now, several types of neurotransmitters, such as amino acids, monoamines, and peptides, have been found, but their function in the brain is not always clear. Exocytotic events occur on a millisecond timescale with transmitter release proportions varying from zepto- to femtomole amounts per vesicle,[5] making them experimentally challenging to monitor. Several different kinds of bioanalytical techniques have been developed to measure chemical messengers in the extracellular fluid following exocytosis from tissue in vivo and to measure individual exocytotic events at single cells under in vitro experimental conditions with biological or artificial models.[6–21]

Electrochemical methods utilizing carbon-fiber microelectrodes have been extremely useful in the detection of neurotransmitters released from single vesicles.[7,8] In a typical experiment, a carbon-fiber microdisk electrode is placed on the cell surface. The electrode potential is held constant (in amperometry) or scanned (in cyclic voltammetry) with respect to a reference electrode placed in the extracellular media. Quantitative and qualitative information about neurotransmission can be obtained from i-t and i-V responses.[22] Experiments using a single carbon-fiber microelectrode provide information including insights into chemical identity, amount of neurotransmitter released from a single vesicle, event frequency, and kinetic information relating to fusion pore opening.[23] However, the specialized protein machineries and lipid domains in the cell membrane lead to spatial variations in the cell membranes as well as the nature and location of exocytotic release.[24,25] For example, the distribution of exocytotic activity has been found to be spatially heterogeneous at the surface of a single cell.[26–30] Such information can be useful in understanding the molecular mechanisms and the chemical basis for the regulation of neural secretion. Due to the difficulties in micromanipulation, it is extremely challenging to record from more than one microelectrode concurrently at a single cell.

In this chapter, recent advances in simultaneously monitoring exocytotic events from multiple subcellular sites of a single cell with microelectrode arrays (MEAs) containing subcellular microelectrodes will be presented. As part of this chapter, we also present the recent advances in exocytosis studies using artificial cell mimic systems and electrochemical monitoring. We do not attempt to be comprehensive surrounding exocytosis methods and do not discuss many of the other elegant electrochemical methods used to measure this process.

14.2 SPATIAL AND TEMPORAL RESOLUTION OF CELL EXOCYTOSIS STUDIED BY INDIVIDUALLY ADDRESSABLE SUBCELLUAR SIZE MEAs

MEAs are devices that contain multiple microelectrodes. These have been used in a variety of applications in electrophysiological recording and stimulation because of their advantages when used with long-term culturing and because they can be used for multisite extracellular recording or stimulation. These include two kinds of MEAs: individually addressable MEAs and nonindividually addressable MEAs. Individually addressable MEAs offer many advantages, such as high spatial resolution, the possibility of sensing multiple analytes using different microelectrodes in the array, and the ability to probe signal transmission in a network of biological cells.[31–36] Considering the patterning of the microelectrodes and relative size of the array compared to the cell, most MEAs contain electrodes larger than a typical mammalian cell, and the electrodes in the whole MEA are not adapted to multisite single-cell membrane studies. In 2005, Lindau and coworkers fabricated an array of four platinum microelectrodes on a chip and recorded exocytotic events simultaneously from these MEAs positioned under a single chromaffin cell.[37] However, due to the patterning of the microelectrodes and relative size of the array compared to the cell, a large fraction of the cell membrane was not directly exposed to the electrode surfaces. And in this experiment, cells were not directly cultured on the electrodes. The cells were detached from the culture surface and deposited on the electrode area by the use of a patch pipette, with one disadvantage that it was easy to disrupt the cell. Nevertheless, this was a significant advance in this area. This chapter considers more recent advances to take the technology to a new level providing high spatial and temporal resolution for the study of cell exocytosis across single-cell membranes by the use of individually addressable MEAs.[38–40]

14.2.1 SPATIALLY AND TEMPORALLY RESOLVED SINGLE-CELL EXOCYTOSIS MEASURED WITH INDIVIDUALLY ADDRESSABLE CARBON-FIBER MEAs

Considering the high temporal resolution of the amperometric technique, it is possible to use closely packed individually addressable carbon-fiber MEAs to characterize the spatial properties of exocytosis at a single cell. SEM pictures of three kinds of carbon-fiber MEAs reported are shown in Figure 14.2. The MEAs contain two, three, and seven carbon-fiber microdisk electrodes. The electrodes are structurally well defined, and the overall diameters of the three-fiber and seven-fiber MEAs are between 15 and 20 μm, respectively. The 2.5 μm radius carbon disks are tightly packed together and surrounded by a thin layer of glass (~1–2 μm). The carbon-fiber MEAs have a total tip dimension of ~20 μm, which indicates that the interelectrode distance in the array is ~7 μm. As the diameter of each microelectrode is 5 μm, the thickness of the glass between adjacent fibers is ~2 μm.

Figure 14.3a through g shows the steady-state voltammetric response of each microelectrode (the inset at the lower left shows the relative position of each microelectrode in the array) measured simultaneously at 20 mV/s in 1 mM FcCH$_2$OH and 0.2 M KCl. The voltammetric response is well defined and has a sigmoidal shape at this scan rate. The steady-state limiting current at each microelectrode is approximately the same, except for electrode G. Close inspection of the voltammogram in Figure 14.3g indicates that the diffusion-limited steady-state current for the center electrode is ~40% smaller than that observed for the edge electrodes (A through F). The limiting current for the surrounding electrodes is ~610 pA, which is 48% smaller than the predicted value. This difference is apparently due to geometrically hindered diffusion combined with depletion of the analyte by the surrounding electrodes. This is more evident for the center electrode as the limiting current at this electrode is somewhat shielded by the collection of microelectrodes surrounding it.

Amperometric detection of neurotransmitter release from single PC12 cells has been carried out with these carbon-fiber MEAs to measure exocytosis. To measure exocytosis, a carbon-fiber MEA is typically placed close to the cell membrane to detect released transmitter. Released transmitters are oxidized locally and free diffusion is minimized by the small electrode/cell separation.

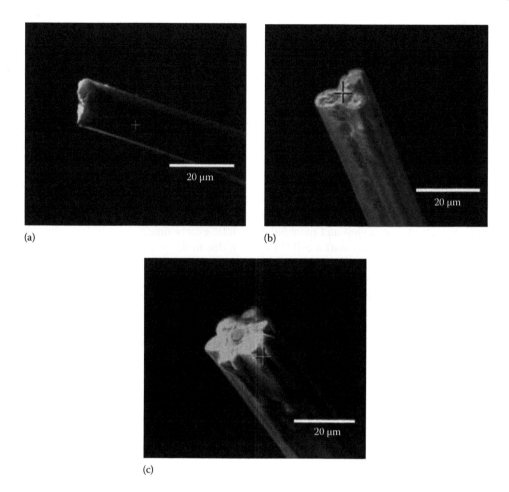

(a)

(b)

(c)

FIGURE 14.2   Scanning electron microscopy pictures of carbon-fiber MEAs having (a) two, (b) three, and (c) seven microdisk microelectrodes. (From Zhang, B. et al., *Anal. Chem.*, 80, 1394, 2008. With permission.)

Electrochemical recordings using carbon-fiber MEAs provide excellent temporal resolution and allow simultaneous examination of different membrane areas with subcellular resolution. Figure 14.4a shows an optical micrograph of a carbon MEA placed on a single PC12 cell. A glass micropipette containing high K^+ solution (100 mM) was positioned ~100 μm away to stimulate secretion. Figure 14.4b displays a 16 min amperometric recording of exocytotic events at a single PC12 cell. Each current transient corresponds to the electrochemical oxidation of dopamine molecules secreted from a single intracellular vesicle. Noticeable in Figure 14.4b is the subcellular heterogeneity observed in exocytosis across the cell surface. For example, the area of cell membrane under electrode F shows fewer events than the others during the first 8 min and appears to be a *cold spot*. More interestingly, spots may be *hot* during a specific period, but then change to be *cold* (or vice versa) after another stimulus. Close inspection of the response from electrode A shows there are fewer events detected after the second and the fifth stimuli than after the third and fourth. In addition, electrodes F and G clearly detected more events in the last 8 min than in the first period of time. Thus, it appears that the array electrode format allows detection of localized membrane function in terms of exocytosis at a single cell. The detection probability for each channel was compared from 16 different PC12 cells ($n = 8$ electrode arrays) to ensure that the differences detected with each channel do not arise from systematic errors, such as electrode placement or response. The number of events at each channel was normalized to the total number of events at a cell before averaging.

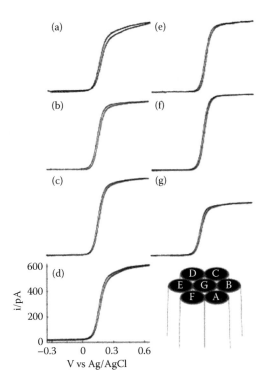

FIGURE 14.3   Steady-state voltammetric response at 20 mV/s of a seven-fiber MEA in 1 mM FcCH$_2$OH and 0.2 M KCl. (a) through (g) are the voltammetric response of individual microelectrodes A through G as shown in the schematic of the microelectrode assemble shown in the bottom right. (From Zhang, B. et al., *Anal. Chem.*, 80, 1394, 2008. With permission.)

Figure 14.4c displays a comparison of the averaged value of normalized number of events for each channel recorded from 16 PC12 cells. From the averaged data, there are no significant differences among the channels (p value = 0.35, one-way ANOVA), indicating that differences seen in the number of detected exocytotic events at a given cell reflect the subcellular spatial heterogeneity in the exocytosis process.

14.2.2   SPATIALLY AND TEMPORALLY RESOLVED SINGLE-CELL EXOCYTOSIS UTILIZING INDIVIDUALLY ADDRESSABLE CARBON-RING MEAs

Regularly patterned addressable seven-barrel carbon-fiber MEAs are suitable for acquiring molecular images of small areas with high spatial and temporal resolution. In order to make more electrodes in an MEA probe, a new and facile method to produce carbon-ring MEAs has been developed. This method offers a very flexible fabrication method, without using commercial multibarrel glass capillaries. The number of electrodes in the array is tunable, and arrays feature from 8 to 16 electrodes in a single tip.

Photos in Figure 14.5 show the simple fabrication steps for electrode construction. The procedure involves pyrolysis of hydrocarbon inside clusters of quartz capillaries. Fused silica capillaries are rinsed with pure water, dried in air, and then cut into pieces approximately 8 cm long. Several 8 cm long capillaries are initially packed and held together at one end with tape to form multibarrel capillaries of 8–16 barrels. The capillaries were then twisted and bound together on the other end of the array with tape (Figure 14.5a). This multibarrel capillary is manually pulled into two conical multibarrel capillaries with a very sharp tip by heating the middle of the multibarrel capillary with

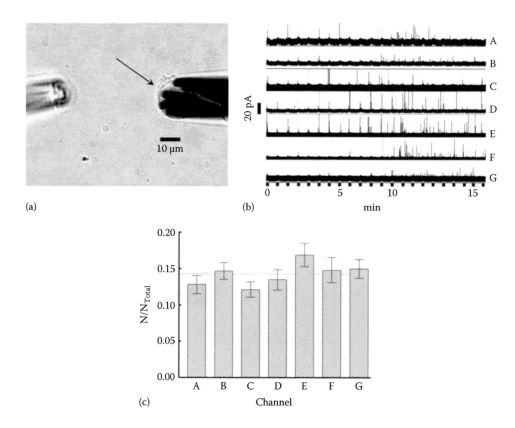

FIGURE 14.4 (a) Optical image showing an MEA positioned over a single PC12 cell (right) and a stimulation pipette (left). The cell is denoted with an arrow. (b) Amperometric traces of exocytotic release from a PC12 cell recorded using an MEA. Thick black lines along the time axis indicate exposure to high-potassium stimuli (100 mM, 5 s pulse every 45 s). (c) The average value of normalized number of exocytotic events recorded with each electrode in the array. Eight different MEAs were used to examine 16 PC12 cells. Error bars are SEM. (From Zhang, B. et al., *Anal. Chem.*, 80, 1394, 2008. With permission.)

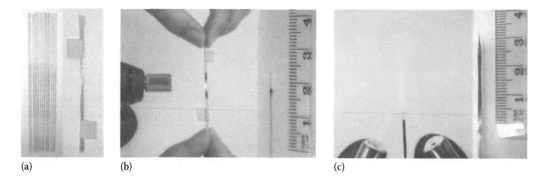

FIGURE 14.5 Pictures of the MEA fabrication process, (a) cleaned fused quartz capillaries are (b) heated and pulled together into two bundles of capillary arrays with an elongated tip, (c) which is used to deposit a layer of carbon on the inner surface of the capillary. (From Lin, Y. et al., *Anal. Chem.*, 84, 2949, 2012. With permission.)

FIGURE 14.6 Scanning electron microscopy of carbon-ring MEAs having (a) 8, (b) 10, (c) 12, and (d) 15 microring electrodes. Scale bars indicate 5 μm. (From Lin, Y. et al., *Anal. Chem.*, 84, 2949, 2012. With permission.)

a butane/oxygen torch flame (Figure 14.5b). Finally, the sharp tip of the conical multibarrel capillary is cut off with a scalpel under an optical microscope to remove the heat soldered part and fabricate a multibarrel capillary (Figure 14.5c).

Typical SEM images of carbon-ring MEAs containing 8, 10, 12, and 15 microelectrodes are shown in Figure 14.6. These carbon-ring MEAs are structurally well defined, and the overall diameters of these carbon-ring MEAs are between 15 and 25 μm. The carbon-deposited microelectrodes in these arrays are tightly packed together and surrounded by a thin layer of fused silica (~1–2 μm) and an epoxy matrix. This diameter can be adjusted during the cutting and beveling process. Due to the conical shape of each pulled capillary and the tapered tip of the whole pulled multibarreled capillary, the electrode radii and the whole tip size of the array are partially controlled by polishing. As more electrode material is polished away from the tip, the interelectrode distance and the total tip size are increased, achieving carbon-ring MEAs with tip sizes ranging from 10 to 50 μm. Interelectrode insulation is large enough to prevent current leakage, even when adjacent carbon-deposited microring electrodes are separated by submicrometer-thick glass, due to the high electrical resistivity of silica, thus limiting the probability of cross talk.

Arrays of micro- or nanometer electrodes can again be used to electrochemically map the change in easily oxidized substances released from different single secretory vesicles across the surface of cells, with even greater spatial resolution. Optical microscopy images in Figure 14.7a and b show carbon-ring MEAs containing eight microelectrodes with a total tip dimension of ~20 μm placed on a single PC12 cell. A glass micropipette containing high K+ solution (100 mM) was positioned ~60 μm away to stimulate secretion from the cell. Figure 14.7c displays a 180 s amperometric recording of exocytotic events detected with eight different microelectrodes at a single PC12 cell.

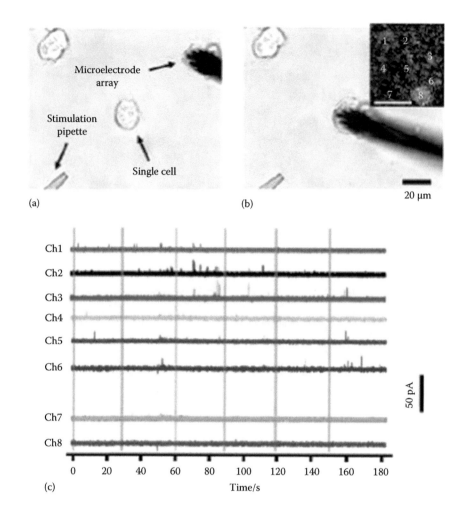

FIGURE 14.7 (a and b) Optical images showing an eight-electrode carbon-ring MEA (a) before and (b) after positioning on a single PC12 cell. The carbon-ring MEA, the cell, and the stimulation pipette are denoted with arrows. The insert in (b) is the image of the used carbon-ring MEA, showing the relative position of each microelectrode in the array (scale bar, 10 μm). (c) Representative amperometric traces of exocytotic release from a PC12 cell recorded using an eight-electrode carbon-ring MEA. Blue stripes indicate high-potassium stimuli (100 mM, 1 s pulse every 30 s). (From Lin, Y. et al., *Anal. Chem.*, 84, 2949, 2012. With permission.)

Each current transient corresponds to the electrochemical oxidation of catecholamine (dopamine in these cells) molecules released from a single intracellular vesicle. No event was recorded over 180 s from Electrode 8, which apparently represents a *cold spot* on the cell, indicating that differences seen in different electrodes at a given cell reflect the subcellular spatial heterogeneity in the exocytosis process.

14.2.3 SPATIALLY AND TEMPORALLY RESOLVED SINGLE-CELL EXOCYTOSIS UTILIZING INDIVIDUALLY ADDRESSABLE THIN-FILM MEAS

Simultaneous recordings of exocytotic events from single cells or cell clusters can be accomplished by microfabricated MEAs with individually addressable microelectrodes, which can be fabricated by the most widely used microelectromechanical system (MEMS) techniques, involving the use of

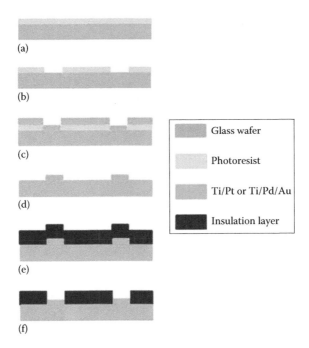

FIGURE 14.8 Process steps for the preparation of thin-film MEAs by photolithography, thin-film deposition, and reaction ion etching techniques. (a) Photoresist spin coating on glass wafer; (b) photolithography; (c) thin-film metal deposition; (d) liftoff; (e) Si$_3$N$_4$ insulation layer deposition; (f) reactive ion etching to prepare recording sites of microelectrodes. (From Wang, J. et al., *Anal. Chem.*, 85, 5600, 2013. With permission.)

photolithography, thin-film metal deposition, reactive ion etching, and other semiconductor techniques.[41–44] MEAs with individual electrodes smaller than 5 μm (typical diameter of carbon-fiber microelectrode) and tightly packed microelectrodes in thin-film planar MEAs have been reported. These thin-film MEAs with microelectrodes at the subcellular scale can be used not only for spatial measurements of neurotransmitter release across single cells but also at cells cultured in clusters.

These thin-film MEAs were fabricated on a 4 in. Borofloat glass wafer by using photolithography, thin-film metal deposition, reactive ion etching, and other techniques. The low cost, high optical transparency, and durability to solvents of the Borofloat glass wafer make it ideally suited for microfabricating fabrication. Figure 14.8 shows the schematic process for the fabrication processes of MEAs. Glass wafers are spin coated with s1813 photoresist and patterned by UV lithography with a mask showing the MEAs design. They are then developed with a positive developer MF319, so that the wafers are coated with photoresist except where the metal is deposited. Then the wafer is coated with 5 nm Ti and 45 nm Au or 45 nm Pt by electron beam evaporation. The wafers are then placed in a beaker containing Microposit remover 1165 and left overnight for the photoresist to come off (metal liftoff). The wafers are then rinsed with isopropanol and distilled water to remove residual metal sheets and the remover. The wafers are then coated with a 420 nm low-stress Si$_3$N$_4$ film using a plasma-enhanced chemical vapor deposition (PECVD) system, which is maintained at a constant temperature of 300°C. The wafers are spin coated with S1813 photoresist. A second mask is used for the second lithography to selectively expose UV light and thereby define the insulation layer on the wafer. The exposed Si$_3$N$_4$ for the recording ultramicroelectrode sites and the connection pads are then etched using CF$_4$ plasma, and the MEAs are checked with optical microscopy and SEM. Nine arrays with three kinds of MEAs (containing 16, 25, and 36 exposed ultramicroelectrodes, respectively) are typically fabricated in the center of each wafer.

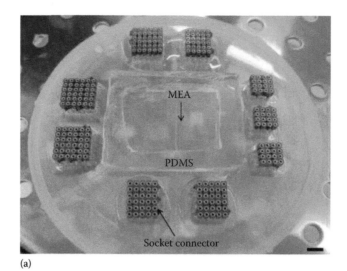

(a)

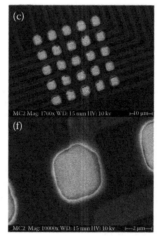

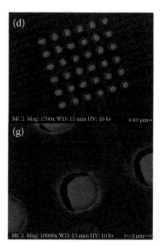

FIGURE 14.9 Optical photo of the whole MEA device and SEM pictures of MEAs containing 16, 25, and 36 microelectrodes, respectively. (a) Typical devices containing PDMS chamber, pin connectors, and the MEAs (scale bar, 10 μm) (one arrow points to one of the nine MEAs and the other arrow points to one of the nine socket connectors); (b–d) SEM pictures of MEAs containing 16, 25, and 36 microelectrodes (scale bar, 10 μm); (e–g) SEM pictures of single microelectrode in the corresponding MEAs of 16, 25, and 36 microelectrodes (scale bar, 2 μm). (From Wang, J. et al., *Anal. Chem.*, 85, 5600, 2013. With permission.)

The MEA-based device for cell culture and amperometry detection is shown in Figure 14.9a. It contains a polydimethylsiloxane (PDMS) well for cell culture, nine socket connectors for outside wire connection, and nine thin-film MEAs in the middle of a 4 in. glass wafer. SEM images of the three kinds of MEAs (i.e., containing 16, 25, and 36 ultramicroelectrodes) are shown in Figure 14.9b through d, respectively. The corresponding recessed square ultramicroelectrodes of different sizes (4, 3, and 2 μm) in the 16, 25, and 36 MEAs are shown in Figure 14.9e through g, respectively. The electrodes in each MEA are tightly defined in a 30 × 30 μm square, which is potentially useful to measure exocytosis across a single cell or clusters of single cells. Each individual microelectrode in the MEAs used in these experiments is formed in slightly recessed form with a 375 nm recess depth. These recessed microelectrodes are obtained by depositing a thin film of silicon nitride

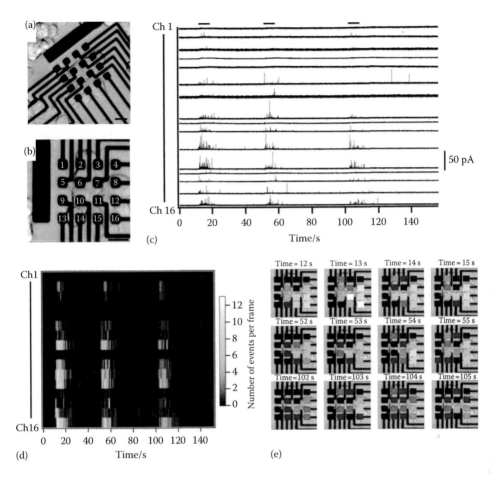

FIGURE 14.10 Electrochemical imaging of a layer of PC12 cells partially covering the MEA. (a) Micrograph of the setup, showing the 16-electrode array partially covered by a small population of PC12 cells (scale bar, 10 μm); (b) expanded view of the electrode array showing the cell population and the labeling of the electrodes (scale bar, 10 μm); (c) traces obtained for 5 s stimulations of the cell layer (the stimulations are indicated by a black bar at the top of the plot); (d) frequency plots showing the release frequency obtained for each channel, for 1 s frames; (e) imaging of the release frequency at each of the 16 electrodes (the pictures shown correspond to typical event sequences for the three exposures to K$^+$). (From Wang, J. et al., *Anal. Chem.*, 85, 5600, 2013. With permission.)

insulation layer over a metal microelectrode. This thin-film silicon nitride is sufficient to insulate the microelectrodes at low applied voltages.

The imaging ability of the electrochemical array is demonstrated in Figure 14.10. The surface of the 16 MEAs has been covered with an incomplete layer of cells (Figure 14.10a). Each electrode is identified with a specific number, and these notations are conserved for the rest of the study (Figure 14.10b). Electrodes 1, 4, 5, and 9 were not covered by cells, and no exocytotic events were observed at these electrodes upon stimulation (Figure 14.10c). In this data set, each current transient corresponds to the electrochemical oxidation of dopamine molecules released from a single intracellular vesicle across this small array of cells. Figure 14.10d shows the release frequency observed for each electrode (1 s time bins). The green and white colors show a high release frequency, while the dark green and black colors show a low release frequency. Several locations showing high release frequency are easily identified. The electrochemical imaging results shown in Figure 14.10e have been correlated to the precise geometry of the array provided by lithography to achieve electrochemical imaging of the exocytotic activity of the PC12 clusters. The pictures shown here are focused on the three stimulations performed during the course of the experiment, thus showing the spatial release of dopamine and the exocytotic activity. Furthermore, as shown on the frequency plots (Figure 14.10d and e), the electrodes showing the highest release frequency are also the ones located at the places of highest cell density (electrodes 8, 11, 12, and 16). This could indicate a higher level of diffusional cross talk with neighboring electrodes or increased activity of PC12 cells embedded in a layer of cells. And the different release frequency between electrodes 13 and 14 across a single cell might be due to the different part of single-cell membrane, indicating that the differences seen in different electrodes at a given cell reflect the subcellular spatial heterogeneity in the exocytosis process. The further study of single-cell exocytosis by 6 × 6 MEAs will be quite interesting for high spatial and temporal heterogeneity in the exocytosis process.

14.2.4 CONCURRENT EVENTS DETECTED AT SINGLE CELLS WITH DIFFERENT ELECTRODES OF THE MEA

MEAs have been used to examine the incidence of concurrent exocytotic events on the same cell. When more than one event occurs simultaneously, it is challenging to resolve them using a single microelectrode. In the absence of spatial resolution, these events will overlap and result in a large, broad current spike. Simultaneous, parallel recordings using multiple microelectrodes allow these events to be resolved based on spatial identification. Figure 14.11 shows a 1 s amperometric recording on a single PC12 cell using a seven-fiber carbon-fiber MEA. The red arrows indicate three different exocytotic events detected from electrodes B, C, and G, labeled as in Figure 14.3.

Numerical simulations have been utilized to understand the temporal and spatial resolution in electrochemical imaging at single cells with multiple electrodes. To analyze the potential for cross talk in amperometric imaging, the position of the release site (the star in the insets of Figure 14.12) was varied between the microelectrode in the center (black circle) and two adjacent electrodes (red and blue circles), as shown in Figure 14.12. Figure 14.12a through c shows the amperometric response at three adjacent electrodes as a function of the position of the release site. Figure 14.12a shows the responses at each microelectrode when the release site is present at the midpoint of the center electrode. One can see that there is no response at any other electrodes in the same array

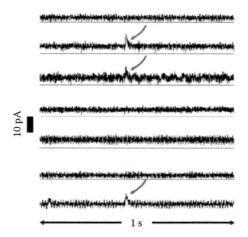

FIGURE 14.11 A 1 s time period of an exocytotic response of a PC12 cell after potassium stimulation showing simultaneous detection of concurrent events at different locations on the same cell. Red arrows indicate these events. (From Zhang, B. et al., *Anal. Chem.*, 80, 1394, 2008. With permission.)

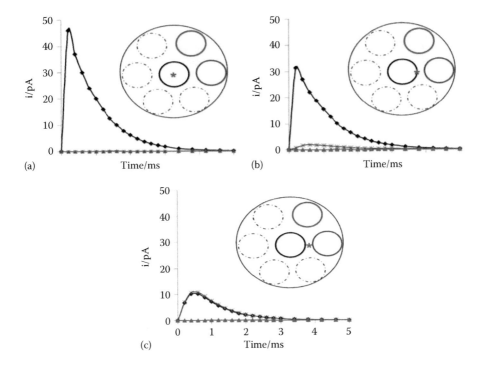

FIGURE 14.12 Simulated amperometric responses at three adjacent microelectrodes in a MEA when the site of exocytotic release is localized at three different locations under the center carbon surface. (a) Exocytotic release site present at mid-point of the center electrode. (b) Exocytotic release site present ~50-nm away from the edge of center electrode. (c) Exocytotic release site present between the center electrode (black) and the edge electrode (blue). The "*" in each figure indicates the location of release. The amperometric response is coded to the electrode by color. (From Zhang, B. et al., *Anal. Chem.*, 83, 571, 2011. With permission.)

when release occurs at the center of a given microelectrode in the array. Figure 14.12b shows the response of three electrodes when release occurs under the center electrode but ~50 nm away from the edge. The response at the adjacent electrode closest to the location of release shows about 2% the peak current observed under the center electrode, and no current is observed at the other adjacent electrode on this current scale. Thus, there is very little cross talk in amperometric imaging when the release site is located underneath one microelectrode in the same array. When release occurs under the insulating material in between two microelectrodes, the simulation shows that the current at each electrode is dependent on their distances to the release site. Figure 14.12c displays the simulated amperometric response at three electrodes when the release site is present in between the center electrode (black) and the edge (blue) one. The currents at the center electrode (black) and the edge electrode are comparable when release occurs in the middle of the two microelectrodes. The current at the other edge electrode (red) is negligible as compared to the other two. Another important fact is that the two signals are comparable only when the release location is very close to the middle of the two microelectrodes. Therefore, cross talk only occurs when the release occurs right at the middle of two adjacent electrodes. When the location of release is close to the center of three adjacent microelectrodes, the simulated signals are comparable but are much smaller than the signals recorded when the vesicle is underneath the electrodes. In fact, the average peak current in this case is only ~10% of the current when the vesicle is present under a single microelectrode.

Based on simulation and the experiment, the concurrent events in Figure 14.11 appear to be from different vesicles docked in different membrane parts of single cell. Although rare, these overlapping events can be spatially resolved with the MEA. The multiple detection sites present allow the temporally synchronous release events present in single cells at different electrodes to be resolved.

14.3   CELL MIMIC SYSTEMS FOR EXOCYTOSIS

Exocytosis is the fundamental process underlying neuronal communication; it involves fusion of a neurotransmitter-containing vesicle (ranging from 50 nm to 1 µm in diameter). The extremely small size and high chemical diversity of exocytotic events have made it difficult to piece together all of the molecular and biophysical mechanisms involved during the release process. Nanoscale measurements[8–13,46–49] and molecular biology applied to cellular model systems have been used to provide insights into exocytotic release and strong evidence that protein networks are involved in the formation of the initial fusion pore.[50,51] In contrast, leakage of transmitter through the nanometer fusion pore and expansion of this pore to the final stage of exocytosis is poorly understood, and new cell models are needed to differentiate mechanisms involving proteins and membrane mechanics in driving the various stages of the exocytotic process.

Artificial cell models have been developed to help understand the membrane fusion process. A common model that involves vesicles containing channel proteins that are driven by osmotic pressure to fuse with a planar lipid bilayer was reported.[52,53] A protein-free model has also been used to demonstrate transient opening of fusion pores.[54] Liposomes have been described as artificial cells and have also been used to examine membrane fusion.[55–58] Recently, an electroinjection technique has been developed, which makes it possible to form lipid nanotubes and networks between liposome reservoirs.[59–61] Here, we briefly present two models, recent advances in the development of a liposome–lipid nanotube network as an artificial cell model that mimics the latter stages of exocytosis for cell study and fusion of vesicles inside liposomes with a DNA zipper.

14.3.1   STUDY OF EXOCYTOSIS WITH AN ARTIFICIAL CELL MODEL USING LIPOSOMES AND LIPID NANOTUBES

An artificial cell system to model the opening of the fusion pore in exocytosis has been illustrated with a liposome–lipid nanotube network.[62] The basic structure and setup of this kind artificial cell model using liposomes and lipid nanotubes for mimicking the complex dynamic cellular membrane process of exocytosis are shown in Figure 14.13. A vesicle is formed inside a surface-immobilized liposome via electroinjection (Figure 14.13a through d). The vesicle is connected to the artificial cell membrane by a lipid nanotube. This nanotube initially resembles an elongated fusion pore. Microinjection of fluid into the pulled lipid nanotube is used to inflate a daughter vesicle inside the artificial cell (Figure 14.13d). Inflation of the pulled nanotube leads to a local increase in membrane tension. To reduce this tension difference, lipid material flows from regions of lower tension (outer membrane) along the nanotube toward higher tension, forming the membrane of the small vesicle. Figure 14.13e shows a Nomarski image of a vesicle formed inside an artificial cell under pressure from the pipette. Because the distance between the injection tip and the outer membrane is fixed, the lipid nanotube shortens when the vesicle is expanded. As the vesicle membrane approaches the artificial cell surface, a transition from a tube of cylindrical geometry to a toroid-shaped fusion pore takes place. At this stage, the system directly mimics a cell before undergoing the later stages of exocytosis (pore opening), and the distance between the pipette tip and the membrane of the artificial cell determines the vesicle diameter for exocytosis. Provided that the pore radius is large enough, it will expand in size exponentially driven by total tension in the membrane system. Thus, in a system held at high surface tension, the vesicle membrane will be rapidly integrated with the outer membrane, leading to the final stage of exocytosis (Figure 14.13f through i). This artificial exocytosis resembles that for cellular release of large dense core vesicles in many ways. The nanotube of the artificial cell model has an estimated diameter of 100–300 nm.[63,64]

Artificial exocytosis with an expanding vesicle and a nanotube attached to a micropipette allows two significant advantages in these experiments. First, the control of solution pressure in the micropipette allows variation of solution flow rate and, therefore, the rate of vesicle expansion.

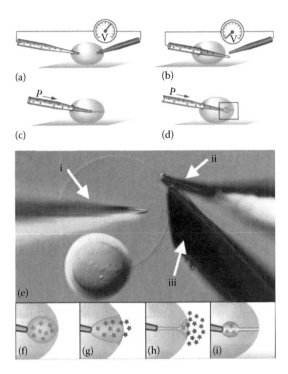

FIGURE 14.13 Formation and release of vesicles in an artificial cell. (a–d) Schematics of a microinjection pipette electroinserted into the interior of a unilamellar liposome and then through the opposing wall, pulled back in to the interior, followed by spontaneous formation of a lipid nanotube and formation of a vesicle from flow out of the tip of the micropipette. (e) Optical image of a unilamellar liposome, with a multilamellar liposome attached as a reservoir of lipid, microinjection pipette (i), electrode for electroinsertion (ii), and 30 μm diameter amperometric electrode beveled to a 45° angle (iii). A small red line depicts the location of the lipid nanotube, which is difficult to observe in the computer image with a 20× objective, illustrating a vesicle with connecting nanotube inside a liposome. (f–i) Fluid injection at a constant flow rate results in the growth of the newly formed vesicle with a simultaneous shortening of the nanotube until the final stage of exocytosis takes place spontaneously and a new vesicle is formed with the attached nanotube. (From Cans, A.-S. et al., *Proc. Natl. Acad. Sci. USA*, 100, 400, 2003. With permission.)

Second, as the lipid nanotube remains after exocytosis, new vesicles are continually formed; thus, controlled release of a vesicle can be repeatedly carried out while maintaining membrane integrity.

It has been confirmed that secretion of neurotransmitters can occur, prior to full exocytosis, through the fusion pore in adrenal chromaffin and beige mouse mast cells by using patch clamp techniques.[10,11] This type of secretion is observed as a prespike increase in oxidation current in the first and second transients from PC12 cells, indicating that transmitter is leaking out of the fusion pore and is oxidized before the later stage of exocytosis. Exocytosis in the artificial model presented in Figure 14.13 can be controlled to demonstrate this prespike amperometric *foot* as shown in Figure 14.14a. A schematic correlating the stages of exocytosis to an amperometric trace is shown in Figure 14.14b. Using higher flow rates in the injection pipette during artificial vesicle formation and expansion, a foot is consistently observed, and its duration and amplitude can be characterized from the amperometric trace. Although it is now apparent that the final stage of exocytosis involves vesicle closing and not full expansion (see Section 14.3.3), the artificial cell model makes it possible to examine fundamental aspects of exocytosis with easily controlled parameters, such as membrane and solution composition, differential pH values across vesicle membranes, temperature, and vesicle size.

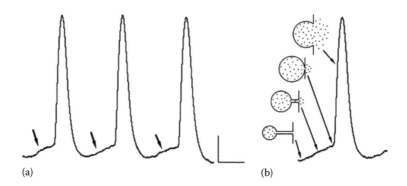

FIGURE 14.14 Amperometric monitoring of release via an artificial fusion pore. (a) Amperometric detection of release from a 5 μm radius vesicle showing prespike feet (arrows), indicating catechol transport through the lipid nanotube or fusion pore (scale bar, 80 pA × 500 ms). (b) Time correlation of vesicle growth, transport of transmitter through the lipid nanotube, and the final stage of exocytosis with amperometric detection. (From Cans, A.-S. et al., *Proc. Natl. Acad. Sci. USA*, 100, 400, 2003. With permission.)

14.3.2 STUDY OF EXOCYTOSIS WITH AN ARTIFICIAL CELL MODEL USING A DNA ZIPPER AS A SNARE-PROTEIN MIMIC

Exocytosis is a complex process involving a wide variety of lipids, proteins (the most important being the soluble N-ethylmaleimide-sensitive factor attachment protein receptor (SNARE) proteins), the molecules to be secreted, and Ca^{2+} ions. A free-protein artificial secretory cell with a DNA zipper mechanism has been designed to model the actions of the SNARE proteins in exocytosis. To replace the SNARE complex, the cell model is equipped with an analogue composed of complimentary DNA constructs. The DNA constructs hybridize in a zipper-like fashion to pull the membranes together causing docking and fusion of the artificial secretory vesicles (containing one single-stranded DNA) with the outer artificial cell membrane (containing corresponding complementary DNA). Interestingly, fusion appears to be controlled by the presence of Ca^{2+} even for the DNA zipper model.[65]

Schematics of this artificial cell model with liposomes, lipid nanotubes, and DNA zipper are shown in Figure 14.15. Artificial secretory vesicles are made from large unilamellar vesicles (LUVs, mean diameter of 200 nm) and filled with catechol as a neurotransmitter analogue. Cholesterol–DNA strands complementary to the t-CH-DNA (labeled a/a′, in Figure 14.15a) are incorporated into the outer leaflet of the vesicle compartment and served as vesicle-SNARE (v-CH-DNA) synaptobrevin analogues. The artificial secretory vesicles are electroinjected into the giant unilamellar vesicles (GUV) where they presumably dock to the DNA-labeled GUV membrane. Vesicle fusion is then initiated by Ca^{2+} with a subsequent injection into the system, hence simulating the critical aspects of exocytosis in live cells. A 5 μm carbon-fiber microelectrode placed against the outer leaflet of the GUV membrane (Figure 14.15a and b) is used to detect the released catecholamine and completes the artificial synapse with an estimated spacing of <300 nm. Upon the addition of Ca^{2+} to the interior of the GUV, a train of amperometric spikes is observed (Figure 14.15c), where each amperometric spike corresponds to an individual vesicular release event. The apparent delay of the release events in Figure 14.15c is attributed to a repositioning of the electrode following the addition of calcium. A negative control using LUVs lacking the v-CH-DNA does not show any amperometric spikes upon Ca^{2+} addition (lower trace in Figure 14.15c). Accordingly, it has been confirmed that the combination of a SNARE analogue and the presence of calcium are required for this functioning artificial secretory system.

A comparison of the normalized average spikes measured at this secretory cell model and at a PC12 cell is shown in Figure 14.16a. This comparison reveals that the peaks are quite similar in

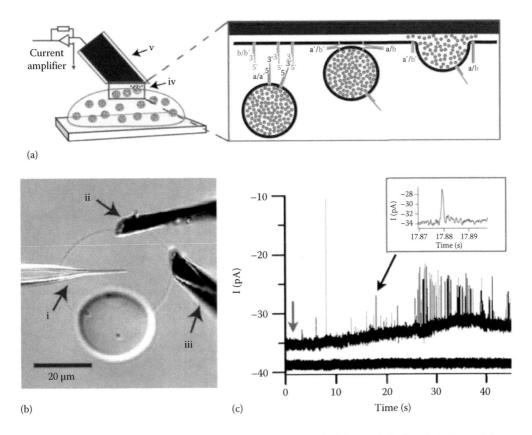

FIGURE 14.15 (a) Schematics of a GUV containing LUVs filled with catechol. Catechol release (iv) upon LUV fusion to the GUV membrane is detected by a carbon-fiber electrode (v). The expanded view provides details of DNA-mediated exocytosis. Injection and self-insertion of CH-DNA b/b′ (red) into the GUV membrane is followed by a second injection of catechol-filled LUVs decorated with the complementary CH-DNA a/a′ (black). The DNA strands hybridize in a zipper-like fashion and bring about fusion of the two membranes. Cartoons are not drawn to scale. (b) Optical image of a GUV, with a multilamellar bleb attached as a lipid reservoir, a microinjection pipette electrode inserted into the interior of a GUV (i), electrode for electroinsertion (ii), and a 5 μm diameter amperometric electrode, beveled to a 45° angle (iii). (c) Amperometric recording of repeated exocytotic events at the secretory cell model (the apparent delay in release is due to repositioning of the electrode [upper trace]). The lower trace represents amperometric monitoring after injection of vesicles that do not display the CH-DNA. The arrow indicates the start of Ca²⁺ addition. The inset shows a single exocytotic event spike at an expanded timescale. (From Simonsson, L. et al., *Sci. Rep.*, 2, 824, 2012. With permission.)

terms of shape characteristics. The temporal statistics for the artificial cell model and comparative data from measurements at a PC12 cell are shown in Figure 14.16b and c. As seen in Figure 14.16a and in the histograms (Figure 14.16b and c), fusion events recorded from the secretory cell model are slightly faster in all respects when compared to the events measured at living PC12 cells. Diffusion during exocytosis in living cells is apparently restricted to a greater degree than that observed in the lipid-based cell model.

Using this artificial secretory cell model combined with amperometry, a well-established method for studies of exocytosis, it is possible to control the kinetics of fusion pore opening and vesicle secretion, which might give some insight into observations made for living cells. The finding that lipids seem critical for the process relates to previous studies of live cells, which upon incubation with different lipid species can lead to altered kinetics during exocytosis. This bottom-up cell model, where it is possible to increase complexity in a stepwise manner, will constitute a powerful

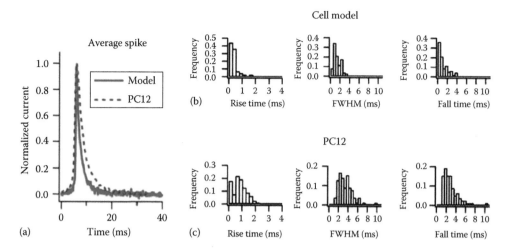

FIGURE 14.16 (a) Normalized average exocytotic event spikes for the artificial cell model (red) and for a PC12 cell (blue). (b and c) Histograms showing distributions of spike kinetics for the artificial cell model (b, n_{model} = 113) and a PC12 cell (c, n_{PC12} = 578). The displayed spike features are the rise time, calculated as the time it takes for the current to rise from 25% of its maximum value to 75%, the full width at half maximum (FWHM) current value, and the fall time, calculated as the time it takes for the current to fall from 75% of its maximum value to 25%. (From Simonsson, L. et al., *Sci. Rep.*, 2, 824, 2012. With permission.)

tool for studies of exocytosis on the molecular level. And also, this artificial secretory cell has been constructed from the minimal components needed to mimic exocytosis. This could be an important step toward developing a minimal synthetic living system.

14.3.3 STUDY OF EXOCYTOSIS WITH AN ARTIFICIAL CELL MODEL CONSTRUCTED FROM PC12 CELL PLASMA MEMBRANE VESICLES (BLEBS)

Much knowledge of the exocytotic process can be gained through the exploration of artificial cell models; however, present models of vesicle fusion and exocytotic release have all only mimicked the full distension mode with complete release of the vesicle content. It has become apparent recently that this mode of release is not as prevalent as open and closed exocytosis or partial release.[66–69]

Recently, artificial cells constructed from PC12 cell plasma membrane vesicles (also referred to as blebs) have been used to directly observe neurotransmitter release in two distinct modes: full and partial distensions of the fusion pore.[66] Blebs are thought to possess a lipid composition representative of the plasma membrane. A schematic of the experimental design and a picture of a plasma membrane vesicle under manipulation are shown in Figure 14.17. The small size of the plasma membrane vesicle inherently makes the daughter vesicles smaller than the vesicles that are formed when using GUVs. The plasma membrane vesicles have an average diameter of approximately 10 μm, whereas the daughter vesicles formed range from 1 to 4 μm in diameter. In comparison to chemical release from vesicles in the GUV model, release from the smaller plasma membrane vesicles results in considerably faster release kinetics. The small-scale system and the fast kinetics make the visual observation of the membrane during release difficult. However, a fraction of the events can be visually inspected. These observations lead to the finding that release in this model occurs through two distinct mechanisms. One mode is referred to as full distension that occurs when the nanotube connecting the daughter vesicle to the artificial cell membrane is completely extinguished, as the vesicle grows larger under pressure (Figure 14.17a). This results in the complete distension of the nanotube until it forms a frustum-shaped connection between the pipette and the cell membrane. The frustum

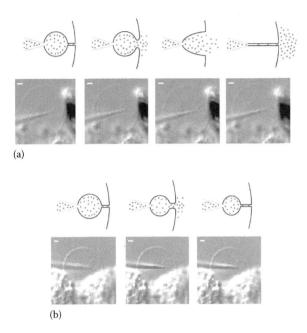

(a)

(b)

FIGURE 14.17 Schematic and micrographs of the two release modes. (a) The full distension mode, where the vesicle opens up to a frustum, which then collapses into a nanotube resulting in the complete release of the vesicle content. (b) The partial distension mode where the nanotube opens up to a larger pore followed by its reclosing to a nanotube. The partial distension results in incomplete release of the vesicle content. Scale bars, 1 µm. (From Mellander, L.J. et al., *Sci. Rep.*, 4, 3847, 2014. With permission.)

then reforms into a nanotube, leading to the expulsion of the entire vesicle content. The nanotube is again filled from the pipette tip end and fuses repeatedly in this manner. The second mode of release observed has apparently not been observed in previous model systems of exocytosis. In these events, the spherical part of the vesicle never reaches the cell membrane but instead releases some of its content in a burst before it can accomplish full distension. This release mode is referred to as partial distension since the nanotube does not distend completely. Instead, a wider tube is transiently formed, allowing partial release of the vesicle content. The result, as the pore again collapses into a nanotube, is a smaller vesicle still present at the tip of the micropipette, and importantly, only part of the content is released. This smaller vesicle is then refilled and the system fuses repeatedly this way (Figure 14.17b).

The release of dopamine from the artificial cell has been monitored with carbon-fiber amperometry, and interestingly, complete and partial distension modes display distinct release kinetics. As discussed earlier, analyzing the release events based on visual observations allows the assignment of two groups of events: one type that releases through full distention of the nanotube and one that releases through partial distension. The visual event assignments can then be used to categorize the collected amperometric traces. In general, it is rare to observe a full distension event in a trace containing partial events and likewise a partial distension event in a full release trace. Representative amperometric traces as well as the average amperometric peaks of the two release modes based on the visual characterization are shown in Figure 14.18. The amperometric peaks, in addition to revealing the number of molecules released, provide detailed information on the kinetics of the release events with millisecond time resolution. Thus, the peaks have been analyzed for peak current: half width ($t_{1/2}$), identified as the width of the peak at half its maximum; rise time, defined as the time of rise from 25% to 75% of the peak height; and fall time, defined as the time between 75% and 25% of the height of the peak. The average peaks for full versus partial opening display

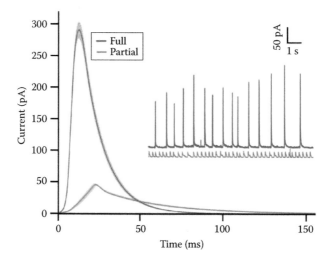

FIGURE 14.18 Representative traces and average peaks from the two modes of release. The average ampero-metric peaks for release though full (red) and partial (blue) distension plotted together (the standard errors are shown in the shaded regions). The inset shows representative amperometric traces for the two modes of release. Data were collected from 11 plasma membrane vesicles where recordings from two of these vesicles were defined as full release and nine of them as partial release. (From Mellander, L.J. et al., *Sci. Rep.*, 4, 3847, 2014. With permission.)

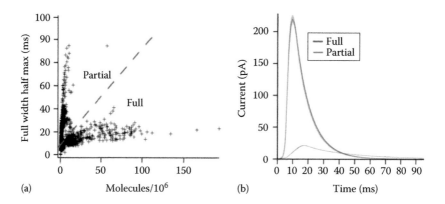

FIGURE 14.19 Distribution of release kinetics for all events and reassigned average peaks. (a) The FWHM plotted against the number of molecules released for all detected peaks. The two distributions appear to result from partial and full distension modes of release. (b) Average peaks of the two modes with assignments based on amount and kinetics of release (the standard errors are shown in the shaded regions). Data were collected from 11 plasma membrane vesicles where five traces were defined as full release and six as partial release. (From Mellander, L.J. et al., *Sci. Rep.*, 4, 3847, 2014. With permission.)

distinctly different kinetics and amounts of release with partial distension events releasing less neurotransmitter and having slower release kinetics when compared to the full distension events.

The half width of the release peak provides information about the duration of the event, whereas the number of molecules released relates to the size of the vesicle. These data are plotted in Figure 14.19a showing two distinct populations: one with large half widths and small release amounts and one with small half widths and large amounts released. It appears that these two peak populations represent partial and full distensions, respectively. In support of this hypothesis, the initial assignments on an individual basis have been examined and confirmed that these are generally accurate. However, these data have also been used to reassign some traces to the other mode. It is clearly difficult to define the event mode visually with 100% accuracy given the small sizes of the vesicles. The reassigned average peaks are displayed in Figure 14.19b showing an even greater distinction between the partial and full release modes.

Several characteristics of the partial release events that have been measured here can be used to draw conclusions about the hypothesized mechanism of open and closed or partial exocytosis. The most exciting finding is that the peak height of the partial release events is fairly uniform. The peak height represents the maximum flux of neurotransmitter through the pore, and it follows that the partial release events have a uniform pore size. This finding suggests that some independent factor determines the pore size, and it has been suggested that lipid composition might play a role here.[19,21,70] Furthermore, the kinetics of the events seems to be coupled to the size of the vesicle, specifically the fall time of the peak. The larger vesicles take a longer time to close than the smaller vesicles. Thus, larger vesicles will release more cargo by extending the time of release in the partial distension mode, while the full release mechanism increases flux as indicated by the increase in peak current. Finally, only a weak connection between the rise time and released amount indicates again that some independent factor might determine the rate of pore expansion.

The observation of a second mode of release in this model is exciting as it shows the simplicity of this event taking place in the biological system and it can be used to study the parallel process of kiss-and-run release. This model has the ability to mimic the expansion and subsequent restoration of the initial fusion pore and the resultant partial release of the vesicle content. The apparent switching mechanism for determining the mode of release demonstrated to be related to membrane tension that can be differentially induced during artificial exocytosis. These results suggest that the partial distension mode might correspond to an open and closed or partial mechanism of release from secretory cells, which has been proposed as a major pathway of exocytosis in neurons and neuroendocrine cells.

14.4 FUTURE PERSPECTIVES

Subcellular cell exocytosis study by using tightly packed MEAs to date has been aimed on evaluation of PC12 cells and confirmed the subcellular heterogeneity of single-cell membrane. It is important to examine exocytosis in different cell systems to verify the importance of the heterogeneity of the cell membrane in the chemical communication process. Additionally, it will be interesting to study the spatial differences in drug action across cells or clusters of cells. Further study by the use of MEAs might be improved by combining other methods, such as microfluidics and fluorescence to the measurement repertoire.

The artificial cell model for the study of cell exocytosis provides a powerful tool in which it is possible to increase the complexity of the system in a stepwise manner. This model is potentially an important step toward developing a minimal synthetic living system in the future.

Applications of the methods described here will likely include investigations of disease states like Parkinson's disease models and perhaps more. It will also be useful to determine the effect of drugs on vesicular content and therefore provide novel targets for pharmaceuticals related to vesicles and disease.

ACKNOWLEDGMENTS

We acknowledge the many coworkers that have come before and whose works we cite in this chapter. We acknowledge the support from the European Research Council (ERC Advanced Grant), the Knut and Alice Wallenberg Foundation in Sweden, the Swedish Research Council (VR), and the U.S. National Institutes of Health (NIH).

REFERENCES

1. Valtorta, F., R. Fesce, F. Grohovaz et al. 1990. Neurotransmitter release and synaptic vesicle recycling. *Neuroscience* 35:477–489.
2. Edwards, R.H. 2007. The neurotransmitter cycle and quantal size. *Neuron* 55:835–858.
3. Sudhof, T.C. 2004. The synaptic vesicle cycle. *Annu. Rev. Neurosci.* 27:509–547.
4. Rahamimoff, R. and J.M. Fernandez. 1997. Pre- and postfusion regulation of transmitter release. *Neuron* 18:17–27.
5. Garris, P.A., E.L. Ciolkowski, P. Pastore, and R.M. Wightman. 1994. Efflux of dopamine from the synaptic cleft in the nucleus-accumbens of the rat-brain. *J. Neurosci.* 14:6084–6093.
6. Bruns, D. and R. Jahn. 1995. Real-time measurement of transmitter release from single synaptic vesicles. *Nature* 377:62–65.
7. Chen, T.K., G.A. Luo, and A.G. Ewing. 1994. Amperometric monitoring of stimulated catecholamine release from rat pheochromocytoma (PC12) cells at the zeptomole level. *Anal. Chem.* 66:3031–3035.
8. Wightman, R.M., J.A. Jankowski, R.T. Kennedy et al. 1991. Temporally resolved catecholamine spikes correspond to single vesicle release from individual chromaffin cells. *Proc. Natl. Acad. Sci. USA* 88:10754–10758.
9. Leszczyszyn, D.J., J.A. Jankowski, O.H. Viveros, E.J. Diliberto, J.A. Near, and R.M. Wightman. 1990. Nicotinic receptor-mediated catecholamine secretion from individual chromaffin cells—Chemical evidence for exocytosis. *J. Biol. Chem.* 265:14736–14737.
10. Chow, R.H., L. Vonruden, and E. Neher. 1992. Delay in vesicle fusion revealed by electrochemical monitoring of single secretory events in adrenal chromaffin cells. *Nature* 356:60–63.
11. Detoledo, G.A., R. Fernandezchacon, and J.M. Fernandez. 1993. Release of secretory products during transient vesicle fusion. *Nature* 363:554–558.
12. Chen, G.Y., P.F. Gavin, G.A. Luo, and A.G. Ewing. 1995. Observation and quantitation of exocytosis from the cell body of a fully-developed neuron in planorbis-corneus. *J. Neurosci.* 15:7747–7755.
13. Wightman, R.M., T.J. Schroeder, J.M. Finnegan, E.L. Ciolkowski, and K. Pihel. 1995. Time-course of release of catecholamines from individual vesicles during exocytosis at adrenal-medullary cells. *Biophys. J.* 68:383–390.
14. Finnegan, J.M., K. Pihel, and P.S. Cahill et al. 1996. Vesicular quantal size measured by amperometry at chromaffin, mast, pheochromocytoma, and pancreatic beta-cells. *J. Neurochem.* 66:1914–1923.
15. Anderson, B.B., G.Y. Chen, D.A. Gutman, and A.G. Ewing. 1999. Demonstration of two distributions of vesicle radius in the dopamine neuron of *Planorbis corneus* from electrochemical data. *J. Neurosci. Methods* 88:153–161.
16. Colliver, T.L., S.J. Pyott, M. Achalabun, and A.G. Ewing. 2000. VMAT-mediated changes in quantal size and vesicular volume. *J. Neurosci.* 20:5276–5282.
17. Westerink, R.H., A. de Groot, and H.P. Vijverberg. 2000. Heterogeneity of catecholamine-containing vesicles in PC12 cells. *Biochem. Biophys. Res. Commun.* 270:625–630.
18. Sombers, L.A., M.M. Maxson, and A.G. Ewing. 2005. Loaded dopamine is preferentially stored in the halo portion of PC12 cell dense core vesicles. *J. Neurochem.* 93:1122–1131.
19. Amatore, C., S. Arbault, Y. Bouret, M. Guille, F. Lemaitre, and Y. Verchier. 2006. Regulation of exocytosis in chromaffin cells by trans-insertion of lysophosphatidylcholine and arachidonic acid into the outer leaflet of the cell membrane. *Chembiochem* 7:1998–2003.
20. Amatore, C., S. Arbault, I. Bonifas, F. Lemaitre, and Y. Verchier. 2007. Vesicular exocytosis under hypotonic conditions shows two distinct populations of dense core vesicles in bovine chromaffin cells. *Chemphyschem* 8:578–585.
21. Uchiyama, Y., M.M. Maxson, T. Sawada, A. Nakano, and A.G. Ewing. 2007. Phospholipid mediated plasticity in exocytosis observed in PC12 cells. *Brain Res.* 1151:46–54.
22. Wightman, R.M. 2006. Probing cellular chemistry in biological systems with microelectrodes. *Science* 311:1570–1574.
23. Mosharov, E. and D. Sulzer. 2005. Analysis of exocytotic events recorded by amperometry. *Nat. Methods* 2:651–658.
24. Walch-Solimena, C., R. Jahn, and T.C. Sudhof. 1993. Synaptic vesicle proteins in exocytosis: What do we know. *Curr. Opin. Neurobiol.* 3:329–336.
25. Kelly, R.B. 1988. The cell biology of the nerve terminal. *Neuron* 1:431–438.
26. Bonner-Weir, S. 1988. Morphological evidence for pancreatic polarity of beta-cell within islets of Langerhans. *Diabetes* 37:616–621.

27. Bokvist, K., L. Eliasson, C. Ammala, E. Renstrom, and P. Rorsman. 1995. Co-localization of L-type Ca^{2+} channels and insulin-containing secretory granules and its significance for the initiation of exocytosis in mouse pancreatic B-cells. *EMBO J.* 14:50–57.

28. Schroeder, T.J., J.A. Jankowski, J. Senyshyn, R.W. Holz, and R.M. Wightman. 1994. Zones of exocytotic release on bovine adrenal medullary cells in culture. *J. Biol. Chem.* 269:17215–17220.

29. Robinson, I.M., J.M. Finnegan, J.R. Monck, R.M. Wightman, and J.M. Fernandez. 1995. Colocalization of calcium entry and exocytotic release sites in adrenal chromaffin cells. *Proc. Natl. Acad. Sci. USA* 92:2474–2478.

30. Paras, C.D., W. Qian, J.R. Lakey, W. Tan, and R.T. Kennedy. 2000. Localized exocytosis detected by spatially resolved amperometry in single pancreatic beta-cells. *Cell Biochem. Biophys.* 33(3):227–240.

31. Stefan, R.I., J.F. Staden, and H.Y. Aboul-Ensin. 1999. Electrochemical sensor arrays. *Crit. Rev. Anal. Chem.* 29:133–153.

32. Droge, M.H., G.W. Gross, M.G. Hightower, and L.E. Czisny. 1986. Multielectrode analysis of coordinated, multisite, rhythmic bursting in cultured CNS monolayer networks. *J. Neurosci.* 6:1583–1592.

33. Maher, M.P., J. Pine, J. Wright, and Y.C. Tai. 1999. The neurochip: A new multielectrode device for stimulating and recording from cultured neurons. *J. Neurosci. Methods* 87:45–56.

34. Suzuki, I., Y. Sugio, Y. Jimbo, and K. Yasuda. 2005. Stepwise pattern modification of neuronal network in photo-thermally-etched agarose architecture on multi-electrode array chip for individual-cell-based electrophysiological measurement. *Lab Chip* 5:241–247.

35. Yakushenko, A., E. Kätelhön, and B. Wolfrum. 2013. Parallel on-chip analysis of single vesicle neurotransmitter release. *Anal. Chem.* 85(11):5483–5490.

36. Meunier, A., R. Fulcrand, F. Darchen, C.M. Guille, F. Lemaître, C. Amatore. 2012. Indium tin oxide devices for amperometric detection of vesicular release by single cells. *Biophys. Chem.* 162:14–21.

37. Hafez, I., K. Kisler, and K. Berberian et al. 2005. Electrochemical imaging of fusion pore openings by electrochemical detector arrays. *Proc. Natl. Acad. Sci. USA* 102:13879–13884.

38. Zhang, B., K.L. Adams, S.J. Luber, D.J. Eves, M.L. Heien, and A.G. Ewing. 2008. Spatially and temporally resolved single-cell exocytosis utilizing individually addressable carbon microelectrode arrays. *Anal. Chem.* 80:1394–1400.

39. Lin, Y., R. Trouillon, M.I. Svensson, J.D. Keighron, A.-S. Cans, and A.G. Ewing. 2012. Carbon-ring microelectrode arrays for electrochemical imaging of single cell exocytosis: Fabrication and characterization. *Anal. Chem.* 84:2949–2954.

40. Wang, J., R. Trouillon, Y. Lin, M.I. Svensson, and A.G. Ewing. 2013. Individually addressable thin-film ultramicroelectrode array for spatial measurements of single vesicle release. *Anal. Chem.* 85(11):5600–5608.

41. Carabelli, V., S. Gosso, and A. Marcantoni et al. 2010. Nanocrystalline diamond microelectrode arrays fabricated on sapphire technology for high-time resolution of quantal catecholamine secretion from chromaffin cells. *Biosens. Bioelectron.* 26:92–98.

42. Chen, X.H., Y.F. Gao, M. Hossain, S. Gangopadhyay, and K.D. Gillis. 2008. Controlled on-chip stimulation of quantal catecholamine release from chromaffin cells using photolysis of caged Ca2+ on transparent indium-tin-oxide microchip electrodes. *Lab Chip* 8:161–169.

43. Dittami, G.M. and R.D. Rabbitt. 2010. Electrically evoking and electrochemically resolving quantal release on a microchip. *Lab Chip* 10:30–35.

44. Berberian, K., K. Kisler, Q. Fang, and M. Lindau. 2009. Improved surface-patterned platinum microelectrodes for the study of exocytotic events. *Anal. Chem.* 81:8734–8740.

45. Zhang, B., M.L. Heien, M.F. Santillo, L. Mellander, and A.G. Ewing. 2011. Temporal resolution in electrochemical imaging on single PC12 cells using amperometry and voltammetry at microelectrode arrays. *Anal. Chem.* 83(2):571–577.

46. Chandler, D.E. and J.E. Heuser. 1980. Arrest of membrane fusion events in mast cells by quick-freezing. *J. Cell Biol.* 86:666–674.

47. Neher, E. and A. Marty. 1982. Discrete changes of cell membrane capacitance observed under conditions of enhanced secretion in bovine adrenal chromaffin cells. *Proc. Natl. Acad. Sci. USA* 79:6712–6716.

48. Steyer, J.A., H. Horstmann, and W. Almers. 1997. Transport, docking and exocytosis of single secretory granules in live chromaffin cells. *Nature* 388:474–478.

49. Angleson, J.K., A.J. Cochilla, G. Kilic, I. Nussinovitch, and W.J. Betz. 1999. Regulation of dense core release from neuroendocrine cells revealed by imaging single exocytic events. *Nat. Neurosci.* 2:440–446.

50. Calakos, N. and R.H. Scheller. 1996. Synaptic vesicle biogenesis, docking, and fusion: A molecular description. *Physiol. Rev.* 76:1–29.
51. Monck, J.R. and J.M. Fernandez. 1996. The fusion pore and mechanisms of biological membrane fusion. *Curr. Opin. Cell Biol.* 8:524–533.
52. Zimmerberg, J., F.S. Cohen, and A. Finkelstein. 1980. Fusion of phospholipid vesicles with planar phospholipid bilayer membranes. I. Discharge of vesicular contents across the planar membrane. *J. Gen. Physiol.* 75:241–250.
53. Woodbury, D.J. 1999. Building a bilayer model of the neuromuscular synapse. *Cell Biochem. Biophys.* 30:303–329.
54. Chanturiya, A., L.V. Chernomordik, and J. Zimmerberg. 1997. Flickering fusion pores comparable with initial exocytotic pores occur in protein-free phospholipid bilayers. *Proc. Natl. Acad. Sci. USA* 94:14423–14428.
55. Evans, E., H. Bowman, A. Leung, D. Needham, and D. Tirrell. 1996. Biomembrane templates for nanoscale conduits and networks. *Science* 273:933–935.
56. Hoekstra, D. 1990. Fluorescence assays to monitor membrane fusion: Potential application in biliary lipid secretion and vesicle interactions. *Hepatology* 12:61S–66S.
57. Takei, K., V. Haucke, V. Slepnev, K. Farsad, M. Salazar, H. Chen and P. De Camilli. 1998. Generation of coated intermediates of clathrin-mediated endocytosis on protein-free liposomes. *Cell* 94:131–141.
58. Kahya, N., E.I. Pecheur, W.P. De Boeij, D.A. Wiersma, and D. Hoekstra. 2001. Reconstitution of membrane proteins into giant unilamellar vesicles via peptide-induced fusion. *Biophys. J.* 81:1464–1474.
59. Karlsson, A., R. Karlsson, M. Karlsson, A-S. Cans, A. Stromberg, F. Ryttsen, and O. Orwar. 2001. Networks of nanotubes and containers. *Nature* 409:150–152.
60. Karlsson, M., K. Nolkrantz, M.J. Davidson, A. Stromberg, F. Ryttsen, B. Akerman, and O. Orwar. 2000. Electroinjection of colloid particles and biopolymers into single unilamellar liposomes and cells for bioanalytical applications. *Anal. Chem.* 72:5857–5862.
61. Wittenberg, N.J., L. Zheng, N. Winograd, and A.G. Ewing. 2008. Short-chain alcohols promote accelerated membrane distention in a dynamic liposome model of exocytosis. *Langmuir* 24:2637–2642.
62. Cans, A-S., N. Wittenberg, R. Karlsson, L. Sombers, M. Karlsson, O. Orwar, and A.G. Ewing. 2003. Artificial cells: Unique insights into exocytosis using liposomes and lipid nanotubes. *Proc. Natl. Acad. Sci. USA* 100(2):400–404.
63. Karlsson, R., M. Karlsson, and A. Karlsson et al. 2002. Discovery of potent and cell-active allosteric dual Akt 1 and 2 inhibitors. *Langmuir* 18:4186–4190.
64. Cans, A.-S., N.J. Wittenberg, D. Eves, R. Karlsson, A. Karlsson, O. Orwar, and A.G. Ewing. 2003. Amperometric detection of exocytosis in an artificial synapse. *Anal. Chem.* 75:4168–4175.
65. Simonsson, L., M.E. Kurczy, R. Trouillon, F. Hook, and A-S. Cans. 2012. A functioning artificial secretory cell. *Sci. Rep.* 2:824.
66. Mellander, L.J., M.E. Kurczy, N. Najafinobar, J. Dunevall, A.G. Ewing, and A.-S. Cans. 2014. Two modes of exocytosis in an artificial cell. *Sci. Rep.* 4:3847.
67. Omiatek, D.M., Y. Dong, M.L. Heien, and A.G. Ewing. 2010. Only a fraction of quantal content is released during exocytosis as revealed by electrochemical cytometry of secretory vesicles. *ACS Chem. Neurosci.* 1:234–245.
68. Amatore, C., A.I. Oleinick, and I. Svir. 2010. Reconstruction of aperture functions during full fusion in vesicular exocytosis of neurotransmitters. *Chemphyschem* 11:159–174.
69. Mellander, L.J., R. Trouillon, M.I. Svensson, and A.G. Ewing. 2012. Amperometric post spike feet reveal most exocytosis is via extended kiss-and-run fusion. *Sci. Rep.* 2:907.
70. Amatore, C., S. Arbault, I. Bonifas, M. Guille, F. Lemaître, and Y. Verchier. 2007. Relationship between amperometric pre-spike feet and secretion granule composition in chromaffin cells: An overview. *Biophys. Chem.* 129:181–189.

Section III

Nanoelectrochemical Methods

Michael V. Mirkin

CONTENTS

15.1 INTRODUCTION

To perform electrochemical experiments on the nanoscale and probe nanometer-sized objects, one needs comparably sized electrochemical tools.[1] A number of such tools—solid nanoelectrodes and nanopipette-supported liquid/liquid interfaces—have been developed since the late 1980s.[2,3] This review is concerned with new challenges and opportunities stemming from the use of nanoelectrochemical approaches. While nanoprobes offer important advantages and allow one to study numerous phenomena that cannot be observed at macroscopic electrodes, the visualization of their surfaces remains challenging, and the interpretation of the electrochemical response relies on assumptions about the electrode size and geometry. Here, we discuss methodologies that have been used to fabricate and characterize electrochemical nanoprobes and some typical pitfalls encountered in nanoelectrochemical experiments.

A wide variety of nanoelectrode shapes and features have recently been reported. In this chapter, we will only discuss several representative geometries as well as general concepts and approaches to nanoelectrochemical experiments. The survey of other geometries (such as a nanoring or a cylindrically shaped nanotube/nanowire) as well as arrays of nanoelectrodes can be found in recent review articles.[1,4] The nanoelectrode theory is not discussed here. As long as a nanoelectrode is not too small, its behavior follows classical microelectrode theory; and the theoretical description of nanosize-related effects on electrochemical processes is reviewed in Chapters 1 and 2. The applications of nanoelectrodes and nanopipettes range from the studies of biological and artificial membranes[5] to single-nanoparticle electrocatalysis[6] to nucleation/growth of metals.[7] To avoid overlap with other chapters, the review of applications is limited to studies of charge-transfer (CT) reactions, and the works employing nanoelectrodes/pipettes as scanning electrochemical microscope (SECM) and scanning electrochemical cell microscopy (SECCM) tips will be surveyed in Chapters 18 and 19.

15.2 FABRICATION AND CHARACTERIZATION OF METAL NANOELECTRODES

The fabrication of a nanoelectrode can be deceptively easy: by using a laser pipette puller, it takes only a few minutes to seal a commercially available metal wire into glass and obtain an electrode with an effective radius on the nanometer scale. However, most of those electrodes are likely to be unsuitable for quantitative measurements. In this section, we survey methodologies developed for making and characterizing different types of nanoelectrodes as well as some issues affecting the reliability of nanoelectrochemical experiments.

15.2.1 FABRICATION AND BASIC FEATURES OF NANOELECTRODES WITH DIFFERENT GEOMETRIES

Several representative nanoelectrode geometries are shown schematically in Figure 15.1. Historically, the first submicrometer-sized electrochemical electrode was a nanoband (Figure 15.1a) produced by the Wightman group.[2a] Other geometries, including (b) conical electrode, (c) spherical cap, (d) inlaid and (e) recessed disks, and (f) nanopore, have been reported. Nanoelectrodes of different kinds offer specific advantages (and disadvantages) and are suitable for different applications.

15.2.1.1 Nanoband Electrodes

These electrodes can be fabricated by forming a thin film of the electrode material between two insulating layers.[2] Such a film can be either obtained commercially or produced on an insulating substrate by a sputtering technique and then covered by an insulating overlayer. The edge of this assembly exposed to the solution serves as a voltammetric electrode with a nanoscopic width (e.g., 5–2300 nm[2a]) and a macroscopic length (mm or cm). The high mass-transfer rate and quasi-steady-state response are due to the nanoscale bandwidth, while the macroscopic length was initially seen

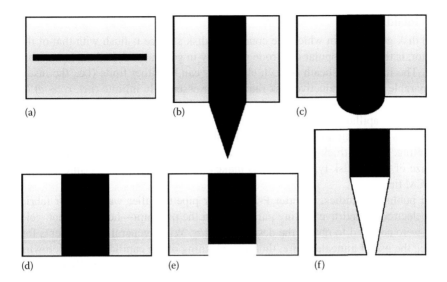

FIGURE 15.1 Examples of nanoelectrode geometries: (a) band, (b) conical electrode, (c) spherical cap, (d) inlaid disk, (e) recessed disk, and (f) nanopore.

as an advantage because of significantly higher Faradaic currents flowing at band electrodes as compared to disks or cones with similar characteristic dimensions (i.e., the radius comparable to the half width of the band). The macroscopic length eventually limited applications of band electrodes, which cannot be used as scanning probes and are not suitable for experiments in small volumes that require small physical size.

15.2.1.2 Conical Nanoelectrodes

The methodology for fabricating conical nanoelectrodes was originally developed to produce scanning tunneling microscope (STM) tips.[8,9] A micrometer-diameter metal wire (e.g., a 125 μm Pt/Ir rod[8]) was etched electrochemically and insulated with molten Apiezon wax,[9] glass,[10] polymer,[10,11] or electrophoretic paint.[12] Similar approaches were used to produce conical carbon nanoelectrodes.[13] The very end of a conical tip was exposed by placing it in an STM and applying voltage (e.g., 10 V) between the tip and a conductive substrate (e.g., a Pt disk). The onset of current flow produced a hole in the tip insulation at the point of closest approach of the tip to the substrate, while leaving most of the tip still insulated.[14] Other methods, such as heat shrinkage of the coating film,[12] have also been used to expose the tip.

The sharp tip of conical nanoelectrodes rendered them useful for penetrating thin films.[14b,c] They have also been used for SECM imaging (see Chapter 18). However, they are less useful for quantitative applications because of the intrinsically imperfect shape and nonpolishable surface.

15.2.1.3 Spherical-Cap Electrodes

Quasi-hemispherical and spherical-cap electrodes (Figure 15.1c) produced by etching metal wires or carbon fibers are not very different from conical nanoelectrodes. Because the exact shape characterization is difficult for both types of electrodes, no clear distinction was made between them in several publications.[10a,12,15]

Spherical-cap electrodes can also be produced by electrodeposition. Three-dimensional nucleation/growth produces metal crystals shaped as a quasi-hemisphere or a spherical cap.[16] The deposit shape is nearly perfect for a liquid metal, for example, Hg. This process was used to produce Hg microelectrodes[17a] and—more recently—nanoelectrodes.[17b] Such electrodes are suitable for electroanalysis and can be employed as SECM tips.[17a,c]

15.2.1.4 Inlaid Disk

The inlaid disk geometry, in which the conductive disk surface is flush with that of the surrounding insulator, is the most popular electrode geometry in general and the hardest to fabricate on the nanoscale. The insulating sheath of such electrodes can be either finite (i.e., the insulator radius, r_g, is comparable to the conductive disk radius, a) or essentially infinite (i.e., $r_g \gg a$). The fabrication procedures are different: the thin-sheath electrodes have been produced by pulling a metal wire into a glass capillary using a micropipette puller,[18] while the first step in the preparation of embedded disks is sealing an electrochemically sharpened metal microwire into a glass capillary without pulling.[19] While thick-glass electrodes are robust and relatively easy to fabricate, the small physical size of pulled disk-type electrodes makes them suitable for experiments in small spaces and as SECM tips.

In most published studies, a Sutter P-2000 laser pipette puller was used for fabricating thin-glass nanoelectrodes, and five pulling parameters in the program—heat, filament, velocity, delay, and pull—were adjusted to obtain the desired a and r_g. While general requirements for successful pulling (i.e., the use of annealed wire, thorough cleaning of the capillary and microwire, and good vacuum) are straightforward, the parameter values may not be the same for different P-2000 pullers, and even for the same instrument, they have to be adjusted occasionally to produce electrodes with the desired size and shape. The procedures were developed for fabricating different metal nanoelectrodes (Pt,[20] Au,[21a] and Ag[21b]). Producing high-quality gold and silver electrodes was found to be harder, especially for Ag, whose low melting point necessitated the use of the pulling sequence consisting of three different programs. A modified pulling procedure was also developed to produce thick-glass nanoelectrodes.[22]

The next step is to expose the conductive surface to solution; in most cases, this was done by mechanical polishing. Thick-glass electrodes were polished manually on the felt polishing cloth, which was wetted with a KCl solution containing 50 nm alumina particles and connected to the external circuit with a metal clip. The effective radius of the exposed metal surface was monitored using a high-input impedance field effect transistor (MOSFET)-based circuit.[19] This protocol yielded electrodes with a very small effective radius of ~10 nm[19] and even smaller.[22]

Polishing sharp, pulled nanoelectrodes is not straightforward because of the tip fragility. The original procedures based on glass etching and micropolishing were only partially successful.[18] Flat disk-type Pt nanoelectrodes suitable for kinetic measurements and quantitative SECM experiments were prepared by polishing on a 50 nm lapping tape under video microscopic control.[20] A micromanipulator was used to move the pipette vertically toward the slowly rotating disk of the micropipette beveller. The choice of the correct separation distance such that the nanoelectrode gets polished without being broken is the most challenging part of this procedure. To prepare a nanoelectrode that can be used as an SECM tip, one has to ensure that its axis is strictly perpendicular to the polishing surface. Several modifications of the aforementioned pulling/polishing procedures have been reported. For instance, a quickly rotating quartz-sealed, pulled nanoelectrode was slowly lowered toward the surface of a fixed polishing plate.[23]

An interesting alternative to mechanical polishing of sharp nanoelectrodes is the use of the focused ion beam (FIB).[24] In Ref. [24a], a pulled capillary containing a glass-sealed Pt wire was heat annealed by using a microforge and then milled by the FIB, producing a well-shaped, smooth disk-type nanoelectrode with a thin-glass sheath. Although, this procedure has only been reported for relatively large ($a \geq 100$ nm) nanoelectrodes, the fabrication of smaller electrodes may also be feasible.

A wider class of disk-type nanoprobes can be prepared by electrodeposition, including metals and other materials not suitable for pulling/polishing. To produce an essentially flat (rather than hemispherical) electrode, one can etch a disk-type Pt nanoelectrode and fill the resulting cavity with a metal of choice. Both sharp[17b] and thick-wall[25] nanoelectrodes have been prepared in this way. To control the size of the deposited electrode, one can either try to stop the deposition process at the moment when the nanocavity is completely filled[17b,26] or polish away the excess metal.[25] The moment when the nanocavity is filled can be detected from the current transient.[26] The deposition current increases slowly with time as the cavity gets filled with metal; and the current–time curve becomes much steeper when the cavity depth becomes smaller than its radius. This method, however, is not exact unless the results of the deposition process can be checked by SECM or AFM (see Sections 15.2.2 and 15.2.3). Removing the excess of deposited metal by polishing is not straightforward. Extensive polishing is likely to remove the deposited metal completely, while slight polishing may result in the electrode radius significantly larger than the original disk before etching[25] and apparently nonflat geometry. It is also difficult to ensure the consistency of the metal/glass seal and the absence of solution leakage. Overall, the electrodeposited probes should be more useful for experiments that do not require perfect electrode geometry (e.g., potentiometric measurements[27] or nanoparticle attachment[25]).

15.2.1.5 Recessed Disk and Nanopore

The possibility of a disk-type nanoelectrode surface becoming recessed into the insulator was first deduced[28] from the analysis of extremely fast electron-transfer (ET) rates measured in early voltammetric experiments at nanoelectrodes.[10b] A nanoelectrode can also become recessed in the process of polishing.[29]

In several publications, different kinds of recessed electrodes were prepared purposefully. For instance, a nanocavity formed within insulating wax was used to trap and detect single molecules.[30] Two extreme examples of recessed nanodisks are *slightly recessed* nanoelectrodes with the recess depth less than or comparable to the disk radius[31] and nanopore electrodes whose recess depth is much larger than the radius.[32] The former were prepared by controlled etching of nanometer-sized, flat Pt electrodes. By using high-frequency (e.g., 2–20 MHz) ac voltage, the layer of Pt as thin as ≥3 nm was removed to produce a cylindrical cavity inside the insulating glass sheath. The recess depth was evaluated from steady-state voltammetry and SECM approach curves,[31] and the possibility of using AFM to measure it more accurately was shown later.[26] Slightly recessed electrodes were used for electrodeposition (see above) and to form ultrathin layer electrochemical cells.[33] Glass nanopore electrodes were created by first fabricating a Pt nanodisk electrode and then etching it with a low-frequency ac current to obtain a Pt disk embedded at the bottom of a conical[32a,b] or cylindrical[32c] pore.

15.2.2 CHARACTERIZATION OF NANOELECTRODES: VOLTAMMETRY AND SECM

The characterization of a nanoelectrode includes the evaluation of its effective radius, true surface area exposed to solution, and the thickness of the insulting sheath, as well as the determination of the electrode geometry. In many published studies, the electrode radius was evaluated from the steady-state diffusion limiting current assuming either hemispherical, or conical, or planar disk geometry. For a nonflat electrode, this assumption is often problematic because of essentially unavoidable imperfections of its geometry.

Steady-state voltammetry is the simplest and most popular technique employed for characterizing nanoelectrodes. The goals here are to check that the electrode response follows the basic electrochemical theory and to determine the effective radius value. Assuming that a nanoelectrode is sufficiently large to avoid deviations from conventional laws of diffusion (see Chapter 2), one can analyze the shape of steady-state voltammograms using the theory developed for micrometer-sized electrodes.[34] The shape of the reversible steady-state voltammogram is independent of the electrode geometry and determined by the Nernst equation.[34a] In earlier publications, the effective radius of a nanoelectrode was typically evaluated from the diffusion limiting steady-state current assuming that it is shaped either as a hemisphere (Equation 15.1) or a disk (Equation 15.2):

$$i_{hs} = 2\pi nFDac* \tag{15.1}$$

$$i_{disk} = 4nFDac* \tag{15.2}$$

where
   n is the number of transferred electrons
   F is the Faraday constant
   a is the electrode radius
   D and $c*$ are the diffusion coefficient and bulk concentration of the reactant, respectively

Nearly perfect steady-state voltammograms of a well-behaved redox mediator (e.g., ferrocene in acetonitrile or ferrocenemethanol in aqueous solutions) at nanometer-sized electrodes have been reported by a number of research groups. A sigmoidal, retraceable curve with an essentially flat diffusion plateau and very low charging current (at potential scan rates up a few hundred millivolts per seconds; Figure 15.2a, red curve) is a good starting point in the nanoelectrode characterization; however, it does not provide any information about electrode geometry or potential problems, such as solution leakage or surface recess.

One approach to leakage detection is based on the comparison of fast-scan (e.g., ≥10 V/s) and slow-scan (e.g., ≤200 mV/s) voltammograms of the dissolved species. In the case of significant leakage, the presence of a thin layer of solution containing electroactive species inside the insulating sheath should result in peak-like features in fast-scan voltammograms, with a peak height proportional to the scan rate. By contrast, sigmoidal fast-scan voltammograms similar to the slow-scan, steady-state voltammograms—except for a moderate charging current contribution—point to the consistent seal and no solution leakage (cf. blue and red curves in Figure 15.2a). However, the apparent capacitance of a nanoelectrode is much larger than the capacitance of the metal/solution nanointerface due to *stray capacitance* of its insulated portion and wiring. Thus, detecting the leakage from capacitive current is not straightforward.

Another approach to detecting the leakage is to use fast-scan voltammetry of adsorbed species to evaluate the effective surface area of the electrode.[15] In principle, one can eliminate the leakage possibility by showing the agreement between the area value and the effective radius obtained from steady-state voltammetry of dissolved redox species. However, it was difficult to use the same electrode for area measurements and kinetic experiments and to accurately measure the amount of adsorbed species for electrodes smaller than ~60 nm radius.[15]

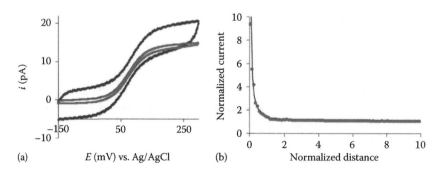

FIGURE 15.2   Characterization of a nanoelectrode by voltammetry and SECM. (a) Slow (red curve) and fast (blue curve) scan voltammograms of 1 mM FcCH₂OH at the 52 nm polished Pt electrode. $\nu = 50$ mV/s (red) and 50 V/s (blue). (b) Experimental (symbols) and theoretical (solid line) current–distance curves obtained with the same electrode as in (a) approaching an evaporated Au substrate.

More reliable characterization can be achieved by combining voltammetry with SECM, where the nanoelectrode serves as a tip (see Chapter 18 for discussion). The SECM is most useful for characterizing planar electrodes because high positive (or negative) feedback can only be obtained with a flat, well-polished tip whose entire surface can be brought close to the flat substrate. No other electrode geometry discussed earlier (i.e., conical, spherical-cap, cylindrical, or recessed disk) can yield either high positive (e.g., $I = i_T/i_{T,\infty} > 7$, where i_T is the tip current in a close proximity of the substrate surface and $i_{T,\infty}$ is its value far away from the substrate) or negative (e.g., $I \leq 0.1$) feedback. In Figure 15.2b, an SECM approach curve is fitted to the theory with $a = 52$ nm. The same a value was obtained independently from the diffusion limiting current at the same Pt electrode (Figure 15.2a). The radius value is reliable because of the high positive feedback current (up to a normalized current of ~9.5, which corresponds to distance of the closest approach, $d < 5$ nm). While well-shaped SECM approach curves showing high positive (and negative) feedback in good agreement with the theory can provide strong evidence that a nanoelectrode is essentially flat, well polished, and not leaky, lower feedback often observed in current–distance curves is hard to interpret. Possible origins of such a response include either recessed or protruding tip geometry, surface contamination, or poor tip/substrate alignment. The ambiguity in interpretation of low-feedback approach curves complicates the detection of the nanometer-scale damage to the electrode,[24a] as discussed in Section 15.2.3.

Both positive and negative feedback currents depend strongly on the height of the convex tip (i.e., one shaped as a cone or a spherical cap) or the depth of the recessed tip. Thus, the effective radius and the height of a conical/spherical nanoelectrode can be determined by fitting the experimental current vs. distance curves to the theory.[14] Similarly, the depth of the recess of the conductive surface into the insulator can be evaluated from the best fit of the experimental approach curve to the theory and compared to the value obtained from steady-state voltammograms.[31,33] Negative feedback current is more sensitive to the insulator thickness, and the r_g value can be evaluated from approach curves obtained at an insulating substrate (see Section 15.3.2). An advantage of the SECM characterization is that it is based not on the appearance of a nanoelectrode but on its current response, which is much more relevant to electrochemical measurements. However, if the tip geometry is imperfect, fitting an experimental approach curve to the theory may be problematic. One should also notice that only a sharp nanoelectrode with a thin insulating sheath can be used as an SECM tip and the reliability of the electrode characterization is largely determined by the attainable distance of the closest approach.

Presently, good-quality SECM approach curves can only be obtained using nanoelectrodes with the radii $\gtrsim 10$ nm. Several recent attempts to employ extremely small ($a < 5$ nm) electrodes for kinetic experiments,[22] transport,[35] and nucleation[13] studies underscore the importance of developing characterization techniques for such electrodes. Small imperfections in the geometry of these electrodes may result is misleading experimental results. For instance, a 1 nm surface recess would cause the current to a 1 nm radius electrode to be only 43% of the current to the equally sized, nonrecessed disk.

15.2.3 CHARACTERIZATION OF NANOELECTRODES: ELECTRON MICROSCOPY AND AFM

Several authors used SEM to evaluate the size and shape of a nanoelectrode.[12,18–20,24,36] The SEM lateral resolution is not sufficiently high to characterize electrodes smaller than ~50 nm radius. Moreover, even for relatively large electrodes (e.g., $a > 50$ nm), SEM micrographs provide mostly qualitative information about electrode shape that cannot be used for quantitative modeling of its response, and insufficient z-axis resolution makes it hard to distinguish between flat, recessed, and protruding nanoelectrodes. Nevertheless, SEM images can provide important information about nanoelectrode geometry and facilitate the detection of damage to its surface. Figure 15.3 shows SEM images of the same submicrometer-sized Pt disk electrode (a) before and (b) after it was damaged by an electrostatic discharge.[24a] The Amemiya group showed that such a damage can be caused

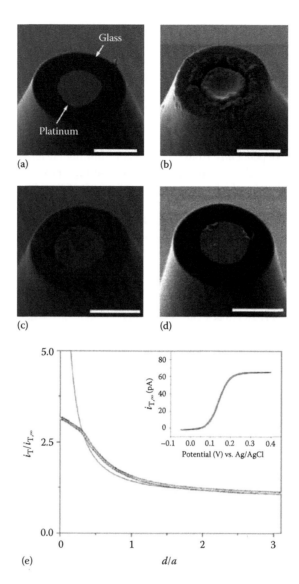

FIGURE 15.3 SEM images and electrochemical responses of a damaged glass-sealed submicrometer Pt electrode. SEM images were obtained (a) before and (b) after electrostatic discharge damage. Panels (c) and (d) show moderately and weakly damaged Pt electrodes, respectively. Scale bars, 1 μm. (e) Experimental (circles) and theoretical (solid line) SECM current–distance curves obtained with the damaged tip (panel b) approaching an unbiased Au substrate. The inset in (e) shows a steady-state voltammogram obtained at the same damaged electrode; $v = 20$ mV/s. Solution contained 0.5 mM FcMeOH in 0.1 M KCl. (Reprinted with permission from Nioradze, N., Chen, R., Kim, J., Shen, M., Santhosh, P., and Amemiya, S., Origins of nanoscale damage to glass-sealed platinum electrodes with submicrometer and nanometer size, *Anal. Chem.*, 2013, 85, 6198–6202. Copyright 2013 American Chemical Society.)

by touching an electrode or by voltage spikes produced by the potentiostat. Surprisingly, a severely damaged electrode (Figure 15.3b) yielded a nearly perfect steady-state voltammogram (inset in Figure 15.3e)—a strong indication that quantitative (especially kinetic) experiments performed at nanoelectrodes without proper characterization are likely to produce erroneous results. An attempt to detect this damage by SECM was only partially successful: the experimental approach curve in Figure 15.3e followed the conventional theory for positive feedback up to $I \approx 2.5$ and then leveled off. Clearly, detecting less profound damage in nanoelectrodes shown in Figures 15.3c and d by SECM and voltammetry could be difficult.

TEM, which offers significantly higher resolution than SEM, is potentially useful for characterizing smaller nanoelectrodes. For instance, Li et al.[22] presented TEM side views of ≤3 nm radius Pt wires inside the insulating sheath of their nanoelectrodes. These impressive micrographs unambiguously demonstrated that the radius of the conductive metal core is comparable to the effective value of the electrode radius obtained from the diffusion limiting current. However, to characterize the geometry of ultrasmall nanoelectrodes, one needs high-resolution images of the metal surface exposed to solution.

The possibility of AFM imaging of laser-pulled, polished nanoelectrodes was shown recently.[29] Although a needle-shaped electrode may not look like a suitable AFM substrate (Figure 15.4a), imaging polished Pt and Au electrodes as small as ~20 nm radius both in air and in liquids is

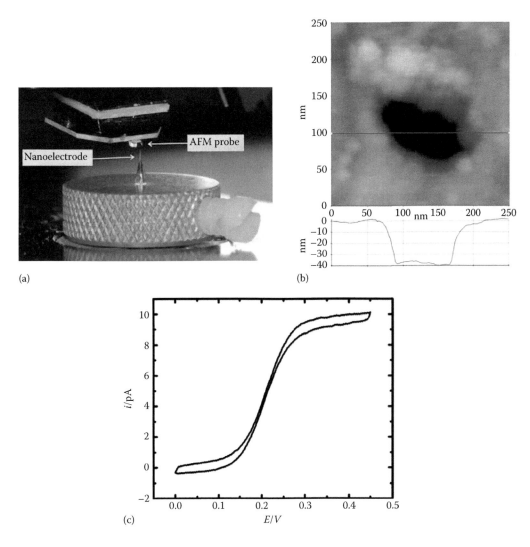

(a)

(b)

(c)

FIGURE 15.4   (a) Experimental setup used for AFM imaging of nanoelectrodes in air. A glass-sealed, polished nanoelectrode is positioned under the AFM probe. (b) Noncontact topographic image of a recessed Pt nanoelectrode in air. The scan rate was 0.5 Hz. The red line corresponds to the shown cross section. (c) Steady-state voltammogram of 1.2 mM FcCH$_2$OH obtained at the same electrode. $v = 50$ mV/s.

relatively easy. Because of its high (subnanometer) z-axis resolution and capacity for imaging in solution, AFM can provide detailed information about nanoelectrode geometry and surface reactivity that would be hard to obtain by any other technique. For example, a noncontact mode AFM image in Figure 15.4b shows a 50–55 nm radius significantly recessed electrode (~40 nm recess depth). Using the available theory (Equation 9b in Ref. [31]), one can obtain the effective value of $a = 54$ nm from the diffusion limiting current measured in 1.2 mM $FcCH_2OH$ (Figure 15.4c) in good agreement with the radius found from Figure 15.4b. Without an AFM image, one would not be able to tell that this electrode is recessed from the voltammogram in Figure 15.4c. The effective radius calculated from this voltammogram without taking into account the recessed geometry would have been as small as 20 nm. Kinetic experiments (and other geometry-sensitive experiments) at such an electrode could yield highly inaccurate results. The cracks in the insulating sheath, which may result is solution leakage, can also be detected in noncontact AFM images.[29]

Electrodes characterized by AFM (especially by noncontact mode imaging) are not damaged and can be employed in electrochemical experiments. The capacity of the AFM for in situ monitoring of surface reactions at nanoelectrodes was used to study nucleation/growth of Ag nanoclusters[7] and dissolution of Pt nanoelectrodes during oxygen reduction[37] and to fabricate nanosensors for reactive oxygen and nitrogen species with well-defined geometry.[26] The main point was to visualize the changes resulting from the deposition process and compare them to the corresponding electrochemical data. The electrodeposition of Pt black under the AFM control is illustrated in Figure 15.5. A noncontact topographic image of an etched Pt electrode (Figure 15.5a) in solution before the platinization shows the effective radius, $a \approx 70$ nm, and the etched cavity depth of ≥ 16 nm. The deposition of Pt black was done by stepping the electrode potential to -100 mV vs. Ag/AgCl, while the AFM tip, immersed in the platinization solution, was scanned in x-direction above the electrode surface. Figure 15.5b shows a stack of 60 consecutive topographic 1D scans obtained over a 60 s period. The deposition was stopped by stepping the electrode potential to 0 mV after Pt black

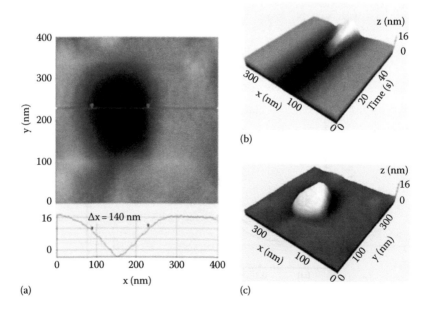

(a)   (c)

FIGURE 15.5 AFM topographic images of an etched Pt nanoelectrode in solution (a) before and (c) after the deposition of Pt black and a time evolution of a line scan (b) during the electrodeposition process. The red line in (a) corresponds to the shown cross section. (b) The tip was scanned along the x-axis with the scan rate of 1 Hz. The position of the line scan approximately corresponded to the red line in (a).

completely filled the cavity and slightly protruded (by ~15 nm) from the glass sheath, as can be seen from the image of the same electrode obtained after the platinization (Figure 15.5c). In another deposition experiment, the initial depth of the nanocavity was only ~2 nm, and the protrusion height after the deposition of Pt black was ~3 nm.[26]

15.3 NANOPIPETTE-SUPPORTED ITIES

15.3.1 ELECTROCHEMISTRY AT NANO-ITIES

Similar to solid/liquid electrochemistry, important advantages can be obtained by replacing a macroscopic interface between two immiscible electrolyte solutions (ITIES) with a liquid/liquid nanointerface. In 1986, Taylor and Girault introduced micrometer-sized liquid/liquid interface supported at the tip of a pulled glass pipette.[38] Nanoscale ITIES and their arrays have later been formed by using nanopipettes, nanopores, and porous membranes.[39] Electrochemistry of nanopores and porous membranes is surveyed in Chapter 11; our focus here is on the ITIES supported at the tip of a nanometer-sized pipette.

In addition to ET reactions occurring also at the metal/solution interfaces, CT reactions at the ITIES include simple and facilitated ion-transfer (IT) processes. All CT processes occurring at macroscopic ITIES can also be observed at a nanopipette-supported ITIES, including simple IT,[40-44] facilitated IT,[3,42,45] and ET reactions.[46] A simple IT process is a one-step reaction in which a cation I^{n+}(or an anion) is transferred directly from one phase (e.g., water) to the second phase (e.g., organic):

$$I^{n+}(W) \rightleftharpoons I^{n+}(O) \qquad (15.3)$$

Facilitated IT reactions require a ligand (L^{m-}) in the organic phase (e.g., 1,2-dichloroethane, DCE), which can react with I^{n+} to form a complex, resulting in the transfer of I^{n+}:

$$I^{n+}(W) + L^{m-}(O) \rightleftharpoons IL^{n-m}(O) \qquad (15.4)$$

The ET reaction between redox molecules confined to two immiscible liquid phases can be described as

$$O_1(W) + R_2(O) \rightleftharpoons R_1(W) + O_2(O) \qquad (15.5)$$

Because of the negligibly small ohmic potential drop (typically, <1 mV) and low double-layer charging current, a nanopipette is an excellent tool for studying CT processes at the ITIES and nanoscale electrochemical imaging.[39] At first glance, it appears to be an extremely simple device; however, quantitative nanopipette voltammetry is not straightforward. It requires a well-defined nanoscopic ITIES formed at the pipette tip and thorough characterization of its geometry.

15.3.2 FABRICATION AND CHARACTERIZATION OF NANOPIPETTES

15.3.2.1 Pulling a Nanopipette

By pulling borosilicate glass or quartz capillaries with a laser pipette puller (e.g., P-2000, Sutter Instrument Co.), a pair of nanopipettes with the same orifice radius can be obtained. When choosing capillaries, one needs to consider several factors, including the material (quartz or glass) and properties of a specific capillary (e.g., the wall thickness or the presence of a filament).[47] Borosilicate glass has a relatively low melting point and requires heat (one of the P-2000 parameters) between 300 and 400, while a significantly higher heat value—between 550 and 900—is used to pull quartz capillaries. Borosilicate glass is much less expensive than quartz and easier to use for pulling relatively large (micrometer- or submicrometer-sized) pipettes. However, ultrasmall nanopipettes with a relatively short taper, which is essential for minimizing the solution resistance, have to be fabricated from quartz. Quartz is very sensitive to uneven heating, which may result in asymmetrical pipettes. In this case, using quartz capillaries with a thicker wall (≥0.5 mm) can help. The ratio of the outer and inner diameters of the capillary largely determines the RG of pulled nanopipettes (RG $= r_g/a$, where r_g is the glass radius at the tip and a is the radius of the orifice).

To support an ITIES, a nanopipette has to be filled with solution. Using capillaries with filaments makes it easier to bring aqueous solution to the end of the pipette tip; otherwise, it is difficult to remove the air and to fill the nanopipette completely. Capillaries without filaments can be used to produce pipettes that will be filled with organic solution. Organic solvents, such as DCE, are relatively easy to inject in a glass or quartz pipettes, and at the same time, the solvent evaporation is slower in the absence of a filament.

The pulling process is controlled by adjusting five parameters in the puller program, which are heat, filament, velocity, delay, and pull. To obtain smaller tips, one can increase the value of heat, velocity, or pull or decrease the value of filament or delay. To control the length of the taper while maintaining the nanometer-scale tip diameter, one can limit the value of velocity and increase pull at the same time. The roughness of the pipette tip can be reduced by polishing[48a] or by FIB milling.[48b] A potential problem is that the pipette orifice can be contaminated by polishing agent.

15.3.2.2 Surface Modification

When a water-filled glass or quartz pipette is immersed in organic solution, a thin aqueous film forms on its hydrophilic outer wall, making the true area of the liquid/liquid interface much larger than the geometrical area of the pipette orifice.[49] The film formation can be avoided by silanizing the outer pipette wall to render it hydrophobic while keeping the interior wall nonsilanized. In most early publications, this was done by dipping the pipette tip into silanizing agent (typically, chlorotrimethylsilane) while passing a flow of argon through the pipette. This procedure is straightforward for micrometer-sized pipettes, but not easy for nanopipettes.[40,46] Silanization of smaller pipettes must be done cautiously to avoid the formation of a film on the inner wall, which can partially block the pipette orifice and induce solvent penetration into its narrow shaft. A recently developed protocol for silanizing pipettes in the vapor phase allows one to avoid oversilanization of relatively small (e.g., ~10 nm radius) pipettes[43]; however, the possibility of silanizing even smaller (e.g., 1–5 nm[42]) pipettes is uncertain.

When the pipette is filled with organic solution and immersed in the aqueous phase, its inner wall needs to be silanized to avoid water penetration into the pipette. This can be done by dipping the tip into chlorotrimethylsilane for 5–7 s.[50] In this case, both the outer and inner walls of the pipette get silanized, but unlike water, organic solution is not likely to form a layer on the outer wall even though it becomes hydrophobic.[49] A more controlled method for vapor silanization was reported recently.[51] The pipettes were fixed in a minivacuum desiccator, which was first evacuated by the pump, and then the vapor of highly pure N-dimethyltrimethyl silylamine was delivered from the flask to the desiccator, where the pipettes were exposed to it for about 15 min.

15.3.2.3 Characterization of Nanopipettes

Checking a nanopipette with an optical microscope is useful for initial evaluation of its properties (i.e., straight or bent, not broken, order of magnitude estimate of the tip diameter). However, the tip of a nanopipette is too small to be quantitatively characterized by optical microscopy. The main parameters defining the pipette geometry are the radius, a; the glass thickness at the tip, r_g (the related dimensionless parameter is RG $= r_g/a$); and the pipette angle, θ_p (Figure 15.6).

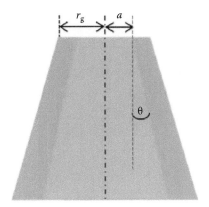

FIGURE 15.6 Schematic diagram of the nanopipette geometry.

The most commonly used methods for characterizing nanopipettes are electron microscopy (SEM and TEM) and electrochemical techniques (cyclic voltammetry and SECM). SEM is a direct way of visualizing the pipette geometry, but it is limited by the resolution of the instrument. For pipettes with diameters smaller than ~50 nm, it is difficult to see the orifice clearly, and a few nanometers thick conductive coating (e.g., Au or Pd) has to be applied to image the insulating nanotip without significant charging. Nevertheless, SEM can provide important information about pipette size and geometry. For instance, both a and r_g of a correctly silanized nanopipette in Figure 15.7a are very similar to those of a nonsilanized pipette (Figure 15.7b), which was pulled from the same quartz capillary; and therefore no orifice blocking occurred during silanization.

Although not as widely used as SEM, TEM can be very useful for characterizing nanopipettes. Unlike metal electrodes, the liquid–liquid interface is supposed to be featureless, and imaging it is not essential for pipette characterization. A high-resolution side view of a nanopipette can be used to evaluate its radius, RG, and θ_p, which are ~55 nm, 1.4, and 6°, respectively, for the pipette imaged in Figure 15.8.

Similar to metal nanoelectrodes, the size and the geometry of a nanopipette can be examined by AFM; however, obtaining high-quality images of small and sharp pipettes with RG < 2 can be difficult because of the stability issues.

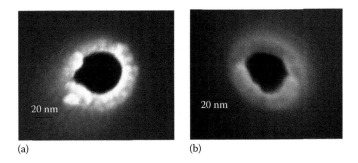

(a) (b)

FIGURE 15.7 Top view of SEM images of (a) silanized and (b) nonsilanized nanopipettes pulled from the same quartz capillary. (Reprinted with permission from Wang, Y., Velmurugan, J., Mirkin, M.V., Rodgers, P.J., Kim, J., and Amemiya, S., Kinetic study of rapid transfer of tetraethylammonium at the 1,2-dichloroethane/water interface by nanopipette voltammetry of common ion, *Anal. Chem.*, 2010, 82, 77–83. Copyright 2010 American Chemical Society.)

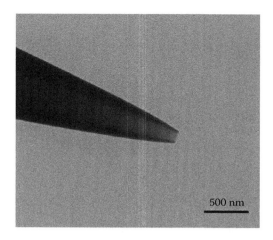

500 nm

FIGURE 15.8 TEM image of a quartz nanopipette.

Steady-state voltammetry of IT from the external liquid phase to the filling solution has often been used to evaluate a from the diffusion limiting current. The radius of a nonsilanized nanopipette can be found from Equation 15.6 proposed by Beattie et al.[52]:

$$i_2 = 3.35\pi z F D_2 c_2^* a \qquad (15.6)$$

where z, D_2, and c_2^* are the charge of the transferred ion, its diffusion coefficient, and bulk concentration in the external solution (phase 2). Sometimes, the background subtraction is necessary to obtain accurate results. One should also keep in mind that Equation 15.6 is an approximate equation obtained empirically for micrometer-sized pipettes. Verifying its accuracy for nanopipettes is difficult.

The equation for the diffusion limiting steady-state current to the orifice of a silanized pipette is more exact:

$$i_2 = 4xzFD_2c_2^*a \qquad (15.7)$$

where x is a function of the dimensionless glass radius, RG, which was tabulated[53a] and expressed by an analytical approximation for disk-shaped interfaces.[53b] Either a or RG can be found from Equation 15.7 if the second parameter is known. The RG value can be found independently from an SECM approach curve by using the nanopipette as a tip[40,48a] (for discussion, see Chapter 18). A good fit of SECM negative feedback approach curve to the theory can also confirm that the ITIES is essentially flat and not recessed. Pipettes as small as ~8 nm radius with RG = 1.6 have been characterized in this way.[48a]

If a transferable ion is initially present in the filling solution, the IT current is determined by diffusion inside the pipette. The geometry of the pipette inside can be approximately described by two parameters, a and θ_p (Figure 15.6), which can be evaluated from the steady-state limiting current of the ion egress[43]:

$$i_1 = 4f(\theta_p)zFD_1c_1^*a \qquad (15.8)$$

where D_1 and c_1^* are the diffusion coefficient and bulk concentration of the transferable ion in the filling solution (phase 1) and $f(\theta_p)$ is a function of the tip inner angle, θ_p, as given by[43,54]

$$f(\theta_p) = 0.0023113912 + 0.013191803\theta_p + 0.000317385960\theta_p^{1.5} - 5.8554625 \times 10^{-5}\theta_p^2 \qquad (15.9)$$

The pipette resistance is another important parameter to be measured if the pipette is to be used in kinetic experiments (and also for resistive-pulse sensing or scanning ion conductance microscopy, SICM).[55] The total pipette resistance ($R_T = R_{int} + R_{ext}$) comprises of two components, that is, the resistances of the inner and outer solutions. It can be obtained from the slope of a current vs. voltage curve recorded with the same solution inside and outside the pipette.[55a] Such curves are normally linear for larger (e.g., $a \geq 100$ nm; the exact limit depends on the ionic strength of solution) pipettes, but nonlinear for smaller pipettes. In the latter case, the slope can be determined from the essentially linear low-voltage portion of the curve (e.g., ±20 mV[56b]). Assuming that the pipette orifice is disk shaped, $R_{ext} = 1/(4\kappa a)$ is entirely determined by its radius and solution conductivity (κ). The internal pipette resistance ($R_{int} = R_T - R_{ext}$) can be related to θ_p by a simple analytical approximation[56a]

$$R_{int} = \frac{1}{\kappa \pi a \tan\theta} \tag{15.10}$$

which can be used to evaluate the pipette angle.[56b]

15.3.3 ELECTROCHEMICAL MEASUREMENTS AT NANOPIPETTES

Unlike macroscopic ITIES, in nanopipette voltammetry, the interfacial ET or IT current is very small (pA range). Therefore, potentiostatic experiments at the nano-ITIES are performed by applying voltage between two reference electrodes, and a four-electrode potentiostat is not required. Typically, the potential gradient and the ohmic potential drop inside a pipette are too small for significant electromigration or electro-osmotic flow along its charged inner wall.[3,43,54] The electrostatic and double-layer effects can be more significant for smaller nanopipettes, for example, $a \leq 5$ nm.[42]

Choosing a proper potential sweep rate (v) in pipette voltammetry is essential for attaining a steady-state and sufficiently low charging current. Computer simulations and experiments showed that ion diffusion on either side of the nano-ITIES reaches a steady state during a potential cycle at a moderate v.[43,54,57] The related dimensionless parameter $\sigma = (a^2/4D_2)(z_i Fv/RT)$ compares a to the diffusion distance in the external solution, $\sqrt{D_2 RT / z_2 Fv}$. It was suggested that the IT process attains a steady state if $\sigma \ll \sim 10^{-4}$.[54] In a typical voltammetric experiment at a nanopipette, $v = 10$ mV/s, $a = 50$ nm, and $D_2 = 10^{-5}$ cm^2/s correspond to a very small σ value of 2.4×10^{-7}, and steady state is readily attained in both liquid phases.

One should notice, however, that the σ value reflects the diffusion only in the external solution. The radial diffusion of ions from the external solution to the tip is much less hindered by the glass wall than the diffusion inside a tapered pipette. The time required to attain a steady state is typically determined by the mass-transfer rate inside the pipette and the geometry of the pipette inside. Thus, besides the tip inner radius (which is included in the dimensionless parameter σ), θ_p can also influence the attainment of a steady state. In practice, the variations in θ for quartz nanopipettes are relatively small, that is, from ~4° to ~20°, and sigmoidal forward and reverse waves that completely retrace each other can be obtained at moderate scan rates (e.g., $v \leq 1$ V/s for $D \approx 10^{-5}$ cm^2/s), thereby confirming the apparent steady state on both sides of the nanopipette tip.

It was noted[43a] that transient cyclic voltammetry is not practical with nanopipettes. A σ value of $>10^{-4}$ is required for obtaining a transient cyclic voltammogram (CV) of simple IT even at a narrow pipette with the taper angle of ~6°. This corresponds to $v > 4$ V/s, assuming $z = 1$, $a = 50$ nm, and $D_2 = 10^{-5}$ cm^2/s. At such a fast potential sweep, a large capacitive current (mostly due to stray capacitance of a nanopipette) would severely distort a voltammogram (Figure 15.9b). Transient CVs can be obtained at larger (micrometer-sized) pipettes.

The preceding discussion assumes that the diffusion coefficients of the transferred ion are similar inside the pipette and in the outer solution. However, a very slow diffusivity (e.g., in ionic liquid [IL][44]) can result in a significantly longer time required for the IT to reach a steady state in the external solution. Figure 15.9a shows two CVs obtained at a 500 nm pipette. At $v = 1$ mV/s, both egress

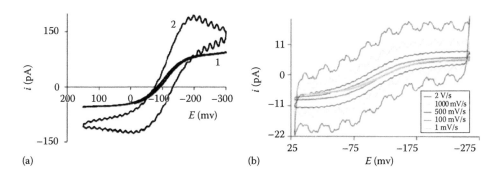

FIGURE 15.9 Effect of the potential sweep rate on CVs of TBA⁺ transfer at the nanopipette-supported water/IL interface. (a) $a = 500$ nm. ν, mV/s = 1 (1) and 1000 (2). (b) $a = 60$ nm. ν was varied between 1 mV/s and 2 V/s, as shown in the color legend. (Reprinted with permission from Wang, Y., Kakiuchi, T., Yasui, Y., and Mirkin, M.V., Kinetics of ion transfer at the ionic liquid/water nanointerface, *J. Am. Chem. Soc.*, 2010, 132, 16945–16952. Copyright 2010 American Chemical Society.)

and ingress currents in curve 1 attain a steady state; however, at $\nu = 1$ V/s (curve 2), the egress wave remains essentially sigmoidal, while the ingress wave is peak shaped. At higher ν, the charging current becomes significant (Figure 15.9b).

15.4 KINETICS OF CHARGE-TRANSFER REACTIONS AT THE NANOINTERFACES

Steady-state voltammetry at nanometer-sized interfaces is one of the best techniques for studying fast electrochemical kinetics. Its advantages over the transient methods include the absence of limitations caused by the charging current and ohmic potential drop, relative insensitivity to low levels of reactant adsorption, and relative simplicity of data acquisition and analysis. The size of a nanointerface is the origin of these advantages and also of numerous technical difficulties, some of which are considered as follows.

15.4.1 Mass-Transfer and Kinetic Measurements at Nanointerfaces

The CT rate constant can be measured only if it is smaller than or comparable to the mass-transfer coefficient, m. For uniformly accessible solid/liquid or liquid/liquid interfaces, the mass-transfer coefficient can be defined as[58]

$$m = \frac{i_d}{(nFAc^*)} \tag{15.11}$$

where
  i_d is the characteristic mass-transfer current for a specific electrochemical method, for example, diffusion limiting current in steady-state voltammetry
  A is the interfacial area
  c^* is the bulk concentration of the reactant

For a disk-type interface (as well as ones shaped as a cone, a ring, or a spherical cap) under steady-state conditions, $m \sim D/a$. Thus, an important feature of nanoelectrodes (and nanopipette-supported liquid interfaces) is a very high steady-state mass-transfer rate. For example, with $D = 10^{-5}$ cm²/s, $a = 10$ nm corresponds to $m = 10$ cm/s. This gives the upper limit for the determinable rate constant of ~50 cm/s.

Assuming Butler–Volmer kinetics, Equation 15.12 describes the shape of a steady-state voltammogram produced by CT at any uniformly accessible electrochemical interface (i.e., when the surface concentrations and diffusion fluxes of electroactive species are uniform over the entire interfacial area)[58]:

$$\frac{i}{i_d} = \frac{1}{\theta + 1/\kappa} \tag{15.12}$$

This equation is applicable to ET at the metal/solution interface and to IT at the nano-ITIES if the diffusion inside the pipette shaft does not have to be taken into account. In the former case, $\theta = 1 + exp[nF(E - E^{0'})/RT]m_O/m_R$ and $\kappa = k^0 exp[-\alpha nF(E - E^{0'})/RT]/m_O$ for the reduction reaction, where m_O and m_R represent mass-transfer coefficients of oxidized and reduced species, respectively, E is the electrode potential, and $E^{0'}$ is the formal potential of the redox couple and k^0 and α are the standard rate constant and the transfer coefficient, respectively. For IT from the external organic solution (phase 2) to the aqueous solution (phase 1), $\theta = 1 + exp[(\Delta_o^w\varphi - \Delta_o^w\varphi^{0'})zF/RT]D_2/D_1$ and $\kappa = k^0 exp[-\alpha(\Delta_o^w\varphi - \Delta_o^w\varphi^{0'})zF/RT]/m_2$, where $\Delta_o^w\varphi$ and $\Delta_o^w\varphi^{0'}$ are the Galvani potential difference across the ITIES and its standard value for the given IT, respectively, and m_2 is the mass-transfer coefficient in the outer solution. Conceptually similar equations were derived for a nonuniformly accessible disk-shaped interface[59] and SECM.[20]

High mass-transfer rates under steady state can also be attained when two electrodes are separated by a nanoscale gap in either a thin-layer cell (TLC) or a SECM. In this case, the mass-transfer rate is a function of the separation distance, d, and $m \sim D/d$ if $d < a$.

In liquid/liquid electrochemistry, the rates of simple and facilitated IT and ET were determined from steady-state voltammograms obtained using nanopipettes filled with an aqueous solution (the resistance of an organic-filled pipette is usually too high to attain the ohmic potential drop of <1 mV required for reliable kinetic measurements[3]). In early experiments, the ion of interest was initially present only in one phase (ether aqueous or organic), and its transfer across the ITIES produced sigmoidal voltammograms, which were used to extract kinetic parameters. In the case of a facilitated IT Equation 15.4, an excess amount of the transferable ion was added to the pipette to deplete a ligand in the external solution. The essentially spherical diffusion of a ligand species to the pipette orifice resulted in the true steady-state voltammogram. The kinetic parameters were extracted either by fitting the entire voltammogram to Equation 15.12 (or to the corresponding equation for quasi-reversible steady-state voltammogram at a disk-shaped interface[59]) or by using the three-point method based on the determination of the half-wave potential, $E_{1/2}$, and two quartile potentials, $E_{1/4}$ and $E_{3/4}$.[60] The same approaches were used for the analysis of steady-state voltammograms of ET obtained at the nano-ITIES.

The asymmetry of the diffusion field at a pipette-based ITIES is important for studies of simple IT processes (Equation 15.3). With a small taper angle, the diffusion inside the narrow shaft is almost linear in contrast to the spherical diffusion of ions to the pipette orifice in the outer solution, which makes the mass transport more complicated.[41,54,57] Depending on experimental conditions, simple IT at a nanopipette may yield either a sigmoidal and retraceable steady-state voltammetric curve or an asymmetrical, transient voltammogram.[40–42] The latter consists of an apparently steady-state, sigmoidal wave corresponding to ingress of an ion into the pipette and a time-dependent, peak-shaped wave produced by egress of the same ion to the external solution.[61]

In earlier studies, sigmoidal waves of simple IT were treated using simple steady-state theory (e.g., Equation 15.12) and assuming that their shape is independent of geometry of the pipette inside. More recent simulations and experiments[54,57] showed that this simplification is not realistic, and the reversible half-wave potential of simple IT from the external solution to the pipette under steady state depends on pipette angle, θ_p. It was suggested that kinetic and thermodynamic parameters of simple ITs determined without taking into account the effects of ion diffusion in the inner shaft of a nanopipette may not be accurate.

Another issue complicating kinetic analysis of rapid CT reactions is a weak dependence of the shape of an almost reversible steady-state voltammogram on kinetic parameters and, consequently, the lack of the unique fit between the theoretical and experimental curves. This problem was addressed by using common ion voltammetry[43] (see Section 15.4.4).

15.4.2 KINETICS OF ELECTRON-TRANSFER REACTIONS AT THE NANOELECTRODES

ET kinetics at nanometer-sized electrodes have been measured either by steady-state voltammetry or by using a nanoelectrode as an SECM tip; the latter approach is discussed in Chapter 18. Voltammetric kinetic experiments have been carried at polished, flat electrodes as well as at nonpolishable conical or spherical-type tips.[15,22,62] Polished electrodes with a RG $\leq$ 10 have been employed as SECM tips for feedback mode kinetic experiments.[20,21a]

Watkins et al.[15] measured the kinetics of ferrocenylmethyltrimethylammonium (TMAFc$^+$) oxidation at 19 quasi-hemispherical Pt electrodes with the effective radius varied between 2 and 150 nm. This work shows how hard it is to make kinetic measurements at nonflat nanoelectrodes. Despite major efforts made to characterize the electrode size and shape (see Section 15.2.2) and a large number of analyzed voltammograms, a significant uncertainty in the determined rate constant ($k^0 = 4.8 \pm 3$ cm/s) was apparently due to imperfect electrode geometry. No strong correlation was found between the electrode size and the measured kinetic parameters even for the radii as small as $a \leq 10$ nm, for which such correlation can be expected from existing theory[63] (see Chapter 2 for discussion). Another important lesson to be learned from Ref. [15] is that an individual kinetic experiment at a nanometer-sized electrode may not be reliable. To ensure that the results are meaningful, one has to treat a number of CVs obtained for a wide range of experimental conditions.

Similar quasi-hemispherical Pt electrodes were used to study kinetics of IrCl$_6^{3-}$ oxidation in 0.5 M KCl.[62] The electrode radii in this case were somewhat larger (48–654 nm), which may be the reason for much smaller uncertainties in the measured kinetic parameters ($k^0 = 2.9 \pm 0.2$; $\alpha = 0.50 \pm 0.01$). The authors have stressed significant deviations of the nanoelectrode responses from the classical theory observed in the absence of the supporting electrolyte and additional complications caused by ion pairing. However, no effect of the electrode size on the measured ET rate with excess KCl was reported, and the k^0 values measured at nanoelectrodes were similar to those obtained at larger electrodes in Ref. [62] and in the literature. One should also notice that the oxidation of IrCl$_6^{3-}$ occurs at potentials sufficiently positive for formation of Pt oxide, which could have affected the ET rate.

Another kinetic study[36] employed nonflat Pt–Ir electrodes with the effective radii ranging from extremely small (e.g., 1.1 nm) to relatively large (e.g., 150 nm). The k^0 values were determined for FcTMA^{2+}/FcTMA$^+$ (1.1–11.9 cm/s) and Fe(CN)$_6^{3-}$/Fe(CN)$_6^{4-}$ (0.12–17.3 cm/s) redox couples. Unlike other published results, the measured rate constants for both ET reactions increased markedly (i.e., by about one and two orders of magnitude, for FcTMA$^{2+/+}$ and Fe(CN)$_6^{3-/4-}$, respectively) with decreasing a value. Moreover, the k^0 values obtained for the former reaction at larger nanoelectrodes were several times lower than those measured in Ref. [15] for the same reaction and for oxidation of either aqueous ferrocenemethanol (FcMeOH)[20,21a] or ferrocenedimethanol (Fc(MeOH)$_2$) in KCl.[64a] In addition to uncertainties in electrode geometry, these results may have been affected by unusual approach to data analysis. The voltammograms were obtained in solution containing both oxidized and reduced forms of redox species, and only a small portion of each curve, corresponding to low (≤ 5 mV) overpotentials, was analyzed. The diffuse double-layer effects were assumed to be negligible at low overpotentials. The validity of this assumption is not obvious because the equilibrium potential is not necessarily close to the potential of zero charge. Moreover, the developed approach did not allow the authors to evaluate α and to check whether experimental curves were in agreement with the voltammetric theory.

No strong correlation between the electrode radius and kinetic parameters was found with glass-sealed, polished, thick-glass nanoelectrodes.[22] The average rate constant values determined for the oxidations of Fc in acetonitrile (3.4–13.4 cm/s), FcMeOH in aqueous 0.1 M NaCl (0.5–18.8 cm/s), and $IrCl_6^{3-}$ in 0.2 M KCl (0.5–13 cm/s) were close to those found by other groups. However, the variation in k^0 was significant—more than an order of magnitude for the oxidation of FcMeOH—and the determined α values (0.72–0.85) were much higher than $\alpha = 0.5$ expected from classical ET theory and also higher than the experimental values reported by others.

In addition to characterizing the electrode geometry (Section 15.2.2), using the SECM, one can further increase the mass-transfer rate by bringing a nanoelectrode tip very close to the surface of the conductive substrate. In this way, the kinetic parameters were determined for several rapid ET reactions—the oxidation of FcMeOH in 0.2 M NaCl ($k^0 = 6.8 \pm 0.7$ cm/s; $\alpha = 0.42 \pm 0.03$), the reduction of 7,7,8,8-tetracyanoquinodimethane (TCNQ; $k^0 = 1.1 \pm 0.04$ cm/s; $\alpha = 0.42 \pm 0.02$), the oxidation of ferrocene in acetonitrile ($k^0 = 8.4 \pm 0.2$ cm/s; $\alpha = 0.47 \pm 0.02$), and the reduction of $Ru(NH_3)_6^{3+}$ in 0.5 M KCl ($k^0 = 17.0 \pm 0.9$ cm/s; $\alpha = 0.45 \pm 0.03$) at Pt.[20] The kinetic parameters were found to be essentially independent of m, which was varied by two orders of magnitude by changing both a and d. In a similar manner, the kinetics of the same ET reactions (plus the oxidation of tetrathiafulvalene in DCE) were measured at Au nanoelectrodes and compared to those studied at Pt tips.[21a] Very similar k^0 and α values were obtained with Pt and Au polished electrodes for all investigated redox species except the reduction of $Ru(NH_3)_6^{3+}$, which was found to be somewhat faster at Pt than at Au with either KCl or KF used as a supporting electrolyte; it was concluded that this reaction is not fully adiabatic.

More recently, the theory was developed[65a] and experiments were carried out[65b] to probe rapid ET kinetics at a macroscopic SECM substrate by quasi-steady-state voltammetry using a submicrometer-sized tip positioned at a nanometer-scale distance from the substrate surface. These results are further discussed in Chapter 18.

The nanoelectrochemical ET experiments surveyed in this section involved a number of redox couples, different supporting electrolytes, and several solvents. Surprisingly, most measured k^0 values (except for a few values obtained at poorly characterized electrodes) are within one order of magnitude range. By contrast, the homogeneous self-exchange rate constants of the same redox species cover the range of more than six orders of magnitude (e.g., from $\sim 10^3$ M/s for $Ru(NH_3)_6^{3/2+}$ to 10^9 M/s for TCNQ/TCNQ$^-$). Moreover, no direct correlation between the homogeneous and heterogeneous rate constants expected from the Marcus formula was found. There are also no striking differences between the rate constants measured by different nanoelectrochemical techniques and at nanoelectrodes of different size. In general, the ET rate constants measured at nanoelectrodes are much larger than the values obtained for the same processes at macroscopic and micrometer-sized interfaces. It is not yet clear whether the observed faster rates are due to the improved capabilities for measuring fast kinetics or fundamental differences between ET processes at nanoscopic and macroscopic interfaces[66] (see Chapters 1 and 2) or experimental issues inherent in nanoelectrochemical kinetic studies.

15.4.3 KINETICS OF CHARGE-TRANSFER REACTIONS AT THE NANO-ITIES

15.4.3.1 Electron-Transfer Kinetics

ET measurements at nanopipette-supported polarizable ITIES are challenging because of interfering IT reactions and/or interfacial precipitation.[39] Steady-state voltammetry was used to investigate ET reactions at the polarizable ITIES formed at the tip of 50–400 nm radius pipettes.[46] Each pipette was filled with an aqueous solution containing a mixture of two forms of redox species (O_1 and R_1) and immersed in organic solution containing water-insoluble redox species (O_2). The application of a sufficiently negative potential to the internal reference electrode with respect to the external (organic) reference resulted in the electric current across the nano-ITIES due to the interfacial ET between R_1 and O_2 species. The condition $c_{R1} \gg c_{O2}$ was maintained in all experiments, so that the diffusion of R_1 species inside the pipette was negligible and did not control the overall current and the aqueous phase showed a metal-like behavior.[67] Steady-state voltammograms obtained by varying the voltage applied across the ITIES were similar to those recorded at the metal/solution nanointerfaces.

By exploring a number of combinations of aqueous and organic redox couples and different supporting electrolytes, two experimental systems—the reduction of TCNQ in DCE by aqueous $Ru(NH_3)_6^{2+}$ and the ET between Fe(EDTA)$^{2-}$ and TCNQ—were shown to be suitable for such studies, while all other systems failed to yield high-quality voltammograms.[46] The kinetic parameters obtained in this way for the former ET reaction ($k_{12}^0 = 2.75$ M/cm/s and $\alpha = 0.53$) were thought to be less reliable because of significant sensitivity of $Ru(NH_3)_6^{2+}$ species to oxygen. An extensive set of data was obtained for the TCNQ reduction by Fe(EDTA)$^{2-}$.

Steady-state voltammograms obtained for this ET reaction were further improved by background subtraction and fitted to theoretical curves calculated either for a microdisk geometry or for a uniformly accessible ITIES (Equation 15.12). Kinetic parameters were obtained for different concentrations of organic and aqueous redox species and for a range of pipette radii (~50 to ~350 nm). While the determined α values were close to 0.5 and essentially independent of a and concentrations of redox species, the bimolecular standard rate constants were much larger than the values previously measured for any ET macroscopic polarized interfaces and at micrometer-sized nonpolarized ITIES. More surprisingly, the apparent k^0 increased markedly with decreasing pipette radius (i.e., from ~0.4 cm/s at $a = 300$ nm to ~1.8 cm/s at $a = 50$ nm). The authors noted that this behavior is at variance with existing ET theories. They eliminated the possibilities of the recessed interface, incorrectly determined pipette radius, and other artifacts by thoroughly characterizing nanopipettes (including SECM experiments with conductive and insulating substrates; see Chapter 18). However, other factors, including the lack of the unique fit between the theoretical and experimental steady-state voltammograms and possible double-layer effects, may have affected the determined kinetic parameters. The scarcity of the available literature data precluded the comparison of the measured ET rates to those determined for the same ET reactions at larger ITIES.

15.4.3.2 Ion-Transfer Kinetics

The rates of most IT processes are too fast to be accurately measured at either macroscopic- or micrometer-sized interfaces. In contrast, the mass-transfer coefficient for a 10 nm radius pipette is ≥ 10 cm/s (assuming $D = 10^{-5}$ cm^2/s), and the corresponding upper limit for the determinable heterogeneous rate constant is ~50 cm/s. The first IT kinetics studied at the nanopipette-supported ITIES was that of potassium transfer from the aqueous filling solution to DCE facilitated by dibenzo-18-crown-6 (DB18C6)[3]:

$$K+(w)+DB18C6(DCE) \rightarrow \left[K^+DB18C6 \right](DCE) \qquad (15.13)$$

The mass-transfer rate was sufficiently high to measure the rate constant of potassium transfer under steady-state conditions using pipettes with $a \leq 250$ nm. Assuming uniform accessibility of the ITIES, k^0 and α values were found by fitting the experimental data to Equation 15.12. Additionally, the kinetic parameters were evaluated by the three-point method.[60] A number of voltammograms obtained at 5–250 nm pipettes yielded $k^0 = 1.3 \pm 0.6$ cm/s and $\alpha = 0.4 \pm 0.1$, and no apparent correlation was found between the measured rate constant and the pipette radius. The determined k^0 was significantly higher than rate constants measured for this reaction at larger interfaces, thus providing the first evidence that the IT rates may be faster than it appeared from earlier experiments.

Yuan and Shao investigated the kinetics of several alkali metal ITs facilitated by DB18C6 at the water/DCE nanointerfaces.[45] Their measurements yielded the rate constant for potassium transfer similar to that reported in Ref. [3]. Well-shaped steady-state voltammograms were also obtained for other alkali metal cations, but the kinetic parameters determined for Li$^+$, Rb$^+$, and Cs$^+$ showed significant correlation with the pipette radius. A similar approach was used by the same group to measure the kinetics of alkali metal transfers across the water/DCE interface facilitated with N-(2-tosylamino)-isopentyl-monoaza-15-crown-5.[68] The association constants were measured for alkali metal complexes in DCE, and the selectivity of this ionophore was shown to follow the sequence $Na^+ > Li^+ > K^+ > Rb^+ > Cs^+$. The standard rate constants determined from steady-state voltammograms were similar for all studied cations (~0.5 cm/s) and somewhat lower than those measured with DB18C6 for K^+ and Na^+.[3,45]

Two sources of error, which may have affected the accuracy of the results reported in Refs. [3,45], were identified in later studies. One of them is the lack of silanization of the outer pipette wall. The formation of a thin aqueous film on the hydrophilic glass surface may have resulted in the true ITIES area significantly larger than that evaluated from the diffusion limiting current (see Section 15.3.2). This should have resulted in overestimated values of the mass-transfer coefficient and standard rate constants calculated from the dimensionless parameter $\lambda = k^0/m$. Another source of error—the uncertainty in fitting experimental IT voltammograms to the theory—is discussed in Section 15.4.4.

The first attempt to measure kinetics of two rapid simple ITs at the water/DCE interface formed at the tip of a nanopipette was reported by Cai et al.[40] Employing 10–300 nm radius pipettes, $k^0 = 2.3$ cm/s was found from quasi-steady-state voltammograms of the TEA^+ transfer from DCE to the aqueous filling solution, and a similar value ($k^0 = 2.1$ cm/s) was obtained by steady-state voltammetry for the reverse reaction. The pipettes were silanized and characterized by SECM and voltammetry. The fit between the theory and experimental curves was very good; however, the corresponding transfer coefficients, $\alpha = 0.70$ and $\beta = 0.60$, were larger than 0.5, and their sum was larger than the theoretically expected value of 1.0. Additionally, a noticeable inverse correlation between the k^0 and the a suggests that the data are not completely reliable. A slightly lower rate constant ($k^0 = 1.5 \pm 0.3$ cm/s) and $\alpha = 0.60 \pm 0.04$ were obtained for the tetramethylammonium transfer. One of the possible sources of error in these measurements was that the diffusion of the transferred ions inside the pipette was not taken into account.

Jing et al. studied IT kinetics at the nanoscopic water/n-octanol (OC) interface, which is often used as a model system to mimic CT processes through biomembranes.[41] Although the potential window (~400 mV) was narrower than that observed with the same supporting electrolytes at the water/DCE interface, it was possible to obtain sufficiently well-defined steady-state voltammograms to determine partition coefficients and standard potentials for the transfers of tetraphenylarsonium, TBA^+, and laurate from OC to water. These results suggest the possibility of a more straightforward approach to investigating the transfers of ionizable drugs through cell membranes. Kinetic parameters were determined for laurate transfer at the water/OC and water/DCE nanointerfaces, and the rate constant measured at the former was about six times slower.

Li et al.[42] measured very large rate constants for simple transfers of TEA^+ (110 ± 23 cm/s) and ClO_4^- (35 ± 8 cm/s) and facilitated transfer of K^+ with DB18C6 (95 ± 31 cm/s) from extremely small (1 nm $\leq a \leq 5$ nm) water-filled pipettes to DCE. It was noticed later that the reported k^0 values may have been significantly overestimated because of problems with the data analysis and lack of pipette silanization.[43] Additional factors that could have increased the apparent IT rate constant are double-layer effects and possible deviations from the conventional theory at ultrasmall pipettes.

15.4.3.3 Some Experimental Issues

A number of experimental problems may plague kinetic experiments at the nano-ITIES. A hard question is whether the phase boundary is flat and located exactly at the pipette tip. In early studies at micrometer-sized pipettes, the assumption was that the water/organic interface is convex (i.e., protruding into the external liquid phase).[52] In situ microscopy showed that a micropipette-supported ITIES tends to be flat when no external pressure is applied[49]; however, a nanopipette tip is too small to be visualized by optical microscopy. The SECM approach curves obtained with nanopipette tips suggested that the ITIES is essentially flat and not recessed (see Section 15.3.2 and Chapter 18).

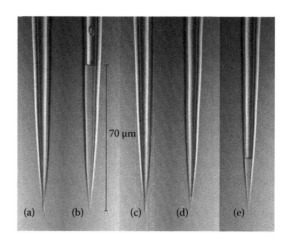

FIGURE 15.10 Sequential ingress/egress of water in a DCE-filled nanopipette. (a) Initial immersion, $E = +600$ mV; (b) ingress of water after the voltage was stepped to -100 mV and then to $+90$ mV; (c) complete egress of water at $E = +600$ mV; (d) same as (b) but with a shorter step time at $E = -100$ mV; (e) the voltage was stepped again to -100 mV and then back to $+90$ mV. The aperture radius was ~300 nm. The pipette was filled with 10 mM THATPBCl in DCE and immersed in 10 mM KF aqueous solution. (Adapted with permission Laforge, F.O., Carpino, J., Rotenberg, S. A., and Mirkin, M.V., Electrochemical attosyringe, *Proc. Natl. Acad. Sci. U.S.A.*, 2007, 104, 11895–11900. Copyright 2007 National Academy of Science, U.S.A.)

The shape and position of very small (e.g., <5 nm) ITIES cannot be characterized at this time, and such interfaces are sometimes assumed to be essentially hemispherical.[42]

It was also noticed that the outer organic solvent can get drawn inside a water-filled pipette and the thickness of the organic phase can vary with the voltage applied across the ITIES (Figure 15.10).[56a] Similarly, one can fill a pipette with water and draw organic solution into it (and then eject it) by applying a suitable interfacial voltage. The movement of a micrometer-sized ITIES that accompanied IT was discussed in detail by Dale and Unwin.[69] The displacement of the liquid phase boundary during a voltammetric experiment can be a source of major distortions. In addition to changing the interfacial area, it may affect the mass-transfer rate and cause a significant increase in pipette resistance (if resistive organic solution is drawn inside the narrow shaft of the pipette). With no obvious way to detect or exclude the motion of a nano-ITIES, one has to assume that a well-shaped, sigmoidal, and retraceable voltammogram with a flat and stable diffusion limiting current is indicative of a stationary phase boundary.

Another source of uncertainty in defining the exact position of the ITIES is the glass roughness. This issue is especially important in the case of ultrasmall (e.g., $a \leq 5$ nm[42]) pipettes, where even miniscule roughness of the glass surface should be comparable to the radius of the orifice. It was shown recently that the roughness of the pipette tip can be reduced by mechanical polishing[48a] or by FIB milling.[48b]

Kinetic analysis of nanopipette voltammograms can be further complicated by electrostatic effects produced by the negatively charged inner glass wall.[43b] The surface charge can influence ion transport along the wall electrostatically and also affect the IT rate at the edge of the nano-ITIES. Various effects of the surface charge and electrical double layer present at the inner wall, including current rectification,[70] accumulation, or depletion of ions near the orifice[71] and electrostatically gated transport[72] have been reported for nanopipettes and glass nanopore electrodes immersed in an aqueous electrolyte solution (i.e., single-phase systems with no liquid/liquid interface).[55a] In voltammetry across the nanoscale ITIES, the interfacial transfer of an ion at a few millimolars bulk concentration produces a pA-range current, which is much lower than that in single-phase experiments. Typically, the potential gradient and the ohmic potential drop inside the pipette are too small for significant electromigration or electro-osmotic flow along its charged inner wall.[43] The electrostatic effects can be more important for smaller nanopipettes, for example, $a < 10$ nm, and more experiments are needed to check whether they can significantly influence kinetic parameters determined from IT voltammograms. Also, double-layer effects at the nanoscopic ITIES, where the diffusion layer thickness is comparable to that of the diffuse double layer, may result in deviations from the conventional electrochemical theory, as discussed for solid nanoelectrodes (see Chapter 2).

15.4.4 COMMON ION VOLTAMMETRY

A recently identified problem in the kinetic analysis of steady-state IT and ET voltammograms is the lack of the unique fit between the experimental and theoretical curves.

It was shown previously that a steady-state voltammogram is quasi-reversible when the dimensionless standard rate constant, $\lambda = k^0 a/D$, is $\lesssim 10$ and kinetic parameters of interfacial CT reaction can, in principle, be extracted by fitting such a curve to the theory.[20,59,60] However, this approach works well only if the shape of the voltammogram depends strongly on the values of kinetic parameters. For near-Nernstian CT processes (i.e., $1 \lesssim \lambda$), the same experimental voltammogram can be fit to the theory using different combinations of kinetic parameters with only minor adjustments in the formal potential value. For example, a satisfactory fit between experimental and theoretical voltammograms was obtained for IT of TEA^+ at a nanopipette-supported ITIES with different k^0 (from 1.2 to 4.3 cm/s) and α (from 0.2 to 0.7) values.[43] The same problem can lead to large uncertainties in kinetic parameters extracted from near-Nernstian steady-state ET voltammograms.[73] It may have compromised the results of previously reported steady-state measurements of rapid ET kinetics at metal electrodes.

A simple modification was shown to essentially eliminate this problem and improve the accuracy and precision of CT kinetic measurements by steady-state voltammetry. For IT at the liquid/liquid interface, this approach—common ion voltammetry—is based on the addition of a transferable ion to both liquid phases, that is, the filling solution inside the pipette and the external solution.[43] The advantages of common ion voltammetry over the conventional protocol, in which a transferable ion is initially present only in one liquid phase, stem from the availability of two waves corresponding to the ingress of the common ion into the pipette and its egress to the external solution (positive and negative waves in Figure 15.11, respectively). If the D_1 and D_2 values are known, geometric parameters can be evaluated from the two limiting currents in the same voltammogram using Equations 15.7 and 15.8. Then, the unique combination of the kinetic parameters can be found by fitting an experimental voltammogram to the theory expressed by the following equation: [43a]

$$\frac{i}{i_2} = \frac{1}{m_2/m_1 + m_2/k_b + k_f/k_b} \left(\frac{k_f}{k_b} - \frac{c_2^*}{c_1^*} \right) \tag{15.14}$$

where i_2 is the diffusion limiting current for ion ingress (Equation 15.7) shown in Figure 15.11; k_f and k_b are the heterogeneous rate constants given by the Butler–Volmer-type model; c_1 and c_2 are the bulk concentrations of the transferred ion in the outer and inner solutions, respectively; and $m_1 = (4f(\theta)D_1)/(\pi a)$ and $m_2 = (4xD_2)/(\pi a)$ are the mass-transfer coefficients representing ingress and egress transfers of the common ion.

The precision is further enhanced by directly determining the formal potential ($\Delta\varphi^{0\prime}$) from the potential of zero current (equilibrium potential, $\Delta\varphi_{eq}$) given by the Nernst equation

$$\Delta\varphi_{eq} = \Delta\varphi^{0\prime} + \frac{RT}{zF} \ln \frac{c_2^*}{c_1^*} \tag{15.15}$$

instead of finding it from the fit of a conventional IT voltammogram to the theory.

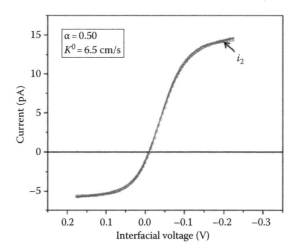

FIGURE 15.11 Steady-state common ion voltammogram of TEA$^+$ transfer across the DCE/water interface obtained with a 19 nm radius pipette (red curve). The best theoretical fit (symbols) to the experimental curve was calculated from Equation 15.14. The scan rate was 10 mV/s. (Reprinted with permission from Wang, Y., Velmurugan, J., Mirkin, M.V., Rodgers, P.J., Kim, J., and Amemiya, S., Kinetic study of rapid transfer of tetraethylammonium at the 1,2-dichloroethane/water interface by nanopipette voltammetry of common ion, *Anal. Chem.*, 2010, 82, 77–83. Copyright 2010 American Chemical Society.)

The unique fit of the experimental steady-state voltammogram to the theory can be obtained when both ingress and egress IT waves are quasi-reversible.[43] The asymmetry of the diffusion field results in different extents of reversibility (i.e., kinetic vs. diffusion control) of the ion ingress and egress processes, which can be assessed using two dimensionless parameters: $\lambda_1 = k^0/m_1$ and $\lambda_2 = k^0/m_2$. If the ratio of diffusion coefficients, D_2/D_1, is not very far from the unity, reliable kinetic parameters can be extracted from a common ion voltammogram if both λ_{ing} and λ_{eg} are smaller than 10. In this way, the unique combination of the kinetic parameters, $\alpha = 0.50$ and $k^0 = 6.5$ cm/s, was obtained for the TEA$^+$ transfer across the water/DCE interface from the best fit shown in Figure 15.11.[43b] Similar values ($k^0 = 6.1 \pm 0.9$ cm/s and $\alpha = 0.49 \pm 0.09$) were determined with various pipettes (9.7 nm $\leq a \leq$ 33 nm) at different TEA$^+$ concentrations and essentially independent of a. These k^0 values are much higher than those determined previously from conventional nanopipette voltammograms[40] ($k^0 \sim 2$ cm/s). In the latter case, the analysis of a nearly reversible voltammogram with λ_{ing} (or λ_{eg}) > 1 did not give a unique combination of kinetic and thermodynamic parameters for rapid IT. An additional source of error was the neglected effect of ion diffusion in the internal solution.

Common ion voltammetry was employed to study IT reactions at the water/IL interface.[44] Kinetic measurements at such interfaces are challenging because of slow mass-transfer rates in IL. For instance, the IL employed in Ref. [44], [THTDP$^+$][C$_4$C$_4$N$^-$], is ~700 times more viscous than water. Slow mass transfer in the IL phase results in a low diffusion current and at the same time necessitates the use of small nanopipettes and very low potential sweep rates to attain a steady state. Kinetic parameters of the TBA$^+$ transfer ($k^0 = 0.12 \pm 0.02$ cm/s and $\alpha = 0.50 \pm 0.06$) were extracted by fitting common ion voltammograms to the theory (Equation 15.14). Because of the large ratio of diffusion coefficients ($D_1/D_2 = 275$), the λ_2 values were much larger than the corresponding λ_1 values; and almost all λ_2 values were $\geq$10. However, unlike water/organic interface, where $D_1/D_2 \approx 1$ and $\lambda_2 \geq 10$ corresponds to an essentially Nernstian IT, the ingress waves at the water/IL interface were quasi-reversible for $\lambda_2 \leq 50$. Several factors that could affect the results of kinetic experiments at the water/IL nanointerface were investigated. Very similar IT rate constants were determined for TBA$^+$ and similarly sized but asymmetric C$_8$ mim$^+$ ion.

This result was taken as an evidence that ionic adsorption is not a major rate-determining factor in the studied system. The comparison of the diffusion currents produced by the egress of cations and anions from the water-filled nanopipettes ($a \geq 11$ nm) to IL showed that the mass transfer inside the pipette shaft is not significantly affected by migration and other electrostatic effects. No correlation was found between the interfacial size and IT kinetics, which would be indicative of double-layer effects.

A conceptually similar approach to measurements of rapid ET kinetics at nanoelectrodes requires both oxidized and reduced forms of redox species to be simultaneously present in solution.[73] An experimental voltammogram comprising steady-state oxidation and reduction waves can be used to determine mass-transfer coefficients of the reduced and oxidized form of electroactive species. The analysis of such curves should improve accuracy and precision of the evaluation of k^0 and α at nanoelectrodes.

15.5 SPECIAL NANOELECTROCHEMICAL PROBES

Several types of nanoelectrochemical probes surveyed in this section differ from those discussed earlier by their geometry and/or are designed for specific applications.

15.5.1 NANO-TLC

Nanometer-sized TLCs fabricated by several groups enabled electrochemical experiments in ultrasmall volumes, in which the total number of redox molecules could be varied between one and a few thousands. The pioneering studies conducted by the Bard group were aimed at observing single molecular events.[30] In those experiments, a Pt nanotip was recessed inside a small compartment formed within the wax insulator. A nano-TLC was produced by pushing such an electrode against a conducting surface. The redox cycling in such a cell is conceptually similar to the positive feedback in SECM: the redox mediator was oxidized (or reduced) at the nanoelectrode surface and regenerated via the reduction (or oxidation at the conductive substrate). With the TLC thickness sufficiently small (a few nanometers), the feedback process can provide sufficient amplification to measure the current produced by oxidation/reduction of a single redox molecule.[30] The voltammograms and current–time dependencies recorded in Ref. [30] exhibited high-amplitude current fluctuations, which were attributed to the migration of single electroactive molecules in and out of the TLC through a tiny hole in the wax insulator (for the discussion of single-molecule/single-nanoparticle electrochemistry, see Chapter 8).

In Ref. [33], nano-TLCs were prepared using recessed disk-type Pt electrodes ($a \geq 5$ nm; recess depth, ≥ 1 nm). A TLC was produced by immersing a recessed nanoelectrode in solution to fill its cavity and then transferring it into the pool of dry mercury. Because solution in this system was present only in the gap between Pt and Hg, there was no possibility of redox molecules escaping the cavity. Thus, unlike experiments described in Ref. [30], the recorded current did not exhibit significant fluctuations. The analysis of steady-state voltammograms yielded information about mass transfer, adsorption, ET kinetics, and double-layer effects on the nanoscale. The radius and the effective thickness of the TLC were determined from voltammetry and SECM. Although a good agreement was found between the determined TLC thickness and the recess depth of the nanoelectrode, more reliable characterization could probably be attained using the AFM (Section 15.2.3).

Several unusual size-related electrochemical phenomena were observed in nano-TLCs, including the current rectification due to nonpolarizability of the Hg electrode, strong dependence of the response on concentration of supporting electrolyte when the number of ionic species inside TLC becomes too small for the formation of two electrical double layers, and an enhanced voltammetric response to one redox species relative to the other.[33]

Another approach to the preparation of nano-TLCs is to fabricate a nanogap device lithographically (see Chapter 16 for detailed discussion).[64] An advantage of this approach is in precise control of the gap width and, therefore, high reproducibility of nanoelectrochemical experiments. A 50 nm thick nanofluidic cell with a well-defined geometry was used to measure kinetics of $Fc(MeOH)_2$ oxidation with different supporting electrolytes.[64a] While the kinetic parameters determined in KCl solutions were reasonably close to those measured by SECM for FcMeOH,[20] the effects of the nature and concentration of electrolyte on k^0 and i_d values have yet to be clarified. All measurements in Ref. [64a] were made using essentially identical devices because, unlike SECM, neither the electrode radius nor the separation distance in the nano-TLC can be varied easily. A well-defined, smaller (30 nm thick) device was recently fabricated and used for amperometric detection of single molecules[64b]; however, the thickness of lithographically prepared TLCs so far remains significantly larger than those attained using recessed nanoelectrodes.[30,33] This resulted in a lower amplification level and small amplitude electrochemical signal (e.g., ~50 fA).

15.5.2 CARBON-BASED NANOELECTRODES AND PIPETTES

Different approaches to preparing carbon nanoelectrodes have been reported over the last decade.[13,24b,27,74–76] Unlike metal microwires, carbon fibers cannot be easily pulled into glass capillaries, and the reported probes were often prepared by etching.[13,74] Although etching can yield electrodes with extremely small effective radius (e.g., ~5 nm[13]), producing a carbon nanoprobe with a well-defined geometry is not straightforward.

Another approach is based on chemical vapor deposition (CVD) of carbon inside quartz nanopipettes,[27,75–77] typically using methane or butane as a carbon source. The Bau and Gogotsi groups[75] developed detailed protocols for depositing different amounts of carbon on the inner pipette wall by varying a number of experimental parameters. The thickness and distribution of carbon layer depend on the CVD time, the composition of the gas mixture, the pipette shape, and tip diameter.[75b] A relatively short CVD time resulted in the pipettes with an open path in the middle (Figure 15.12a). Conversely, by increasing CVD time and using a higher methane to argon ratio, one can increase the carbon layer thickness and close the path, leaving a cavity at the very end of tapered shaft (Figure 15.12b). Even longer deposition time (e.g., ~3 h) produced pipettes completely filled with carbon (Figure 15.12c).

Three types of carbon probes shown in Figure 15.12 have been employed in different experiments. Open carbon pipettes were inserted inside biological cells and used for concurrent intracellular injection and electrical measurements.[78] The pipettes in Ref. [78] had a relatively long (submicrometer to micron) piece of tapered carbon tube exposed to solution. Such a large conductive surface can be useful for potentiometric measurements, but not for amperometric experiments and applications requiring high z-axis resolution. A probe with the carbon layer confined to the inside of the quartz pipette (Figure 15.12a) is more suitable for amperometric applications and can be used as an SECM tip.[77] A possibility of using open carbon pipettes as rectification-based sensors and resistive-pulse sensors has been explored recently.[79]

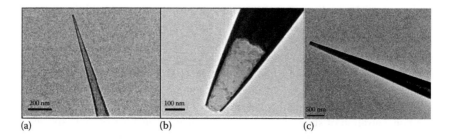

(a)   (b)   (c)

FIGURE 15.12 TEM images of carbon-coated nanopipettes. (a) An open pipette, $a = 4$ nm, $\theta = 2°$; (b) a pipette with a nanocavity, $a = 33$ nm, $\theta = 8°$, and the cavity depth equivalent to ~12a; and (c) a pipette completely filled with carbon, $a = 50$ nm.

A carbon pipette with a nanocavity (Figure 15.12b) can be either prepared directly, as discussed earlier, or obtained by annealing a nanoelectrode for a few seconds in the oven to oxidize carbon in air.[27] The reported cavity depth was equivalent to ~4 to ~200 pipette radii depending on tip preparation conditions. The cavity can be filled with various solid or liquid agents to produce a sensor, for example, a nanosensor for reactive oxygen and nitrogen species was prepared by electrodepositing Pt black.[27] Such a probe can be used for amperometric and potentiometric measurements inside living cells. A carbon pipette with a nanocavity was also used for sampling attoliter-to-picoliter volumes of fluids and determining redox species by voltammetry and coulometry.[77] Very fast mass transport inside the carbon-coated nanocavity allows for rapid exhaustive electrolysis of the sampled material. The signal produced by oxidation/reduction inside the cavity can be significantly higher than the steady-state current to the orifice of the same pipette. The developed device is potentially useful for solution sampling from biological cells, micropores, and other microscopic objects.

An advantage of a flat carbon nanoelectrode (Figure 15.12c) over a wire-in-glass disk-type electrode with a similar radius of the conductive core is in a much smaller RG, which can be as small as that of a pulled quartz pipette (typically, <2). The resulting small physical size of the carbon probe is essential for intracellular measurements and high-resolution imaging. Thus, Takahashi et al.[76a] used carbon tips to simultaneously obtain high-resolution topographical and electrochemical images of living cells. Actis et al.[76b] showed that carbon nanoelectrodes can be inserted into individual cells both in tissue and in isolated cells to perform electrochemical measurements with minimal disruption to cell function. However, producing well-shaped, flat C nanoelectrodes is challenging. Polishing such electrodes is much harder than similarly sized metal nanoelectrodes (Section 15.2.1), and the utility of FIB milling has yet to be demonstrated. A very simple methodology for fabricating nanoelectrodes developed in Ref. [76] is based on the pyrolytic deposition of C from butane, which was passed through the quartz pipette by using a Tygon tube. The taper of the pipette was inserted into another quartz capillary, which was filled with N_2 to prevent oxidation of the carbon structure formed and bending of the capillary by high temperature. To form a pyrolytic carbon layer inside the capillary, the pipette taper was then heated with a Bunsen burner for a few seconds. Although very easy and efficient (takes ~1 min per electrode[76b]), this method is not likely to yield electrodes with well-defined geometry, as can be seen from SECM approach curves presented in Ref. [76]. Multifunctional carbon nanoelectrodes with well-defined geometry were fabricated recently by CVD of parylene, followed by thermal pyrolysis and FIB milling.[24b]

15.5.3 Dual Nanoelectrodes and Pipettes

Electrochemical nanoprobes with two closely separated sensing elements have been fabricated and used either for generation/collection (G/C) experiments or for simultaneous recording of two different signals. Such a probe can consist of either two solid nanoelectrodes (e.g., two disk-type electrodes or a disk and a concentric ring)[27b,80–83] or two liquid–liquid nanointerfaces[84] or a combination of a nanoelectrode and an open nanopipette.[27b,85,86] Dual probes are produced either by using θ-tubing or by forming a concentric conducting ring surrounding a disk electrode. Most dual solid electrodes and nanoelectrode–nanopipette probes were used as tips in SECM or SICM; these publications are surveyed in Chapter 18. Dual pipettes (or θ-pipettes) were also employed in scanning probe experiments, including different versions of reactant delivery in SICM[87] and SECCM (Chapter 19).

Shao et al.[84] developed the θ-pipette-based G/C technique as a tool for studying heterogeneous IT reactions and homogeneous chemical reactions of ionic species in solution. They fabricated submicrometer- and nanometer-sized dual pipettes from borosilicate θ-tubing using a laser puller and formed two independent ITIES at the tip of such a device. Figure 15.13a illustrates the geometry, and Figure 15.13b shows an SEM image of a θ-shaped tip. Typically, both barrels of the dual pipette are filled with water.

If one of the barrels (*generator*) contains a cation, it can be transferred to the outer organic solvent by biasing this pipette at a sufficiently positive potential (E_g). A significant fraction of ejected

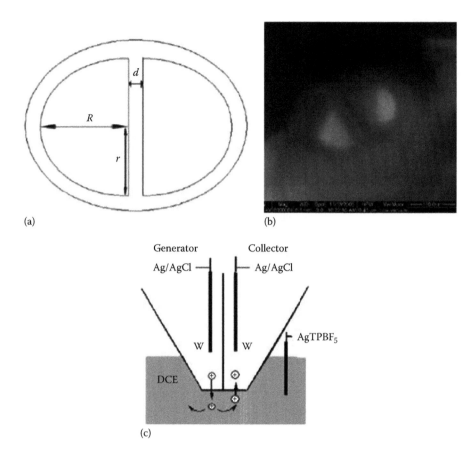

FIGURE 15.13   (a) Schematic representation of the θ-pipette, (b) an SEM image of the pipette with $R = 65$ nm and $d = 36$ nm, and (c) the G/C IT process at two liquid/liquid interfaces supported at the pipette tip. (Reprinted with permission from Hu, S., Xie, X., and Meng, P. et al., Fabrication and characterization of submicrometer- and nanometer-sized double-barrel pipettes, *Anal. Chem.*, 2006, 78, 7034–7039. Copyright 2006 American Chemical Society.)

cations reaches the negatively biased second pipette (*collector*) and gets transferred back into the aqueous phase (Figure 15.13c). The collection efficiency, $\eta = i_c/i_g$ (i_c is the collector current and i_g is the generator current), can be used to investigate CT and chemical reactions occurring in the space between two channels.[84] The η value depends on the collection potential and the geometry of the θ-pipette. In the absence of chemical reactions in solution, its maximum value, η_{max}, is obtained when all ions reaching the opening of the collector pipette are transferred into it. The η_{max} value depends only on the normalized distance between the centers of two barrels. The K$^+$ transfer at the W/DCE interface facilitated by DB18C6 was used as the model system to study the effects of geometric parameters of the pipette on collection efficiency. The larger, submicrometer-sized pipettes showed higher collection efficiency, while nanometer-sized pipettes produced better results for a system without supporting electrolyte.[84d]

The dual-pipette technique allows quantitative separation of different IT and ET processes simultaneously occurring at the liquid/liquid interface (e.g., simple transfer of potassium, facilitated transfer of the same ion with a crown ether, and IT of supporting electrolyte). It can also be used to overcome potential window limitations and study numerous important reactions occurring at high positive or negative voltages applied across an ITIES (e.g., transfers of alkali metals from water to organic media).

15.5.4 Electrochemical Attosyringe

Laforge et al.[56a] showed that a nanopipette can also be used as an *electrochemical attosyringe* for controlled fluid delivery. The prepared nanopipette was filled with an organic solvent and immersed in an aqueous solution. The ITIES at the pipette orifice was shown to move in response to variations in applied voltage. Water entered the pipette when the potential of the inner organic solution was made negative and was expelled at positive potentials (Figure 15.10). This phenomenon was used to sample and deliver attoliter-to-picoliter volumes of fluorescent dyes into human breast cells in culture. The injection volumes could be monitored and evaluated by measuring the pipette resistance and/or current vs. potential curves. Compared to other existing microinjectors, this device is inexpensive and easy to fabricate and use; it can be made very small and used repeatedly. Potential applications are in cell biology, nanolithography, and microfluidics.

Actis et al.[88] used an attosyringe as a scanning probe to sample small amounts of total RNA and mitochondrial DNA from a single cell. This approach, which the authors called *single-cell nanobiopsy*, offers the subcellular resolution and a minimal disruption of the cell function due to a small pipette size. The reported cellular survival rate was higher than 70%.

15.6 SUMMARY AND OUTLOOK

The use of nanometer-sized electrodes and liquid nanointerfaces enabled experimentalists to study a number of systems and processes that would not be accessible by macroscopic electrochemical probes. Some reported results, including incredibly fast rate constants of CT reactions,[10b,42] multiple nucleation of metal on extremely small (~5 nm) electrodes,[13] unusual transport phenomena,[35] and incompletely charged electrical double layer,[33] may be hard to reconcile with the existing theory. High spatial resolution of nanoelectrode experiments enables localized measurements of physicochemical properties that can challenge the consensus based on macroscopic electrochemical measurements.[89] In such extreme cases as well as in more routine kinetic experiments, the thorough characterization of nanoelectrodes and nanopipettes is essential to avoid misleading or highly inaccurate results. Unfortunately, the data obtained at nanoelectrodes characterized only by extracting the effective radius value from a steady-state voltammogram continue to be published. In addition to double-layer effects and other sources of deviations from conventional electrochemical theory discussed in Chapters 1 and 2, a number of experimental factors, such as the effects of the metal/insulator boundary and charged inner wall of a nanopipette on interfacial CT reactions, have to be better understood to ensure meaningful data analysis.

ACKNOWLEDGMENTS

Our research on nanoelectrochemistry has been supported by the National Science Foundation (CHE-1026582, CHE-1300158, and CBET-1251232) and AFOSR MURI (FA9550-14-1-0003).

REFERENCES

1. (a) Murray, R. W. 2008. Nanoelectrochemistry: Metal nanoparticles, nanoelectrodes, and nanopores. *Chem. Rev.* 108:2688–2720. (b) Oja, S. M., Wood, M., and Zhang, B. 2013. Nanoscale electrochemistry. *Anal. Chem.* 85:473–486. (c) Chen, S. and Liu, Y. 2014. Electrochemistry at nanometer-sized electrodes. *Phys. Chem. Chem. Phys.* 16:635–652.
2. (a) Wehmeyer, K. R., Deakin, M. R., and Wightman, R. M. 1985. Electroanalytical properties of band electrodes of submicrometer width. *Anal. Chem.* 57:1913–1916. (b) Bond, A. M. 1986. Theory and experimental characterization of linear gold microelectrodes with submicrometer thickness. *J. Phys. Chem.* 90:2911–2917. (c) Morris, R., Franta, D. J., and White, H. S. 1987. Electrochemistry at platinum bane electrodes of width approaching molecular dimensions: Breakdown of transport equations at very small electrodes. *J. Phys. Chem.* 91:3559–3564.

3. Shao, Y. and Mirkin, M. V. 1997. Fast kinetic measurements with nanometer-sized pipettes. Transfer of potassium ion from water into dichloroethane facilitated by dibenzo-18-crown-6. *J. Am. Chem. Soc.* 119:8103–8104.

4. (a) Arrigan, D. W. M. 2004. Nanoelectrodes, nanoelectrode arrays and their applications. *Analyst* 129:1157–1165. (b) Cox, J. T. and Zhang, B. 2012. Nanoelectrodes: Recent advances and new directions. *Annu. Rev. Anal. Chem.* 5:253–272.

5. Shen, M., Ishimatsu, R., Kim, J., and Amemiya, S. 2012. Quantitative imaging of ion transport through single nanopores by high-resolution scanning electrochemical microscopy. *J. Am. Chem. Soc.* 134:9856–9859.

6. Li, Y., Cox, J. T., and Zhang, B. 2010. Electrochemical response and electrocatalysis at single Au nanoparticles. *J. Am. Chem. Soc.* 132:3047–3054.

7. Velmurugan, J., Noël, J.-M., Nogala, W., and Mirkin, M. V. 2012. Nucleation and growth of metal on nanoelectrodes. *Chem. Sci.* 3:3307–3314.

8. Gewirth, A. A., Craston, D. H., and Bard, A. J. 1989. Fabrication and characterization of microtips for in situ scanning tunneling microscopy. *J. Electroanal. Chem.* 261:477–482.

9. Nagahara, L. A., Thundat, T., and Lindsay, S. M. 1989. Preparation and characterization of STM Tips for electrochemical studies. *Rev. Sci. Instrum.* 60:3128–3130.

10. (a) Penner, R. M., Heben, M. J., and Lewis, N. S. 1989. Preparation and electrochemical characterization of conical and hemispherical ultramicroelectrodes. *Anal. Chem.* 61:1630–1636. (b) Penner, R. M., Heben, M. J., Longin, T.L., and Lewis, N. S. 1990. Fabrication and use of nanometer-sized electrodes in electrochemistry. *Science* 250:1118–1121.

11. Sun, P., Zhang, Z., Guo, J., and Shao, Y. 2001. Fabrication of nanometer-sized electrodes and tips for scanning electrochemical microscopy. *Anal. Chem.* 73:5346–5351.

12. Slevin, C. J., Gray, N. J., Macpherson, J. V., Webb, M. A., and Unwin, P. R. 1999. Fabrication and characterisation of nanometre-sized platinum electrodes for voltammetric analysis and imaging. *Electrochem. Commun.* 1:282–288.

13. Chen, S. L. and Kucernak, A. 2002. Fabrication of carbon microelectrodes with an effective radius of 1 nm. *Electrochem. Commun.* 4:80–85.

14. (a) Mirkin, M. V., Fan, F.-R. F., and Bard, A. J. 1992. Scanning electrochemical microscopy. 13. Evaluation of the tip shapes of nanometer size microelectrodes. *J. Electroanal. Chem.* 328:47–62. (b) Mirkin, M. V., Fan, F.-R. F., and Bard, A. J. 1992. Direct electrochemical measurements inside a 2000 Å—Thick polymer film by scanning electrochemical microscopy. *Science* 257:364–366.

15. Watkins, J. J., Chen, J. Y., White, H. S., Abruna, H. D., Maisonhaute, E., and Amatore, C. 2003. Zeptomole voltammetric detection and electron-transfer rate measurements using platinum electrodes of nanometer dimensions. *Anal. Chem.* 75:3962–3971.

16. (a) Hills, G. J., Schiffrin, D. J., and Thompson, J. 1974. Electrochemical nucleation from molten salts. Part I: Diffusion controlled electrodeposition of silver from alkali molten nitrates. *Electrochim. Acta* 19:657–670. (b) Branco, P. D., Mostany, J., Borrás, C., and Scharifker, B. R. 2009. The current transient for nucleation and diffusion-controlled growth of spherical caps. *J. Solid State Electrochem.* 13:565–571.

17. (a) Wehmeyer, K. R. and Wightman, R. M. 1985. Cyclic voltammetry and anodic stripping voltammetry with mercury ultra-microelectrodes. *Anal. Chem.* 57:1989–1993. (b) Velmurugan, J. and Mirkin, M. V. 2010. Fabrication of nanoelectrodes and metal clusters by electrodeposition. *ChemPhysChem* 11:3011–3017. (c) Mauzeroll, J., Hueske, E. A., and Bard, A. J. 2003. Scanning electrochemical microscopy. 48. Hg/Pt hemispherical ultramicroelectrodes: Fabrication and characterization. *Anal. Chem.* 75:3880–3889.

18. Shao, Y., Mirkin, M. V., Fish, G., Kokotov, S., Palanker, D., and Lewis, A. 1997. Nanometer-sized electrochemical sensors. *Anal. Chem.* 69:1627–1634.

19. Zhang, B., Galusha, J., Shiozawa, P. G., Wang, G. L., Bergren, A. J., Jones, R. M., White, R. J., Ervin E. N., Cauley, C. C., and White, H. S. 2007. Bench-top method for fabricating glass-sealed nanodisk electrodes, glass nanopore electrodes, and glass nanopore membranes of controlled size. *Anal. Chem.* 79:4778–4787.

20. Sun, P. and Mirkin, M. V. 2006. Kinetics of electron transfer reactions at nanoelectrodes. *Anal. Chem.* 78:6526–6534.

21. (a) Velmurugan, J., Sun, P., and Mirkin, M. V. 2009. Scanning electrochemical microscopy with gold nanotips: The effect of electrode material on electron transfer rates. *J. Phys. Chem. C* 113:459–464. (b) Noël, J.-M., Velmurugan, J., Gökmeşe, E., and Mirkin, M. V. 2013. Fabrication, characterization and chemical etching of Ag nanoelectrodes. *J. Solid State Electrochem.* 17:385–389.

22. Li, Y., Bergman, D. and Zhang, B. 2009. Preparation and electrochemical response of 1–3 nm Pt disk electrodes. *Anal. Chem.* 81:5496–5502.
23. (a) Katemann, B. B. and Schuhmann, W. 2002. Fabrication and characterization of needle-type Pt-disk nanoelectrodes. *Electroanalysis* 14:22–28. (b) Mezour, M. A., Morin, M., and Mauzeroll, J. 2011. Fabrication and characterization of laser pulled platinum microelectrodes with controlled geometry. *Anal. Chem.* 83:2378–2382.
24. (a) Nioradze, N., Chen, R., Kim, J., Shen, M., Santhosh, P., and Amemiya, S. 2013. Origins of nanoscale damage to glass-sealed platinum electrodes with submicrometer and nanometer size. *Anal. Chem.* 85:6198–6202. (b) Thakar, R., Weber, A. E., Morris, C. A., Baker, L. A. 2013. Multifunctional carbon nanoelectrodes fabricated by focused ion beam milling. *Analyst* 138:5973–5982.
25. Jena, B. K., Percival, S. J., and Zhang, B. 2010. Au disk nanoelectrode by electrochemical deposition in a nanopore. *Anal. Chem.* 82:6737–6743.
26. Wang, Y., Noel, J.-M., Velmurugan, J., Nogala, W., Mirkin, M. V., Lu, C., Guille Collignon, M., Lemaître, F., and Amatore, C. 2012. Nanoelectrodes for determination of reactive oxygen and nitrogen species inside murine macrophages. *Proc. Natl. Acad. Sci. U. S. A.* 109:11534–11539.
27. Hu, K., Gao, Y., Wang, Y. et al. 2013. Platinized carbon nanoelectrodes as potentiometric and amperometric SECM probes. *J. Solid State Electrochem.* 17:2971–2977.
28. (a) Baranski, A. S. 1991. On possible systematic errors in determinations of charge transfer kinetics at very small electrodes. *J. Electroanal. Chem.* 307:287–292. (b) Oldham, K. B. 1992. A hole can serve as a microelectrode. *Anal. Chem.* 64:646–651.
29. Nogala, W., Velmurugan, J., and Mirkin, M. V. 2012. Atomic force microscopy of electrochemical nanoelectrodes. *Anal. Chem.* 84:5192–5197.
30. (a) Fan, F.-R. F. and Bard, A. J. 1995. Electrochemical detection of single molecules. *Science* 267: 871–874. (b) Fan, F. R. F., Kwak, J., and Bard, A. J. 1996. Single molecule electrochemistry. *J. Am. Chem. Soc.* 118:9669–9675.
31. Sun, P. and Mirkin, M. V. 2007. Scanning electrochemical microscopy with slightly recessed nanotips. *Anal. Chem.* 79:5809–5816.
32. (a) Zhang, B., Zhang, Y. H., and White, H. S. 2004. The nanopore electrode. *Anal. Chem.* 76: 6229–6238. (b) Zhang, B., Zhang, Y. H., and White, H. S. 2006. Steady-state voltammetric response of the nanopore electrode. *Anal. Chem.* 78:477–483. (c) Sun, P. 2010. Cylindrical nanopore electrode and its application to the study of electrochemical reaction in several hundred attoliter volume. *Anal. Chem.* 82:276–281.
33. Sun, P. and Mirkin, M. V. 2008. Electrochemistry of individual molecules in zeptoliter volumes. *J. Am. Chem. Soc.* 130:8241–8250.
34. (a) Bond, A. M., Oldham, K. B., and Zoski, C. G. 1989. Steady-state voltammetry. *Anal. Chim. Acta* 216:177–230. (b) Wightman, R. M. and Wipf, D. O. 1989. Voltammetry at ultramicroelectrodes. In *Electroanalytical Chemistry*, vol. 15, Bard, A. J. (ed.), pp. 267–353. New York: Marcel Dekker. (c) Amatore, C. 1995. Electrochemistry at ultramicroelectrodes. In *Physical Electrochemistry: Principles, Methods, and Applications*, Rubinstein, I. (ed.), pp. 131–208. New York: Marcel Dekker.
35. Sun, Y., Liu, Y., Liang, Z., Xiong, L., Wang, A., and Chen, S. 2009. On the applicability of conventional voltammetric theory to nanoscale electrochemical interfaces. *J. Phys. Chem. C* 113:9878–9883.
36. Zhang, Y., Zhou, J., Lin, L., and Lin, Z. 2008. Determination of electrochemical electron-transfer reaction standard rate constants at nanoelectrodes: Standard rate constants for ferrocenylmethyltrimethylammonium(III)/(II) and hexacyanoferrate(III)/(II). *Electroanalysis* 20:1490–1494.
37. Noël, J.-M., Yu, Y., and Mirkin, M. V. 2013. Dissolution of Pt at moderately negative potentials during oxygen reduction in water and organic media. *Langmuir* 29:1346–1350.
38. Taylor, G. and Girault, H. H. 1986. Ion transfer reactions across a liquid–liquid interface supported on a micropipette tip. *J. Electroanal. Chem.* 208:179–183.
39. Amemiya, S., Wang, Y., and Mirkin, M. V. 2013. Nanoelectrochemistry at the liquid/liquid interfaces. In *Specialist Periodical Reports in Electrochemistry*, vol. 12, Compton, R. and Wadhawan, J. (eds.), pp. 1–43. RSC Publishing.
40. Cai, C. X., Tong, Y. H., and Mirkin, M. V. 2004. Probing rapid ion transfer across nanoscopic liquid-liquid interface. *J. Phys. Chem. B* 108:17872–17878.
41. Jing, P., Zhang, M. Q., Hu, H., Xu, X. D., Liang, Z. W., Li, B., Shen, L., Xie, S. B., Pereira, C. M., and Shao, Y. H. 2006. Ion-transfer reactions at the nanoscopic water/n-octanol interface. *Angew. Chem. Int. Ed.* 45:6861–6864.
42. Li, Q., Xie, S., Liang, Z., Meng, X., Liu, S., Girault, H. H., and Shao, Y. 2009. Fast ion-transfer processes at nanoscopic liquid/liquid interfaces. *Angew. Chem. Int. Ed.* 48:8010–8013.

43. (a) Rodgers, P. J., Amemiya, S., Wang, Y., and Mirkin, M. V. 2010. Nanopipette voltammetry of common ion across a liquid–liquid interface. Theory and limitations in kinetic analysis of nanoelectrode voltammograms. *Anal. Chem.* 82:84–90. (b) Wang, Y., Velmurugan, J., Mirkin, M. V., Rodgers, P. J., Kim, J., and Amemiya, S. 2010. Kinetic study of rapid transfer of tetraethylammonium at the 1,2-dichloroethane/water interface by nanopipette voltammetry of common ion. *Anal. Chem.* 82:77–83.

44. Wang, Y., Kakiuchi, T., Yasui, Y., and Mirkin, M. V. 2010. Kinetics of ion transfer at the ionic liquid/water nanointerface. *J. Am. Chem. Soc.* 132:16945–16952.

45. Yuan, Y. and Shao, Y. H. 2002. Systematic investigation of alkali metal ion transfer across the micro- and nano-water/1,2-dichloroethane interfaces facilitated by dibenzo-18-crown-6. *J. Phys. Chem. B* 106:7809–7814.

46. Cai, C. and Mirkin, M. V. 2006. Electron transfer kinetics at polarized nanoscopic liquid/liquid interfaces. *J. Am. Chem. Soc.* 128:171–179.

47. Brown, K. T. and Flaming, D. G. 1986. *Advanced Micropipette Techniques for Cell Physiology.* New York: Wiley.

48. (a) Elsamadisi, P., Wang, Y., Velmurugan, J., and Mirkin, M. V. 2011. Polished nanopipets: New probes for high-resolution scanning electrochemical microscopy. *Anal. Chem.* 83:671–673. (b) Kim, J., Izadyar, A., Shen, M., Ishimatsu, R., and Amemiya, S. 2014. Ion Permeability of the nuclear pore complex and ion-induced macromolecular permeation as studied by scanning electrochemical and fluorescence microscopy. *Anal. Chem.* 86:2090–2098.

49. Shao, Y. and Mirkin, M. V. 1998. Voltammetry at micropipette electrodes. *Anal. Chem.* 70:3155–3161.

50. Laforge, F. O., Velmurugan, J., Wang, Y., and Mirkin, M. V. 2009. Nanoscale imaging of surface topography and reactivity with the scanning electrochemical microscope (SECM). *Anal. Chem.* 81:3143–3150.

51. Kim, J., Shen, M., Nioradze, N., and Amemiya, S. 2012. Stabilizing nanometer scale tip-to-substrate gaps in scanning electrochemical microscopy using an isothermal chamber for thermal drift suppression. *Anal. Chem.* 84:3489–3492.

52. Beattie, P. D., Delay, A., and Girault, H. H. 1995. Investigation of the kinetics of assisted potassium ion transfer by dibenzo-18-crown-6 at the micro-ITIES by means of steady-state voltammetry. *J. Electroanal. Chem.* 380:167–175.

53. (a) Shoup, D. and Szabo, A. 1984. Influence of insulation geometry on the current at microdisk electrodes. *J. Electroanal. Chem.* 160:27–31. (b) Zoski, C. G. and Mirkin, M. V. 2002. Steady-state limiting currents at finite conical microelectrodes. *Anal. Chem.* 74:1986–1992.

54. Rodgers, P. J. and Amemiya, S. 2007. Cyclic voltammetry at micropipette electrodes for the study of ion-transfer kinetics at liquid/liquid interfaces. *Anal. Chem.* 79:9276–9285.

55. (a) Morris, C., Friedman, A. K., and Baker, L. A. 2010. Applications of nanopipettes in the analytical sciences. *Analyst* 135:2190–2202. (b) Chen, C.-C., Zhou, Y., and Baker, L. A. 2012. Scanning ion conductance microscopy. *Annu. Rev. Anal. Chem.* 5:207–228.

56. (a) Laforge, F. O., Carpino, J., Rotenberg, S. A., and Mirkin, M. V. 2007. Electrochemical Attosyringe. *Proc. Natl. Acad. Sci. USA* 104:11895–11900. (b) Wang, Y., Kececi, K., Mirkin, M. V., Mani, V., Sardesai, N., and Rusling, J. F. 2013. Quantitative resistive-pulse measurements with nanopipettes: Detection of Au nanoparticles and nanoparticle-bound anti-peanut IgY. *Chem. Sci.* 4:655–663.

57. (a) Tsujioka, N., Imakura, S., Nishi, N., and Kakiuchi, T. 2006. Voltammetry of ion transfer across the electrochemically polarized micro liquid-liquid interface between water and a room-temperature ionic liquid, tetrahexylammonium bis(trifluoromethylsulfonyl)imide, using a glass capillary micropipette. *Anal. Sci.* 22:667–671. (b) Nishi, N., Imakura, S., and Kakiuchi, T. 2008. A digital simulation study of steady-state voltammograms for the ion transfer across the liquid–liquid interface formed at the orifice of a micropipette. *J. Electroanal. Chem.* 621:297–303.

58. Bard, A. J. and L. R. Faulkner. 2001. *Electrochemical Methods: Fundamentals and Applications,* 2nd edn. New York: John Wiley & Sons.

59. Oldham, K. B. and Zoski, C. G. 1988. Comparison of voltammetric steady states at hemispherical and disc microelectrodes. *J. Electroanal. Chem.* 256:11–19.

60. Mirkin, M. V. and Bard, A. J. 1992. A simple analysis of quasi-reversible steady-state voltammograms. *Anal. Chem.* 64:2293–2202.

61. Stewart, A. A., Taylor, G., Girault, H. H., and McAleer, J. 1990. Voltammetry at micro ITIES supported at the tip of a micropipette: Part I. Linear sweep voltammetry. *J. Electroanal. Chem.* 296:491–515.

62. Watkins, J. J. and White, H. S. 2004. The role of the electrical double layer and ion pairing on the electrochemical oxidation of hexachloroiridate(III) at Pt electrodes of nanometer dimensions. *Langmuir* 20:5474–5483.

63. (a) Smith, C. P. and White, H. S. 1993. Theory of the voltammetric response of electrodes of submicron dimensions. Violation of electroneutrality in the presence of excess supporting electrolyte. *Anal. Chem.* 65:3343–3353. (b) He, R., Chen, S., Yang, F., and Wu, B. 2006. Dynamic diffuse double-layer model for the electrochemistry of nanometer-sized electrodes. *J. Phys. Chem. B* 110:3262–3270. (c) Liu, Y., He, R., Zhang, Q., and Chen, S. 2010. Theory of electrochemistry at nanometer-sized disk electrodes. *J. Phys. Chem. C* 114:10812–10822.

64. (a) Zevenbergen, M. A. G., Wolfrum, B. L., Goluch, E. D., Singh, P. S., and Lemay, S. G. 2009. Fast electron-transfer kinetics probed in nanofluidic channels. *J. Am. Chem. Soc.* 131:11471–11477. (b) Zevenbergen, M. A. G., Singh, P. S., Goluch, E. D., Wolfrum, B. L., and Lemay, S. G. 2011. Stochastic sensing of single molecules in a nanofluidic electrochemical device. *Nano Lett.* 11, 2881–2886.

65. (a) Amemiya, S., Nioradze, N., Santhosh, P., and Deible, M. J. 2011. Generalized theory for nanoscale voltammetric measurements of heterogeneous electron-transfer kinetics at macroscopic substrates by scanning electrochemical microscopy. *Anal. Chem.* 83:5928–5935. (b) Nioradze, N., Kim, J., and Amemiya, S. 2011. Quasi-steady-state voltammetry of rapid electron transfer reactions at the macroscopic substrate of the scanning electrochemical microscope. *Anal. Chem.* 83:828–835.

66. Garcia-Morales, V. and Krischer, K. 2010. Fluctuation enhanced electrochemical reaction rates at the nanoscale. *Proc. Natl. Acad. Sci. U. S. A.* 107:4528–4532.

67. Geblewicz, G. and Schiffrin, D. J. 1988. Electron transfer between immiscible solutions: The hexacyanoferrate-lutetium biphthalocyanine. *J. Electroanal. Chem.* 244:27–37.

68. Zhan, D. P., Yuan, Y., Xiao, Y. J., Wu, B. L., and Shao, Y. H. 2002. Alkali metal ions transfer across a water/1,2-dichloroethane interface facilitated by a novel monoaza-B15C5 derivative. *Electrochim. Acta* 47:4477–4483.

69. Dale, S. E. C. and Unwin, P. R. 2008. Polarised liquid/liquid micro-interfaces move during charge transfer. *Electrochem. Commun.* 10:723–726.

70. Wei, C., Bard, A. J., and Feldberg, S. W. 1997. Current rectification at quartz nanopipet electrodes. *Anal. Chem.* 69:4627–4633.

71. (a) White, H. S. and Bund, A. 2008. Ion current rectification at nanopores in glass membranes. *Langmuir* 24:2212–2218. (b) Calander, N. 2009. Analyte concentration at the tip of a nanopipette. *Anal. Chem.* 81:8347–8353.

72. Wang, G., Zhang, B., Wayment, J. R., Harris, J. M., and White, H. S. 2006. Electrostatic-gated transport in chemically modified glass nanopore electrodes. *J. Am. Chem. Soc.* 128:7679–7686.

73. Yu, Y., Velmurugan, J., and Mirkin, M. V., unpublished results.

74. Huang, W. H., Pang, D. W., Tong, H., Wang, Z. L., Cheng, J. K. 2001. A method for the fabrication of low-noise carbon fiber nanoelectrodes. *Anal. Chem.* 73:1048–1052.

75. (a) Kim, B. M., Murray, T., and Bau, H. H. 2005. The fabrication of integrated carbon pipes with sub-micron diameters. *Nanotechnology* 16:1317–1320. (b) Vitol, E.; Schrlau, M. G., Bhattacharyya, S., Ducheyne, P., Bau, H. H., Friedman, G., Gogotsi, Y. 2009. Effects of deposition conditions on the structure and chemical properties of carbon nanopipettes. *Chem. Vap. Deposit.* 15:204–208. (c) Singhal, R., Bhattacharyya, S., Orynbayeva, Z., Vitol, E., Friedman, G., and Gogotsi, Y. 2010. Small diameter carbon nanopipettes. *Nanotechnology* 21:015304.

76. (a) Takahashi, Y., Shevchuk, A. I., Novak, P. et al. 2012. Topographical and electrochemical nanoscale imaging of living cells using voltage-switching mode scanning electrochemical microscopy. *Proc. Natl. Acad. Sci. U. S. A.* 109:11540–11545. (b) Actis, P., Tokar, S., Clausmeyer, J. et al. 2014. Electrochemical nanoprobes for single-cell analysis. *ACS Nano* 8:875–884. DOI: 10.1021/nn405612q.

77. Yu, Y., Noël, J.-M., Mirkin, M. V., Gao, Y., Mashtalir, O., Friedman, G., and Gogotsi. Y. 2014. Carbon pipette-based electrochemical nanosampler. *Anal. Chem.* 86:3365–3372.

78. Schrlau, M., Dun, N, and Bau, H. 2009. Cell electrophysiology with carbon nanopipettes. *ACS Nano* 3:563–568.

79. Hu, K., Wang, Y., Cai, H., Mirkin, M. V., Gao, Y., Fridman, G., and Gogotsi, Y. 2014. Open carbon nanopipettes as resistive-pulse sensors, rectification sensors and electrochemical nanoprobes. *Anal. Chem.* 86:8897–8901.

80. Yasukawa, T., Kaya, T., and Matsue, T. 1999. Dual imaging of topography and photosynthetic activity of a single protoplast by scanning electrochemical microscopy. *Anal. Chem.* 71:4637–4641.

81. Yang, C. and Sun, P. 2009. Fabrication and characterization of a dual submicrometer-sized electrode. *Anal. Chem.* 81:7496–7500.

82. McKelvey, K., Nadappuram, B. P., Actis, P. et al. 2013. Fabrication, characterization, and functionalization of dual carbon electrodes as probes for scanning electrochemical microscopy (SECM). *Anal. Chem.* 85:7519–7526.

83. Johnson, L. and Walsh, D. A. 2012. Tip generation–substrate collection–tip collection mode scanning electrochemical microscopy of oxygen reduction electrocatalysts. *J. Electroanal. Chem.* 682:45–52.

84. (a) Shao, Y., Liu, B., and Mirkin, M. V. 1998. Studying ionic reactions by new generation/collection technique. *J. Am. Chem. Soc.* 120:12700–12701. (b) Liu, B., Shao, Y., and Mirkin, M. V. 2000. Dual-pipette techniques for probing ionic reactions. *Anal. Chem.* 72:510–519. (c) Chen, Y., Gao, Z., Li, F. et al. 2003. Studies of electron-transfer and charge-transfer coupling processes at a liquid/liquid interface by double-barrel micropipette technique. *Anal. Chem.* 75:6593–6601. (d) Hu, S., Xie, X., Meng, P. et al. 2006. Fabrication and characterization of submicrometer- and nanometer-sized double-barrel pipettes. *Anal. Chem.* 78:7034–7039.

85. Takahashi, Y., Shevchuk, A. I., Novak, P. et al. 2011. Multifunctional nanoprobes for nanoscale chemical imaging and localized chemical delivery at surfaces and interfaces. *Angew. Chem. Int. Ed.* 50:9638–9642.

86. (a) Comstock, D. J., Elam, J. W., Pellin, M. J., and Hersam, M. C. 2010. Integrated ultramicroelectrode-nanopipet probe for concurrent scanning electrochemical microscopy and scanning ion conductance microscopy. *Anal. Chem.* 82:1270–1276. (b) Takahashi, Y., Shevchuk, A. I., Novak, P. et al. 2010. Simultaneous noncontact topography and electrochemical imaging by SECM/SICM featuring ion current feedback regulation. *J. Am. Chem. Soc.* 132:10118–10126.

87. (a) Rodolfa, K. T., Bruckbauer, A., Zhou, D., Korchev, Y. E., and Klenerman, D. 2005. Two-component graded deposition of biomolecules with a double-barreled nanopipette. *Angew. Chem., Int. Ed.* 44: 6854–6859. (b) Rodolfa, K. T., Bruckbauer, A., Zhou, D., Schevchuk, A. I., Korchev, Y. E., and Klenerman, D. 2006. Nanoscale pipetting for controlled chemistry in small arrayed water droplets using a double-barrel pipette. *Nano Lett.* 6:252–257.

88. Actis, P., Maalouf, M. M., Kim, H. J. et al. 2014. Compartmental genomics in living cells revealed by single-cell nanobiopsy. *ACS Nano* 8:546–553. DOI: 10.1021/nn405097u.

89. (a) Williams, C. G., Edwards, M. A., Colley, A. L., Macpherson, J. V., and Unwin, P. R. 2009. Scanning micropipette contact method for high-resolution imaging of electrode surface redox activity. *Anal. Chem.* 81:2486–2495. (b) Patel, A. N., Guille Collignon, M., O'Connell, M. A. et al. 2012. A new view of electrochemistry at highly oriented pyrolytic graphite. *J. Am. Chem. Soc.* 134:20117–20130.

16 Microfabricated Electrochemical Systems

Shuo Kang and Serge G. Lemay

CONTENTS

16.1 INTRODUCTION

The terms *microfabrication* and *micromachining* represent a broad set of techniques for systematically creating solid-state structures on the micro- and nanometer scales. Primarily developed by the semiconductor industry as an enabler for cheaper and more complex microelectronics circuitry, the resulting capabilities have since been exploited throughout most other areas of science and technology. In particular, microfabrication, having first become a workhorse of solid-state physics research, has become increasingly common in a variety of *wet* fields ranging from biophysics and neuroscience to environmental sensing and bioanalytical applications. Lithographic approaches are particularly well matched to the demands of electroanalytical methods due to the latter's emphasis on solid-state electrodes and electrical signals and a growing interest in micro- and nanoscale systems and processes.

Microfabrication techniques offer several broad benefits when compared to alternative methods for fabricating miniaturized electrochemical measurement systems:

- Harnessing the well-developed, systematic fabrication protocols developed in the context of microelectronics leads in principle to highly reproducible results for the size and geometry of nanostructures. This is notoriously difficult to achieve on the nanometer scale using alternative approaches based on more ad hoc protocols.
- This reproducibility in turn greatly facilitates characterization since a battery of tools can be brought to bear on a series of nominally identical structures, even when some of these tools are mutually exclusive and/or destructive to the structures. This is again in contrast

to approaches where each (nanoscale) system is individually realized and thus needs to be separately characterized; in these cases, electrochemical measurements themselves are often the only source of characterization available.

- Once a measurement system is developed, the marginal costs associated with large-scale production become relatively low. To fully appreciate the full extent of this point, note that standard complementary metal-oxide semiconductor (CMOS) technology allows integrating millions of functional components on a mass-produced chip at a cost of only a few dollars.
- For sufficiently complex geometries, there are often no alternative clever *tricks* available and brute-force lithography-based methods are the only option.
- Individual devices can be easily integrated with each other as well as with other electronic and/or fluid handling components. This is particularly relevant in the context of so-called lab-on-a-chip applications. At the extreme limit of integration, a complete measurement system can be integrated on a single chip with a liquid sample as input and digital data as output.

Offsetting these benefits are several complications and limitations introduced by microfabrication:

- Specialized equipment is required that is not available in all laboratories. This is particularly true of the high-end lithography equipment employed in several common approaches for patterning thin-film materials at the submicron level.
- There are experimental issues to which widely accepted solutions have been developed in conventional systems, but that cannot easily be replicated in microfabricated devices. Probably the best example is the difficulty of polishing most microfabricated electrodes, a common procedure with macro- and ultramicroelectrodes (UMEs).
- The extensive processing involved in microfabrication largely precludes working with advanced materials such as single crystals.

These limitations and some of the approaches that have been explored to mitigate them will be the main focus of this chapter, with a particular focus on concrete examples.

We note that the development of microfabricated electrochemical systems over the last 30 years has largely progressed in an evolutionary rather than revolutionary manner. But whereas many of the basic motivations, principles, and approaches have remained relatively unchanged, their realization has become increasingly sophisticated and their performance has continually improved as a result of new insights and more advanced fabrication methods. This is illustrated in Figure 16.1, which contrasts two setups—one early and one recent—for redox-cycling measurements. Figure 16.1a shows a measurement cell based on microfabricated interdigitated electrodes (IDEs) (discussed in Section 16.4.1). The critical dimension of the microfabricated structure, namely, the spacing between the electrodes, was 50 μm. Figure 16.1b shows the corresponding arrangement for a recent nanofluidic thin-layer cell (discussed in Section 16.4.3). Here, the electrode spacing is 50 nm, leading to a thousandfold increase in diffusive fluxes. Both cells allow for convective transport, with the caveat that this requires a more sophisticated polydimethylsiloxane (PDMS) microfluidic interface in the case of the nanodevice.

The present chapter focuses on summarizing the evolution and the current status of microfabrication-based approaches for the realization of electroanalytical systems. In keeping with the general theme of this book, we focus primarily on nanoscale systems where possible. In areas where little work has reached this level of miniaturization, we instead discuss the state of the art at the micrometer scale. We assume that the reader has some familiarity with basic lithography-based fabrication methods and dwell only briefly on the general methods. For a more general introduction, we refer the uninitiated reader to a recent tutorial overview.[1] Here, we instead concentrate on aspects of direct relevance to electrochemical methods or to the specific works being reviewed.

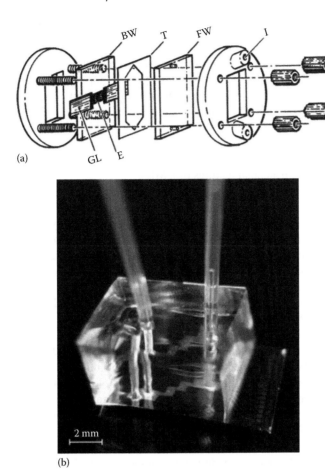

(a)

(b)

FIGURE 16.1 (a) Schematic drawing of assembly of IDEs (E) microfabricated on a quartz substrate with electrical contacts (GL) and liquid chamber (T + BW + FW). BW, back window; FW, front window; GL, gold leaf contact; I, injection port; T, Teflon spacer. (Reprinted with permission from Sanderson, D.G. and Anderson, L.B., Filar electrodes—Steady-state currents and spectroelectrochemistry at twin interdigitated electrodes, *Analytical Chemistry*, 1985, 57, 2388–2393. Copyright 1985 American Chemical Society.) (b) Photograph of a microfabricated electrochemical nanofluidic device; the contact pads and wires to individual electrodes are visible on the bottom right. Microfluidic channels molded in the transparent PDMS block allow delivering fluid to the electrochemical device. (From Mathwig, K. and Lemay, S.G., *Micromachines*, 4, 138, 2013.)

The chapter is further organized in order of increasing complexity of the structures being discussed, starting with methods for the fabrication of individual electrodes and concluding with a brief discussion of systems in which electrochemical probes are fully integrated with microelectronics on the same chip.

16.2   ULTRAMICROELECTRODES AND ULTRAMICROELECTRODE ARRAYS

UMEs[2,3] offer several advantageous features compared to their macroscopic counterparts including a true steady-state diffusion-limited current, small IR drops from solution resistance, and short RC response times. Originally aimed at precise measurements of diffusion coefficients, interest in UMEs was further stoked by attempts at probing electroactive species inside brain tissue, which necessitated small, nonperturbing probes.[4,5] Classical methods for fabricating UMEs were largely based on micrometer-diameter wires that were either selectively insulated or encased in glass micropipettes. These methods were used successfully in producing high-quality monolithic UMEs that were suitable for intra- and extracellular stimulation and recording[6,7]; indeed, similar electrodes are still in use today. It, however, proved more challenging to employ these approaches to fabricate bundles of closely spaced microelectrodes to monitor neural activity at a number of nearby sites simultaneously. In the 1970s, micromachining technology was thus introduced to fabricate arrays of (separately addressable) microelectrodes for both in vitro and in vivo experiments.[8–12] Arrays of identical UMEs connected in parallel can also be beneficial in other applications since faradaic currents at UMEs are relatively small: wiring many electrodes together amplifies the magnitude of the current while retaining the beneficial features of UMEs.[13]

An early work was presented by Thomas et al.[8] who fabricated a miniature microelectrode array to monitor the bioelectric activity of cultured heart cells. A glass coverslip was used as a substrate on which a 200 nm thick nickel film was deposited and then defined by lithography. Afterwards, gold was electroplated onto the nickel pads, and a photoresist layer was coated and patterned to reveal only the gold electrodes. The remaining resist then functioned as a passivation layer. Finally, a glass ring was affixed to the insulated array with bees' wax, creating a culture chamber, and platinum black was electrochemically deposited on the electrodes.

As an example of a miniaturized tool for in vivo neural recordings, a 24-channel microelectrode array fabricated based on thin-film technology was developed by Kuperstein and Whittington.[14] In this work, Mo foil was used as a temporary substrate on which to build structures. KTFR photoresist, Au, and another layer of KTFR resist were deposited and patterned in succession; thereafter, the Mo foil was electrolytically etched away in an aqueous solution of 5% KOH, 5% $K_3Fe(CN)_6$, and 1% liquid Woolite (the latter atypical reagent playing the role of "low foaming, nonionic, water soluble, and alkali resistant surfactant compound"[15]). In this manner, a probe consisting of arrays of Au recording sites sandwiched between two KTFR resist layers was generated, each of the recording site having an area of 120 μm^2 and being separated from neighboring sites by a gap of 85 μm. Finally, platinum black was plated onto the recording sites of the probe.

During the same period, a multicathode polarographic oxygen electrode with several cathodes connected in parallel in a single package was demonstrated by Siu and Cobbold.[16] The device consisted of circular Au cathodes surrounded by a continuous Ag/AgCl anode created with thin-film technology. Electrical contact between the anode and the cathode was maintained via a salt bridge formed by an electrolyte-containing membrane that covered the surface of the electrodes. The membrane also functioned as a protection layer to prevent the electrodes being contaminated in the meantime.

In the following decades, microfabricated UMEs and UME arrays became increasingly widespread, as reviewed by Feeney and Kounaves.[17] An advantage of the added flexibility provided by micromachining started to be exploited by fashioning sets of electrodes from different materials. For example, Glass et al.[18] fabricated a multielement microelectrode array for environmental monitoring including 66 working electrodes on a 2 in. silicon wafer with a variety of electrode materials including Pt, Au, V, Ir, and carbon deposited and defined by separate lithography steps. Different electrode materials displayed somewhat different responses to a given compound in voltammetric measurements, in principle increasing the selectivity compared with using a single electrode material.

In recent years, designs for UMEs and UME arrays continue to evolve. For example, works based on microfabricated diamond UMEs and arrays are increasingly common, motivated by this material's attractive properties as an electrode that include mechanical stability, chemical inertness, low background currents, wide potential window, and resistance to electrode fouling.[19] Individual electrodes fabricated with focused ion beam (FIB)[20] and arrays fabricated with thin-film technology[21,22] were demonstrated.

Instead of exploiting the advantages of a high degree of integration, addressable electrode arrays with each sensing pixel wired via multiplexing circuitry to a potentiostat were developed for sensing and imaging.[23,24] For instance, a multianalyte microelectrode detection platform capable of discriminating between multiple protein and DNA analytes simultaneously was demonstrated.[25] The electrodes were selectively functionalized with enzymes, antibodies, DNA, and peptide probes using an electrically addressable deposition procedure.

A method for fabricating 3D electrode structures was demonstrated by Sanchez-Molas et al.[26] to effectively extend the electrode surface area. In this case, the motivation for creating such structures originated from biofilm-based microbial fuel cell applications. High-aspect-ratio micropillars were formed by micromachining a silicon wafer with deep reactive ion etching (DRIE), the radius of the pillars being 5–10 μm with a separation of 20–100 μm in between and a height of 5–125 μm. A multilayer of Ti/Ni/Au was sputtered onto the structure surface to ensure the metallization of both the vertical walls and the bottom surface between the pillars.

16.3 NANOELECTRODES AND NANOELECTRODE ARRAYS

In recent years, considerable attention has shifted to nanoscale electrodes (as already discussed in Chapter 15) and integrated systems.[27–30] With this further downscaling, the intrinsic advantages of UMEs such as small ohmic drops and fast response times are further amplified. Mass transport also becomes so efficient that even fast electrochemical reactions become increasingly limited by the rate of heterogeneous electron transfer, allowing ultrafast electron-transfer kinetics to be studied. Furthermore, because the electrode size becomes comparable to the thickness of the electrical double layer and to the size of macromolecular analytes, new mass-transport phenomena have been predicted and new analytical applications can be considered, respectively.[31]

The challenge of fabricating and characterizing nanometer-scale electrodes is, however, substantial compared to microelectrodes. In particular, the ability to project a sharp image of a small feature onto the substrate in photolithography is limited by the wavelength of the light that is used and the ability of the reduction lens system to capture enough diffraction orders from the illuminated mask.[32] Even though the most advanced optical immersion lithography tools currently allow features of ~40 nm to be realized in integrated-circuit (IC) processing, the necessary equipment is very specialized and mostly targeted at semiconductor research and manufacturing. Most readily accessible optical-lithography equipment in universities and research laboratories instead has a much more modest practical resolution of ~1 μm. Consequently, a broad range of alternative approaches has been explored for micromachining nanoscale electrochemical systems. These include lithographic methods with higher resolution, such as e-beam, nanoimprint, and nanosphere lithography, electrode materials prepared by bottom-up approaches, and a number of one-of-a-kind solutions for creating specific structures.

16.3.1 TIP-BASED NANOELECTRODES

The bulk of the approaches employed for pioneering studies of nanoelectrodes was evolved from methods for preparing UMEs and/or tips for scanning electrochemical microscopy (SECM).[33,34] Broadly speaking, these methods rely on preparing sharp conducting wires or tips and covering all but the apex with an insulating material, including wax,[35–37] polyimide,[38] electrophoretic paint,[39–46] or glass.[47–53] Because the electrodes are prepared individually, these approaches have historically tended to exhibit limited reproducibility. This prompted some authors to explore micromachining-based approaches for fabricating tip-based electrodes.

Thiébaud et al.[54] developed tip-like electrodes based on a fully controlled lithographic process. Atomically faceted, 47 μm high tips were carved out of a <100> silicon wafer by anisotropic etching in KOH. The silicon was then successively coated with thin films of silicon dioxide, platinum, and silicon nitride. Following a final lithography step, the nitride was etched away from the apex of the tip to leave a Pt tip exposed with a height as small as 2 μm. In an alternative hybrid approach, Qiao et al.[55] first etched tungsten wires to yield tips with diameters below 100 nm and insulated these wires using electrophoretic paint. The FIB technique was then employed to selectively remove the insulating paint and sculpt the Pt tip apex to the desired shape. Tips with dimensions 100–1000 nm were realized in this manner.

Despite their potential benefits in terms of control and characterization, however, these approaches have not proven competitive compared to the more accessible classic approaches for fabricating tip electrodes.

16.3.2 Top-Down Fabrication of Nanoelectrodes

Despite the limited resolution of optical lithography, this method has been employed to create nanoelectrodes by incorporating nonstandard microfabrication steps. For example, Menke et al.[56] combined top-down lithography and electrodeposition to generate band electrodes with a width of 40–50 nm in a process coined lithographically patterned nanowire electrodeposition (LPNE). The process flow for the fabrication is shown in Figure 16.2. By undercutting nickel bands that were covered with a layer of photoresist, a trench was formed, and nanowires were grown by electrodeposition in the trench along the edge of the nickel bands. The height of the nanowires was determined by the thickness of the nickel bands and the width by controlling the deposition process. A hydrogen gas detector consisting of Pd nanowires fabricated using this method was demonstrated,[57] and the method was also improved by adding further processing steps to fabricate arrays of nanowires.[58] To overcome the restrictions imposed on the array density by the limited resolution of photolithography, repeated alternating deposition of nanowire electrodes and nickel bands was performed, the array being generated when all the nickel bands were simultaneously released in a subsequent step.

Another method for beating the resolution limitations of optical lithography was demonstrated by Heo et al.[59] who derived a carbon linear nanoelectrode array from optical-lithography-defined polymer microstructures. Photosensitive polymer SU-8 was coated and patterned on a 6 in. passivated silicon wafer and subsequently pyrolyzed at 900°C in vacuum. During the pyrolysis process, the SU-8 was carbonized and the dimension of the structures shrank by approximately 60% in width and 90% in height, as shown in the scanning electron microscope (SEM) images of Figure 16.3.

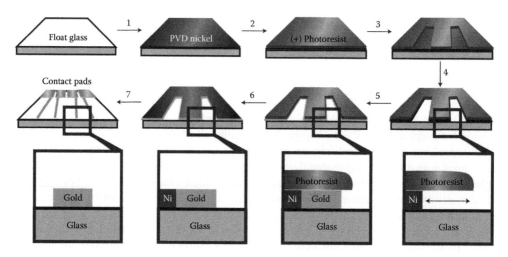

FIGURE 16.2 Process flow for lithographically patterned nanowire electrodeposition. (Reprinted by permission from Macmillan Publishers Ltd. *Nature Materials*, Menke, E.J., Thompson, M.A., Xiang, C., Yang, L.C., and Penner, R.M, Lithographically patterned nanowire electrodeposition, 5, 914–916, 2006. Copyright 2006.)

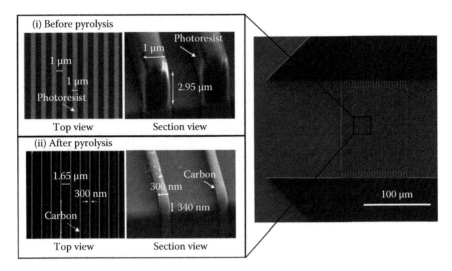

FIGURE 16.3 SEM images of nanoscale carbon electrodes pyrolyzed from SU-8 microstructures. (From Heo, J.I. et al., *J. Electrochem. Soc.*, 158, J76, 2011. By permission of The Electrochemical Society.)

The pyrolysis process was reported to be controllable such that the final dimensions of the carbon electrodes were predictable.

Despite these successes of optical-lithography-based approaches, patterning of nanoscale structures is more typically carried out using a workhorse of nanoscience and nanotechnology, electron-beam lithography (EBL). This tool, which was developed in the early 1970s,[60] employs a focused beam of electrons to write arbitrary 2D patterns on a surface covered with an electron-sensitive resist. Apart from these differences, the whole range of thin-film technologies can be combined with EBL with only minor adjustments to the processes compared with optical lithography. It is a serial patterning technology rather than simultaneous patterning as in optical lithography, rendering the process more time consuming and therefore costly, but this is compensated by the feature that resolutions in the range 10–100 nm can be achieved with EBL, depending on the specific equipment employed.

A variety of nanoelectrochemical systems fabricated with EBL has been demonstrated.[61–67] As an early example, Niwa et al.[68] reported electrode arrays with submicron dimensions. Electrochemical analysis based on EBL-generated individual Au nanowires was reported by Dawson et al.[66] A catalytic signal from fewer than 50 enzyme molecules immobilized on an EBL-patterned nanoelectrode was also reported.[64]

Another technique used to pattern nanostructures from thin films is FIB milling, which operates in a fashion analogous to an SEM except that a finely focused beam of ions (usually gallium) is used instead of electrons. A FIB can be operated at low beam currents for imaging or high beam currents for site-specific sputtering or milling. A disadvantage is that this is also a serial method, individual structures needing to be prepared separately. One way to use FIB to generate electrodes is to first deposit a metal and an insulating layer and then drill holes through the insulating layer to uncover the electrodes.[69–71] With this method, recessed electrodes located at the bottom of truncated conical pores result.[69]

Alternatively, it is also possible to generate electrodes by first creating nanoscale holes through thin insulating membranes and then filling these holes from one side of the membrane with a conducting material to create electrode structures on the other side of the membrane. This approach is conceptually descended from earlier protocols to create nanoelectrode ensembles by depositing metal in a porous host membrane such as polycarbonate (PC).[72] Besides FIB milling, a focused electron beam from a transmission electron microscope can also be used to drill individual nanopores.[73–76] An advantage of the latter approach is that a nanometer-resolution image of each nanopore can be simultaneously obtained. Since the diameter of the finished electrodes is dictated by that of the original pores, this provides an independent characterization of the electrode size. Krapf et al. demonstrated electrodes as small as 2 nm using this approach.[75]

High-throughput, high-resolution lithographic methods have also been developed. Nanoimprint lithography[77] creates patterns by mechanical deformation of a so-called imprint resist that typically consists of a monomer or polymer formulation cured by heat or UV light during the imprinting. A master stamp provides the pattern to be imprinted; while this stamp must first be created using another lithography method, it is not significantly degraded by the imprinting process and can be reused over an extended period of time. A challenge of this technique is that the process is strongly dependent upon the pressure, temperature, time control, and even the geometry of the stamp. Nonetheless, electrode arrays created by nanoimprint lithography have been demonstrated and suggested for low-cost sensor production.[62,78,79]

Another, much simpler and low-cost alternative for fabricating electrodes is nanosphere lithography,[80–84] in which self-assembled monolayers of spheres are used as masks instead of selectively exposed polymer layers. For example, Valsesia et al.[81] spin-coated polystyrene beads with a diameter of 500–1000 nm onto an Au-coated substrate, forming a monolayer of hexagonally packed beads whose surface coverage could be adjusted by tuning the spin-coating acceleration. With a treatment in oxygen plasma, the size of the beads was reduced by half. Afterwards, a layer of silicon oxide was deposited and lifted off by mechanically removing the beads in an ultrasonic bath. The resulting recessed Au spots with dimensions in the range of 50–120 nm and surrounded by silicon oxide were then used as templates to electrochemically grow polypyrrole nanopillar electrodes.

Diamond nanoelectrode ensembles and arrays were created by Hees et al.[83] using nanosphere lithography and EBL, respectively. In the first approach, a substrate coated with a trimethylboron-doped nanocrystalline diamond (NCD) film was immersed in an ultrasonic bath with suspended SiO_2 spheres having a radius of 500 nm and a concentration of $\sim 10^{-7}$ cm^{-3}. The spheres adhered to the surface in a random pattern. An insulating NCD layer was then deposited onto the surface and lifted off by removing the SiO_2 beads with hydrofluoric acid (HF), creating recessed boron-doped diamond electrodes surrounded by an insulating NCD layer. The radius of the electrodes was about 175 nm and the average distance between them was 10 μm. In the second approach, all the process steps were identical except that EBL-patterned plasma-enhanced chemical vapor deposition (PECVD) SiO_2 was used instead of SiO_2 beads for lifting off the passivation NCD film. Electrode arrays following regular hexagonal patterns were formed in this manner. SEM images of the electrodes and arrays fabricated with both methods are shown in Figure 16.4. Based on these arrays, changes in electron-transfer rates were observed to change when switching the NCD surface termination from hydrogen to oxygen; this subtle effect was not observed based on macroscopic planar diamond electrodes.

16.3.3 NANOWIRE-BASED NANOELECTRODES

In the approaches described so far, micro- and nanoscale electrodes were created by patterning thin conductive and/or insulating films into the desired geometry. An alternative bottom-up approach is to first synthesize electrode materials with nanoscopic dimensions, then to interface these materials to external interconnects to enable electrochemical measurements. Wire-shaped objects with nanometer-scale diameters and micrometer-scale lengths are particularly well suited for this approach: the long lengths make it relatively straightforward to pattern interconnects using relatively low-resolution lithography, while the small diameters mean that the materials effectively function as nanoscale band electrodes.[85]

This approach is perhaps best illustrated by the use of single-wall carbon nanotubes (SWNTs) as electrode materials. SWNTs are cylindrically shaped carbon macromolecules. They can be readily deposited on a substrate or, often preferably for device applications, grown by CVD from catalyst

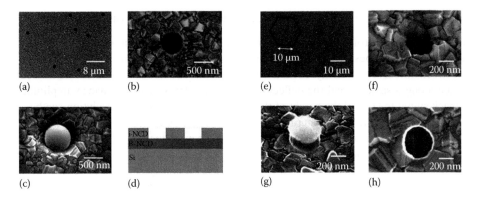

FIGURE 16.4 (a through d) SEM images and schematic cross section of diamond nanoelectrode ensembles fabricated with nanosphere lithography and (e through h) arrays fabricated with EBL. (a) Overview of randomly distributed electrodes. (b) SiO$_2$ sphere after deposition of insulating diamond. (c) Final boron-doped NCD electrode after removal of SiO$_2$. (d) Schematic cross section of fabricated electrodes. (e) Overview of electrodes distributed in hexagonal pattern. (f) Structured SiO$_2$ island on boron-doped NCD layer. (g) Insulating diamond grown around SiO$_2$. (h) Final recessed diamond electrode. (Reprinted with permission from Hees, J., Hoffmann, R., Kriele, A. et al., Nanocrystalline diamond nanoelectrode arrays and ensembles, *ACS Nano*, 2011, 5, 3339–3346. Copyright 2011 American Chemical Society.)

particles that can be deposited according to lithographically defined patterns on a solid substrate. The diameter and length distribution varies substantially depending on the growth method, but diameters of 1–3 nm and lengths of a few micrometers are typical and readily achievable. In a common approach, the nanotubes are first deposited or grown on the substrate, metal interconnects are added to make contact to one or more nanotube, and a passivation layer is deposited and patterned so as to cover the electrodes but leave (part of) the nanotubes exposed. Since the sidewalls are electrochemically active,[86] each individual nanotube functions as a band nanoelectrode. But because the geometry of the nanotube(s) and passivation can be controlled, a greater range of electrode geometries can also be created. In particular, Dumitrescu et al.[87,88] showed that a relatively sparse network of randomly oriented, interconnected SWNTs can effectively function as a 2D array of nanoelectrodes with overlapping diffusion fields: the total diffusion-limited current at a disk-shaped network electrode was shown to be equivalent to that to an UME of the same shape and size, but the current density at the surface of the SWNTs was much higher than at the corresponding UME. Alternatively, exposing only the sidewall of an individual SWNT leads to a near-ideal cylindrical electrode with a radius of ~1 nm.[89] Finally, exposing only the end allows forming a point-like electrode with the same radius.[90] In cases where a different electrode material is needed, it was also shown that SWNTs can also be modified with metal nanoparticles by electrodeposition. In these applications, the SWNTs serve both as a template for deposition and as interconnects between the nanoparticles and external wiring. Paralleling the work on bare SWNTs, such deposition has been employed to create 2D (networks), 1D (wires), and 0D (single particles) nanoparticle electrodes.[90–92]

Similar approaches have been applied to a broad range of other 1D nanostructures. For example, Dawson et al.[93] demonstrated electrodes based on Au nanowires with a rectangular (~210×250 nm) cross section created by nanoskiving.[94] This method is based on first forming a block consisting of thick epoxy layers separated by an Au film. Thin slices of this block are then sectioned off in a plane perpendicular to the layers. Finally, the epoxy is dissolved, leaving only Au nanowires available for contacting via lithographically defined external wires. Other examples of individual nanowires that have been investigated as electrochemical nanoelectrodes include multiwalled carbon nanotubes,[95] carbon nanofibers,[96] vanadium oxide nanowires and Si/amorphous–Si core/shell nanowires,[97] mesoporous ZnO nanofibers,[98] and platinum nanowires prepared by laser pulling.[99]

16.3.4 ELECTRODES FOR ELECTROCHEMICAL ATOMIC FORCE MICROSCOPY

Another area where microfabricated electrodes have played a significant role is in the preparation of advanced scanning probes, in particular modified cantilevers for atomic force microscopy (AFM) with electrochemical functionality. In AFM, a sharp point mounted at the end of a flexible cantilever is scanned along a surface, and the deflection of the cantilever or its resonance amplitude is used as feedback signal to control the height of the cantilever. Subnanometer resolution can be achieved in the height direction, while the lateral resolution is largely determined by the sharpness of the tip being employed; micromachining is commonly used for manufacturing sharp, reproducible cantilever and tip structures. Several authors have explored the possibility of modifying cantilevers to incorporate one or more electrodes in AFM tips.[100–104] In this way, local electrochemical measurements can be performed while AFM feedback is employed for imaging and tip positioning.

As an early example, silicon nitride cantilevers were modified by patterning a ring electrode immediately around the apex of the AFM tip.[100,101] This was achieved by coating the original silicon nitride cantilever with Au and an insulating silicon nitride layer, then milling the apex of the tip to create a sharp silicon nitride point (made from the original cantilever material) surrounded by a ring of exposed gold. The sharp nitride tip provides imaging capabilities and stability comparable to those of the original cantilever, while the ring electrode, contacted via the Au film, permits electrochemical measurements. In an alternative approach, Burt et al.[102] attached a metal nanowire to the end of an AFM tip. The wire, which was fabricated by coating an SWNT template, was insulated and then cut to create an Au disk nanoelectrode. This geometry results in a flat tip that reduces AFM resolution but has the benefit of allowing SECM measurements with simultaneous AFM imaging. More recent developments in this area include needle-shaped, individually addressable dual tips[103] and insulating diamond tips with integrated boron-doped diamond electrodes.[104] In most approaches to AFM tip modification, the FIB technique has been the method of choice to precisely sculpt the complex geometry of the critical region near and at the apex of the tip.

16.4 REDOX-CYCLING AND GENERATOR–COLLECTOR ELECTRODES

In the micro- and nanoelectrode arrays discussed earlier, the motivation for creating a multielectrode system is most often to amplify the faradaic current while retaining the beneficial properties of the individual miniature electrodes. The constituting electrodes thus function essentially independent of each other. Redox-cycling and generator–collector approaches instead exploit the interplay between redox reactions taking place at two or more electrodes. Establishing an effective coupling between electrodes requires careful control of electrode geometry and placement, a challenge that plays to the strengths of microfabrication techniques.

In generator–collector systems, the product of a reaction taking place at a generator electrode is detected at a second, so-called collector electrode. A natural figure of merit is the collection efficiency, which corresponds to the fraction of generated molecules that are collected. In redox cycling, both electrodes instead share both roles of generator and collector, as chemically reversible species are repeatedly reduced at one electrode and oxidized at the other. The geometries required for efficient redox cycling tend to be more restrictive than for generation–collection, since in this case the collection efficiency should be high for both halves of the cycle. A common figure of merit in redox cycling is the amplification factor, which essentially corresponds to the average number of times that each molecule is cycled before it exits the detection domain. Consistent with intuition, both the collection efficiency and the amplification factor tend to increase as the distance between the electrodes is reduced due to more effective mass transport.[3] Generator–collector and redox-cycling systems are thus natural candidates for miniaturization to the nanoscale.

At this time, three main classes of devices are undergoing the most extensive development toward nanoscale applications: IDEs,[59,68,105–111,185] recessed ring–disk (RRD) electrodes,[84,112–117] and nanogaps,[118–130] as summarized in Figure 16.5.

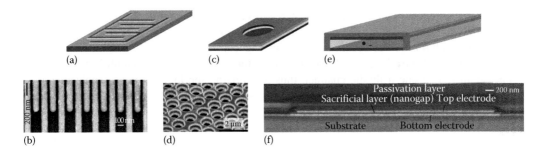

FIGURE 16.5 (a, c, e) Schematic drawings and (b, d, f) SEM images of IDEs, RRD electrodes, and nanogaps, respectively. (b) Top view of IDEs. (Reprinted from Ueno, K., Hayashida, M., Ye, J. and Misawa, H. Fabrication and electrochemical characterization of interdigitated nanoelectrode arrays, *Electrochemistry Communications*, 7, 161–165, Copyright 2005, with permission from Elsevier.) (d) View from an angle of an RRD electrode array. (Reprinted with permission from Ma, C., Contento, N.M., Gibson, L.R., 2nd, and Bohn, P.W., Recessed ring-disk nanoelectrode arrays integrated in nanofluidic structures for selective electrochemical detection, *Analytical Chemistry*, 2013, 85, 9882–9888. Copyright 2013 American Chemical Society.) (f) View from an angle of the cross-section of a nanogap. (Reprinted with permission from Kang, S., Nieuwenhuis, A.F., Mathwig, K., Mampallil, D., and Lemay, S.G., Electrochemical single-molecule detection in aqueous solution using self-aligned nanogap transducers, *ACS Nano*, 2013, 7, 10931–10937. Copyright 2013 American Chemical Society.)

16.4.1 INTERDIGITATED ELECTRODES

The most widely reported redox-cycling device configuration, illustrated in Figure 16.5a and b, is the IDE or, equivalently, interdigitated array (IDA).[59,68,105–111,185] It consists of two coplanar, inter-penetrating comb-shaped electrodes. Because the two electrodes can be realized simultaneously by patterning a single layer of conducting material, this geometry is conceptually straightforward from a fabrication point of view. By the same token, the smallest achievable electrode spacing is set by the lateral resolution of the lithographic process employed. IDEs with electrode spacing ranging from microns down to tens of nanometers were correspondingly demonstrated using optical,[107,131,132] e-beam,[61] and nanoimprint lithography.[78,79] Amplification factors up to ~10^2 are typically reported with these structures.

In a pioneering article, Sanderson and Anderson[105] reported coplanar IDEs fabricated by depositing and defining a layer of Au (1000–2000 Å) with 200–400 Å Cr as an adhesion layer on a quartz substrate with photolithography and subsequent wet etching. Each electrode was 0.5 cm long and 50 μm wide, separated from the adjacent electrodes by a gap of 50 μm. Two strips of gold leaves were placed onto the metal pads to make electrical contacts. A liquid cell was formed by clamping the quartz substrate and a Teflon spacer between two quartz windows with quick-tightened screws, as indicated in the schematic drawing shown in Figure 16.1a; amplification of the faradaic current by redox cycling was successfully observed in this system. Several years later, further downscaled electrode arrays with feature sizes ranging from 0.75 to 10 μm fabricated using both optical and EBL were reported by Niwa et al.[68] A layer of spin-on glass was coated onto wafers as passivation, and the electrodes and contact pads were uncovered by etching through this passivation layer using reactive ion etching (RIE).

Besides electrode spacing, the signal amplification provided by IDEs also depends on the width and the aspect ratio of the electrodes.[133] Electrodes with a relatively large height-to-width ratio were shown to generate a higher amplification factor than planar electrodes, as the short linear diffusion path created between the electrode sidewalls increases the diffusive flux. Dam et al.[108] reported intentionally vertically faced IDEs. Trenches were first created by DRIE on a silicon wafer, after which the electrode material (Pt together with a Ti adhesion layer) was deposited onto the sidewalls of the trenches by evaporation under a 45° incident angle. While the minimum separation between the electrodes was only 2 μm, a relatively high amplification factor of 60–70 was nonetheless achieved with this device because of the advantageous 3D geometry.

Another method introduced by Goluch et al.[109] for achieving higher amplification factors is to encase an IDE inside a fluidic channel, thus minimizing the loss of analyte molecules to the bulk solution above the IDE and increasing the average number of cycles undergone per molecule. Calculations indicate that the increase becomes most pronounced once the height of the channel becomes comparable to or smaller than the lateral electrode finger spacing. By embedding an IDE with a finger spacing of 250 nm in a series of parallel, 75 nm tall fluidic channels, an amplification factor of 110 was obtained. This was used to show that the confined IDE was capable of detecting paracetamol, a chemically reversible species, in the presence of a large excess of (irreversible) ascorbic acid. More recently, Heo et al. reported an amplification factor of 1100 in devices combining vertical face and confinement in a microchannel.[133]

At a higher degree of parallelization (albeit not of miniaturization), an addressable IDA fabricated on a single glass substrate and consisting of 32 rows and 32 columns of electrodes forming 1024 addressable sensing pixels was reported by Ino et al.[110,111] The electrodes were defined by sputtered Ti/Pt and the gap between the fingers was 12 μm; each sensing pixel was located at the bottom of a microwell that was formed by photoresist SU-8 and had a dimension of $100 \times 100 \times 7$ μm. Redox signals at each of the 1024 pixels could be acquired within 1 min, based on which a 2D map of the distributions of electrochemical species could be obtained.

16.4.2 Recessed Ring–Disk Electrodes and Arrays

An alternative to the IDE is the coplanar ring–disk electrode, which consists of a central disk-shaped electrode surrounded by a second, ring-shaped electrode.[134–136] A further refinement of this structure that is particularly suitable for microfabrication is the RRD electrode,[84,112–117] (Figure 16.5c and d), in which the two electrodes are placed on different planes. That is, a disk-shaped electrode forms the bottom of a recessed pit, while the ring electrode is located at the rim, also forming part of the sidewalls of the pit. Most such devices are fabricated by etching cylindrical cavities through the first two layers of a metal/insulator/metal stack, the two metal layers thus becoming the electrodes. An important advantage of this approach compared to IDEs is that the size of the gap between the two electrodes is determined by the thickness of the insulating layer, which does not depend on the resolution of the lithographic method employed and which can be straightforwardly controlled down to nanometer resolution.

A theoretical analysis focusing on the current collection efficiency and the transient response for this device geometry was provided by Menshykau et al.[115,116] It was concluded, with support from some experiments, that, in the operation mode where the disk acted as generator electrode and the ring as collector electrode, the current collecting efficiency, which depends on the recess depth and size of the collector ring, could reach 90%.

An interesting work in which RRD electrodes were characterized by both cyclic voltammetry and SECM was provided by Neugebauer et al.[114] Structures with a vertical space between the bottom and rim electrodes of about 200 nm and ring-electrode diameters varying between 200 and 800 nm were created with nanosphere lithography. Electrochemical activity images of single RRD electrodes in good agreement with the ring dimensions were captured, and it was demonstrated how the potential of the unbiased top electrode was influenced by the ratio of the oxidized and reduced form of the redox couples.

In a recent proof of concept for sensor applications, Ma et al.[84,117] reported an RRD electrode array in which the distance between the two electrodes was ~100 nm. Cavities were created with nanosphere lithography through deposited layers of $Au/SiN_x/Au/SiO_2$. The cavities had a radius of about 230 nm, as defined by the size of the polystyrene spheres; an SEM image of the array is demonstrated in Figure 16.5d. The collection efficiency was 98%. The arrays were also confined in a nanochannel; as a result, the detection selectivity for $Ru(NH_3)_6^{3+}$ in the presence of ascorbic acid was increased by a factor of 7 compared to an array in the absence of confinement.

16.4.3 NANOGAPS

Collection efficiency is further improved in a nanogap consisting of two parallel micrometric metal electrodes separated by a thin liquid layer,[118–130] as illustrated in Figure 16.5e and f. Conceptually, this configuration represents a direct downscaling of classic thin-layer cells. But whereas thin-layer cells with micron-scale spacing can be fabricated simply by sandwiching a thin spacer material between two flat electrodes, microfabrication mostly relies on a so-called sacrificial layer approach. A bottom electrode, a sacrificial layer made of a different material, and a top electrode are deposited and patterned on top of each other and passivated with an insulating layer. The resulting structure is illustrated in Figure 16.5f, which shows an SEM image of the cross section of a nanogap device from the authors' laboratory. At least one access hole is then opened through the insulating layer to make contact to the sacrificial layer. In a final step, the sacrificial layer is selectively etched away via the access hole(s) using a wet or isotropic dry etch, leaving a thin-layer cell structure with an electrode spacing determined by the thickness of the sacrificial layer before its removal. The nanogap geometry thus shares with RRDs the benefit that the electrode spacing is set by the thickness of a thin film, which can be accurately controlled, rather than by the resolution of the lithographic method employed. Indeed, nanogaps with spacing 40–65 nm have been demonstrated using micron-resolution optical lithography.[122,125,127] A potential pitfall of this geometry is that any residual strain in the top electrode can cause it to deform slightly; because of the small spacing between the electrodes, even minor buckling can result in a significant relative change in the electrode spacing. This problem was encountered in some early designs in which the electrodes had a square geometry,[118] but was later alleviated through the use of a thin rectangular electrodes[120] or judicious choices of materials.[122,125]

Because of the confined geometry of nanogap devices, the collection efficiency can in certain cases approach 100%, corresponding to a lower bound of ~10^4 for the amplification factor.[120] Largely thanks to this high degree of amplification, the detection of single molecules by redox-cycling electrochemistry was realized in nanogap devices.[123,127] The ability to form arrays of separately addressable nanogap detectors was further exploited in a chip-based recording system enabling in vitro measurements of individual neurotransmitter release events from neurons cultured directly on the chip.[137]

A strategy for further downscaling nanogaps to the sub–10 nm range was demonstrated by McCarty et al.[130] who employed a combination of optical and molecular lithography to minimize the gap size. In their approach, a single- or multilayered molecular film was grown selectively on a first electrode followed by the deposition and patterning of a second electrode, so that the space between the two electrodes was controlled by the thickness of the molecular resist. This molecular layer thus fulfilled the function of sacrificial layer described earlier. The resulting nanogaps took the form of 2 µm long, 50 nm deep crevices between the two electrodes. Gap sizes as small as 4 nm were reported and successfully employed in redox-cycling experiments. The crevice geometry is open to bulk solution in a manner reminiscent of IDEs, for which these devices could provide a higher-performance substitute; the formation of sealed channel structures with higher collection efficiencies can also be envisioned with additional processing steps.

An alternative strategy for further downscaling nanogaps is to decrease the spacing between the electrodes by controlled electrodeposition of additional material on the electrode surfaces.[124] This approach has been employed extensively to create closely separated electrodes, in particular with the aim of measuring the electronic properties of molecules trapped between the electrodes.[138–140] These applications, however, tend to focus on sharp, point-like electrodes that lead to low collection efficiencies. Applying it to parallel planar electrodes would require electrodeposition under conditions in which mass transport is not limiting in order to achieve a uniform decrease of the electrode spacing. To our knowledge, this has not been realized to date, however.

16.5 ELECTROCHEMISTRY AND MICROFLUIDIC INTEGRATION

Microfluidic systems,[141–143] also referred to as *lab-on-chip* or *micro-total-analysis systems*, consist of fluid handling elements such as valves, mixers, and pumps integrated on a microchip. In general, such miniaturized platforms offer several advantages including the ability to analyze small-volume samples, reduction in reagent consumption and a consequent reduction in the amount of waste to be disposed, and increased speed of analysis as well as potential for parallelization. Electrochemical detection is well suited for these applications as it is more readily integrated with fluidic elements than, for example, optical systems. Indeed, numerous integrated microfluidic electrochemical analytical systems have been reported in the last decade.[144–154]

In the early stages of development, the use of relatively complex silicon- and glass-based micromachining technology developed for ICs was explored to fabricate microfluidic chips.[143] More recently, the focus has shifted toward simpler techniques, micro-/nanofluidic channels being created directly by lithography or molding using low-cost polymers, such as PC,[155] PDMS,[145,156,157] olefin copolymer (COC),[158] and SU-8. [159,160] Among these materials, PDMS has been the most employed for its gas permeability, deformability, and ability to quickly produce prototype devices. It, however, has important drawbacks including in particular analyte absorption and low solvent resistance. COC is a popular alternative for environmental lab-on-a-chip applications due to its high chemical resistance and minimal water adsorption. SU-8, a form of photoresist, is available in a wide range of viscosities, making it suitable to form thick layers and high-aspect-ratio structures. It can be directly spin-coated onto substrates and patterned lithographically, making it particularly convenient for integration with electronic components. Depending on the choice of materials, either the microfluidics are fabricated directly onto a substrate on which electrochemical components have already been defined or the fluidic and electrochemical structures are formed on independent substrates that are bonded together afterwards.

In vitro experiments on living cells have benefited directly from fluidic integration. As discussed in Section 16.2, early approaches relied on glass rings or pierced petri dishes being glued onto electrode substrates to create culture chambers. It is now instead relatively straightforward to build arrays of independent chambers addressable with individual electrodes. For instance, a microwell device fabricated with SU-8 and PDMS for targeting single cells to detect quantal exocytosis—the burst release of intracellular transmitter molecules—was reported by Liu et al.[161] Transparent nitrogen-doped diamond-like-carbon (DLC:N)/indium-tin-oxide (ITO) films were defined on glass slides using photolithography and thin-film etching as electrodes in order to allow visualization of cells immobilized on the electrodes using a conventional inverted microscope. DLC:N was reported to promote cell adhesion and to exhibit good electrochemical properties. SU-8 was then coated and patterned on the slides to form microwells as well as to insulate inactive areas of the conductive film, following which a poly(ethylene glycol) film was grafted to the surface of the SU-8 to inhibit protein adsorption and cell adhesion. Finally, a PDMS gasket was cut and bonded to the substrate to confine a drop of solution containing cells to the middle of the device where 40 working electrodes were located. Single cells were targeted to the electrodes by functionalizing the electrodes with poly(L-lysine). Amperometric responses from individual cells could be recorded unambiguously without interference from nearby extraneous cells, and multiple recordings from the same electrode demonstrated that the device can be cleaned and reused without significant degradation of performance. This showed the potential of this platform as an alternative to carbon-fiber microelectrodes, which are extensively used to study quantal exocytosis of electroactive transmitters, with the additional advantage of increased throughput.

As another example, a disposable polymer-based protein immunosensor was demonstrated by Zou et al.[158] Images, a schematic sketch, and the fabrication process of the device are illustrated in

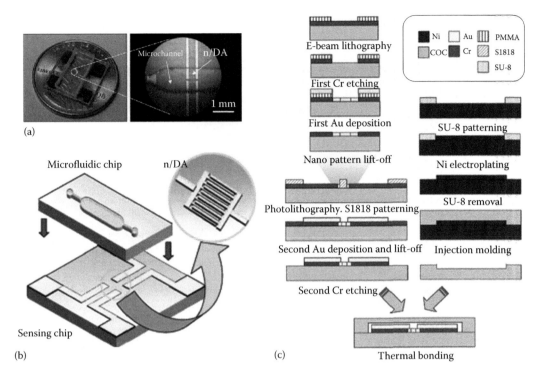

(a)

(b)

Microfluidic chip   n/DA

Sensing chip

(c)

Thermal bonding

E-beam lithography

First Cr etching

First Au deposition

Nano pattern lift-off

Photolithography. S1818 patterning

Second Au deposition and lift-off

Second Cr etching

SU-8 patterning

Ni electroplating

SU-8 removal

Injection molding

Ni   Au   PMMA
COC   Cr   S1818
SU-8

FIGURE 16.6 (a) Optical images, (b) schematic sketch of assembly, and (c) fabrication process for a micro-fluidic protein immunosensor based on nanoscale IDEs. (Reprinted from *Sens. Actuators A*, 136, Zou, Z.W., Kai, J.H., Rust, M.J., Han, J., and Ahn, C.H., Functionalized nano interdigitated electrodes arrays on polymer with integrated microfluidics for direct bio-affinity sensing using impedimetric measurement, 518–526, Copyright 2006, with permission from Elsevier.)

Figure 16.6. A 3 in. blank cyclic COC wafer with an ultrasmooth surface prepared by plastic injection molding was used as substrate. A gold IDE and contact pads were defined with e-beam lithography and lift-off after the COC wafer was coated with 10 nm Cr layer to render it compatible with lithography. Microfluidic channels were also fabricated in a second COC substrate using the same technique, except that here a Ni mold defined by a combination of lithography and electroplating was used, as shown in the right column of Figure 16.6c. After drilling holes for fluidic connections in the microfluidic chip using a microdrill and growing a self-assembled monolayer of alkanethiols on the gold electrode surfaces, the two substrates were thermally bonded, generating a reaction chamber with a volume of 0.2 µL.

An example of a higher level of multifunctional integration was reported by Ferguson et al.[162] who combined in a microfluidic electrochemical DNA sensor the functionalities of polymerase chain reaction (PCR), single-stranded DNA generation, and sequence-specific electrochemical detection. The architecture and fabrication process of the device are shown in Figure 16.7. The detection system incorporated counter, (quasi-) reference, and working electrodes that were defined by photolithography and lift-off. DNA probes were immobilized on the gold working electrodes via thiol chemistry. In parallel, a liquid chamber was fabricated by bonding a glass chip to a UV–ozone-treated PDMS sheet in which fluidic channels had been cut, and fluidic vias were generated by drilling through the glass chip with a mill equipped with a diamond bit. The exposed side of the PDMS was then bonded to the chip to complete the integration. During use, liquid was pumped into the chamber through eyelets affixed to the vias with epoxy. Comparing with traditional methods, this disposable device was argued to minimize both the sample loss and the likelihood of contamination as the fluid pathways were contained within a sterile system.

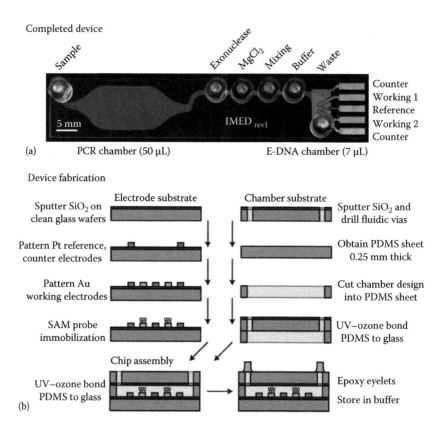

FIGURE 16.7 (a) Architecture and (b) fabrication process of a microfluidic electrochemical DNA sensor. (Reprinted with permission from Ferguson, B.S., Buchsbaum, S.F., Swensen, J.S. et al., Integrated microfluidic electrochemical DNA sensor, *Analytical Chemistry*, 2009, 81, 6503–6508. Copyright 2009 American Chemical Society.)

Fragoso and coworkers reported a system to electrochemically detect breast cancer markers[155] that was realized by high-precision milling of PC sheets forming two distinct sections, a detection zone incorporating an electrode array and a fluid storage zone. The detection zone was divided into separate microfluidic chambers for samples and calibrators, and the fluidic storage zone was split into five reservoirs to store the reagents and sample. The solutions in the separate reservoirs were actuated by applying pressure through a syringe pump and steered to the detection zone via two integrated valves. The detection of protein cancer markers in patient serum samples was demonstrated with detection limits below 10 ng/mL.

Nanogap devices can also be interfaced to fluidic systems. Under typical conditions, the fluidic resistance of the nanochannel is so high that negligible flow takes place within the device; in this case, the fluidics merely serve to bring a sample to the device, but analyte mass transport inside the device remains purely diffusive.[122,163] When sufficient pressure is applied between access points to the detection region, on the other hand, advective flows can develop along the surface of the electrodes. Figure 16.1b shows such a device reported by Mathwig and Lemay[128] in which microfluidic channels created in PDMS were interfaced to a nanochannel containing two separately addressable nanogap transducers. Record-low flow rates at the pL/min level could be measured from analyte time-of-flight measurements between the two electrochemical transducers.

Besides the benefits such as low consumption resulting from miniaturizing the fluidic components, phenomena specific to micro- and nanofluidic systems can also be harnessed to enhance electrochemical response.[1] For example, Wang et al. introduced a preconcentrating device that could be integrated with an electrochemical detector to study homogeneous enzyme reaction kinetics.[164] The negatively charged enzymes were concentrated via the exclusion-enrichment effect in a nanochannel[165] before being detected electrochemically near the outlet of the channel. In another example, Branagan et al. induced an electroosmotic flow (EOF) through a nanocapillary array membrane to enhance the delivery rate of analyte to annular nanoband electrodes embedded in the membrane.[166] An array of cylindrical nanochannels was created by FIB milling through Au/polymer/Au/polymer membranes and subsequently sandwiched between two axially separated microchannels. The generated EOF enhanced the steady-state current by a factor >10 compared to a comparable structure without convective transport.

It is important to note that many electrochemical micro-/nanofluidic systems still rely on macroscopic reference electrodes that are inserted in a solution reservoir external to the microfluidic system. This is because integrating a reliable, long-lived, and stable microfabricated reference electrode in miniaturized fluidic systems remains a challenge.[167] The main problem is the rapid dissolution of the (small) electrode volume, which leads to short lifetimes. Though pseudoreference electrodes—usually in the form of patterned metal thin films—can be used as a replacement, a true reference is often highly desirable. Analogues to conventional macroscopic liquid-junction reference electrodes have been demonstrated[168–170] in the form of encapsulated thin-film Ag/AgCl electrodes located in a dedicated compartment filled with reference electrolyte of constant Cl⁻ activity. Incorporating such a device, however, represents significant added complexity of design and fabrication. To form Ag/AgCl layers, an Ag film is normally deposited on an Au or Pt backbone layer in a first step, after which AgCl is formed by passing a current through the Ag layer in a solution with Cl^-.[170] Suzuki et al.[168] demonstrated an approach to fabricate a liquid-junction Ag/AgCl reference electrode using a resin sheet mainly formed by poly(ethylene glycol) as the liquid junction and screen-printed paste prepared from a mixture of KCl and 2-propanol as the electrolyte layer. Poly(vinylpyrrolidone) was added into the electrolyte layer to suppress the dissolution of AgCl, after which the electrode could maintain a stable potential level within ±1 mV for longer than 100 h. In another work Huang et al.[169] demonstrated a gel-coated Ti/Pd/Ag/AgCl electrode in which an agarose-stabilized KCl-gel membrane was introduced to serve both as a polymer-supported solid reference electrolyte and as ionic bridge for the electrode. The variation of the cell potential was less than 2 mV over pH 4–10 and insensitive to changes in the concentration of Cl^- (about 0.02–0.25 mV/pKCl).

16.6 INTEGRATION OF ELECTROCHEMICAL SYSTEMS WITH CMOS ELECTRONICS

By virtue of being inherently electrical in nature, electrochemical sensors are particularly well suited for integration with microelectronics compared to sensors based on other detection principles. While still in relatively early stages of development, such integration could open significant opportunities in applications such as high-throughput screening, point-of-care (POC) diagnosis, and implantable devices with flexibility, scalability, and low cost.

CMOS electronics provide the backbone of most commercial ICs, including microprocessors, microcontrollers, and image sensors. Several CMOS-based potentiostats have been reported. A great deal of flexibility in circuit topology is provided by CMOS, such that it is possible to design integrated circuitry with full potentiostat functionality and a range of operation modes approaching that of tabletop instruments.[171] For particular applications, however, it is often more practical to design more specialized electronics that implement a single electrochemical measurement technique of interest.[172–174] For example, Martin et al.[175] reported a custom integrated system for anodic stripping voltammetry (ASV) in which the detection circuit architecture was cooptimized with the electrode design to minimize parasitics, cancel solution matrix effects, and improve the dynamic range of the system. Alternatively, the relative ease with which addressable arrays can be implemented using CMOS electronics provides an ideal platform for parallelized assays.

This is dramatically illustrated by multiplexed microarrays functionalized with user-dialed probes via local, electronically controlled functionalization[176,177] or DNA synthesis.[178,179] Such systems have been used, for example, for DNA hybridization,[176,179–181] protein arrays,[178] and a range of immunoassays.[177,180]

The electrode materials most commonly employed in electrochemistry span a wide range including gold, platinum, palladium, carbon, graphite, and silver, all of which share the feature of being incompatible with CMOS manufacturing equipment and processes. Electrodes must therefore be formed subsequently to the completion of any CMOS circuitry, in the so-called post-CMOS processing stage. During this stage, high temperatures or intense plasmas that can destroy the circuits must be avoided. Martin et al.[175,182] reported integrated sensing systems for environmental monitoring with two sets of seven Au working electrodes that were selectable via an electronic multiplexer, in addition to two sets of Pt auxiliary electrodes and Ag/AgCl reference electrodes. The post-CMOS processing started with the deposition and lift-off of a Ti/TiN/Ti/Pt layer stack in which the two Ti layers were applied as adhesion promoters and the TiN layer was used as a diffusion barrier between the top-level CMOS metallization and the Pt sensing electrode. Cr/Au and Ti/Ag electrodes were deposited and defined separately in the following steps. Subsequently, the Ag/AgCl reference electrodes were created from the Ag surfaces by submersion of the chip in 1 mM $FeCl_3$ for 2 min.

Another consideration is that it is in general more demanding to package CMOS-based chemical sensors. The chips must be packaged in such a way that the electrical components and interconnects are protected from contact with liquid; otherwise, contaminants from solution may cause the properties of the transistors to drift over time. In specific cases such as protein-based sensors, aggressive electrode cleaning by piranha following organic solutions is required for reliable self-assembly of nanostructured biointerfaces, so the packaging material must withstand this strong corrosiveness. SU-8, polyimide, epoxy, and parylene are the most commonly used passivation material in the reported CMOS electrochemical microsystems.[173] In a microsystem for in situ detection of heavy metals in rainwater,[175] SU-8 was used to form a dam-like structure between the bonding pads and the sensor sites. Packaging was accomplished by fixing the device onto a printed circuit board (PCB) by epoxy, making electrical connections by wire bonding and then encapsulating the wires in a two-coat epoxy process. It was concluded that, using this packaging strategy, the passivation had a lifetime greater than 100 days in saturated salt solutions and the properties of the electronics exhibited only minor drift after soaking in a 100 mM NaCl solution for more than 35 days.

A long-lasting, parylene-packaged, wire-bonded chip that survived a harsh piranha electrode cleaning process was demonstrated by Li et al.[173] The post-CMOS fabrication began with the evaporation and wet etching of Ti/Au to form electrodes, and in the following step, polyimide was spin-coated on the chip and patterned to uncover the electrodes and bonding pads. Afterwards, the chip was wire-bonded to a packaging board and the assembly was coated with 5 μm parylene, following which the parylene was patterned to uncover the electrode sites using RIE in oxygen with a layer of crystal adhesive as the mask. This highlights the strengths of parylene as passivation material for CMOS-integrated biosensors: high chemical resistance in addition to biocompatibility, biostability, low cytotoxicity, relatively simple chemical vapor deposition methods with low process temperatures, and easy etching in O_2 plasmas.[173]

To achieve a higher level of integration by incorporating microfluidics with CMOS electrochemical sensors and realize a complete lab-on-CMOS system, problems caused by topographical conflicts also need to be solved. The first issue is the size disparity between conventional CMOS chips and microfluidic components, such as channels, valves, and pumps: the former typically occupy a few square millimeters, while microfluidic structures require significantly more area. The other inconvenience is the nonflat morphologies formed through the use of wire bonding or flip-chip bonding, which are the standard packaging techniques employed in the semiconductor industry to form electrical interconnections between CMOS chips and PCBs.

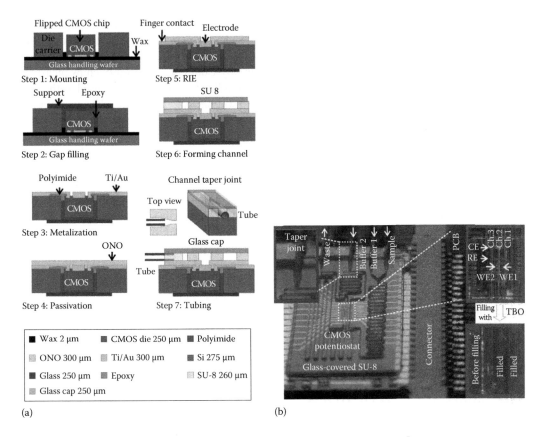

FIGURE 16.8 (a) Schematics of process flow for integrating microfluidics and CMOS electrochemical sensors. (b) The integrated device and schematic of the microfluidic circuits. The insets show (upper left) the taper joint, (upper right) the three microfluidic detection channels across the CMOS potentiostat with on-chip electrodes, and (lower right) the microfluidic channels being filled by TBO. (From Huang, Y. and Mason, A.J., Lab-on-CMOS integration of microfluidics and electrochemical sensors, *Lab Chip*, 13, 3929–3934, 2013. Reproduced by permission of The Royal Society of Chemistry.)

These methods lead to uneven surfaces due to wires protruding out of the surface of the chip, which is itself at a different height than the surrounding board. However, a smooth surface without bumps or steps is usually a necessary starting point to create the microfluidic systems described earlier.

To address these issues, Huang and Mason[183] recently introduced an integration scheme in which the CMOS chip was embedded into a micromachined silicon carrier as the packaging board, as shown in Figure 16.8. Both the CMOS chip and the carrier were first pressed onto a wax-coated glass handling wafer with the front side of the chip facing the handling wafer. The assembly was then placed in a 150°C chamber to allow the wax to melt, evening out the vertical position of the chip and carrier and attaching the chip to the handling wafer. Following this procedure, epoxy was applied to fill the gap between the chip and the carrier. Afterwards, the glass handling wafer was released by softening the wax at 100°C, and the wax remaining on the assembly was cleaned off. Polyimide was then coated onto the surface to smoothen it, metal wires for electrical interconnections were added by thin-film deposition and lithography, and a passivation layer consisting of silicon oxide/nitride/oxide was deposited at 100°C using PECVD. Finally, microfluidic structures with open channels made of SU-8 resist and covered by a glass cap were incorporated. Tubing was inserted laterally into the taper joint located at the sidewall of the SU-8 layer to enable high-density world-to-chip microfluidic interconnections. The integrated device and schematic of the microfluidic circuits are shown in Figure 16.8b. The simultaneous fluidic and electrical operation of the lab-on-CMOS device was demonstrated by detecting a diluted toluidine blue O (TBO) sample.

A method based on a similar concept was reported by Uddin et al.[184] CMOS chips measuring $3 \text{ mm} \times 3 \text{ mm} \times 260 \text{ μm}$ were placed on a resist-coated oxidized silicon wafer and used as the mask to pattern the resist. The pattern was then transferred to the oxide layer and subsequently to the silicon wafer by RIE and DRIE, respectively, so that cavities with approximately the same size as the CMOS chips were generated in this wafer. The wafer was then placed on a handle substrate, and the CMOS chips were placed facedown inside the cavities of the wafer with the help of a flip-chip bonder. Another wafer coated with benzocyclobutene (BCB) was then placed on top of the CMOS chips, and the cavity wafer and the whole stack were placed in a wafer bonding machine with pressure and elevated temperature (250°C). As a result, the CMOS chips and the cavity wafer were bonded to the wafer coated with BCB. In the next step, the handle substrate was removed from the front side of the CMOS chips, and spin-on glass was coated onto the surface to fill in the gap between the chip and the cavity wafer. Access to the contact pads was opened by RIE through the spin-on glass, and metal interconnects between the chip and the carrier were created by evaporation and lift-off. Up to this step, the process was done on the wafer scale, after which the wafer was diced into individual chips. Measurements with the packaged chip showed that the postintegration processing did not affect the CMOS device parameters. A hybrid CMOS/microfluidic system was completed by placing the embedded chip on an acrylic stage and securing it mechanically by fastening an acrylic microfluidic channel on top with screws. A PDMS gasket was used to achieve a tight seal between chip and fluidic channel.

16.7   SUMMARY AND OUTLOOK

This chapter reviewed the development of electrochemical measurement systems fabricated with micromachining technology. This set of techniques enables the systematic downscaling of the dimensions of experimental elements to explore electrochemistry in new regimes and to enhance sensitivity and selectivity in sensor applications. Lithography-based techniques provide the opportunity to build arrays of components with high controllability and repeatability. Additionally, the flexibility to integrate detecting electrodes, microfluidics, and even ICs onto a single chip that includes the functionalities of sensing, fluidic handling, and signal processing potentially creates new opportunities: while not going to replace existing instrumentation for classic measurements, the low costs, low sample and reagent volumes, low power consumptions, and possibility of massive parallelization open the door to new classes of electrochemical analytical methods. One can envision fully integrated microfluidic-based electrochemical measurement systems implemented on top of CMOS electronics to provide high-throughput biomedical analytical platforms, POC diagnostic tools, implantable devices, and portable and disposable food- and environment-monitoring sensors. Although a variety of difficulties remain, such as integration of reference electrode and effective packaging of the compact systems, the rapid pace of development means that such systems could become a practical reality on a relatively short timescale.

REFERENCES

1. Rassaei, L., Singh, P. S., and Lemay, S. G. 2011. Lithography-based nanoelectrochemistry. *Anal. Chem.* 83: 3974–3980.
2. Wightman, R. M. 1981. Microvoltammetric electrodes. *Anal. Chem.* 53: 1125A–1134A.
3. Bard, A. J. and Faulkner, L. R. 2001. *Electrochemical Methods: Fundamentals and Applications.* New York: John Wiley & Sons.
4. Adams, R. N. 1976. Probing brain chemistry with electroanalytical techniques. *Anal. Chem.* 48: 1126A–1138A.

5. Ponchon, J. L., Cespuglio, R., Gonon, F., Jouvet, M., and Pujol, J. F. 1979. Normal pulse polarography with carbon-fiber electrodes for in vitro and in vivo determination of catecholamines. *Anal. Chem.* 51: 1483–1486.

6. McCreery, R. L., Dreiling, R., and Adams, R. N. 1974. Voltammetry in brain-tissue—Quantitative studies of drug interactions. *Brain Res.* 73: 23–33.

7. Adams, R. N. 1978. In vivo electrochemical recording—A new neurophysiological approach. *Trends Neurosci.* 1: 160–163.

8. Thomas, C. A., Jr., Springer, P. A., Loeb, G. E., Berwald-Netter, Y., and Okun, L. M. 1972. A miniature microelectrode array to monitor the bioelectric activity of cultured cells. *Exp. Cell Res.* 74: 61–66.

9. May, G. A., Shamma, S. A., and White, R. L. 1979. Tantalum on sapphire micro-electrode array. *IEEE Trans. Electron Dev.* 26: 1932–1939.

10. Prohaska, O., Olcaytug, F., Womastek, K., and Petsche, H. 1977. Multielectrode for intracortical recordings produced by thin-film technology. *Electroencephalogr. Clin. Neurophysiol.* 42: 421–422.

11. Gross, G. W. 1979. Simultaneous single unit recording in vitro with a photoetched laser deinsulated gold multimicroelectrode surface. *IEEE Trans. Biomed. Eng.* 26: 273–279.

12. Pine, J. 1980. Recording action-potentials from cultured neurons with extracellular micro-circuit electrodes. *J. Neurosci. Methods* 2: 19–31.

13. Thormann, W., Vandenbosch, P., and Bond, A. M. 1985. Voltammetry at linear gold and platinum microelectrode arrays produced by lithographic techniques. *Anal. Chem.* 57: 2764–2770.

14. Kuperstein, M. and Whittington, D. A. 1981. A practical 24 channel microelectrode for neural recording in vivo. *IEEE Trans. Biomed. Eng.* 28: 288–293.

15. Kern, W. and Shaw, J. M. 1971. Electrochemical delineation of tungsten films for microelectronic devices. *J. Electrochem. Soc.* 118: 1699–1704.

16. Siu, W. and Cobbold, R. S. C. 1976. Characteristics of a multicathode polarographic oxygen electrode. *Med. Biol. Eng.* 14: 109–121.

17. Feeney, R. and Kounaves, S. P. 2000. Microfabricated ultramicroelectrode arrays: Developments, advances, and applications in environmental analysis. *Electroanalysis* 12: 677–684.

18. Glass, R. S., Perone, S. P., and Ciarlo, D. R. 1990. Application of information-theory to electroanalytical measurements using a multielement, microelectrode array. *Anal. Chem.* 62: 1914–1918.

19. Compton, R. G., Foord, J. S., and Marken, F. 2003. Electroanalysis at diamond-like and doped-diamond electrodes. *Electroanalysis* 15: 1349–1363.

20. Hu, J. P., Holt, K. B., and Foord, J. S. 2009. Focused ion beam fabrication of boron-doped diamond ultramicroelectrodes. *Anal. Chem.* 81: 5663–5670.

21. Soh, K. L., Kang, W. P., Davidson, J. L. et al. 2004. Diamond-derived microelectrodes array for electrochemical analysis. *Diam. Relat. Mater.* 13: 2009–2015.

22. Simm, A. O., Banks, C. E., Ward-Jones, S. et al. 2005. Boron-doped diamond microdisc arrays: Electrochemical characterisation and their use as a substrate for the production of microelectrode arrays of diverse metals (Ag, Au, Cu) via electrodeposition. *Analyst* 130: 1303–1311.

23. Zoski, C. G., Simjee, N., Guenat, O., and Koudelka-Hep, M. 2004. Addressable microelectrode arrays: Characterization by imaging with scanning electrochemical microscopy. *Anal. Chem.* 76: 62–72.

24. Lin, Z., Takahashi, Y., Kitagawa, Y. et al. 2008. An addressable microelectrode array for electrochemical detection. *Anal. Chem.* 80: 6830–6833.

25. Polsky, R., Harper, J. C., Wheeler, D. R., and Brozik, S. M. 2008. Multifunctional electrode arrays: Towards a universal detection platform. *Electroanalysis* 20: 671–679.

26. Sanchez-Molas, D., Esquivel, J. P., Sabate, N., Munoz, F. X., and del Campo, F. J. 2012. High aspect-ratio, fully conducting gold micropillar array electrodes: Silicon micromachining and electrochemical characterization. *J. Phys. Chem. C* 116: 18831–18846.

27. Wehmeyer, K. R., Deakin, M. R., and Wightman, R. M. 1985. Electroanalytical properties of band electrodes of submicrometer width. *Anal. Chem.* 57: 1913–1916.

28. Arrigan, D. W. M. 2004. Nanoelectrodes, nanoelectrode arrays and their applications. *Analyst* 129: 1157–1165.

29. Huang, X. J., O'Mahony, A. M., and Compton, R. G. 2009. Microelectrode arrays for electrochemistry: Approaches to fabrication. *Small* 5: 776–788.

30. Oja, S. M., Wood, M., and Zhang, B. 2013. Nanoscale electrochemistry. *Anal. Chem.* 85: 473–486.

31. Murray, R. W. 2008. Nanoelectrochemistry: Metal nanoparticles, nanoelectrodes, and nanopores. *Chem. Rev.* 108: 2688–2720.

32. Okazaki, S. 1991. Resolution limits of optical lithography. *J. Vac. Sci. Technol. B* 9: 2829–2833.

33. Zoski, C. G. 2002. Ultramicroelectrodes: Design, fabrication, and characterization. *Electroanalysis* 14: 1041–1051.

34. Cox, J. T. and Zhang, B. 2012. Nanoelectrodes: Recent advances and new directions. *Ann. Rev. Anal. Chem.* 5: 253–272.

35. Nagahara, L. A., Thundat, T., and Lindsay, S. M. 1989. Preparation and characterization of STM tips for electrochemical studies. *Rev. Sci. Instrum.* 60: 3128–3130.

36. Mirkin, M. V., Fan, F. R. F., and Bard, A. J. 1992. Scanning electrochemical microscopy part 13. Evaluation of the tip shapes of nanometer size microelectrodes. *J. Electroanal. Chem.* 328: 47–62.

37. Fan, F. R. F., Kwak, J., and Bard, A. J. 1996. Single molecule electrochemistry. *J. Am. Chem. Soc.* 118: 9669–9675.

38. Sun, P., Zhang, Z. Q., Guo, J. D., and Shao, Y. H. 2001. Fabrication of nanometer-sized electrodes and tips for scanning electrochemical microscopy. *Anal. Chem.* 73: 5346–5351.

39. Slevin, C. J., Gray, N. J., Macpherson, J. V., Webb, M. A., and Unwin, P. R. 1999. Fabrication and characterisation of nanometre-sized platinum electrodes for voltammetric analysis and imaging. *Electrochem. Commun.* 1: 282–288.

40. Conyers, J. L. and White, H. S. 2000. Electrochemical characterization of electrodes with submicrometer dimensions. *Anal. Chem.* 72: 4441–4446.

41. Macpherson, J. V. and Unwin, P. R. 2000. Combined scanning electrochemical-atomic force microscopy. *Anal. Chem.* 72: 276–285.

42. Gray, N. J. and Unwin, P. R. 2000. Simple procedure for the fabrication of silver/silver chloride potentiometric electrodes with micrometre and smaller dimensions: Application to scanning electrochemical microscopy. *Analyst* 125: 889–893.

43. Chen, S. L. and Kucernak, A. 2002. Fabrication of carbon microelectrodes with an effective radius of 1 nm. *Electrochem. Commun.* 4: 80–85.

44. Chen, S. L. and Kucernak, A. 2002. The voltammetric response of nanometer-sized carbon electrodes. *J. Phys. Chem. B* 106: 9396–9404.

45. Abbou, J., Demaille, C., Druet, M., and Moiroux, J. 2002. Fabrication of submicrometer-sized gold electrodes of controlled geometry for scanning electrochemical-atomic force microscopy. *Anal. Chem.* 74: 6355–6363.

46. Watkins, J. J., Chen, J., White, H. S. et al. 2003. Zeptomole voltammetric detection and electron-transfer rate measurements using platinum electrodes of nanometer dimensions. *Anal. Chem.* 75: 3962–3971.

47. Penner, R. M., Heben, M. J., Longin, T. L., and Lewis, N. S. 1990. Fabrication and use of nanometer-sized electrodes in electrochemistry. *Science* 250: 1118–1121.

48. Pendley, B. D. and Abruna, H. D. 1990. Construction of submicrometer voltammetric electrodes. *Anal. Chem.* 62: 782–784.

49. Shao, Y. H., Mirkin, M. V., Fish, G. et al. 1997. Nanometer-sized electrochemical sensors. *Anal. Chem.* 69: 1627–1634.

50. Katemann, B. B. and Schuhmann, T. 2002. Fabrication and characterization of needle-type Pt-disk nanoelectrodes. *Electroanalysis* 14: 22–28.

51. Zhang, B., Zhang, Y. H., and White, H. S. 2004. The nanopore electrode. *Anal. Chem.* 76: 6229–6238.

52. Sun, P. and Mirkin, M. V. 2006. Kinetics of electron-transfer reactions at nanoelectrodes. *Anal. Chem.* 78: 6526–6534.

53. Zhang, B., Galusha, J., Shiozawa, P. G. et al. 2007. Bench-top method for fabricating glass-sealed nanodisk electrodes, glass nanopore electrodes, and glass nanopore membranes of controlled size. *Anal. Chem.* 79: 4778–4787.

54. Thiébaud, P., Beuret, C., de Rooij, N. F., and Koudelka-Hep, M. 2000. Microfabrication of Pt-tip microelectrodes. *Sens. Actuators, B* 70: 51–56.

55. Qiao, Y., Chen, J., Guo, X. L. et al. 2005. Fabrication of nanoelectrodes for neurophysiology: Cathodic electrophoretic paint insulation and focused ion beam milling. *Nanotechnology* 16: 1598–1602.

56. Menke, E. J., Thompson, M. A., Xiang, C., Yang, L. C., and Penner, R. M. 2006. Lithographically patterned nanowire electrodeposition. *Nat. Mater.* 5: 914–919.

57. Yang, F., Taggart, D. K., and Penner, R. M. 2009. Fast, sensitive hydrogen gas detection using single palladium nanowires that resist fracture. *Nano Lett.* 9: 2177–2182.

58. Hujdic, J. E., Sargisian, A. P., Shao, J. R., Ye, T., and Menke, E. J. 2011. High-density gold nanowire arrays by lithographically patterned nanowire electrodeposition. *Nanoscale* 3: 2697–2699.

59. Heo, J. I., Shim, D. S., Teixidor, G. T. et al. 2011. Carbon interdigitated array nanoelectrodes for electrochemical applications. *J. Electrochem. Soc.* 158: J76–J80.

60. Varnell, G. L., Spicer, D. F., and Rodger, A. C. 1973. E-beam writing techniques for semiconductor-device fabrication. *J. Vac. Sci. Technol.* 10: 1048–1051.

61. Finot, E., Bourillot, E., Meunier-Prest, R. et al. 2003. Performance of interdigitated nanoelectrodes for electrochemical DNA biosensor. *Ultramicroscopy* 97: 441–449.

62. Sandison, M. E. and Cooper, J. M. 2006. Nanofabrication of electrode arrays by electron-beam and nano-imprint lithographies. *Lab Chip* 6: 1020–1025.

63. Naka, K., Hayashi, H., Senda, M., Shiraishi, H., and Konishi, S. 2007. Effect of nano stripe carbonized-polymer electrode on high S/N ratio in electrochemical detection. *Proceedings of the IEEE Twentieth Annual International Conference on Micro Electro Mechanical Systems*, Hyogo, Japan 1/2: 374–377.

64. Hoeben, F. J. M., Meijer, F. S., Dekker, C. et al. 2008. Toward single-enzyme molecule electrochemistry: [NiFe]-hydrogenase protein film voltammetry at nanoelectrodes. *ACS Nano* 2: 2497–2504.

65. Moretto, L. M., Tormen, M., De Leo, M., Carpentiero, A., and Ugo, P. 2011. Polycarbonate-based ordered arrays of electrochemical nanoelectrodes obtained by e-beam lithography. *Nanotechnology* 22: 185305.

66. Dawson, K., Wahl, A., Murphy, R., and O'Riordan, A. 2012. Electroanalysis at single gold nanowire electrodes. *J. Phys. Chem. C* 116: 14665–14673.

67. Kleijn, S. E. F., Yanson, A. I., and Koper, M. T. M. 2012. Electrochemical characterization of nano-sized gold electrodes fabricated by nano-lithography. *J. Electroanal. Chem.* 666: 19–24.

68. Niwa, O., Morita, M., and Tabei, H. 1990. Electrochemical-behavior of reversible redox species at inter-digitated array electrodes with different geometries—consideration of redox cycling and collection efficiency. *Anal. Chem.* 62: 447–452.

69. Lanyon, Y. H. and Arrigan, D. W. M. 2007. Recessed nanoband electrodes fabricated by focused ion beam milling. *Sens. Actuators B* 121: 341–347.

70. Lanyon, Y. H., De Marzi, G., Watson, Y. E. et al. 2007. Fabrication of nanopore array electrodes by focused ion beam milling. *Anal. Chem.* 79: 3048–3055.

71. Rauf, S., Shiddiky, M. J. A., Asthana, A., and Dimitrov, K. 2012. Fabrication and characterization of gold nanohole electrode arrays. *Sens. Actuators B* 173: 491–496.

72. Menon, V. P. and Martin, C. R. 1995. Fabrication and evaluation of nanoelectrode ensembles. *Anal. Chem.* 67: 1920–1928.

73. Penner, R. M. and Martin, C. R. 1987. Preparation and electrochemical characterization of ultramicro-electrode ensembles. *Anal. Chem.* 59: 2625–2630.

74. Lemay, S. G., van den Broek, D. M., Storm, A. J. et al. 2005. Lithographically fabricated nanopore-based electrodes for electrochemistry. *Anal. Chem.* 77: 1911–1915.

75. Krapf, D., Wu, M. Y., Smeets, R. M. M. et al. 2006. Fabrication and characterization of nanopore-based electrodes with radii down to 2 nm. *Nano Lett.* 6: 105–109.

76. Yang, M., Qu, F., Lu, Y. et al. 2006. Platinum nanowire nanoelectrode array for the fabrication of biosensors. *Biomaterials* 27: 5944–5950.

77. Chou, S. Y., Krauss, P. R., and Renstrom, P. J. 1996. Imprint lithography with 25-nanometer resolution. *Science* 272: 85–87.

78. Beck, M., Persson, F., Carlberg, P. et al. 2004. Nanoelectrochemical transducers for (bio-) chemical sensor applications fabricated by nanoimprint lithography. *Microelectron. Eng.* 73–4: 837–842.

79. Huang, C. W. and Lu, M. S. C. 2011. Electrochemical detection of the neurotransmitter dopamine by nanoimprinted interdigitated electrodes and a CMOS circuit with enhanced collection efficiency. *IEEE Sens. J.* 11: 1826–1831.

80. Haynes, C. L. and Van Duyne, R. P. 2001. Nanosphere lithography: A versatile nanofabrication tool for studies of size-dependent nanoparticle optics. *J. Phys. Chem. B* 105: 5599–5611.

81. Valsesia, A., Lisboa, P., Colpo, P., and Rossi, F. 2006. Fabrication of polypyrrole-based nanoelectrode arrays by colloidal lithography. *Anal. Chem.* 78: 7588–7591.

82. Lohmuller, T., Muller, U., Breisch, S. et al. 2008. Nano-porous electrode systems by colloidal lithography for sensitive electrochemical detection: Fabrication technology and properties. *J. Micromech. Microeng.* 18: 115011.

83. Hees, J., Hoffmann, R., Kriele, A. et al. 2011. Nanocrystalline diamond nanoelectrode arrays and ensembles. *ACS Nano* 5: 3339–3346.

84. Ma, C., Contento, N. M., Gibson, L. R., 2nd, and Bohn, P. W. 2013. Recessed ring-disk nanoelectrode arrays integrated in nanofluidic structures for selective electrochemical detection. *Anal. Chem.* 85: 9882–9888.

85. Morris, R. B., Franta, D. J., and White, H. S. 1987. Electrochemistry at Pt band electrodes of width approaching molecular dimensions—Breakdown of transport-equations at very small electrodes. *J. Phys. Chem.* 91: 3559–3564.

86. Dumitrescu, I., Unwin, P. R., and Macpherson, J. V. 2009. Electrochemistry at carbon nanotubes: Perspective and issues. *Chem. Commun.* 7: 6886–6901.

87. Dumitrescu, I., Unwin, P. R., Wilson, N. R., and Macpherson, J. V. 2008. Single-walled carbon nanotube network ultramicroelectrodes. *Anal. Chem.* 80: 3598–3605.

88. Dumitrescu, I., Edgeworth, J. P., Unwin, P. R., and Macpherson, J. V. 2009. Ultrathin carbon nanotube mat electrodes for enhanced amperometric detection. *Adv. Mater.* 21: 3105–3109.

89. Heller, I., Kong, J., Heering, H. A. et al. 2005. Individual single-walled carbon nanotubes as nanoelectrodes for electrochemistry. *Nano Lett.* 5: 137–142.

90. Quinn, B. M. and Lemay, S. G. 2006. Single-walled carbon nanotubes as templates and interconnects for nanoelectrodes. *Adv. Mater.* 18: 855–859.

91. Day, T. M., Unwin, P. R., Wilson, N. R., and Macpherson, J. V. 2005. Electrochemical templating of metal nanoparticles and nanowires on single-walled carbon nanotube networks. *J. Am. Chem. Soc.* 127: 10639–10647.

92. Quinn, B. M., Dekker, C., and Lemay, S. G. 2005. Electrodeposition of noble metal nanoparticles on carbon nanotubes. *J. Am. Chem. Soc.* 127: 6146–6147.

93. Dawson, K., Strutwolf, J., Rodgers, K. P. et al. 2011. Single nanoskived nanowires for electrochemical applications. *Anal. Chem.* 83: 5535–5540.

94. Xu, Q., Rioux, R. M., Dickey, M. D., and Whitesides, G. M. 2008. Nanoskiving: A new method to produce arrays of nanostructures. *Acc. Chem. Res.* 41: 1566–1577.

95. Campbell, J. K., Sun, L., and Crooks, R. M. 1999. Electrochemistry using single carbon nanotubes. *J. Am. Chem. Soc.* 121: 3779–3780.

96. Guillorn, M. A., McKnight, T. E., Melechko, A. et al. 2002. Individually addressable vertically aligned carbon nanofiber-based electrochemical probes. *J. Appl. Phys.* 91: 3824–3828.

97. Mai, L., Dong, Y., Xu, L., and Han, C. 2010. Single nanowire electrochemical devices. *Nano Lett.* 10: 4273–4278.

98. Zhao, M., Huang, J., Zhou, Y. et al. 2013. A single mesoporous ZnO/chitosan hybrid nanostructure for a novel free nanoprobe type biosensor. *Biosens. Bioelectron.* 43: 226–230.

99. Percival, S. J. and Zhang, B. 2013. Electrocatalytic reduction of oxygen at single platinum nanowires. *J. Phys. Chem. C* 117: 13928–13935.

100. Kranz, C., Friedbacher, G., Mizaikoff, B. et al. 2001. Integrating an ultramicroelectrode in an AFM cantilever: Combined technology for enhanced information. *Anal. Chem.* 73: 2491–2500.

101. Lugstein, A., Bertagnolli, E., Kranz, C., and Mizaikoff, B. 2002. Fabrication of a ring nanoelectrode in an AFM tip: Novel approach towards simultaneous electrochemical and topographical imaging. *Surf. Interface Anal.* 33: 146–150.

102. Burt, D. P., Wilson, N. R., Weaver, J. M. R., Dobson, P. S., and Macpherson, J. V. 2005. Nanowire probes for high resolution combined scanning electrochemical microscopy—Atomic force microscopy. *Nano Lett.* 5: 639–643.

103. Bai, S. J., Fabian, T., Prinz, F. B., and Fasching, R. J. 2008. Nanoscale probe system for cell-organelle analysis. *Sens. Actuators B* 130: 249–257.

104. Smirnov, W., Kriele, A., Hoffmann, R. et al. 2011. Diamond-modified AFM probes: From diamond nanowires to atomic force microscopy-integrated boron-doped diamond electrodes. *Anal. Chem.* 83: 4936–4941.

105. Sanderson, D. G. and Anderson, L. B. 1985. Filar electrodes—steady-state currents and spectroelectrochemistry at twin interdigitated electrodes. *Anal. Chem.* 57: 2388–2393.

106. Aoki, K., Morita, M., Niwa, O., and Tabei, H. 1988. Quantitative-analysis of reversible diffusion-controlled currents of redox soluble species at interdigitated array electrodes under steady-state conditions. *J. Electroanal. Chem.* 256: 269–282.

107. Niwa, O., Xu, Y., Halsall, H. B., and Heineman, W. R. 1993. Small-volume voltammetric detection of 4-Aminophenol with interdigitated array electrodes and its application to electrochemical enzyme-immunoassay. *Anal. Chem.* 65: 1559–1563.

108. Dam, V. A. T., Olthuis, W., and van den Berg, A. 2007. Redox cycling with facing interdigitated array electrodes as a method for selective detection of redox species. *Analyst* 132: 365–370.

109. Goluch, E. D., Wolfrum, B., Singh, P. S., Zevenbergen, M. A. G., and Lemay, S. G. 2009. Redox cycling in nanofluidic channels using interdigitated electrodes. *Anal. Bioanal. Chem.* 394: 447–456.

110. Ino, K., Saito, W., Koide, M. et al. 2011. Addressable electrode array device with IDA electrodes for high-throughput detection. *Lab Chip* 11: 385–388.

111. Ino, K., Nishijo, T., Arai, T. et al. 2012. Local redox-cycling-based electrochemical chip device with deep microwells for evaluation of embryoid bodies. *Angew. Chem. Int. Ed.* 51: 6648–6652.

112. Henry, C. S. and Fritsch, I. 1999. Microcavities containing individually addressable recessed microdisk and tubular nanoband electrodes. *J. Electrochem. Soc.* 146: 3367–3373.

113. Vandaveer, W. R., Woodward, D. J., and Fritsch, I. 2003. Redox cycling measurements of a model compound and dopamine in ultrasmall volumes with a self-contained microcavity device. *Electrochim. Acta* 48: 3341–3348.

114. Neugebauer, S., Muller, U., Lohmuller, T. et al. 2006. Characterization of nanopore electrode structures as basis for amplified electrochemical assays. *Electroanalysis* 18: 1929–1936.

115. Menshykau, D., O'Mahony, A. M., del Campo, F. J., Munoz, F. X., and Compton, R. G. 2009. Microarrays of ring-recessed disk electrodes in transient generator-collector mode: Theory and experiment. *Anal. Chem.* 81: 9372–9382.

116. Menshykau, D., del Campo, F. J., Munoz, F. X., and Compton, R. G. 2009. Current collection efficiency of micro- and nano-ring-recessed disk electrodes and of arrays of these electrodes. *Sens. Actuators B* 138: 362–367.

117. Ma, C. X., Contento, N. M., Gibson, L. R., and Bohn, P. W. 2013. Redox cycling in nanoscale-recessed ring-disk electrode arrays for enhanced electrochemical sensitivity. *ACS Nano* 7: 5483–5490.

118. Zevenbergen, M. A. G., Krapf, D., Zuiddam, M. R., and Lemay, S. G. 2007. Mesoscopic concentration fluctuations in a fluidic nanocavity detected by redox cycling. *Nano Lett.* 7: 384–388.

119. Wolfrum, B., Zevenbergen, M., and Lemay, S. 2008. Nanofluidic redox cycling amplification for the selective detection of catechol. *Anal. Chem.* 80: 972–977.

120. Zevenbergen, M. A. G., Wolfrum, B. L., Goluch, E. D., Singh, P. S., and Lemay, S. G. 2009. Fast electron-transfer kinetics probed in nanofluidic channels. *J. Am. Chem. Soc.* 131: 11471–11477.

121. Zevenbergen, M. A. G., Singh, P. S., Goluch, E. D., Wolfrum, B. L., and Lemay, S. G. 2009. Electrochemical correlation spectroscopy in nanofluidic cavities. *Anal. Chem.* 81: 8203–8212.

122. Katelhon, E., Hofmann, B., Lemay, S. G. et al. 2010. Nanocavity redox cycling sensors for the detection of dopamine fluctuations in microfluidic gradients. *Anal. Chem.* 82: 8502–8509.

123. Zevenbergen, M. A. G., Singh, P. S., Goluch, E. D., Wolfrum, B. L., and Lemay, S. G. 2011. Stochastic sensing of single molecules in a nanofluidic electrochemical device. *Nano Lett.* 11: 2881–2886.

124. Kim, J. H., Moon, H., Yoo, S., and Choi, Y. K. 2011. Nanogap electrode fabrication for a nanoscale device by volume-expanding electrochemical synthesis. *Small* 7: 2210–2216.

125. Kang, S., Mathwig, K., and Lemay, S. G. 2012. Response time of nanofluidic electrochemical sensors. *Lab Chip* 12: 1262–1267.

126. Mathwig, K., Mampallil, D., Kang, S., and Lemay, S. G. 2012. Electrical cross-correlation spectroscopy: Measuring picoliter-per-minute flows in nanochannels. *Phys. Rev. Lett.* 109: 118302.

127. Kang, S., Nieuwenhuis, A. F., Mathwig, K., Mampallil, D., and Lemay, S. G. 2013. Electrochemical single-molecule detection in aqueous solution using self-aligned nanogap transducers. *ACS Nano* 7: 10931–10937.

128. Mathwig, K. and Lemay, S. G. 2013. Pushing the limits of electrical detection of ultralow flows in nanofluidic channels. *Micromachines* 4: 138–148.

129. Mampallil, D., Mathwig, K., Kang, S., and Lemay, S. G. 2013. Redox couples with unequal diffusion coefficients: Effect on redox cycling. *Anal. Chem.* 85: 6053–6058.

130. McCarty, G. S., Moody, B., and Zachek, M. K. 2010. Enhancing electrochemical detection by scaling solid state nanogaps. *J. Electroanal. Chem.* 643: 9–14.

131. Van Gerwen, P., Laureyn, W., Laureys, W. et al. 1998. Nanoscaled interdigitated electrode arrays for biochemical sensors. *Sens. Actuators B* 49: 73–80.

132. Yang, L. J., Li, Y. B., and Erf, G. F. 2004. Interdigitated array microelectrode-based electrochemical impedance immunosensor for detection of *Escherichia coli* O157: H7. *Anal. Chem.* 76: 1107–1113.

133. Heo, J. I., Lim, Y., and Shin, H. 2013. The effect of channel height and electrode aspect ratio on redox cycling at carbon interdigitated array nanoelectrodes confined in a microchannel. *Analyst* 138: 6404–6411.

134. Albery, W. J. H. and Hitchman, M. L. 1971. *Ring-Disc Electrodes*. Lodon, U.K.: Oxford University Press.

135. Zhao, G., Giolando, D. M., and Kirchhoff, J. R. 1995. Carbon ring disk ultramicroelectrodes. *Anal. Chem.* 67: 1491–1495.

136. Liljeroth, P., Johans, C., Slevin, C. J., Quinn, B. M., and Kontturi, K. 2002. Disk-generation/ring-collection scanning electrochemical microscopy: Theory and application. *Anal. Chem.* 74: 1972–1978.

137. Yakushenko, A., Kaetelhoen, E., and Wolfrum, B. 2013. Parallel on-chip analysis of single vesicle neurotransmitter release. *Anal. Chem.* 85: 5483–5490.

138. Reed, M. A., Zhou, C., Muller, C. J., Burgin, T. P., and Tour, J. M. 1997. Conductance of a molecular junction. *Science* 278: 252–254.
139. Joachim, C., Gimzewski, J. K., and Aviram, A. 2000. Electronics using hybrid-molecular and mono-molecular devices. *Nature* 408: 541–548.
140. Akkerman, H. B., Blom, P. W. M., de Leeuw, D. M., and de Boer, B. 2006. Towards molecular electronics with large-area molecular junctions. *Nature* 441: 69–72.
141. Manz, A., Fettinger, J. C., Verpoorte, E. et al. 1991. Micromachining of monocrystalline silicon and glass for chemical-analysis systems—A look into next century technology or just a fashionable craze. *Trends Anal. Chem.* 10: 144–149.
142. Kovacs, G. T. A., Petersen, K., and Albin, M. 1996. Silicon micromachining—Sensors to systems. *Anal. Chem.* 68: A407–A412.
143. Erickson, D. and Li, D. Q. 2004. Integrated microfluidic devices. *Anal. Chim. Acta* 507: 11–26.
144. Lin, Y. H., Wu, H., Timchalk, C. A., and Thrall, K. D. 2002. Integrated microfluidics/electrochemical sensor system for monitoring of environmental exposures to toxic chemicals. *Abstr. Pap. Am. Chem. Soc.* 223: U77.
145. Wang, J. 2002. Electrochemical detection for microscale analytical systems: A review. *Talanta* 56: 223–231.
146. Rossier, J., Reymond, F., and Michel, P. E. 2002. Polymer microfluidic chips for electrochemical and biochemical analyses. *Electrophoresis* 23: 858–867.
147. Wang, J. 2005. Electrochemical detection for capillary electrophoresis microchips: A review. *Electroanalysis* 17: 1133–1140.
148. Bange, A., Halsall, H. B., and Heineman, W. R. 2005. Microfluidic immunosensor systems. *Biosens. Bioelectron.* 20: 2488–2503.
149. Goral, V. N., Zaytseva, N. V., and Baeumner, A. J. 2006. Electrochemical microfluidic biosensor for the detection of nucleic acid sequences. *Lab Chip* 6: 414–421.
150. Sadik, O. A., Aluoch, A. O., and Zhou, A. L. 2009. Status of biomolecular recognition using electro-chemical techniques. *Biosens. Bioelectron.* 24: 2749–2765.
151. Swensen, J. S., Xiao, Y., Ferguson, B. S. et al. 2009. Continuous, real-time monitoring of cocaine in undiluted blood serum via a microfluidic, electrochemical aptamer-based sensor. *J. Am. Chem. Soc.* 131: 4262–4266.
152. Kraly, J. R., Holcomb, R. E., Guan, Q., and Henry, C. S. 2009. Review: Microfluidic applications in metabolomics and metabolic profiling. *Anal. Chim. Acta* 653: 23–35.
153. Dungchai, W., Chailapakul, O., and Henry, C. S. 2009. Electrochemical detection for paper-based micro-fluidics. *Anal. Chem.* 81: 5821–5826.
154. Jang, A., Zou, Z. W., Lee, K. K., Ahn, C. H., and Bishop, P. L. 2011. State-of-the-art lab chip sensors for environmental water monitoring. *Meas. Sci. Technol.* 22: 032001.
155. Fragoso, A., Latta, D., Laboria, N. et al. 2011. Integrated microfluidic platform for the electrochemical detection of breast cancer markers in patient serum samples. *Lab Chip* 11: 625–631.
156. Hebert, N. E., Kuhr, W. G. and Brazill, S. A. 2002. Microchip capillary electrophoresis coupled to sinu-soidal voltammetry for the detection of native carbohydrates. *Electrophoresis* 23: 3750–3759.
157. Zhan, W., Alvarez, J., and Crooks, R. M. 2003. A two-channel microfluidic sensor that uses anodic electrogenerated chemiluminescence as a photonic reporter of cathodic redox reactions. *Anal. Chem.* 75: 313–318.
158. Zou, Z. W., Kai, J. H., Rust, M. J., Han, J., and Ahn, C. H. 2007. Functionalized nano interdigitated elec-trodes arrays on polymer with integrated microfluidics for direct bio-affinity sensing using impedimetric measurement. *Sens. Actuators A* 136: 518–526.
159. Lorenz, H., Despont, M., Fahrni, N. et al. 1997. SU-8: A low-cost negative resist for MEMS. *J. Micromech. Microeng.* 7: 121–124.
160. Castano-Alvarez, M., Fernandez-Abedul, M. T., Costa-Garcia, A. et al. 2009. Fabrication of SU-8 based microchip electrophoresis with integrated electrochemical detection for neurotransmitters. *Talanta* 80: 24–30.
161. Liu, X., Barizuddin, S., Shin, W. et al. 2011. Microwell device for targeting single cells to electrochemi-cal microelectrodes for high-throughput amperometric detection of quantal exocytosis. *Anal. Chem.* 83: 2445–2451.
162. Ferguson, B. S., Buchsbaum, S. F., Swensen, J. S. et al. 2009. Integrated microfluidic electrochemical DNA sensor. *Anal. Chem.* 81: 6503–6508.
163. Rassaei, L., Mathwig, K., Goluch, E. D., and Lemay, S. G. 2012. Hydrodynamic voltammetry with nanogap electrodes. *J. Phys. Chem. C* 116: 10913–10916.

164. Wang, C., Li, S. J., Wu, Z. Q. et al. 2010. Study on the kinetics of homogeneous enzyme reactions in a micro/nanofluidics device. *Lab Chip* 10: 639–646.

165. Plecis, A., Schoch, R. B., and Renaud, P. 2005. Ionic transport phenomena in nanofluidics: Experimental and theoretical study of the exclusion-enrichment effect on a chip. *Nano Lett.* 5: 1147–1155.

166. Branagan, S. P., Contento, N. M., and Bohn, P. W. 2012. Enhanced mass transport of electroactive species to annular nanoband electrodes embedded in nanocapillary array membranes. *J. Am. Chem. Soc.* 134: 8617–8624.

167. Shinwari, M. W., Zhitomirsky, D., Deen, I. A. et al. 2010. Microfabricated reference electrodes and their biosensing applications. *Sensors* 10: 1679–1715.

168. Suzuki, H., Shiroishi, H., Sasaki, S., and Karube, I. 1999. Microfabricated liquid junction Ag/AgCl reference electrode and its application to a one-chip potentiometric sensor. *Anal. Chem.* 71: 5069–5075.

169. Huang, I. Y., Huang, R. S., and Lo, L. H. 2003. Improvement of integrated Ag/AgCl thin-film electrodes by KCl-gel coating for ISFET applications. *Sens. Actuators B* 94: 53–64.

170. Zhou, J. H., Ren, K. N., Zheng, Y. Z. et al. 2010. Fabrication of a microfluidic Ag/AgCl reference electrode and its application for portable and disposable electrochemical microchips. *Electrophoresis* 31: 3083–3089.

171. Hassibi, A. and Lee, T. H. 2006. A programmable 0.18-mu m CMOS electrochemical sensor microarray for biomolecular detection. *IEEE Sens. J.* 6: 1380–1388.

172. Zhu, X. S. and Ahn, C. H. 2006. On-chip electrochemical analysis system using nanoelectrodes and bioelectronic CMOS chip. *IEEE Sens. J.* 6: 1280–1286.

173. Li, L., Liu, X. W., Qureshi, W. A., and Mason, A. J. 2011. CMOS amperometric instrumentation and packaging for biosensor array applications. *IEEE Trans. Biomed. Circuits Syst.* 5: 439–448.

174. Huang, Y., Liu, Y., Hassler, B. L., Worden, R. M., and Mason, A. J. 2013. A protein-based electrochemical biosensor array platform for integrated microsystems. *IEEE Trans. Biomed. Circuits Syst.* 7: 43–51.

175. Martin, S. M., Gebara, F. H., Larivee, B. J., and Brown, R. B. 2005. A CMOS-integrated microinstrument for trace detection of heavy metals. *IEEE J. Solid-State Circ.* 40: 2777–2786.

176. Swanson, P., Gelbart, R., Atlas, E. et al. 2000. A fully multiplexed CMOS biochip for DNA analysis. *Sens. Actuators B* 64: 22–30.

177. Dill, K., Montgomery, D. D., Wang, W., and Tsai, J. C. 2001. Antigen detection using microelectrode array microchips. *Anal. Chim. Acta.* 444: 69–78.

178. Oleinikov, A. V., Gray, M. D., Zhao, J. et al. 2003. Self-assembling protein arrays using electronic semiconductor microchips and in vitro translation. *J. Proteome Res.* 2: 313–319.

179. Maurer, K., Cooper, J., Caraballo, M. et al. 2006. Electrochemically generated acid and its containment to 100 micron reaction areas for the production of DNA microarrays. *PLoS ONE* 1: e34.

180. Dill, K., Montgomery, D. D., Ghindilis, A. L., and Schwarzkopf, K. R. 2004. Immunoassays and sequence-specific DNA detection on a microchip using enzyme amplified electrochemical detection. *J. Biochem. Biophys. Methods* 59: 181–187.

181. Maurer, K., Yazvenko, N., Wilmoth, J. et al. 2010. Use of a multiplexed CMOS microarray to optimize and compare oligonucleotide binding to DNA probes synthesized or immobilized on individual electrodes. *Sensors* 10: 7371–7385.

182. Martin, S. M., Strong, T. D., and Brown, R. B. 2004. Design, implementation, and verification of a CMOS-integrated chemical sensor system. *Proceedings. 2004 International Conference on MEMS, NANO and Smart Systems, 2004*, pp. 379–385.

183. Huang, Y. and Mason, A. J. 2013. Lab-on-CMOS integration of microfluidics and electrochemical sensors. *Lab Chip* 13: 3929–3934.

184. Uddin, A., Milaninia, K., Chen, C. H., and Theogarajan, L. 2011. Wafer scale integration of CMOS chips for biomedical applications via self-aligned masking. *IEEE Trans. Compon., Packag., Manuf. Technol.* 1: 1996–2004.

185. Ueno, K., Hayashida, M., Ye, J., and Misawa, H. 2005. Fabrication and electrochemical characterization of interdigitated nanoelectrode arrays. *Electrochemistry Communications* 7: 161–165.

17 Electrodeposition at the Nanoscale

Jay A. Switzer

CONTENTS

17.1 INTRODUCTION

In this chapter, we will review the production of nanostructured materials by electrodeposition. We will first give a general introduction to the advantages and limitations of electrodeposition. We will then discuss the electrodeposition of metals, semiconductors, and oxides, followed by a survey of some of the techniques used to electrodeposit nanostructures of these materials.

Electrodeposition is a solution processing method that assembles solid materials from molecules, ions, or complexes in solution.[1] The reactions occur on solid surfaces to produce polycrystalline, epitaxial, and ultrathin films, porous networks, nanorods, superlattices, and composites.[2] Because of the low processing temperatures (often near room temperature), the technique is ideal for producing nanostructured materials and interfaces because interdiffusion and sintering can be minimized. Electrodeposition is not only inexpensive and relatively simple, but it can also often produce materials and nanostructures that cannot be accessed in ultrahigh vacuum (UHV). For example, the shape and orientation can be tuned by controlling the pH or through solution additives. In this regard, electrodeposition is similar to biomineralization—the solution method utilized by *nature* to assemble very elegant, sophisticated, and highly functional structures from solution precursors.

What really distinguishes electrodeposition from other deposition techniques is the applied potential. The applied potential controls the departure from equilibrium and, therefore, the rate of the reaction. The electrodeposition of metals only requires that the electrode potential be driven negative of the equilibrium potential. The difference between the applied potential and the equilibrium potential is called the overpotential. Because the electrode must be poised at a potential for deposition to occur, the substrate must be a conductor or semiconductor. In contrast, vapor phase deposition does not require a conducting substrate. This substrate constraint is not a limitation in the electrodeposition of catalysts for fuel cells, batteries, and photoelectrochemical solar cells, however, where electrical continuity from the substrate to the surface of the film is a requirement for the device. There is also some limitation as to what materials can be deposited, based on the electrochemical window of the solvent that is used to prepare the deposition solution. As we discuss later, however, there are often clever chemical tricks that can be used to circumvent this issue. For example, we discuss the electrodeposition of elemental Si and Ge from aqueous solution, a process that should not be possible according to simple thermodynamic arguments.

Electrodeposition can be used to produce metals[1,3–13] or to induce the deposition of compound materials, such as metal oxide ceramics,[14–43] biominerals,[44] and metal chalcogenide (e.g., CdS) semiconductors.[45–58] Deposition is not a line-of-sight method, so conformal films can be grown on complex shapes. This aspect of electrodeposition is exploited, for instance, in the on-chip deposition of copper interconnects into submicron-sized features of semiconductor devices.[3] The bottom-up filling of vias is controlled by solution additives, which promote the growth of copper at the bottom and sides of the via but inhibit growth on the shoulders. The use of solution additives to control the morphology of electrodeposits is shown in Figure 17.1, in which the Choi group deposits Cu_2O in the presence of various additives.[59] The basic concept that is used in controlling morphology through additive addition is that the final morphology of a crystal is determined by the slowest growing crystal face. Hence, if molecules or ions can be selectively adsorbed on crystal faces that inhibit growth, those slow-growing faces will be the dominant surfaces in the final microstructure. This is shown schematically in Figure 17.2. If chiral ligands are used, the chiral ligand can template the growth of materials with chiral surfaces or morphologies.[41,42,44,60] In the case of single-crystal substrates, it is possible to electrodeposit epitaxial films, ultrathin films, and self-ordered nano-structures.[12,13] The stoichiometry of compounds and metal alloys can often be controlled through the applied potential, leading to the possibility of electrodepositing layered nanostructures such as superlattices.[15,31,33,34,61] A final advantage of electrodeposition over other deposition methods is that

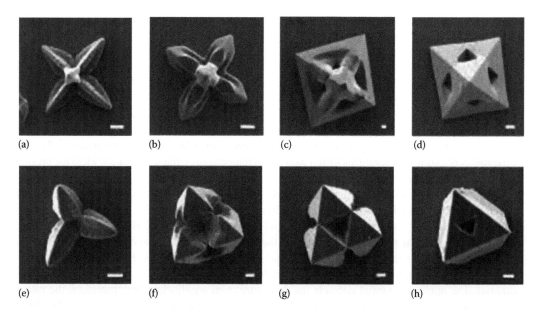

(a)   (b)   (c)   (d)

(e)   (f)   (g)   (h)

FIGURE 17.1 Use of solution additives to control the morphology of electrodeposited Cu_2O. SEM images of deposited octahedral Cu_2O crystals that display systematically varying degrees of branching with (100) planes parallel to the substrate (a–d) and with (111) planes parallel to the substrate (e–h; scale bar = 1 μm). These crystals were obtained at a constant temperature (60°C) and concentration of Cu^{2+} (0.02 M) and by applying deposition conditions of 0.10 mA $cm^{-2} \leq I \leq 0.12$ mA cm^{-2} and 0.08 V $\leq E \leq 0.12$ V. (From Siegfried M.J. and Choi, K.-S., Directing the architecture of cuprous oxide crystals during electrochemical growth, *Angew. Chem. Int. Ed.*, 44, 3218–3223, 2005. Copyright Wiley-VCH Verlag GmbH & Co. KGaA. Reprinted with permission.)

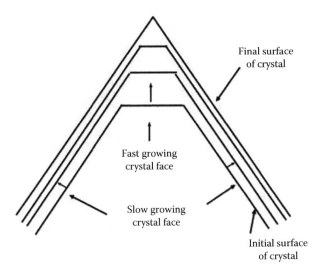

FIGURE 17.2 The control of crystal morphology by controlling the crystal face growth rates. If additives are adsorbed on a face that inhibit growth, the final surface of the crystal will be dominated by the slowest growing crystal face.

the current-time transient following a step in applied potential provides a real-time snapshot of the nucleation and growth of electrodeposited nanostructures.[62,63]

17.2 ELECTRODEPOSITION

17.2.1 METAL AND ELEMENTAL SEMICONDUCTOR ELECTRODEPOSITION

The electrodeposition of bulk metal films only requires that the electrode potential, U, be driven negative of the equilibrium potential, ϕ_{Nernst}, for the metal ion or metal complex. The difference between the applied potential and the equilibrium potential (i.e., the departure from equilibrium) is known as the overpotential, η. The morphology, grain size, and crystallographic orientation of electrodeposited metals are strongly dependent on the overpotential.[1,3–13] For example, the nucleation density typically increases as the overpotential is increased, leading to smaller grain sizes. At very large overpotentials, at which the deposition is mass transport limited, dendritic growth is often observed. It is also possible by underpotential deposition (UPD) to deposit a highly ordered monolayer or submonolayer of a metal on single-crystal substrates at potentials positive of the equilibrium potential.[10]

The electrodeposition of elemental semiconductors such as Si and Ge from aqueous solution is not thermodynamically allowed, because the equilibrium potentials for the deposition of these materials are negative of the hydrogen evolution reaction. Hence, according to thermodynamics, hydrogen gas should evolve before the equilibrium potential for the elemental semiconductors is reached. Maldonado and coworkers have circumvented this thermodynamic constraint by a kinetic trick that they call electrochemical liquid–liquid–solid (ec-LLS) crystal growth.[64–66] They use a low-melting metal such as Hg, Ga, or In as a metallic liquid *flux* that acts both as a traditional electrode and as a solvent for crystallization of the semiconductors. The liquid metals also have a very large overpotential for hydrogen gas evolution, so the hydrogen evolution reaction is driven more negative in potential (due to the kinetic overpotential) than the equilibrium potential for deposition of the semiconductors. The Maldonado group has produced films and nanowires of Si, Ge, and GaAs by this route. Figure 17.3 shows a cartoon of the growth of crystalline Si from liquid Ga (a) and a scanning electron micrograph (SEM) and transmission electron micrograph (TEM) diffraction

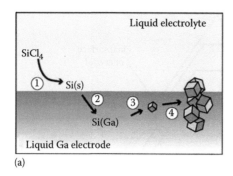
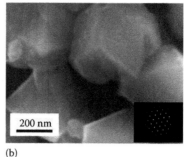

(a)   (b)

FIGURE 17.3 Cartoon showing the electrodeposition of Si from a liquid Ga electrode in aqueous solution (a) and an SEM micrograph and TEM diffraction pattern of the crystalline electrodeposited Si (b). (Reprinted with permission from Gu, J., Fahrenkrug, E., and Maldonado, S., Direct electrodeposition of crystalline silicon at low temperatures, *J. Am. Chem. Soc.*, 135, 1684–1687, 2013. Copyright 2013 American Chemical Society.)

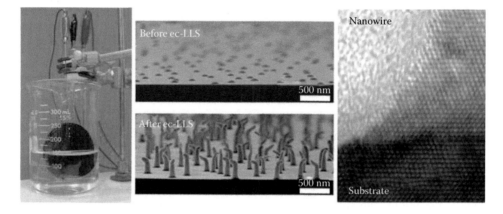

FIGURE 17.4 Electrodeposition of Ge nanowires from liquid Ga nanodots on single-crystal Ge by ec-LLS deposition. (Reprinted with permission from Fahrenkrug, E., Gu, J., Jeon, S., Veneman, P.A., Goldman, R.S., and Maldonado, S., Room-temperature epitaxial electrodeposition of single-crystalline germanium nanowires at the wafer scale from an aqueous solution, *Nano Lett.*, 14, 847–852, 2014. Copyright 2014 American Chemical Society.)

pattern of the electrodeposited Si (b). Figure 17.4 shows the SEM micrographs of Ge nanowires electrodeposited from liquid Ga nanodots on single-crystal Ge substrates by ec-LLS crystal growth.

17.2.2 COMPOUND ELECTRODEPOSITION

For the electrodeposition of compounds such as metal oxides and metal chalcogenides, the process involves both electrochemical and chemical reactions. That is, electrodeposition of compounds can be considered to occur through electrochemical–chemical (or EC) mechanisms in which chemical reactions follow an initial electron transfer. To use the electrodeposition of metal oxides as an example, the oxides can be produced either by the oxidation/reduction of a metal ion or complex, followed by hydrolysis/condensation to form the oxide or by the electrochemical generation of acid or base at the electrode surface to induce the hydrolysis/condensation reactions. In both cases, the source of oxygen in the final oxide is from H_2O/OH^-. The oxidation/reduction method has been used to produce oxides such as Fe_3O_4,[14,15] Cu_2O,[18–24,26,27,29] Bi_2O_3,[30] CeO_2,[67] and PbO_2/Tl_2O_3.[31–34] The base or acid generation method has been used to produce materials such as ZnO,[25,35–40] CuO,[41,42,68] and

CeO_2.[43] To use the electrodeposition of Fe_3O_4 as an example of the electrodeposition of metal oxides using the redox method, the material can be produced by electrochemically reducing a Fe(III) complex of triethanolamine (TEA) as shown in Equations 17.1 and 17.2.[14,15]

$$Fe(TEA)^{3+} + e^- \rightarrow Fe^{2+} + TEA \tag{17.1}$$

$$Fe^{2+} + 2Fe(TEA)^{3+} + 8OH^- \rightarrow Fe_3O_4 + 2TEA + 4H_2O \tag{17.2}$$

The production of metal oxide films by the electrochemical generation of acid or base method is similar to sol–gel processing, except that the pH change is at the electrode surface, inducing heterogeneous nucleation on the electrode surface with little or no homogeneous nucleation in the solution. The most obvious way to change the pH is to electrochemically generate base to induce hydrolysis.[25,35–39] This can be done by reducing water to hydrogen gas, or by reducing other species such as nitrate ion or molecular oxygen. The local pH at the electrode surface is increased, but the pH of the bulk solution is not changed. With amphoteric ions such as Zn(II), Cr(II), and Al(III), which can function as both an acid or a base, it is possible to deposit the oxides by electrochemically generating acid in a saturated alkaline solution of the metal ion.[40] For example, Figure 17.5 shows the speciation diagram (a) and solubility (a) of Zn(II) as a function of pH at 65°C. The minimum in solubility

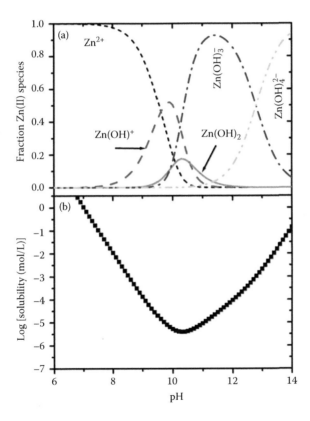

FIGURE 17.5 Speciation diagram (a) and solubility (b) of Zn(II) as a function of pH at 65°C. The minimum solubility occurs at pH 10.3. For solutions at a pH less than 10.3, ZnO can be produced by electrochemically generating base. For solutions at a pH greater than 10.3, ZnO can be produced by electrochemically generating acid. At a pH of 13.3 (the pH used to generate ZnO in Figures 17.6 and 17.7), the solution consists of 77% $Zn(OH)_4^{2-}$ and 23% $Zn(OH)_3^-$. (Reprinted with permission from Limmer, S.J., Kulp, E.A., Switzer, J.A., Epitaxial electrodeposition of ZnO on Au(111) from alkaline solution: Exploiting amphoterism in Zn(II), *Langmuir*, 22, 10535–10539, 2006. Copyright 2006 American Chemical Society.)

FIGURE 17.6 Epitaxial ZnO crystals that were electrodeposited onto single-crystal Au(111) by electro-chemically generating acid in the solution of Zn(II) at pH 13.3. (Reprinted from Limmer, S.J., Kulp, E.A., and Switzer, J.A. Epitaxial electrodeposition of ZnO on Au(111) from alkaline solution: Exploiting amphoterism in Zn(II), *Langmuir*, 22, 10535–10539, 2006. Copyright 2006 American Chemical Society.)

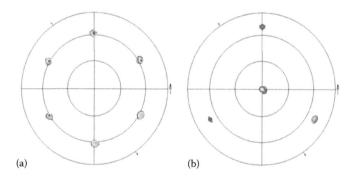

(a)   (b)

FIGURE 17.7 X-ray pole figures of epitaxial electrodeposited ZnO (a) and the Au(111) single-crystal sub-strate (b). The epitaxial ZnO was electrodeposited onto single-crystal Au(111) by electrochemically generating acid in the solution of Zn(II) at pH 13.3. The radial grid lines on the pole figures correspond to 30° incre-ments of the tilt angle, χ. The pole figures show that the out-of-plane and in-plane orientation of the epitaxial ZnO is determined by the Au(111) single-crystal substrate. The orientation relationship is ZnO(0001)[10$\bar{1}$1]// Au(111)[1$\bar{1}$0]. (Reprinted from Limmer, S.J., Kulp, E.A., and Switzer, J.A. Epitaxial electrodeposition of ZnO on Au(111) from alkaline solution: Exploiting amphoterism in Zn(II), *Langmuir*, 22, 10535–10539, 2006. Copyright 2006 American Chemical Society.)

occurs at pH 10.3. For solutions at a pH less than 10.3, ZnO can be produced by electrochemically generating base. For solutions at a pH greater than 10.3, ZnO can be produced by electrochemically generating acid. At a pH of 13.3 (the pH used to generate ZnO in Figures 17.6 and 17.7), the solution consists of 77% $Zn(OH)_4^{2-}$ and 23% $Zn(OH)_3^{2-}$. Figure 17.6 shows an SEM micrograph of epitaxial ZnO crystals that were produced in alkaline solution by electrochemically generating acid at an Au(111) surface.[40] Figure 17.7 shows x-ray pole figures of an epitaxial electrodeposited ZnO film and the Au(111) single-crystal substrate. The epitaxial ZnO was electrodeposited onto single-crystal Au(111) by electrochemically generating acid in the solution of Zn(II) at pH 13.3. The radial grid lines on the pole figures correspond to 30° increments of the tilt angle, χ. The pole figures show that the out-of-plane and in-plane orientation of the ZnO is determined by the Au(111). The orientation relationship is ZnO(0001)[10$\bar{1}$1]//Au(111)[$\bar{1}$10].[40] The base/acid generation method does not require that the metal ion is redox-active. For example, it has also been used to induce the electrochemical

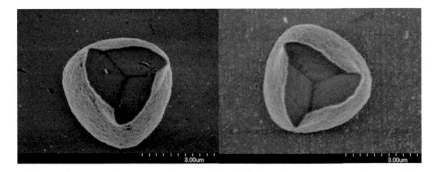

FIGURE 17.8 Chiral morphologies of calcite electrodeposited by electrochemically generating base in a solution of calcium bicarbonate. The chiral morphology was developed by depositing the calcite in the presence of chiral molecules. The structure on the left was grown in the presence of L-tartrate, whereas the structure on the right was grown in the presence of D-tartrate. (Reprinted from Kulp, E.A. and Switzer, J.A., Electrochemical biomineralization: The deposition of calcite with chiral morphologies, *J. Am. Chem. Soc.*, 129, 15120–15121. Copyright 2007 American Chemical Society.)

nucleation and growth of biominerals such as calcite on an electrode surface.[44] Figure 17.8 shows SEM micrographs of chiral calcite architectures that were grown by generating base in a solution of calcium bicarbonate. The chiral morphologies were obtained by depositing the calcite in the presence of L- or D-tartrate ion.[44]

Metal compound semiconductors such as CdS, CdSe, $Hg_{(1-x)}Cd_xTe$, Bi_2Se_3, and InAs can also be produced by electrodeposition. Films of these materials have been produced by codeposition of the elements,[45–51] or by a technique known as electrochemical atomic layer epitaxy (ECALE).[52–58] In codeposition, the films or nanostructures are produced from solution precursors by EC reactions, such as those shown in Equations 17.3 through 17.5. ECALE is the electrochemical equivalent of atomic layer epitaxy (ALE), because the material is assembled monolayer-by-monolayer using surface-limited reactions.[52–58] The alternating layers in ECALE are deposited by UPD.

$$Cd^{2+} + 2e^- \rightarrow Cd \tag{17.3}$$

$$HTeO_2^+ + 4e^- + 3H^+ \rightarrow Te + 2H_2O \tag{17.4}$$

$$Cd + Te \rightarrow CdTe \tag{17.5}$$

17.3 ELECTRODEPOSITION OF NANOSTRUCTURES

Electrodeposition is a great way to produce nanostructures, because the low processing temperatures allow the production of nanoscale materials or interfaces that may not be accessible to higher-temperature processing methods due to interdiffusion. In addition, because the nanostructures grow from the bottom up, they maintain electrical contact with the substrate. This is particularly important for the production of devices for the conversion (e.g., solar cells) or storage (e.g., batteries) of energy. Techniques for electrodepositing nanostructures include, but are not limited to, deposition into nanoscale templates,[69–71] self-assembly,[72–74] growth of ordered nanostructures and ultrathin films on single-crystal substrates,[13,75] electrodeposition of quantum dots,[76,77] scanning probe nanolithography,[7,78–81] optical or electron beam lithography of patterns on nonconducting substrates,[82,83] electrodeposition of nanostructured dendrites,[9] assembly of nanostructures by electrochemical oscillations,[9,19,84–86] and the growth of superlattices by pulsing the applied potential or current.[15,31–34,87–89] Some of these methods will be discussed in more detail in the following sections.

17.3.1   UPD Layers and Ultrathin Films

Although the deposition of thick films is straightforward, the assembly of ultrathin layers (i.e., up to several monolayers) is a different story. The growth of ultrathin films of materials is hindered by roughening and three-dimensional mound formation. This roughening can be eliminated or minimized by atomic layer deposition (ALD), in which atomic layer control and conformal growth are achieved using sequential, self-limiting surface reactions.[90] ALD has been used in the electronics field to produce high dielectric gate oxides in metal oxide field-effect transistors (MOSFETs) and to deposit copper diffusion barriers in back-end interconnects. Another application of ALD is to deposit ultrathin layers of expensive metals such as Pt that are used, for example, as the catalyst in proton exchange membrane fuel cells.[91] The cost of the Pt oxygen reduction catalyst in a fuel cell is a major impediment to commercialization.[92] Besides the economic incentives to produce ultrathin films, there are also scientific payoffs. Ultrathin films of materials often have catalytic, electronic, or magnetic properties that are not found in the bulk material.[93–95]

The electrochemical equivalent of ALD is UPD. It is possible by UPD to produce a highly ordered monolayer or submonolayer of a metal on a foreign substrate at potentials positive of the equilibrium potential.[96,97] The substrate can be single crystalline or polycrystalline. The UPD occurs because the binding of the monolayer to the foreign substrate is stronger than the binding of the monolayer to a substrate of the same material as the monolayer. This phenomenon is a surface-limited reaction, because only a monolayer will be deposited, no matter how long the UPD potential is held. It is the self-limiting action of metal UPD that allows electrochemical deposition to function as an ALD method. Electrochemical ALD has been used, for instance, to grow compound semiconductors by sequentially depositing each element in a UPD cycle.[98]

One approach to ALD of metals is a process known as surface-limited redox replacement or galvanic displacement. This method has been used to produce ultrathin layers of metals such as Pt, Pd, and Ag.[99–102] The ALD is brought about by the galvanic replacement of underpotentially deposited metal monolayers of less noble metals such as Pb or Cu. The surface-limited redox replacement occurs spontaneously, because the reduction potential of the less noble metal is more negative than that of the subsequent ALD layer. The two deposition steps in the surface-limited redox replacement of metals must be performed in separate solutions, so it is necessary to exchange reactants—much like the case of ALD by vapor deposition methods.

Conventional wisdom would suggest that the best way to electrodeposit ultrathin metal films would be to apply either an underpotential or a very small overpotential. An unexpected result in the electrochemical ALD reported by Moffat and coworkers is that only a monolayer of Pt is deposited at an overpotential as large as 1 V.[103,104] At this overpotential, the deposition rate of Pt should be very large, because electrodeposition rates increase exponentially as the overpotential is increased. It turns out that at this high overpotential for Pt deposition, a monolayer of hydrogen (H_{UPD}) is formed on the Pt surface by UDP. The H_{UPD} adlayer completely blocks the deposition of additional Pt, making the process self-limiting. Liu et al. attribute this blocking to the disruption of the electrical double layer by the H_{UPD}.[103] Multilayers of Pt can be produced, because the hydrogen that is underpotentially deposited at -0.8 V_{SSCE} can be desorbed at a potential of $+0.4_{SSCE}$. By pulsing the potential between -0.8 and $+0.4$ V_{SSCE} in a single plating bath, Pt is deposited on the surface monolayer-by-monolayer in a digital manner (see Figure 17.9).[104] This new deposition method is potentially much faster than both electrochemical surface-limited redox replacement and vapor phase ALD, because it is not necessary to exchange reactants. The deposition of a Pt monolayer is complete in 1 s. The new process may also lead to less carryover of contaminants that could occur when the reactants are exchanged in the surface-limited redox replacement scheme.

The layer-by-layer electrochemical ALD described by Liu et al.[103] offers an excellent platform to study the evolution of the catalytic, electronic, and magnetic properties of ultrathin films as a function of thickness.[93–95] If the self-limiting growth is observed in alloys of Pt, magnetism could be imparted in the films by incorporation of magnetic metals. It would also be intriguing to digitally

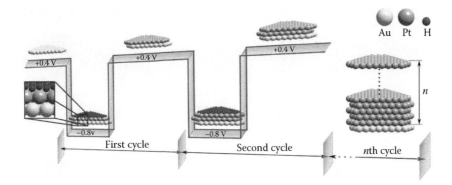

FIGURE 17.9 Electrochemical atomic layer deposition (ALD) of ultrathin platinum films. The films are deposited one monolayer at a time by simply pulsing the electrode potential between +0.4 and −0.8 V. A capping layer of underpotential hydrogen (H_{UPD}) is produced at −0.8 V that blocks the deposition of more than one monolayer of Pt. When the potential is stepped to +0.4 V, the H_{UPD} blocking layer is desorbed, and the cycle can begin again. The self-limiting, digital processing method is fast because it is performed in a single plating bath, so it is not necessary to exchange reactants. The ultrathin Pt films could lower the costs of the Pt catalyst in fuel cells, and they could provide a platform to study how the catalytic, electronic, and magnetic properties of ultrathin films evolve with film thickness. (Reprinted from Switzer, J.A., Atomic layer electrodeposition, *Science*, 338, 1300–1301, 2012. With permission from the American Association for the Advancement of Science.)

deposit compositionally modulated superlattices[15] one monolayer at a time. How general is this digital processing method? Other metals have well-defined H_{UPD} adlayers, so the prospects are encouraging. What about deposition of materials besides metals? Is there some potential-controlled process that can make the growth of metal oxides or semiconductors self-limiting? Ultrathin films deposited layer-by-layer onto single-crystalline substrates should be epitaxial. Electrochemical ALD would be ideal for in situ studies of the effect of lattice mismatch on the transition from two-dimensional to three-dimensional growth in epitaxial films. The beauty of this research is that it blends basic electrochemistry and surface science to unlock an important new technology. The future looks very bright for this new electrochemical approach to ALD.

17.3.2 TEMPLATE SYNTHESIS BY DEPOSITION INTO NANOPOROUS MEMBRANES

The deposition of nanowires and nanotubes into templates was pioneered by Martin.[69] In template deposition, the materials are deposited into nanoporous membranes, such as anodized aluminum or track-etch polymers.[69,70] The nanoporous membranes function as nanosized beakers that constrain the crystal growth. Figure 17.10 shows TEM micrographs of Au nanowires with a diameter of 70 nm and polypyrrole nanotubes with an outside diameter of 90 nm and an inside diameter of 20–30 nm that were electrodeposited into an alumina nanoporous template.[70]

An interesting variation on template deposition is to self-assemble ordered nanostructures (e.g., surfactants) and microstructures (e.g., polystyrene or SiO_2 beads) on the surface of an electrode and then electrodeposit into the self-assembled pores.[72–74,105] The order in the resulting nanostructure is imposed by the self-assembled layer, not by the substrate. Schwartz and coworkers have extended this idea to the use of crystalline protein masks to produce ordered nanostructures of metals (such as Ni, Pt, Pd, and Co) and metal oxides (such as Cu_2O).[6,29,106] Braun and coworkers have used the electrodeposition of materials into self-assembled colloidal crystals or silica or polymer opals. The template is then removed (see Figure 17.11) to produce an inverse opal.[105] This type of templating produces periodic microstructures that can be used to produce functional photonics.[105] Figure 17.11 shows the production of CdSe and Ni inverse opals by electrodeposition into a colloidal crystal with subsequent removal of the colloidal crystal template.[105]

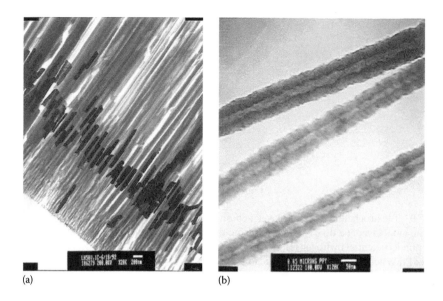

FIGURE 17.10 Nanowires and nanotubes produced by the template method in an alumina template. (a) TEM micrograph of Au nanowires with a 70 nm diameter. (b) TEM micrograph of polypyrrole nanotubes. The outside diameter is about 90 nm, and the inside diameter is 20–30 nm. (Reprinted with permission from Martin, C.R., Membrane-based synthesis of nanomaterials, *Chem. Mater.*, 8, 1739–1746, 1994. Copyright 1996 American Chemical Society.)

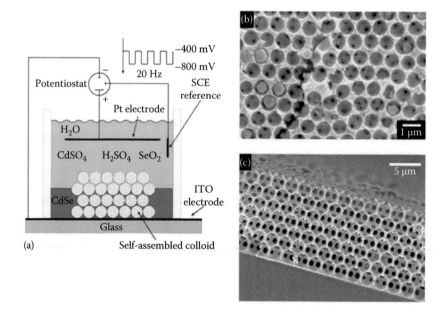

FIGURE 17.11 Electrodeposition scheme (a) for the production of CdSe (b) and Ni (c) reverse opals by electrodeposition onto an ordered colloidal crystal with subsequent removal of the colloidal crystal template. (Reprinted with permission from Braun, P.V., Materials chemistry in 3D templates for functional photonics, *Chem. Mater.*, 26, 277–286. Copyright 2013 American Chemical Society.)

17.3.3 Epitaxial Nanostructures on Single Crystal Surfaces

In contrast with template deposition, it is also possible to start with single-crystal substrates and deposit ultrathin films and ordered nanostructures onto patterns that are determined by the substrate. Although the use of single-crystal substrates is usually constrained for use in UHV, there are numerous examples of ordered nanostructures that have been produced on single-crystal surfaces by electrodeposition. For example, Ni nanowires have been formed by direct electrodeposition on Ag(111) monatomic steps,[75] and self-ordered Co/Au dots have been grown on a vicinal H-terminated Si(111) surface.[13] Rubinstein, Hodes, and coworkers have produced epitaxially oriented CdSe quantum dots with diameters of about 5 nm and controllable spatial distribution[76] and size[77] by electrodepositing the nanocrystals on Au(111) substrates, as well as modifying the substrate by either alloying[107] or mechanical bending[108] to obtain the epitaxial rock-salt phase of CdSe, normally only stable at high pressure. The size of the quantum dots is believed to be controlled by the strain that is induced by the CdSe/Au lattice mismatch. Figure 17.12 shows an SEM micrograph of epitaxial Cu_2O nanocubes that were electrodeposited onto single-crystal InP in the author's laboratory.[24]

17.3.4 Scanning Probe Lithography

The scanning probe microscope can be used to write nanostructures onto surfaces.[7] Penner has used the scanning tunneling microscope (STM) to modify a surface with nanometer-scale defects, so as to induce nucleation of the deposited material at these defect sites.[78,79] In another method, pioneered by Dieter Kolb, metal nanostructures are produced by electrochemically depositing the metal onto the STM tip and then transferring the material to the surface during the tip approach.[7,80,81] An example of a corral of Fe nanoclusters on Au(111) produced by this method is shown in Figure 17.13.[81]

17.3.5 Lithographically Patterned Nanowire Deposition

In a method called lithographically patterned nanowire deposition (LPNE), developed by Penner and coworkers, a template for the growth of nanowires is temporarily fabricated on the surface of a dielectric using standard photolithographic methods.[83,109,110] The nanowire is grown by

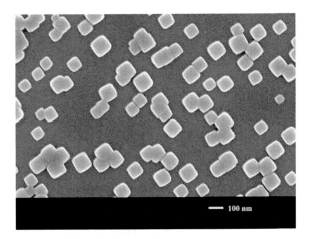

FIGURE 17.12 Scanning electron micrograph of Cu_2O nanocubes that were electrodeposited epitaxially onto single-crystal InP(100). The scale marker is 100 nm. (Reprinted with permission from Liu, R., Oba, F., Bohannan, E.W., Ernst, F., and Switzer, J.A., Shape control in epitaxial electrodeposition: Cu_2O nanocubes on InP(001), *Chem. Mater.*, 15, 4882–4885. Copyright 2003 American Chemical Society.)

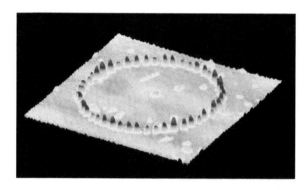

FIGURE 17.13 A corral of Fe nanoclusters on Au(111) produced by scanning tunneling microscopy (STM). The Fe was electrochemically deposited onto the STM tip and then transferred to the surface during the tip approach. (From Wei, Y.-M., Zhou, X.-S., Wang, J.-G., Tang, J., Mao, B.-W., and Kolb, D.M. The creation of nanostructures on an Au(111) electrode by tip-induced iron deposition from an ionic liquid, *Small*, 2008, 4, 1355–1358. Copyright Wiley-VCH Verlag GmbH & Co. KGaA. Reproduced with permission.)

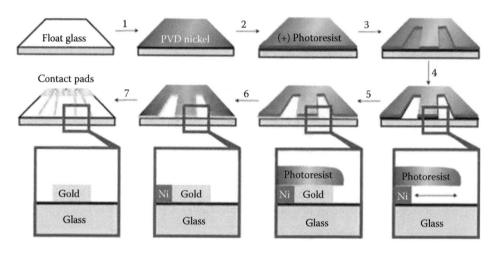

FIGURE 17.14 Scheme used to produce patterned noble metal nanowires on insulating substrates using lithographically patterned nanowire electrodeposition (LPNE). (Reprinted by permission from Macmillan Publishers Ltd. *Nat. Mater.*, Menke, E.J., Thompson, M.A., Xiang, C., Yang, L.C., and Penner, R.M., Lithographically patterned nanowire electrodeposition, 5, 914–919, 2006. Copyright 2006.)

electrodeposition from an aqueous solution, and the template is then removed without disrupting the deposited nanowires. An advantage of this method is that nanowires can be produced on nonconducting substrates. This circumvents the normal requirement for conducting substrates, and opens up device capabilities.

The LPNE deposition scheme is outlined in Figure 17.14. It combines traditional *top-down* lithographic techniques with *bottom-up* electrodeposition. It involves the preparation of a sacrificial nickel nanoband electrode on glass or oxidized silicon surfaces. The Ni is overetched in step 4 to produce the Ni nanoband onto which a noble metal such as Au, Pt, or Pd is electrodeposited in step 5. The electrodeposition into this nanoform that is produced by the overetching produces a wire with a rectangular cross section, with a size that can be as low as 18 nm in height and 40 nm in width. The Ni and photoresist are then removed to produce the noble metal wires on an insulating substrate. Figure 17.15 shows examples of gold and platinum patterned nanowires that were deposited by LPNE.

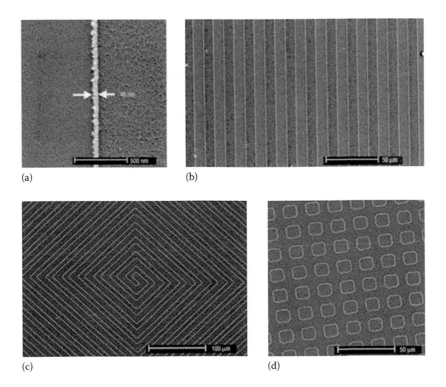

(a)   (b)

(c)   (d)

FIGURE 17.15 Examples of patterned nanowires that were produced by lithographically patterned nanowire electrodeposition (LPNE). (a) A platinum nanowire with a width of 46 nm and a thickness of 39 nm. The wire was continuous for more than one cm. (b) Parallel gold nanowires, 1 cm in length and spaced by 9 μm. (c) A coiled nanowire with a total length of 2.7 cm. (d) Nanowire loops. (Reprinted by permission from Macmillan Publishers Ltd. *Nat. Mater.*, Menke, E.J., Thompson, M.A., Xiang, C., Yang, L.C., and Penner, R.M., Lithographically patterned nanowire electrodeposition, 5, 914–919, 2006. Copyright 2006.)

17.3.6 NANOLAYERED MATERIALS SUCH AS MULTILAYERS AND SUPERLATTICES

The intense interest in nanomaterials stems from the fact that the optical, electrical, magnetic, or mechanical properties can be tuned by changing the physical dimensions of the material. Superlattices and multilayers are especially well suited for device applications, because the confinement dimensions (i.e., the individual layer thickness) can be kept in the nanometer range even for films that have an overall thickness that is quite large. A schematic for both multilayers and superlattices is shown in Figure 17.16. They are composed of alternating layers of materials A and B, with a bilayer thickness known as the modulation wavelength, Λ. Superlattices are a particular type of crystalline multilayer, in which there is coherent stacking of atomic planes, and periodic modulation of the structure, or composition, or both. That is, multilayers and superlattices are both modulated materials, but superlattices have the additional constraint that they are crystallographically coherent. Because of this constraint, superlattices are usually produced with alternating layers of materials with very low lattice mismatch, whereas multilayers can be produced using even amorphous materials. The substrate in Figure 17.16 is typically a single crystal for a superlattice, and the crystallographic orientation of the superlattice is determined by the orientation of the substrate (i.e., superlattices are epitaxial).

The best evidence for a superlattice is the observation of satellites around the Bragg reflection in x-ray diffraction (XRD).[111] The satellites are caused by the superperiodicity in the system, because the x-ray pattern is the Fourier transform of the product of the lattice and modulation functions convoluted with the basis. For composition waveforms that vary sinusoidally, only first order satellites are expected, because only one Fourier term is needed to describe a sine wave. As the waveform

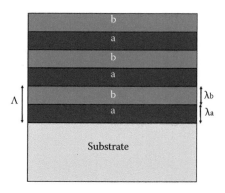

FIGURE 17.16 Schematic of a multilayer or superlattice. The alternating layers can be electrochemically deposited by pulsing either the applied current or potential.

becomes more square, higher order Fourier terms are necessary, and higher order satellites are observed. It is straightforward to determine the modulation wavelength, Λ, from the XRD pattern from the satellite spacing. In a superlattice, the modulation causes an artificial spacing that is not present in bulk materials. These planes scatter x-rays similar to a bulk crystal, where scattering from the ith plane can be represented by Equation 17.6, a modification of Bragg's law:

$$L_i\lambda = 2\Lambda\sin\theta_i \tag{17.6}$$

For two planes i and j separated by a distance Λ, the path distance is

$$(L_1 - L_2)\lambda = 2\Lambda(\sin\theta_i - \sin\theta_j), \tag{17.7}$$

or, in terms of modulation wavelength,

$$\Lambda = (L_i - L_j)\lambda/2(\sin\theta_1 - \sin\theta_2), \tag{17.8}$$

where
  L_i and L_j are the satellite orders
  θ_i and θ_j are the angles corresponding to the positions of the satellites
  λ is the wavelength of the x-ray radiation

Figure 17.17 shows the XRD of a Pb–Tl–O superlattice from the author's laboratory that shows satellites out to seventh order around the (400) Bragg reflection.[34] The modulation wavelength of the superlattice is 18.9 nm.

There are two general techniques for electrodepositing superlattices: dual bath and single bath.[111] Dual bath deposition is the easier of the two to design, but more difficult to implement for superlattices with a large number of layers or with very small modulation wavelengths. Also, the electrode is exposed to the atmosphere between the deposition steps. The electrode is simply alternated between two deposition solutions containing the precursors for the two layers. The main advantage of dual bath deposition is that it is possible to deposit pure alternating layers of different materials. Most superlattices (and multilayers) are produced from a single bath. The usual method for single bath deposition is to use a very low concentration of the precursor for the layer that is deposited at low overpotential (layer A) and a high concentration for the layer that is deposited at higher overpotential (layer B). Therefore, at low overpotentials, pure material A is deposited, whereas at higher overpotentials, an alloy or solid solution, which is predominately material B, is deposited. The higher the applied current density during the high overpotential pulse, the higher the concentration of B in that layer. The superlattice or multilayer is deposited by pulsing either the applied potential or current,

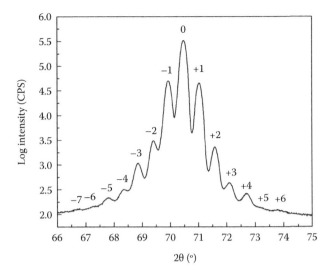

FIGURE 17.17 XRD scan for a Pb–Tl–O superlattice that was deposited by pulsing the applied current density between 0.05 (150 s) and 5 (1.5 s) mA/cm². The modulation wavelength calculated from the satellite spacing is 18.9 nm. (Reprinted with permission from Kothari, H.M., Vertegel, A.A., Bohannan, E.W., and Switzer, J.A., Epitaxial electrodeposition of Pb-Tl-O superlattices on single-crystal Au(100), *Chem. Mater.*, 14, 2750–2756, 2002. Copyright 2002 American Chemical Society.)

and the thickness of each layer is determined by the current and dwell time used to deposit that layer.[111] Typically, square waves are applied to the electrochemical cell to produce the superlattice.

Superlattices of metals,[87,88] semiconductors,[57] and ceramics[15,31–34,61] can be produced by single bath electrodeposition.[111] In addition to superlattices in which the composition is modulated, it is also possible to alternate layers of metal oxides with varying defect chemistry.[33] The defect chemistry superlattices are produced by simply pulsing the applied potential during deposition. A cross-sectional SEM image of a compositional superlattice based on magnetite is shown in Figure 17.18.[15]

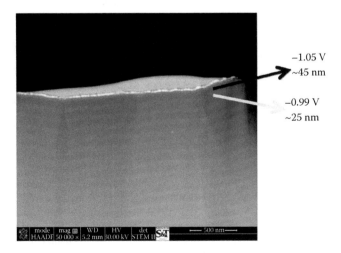

FIGURE 17.18 Cross-sectional STEM-high angle annular dark field image of an electrodeposited zinc ferrite superlattice with a modulation wavelength of 70 nm. (Reprinted from Switzer, J.A., Gudavarthy, R.V., Kulp, E.A., Mu, G., He, Z., and Wessel, A.J., Resistance switching in electrodeposited magnetite superlattices, *J. Am. Chem. Soc.*, 132, 1258–1260, 2010. Copyright 2010 American Chemical Society.)

These superlattices show multistate resistance switching that may be applicable to the production of resistance random access memory (RRAM).[15]

ACKNOWLEDGMENTS

The author acknowledges financial support by the U.S. Department of Energy, Office of Basic Energy Sciences under Grant No. DE-FG02-08ER46518 and by the National Science Foundation under Grant No. DMR-1104801.

REFERENCES

1. Paunovic, M. and M. Schlesinger. 1998. *Fundamentals of Electrochemical Deposition*. New York: Wiley-Interscience.
2. Switzer, J. A. and G. Hodes. 2010. Electrodeposition and chemical bath deposition of functional nano-materials. *MRS Bull.* 35: 743–751.
3. Andricacos, P. C. 1999. Copper on-chip interconnections: A breakthrough in electrodeposition of make better chips. *Interface* 8: 32–37.
4. Guo, L. and P. C. Searson. 2008. Anisotropic island growth: A new approach to thin film electrocrystallization. *Langmuir* 24: 10557–10559.
5. Hautier, G., J. D'Haen, K. Maex, and P. M. Vereecken. 2008. Electrodeposited free-standing single-crystal indium nanowires. *Electrochem. Solid-State Lett.* 11: K47–K49.
6. Grosh, C. D., D. T. Schwartz, and F. Baneyx. 2009. Protein-based control of silver growth habit using electrochemical deposition. *Cryst. Growth Des.* 9: 4401–4406.
7. Kolb, D. M. and F. C. Simeone. 2005. Electrochemical nanostructuring with an STM: A status report. *Electrochim. Acta* 50: 2989–2996.
8. O'Brien, B., M. Plaza, L. Y. Zhu, L. Perez, C. L. Chien, and P. C. Searson. 2008. Magnetotransport properties of electrodeposited bismuth films. *J. Phys. Chem. C* 112: 12018–12023.
9. Fukami, K., S. Nakanishi, H. Yamasaki et al. 2007. General mechanism for the synchronization of electrochemical oscillations and self-organized dendrite electrodeposition of metals with ordered 2D and 3D microstructures. *J. Phys. Chem. C* 111: 1150–1160.
10. Budevski, E., G. Staikov, W. J. Lorenz (Eds.). 1996. *Electrochemical Phase Formation and Growth: An Introduction to the Initial Stages of Metal Deposition*. Chapter 3. Weinhiem, Germany: VCH Publishers.
11. Prod'homme, P., F. Maroun, R. Cortes, and P. Allongue. 2008. Electrochemical growth of ultraflat Au(111) epitaxial buffer layers on H-Si(111). *Appl. Phys. Lett.* 93: 171901.
12. Allongue, P. and F. Maroun. 2006. Metal electrodeposition on single crystal metal surfaces mechanisms, structure and applications. *Curr. Opin. Solid State Mater. Sci.* 10: 173–181.
13. Allongue, P. and F. Maroun. 2006. Self-ordered electrochemical growth on single-crystal electrode surfaces. *J. Phys.: Cond. Matter.* 18: S97–S114.
14. Kulp, E. A., H. M. Kothari, S. J. Limmer et al. 2009. Electrodeposition of epitaxial magnetite films and ferrihydrite nanoribbons on single-crystal gold. *Chem. Mater.* 21: 5022–5031.
15. Switzer, J. A., R. V. Gudavarthy, E. A. Kulp, G. Mu, Z. He, and A. J. Wessel. 2010. Resistance switching in electrodeposited magnetite superlattices. *J. Am. Chem. Soc.* 132: 1258–1260.
16. Koza, J. A., C. M. Hull, Y.-C. Liu, and J. A. Switzer. 2013. Deposition of β-Co(OH)$_2$ films by electrochemical reduction of tris(ethylenediamine)cobalt(III) in alkaline solution. *Chem. Mater.* 25: 1922–1926.
17. Koza, J. A., E. W. Bohannan, and J. A. Switzer. 2013. Superconducting filaments formed during nonvolatile resistance switching in electrodeposited δ-Bi$_2$O$_3$. *ACS Nano* 7: 9940–9946.
18. Golden, T. D., M. G. Shumsky, Y. Zhou, R. A. VanderWerf, R. A. Van Leeuwen, and J. A. Switzer. 1996. Electrochemical deposition of copper(I) oxide films. *Chem. Mater.* 8: 2499–2504.
19. Switzer, J. A., C.-J. Hung, L.-Y. Huang et al. 1998. Electrochemical self-assembly of copper/cuprous oxide layered nanostructures. *J. Am. Chem. Soc.* 120: 3530–3531.
20. Switzer, J. A., H. M. Kothari, and E. W. Bohannan. 2002. Thermodynamic to kinetic transition in epitaxial electrodeposition. *J. Phys. Chem. B* 106: 4027–4031.
21. Switzer, J. A., R. Liu, E. W. Bohannan, and F. Ernst. 2002. Epitaxial electrodeposition of a crystalline metal oxide onto single-crystalline silicon. *J. Phys. Chem. B* 106: 12369–12372.
22. Liu, R., E. W. Bohannan, J. A. Switzer, F. Oba, and F. Ernst. 2003. Epitaxial electrodeposition of Cu$_2$O films onto InP(001). *Appl. Phys. Lett.* 83: 1944–1946.

23. Liu, R., E. A. Kulp, F. Oba, E. W. Bohannan, F. Ernst, and J. A. Switzer. 2005. Epitaxial electrodeposition of high-aspect-ratio Cu$_2$O(110) nanostructures on InP(111). *Chem. Mater.* 17: 725–729.

24. Liu, R., F. Oba, E. W. Bohannan, F. Ernst, and J. A. Switzer. 2003. Shape control in epitaxial electrodeposition: Cu$_2$O nanocubes on InP(001). *Chem. Mater.* 15: 4882–4885.

25. Liu, R., A. A. Vertegel, E. W. Bohannan, T. A. Sorenson, and J. A. Switzer. 2001. Epitaxial electrodeposition of zinc oxide nanopillars on single-crystal gold. *Chem. Mater.* 13: 508–512.

26. Oba, F., F. Ernst, Y. Yu, R. Liu, H. M. Kothari, and J. A. Switzer. 2005. Epitaxial growth of cuprous oxide electrodeposited onto semiconductor and metal substrates. *J. Am. Ceram. Soc.* 88: 253–270.

27. Choi, K.-S. 2008. Shape control of inorganic materials via electrodeposition. *Dalton Trans.* 2008(40): 5432–5438.

28. McShane, C. M. and K.-S. Choi. 2009. Photocurrent enhancement of n-type Cu$_2$O electrodes achieved by controlling dendritic branching growth. *J. Am. Chem. Soc.* 131: 2561–2569.

29. Allred, D. B., A. Cheng, M. Sarikaya, F. Baneyx, and D. T. Schwartz. 2008. Three-dimensional architecture of inorganic nanoarrays electrodeposited through a surface-layer protein mask. *Nano Letts.* 8: 1434–1438.

30. Switzer, J. A., M. G. Shumsky, and E. W. Bohannan. 1999. Electrodeposited ceramic single crystals. *Science* 284: 293–296.

31. Switzer, J. A., M. J. Shane, and R. J. Phillips. 1990. Electrodeposited ceramic superlattices. *Science* 247: 444–446.

32. Switzer, J. A., R. P. Raffaelle, R. J. Phillips, C. J. Hung, and T. D. Golden. 1992. Scanning tunneling microscopy of electrodeposited ceramic superlattices. *Science* 258: 1918–1921.

33. Switzer, J. A., C. J. Hung, B. E. Breyfogle, M. G. Shumsky, R. Van Leeuwen, and T. D. Golden. 1994. Electrodeposited defect chemistry superlattices. *Science* 264: 1573–1576.

34. Kothari, H. M., A. A. Vertegel, E. W. Bohannan, and J. A. Switzer. 2002. Epitaxial electrodeposition of Pb-Tl-O superlattices on single-crystal Au(100). *Chem. Mater.* 14: 2750–2756.

35. Peulon, S. and D. Lincot. 1996. Cathodic electrodeposition of dense or open-structured zinc oxide films from aqueous solution. *Adv. Mater.* 8: 166–170.

36. Peulon, S. and D. Lincot. 1998. Mechanistic study of cathodic electrodeposition of zinc oxide and zinc hydroxychloride films from oxygenated aqueous zinc chloride solutions. *J. Electrochem. Soc.* 145: 864–874.

37. Izaki, M. and T. Omi. 1996. Transparent zinc oxide films prepared by electrochemical reaction. *Appl. Phys. Lett.* 68: 2439–2440.

38. Pauporte, T. and D. Lincot. 1999. Heteroepitaxial electrodeposition of zinc oxide films on gallium nitride. *Appl. Phys. Lett.* 75: 3817–3819.

39. Voss, T., C. Bekeny, J. Gutowski et al. 2009. Localized versus delocalized states: Photoluminescence from electrochemically synthesized ZnO nanowires. *J. Appl. Phys.* 106: 054304.

40. Limmer, S. J., E. A. Kulp, and J. A. Switzer. 2006. Epitaxial electrodeposition of ZnO on Au(111) from alkaline solution: Exploiting amphoterism in Zn(II). *Langmuir* 22: 10535–10539.

41. Switzer, J. A., H. M. Kothari, P. Poizot, S. Nakanishi, and E. W. Bohannan. 2003. Enantiospecific electrodeposition of a chiral catalyst. *Nature* 425: 490–493.

42. Bohannan, E. W., H. M. Kothari, I. M. Nicic, and J. A. Switzer. 2004. Enantiospecific electrodeposition of chiral CuO films on single-crystal Cu(111). *J. Am. Chem. Soc.* 126: 488–489.

43. Switzer, J. A. 1987. Electrochemical synthesis of ceramic films and powders. *Am. Ceram. Soc. Bull.* 66: 1521–1524.

44. Kulp, E. A. and J. A. Switzer. 2007. Electrochemical biomineralization: The deposition of calcite with chiral morphologies. *J. Am. Chem. Soc.* 129: 15120–15121.

45. Kroger, F. A. 1978. Cathodic deposition and characterization of metallic or semiconducting binary alloys or compounds. *J. Electrochem. Soc.* 125: 2028–2034.

46. Miller, B., S. Menezes, and A. Heller. 1978. Anodic formations of semiconductive sulfide films at cadmium and bismuth. Rotating ring-disk electrode studies. *J. Electroanal. Chem. Interface Electrochem.* 94: 85–97.

47. Rajeshwar, K. 1992. Electrosynthesized thin films of group II-VI compound semiconductors, alloys and superstructures. *Adv. Mater.* 4: 23–29.

48. Pandey, R. K., S. N. Sahu, S. Chandra (Eds.). 1996. *Handbook of Semiconductor Electrodeposition.* New York: Marcel Dekker. 312 pp.

49. Ruach-Nir, I., Y. Zhang, R. Popovitz-Biro, I. Rubinstein, and G. Hodes. 2003. Shape control in electrodeposited, epitaxial CdSe nanocrystals on (111) gold. *J. Phys. Chem. B* 107: 2174–2179.

50. Brownson, J. R. S., C. Georges, and C. Levy-Clement. 2006. Synthesis of a δ-SnS polymorph by electrodeposition. *Chem. Mater.* 18: 6397–6402.

51. Ham, S., S. Jeon, M. Park et al. 2010. Electrodeposition and stripping analysis of bismuth selenide thin films using combined electrochemical quartz crystal microgravimetry and stripping voltammetry. *J. Electroanal. Chem.* 638: 195–203.

52. Stickney, J. L. 1999. Electrochemical atomic layer epitaxy. *Electroanal. Chem.* 21: 75–209.

53. Boone, B. E. and C. Shannon. 1996. Optical properties of ultrathin electrodeposited CdS films probed by resonance raman spectroscopy and photoluminescence. *J. Phys. Chem.* 100: 9480–9484.

54. Colletti, L. P., D. Teklay, and J. L. Stickney. 1994. Thin-layer electrochemical studies of the oxidative underpotential deposition of sulfur and its application to the electrochemical atomic layer epitaxy deposition of CdS. *J. Electroanal. Chem.* 369: 145–152.

55. Huang, B. M., L. P. Colletti, B. W. Gregory, J. L. Anderson, and J. L. Stickney. 1995. Preliminary studies of the use of an automated flow-cell electrodeposition system for the formation of cdte thin films by electrochemical atomic layer epitaxy. *J. Electrochem. Soc.* 142: 3007–3016.

56. Villegas, I. and J. L. Stickney. 1992. Preliminary studies of gallium arsenide deposition on gold (100), (110), and (111) surfaces by electrochemical atomic layer epitaxy. *J. Electrochem. Soc.* 139: 686–694.

57. Vaidyanathan, R., S. M. Cox, U. Happek, D. Banga, M. K. Mathe, and J. L. Stickney. 2006. Preliminary studies in the electrodeposition of PbSe/PbTe superlattice thin films via electrochemical atomic layer deposition (ALD). *Langmuir* 22: 10590–10595.

58. Venkatasamy, V., N. Jayaraju, S. M. Cox, C. Thambidurai, and J. L. Stickney. 2007. Studies of $Hg_{(1-x)}$$Cd_x$Te formation by electrochemical atomic layer deposition and investigations into bandgap engineering. *J. Electrochem. Soc.* 154: H720–H725.

59. Siegfried, M. J. and K.-S. Choi. 2005. Directing the architecture of cuprous oxide crystals during electrochemical growth. *Angew. Chemie, Int. Ed.* 44: 3218–3223.

60. Sarkar, S. K., N. Burla, E. W. Bohannan, and J. A. Switzer. 2007. Enhancing enantioselectivity of electrodeposited CuO films by chiral etching. *J. Am. Chem. Soc.* 129: 8972–8973.

61. He, Z., J. A. Koza, G. Mu, A. S. Miller, E. W. Bohannan, and J. A. Switzer. 2013. Electrodeposition of $Co_xFe_{(3-x)}O_4$ epitaxial films and superlattices. *Chem. Mater.* 25: 223–232.

62. Velmurugan, J., J.-M. Noel, and M. V. Mirkin. 2014. Nucleation and growth of mercury on Pt nanoelectrodes at different overpotentials. *Chem. Sci.* 5: 189–194.

63. Velmurugan, J., J.-M. Noel, W. Nogala, and M. V. Mirkin. 2012. Nucleation and growth of metal on nanoelectrodes. *Chem. Sci.* 3: 3307–3314.

64. Fahrenkrug, E., J. Gu, S. Jeon, P. A. Veneman, R. S. Goldman, and S. Maldonado. 2014. Room-temperature epitaxial electrodeposition of single-crystalline germanium nanowires at the wafer scale from an aqueous solution. *Nano Letts.* 14: 847–852.

65. Gu, J., S. M. Collins, A. I. Carim, X. Hao, B. M. Bartlett, and S. Maldonado. 2012. Template-free preparation of crystalline Ge nanowire film electrodes via an electrochemical liquid-liquid-solid process in water at ambient pressure and temperature for energy storage. *Nano Letts.* 12: 4617–4623.

66. Gu, J., E. Fahrenkrug, and S. Maldonado. 2013. Direct electrodeposition of crystalline silicon at low temperatures. *J. Am. Chem. Soc.* 135: 1684–1687.

67. Kulp, E. A., S. J. Limmer, E. W. Bohannan, and J. A. Switzer. 2007. Electrodeposition of nanometer-thick ceria films by oxidation of cerium(III)-acetate. *Solid State Ionics* 178: 749–757.

68. Poizot, P., C.-J. Hung, M. P. Nikiforov, E. W. Bohannan, and J. A. Switzer. 2003. An electrochemical method for CuO thin film deposition from aqueous solution. *Electrochem. Solid-State Lett.* 6: C21–C25.

69. Martin, C. R. 1994. Nanomaterials: A membrane-based synthetic approach. *Science* 266: 1961–1966.

70. Martin, C. R. 1996. Membrane-based synthesis of nanomaterials. *Chem. Mater.* 8: 1739–1746.

71. Liu, Z., G. Xia, F. Zhu et al. 2008. Exploiting finite size effects in a novel core/shell microstructure. *J. Appl. Phys.* 103: 064313.

72. Attard, G. S., P. N. Bartlett, N. R. B. Coleman, J. M. Elliott, J. R. Owen, and J. H. Wang. 1997. Mesoporous platinum films from lyotropic liquid crystalline phases. *Science* 278: 838–840.

73. Attard, G. S., S. A. A. Leclerc, S. Maniguet, A. E. Russell, I. Nandhakumar, and P. N. Bartlett. 2001. Mesoporous Pt/Ru alloy from the hexagonal lyotropic liquid crystalline phase of a nonionic surfactant. *Chem. Mater.* 13: 1444–1446.

74. Abdelsalam, M. E., P. N. Bartlett, J. J. Baumberg, and S. Coyle. 2004. Preparation of arrays of isolated spherical cavities by self-assembly of polystyrene spheres on self-assembled pre-patterned macroporous films. *Adv. Mater.* 16: 90–93.

75. Morin, S., A. Lachenwitzer, O. M. Magnussen, and R. J. Behm. 1999. Potential-controlled step flow to 3D step decoration transition: Ni electrodeposition on Ag(111). *Phys. Rev. Lett.* 83: 5066–5069.

76. Golan, Y., L. Margulis, I. Rubinstein, and G. Hodes. 1992. Epitaxial electrodeposition of cadmium selenide nanocrystals on gold. *Langmuir* 8: 749–752.

77. Golan, Y., A. Hatzor, J. L. Hutchinson, I. Rubinstein, and G. Hodes. 1997. Electrodeposited quantum dots. 6. Epitaxial size control in Cd(Se,Te) nanocrystal on {111} gold. *Israel J. Chem.* 37: 303–313.
78. Li, W. J., J. A. Virtanen, and R. M. Penner. 1992. Nanometer-scale electrochemical deposition of silver on graphite using a scanning tunneling microscope. *Appl. Phys. Lett.* 60: 1181–1183.
79. Li, W., J. A. Virtanen, and R. M. Penner. 1992. A nanometer-scale galvanic cell. *J. Phys. Chem.* 96: 6529–6532.
80. Kolb, D. M., R. Ullmann, and T. Will. 1997. Nanofabrication of small copper clusters on gold(111) electrodes by a scanning tunneling microscope. *Science* 275: 1097–1099.
81. Wei, Y.-M., X.-S. Zhou, J.-G. Wang, J. Tang, B.-W. Mao, and D. M. Kolb. 2008. The creation of nanostructures on an Au(111) electrode by tip-induced iron deposition from an ionic liquid. *Small* 4: 1355–1358.
82. Xiang, C., S.-C. Kung, D. K. Taggart et al. 2008. Lithographically patterned nanowire electrodeposition: A method for patterning electrically continuous metal nanowires on dielectrics. *ACS Nano* 2: 1939–1949.
83. Yang, Y., S. C. Kung, D. K. Taggart et al. 2008. Synthesis of PbTe nanowire arrays using lithographically patterned nanowire electrodeposition. *Nano Lett.* 8: 2447–2451.
84. Switzer, J. A., C. J. Hung, E. W. Bohannan, M. G. Shumsky, T. D. Golden, and D. C. Van Aken. 1997. Electrodeposition of quantum-confined metal/semiconductor nanocomposites. *Adv. Mater.* 9: 334–338.
85. Switzer, J. A., B. M. Maune, E. R. Raub, and E. W. Bohannan. 1999. Negative differential resistance in electrochemically self-assembled layered nanostructures. *J. Phys. Chem. B* 103: 395–398.
86. Switzer, J. A., C.-J. Hung, L.-Y. Huang et al. 1998. Potential oscillations during the electrochemical self-assembly of copper/cuprous oxide layered nanostructures. *J. Mater. Res.* 13: 909–916.
87. Lashmore, D. S. and M. P. Dariel. 1988. Electrodeposited copper-nickel textured superlattices. *J. Electrochem. Soc.* 135: 1218–1221.
88. Moffat, T. P. 1995. Electrochemical production of single-crystal Cu-Ni strained-layer superlattices on Cu(100). *J. Electrochem. Soc.* 142: 3767–3770.
89. Ross, C. A. 1994. Electrodeposited multilayer thin films. *Annu. Rev. Mater. Sci.* 24: 159–188.
90. George, S. M. 2010. Atomic layer deposition: An overview. *Chem. Rev.* 110: 111–131.
91. Baker, L., A. S. Cavanagh, J. Yin, S. M. George, A. Kongkanand, and F. T. Wagner. 2012. Growth of continuous and ultrathin platinum films on tungsten adhesion layers using atomic layer deposition techniques. *Appl. Phys. Lett.* 101: 111601.
92. Debe, M. K. 2012. Electrocatalyst approaches and challenges for automotive fuel cells. *Nature* 486: 43–51.
93. Chen, M. S. and D. W. Goodman. 2004. The structure of catalytically active gold on titania. *Science* 306: 252–255.
94. Jiang, X., T. M. Gur, F. B. Prinz, and S. F. Bent. 2010. Atomic layer deposition (ALD) co-deposited Pt-Ru binary and Pt skin catalysts for concentrated methanol oxidation. *Chem. Mater.* 22: 3024–3032.
95. Allongue, P. and F. Maroun. 2010. Electrodeposited magnetic layers in the ultrathin limit. *MRS Bull.* 35: 761–770.
96. Kolb, D. M., M. Przasnyski, and H. Gerischer. 1974. Underpotential deposition of metals and work function differences. *J. Electroanal. Chem. Interface Electrochem.* 54: 25–38.
97. Herrero, E., L. J. Buller, and H. D. Abruna. 2001. Underpotential deposition at single crystal surfaces of Au, Pt, Ag and other materials. *Chem. Rev.* 101: 1897–1930.
98. Gregory, B. W. and J. L. Stickney. 1991. Electrochemical atomic layer epitaxy (ECALE). *J. Electroanal. Chem. Interface Electrochem.* 300: 543–561.
99. Brankovic, S. R., J. X. Wang, and R. R. Adzic. 2001. Metal monolayer deposition by replacement of metal adlayers on electrode surfaces. *Surf. Sci.* 474: L173–L179.
100. Jayaraju, N., D. Vairavapandian, Y. G. Kim, D. Banga, and J. L. Stickney. 2012. Electrochemical atomic layer deposition (e-ALD) of Pt nanofilms using SLRR cycles. *J. Electrochem. Soc.* 159: D616–D622.
101. Fayette, M., Y. Liu, D. Bertrand, J. Nutariya, N. Vasiljevic, and N. Dimitrov. 2011. From Au to Pt via surface limited redox replacement of Pb UPD in one-cell configuration. *Langmuir* 27: 5650–5658.
102. Sheridan, L. B., D. K. Gebregziabiher, J. L. Stickney, and D. B. Robinson. 2013. Formation of palladium nanofilms using electrochemical atomic layer deposition (e-ALD) with chloride complexation. *Langmuir* 29: 1592–1600.
103. Liu, Y., D. Gokcen, U. Bertocci, and T. P. Moffat. 2012. Self-terminating growth of platinum films by electrochemical deposition. *Science* 338: 1327–1330.
104. Switzer, J. A. 2012. Atomic layer electrodeposition. *Science* 338: 1300–1301.
105. Braun, P. V. 2014. Materials chemistry in 3D templates for functional photonics. *Chem. Mater.* 26: 277–286.

106. Allred, D. B., M. Sarikaya, F. Baneyx, and D. T. Schwartz. 2005. Electrochemical nanofabrication using crystalline protein masks. *Nano Letts.* 5: 609–613.

107. Zhang, Y., G. Hodes, I. Rubinstein, E. Grunbaum, R. R. Nayak, and J. L. Hutchison. 1999. Electrodeposited quantum dots. Metastable rocksalt CdSe nanocrystals on {111} gold alloys. *Adv. Mater.* 11: 1437–1441.

108. Ruach-Nir, I., H. D. Wagner, I. Rubinstein, and G. Hodes. 2003. Structural effects in the electrodeposition of CdSe quantum dots on mechanically strained gold. *Adv. Funct. Mater.* 13: 159–164.

109. Yang, F., D. K. Taggart, and R. M. Penner. 2009. Fast, sensitive hydrogen gas detection using single palladium nanowires that resist fracture. *Nano Letts.* 9: 2177–2182.

110. Menke, E. J., M. A. Thompson, C. Xiang, L. C. Yang, and R. M. Penner. 2006. Lithographically patterned nanowire electrodeposition. *Nat. Mater.* 5: 914–919.

111. Switzer, J. A. 2001. Electrodeposition of superlattices and multilayers. In *Electrochemistry of Nanomaterials*, ed. G. Hodes, Chapter 3, 67–101. Weinheim, Germany: Wiley-VCH.

18 Scanning Electrochemical Microscopy of Nanopores, Nanocarbons, and Nanoparticles

Shigeru Amemiya

CONTENTS

18.1 INTRODUCTION

In this chapter, the applications of scanning electrochemical microscopy (SECM) to the study of nanopores, nanocarbons, and nanoparticles are discussed. Electrochemistry of these nanosystems is an important topic and is treated thoroughly in Chapters 3, 9, and 11. Moreover, the fundamentals and applications of SECM are summarized in many reviews[1,2] and a couple of comprehensive monographs.[3,4] By contrast, this chapter is uniquely and timely devoted to the recent evolution of SECM as a powerful electrochemical method to quantitatively characterize the (electro)chemical reactivity of these nanosystems. In the late 1980s,[5–7] Bard and coworkers invented SECM as the unique scanning probe microscopy technique by employing an ultramicroelectrode (UME) as a probe to quantitatively image and measure dynamic (electro)chemical processes at and near the interfaces. This invention was followed quickly by the establishment, wide acceptance, and commercialization of SECM by the late 1990s to find a wide range of applications from electrochemistry to various fields of chemistry, biology, and materials science.[3] This success is due to the versatility of the SECM principle, which allows for the study of various types of interfaces, including solid–liquid, liquid–liquid, and gas–liquid interfaces as well as biological cell and membranes. More recently, the successful applications of SECM were demonstrated in the numerous studies of the ensembles of nanoscale objects formed at different interfaces as discussed in this chapter. The micrometer spatial resolution of the standard SECM setup, however, limits the study of individual nanostructures. This limitation will be overcome by developing nanoscale SECM based on nanoelectrode tips (see also Chapter 15). The exciting future avenue of nanoscale SECM of individual nanostructures is overviewed in the perspectives of this chapter, while another chapter by this author was devoted to recent progresses in the development of nanoscale SECM.[8]

18.2 NANOPORES

The SECM studies of nanoporous membranes illustrate the power of this electrochemical method in spatial resolution, kinetic resolution, sensitivity, and versatility in comparison to other electrochemical methods.[9] This section is focused on the recent SECM studies of two nanoporous membranes, that is, the nuclear envelope (NE) and porous nanocrystalline silicon (pnc-Si) membranes. Commonly, these nanoporous membranes possess nanometer thickness and self-stand in solution to mediate fast molecular transport. The resultant high permeability of the ultrathin nanoporous membranes can be reliably and quantitatively measured by using SECM without a limitation due to molecular diffusion between the membrane and the solution. The quantitative data allow for the assessment of pore dimensions and pore–molecule interactions. Interestingly, SECM is highly sensitive to the kinetics of molecular transport through the nanoporous membranes with a low porosity of <0.1 because of efficient diffusional interactions between nanopores.[10] In fact, SECM has been successfully applied to the study of the substrates that are partially covered with nanometer-sized active spots, for example, the defects of self-assembled monolayers of alkylthiolates on the gold electrodes,[11–13] the defects of TiO_2 films based on atomic layer deposition,[14] and the boron-doped sites of diamond electrodes.[15] By contrast, single nanopores can be imaged by using nanoscale SECM with the unprecedentedly high spatial resolution.

18.2.1 Nuclear Pore Complex

The nuclear pore complex (NPC) is the proteinaceous nanopore that solely mediates molecular transport across the NE between the nucleus and cytoplasm of a eukaryotic cell. Recently, SECM was used to determine the high ion permeability and the corresponding dimensions of the NPC nanopores.[16,17] In these studies, several key experimental strategies were developed to form a nanoscale gap between the tip and the NE. The nanogap was required for the precise determination of the high ion permeability of the NE with multiple NPCs under high mass transport conditions. Specifically, the large nucleus (~400 μm in diameter) was isolated from the *Xenopus laevis* oocyte and swollen in the hypotonic buffer solution to obtain the smooth patch of the NE that is detached from the nucleoplasm and stabilized by the hole of the Si membrane frame (Figure 18.1a).[16] Moreover, ~1 μm Pt tips were insulated with a thin glass sheath ($RG = r_g/a = 2$ in Figure 18.1b, where r_g and a are the outer and inner tip radii) and milled and smoothened by focused ion beam (FIB) technology.[18] The small Pt tips were protected from severe nanoscale damage by electrostatic discharge (ESD) from an operator as well as transient current flow from a potentiostat.[19] Previously, the reliable measurement of the NE permeability was hampered by using a ~2 μm Pt tip with a thick glass sheath ($RG = 10$) without any protection from electrostatic and electrochemical damages.[20] High-quality SECM tips as required for the NPC studies were also obtained by filling FIB-milled micropipets (Figure 18.1c) with the electrolyte solution of 1,2-dichloroethane. The organic-filled micropipets were immersed in the aqueous solution to form the interface between two immiscible electrolyte solutions (ITIES). The ITIES-based tips served as the ion-selective SECM tips that are surrounded by the thin wall

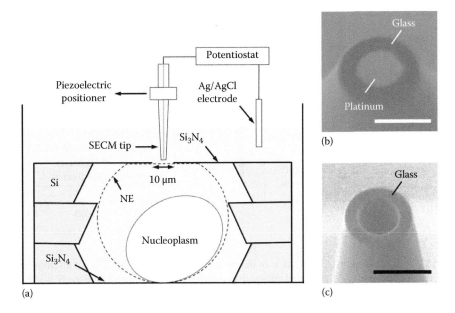

FIGURE 18.1 (a) The SECM cell with the nucleus of a *Xenopus laevis* oocyte. The nucleus was swollen to detach the NE from the nucleoplasm for smoothening. (Reprinted with permission from Kim, J., Izadyar, A., Nioradze, N., and Amemiya, S., Nanoscale mechanism of molecular transport through the nuclear pore complex as studied by scanning electrochemical microscopy, *J. Am. Chem. Soc.*, 135, 2321–2329, 2013. Copyright 2013 American Chemical Society.) SEM images of (b) Pt and (c) micropipet tips. The tip ends were FIB-milled. Scale bar, 1 μm. (Panel b was reprinted with permission from Nioradze et al., Origins of nanoscale damage to glass-sealed platinum electrodes with submicrometer and nanometer size, *Anal. Chem.*, 85, 6198–6202, 2013. Copyright 2013 American Chemical Society; Panel c was adapted with permission from Kim, J., Izadyar, A., Shen, M., Ishimatsu, R., and Amemiya, S., *Anal. Chem.*, Ion permeability of the nuclear pore complex and ion-induced macromolecular permeation as studied by scanning electrochemical and fluorescence microscopy, 86, 2090–2098, 2014. Copyright 2014 American Chemical Society.)

of the glass micropipets ($RG = 1.7$) and are free from electrostatic and electrochemical damages, thereby enabling the formation of a nanogap over the NE (see the following).

The SECM-induced transfer mode was employed to determine the high ion permeability of the NPCs by using the high-quality Pt and ITIES-based tips. In this operation mode, a probe ion was amperometrically transferred at the ITIES tip (or electrolyzed at the Pt tip) to deplete the probe ion between the tip and the NE (Figure 18.2a). The resultant concentration gradient drove the flux of the probe ion through the NPCs from the opposite side of the NE to enhance the amperometric tip current. Figure 18.2b shows plots of tip current, i_T, based on the transfer of tetraphenylarsonium (TPhAs$^+$) as a probe ion when 0.9 μm diameter micropipet-supported ITIES tips approached the NE and the Si wafer as substrates. In both approach curves, the ionic current at the tip near the substrates decreased from the diffusion-limited current in the bulk solution, $i_{T,\infty}$, as given by

$$i_{T,\infty} = 4xnFD_w ca, \tag{18.1}$$

where
  x is a function of RG
  n is the charge of a probe ion transferred at an ITIES-based tip (or the number of electrons accepted or donated by a probe ion at a Pt tip)
  c is the concentration of the probe ion in the bulk solution

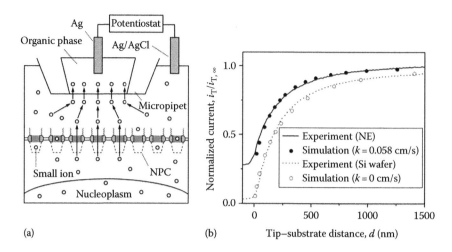

(a)   (b)   Tip–substrate distance, d (nm)

FIGURE 18.2 (a) Measurement illustration of the ion permeability of the NE using a micropipet-supported ITIES tip. (b) Experimental and simulated approach curves for TPhAs$^+$ at the NE and the Si wafer as obtained with micropipet-supported ITIES tips. Tip approach rate, 0.30 μm/s. (Adapted with permission from Kim, J., Izadyar, A., Shen, M., Ishimatsu, R., and Amemiya, S., *Anal. Chem.*, Ion permeability of the nuclear pore complex and ion-induced macromolecular permeation as studied by scanning electrochemical and fluorescence microscopy, 86, 2090–2098, 2014. Copyright 2014 American Chemical Society.)

The approach curve at the NE was more positive than the purely negative approach curve at the inert Si wafer (solid and dashed lines, respectively). The higher tip current at the NE is due to the SECM-induced transport of TPhAs$^+$ through the NPCs from the nucleus to the tip. The experimental approach curve at the NE was fitted with a theoretical curve as obtained by employing the finite element method with NE permeability, tip inner and outer radii, and tip position at the zero tip–NE distance as fitting parameters. A good fit was obtained by using a high permeability value, k, of 0.058 cm/s and tip inner and outer radii of 0.45 and 0.77 μm, respectively. In contrast, the experimental approach curve at the Si wafer was fitted with a theoretical negative approach curve, that is, $k = 0$ cm/s, with tip inner and outer radii of 0.47 and 0.80 μm. The tip radii determined by the theoretical analysis were consistent with those determined by SEM (e.g., Figure 18.1c). Remarkably, this analysis also showed that the smooth ~0.9 μm diameter tips approached as close as 22 nm from the NE and even closer to the Si wafer, down to a separation of 9 nm.

The passive permeability of the NPCs to various probe ions thus determined by SECM was quantitatively analyzed to assess the dimensions of the NPCs and the strength of NPC–probe interactions. Notably, the permeability values were proportional to ion diffusion coefficients in the aqueous phase (Figure 18.3). This proportionality is expected theoretically when the probe ions freely diffuse through nanopores as established by the SECM study of a pnc-Si membrane as a model (see Section 18.2.2). Theoretically, the permeability of a nanopore membrane to a probe ion, k, is based on three diffusion steps: the probe ion accesses from the aqueous solution to the pore, moves through the pore, and escapes from the pore to the aqueous solution.[21] This theoretical analysis gives[16,17]

$$k = \frac{2rN}{2l/\pi r + 1/f(\sigma)} D_{\mathrm{w}},$$ (18.2)

where
  N is pore density
  r and l are the radius and length of a cylindrical nanopore, respectively
  $\sigma (=\pi N r^2)$ is membrane porosity
  $f(\sigma) \approx 1$ for $\sigma < 0.1$[22]

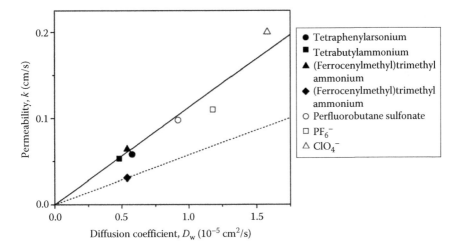

FIGURE 18.3   Plot of the ion permeability of the NE, k, versus ion diffusion coefficient, D_w. Equation 18.2 was fitted to experimental k values in the absence and presence of WGA to yield the solid and dotted lines, respectively. (Adapted with permission from Kim, J., Izadyar, A., Shen, M., Ishimatsu, R., and Amemiya, S., Ion permeability of the nuclear pore complex and ion-induced macromolecular permeation as studied by scanning electrochemical and fluorescence microscopy, *Anal. Chem.*, 86, 2090–2098, 2014. Copyright 2014 American Chemical Society.).

The best fit of a k versus D_w plot for six probe ions with Equation 18.2 (solid line in Figure 18.3) gave a slope that was consistent with typical geometrical parameters for the *Xenopus* oocyte NPCs, that is, $N=40$ NPCs/μm^2, $r=24$ nm, $l=35$ nm, and $\sigma=0.07$.[23,24] Importantly, lower permeability was obtained when the NPCs were blocked by wheat germ agglutinin (WGA) to give $r=17$ nm in Equation 18.2 (dotted line in Figure 18.3b), where the other parameters were the same. This smaller radius was ascribed to that of the central zone of the NPC because WGA blocks the peripheral zone.[25] In fact, a difference of ~7 nm between pore radii with and without WGA is similar to the diameter of WGA (~5 nm).[26] Interestingly, WGA completely blocks the nuclear import of large proteins (>40 kDa) with nuclear localization signal (NLS) peptides,[16,20] which are chaperoned by nuclear transport receptors, for example, importins. Overall, this SECM study demonstrated that the NPC possesses central and peripheral routes with distinct permeability to passive transport of small probe ions and importin-facilitated transport of NLS-tagged proteins.[16]

Advantageously, ITIES-based tips can be miniaturized to nanometer scale by using nanopipets (see Chapter 15), which are attractive for the study of single NPC permeability.[17] Unfortunately, this nanoscale SECM study was unsuccessful because the nano-ITIES was readily blocked by the adsorption of several protein molecules leaching from the nucleoplasm. This study, however, led to an interesting finding that the NPC can be permeabilized to naturally impermeable macromolecules by highly lipophilic probe ions such as TPhAs⁺ and perfluorobutane sulfonate. The detailed mechanism of the ion-induced macromolecular permeation was studied by using fluorescence microscopy to ascribe the permeabilization of the NPCs to cooperative hydrophobic interactions between the NPC-permeabilizing ions and the hydrophobic transport barriers of the NPC based on the repeats of phenylalanine and glycine. The permeabilization of the NPC by the hydrophobic ions was pathway-selective, which supports a new transport model based on different transport barriers through the central and peripheral zones of the NPC. The use of the nonphysiological probe ions in the SECM study provided unprecedented insights into the gating mechanism of molecular transport through the NPC.

18.2.2 Porous Nanocrystalline Silicon Membrane

Ion transport through pnc-Si membranes was studied by using SECM at both multiple[10,27] and single[28] pore levels. This ultrathin nanoporous membrane was reported first in 2007[29] and quickly found various applications including the separation of macromolecules[29] and nanoparticles[30] as well as the development of artificial kidney[31] and electro-osmotic pump.[32] Moreover, the dimensions and density of the silicon nanopores can be tuned to be comparable to those of the NPCs, thereby serving as an excellent model system for the SECM study of the NPCs.[10] Specifically, the SECM-induced transfer mode was employed to determine the permeability of pnc-Si membranes to various probe ions by using micrometer-sized Pt[10] and ITIES[27] tips. In the latter study, the tip of a micropipet was not only surrounded by a thin glass wall, but also smoothened by FIB milling to be positioned close to the membrane surface for precise permeability measurement. Permeability values for small probe ions were proportional to their diffusion coefficients in the aqueous solution (Figure 18.4). The slope agreed with Equation 18.2 with the parameters determined by using transmission electron microscopy, that is, $N = 67$ pores/μm^2, $r = 5.6$ nm as an average, $l = 16$ nm, and $\sigma = 0.007$. The excellent agreement between experimental and theoretical permeability values confirmed that these small probe ions freely diffuse through the water-filled nanopores. By contrast, these nanopores were small enough to slow down the transport of polycationic peptide protamines (4.5 kDa and +20 charges) with a Stokes radius of 2.0 nm. The corresponding permeability value was lower than expected from Equation 18.2 (Figure 18.4) despite attractive electrostatic interactions of the polycations with the negatively charged wall of the SiO_2-coated silicon nanopore. Noticeably, electrostatic repulsion from the negatively charged pore wall was strong enough to slow down the transport of polyanionic pentasaccharide (Arixtra; 1.5 kDa and −10 charges) as ionic strength decreased (from 0.60 to 0.06 M in Figure 18.4).

Ion transport through the single nanopores of a pnc-Si membrane was imaged by using nanoscale SECM to quantitatively demonstrate an unprecedentedly high spatial resolution of ~30 nm.[28] In this nanoscale SECM imaging, a nanopipet-supported ITIES tip approached to a distance of down to

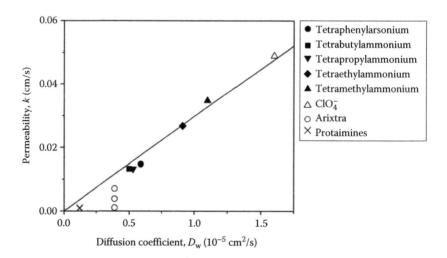

FIGURE 18.4 The permeability of a pnc-Si membrane, k, plotted against the diffusion coefficients of probe ions in the aqueous phase, D_w. The solid line represents the best fit of Equation 18.2 for monovalent ions. The permeability to Arixtra and protamine was measured with 0.10, 0.03, and 0.01 M phosphate buffer at pH 7.0 while the buffer concentration was 0.10 M for monovalent probe ions. (Adapted with permission from Ishimatsu, R., Kim, J., Jing, P. et al., Ion-selective permeability of a ultrathin nanopore silicon membrane as probed by scanning electrochemical microscopy using micropipet-supported ITIES tips, *Anal. Chem.*, 82, 7127–7134, 2010. Copyright 2010 American Chemical Society.)

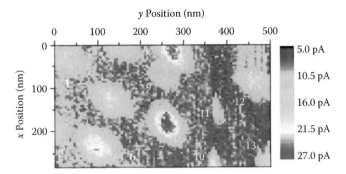

FIGURE 18.5   Constant-height SECM image of a pnc-Si membrane based on an amperometric response to tetrabutylammonium at a 34 nm diameter quartz pipet filled with a 1,2-dichloroethane solution of organic supporting electrolytes. The external aqueous solution contained 10 mM tetrabutylammonium chloride and 0.3 M KCl. (Reprinted with permission from Shen, M., Ishimatsu, R., Kim, J., and Amemiya, S., Quantitative imaging of ion transport through single nanopores by high-resolution scanning electrochemical microscopy, *J. Am. Chem. Soc.*, 134, 9856–9859, 2012. Copyright 2012 American Chemical Society.)

1.3 nm from the membrane in water by monitoring the tip current response to tetrabutylammonium as a probe ion. Then, the tip was scanned laterally above the flat membrane at a constant height to obtain a 275 nm × 500 nm image (Figure 18.5). The tip current was lowered when the nanopipet was positioned over the impermeable region of the membrane, which hindered the diffusion of the probe ion to the nanoscale ITIES (i.e., negative feedback). A higher tip current was obtained when a tip was positioned at the same distance over the nanopore, which supplied the probe ion. Higher current responses were observed over 13 nanopores in the image as expected from a density of 90 nanopores/μm^2. The SECM image of nanopore 7 agreed very well with the theoretical image simulated by using the finite element method to provide major and minor axes of 53 and 41 nm and a depth of 30 nm for a nanopore as an elliptical cylinder. These pore dimensions are similar to those of the NPCs (see preceding text). This finite element analysis also confirmed that the spatial resolution was limited by the tip diameter (34 nm) and high enough to resolve the dense silicon nanopores with short pore–pore separations of ~100 nm. Such high-density nanopores have not been resolved by using other electrochemical imaging methods, that is, SECM,[9] scanning ion-conductance microscopy (SICM),[33–38] SECM–SICM,[39,40] and SECM–AFM,[41–43] where the shortest separations between two resolvable nanopores were limited to >250 nm and ~1.5 μm for SECM–AFM[42] and SICM,[38] respectively.

18.3   NANOCARBONS

In this section, the advantages of SECM for the kinetic study of carbon nanotube and graphene are discussed. SECM is an attractive method to investigate the spatially resolved electrochemical activity of the nanocarbon materials, which is not addressable by standard electrochemical methods.[44] The high spatial resolution of SECM enables us to selectively study the electroactivity of the sidewall of individual single-walled carbon nanotubes (SWCNTs). The high total electroactivity of SWCNT networks is also measurable by using SECM owing to its high mass transport conditions. Importantly, SWCNTs can be grown on insulating substrates by chemical vapor deposition (CVD) to eliminate a background electrochemical response from the substrates. In addition, an advantage of SECM for the kinetic study of CVD-grown graphene is a negligibly small iR drop in comparison to cyclic voltammetry with a macroscopic graphene electrode. SECM can be also used to measure the lateral conductivity of the graphene oxide film deposited on an insulating substrate.

18.3.1 Individual Carbon Nanotube

Carbon nanotube is attractive as a nanoscale electrode material with well-defined, structure-dependent electrical properties. The origins of its electroactivity, however, have been poorly understood and often ascribed to defects, contaminants, and edges.[44,45] Recently, the feedback mode of SECM was employed to find that the sidewall of SWCNT was highly electroactive (Figure 18.6a).[46] In this study, ultralong SWCNTs (~1.6 nm in diameter and ~2 mm in length) were grown horizontally and sparsely on a SiO$_2$ surface by CVD. The SWCNTs were so widely separated that each nanotube was detected by SECM imaging with a 10 µm diameter Pt tip in the feedback mode (Figure 18.6b). Specifically, (ferrocenylmethyl)trimethyl ammonium (FcTMA$^+$) was oxidized at the Pt tip to generate the ferrocenium species (FcTMA^{2+}), which diffused to the substrate. When the tip was scanned over a SWCNT at a constant height, tip-generated FcTMA^{2+} was reduced at the sidewall of the nanotube to regenerate FcTMA$^+$, which was detected again at the tip to enhance tip current. Importantly, mediator regeneration (step i in Figure 18.6a) at the unbiased SWCNT on the

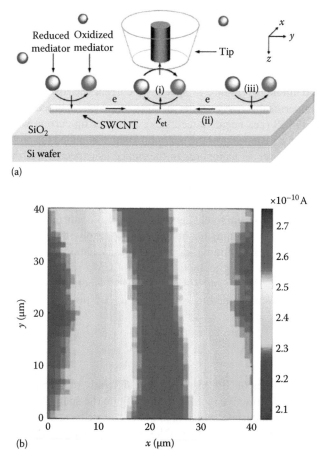

(b)

FIGURE 18.6 (a) Scheme of an SECM feedback experiment with a disk Pt tip positioned above an individual SWCNT as grown on an insulating substrate. (b) SECM image of an individual SWCNT in the feedback mode based on the oxidation of FcTMA$^+$ at a 10 µm diameter Pt tip with a tip–substrate distance of 5.2 µm. Probe scan rate, 15 µm/s. (Reprinted with permission from Kim, J., Xiong, H., Hofmann, M., Kong, J., and Amemiya, S., Scanning electrochemical microscopy of individual single-walled carbon nanotubes, *Anal. Chem.*, 82, 1605–1607, 2010. Copyright 2010 American Chemical Society.)

SiO$_2$ surface was coupled with the reverse reaction (step iii) at the exterior surface of the ultralong nanotube by efficient electron transport (step ii) along the nanotube. By contrast, tip current was lower due to the negative feedback effect when the tip was positioned over the insulating SiO$_2$ surface. The tip current based on the feedback effect from the individual SWCNT was analyzed to estimate a high electron-transfer (ET) rate constant of >4 cm/s for its sidewall. This high ET rate constant was measurable because of a high mass transport rate at the 1.6 nm diameter nanotube. Importantly, efficient mass transport and facile ET at the nanotube sidewall enabled the detection of the individual SWCNT by using the 10 μm diameter tip, which is ~6000 times larger than the nanotube diameter. The spatial resolution of the SECM image was determined by the diameter of the disk-shaped tip rather than by the nanotube diameter and was improved by using a 1.5 μm diameter tip. More recently, SECM was combined with AFM by using a nanofabricated cantilever based on a triangular-frame electrode to obtain the topography and reactivity images of individual SWCNTs in their network.[47] The spatial resolution with the triangular-frame-shaped tip was limited by its edge length of ~300 nm, which were ~200 times larger than the nanotube diameter. Noticeably, a disk-shaped conductive spot must be as large as at least 10%–20% of a tip diameter for the SECM feedback detection.[48]

Originally, the feasibility of the SECM-based detection of individual SWCNTs was suggested by using nanoband electrodes as model substrates.[49,50] For instance, the SECM images of a 100 nm wide Pt nanoband in an insulating matrix were obtained by using $Ru(NH_3)_6^{3+}$ as a redox mediator in the feedback mode. Higher tip current was observed when the tip was scanned over the Pt nanoband. The width of the nanoband in the image was determined by tip diameter (2–25 μm) as confirmed by three-dimensional numerical simulations. Importantly, this quantitative theory predicted that a feedback effect from a nanoband electrode weakly depended on its nanometer width because of a hemicylindrical diffusion pattern of the redox mediator at the nanoband surface. The high sensitivity of SECM to an anisotropic one-dimensional material with small nanoscale dimensions enabled the detection of an individual SWCNT using micrometer-sized tips (see earlier text). In addition, the damaged region of a nanoband electrode gave an image that was broader than expected from the tip diameter. The tip current over the damaged region was lower due to the lower electroactivity of the damaged surface. Interestingly, the nanometer-wide gap formed in an Au nanoband was also detectable by using the feedback mode of SECM.[51] Specifically, a peak-shaped tip current response was obtained by scanning a 2 μm diameter Pt tip over a 37 nm wide gap formed near the middle of an Au nanoband with a length of 50 μm although the SiO$_2$ surface exposed under the gap was insulating. The tip current was enhanced above the nanogap, where both fragments of the unbiased nanoband exert a feedback effect on tip current. This mechanism was confirmed quantitatively by using three-dimensional finite element simulation.

18.3.2 CARBON NANOTUBE NETWORK

The feedback mode of SECM was employed to reveal the intrinsically high electrochemical activity of SWCNT network under unbiased conditions.[52] A uniform film or patterned sample of a two-dimensional random SWCNT network was directly grown on a SiO$_2$-covered Si wafer by CVD with iron nanoparticles as catalysts. In the SECM imaging of a patterned sample, positive feedback responses were obtained over the regions of high-density SWCNTs separated by low-density regions. The tip current over the lower-density region was still higher than the negative feedback current over an insulating surface. Remarkably, the significant feedback effect was obtained from a SWCNT film with a coverage of only ~1% on the underlying SiO$_2$ surface owing to diffusional overlap between neighboring SWCNTs. In fact, the low-density SWCNT network exerted a positive feedback effect on the tip as expected for a uniformly active film. The high feedback effect from the SWCNT network indicates that individual SWCNTs in the sparse film must be highly electroactive to support the extremely high fluxes that correspond to >100 times the flux at a uniformly active surface. Noticeably, the feedback mode of SECM was also employed to image and measure the significant electrochemical activities of multiwalled carbon nanotubes covalently patterned on Si(111) surfaces.[53]

The high electroactivity of a SWCNT network was quantitatively measured under biased conditions to determine standard ET rate constant, k^0, and transfer coefficient, α.[54] An UME based on a SWCNT network was fabricated as a substrate to enable steady-state SECM measurement in the substrate generation/tip collection (SG/TC) mode. A SWCNT network was grown on the SiO_2/Si substrate and insulated with a photoresist to expose the 25 μm radius disk-shaped surface of the network as an UME. FcTMA$^+$ was voltammetrically oxidized at the SWCNT-network UME positioned under a 50 μm radius disk Pt tip held at a potential of 0.0 V for the diffusion-limited reduction of substrate-generated FcTMA^{2+} (Figure 18.7a). The plots of substrate current versus substrate potential became less reversible owing to a higher mass transport condition as the tip–substrate distance decreased from 18 to 7 μm (Figure 18.7b). The substrate voltammograms were analyzed to yield $k^0 = 0.010$ cm/s and $\alpha = 0.31$. This k^0 value was estimated to be <1% of an average k^0 value at individual SWCNTs, which cover <1% of the 50 μm diameter UME. An minimum average k^0 value of >1 cm/s thus assigned to individual SWCNTs is comparable to the large k^0 values of the

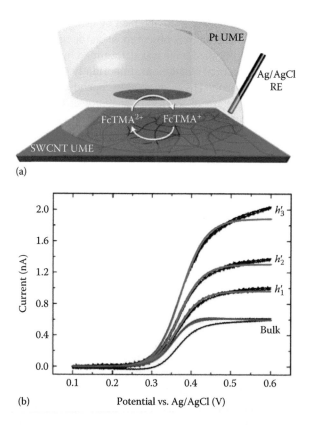

(a)

(b)

FIGURE 18.7 (a) Experimental setup for the kinetic measurements of FcTMA$^+$ oxidation at a 50 μm diameter SWCNT-network UME as a substrate under a 100 μm diameter Pt tip in the SG/TC mode. (b) Experimental current–voltage curves for the oxidation of 0.1 mM FcTMA$^+$ at the SWCNT substrate in 0.1 M NaCl and corresponding simulations based on the Butler–Volmer model with tip–substrate distances of $h_1' = 18\,\mu m$, $h_2' = 12\,\mu m$, and $h_3' = 7\,\mu m$ (red lines). The blue line represents a simulated reversible wave. (Adapted with permission from Dumitrescu, I., I., Dudin, P.V., Edgeworth, J.P., Macpherson, J.V., and Unwin, P.R., Electron transfer kinetics at single-walled carbon nanotube electrodes using scanning electrochemical microscopy, *J. Phys. Chem. C*, 114, 2633–2639, 2010. Copyright 2010 American Chemical Society.)

same redox couple at individual SWCNTs as obtained by steady-state voltammetry (7.6 cm/s)[55] and scanning electrochemical cell microscopy (9 cm/s).[56] The large k^0 values at SWCNTs were ascribed to the high electroactivity of the sidewall of SWCNTs despite its sp^2-carbon character. This high electroactivity cannot be explained if only defects on the sidewall are reactive.

18.3.3 GRAPHENE

The relatively low electroactivity of graphene was demonstrated systematically for various aqueous and nonaqueous redox mediators by using SECM.[57] In this study, centimeter-sized monolayer graphene was grown on a copper foil by CVD, transferred to a SiO_2-coated Si wafer, and attached to an indium wire for electrical connection. An area of 0.5 cm^2 was exposed to an aqueous or nonaqueous solution for the SECM characterization of various redox mediators in the feedback mode by using a 15 μm diameter Pt tip (Figure 18.8a). Figure 18.8b and c shows CV and SECM approach curves for $Fe(CN)_6^{4-}$. While the CV was very broad with a wide peak separation, well-defined approach curves were obtained for the graphene electrode at different potentials. These approach curves fitted well with theoretical curves to yield the corresponding ET rate constants. Surprisingly, this analysis showed that the approach curve at a large overpotential of −0.98 V was still limited by the kinetics of $Fe(CN)_6^{4-}$ regeneration at the graphene surface, not by the diffusion of the redox mediator between the tip and the graphene surface. This result was attributed to the

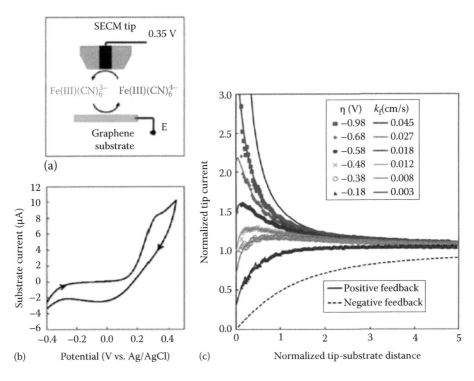

FIGURE 18.8 (a) Scheme of SECM feedback measurements of $Fe(CN)_6^{3-}$ reduction kinetics at a graphene substrate electrode in 1 mM $Na_4[Fe(CN)_6]$ and 0.2 M phosphate buffer at pH 7. (b) CV of $Fe(CN)_6^{4-}$ at a graphene electrode. Sweep rate, 20 mV/s. (c) Experimental (dots) and theoretical (lines) approach curves for the reduction of $Fe(CN)_6^{3-}$ at the graphene substrate as obtained by using a 15 μm diameter Pt tip. Overpotential and ET rate constant for $Fe(CN)_6^{3-}$ reduction are given by η and k_f, respectively. (Reprinted with permission from Ritzert, N.L., Rodriguez-Lopez, J., Tan, C., and Abruna, H.D., Kinetics of interfacial electron transfer at single-layer graphene electrodes in aqueous and nonaqueous solutions, *Langmuir*, 29, 1683–1694, 2013. Copyright 2013 American Chemical Society.)

very weak dependence of the ET rate on overpotential. In fact, an extremely low transfer coefficient, α, of 0.1 was determined from a plot of the measured rate constant versus overpotential, that is, the Tafel plot. In addition, the anodic branch of the Tafel plot was measured by using $Fe(CN)_6^{3-}$ in the feedback mode to observe the weak overpotential-dependence of ET rate as represented by an extremely large transfer coefficient, α, of 0.9. In contrast to the inconsistent α values, the cathodic and anodic Tafel plots gave relatively similar standard ET rate constants, k^0, of 9.5×10^{-4} and 1.9×10^{-3} cm/s, respectively.

The SECM kinetic study of the CVD-grown graphene electrode manifested the limitation of the traditional feedback mode,[57] where overpotential must be large enough to achieve the steady states based on the irreversible mediator regeneration at the macroscopic substrate electrode.[18] Small overpotentials result in the development of a time-dependent diffusion profile at the whole graphene surface. Approach curves at large overpotentials were diffusion-limited for fast redox mediators, that is, methyl viologen, $Ru(NH_3)_6^{3+}$, and tris(2,2'-bipyridyl)ruthenium(II), to estimate k^0 values of $>2 \times 10^{-2}$ cm/s. By contrast, kinetically limited approach curves were obtained for slow mediators even with large overpotentials to yield k^0 values in the range between 2×10^{-2} cm/s and 5.4×10^{-4} cm/s with the order of $FcMeOH > Co(III)$ sepulchrate $> Mo(CN)_8^{4-} > Fe(CN)_6^{3-} > Fe(II)$ EDTA. This finite kinetics at large overpotentials was due to the weak overpotential-dependence of ET rates, thereby yielding extremely high or low α values of 0.9 or 0.1 for oxidation and reduction at the graphene surface, respectively. These anomalous α values were ascribed to the presence of defects as active sites or the reflection of the electronic structure of the graphene. Noticeably, another SECM study demonstrated that the electroactivity of the graphene surface was enhanced by introducing mechanical and chemical damages.[58] The damages were made by locally scratching graphene surface or by reaction with an oxidizer, NaOCl. The local damages were imaged by SECM to show their higher reactivity to the reduction of $Fe(CN)_6^{3-}$ as a tip-generated redox mediator in the feedback mode.

Interestingly, the redox molecules adsorbed on the graphene surface are highly electroactive[59] and efficiently mediate electron transfer between graphene and solution species, as demonstrated by using SECM.[57,60] For instance, a tripodal compound with a cobalt bis-terpyridine group (**1·2PF₆**) was noncovalently adsorbed on single-layer graphene by π–π interactions with three pyrene groups (Figure 18.9a). The CV of Co(II) complex **1** gave well-defined surface waves to yield a high k^0 value of 13.5 s^{-1}.[59] In this case, the anodic and cathodic branches of the surface waves were symmetric and as wide as expected for a normal transfer coefficient, α, of 0.5. This behavior contrasts to the anomalously weak potential dependence of diffusing redox species at graphene (see previous text). In addition, the surface bound Co(II) complex can mediate the oxygen reduction reaction (ORR) as shown by using the SG/TC mode.[60] In this experiment, the ORR at the graphene substrate with and without the Co(II) complex was voltammetrically studied to yield similar substrate current (the bottom panel of Figure 18.9b). By contrast, the significantly higher tip current based on the oxidation of hydrogen peroxide at a 25 μm diameter Pt tip was observed when the tip was positioned over the graphene substrate covered with the Co(II) complex (the top panel of Figure 18.9b). The tripodal compound also mediates the oxidation of $Fe(CN)_6^{4-}$, which was generated from $Fe(CN)_6^{3-}$ at the SECM tip in the feedback mode. The resultant activity of the graphene surface coated with complex **1** was higher than the surrounding bare graphene surface as spatially resolved by SECM imaging. Importantly, the tip current response to either hydrogen peroxide or $Fe(CN)_6^{4-}$ decreased after the graphene electrode was locally modified with spots of the Co(II) complex and was kept in the THF solution. Since Co(II) complex **1** was nearly insoluble in THF, the decrease of the tip current was attributed to the loss of Co(II) complex **1** by its lateral diffusion from the deposited spot to the surrounding bare graphene surface. The time dependence of tip current was analyzed quantitatively to determine a lateral diffusion coefficient of $\sim 1.5 \times 10^{-9}$ cm^2/s for Co(II) complex **1** on the graphene surface. More recently, a redox-active species, $[Os(bpy)_2(dipy)Cl]PF_6$(bpy, 2,2'-bipyridine; dipy, 4,4'-trimethylenedipyridine), was also adsorbed onto the graphene surface and oxidized to mediate the oxidation of Fe(II)EDTA and $Fe(CN)_6^{4-}$ generated at a 15 μm diameter Pt tip in the feedback

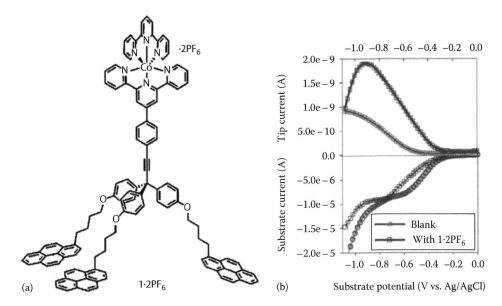

FIGURE 18.9 (a) Structure of a tripodal graphene binder bearing a redox-active Co(II) bis-terpyridyl complex, **1·2PF$_6$**. (Reprinted with permission from Mann, J.A., Rodríguez-López, J., Abruña, H.D., and Dichtel, W.R., Multivalent binding motifs for the noncovalent functionalization of graphene, *J. Am. Chem. Soc.*, 133, 17614–17617, 2011. Copyright 2011 American Chemical Society.) (b) Substrate and tip current responses during the ORR at the graphene substrate electrode with a coverage of 23 pmol/cm^2 by **1·2PF$_6$** in an air-saturated 0.2 M phosphate buffer at pH = 7. H$_2$O$_2$ were generated voltammetrically at the substrate and amperometrically collected at a 25 μm diameter Pt SECM tip (*RG* = ~7) with a tip–substrate distance of 10 μm. Tip potential, 0.6 V versus Ag/AgCl. Sweep rate of the substrate potential, 10 mV/s. (Reprinted with permission from Rodriguez-Lopez, J., Ritzert, N.L., Mann, J.A., Tan, C., Dichtel, W.R., and Abruna, H.D., Quantification of the surface diffusion of tripodal binding motifs on graphene using scanning electrochemical microscopy, *J. Am. Chem. Soc.*, 134, 6224–6236, 2012. Copyright 2012 American Chemical Society.)

mode.[57] Remarkably, the oxidation of both mediators was significantly accelerated in the presence of only one-hundredth of a monolayer of the osmium complex.

The feedback mode of SECM was also used to quantitatively probe the conductivity of the reduced film of graphene oxide under unbiased conditions.[61] In this study, graphene oxide was deposited on glass by the bubble deposition method[62] and was reduced by exposition to an aqueous solution of 0.1 M KOH. Approach curves with FcMeOH changed from positive to negative as the mediator concentration increased. This result indicates that the feedback effect was limited by the conductivity of the graphene oxide film rather than by the electroactivity of its surface, where the regeneration of FcMeOH should be independent of the mediator concentration. In SECM imaging, the feedback tip current varied over the graphene oxide film before exposition to 0.1 M KOH (Figure 18.10a). The normalized feedback current was consistently less than 1, thereby indicating a negative feedback effect due to the limited conductivity of the graphene oxide film. Figure 18.10b shows the SECM image of the graphene oxide film reduced by exposition to 0.1 M KOH solution at room temperature for 30 s. The reduced graphene oxide film showed similar positive feedback current for FcMeOH over the substrate, which corresponds to higher and more uniform conductivity of the reduced film. A theory was developed to calculate local film conductivity from tip current and, subsequently, obtain a conductivity map of the graphene oxide film. Similarly, the feedback mode of SECM was used to study reduced graphene oxide on polyester fabrics.[63] An increase in electroactivity was observed after the inactive graphene oxide film was reduced. Moreover, an increase in electroactivity was observed with an increasing number of the reduced graphene oxide layers.

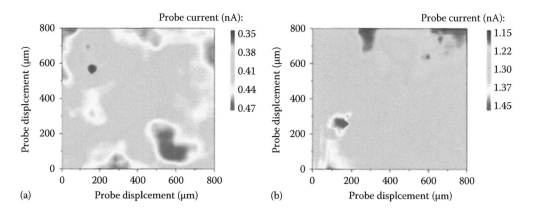

FIGURE 18.10 SECM feedback images of graphene oxide (a) before and (b) after reduction by 30 s exposition to a 0.1 M KOH solution as obtained for FcMeOH at a 24 μm diameter Pt tip ($RG=3$) with a tip–substrate distance of 12 μm. The scanned zones are not the same. (Adapted with permission from Azevedo, J., Bourdillon, C., Derycke, V., Campidelli, S., Lefrou, C., and Cornut, R., Contactless surface conductivity mapping of graphene oxide thin films deposited on glass with scanning electrochemical microscopy, *Anal. Chem.*, 85, 1812–1818, 2013. Copyright 2013 American Chemical Society.)

18.4 NANOPARTICLES

The SECM studies of nanoparticles are of great interest and diversity. Metal nanoparticles can serve as redox mediators in solution to investigate their redox activities at solid–liquid[64,65] and liquid–liquid[66,67] interfaces by using SECM as discussed in Chapter 3. By contrast, this section is focused on the SECM studies of monolayers and multilayers of metal nanoparticles formed at various interfaces. In these studies, SECM was employed to quantitatively investigate the lateral conductivity and interfacial electroactivity of nanoparticle films. In addition, new experimental setups were developed to address the electrocatalytic and photocatalytic activities of nanoparticle films. Moreover, significant progresses were made to deposit and pattern nanoparticles on various substrates by using SECM.

18.4.1 CONDUCTIVITY AND ELECTROACTIVITY

SECM has been used to study the conductivity and electroactivity of nanoparticle monolayers formed at various interfaces. For instance, conductivity measurement was reported by Bard and coworkers to observe the metal–insulator transition of the Langmuir film of hexanethiolate-protected Ag nanoparticles with ~4 nm diameter at the air–water interface (Figure 18.11a).[68] Advantageously, this setup allows for the precise control of interparticle distances by moving a barrier to compress the nanoparticle film for in situ SECM measurement. In addition, the formation of a stable monolayer can be confirmed by a surface pressure–trough area isotherm. Specifically, a *submarine* Pt SECM tip approached from the aqueous side of the Langmuir film to obtain a family of approach curves for FcMeOH in the feedback mode (Figure 18.11b). As the barrier was moved to compress the film, the approach curve gradually changed from a negative one (curve 1) to a positive one (curve 5), which demonstrates a transition from an insulating film to a metallic film with high conductivity. This gradual transition was ascribed to disorder in the nanoparticle film. In fact, the intermediate feedback responses (curves 2–4) were limited by electron transport between nanoparticles. Noticeably, feedback responses were obtainable without externally biasing the nanoparticle film. The reductive regeneration of FcMeOH at the nanoparticle film under the tip was coupled with the oxidation of FcMeOH at other points on the film far away from the tip. These interfacial ET processes were coupled by electron transport between nanoparticles. Similarly, SECM was applied to the recent study of gold nanoparticles adsorbed at the ITIES.[69] In this work, a 10 μm diameter Pt SECM tip approached from the upper solution of

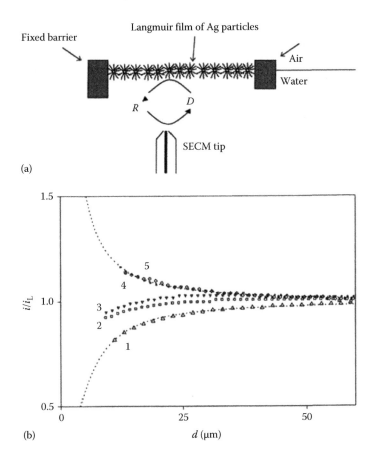

FIGURE 18.11 (a) Schematic representation of positive feedback at an inverted SECM tip with a Langmuir film of Ag nanoparticles at the air–water interface (not to scale). (b) In situ SECM approach curves to the nanoparticle film at the air–water interface with a surface pressure of (1) 0 mN/m, (2) 11 mN/m, (3) 22 mN/m, (4) 42 mN/m, and (5) 56 mN/m. Dotted lines represent the theoretical approach curves to an insulator (lower) and a conductor (upper). (Reprinted with permission from Quinn, B.M., Prieto, S.K., Haram, S.K., Bard, A.J., Electrochemical observation of a metal/insulator transition by scanning electrochemical microscopy, *J. Phys. Chem. B*, 105, 7474–7476, 2001. Copyright 2001 American Chemical Society.)

1,2-dichloroethane and heptanes containing decamethylferrocene as a redox mediator. Approach curves based on this redox mediator became more positive as the interfacial coverage with gold nanoparticles increased. The family of approach curves was quantitatively analyzed by using the finite element method[70] to determine film conductivity as a function of surface coverage. This analysis demonstrated a gradual change in the electric conductance in contrast to a sharp change in the optical reflectance of the nanoparticle film.

The Langmuir film of monolayer-protected gold nanoparticles was transferred to an insulating solid substrate for the SECM study of film conductivity and electroactivity.[71] In this SECM measurement (Figure 18.12a),[72] a Pt tip approached the solid-supported monolayer to drive electron transport through the film. The rates of interfacial and interparticle ET processes (steps 1 and 2, respectively, in Figure 18.12a) were determined separately by adjusting the concentration of redox mediators. Steady-state current through the monolayer was maintained by transferring electrons from a redox mediator, FcMeOH, to the region of the nanoparticle film exposed to the bulk solution (step 3). Specifically, approach curves were measured at the monolayer of dodecanethiolate-protected Au nanoparticles (Figure 18.12b), which were heat-annealed to achieve uniform particle

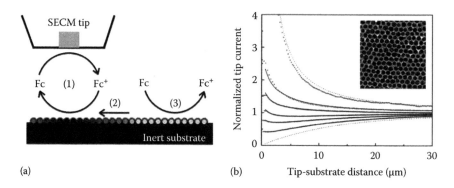

(a)   (b)   Tip-substrate distance (µm)

FIGURE 18.12 (a) The feedback mechanism of the SECM response to FcMeOH (Fc) at the monolayer of Au nanoparticles on the insulating substrate. See main text for steps 1–3. (b) Experimental approach curves (dots) and the corresponding fits to the theory (solid lines) at the monolayer of dodecanethiolate-protected Au particles as obtained by using a 25 µm diameter Pt tip ($RG = 5$) with FcMeOH concentrations of 1070, 450, 280, 170, 60, and 1.3 µM (from bottom to top). The lower and upper dotted lines are the theoretical responses for negative and positive feedbacks, respectively. The Langmuir monolayer was transferred to a microscope slide at a surface pressure of 10 mN/m. The inset shows a TEM image of the nanoparticle film transferred at a surface pressure of 20 mN/m. (Reprinted with permission from Liljeroth, P., Vanmaekelbergh, D., Ruiz, V. et al., Electron transport in two-dimensional arrays of gold nanocrystals investigated by scanning electrochemical microscopy, *J. Am. Chem. Soc.*, 126, 7126–7132, 2004. Copyright 2004 American Chemical Society.)

size (core diameter 6.6 nm; inset of Figure 18.12b). The approach curves were more positive at a lower FcMeOH concentration, which drives a lower flux of electrons through the resistive film. Eventually, an almost purely positive curve was obtained at the lowest FcMeOH concentration of 1.3 µM, where the reductive regeneration of FcMeOH at the surface of the nanoparticle film was diffusion limited. This fast reduction is due to either intrinsically high electroactivity of the nanoparticles or large overpotential of the unbiased nanoparticle film, which is much larger than the tip.[73] On the other hand, approach curves were saturated at low concentrations of $Fe(CN)_6^{4-}$ as a redox mediator to be kinetically limited.[74] The approach curves were analyzed to yield both monolayer conductivity (0.018 Ω/cm) and standard ET rate constant, k^0 (1.1×10^{-3} cm/s). To determine the standard ET rate constant, the potential of the unbiased monolayer was fixed by adding both $Fe(CN)_6^{4-}$ and $Fe(CN)_6^{3-}$ to the solution. This approach was also applied to determine the potential dependence of monolayer conductivity.[75] More recently, SECM was used to study the Langmuir film of *Janus* gold nanoparticles stabilized asymmetrically by hydrophobic tetraoctylammonium and hydrophilic tryptophan at the air and aqueous sides of the interface, respectively.[76] The Langmuir film was transferred to a silanized Si(100) substrate at different surface pressures. Approach curves became more positive as the surface pressure increased from 0 to 23 mN/m to obtain higher film conductivity due to a shorter distance between nanoparticles.

Noticeably, metal nanoparticles were also attached to solid electrodes to study their conductivity and electroactivity by using SECM with or without potential control. For instance, diazonium-substituted polyphenyl acetate was electrografted to a glassy carbon electrode to electrostatically immobilize Ag nanoparticles.[77] Since the carbon electrode was not electronically connected, the regeneration of FcMeOH at the Ag nanoparticles under the tip was coupled with the conduction of electrons through the carbon electrode or between nanoparticles. The strong solvent effect on the SECM feedback response was ascribed to the solvent-dependent distance between nanoparticles. In another study, the feedback mode of SECM was employed to study the electroactivity of the Ag particles electrodeposited on the monolayer of dodecanethiolate-protected Au nanoparticles at highly oriented pyrolytic graphite (HOPG).[78] Polyhedral and plate Ag particles were obtained by controlling temperature and applied potential during deposition. The anisotropic growth of the plate particle was ascribed to the preferential adsorption of dodecanethiolate molecules from the Au nanoparticles to the (111) plane of the deposited Ag particles, where Ag^+ reduction was blocked. The Ag particles were electronically connected to the underlying Au nanoparticles and HOPG such that the feedback effect at a Pt SECM tip was limited by the reduction of $Fe(CN)_6^{3-}$ at the substrate surface rather than by substrate conductivity. Approach curves were quantitatively analyzed to demonstrate the higher reactivity of the polyhedral particle than the plate particle, which was covered by dodecanethiolate molecules. In another study, SECM was used to characterize the multilayers of gold nanoparticles assembled on the gold surface by using 5,15-di-[p-(6-mercaptohexyl)-phenyl]-10,20-diphenylporphyrin as a linker.[79] A gold electrode was modified with the multilayers by using the layer-by-layer (LBL) technique, where porphyrin and nanoparticle layers were deposited alternatively. Approach curves with $Fe(CN)_6^{4-}$ became more negative with the deposition of the porphyrin layer and then became more positive with the deposition of the nanoparticle layer. In either case, approach curves became more negative as more layers were deposited. This result indicates that the feedback effect was controlled by electron transport through the multilayers.

18.4.2 Electrocatalysis

The electrocatalytic activity of various nanoparticles was characterized by using SECM. Uniquely, SECM enables the study of the intrinsic electrocatalytic activity of nanoparticles on insulating substrates without electronic interactions between the nanoparticles and the substrates. This powerful SECM approach was employed in the studies of the hydrogen evolution reaction (HER) mediated by gold nanoparticles on glass[80] and palladium nanoparticles on the polylysine-coated mica[81] and on nafion-coated glass[82] as well as the catalytic decomposition of hydrogen peroxide at platinum nanoparticles on glass.[83] For instance, SECM was used to study the HER coupled with the oxidation of a tip-generated redox species as a discharging step (Figure 18.13a) to yield the positive feedback tip response at steady states.[81] A higher positive feedback effect was observed as the density of Pt nanoparticles on the polylysine-coated mica increased from 4.3×10^9 to 2.4×10^{10} cm^{-2} (Figure 18.13b). The approach curves were fitted well with simulated curves as obtained by assuming the

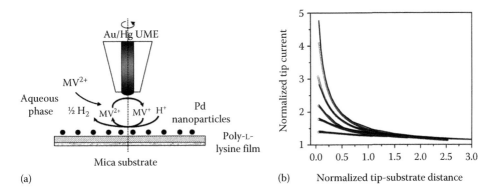

(a)   (b)   Normalized tip-substrate distance

FIGURE 18.13 (a) Scheme of the SECM feedback experiment (not to scale) for probing the HER catalyzed by citrate-stabilized Pd nanoparticles adsorbed on a mica surface. (b) Experimental (circles) and simulated (solid curves) approach curves for the mica surfaces modified with Pd nanoparticles in a pH 3 buffered solution containing 0.1 mM methyl viologen (MV^{2+}) and 0.05 M KNO$_3$. The nanoparticle number density was $(4.3 \pm 0.3) \times 10^9$, $(7.8 \pm 0.3) \times 10^9$, $(1.2 \pm 0.1) \times 10^{10}$, $(2.1 \pm 0.1) \times 10^{10}$, and $(2.4 \pm 0.1) \times 10^{10}$ from bottom to top. The highest curve corresponds to diffusion-controlled positive feedback. The remaining curves fit well to a transfer coefficient, α, of 0.5 and kinetic rate constants of 3.5×10^{-10}, 6.5×10^{-10}, 1.0×10^{-9}, and 2.0×10^{-9} from bottom to top. (Reprinted from Li, F., Ciani, I., Bertoncello, P. et al., Scanning electrochemical microscopy of redox-mediated hydrogen evolution catalyzed by two-dimensional assemblies of palladium nanoparticles, *J. Phys. Chem. C*, 112, 9686–9694, 2008. Copyright 2008 American Chemical Society.)

Volmer–Heyrovsky mechanism with an apparent transfer coefficient, α, of 0.5 for the HER although this α value was also expected when the discharging step was the rate-determining step. By contrast, α = 1.5 or 2.0 did not give a good fit between experimental and simulated approach curves. The numerical simulation also demonstrated that an overpotential at the unbiased nanoparticles exceeded 250 mV as the tip approached the substrate. In fact, the Volmer–Heyrovsky mechanism was expected at high overpotentials. In contrast to the HER study, no additional redox couple was required when the feedback mode of SECM was employed to study the decomposition of hydrogen peroxide at Pt nanoparticles on glass.[83] Specifically, a 25 μm diameter Au tip was coated with mercury to reduce oxygen as

$$O_2 + 2e + H_2O \rightarrow HO_2^- + OH^- \qquad (18.3)$$

Hydrogen peroxide was decomposed at Pt nanoparticles on glass to regenerate oxygen as

$$HO_2^- \rightarrow 1/2 O_2 + OH^- \qquad (18.4)$$

Importantly, the reverse reaction of Equation 18.3 was prevented at the Pt nanoparticles electronically isolated from each other. The resultant asymmetric stoichiometries of the tip and substrate reactions were treated by the finite element simulation for the quantitative analysis of experimental approach curves as measured at various pH values. This analysis demonstrated that the corresponding heterogeneous rate constants were highest at pH ~ 12.

Various operation modes of SECM were systematically developed for the study of electrocatalysis at conductive substrates by Bard and coworkers[84] (Figure 18.14) and recently applied to the

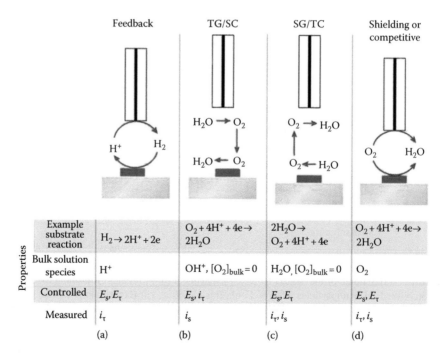

FIGURE 18.14 Schematic representation of different operation modes of SECM for screening electrocatalysts. E_T, tip potential; i_T, tip current; E_S, substrate potential; and i_S, substrate current. (Reprinted with permission from Amemiya, S., Bard, A.J., Fan, F.-R.F., Mirkin, M.V., and Unwin, P.R., Scanning electrochemical microscopy, *Ann. Rev. Anal. Chem.*, 1, 95–131, 2008. Copyright 2008 Annual Reviews.)

study of nanoparticle systems by various groups. For instance, the feedback mode[85] (Figure 18.14a) was employed to image the activity of the platinum-loaded carbon nanoparticles as spin-coated on HOPG for the hydrogen oxidation reaction.[86] The tip-generated hydrogen molecules were oxidized at the nanoparticle-coated regions of the unbiased substrate to enhance the tip current based on the HER.

The tip generation/substrate collection (TG/SC) mode was developed (Figure 18.14b)[87] and applied to the study of the ORR at various nanoparticles and nanorods. In this operation mode, tip-generated oxygen molecules were reduced at the biased substrate to evaluate the rate of the ORR by monitoring substrate current. For instance, the arrays of platinum nanoparticles with spherical, cubic, hexagonal, and tetrahedral–octahedral shapes were prepared on a carbon plate and imaged in the TG/SC mode.[88] The highest substrate current was obtained when the tip was positioned over the spot of hexagonal platinum nanoparticles, which were most active. Similarly, the nanometer-sized spheres, cubes, and rods of gold were studied in the TG/SC mode to find the highest ORR activity of the cubic gold nanoparticles.[89] The TG/SC mode was also employed to image the ORR activity of the Pt(111) single crystal electrode decorated with the spots of a nafion film entrapping commercial carbon-supported platinum nanoparticles.[90] A higher activity of nanoparticle-coated regions was ascribed to the larger surface area of the platinum nanoparticles and the lower intrinsic activity of the (111) surface due to the specific adsorption of bisulfate in 0.5 M H_2SO_4 solution.

The oxygen evolution reaction (OER) can be studied by detecting the substrate-generated oxygen molecules at the tip (Figure 18.14c).[91] The SG/TC mode was used to study the OER at TiO_2-doped RuO-nanoparticle-coated electrodes.[92] Moreover, the oxidation of hydrogen peroxide at Au nanoparticles was also studied by using the SG/TC mode.[93] Noticeably, these SG/TC studies of nanoparticles were limited to qualitative imaging, because the quantitative analysis of the time-dependent tip current was difficult when the catalytic reactions occur at the macroscopic substrates. By contrast, the generation of hydrogen peroxide during the ORR was quantitatively studied in the steady-state SG/TC mode by employing the UMEs of pure metals as substrates.[94]

Finally, the shielding mode of SECM was developed as a variation of the feedback mode (Figure 18.14d).[95] This operation mode was also called the redox competition (RC) mode when it was applied to the characterization of electrocatalysts,[96] including nanomaterials. In the RC mode, the ORR was driven competitively at both tip and substrate (Figure 18.15d) so that a higher substrate activity resulted in a lower tip current. The applications of the RC mode include the ORR at gold nanoparticles,[93] Pt–Ag nanoparticles,[97,98] and Ru–Au nanoparticles[99] as well as the chlorine evolution reaction at $Ru_{0.3}Sn_{0.7}O_2$ nanocatalysts.[100] Interestingly, the reduction of hydrogen peroxide at Prussian blue nanoparticles was studied in the RC mode by oxidizing hydrogen peroxide at the tip.[101] The versatile RC mode has been also applied for the study of electrocatalysis at nanoparticle-modified carbon nanotubes (see Section 18.5.1). In the RC mode, a transient reaction is driven at a macroscopic substrate to yield the time-dependent tip current, while a series of pulse potentials are applied to the tip to prevent complete O_2 depletion during imaging.[96]

18.4.3 PHOTOCATALYSIS

The photoelectrochemical activities of nanoparticles were studied under illumination by using SECM. For instance, Bard and coworkers employed an SECM tip based on an optical fiber for the photoactivation of catalytic substrates during the simultaneous electrochemical detection of oxygen as the product of photoelectrochemical water oxidation in the SG/TC mode.[102] This approach was employed to demonstrate the enhancement of photoelectrochemical water oxidation at the $BiVO_4$ photoelectrodes doped with W by addition and decomposition of H_2WO_4 nanoparticles.[103] Specifically, the amount of evolved oxygen was amperometrically monitored by a platinum-coated Au ring electrode formed at the tip of an optical fiber while the fiber was coupled with the Xe lamp as a light source to locally illuminate the substrate surface (Figure 18.15a).

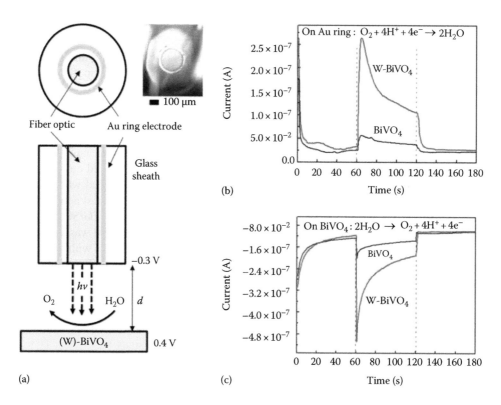

FIGURE 18.15 (a) Schematic diagram of the SG/TC mode of SECM for the detection of oxygen as generated by the photoelectrochemical water oxidation at the $BiVO_4$-based substrates under irradiation. Chronoamperograms for (b) oxygen reduction at the Pt–Au ring electrode tip and (c) water oxidation at $BiVO_4$ (black) and W-doped $BiVO_4$ (red) electrodes in 0.2 M sodium phosphate buffer at pH 7. The measurements started in the dark. UV–vis light was irradiated from 60 s to 120 s. (Reprinted with permission from Cho, S.K., Park, H.S., Lee, H.C., Nam, K.M., and Bard, A.J., Metal doping of $BiVO_4$ by composite electrodeposition with improved photoelectrochemical water oxidation, *J. Phys. Chem. C*, 117, 23048–23056, 2013. Copyright 2013 American Chemical Society.)

Chronoamperograms of the Pt–Au ring tip and $BiVO_4$ substrate were monitored (Figure 18.15b and c, respectively) during the illumination of the substrate. The substrate water oxidation current increased sharply and immediately upon illumination from 60 to 120 s. The increment of photocurrent at W-doped $BiVO_4$ was about three times larger than that of pure $BiVO_4$. An increase in the oxygen reduction current at the tip was synchronized with the rising oxidation photocurrent at the substrate. The tip current with W-doped $BiVO_4$ was larger than that with $BiVO_4$, thereby confirming the enhanced water oxidation at the former substrate. The ORR current at the tip also showed an instant response to the irradiation, because the diffusion of produced oxygen from the substrate to the tip only took a few tens of milliseconds. Noticeably, such optical fiber tips were used also successfully for the study of the photoelectrochemical activities of Ta_3N_5 nanotubes[104] and ZnO nanorods.[105]

Alternatively, a photocatalytic substrate was illuminated from behind while the reactant or product of the resultant photoelectrochemical reaction was monitored by a conventional metal SECM tip above the substrate.[106] This approach was recently employed to study the photodegradation of 4-chlorophenol at Fe-doped TiO_2 nanoparticles.[107] The progression of this reaction was coupled with the consumption of oxygen at the nanoparticle-coated ITO electrode, which was monitored by using a Pt tip.

18.4.4 DEPOSITION

Metal nanoparticles were deposited and patterned on various substrates by using the TG/SC mode of SECM. In this operation mode, either Au or Ag tip was used to oxidatively generate metal ions as

$$Au\ (tip) + 4Cl^- \rightarrow AuCl_4^- + 3e \qquad (18.5)$$

or

$$Ag\ (tip) \rightarrow Ag^+ + e \qquad (18.6)$$

The tip-generated metal ions were reduced at the substrate surface under the tip to locally deposit the corresponding metal nanoparticles. Figure 18.16 shows anisotropic gold nanoparticles thus deposited on the ITO electrode.[108] While the tip potential of +0.67 V was applied to drive the dissolution of a gold tip (Equation 18.5), the substrate potential was stepped twice to quickly form small gold seeds at −0.2 V for 1 s and then slowly grow gold nanorods (GNRs) at 0 V. The anisotropy was induced by a surfactant, cetyltrimethylammonium, and halide ions (Cl^- and Br^-), which adsorbed preferentially on different facets of the gold seeds to break growth symmetry. The versatility of the TG/SC mode was also demonstrated by locally depositing Au nanoparticles on polyanilines[109,110] and poly(3,4-ethylenedioxythiophene) (PEDOT).[111] Similarly, Ag tips were used to locally generate Ag^+ (Equation 18.6) and deposit Ag nanoparticles on PEDOT[111] and on the monolayer of ω-mercaptoalkanoic acids self-assembled on gold.[112] Ag nanoparticles were also deposited at the interface between 1,2-dichloroethane and water.[113] In this case, nanoparticles were formed at the ITIES, where decamethylferrocene in the 1,2-dichloroethane solution reduced Ag^+ as generated at an Ag tip in the aqueous phase. Alternatively, Ag^+ was electrochemically delivered from the tip of a glass micropipet filled with an aqueous solution of Ag^+.[114] Advantageously, the pipet-supported ITIES tips were miniaturized and positioned within nanometer distances from an Au substrate to locally deposit Ag nanoparticles with higher lateral resolution.[115]

Various strategies were developed to enable the local electroless deposition of metal nanoparticles on the insulating or unbiased substrates by using SECM. For instance, the deposition of

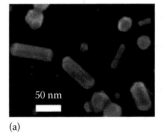 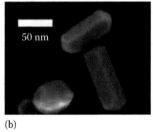 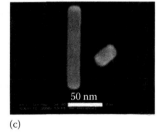

(a)   (b)   (c)

FIGURE 18.16 SEM images of (a) mixture of different GNRs, (b) GNR with orthogonal cross section, and (c) GNR with octagonal cross section as deposited locally on the ITO electrode by using SECM. The deposition was performed by double-pulse chronoamperometry. First pulse: E_T=0.67 V, E_S=−0.2 V, t=1 s. Second pulse: E_T=0.67 V, E_S=0 V, t=200 s. The solution consisted of 50 mM cetyltrimethylammonium chloride, 50 mM KCl, and 50 mM KBr. (Reproduced from Fedorov, R.G. and Mandler, D., Local deposition of anisotropic nanoparticles using scanning electrochemical microscopy (SECM), *Phys. Chem. Chem. Phys.*, 15, 2725–2732, 2013. Reproduced by permission of The Royal Society of Chemistry.)

Au nanoparticles was catalytically coupled with the oxidation of hydroquinone (HQ) at the substrate surface as given by[116]

$$3HQ + 2AuCl_4^- \rightarrow 2Au + 3BzQ + 6H^+ + 8Cl^- \tag{18.7}$$

where BzQ represents benzoquinone. By this way, Au nanoparticles were successfully deposited on the Pd-/Sn-activated SiO_2 substrate to form a ring-shaped spot under the tip. The formation of the ring was ascribed to the slow reduction of $AuCl_4^-$ on the sparsely active substrate (Figure 18.17). Subsequently, $AuCl_4^-$ diffused laterally from the tip toward the edge of the surrounding glass sheath of the tip to be reduced by HQ from the bulk solution, thereby depositing Au nanoparticles under the edge. In contrast, Au nanoparticles were deposited on the Pd substrate to form a disk-shaped deposit under the tip. In this case, $AuCl_4^-$ was generated at the tip to be immediately reduced by HQ at the highly active Pd surface under the tip. Alternatively, cellobiose dehydrogenase was used as a surface catalyst for the electroless deposition of Au nanoparticles coupled with the oxidation of β-D-lactose.[117] In another study, $AuCl_4^-$ was generated at an Au tip to enable the local electroless deposition of Au nanoparticles at the CdTe films and the CdTe nanoparticles embedded in a polymer film, as given by[118] the reaction:

$$3CdTe + 2AuCl_4^- \rightarrow 2Au + 3Cd^{2+} + 3Te + 8Cl^- \tag{18.8}$$

A chronoamperometric response at the Au tip during nanoparticle deposition was examined to demonstrate the nucleation and growth mechanism of nanoparticle deposition in the polymer film. Similarly, a galvanic displacement process was employed to enable the electroless deposition of Au nanoparticles on a copper substrate in the TG/SC mode as given by[119] the reaction:

$$3Cu + 2AuCl_4^- \rightarrow 2Au + 3Cu^{2+} + 8Cl^- \tag{18.9}$$

By this way, bimetallic nanomaterials were prepared without externally biasing the copper substrate.

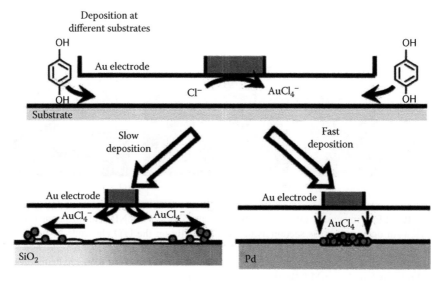

FIGURE 18.17 Schematic presentation of the localized electroless deposition of Au nanoparticles on (left) Pd-activated SiO_2 and (right) Pd surfaces. (From Malel, E. and Mandler, D., *J. Electrochem. Soc.*, 155, D459, 2008. Copyright 2008 The Electrochemical Society.)

SECM enables the deposition of Au nanoparticles without dissolving an Au tip. For instance, chitosan-stabilized Au nanoparticles were provided from solution and deposited by changing local pH near the substrate under an SECM tip.[120] In this case, a Pt tip served as a counter electrode to locally reduce water to OH⁻ at a conductive substrate under the tip. Subsequently, the amino groups of chitosan were deprotonated in the OH⁻ rich solution near the substrate to deposit the gold nanoparticles on the substrate surface. The versatility of this approach was demonstrated by using stainless steel, indium tin oxide, and HOPG as substrates. In another approach, a Pt tip was used to generate Au nanoparticles by the electrochemical reduction of $AuCl_4^-$ dissolved in the solution.[121] Tip-generated gold nanoparticles were directly deposited on the gold substrate, which was modified with a self-assembled monolayer of biphenyl dithiol as a linker.

18.5 NANOCOMPOSITES

In this section, the applications of SECM for the electrochemical characterization of nanocomposites are discussed. Synergy between two nanomaterials or between a nanomaterial and a macroscopic material is highly attractive for various electrochemical applications. SECM enables the screening of such nanocomposite systems with various mixing ratios to find the best composition. SECM has been also used to study the interactions of nanoparticles with biological macromolecules. The resultant biological nanocomposites are crucial to the highly sensitive detection of DNA and proteins.

18.5.1 COMPOSITE NANOMATERIALS

The electroactivity of composites between carbon nanotubes and metal nanoparticles has been studied by using SECM for sensing and electrocatalysis applications. Yogeswaran et al. electrodeposited gold nanoparticles on multiwalled carbon nanotubes by biasing the underlying glassy carbon electrode.[122] The electroactivity of the nanocomposite-modified electrode was imaged by using $Fe(CN)_6^{3-}$ as a mediator in the feedback mode. In another study,[123] multiwalled carbon nanotubes on a glassy carbon electrode were coated with a nafion film and modified with platinum and gold nanoparticles by electrodeposition. Epinephrine was employed as a redox mediator for SECM imaging in the feedback mode to investigate the utility of the modified electrode for bioanalytical sensing applications. Schumann and coworkers employed the RC mode of SECM to study the ORR at the nanoparticles of various metals (Pt, Au, Ru, Rh, and their codeposits) on carbon nanotubes.[124] SECM images clearly demonstrated the lowest tip current over an Au spot against the spots of other pure metals. The higher electrocatalytic activity of the Au spot was ascribed to higher loading of Au nanoparticles, as confirmed by SEM. In fact, the Ru spot turned out to be more reactive than the Au spot when their reactivity was semiquantitatively normalized against the amount of metal loading. The higher mass specific activity of the Ru spot is likely due to smaller nanoparticle size. The RC mode was also employed to study the ORR at the Pt nanoparticles deposited on carbon nanotubes by CVD.[125] Uniquely, the carbon nanotubes were grown on carbon microfibers, which were originally grown on a carbon cloth. Advantageously, the resultant three-dimensional hierarchical structure of carbon nanotubes and microfibers possesses a large surface area. SECM imaging confirmed a higher activity of Pt nanoparticles deposited on the hierarchical structure than those deposited on a carbon cloth. The RC mode of SECM was also employed to demonstrate that the ORR activity of carbon nanotubes is higher than that of a glassy carbon electrode but is lower than that of Pt-nanoparticle–carbon nanotube composites.[126] Other applications of the RC mode include the study of the ORR at nitrogen-containing carbon nanotubes[127] and the composite of multiwalled carbon nanotubes with cobalt protoporphyrins.[128]

The composites of metal nanoparticles with polymer films have also been characterized by using SECM. Of particular interest is the composite of metal nanoparticles with conducting polymers as

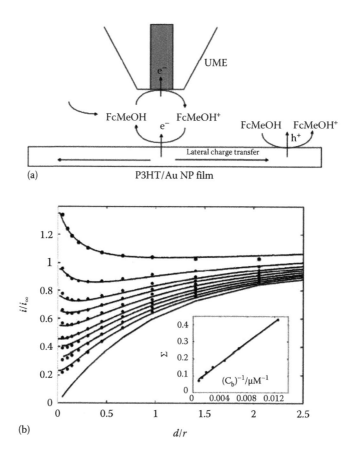

FIGURE 18.18 (a) Schematic of the SECM setup for measuring the lateral conductivity of P3HT-Au nanoparticle films on an inert solid support. (b) Experimental approach curves (solid lines) and corresponding fits to theory (dots) for the composite film with concentrations of FcMeOH at 3300, 1700, 1170, 770, 520, 340, 210, 150, and 80 μM (from bottom to top). In the approach curves, tip current, i, and tip–substrate distance, d, are normalized against tip limiting current in the bulk solution, i_∞, and tip radius, r, respectively. The lowest curve represents the theoretical approach curve at an insulating substrate. The insets show the normalized conductivity (Σ) as a function of the inverse of the solution redox mediator concentration (C_b). (Reprinted with permission from Ruiz, V., Nichiolson, S., Thomas, P.A., Macpherson, J.V., Unwin, P.R., Molecular ordering and 2D conductivity in ultrathin poly(3-hexylthiophene)/gold nanoparticle composite films, *J. Phys. Chem. B,* 109, 19335–19344, 2005. Copyright 2005 American Chemical Society.)

studied by Unwin and coworkers.[129,130] In their studies, the conductivity of the regioregular poly(3-hexyl thiophene) (P3HT) film incorporating Au nanoparticles was measured by using the feedback mode (Figure 18.18a). Approach curves to the unbiased polymer–nanoparticle composite film became negative as the concentration of FcMeOH in the solution increased (Figure 18.18b), thereby indicating that the feedback process was limited by electron transport through the composite film. All experimental approach curves fitted well with simulated curves to yield consistent film conductivity. Similarly, the conductivity of polyaniline films with and without incorporating Pt nanoparticles was studied by using SECM.[131] An approach curve with Pt nanoparticles was more positive than that without the particles when the polyaniline film was deprotonated at pH 5.4. This difference was observable because the deprotonated polyaniline film was not only insulating but also blocking mediator diffusion to the underlying Pt substrate. By contrast, a purely positive feedback effect was obtained even without Pt nanoparticles at pH 2.3, where the protonated polyaniline film was sufficiently conductive and electroactive.

The LBL deposition of polymer layers incorporating metal nanoparticles was employed to study the electrochemical properties of the resultant composite films by using SECM. Wittstock and coworkers electrodeposited Pd and Pt nanoparticles in the matrix of LBL-deposited multilayers of polyelectrolytes, that is, poly(diallyldimethylammonium) and poly(4-stylene sulfonate).[132] The production of hydrogen peroxide during the ORR at the nanoparticle-incorporated LBL films was monitored in the SG/TC mode. The transient current at a 25 μm diameter Pt SECM tip was measured at various tip–substrate distances and quantitatively analyzed to determine effective rate constants k_1, k_2, and k_3, for the following reactions at the film:

$$O_2 + 2H^+ + 2e \xrightarrow{k_1} H_2O_2 \tag{18.10}$$

$$O_2 + 4H^+ + 4e \xrightarrow{k_2} 2H_2O_2 \tag{18.11}$$

$$H_2O_2 + 2H^+ + 2e \xrightarrow{k_3} 2H_2O \tag{18.12}$$

Good fits between experimental and theoretical amperometric responses at the tip were obtained for the composite films incorporating Pd nanoparticles to demonstrate that the generation of hydrogen peroxide (Equation 18.10) was accelerated from $k_1 = 0.0014$ cm/s to $k_1 = 0.00534$ cm/s as more Pd nanoparticles were loaded. By contrast, data for Pt composite films did not fit well with theory, which was ascribed to the formation of an oxide layer on Pt nanoparticles. In another study,[133] LBL approach was employed to prepare the composite of gold nanoparticles, DNA, and polyethylenimine (PEI). Specifically, the SAM of 3-mercapto-1-propanesulfonic acid was formed on the gold electrode to deposit a cationic PEI film, which was followed by the deposition of double-stranded DNA or citrate-stabilized Au nanoparticles and PEI. Approach curves were measured to demonstrate that k^0 values for both FcMeOH and ferrocenecarboxylic acid were highly dependent on the type of the outmost layer.

18.5.2 BIOLOGICAL NANOCOMPOSITES

Several groups reported the use of metal nanoparticles as a label of biological macromolecules for their sensitive detection by using SECM. For instance, Zhou and coworkers employed SECM for highly sensitive monitoring of DNA hybridization on the microarray surface.[134] With this approach, a single-stranded DNA molecule on the array surface was hybridized with a complementary DNA probe with a biotin residue. Once the duplex was formed, a streptavidin–gold nanoparticle conjugate was attached to the biotin residue for the electroless deposition of silver nanoparticles. Subsequently, the microarray surface became conductive to exert a positive feedback effect on $Ru(NH_3)_6^{3+}$ as a redox mediator for SECM imaging. This approach enabled the highly sensitive detection of 30 amol (0.1 nM) of a 17-mer DNA probe on the microarray. More recently, the silver-staining approach was also applied to the SECM-based detection of proteins in the feedback mode.[135] In this study, proteins, such as β-lactoglobulin A and myoglobin, were separated by SDS-PAGE, electroblotted to the poly(vinylidenedifluoride) (PVDF) membrane, and stained by silver nanoparticles in $AgNO_3$ solution. The stained bands of the proteins were resolved in the SECM image as obtained by using $IrCl_6^{3-}$ as a redox mediator (Figure 18.19a). Sufficiently strong oxidants, that is, $IrCl_6^{2-}$, were generated at the tip and then reduced by the silver nanoparticles to enhance the tip current while the silver nanoparticles were oxidatively dissolved. Interestingly, the soft and porous PVDF membrane allowed for the constant-height imaging of large millimeter-sized bands, where the tip touched and squeezed the flexible membrane to bring the silver nanoparticles within a feedback distance from the tip (Figure 18.19b). The short tip–particle distances were advantageous for the highly sensitive detection of a spot of down to 0.5 ng of bovine serum albumin, which was not detectable by optical

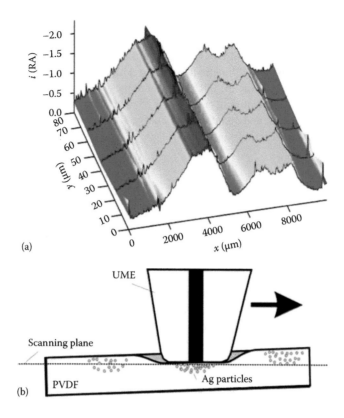

FIGURE 18.19 (a) SECM image of β-lactoglobulin A (left band) and myoglobin (right band) electroblotted from an electrophoresis gel as obtained by detecting $IrCl_6^{3-}$ as a redox mediator at a 20 μm diameter Pt tip in the feedback mode. Scan speed: 50 μm/s. 10 μL of 0.1 mg/mL β-lactoglobulin A and 10 μL of 0.1 mg/mL myoglobin were separated by SDS-PAGE. (b) Schematic representation of the SECM tip indenting the flexible PVDF membrane and keeping a short working distance to Ag nanoparticles. (Reprinted with permission from Zhang, M.Q., Wittstock, G., Shao, Y.H., Girault, H.H. Scanning electrochemical microscopy as a readout tool for protein electrophoresis, *Anal. Chem.*, 79, 4833–4839, 2007. Copyright 2007 American Chemical Society.)

microscopy. Noticeably, the feedback mode of SECM was also applied to detect the adsorption of silver nanoparticles on HeLa cells.[136]

The electroactivity of enzyme-modified nanoparticles was studied by using SECM for various applications. The RC mode (Figure 18.11d) was applied to study the ORR at the carbon nanoparticles modified with laccase[137] or bilirubin oxidase.[138] Specifically, carbon nanoparticles (~8 nm in diameter) were coated with a sol–gel film incorporating the enzymes and were deposited on an ITO electrode. A potential pulse was applied to the ITO electrode for the reduction of oxygen molecules, which were competitively reduced at a ~25 μm diameter Pt tip. Moreover, Bracamonte et al. studied the oxidation of glucose at the gold nanoparticles modified with glucose oxidase (GOD) for sensor applications.[139] The nanoparticles were attached to a gold substrate and were investigated by using SECM in the SG/TC mode (Figure 18.11c). In this case, GOD coupled the oxidation of glucose with the reduction of oxygen to hydrogen peroxide as given by the following equation:

$$O_2 + glucose \xrightarrow{\text{GOD}} H_2O_2 + gluconolactone \qquad (18.13)$$

Hydrogen peroxide thus generated at the substrate was detected at a 10 μm diameter carbon fiber tip as a measure of the rate of glucose oxidation. Another application of enzyme-modified nanoparticles is the enzyme-linked immunosorbent assay based on SECM.[140] Specifically, CD10, that is, neprily-sin, was captured on an antibody-modified gold electrode for sandwich immunoassay to be further recognized by the second antibody attached to gold nanoparticles. The gold nanoparticles were also modified with horseradish peroxidase, which catalytically coupled the reduction of hydrogen per-oxide to water with the oxidation of Fe^{2+}–Fe^{3+}. Subsequently, Fe^{3+} was detected at a 10 μm diameter Pt tip during imaging in the SG/TC mode. More recently, SiO_2 nanoparticles were modified with horseradish peroxidase and single-stranded DNA to serve as a marker for the SECM-based detec-tion of DNA hybridization on a microarray in the SG/TC mode.[141,142] In these studies, the reduction of hydrogen peroxide to water was enzymatically coupled with the oxidation of hydroquinone to benzoquinone. The generation of benzoquinone was monitored by using a 10 or 25 μm diameter Pt tip to determine the amount of probe DNA.

18.6 PERSPECTIVES

Exciting opportunities are envisaged for the nanometer-scale SECM studies of individual nano-pores, nanocarbons, and nanoparticles, which will be built upon micrometer-scale studies of their ensembles as discussed in this chapter. As thoroughly discussed in a recent chapter,[8] significant progresses have been made to enable truly nanoscale SECM measurement by developing reliable nanometer-sized tips (see Chapter 15) as well as by identifying and overcoming the fundamental problems associated with the use of a nanometer-sized tip,[19] the control of a nanometer-wide tip–substrate gap,[143] etc. In fact, molecular transport through single artificial nanopores has been imaged by using SECM (see Section 18.2.2) or SECM combined with AFM.[9] There, however, is no SECM study of single biological nanopores. Moreover, micrometer-scale SECM studies of SWCNT and graphene raised a question as to why the former is much more electroactive than the latter. The effects of defects and impurities on the electroactivity of these nanocarbons will be assessable by using nanoscale SECM with higher spatial resolution under more comprehensive voltammetric con-ditions.[18,144] Finally, SECM imaging or detection of individual nanoparticles has not been reported. Such a study is highly attractive to assess how the reactivity of nanoparticle is related to their size and structure.

ACKNOWLEDGMENTS

This work was supported by the National Institutes of Health (GM073439) and the National Science Foundation (CHE-1213452).

REFERENCES

1. Amemiya, S., A. J. Bard, F.-R. F. Fan, M. V. Mirkin, and P. R. Unwin. 2008. Scanning electrochemical microscopy. *Annu. Rev. Anal. Chem.* 1: 95–131.
2. Mirkin, M. V., W. Nogala, J. Velmurugan, and Y. Wang. 2011. Scanning electrochemical microscopy in the 21st century. Update 1: Five years after. *Phys. Chem. Chem. Phys.* 13: 21196–21212.
3. Bard, A. J. and M. V. Mirkin, eds. 2001. *Scanning Electrochemical Microscopy*. New York: Marcel Dekker.
4. Bard, A. J. and M. V. Mirkin, eds. 2012. *Scanning Electrochemical Microscopy*. 2nd edn. Boca Raton, FL: CRC Press.
5. Liu, H.-Y., F.-R. F. Fan, C. W. Lin, and A. J. Bard. 1986. Scanning electrochemical and tunneling ultrami-croelectrode microscope for high-resolution examination of electrode surfaces in solution. *J. Am. Chem. Soc.* 108: 3838–3839.
6. Bard, A. J., F.-R. F. Fan, J. Kwak, and O. Lev. 1989. Scanning electrochemical microscopy. Introduction and principles. *Anal. Chem.* 61: 132–138.
7. Kwak, J. and A. J. Bard. 1989. Scanning electrochemical microscopy. Apparatus and two-dimensional scans of conductive and insulating substrates. *Anal. Chem.* 61: 1794–1799.

8. Amemiya, S. Nanoscale scanning electrochemical microscopy. In *Electroanalytical Chemistry*, eds. A. J. Bard and C. G. Zoski, in press. Boca Raton, FL: Taylor & Francis.

9. White, H. S. and F. Kanoufi. 2012. Imaging molecular transport across membranes. In *Scanning Electrochemical Microscopy*, eds. A. J. Bard and M. V. Mirkin, pp. 233–273. Boca Raton, FL: CRC Press.

10. Kim, E., H. Xiong, C. C. Striemer et al. 2008. A structure-permeability relationship of ultrathin nanoporous silicon membrane: A comparison with the nuclear envelope. *J. Am. Chem. Soc.* 130: 4230–4231.

11. Forouzan, F., A. J. Bard, and M. V. Mirkin. 1997. Voltammetric and scanning electrochemical microscopic studies of the adsorption kinetics and self-assembly of *n*-alkanethiol monolayers on gold. *Isr. J. Chem.* 37: 155–163.

12. Liu, B., A. J. Bard, M. V. Mirkin, and S. E. Creager. 2004. Electron transfer at self-assembled monolayers measured by scanning electrochemical microscopy. *J. Am. Chem. Soc.* 126: 1485–1492.

13. Kiani, A., M. A. Alpuche-Aviles, P. K. Eggers et al. 2008. Scanning electrochemical microscopy. 59. Effect of defects and structure on electron transfer through self-assembled monolayers. *Langmuir* 24: 2841–2849.

14. Satpati, A. K., N. Arroyo-Curras, L. Ji, E. T. Yu, and A. J. Bard. 2013. Electrochemical monitoring of TiO_2 atomic layer deposition by chronoamperometry and scanning electrochemical microscopy. *Chem. Mater.* 25: 4165–4172.

15. Holt, K. B., A. J. Bard, Y. Show, and G. M. Swain. 2004. Scanning electrochemical microscopy and conductive probe atomic force microscopy studies of hydrogen-terminated boron-doped diamond electrodes with different doping levels. *J. Phys. Chem. B* 108: 15117–15127.

16. Kim, J., A. Izadyar, N. Nioradze, and S. Amemiya. 2013. Nanoscale mechanism of molecular transport through the nuclear pore complex as studied by scanning electrochemical microscopy. *J. Am. Chem. Soc.* 135: 2321–2329.

17. Kim, J., A. Izadyar, M. Shen, R. Ishimatsu, and S. Amemiya. 2014. Ion permeability of the nuclear pore complex and ion-induced macromolecular permeation as studied by scanning electrochemical and fluorescence microscopy. *Anal. Chem.* 86: 2090–2098.

18. Nioradze, N., J. Kim, and S. Amemiya. 2011. Quasi-steady-state voltammetry of rapid electron transfer reactions at the macroscopic substrate of the scanning electrochemical microscope. *Anal. Chem.* 83: 828–835.

19. Nioradze, N., R. Chen, J. Kim, M. Shen, P. Santhosh, and S. Amemiya. 2013. Origins of nanoscale damage to glass-sealed platinum electrodes with submicrometer and nanometer size. *Anal. Chem.* 6198–6202.

20. Guo, J. and S. Amemiya. 2005. Permeability of the nuclear envelope at isolated *Xenopus* oocyte nuclei studied by scanning electrochemical microscopy. *Anal. Chem.* 77: 2147–2156.

21. Berg, H. C. 1993. *Random Walks in Biology*. Princeton, NJ: Princeton University Press.

22. Makhnovskii, Y. A., A. M. Berezhkovskii, and V. Y. Zitserman. 2005. Homogenization of boundary conditions on surfaces randomly covered by patches of different sizes and shapes. *J. Chem. Phys.* 122: 236102.

23. Wang, H. and D. E. Clapham. 1999. Conformational changes of the in situ nuclear pore complex. *Biophys. J.* 77: 241–247.

24. Frenkiel-Krispin, D., B. Maco, U. Aebi, and O. Medalia. 2010. Structural analysis of a metazoan nuclear pore complex reveals a fused concentric ring architecture. *J. Mol. Biol.* 395: 578–586.

25. Loschberger, A., S. van de Linde, M. C. Dabauvalle et al. 2012. Super-resolution imaging visualizes the eightfold symmetry of gp210 proteins around the nuclear pore complex and resolves the central channel with nanometer resolution. *J. Cell Sci.* 125: 570–575.

26. Schwefel, D., C. Maierhofer, J. G. Beck et al. 2010. Structural basis of multivalent binding to wheat germ agglutinin. *J. Am. Chem. Soc.* 132: 8704–8719.

27. Ishimatsu, R., J. Kim, P. Jing et al. 2010. Ion-selective permeability of a ultrathin nanoporous silicon membrane as probed by scanning electrochemical microscopy using micropipet-supported ITIES tips. *Anal. Chem.* 82: 7127–7134.

28. Shen, M., R. Ishimatsu, J. Kim, and S. Amemiya. 2012. Quantitative imaging of ion transport through single nanopores by high-resolution scanning electrochemical microscopy. *J. Am. Chem. Soc.* 134: 9856–9859.

29. Striemer, C. C., T. R. Gaborski, J. L. McGrath, and P. M. Fauchet. 2007. Charge- and size-based separation of macromolecules using ultrathin silicon membranes. *Nature* 445: 749–753.

30. Gaborski, T. R., J. L. Snyder, C. C. Striemer et al. 2010. High-performance separation of nanoparticles with ultrathin porous nanocrystalline silicon membranes. *ACS Nano* 4: 6973–6981.

31. Johnson, D. G., T. S. Khire, Y. L. Lyubarskaya et al. 2013. Ultrathin silicon membranes for wearable dialysis. *Adv. Chronic Kidney Dis.* 20: 508–515.

32. Snyder, J. L., J. Getpreecharsawas, D. Z. Fang et al. 2013. High-performance, low-voltage electroosmotic pumps with molecularly thin silicon nanomembranes. *Proc. Natl. Acad. Soc. U.S.A.* 110: 18425–18430.

33. Proksch, R., R. Lal, P. K. Hansma, D. Morse, and G. Stucky. 1996. Imaging the internal and external pore structure of membranes in fluid: Tapping mode scanning ion conductance microscopy. *Biophys. J.* 71: 2155–2157.

34. Böcker, M., S. Muschter, E. K. Schmitt, C. Steinem, and T. E. Schäffer. 2009. Imaging and patterning of pore-suspending membranes with scanning ion conductance microscopy. *Langmuir* 25: 3022–3028.

35. Chen, C.-C., M. A. Derylo, and L. A. Baker. 2009. Measurement of ion currents through porous membranes with scanning ion conductance microscopy. *Anal. Chem.* 81: 4742–4751.

36. Chen, C.-C. and L. A. Baker. 2011. Effects of pipette modulation and imaging distances on ion currents measured with scanning ion conductance microscopy (SICM). *Analyst* 136: 90–97.

37. Chen, C.-C., Y. Zhou, and L. A. Baker. 2011. Single-nanopore investigations with ion conductance microscopy. *ACS Nano* 5: 8404–8411.

38. Zhou, Y., C.-C. Chen, and L. A. Baker. 2012. Heterogeneity of multiple-pore membranes investigated with ion conductance microscopy. *Anal. Chem.* 84: 3003–3009.

39. Takahashi, Y., A. I. Shevchuk, P. Novak et al. 2011. Multifunctional nanoprobes for nanoscale chemical imaging and localized chemical delivery at surfaces and interfaces. *Angew. Chem. Int. Ed.* 50: 9638–9642.

40. Morris, C. A., C.-C. Chen, and L. A. Baker. 2012. Transport of redox probes through single pores measured by scanning electrochemical-scanning ion conductance microscopy (SECM-SICM). *Analyst* 137: 2933–2938.

41. Macpherson, J. V. and P. R. Unwin. 2000. Combined scanning electrochemical-atomic force microscopy. *Anal. Chem.* 72: 276–285.

42. Macpherson, J. V., C. E. Jones, A. L. Barker, and P. R. Unwin. 2002. Electrochemical imaging of diffusion through single nanoscale pores. *Anal. Chem.* 74: 1841–1848.

43. Gardner, C. E., P. R. Unwin, and J. V. Macpherson. 2005. Correlation of membrane structure and transport activity using combined scanning electrochemical–atomic force microscopy. *Electrochem. Commun.* 7: 612–618.

44. Dumitrescu, I., P. R. Unwin, and J. V. Macpherson. 2009. Electrochemistry at carbon nanotubes: Perspective and issues. *Chem. Commun.* 6886–6901.

45. McCreery, R. L. 2008. Advanced carbon electrode materials for molecular electrochemistry. *Chem. Rev.* 108: 2646–2687.

46. Kim, J., H. Xiong, M. Hofmann, J. Kong, and S. Amemiya. 2010. Scanning electrochemical microscopy of individual single-walled carbon nanotubes. *Anal. Chem.* 82: 1605–1607.

47. Lee, E., M. Kim, J. Seong, H. Shin, and G. Lim. 2013. An L-shaped nanoprobe for scanning electrochemical microscopy-atomic force microscopy. *Phys. Status Solidi (RRL)* 7: 406–409.

48. Bard, A. J., M. V. Mirkin, P. R. Unwin, and D. O. Wipf. 1992. Scanning electrochemical microscopy. 12. Theory and experiment of the feedback mode with finite heterogeneous electron-transfer kinetics and arbitrary substrate size. *J. Phys. Chem.* 96: 1861–1868.

49. Xiong, H., D. A. Gross, J. Guo, and S. Amemiya. 2006. Local feedback mode of scanning electrochemical microscopy for electrochemical characterization of one-dimensional nanostructure: Theory and experiment with nanoband electrode as model substrate. *Anal. Chem.* 78: 1946–1957.

50. Xiong, H., J. Kim, E. Kim, and S. Amemiya. 2009. Scanning electrochemical microscopy of one-dimensional nanostructure: Effects of nanostructure dimensions on the tip feedback current under unbiased conditions. *J. Electroanal. Chem.* 629: 78–86.

51. Kim, E., J. Kim, and S. Amemiya. 2009. Spatially resolved detection of a nanometer-scale gap by scanning electrochemical microscopy. *Anal. Chem.* 81: 4788–4791.

52. Wilson, N. R., M. Guille, I. Dumitrescu et al. 2006. Assessment of the electrochemical behavior of two-dimensional networks of single-walled carbon nanotubes. *Anal. Chem.* 78: 7006–7015.

53. Fabre, B., F. Hauquier, C. Herrier et al. 2008. Covalent assembly and micropatterning of functionalized multiwalled carbon nanotubes to monolayer-modified Si(111) surfaces. *Langmuir* 24: 6595–6602.

54. Dumitrescu, I., P. V. Dudin, J. P. Edgeworth, J. V. Macpherson, and P. R. Unwin. 2010. Electron transfer kinetics at single-walled carbon nanotube electrodes using scanning electrochemical microscopy. *J. Phys. Chem. C* 114: 2633–2639.

55. Heller, I., J. Kong, H. A. Heering, K. A. Williams, S. G. Lemay, and C. Dekker. 2005. Individual single-walled carbon nanotubes as nanoelectrodes for electrochemistry. *Nano Lett.* 5: 137–142.

56. Guell, A. G., N. Ebejer, M. E. Snowden, K. McKelvey, J. V. Macpherson, and P. R. Unwin. 2012. Quantitative nanoscale visualization of heterogeneous electron transfer rates in 2D carbon nanotube networks. *Proc. Natl. Acad. Soc. U.S.A.* 109: 11487–11492.

57. Ritzert, N. L., J. Rodriguez-Lopez, C. Tan, and H. D. Abruna. 2013. Kinetics of interfacial electron transfer at single-layer graphene electrodes in aqueous and nonaqueous solutions. *Langmuir* 29: 1683–1694.

58. Tan, C., J. Rodriguez-Lopez, J. J. Parks, N. L. Ritzert, D. C. Ralph, and H. D. Abruna. 2012. Reactivity of monolayer chemical vapor deposited graphene imperfections studied using scanning electrochemical microscopy. *ACS Nano* 6: 3070–3079.

59. Mann, J. A., J. Rodriguez-Lopez, H. D. Abruna, and W. R. Dichtel. 2011. Multivalent binding motifs for the noncovalent functionalization of graphene. *J. Am. Chem. Soc.* 133: 17614–17617.

60. Rodriguez-Lopez, J., N. L. Ritzert, J. A. Mann, C. Tan, W. R. Dichtel, and H. D. Abruna. 2012. Quantification of the surface diffusion of tripodal binding motifs on graphene using scanning electrochemical microscopy. *J. Am. Chem. Soc.* 134: 6224–6236.

61. Azevedo, J., C. Bourdillon, V. Derycke, S. Campidelli, C. Lefrou, and R. Cornut. 2013. Contactless surface conductivity mapping of graphene oxide thin films deposited on glass with scanning electrochemical microscopy. *Anal. Chem.* 85: 1812–1818.

62. Azevedo, J. L., C. Costa-Coquelard, P. Jegou, T. Yu, and J.-J. Benattar. 2011. Highly ordered monolayer, multilayer, and hybrid films of graphene oxide obtained by the bubble deposition method. *J. Phys. Chem. C* 115: 14678–14681.

63. Molina, J., J. Fernandez, J. C. Ines, A. I. del Rio, J. Bonastre, and F. Cases. 2013. Electrochemical characterization of reduced graphene oxide-coated polyester fabrics. *Electrochim. Acta* 93: 44–52.

64. Peterson, R. R. and D. E. Cliffel. 2006. Scanning electrochemical microscopy determination of organic soluble MPC electron-transfer rates. *Langmuir* 22: 10307–10314.

65. Toikkanen, O., V. Ruiz, G. Ronholm, N. Kalkkinen, P. Liljeroth, and B. M. Quinn. 2008. Synthesis and stability of monolayer-protected Au38 clusters. *J. Am. Chem. Soc.* 130: 11049–11055.

66. Georganopoulou, D. G., M. V. Mirkin, and R. W. Murray. 2004. SECM measurement of the fast electron transfer dynamics between Au_{38}^{1+} nanoparticles and aqueous redox species at a liquid/liquid interface. *Nano Lett.* 4: 1763–1767.

67. Quinn, B. M., P. Liljeroth, and K. Kontturi. 2002. Interfacial reactivity of monolayer-protected clusters studied by scanning electrochemical microscopy. *J. Am. Chem. Soc.* 124: 12915–12921.

68. Quinn, B. M., I. Prieto, S. K. Haram, and A. J. Bard. 2001. Electrochemical observation of a metal/insulator transition by scanning electrochemical microscopy. *J. Phys. Chem. B* 105: 7474–7476.

69. Fang, P. P., S. Chen, H. Q. Deng et al. 2013. Conductive gold nanoparticle mirrors at liquid/liquid interfaces. *ACS Nano* 7: 9241–9248.

70. Whitworth, A. L., D. Mandler, and P. R. Unwin. 2005. Theory of scanning electrochemical microscopy (SECM) as a probe of surface conductivity. *Phys. Chem. Chem. Phys.* 7: 356–365.

71. Liljeroth, P., B. M. Quinn, V. Ruiz, and K. Kontturi. 2003. Charge injection and lateral conductivity in monolayers of metallic nanoparticles. *Chem. Commun.* 1570–1571.

72. Liljeroth, P., D. Vanmaekelbergh, V. Ruiz et al. 2004. Electron transport in two-dimensional arrays of gold nanocrystals investigated by scanning electrochemical microscopy. *J. Am. Chem. Soc.* 126: 7126–7132.

73. Xiong, H., J. Guo, and S. Amemiya. 2007. Probing heterogeneous electron transfer at an unbiased conductor by scanning electrochemical microscopy in the feedback mode. *Anal. Chem.* 79: 2735–2744.

74. Liljeroth, P. and B. M. Quinn. 2006. Resolving electron transfer kinetics at the nanocrystal/solution interface. *J. Am. Chem. Soc.* 128: 4922–4923.

75. Ahonen, P., V. Ruiz, K. Kontturi, P. Liljeroth, and B. M. Quinn. 2008. Electrochemical gating in scanning electrochemical microscopy. *J. Phys. Chem. C* 112: 2724–2728.

76. Biji, P., N. K. Sarangi, and A. Patnaik. 2010. One pot hemimicellar synthesis of amphiphilic janus gold nanoclusters for novel electronic attributes. *Langmuir* 26: 14047–14057.

77. Noël, J. M., D. Zigah, J. Simonet, and P. Hapiot. 2010. Synthesis and immobilization of Ag^0 nanoparticles on diazonium modified electrodes: SECM and cyclic voltammetry studies of the modified interfaces. *Langmuir* 26: 7638–7643.

78. Taleb, A., Y. P. Xue, S. Munteanu, F. Kanoufi, and P. Dubot. 2013. Self-assembled thiolate functionalized gold nanoparticles template toward tailoring the morphology of electrochemically deposited silver nanostructure. *Electrochim. Acta* 88: 621–631.

79. Lu, X. Q., F. P. Zhi, H. Shang, X. Y. Wang, and Z. H. Xue. 2010. Investigation of the electrochemical behavior of multilayers film assembled porphyrin/gold nanoparticles on gold electrode. *Electrochim. Acta* 55: 3634–3642.

80. Zhang, J., R. M. Lahtinen, K. Kontturi, P. R. Unwin, and D. J. Schiffrin. 2001. Electron transfer reactions at gold nanoparticles. *Chem. Commun.* 1818–1819.

81. Li, F., I. Ciani, P. Bertoncello et al. 2008. Scanning electrochemical microscopy of redox-mediated hydrogen evolution catalyzed by two-dimensional assemblies of palladium nanoparticles. *J. Phys. Chem. C* 112: 9686–9694.

82. Li, F., P. Bertoncello, I. Ciani, G. Mantovani, and P. R. Unwin. 2008. Incorporation of functionalized palladium nanoparticles within ultrathin nafion films: A nanostructured composite for electrolytic and redox-mediated hydrogen evolution. *Adv. Funct. Mater.* 18: 1685–1693.

83. Fernández, J. L., C. Hurth, and A. J. Bard. 2005. Scanning electrochemical microscopy #54. Application to the study of heterogeneous catalytic reactions—Hydrogen peroxide decomposition. *J. Phys. Chem. B* 109: 9532–9539.

84. Bard, A. J. 2010. Inner-sphere heterogeneous electrode reactions. Electrocatalysis and photocatalysis: The challenge. *J. Am. Chem. Soc.* 132: 7559–7567.

85. Zhou, J., Y. Zu, and A. J. Bard. 2000. Scanning electrochemical microscopy part 39. The proton/hydrogen mediator system and its application to the study of the electrocatalysis of hydrogen oxidation. *J. Electroanal. Chem.* 491: 22–29.

86. Nicholson, R., S. Zhou, G. Hinds, A. J. Wain, and A. Turnbull. 2009. Electrocatalytic activity mapping of model fuel cell catalyst films using scanning electrochemical microscopy. *Electrochim. Acta* 54: 4525–4533.

87. Fernandez, J. L. and A. J. Bard. 2003. Scanning electrochemical microscopy. 47. Imaging electrocatalytic activity for oxygen reduction in an acidic medium by the tip generation-substrate collection mode. *Anal. Chem.* 75: 2967–2974.

88. Sanchez-Sanchez, C. M., J. Solla-Gullon, F. J. Vidal-Iglesias, A. Aldaz, V. Montiel, and E. Herrero. 2010. Imaging structure sensitive catalysis on different shape-controlled platinum nanoparticles. *J. Am. Chem. Soc.* 132: 5622–5624.

89. Sanchez-Sanchez, C. M., F. J. Vidal-Iglesias, J. Solla-Gullon et al. 2010. Scanning electrochemical microscopy for studying electrocatalysis on shape-controlled gold nanoparticles and nanorods. *Electrochim. Acta* 55: 8252–8257.

90. Sanchez-Sanchez, C. M., J. Souza-Garcia, A. Saez et al. 2011. Imaging decorated platinum single crystal electrodes by scanning electrochemical microscopy. *Electrochim. Acta* 56: 10708–10712.

91. Minguzzi, A., M. A. Alpuche-Aviles, J. R. López, S. Rondinini, and A. J. Bard. 2008. Screening of oxygen evolution electrocatalysts by scanning electrochemical microscopy using a shielded tip approach. *Anal. Chem.* 80: 4055–4064.

92. Naslund, L. A., C. M. Sanchez-Sanchez, A. S. Ingason et al. 2013. The role of TiO_2 doping on RuO_2-coated electrodes for the water oxidation reaction. *J. Phys. Chem. C* 117: 6126–35.

93. Wain, A. J. 2013. Imaging size effects on the electrocatalytic activity of gold nanoparticles using scanning electrochemical microscopy. *Electrochim. Acta* 92: 383–391.

94. Sanchez-Sanchez, C. M., J. Rodriguez-Lopez, and A. J. Bard. 2008. Scanning electrochemical microscopy. 60. Quantitative calibration of the SECM substrate generation/tip collection mode and its use for the study of the oxygen reduction mechanism. *Anal. Chem.* 80: 3254–3260.

95. Zoski, C. G., J. C. Aguilar, and A. J. Bard. 2003. Scanning electrochemical microscopy. 46. Shielding effects on reversible and quasi reversible reactions. *Anal. Chem.* 75: 2959–2966.

96. Eckhard, K., X. X. Chen, F. Turcu, and W. Schuhmann. 2006. Redox competition mode of scanning electrochemical microscopy (RC-SECM) for visualisation of local catalytic activity. *Phys. Chem. Chem. Phys.* 8: 5359–5365.

97. Nagaiah, T. C., A. Maljusch, X. X. Chen, M. Bron, and W. Schuhmann. 2009. Visualization of the local catalytic activity of electrodeposited Pt–Ag catalysts for oxygen reduction by means of SECM. *Chemphyschem* 10: 2711–2718.

98. Maljusch, A., T. C. Nagaiah, S. Schwamborn, M. Bron, and W. Schuhmann. 2010. Pt–Ag catalysts as cathode material for oxygen-depolarized electrodes in hydrochloric acid electrolysis. *Anal. Chem.* 82: 1890–1896.

99. Kulp, C., X. X. Chen, A. Puschhof et al. 2010. Electrochemical synthesis of core-shell catalysts for electrocatalytic applications. *ChemPhysChem* 11: 2854–2861.

100. Chen, R. Y., V. Trieu, A. R. Zeradjanin et al. 2012. Microstructural impact of anodic coatings on the electrochemical chlorine evolution reaction. *Phys. Chem. Chem. Phys.* 14: 7392–7399.

101. Song, Y. Y., W. Z. Jia, Y. Li et al. 2007. Synthesis and patterning of Prussian blue nanostructures on silicon wafer via galvanic displacement reaction. *Adv. Funct. Mater.* 17: 2808–2814.

102. Lee, J. W., H. C. Ye, S. L. Pan, and A. J. Bard. 2008. Screening of photocatalysts by scanning electrochemical microscopy. *Anal. Chem.* 80: 7445–7450.

103. Cho, S. K., H. S. Park, H. C. Lee, K. M. Nam, and A. J. Bard. 2013. Metal doping of $BiVO_4$ by composite electrodeposition with improved photoelectrochemical water oxidation. *J. Phys. Chem. C* 117: 23048–23056.

104. Cong, Y. Q., H. S. Park, S. J. Wang et al. 2012. Synthesis of Ta_3N_5 nanotube arrays modified with electrocatalysts for photoelectrochemical water oxidation. *J. Phys. Chem. C* 116: 14541–14550.

105. Chang, C. J., M. H. Hsu, Y. C. Weng, C. Y. Tsay, and C. K. Lin. 2013. Hierarchical ZnO nanorod-array films with enhanced photocatalytic performance. *Thin Solid Films* 528: 167–174.

106. Fonseca, S. M., A. L. Barker, S. Ahmed, T. J. Kemp, and P. R. Unwin. 2003. Direct observation of oxygen depletion and product formation during photocatalysis at a TiO_2 surface using scanning electrochemical microscopy. *Chem. Commun.* 1002–1003.

107. Alexandrescu, R., I. Morjan, M. Scarisoreanu et al. 2010. Development of the IR laser pyrolysis for the synthesis of iron-doped TiO_2 nanoparticles: Structural properties and photoactivity. *Infrared Phys. Technol.* 53: 94–102.

108. Fedorov, R. G. and D. Mandler. 2013. Local deposition of anisotropic nanoparticles using scanning electrochemical microscopy (SECM). *Phys. Chem. Chem. Phys.* 15: 2725–2732.

109. Sheffer, M., V. Martina, R. Seeber, and D. Mandlera. 2008. Deposition of gold nanoparticles on thin polyaniline films. *Isr. J. Chem.* 48: 349–357.

110. Sheffer, M. and D. Mandler. 2009. Control of locally deposited gold nanoparticle on polyaniline films. *Electrochim. Acta* 54: 2951–2956.

111. Danieli, T., J. Colleran, and D. Mandler. 2011. Deposition of Au and Ag nanoparticles on PEDOT. *Phys. Chem. Chem. Phys.* 13: 20345–20353.

112. Malel, E., J. Colleran, and D. Mandler. 2011. Studying the localized deposition of Ag nanoparticles on self-assembled monolayers by scanning electrochemical microscopy (SECM). *Electrochim. Acta* 56: 6954–6961.

113. Li, F., M. Edwards, J. Guo, and P. R. Unwin. 2009. Silver particle nucleation and growth at liquid/liquid interfaces: A scanning electrochemical microscopy approach. *J. Phys. Chem. C* 113: 3553–3565.

114. Yatziv, Y., I. Turyan, and D. Mandler. 2002. A new approach to micropatterning: Application of potential-assisted ion transfer at the liquid-liquid interface for the local metal deposition. *J. Am. Chem. Soc.* 124: 5618–5619.

115. Turyan, I., M. Etienne, D. Mandler, and W. Schuhmann. 2005. Improved resolution of local metal deposition by means of constant distance mode scanning electrochemical microscopy. *Electroanalysis* 17: 538–542.

116. Malel, E. and D. Mandler. 2008. Localized electroless deposition of gold nanoparticles using scanning electrochemical microscopy. *J. Electrochem. Soc.* 155: D459–D467.

117. Malel, E., R. Ludwig, L. Gorton, and D. Mandler. 2010. Localized deposition of Au nanoparticles by direct electron transfer through cellobiose dehydrogenase. *Chem. Eur. J.* 16: 11697–11706.

118. Danieli, T., N. Gaponik, A. Eychmuller, and D. Mandler. 2008. Studying the reactions of CdTe nanostructures and thin CdTe films with Ag^+ and $AuCl_4^-$. *J. Phys. Chem. C* 112: 8881–8889.

119. O'Mullane, A. P., S. J. Ippolito, A. M. Bond, and S. K. Bhargava. 2010. A study of localised galvanic replacement of copper and silver films with gold using scanning electrochemical microscopy. *Electrochem. Commun.* 12: 611–615.

120. Danieli, T. and D. Mandler. 2013. Local surface patterning by chitosan-stabilized gold nanoparticles using the direct mode of scanning electrochemical microscopy (SECM). *J. Solid State Electrochem.* 17: 2989–2997.

121. Abad, J. M., A. Y. Tesio, F. Pariente, and E. Lorenzo. 2013. Patterning gold nanoparticle using scanning electrochemical microscopy. *J. Phys. Chem. C* 117: 22087–22093.

122. Yogeswaran, U., S. Thiagarajan, and S. M. Chen. 2007. Pinecone shape hydroxypropyl-β-cyclodextrin on a film of multi-walled carbon nanotubes coated with gold particles for the simultaneous determination of tyrosine, guanine, adenine and thymine. *Carbon* 45: 2783–2796.

123. Yogeswaran, U., S. Thiagarajan, and S. M. Chen. 2007. Nanocomposite of functional multiwall carbon nanotubes with nafion, nano platinum, and nano biosensing film for simultaneous determination of ascorbic acid, epinephrine, and uric acid. *Anal. Biochem.* 365: 122–131.

124. Chen, X. X., K. Eckhard, M. Zhou, M. Bron, and W. Schuhmann. 2009. Electrocatalytic activity of spots of electrodeposited noble-metal catalysts on carbon nanotubes modified glassy carbon. *Anal. Chem.* 81: 7597–7603.

125. Kundu, S., T. C. Nagaiah, X. X. Chen et al. 2012. Synthesis of an improved hierarchical carbon-fiber composite as a catalyst support for platinum and its application in electrocatalysis. *Carbon* 50: 4534–4542.

126. Schwamborn, S., L. Stoica, X. X. Chen et al. 2010. Patterned CNT arrays for the evaluation of oxygen reduction activity by SECM. *ChemPhysChem* 11: 74–78.

127. Kundu, S., T. C. Nagaiah, W. Xia et al. 2009. Electrocatalytic activity and stability of nitrogen-containing carbon nanotubes in the oxygen reduction reaction. *J. Phys. Chem. C* 113: 14302–14310.

128. Dobrzeniecka, A., A. Zeradjanin, J. Masa et al. 2013. Application of SECM in tracing of hydrogen peroxide at multicomponent non-noble electrocatalyst films for the oxygen reduction reaction. *Catal. Today* 202: 55–62.

129. Ruiz, V., P. G. Nicholson, S. Jollands, P. A. Thomas, J. V. Macpherson, and P. R. Unwin. 2005. Molecular ordering and 2D conductivity in ultrathin poly(3-hexylthiophene)/gold nanoparticle composite films. *J. Phys. Chem. B* 109: 19335–19344.

130. Nicholson, P. G., V. Ruiz, J. V. Macpherson, and P. R. Unwin. 2006. Effect of composition on the conductivity and morphology of poly (3-hexylthiophene)/gold nanoparticle composite Langmuir–Schaeffer films. *Phys. Chem. Chem. Phys.* 8: 5096–5105.

131. Molina, J., J. Fernandez, A. I. del Rio, J. Bonastre, and F. Cases. 2012. Characterization of azo dyes on Pt and Pt/polyaniline/dispersed Pt electrodes. *Appl. Surf. Sci.* 258: 6246–6256.

132. Shen, Y., M. Trauble, and G. Wittstock. 2008. Electrodeposited noble metal particles in polyelectrolyte multilayer matrix as electrocatalyst for oxygen reduction studied using SECM. *Phys. Chem. Chem. Phys.* 10: 3635–3644.

133. Ferreyra, N. F., S. Bollo, and G. A. Rivas. 2010. Self-assembled multilayers of polyethylenimine, DNA and gold nanoparticles. A study of electron transfer reaction. *J. Electroanal. Chem.* 638: 262–268.

134. Wang, J., F. Y. Song, and F. M. Zhou. 2002. Silver-enhanced imaging of DNA hybridization at DNA microarrays with scanning electrochemical microscopy. *Langmuir* 18: 6653–6658.

135. Zhang, M. Q., G. Wittstock, Y. H. Shao, and H. H. Girault. 2007. Scanning electrochemical microscopy as a readout tool for protein electrophoresis. *Anal. Chem.* 79: 4833–4839.

136. Chen, Z., S. B. Xie, L. Shen et al. 2008. Investigation of the interactions between silver nanoparticles and HeLa cells by scanning electrochemical microscopy. *Analyst* 133: 1221–1228.

137. Szot, K., W. Nogala, J. Niedziolka-Jonsson et al. 2009. Hydrophilic carbon nanoparticle-laccase thin film electrode for mediator less dioxygen reduction SECM activity mapping and application in zinc-dioxygen battery. *Electrochim. Acta* 54: 4620–4625.

138. Nogala, W., A. Celebanska, K. Szot, G. Wittstock, and M. Opallo. 2010. Bioelectrocatalytic mediatorless dioxygen reduction at carbon ceramic electrodes modified with bilirubin oxidase. *Electrochim. Acta* 55: 5719–5724.

139. Bracamonte, M. V., S. Bollo, P. Labbe, G. A. Rivas, and N. F. Ferreyra. 2011. Quaternized chitosan as support for the assembly of gold nanoparticles and glucose oxidase: Physicochemical characterization of the platform and evaluation of its biocatalytic activity. *Electrochim. Acta* 56: 1316–1322.

140. Song, W. L., Z. Y. Yan, and K. C. Hu. 2012. Electrochemical immunoassay for CD10 antigen using scanning electrochemical microscopy. *Biosens. Bioelectron.* 38: 425–429.

141. Fan, H. J., X. L. Wang, F. Jiao et al. 2013. Scanning electrochemical microscopy of DNA hybridization on DNA microarrays enhanced by HRP-modified SiO_2 nanoparticles. *Anal. Chem.* 85: 6511–6517.

142. Fan, H. J., F. Jiao, H. Chen et al. 2013. Qualitative and quantitative detection of DNA amplified with HRP-modified SiO_2 nanoparticles using scanning electrochemical microscopy. *Biosens. Bioelectron.* 47: 373–378.

143. Kim, J., M. Shen, N. Nioradze, and S. Amemiya. 2012. Stabilizing nanometer scale tip-to-substrate gaps in scanning electrochemical microscopy using an isothermal chamber for thermal drift suppression. *Anal. Chem.* 84: 3489–3492.

144. Amemiya, S., N. Nioradze, P. Santhosh, and M. J. Deible. 2011. Generalized theory for nanoscale voltammetric measurements of heterogeneous electron-transfer kinetics at macroscopic substrates by scanning electrochemical microscopy. *Anal. Chem.* 83: 5928–5935.

19 Scanning Electrochemical Cell Microscopy

Mapping, Measuring, and Modifying Surfaces and Interfaces at the Nanoscale

Barak D.B. Aaronson, Aleix G. Güell, Kim McKelvey,
Dmitry Momotenko, and Patrick R. Unwin

CONTENTS

19.1   INTRODUCTION: ELECTROCHEMICAL IMAGING AND DROPLET CELL TECHNIQUES

The acquisition of spatially resolved functional and structural information on surfaces and interfaces is a major theme in contemporary microscopy, with applications spanning materials science and technology, biology, medicine, and nanotechnology generally. However, access to local chemical information (especially concerning dynamics at condensed phase interfaces) from mainstream microscopy techniques is often challenging, and structure–function relationships remain obscured. To expand the capabilities of microscopy, significant efforts have been invested in the development of scanning probe microscopy (SPM) techniques[1] that facilitate direct measurements of various types of processes at a wide range of interfaces. The purpose of this chapter is to highlight scanning droplet-based techniques, with a particular focus on scanning electrochemical cell microscopy (SECCM).[2] This is a very recently developed methodology within the family of more established SPMs, but one that has already demonstrated considerable versatility for probing and visualizing interfacial function, opening up new opportunities for understanding electrochemical processes, and as a tool for the modification and patterning of surfaces.

When considering electrochemical imaging,[3–5] the most prominent technique is scanning electrochemical microscopy (SECM),[6–9] which has been used to investigate (electro)chemical properties and reactions at a variety of interfaces. The translation of SECM into diverse fields (biology,[10–13] energy research,[14,15] and materials science[16,17]) has been facilitated by the development of different operation modes, the wide range of tips (probes), and a well-established theory and models for quantitative and qualitative data analyses.[3,18–21] However, there are some limitations of SECM experiments, particularly the fact that the tip electrode response (and spatial resolution) depends on the probe-to-substrate separation and reactivity. This arises because the SECM probe is a remote sensor of reactivity, detecting products or intermediates of the surface process or competing with the surface for a particular reagent. The tip–substrate distance thus needs to be known precisely, and, particularly for nanoscale measurements, there is a need for additional instrumentation for reliable tip positioning. Although various positioning control mechanisms have been explored, including the detection of shear forces,[22–25] coupling SECM with scanning ion conductance microscopy (SICM),[26–30] impedance measurements,[31,32] the combination of SECM with atomic force microscopy (AFM),[33–36] the use of tip position modulation,[37–39] and intermittent contact techniques,[40–42] none of these methods has emerged as definitive, and while each has its merits, there are also some issues related to tip design and reliability.[3] Another important consideration for the functional characterization of interfaces with SECM arises from the requirement to immerse the entire sample into electrolyte solution and typically maintain a reaction at the tip and substrate throughout the duration of an image, which may require a period of upward of an hour, with implications for the stability of the tip and sample, especially from fouling or contamination. This problem can be circumvented to some extent through the use of multielectrode SECM probes for parallel imaging,[43–48] which speed up the process and allow access to wider areas of a sample.

SECCM[2,49] overcomes some of the aforementioned issues and allows the direct measurement of charge transfer and other processes at a variety of electrodes, and interfaces generally, in many cases at the nanoscale. The intrinsic principle of scanning droplet cell methods is the confinement of electrochemical measurements in a small area of the surface allowing direct, localized spatially resolved surface investigations of (semi)conductive electrode substrates (and, for SECCM, insulating surfaces where there are ion fluxes). A small mobile droplet is typically created as a liquid meniscus between a (scanned) probe and the specimen surface, which serves as a working electrode. The working electrode size is defined by the wetted area of the surface, while integrated counter-reference electrodes inside the probe, filled with electrolyte, result in a dynamic electrochemical cell setup. In practice, this experimental arrangement is particularly advantageous because of the possibility of examining only small areas of the substrate. It is especially beneficial for the investigation of delicate samples, the characterization of dry surfaces, and the avoidance of complexity in sample preparation (encapsulation). Furthermore, only small amounts of electrolyte solution are needed. This approach yields relatively low background signals on the working electrode, enabling direct and sensitive measurements.

The conventional setup for a scanning droplet cell microscope typically utilizes a pipette or a capillary with a small opening (usually with a tip diameter of the order of 1–1000 μm), mounted on a microscope stage for positioning of the probe over the sample.[50–56] In most cases, the capillary has a small silicon or rubber seal at the tip to prevent solvent evaporation and to provide a working electrode area of well-defined dimensions (Figure 19.1a and b). The implementation of this type of scanning droplet cell design has been demonstrated with single-point (static) measurements as well as in a scanning mode for corrosion research, for example, in local pit initiation on a stainless steel,[50] local surface modification through direct lithographic writing of surface oxides,[51] and for the characterization of individual grains and grain boundaries on alloys.[52–54,56] Staemmler et al.[55] attempted

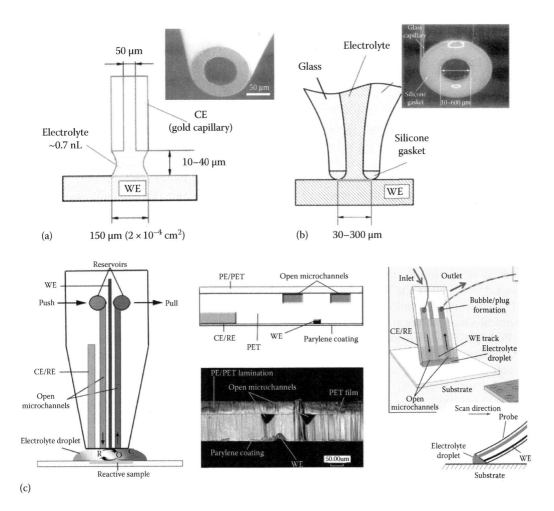

FIGURE 19.1 Examples of droplet confinement. (a) Micron-sized openings at glass pipettes employed to confine measurements at a working electrode (WE) to the microscale. (Adapted from Suter, T. and Böhni, H., *Electrochim. Acta*, 42, 3275, 1997.) (b) Incorporating a gasket to avoid evaporation and delimit an area of interrogation. (Adapted from Lohrengel, M.M. et al., *Electrochim. Acta*, 47, 137, 2001.) (c) Flexible multichannel push–pull probes, with integrated electrodes and connected to microfluidic systems, as applied to nonconductive surfaces. (Adapted from Momotenko, D. et al., *Anal. Chem.*, 83, 5275, 2011; Momotenko, D. et al., *Anal. Chem.*, 84, 6630, 2012.) *(Continued)*

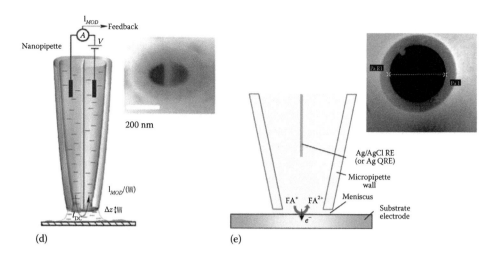

(d)   (e)

FIGURE 19.1 (CONTINUED) Examples of droplet confinement. (d) Confined measurements at the submicron level, employing double barrel. (Adapted from Rodolfa, K. et al., *Angew. Chem., Int. Ed.*, 44, 6854, 2005.) (e) Single-barrel glass pipettes. (Adapted from Williams, C.G. et al., *Anal. Chem.*, 81, 2486, 2009.)

to bring the capabilities of this technique to the nanoscale through the electrochemical surface modification of gold and silicon substrates. Nanolithographic patterning with an electrochemical scanning capillary microscope was demonstrated through the elecrodeposition of thin copper lines and dots using a pipette with 150 nm opening and shear-force positional feedback control.[55]

A flow-through cell design has also been introduced to the scanning droplet cell concept, where a constant electrolyte flow passes over the wetted surface area.[62] This approach has been used for photoelectrochemical analysis of semiconductors and solar cell materials,[63] but the versatility of such a strategy is the possibility of adding additional complements to electrochemical microscopy and point measurements through integration with other techniques for the downstream analysis of products of (electro)chemical reactions. Electrochemical studies of the anodic dissolution of Zn alloys,[64–66] as well as Cu[65,67] and Fe,[68] were combined with spectroscopic and mass-spectrometric (MS) techniques for online detection, demonstrating the efficacy of this approach for integrated chemical–electrochemical surface analysis.

Push–pull[58] and fountain pen[69] soft SECM probes have also been introduced for the analysis of dry surfaces (Figure 19.1c). These microfluidic SECM probes were developed following the concept of soft probes for topography-tolerant and high-throughput SECM investigations of samples with extended surface area.[44–47,69] Probes fabricated from soft polymeric materials (e.g., polyethylene-terephthalate [PET]) have allowed SECM experiments in contact mode, providing close to constant probe-to-substrate separation, without the need for additional positional feedback and also providing a versatile platform for integration with microfluidic systems. Coupling the resolving power of SECM with highly sensitive MS techniques within microfluidic push–pull probes has opened up the possibility of simultaneous electrochemical imaging and MS analysis, as illustrated by studies of latent fingerprint samples and online ESI-MS and SECM characterization of immobilized enzyme reactivity.[59]

Although the techniques described earlier are capable of local electrochemical measurements, they typically operate at a scale of tens of microns or larger. The ability to readily pull glass or quartz pipettes with tip diameters in the range of several nanometers to hundreds of nanometers has led to a breakthrough in electrochemical imaging resolution, coupled with extremely simple, fast, and cheap probe preparation protocols, which have high success rates. SICM, which was introduced in late 1980s, has been proven to be a powerful high-resolution non-contact pipette-based technique for the visualization of substrate topography.[70] The positioning of a mobile nanopipette is achieved through the detection of ion currents arising from the migration of ionic species between two biased quasi-reference counter electrodes (QRCEs)—one located within a sharp pipette filled with electrolyte solution and another one in the bulk of a bathing solution. The feedback mechanism relies on the moderation of the ion flow between the QRCEs (governed by a distance-dependent resistance of the electrolyte solution around the tip opening)[71] when the nanopipette is brought into the vicinity of a substrate surface. The vertical position of the probe can be controlled using either the direct (ion conductance) current (DC) magnitude or the amplitude of the alternating current (AC) component of a modulated ion current generated by a physical oscillation of the probe around its vertical (z) position.[72,73] The latter provides a more sensitive and stable mechanism for the control of tip–substrate separation distance as the ion current can be detected using phase-sensitive techniques and the AC ion current is also less susceptible to bulk changes in the solution conductivity. The evident advantages and capabilities of the SICM technique have been successfully demonstrated in nanobioscience for the imaging of living cells,[74–77] cell membrane structure down to the protein level,[78] studies of ion fluxes across biological and artificial membranes,[28,79,80] the patch clamp technique in electrophysiology,[75,79] and the modification of surfaces with biomaterials using a double-barrel pipette (Figure 19.1d).[60,81–83] At the same time, as highlighted earlier, SICM has recently been used in combination with other methods, such as SECM, as an independent method for maintaining constant probe-to-substrate distance.[26–30]

It has also been shown recently that pulled nanopipettes can also be employed as multipurpose tools for concurrent topographical and electrochemical reactivity imaging in a novel electron transfer/ion transfer mode of SECM.[84] The tip of a silanized nanopipette was filled with an organic electrolyte solution (immiscible with the external aqueous solution) to make a nanoscale liquid/liquid interface. Controlling the bias across this boundary allowed ion transfer to be driven, enabling tracking of the substrate topography, similar to SICM, as well as the localized reactivity, similar to SECM, depending on the nanopipette bias. For example, with a positively polarized nanopipette, the transfer of PF_6^- anions from aqueous solution into the nanopipette was used to track the probe-to-substrate distance. The organic phase within the nanopipette could also contain a neutral redox probe (e.g., a ferrocene derivative) that would diffuse away from the nanopipette opening regardless of the bias and could undergo electrochemical transformation into its ionic form (ferrocenium derivative) at reactive sites on the specimen surface. The ion transfer current into a negatively biased pipette was then used to record local reactive properties of the surface. This sequential sensing of topography and reactivity with a single probe (pipette orifice) is a good example of the exploitation of the mass-transport properties of an interface between two immiscible electrolyte solutions supported at the tip of a nanopipette.[85,86]

In order to bring the high-resolution capacity of pulled pipettes into scanning droplet cell methodology for the analysis of dry substrates, a scanning micropipette contact method (SMCM) was recently developed (Figure 19.1e).[61] The probe was a single-barrel pipette with dimensions down to a few hundred nanometers, filled with electrolyte solution containing redox-active species and a QRCE. SMCM has been successfully demonstrated for the electrochemical interrogation of redox activity at highly oriented pyrolytic graphite (HOPG) and investigations of heterogeneities in the electroactivity of aluminum alloys using point-by-point measurements. Typically, the pipette was operated in a hopping mode, whereby the electrode substrate was approached, then paused upon the contact of the meniscus with the substrate (detected as a current flow). The steady-state current value or a current–voltage curve was recorded before the pipette was retracted back from the substrate for further repositioning over different locations at the substrate. The implementation of the hopping mode enabled images of reactivity to be built up across the substrate of interest. Furthermore, the simplified probe design avoided the need for a rubber gasket/seal at the pipette tip, yet allowed precise control of the contact meniscus and the area of the working electrode with high reproducibility. Similar techniques, but with pipettes that are tens of microns in size, have been implemented for voltammetric studies at single-walled carbon nanotubes (SWNTs) and metal nanowires[87] and for the analysis of the parameters controlling the electrodeposition of metal nanoparticles on SWNTs.[88]

SECCM represents a new generation of scanning droplet cells, making use of a double-barrel pipette to create a probe that functions as both a conductivity cell between QRCEs in the two nanochannels of the probe and an amperometric–voltammetric cell with a (semi)conducting working electrode surface with which the meniscus of the nanopipette makes contact. The technique provides exquisite control of a droplet with a surface (controllable on the ms timescale or shorter) and opens up the possibility of controlling and measuring electrochemical fluxes at interfaces, while simultaneously exploring ionic mass transport between the pipette barrels and measuring substrate topography.[2] Herein, the considerable capabilities of the SECCM technique for nanoscale (electro)chemical analysis are presented, along with a detailed description of instrumentation, application modes, and the theoretical basis of the functioning and operation of SECCM. It will be shown how SECCM allows a myriad of important structure–function relationships to be elucidated, enabling important questions in electrochemistry to be addressed and providing a platform for the future development and application of the technique.

19.2   SECCM EXPERIMENTAL AND INSTRUMENTATION DETAILS

19.2.1   Probe Fabrication

SECCM probes are fabricated from either borosilicate or quartz double-barrel pipettes, pulled to a sharp point in a laser puller (typically a P-2000 from *Sutter Instruments*). The size of the probe can be controlled by adjusting the pulling parameters, with probes between 100 nm and tens of microns across at the tip end fabricated easily and quickly. A field emission scanning electron microscopy (FE-SEM) image of a typical (relatively large) SECCM probe is shown in Figure 19.2a. The pulling parameters differ between laser pullers, but for a typical 500 nm borosilicate probe on a P-2000 laser puller, the parameters are as follows: line 1: heat = 600, filament = 4, velocity = 30, delay = 150. pull = 20; line 2: heat = 500, filament = 4, velocity = 30, delay = 150. pull = 60; line 3: heat = 500, filament = 3, velocity = 30, delay = 135. pull = 60.

The sharp double-barrel pipettes are then filled with electrolyte solution (typically aqueous salt solutions between 1 and 100 mM ionic strength), which can also contain other species (such as redox mediators or a nanoparticle dispersion).[2] We have also successfully employed room temperature ionic liquids (RTILs) in SECCM.[89] The electrolyte solution naturally forms a small liquid meniscus over the end of the pipette, connecting the two barrels.[49,90] QRCEs are inserted into each barrel, and a conductance cell is formed between the QRCEs and across the liquid meniscus. Ag/AgCl[91] and Pd-H$_2$[92,93] electrodes have been used as QRCEs. To help confine an aqueous meniscus, the outside walls of the pipette are often salinized using dimethyldichlorosilane [Si(CH$_3$)$_2$Cl$_2$].[49]

19.2.2   Instrumentation

The SECCM probe needs to be moved both laterally across and vertically away from and toward a sample of interest, and this is typically achieved using three piezoelectric positioners. Two different configurations have been used: tip scanning or sample scanning. In the tip scanning configuration, the tip is mounted on a stack of three piezoelectric positioners and moved in three (x, y, z) dimensions while the sample is held stationary. This configuration is used when the sample is large or bulky, and therefore impractical to move (scan). In the sample scanning configuration, the sample is mounted on two piezoelectric positioners, which allow lateral (x, y) movement of the sample, while the SECCM probe is mounted on a separate piezoelectric positioner that moves the probe vertically (z) away from and toward the surface. This configuration is best suited to high-resolution applications, where the tip is small and there is thus a need to reduce any lateral movement of the z-piezoelectric positioner that could be caused by cross talk between the different positioners.

The piezoelectric positioners are controlled through a servo/amplifier from a personal computer (PC) with a data acquisition (DAQ) or field programmable gate array (FPGA) card. The tip position

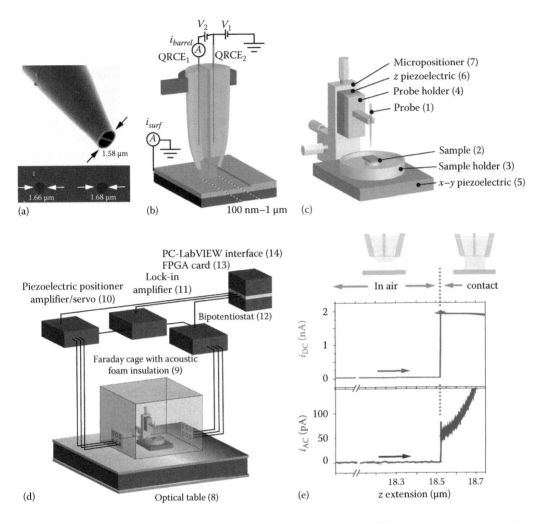

FIGURE 19.2 Scanning electrochemical cell microscopy (SECCM). (a) SEM image of a typical double-barrel probe. Shown in the following are example footprints resulting from meniscus contact with a substrate transmission electron microscopy (TEM) grid, demonstrating a consistency in size that is commensurate with the probe dimensions. Schematics of the setup of a typical SECCM platform: (b) electrochemical cell configuration; (c) probe, positioners, and sample holders; (d) overview of full setup. (e) Ion conductance current between the barrels of the pipette (i_{barrel}) as a function of tip–substrate position (z-piezoelectric positioner extension) showing the i_{DC} and i_{AC} (current) responses.

is typically oscillated normal to the surface (see working principles in the following text) using a lock-in amplifier. The two QRCEs in the SECCM probe are connected to a bipotentiostat, and a small (typically 100 mV) constant potential is applied between the barrels, V_2, as shown in Figure 19.2b. The ion current is measured, and the oscillating components of this current (due to the tip position modulation) are extracted using the lock-in amplifier and used as a feedback signal to control the movement of the tip with respect to the sample.

The contact area between the liquid meniscus and the surface constitutes a working electrode on (semi)conducting surfaces, and so any redox-active species (molecules or ions) present in the solution can be oxidized or reduced at the substrate, causing a current flow that is measured. Additionally, as we describe later, the SECCM nanopipette can be used to deliver nanoparticles to a surface.[94] Typically, the potential of the (semi)conducting sample is held at ground, and the potential of the two QRCEs in the barrels of the probe is floated to drive electrochemical reactions at the substrate (as shown in Figure 19.2b, V_1 is changed, which floats the potential of the QRCEs with respect to the substrate). In this configuration, the effective potential of the working electrode is approximately $-(V_1 + V_2/2)$.[95] This needs to be controlled and checked carefully as asymmetries in the pipette, differences in the two QRCEs, or the vertical alignment of the probe could influence the effective potential of the substrate electrode.[95]

A labeled diagram of a typical sample scanning instrument used for SECCM is shown in Figure 19.2c and d. The sample (2 in Figure 19.2c) is mounted in a sample holder (3 in Figure 19.2c) that is frequently made from a thin piece of polytetrafluoroethylene (PTFE) or other lightweight material. The sample holder can also contain a moat, which is filled with electrolyte solution to create a humidity cell. The sample holder is, in turn, mounted on x, y piezoelectric positioners (typically P-622.1CD, Physik Instrumente, 5 in Figure 19.2c) for lateral movement of the sample. The SECCM probe (1 in Figure 19.2c) is mounted on a z-piezoelectric positioner (typically P-753.31C, Physik Instrumente, 6 in Figure 19.2c) that allows movement of the pipette toward or away from the sample surface. This positioner is mounted, in turn, on a manual x, y, z stage to enable coarse positioning of the pipette probe (7 in Figure 19.2c). The pipette is held in position, on a tip holder, using a v-shaped groove and PTFE screw (4 in Figure 19.2c).

The piezoelectric positioners are controlled through amplifiers/servos (10 in Figure 19.2d). A bipotentiostat (custom built, 12 in Figure 19.2d) is used to measure electrochemical signals at the probe. The probe is oscillated normal to the surface using the oscillating signal that is generated by the lock-in amplifier, and the resulting oscillating current signal is extracted at the same frequency using the same lock-in amplifier (SR830, Stanford Research Systems, 11 in Figure 19.2d). An FPGA card (7852R, National Instruments, 13 in Figure 19.2d) mounted directly into the motherboard of the PC is used to collect all the data and control the instrument. The use of an FPGA card allows complex calculations, such as data filtering and probe position control logic, to be completed quickly. However, a standard DAQ card can also be used. LabVIEW (LabVIEW2011, National Instruments) is used (14 in Figure 19.2d) to control the FPGA card on the PC.

Electric, acoustic, and vibration isolation is essential for high-resolution current measurements and positional control. All instruments are mounted on vibration isolation tables (8 in Figure 19.2d), within a Faraday cage (custom made) with acoustic foam to reduce vibrations (9 in Figure 19.2d). Cameras or an optical microscope (not shown) can be used to aid positioning of the probe.

19.2.3 Working Principles

The liquid meniscus that forms at the end of an SECCM probe is brought into contact with a sample, and localized (electrochemical) measurements are carried out. Importantly, the response only depends on the area of the surface that is in contact with the liquid meniscus. The localized electrochemical measurements can be the electrochemical current that is measured through the surface,[49] and/or the conductance current between the QRCEs of the SECCM probe.[96] The liquid meniscus can also be used to modify the surface, which can be assessed afterward using complementary techniques such as AFM or FE-SEM.

A feedback response is needed to control the distance between the end of the SECCM probe and the surface (i.e., to control the meniscus thickness), to prevent the pipette from crashing into the surface and to maintain a constant distance between the end of the probe and the surface during imaging. This is achieved by monitoring the ion current between the QRCEs in the barrels of the SECCM probe. As discussed in the following text, the ion current depends on the size and shape of the liquid meniscus.

Practically, as in SICM,[5] a distance-modulated protocol is used to maintain a very stable feedback response. The probe position is modulated perpendicular to the surface, and once the liquid meniscus is in contact with the surface, this produces an AC component at the same frequency as the probe oscillation due to the changes in the probe–surface distance. The AC component is extracted using a lock-in amplifier, and the magnitude is used as a feedback signal (see the preceding text). Oscillation frequencies between 70 and 400 Hz are typically used, with an oscillation amplitude between 20 and 300 nm (usually ca. 10%–15% of the meniscus height or, similarly, the probe diameter).

The response of the SECCM probe as it is approached toward a surface illustrates the high precision with which meniscus contact can be controlled and maintained with the surface. Using the data in Figure 19.2e as an example, when the liquid meniscus is in air, the ionic current is constant at a stable finite value. This is relatively low, indicating a very thin meniscus at the end of the pipette. The AC component is nonexistent, as is shown in Figure 19.2e, because the modulation does not appreciably alter the meniscus dimensions when the probe is in air (or other atmosphere). When the liquid meniscus makes contact with the surface, the meniscus size (and shape) changes and this can easily be observed as a sudden increase in the ionic current (i_{DC}) and also the oscillation current amplitude (i_{AC}). This *jump-to-contact* with the surface is a very obvious signal and indicates that the liquid meniscus has made contact with the surface (without contact from the pipette). The contact area of the liquid meniscus is typically the size of the end of the probe (as is shown in Figure 19.2a). However, the chemistry of the surface can have an impact on the contact area, with very hydrophobic surfaces tending to repel aqueous droplets or very hydrophilic surfaces causing leaking of aqueous electrolyte solution over the surface. In general, however, a wide range of surfaces can be investigated with aqueous solutions, as outlined in this contribution, and it may be possible to expand the range of surfaces and solvents by controlling the pressure of the liquid in the pipette.

Once the meniscus is in contact with the surface, the magnitude of the DC and AC depends on the distance between the end of the probe and the surface. As the meniscus is squeezed (the probe is moved closer to the surface), the dimensions of the conductance cell decrease, and so the resistance increases (decreasing i_{DC}) as in SICM.[5,49] The AC value (i_{AC}), however, essentially measures the derivative of i_{DC} with respect to the distance, and so this value increases with decreasing distance and is particularly sensitive to the tip–substrate separation as shown in Figure 19.2e.

19.2.4 POINT MEASUREMENTS

The SECCM meniscus probe can be brought into contact with the surface of interest at a number of discrete points. Since only a tiny fraction of the surface is in contact with the liquid meniscus at any time, this allows many repeat measurements to be made over a surface, which can provide large data sets for robust statistical analysis.

Typically, the SECCM probe is approached toward a surface until the *jump-to-contact* signal is observed (see earlier text), at which point the movement of the probe is halted. The probe is then held stationary, with the liquid meniscus in contact with the surface, for a user-defined period of time. The electrochemical current at a (semi)conducting surface or the change in conductance current between the barrels is measured during the period that the probe is held stationary on the surface. Measurements are made as a function of time, and the potential of the surface can be changed in a user-defined manner to provide voltammetric control (by adjustments of V_1 in Figure 19.2b).

19.2.5 LATERAL SCANNING AND IMAGING

Two-dimensional maps of the surface of interest can be constructed by moving the SECCM probe meniscus laterally over the surface, following the topography and collecting the spatially resolved response. The probe is first approached toward the surface until the liquid meniscus makes contact with it, as described earlier. The probe is then moved laterally across the surface and a feedback loop is used to maintain a user-defined oscillating ion current magnitude, which corresponds to a constant probe–surface distance. Two different methods to maintain a constant set point have been reported: a distance-based method[49,92,93] and a time-based method.[91,97] In the distance-based method, the height of the probe is adjusted based on the lateral distance the probe has moved (i.e., the probe is moved laterally a set distance, the movement is paused, the oscillation amplitude is measured, and the height is adjusted to maintain the set point). In the time-based method, the probe is moved laterally at a constant speed and the height of the probe updated after a set period of time has passed. This allows the probe to be moved constantly and thus reduces the time required to generate a map.

Maps of the surface properties are typically constructed from a series of parallel line scans, although other scan patterns can be employed.[98,99] In a number of reports, the SECCM probe is moved forward and then back over the same line, before being moved laterally to the start of the next line.[91,100] Data are collected during both the forward and reverse movements, and therefore, two maps (a forward and a reverse map) of the surface properties can be constructed from every scan. This allows the consistency of the data to be checked (with the electrode surface at the same potential during both scans), or protocols can be used where a different potential is applied during the forward and reverse scans.

During a scan, the ionic current, magnitude (and phase) of the oscillating ionic current, substrate current, and the position are recorded. This allows 4 (or 5) maps for each direction (forward and reverse) to be generated. This wealth of information can then be used to identify variations in surface activity and local topography.

The ionic current map reveals changes in the conductance between the QRCEs, and is typically relatively featureless. However, changes in the shape or size of the liquid meniscus can affect the ion current signal, and so this map can be used to determine if the meniscus is stable during the scan. Additionally, if the surface process of interest leads to changes in the ionic composition of the electrolyte (interfacial ion flux processes or electrochemical reactions leading to a change in the ionic strength), this can be observed in this map.[49]

The magnitude of the oscillating ion current is used as a feedback signal, and so ideally maps of this signal should be featureless. However, the feedback circuit has a finite response time and therefore, maps of this signal can reveal the edges of topographical features (in the same way as the error signal in AFM).[101] The laterally resolved map of the z position of the probe reveals the apparent topography of the sample. The maps of the surface current (for (semi)conducting substrates) identify the local surface activity. The analysis and interpretation of the surface current are dependent on the experiment, but using appropriate finite element method (FEM) modeling techniques (see the following text), the surface current response can be analyzed quantitatively and used to extract surface kinetic information.

19.2.6 SURFACE MODIFICATION

SECCM has been typically used to map the local electrochemical response and can also be used to modify a surface. Active species can be dissolved or suspended in the electrolyte solution and the contact between the liquid meniscus and the surface used to deliver the active species to a specific location on the surface.[97–99,102] The SECCM probe can be used to generate spots on a surface by approaching the probe to discrete points on the surface[97,98] or generate patterns by moving the probe laterally across the surface while using a feedback loop to follow the topography of the surface and maintain the contact between the liquid meniscus and the surface.[98,99,102] Applications of these methods are discussed herein.

19.3 THEORY AND SIMULATIONS

An important feature of SECCM for the characterization of interfaces is that qualitative and quantitative information can be extracted from experimental data, making the technique particularly beneficial for understanding and interpreting (electro)chemical processes at the nanoscale. This requires a careful analysis of electrochemical and mass-transport phenomena occurring at the substrate as well as within the barrels of the double-barrel pipette, which necessitates the solution of continuum flux equations describing processes of diffusion, migration in the electric field, and convective motion, given by the Nernst–Planck relation:

$$\frac{\partial c_i}{\partial t} + \nabla(-D_i\nabla c_i - \frac{F}{RT}z_iD_ic_i\nabla\phi) + \mathbf{u}\nabla c_i = R_i. \tag{19.1}$$

Here, c_i, D_i, and z_i denote the concentration, diffusion coefficient, and the charge number of species i, while F, R, T, $\mathbf{u}$, ϕ and R_i specify the Faraday and gas constants, temperature, fluid velocity, the value of electrical potential, and the rate of any homogeneous reactions involving species i, respectively. Under certain assumptions, the resolution of the Nernst–Planck equation can be achieved analytically for a freely suspended pipette in a bulk solution; however, due to the overall complexity of the homogeneous and heterogeneous (electro)chemical processes for the typical SECCM configuration, the calculation of mass-transport and chemical phenomena requires access to numerical techniques, such as the FEM. The latter is a powerful methodology for numerical modeling of phenomena involving the coupling of different physical phenomena in 1D, 2D, or 3D environments. The simulation is exceptionally helpful for understanding and interpreting SECCM experimental data, for designing experiments, optimizing experimental conditions and parameters, and for the exploration of any method limitations.

In order to access the response of the SECCM probe and to assess the capabilities of the technique, an FEM model should mimic the geometrical arrangement of the double-barrel pipette, the meniscus, and the substrate. As shown in Figure 19.3a and b, the double-barrel pipette is reasonably approximated to a capped, truncated cone, where the cap represents the opening, and the meniscus is typically represented with a cylinder; variations of this geometry have been explored in some detail.[95] The key geometrical parameters to consider when describing the ion conductance across the meniscus between the QRCEs, and mass transport to the substrate, are the internal radius of the pipette (r_p), the semiangle (θ), the thickness of the central septum (t_w), the tip–substrate separation (or, equivalently, the meniscus height m_h), and the effective bias in the simulated domain (V_2). Experimentally, r_p, θ, and t_w can be measured accurately using FE-SEM and then introduced into a model.[91,95,100,103] Once the simulation domain is set up, that is, the geometrical arrangement of the probe and its position with respect to the substrate are given, one can formulate the set of Nernst–Planck equations (Equation 19.1) for chemical species, complemented with boundary conditions.

With the aim of reducing computational efforts, a few assumptions can be introduced into the SECCM model. For the sake of simplicity, the convective contribution to mass transport, $\mathbf{u}$, can be ignored since the contribution to the overall transport of chemical species caused by electro-osmotic fluid flow is relatively small and the liquid movement due to pipette oscillations and lateral translation (during imaging) is negligible compared to diffusion and migration.[95] Thus, in the absence of any homogeneous reactions of chemical species, the Nernst–Planck equation reads

$$\frac{\partial c_i}{\partial t} + \nabla\left(-D_i\nabla c_i - \frac{F}{RT}z_iD_ic_i\nabla\phi\right) = 0. \tag{19.2}$$

Contributions from double layer effects to the distribution of ionic species and electrical potential are also typically disregarded in the SECCM model as the thickness of the Gouy–Chapman layer in the presence of relatively high electrolyte concentrations is very small compared to the simulation domain (meniscus and probe) size. This allows the application of the electroneutrality condition throughout the simulation domain:

$$\sum_i z_ic_i = 0. \tag{19.3}$$

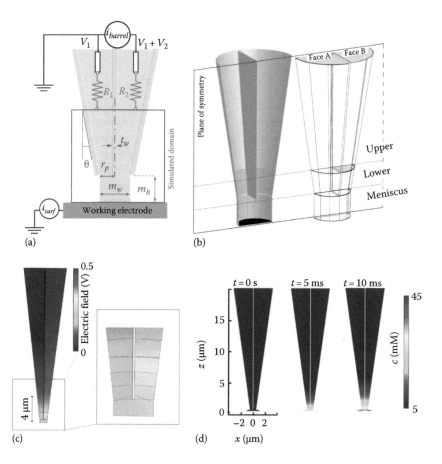

FIGURE 19.3 FEM simulation of SECCM. (a) A schematic of the SECCM setup showing the key geometric dimensions and electronic circuits. (b) Simulation domain. (c) Simulation of the electric field within a typical pipette. (d) Concentration of NaCl within a dual-barrel conductance micropipette at different dissolution times. (Adapted from Snowden, M.E. et al., *Anal. Chem.*, 84, 2483, 2012; Kinnear, S.L. et al., *Langmuir*, 29, 15565, 2014.)

For low electrolyte concentrations and small tips, the charge at the pipette wall and substrate will become important, and these issues can be addressed theoretically.[104,105] Therefore, the equation system, including a set of Nernst–Planck equations (Equation 19.2) and assuming the absence of free charges (Equation 19.3), has to be solved. In general, the calculation is performed over the spatial coordinates and time variable; however, in certain cases, a steady-state formulation may represent the electrical and electrochemical properties of the SECCM experiment in sufficient detail.

19.3.1 Steady-State SECCM Model

A steady-state formulation of the numerical problem considers a flux conservation equation:

$$\nabla\left(-D_i\nabla c_i - \frac{F}{RT}z_i D_i c_i \nabla\phi\right) = 0. \tag{19.4}$$

In the absence of (electro)chemical reactions at the substrate, the solution of a set of equations (Equation 19.4) gives the access to electrical properties of the pipette and the liquid droplet (meniscus) between the probe and the substrate. This is essential for modeling the overall mass-transport/electrical resistance characteristics, which is of primary importance for calculation of the i_{DC} and i_{AC} ionic current components flowing between the two QRCEs in the pipette barrels. In fact, the probe behaves as an ohmic resistor, composed of the internal resistance of the electrolyte in the pipette and the resistance in the liquid meniscus connecting the two barrels.[95] The former is determined primarily by the geometrical parameters of the pipette (i.e., tip radius and the half-cone angle) because the electric field is localized at the most resistive region of the pipette, that is, at the end of the tip (see Figure 19.3c). The resistance of the meniscus is determined by its height and shape, allowing estimation of the probe–substrate separation distance from both the i_{DC} and i_{AC} currents with the high degree of accuracy,[95,100] as well as the prediction of the probe behavior under different experimental conditions.

Most importantly, FEM computations can be used for quantitative characterization of reactivity at heterogeneous substrates. The key advantage of modeling in this case is the possibility to extract kinetic information, particularly of heterogeneous electron transfer (HET), occurring at microscopic and/or nanoscopic interfaces. This task requires thorough characterization of the probe, as well as careful examination of meniscus parameters, followed by the simulation of heterogeneous reactions of redox-active species at reactive sites of the substrate and ionic mass transport within the SECCM tip. For an elementary one-electron process,

$$Red - e^- \underset{k_b}{\overset{k_f}{\rightleftarrows}} Ox,$$

the variation of interfacial kinetics can be encoded in boundary conditions at the reactive areas of the working electrode as a boundary condition specifying the flux, J, of redox species through the interface

$$J = k_f c_{Red} - k_b c_{Ox} \tag{19.5}$$

while specifying rate constants for the forward, k_f, and backward, k_b, reactions within the frameworks of an electrode kinetics model, such as the Butler–Volmer formulation.[91,95,100,103] With this approach, working curves defining the change in the working electrode current as a function of the standard rate constant, k_0, for particular conditions (applied potential, SECCM parameters) are calculated to enable the quantitative analysis of experimental data. For instance, Güell et al.[100] investigated the intrinsic reactivity of a complex electrode, comprising a random network of interconnected SWNTs on an Si/SiO$_2$ substrate. The SECCM currents measured as a function of a probe position at the formal potential of two different redox couples were analyzed to show that the majority of SWNTs exhibited fairly uniform reactivity. The FEM analysis allowed an estimation of the values of kinetic rate constants for ruthenium hexamine Ru(NH$_3$)$_6^{2+/3+}$ and trimethyl (ferrocenylmethyl) ammonium (FcTMA$^{+/2+}$) using different activity models, which indicated that electron transfer (ET) occurred predominantly at the sidewalls of the SWNTs for these outer-sphere redox couples. This approach has also been used to investigate ET at metallic and semiconducting SWNTs at different electrode potentials.[91]

19.3.2 Transient SECCM Model

Despite the robustness and computational simplicity of a steady-state formulation, there is often a demand to explore the dynamic evolution of (electro)chemical systems. In such a case, the numerical problem comprising the Nernst–Planck equation is solved in transient form (Equation 19.1), that is, incorporating the time variable in the equation system. For the simulation of voltammetry, two approaches can be used to treat reactions at the substrate working electrode surface, carefully balancing a trade-off between accuracy and computational resources: the dynamic field and static field approaches. In the dynamic field approach, the electric field that controls the migration of ions between the barrels of the pipette is recalculated for the changing conductivity within the SECCM probe arising from the working electrode reaction. In the static field approach, the potential field is calculated for the initial conditions (with no reaction occurring at the substrate) and kept constant when a substrate reaction is introduced, that is, neglecting any conductivity changes caused by dynamically changing ionic species concentrations. For the sake of simplicity, the latter is often acceptable when high electrolyte concentrations (with respect to the redox species) are used. However, special care is needed when simulating SECCM experiments with low supporting electrolyte concentrations.

Another illustration of time-dependent FEM analysis is the use of SECCM (double-barrel ion conductance probe) to investigate ionic crystal dissolution.[96] A transient Fick's diffusion equation,

$$\frac{\partial c_i}{\partial t} = D_i \nabla c_i, \tag{19.6}$$

was used to calculate ionic concentration profiles due to crystal dissolution from the substrate (see Section 19.4.4). The flux of ions from the substrate was defined, assuming dissolution to be first-order with respect to the degree of undersaturation at the crystal-solution interface

$$J_i = k_{diss}(c_{sat} - c_i), \tag{19.7}$$

where k_{diss} and c_{sat} are the dissolution rate constant and the saturation concentration, respectively. The calculation of the electric field and the total current flowing between the QRCEs in the two barrels was performed at each solver time step (solution of Equation 19.6), by solving the current conservation equation

$$\nabla \mathbf{J} = 0 \tag{19.8}$$

under consideration of a differential form of Ohm's law

$$\mathbf{J} = \sigma \nabla \phi, \tag{19.9}$$

where the value of ionic conductivity σ is determined by the concentration of ionic species.

The developed model allowed quantitative analysis of experimental NaCl crystal dissolution and determination of interfacial dissolution kinetics. Numerical modeling also allowed visualization of concentration profiles. For example, Figure 19.3d shows the NaCl concentrations due to dissolution (first-order dissolution rate constant of $7.5 \cdot 10^{-5}$ cm·s^{-1}) into a pipette containing 5 mM NaCl over the first 10 ms after landing the SECCM pipette meniscus on an NaCl crystal surface.

Numerical resolution of mass-transport equations, along with Laplace formulation of the electrical properties of the SECCM system, is an efficient and reliable strategy for the characterization of SECCM mass transport, technique optimization, and the analysis of experimental data with the aim of extracting quantitative information. This is an indispensable tool not only for detailed examination of experimental data but also for developing novel experiments and exploiting new phenomena that can be examined with SECCM.

19.4 VISUALIZING AND QUANTIFYING HETEROGENEOUS ELECTRON TRANSFER AT ELECTRODE SURFACES

In this section, illustrative example applications of SECCM are presented. These examples are divided into three main sections, reflecting the scale of the measurement and the application: nanoscale mapping of macroscopic samples, nanoscale mapping of nanoscopic samples, and finally, nanoscale modification of substrates. To conclude, the further use of SECCM, which expands the scope of the technique beyond surface electrochemistry, is outlined. This provides an indication of future applications of SECCM-related techniques in other areas.

19.4.1 Nanoscale Mapping of Macroscopic Samples

The capabilities of SECCM are demonstrated most powerfully when it is used in the imaging mode, where a surface of interest is scanned with the SECCM probe meniscus to produce x–y maps of surface reactivity, simultaneously with topography and ion conductance (between the two QRCEs). SECCM images can be complemented with a variety of structural characterization techniques, applied to the same region of the surface, such as FE-SEM, micro-Raman, AFM, and electron backscatter diffraction (EBSD). This type of *multimicroscopy* approach provides a rich and powerful platform for elucidating how local reactivity is influenced by the properties of the surface under investigation.

19.4.1.1 Polycrystalline Platinum

Platinum is a widely used electrode for energy applications.[106,107] The electrochemical activity of platinum is known to be dependent on crystallographic orientation for many reactions, especially those involving electrocatalysis, and the correlation of structure and activity is thus of paramount importance for understanding and optimizing these systems.[108] Although the use of single crystal electrodes is a particularly effective means of correlating structure and activity,[109] the cost and time of manufacturing high-quality single-crystal facets, together with the instability of the electrode structure at high over potentials, make the use of single crystals challenging for certain electrochemical reactions. It is particularly difficult to fabricate and maintain single crystals with high-index facets. Furthermore, single-crystal surfaces do not give access to grain boundaries between facets on a polycrystalline surface, which may be key active sites.[110,111] SECCM provides a *pseudo-single-crystal* approach to the investigation of polycrystalline metal samples if the SECCM probe dimensions are smaller than the individual grains. As shown in Figure 19.4, this is the case for polycrystalline platinum, opening up the possibility of directly correlating electrochemical images with the surface crystallographic orientation obtained with EBSD.[112]

This approach was used to investigate the role of crystallographic orientation of high-index platinum facets on the electrochemical activity of the $Fe^{2+/3+}$ redox couple on polycrystalline platinum foil in two different aqueous media (perchloric acid and sulfuric acid).[93] The electrochemical maps taken at different overpotentials were complemented with EBSD, and the overall activity was compared and verified with single-crystal measurements. Figure 19.4 shows examples of the reactivity patterns across the platinum surface in the cases of the weakly adsorbing perchlorate anion (Figure 19.4b) and the strongly adsorbing sulfate anion (Figure 19.4c). In perchlorate medium, there were clear variations between individual grains, but within a particular grain, more or less uniform activity was observed. Grains with a higher contribution of the (110) orientation exhibited higher activity compared to those with (100) and (111), consistent with single-crystal studies.[93] These variations were found to correlate with the shift of the potential of total zero charge (ptzc) at the same pH (pH = 2) for basal plane platinum.[113] In contrast, in sulfate medium, little variation in activity between different crystallographic orientations was seen, but significantly enhanced activity was evident at the grain boundaries. The enhanced activity at the grain boundaries was estimated to be at least two orders of magnitude higher than in the areas within the grain, revealing that anion adsorption phenomena at grain boundaries may play an important role in electrocatalysis. The uniformity of activity between different facet orientations was again consistent with single-crystal cyclic voltammetry (CV) measurements at basal platinum surfaces.[93]

The triioidode (I_3^-)/iodide (I^-) redox couple in 1-butyl-3-methylimidazolium tetrafluoroborate ([BMIm][BF$_4$]) RTIL,[89] the redox shuttle typically employed in dye-sensitized solar cells (DSSCs),[114] has also been investigated at polycrystalline platinum surfaces. RTILs have generally

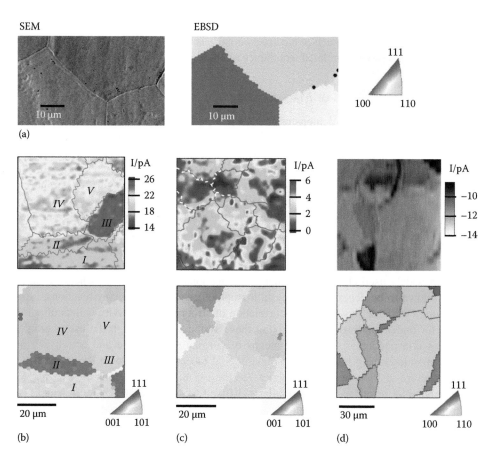

FIGURE 19.4 SECCM studies on polycrystalline platinum. (a) Polycrystalline platinum foil surfaces visualized with FE-SEM and electron backscatter diffraction (EBSD). Oxidation of 1 mM Fe^{2+} to Fe^{3+} in two electrolytes: (b) SECCM image of the activity in 10 mM $HClO_4$ at 0.8 V relative to Pd–H_2 with corresponding EBSD image of the platinum surface and (c) SECCM image of the activity in 10 mM H_2SO_4 at 0.8 V relative to Pd–H_2 with the corresponding EBSD image. Grain boundaries (from EBSD) are marked with either black lines (boundaries at which an enhanced current was observed) or white dotted lines (with no enhanced current) to guide the eye. (Adapted from Aaronson, B.D.B. et al., *J. Am. Chem. Soc.*, 135, 3873, 2013.) (d) SECCM image of activity for the reduction of 10 mM I_3^- in [BMIm][BF_4] at 0.3 V relative to Ag QRCE with the corresponding EBSD image in the following. (Adapted from Aaronson, B.D.B. et al., *Langmuir*, 30, 1915, 2014.)

proven challenging for probe imaging techniques, due to the high viscosity and (for electrochemical probe methods) low (and widely different) diffusion coefficients of the redox species.[115] SECCM has opened up spatially resolved measurements of fundamental electrochemical and electrocatalytic processes in RTILs, which are of considerable (and growing) interest for many important applications in energy technologies, electrosynthesis, and sensing. For the purpose of this study, the SECCM tip and sample were placed in an environmental cell and purged with dry N_2 to assist in the deaeration and drying of the RTIL meniscus.[116] Similar to the structure–activity dependency found for Fe^{2+} oxidation in aqueous perchlorate medium (Figure 19.4b and c), striking correlation between the SECCM images for the reduction of 10 mM I_3^- in [BMIm][BF_4] and the corresponding EBSD was observed (Figure 19.4d), where grains with a higher contribution of the (110) orientation exhibited higher activity for the reduction reaction and those with a higher contribution of (100) planes exhibited the least activity. The strong surface adsorption of the

$[BF_4]^-$ anion to (100) surfaces has been suggested to impede electrocatalysis, as demonstrated in single-crystal studies of other electrochemical processes in imidazolium ionic liquids.[117] In the context of DSSCs, this study reveals that the rate of the I_3^- reduction process at the (polycrystalline) counter electrode (CE) is structure dependent and is strongly influenced by subtle variations in crystallographic orientation.

19.4.1.2 Carbon Materials

SECCM has found particular application in mapping the electrochemical activity of a plethora of carbon materials for which there has been uncertainty and controversy over the nature of the active sites for ET. As highlighted herein, SECCM allows access to key features on surfaces and, through complementary microscopy techniques applied in the same area, enables activity to be linked to local structure or other properties, greatly advancing understanding of structural control of electrochemical activity. A further benefit of SECCM is the possibility of mapping materials with minimal processing or lithography, enabling studies of materials in (close to) pristine form.

19.4.1.2.1 Polycrystalline Boron-Doped Diamond

Polycrystalline boron-doped diamond (pBDD) is inherently heterogeneous due to variations in dopant density across different facets. Figure 19.5a shows an example of an SECCM reactivity map of oxygen-terminated pBDD for a complex multistage reaction, the oxidation of the neurotransmitter serotonin (5-hydroxytryptamine) in aqueous solution.[118] The advantage of a confined electrochemical

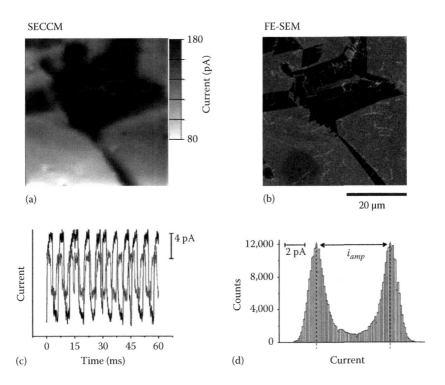

FIGURE 19.5 Polycrystalline boron-doped diamond (pBDD) investigated with SECCM. (a) SECCM map (45 μm×45 μm) of 2 mM serotonin oxidation in 5 mM 4-(2-hydroxyethyl)-1-piperazineethanesulfonic acid (HEPES) and 0.1 M NaCl at +0.65 V vs. Ag/AgCl QRCE on oxygen-terminated pBDD. (b) FE-SEM image of the same area as in (a). (c) Typical capacitance current–time data in high (black) and low (light) doped regions of a pBDD substrate. The triangular wave applied was centered around 0.0 V (vs. substrate at ground) with a peak to peak amplitude of 0.15 V and a scan rate of 30 V s^{-1}. (d) Histograms of the current amplitudes recorded during one typical capacitance measurement on a low-doped facet.

cell as a scanning probe for studying reactions in which the products block the electrode surface was well illustrated in this study, which is notorious for electrode fouling.[119–122] By using SECCM, the effect of electrode blocking was minimized as the reaction product was left behind due to tip movement during imaging and a reactivity map was readily obtained. Moreover, the deposited product served as a marker of the reaction location to enable use of other microscopy techniques. The map in Figure 19.5a identifies patterns of electrochemical activity that strikingly correspond to the local dopant density obtained quantitatively with FE-SEM (in Figure 19.5b, darker zones in the FE-SEM image correspond to higher boron concentrations). Furthermore, no increase in current was observed at grain boundaries or at other hot spots. Similar results (but with more subtle differences) were found for the outer-sphere couple FcTMA$^{+/2+}$, and strikingly, distinct patterns of activity were also found for the simple inner-sphere couple Fe$^{2+/3+}$.[123]

SECCM has also been used to make localized capacitance measurements from which the local density of states (LDOS) in particular grains of pBDD were deduced.[123] For these measurements, a triangular wave form (around 0.0 V and characterized by the scan rate, v, in V s^{-1}) was applied to the QRCEs in the SECCM pipette and the corresponding (capacitive) current–time response was recorded at high and low dopant density areas at the substrate (Figure 19.5c).[118] The amplitude of the current–time response (i_{amp}, defined as the difference in the modal values of the current amplitude histogram shown in Figure 19.5d) reflects the different doping levels of the facets. The measured capacitance (C_{meas}) was extracted as $C_{meas}=i_{amp}/2vA$, where A is the meniscus contact area with the substrate electrode, fully characterized with FE-SEM from an imprint left on the surface. The relationship between C_{meas}, the capacitance of the space charge region (C_{SC}), the Helmholtz capacitance (C_H), and the diffuse layer capacitance (C_{diff}), for carbon materials with a low DOS, is reasonably given by[124]

$$C_{meas}^{-1} = C_H^{-1} + C_{diff}^{-1} + C_{SC}^{-1}. \qquad (19.10)$$

At high electrolyte concentration, C_{diff}^{-1} is negligible and C_H can be estimated reasonably, enabling the determination of C_{SC}. The LDOS at the Fermi level, $D(E_f)$, can then be calculated from[124,125]

$$C_{SC} = \sqrt{e_0 \cdot \varepsilon\varepsilon_0 D(E_f)}, \qquad (19.11)$$

where

e_0 is the electronic charge
ε is the dielectric constant of pBDD
ε_0 is the vacuum permittivity

With this approach, the LDOS in characteristic high- and low-doped grains of pBDD was estimated and correlated with local ET kinetics.[123]

19.4.1.2.2 Graphene

Different types of graphene, such as mechanically exfoliated graphene (Figure 19.6a) and chemical vapor deposition (CVD)–grown graphene (Figure 19.6b), have been examined by SECCM and the response related to information from complementary microscopy techniques. For these studies, graphene was transferred to silicon/silicon oxide substrates, allowing the determination of the number of layers of graphene with optical microscopy.[126] Both types of graphene were employed as working electrode in the one-electron oxidation of the outer-sphere mediator FcTMA$^+$.

CVD graphene was grown on nickel substrates, yielding a heterogeneous, continuous layer of single-layered and multilayered micrometer-sized graphene flakes.[127] The considerable power of SECCM is well illustrated by studies carried out on this type of graphene, as the flakes have

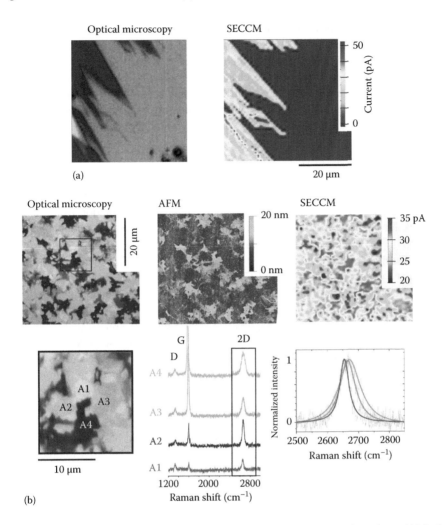

FIGURE 19.6 SECCM interrogation of the electrochemistry of different forms of graphene. (a) Mechanically exfoliated graphene on Si/SiO$_2$ substrates showing an optical image and corresponding SECCM activity map obtained at the formal potential of FcTMA$^{+/2+}$ (oxidation of 2 mM FcTMA$^+$ in 30 mM KCl). (b) CVD-grown graphene showing characterization of the same region with optical microscopy, AFM, and SECCM acquired at the formal potential of FcTMA$^{+/2+}$ (oxidation of 2 mM FcTMA$^+$ in 30 mM KCl). Complementary micro-Raman (633 nm laser) single-point measurements of four regions (A1–A4) of a zoomed-in area in the panel under the optical microscopy image were taken to enable graphene structural characterization in greater detail. The most prominent features in the spectra are the so-called D, G, and 2D bands. The FWHM of the 2D band can be used to determine the number of graphene layers. (Adapted from Güell, A.G. et al., *J. Am. Chem. Soc.*, 134, 7258, 2012; Ebejer, N. et al., *Annu. Rev. Anal. Chem.*, 6, 329, 2013.)

random orientation, shapes, and distribution, which would make their individual interrogation extremely challenging with other technique. The size of the double-barrel pipettes employed was fine-tuned, based on the average size of the flakes (see optical image in Figure 19.6b), so that single flakes could be investigated. The resulting maps were then complemented with AFM, optical microscopy, and Raman spectroscopy to reveal the correlation between structure and electrochemical activity. In Figure 19.6b, a typical electrochemical map for FcTMA$^+$ oxidation obtained via SECCM is correlated to a corresponding AFM image, an optical micrograph, and micro-Raman spectra of the same scanned area.[103] The three primary Raman peaks (D, G, and 2D)[128] are used to assess the number of layers, doping, functionalization, lattice defects, and physical damage of graphene.[128] The Raman spectra for four different flakes confirmed that the darker regions in the optical microscopy images corresponded to thicker graphene flakes. From these data sets, a strong positive correlation between the electrochemical activity of graphene and the number of graphene layers was revealed, where multilayered flakes exhibited the highest electrochemical activity. These findings have important implications for the design and optimization of new graphene-based technologies, particularly for electrochemical applications.

19.4.1.2.3 Highly Oriented Pyrolytic Graphite

HOPG is of intrinsic interest in its own right,[129] and as a comparison to graphene. By studying the HOPG basal surface, with a step-spacing significantly larger than the pipette diameter (Figure 19.7), the HOPG basal plane could be investigated electrochemically, in isolation from the response on the step edges, for the first time.[99,130,131] Figure 19.7 shows SECCM maps for the one-electron reduction of $Ru(NH_3)_6^{3+}$ at the reversible half-wave potential. The topographical response of SECCM is sufficient to distinguish between step edges on the HOPG basal surface (Figure 19.7a). Strikingly, surface current (activity) maps (Figure 19.7b) show that the entire surface is electrochemically active. SECCM provides irrefutable evidence that the HOPG basal plane supports fast ET (close to reversible) for two commonly used redox mediators, $Ru(NH_3)_6^{3+/2+}$ and $Fe(CN)_6^{4-/3-}$,[131] and other more complex electrochemical processes,[97,99,102] even under the relatively high mass-transport rates delivered by the SECCM setup (see Section 19.3). Importantly, SECCM provides unambiguous information as to the location of spatially resolved measurements, because maps of topography (Figure 19.7a), electrochemical activity (Figure 19.7b), and conductance (Figure 19.7c) can be correlated. Detailed comparison of the conductance current and surface electrochemical activity reveals extensive activity across the basal surface and a small increase in redox current at steps edges, which appears to correlate with an increase in the conductance current, attributed to a disturbance of the SECCM droplet as it moves from the hydrophobic basal surface over terminated step edges.[99,131] The findings, which are consistent with those from other microscopic and nanoscale techniques,[132–135] oppose previous models derived from conventional macroscopic CV measurements where the basal surface was considered to be completely or largely inert.[129] SECCM is able to test such surface activity models directly and provides a powerful tool for the development of new models. The reasons for the difference between recent studies[99,102,131–135] and earlier research have been assessed and discussed fully elsewhere.[131] In Section 19.4.3, HOPG is used as a substrate for surface modification by SECCM studies, which also throw considerable light on the intrinsic microscopic activity of HOPG.

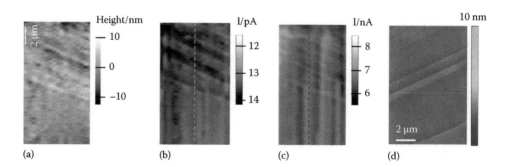

FIGURE 19.7 SECCM co-location maps of $Ru(NH_3)_6^{3+}$ reduction at HOPG. SECCM maps of (a) topography, (b) surface activity, and (c) ion conductance. (d) AFM image showing a typical area of HOPG used for SECCM studies. (Adapted from Patel, A.N. et al., *J. Am. Chem. Soc.*, 134, 20117, 2012.)

19.4.2 Nanoscopic Samples

19.4.2.1 Carbon Nanotube Forests and Networks

The origin of the electrochemistry of carbon nanotubes, and particularly SWNTs, has led to considerable debate and discussion.[136] SECCM has been able to resolve several key issues and identify valid models for SWNT electrochemistry. Early macroscopic studies on rather ill-defined material suggested that electrochemistry, even for outer-sphere redox couples, only occurred at open ends of SWNTs, with the sidewall being inert, a model which was widely accepted and adopted in the field despite studies on individual SWNTs,[87,91,137,138] and well-defined 2D networks of SWNTs showing high sidewall activity.[100,139] Carbon nanotube forests (vertically aligned SWNTs), which are particularly interesting for various electrochemical applications, were also interpreted as showing HET occurring more readily at the open ends of SWNTs than on the sidewalls.[140–142] In the case of carbon nanotube forests, the use of SECCM to confine electrochemical measurements to small regions of a sample allowed the independent investigation of SWNT sidewalls and SWNT ends, avoiding the need for any encapsulation of the electrode material.[143] These studies accessed SWNTs in their pristine state. Thus, Miller et al.[144] showed that SWNTs with closed ends, as well as SWNT sidewalls, promoted fast ET for outer-sphere redox processes, without the requirement of any activation or processing of the nanotubes. The same team has built on this work to further study SWNTs as discussed herein.

Since the SECCM feedback mechanism for the tip–substrate separation does not rely on the electrical nature of the substrate, a further advantage of SECCM is that it enables the imaging of substrates that are (partially) insulating. This feature allowed the study of 2D networks of SWNTs grown by CVD on Si/SiO_2 (Figure 19.8a). In this study, the network of electrically connected SWNTs constituted only 0.3% of an otherwise nonelectrochemically active surface. The SECCM double-barrel pipettes were fine-tuned to a size smaller than the typical spacing between nanotubes (~250 nm),[100] allowing for the acquisition of maps of electrochemical activity with high spatial resolution and for the extraction of information from individual nanotubes. Outer-sphere redox probes such as $FcTMA^{+/2+}$ and $Ru(NH_3)_6^{3+/2+}$ were employed to study the electrochemical activity of the SWNTs. Electrochemical maps (Figure 19.8b and c) revealed that the nanotubes had substantial electrochemical activity along the sidewalls. This finding, in addition to the lack of current enhancement at hot spots across the surface (e.g., at nanotube ends), suggests a model for SWNT electrochemical activity, in which the sidewalls are active components, at least for outer-sphere redox processes.

To further confirm the findings of SECCM imaging studies and to eliminate the possibility that only point defects in the SWNT sidewalls are responsible for HET, scan profiles across individual SWNTs were analyzed, highlighting a further powerful feature of SECCM for kinetic mapping.[100] A gradual increase (and then decrease) in the working electrode current is measured as the SECCM meniscus moves from the inert SiO_2 surface over a nanotube (aligned perpendicular to the scan direction). This is due to a gradual increase (and then decrease) in the area of the SWNT being contacted by the meniscus. In a line scan, after initial contact of the meniscus with the sidewall of the SWNT, the electrochemical contact area of the SWNT increases in small increments (typically in 5 nm steps in the data of Figure 19.8c), exposing a new area of SWNT in each step that is essentially of the same electrochemical activity, as validated by FEM model simulations.[100] FEM modeling allowed scan profiles to be simulated for the case of a fully active SWNT, together with the cases where SWNT activity was determined by either one- or three-point defects (equally spaced) in the SECCM area on an otherwise inactive SWNT, representing a typical defect density and maximum defect density on this type of SWNT.[145] For the cases in which only point defects were considered to be electrochemically active (even when unrealistically high standard ET rate constants were assigned, $k_0 > 10^3$ cm s^{-1}, to the point defects of 1 nm size), the experimentally measured current could not be matched to the FEM simulations, which predicted a much lower current in these situations. This work demonstrated SECCM line profile mapping as a powerful means of distinguishing

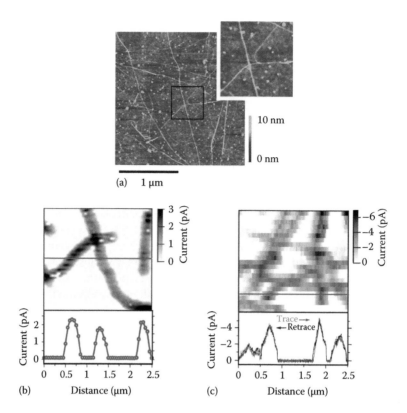

FIGURE 19.8 Electrochemistry of SWNT network electrodes. (a) AFM image of an area of a typical network. SECCM electrochemical activity maps of: (b) FcTMA$^+$ oxidation and (c) $Ru(NH_3)_6^{3+}$ reduction at the corresponding formal potentials of these redox couples. (Adapted from Güell, A.G. et al., *Proc. Natl. Acad. Sci. USA*, 109, 11487, 2012.)

between different models for characteristic activities in nanoscale materials, an aspect which we consider further in Section 19.4.2.2.

19.4.2.2 Platform for the Investigation of Individual SWNTs

The electrochemical response of SWNTs can be affected by the type of nanotube (semiconducting or metallic) and, with reference to Section 19.4.2.1, overlapping (or interconnection) of SWNTs in a network (or other assembly), or from the possibility that residues of the metallic catalyst are left on the surface from the growth stage. In order to fully distinguish between these different possible contributions to the electrochemical response, a versatile experimental platform has been developed, which enables the electrochemical study of truly isolated SWNTs and the utilization of multiple characterization measurements on the same SWNT.[91] Moreover, the combination of the nanoscale meniscus droplet and nanoscale SWNT dimensions results in extremely high mass transport to and from the SWNT electrode surface, allowing access to extremely high reaction kinetics (as demonstrated in Section 19.4.2.1).

Figure 19.9 shows schematics of the platform and an overview of the experimental capabilities. Briefly, aligned pristine SWNTs are grown on Si/SiO$_2$ substrates via catalytic chemical vapor deposition (cCVD)[146] and a palladium contact terminal evaporated at one end of the SWNTs (Figure 19.9a). To locate a particular SWNT and allow thorough investigation with a range of complementary high-resolution techniques, including SECCM, the SWNTs are marked by means of silver electrodeposition. This marking delineates a portion of individual SWNT to enable visualization with optical microscopy. At this point, a truly isolated SWNT can then be electrochemically interrogated

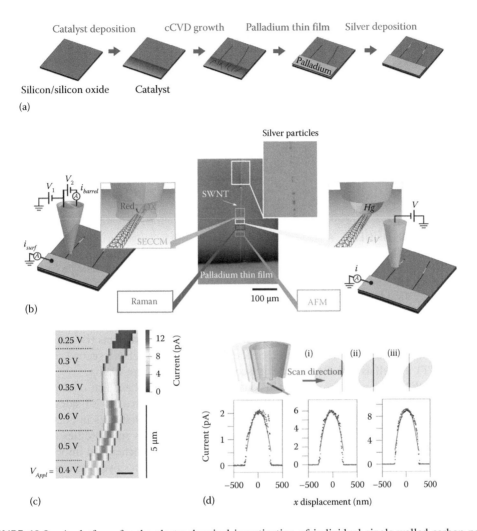

FIGURE 19.9 A platform for the electrochemical investigation of individual single-walled carbon nanotubes (a) Sequence of steps for sample fabrication. Flow-aligned SWNTs are grown on a silicon/silicon oxide wafer by cCVD, depositing the iron catalyst only on one portion of the sample, so that SWNTs grow across the surface extending into the region that is catalyst free. A macroscopic electrical contact to the SWNTs is achieved by thermal evaporation of a palladium thin film in the region of the catalyst layer. One end of each SWNT is decorated by Ag particles via electrodeposition to make their location visible by optical microscopy. (b) SEM image of a final device, showing an as-grown SWNT, delineated by a macroscopic Pd thin film contact at one end and silver decoration at the other (inset optical microscopy image of an ensemble of silver particles decorating an SWNT). The region between is pristine for studies by SECCM and other techniques. Schematics (not to scale) of the experimental setup for SECCM imaging (left-hand side) and for mercury drop contact conductance measurements (right-hand side). Raman spectroscopy and AFM (not shown) can further be deployed to characterize individual SWNT devices. (c) SECCM electrochemical map along an individual metallic SWNT for the oxidation of 2 mM FcTMA$^+$ in 25 mM KCl and phosphate buffer (pH$=7.2$) at a set of distinct applied potentials vs. Ag/AgCl QRCE. Scale bar on the x-axis is 500 nm. Color values are assigned when the signal is detectable at three times above the background current (10 fA; otherwise, the current is assigned to the background (gray). (d) Schematics of the SECCM probe meniscus scanning across an SWNT, and the corresponding FEM simulated profiles, at three different potentials of 0.25 V (left), 0.3 V (center), and 0.35 V (right) vs. Ag/AgCl QRCE. *(Continued)*

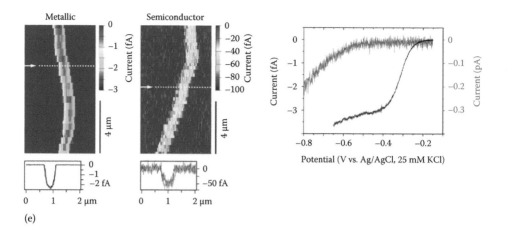

(e)

FIGURE 19.9 (CONTINUED)   A platform for the electrochemical investigation of individual single-walled carbon nanotubes (e) Electrochemical maps of $Ru(NH_3)_6^{3+}$ reduction at individual metallic and semiconducting SWNTs, at -0.35 and -0.65 V, respectively, vs. Ag/AgCl QRCE, together with static-probe cyclic voltammograms (black, left y-axis conducting; red, right y-axis, semiconducting). (Adapted from Güell, A.G. et al., *Nano Lett.*, 14, 220, 2014.)

with SECCM and complemented with electrical conductance measurements (by introducing a mercury droplet at the end of an ultra micro electrode (UME) probe to act as a second contact terminal to the SWNT), as well as with Raman spectroscopy and AFM (Figure 19.9b).

Isolated SWNTs were interrogated with outer-sphere redox probes. The data in Figure 19.9c recorded for FcTMA$^{+/2+}$ at a range of applied potentials demonstrate that SWNT sidewalls are essentially uniformly active for this process, consistent with the SECCM studies in Section 19.4.2.1.[100] An important feature of SECCM is the moveable probe that allows scan profiles to be obtained, enabling the extraction of a vast amount of spatially resolved electrochemical information (Figure 19.9d). Furthermore, comparison of forward and reverse scan profiles over the same feature provides direct information on the stability of the electrode materials. Significantly, a detailed analysis of the scan profiles confers nanoscale spatial resolution to SECCM since the redox current is collected at nanometer intervals (6 nm) along each line scanned by the pipette meniscus, as discussed in Section 19.4.2.1. Knowing the geometry of the pipette tip and footprint of the meniscus (extracted from the profile width), detailed FEM simulations can be carried out. The data in Figure 19.9d are for the oxidation of FcTMA$^{+/2+}$ at different applied potentials, together with a model for uniformly active SWNT sidewalls. The excellent agreement between experimental and simulated scan profiles confirms the highly uniform activity of SWNTs.

The use of complementary techniques on this platform ensures that the wide range of electronic properties of SWNTs that may influence the electrochemistry can be fully accounted for. This is particularly important for carbon nanotubes, considering that 2/3 of SWNTs in an ensemble have a semiconductor behavior.[147] This was highlighted using $Ru(NH_3)_6^{3+/2+}$ as electrochemical mediator, as its potential lies in the charge depletion region of semiconductor SWNTs,[148] and a vast difference in electrochemical activity was seen between metallic and semiconductor SWNTs, as shown in Figure 19.9e.

19.4.2.3   Metal NP Electrocatalysts on SWNTs

The ability to use SECCM on (partly) nonelectrochemically active substrates was further exploited to study the activity of discreet individual platinum nanoparticles (PtNPs).[92] An isolated SWNT on an SiO$_2$ substrate acted as an electrically conducting wire on which single PtNPs of ~100-nm size were electrodeposited with spacing between particles in the micrometer range (Figure 19.10a). SECCM was used to locate and investigate the potential-dependent activity of these individual

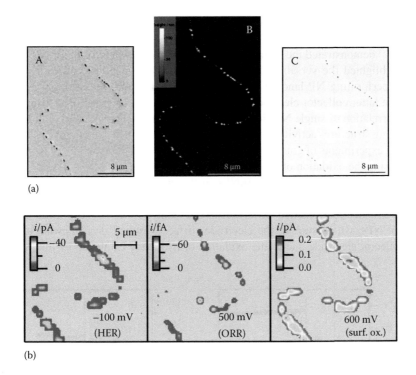

(a)

(b)

FIGURE 19.10 SWNTs as a platform to study individual NPs by SECCM. (a) Optical (A) AFM (B) and (C) SEM images of an SWNT decorated with electrochemically deposited PtNPs. (b) SECCM maps of the electrochemical activity of the same region at different potentials (vs. Pd/H$_2$ QRCE) corresponding to the hydrogen evolution reaction (HER), oxygen reduction reaction (ORR), and Pt transient surface oxidation (surf. ox). (Adapted from Lai, S.C.S. et al., *J. Am. Chem. Soc.*, 133, 10744, 2011.)

PtNPs (Figure 19.10b). These particles were subsequently characterized with AFM and FE-SEM, enabling a direct correlation between the activity of PtNPs, size, and shape. It was demonstrated that subtle variations in NP morphology can lead to dramatic changes in (potential-dependent) activity, which has important implications for the design and optimization of NPs in electrocatalytic applications. This study also highlighted how the SECCM setup can be employed to study events with very low currents (down to ~10 fA during a 40 ms meniscus residence period, which corresponds to only ~2500 molecules in the case of the 4 e$^-$ reduction of O$_2$). These investigations demonstrate the significant potential of SECCM for investigations of electrode reactions at the single NP level by drawing on the excellent spatial and current measurement resolution. Such studies serve to enhance fundamental understanding of electrocatalytic processes at the nanoscale.

19.4.2.4 Single Nanoparticle Collisions on a TEM Grid

To further investigate the structure–activity relationship of NPs, an SECCM platform for single NP electrochemical investigation, complemented with TEM, was developed.[94] In NP collision experiments, NPs in solution collide with a collector electrode, held at a potential where there is no electrode reaction, except when an NP colliding with the collector establishes such a reaction.[149] Electrodes with small areas are required in order to (1) capture a limited number of NP collisions (thus enabling the detection of individual landing events) and (2) minimize the background current on the collector electrode. The use of meniscus-confined pipette-based methods to define the electrode area circumvents the need to prepare UMEs, which is the typical approach to decrease the area of the collector electrode, but limits the range of NPs and electrode materials that can be explored (due to the difficulty of fabricating such electrodes in certain cases).[149–152] An additional advantage of SECCM stems from the level of control over the contact time of the meniscus with the substrate that is possible (<ms control).

Kleijn et al.[94] demonstrated the suitability of SECCM as a new approach for NP landing experiments and highlighted the versatility of confining the electrochemical cell to different types of materials, by performing NP landing experiments on HOPG and, interestingly, a carbon-coated TEM grid. The latter collector electrode, employed for the first time in NP collision experiments, allowed the correlation of single NP electrochemical measurements with TEM studies of the same particle, enabling structure–activity measurements at the single NP level. Figure 19.11 shows examples of landing experiments of gold nanoparticles (AuNPs) on HOPG and on a carbon-coated TEM grid by measuring the oxidation of 2 mM hydrazine in a 50 mM citrate buffer. The SECCM probe was filled with a colloidal solution of AuNPs in the electrolyte solution, and the meniscus was brought into contact with the TEM grid, hooked up as a collector electrode, and held at a potential where the electrochemical reaction occurred on the catalytic NP but not on the electrode itself. The arrival of NPs, which stick at the electrode surface, resulted in a stepwise increase in current at the collector electrode due to the oxidation of hydrazine catalyzed by the (now wired) NP

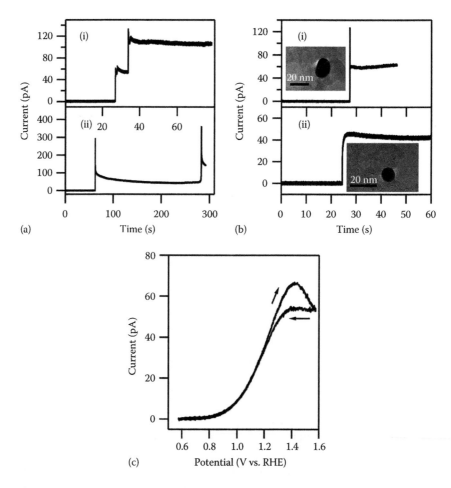

FIGURE 19.11 SECCM in the study of single NP landing events. (a) Landing and sticking of AuNPs on an HOPG substrate held at 1.25 V vs. Pd/H$_2$ QRCE in the presence of 2 mM N$_2$H$_4$. (b) Landing events of two individual AuNPs on a TEM grid electrode, with the same AuNPs imaged by TEM afterward. (c) CV measured for the electro-oxidation of N$_2$H$_4$ on an individual AuNP (as in b(i)). (Adapted from Kleijn, S.E.F. et al., *J. Am. Chem. Soc.*, 134, 18558, 2012.)

(Figure 19.11a). Figure 19.11b presents two separate single AuNP landing experiments that give rise to diffusion-limited current steps of 40 and 60 pA. These two individual AuNPs were subsequently characterized by TEM, and correlation between the NP size and current was found.

One of the key advantages of this SECCM-based approach is the ease with which the nanopipette size can be tuned and that very small pipettes can be produced, thus allowing very small areas of a substrate to be targeted, much smaller than those of a typical UME. This results in much lower background currents, thus allowing the detection of much smaller electrocatalytic currents. Thus, single NP electrocatalysis experiments have been carried out at a range of different (kinetic) potentials, rather than just the mass-transport-limited potential.[152] This also made possible the recording of a full CV of a single AuNP (Figure 19.11c).

19.4.3 Nanoscale Modification of Substrates

In addition to the investigation of surface reactions, SECCM can also be used as a modification and patterning device. Kirkman et al.[97] employed the SECCM platform to locally modify sp^2 carbon surfaces, via diazonium molecules, under electrochemical control.[153] This reaction is attracting considerable attention since, besides being a recognized approach for tailoring the chemical functionality of surfaces,[154] it also provides a potential route to generate a bandgap in graphene by introducing sp^3 defects into the graphene sp^2 carbon lattice.[155,156] Electrochemistry is advantageous compared to other modification methods in both the level of control that can be achieved and the shorter timescales required for modification. A multiple-point SECCM approach was utilized to perform micron-scale modification at spots across an HOPG sample, allowing many measurements with different parameters and facilitating the characterization of the deposits with complementary techniques. Arrays of spots, combining different meniscus hold times in contact with the surface, and different potential values (driving forces) were produced that were subsequently characterized with AFM and Raman spectroscopy, providing information on the diazonium layer thickness and the level of diazonium modification of the HOPG surface (Figure 19.12a and b). As the diazonium modification was driven electrochemically, current–time transients and the electrochemical charge associated with the grafting process for each of the spot depositions could also be examined to gain further insight on film density (Figure 19.12c). Two different growth regimes were identified depending on the potential employed: at high driving forces, a polyaryl multilayer growth process in which the film thickness increased with deposition time; at low driving forces, there was an increase in the concentration of diazonium moieties in a film of more or less constant thickness.

The suitability of meniscus-confined methods to precisely control surface modification with a high degree of reproducibility has been demonstrated, using both single-barrel pipettes[157–160] and double-barreled pipettes.[60,161] However, the positional feedback inherent in the SECCM design allows enhanced control and quantitative analysis of the patterning process. Thus, electrochemical patterning has been used as a means to understand local surface activity, as demonstrated by Patel et al.[99,102] In two separate studies, the residues derived from the electro-oxidation of two highly studied redox-active neurotransmitters, dopamine[99] and epinephrine,[102] were used to mark the location of electrochemical measurements and unambiguously demonstrate the high electrochemical activity of basal plane HOPG. As the SECCM probe moves on to a new location, it leaves the blocking products behind as place markers of the electrochemical reaction. The surface structures at these markers were then analyzed by complementary techniques such as FE-SEM and AFM, enabling structure–electrochemical activity correlations with high resolution (see Figure 19.13). In particular, the SECCM reactive patterning of dopamine oxidation[99] revealed HOPG basal plane to be highly electrochemically active, as confirmed by subsequent macroscopic measurements.[135,162]

The patterning capabilities of SECCM have been further exploited in the creation of 3D structures by McKelvey et al.[98] In contrast to previous meniscus-based fabrication techniques, which tended to use single-channel devices to make contact with a substrate at a limited number of points,[60,161] SECCM allows the fabrication of structures on conducting substrates, across

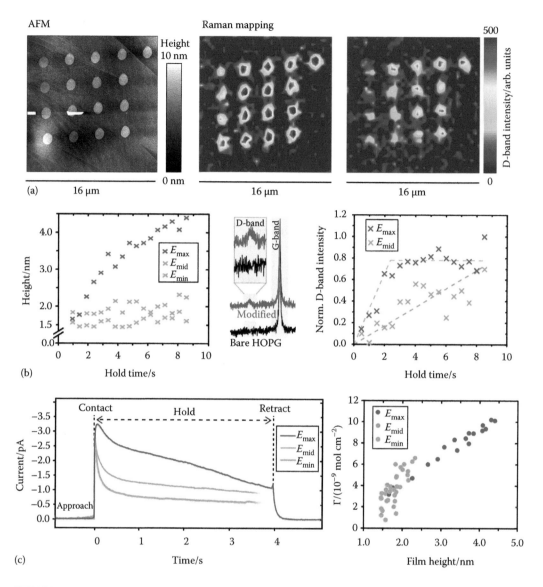

FIGURE 19.12 SECCM diazonium spot modification of HOPG substrates. (a, left) AFM topography image of an array of diazonium modifications on HOPG created via SECCM using various deposition times at a fixed electrochemical potential. (a, right) Raman maps plotted as D-band intensity over the surface of two different arrays created at two different electrochemical potentials (lower driving force on the right). (b) AFM heights and Raman D-band intensity exhibit a dependence on electrochemical potential and deposition (hold) time. (c) Typical current–time transients obtained during spot depositions, one for each characteristic potential. Diazonium coverage (Γ) vs. film height for the different modification potentials. Depositions for low ($E_{min} = 0.4$ V), medium ($E_{mid} = 0.3$ V), and high ($E_{max} = 0.15$ V) driving forces, all potentials vs. Pd/H$_2$ QRCE. (Adapted from Kirkman, P.M. et al., *J. Am. Chem. Soc.*, 136, 36, 2014.)

insulating substrates and also in the form of 3D pillars.[98,157] This was proven by patterning structures of the conducting polymer polyaniline (PANI), constructed by controlling the position at which a liquid meniscus containing aniline made contact with a surface (at a potential that drove the aniline oxidation reaction), or by controlling the potential of the surface so that the reaction was switched between *on* and *off* while moving the probe laterally across the surface with the meniscus always in contact. The continuous recording of the tip position (probe height), substrate

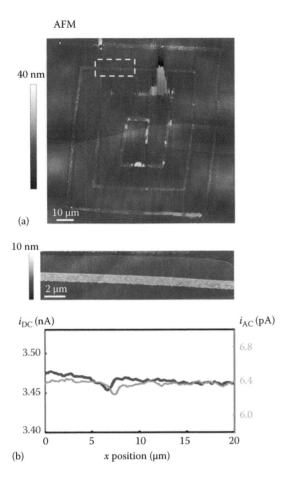

AFM

40 nm

10 μm

(a)

10 nm

2 μm

i_{DC} (nA) i_{AC} (pA)

3.50   6.8

                                                                        6.4
3.45

                                                                        6.0
3.40

        0   5   10   15   20

(b)   x position (μm)

FIGURE 19.13 SECCM reactive patterning: correlation of HOPG topography and activity at the nanoscale. (a) AFM image of an HOPG surface showing an SECCM-deposited pattern produced during the electro-oxidation of dopamine. (b) AFM image (20 μm × 5 μm) of the section of the line pattern marked in white in (a) and an overlay of the corresponding surface electrochemical activity (green) and DC component of the conductance current (blue) recovered along the patterned line, which is entirely the basal surface except for one step at ca. 7 μm (x position). (Adapted from Patel, A.N. et al., *J. Am. Chem. Soc.*, 134, 20246, 2012.)

current, and both DC and AC components of the barrel current during deposition provided exquisite control over electropolymerization, leading to highly consistent patterns. Furthermore, the use of two feedback loops, one to control the contact of the meniscus with the surface and another to control the extent of PANI deposition on the surface, allowed high precision modification. Both potential control and current (galvanostatic) control (Figure 19.14a) of the deposition pattern were demonstrated. SECCM also allowed the construction of multidimensional structures across conducting and insulating substrates, Figure 19.14b. In this case, the probe was scanned from gold, where electrochemical polymerization was initiated, to SiO_2, followed by the growth of a PANI tower by maintaining the AC conductance current feedback response while retracting the SECCM pipette.

19.4.4 BEYOND SURFACE ELECTROCHEMISTRY

As shown in the previous section (19.4.3), an important feature of meniscus-confined pipette-based methods is that the electrochemical cell can be created and detached readily at a specific

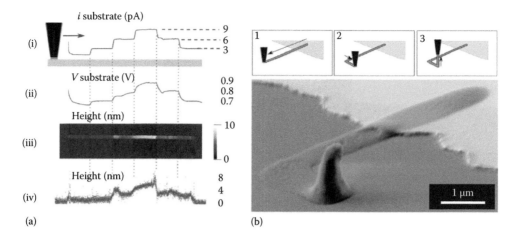

FIGURE 19.14 SECCM surface patterning with polyaniline (PANI). (a) Patterning of PANI on a gold surface with galvanostatic control as a meniscus was moved laterally across the surface. (i) Measured substrate current stepped through values of 1, 3, 6, and 9 pA, and corresponding applied surface potential (ii) to drive the surface current. The AFM image of the resulting pattern (iii) and cross-sectional height (iv) showing control of the extent of PANI deposition. (b) SEM (false color) of a three-dimensional PANI structure created across a conducting gold and nonconducting SiO_2 surface. The probe movement steps are shown in the top figure schematics. (Adapted from McKelvey, K. et al., *Chem. Commun.*, 49, 2986, 2013.)

location on the surface of interest at an ms timescale by simply moving the pipette toward or away from the substrate. This capability was used in the study of ionic crystal growth and dissolution kinetics,[49,96] extending SECCM methodology beyond electrochemistry. Dissolution experiments of 30 ms duration or less (precisely controlled) were carried out on NaCl single crystals in aqueous solution. In contrast, due to the high solubility of NaCl and relatively large dissolution fluxes, conventional studies requiring immersion of the solid in a bulk solvent are somewhat challenging.

The double-barrel SECCM platform allowed conductance measurements to be confined to micro-sized areas. The experimental procedure was to approach the pipette toward the crystal surface at a low speed until the meniscus at the end of the pipette wetted the substrate surface, without the pipette itself making contact. This event was readily detected as an abrupt change in conductance current (*jump-to-contact*) (Figure 19.15a), used to automatically stop the motion of the pipette and maintain meniscus contact with the crystal surface for a predetermined etch time. Dissolution of the crystal was promoted by the use of a greatly undersaturated solution (5 mM NaCl) in the pipette. The resulting ion dissolution flux from the crystal surface into the solution in the meniscus and pipette decreased the resistance between the two QRCEs and hence increased the current on a longer timescale (Figure 19.15b).

Once the set dissolution time had elapsed, the tip was withdrawn at a high speed to ensure that the contact between meniscus and substrate was broken abruptly, resulting in a rapid ionic conductance current decay as the meniscus shrank in size (Figure 19.15b). This variation of the ionic conductance with respect to time, together with an accurate measurement of the pipette shape and dimensions (by FE-SEM), and footprint of the meniscus on the surface (by AFM), permitted quantitative analysis of experimental data by FEM modeling.[95,96]

As a localized microscale surface approach, it is possible to perform many repeat measurements over one macroscopic sample by laterally repositioning the tip over new sections of the surface. The data in Figure 19.15 highlight the high reproducibility of meniscus landing and localized measurements attainable with SECCM techniques.[94,96,97]

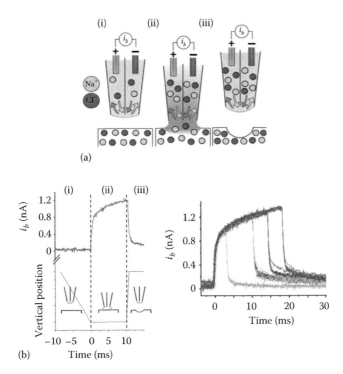

FIGURE 19.15 Crystal dissolution studies with SECCM. (a) Schematic of the experimental procedure for microscale dissolution: (i) pipette held in air; (ii) dissolution occurs upon contact of the meniscus with the crystal for a defined short period; (iii) tip is retracted, breaking meniscus contact with the surface. (b, left) Plot of barrel conductance current–time above a schematic of the corresponding vertical pipette position during a dissolution experiment, as described in (a). (b, right) Current–time plots of multiple repeat measurements of dissolution pits at different spots on an NaCl crystal, for different meniscus contact times (as short as 3 ms), showing consistency in the transients. (Adapted from Kinnear, S.L. et al., *Langmuir*, 29, 15565, 2014.)

19.5 CONCLUSIONS AND OUTLOOK

With the advent of SECCM, scanning droplet cell methodology has undergone a step change in capability, with multidimensional imaging, and the modification of surfaces and interfaces at the microscale and nanoscale, readily achievable for a wide variety of processes.

SECCM produces an electrochemical cell on a surface of interest, via simple (and well-defined) meniscus contact from a double-barrel nanopipette containing electrolyte solution and QRCEs. A bias between the two electrodes produces an ion-migration current that is sensitive to the meniscus dimensions and informs on meniscus contact with the surface of interest, particularly when a small amplitude positional modulation is applied to the pipette in the direction normal to the surface and an AC current is generated that can be detected with a lock-in amplifier. This can be used as a set point in imaging with a scanned pipette, so that the meniscus traces the contours of the surface, producing topographical maps. Simultaneously, the potentials of the QRCEs with respect to a (semi)conducting substrate can be used to drive electrochemical processes at the substrate, so that the SECCM configuration represents a dynamic electrochemical (amperometric–voltammetric) cell. A broad family of dynamic electrochemical techniques can thus be applied at the micro-/nanoscale with SECCM and, furthermore, direct electrochemical images of surfaces and interfaces can be obtained.

Alongside surface electrochemical and topographical maps, SECCM also records the DC conductive current, AC set point (amplitude), and AC phase, and each of these data sets can provide complementary information on surface processes and inform on the quality of SECCM images and the stability of the meniscus. The DC conductance current is particularly informative if the surface process of interest involves interfacial ion transfer or changes in the local ionic strength in the meniscus. Such effects are naturally involved in all electrochemical processes, but other processes such as adsorption, desorption, dissolution, and growth of insulating surfaces are also open to study as highlighted in the latter part of this chapter. We envisage myriad possibilities of SECCM-related techniques in materials science, geoscience, and life sciences.

SECCM has found particular application in the study of carbon electrode materials, enabling microscopic models for activity, derived from macroscopic electrochemical measurements, to be rigorously tested at the microscale and nanoscale. Notably, SECCM studies have provided major new and unambiguous findings on the activity of basal plane graphite, conducting diamond, SWNTs, and graphene. Importantly, SECCM data and maps can often be combined with complementary *multimicroscopy* information obtained in the same area of the surface of interest, providing clear and definitive views of local electrochemical activity. This new information is not only important in terms of fundamental understanding of structural and electronic controls of HET, but is also hugely valuable in selecting and tuning the most appropriate electrode materials for practical electrocatalysis and electroanalysis.

Noble metal electrocatalysts are also amenable to study with SECCM, spanning localized measurements on polycrystalline materials (*pseudo-single-crystal* approach) to individual nanoparticles. At both scales, the electrochemical activity can be correlated with the structure of the catalyst (e.g., complementing SECCM maps with the use of EBSD to determine the crystallographic orientation of polycrystalline materials, and visualizing individual NPs by SEM or TEM that have been studied electrochemically). Thus, SECCM opens up new avenues in the study of electrocatalysis which is of huge importance from a fundamental standpoint and with a view to discovering new electrocatalytic materials and improving electrocatalyst usage.

The double-barrel pipette deployed in SECCM can further be used to deliver material locally, whether it be a preformed NP or a precursor for an electrochemical process (e.g., electropolymerization or electrochemical grafting). The high level of control over the position of the meniscus is particularly powerful for patterning and modifying substrates, and we anticipate much further use of SECCM in this direction, especially as meniscus contact can be made with excellent time control (sub-ms resolution).

Looking forward, although SECCM methodology has already proven to be extremely powerful, there are several options to further improve its capability. Enhancement in imaging speed, and the incorporation of complementary sensors into the pipette probe (creating a *lab on a tip*) should be possible, and work in these directions is underway in our laboratory. Furthermore, it should be possible to shrink the double-barrel pipette design to the sub-100 nm scale to open up the possibility of solving new problems in electrochemical science and technology, and allow fundamental aspects of electrochemical interfaces to be probed, such as local charge, potential, and double layer properties.

ACKNOWLEDGMENTS

We are grateful to the European Research Council for generously supporting the development of SECCM in our group through the Advanced Investigator Grant ERC-2009-AdG 247143, *QUANTIF*. A large number of current and past colleagues at the University of Warwick have contributed to the development and application of SECCM, as described in this chapter, and we particularly thank Neil Ebejer, Stanley Lai, Anisha Patel, Sophie Kinnear, Mike Snowden, Massimo Peruffo, Paul Kirkman, Chang-Hui Chen, Josh Byers, Petr Dudin, Yangrae Kim, Kate Meadows, Hollie Patten, Anatolii Cuharuc, Minkyung Kang, and Guohui Zhang for their expert contributions. Finally, our SECCM program has been greatly enhanced by significant contributions from Alex Colburn on current measurement instrumentation development and from Lee Butcher and Marcus Grant who fabricated various mechanical components for our instruments.

REFERENCES

1. Bhushan, B.; Fuchs, H. 2007. *Applied Scanning Probe Methods VII: Biomimetics and Industrial Applications*; Springer, Berlin, Germany.
2. Ebejer, N.; Güell, A. G.; Lai, S. C. S.; McKelvey, K.; Snowden, M. E.; Unwin, P. R. 2013. Scanning electrochemical cell microscopy: A versatile technique for nanoscale electrochemistry and functional imaging. *Annu. Rev. Anal. Chem.*, *6*, 329–351.
3. Kranz, C. 2014. Recent advancements in nanoelectrodes and nanopipettes used in combined scanning electrochemical microscopy techniques. *Analyst*, *139*, 336–352.
4. Bandarenka, A. S.; Ventosa, E.; Maljusch, A.; Masa, J.; Schuhmann, W. 2014. Techniques and methodologies in modern electrocatalysis: Evaluation of activity, selectivity and stability of catalytic materials. *Analyst*, *139*, 1274–1291.
5. Chen, C.-C.; Zhou, Y.; Baker, L. A. 2012. Scanning ion conductance microscopy. *Annu. Rev. Anal. Chem.*, *5*, 207–228.
6. Engstrom, R. C.; Weber, M.; Wunder, D. J.; Burgess, R.; Winquist, S. 1986. Measurements within the diffusion layer using a microelectrode probe. *Anal. Chem.*, *58*, 844–848.
7. Bard, A. J.; Fan, F. R. F.; Kwak, J.; Lev, O. 1989. Scanning electrochemical microscopy. Introduction and principles. *Anal. Chem.*, *61*, 132–138.
8. Bard, A. J.; Fan, F. R. F.; Pierce, D. T.; Unwin, P. R.; Wipf, D. O.; Zhou, F. 1991. Chemical imaging of surfaces with the scanning electrochemical microscope. *Science*, *254*, 68–74.
9. Amemiya, S.; Bard, A. J.; Fan, F.-R. F.; Mirkin, M. V.; Unwin, P. R. 2008. Scanning electrochemical microscopy. *Annu. Rev. Anal. Chem.*, *1*, 95–131.
10. Beaulieu, I.; Kuss, S.; Mauzeroll, J.; Geissler, M. 2011. Biological scanning electrochemical microscopy and its application to live cell studies. *Anal. Chem.*, *83*, 1485–1492.
11. Amemiya, S.; Guo, J.; Xiong, H.; Gross, D. A. 2006. Biological applications of scanning electrochemical microscopy: Chemical imaging of single living cells and beyond. *Anal. Bioanal. Chem.*, *386*, 458–471.
12. Schulte, A.; Nebel, M.; Schuhmann, W. 2010. Scanning electrochemical microscopy in neuroscience. *Annu. Rev. Anal. Chem.*, *3*, 299–318.
13. Edwards, M. A.; Martin, S.; Whitworth, A. L.; Macpherson, J. V.; Unwin, P. R. 2006. Scanning electrochemical microscopy: Principles and applications to biophysical systems. *Physiol. Meas.*, *27*, R63.
14. Bertoncello, P. 2010. Advances on scanning electrochemical microscopy (SECM) for energy. *Energy Environ. Sci.*, *3*, 1620–1633.
15. Lai, S. C. S.; Macpherson, J. V.; Unwin, P. R. 2012. In situ scanning electrochemical probe microscopy for energy applications. *MRS Bull.*, *37*, 668–674.
16. Amemiya, S., 2002. *Scanning Electrochemical Microscopy*; John Wiley & Sons, Inc., New York.
17. Szunerits, S.; Pust, S. E.; Wittstock, G. 2007. Multidimensional electrochemical imaging in materials science. *Anal. Bioanal. Chem.*, *389*, 1103–1120.
18. Bard, A. J.; Mirkin, M. V. 2012. *Scanning Electrochemical Microscopy*; 2nd edn.; CRC Press, Boca Raton, FL.
19. Wittstock, G.; Burchardt, M.; Pust, S. E.; Shen, Y.; Zhao, C. 2007. Scanning electrochemical microscopy for direct imaging of reaction rates. *Angew. Chem., Int. Ed.*, *46*, 1584–1617.
20. Mirkin, M. V.; Nogala, W.; Velmurugan, J.; Wang, Y. 2011. Scanning electrochemical microscopy in the 21st century. Update 1: Five years after. *Phys. Chem. Chem. Phys.*, *13*, 21196–21212.
21. Guo, S. X.; Unwin, P. R.; Whitworth, A. L.; Zhang, T. 2004. Microelectrochemical techniques for probing kinetics at liquid/liquid interfaces. *Prog. React. Kinet. Mech.*, *29*, 125.
22. Nebel, M.; Eckhard, K.; Erichsen, T.; Schulte, A.; Schuhmann, W. 2010. 4D shearforce-based constant-distance mode scanning electrochemical microscopy. *Anal. Chem.*, *82*, 7842–7848.
23. Takahashi, Y.; Shiku, H.; Murata, T.; Yasukawa, T.; Matsue, T. 2009. Transfected single-cell imaging by scanning electrochemical optical microscopy with shear force feedback regulation. *Anal. Chem.*, *81*, 9674–9681.
24. Yamada, H.; Ogata, M.; Koike, T. 2006. Scanning electrochemical microscope observation of defects in a hexadecanethiol monolayer on gold with shear force-based tip−substrate positioning. *Langmuir*, *22*, 7923–7927.
25. Etienne, M.; Layoussifi, B.; Giornelli, T.; Jacquet, D. 2012. SECM-based automate equipped with a shear-force detection for the characterization of large and complex samples. *Electrochem. Commun.*, *15*, 70–73.
26. Takahashi, Y.; Shevchuk, A. I.; Novak, P.; Murakami, Y.; Shiku, H.; Korchev, Y. E.; Matsue, T. 2010. Simultaneous noncontact topography and electrochemical imaging by SECM/SICM featuring ion current feedback regulation. *J. Am. Chem. Soc.*, *132*, 10118–10126.

27. Comstock, D. J.; Elam, J. W.; Pellin, M. J.; Hersam, M. C. 2010. Integrated ultramicroelectrode-nanopipet probe for concurrent scanning electrochemical microscopy and scanning ion conductance microscopy. *Anal. Chem.*, *82*, 1270–1276.

28. Morris, C. A.; Chen, C. C.; Baker, L. A. 2012. Transport of redox probes through single pores measured by scanning electrochemical-scanning ion conductance microscopy (SECM-SICM). *Analyst*, *137*, 2933–2938.

29. Nadappuram, B.P.; McKelvey, K.; Al-Botros, R.; Colburn, A. W.; Unwin, P. R. 2013. Fabrication and characterization of dual function nanoscale pH-scanning ion conductance microscopy (SICM) probes for high resolution pH mapping. *Anal. Chem.*, *85*, 8070–8074.

30. Takahashi, Y.; Shevchuk, A. I.; Novak, P.; Zhang, Y.; Ebejer, N.; Macpherson, J. V. et al. 2011. Multifunctional nanoprobes for nanoscale chemical imaging and localized chemical delivery at surfaces and interfaces. *Angew. Chem., Int. Ed.*, *50*, 9638–9642.

31. Kurulugama, R. T.; Wipf, D. O.; Takacs, S. A.; Pongmayteegul, S.; Garris, P. A.; Baur, J. E. 2005. Scanning electrochemical microscopy of model neurons: Constant distance imaging. *Anal. Chem.*, *77*, 1111–1117.

32. Alpuche-Aviles, M. A.; Wipf, D. O. 2001. Impedance feedback control for scanning electrochemical microscopy. *Anal. Chem.*, *73*, 4873–4881.

33. Macpherson, J. V.; Unwin, P. R.; Hillier, A. C.; Bard, A. J. 1996. In-situ imaging of ionic crystal dissolution using an integrated electrochemical/AFM probe. *J. Am. Chem. Soc.*, *118*, 6445–6452.

34. Kranz, C.; Friedbacher, G.; Mizaikoff, B. 2001. Integrating an ultramicroelectrode in an AFM cantilever: Combined technology for enhanced information. *Anal. Chem.*, *73*, 2491–2500.

35. Salomo, M.; Pust, S. E.; Wittstock, G.; Oesterschulze, E. 2010. Integrated cantilever probes for SECM/AFM characterization of surfaces. *Microelectron. Eng.*, *87*, 1537–1539.

36. Macpherson, J. V.; Unwin, P. R. 1999. Combined scanning electrochemical–atomic force microscopy. *Anal. Chem.*, *72*, 276–285.

37. Wipf, D. O.; Bard, A. J. 1992. Scanning electrochemical microscopy. 15. Improvements in imaging via tip-position modulation and lock-in detection. *Anal. Chem.*, *64*, 1362–1367.

38. Wipf, D. O.; Bard, A. J.; Tallman, D. E. 1993. Scanning electrochemical microscopy. 21. Constant-current imaging with an autoswitching controller. *Anal. Chem.*, *65*, 1373–1377.

39. Edwards, M. A.; Whitworth, A. L.; Unwin, P. R. 2011. Quantitative analysis and application of tip position modulation-scanning electrochemical microscopy. *Anal. Chem.*, *83*, 1977–1984.

40. McKelvey, K.; Edwards, M. A.; Unwin, P. R. 2010. Intermittent contact–scanning electrochemical microscopy (IC–SECM): A new approach for tip positioning and simultaneous imaging of interfacial topography and activity. *Anal. Chem.*, *82*, 6334–6337.

41. McKelvey, K.; Snowden, M. E.; Peruffo, M.; Unwin, P. R. 2011. Quantitative visualization of molecular transport through porous membranes: Enhanced resolution and contrast using intermittent contact-scanning electrochemical microscopy. *Anal. Chem.*, *83*, 6447–6454.

42. Lazenby, R. A.; McKelvey, K.; Unwin, P. R. 2013. Hopping intermittent contact-scanning electrochemical microscopy (HIC-SECM): Visualizing interfacial reactions and fluxes from surfaces to bulk solution. *Anal. Chem.*, *85*, 2937–2944.

43. Cortés-Salazar, F.; Momotenko, D.; Lesch, A.; Wittstock, G.; Girault, H. H. 2010. Soft microelectrode linear array for scanning electrochemical microscopy. *Anal. Chem.*, *82*, 10037–10044.

44. Cortes-Salazar, F.; Momotenko, D.; Girault, H. H.; Lesch, A.; Wittstock, G. 2011. Seeing big with scanning electrochemical microscopy. *Anal. Chem.*, *83*, 1493–1499.

45. Lesch, A.; Momotenko, D.; Cortes-Salazar, F.; Wirth, I.; Tefashe, U. M.; Meiners, F.; Vaske, B.; Girault, H. H.; Wittstock, G. 2012. Fabrication of soft gold microelectrode arrays as probes for scanning electrochemical microscopy. *J. Electroanal. Chem.*, *666*, 52–61.

46. Lesch, A.; Vaske, B.; Meiners, F.; Momotenko, D.; Cortes-Salazar, F.; Girault, H. H.; Wittstock, G. 2012. Parallel imaging and template-free patterning of self-assembled monolayers with soft linear microelectrode arrays. *Angew. Chem., Int. Ed.*, *51*, 10413–10416.

47. Lesch, A.; Momotenko, D.; Cortés-Salazar, F.; Roelfs, F.; Girault, H. H.; Wittstock, G. 2013. High-throughput scanning electrochemical microscopy brushing of strongly tilted and curved surfaces. *Electrochim. Acta*, *110*, 30–41.

48. Lesch, A.; Chen, P.-C.; Roelfs, F.; Dosche, C.; Momotenko, D.; Cortés-Salazar, F.; Girault, H. H.; Wittstock, G. 2013. Finger probe array for topography-tolerant scanning electrochemical microscopy of extended samples. *Anal. Chem.*, *86*, 713–720.

49. Ebejer, N.; Schnippering, M.; Colburn, A. W.; Edwards, M. A.; Unwin, P. R. 2010. Localized high resolution electrochemistry and multifunctional imaging: Scanning electrochemical cell microscopy. *Anal. Chem.*, *82*, 9141–9145.

50. Böhni, H.; Suter, T.; Schreyer, A. 1995. Micro- and nanotechniques to study localized corrosion. *Electrochim. Acta*, *40*, 1361–1368.

51. Mardare, A. I.; Wieck, A. D.; Hassel, A. W. 2007. Microelectrochemical lithography: A method for direct writing of surface oxides. *Electrochim. Acta*, *52*, 7865–7869.

52. Lohrengel, M. M.; Moehring, A.; Pilaski, M. 2001. Capillary-based droplet cells: Limits and new aspects. *Electrochim. Acta*, *47*, 137–141.

53. Lill, K. A.; Hassel, A. W.; Frommeyer, G.; Stratmann, M. 2005. Scanning droplet cell investigations on single grains of a FeAlCr light weight ferritic steel. *Electrochim. Acta*, *51*, 978–983.

54. Hassel, A. W.; Seo, M. 1999. Localised investigation of coarse grain gold with the scanning droplet cell and by the Laue method. *Electrochim. Acta*, *44*, 3769–3777.

55. Staemmler, L.; Suter, T.; Bohni, H. 2004. Nanolithography by means of an electrochemical scanning capillary microscope. *J. Electrochem. Soc.*, *151*, G734–G739.

56. Woldemedhin, M. T.; Raabe, D.; Hassel, A. W. 2011. Grain boundary electrochemistry of β-type Nb–Ti alloy using a scanning droplet cell. *Phys. Status Solidi A*, *208*, 1246–1251.

57. Suter, T.; Böhni, H. 1997. A new microelectrochemical method to study pit initiation on stainless steels. *Electrochim. Acta*, *42*, 3275–3280.

58. Momotenko, D.; Cortes-Salazar, F.; Lesch, A.; Wittstock, G.; Girault, H. H. 2011. Microfluidic push-pull probe for scanning electrochemical microscopy. *Anal. Chem.*, *83*, 5275–5282.

59. Momotenko, D.; Qiao, L.; Cortes-Salazar, F.; Lesch, A.; Wittstock, G.; Girault, H. H. 2012. Electrochemical push-pull scanner with mass spectrometry detection. *Anal. Chem.*, *84*, 6630–6637.

60. Rodolfa, K. T.; Bruckbauer, A.; Zhou, D. J.; Korchev, Y. E.; Klenerman, D. 2005. Two-component graded deposition of biomolecules with a double-barreled nanopipette. *Angew. Chem., Int. Ed.*, *44*, 6854–6859.

61. Williams, C. G.; Edwards, M. A.; Colley, A. L.; Macpherson, J. V.; Unwin, P. R. 2009. Scanning micropipet contact method for high-resolution imaging of electrode surface redox activity. *Anal. Chem.*, *81*, 2486–2495.

62. Vogel, A.; Schultze, J. W. 1999. A new microcell for electrochemical surface analysis and reactions. *Electrochim. Acta*, *44*, 3751–3759.

63. Kollender, J. P.; Mardare, A. I.; Hassel, A. W. 2013. Photoelectrochemical scanning droplet cell microscopy (PE-SDCM). *ChemPhysChem*, *14*, 560–567.

64. Klemm, S. O.; Schauer, J.-C.; Schuhmacher, B.; Hassel, A. W. 2011. High throughput electrochemical screening and dissolution monitoring of Mg–Zn material libraries. *Electrochim. Acta*, *56*, 9627–9636.

65. Klemm, S. O.; Schauer, J.-C.; Schuhmacher, B.; Hassel, A. W. 2011. A microelectrochemical scanning flow cell with downstream analytics. *Electrochim. Acta*, *56*, 4315–4321.

66. Klemm, S.; Pust, S.; Hassel, A.; Hüpkes, J.; Mayrhofer, K. J. 2012. Electrochemical texturing of Al-doped ZnO thin films for photovoltaic applications. *J. Solid State Electrochem.*, *16*, 283–290.

67. Klemm, S. O.; Topalov, A. A.; Laska, C. A.; Mayrhofer, K. J. J. 2011. Coupling of a high throughput microelectrochemical cell with online multielemental trace analysis by ICP-MS. *Electrochem. Commun.*, *13*, 1533–1535.

68. Lohrengel, M. M.; Rosenkranz, C.; Klüppel, I.; Moehring, A.; Bettermann, H.; Van den Bossche, B.; Deconinck, J. 2004. A new microcell or microreactor for material surface investigations at large current densities. *Electrochim. Acta*, *49*, 2863–2870.

69. Cortes-Salazar, F.; Lesch, A.; Momotenko, D.; Busnel, J. M.; Wittstock, G.; Girault, H. H. 2010. Fountain pen for scanning electrochemical microscopy. *Anal. Methods*, *2*, 817–823.

70. Hansma, P. K.; Drake, B.; Marti, O.; Gould, S. A.; Prater, C. B. 1989. The scanning ion-conductance microscope. *Science*, *243*, 641–643.

71. Edwards, M. A.; Williams, C. G.; Whitworth, A. L.; Unwin, P. R. 2009. Scanning ion conductance microscopy: A model for experimentally realistic conditions and image interpretation. *Anal. Chem.*, *81*, 4482–4492.

72. Shevchuk, A. I.; Gorelik, J.; Harding, S. E.; Lab, M. J.; Klenerman, D.; Korchev, Y. E. 2001. Simultaneous measurement of Ca^{2+} and cellular dynamics: Combined scanning ion conductance and optical microscopy to study contracting cardiac myocytes. *Biophys. J.*, *81*, 1759–1764.

73. Pastre, D.; Iwamoto, H.; Liu, J.; Szabo, G.; Shao, Z. 2001. Characterization of AC mode scanning ion-conductance microscopy. *Ultramicroscopy*, *90*, 13–19.

74. Novak, P.; Li, C.; Shevchuk, A. I.; Stepanyan, R.; Caldwell, M.; Hughes, S. et al. 2009. Nanoscale live-cell imaging using hopping probe ion conductance microscopy. *Nat. Methods*, *6*, 279–281.

75. Sanchez, D.; Arand, U.; Gorelik, J.; Benham, C. D.; Bountra, C.; Lab, M.; Klenerman, D.; Birch, R.; Arland, P.; Korchev, Y. 2007. Localized and non-contact mechanical stimulation of dorsal root ganglion sensory neurons using scanning ion conductance microscopy. *J. Neurosci. Methods.*, *159*, 26–34.

76. Gorelik, J.; Zhang, Y. J.; Shevchuk, A. I.; Frolenkov, G. I.; Sanchez, D.; Lab, M. J.; Vodyanoy, I.; Edwards, C. R. W.; Klenerman, D.; Korchev, Y. E. 2004. The use of scanning ion conductance microscopy to image A6 cells. *Mol. Cell. Endocrinol.*, *217*, 101–108.

77. Happel, P.; Hoffmann, G.; Mann, S. A.; Dietzel, I. D. 2003. Monitoring cell movements and volume changes with pulse-mode scanning ion conductance microscopy. *J. Microsc.-Oxford*, *212*, 144–151.

78. Shevchuk, A. I.; Frolenkov, G. I.; Sanchez, D.; James, P. S.; Freedman, N.; Lab, M. J.; Jones, R.; Klenerman, D.; Korchev, Y. E. 2006. Imaging proteins in membranes of living cells by high-resolution scanning ion conductance microscopy. *Angew. Chem., Int. Ed.*, *45*, 2212–2216.

79. Korchev, Y. E.; Negulyaev, Y. A.; Edwards, C. R.; Vodyanoy, I.; Lab, M. J. 2000. Functional localization of single active ion channels on the surface of a living cell. *Nat. Cell Biol.*, *2*, 616–619.

80. Proksch, R.; Lal, R.; Hansma, P. K.; Morse, D.; Stucky, G. 1996. Imaging the internal and external pore structure of membranes in fluid: Tapping mode scanning ion conductance microscopy. *Biophys. J.*, *71*, 2155–2157.

81. Bruckbauer, A.; Ying, L. M.; Rothery, A. M.; Zhou, D. J.; Shevchuk, A. I.; Abell, C.; Korchev, Y. E.; Klenerman, D. 2002. Writing with DNA and protein using a nanopipet for controlled delivery. *J. Am. Chem. Soc.*, *124*, 8810–8811.

82. Bruckbauer, A.; Zhou, D.; Kang, D. J.; Korchev, Y. E.; Abell, C.; Klenerman, D. 2004. An addressable antibody nanoarray produced on a nanostructured surface. *J. Am. Chem. Soc.*, *126*, 6508–6509.

83. Bruckbauer, A.; Zhou, D.; Ying, L.; Korchev, Y. E.; Abell, C.; Klenerman, D. 2003. Multicomponent submicron features of biomolecules created by voltage controlled deposition from a nanopipet. *J. Am. Chem. Soc.*, *125*, 9834–9839.

84. Wang, Y.; Kececi, K.; Velmurugan, J.; Mirkin, M. V. 2013. Electron transfer/ion transfer mode of scanning electrochemical microscopy (SECM): A new tool for imaging and kinetic studies. *Chem. Sci.*, *4*, 3606–3616.

85. Zhan, D. P.; Li, X.; Zhan, W.; Fan, F. R. F.; Bard, A. J. 2007. Scanning electrochemical microscopy. 58. Application of a micropipet-supported ITIES tip to detect Ag+ and study its effect on fibroblast cells. *Anal. Chem.*, *79*, 5225–5231.

86. Yatziv, Y.; Turyan, I.; Mandler, D. 2002. A new approach to micropatterning: Application of potential-assisted ion transfer at the liquid-liquid interface for the local metal deposition. *J. Am. Chem. Soc.*, *124*, 5618–5619.

87. Dudin, P. V.; Snowden, M. E.; Macpherson, J. V.; Unwin, P. R. 2011. Electrochemistry at nanoscale electrodes: Individual single-walled carbon nanotubes (SWNTs) and SWNT-templated metal nanowires. *ACS Nano*, *5*, 10017–10025.

88. Day, T. M.; Unwin, P. R.; Macpherson, J. V. 2007. Factors controlling the electrodeposition of metal nanoparticles on pristine single walled carbon nanotubes. *Nano Lett.*, *7*, 51–57.

89. Aaronson, B. D. B.; Lai, S. C. S.; Unwin, P. R. 2014. Spatially resolved electrochemistry in ionic liquids: Surface structure effects on triiodide reduction at platinum electrodes. *Langmuir*, *30*, 1915–1919.

90. Liu, B.; Shao, Y.; Mirkin, M. V. 1999. Dual-pipet techniques for probing ionic reactions. *Anal. Chem.*, *72*, 510–519.

91. Güell, A. G.; Meadows, K. E.; Dudin, P. V.; Ebejer, N.; Macpherson, J. V.; Unwin, P. R. 2014. Mapping nanoscale electrochemistry of individual single-walled carbon nanotubes. *Nano Lett.*, *14*, 220–224.

92. Lai, S. C. S.; Dudin, P. V.; Macpherson, J. V.; Unwin, P. R. 2011. Visualizing zeptomole (electro)catalysis at single nanoparticles within an ensemble. *J. Am. Chem. Soc.*, *133*, 10744–10747.

93. Aaronson, B. D. B.; Chen, C.-H.; Li, H.; Koper, M. T. M.; Lai, S. C. S.; Unwin, P. R. 2013. Pseudo-single-crystal electrochemistry on polycrystalline electrodes: Visualizing activity at grains and grain boundaries on platinum for the Fe^{2+}/Fe^{3+} redox reaction. *J. Am. Chem. Soc.*, *135*, 3873–3880.

94. Kleijn, S. E. F.; Lai, S. C. S.; Miller, T. S.; Yanson, A. I.; Koper, M. T. M.; Unwin, P. R. 2012. Landing and catalytic characterization of individual nanoparticles on electrode surfaces. *J. Am. Chem. Soc.*, *134*, 18558–18561.

95. Snowden, M. E.; Güell, A. G.; Lai, S. C. S.; McKelvey, K.; Ebejer, N.; O'Connell, M. A.; Colburn, A. W.; Unwin, P. R. 2012. Scanning electrochemical cell microscopy: Theory and experiment for quantitative high resolution spatially-resolved voltammetry and simultaneous ion-conductance measurements. *Anal. Chem.*, *84*, 2483–2491.

96. Kinnear, S. L.; McKelvey, K.; Snowden, M. E.; Peruffo, M.; Colburn, A. W.; Unwin, P. R. 2014. Dual-barrel conductance micropipet as a new approach to the study of ionic crystal dissolution kinetics. *Langmuir*, *29*, 15565–15572.

97. Kirkman, P. M.; Güell, A. G.; Cuharuc, A. S.; Unwin, P. R. 2014. Spatial and temporal control of the diazonium modification of sp^2 carbon surfaces. *J. Am. Chem. Soc.*, *136*, 36–39.

98. McKelvey, K.; O'Connell, M. A.; Unwin, P. R. 2013. Meniscus confined fabrication of multidimensional conducting polymer nanostructures with scanning electrochemical cell microscopy (SECCM). *Chem. Commun.*, *49*, 2986–2988.

99. Patel, A. N.; McKelvey, K.; Unwin, P. R. 2012. Nanoscale electrochemical patterning reveals the active sites for catechol oxidation at graphite surfaces. *J. Am. Chem. Soc.*, *134*, 20246–20249.

100. Güell, A. G.; Ebejer, N.; Snowden, M. E.; McKelvey, K.; Macpherson, J. V.; Unwin, P. R. 2012. Quantitative nanoscale visualization of heterogeneous electron transfer rates in 2D carbon nanotube networks. *Proc. Natl. Acad. Sci. USA*, *109*, 11487–11492.

101. Putman, C. A.; van der Werf, K. O.; de Grooth, B. G.; van Hulst, N. F.; Greve, J.; Hansma, P. K. New imaging mode in atomic-force microscopy based on the error signal. In *Scanning Probe Microscopies*, S. Manne (ed.), SPIE, Los Angeles, CA, 1992; Vol. 1639, pp. 198–204.

102. Patel, A. N.; Tan, S.; Unwin, P. R. 2013. Epinephrine electro-oxidation highlights fast electrochemistry at the graphite basal surface. *Chem. Commun.*, *49*, 8776–8778.

103. Güell, A. G.; Ebejer, N.; Snowden, M. E.; Macpherson, J. V.; Unwin, P. R. 2012. Structural correlations in heterogeneous electron transfer at monolayer and multilayer graphene electrodes. *J. Am. Chem. Soc.*, *134*, 7258–7261.

104. Sa, N.; Lan, W.-J.; Shi, W.; Baker, L. A. 2013. Rectification of ion current in nanopipettes by external substrates. *ACS Nano*, *7*, 11272–11282.

105. Momotenko, D.; Cortes-Salazar, F.; Josserand, J.; Liu, S.; Shao, Y.; Girault, H. H. 2011. Ion current rectification and rectification inversion in conical nanopores: A perm-selective view. *Phys. Chem. Chem. Phys.*, *13*, 5430–5440.

106. Marković, N. M.; Ross Jr., P. N. 2002. Surface science studies of model fuel cell electrocatalysts. *Surf. Sci. Rep.*, *45*, 117–229.

107. Koper, M. T. M. 2009. *Fuel Cell Catalysis: A Surface Science Approach*; Wiley, Hoboken, NJ.

108. Koper, M. T. M. 2011. Structure sensitivity and nanoscale effects in electrocatalysis. *Nanoscale*, *3*, 2054–2073.

109. Climent, V.; Feliu, J. M. 2011. Thirty years of platinum single-crystal electrochemistry. *J. Solid State Electrochem.*, *15*, 1297–1315.

110. Cherstiouk, O. V.; Gavrilov, A. N.; Plyasova, L. M.; Molina, I. Y.; Tsirlina, G. A.; Savinova, E. R. 2008. Influence of structural defects on the electrocatalytic activity of platinum. *J. Solid State Electrochem.*, *12*, 497–509.

111. Maillard, F.; Savinova, E. R.; Stimming, U. 2007. CO monolayer oxidation on Pt nanoparticles: Further insights into the particle size effects. *J. Electroanal. Chem.*, *599*, 221–232.

112. Wilkinson, A. J.; Britton, T. B. 2012. Strains, planes, and EBSD in materials science. *Mater. Today*, *15*, 366–376.

113. Garcia-Araez, N.; Climent, V.; Feliu, J. 2009. Potential-dependent water orientation on Pt(111), Pt(100), and Pt(110), as inferred from laser-pulsed experiments. Electrostatic and chemical effects. *J. Phys. Chem. C*, *113*, 9290–9304.

114. Hagfeldt, A.; Boschloo, G.; Sun, L. C.; Kloo, L.; Pettersson, H. 2010. Dye-sensitized solar cells. *Chem. Rev.*, *110*, 6595–6663.

115. Walsh, D. A.; Lovelock, K. R. J.; Licence, P. 2010. Ultramicroelectrode voltammetry and scanning electrochemical microscopy in room-temperature ionic liquid electrolytes. *Chem. Soc. Rev.*, *39*, 4185–4194.

116. Zhao, C.; Bond, A. M.; Compton, R. G.; O'Mahony, A. M.; Rogers, E. I. 2010. Modification and implications of changes in electrochemical responses encountered when undertaking deoxygenation in ionic liquids. *Anal. Chem.*, *82*, 3856–3861.

117. Navarro-Suárez, A. M.; Hidalgo-Acosta, J. C.; Fadini, L.; Feliu, J. M.; Suárez-Herrera, M. F. 2011. Electrochemical oxidation of hydrogen on basal plane platinum electrodes in imidazolium ionic liquids. *J. Phys. Chem. C*, *115*, 11147–11155.

118. Patten, H. V.; Lai, S. C. S.; Macpherson, J. V.; Unwin, P. R. 2012. Active sites for outer-sphere, inner-sphere, and complex multistage electrochemical reactions at polycrystalline boron-doped diamond electrodes (pBDD) revealed with scanning electrochemical cell microscopy (SECCM). *Anal. Chem.*, *84*, 5427–5432.

119. Güell, A. G.; Meadows, K. E.; Unwin, P. R.; Macpherson, J. V. 2010. Trace voltammetric detection of serotonin at carbon electrodes: Comparison of glassy carbon, boron doped diamond and carbon nanotube network electrodes. *Phys. Chem. Chem. Phys.*, *12*, 10108–10114.

120. Sarada, B. V.; Rao, T. N.; Tryk, D. A.; Fujishima, A. 2000. Electrochemical oxidation of histamine and serotonin at highly boron-doped diamond electrodes. *Anal. Chem.*, *72*, 1632–1638.

121. Wrona, M. Z.; Dryhurst, G. 1987. Oxidation chemistry of 5-hydroxytryptamine. 1. Mechanism and products formed at micromolar concentrations. *J. Org. Chem.*, *52*, 2817–2825.

122. Wrona, M. Z.; Dryhurst, G. 1990. Oxidation chemistry of 5-hydroxytryptamine: Part II. Mechanisms and products formed at millimolar concentrations in acidic aqueous solution. *J. Electroanal. Chem. Interf. Electrochem.*, *278*, 249–267.

123. Patten, H. V.; Meadows, K. E.; Hutton, L. A.; Iacobini, J. G.; Battistel, D.; McKelvey, K.; Colburn, A. W.; Newton, M. E.; Macpherson, J. V.; Unwin, P. R. 2012. Electrochemical mapping reveals direct correlation between heterogeneous electron-transfer kinetics and local density of states in diamond electrodes. *Angew. Chem., Int. Ed.*, *51*, 7002–7006.

124. Gerischer, H. 1985. An interpretation of the double layer capacity of graphite electrodes in relation to the density of states at the fermi level. *J. Phys. Chem.*, *89*, 4249–4251.

125. Hahn, M.; Baertschi, M.; Barbieri, O.; Sauter, J.-C.; Kötz, R.; Gallay, R. 2004. Interfacial capacitance and electronic conductance of activated carbon double-layer electrodes. *Electrochem. Solid-State Lett.*, *7*, A33–A36.

126. Blake, P.; Hill, E. W.; Castro Neto, A. H.; Novoselov, K. S.; Jiang, D.; Yang, R.; Booth, T. J.; Geim, A. K. 2007. Making graphene visible. *Appl. Phys. Lett.*, *91*, 063124-1–063124-3.

127. Reina, A.; Jia, X.; Ho, J.; Nezich, D.; Son, H.; Bulovic, V.; Dresselhaus, M. S.; Kong, J. 2008. Large area, few-layer graphene films on arbitrary substrates by chemical vapor deposition. *Nano Lett.*, *9*, 30–35.

128. Ferrari, A. C.; Meyer, J. C.; Scardaci, V.; Casiraghi, C.; Lazzeri, M.; Mauri, F.; Piscanec, S.; Jiang, D.; Novoselov, K. S.; Roth, S.; Geim, A. K. 2006. Raman spectrum of graphene and graphene layers. *Phys. Rev. Lett.*, *97*, 187401.

129. McCreery, R. L. 2008. Advanced carbon electrode materials for molecular electrochemistry. *Chem. Rev.*, *108*, 2646–2687.

130. Lai, S. C. S.; Patel, A. N.; McKelvey, K.; Unwin, P. R. 2012. Definitive evidence for fast electron transfer at pristine basal plane graphite from high-resolution electrochemical imaging. *Angew. Chem., Int. Ed.*, *51*, 5405–5408.

131. Patel, A. N.; Collignon, M. G.; O'Connell, M. A.; Hung, W. O. Y.; McKelvey, K.; Macpherson, J. V.; Unwin, P. R. 2012. A new view of electrochemistry at highly oriented pyrolytic graphite. *J. Am. Chem. Soc.*, *134*, 20117–20130.

132. Anne, A.; Bahri, M. A.; Chovin, A.; Demaille, C.; Taofifenua, C. 2014. Probing the conformation and 2D-distribution of pyrene-terminated redox-labeled poly(ethylene glycol) chains end-adsorbed on HOPG using cyclic voltammetry and atomic force electrochemical microscopy. *Phys. Chem. Chem. Phys*, *16*, 4642–4652.

133. Patrick, L. T. M. F.; Patrick, D. B.; Terunobu, A.; Mohamed, C.; Maurizio, R. G.; Jason, J. B.; Karin, D.; Nico, F. d. R.; Urs, S.; Andreas, E. 2008. Conductive supports for combined AFM–SECM on biological membranes. *Nanotechnology*, *19*, 384004.

134. Anne, A.; Cambril, E.; Chovin, A.; Demaille, C.; Goyer, C. 2009. Electrochemical atomic force microscopy using a tip-attached redox mediator for topographic and functional imaging of nanosystems. *ACS Nano*, *3*, 2927–2940.

135. Lhenry, S.; Leroux, Y. R.; Hapiot, P. 2012. Use of catechol as selective redox mediator in scanning electrochemical microscopy investigations. *Anal. Chem.*, *84*, 7518–7524.

136. Dumitrescu, I.; Unwin, P. R.; Macpherson, J. V. 2009. Electrochemistry at carbon nanotubes: Perspective and issues. *Chem. Commun.*, *23* (45), 6886–6901.

137. Heller, I.; Kong, J.; Heering, H. A.; Williams, K. A.; Lemay, S. G.; Dekker, C. 2004. Individual single-walled carbon nanotubes as nanoelectrodes for electrochemistry. *Nano Lett.*, *5*, 137–142.

138. Kim, J.; Xiong, H.; Hofmann, M.; Kong, J.; Amemiya, S. 2010. Scanning electrochemical microscopy of individual single-walled carbon nanotubes. *Anal. Chem.*, *82*, 1605–1607.

139. Bertoncello, P.; Edgeworth, J. P.; Macpherson, J. V.; Unwin, P. R. 2007. Trace level cyclic voltammetry facilitated by single-walled carbon nanotube network electrodes. *J. Am. Chem. Soc.*, *129*, 10982–10983.

140. Li, J.; Cassell, A.; Delzeit, L.; Han, J.; Meyyappan, M. 2002. Novel three-dimensional electrodes: Electrochemical properties of carbon nanotube ensembles. *J. Phys. Chem. B*, *106*, 9299–9305.

141. Chou, A.; Bocking, T.; Singh, N. K.; Gooding, J. J. 2005. Demonstration of the importance of oxygenated species at the ends of carbon nanotubes for their favourable electrochemical properties. *Chem. Commun.*, 842–844.

142. Holloway, A. F.; Toghill, K.; Wildgoose, G. G.; Compton, R. G.; Ward, M. A. H.; Tobias, G.; Llewellyn, S. A.; Ballesteros, B.; Green, M. L. H.; Crossley, A. 2008. Electrochemical opening of single-walled carbon nanotubes filled with metal halides and with closed ends. *J. Phys. Chem. C*, *112*, 10389–10397.

143. Gong, K.; Chakrabarti, S.; Dai, L. 2008. Electrochemistry at carbon nanotube electrodes: Is the nanotube tip more active than the sidewall? *Angew. Chem., Int. Ed.*, *47*, 5446–5450.

144. Miller, T. S.; Ebejer, N.; Güell, A. G.; Macpherson, J. V.; Unwin, P. R. 2012. Electrochemistry at carbon nanotube forests: Sidewalls and closed ends allow fast electron transfer. *Chem. Commun.*, *48*, 7435–7437.

145. Fan, Y.; Goldsmith, B. R.; Collins, P. G. 2005. Identifying and counting point defects in carbon nanotubes. *Nat. Mater.*, *4*, 906–911.

146. Li, S.; Yu, Z.; Rutherglen, C.; Burke, P. J. 2004. Electrical properties of 0.4 cm long single-walled carbon nanotubes. *Nano Lett.*, *4*, 2003–2007.

147. Biercuk, M.; Ilani, S.; Marcus, C.; McEuen, P. 2008. *Electrical Transport in Single-Wall Carbon Nanotubes*; Springer, Berlin, Germany.

148. Day, T. M.; Wilson, N. R.; Macpherson, J. V. 2004. Electrochemical and conductivity measurements of single-wall carbon nanotube network electrodes. *J. Am. Chem. Soc.*, *126*, 16724–16725.

149. Zhou, H.; Fan, F.-R. F.; Bard, A. J. 2010. Observation of discrete Au nanoparticle collisions by electrocatalytic amplification using Pt ultramicroelectrode surface modification. *J. Phys. Chem. Lett.*, *1*, 2671–2674.

150. Wakerley, D.; Güell, A. G.; Hutton, L. A.; Miller, T. S.; Bard, A. J.; Macpherson, J. V. 2013. Boron doped diamond ultramicroelectrodes: A generic platform for sensing single nanoparticle electrocatalytic collisions. *Chem. Commun.*, *49*, 5657–5659.

151. Kwon, S. J.; Fan, F.-R. F.; Bard, A. J. 2010. Observing iridium oxide (IrOx) single nanoparticle collisions at ultramicroelectrodes. *J. Am. Chem. Soc.*, *132*, 13165–13167.

152. Xiao, X.; Bard, A. J. 2007. Observing single nanoparticle collisions at an ultramicroelectrode by electrocatalytic amplification. *J. Am. Chem. Soc.*, *129*, 9610–9612.

153. Delamar, M.; Hitmi, R.; Pinson, J.; Saveant, J. M. 1992. Covalent modification of carbon surfaces by grafting of functionalized aryl radicals produced from electrochemical reduction of diazonium salts. *J. Am. Chem. Soc.*, *114*, 5883–5884.

154. Bahr, J. L.; Yang, J.; Kosynkin, D. V.; Bronikowski, M. J.; Smalley, R. E.; Tour, J. M. 2001. Functionalization of carbon nanotubes by electrochemical reduction of aryl diazonium salts: A bucky paper electrode. *J. Am. Chem. Soc.*, *123*, 6536–6542.

155. Paulus, G. L. C.; Wang, Q. H.; Strano, M. S. 2012. Covalent electron transfer chemistry of graphene with diazonium salts. *Acc. Chem. Res.*, *46*, 160–170.

156. Huang, P.; Jing, L.; Zhu, H.; Gao, X. 2012. Diazonium functionalized graphene: Microstructure, electric, and magnetic properties. *Acc. Chem. Res.*, *46*, 43–52.

157. Hu, J.; Yu, M. F. 2010. Meniscus-confined three-dimensional electrodeposition for direct writing of wire bonds. *Science*, *329*, 313–316.

158. Laslau, C.; Williams, D. E.; Kannan, B.; Travas-Sejdic, J. 2011. Scanned pipette techniques for the highly localized electrochemical fabrication and characterization of conducting polymer thin films, microspots, microribbons, and nanowires. *Adv. Funct. Mater.*, *21*, 4607–4616.

159. Kim, J. T.; Seol, S. K.; Pyo, J.; Lee, J. S.; Je, J. H.; Margaritondo, G. 2011. Three-dimensional writing of conducting polymer nanowire arrays by meniscus-guided polymerization. *Adv. Mater.*, *23*, 1968–1970.

160. Yang, D.; Han, L.; Yang, Y.; Zhao, L.-B.; Zong, C.; Huang, Y.-F.; Zhan, D.; Tian, Z.-Q. 2011. Solid-state redox solutions: Microfabrication and electrochemistry. *Angew. Chem., Int. Ed.*, *50*, 8679–8682.

161. Rodolfa, K. T.; Bruckbauer, A.; Zhou, D. J.; Schevchuk, A. I.; Korchev, Y. E.; Klenerman, D. 2006. Nanoscale pipetting for controlled chemistry in small arrayed water droplets using a double-barrel pipet. *Nano Lett.*, *6*, 252–257.

162. Patel, A. N.; Tan, S.; Miller, T. S.; Macpherson, J. V.; Unwin, P. R. 2013. Comparison and reappraisal of carbon electrodes for the voltammetric detection of dopamine. *Anal. Chem.*, *85*, 11755–11764.

20 In Situ Atomic Resolution Studies of the Electrode/ Solution Interface by Electrochemical Scanning Tunneling Microscopy

Scott N. Thorgaard and Philippe Bühlmann

CONTENTS

20.1 INTRODUCTION

Few scientific developments in the last 50 years have revolutionized the understanding of the solid/liquid interface as much as the invention of the scanning tunneling microscope (STM).[1] The emergence of electrochemical scanning tunneling microscopy (EC-STM) in the late 1980s and early 1990s,[2–5] coupled with the rapid advancement of nanofabrication techniques and the increasing use of ultramicroelectrodes and nanoelectrodes,[6–8] has dramatically expanded capabilities to relate observed electrochemistry to the properties of single atoms and molecules at the electrode surface. The ability of EC-STM to act as an atomic scale, in situ probe of electrochemical systems affords it applications in nanoscale electrochemistry that would be inaccessible to techniques such as cyclic voltammetry or surface sensitive spectroscopies, where experimental data are averaged over a large fraction of the working electrode area.

While there has also been major progress recently toward atomic scale imaging in liquids by atomic force microscopy (AFM) and the analogous electrochemical AFM (EC-AFM),[9–11] EC-STM is still the only experimental technique readily capable of providing atomic resolution images in electrolyte solutions. EC-STM has been utilized extensively in atomic imaging studies of bare electrodes[4,12–17] and electrode surfaces modified with adsorbed species[14,18–23] as well as the characterization of single nanoparticles and nanostructures.[24–27] Beyond pure imaging, EC-STM has been used to observe numerous potential-dependent electrochemical processes at the nanoscale, including adsorption, surface reconstruction, and electron transfer reactions of surface species.[20,28–31] Moreover, EC-STM has been used as a platform for numerous nonimaging experiments that exploit the instrument's ability to control an electrical probe in situ over atomic scale distances, including studies of single molecule trapping and conduction,[32,33] and also as a tool for nanofabrication.[26,34,35]

Several review articles and books have already been published that cover in detail the basic principles of (nonelectrochemical) STM, covering also theoretical aspects of quantum mechanical tunneling, the design of STM scanners and vibration isolation, and the application of STM to imaging studies under ultrahigh vacuum (UHV) conditions.[36–44] For this chapter, preference is given instead to information specific to EC-STM. Section 20.2 briefly describes the development of STM leading up to the first EC-STM experiments, while Section 20.3 describes practical considerations for the use of EC-STM, including the preparation of insulated tips and metal single crystal samples. The remaining sections of this chapter are devoted to experimental results, with Section 20.4 describing the use of EC-STM to characterize bare and anion-covered substrates, and Section 20.5 describing EC-STM studies of organic self-assembled monolayers (SAMs). Finally, Section 20.6 briefly covers a key nonimaging application of EC-STM, which is the use of EC-STM break junctions for trapping single molecules in electrolyte solutions.

20.2 DEVELOPMENT OF EC-STM

20.2.1 Brief Introduction to STM

Following the first observations of electron tunneling in p–n junctions by Esaki,[45] as well as the realization of the basic components of probe microscopy in the field ion microscope (FIM) by

Müller and Bahadur[46] and later the topografiner by Young and coworkers,[47] Binnig and coworkers developed the first STM in 1981.[1] For this work, Gerd Binnig and Heinrich Rohrer were awarded the Nobel Prize in physics in 1986. The first use of STM to produce images showing lateral atomic resolution on a surface occurred in 1983 with imaging studies of the 7×7 reconstruction of Si(111), again by Binnig and coworkers.[48] The essential elements of the STM technique are shown in the schematic in Figure 20.1.

In STM, a sharp metal tip is positioned sufficiently close to a conductive sample that a measurable current (pA–nA) due to quantum mechanical tunneling of electrons between states in the tip and sample flows when a bias voltage (usually less than 1 V) is applied to the gap.[1,36,39] The magnitude of the tunneling current is related to the densities of electronic states near the Fermi level in the tip/sample gap, and depends exponentially on the tip/sample separation distance as follows[49]:

$$i_t = aVe^{-2\beta z}$$

where
  i_t is the tunneling current
  a is a term related to the overlap of electronic states in the tip and sample
  V is the bias voltage
  ß is the barrier height between the tip and the sample
  z is the tip/sample separation distance

Due to this exponential dependence on z, the tunneling current typically falls off more than an order of magnitude per ångström of tip/sample separation. It is this dependence that gives rise to the extraordinary resolving power of STM, as nearly all electron tunneling occurs through the terminating atom on the tip apex.

Positioned within tunneling distance of the sample ($z < 1$ nm) using a piezoelectric positioner, the STM tip is then raster-scanned over the sample surface to produce images. In the constant-height mode of imaging, the images consist of contour maps of the recorded tunneling current. In the more commonly used constant-current mode of imaging, images instead consist of contour maps of the response of a feedback loop that either extends or retracts the tip during scanning in order to keep the tunneling current at a chosen setpoint. In either mode, STM images may be interpreted as plots of the density of electronic states near the Fermi level directly beneath the tip, and therefore contain

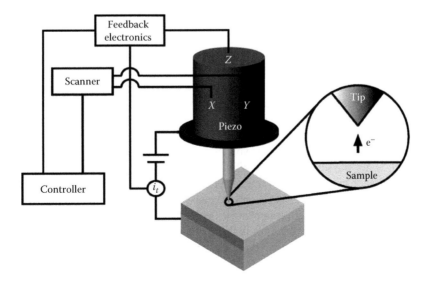

FIGURE 20.1 Schematic showing the basic components of conventional STM.

both chemical and topographic information superimposed. This convolution presents one of the fundamental limitations of the technique. Because topographic information may not be distinguished from molecular effects in images, it is often difficult or impossible to extract chemically specific information directly using STM imaging. On the other hand, when supplemented using chemical or potential sensitive techniques such as spectroscopy or voltammetry, STM has enormous capability for elucidating the atomic structure of surfaces. In contrast to techniques that require operation in vacuum, such as electron microscopy, STM has been used for atomic resolution imaging not only in vacuum but also in air and liquids. A key requirement for STM is that the sample be conductive to allow for electron tunneling, although a limited number of studies have attempted STM imaging of insulating substrates using either alternating currents[50,51] or hybrid modes of STM using conduction through ultrathin water layers in contact with the substrate.[52,53]

The STM instrument can also be used to probe the electronic properties of a surface as a function of the applied bias voltage, permitting scanning tunneling spectroscopy (STS).[39,40,54] In STS, the tip's position is fixed in the plane of the substrate, and the tunneling current is recorded as the bias voltage (or the tip/substrate spacing) is varied. STS enables local characterization of the electronic states immediately beneath the tip, and has been used as a means of interpreting the contrast in STM images for clean[55,56] and adsorbate-covered surfaces.[57–59] STS has seen much more limited application in electrochemical systems due to the convolution of the tunneling current with electrochemical processes occurring at the tip surface.[60,61] One particularly significant use of STS in electrochemically active phases has been for the electronic characterization of single redox proteins.[62,63]

20.2.2 Invention of EC-STM

The first example of STM applied in an electroactive phase appeared with Sonnenfeld and Hansma's 1986 report on the imaging of graphite and Au surfaces in aqueous solution.[2] However, to perform imaging, this work used a conventional STM instrument that did not offer independent control of the sample electrochemistry, and thus was not an example of a true EC-STM experiment as it is now understood. Instead, the authors chose a low bias voltage (100 mV) to limit electrochemical current at the tip and sample, while simultaneously positioning the tip at a high tunneling setpoint (31 nA tunneling current) during imaging. In this way, the fraction of the tip's current due to tunneling (as opposed to electrochemical processes) was sufficient to allow atomic resolution imaging despite the electroactive phase.

This early work illustrates the two principal challenges of STM imaging within electroactive environments. In an electroactive phase, the bias voltage applied for imaging inevitably gives rise to electrochemical processes (due to both electron transfer and charging of the interface) occurring at both the tip and sample. The first challenge here is that the current associated with these processes is superimposed onto the tunneling current monitored for STM imaging. If the electrochemical current is much larger than the tunneling current (which is very likely if the solution contains readily oxidized or reduced species), imaging capability is destroyed. Closely related is the second challenge, which is that the sample electrochemistry (and thereby, state of the surface on the atomic scale) depends entirely on the bias voltage. Without independent control of the sample electrochemistry, it is not possible to use the STM to investigate potential-dependent processes (such as adsorption or reconstruction) occurring at the sample surface. Moreover, interpretation of any STM images becomes inherently problematic because the chemical state of the surface cannot be decoupled from the choice of the tip/sample bias (which further influences contrast in images, which is dependent on the local density of states in the tip/sample gap).

These challenges were resolved by three research groups nearly concurrently in 1988.[3–5] In order to reduce the electrochemical contribution to the tip current during imaging, tips were used whose surface had been electrically isolated from the solution (save the last few nanometers of the tip apex to allow electron tunneling) using an insulating coating. To achieve control over the tip and sample electrochemistry, the two-electrode setup of conventional STM (with one controlled bias between the tip and sample) was replaced with a bipotentiostat design (with a tip, sample, reference electrode, and auxiliary electrode). Using this setup, the tip and sample potentials may be controlled independently relative to the reference electrode, enabling control over the sample electrochemistry during imaging. Moreover, the use of an auxiliary electrode minimizes the current through the reference electrode. A representative set of EC-STM images showing potential-dependent imaging is given in Figure 20.2. Here, Yoshimoto and coworkers have elucidated potential-dependent structural changes in a mixed monolayer of copper octaethylporphyrin and cobalt phthalocyanine on an Au(111) surface in dilute $HClO_4$ solution.[64] Highly ordered domains of the monolayer were observed when the sample electrode potential was set at +0.8 V versus reversible hydrogen electrode (RHE) and were found to become unstable and eventually disappear when the electrode potential was made progressively less positive. The following section describes specific considerations for the tip, sample, and electrochemical cells as they are required for EC-STM.

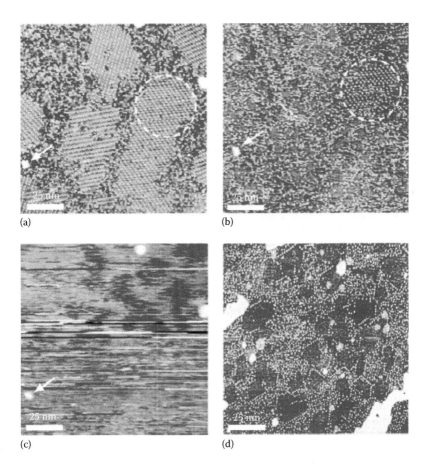

FIGURE 20.2 Potential-dependent EC-STM images of a mixed monolayer of copper octaethylporphyrin and cobalt phthalocyanine on an Au(111) single crystal in 0.1 M $HClO_4$. Sample potentials were +0.8 V (a), +0.6 V (b), +0.35 V (c), and +0.85 V (d), respectively, where the last potential of +0.85 V was applied by stepping from +0.3 V. The potential of the tip and the tunneling current were +0.47 V and 1.0 nA (a–c) and +0.46 V and 0.75 nA (d), respectively. (Reprinted with permission from Yoshimoto, S., Higa, N., and Itaya, K., Two-dimensional supramolecular organization of copper octaethylporphyrin and cobalt phthalocyanine on Au(111): Molecular assembly control at an electrochemical interface, *J. Am. Chem. Soc.*, 126, 8540–8545, 2004. Copyright 2004 American Chemical Society.)

20.3   EXPERIMENTAL CONSIDERATIONS

20.3.1   PREPARATION OF INSULATED TIPS FOR EC-STM

The preparation of sharp metal tips for conventional STM or other microscopies has been covered extensively in the literature.[65–67] Ideally, a tip should have as low an aspect ratio as possible to improve stability during imaging while still presenting a single terminating atom at its apex through which electron tunneling may occur. Sharp metal tips have been prepared by numerous methods, including electrochemical etching, mechanical cutting, and ion beam milling. One advantage of STM is that only the last few atoms along the tip's length participate in the experiment due to the extremely steep distance dependence of the electron tunneling interaction. This makes tip preparation in STM somewhat simpler than for AFM, where the entirety of the probe shape must be considered. On the other hand, if the STM tip apex consists not of one atom but of two or more atoms positioned at roughly the same distance from the sample, this may result in a loss of atomic resolution and imaging artifacts due to this multitip.[68]

To be usable for EC-STM, a metal tip's surface must be isolated from the electrolyte solution in order to reduce the electrochemical contribution to the tip current. At the same time, the tip apex must remain exposed as the presence of an insulator in the tip/sample gap would interfere with the electron tunneling required for STM imaging. Numerous materials have been used to insulate tips for EC-STM, including glass, wax, epoxy resin, oxide layers, nail polish, and electrophoretic paint.[3,66,69–74] Some early papers applied the coating manually by extending tips through reservoirs of liquid or molten material. In principle, the sharp metal apex should pierce the liquid material while the insulator adheres to the remaining shank (length) of the tip. An important example is the 1989 work by Nagahara et al., where the authors applied molten Apiezon wax to Pt/Ir tips using the apparatus shown in Figure 20.3.[66] This method produces tips that exhibit electrochemical leakage currents in the low pA range, allowing STM imaging using tunneling setpoints in the nA range. For the Apiezon wax method or other molten material, proper insulation of the tip depends on multiple factors, including the speed at which the tip is translated through the material and the viscosity of the material. Consequently, multiple authors reported either difficulties with this technique or a low rate of successfully producing usable tips.[61,73]

More common in the field have been dissolved polymer or varnish (specifically, nail polish) coatings, which are applied to tips by dipping methods.[71,75,76] Allowing the coating to dry with the tip in an inverted position causes it to recede and expose the apex. These methods have the distinct advantage of not requiring careful translation of the tip through molten material, and have been the prevailing choice for many research groups in the decades since EC-STM was developed. However, dissolved coatings applied by dipping or dropping do present a few unique challenges in that effective application can be somewhat sensitive to the amount of solvent present in the coating mixture as well as the mass of material applied to a tip. These difficulties can also be compounded by poor wetting of the metal by the varnish during application.

An alternative insulation material that has become increasingly utilized is electrophoretic paint. The use of electrophoretic paint to insulate STM tips was first demonstrated in the 1992 work by Mao et al.,[77] although the 1993 work by Bach et al. first reported what has become the established general procedure for applying the paint to tips,[73] with a handful of additional papers presenting improvements to the method in the mid-2000s.[61,78,79] The original two-electrode cell used by Bach is shown in Figure 20.4. This first work used an anodic electropaint, requiring the tip to be polarized at a positive potential.[73] In our experience, we have instead used a cathodic paint for tip insulation, as described in the following text.

Electrophoretic paints consist of ionic oligomers dispersed in an organic solvent and usually diluted with water to form an emulsion used for deposition. When cathodic current (for a cationic paint) is driven through a tip connected to a two-electrode cell containing the electropaint solution, the oligomers migrate to the tip and deposit on the surface due to reactions with the reduction

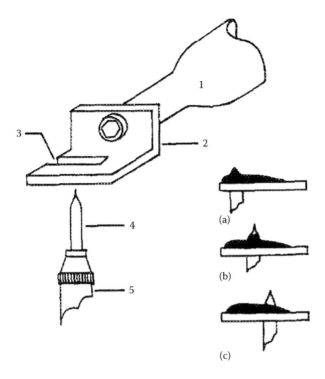

(a)

(b)

(c)

FIGURE 20.3 Illustration of the apparatus used by Nagahara and coworkers for preparing insulated tips. (1) Soldering iron, (2) copper plate, (3) 1 mm wide slit, (4) STM tip, and (5) holder for manipulator. Where the tip penetrates the Apiezon wax (black) determines how much of the tip is insulated. (a) When translated through a region that is too cold, the tip is completely covered with wax. (b) The optimum region allows only the apex of the tip to be exposed. (c) When translated through a region that is too hot, the tip receives little insulation. (Reprinted with permission from Nagahara, L.A., Thundat, T., Lindsay, S.M., Preparation and characterization of STM tips for electrochemical studies, *Rev. Sci. Instrum.*, 60, 3128–3130, 1989. Copyright 1989, American Institute of Physics.)

products nearby (usually OH⁻). After electrophoretic deposition of the paint, tips are transferred to an oven to cure the polymer film. This curing period strengthens the coating and also causes it to shrink slightly, which for a sharp tip results in exposure of the apex. Because no part of this procedure requires manual application of material to tips, this electropainting method is far less effort intensive than either the application of molten materials (wax and glass) or dipped varnishes. Most published reports have deposited the electropaint at a large applied potential (–7 to –55 V, for cathodic electropaint),[61,78] which is required to drive the deposition in a standard water emulsion of the paint. In the authors' experience, we have found that applying the electropaint instead in multiple deposition cycles (4 or 5) at a smaller applied potential (–2 V) and in an organic solution (not an emulsion) resulted in a more even coating of the paint with a higher fraction of tips exhibiting low electrochemical leakage currents (<100 pA) under typical imaging conditions.[79] Under these conditions, the circular counter electrode commonly employed for electropaint application was not needed, enabling batches of 8–12 tips to be connected in parallel for the deposition, with a length of Pt wire as the counter electrode immersed a few centimeters away.

A second important consideration related to the use of insulated tips is the potential of the tip relative to the reference electrode during EC-STM operation. Even with the use of a bipotentiostat to independently control the electrochemistry at the imaged sample surface, the bias voltage used for imaging still affects electrochemical processes occurring at the tip surface. If the desired tip–sample bias voltage causes the tip potential to be set to a region where a redox process may occur,

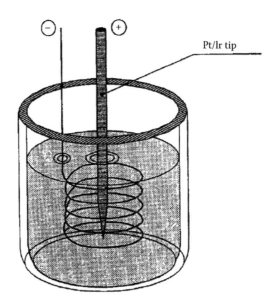

FIGURE 20.4   Cell used for electropainting STM tips. (Adapted from Bach, C.E. et al., Microscopy tips for electrochemical studies using an electropainting method, *J. Electrochem. Soc.*, 140, 1281, 1993. Copyright 1993, The Electrochemical Society. With permission.)

the magnitude of the electrochemical leakage current will be much greater than if the tip potential is shifted to a region mostly devoid of redox processes. Thus, tips with superior insulation (i.e., with a very low exposed surface area) may be used with a wider range of bias voltages and sample potentials than those with poor insulation. The potential-dependent surface chemistry of the tip metal must also be considered when selecting a bias voltage. For example, the surface of tungsten tips (often chosen for their superior hardness and ease of preparation compared to Pt/Ir) may become oxidized at positive potentials in the EC-STM cell. Thus, bias voltages during EC-STM with these tips must be chosen such that the tip potential remains negative of the region where an insulating oxide layer can form at the tip's surface.

20.3.2 Preparation of Metal Surfaces for EC-STM

By far, the majority of EC-STM studies have been carried out on metal surfaces. Metals, especially noble metal single crystals, have been obvious first choices for many types of electrochemical surface studies due to their wide potential windows of polarizability and their highly reproducible kinetics.[80–82] For EC-STM, in contrast to conventional STM, local images from in situ STM on metal single crystals may be interpreted in concert with voltammetry recorded using larger area single-crystal electrodes, enabling elucidation of potential-dependent surface processes. In order to function adequately as a sample for EC-STM, a metal substrate must exhibit large (>100 nm diameter) areas of atomic flatness while also being durable enough to survive introduction to the electrochemical cell. Also of major importance are the methods available for cleaning a particular substrate. Given the very high susceptibility of metal surfaces to contamination from air, scrupulous cleaning of the surface (either by flame annealing or by electrochemical methods) is generally required immediately before introduction to the EC-STM cell. In general, metal substrates used for EC-STM have been either epitaxially grown thin films or—more commonly—monolithic single crystals, in particular bead-type (Clavilier-type) single crystals.

Thin films grown on either mica or Cr-modified glass have been used on a limited basis as substrates in EC-STM for a small number of metals including Au and Pt.[83–86] The preparation of epitaxially grown metal thin films, in particular for the case of Au, has been described extensively in the literature, as such films have been the predominant choice among metal substrates for many types of nonelectrochemical surface studies.[87–91] One refinement to the general procedure of film deposition by thermal evaporation has been the use of template stripping, in which the film is inverted and the underlying substrate is removed to present a metal surface exhibiting decreased roughness.[92] Epitaxially grown thin films may also be thermally annealed and cleaned by careful heating of the substrate (generally indirectly and in an inert atmosphere) immediately prior to the STM experiment.[88,93] Epitaxially grown metal thin films do possess some desirable characteristics for EC-STM in that they are relatively straightforward to introduce to a liquid STM cell (usually by assembling the entire liquid cell atop a sealed metal thin film) and that they are also cheaper on a per-sample basis than monolithic single crystals (although a large initial investment in high vacuum instrumentation is required to produce them).

On the other hand, metal thin films present several major limitations as compared to monolithic single crystals. First, the thermal annealing used to prepare such metal films will favor the formation of grains having the lowest energy crystal face, such as the (111) face for the case of Au. As a result, surfaces showing higher energy crystal faces (such as Au(100)) cannot be readily obtained by this method. Furthermore, the grain size and topography (i.e., the formation of useful flat terraces rather than small, rounded grains) depend on numerous parameters, including the vacuum chamber pressure, metal deposition rate, substrate temperature, and annealing time, and thus, extensive optimization may be required to produce films usable for STM.[88,91]

Metal thin films also present some additional difficulties specific to EC-STM and other electrochemical experiments. Due to the sensitivity of EC-STM (and electrochemistry in general) to contaminants on the electrode surface arriving from ambient air, it is often necessary to clean metal substrates such as Au and Pt by flame annealing immediately prior to the experiment. Unlike a monolithic single crystal, which can be easily heated to near its melting point for optimal annealing, metal films may become unstable due to decomposition of the mica or glass substrate if heated similarly. Therefore, more complex or indirect annealing methods are often required for metal films.[93] Compared to monolithic single crystals, thin films are also much less straightforward to use with supporting electrochemical techniques such as cyclic voltammetry due to their large surface area and the contribution of rough grain boundaries to the observed electrochemistry. Lastly, use of metal thin films for electrochemistry is often complicated by delamination of the metal film from the mica or glass substrate due to the hydrophilicity of the underlying material. In our experience using Au(111) films for EC-STM, experiments that lasted longer than a couple of hours or involved exchange of the electrolyte solution often saw the Au film peeling away from the substrate, usually beginning at cracks near the edge of the film where it was sealed using an o-ring. This delamination problem could be reduced by depositing much thicker Au films than normal, but doing so had the counterproductive side effect of also reducing the grain size on the surface.

In contrast to thin films, monolithic single crystals present improved resilience and ease of cleaning when used for electrochemical experiments including EC-STM. Due to these advantages, such single crystals have been by far the predominant choice for EC-STM investigations by major groups in the field.[19,23,29,76,94] The 1980 work by Clavilier and coworkers found that high-quality monolithic Pt single-crystal substrates could be prepared by first melting high-purity Pt wire in a flame to produce a single-crystal bead and then cutting the bead along crystallographic directions to yield a single-crystal face with a large surface area.[95] The generation of spherical Pt single crystals by melting high-purity wire was first noted by Kaischew in 1955,[96] but Clavilier's refinement to include cutting of the bead (and thus, the ability to create different crystal face electrodes with large, uniform surfaces) massively expanded the applicability of the technique within electrochemistry. Additional refinements by Hamelin allowed for more sophisticated alignment and cutting of the beads.[97] Since Clavilier's work with Pt single crystals, similar techniques have been used to prepare bead crystals of several metals, including Au, Rh, Ir, and Pd, among others.[81,98–100] More recently, Voigtländer and coworkers have modified the bead formation procedure to use electron beam heating in vacuum, which has enabled the generation of bead single crystals of

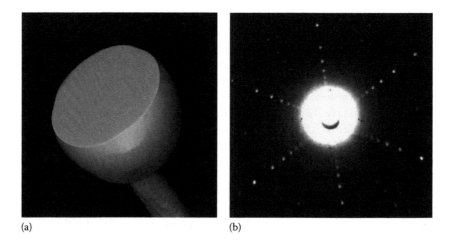

(a)   (b)

FIGURE 20.5   (a) SEM image and (b) Laue x-ray diffraction pattern of a high-quality Pd(111) bead electrode. Diameter of the polished surface: 2.5 mm. (Reprinted from *Electrochim. Acta*, 52, Hara, M., Linke, U., and Wandlowski, T., Preparation and electrochemical characterization of palladium single crystal electrodes in 0.1 M H_2SO_4 and $HClO_4$: Part I. Low-index phases, 5733–5748, Copyright 2007, with permission from Elsevier.)

more reactive metals, such as Ag, Cu, and W.[101] An scanning electron microscope (SEM) image and a Laue x-ray diffraction pattern for a 2.5 mm diameter Clavilier-type Pd electrode prepared by Hara and coworkers are shown in Figure 20.5.[102]

In our own lab, we have had success forming Au(111) bead electrodes by melting high-purity Au wire in a hydrogen/oxygen flame. By heating the wire in air and allowing molten gold droplets to roll up at the wire end, single-crystal beads having diameters between 2 and 3 mm can be routinely formed. The single-crystal nature of the beads can be later verified by short etches in heated aqua regia, which will reveal distinct grain boundaries for polycrystals. Allowing the molten Au to cool slowly to room temperature in air causes the formation of facets on the bead surface. In our experience, facets could be grown to roughly 0.5–1 mm in diameter by repeated melting and solidification. We were able to perform atomic resolution EC-STM imaging directly on these facets in dilute electrolyte solutions and found them to show the hexagonally close-packed atomic structure characteristic of the Au(111) face.[30]

Such as-formed Au(111) and Pt(111) facet electrodes have been used by numerous groups for EC-STM.[19,20,29,76,99] As they are generated through recrystallization rather than any mechanical cutting or polishing, these facets are perhaps the most pristine metal single-crystal surfaces possible. However, the facets can also be problematic to isolate for nonlocal techniques such as cyclic voltammetry, and thus, beads cut using techniques similar to Clavilier's method are more often applied for these larger area experiments. For Clavilier-type electrodes, single-crystal beads are prepared as mentioned earlier, and then the recrystallized facets are used as a guide to align the crystal along a desired crystallographic direction.[81,95] Once aligned, the crystal can be ground or cut parallel to the desired crystal face using a lapping device. After cutting, the crystal must be subjected to many cycles of polishing and high-temperature annealing to return an atomically flat surface. This type of cut electrode is advantageous as the larger crystal face is much easier to isolate for supporting experiments using techniques such as the hanging meniscus contact, in which the cut face of the bead electrode is contacted to the surface of the electrolyte solution and then drawn up to create a meniscus that isolates it for voltammetry.[103] Also, while the initial recrystallized facets favor the formation of low-index crystal faces due to their lower surface area, Clavilier-type electrodes may be cut to expose higher-index faces for characterization by either voltammetry or EC-STM.[95,97]

An advantage of using Au single crystals is that the metal surface is more thermodynamically stable than its oxide.[98] This enables straightforward annealing and cleaning of the surface by exposure of the crystal to a hydrogen or hydrogen/oxygen flame in air (i.e., flame annealing). For more active metals such as Ag or Cu, high-temperature annealing must be performed in an inert atmosphere.[98] We have had success cleaning and annealing our Au single crystals by heating them to a red/orange glow in a pure hydrogen flame for 5 min in air, followed by cooling for 10 s in air and then immersion in pure water.[104] By *sealing* the freshly prepared surface in a droplet of pure water, it is protected from further contamination from the air during transfer to the EC-STM cell. The effectiveness of this cleaning procedure can be verified using either cyclic voltammetry (which will show peak shapes characteristic of the (111) surface)[81] or atomic resolution EC-STM imaging.

During flame annealing of a Clavilier-type electrode, the crystal must not be heated past its melting point (evident by a bright yellow glow for Au), as this will destroy the cut surface. It is important to recognize that there is no bulk flow of molten material during the annealing procedure, and flame annealing will not remove deep scratches on the surface visible to the naked eye. Instead, the objective is to promote the formation of large atomic terraces on the already polished surface. A point of contention in the field has been quenching of single-crystal electrodes by rapid immersion of the still hot crystal in water immediately following flame annealing.[30,105] Some authors have performed this quenching procedure using water saturated with hydrogen to limit oxide formation on the metal surface.[12] In principle, immediate transfer into a liquid should reduce the amount of surface contamination arriving from ambient air. However, it has been demonstrated that such rapid cooling procedures can have a major impact on the observed surface structure and result in the formation of defects.[30,106]

20.3.3 EC-STM Cell

Many of the operating requirements for EC-STM cells and sample stages largely mirror those for conventional liquid phase or vacuum STM, which have been addressed in the literature.[36,39,40] Specific to EC-STM are considerations related to compatibility of the liquid cell parts with the electrolyte solution, insertion of the reference and auxiliary electrodes into the EC-STM cell, and the ability of the cell to accommodate different types of substrate electrodes (e.g., thin films or monolithic single crystals). Figure 20.6 shows a schematic of the EC-STM cell we have used routinely in our lab. This cell closely resembles similar cells that have been described by Itaya and Wang.[5,18]

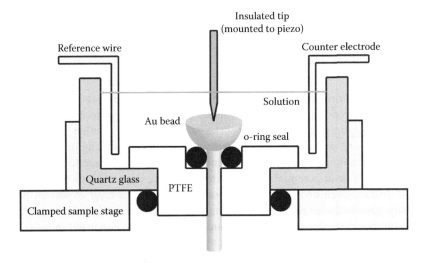

FIGURE 20.6 Cross section illustration of an EC-STM cell constructed to accommodate a Clavilier-type Au bead crystal as the sample. The volume of liquid used with a cell of this design is around 0.8 mL.

Our design differs from these slightly by the long stem wire attached to the Au crystal, which we are able to accommodate because of the top-down design of our STM (with the piezo positioner connected to the tip rather than the sample stage).[79] More commonly, EC-STM cells using bead single crystals find those beads spot-welded to an Au plate that creates the cell bottom.[18] In our case, the longer stem wire enables easy handling of the crystal during flame annealing and also reduces difficulties in transferring the electrode to a hanging meniscus cell for supplemental characterization by voltammetry. We have used both homemade, spherical Au bead crystals and cut, Clavilier-type Au(111) electrodes for EC-STM imaging with this cell. We have also occasionally performed EC-STM imaging using Au thin films evaporated onto mica slides and held in a more conventional liquid STM cell (with the cell walls clamped to the substrate), but difficulties with the film preparation and its stability in electrolyte solution have caused us to instead select Au bead single crystals for nearly all experiments.

Glass and Teflon have been the preferred materials for constructing EC-STM cells due to their high resilience in aqueous solution and, thereby, the reduced possibility of contaminants leaching into (or out of) the cell material. In our own experiments, the glass and polytetrafluoroethylene (PTFE) parts of the cell were washed in heated piranha solution (3:1 mixture of concentrated H_2SO_4 and 30% H_2O_2 solution; CAUTION: This solution is very oxidizing and should never be stored in closed containers) and then thoroughly rinsed using pure water immediately prior to assembling the EC-STM cell. The perfluoroelastomer o-ring sealing the Au bead crystal to the cell bottom is a necessary liability with this design; the hydrophilicity of the freshly cleaned Au causes the cell to leak easily without an o-ring, but the o-ring itself cannot withstand repeated cleaning with piranha solution and so becomes a potential source of organic contaminants. To mitigate this issue, we have stored o-rings for the EC-STM cell in pure water (exchanged periodically) for several days before each experiment.

Because the EC-STM is usually operated with the tip and sample at fixed potentials in dilute electrolyte solutions, the overall currents are low and, therefore, requirements for the reference and counter electrodes are modest. Several authors have used simple Pt wire counter electrodes with a second Pt or Ag wire serving as a quasi-reference electrode.[18,19,29] The potential of the quasi-reference electrode may later be checked against a proper reference such as the Ag/AgCl electrode using voltammetry. Some significant additions to the design of EC-STM cells since the early 1990s have been the incorporation of gas-handling systems to allow transfer of the sample electrode to an UHV system for follow-up characterization without exposing the electrode to laboratory air,[107] as well as the development of liquid flow cells for EC-STM.[108]

20.4   EC-STM STUDIES OF BARE AND ANION-MODIFIED METAL ELECTRODES

20.4.1   METAL ELECTRODES FREQUENTLY USED AS EC-STM SUBSTRATES

Unmodified (*clean* or *bare*) electrode surfaces have been extensively studied by EC-STM, dating back to the very first reports utilizing STM under potential control to characterize Au(111) and highly oriented pyrolytic graphite (HOPG) surfaces in electrolyte solution.[3–5] In this context, the terms *unmodified, bare,* and *clean* all indicate that the electrode is being imaged either in a solution containing only anions that do not specifically adsorb to the surface to an appreciable extent (such as F^- or ClO_4^-) or at potentials at which more surface-active anions (such as Cl^- or SO_4^{2-}) are desorbed. In fact, even under these conditions, there is still a layer of solvent molecules and nonspecifically adsorbed electrolyte ions in contact with the electrode.[49] Under potential control in liquids, many metal single-crystal surfaces are found to be reconstructed in order to maximize favorable atomic interactions at the metal/electrolyte interface.[30] This process of metal surface reconstruction depends strongly on both the applied potential and the nature of the electrolyte solution (i.e., the possibility of specific anion adsorption).[109,110] The phenomena of anion adsorption and potential-dependent surface reconstruction are discussed further in Section 20.4.2. In this section, we describe EC-STM work related to the characterization of unmodified metal electrode surfaces, with emphasis placed on Au and Pt, which have been some of the most frequently employed substrates in the field of EC-STM due to their relative ease of preparation and reproducible surface characteristics.

20.4.1.1 Au(111) and Pt(111)

The (111) faces of Au and Pt have been particularly popular metal substrates in EC-STM, at least in part because they are the lowest energy crystal faces of these metals, making it possible to produce high-quality (111) substrates directly from a melt without any further cutting or polishing of a single crystal. Au and Pt also present relatively wide potential windows in aqueous solution as compared to more reactive metals (such as Ag or Cu),[49] making them ideal substrates for the characterization of surface-adsorbed redox species. These faces of Au and Pt have historically been some of the most characterized in the field of EC-STM, starting with early potential controlled observations of singe atom steps[12,111] followed by atomic resolution imaging of the (111) terraces themselves.[14,112–114]

Several studies carried out in the 1990s by Itaya's group illustrate the usefulness of Pt(111) as a substrate for EC-STM.[14,111,114–116] The 1999 work by Kim and coworkers provides a representative example of atomic resolution EC-STM imaging of the unmodified Pt(111) surface.[116] An image of a Pt(111)−(1×1) surface exposed to 0.1 M HClO$_4$ and at a substrate potential set to +0.3 V versus RHE is shown in Figure 20.7. This work found the expected hexagonally close-packed structure of the Pt(111)−(1×1) surface with an interatomic spacing of 0.28 nm. Even at this significantly positive applied potential, the Pt(111) atomic structure remains visible due to the very low amount of specific ClO$_4^-$ adsorption. In the presence of a more surface-active anion such as Cl$^-$, the structure of an adsorbed anion layer may be found instead of the Pt(111) substrate (a phenomenon described further in Section 20.4.2).[109] A key characteristic of Pt(111) is that its surface does not undergo

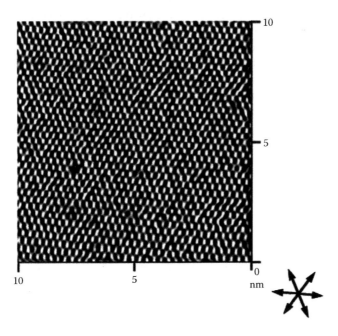

FIGURE 20.7   In situ STM image of a Pt (111) substrate at +0.3 V versus RHE in pure 0.1 M HClO$_4$. The potential of the tip was +0.35 V versus RHE, and the tunneling current was 10 nA. The arrows refer to the close-packed directions of the Pt (111) substrate. The image was 2D Fourier transform filtered. (Reprinted with permission from Kim, Y.-G., Yau, S.-L., and Itaya, K., In situ scanning tunneling microscopy of highly ordered adlayers of aromatic molecules on well–defined Pt(111) electrodes in solution: Benzoic acid, terephthalic acid, and pyrazine, *Langmuir*, 15, 7810–7815, 1999. Copyright 1999 American Chemical Society.)

potential-dependent reconstruction—the hexagonally close-packed (111)–(1 × 1) structure persists irrespective of anion adsorption or the applied potential.

In the work that produced the image shown in Figure 20.7, the authors went on to introduce benzoic acid to the EC-STM cell, which then became specifically adsorbed to the surface and was monitored by EC-STM imaging.[116] This illustrates an important principle for EC-STM experimental design when working with adsorbing species; the bare substrate is first imaged under potential control as a means of calibrating the observation, and then an adsorbing species is introduced to enable imaging of an adsorbed layer generated in situ. One advantage of EC-STM is that adsorption can often be controlled by the applied potential, so that the experimenter may characterize either the bare substrate or an adsorbed monolayer within the same cell. This type of experiment is of particular importance for elucidating commensurate structures of adsorbed species on metal surfaces.

Au(111) has been the single most commonly used substrate for EC-STM, starting from the first EC-STM studies in the early 1990s through to many recent examples.[23,64,75,76,117] In contrast to Pt(111), Au(111) surfaces can be found in aqueous solutions both reconstructed (as the ($\sqrt{3} \times 22$) structure) and unreconstructed (as the hexagonally close-packed (1 × 1) structure) depending on the applied potential and the presence of adsorbing anions.[109,118] In the absence of stabilizing interactions with specific adsorbates, the surface structure of Au(111) contracts along the [110] direction, with every 23rd surface atom retaining the position expected from the underlying bulk. A drawing showing the overlaid geometries for both the compressed ($\sqrt{3} \times 22$) reconstruction and the underlying (1 × 1) structure is shown in Figure 20.8.

The reconstruction of Au(111) results in the formation of raised ridges on the surface with dimensions much larger than the unreconstructed unit cell, resulting in the so-called herringbone pattern of Au(111).[30,119,120] While the ($\sqrt{3} \times 22$) structure has been routinely imaged using STM in vacuum, the herringbone pattern has also been imaged using EC-STM in electrolyte solutions. An EC-STM image of the reconstructed Au(111) surface recorded in 0.1 M HClO$_4$ with an applied sample potential of +0.1 V versus Ag/AgCl is shown in panel (a) of Figure 20.9. This image shows both single atom steps on the Au(111) surface and the raised ridges characteristic of the reconstruction, which appear as faint lines on the Au(111) terraces. The cross sections in panel (b) of the figure indicate a height of roughly 100 pm for the raised ridges, which is less than half of that expected for a monoatomic step on the Au(111) surface.

The observed structure for Au(111) in electrolyte solutions depends on both the applied potential and the electrolyte solution; with the electrode potential set negative of the potential of zero charge (PZC), the amount of specific adsorption by anions is low, and the reconstructed ($\sqrt{3} \times 22$) surface is observed. When the electrode potential is moved positive of the PZC, the ($\sqrt{3} \times 22$) reconstruction lifts to return the (1 × 1) structure. This generally coincides with the onset of a larger degree of specific anion adsorption.[109,110] For electrolyte systems containing strongly adsorbing anions, EC-STM

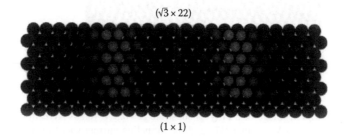

FIGURE 20.8 Structure of the reconstructed surface of Au(111). (Reprinted from *Prog. Surf. Sci.*, 51, Kolb, D.M., Reconstruction phenomena at metal-electrolyte interfaces, 109–173, Copyright 1996, with permission from Elsevier.)

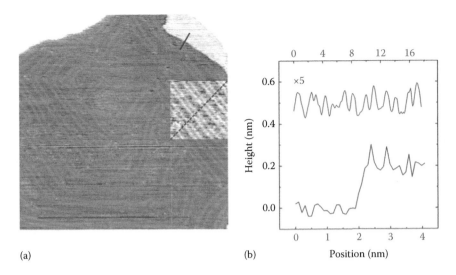

(a)   (b)   Position (nm)

FIGURE 20.9 In situ STM images of a clean Au(111) electrode surface in 0.1 M HClO$_4$, showing (a) an image of the characteristic surface topography, exhibiting steps and the herringbone surface reconstruction (69×69 nm^2, tunneling current = 0.5 nA, sample potential = +0.1 V versus Ag/AgCl, tip potential = −0.11 V), (b) cross sections along the lines indicated in panel (a). (Reprinted with permission from Sripirom, J., Kuhn, S., Jung, U., Magnussen, O., and Schulte, A., Pointed carbon fiber ultramicroelectrodes: A new probe option for electrochemical scanning tunneling microscopy, *Anal. Chem.*, 85, 837–842, 2013. Copyright 2013 American Chemical Society.)

imaging of the (1×1) structure is often convoluted with imaging of the adsorbed anion layer.[75,109] Likewise, reappearance of the herringbone pattern in EC-STM images has often been found to accompany the potential-dependent desorption of an adsorbed species.[20,121] On the other hand, if the Au(111) surface is imaged in a solution containing anions that exhibit very low specific adsorption, such as ClO$_4^-$, the (1×1) structure can be observed.[122] Representative EC-STM images of the (1×1) structure of Au(111) recorded in HClO$_4$ solution are shown in Figure 20.10. Further discussion of EC-STM experiments examining Au(111) surface reconstruction can be found in Section 20.4.2.

20.4.1.2 Other Metal Electrodes Used in EC-STM

Although Au(111) and Pt(111) have been by far the most commonly used Au and Pt substrates for EC-STM, other crystal faces of both metals, such as Au(100), Au(110), Pt(100), and Pt(110), have also seen application in the field.[123–126] A representative EC-STM image of the Pt(110)–(1×1) surface recorded for a clean Pt electrode in 0.1 M HClO$_4$ is shown in Figure 20.11.[123] The (110)–(1×1) surfaces for both metals are capable of reconstructing to give the ridged (110)–(2×1) structure.[30]

Single-crystal surfaces of other noble metals, including Rh and Pd, have also been used as substrates for EC-STM. As with Au and Pt, these metals possess electrochemical windows and surface characteristics suitable for EC-STM studies of adsorbed species.[14,99,127] Similar to Au, the most frequently employed crystal faces for Rh and Pd electrodes have been the (111) faces, which present a hexagonally close-packed surface structure.

Single crystals of reactive metals such as Cu,[128,129] Ag,[21,72] Fe,[130,131] and Ni[132] have also seen application in EC-STM. These metals have more limited potential windows due to their less positive reduction potentials[49] and, thereby, the ease of irreversible oxide formation on the electrode surface at positive applied potentials. As a representative example of atomic resolution EC-STM imaging of a reactive metal surface, Figure 20.12 shows an image recorded for an Fe(110) electrode in 0.1 M NaClO$_4$. The large negative applied sample potential of −1.62 V versus a Pt quasi-reference

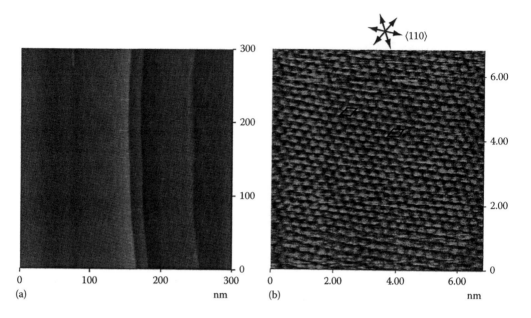

FIGURE 20.10 EC-STM images of Au(111) in 0.1 M HClO₄. (a) Large-scale EC-STM image obtained at a sample potential of +0.13 V versus standard calomel electrode (SCE), showing a clean, atomically flat Au(111) surface, and (b) the Au(111)–(1 × 1) structure obtained at +0.29 V versus SCE. Scanning rates were 8 Hz for (a) and 12 Hz for (b), respectively. Tunneling currents were (a) 1.0. and (b) 3.2 nA. A set of arrows indicates the [110] directions of Au(111). (Reprinted from *Electrochim. Acta*, 48, Kong, D.-S., Wan, L.-J., Han, M.-J., Pan, G.-B., Lei, S.-B., Bai, C.-L., and Chen, S.-H., Self-assembled monolayer of a Schiff base on Au(111) surface: Electrochemistry and electrochemical STM study, 303–309, Copyright 2002, with permission from Elsevier.)

electrode used during the recording of this image was required to prevent the formation of a surface oxide layer, the onset of which was found to occur at roughly −1.30 V.[130]

20.4.2 EC-STM Studies of Anion-Modified Electrodes and Surface Reconstruction

Adsorbed monolayers of anions have been some of the most extensively studied systems by EC-STM due to their major influence on the structure of electrode/solution interface. As a working electrode's potential is moved positive of its PZC, there will nearly always be some degree of specific adsorption of anions, with a range from those anions with a strong affinity for metal electrode surfaces that adsorb very readily (such as I⁻) to those that adsorb appreciably only with a larger positive applied potential (such as SO_4^{2-}).[109,110] At potentials sufficient to allow the specific adsorption of anions, EC-STM may be used to produce atomic resolution images of the adsorbed anion monolayer. In this section, EC-STM studies of anion adsorption will be discussed briefly, again with some emphasis placed on the most extensively used Au(111) electrodes. We will also discuss here the use of EC-STM to monitor the reconstruction of metal electrode surfaces, as anion adsorption plays a critical role in the potential dependence and dynamics of these processes. An especially thorough review article on the topic of anion monolayers at metal electrodes has been written by Magnussen.[109]

It has been shown that there is significant partial charge transfer between anions and the electrode surface upon adsorption.[109] Therefore, there has been some ambiguity as to whether such an adsorbed layer should be thought of as an anion monolayer (e.g., a *halide-modified* or *sulfide-modified electrode*) or a monolayer of a neutral species (e.g., a *halogen-modified* or *sulfur-modified* electrode), given that the true state of the surface species is somewhere in between these two. To

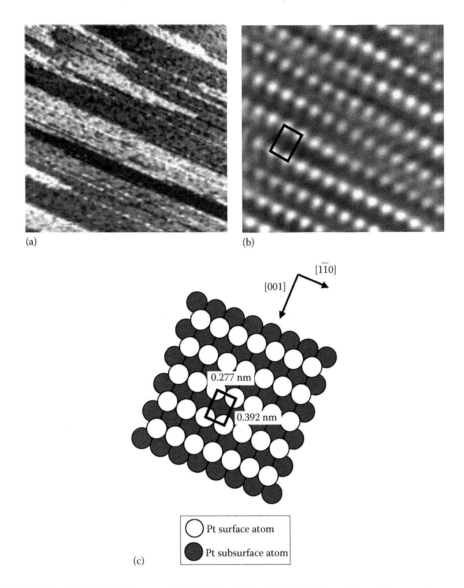

FIGURE 20.11 (a) EC-STM image of Pt(110)–(1×1) in 0.1 M HClO$_4$ solution. The scanning area was 100×100 nm^2. The electrode potential (E), bias voltage ($V_b = E_{tip} - E_{sample}$), and tunneling current (I_t) were +0.1 V versus RHE, +0.4 V, and 9.7 nA, respectively (no filtering used). (b) Atomic resolution image of Pt(110) in 0.1 M HClO$_4$ solution. The scanning area was 3.3×3.3 nm^2 ($E = +0.1$ V vs. RHE, $V_b = +0.03$ V, $I_t = 8.8$ nA, low pass filtered). (c) Ball model of the Pt(110)–(1 × 1) substrate. (Reprinted with permission from Wakisaka, M., Asizawa, S., Yoneyama, T., Uchida, H., and Watanabe, M., In situ STM observation of the CO adlayer on a Pt(110) electrode in 0.1 M HClO$_4$ solution, *Langmuir*, 26, 9191–9194, 2010. Copyright 2010 American Chemical Society.)

minimize inconsistency with existing literature, we will refer to monolayers generated by adsorbing halides as *halogen monolayers*, while for the polyatomic species we will use the anion name throughout.

20.4.2.1 Halides

Since the first atomic resolution images of the iodine adlayer on Pt(111) were produced by Yau et al.,[133] numerous halide/metal systems have been characterized by EC-STM. In particular, adlayers

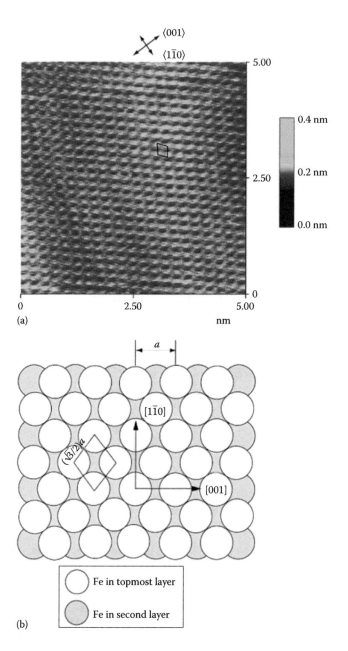

FIGURE 20.12   (a) Atomically resolved STM image of the Fe(110) surface showing a (1×1) structure. The image was obtained in 0.1 M NaClO$_4$ at −1.62 V versus a Pt quasi-reference with a scan rate of 12 Hz and a tunneling current of 28 nA. (b) Schematic top view of the Fe(110)–(1×1) structure. The rhombus indicates the two-dimensional unit cell of the surface lattice, where a is the lattice constant (a=2.87 Å). (Reprinted with permission from Kim, J.-W., Lee, J.-Y., and Park, S.-M., Effects of organic addititves on zinc electrodeposition at iron electrodes studied by EQCM and in situ STM, *Langmuir*, 20, 459–466, 2004. Copyright 2004 American Chemical Society.)

of I and Br on both Au(111) and Pt(111) have been investigated by multiple groups.[75,114,133,134] Cl⁻ adsorption on Au has also been studied using EC-STM, although to a lesser extent due to anodic dissolution of the metal.[135] There have also been reports for halogen monolayers on other crystal faces of Au and Pt[136,137] as well as some studies of halogen monolayers on more reactive metal surfaces such as Cu[128,129,138] and Ag.[139] A well-resolved image of a Br adlayer on Cu(110) recorded in 0.1 M HClO₄ containing 1 mM KBr is shown in Figure 20.13. Here, the adlayer adopts a hexagonally close-packed structure, with each bromine occupying a fourfold hollow site on the ridged (110) substrate.[138] In this work, EC-STM imaging of the bare Cu(110) substrate was first performed at potentials negative of those where Br⁻ adsorbs to allow for elucidation of the correspondence between the adlayer and substrate structures.

Several of these monolayers present potential-dependent structures on the metal electrode surface, making them ideally suited for potentiodynamic characterization using EC-STM. For example, both I and Br adlayers undergo potential-dependent compression and rotation on Au(111).[75,134] Figure 20.14 shows atomic resolution EC-STM images recorded at different potentials for an Au(111) substrate immersed in 1 mM NaBr with a 0.1 M HClO₄ background.[134] The first image in panel (a), recorded before the formation of an ordered Br adlayer, shows the underlying Au(111) substrate, visible despite submonolayer coverage of Br expected at this potential. As the potential is gradually moved from +0.44 V versus Ag/AgCl to +0.74 V over the course of the remaining images, the ordered Br⁻ adlayer becomes visible and is observed to change its lattice parameters with the successive potential steps. The larger bright spots visible in panels (c) and (d) are a Moiré pattern generated by size mismatch between the Br layer and the underlying Au(111). Using the size of this Moiré pattern, the Br interatomic spacing was estimated to decrease from 4.17 ± 0.04 Å at +0.59 V versus Ag/AgCl to 4.01 ± 0.04 Å at +0.74 V.

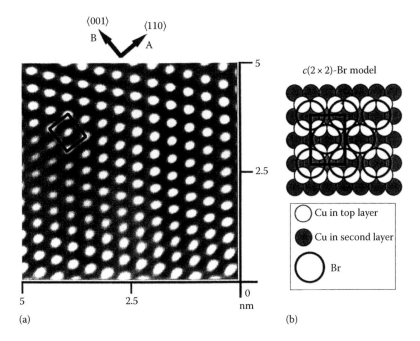

FIGURE 20.13 (a) STM top view of a bromide adlayer on the Cu(110) surface, recorded with the sample potential at −0.35 V versus RHE in 0.1 M HClO₄ with 1 mM KBr. The scan rate and tunneling current were 20.35 Hz and 15 nA, respectively. (b) Schematic representation of the Cu(110)–$c(2 \times 2)$–Br structure. (Adapted from *J. Electroanal. Chem.*, 473, Wan, L.-J. and Itaya, K., In situ scanning tunneling microscopy of Cu(110): Atomic structures of halide layers and anodic dissolution, 10–18, Copyright 1999, with permission from Elsevier.)

(a)   (b)

(c)   (d)

FIGURE 20.14   STM images of (a, c, d) 80×80 Å² and (b) 65×65 Å² surface areas on Au(111) in 0.1 M HClO₄ + 1 mM NaBr. (a) At +0.44 V versus Ag/AgCl, only the hexagonal Au substrate lattice is visible. (b) Upon increasing the potential with a sweep rate of 5 mV/s from +0.48 V (upper edge) to +0.59 V (lower edge), the rotated-hexagonal bromide adlayer is formed (center). Due to the different mismatch of the adlattice and the Au substrate at (c) +0.59 V and (d) +0.74 V, a slightly different long-range modulation pattern is observed. (Reprinted with permission from Magnussen, O.M., Ocko, B.M., Wang, J.X., and Adzic, R.R., In-situ x-ray diffraction and STM studies of bromide adsorption on Au(111) electrodes, *J. Phys. Chem.*, 100, 5500–5508, 1996. Copyright 1996 American Chemical Society.)

20.4.2.2   Sulfate and Phosphate

Monolayers of oxoanions, such as sulfate and phosphate, formed on metal single crystals have been characterized using EC-STM.[140–142] For Au(111), these anions show less affinity for the metal surface than the halides, and may not be easily imaged without applying a sufficiently positive potential. At low applied potentials, the surface coverage of sulfate is so low[110] that it may be deemed nearly *nonadsorbing*, allowing straightforward imaging of the metal substrate (or of more readily adsorbing species). In this way, sulfate is similar to perchlorate in that it is often chosen as a supporting electrolyte for EC-STM studies in which specific anion adsorption is not desired. However, sulfate has been shown to adopt adlayer structures on Au(111) in H_2SO_4 solution. Their structure is more complex than in the case of halogen adlayers and consists of rows of co-adsorbed sulfate anions and water molecules.[142] Figure 20.15 shows a set of EC-STM images of sulfate/water monolayers recorded in 0.5 M H_2SO_4 with an applied potential of +0.9 V versus RHE and a structural

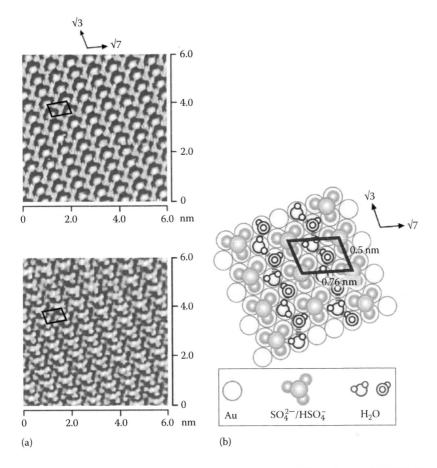

FIGURE 20.15 (a) High-resolution STM images (6×6 nm²) of Au(111) surface in 0.5 M H₂SO₄ acquired with a sample potential of +0.9 V versus RHE. The tunneling currents were 3.0 and 1.0 nA, respectively. (b) Structural model of the ($\sqrt{3}\times\sqrt{7}$) adlayer on Au(111)–(1×1). (Adapted from *Electrochem. Commun.*, 8, Sato, K., Yoshimoto, S., Inukai, J., and Itaya, K., Effect of sulfuric acid concentration on the structure of sulfate adlayer on Au(111) electrode, 725–730, Copyright 2006, with permission from Elsevier.)

model of the adlayer from the work by Sato and coworkers. Sulfate adlayers have also been imaged on other metal single-crystal surfaces, including Rh(111) and Cu(111).[99,143]

20.4.2.3 Cyanide and Sulfide

In contrast to the oxoanions, cyanide and sulfide strongly adsorb to metal surfaces, producing an adlayer with bonds to the substrate that have a strongly covalent character.[144,145] Highly ordered monolayers of both adsorbed cyanide and sulfide have been characterized in aqueous solution using EC-STM.[115,132,144–148] The CN⁻ adlayer on Pt(111) has been found to adopt a potential-dependent structure consisting of hollow, hexagonal rings,[115] as shown in the image and structure drawing in Figure 20.16. Kim and coworkers went on to show that within a limited potential window K⁺ ions may be complexed in the hexagonal hollow sites in the CN⁻ adlayer.

20.4.2.4 Monitoring Surface Reconstruction in the Presence of Various Anions

Adsorbing anions have a major effect on the potentials at which metal surfaces are found to be reconstructed.[110] For the case of halides and oxoanions such as sulfate on Au(111), the potential at which the ($\sqrt{3}\times22$) reconstruction lifts to give the (1×1) surface is found to shift by several hundred millivolts depending on the identity of the adsorbing anion. This is evident from single-crystal

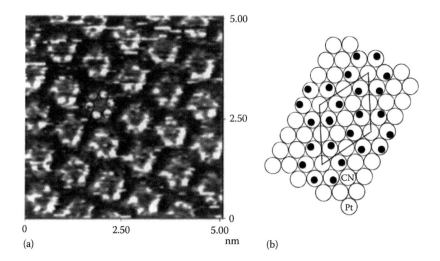
(a)   (b)

FIGURE 20.16 (a) Unfiltered, high-resolution STM image of the hollow CN^- hexagonal arrangement. The image was obtained in 0.1 mM NaCN + 0.1 M $NaClO_4$ (pH 9.5). The electrode potentials of Pt(111) and the tip were +0.6 and +0.5 V versus RHE, respectively. The tunneling current was 20 nA. (b) Ball model of the $(2\sqrt{3}\times2\sqrt{3})R30°$ surface. (Reprinted with permission from Kim, Y.-G., Yau, S.-L., and Itaya, K., Direct observation of complexation of alkali cations on cyanide-modified Pt(111) by scanning tunneling microscopy, *J. Am. Chem. Soc.*, 118, 393–400. Copyright 1996 American Chemical Society.)

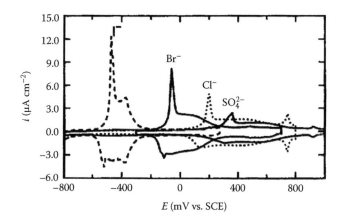

FIGURE 20.17 Comparison cyclic voltammograms for Au(111) electrodes in 0.1 M $HClO_4 + 10^{-3}$ M K_2SO_4 and 0.1 M $KClO_3 + 10^{-3}$ M KCl, KBr, and KI solutions. The sweep rate was 10 mV s^{-1}. (Adapted from *Electrochim. Acta*, 43, Lipkowski, J., Shi, Z., Chen, A., Pettinger, B., and Bilger, C., Ionic adsorption at the Au(111) electrode, 2875–2888, 1996, Copyright 1998, with permission from Elsevier.)

voltammograms recorded in dilute electrolyte solutions (Figure 20.17), where the lifting of the reconstruction is marked by a sharp anodic peak.[110] The current flowing in this region corresponds to both charging current (due to a shift in the PZC of the electrode) as well as partial charge transfer to adsorbing ions, the surface coverage of which has been shown to rise dramatically at potentials just positive of the PZC.[109] Lifting of the reconstruction has also been reported in ClO_4^- solution, although at a more positive potential than for SO_4^{2-}.[150]

The reconstruction process of Au(111) in electrolyte solutions has also been monitored using EC-STM imaging.[149–151] Figure 20.18 shows a series of EC-STM images recorded at multiple potentials for an Au(111) electrode in 0.1 M H_2SO_4.[149] The applied potential of the Au(111) electrode

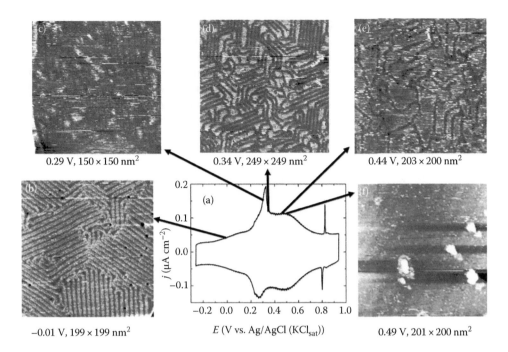

FIGURE 20.18 Cyclic voltammogram at 10 mV s^{-1} (a) and surface morphologies observed using in situ STM on an Au(111) electrode at −0.01 V versus Ag/AgCl (b), +0.29 V (c), +0.34 V (d), +0.44 V (e), and +0.49 V (f) in 0.1 M H$_2$SO$_4$. The tip was kept at −0.105 V. (Reprinted with permission from Vaz-Domínguez, C., Aranzábal, A., and Cuesta, A., In situ STM observation of stable dislocation networks during the initial stages of the lifting of the reconstruction on Au(111) electrodes, *J. Phys. Chem. Lett.*, 1, 2059–2062, 2010. Copyright 2010 American Chemical Society.)

for each image is given by arrows connected to the corresponding potential in a cyclic voltammogram, which appears similar to that shown in Figure 20.17. At potentials far negative of that required to lift the reconstruction, the characteristic herringbone pattern of the Au(111)–($\sqrt{3}\times22$) structure is seen. As the electrode potential is made more positive, passing the large peak near +0.3 V versus Ag/AgCl, the herringbone pattern becomes disrupted before finally disappearing (indicating that the reconstruction has fully lifted) when the electrode is imaged at +0.49 V. This visibly demonstrates that the lifting of the reconstruction occurs over a potential window of several tens of mV in the region where sulfate is gradually reaching its maximum surface coverage, rather than immediately upon passing the tall peak in the CV, where the majority of sulfate adsorption occurs. This result suggests that the lifting of the reconstruction in Au(111) is not due solely to anion adsorption, but rather the positive surface charge of the electrode at this potential. The negatively charged anions instead act to stabilize the nonreconstructed surface.[150] Other metal surfaces for which reconstruction phenomena have been studied using EC-STM include Au(100),[124] Au(110),[125] Pt(100),[126] Ag(111),[16] and Cu(111).[152]

20.4.3 EC-STM Investigations of Metal Deposition

One of the most important applications of EC-STM has been for investigations of metal electrodeposition, due to the great importance of the latter in materials science fields. Both underpotential deposition (UPD),[153] in which a monolayer of metal adatoms is generated at a lower work function metal substrate,[49] and bulk deposition[154–156] have been investigated using EC-STM for numerous metal/substrate systems. EC-STM studies of UPD on Au surfaces include observations of the deposition of Ag,[157] Cu,[83] Cd,[158] Pb,[159] and Hg[160] on Au(111). The strength of EC-STM as a tool for characterizing metal UPD is clear: the UPD monolayer may be generated in situ during EC-STM imaging, allowing for simultaneous atomic resolution characterization of the underlying metal substrate, the UPD monolayer, and the dynamics of the UPD process itself.

Representative of much of the early EC-STM work focused on UPD on Au are the early/mid-1990s reports from Kolb's group.[23,154,161–163] Among these, the 1992 report on Cu UPD at Au(111) highlights the usefulness of performing in situ EC-STM imaging at different potentials.[83] Figure 20.19 shows EC-STM images recorded in this work for an Au(111) substrate in contact with a

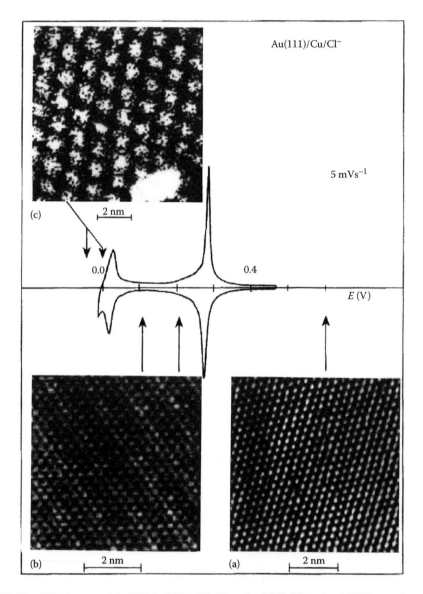

FIGURE 20.19   STM images of Au(111) in 0.05 M H_2SO_4 + 1 mM $CuSO_4$ + 1 mM HCl at various electrode potentials (see arrows in the cyclic voltammogram). (a) Bare Au(111) at +600 mV versus Cu/Cu^{2+}. The tunneling current (I_t) was 120 nA. (b) Au(111) covered with Cu adatoms arranged in a (5×5) superstructure. The Cu next neighbor distance varies with potential from 0.40 to 0.36 nm (I_t = 11 nA). (c) Au(111) with (5×5)–Cu, but now the long-range corrugation of 1.5 nm length is particularly clearly visible. (From Batina, N., Will, T., and Kolb, D.M., Study of the initial stages of copper deposition by in situ scanning tunnelling microscopy, *Faraday Discuss.*, 94, 93–106, 1992. Reproduced by permission of The Royal Society of Chemistry.)

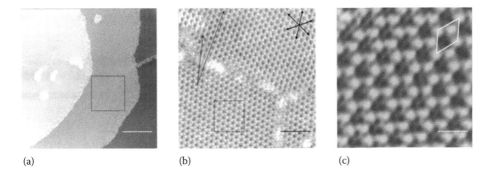

(a) (b) (c)

FIGURE 20.20 In situ STM images with successively finer resolution acquired at −0.55 V in 0.1 M K_2SO_4 with 1 mM H_2SO_4, 1 mM HCl, and 1 mM $In_2(SO_4)_3$. The squares in (a) and (b) indicate regions of zoom-in. The smooth terraces seen in panel (a) are separated by rough edges. Two rotational domains of the ordered structure can be seen on a terrace (b). These two domains enclose an angle of 15° with their molecule rows. This $(\sqrt{43} \times \sqrt{43})R7.6°$ ordered array has a rhombus unit cell marked in (c). Both (b) and (c) were treated with a 2D FT (Fourier transformation) filtering method. Scale bars are (a) 30, (b) 10, and (c) 2 nm. (Reprinted with permission from Pao, T., Chen, Y.-Y., Chen, S., and Yau, S.-L., In situ scanning tunneling microscopy of electrodeposition of indium on a copper thin film electrode predeposited on Pt(111) Eelectrode, *J. Phys. Chem. C*, 117, 26659–26666, 2013. Copyright 2013 American Chemical Society.)

solution containing 0.05 M H_2SO_4, 1 mM $CuSO_4$, and 1 mM HCl. From cyclic voltammetry, it is apparent that the Cu UPD in this solution occurs in two steps at roughly +0.300 and +0.020 V versus Cu/Cu^{2+}. At potentials positive of the redox peaks around +0.320 V, the bare Au(111) surface structure is observed (image A in the figure). At potentials between the two sets of redox peaks, a partial Cu monolayer incorporating adsorbed Cl^- with a larger adlattice is observed (image b in the figure). Finally, at potentials negative of the redox peaks near +0.020 V, a full Cu monolayer is observed, which in this case gave rise to a Moiré pattern due to size mismatch with the underlying Au (image c in the figure). This work also illustrates the large influence of anions on the process of Cu UPD on Au(111), which has been similarly recognized for other metal/substrate systems using EC-STM.[153,164]

UPD as well as overpotential deposition have also been investigated using EC-STM on other metal surfaces, including Pt,[165,166] Ag,[167] and Pd.[168] In recent years, increasingly complex studies of metal deposition using EC-STM have been performed, including studies of further adsorption atop UPD metal adlayers.[169,170] For example, Pao et al. have elucidated the structure of an In adlayer generated by UPD atop a Cu(111) film that was itself generated by overpotential deposition on a Pt(111) single crystal.[170] It was found that the In adlayer adopted multiple potential-dependent surface geometries as the electrode potential was progressively moved through the two-step sequence of In UPD, which occurs in a potential region from between roughly −0.3 and −0.5 V versus Ag/AgCl for a solution containing 0.1 M K_2SO_4 with 1 mM H_2SO_4, 1 mM HCl, and 1 mM $In_2(SO_4)_3$. These potential-dependent structures were also found to be influenced by the presence of adsorbed sulfate/bisulfate or chloride, which are capable of co-adsorbing with the In adlayer during UPD. Figure 20.20 depicts an EC-STM image acquired for an In adlayer atop the Cu(111) substrate, recorded at −0.55 V versus AgCl, showing the geometry of the monolayer at multiple length scales on the Cu(111) terraces.

20.4.4 EC-STM Investigations of Metal Corrosion and Oxidation

EC-STM monitoring has been used to study the dissolution of metal surfaces and the formation of passivating oxide films—both areas of critical importance to corrosion-related fields.[171] Significant metal dissolution processes characterized with EC-STM using single-crystal substrates include

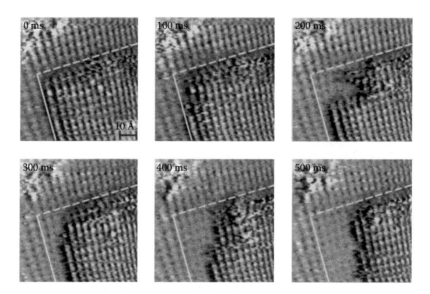

FIGURE 20.21 Series of in situ STM images recorded on Cu(100) in 0.01 M HCl after a potential step from −0.23 to −0.17 V. The images show the progressing dissolution of a Cu terrace starting at the outer terrace corner. The strongly different dissolution behavior at the active and at the stable step (initial positions marked by a solid line and dashed white line, respectively) is clearly visible. (Reprinted from *Electrochim. Acta*, 46, Magnussen, O.M., Zitzler, L., Gleich, B., Vogt, M.R., and Behm, R.J., In-situ atomic-scale studies of the mechanisms and dynamics of metal dissolution by high-speed STM, 3725–3733, Copyright 2001, with permission from Elsevier.)

the anodic dissolutions of Au and Ag,[16,21,135] the dissolution of Cu and Ni in sulfuric acid solution,[138,172–174] and several studies of dissolution at metal single crystals previously modified with electrodeposited metal thin films.[155,175,176] Because the extent of dissolution of a metal substrate can be controlled by the applied potential, sequences of time-dependent EC-STM images can be used to elucidate the dynamics of the dissolution process. Magnussen et al. successfully imaged the dissolution of Cu(100) in HCl solution.[177] In their work, a very high scan rate was used for EC-STM imaging (*video STM*), which has enabled the time-dependent dissolution process at single Cu(100) terraces to be captured with atomic resolution. Figure 20.21 shows a series of time-dependent EC-STM images for a Cu(100) substrate immersed in 0.01 M HCl solution recorded immediately after a potential step from −0.23 V versus SCE to −0.17 V. At this potential, etching of the Cu(100) terraces in the presence of chloride occurs preferentially along the more reactive [010] direction step edges as opposed to the less reactive [001] direction step edges, as illustrated by changes to the terrace highlighted by white lines in Figure 20.21. Closely linked with studies of metal dissolution in corrosion research are investigations of the formation of passivating oxide films on metals.[171] Some electrochemical oxide formation processes that have been characterized using EC-STM include those on single crystals of Ag,[178] Au,[179] Cr,[180] Cu[181], and Ni[174].

20.4.5 EC-STM OF OTHER ELECTRODE MATERIALS: HOPG AND SEMICONDUCTORS

Throughout this section, the discussion focused on the characterization of metal electrodes, as these have been by far the most frequently employed substrates in EC-STM. However, in attempting to provide an overview of substrates used in the field, it is important to recognize significant work that has been done using other electrode materials, chief among these being HOPG, single-crystal Si, and a limited number of other semiconductors. The first use of HOPG as a substrate for EC-STM occurred with the very first paper from Itaya on the implementation of potential control for STM.[5] Compared to metal single crystals, a much smaller number of reports describe EC-STM imaging of bare HOPG substrates or of molecular adsorbates on HOPG.[182,183] More commonly, HOPG has been used as a substrate for the deposition of either metal[184–186] or semiconducting[187] thin films, with subsequent characterization by EC-STM. Multiple groups have performed EC-STM imaging studies on single crystals of n-type Si(111) in aqueous solution,[17,188–190] and one report from Szklarczyk et al. describes p-type Si(111) in a propylene carbonate solution.[191] Due to the importance of Si etching in semiconductor device manufacturing, much of this work has focused on using EC-STM to characterize etching processes at the Si(111)/solution interface, although EC-STM has also been used to characterize films deposited on Si(111) substrates.[192,193] Finally, a small group of binary semiconductors have been used as substrates for EC-STM imaging (among these are GaAs,[194] ZnO,[195] $MoSe_2$,[196] MoS_2,[197] and PbS),[198] and furthermore, there have also been several EC-STM investigations of semiconductors deposited on underlying metal substrates, such as thin films of Si, Ge, Te, and CdTe on Au.[199–203]

20.5   EC-STM OBSERVATION OF ORGANIC ADSORBATES

A wide range of organic adsorbates have been observed with EC-STM under potential control. While initial studies primarily took advantage of the fact that EC-STM is able to image surfaces with submolecular resolution under well-controlled conditions not accessible with any other technique, more recent contributions increasingly focus on the fundamental understanding of surface electrochemistry, the design of novel supramolecular structures at the electrolyte–electrode interface, and the contribution of EC-STM toward the design of functional surfaces.[164,204–209]

20.5.1   SELF-ASSEMBLED MONOLAYERS OF ALKANES

Several thousand publications have been published over the past 30 years on SAMs of alkanethiols on metal surfaces. It would be well outside of the scope of this chapter to comprehensively review all the work that has been performed to characterize such SAMs, even when focusing only on STM studies. Therefore, we will concentrate here on the formation of SAMs and the effect of the applied potential on the conformation and stability of such SAMs as investigated by EC-STM.

The first study of ethanethiol SAMs on Au(111) was only reported in 1999.[210] While oxidative desorption in 0.1 M H_2SO_4 occurs at +1.15 V versus SCE, potentials below 0 V cause a structural transformation, resulting slowly in pit and island formation, but only at −0.31 V is reductive desorption observed. Two differently ordered types of domains, a pinstripe $(p \times \sqrt{3})$ structure and an oblique primitive (4×3) superstructure, are simultaneously observed in the range between +0.75 and 0 V, a range slightly narrower than the one defined by the desorption reactions.

An extremely thorough study of propanethiol SAMs on Au(111) involved not only the determination of the structure of stable, ordered adlayers, but also the process of self-assembly.[211] The initial phase of assembly involves lifting of the Au(111) surface reconstruction, while the following stage results in the formation of a denser adlayer and appearance of triangular pits covering approximately 4% of the total surface. The self-assembly results in a $(2\sqrt{3} \times 3)R30°$ surface lattice consisting of unit cells with four individual thiols. The brighter appearance of one of these was interpreted as arising from different conformations of the adsorbed thiolate (see Figure 20.22).

EC-STM is also useful to elucidate electrochemical adlayer desorption, as demonstrated, for example, in a study of hexanethiol SAMs on Au(111).[121] As may be expected, desorption was confirmed to start at surface defects. Interestingly, images following the reductive desorption in H_2SO_4 showed surfaces with aggregates of desorbed thiols, which is consistent with the low solubility of hexanethiol in acidic solution. No such aggregates were observed in KOH solution, where the deprotonation of the thiol results in a much higher solubility.

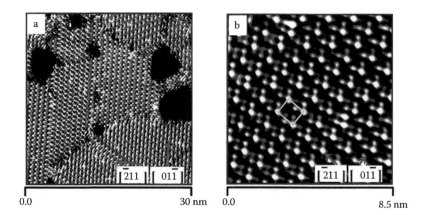

FIGURE 20.22 EC-STM images of a 1-propanethiol SAM on Au(111) in NH$_4$OAc buffer (pH 4.6). Scan area: (a) 30×30 nm^2 and (b) 8.5×8.5 nm^2. (a) $I_t = 0.30$ nA, $V_{bias} = -0.13$ V, sample potential $E_w = -0.07$ V versus SCE; and (b) $I_t = 0.30$ nA, $V_{bias} = 0.07$ V, sample potential $E_w = -0.15$ V. (Reprinted with permission from Zhang, J.D., Chi, Q.J., and Ulstrup, J., Assembly dynamics and detailed structure of 1-propanethiol monolayers on Au(111) surfaces observed real time by in situ STM, *Langmuir*, 22, 6203–6213, 2006. Copyright 2006 American Chemical Society.)

In situ alkanethiol desorption followed by the assembly of another type of thiol offers the possibility to engineer surface properties. This is exemplified by a study from Hobara et al., who first self-assembled mixed monolayers of the short-chain compound 3-mercaptopropionic acid and the long-chain compound 1-hexanethiol on Au(111).[212] Because of spontaneous phase separation in this type of mixed monolayer and the different desorption potential for the two types of thiols, selective desorption of 3-mercaptopropionic acid can be performed. This results in bare patches on the Au(111) surface, in which another thiol can self-assemble.

Evidently, this type of work is not limited to gold surfaces. For example, Scherer et al. studied the corrosion of Cu(100) modified with an adlayer of octanethiol or hexadecanethiol in HCl solution.[213] On one hand, they found that the SAM modification inhibited the formation of corrosion sites and the enlargement of surface defects. On the other hand, they observed surface roughening in a potential range in which no net thiol desorption occurred, presumably due to an exchange of Cu and Cu thiolates with the solution.

20.5.2 Self-Assembled Monolayers of Aromatic Compounds

A key reason for the interest in adlayers of aromatic thiols arises from their molecular level conductivity, which is much higher than in the case of alkanethiols. This has made Au electrodes modified with SAMs of aromatic thiols attractive as substrates for modification with redox-active enzymes.

The first EC-STM study of Au(111) modified with 4-mercaptopyridine in 0.05 M HClO$_4$ showed the adlayer to be highly ordered, taking a ladder like structure.[76] Individual molecules exhibited a molecular plane perpendicular to the surface, with a tilt of the molecular axis with respect to the surface, and the nitrogen atom directed away from the metal surface. In contrast, 2-mercaptopyridine binds to Au(111) both through the nitrogen and sulfur atoms, and thiophenol forms disordered SAMs, explaining why among these three types of SAMs, only the 4-mercaptopyridine adlayer facilitates electron transfer to cytochrome c.[214] Incidentally, the conformation of the 4-mercaptopyridine adlayer depends strongly on the electrolyte.[215] When 0.1 M H$_2$SO$_4$ is used rather than HClO$_4$, a potential-dependent phase transition at +0.35 V versus SCE is observed. This has been proposed to be consistent with HSO$_4^-$ adsorption to the adlayer, forming a hydrogen bond network with the pyridinium moieties of the adlayer. In the even more complicated case of 4-methylbenzenethiol SAMs, multiple phases are observed as a function of the applied potential, modulated by the strength of the adsorbate–substrate interaction.[216] Similarly, a conformational transition has also been found for SAMs of 4-pyridylethanethiol on Au(111), but driven in this case primarily by protonation/deprotonation.[217]

20.5.3 PORPHYRINS AND PHTHALOCYANINES AS ADSORBATES

More EC-STM studies have been devoted to porphyrin and phthalocyanine adlayers on single-crystal metal electrodes than to any other type of organic adlayers. Because extensive reviews have discussed this particular topic,[218–221] we will limit the discussion of this subject here to a few examples that emphasize the particular features of porphyrin and phthalocyanine adlayers and emphasize the versatility that they offer for the design of functional surfaces under electrochemical control.

Different methods of adlayer formation have to be used depending on whether the porphyrin or phthalocyanine adlayer has a low or high solubility in the electrolyte solution in which EC-STM imaging is performed. To enable spontaneous adsorption from aqueous solutions, water-soluble porphyrins that have, for example, four pyridine,[222] methylpyridinium,[19] or sulfonate groups have been employed. Alternatively, porphyrin adlayers have been formed by self-assembly from organic solutions of the adsorbate prior to insertion into the EC-STM cell.[223–225] A nice example of the control of the adlayer structure during the preassembly in an organic solvent was demonstrated by Yoshimoto et al. when they used hydrogen bonds to form highly ordered square arrays of cobalt(II) porphyrins with one or four 4-carboxyphenyl substituents.[225] Later work visualized for adlayers of the same porphyrin the porphyrin reduction as a function of the pH, showing in the same image area the simultaneous presence of readily distinguishable oxidized and reduced porphyrins.[226]

With their large molecular footprint, porphyrins and phthalocyanines tend to adsorb quite strongly to metal surfaces, which can inhibit the lateral mobility needed for the formation of a highly ordered adlayer. To circumvent this problem, Itaya and coworkers introduced the use of iodine-modified single-crystal metal electrodes as the substrate for the self-assembly and EC-STM investigation of porphyrins, crystal violet, phthalocyanines, and other organic species.[19,227] The approach is not only suitable for metal-free but also for metalloporphyrins.[228] While much of this type of work has been performed with iodine-modified Au(111), other metal surfaces have been used too, such as iodine-modified Ag(111).[139] Alternatives to iodine exist as well. While bromine adlayers offer the advantage of a wider electrochemical window than iodine adlayers on Au(111),[104] hydrogen bonding appears to play an important role in porphyrin monolayer formation on Au(111) modified with a sulfate/bisulfate adlayer.[229]

An alternative approach to modulate the strength of the interaction between the organic adlayer and the underlying electrode is based on the control of the electrode potential.[206,222] For example, Borguet and coworkers showed that in 0.1 M H_2SO_4 and in the potential range between −0.2 and +0.2 V versus SCE, a porphyrin derivative with four 4-pyridyl substituents spontaneously forms well-ordered adlayers on Au(111).[222] At lower potentials, the adsorbed porphyrin molecules are too mobile to be imaged by EC-STM, and at higher potentials, the adsorbed molecules interact so strongly with the surface that they lack the mobility to form an ordered adlayer. Interestingly, once an ordered monolayer has been formed, the potential can be raised above the +0.2 V threshold without perturbing the monolayer order. The potential applied to the substrate can also have a crucial effect determining whether adsorption results in a monolayer or a multilayer, as shown by Borguet and coworkers for a metal-free porphyrin with four 4-pyridyl substituents.[230] While the oxidized porphyrin only forms monolayers, a more negative potential results in porphyrin reduction and multilayer formation.

Porphyrin adlayers of the aforementioned type gain particular interest if they perform a function. Examples include the catalysis of O_2 reduction, as in the case of a cobalt(II) tetraphenyl-porphyrin adlayer on Au(111).[223] Another interesting role of underlying porphyrin adlayers is the modulation of the electroactivity of a co-adsorbate, as shown by Yoshimoto for a C_{60} fullerene with a covalently attached ferrocene pendant.[231] When the fullerene was allowed to form an adlayer directly on Au(111) by self-assembly from a benzene solution, a clear redox reaction could not be observed. However, when the Au(111) surface was first modified with a Zn(II), Co(II), or Cu(II) octaethylporphyrin, distinct redox activity was observed for the ferrocene group at +0.78 V versus RHE. This can be correlated to the highly ordered binary adlayers on Au(111), in which the porphyrin makes direct contact to the Au(111) surface while the fullerenes sit atop the porphyrins (see Figure 20.23).

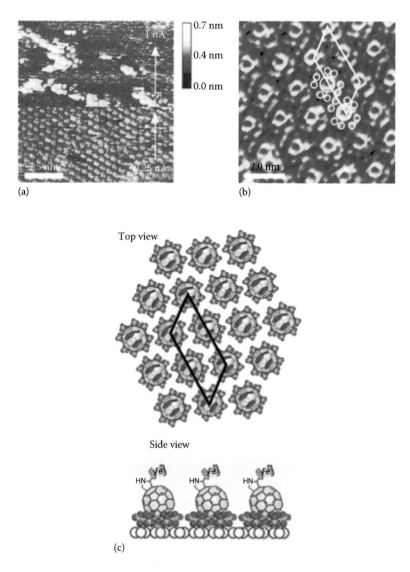

FIGURE 20.23 (a) Composite EC-STM image (30×30 nm²) and (b) high-resolution EC-STM image (8×8 nm²) of the underlying layer of a $C_{60}Fc/ZnOEP$ supramolecular assembled layer on Au(111) in 0.1 M $HClO_4$ acquired at +0.75 V. $I_t = 0.25$ nA, $V_{bias} = +0.35$ V for lower half of (a); $I_t = 1.0$ nA, $V_{bias} = +0.35$ V for upper half of (a); $I_t = 1.8$ nA, $V_{bias} = +0.46$ V for (b). (c) Structural model for supramolecular adlayer consisting of the C_{60} with a ferrocene pendant and zinc(II) octaethylporphyrin. (Reprinted with permission from Yoshimoto, S., Saito, A., Tsutsumi, E., D'Souza, F., Ito, O., and Itaya, K., Electrochemical redox control of ferrocene using a supramolecular assembly of ferrocene-linked C-60 derivative and metallooctaethylporphyrin array on a Au(111) electrode, *Langmuir*, 20, 11046–11052, 2004. Copyright 2004 American Chemical Society.)

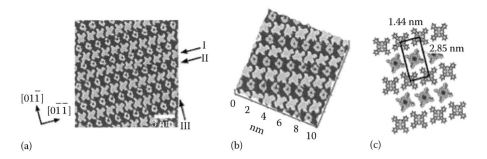

FIGURE 20.24 High-resolution EC-STM image (15×15 nm^2) of a cobalt(II) phthalocyanine/copper(II) octaethylporphyrin mixed adlayer formed on a Au(100)–(hex) surface in 0.1 M HClO$_4$ acquired at +0.75 V versus RHE (a), height-shaded view (10×10 nm^2) (b), and structural model of the mixed adlayer (c). $I_t = 1.2$ nA, $V_{bias} = +0.45$ V. (Reprinted with permission from Suto, K., Yoshimoto, S., and Itaya, K., Two-dimensional self-organization of phthalocyanine and porphyrin: Dependence on the crystallographic orientation of Au, *J. Am. Chem. Soc.*, 125, 14976–14977, 2003. Copyright 2003 American Chemical Society.)

A new level of surface design is achieved when multicomponent monolayers are formed. An excellent surface order with alternating rows of a cobalt(II) phthalocyanine and a copper(II) tetraphenylporphyrin was formed on Au(100)–(hex) in 0.1 M HClO$_4$ (see Figure 20.24).[31,232] The two types of molecules can be readily distinguished by their shape since the phthalocyanine, with its bright spot for the cobalt center, gives a propeller-shaped image while the porphyrins are seen as rings with a dark center. Similar results were also obtained with monolayers consisting of mixtures of a cobalt(II) phthalocyanine and a copper(II) octaethylporphyrin[64] or zinc(II) phthalocyanine and zinc(II) porphyrins.[233]

20.5.4 FULLERENES

The unique capabilities of EC-STM are demonstrated by the formation and observation of fullerene monolayers on Au(111) surfaces. On the one hand, as shown also in several STM studies without electrochemical control, when C$_{60}$ monolayers are formed directly on Au(111), the lateral mobility of the C$_{60}$ molecules is low, and a lot of attention to experimental details is required to avoid the formation of disordered monolayers. For example, using sublimation techniques, the C$_{60}$ deposition speed and substrate temperature have to be controlled very carefully. On the other hand, if C$_{60}$ monolayers are formed on an iodine-modified Au(111) surface, the C$_{60}$ molecules have such a high lateral mobility that imaging does not even reveal direct evidence for the presence of C$_{60}$. Uemura et al. have avoided these problems by first modifying the Au(111) surface with iodine, transferring a monolayer of C$_{60}$ formed with the Langmuir–Blodgett technique at the air–water interface onto the iodine-modified Au(111), and then electrochemically removing the iodine with an applied potential.[234] Key to the formation of high-quality films is a slow removal of the iodine, allowing the C$_{60}$ molecules to form an ordered monolayer on the Au(111) (see Figure 20.25).

In this process, the nature of the underlying metal is crucial, as shown by attempts to replace the Au(111) by Pt(111). Because the interaction of C$_{60}$ is much stronger with both the Pt(111) and the iodine-modified Pt(111) than with the corresponding Au(111) surfaces, ordered monolayers cannot be obtained with either of these Pt substrates.[235] However, as the result of the much stronger interaction of the C$_{60}$ molecules with the Pt(111) substrates, intramolecular features of C$_{60}$ could be observed, which were not visible in the case of Au(111) case, on which the C$_{60}$ exhibited enough rotational freedom to make the C$_{60}$ molecules appear as featureless balls.

Iodine-modified Au(111) was more recently also used to form mixed monolayers of C$_{60}$ and C$_{70}$.[236] The latter exhibited the shape of an ellipsoid, with a long axis and a short axis. In the hexagonal

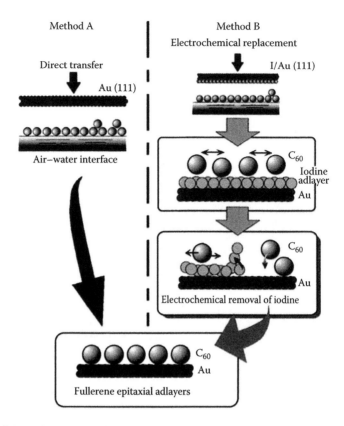

FIGURE 20.25 Schematic representation of the direct transfer method and the electrochemical replacement method. (Reprinted with permission from Uemura, S., Sakata, M., Taniguchi, I., Kunitake, M., and Hirayama, C., Novel "Wet process" technique based on electrochemical replacement for the preparation of fullerene epitaxial adlayers, *Langmuir*, 17, 5–7, 2001. Copyright 2001 American Chemical Society.)

arrays of C_{70} on Au(111), the C_{70} predominantly stand on the surface, with their long axis perpendicular to the surface. Apparently as a result of the closely matching footprints of C_{60} and standing C_{70}, there is only limited evidence for phase separation in the mixed monolayers. As a result of the standing conformation of C_{70}, mixed monolayers with a small fraction of C_{70} show bright spots for protruding C_{70}, while mixed monolayers with a small fraction of C_{60} show dark spots. It is not possible though to distinguish individual C_{60} from C_{70} molecules lying on their side, as they have very similar heights.

20.5.5 INTERPLAY OF DIFFERENT INTERACTION FORCES IN DETERMINING THE STRUCTURE OF ADSORBATE MONOLAYERS

Melem

A number of adsorbed monolayers with a structure controlled by intermolecular hydrogen bonds have been investigated by EC-STM. In these experiments, the choice of EC-STM over STM is particularly important because the overlying solvent strongly affects surface adsorption energies and the individual forces that control the geometry of the adsorbed layer. Moreover, the potential applied to the surface with respect to the reference electrode can have a decisive effect on the structure of the electrode itself, the adlayer structure, the mobility of the adsorbed species, and, in some cases, even the redox state of the adsorbed monolayer.

A nice example that shows how the structure of an adsorbate monolayer depends on the substrate potential is given by the adsorption of 1,3,5-benzene-tricarboxylic acid (also known as trimesic acid) onto Au(111).[237] While a close-packed monolayer is observed at +0.6 V versus RHE, a planar, hexagonal 2D network is formed at +0.2 to +0.4 V, with each molecule interacting with three neighboring molecules through hydrogen bonding. Both surface structures represent a thermodynamic equilibrium, and by cycling the potential, one can flip forth and back between the two surface structures. However, when the potential is brought to about +0.8 V, the adsorbate–surface interactions become so strong that a well-ordered monolayer is no longer observed. A later follow-up study that used sulfuric acid rather than perchloric acid identified as many as five differently organized adlayers and also showed that application of a positive potential to the electrode can cause partial deprotonation of adsorbed species.[238]

Hydrogen-bonded 2D networks were also observed for two triamines, melamine and melem.[239] Here too, depending on the applied potential, either disorganized monolayers or highly ordered monolayers of differing surface density were observed (Figure 20.26).

Nucleobases represent a classical system for hydrogen bonding. While cytosine, guanine, and adenine all adsorb spontaneously onto Au(111), thymine has only been found to adsorb after the

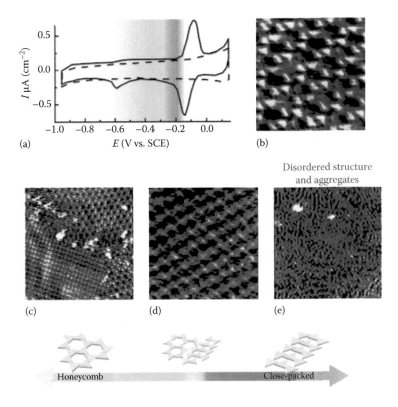

FIGURE 20.26 (a) Cyclic voltammogram of a melem aqueous solution using the Au(111) electrode. The dotted line in the CV is for a melem-free solution, and the solid line is for a 1×10^{-5} M melem aqueous solution. (b–e) Corresponding STM images. (b) $E = -0.60$ V, (c) $E = -0.30$ V, (d) $E = -0.15$ V, and (e) $E = -0.05$ V. (Reprinted with permission from Uemura, S., Aono, M., Komatsu, T., and Kunitake, M., Novel "Wet process" technique based on electrochemical replacement for the preparation of fullerene epitaxial adlayers, *Langmuir*, 27, 1336–1340, 2011. Copyright 2011 American Chemical Society.)

substrate potential is raised sufficiently high to oxidize thymine.[240] Interestingly, π stacking rather than hydrogen bonding is the dominating force that controls the structure of guanine and adenine adsorbate monolayers, while cytosine forms a planar hydrogen-bonded network. A recent reinvestigation of adenine adsorption onto Au(111) confirms that the formation of an adenine monolayer on Au(111) lifts the surface reconstruction. Since in the range from pH 1 to 7, the pH has no effect on the adsorbed monolayer, it was concluded that, unlike in solution, the adsorbed adenine is not protonated.[241] While surface pK values differing from those in solution are well known, there is still only a limited number of cases where this effect has been directly observed with EC-STM.

The adsorption of surfactants is another example for the intricate interplay of numerous factors affecting the structure of an adsorbent layer. At moderate potentials, dodecyl sulfate adsorbs onto Au(111) in the form of hemi-micelles.[242] The surfactant molecules in direct contact with the Au(111) lie flat on the surface in lamella forming hydrogen bonds to water molecules that bridge the lamella (see Figure 20.27, panels a and b). Additional surfactant molecules are tilted with respect to the surface, forming the hemi-micelles that have the shape of a cylinder cut in half along its axis (Figure 20.27, panel c). The hemi-micelles collapse and give way to a condensed, melted film when the positive charge approaches or exceeds the charge of the hemi-micelle sulfate groups.

In view of biological structures, the use of EC-STM to observe phospholipid layers on octanethiol modified gold, mimicking biological bilayers, is of much interest. While a study of Matsunaga et al. using a phosphocoline derivative as the layer on top of SAM-modified gold did not reveal high

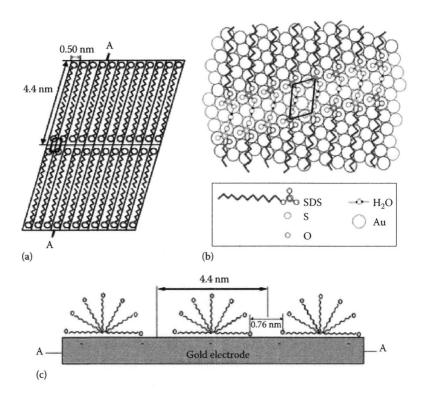

FIGURE 20.27 (a) Model of the long-range ordering of sodium dodecyl sulfate (SDS) aggregates built around a unit cell having vectors of 4.4 and 0.5 nm orientated at an angle of 70°. (b) Model of the water-bridged sulfate groups arranged into ($\sqrt{3} \times \sqrt{7}$) cells. (c) Model of the cross section of the hemi-micelle along the direction of the longer vector of the unit cell. (Reprinted with permission from Burgess, I., Jeffrey, C.A., Cai, X., Szymanski, G., Galus, Z., and Lipkowski, J., Direct visualization of the potential-controlled transformation of hemimicellar aggregates of dodecyl sulfate into a condensed monolayer at the Au(111) electrode surface, *Langmuir*, 15, 2607–2616, 1999. Copyright 1999 American Chemical Society.)

resolution, it demonstrated the possibility of reversibly switching between a striped/grainy lipid layer in which the lipids lie parallel to the surface and a layer with the character of a 2D liquid.[243]

The incorporation of coordinating metal cations offers an additional opportunity to control the structure of adsorbates. Nishiyama et al. observed the monolayer structure of a compound with two terminal metal cation coordinating groups with three pyridine groups in the presence of different metal cations in solution.[244] As may be expected, a very strong effect of the character of the metal cation was observed. While in the presence of Pt^{2+}, Ag^+, and Cu^{2+} no well-organized monolayers could be observed, coordination of Fe^{2+} to four pyridine ligands was seen to give rise to monolayers with a disordered structure. Use of $AuBr_4^-$ gave rise to the most organized monolayers, but it was not possible to conclude whether in those monolayers the gold atoms were still present in the form of $AuBr_4^-$ or whether the bromide ligands had been lost in favor of direct metal–pyridine interactions. A particular interest in structures of this type lies in the electrocatalytic properties of such surface adsorbate layers, which may be tuned by the selection of the metal cation.

20.5.6 MACROCYCLIC ADSORBATES

Macrocyclic compounds with multiple hydroxyl groups on one rim, as they are common for many cyclodextrins and calixarenes, have a tendency to form barrel-shaped dimers that, due to their overall dimensions, often lie sideways onto the electrode surface. In the case of α-, β-, and γ-cyclodextrin, which are macrocyclic oligosaccharides that have 10, 12, or 14 hydroxyl groups on the upper rim and 5, 6, or 7 hydroxyl groups on the lower rim, not only dimers but whole tubes can be formed on Au(111) (Figure 20.28).[245] The specific range of potentials within which these

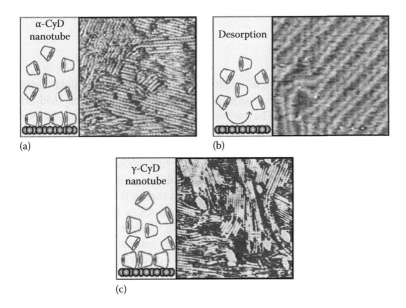

FIGURE 20.28 EC-STM images for selective adsorption of γ-cyclodextrin (in red) from an α-cyclodextrin (in blue) and γ-cyclodextrin mixture solution. The α-*CyD-nanotube* image ((a) 25 nm×25 nm, $I_t = 2.5$ nA, and $E_t = 0.00$ V) was collected at −0.20 V in 10 mM NaClO$_4$ in the presence of 2.0 μM α-CyD. The reconstructed surface image ((b) 25 nm×25 nm, $I_t = 2.0$ nA, and $E_t = -0.05$ V) was obtained 6 min after the potential jump from −0.20 to −0.40 V to desorb α-CyD molecules. The typical γ-*CyD-nanotube* image ((c) 34 nm×34 nm, $I_t = 2.0$ nA, and $E_t = +0.03$ V) was obtained 5 min after the injection of the γ-CyD stock solution at −0.40 V. The final concentration of the α-CyD and γ-CyD solution was approximately 2.0 μM. (Reprinted with permission from Ohira, A., Sakata, M., Taniguchi, I., Hirayama, C., and Kunitake, M., Comparison of nanotube structures constructed from alpha-, beta-, and gamma-cyclodextrins by potential-controlled adsorption, *J. Am. Chem. Soc.*, 125, 5057–5065, 2003. Copyright 2003 American Chemical Society.)

cyclodextrin nanotubes are formed depends on the type of cyclodextrin, with the width of that range increasing from α- to γ-cyclodextrin from +0.05 to +0.23 V. EC-STM allows not only the in situ observation of the monolayer structure but also of transitions between disorder and order and vice versa.

To avoid sideways adsorption, Yoshimoto et al. used a p-tert-butylcalix[4]arene with two thiol substituents for attachment to gold electrodes.[246] While this molecule was apparently not compatible with Au(111), it formed well-ordered monolayers on Au(100)–(1 × 1), emphasizing the importance, in this case, of the underlying substrate. Many other STM and EC-STM studies of macrocycle monolayers have found it difficult to obtain very high resolution, but the images of Yoshimoto et al. exhibited distinct submolecular resolution, showing even the central cavity of the calixarene.

Further intricacy is added if the macrocycle has the ability to bind a guest, such as in the case of dibenzo-18-crown-6, which bind potassium ions. As shown by Ohira et al., the crown ether forms disordered monolayers of strongly bound neutral molecules at the open circuit potential, but raising the potential by 0.2 to +0.66 V versus SCE reduces the adsorbate–substrate interaction and results in the formation of ordered domains.[247] Much more negative potentials are needed for the observation of monolayers of the potassium ion complexes of this crown ether, which were prepared by adsorption from a very dilute KI solution. While in the positive voltage range only an iodine adlayer could be observed, voltages below +0.01 V permitted the observation of a well-ordered monolayer with ball-like protrusions at the center of the crown ethers. The authors did not want to assign those protrusions with certainty to the potassium ions but left open the possibility of their association with the iodide counterions. As often in STM, the observation of these very distinct protrusions was highly dependent on the selection of an appropriate tunneling current.

20.5.7 CONJUGATED POLYMERS

Adsorption of conducting polymers to conducting surfaces is of particular interest because of applications of conducting polymers in organic electronic devices, such as field-effect transistors, light-emitting devices, or sensors. The knowledge of the conformation of adsorbed polymer molecules is crucial for the interpretation of the (conformation-dependent) electron conduction in conducting polymers and the design of devices based on such compounds. A number of STM studies on the conformation of, for example, oligothiophenes have shown often highly ordered adlayers. Unique to EC-STM is the ability to determine the stability of such adlayers as a function of the substrate potential. For example, Tongol et al. showed that in 0.10 M $HClO_4$ and in a range from +0.05 to +0.50 V versus RHE, oligothiophenes with 24, 48, or 96 monomer units formed stable, well-ordered monolayers on Au(111) (see Figure 20.29).[248] Beyond this range, the oligothiophene chains were either too mobile to be imaged, desorbed, or underwent oxidation.

Another unique feature of EC-STM in the context of conjugated polymers is the possibility of synthesizing the polymers in situ, taking advantage of the preorganizing influence of surface adsorption (resulting in surface polymerization) and, in the case of in situ electropolymerization,[249] the potential applied to the substrate. Adlayers of conjugated polymers can also take advantage of the halogen modification of Au(111) to increase the mobility of the adsorbents and promote the formation of adlayers with a higher level of order. Where desired, the iodine layer can be removed subsequently at a sufficiently negative potential. For example, for poly(3-hexylthiophene) on iodine-modified Au(111), it was shown that the iodine can be removed without affecting the order of the polymer adlayer.[250]

In a 2012 EC-STM investigation, Tanoue et al. prepared highly conjugated polymer adlayers in situ by polycondensation of combinations of aromatic diamines and dialdehydes on Au(111), some of which polymerized only on the surface but not in solution (see Figure 20.30).[251] Highly ordered arrays were only observed in a relatively narrow range between +0.70 and +0.90 V versus RHE,

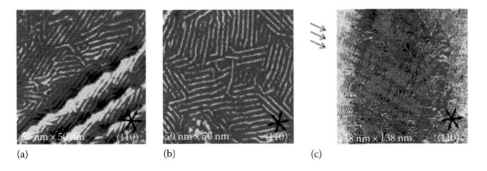

(a)   (b)   (c)

FIGURE 20.29 STM images of oligothiophenes consisting of (a) 24, (b) 48, and (c) 96 monomers, adsorbed on Au(111) electrodes at $E = +0.25$ V (vs. RHE) in 0.10 M $HClO_4$. Tunneling parameters are $V_B = -107$ mV and $I_T = 300$ pA for (a), $V_B = 165$ mV and $I_T = 200$ pA for (b), and $V_B = -150$ mV and $I_T = 2$ nA for (c). The three arrows at the bottom right indicate the close-packed directions of the Au(111) substrate. The black arrows at the left of (c) indicate the reconstruction lines of Au(111). (Reprinted with permission from Tongol, B.J.V., Wang, L., Yau, S.L., Otsubo, T., and Itaya, K., Nanostructures and molecular assembly of beta-blocked long oligothiophenes up to the 96-Mer on Au(111) as probed by in situ electrochemical scanning tunneling microscopy, *J. Phys. Chem. C*, 113, 13819–13824, 2009. Copyright 2009 American Chemical Society.)

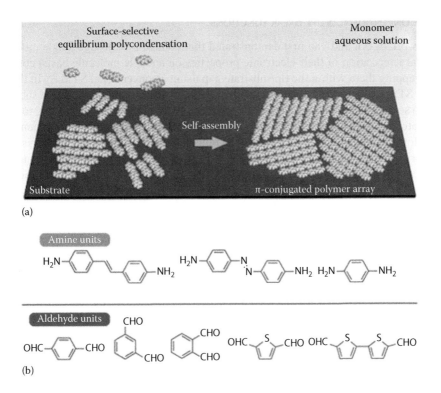

FIGURE 20.30 (a) Schematic representation of surface-induced, on-site polycondensation of a π-conjugated polymer and (b) aromatic amine and aromatic aldehyde units used for *on-site* polycondensation. (Reprinted with permission from Liu, Y.F., Krug, K., Lee, Y.L., Self-organization of two-dimensional poly(3-hexylthiophene) crystals on Au(111) surfaces, *Nanoscale*, 5, 7936–7941, 2013. Copyright 2013 American Chemical Society.)

with more negative potentials resulting in weaker adsorption and more positive potentials causing Au(111) oxidation. While the first observations of conjugated polymers with EC-STM have only been made a few years ago, it appears that this emerging approach to study conjugated polymers has great promise to provide device fabrication with crucial design information.

20.5.8 Carbon Nanotubes

Despite the advantage given by the independent control of the tip–sample voltage bias and the bias between the sample and an external reference, EC-STM has only recently been used to study single carbon nanotubes.[24] While the observed resolution was not high enough to determine the exact diameter and the helicity of the nanotubes, STS under electrochemically controlled conditions made it possible to distinguish semiconducting nanotubes from metallic ones. It was also suggested that the slightly larger than expected bandgap was evidence for the presence of water within the nanotube.

20.6 APPLICATION OF EC-STM TO SINGLE MOLECULE TRAPPING

While the primary application of EC-STM has been for atomic resolution imaging of electrode surfaces under potential control, additional applications make use of the instrument's ability to position an insulated electrochemical probe within tunneling distance of a sample under potential control. Perhaps most significant among these have been investigations directed at trapping or characterizing single molecules in electrolyte solutions using a break junction approach. In this section, we briefly discuss developments in this area.

20.6.1 Conventional STM Break Junctions

The 2003 work by Xu and Tao first demonstrated that conduction through single molecules (and thereby characterization of their electronic properties on a single molecule basis) could be measured by trapping them within the tip/substrate gap using a conventional STM.[252] In these experiments, the STM tip is first brought into gentle physical contact with a conductive substrate. This is done in a solution that contains a molecule of interest capable of strongly binding to both the tip and substrate (e.g., an alkane dithiol). Following contact, the tip is retracted until conductance in the gap is lost (i.e., until the tip is far out of the tunneling range). As this retraction is performed, there is a brief period when the size of the tip/substrate gap is nearly the same size as the molecule of interest, in which case conduction in the gap may be modified by the trapped molecule, and plots of conductance versus tip movement for the retraction may be punctuated by step features corresponding to conduction through single molecules. Although the conduction plots obtained will generally have a high degree of variation for successive tip retractions, if a large number of tip retractions (hundreds to thousands) are performed for a system of interest, distributions of the observed step features in the conductance plots can be interpreted to yield measurements of single molecule conduction. In the initial report by Xu and Tao, single molecule conduction was measured for hexanedithiol, octanedithiol, decanedithiol, and 4,4′-bipyridine molecules that were trapped between a gold tip and a gold substrate. Figure 20.31 summarizes this experiment and presents single molecule conductance data for 4,4′-bipyridine. Panels c and e of the figure show representative raw data from tip retraction experiments in the presence (c) or absence (e) of 4,4′-bipyridine, while panels d and f in the figure show conductance measurement distributions from many such retractions. For comparison, panels a and b show the much larger conductance measured for a metal/metal contact. The units of conductance shown here are fractions of the quantum conductance, $G_0 = 2e^2/h$. In this experiment, single molecule conduction is ultimately reflected in peaks in the distributions of measured conductances in the distance regime immediately following the breaking of metal/metal contact during the tip retraction, as shown in panel d of the figure.

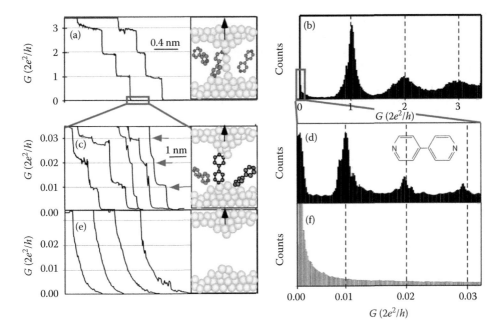

FIGURE 20.31 (a) Conductance of a gold contact formed between a gold STM tip and a gold substrate decreases in quantum steps near multiples of G_0 ($=2e^2/h$) as the tip is pulled away from the substrate. (b) A corresponding conductance histogram constructed from 1000 conductance curves as shown in (a) shows well-defined peaks near 1 G_0, 2 G_0, and 3 G_0 due to conductance quantization. (c) When the contact shown in (a) is completely broken, corresponding to the collapse of the last quantum step, a new series of conductance steps appears if molecules such as 4,4'-bipyridine are present in solution. These steps are due to the formation of a stable molecular junction between the tip and substrate electrodes. (d) A conductance histogram obtained from 1000 measurements as shown in (c) shows peaks near 1×, 2×, and 3×0.01 G_0 that are ascribed to one, two, and three molecules, respectively. (e and f) In the absence of molecules, no such steps are observed within the same conductance range. (From Xu, B. and Tao, N., Measurement of single-molecule resistance by repeated formation of molecular junctions, *Science*, 301, 1221–1223, 2003. Reprinted with permission of American Association for the Advancement of Science.)

This general approach draws from earlier efforts to measure single molecule conduction for trapped molecules using an AFM tip,[253] and also trapping experiments performed with mechanically controllable break junctions (MCBJs).[254] The real strength of the STM-based approach by Tao (often referred to as an STM break junction [STM-BJ])[255] is the use of the STM hardware to easily bring a metal tip into tunneling distance and then rapidly regenerate an ångström-scale gap by using the measured tip current as a guide. This is in contrast to MCBJ experiments, wherein a very thin (nm dimensions) piece of conductive material is broken perpendicular to its length to yield a gap for molecule trapping.[254] Besides studies of single molecule conductance, the STM-BJ technique has also been utilized as a strategy for nanofabrication, as in the 2008 work by Zhou et al. in which arrays of Cu, Fe, and Pd nanowires were fabricated with the break junction technique by using electro-deposition in the tip/substrate gap (although not while under bipotentiostatic control, as in EC-STM).[256]

20.6.2 EC-STM Break Junctions

A natural extension of the STM break junction technique was to perform analogous experiments using an EC-STM setup, allowing conductance measurements and single molecule trapping in electrolyte solutions. Haiss et al. were the first to accomplish this with their 2003 report on single molecule conductance measurements in a phosphate buffer solution for a 4,4'- bipyridinium derivative with two alkanethiol substituents attached to the nitrogen atoms on opposite ends to enable tethering of the molecule between the tip and substrate.[257] Tao's group followed in 2004 with experiments using an EC-STM break junction to perform trapping measurements of benzenedithiol and benzenedimethanethiol in NaClO$_4$ solution.[258] In the benzenedithiol experiments, single molecule conductance peaks were found only at applied potentials between −1.1 V and +0.4 V versus Ag/AgCl, which is consistent with the interpretation that conduction through a benzenedithiol SAM is limited by reductive desorption of the thiol and oxidation of the Au surface at the low and high potential limits, respectively.

Beyond the ability to trap single molecules in electrolyte solutions, a major advantage of break junction experiments with an EC-STM is the possibility of forming a single molecule electrochemical gate, which makes it possible to modulate the single molecule conductance by varying the working electrode potential relative to a reference electrode. An illustrative example is given in the 2005 work by Xiao et al.[259] Here, the conductance of a tip/substrate gap spanned by a trapped oligo (phenylene ethynylene) was found to depend on the electrode potential in 0.1 M NaOH solution due to the changing redox state of the oligomer. Conductance histograms from this work collected at various values of the electrode potential are shown in Figure 20.32. The electrochemical gating effect is demonstrated by the single molecular conductance peak (marked with a black arrow), which is

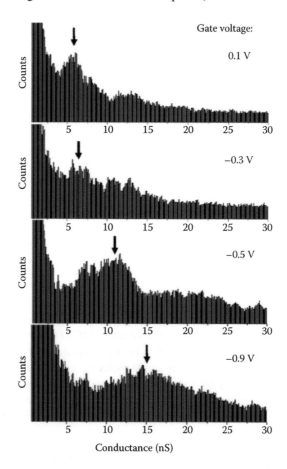

FIGURE 20.32 Conductance histograms for a nitro-substituted oligo(phenylene ethynylene) at different electrode potentials in 0.1 M NaOH solution. The single molecular conductance value is marked by black arrows. (Reprinted with permission from Xiao, X., Nagahara, L.A., Rawlett, A.M., and Tao, N., Electrochemical gate-controlled conductance of single oligo(phenylene ethynylene)s, *J. Am. Chem. Soc.*, 127, 9235–9240, 2005. Copyright 2005 American Chemical Society.)

found to move with the applied potential. Due to major interest in single molecule electrochemical gates, as they relate to molecular electronics applications, further experiments using EC-STM tip/substrate gaps to trap either small molecules or short molecular wires have seen continued attention from multiple groups.[260–264]

20.6.3 Application of EC-STM Break Junctions to Single Protein Measurements

An emerging application of EC-STM related to molecular trapping has been single molecule measurements of biomolecules. Such work includes break junction studies of proteins using the tip retraction technique described earlier,[265] as well as experiments in which the protein is attached to the EC-STM tip to allow for characterization with a fixed tip/substrate gap.[63,266] In a 2011 report by Artés et al., single azurin proteins were trapped and characterized using an EC-STM break junction approach in an ammonium acetate solution.[265] This demonstrated the first use of the EC-STM break junction method to generate single protein tip/substrate junctions, allowing for measurements of single protein conductance. Moreover, the authors performed a further experiment wherein the EC-STM tip was approached into tunneling with the sample and then the feedback loop turned off to produce a tunneling gap with a fixed width for the monitoring of single proteins. This experiment is summarized with representative data in Figure 20.33. When individual proteins spontaneously move into the tip/substrate gap, so-called blink features (jumps/steps) in plots of the tip current versus time denote conduction through a protein bridging the gap. This work demonstrated that the observed blinks not only had heights consistent with conduction through single proteins, but also exhibited dependence on the applied electrode potential, and thus behave as electrochemical gates as described in Section 20.6.2. Azurin has also been characterized using EC-STM imaging in earlier work.[267] There, the redox state of azurin, and moreover a redox-gated tunneling effect for azurin proteins interposed between the tip and sample, was ascertained by interpreting contrast in molecular resolution EC-STM images of azurin proteins immobilized on an Au surface.

20.7 CONCLUSIONS

From the time of its introduction nearly 30 years ago, EC-STM has had the unique capability to image in situ the electrolyte–electrode interface under ambient conditions with submolecular and often atomic resolution. The ability to uncouple, with the use of a bipotentiostat, the conditions necessary for imaging from the electrochemical control of the substrate potential introduced a new level of sophistication, not unlike the advances from single wavelength spectrometry to full spectrum spectroscopy or from one- to two-dimensional nuclear magnetic resonance (NMR) spectroscopy. Early studies have focused primarily on comparatively simple surfaces with fundamental interest to surface science, electrochemistry, sensing, catalysis, and energy research. Nevertheless, the impact of that work was considerable since it enabled visualization of surface chemistry never seen before. Over the years, the complexity of the surface chemistry studied by EC-STM has grown substantially, as exemplified by work with multicomponent adlayers and catalytic surfaces. Crucial to the success of EC-STM has been not only its unparalleled ability to provide insight into the atomic structure of the electrolyte–electrode interface but also the possibility to observe the kinetics of surface reactions, such as in the case of electrochemical metal deposition, the self-assembly of monolayers, or corrosion. Likely because of the added experimental complexity associated with the bipotentiostat, the need for tip insulation, the level of cleanliness required when dealing with electrolyte solutions, and the parameter space increased by all the variables of an electrochemical experiment, EC-STM has not become as widely popular as STM. Indeed, the number of reports describing the use of EC-STM in recent years is only a few percent of the number of publications on research involving STM, and the number of laboratories performing EC-STM work is considerably smaller than in the case of STM. However, the initial burst

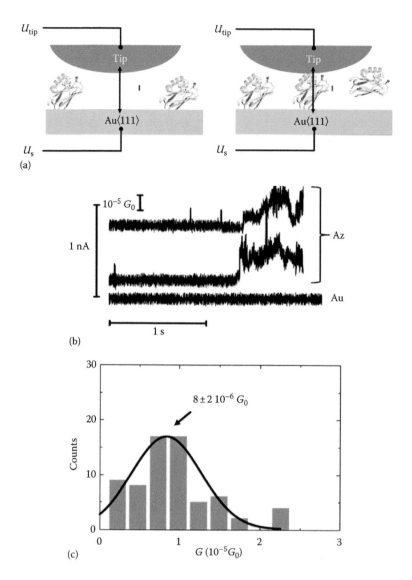

FIGURE 20.33 (a) Diagram illustrating a *blinking* experiment. (b) Example of current versus time traces showing spontaneous single protein junction formation as jumps (blink) in the current traces after the feed-back loop is turned off. The lower trace corresponds to a control performed on a clean gold surface. Traces have been vertically offset for clarity. (c) Conductance histogram built from 75 *blinks* traces. The black line is a Gaussian fit centered at $(8.5 \pm 2.4) \times 10^{-6} \, G_0$. The experiments were performed in 50 mM ammonium acetate solution (pH 4.5) with a tip potential (U_p) of −0.1 V versus SSC, a sample potential (U_s) of +0.2 V versus SSC, and a bias voltage ($U_p - U_s$) of −0.3 V. (Reprinted with permission from Artés, J.M., Díez-Pérez, I., and Gorostiza, P., Transistor-like behavior of single metalloprotein junctions, *Nano Lett.*, 12, 2679–2684, 2012. Copyright 2012 American Chemical Society.)

of activity in the late 1990s has not been followed by a decline in the number of EC-STM studies, but it has led to a nearly constant level of publication activity for almost 20 years. This shows that the field of EC-STM is alive and well. With the recent trend in academia to focus more and more on problems with real-life applications, which can rarely avoid the difficulty of complex systems, the future for EC-STM looks bright, in particular in view of the key insights that EC-STM can give to further nanosciences and nanotechnology.

REFERENCES

1. Binnig, G.; Rohrer, H.; Gerber, C.; Weibel, E. 1982. Surface studies by scanning tunneling microscopy. *Phys. Rev. Lett.* 42: 57–60.
2. Sonnenfeld, R.; Hansma, P. K. 1986. Atomic-resolution microscopy in water. *Science* 232: 211–213.
3. Lustenberger, P.; Rohrer, H. 1988. Scanning tunneling microscopy at potential controlled electrode surfaces in electrolytic environment. *J. Electroanal. Chem.* 243: 225–235.
4. Wiechers, J.; Twomey, T.; Kolb, D. M. 1988. An in-situ scanning tunneling microscopy study of Au (111) with atomic scale resolution. *J. Electroanal. Chem.* 248: 451–460.
5. Itaya, K.; Tomita, E. 1988. Scanning tunneling microscope for electrochemistry—A new concept for the in situ scanning tunneling microscope in electrolyte solutions. *Surf. Sci.* 201: L507–L512.
6. Oja, S. M.; Wood, M.; Zhang, B. 2013. Nanoscale electrochemistry. *Anal. Chem.* 85: 473–486.
7. Murray, R. W. 2008. Nanoelectrochemistry: Metal nanoparticles, nanoelectrodes, and nanopores. *Chem. Rev.* 108: 2688–2720.
8. Nagy, G.; Nagy, L. 2007. Electrochemical sensors developed for gathering microscale chemical information. *Anal. Lett.* 40: 3–38.
9. Fukuma, T.; Kobayashi, K.; Matsushige, K.; Yamada, H. 2005. True atomic resolution in liquid by frequency-modulation atomic force microscopy. *Appl. Phys. Lett.* 87: 034101–034103.
10. Umeda, K.; Fukui, K. 2010. Observation of redox-state-dependent reversible local structural change of ferrocenyl-terminated molecular island by electrochemical frequency modulation AFM. *Langmuir* 26: 9104–9110.
11. Giessibl, F. J. 2005. AFM's path to atomic resolution. *Mater. Today* 8: 32–41.
12. Honbo, H.; Sugawara, S.; Itaya, K. 1990. Detailed in situ scanning tunneling microscopy of single crystal planes of gold(111) in aqueous solutions. *Anal. Chem.* 62: 2424–2429.
13. Gao, X.; Edens, G. J.; Hamelin, A.; Weaver, M. J. 1994. Charge-dependent atomic-scale structures of high-index and (110) gold electrode surfaces as revealed by scanning tunneling microscopy. *Surf. Sci.* 318: 1–20.
14. Yau, S.-L.; Kim, Y.-G.; Itaya, K. 1996. In situ scanning tunneling microscopy of benzene adsorbed on Rh(111) and Pt(111) in HF solution. *J. Am. Chem. Soc.* 118: 7795–7803.
15. Xu, Q.; He, T.; Wipf, D. O. 2007. In situ electrochemical STM study of the coarsening of platinum islands at double-layer potentials. *Langmuir* 23: 9098–9103.
16. Maurice, V.; Klein, L. H.; Strehblow, H. H.; Marcus, P. 2007. In situ STM study of the surface structure, dissolution, and early stages of electrochemical oxidation of the Ag(111) electrode. *J. Phys. Chem. C* 111: 16351–16361.
17. Yau, S.-L.; Fan, F.-R. F.; Bard, A. J. 1992. In situ STM imaging of silicon(111) in HF under potential control. *J. Electrochem. Soc.* 139: 2825–2829.
18. Wang, D.; Wan, L.-J. 2007. Electrochemical scanning tunneling microscopy: Adlayer structure and reaction at solid/liquid interface. *J. Phys. Chem. C* 111: 16109–16130.
19. Kunitake, M.; Batina, N.; Itaya, K. 1995. Self-organized porphyrin array on iodine-modified Au(111) in electrolyte solutions: In situ scanning tunneling microscopy study. *Langmuir* 11: 2337–2340.
20. Ohira, A.; Ishizaki, T.; Sakata, M.; Taniguchi, I.; Hirayama, C.; Kunitake, M. 2000. Formation of the 'nanotube' structure of β-cyclodextrin on Au(III) surfaces induced by potential controlled adsorption. *Colloids Surf. A* 169: 27–33.
21. Teshima, T.; Ogaki, K.; Itaya, K. 1997. Effect of adsorbed iodine on the dissolution and deposition reactions of Ag(100): Studies by in situ STM. *J. Phys. Chem. B* 101: 2046–2053.
22. Zhou, W.; Baunach, T.; Ivanova, V.; Kolb, D. M. 2004. Structure and electrochemistry of 4–4′-dithiodipyridine self-assembled monolayers in comparison with 4-mercaptopyridine self-assembled monolayers on Au(111). *Langmuir* 20: 4590–4595.
23. Magnussen, O. M.; Hotlos, J.; Nichols, R. J.; Kolb, D. M.; Behm, R. J. 1990. Atomic structure of Cu adlayers on Au(100) and Au(111) electrodes observed by in situ scanning tunneling microscopy. *Phys. Rev. Lett.* 64: 2929–2932.
24. Yasuda, S.; Ikeda, K.; Yu, L.; Murakoshi, K. 2012. Characterization of isolated individual single-walled carbon nanotube by electrochemical scanning tunneling microscopy. *Jpn. J. Appl. Phys.* 51: 08KB06-1–08KB06-4.
25. Boxley, C. J.; White, H. S. 2003. Electrochemical deposition and reoxidation of Au at highly oriented pyrolytic graphite. Stabilization of Au nanoparticles on the upper plane of step edges. *J. Phys. Chem. B* 107: 451–458.
26. Jakob, S.; Schindler, W. 2013. Electric field assisted electrochemical "writing" of Co nanostructures onto n-Si(111):H surfaces. *Surf. Sci.* 612: L1–L4.

27. Li, Z.; Liu, Y.; Mertens, S. F. L.; Pobelov, I. V.; Wandlowski, T. 2010. From redox gating to quantized charging. *J. Am. Chem. Soc.* 132: 8187–8193.
28. Noda, H.; Uehara, H.; Abe, M.; Michi, T.; Osawa, M.; Uosaki, K.; Sasaki, Y. 2009. In situ scanning tunneling microscopy observation of metal-cluster redox interconversion and CO dissociation reactions at a solution/Au(111) interface. *Bull. Chem. Soc. Jpn.* 82: 1227–1231.
29. He, Y.; Borguet, E. 2007. Dynamics of porphyrin electron-transfer reactions at the electrode-electrolyte interface at the molecular level. *Angew. Chem. Int. Ed.* 46: 6098–6101.
30. Kolb, D. M. 1996. Reconstruction phenomena at metal-electrolyte interfaces. *Prog. Surf. Sci.* 51: 109–173.
31. Suto, K.; Yoshimoto, S.; Itaya, K. 2006. Electrochemical control of the structure of two-dimensional supramolecular organization consisting of phthalocyanine and porphyrin on a gold single-crystal surface. *Langmuir* 22: 10766–10776.
32. Tao, N. J. 1996. Probing potential-tuned resonant tunneling through redox molecules with scanning tunneling microscopy. *Phys. Rev. Lett.* 76: 4066–4069.
33. Chen, F.; He, J.; Nuckolls, C.; Roberts, T.; Klare, J. E.; Lindsay, S. 2005. A molecular switch based on potential-induced changes of oxidation state. *Nano Lett.* 5: 503–506.
34. Kolb, D. M.; Ullmann, R.; Will, T. 1997. Nanofabrication of small copper clusters on gold(111) electrodes by a scanning tunneling microscope. *Science* 21: 1097–1099.
35. Valov, I.; Staikov, G. 2013. Nucleation and growth phenomena in nanosized electrochemical systems for resistive switching memories. *J. Solid State Electrochem.* 17: 365–371.
36. Güntherodt, H.-J.; Wiesendanger, R., Eds. 1992. *Scanning Tunneling Microscopy I: General Principles and Applications to Clean and Adsorbate-Covered Surfaces.* Springer Series in Surface Sciences: Vol. 20. Berlin, Germany: Springer-Verlag.
37. Wiesendanger, R.; Güntherodt, H.-J., Eds. 1992. *Scanning Tunneling Microscopy II: Further Applications and Related Scanning Techniques.* Springer Series in Surface Sciences: Vol. 28. Berlin, Germany: Springer-Verlag.
38. Wiesendanger, R.; Güntherodt, H.-J., Eds. 1993. *Scanning Tunneling Microscopy III: Theory of STM and Related Scanning Techniques.* Springer Series in Surface Sciences: Vol. 29. Berlin, Germany: Springer-Verlag.
39. Bai, C. 2000. *Scanning Tunneling Microscopy and Its Applications.* Springer Series in Surface Sciences: Vol. 32. Berlin, Germany: Springer-Verlag.
40. Stroscio, J. A.; Kaiser, W. J., Eds. 1993. *Scanning Tunneling Microscopy.* San Diego, CA: Academic Press.
41. Kuk, Y.; Silverman, P. J. 1989. Scanning tunneling microscope instrumentation. *Rev. Sci. Instrum.* 60: 165–180.
42. Vang, R. T.; Lauritsen, J. V.; Lægsgaard, E.; Besenbacher, F. 2008. Scanning tunneling microscopy as a tool to study catalytically relevant model systems. *Chem. Soc. Rev.*: 2191–2203.
43. Hansma, P. K.; Tersoff, J. 1987. Scanning tunneling microscopy. *J. Appl. Phys.* 61: R1–R23.
44. Chiang, S. 2011. Imaging atoms and molecules on surfaces by scanning tunnelling microscopy. *J. Phys. D: Appl. Phys.* 44: 464001.
45. Esaki, L. 1958. New phenomenon in narrow germanium p-n junctions. *Phys. Rev.* 109: 603–604.
46. Müller, E. W.; Bahadur, K. 1956. Field ionization of gases at a metal surface and the resolution of the field ion microscope. *Phys. Rev.* 102: 624–631.
47. Young, R.; Ward, J.; Scire, F. 1972. The topografiner: An instrument for measuring surface microtopography. *Rev. Sci. Instrum.* 43: 999–1011.
48. Binnig, G.; Rohrer, H.; Gerber, C.; Weibel, E. 1983. 7 × 7 reconstruction on Si(111) resolved in real space. *Phys. Rev. Lett.* 50: 120–123.
49. Bard, A. J.; Faulkner, L. R. 2001. *Electrochemical Methods: Fundamentals and Applications*; 2nd edn.: New York: John Wiley & Sons.
50. Kochanski, G. P. 1989. Nonlinear alternating-current tunneling microscopy. *Phys. Rev. Lett.* 62: 2285–2288.
51. Stranick, S. J.; Weiss, P. S. 1994. Alternating current scanning tunneling microscopy and nonlinear spectroscopy. *J. Phys. Chem.* 98: 1762–1764.
52. Guckenberger, R.; Heim, M.; Cevc, G.; Knapp, H. F.; Wiegrabe, W.; Hillebrand, A. 1994. Scanning tunneling microscopy of insulators and biological specimens based on lateral conductivity of ultrathin water films. *Science* 266: 1538–1540.
53. Patel, N.; Davies, M. C.; Lomas, M.; Roberts, C. J.; Tendler, S. J. B.; Williams, P. M. 1997. STM of insulators with the probe in contact with an aqueous layer. *J. Phys. Chem. B* 101: 5138–5142.

54. Chen, C. J. 1988. Theory of scanning tunneling spectroscopy. *J. Vac. Sci. Technol. A* 6: 319–322.
55. Tersoff, J.; Hamann, D. R. 1983. Theory and application for the scanning tunneling microscope. *Phys. Rev. Lett.* 50: 1998–2001.
56. Tersoff, J.; Hamann, D. R. 1985. Theory of the scanning tunneling microscope. *Phys. Rev. B* 31: 805–813.
57. Mårtensson, P.; Feenstra, R. M. 1989. Geometric and electronic structure of antimony on the GaAs(110) surface studied by scanning tunneling microscopy. *Phys. Rev. B* 39: 7744–7753.
58. Stipe, B. C.; Rezaei, M. A.; Ho, W. 1998. Single-molecule vibrational spectroscopy and microscopy. *Science* 280: 1732–1735.
59. Overgaag, K.; Liljeroth, P.; Grandidier, B.; Vanmaekelbergh, D. 2008. Scanning tunneling spectroscopy of individual PbSe quantum dots and molecular aggregates stabilized in an inert nanocrystal matrix. *ACS Nano* 2: 600–606.
60. Tomita, E.; Matsuda, N.; Itaya, K. 1990. Surface electronic structure of semiconductor (*p*- and *n*-Si) electrodes in electrolyte solution. *J. Vac. Sci. Technol. A* 8: 534–538.
61. Güell, A. G.; Díez-Pérez, I.; Gorostiza, P.; Sanz, F. 2004. Preparation of reliable probes for electrochemical tunneling spectroscopy. *Anal. Chem.* 76: 5218–5222.
62. Artés, J. M.; Díez-Pérez, I.; Sanz, F.; Gorostiza, P. 2011. Direct measurement of electron transfer distance decay constants of single redox proteins by electrochemical tunneling spectroscopy. *ACS Nano* 5: 2060–2066.
63. Artés, J. M.; López-Martínez, M.; Giraudet, A.; Díez-Pérez, I.; Sanz, F.; Gorostiza, P. 2012. Current-voltage characteristics and transition voltage spectroscopy of individual redox proteins. *J. Am. Chem. Soc.* 134: 20218–20221.
64. Yoshimoto, S.; Higa, N.; Itaya, K. 2004. Two-dimensional supramolecular organization of copper octa-ethylporphyrin and cobalt phthalocyanine on Au(111): Molecular assembly control at an electrochemical interface. *J. Am. Chem. Soc.* 126: 8540–8545.
65. Melmed, A. J. 1991. The art and science and other aspects of making sharp tips. *J. Vac. Sci. Technol. B* 9: 601–608.
66. Nagahara, L. A.; Thundat, T.; Lindsay, S. M. 1989. Preparation and characterization of STM tips for electrochemical studies. *Rev. Sci. Instrum.* 60: 3128–3130.
67. Gingery, D.; Bühlmann, P. 2007. Single-step electrochemical method for producing very sharp Au scanning tunneling microscopy tips. *Rev. Sci. Instrum.* 78: 113703-1–113703-4.
68. Mizes, H. A.; Park, S.; Harrison, W. A. 1987. Multiple-tip interpretation of anomalous scanning-tunneling-microscopy images of layered materials. *Phys. Rev. B* 36: 4491–4494.
69. Wang, E.; Zhang, B. 1994. Fabrication of STM tips with controlled geometry by electrochemical etching and ECSTM tips coated with paraffin. *Electrochim. Acta* 39: 103–106.
70. Heben, M. J.; Dovek, M. M.; Lewis, N. S.; Penner, R. M.; Quate, C. F. 1988. Preparation of STM tips for in-situ characterization of electrode surfaces. *J. Microsc.* 152: 651–661.
71. Gewirth, A. A.; Craston, D. H.; Bard, A. J. 1989. Fabrication and characterization of microtips for in situ scanning tunneling microscopy. *J. Electroanal. Chem.* 261: 477–482.
72. Christoph, R.; Siegenthaler, H.; Rohrer, H.; Wiese, H. 1989. In situ scanning tunneling microscopy at potential controlled Ag(100) substrates. *Electrochim. Acta* 34: 1011–1022.
73. Bach, C. E.; Nichols, R. J.; Bechmann, W.; Meyer, H.; Schulte, A.; Besenhard, J. O.; Junnakoudakis, P. D. 1993. Microscopy tips for electrochemical studies using an electropainting method. *J. Electrochem. Soc.* 140: 1281–1284.
74. Salerno, M. 2010. Coating of tips for electrochemical scanning tunneling microscopy by means of silicon, magnesium, and tungsten oxides. *Rev. Sci. Instrum.* 81: 093703-1–093703-7.
75. Batina, N.; Yamada, T.; Itaya, K. 1995. Atomic level characterization of the iodine-modified Au(111) electrode surface in perchloric acid solution by in-situ STM and ex-situ LEED. *Langmuir* 11: 4568–4576.
76. Sawaguchi, T.; Mizutani, F.; Taniguchi, I. 1998. Direct observation of 4-mercaptopyridine and bis(4-pyridyl) disulfide monolayers on Au(111) in perchloric acid solution using in situ scanning tunneling microscopy. *Langmuir* 14: 3565–3569.
77. Mao, B. W.; Ye, X. D.; Zhou, J. Q.; Fen, Z. D.; Tian, Z. W. 1992. A new method of STM tip fabrication for in-situ electrochemical studies. *Ultramicroscopy* 42: 464–467.
78. Zhu, L.; Claude-Montigny, B.; Gattrell, M. 2005. Insulating method using cataphoretic paint for tungsten tips for electrochemical scanning tunneling microscopy (ECSTM). *Appl. Surf. Sci.* 252: 1833–1845.
79. Thorgaard, S.; Bühlmann, P. 2007. Cathodic electropaint insulated tips for electrochemical scanning tunneling microscopy. *Anal. Chem.* 79: 9224–9228.
80. Motoo, S.; Furuya, N. 1984. Electrochemistry of platinum single crystal surfaces part I. Structural change of the Pt (111) surface followed by an electrochemical method. *J. Electroanal. Chem.* 172: 339–358.

81. Hamelin, A. 1996. Cyclic voltammetry at gold single-crystal surfaces. Part 1. Behavior at low-index faces. *J. Electroanal. Chem.* 407: 1–11.
82. Zurawski, D.; Rice, L.; Hourani, M.; Wiekowski, A. 1987. The in-situ preparation of well-defined, single crystal electrodes. *J. Electroanal. Chem.* 230: 221–231.
83. Batina, N.; Will, T.; Kolb, D. M. 1992. Study of the initial stages of copper deposition by in situ scanning tunnelling microscopy. *Faraday Discuss.* 94: 93–106.
84. Cavalleri, O.; Gilbert, S. E.; Kern, K. 1997. Electrochemical Cu deposition on thiol covered Au(111) surfaces. *Surf. Sci.* 377–379: 931–936.
85. Tang, L.; Han, B.; Persson, K.; Friesen, C.; He, T.; Sieradzki, K.; Ceder, G. 2010. Electrochemical stability of nanometer-scale Pt particles in acidic environments. *J. Am. Chem. Soc.* 132: 596–600.
86. Xu, Q.; He, T.; Wipf, D. O. 2007. In situ electrochemical STM study of the coarsening of platinum islands at double-layer potentials. *Langmuir* 17: 9098–9103.
87. Chidsey, C. E. D.; Loiacono, D. N.; Sleator, T.; Nakahara, S. 1988. STM study of the surface morphology of gold on mica. *Surf. Sci.* 200: 45–66.
88. Dishner, M. H.; Ivey, M. M.; Gorer, S.; Hemminger, J. C.; Feher, F. J. 1998. Preparation of gold films by epitaxial growth on mica and the effect of flame annealing. *J. Vac. Sci. Technol. A* 16: 3295–3300.
89. Kang, J.; Rowntree, P. A. 2007. Gold film surface preparation for self-assembled monolayer studies. *Langmuir* 23: 509–516.
90. Goss, C. A.; Charych, D. H.; Majda, M. 1991. Application of (3-mercaptopropyl) trimethoxysilane as a molecular adhesive in the fabrication of vapor-deposited gold electrodes on glass substrates. *Anal. Chem.* 63: 85–88.
91. Lüssem, B.; Karthäuser, S.; Haselier, H.; Waser, R. 2005. The origin of faceting of ultraflat gold films epitaxially grown on mica. *Appl. Surf. Sci.* 249: 197–202.
92. Wagner, P.; Hegner, M.; Güntherodt, H.-J.; Semenza, G. 1995. Formation and in situ modification of monolayers chemisorbed on ultraflat template-stripped gold surfaces. *Langmuir* 1995: 3867–3875.
93. Nogues, C.; Wanunu, M. 2004. A rapid approach to reproducible, atomically flat gold films on mica. *Surf. Sci.* 573: L383–L389.
94. Wano, H.; Uosaki, K. 2005. In situ dynamic monitoring of electrochemical oxidative adsorption and reductive desorption processes of a self-assembled monolayer of hexanethiol on a Au(111) surface in KOH ethanol solution by scanning tunneling microscopy. *Langmuir* 21: 4024–4033.
95. Clavilier, J.; Faure, R.; Guinet, G.; Durand, R. 1980. Preparation of monocrystalline Pt microelectrodes and electrochemical study of the plane surfaces cut in the direction of the {111} and {110} planes. *J. Electroanal. Chem.* 107: 205–209.
96. Kaischew, R.; Mutaftschiew, B. 1955. Elektrolytische Keimbildung auf kugelförmigen Pt-Einkristallelekctroden. *Z. Phys. Chem. Leipzig* 204: 334–347.
97. Hamelin, A.; Morin, S.; Richer, J.; Lipkowski, J. 1990. Adsorption of pyridine on the (311) face of silver. *J. Electroanal. Chem.* 285: 249–262.
98. Cuesta, A.; Kibler, L. A.; Kolb, D. M. 1999. A method to prepare single crystal electrodes of reactive metals: Application to Pd(hkl). *J. Electroanal. Chem.* 466: 165–168.
99. Wan, L.-J.; Yau, S.-L.; Itaya, K. 1995. Atomic structure of adsorbed sulfate on Rh(111) in sulfuric acid solution. *J. Phys. Chem.* 99: 9507–9513.
100. Furuya, N.; Koide, S. 1990. Hydrogen adsorption on iridium single-crystal surfaces. *Surf. Sci.* 226: 221–225.
101. Voigtländer, B.; Linke, U.; Stollwerk, H.; Brona, J. 2005. Preparation of bead metal single crystals by electron beam heating. *J. Vac. Sci. Technol. A* 23: 1535–1537.
102. Hara, M.; Linke, U.; Wandlowski, T. 2007. Preparation and electrochemical characterization of palladium single crystal electrodes in 0.1 M H_2SO_4 and $HClO_4$: Part I. Low-index phases. *Electrochim. Acta* 52: 5733–5748.
103. Welford, P. J.; Brookes, B. A.; Climent, V.; Compton, R. G. 2001. The hanging meniscus contact: Geometry induced diffusional overpotential. The reduction of oxygen in dimethylsulphoxide at Au(111). *J. Electroanal. Chem.* 513: 8–15.
104. Thorgaard, S. N.; Bühlmann, P. 2010. Bromine-passivated Au(111) as a platform for the formation of organic self-assembled monolayers under electrochemical conditions. *Langmuir* 26: 7133–7137.
105. Hamelin, A.; Katayama, A. 1981. Lead underpotential deposition on gold single-crystal surfaces: The (100) face and its vicinal faces. *J. Electroanal. Chem.* 117: 221–232.
106. Batina, N.; Dakkouri, A. S.; Kolb, D. M. 1994. The surface structure of flame-annealed Au(100) in aqueous solution: An STM study. *J. Electroanal. Chem.* 370: 87–94.

107. Yamada, T.; Batina, N.; Itaya, K. 1995. Structure of electrochemically deposited iodine adlayer on Au(111) studied by ultrahigh-vacuum instrumentation and in situ STM. *J. Phys. Chem.* 99: 8817–8823.

108. Lay, M. D.; Sorenson, T. A.; Stickney, J. L. 2003. High-resolution electrochemical scanning tunneling microscopy (EC-STM) flow-cell studies. *J. Phys. Chem. B* 107: 10598–10602.

109. Magnussen, O. M. 2002. Ordered anion adlayers of metal electrode surfaces. *Chem. Rev.* 102: 679–726.

110. Lipkowski, J.; Shi, Z.; Chen, A.; Pettinger, B.; Bilger, C. 1998. Ionic adsorption at the Au(111) electrode. *Electrochim. Acta* 43: 2875–2888.

111. Itaya, K.; Sugawara, S.; Sashikata, K.; Furuya, N. 1990. In situ scanning tunneling microscopy of platinum (111) surface with the observation of monatomic steps. *J. Vac. Sci. Technol. A* 8: 515–519.

112. Magnussen, O. M.; Hotlos, J.; Beitel, G.; Kolb, D. M.; Behm, R. J. 1991. Atomic structure of ordered copper adlayers on single-crystalline gold electrodes. *J. Vac. Sci. Technol. B* 9: 969–975.

113. Hachiya, T.; Honbo, H.; Itaya, K. 1991. Detailed underpotential deposition of copper on gold(111) in aqueous solutions. *J. Electroanal. Chem.* 315: 275–291.

114. Tanaka, S.; Yau, S.-L.; Itaya, K. 1995. In-situ scanning tunneling microscopy of bromine adlayers on Pt(111). *J. Electroanal. Chem.* 396: 125–130.

115. Kim, Y.-G.; Yau, S.-L.; Itaya, K. 1996. Direct observation of complexation of alkali cations on cyanide-modified Pt(111) by scanning tunneling microscopy. *J. Am. Chem. Soc.* 118: 393–400.

116. Kim, Y.-G.; Yau, S.-L.; Itaya, K. 1999. In situ scanning tunneling microscopy of highly ordered adlayers of aromatic molecules on well-defined Pt(111) electrodes in solution: Benzoic acid, terephthalic acid, and pyrazine. *Langmuir* 15: 7810–7815.

117. Sripirom, J.; Kuhn, S.; Jung, U.; Magnussen, O.; Schulte, A. 2013. Pointed carbon fiber ultramicro-electrodes: A new probe option for electrochemical scanning tunneling microscopy. *Anal. Chem.* 85: 837–842.

118. Wang, J.; Ocko, B. M.; Davenport, A. J.; Isaacs, H. S. 1992. In situ x-ray-diffraction and -reflectivity studies of the Au(111)/electrolyte interface: Reconstruction and anion adsorption. *Phys. Rev. B* 46: 10321–10337.

119. Poirier, G. E. 1997. Characterization of organosulfur molecular monolayers on Au(111) using scanning tunneling microscopy. *Chem. Rev.* 97: 1117–1127.

120. Narasimhan, S.; Vanderbilt, D. 1992. Elastic stress domains and the herringbone reconstruction on Au(111). *Phys. Rev. Lett.* 69: 1564–1567.

121. Wano, H.; Uosaki, K. 2001. In situ, real-time monitoring of the reductive desorption process of self-assembled monolayers of hexanethiol on Au(111) surfaces in acidic and alkaline aqueous solutions by scanning tunneling microscopy. *Langmuir* 17: 8224–8228.

122. Kong, D.-S.; Wan, L.-J.; Han, M.-J.; Pan, G.-B.; Lei, S.-B.; Bai, C.-L.; Chen, S.-H. 2002. Self-assembled monolayer of a Schiff base on Au(111) surface: Electrochemistry and electrochemical STM study. *Electrochim. Acta* 48: 303–309.

123. Wakisaka, M.; Asizawa, S.; Yoneyama, T.; Uchida, H.; Watanabe, M. 2010. In situ STM observation of the CO adlayer on a Pt(110) electrode in 0.1 M HClO$_4$ solution. *Langmuir* 26: 9191–9194.

124. Gao, X.; Hamelin, A.; Weaver, M. J. 1991. Potential-dependent reconstruction at ordered Au(100)-aqueous interfaces probed by atomic-resolution scanning tunneling microscopy. *Phys. Rev. Lett.* 67: 618–621.

125. Magnussen, O.; Wiechers, J.; Behm, R. J. 1993. In situ scanning tunneling microscopy observations of the potential-dependent (1 × 2) reconstruction on Au(110) in acidic electrolytes. *Surf. Sci.* 289: 139–151.

126. Vitus, C. M.; Chang, S. C.; Schardt, B. C.; Weaver, M. J. 1991. In situ scanning tunneling microscopy as a probe of adsorbate-induced reconstruction at ordered monocrystalline electrodes: Carbon monoxide on Pt(100). *J. Phys. Chem.* 95: 7559–7563.

127. Kim, Y.-G.; Baricuatro, J. H.; Soriaga, M. P. 2006. Molecular adsorption at well-defined electrode surfaces: Hydroquinone on Pd(111) studied by EC-STM. *Langmuir* 22: 10762–10765.

128. Bae, S.-E.; Gewirth, A. A. 2006. In situ EC-STM studies of MPS, SPS, and chloride on Cu(100): Structural studies of accelerators for dual damascene electrodeposition. *Langmuir* 22: 10315–10321.

129. Li, W.-H.; Wang, Y.; Ye, J. H.; Li, S. F. Y. 2001. In situ STM study of chloride adsorption on Cu(110) electrode in hydrochloric acid aqueous solution. *J. Phys. Chem. B* 105: 1829–1833.

130. Kong, D.-S.; Chen, S.-H.; Wan, L.-J.; Han, M.-J. 2003. The preparation and in situ scanning tunneling microscopy study of Fe(110) surface. *Langmuir* 19: 1954–1957.

131. Kim, J.-W.; Lee, J.-Y.; Park, S.-M. 2004. Effects of organic addititves on zinc electrodeposition at iron electrodes studied by EQCM and in situ STM. *Langmuir* 20: 459–466.

132. Suzuki, T.; Yamada, T.; Itaya, K. 1996. In situ electrochemical scanning tunneling microscopy of Ni(111), Ni(100), and sulfur-modified Ni(100) in acidic solution. *J. Phys. Chem.* 100: 8954–8961.

133. Yau, S.-L.; Vitus, C. M.; Schardt, B. C. 1990. In situ scanning tunneling microscopy of adsorbates on electrode surfaces: Images of the ($\sqrt{3} \times \sqrt{3}$)R30°-iodine adlattice on platinum(111). *J. Am. Chem. Soc.* 112.

134. Magnussen, O. M.; Ocko, B. M.; Wang, J. X.; Adzic, R. R. 1996. In-situ X-ray diffraction and STM studies of bromide adsorption on Au(111) electrodes. *J. Phys. Chem.* 100: 5500–5508.

135. Ye, S.; Ishibashi, C.; Uosaki, K. 1999. Anisotropic dissolution of an Au(111) electrode in perchloric acid solution containing chloride anion investigated by in situ STM–The important role of adsorbed chloride anion. *Langmuir* 15: 807–812.

136. Cruesta, A.; Kolb, D. M. 2000. The structure of bromide and chloride adlayers on Au(100) electrodes: An in situ STM study. *Surf. Sci.* 465: 310–316.

137. Sashikata, K.; Sugata, T.; Sugimasa, M.; Itaya, K. 1998. In situ scanning tunneling microscopy observation of a porphyrin adlayer on an iodine-modified Pt(100) electrode. *Langmuir* 14: 2896–2902.

138. Wan, L.-J.; Itaya, K. 1999. In situ scanning tunneling microscopy of Cu(110): Atomic structures of halide layers and anodic dissolution. *J. Electroanal. Chem.* 473: 10–18.

139. Ogaki, K.; Batina, N.; Kunitake, M.; Itaya, K. 1996. In situ scanning tunneling microscopy of ordering processes of adsorbed porphyrin on iodine-modified Ag(111). *J. Phys. Chem.* 100: 7185–7190.

140. Cuesta, A.; Kleinert, M.; Kolb, D. M. 2000. The adsorption of sulfate and phosphate on Au(111) and Au(100) electrodes: An in situ STM study. *Phys. Chem. Chem. Phys.* 2: 5684–5690.

141. Schlaup, C.; Horch, S. 2013. In-situ STM study of phosphate adsorption on Cu(111), Au(111) and Cu/Au(111) electrodes. *Surf. Sci.* 608: 44–54.

142. Sato, K.; Yoshimoto, S.; Inukai, J.; Itaya, K. 2006. Effect of sulfuric acid concentration on the structure of sulfate adlayer on Au(111) electrode. *Electrochem. Commun.* 8: 725–730.

143. Broekmann, P.; Wilms, M.; Wandelt, K. 1999. Atomic structures of a Cu(111) surface under electrochemical conditions: An in-situ STM study. *Surf. Rev. Lett.* 6: 907–916.

144. Sawaguchi, T.; Yamada, T.; Okinaka, Y.; Itaya, K. 1995. Electrochemical scanning tunneling microscopy and ultrahigh-vacuum investigation of gold cyanide adlayers on Au(111) formed in aqueous solution. *J. Phys. Chem.* 1995: 14149–14155.

145. Wan, L.-J.; Shundo, S.; Inukai, J.; Itaya, K. 2000. Ordered adlayers of organic molecules on sulfur-modified Au(111): In situ scanning tunneling microscopy study. *Langmuir* 16: 2164–2168.

146. Vericat, C.; Vela, M. E.; Andreasen, G.; Salvarezza, R. C. 2001. Sulfur-substrate interactions in spontaneously formed sulfur adlayers on Au(111). *Langmuir* 17: 4919–4924.

147. Stuhlmann, C.; Villegas, I.; Weaver, M. J. 1994. Scanning tunneling microscopy and infrared spectroscopy as combined in situ probes of electrochemical adlayer structure. Cyanide on Pt(111). *Chem. Phys. Lett.* 219: 319–324.

148. Gao, X.; Zhang, Y.; Weaver, M. J. 1992. Observing surface chemical transformations by atomic resolution scanning tunneling microscopy: Sulfide electrooxidation on gold(111). *J. Phys. Chem.* 96: 4156–4159.

149. Vaz-Domínguez, C.; Aranzábal, A.; Cuesta, A. 2010. In situ STM observation of stable dislocation networks during the initial stages of the lifting of the reconstruction on Au(111) electrodes. *J. Phys. Chem. Lett.* 1: 2059–2062.

150. He, Y.; Borguet, E. 2011. Metastable phase of the Au(111) surface in electrolyte revealed by STM and asymmetric potential pulse perturbation. *J. Phys. Chem. C* 115: 5726–5731.

151. Gao, X.; Hamelin, A.; Weaver, M. J. 1991. Atomic relaxation at ordered electrode surfaces probed by scanning tunneling microscopy: Au(111) in aqueous solution compared with ultrahigh-vacuum environments. *J. Chem. Phys.* 95: 6993–6996.

152. Broekmann, P.; Wilms, M.; Kruft, M.; Stuhlmann, C.; Wandelt, K. 1999. In-situ STM investigation of specific anion adsorption on Cu(111). *J. Electroanal. Chem.* 467: 307–324.

153. Herrero, E.; Buller, L. J.; D., A. H. 2001 Underpotential deposition at single crystal surfaces of Au, Pt, Ag, and other materials. *Chem. Rev.* 101: 1897–1930.

154. Nichols, R. J.; Kolb, D. M. 1991. STM observations of the initial stages of copper deposition on gold single-crystal electrodes. *J. Electroanal. Chem.* 313: 109–119.

155. Randler, R. J.; Kolb, D. M.; Ocko, B. M.; Robinson, I. K. 2000. Electrochemical copper deposition on Au(100): A combined in situ STM and in situ surface X-ray diffraction study. *Surf. Sci.* 20: 187–200.

156. Maupai, S.; Zhang, Y.; Schmuki, P. 2003. Nanoscale observation of initial stages of Cd-electrodeposition on Au(111). *Surf. Sci.* 527: L165–L170.

157. Garcia, S.; Salinas, D.; Mayer, C.; Schmidt, E.; Staikov, G.; Lorenz, W. J. 1998. Ag UPD on Au(100) and Au(111). *Electrochim. Acta* 43: 3007–3019.

158. Lay, M. D.; Varazo, K.; Srisook, N.; Stickney, J. L. 2003. Cd underpotential deposition (upd) from a sulfate electrolyte on Au(111): Studies by in situ STM and UHV-EC. *J. Electroanal. Chem.* 554–555: 221–231.

159. Green, M. P.; Hanson, K. J.; Scherson, D. A.; Xing, X.; Richter, M.; Ross, P. N.; Carr, R.; Lindau, I. 1989. In situ scanning tunneling microscopy studies of the underpotential deposition of lead on Au(111). *J. Phys. Chem.* 93: 2181–2184.

160. Inukai, J.; Sugita, S.; Itaya, K. 1996. Underpotential deposition of mercury on Au(111) investigated by in situ scanning tunnelling microscopy. *J. Electroanal. Chem.* 403: 159–168.

161. Hölzle, M. H.; Zwing, V.; Kolb, D. M. 1995. The influence of steps on the deposition of Cu onto Au(111). *Electrochim. Acta* 40: 1237–1247.

162. Baldauf, M.; Kolb, D. M. 1993. A hydrogen adsorption and absorption study with ultrathin Pd overlayers on Au(111) and Au(100). *Electrochim. Acta* 38: 2145–2153.

163. Hölzle, M. H.; Apsel, C. W.; Will, T.; Kolb, D. M. 1995. Copper deposition onto Au(111) in the presence of thiourea. *J. Electrochem. Soc.* 142: 3741–3749.

164. Gewirth, A. A.; Niece, B. K. 1997. Electrochemical applications of in situ scanning probe microscopy. *Chem. Rev.* 97: 1129–1162.

165. Sashikata, K.; Furuya, N.; Itaya, K. 1991. In situ scanning tunneling microscopy of underpotential deposition of copper on platiunum(111) in sufluric acid solutions. *J. Electroanal. Chem.* 316: 361–368.

166. Domke, K. F.; Xaio, X.-Y.; Baltruschat, H. 2009. The formation of two Ag UPD layers on stepped Pt single crystal electrodes and their restructuring by co-adsorption of CO. *Electrochim. Acta* 54: 4829–4836.

167. Carnal, D.; Oden, P. I.; Müller, U.; Schmidt, E.; Siegenthaler, H. 1995. In-situ STM investigation of Tl and Pb underpotential deposition on chemically polished Ag(111) electrodes. *Electrochim. Acta* 40: 1223–1235.

168. Okada, J.; Inukai, J.; Itaya, K. 2001. Underpotential and bulk deposition of copper on Pd(111) in sulfuric acid solution studied by in situ scanning tunneling microscopy. *Phys. Chem. Chem. Phys.* 3: 3297–3302.

169. Domke, K. F.; Xaio, X.-Y.; Baltruschat, H. 2008. Co-adsorption of CO onto a Ag-modified Pt(111)—Restructuring of a Ag UPD layer monitored by EC-STM. *Phys. Chem. Chem. Phys.* 10: 1555–1561.

170. Pao, T.; Chen, Y.-Y.; Chen, S.; Yau, S.-L. 2013. In situ scanning tunneling microscopy of electrodeposition of indium on a copper thin film electrode predeposited on Pt(111) electrode. *J. Phys. Chem. C* 117: 26659–26666.

171. Marcus, P. 1998. Surface science approach of corrosion phenomena. *Electrochim. Acta* 43: 109–118.

172. Vogt, M. R.; Lachenwitzer, A.; Magnussen, O. M. 1998. In-situ STM study of the initial stages of corrosion of Cu(100) electrodes in sulfuric and hydrochloric acid solution. *Surf. Sci.* 399: 49–69.

173. Wilms, M.; Broekmann, P.; Stuhlmann, C.; Wandelt, K. 1998. In-situ STM investigation of adsorbate structures on Cu(111) in sulfuric acid electrolyte. *Surf. Sci.* 416: 121–140.

174. Scherer, J.; Ocko, B. M.; Magnussen, O. M. 2003. Structure, dissolution and passivation of Ni(111) electrodes in sulfuric acid solution: An in situ STM, X-ray scattering, and electrochemical study. *Electrochim. Acta* 48: 1169–1191.

175. Lachenwitzer, A.; Morin, S.; Magnussen, O.; Behm, R. J. 2001. In situ STM study of electrodeposition and anodic dissolution of Ni on Ag(111). *Phys. Chem. Chem. Phys.* 3: 3351–3363.

176. Möller, F. A.; Kintrup, J.; Lachenwitzer, A.; Magnussen, O. M.; Behm, R. J. 1997. In situ STM study of the electrodeposition and anodic dissolution of ultrathin epitaxial Ni films on Au(111). *Phys. Rev. B* 56: 12506–12518.

177. Magnussen, O. M.; Zitzler, L.; Gleich, B.; Vogt, M. R.; Behm, R. J. 2001. In-situ atomic-scale studies of the mechanisms and dynamics of metal dissolution by high-speed STM. *Electrochim. Acta* 46: 3725–3733.

178. Kunze, J.; Strehblow, H. H.; Staikov, G. 2004. In situ STM study of the initial stages of electrochemical oxide formation at the Ag(111)/0.1 M NaOH(aq) interface. *Electrochem. Commun.* 6: 132–137.

179. Schneeweiss, M. A.; Kolb, D. M. 1997. Oxide formation on Au(111) an in situ STM study. *Solid State Ionics* 94: 171–179.

180. Müller, M.; Oechsner, H. 1997. In situ STM and AES studies on the oxidation of Cr(110). *Surf. Sci.* 387: 269–278.

181. Kunze, J.; Maurice, V.; Klien, L. H.; Strehblow, H. H.; Marcus, P. 2004. In situ STM study of the duplex passive films formed on Cu(111) and Cu(001) in 0.1 M NaOH. *Corrosion Sci.* 46: 245–264.

182. Inaba, M.; Siroma, Z.; Funabiki, A.; Ogumi, Z. 1996. Electrochemical scanning tunneling microscopy observation of highly oriented pyrolytic graphite surface reactions in an ethylene carbonate-based electrolyte solution. *Langmuir* 12: 1535–1540.

183. Tanaka, M.; Sawaguchi, T.; Sato, Y.; Yoshida, K.; Niwa, O. 2011. Surface modification of GC and HOPG with diazonium, amine, azide, and olefin derivatives. *Langmuir* 27: 170–178.

184. Pötzschke, R. T.; Gervasi, C. A.; Vinzelberg, S.; Staikov, G.; Lorenz, W. J. 1995. Nanoscale studies of Ag electrodeposition on HOPG (0001). *Electrochim. Acta* 40: 1469–1474.

185. Salinas, D. R.; Cobo, E. O.; Garcia, S. G.; Bessone, J. B. 1999. Early stages of mercury electrodeposition on HOPG. *J. Electroanal. Chem.* 470: 120–125.

186. Gloaguen, F.; Léger, J. M.; Lamy, C.; Marmann, A.; Stimming, U.; Vogel, R. 1999. Platinum electrodeposition on graphite: Electrochemical study and STM imaging. *Electrochim. Acta* 44: 1805–1816.

187. Zhao, X. K.; McCormick, L.; Fendler, J. H. 1991. Electrical and photoelectrochemical characterization of CdS particulate films by scanning electrochemical microscopy, scanning tunneling microscopy, and scanning tunneling spectroscopy. *Chem. Mater.* 3: 922–935.

188. Allongue, P.; Brune, H.; Gerischer, H. 1992. In situ STM observations of the etching of n-Si(111) in NaOH solutions. *Surf. Sci.* 275: 414–423.

189. Kaji, K.; Yau, S.-L.; Itaya, K. 1995. Atomic scale etching processes of n-Si(111) in NH$_4$F solutions: In situ scanning tunneling microscopy. *J. Appl. Phys.* 78: 5727–5733.

190. Itaya, K.; Sugawara, R.; Morita, Y.; Tokumoto, H. 1992. Atomic resolution images of H-terminated Si(111) surfaces in aqueous solutions. *Appl. Phys. Lett.* 60: 2534–2536.

191. Szklarczyk, M.; González-Martín, A.; Bokris, J. O. M. 1991. In situ STM studies of surface states at the p-Si(111)/propylene carbonate(TBAP) interface. *Surf. Sci.* 257: 307–318.

192. Ziegler, J. C.; Reitzle, A.; Bunk, O.; Zegenhagen, J.; Kolb, D. M. 2000. Metal deposition on n-Si(111):H electrodes. *Electrochim. Acta* 45: 4599–4605.

193. Rashkova, B.; Guel, B.; Pötzschke, R. T.; Staikov, G.; Lorenz, W. J. 1998. Electrodeposition of Pb on n-Si(111). *Electrochim. Acta* 43: 3021–3028.

194. Yao, H.; Yau, S.-L.; Itaya, K. 1996. In situ scanning tunneling microscopy of GaAs(001), (111)A, and (111)B surfaces in sulfuric acid solution. *Appl. Phys. Lett.* 68: 1473–1475.

195. Itaya, K.; Tomita, E. 1989. Scanning tunneling microscopy of semiconductor (n-ZnO)/liquid interfaces under potentiostatic conditions. *Surf. Sci.* 219: L515–L520.

196. Ohmori, T.; Castro, R. J.; Cabrera, C. R. 1998. In situ study of silver electrodeposition at MoSe$_2$ by electrochemical scanning tunneling microscopy. *Langmuir* 14: 6755–6760.

197. Sakata, M.; Hinokuma, K.; Hashimoto, K.; Fujishima, A. 1990. Adding H$_2$O to study in situ anisotropically enhanced photo-oxidation at n-MoS$_2$/CH$_3$CN interfaces using a scanning tunneling microscope. *Surf. Sci.* 237: L383–L389.

198. Higgins, S. R.; Hamers, R. J. 1996. Chemical dissolution of the galena (001) surface observed using electrochemical scanning tunneling microscopy. *Geochim. Cosmochim. Acta* 60: 3067–3073.

199. Borisenko, N.; Zein El Abedin, S.; Endres, F. 2006. In situ STM investigation of gold reconstruction and of silicon electrodeposition on Au(111) in the room temperature ionic liquid 1-butyl-1methylpyrrolidinium bis(trifluoromethylsulfonyl)imide. *J. Phys. Chem. B* 110: 6250–6256.

200. Endres, F.; Zein El Abedin, S. 2002. Nanoscale electrodeposition of germanium on Au(111) from an ionic liquid: An in situ STM study of phase formation, Part I. Ge from GeBr$_4$. *Phys. Chem. Chem. Phys.* 4: 1640–1648.

201. Varazo, K.; Lay, M. D.; Sorenson, T. A.; Stickney, J. L. 2002. Formation of the first monolayers of CdTe on Au(111) by electrochemical atomic layer epitaxy (EC-ALE): Studied by LEED, Auger, XPS, and in-situ STM. *J. Electroanal. Chem.* 522: 104–114.

202. Hayden, B. E.; Nandhakumar, I. S. 1998. In situ STM study of CdTe ECALE bilayers on gold. *J. Phys. Chem. B* 102: 4897–4905.

203. Hayden, B. E.; Nandhakumar, I. S. 1997. In-situ STM study of Te UPD layers on low index planes of gold. *J. Phys. Chem. B* 101: 7751–7757.

204. Itaya, K. 1998. In situ scanning tunneling microscopy in electrolyte solutions. *Prog. Surf. Sci.* 58: 121–247.

205. Tao, N. J.; Li, C. Z.; He, H. X. 2000. Scanning tunneling microscopy applications in electrochemistry—Beyond imaging. *J. Electroanal. Chem.* 492: 81–93.

206. Yoshimoto, S. 2006. Molecular assemblies of functional molecules on gold electrode surfaces studied by electrochemical scanning tunneling microscopy: Relationship between function and adlayer structures. *Bull. Chem. Soc. Jpn.* 79: 1167–1190.

207. Yoshimoto, S.; Itaya, K. 2013. Adsorption and assembly of ions and organic molecules at electrochemical interfaces: Nanoscale aspects. *Ann. Rev. Anal. Chem.* 6: 213–235.

208. Tanoue, R.; Higuchi, R.; Enoki, N.; Miyasato, Y.; Uemura, S.; Kimizuka, N.; Stieg, A. Z.; Gimzewski, J. K.; Kunitake, M. 2011. Thermodynamically controlled self-assembly of covalent nanoarchitectures in aqueous solution. *ACS Nano* 5: 3923–3929.

209. Alessandrini, A.; Facci, P. 2008. Electrochemically assisted scanning probe microscopy: A powerful tool in nano(bio)science. In *The New Frontiers of Organic and Composite Nanotechnology,* V. Erokhin, M. Kumar Ram, Ö. Yavuz (Eds.), Elsevier (Amsterdam), pp. 237–286.

210. Hagenstrom, H.; Schneeweiss, M. A.; Kolb, D. M. 1999. Modification of a Au(111) electrode with ethanethiol. 1. Adlayer structure and electrochemistry. *Langmuir* 15: 2435–2443.

211. Zhang, J. D.; Chi, Q. J.; Ulstrup, J. 2006. Assembly dynamics and detailed structure of 1-propanethiol monolayers on Au(111) surfaces observed real time by in situ STM. *Langmuir* 22: 6203–6213.

212. Hobara, D.; Sasaki, T.; Imabayashi, S.; Kakiuchi, T. 1999. Surface structure of binary self-assembled monolayers formed by electrochemical selective replacement of adsorbed thiols. *Langmuir* 15: 5073–5078.

213. Scherer, J.; Vogt, M. R.; Magnussen, O. M.; Behm, R. J. 1997. Corrosion of alkanethiol-covered Cu(100) surfaces in hydrochloric acid solution studied by in-situ scanning tunnelling microscopy. *Langmuir* 13: 7045–7051.

214. Sawaguchi, T.; Mizutani, F.; Yoshimoto, S.; Taniguchi, I. 2000. Voltammetric and in situ STM studies on self-assembled monolayers of 4-mercaptopyridine, 2-mercaptopyridine and thiophenol on Au(111) electrodes. *Electrochim. Acta* 45: 2861–2867.

215. Baunach, T.; Ivanova, V.; Scherson, D. A.; Kolb, D. A. 2004. Self-assembled monolayers of 4-mereaptopyridine on Au(111): A potential-induced phase transition in sulfuric acid solutions. *Langmuir* 20: 2797–2802.

216. Seo, K.; Borguet, E. 2007. Potential-induced structural change in a self-assembled monolayer of 4-methylbenzenethiol on Au(111). *J. Phys. Chem. C* 111: 6335–6342.

217. Nishiyama, K.; Tsuchiyama, M.; Kubo, A.; Seriu, H.; Miyazaki, S.; Yoshimoto, S.; Taniguchi, I. 2008. Conformational change in 4-pyridineethanethiolate self-assembled monolayers on Au(111) driven by protonation/deprotonation in electrolyte solutions. *Phys. Chem. Chem. Phys.* 10: 6935–6939.

218. Yoshimoto, S.; Itaya, K. 2007. Advances in supramolecularly assembled nanostructures of fullerenes and porphyrins at surfaces. *J. Porphyr. Phthalocyanines* 11: 313–333.

219. Yoshimoto, S.; Kobayashi, N. Supramolecular nanostructures of phthalocyanines and porphyrins at surfaces based on the "bottom-up assembly". In *Functional Phthalocyanine Molecular Materials*; Vol. 135, Jiang, J. (ed.), Berlin, Germany: Springer-Verlag, 2010; pp. 137–167.

220. Otsuki, J. 2010. STM studies on porphyrins. *Coord. Chem. Rev.* 254: 2311–2341.

221. Bonifazi, D.; Kiebele, A.; Stohr, M.; Cheng, F. Y.; Jung, T.; Diederich, F.; Spillmann, H. 2007. Supramolecular nanostructuring of silver surfaces via self-assembly of 60 fullerene and porphyrin modules. *Adv. Funct. Mater.* 17: 1051–1062.

222. He, Y.; Ye, T.; Borguet, E. 2002. Porphyrin self-assembly at electrochemical interfaces: Role of potential modulated surface mobility. *J. Am. Chem. Soc.* 124: 11964–11970.

223. Yoshimoto, S.; Tada, A.; Suto, K.; Narita, R.; Itaya, K. 2003. Adlayer structure and electrochemical reduction of O-2 on self-organized arrays of cobalt and copper tetraphenyl porphines on a Au(111) surface. *Langmuir* 19: 672–677.

224. Yoshimoto, S.; Tsutsumi, E.; Suto, K.; Honda, Y.; Itaya, K. 2005. Molecular assemblies and redox reactions of zinc(II) tetraphenylporphyrin and zinc(II) phthalocyanine on Au(111) single crystal surface at electrochemical interface. *Chem. Phys.* 319: 147–158.

225. Yoshimoto, S.; Yokoo, N.; Fukuda, T.; Kobayashi, N.; Itaya, K. 2006. Formation of highly ordered porphyrin adlayers induced by electrochemical potential modulation. *Chem. Commun.* 500–502.

226. Yuan, Q. H.; Xing, Y. J.; Borguet, E. 2010. An STM study of the pH dependent redox activity of a two-dimensional hydrogen bonding porphyrin network at an electrochemical interface. *J. Am. Chem. Soc.* 132: 5054–5060.

227. Batina, N.; Kunitake, M.; Itaya, K. 1996. Highly ordered molecular arrays formed on iodine-modified Au(111) in solution: In situ STM imaging. *J. Electroanal. Chem.* 405: 245–250.

228. Thorgaard, S. N.; Bühlmann, P. 2012. Self-assembled monolayers formed by 5,10,15,20-tetra(4-pyridyl) porphyrin and cobalt 5,10,15,20-tetra(4-pyridyl)-21H,23H-porphine on iodine-passivated Au(111) as observed using electrochemical scanning tunneling microscopy and cyclic voltammetry. *J. Electroanal. Chem.* 664: 94–99.

229. Yoshimoto, S.; Sawaguchi, T. 2008. Electrostatically controlled nanostructure of cationic porphyrin diacid on sulfate/bisulfate adlayer at electrochemical interface. *J. Am. Chem. Soc.* 130: 15944–15949.

230. Ye, T.; He, Y. F.; Borguet, E. 2006. Adsorption and electrochemical activity: An in situ electrochemical scanning tunneling microscopy study of electrode reactions and potential-induced adsorption of porphyrins. *J. Phys. Chem. B* 110: 6141–6147.

231. Yoshimoto, S.; Saito, A.; Tsutsumi, E.; D'Souza, F.; Ito, O.; Itaya, K. 2004. Electrochemical redox control of ferrocene using a supramolecular assembly of ferrocene-linked C-60 derivative and metallooctaethylporphyrin array on a Au(111) electrode. *Langmuir* 20: 11046–11052.

232. Suto, K.; Yoshimoto, S.; Itaya, K. 2003. Two-dimensional self-organization of phthalocyanine and porphyrin: Dependence on the crystallographic orientation of Au. *J. Am. Chem. Soc.* 125: 14976–14977.

233. Yoshimoto, S.; Honda, Y.; Ito, O.; Itaya, K. 2008. Supramolecular pattern of fullerene on 2D bimolecular "chessboard" consisting of bottom-up assembly of porphyrin and phthalocyanine molecules. *J. Am. Chem. Soc.* 130: 1085–1092.

234. Uemura, S.; Sakata, M.; Taniguchi, I.; Kunitake, M.; Hirayama, C. 2001. Novel "Wet process" technique based on electrochemical replacement for the preparation of fullerene epitaxial adlayers. *Langmuir* 17: 5–7.

235. Uemura, S.; Sakata, M.; Hirayama, C.; Kunitake, M. 2004. Fullerene adlayers on various single-crystal metal surfaces prepared by transfer from L films. *Langmuir* 20: 9198–9201.

236. Uemura, S.; Taniguchi, I.; Sakata, M.; Kunitake, M. 2008. Electrochemical STM investigation of C-70, C-60/C-70 mixed fullerene and hydrogenated fullerene adlayers on Au(111) prepared using the electrochemical replacement method. *J. Electroanal. Chem.* 623: 1–7.

237. Ishikawa, Y.; Ohira, A.; Sakata, M.; Hirayama, C.; Kunitake, M. 2002. A two-dimensional molecular network structure of trimesic acid prepared by adsorption-induced self-organization. *Chem. Commun.* 2652–2653.

238. Li, Z.; Han, B.; Wan, L. J.; Wandlowski, T. 2005. Supramolecular nanostructures of 1,3,5-benzenetricarboxylic acid at electrified au(111)/0.05 M H_2SO_4 interfaces: An in situ scanning tunneling microscopy study. *Langmuir* 21: 6915–6928.

239. Uemura, S.; Aono, M.; Komatsu, T.; Kunitake, M. 2011. Two-dimensional self-assembled structures of melamine and melem at the aqueous solution-Au(111) interface. *Langmuir* 27: 1336–1340.

240. Tao, N. J.; Derose, J. A.; Lindsay, S. M. 1993. Self-assembly of molecular superstructures studied by insitu scanning tunneling microscopy—DNA bases on AU(111). *J. Phys. Chem.* 97: 910–919.

241. Vaz-Dominguez, C.; Escudero-Escribano, M.; Cuesta, A.; Prieto-Dapena, F.; Cerrillos, C.; Rueda, M. 2013. Electrochemical STM study of the adsorption of adenine on Au(111) electrodes. *Electrochem. Commun.* 35: 61–64.

242. Burgess, I.; Jeffrey, C. A.; Cai, X.; Szymanski, G.; Galus, Z.; Lipkowski, J. 1999. Direct visualization of the potential-controlled transformation of hemimicellar aggregates of dodecyl sulfate into a condensed monolayer at the Au(111) electrode surface. *Langmuir* 15: 2607–2616.

243. Matsunaga, S.; Yokomori, R.; Ino, D.; Yamada, T.; Kawai, M.; Kobayshi, T. 2007. EC-STM observation on electrochemical response of fluidic phospholipid monolayer on Au(111) modified with 1-octanethiol. *Electrochem. Commun.* 9: 645–650.

244. Nishiyama, K.; Ono, Y.; Taniguchi, I.; Yoshimoto, S. 2012. EC-STM investigation of electrochemically active 2D adlayer consisting of metal ions and a bis(terpyridine) derivative. *Chem. Lett.* 41: 1311–1313.

245. Ohira, A.; Sakata, M.; Taniguchi, I.; Hirayama, C.; Kunitake, M. 2003. Comparison of nanotube structures constructed from alpha-, beta-, and gamma-cyclodextrins by potential-controlled adsorption. *J. Am. Chem. Soc.* 125: 5057–5065.

246. Yoshimoto, S.; Abe, M.; Itaya, K.; Narumi, F.; Sashikata, K.; Nishiyama, K.; Taniguchi, I. 2003. Formation of well-defined p-tent-butylcalix 4 arenedithiolate monolayers on a Au(100)-(1 × 1) surface studied by in situ scanning tunneling microscopy. *Langmuir* 19: 8130–8133.

247. Ohira, A.; Sakata, M.; Hirayama, C.; Kunitake, M. 2003. 2D-supramolecular arrangements of dibenzo-18-crown-6-ether and its inclusion complex with potassium ion by potential controlled adsorption. *Org. Biomol. Chem.* 1: 251–253.

248. Tongol, B. J. V.; Wang, L.; Yau, S. L.; Otsubo, T.; Itaya, K. 2009. Nanostructures and molecular assembly of beta-blocked long oligothiophenes up to the 96-Mer on Au(111) as probed by in situ electrochemical scanning tunneling microscopy. *J. Phys. Chem. C* 113: 13819–13824.

249. Lapitan, L. D. S.; Tongol, B. J. V.; Yau, S. L. 2012. In situ scanning tunneling microscopy imaging of electropolymerized poly(3,4-ethylenedioxythiophene) on an iodine-modified Au(111) single crystal electrode. *Electrochim. Acta* 62: 433–440.

250. Liu, Y. F.; Krug, K.; Lee, Y. L. 2013. Self-organization of two-dimensional poly(3-hexylthiophene) crystals on Au(111) surfaces. *Nanoscale* 5: 7936–7941.

251. Tanoue, R.; Higuchi, R.; Ikebe, K.; Uemura, S.; Kimizuka, N.; Stieg, A. Z.; Gimzewski, J. K.; Kunitake, M. 2012. In situ STM investigation of aromatic poly(azomethine) arrays constructed by "on-site" equilibrium polymerization. *Langmuir* 28: 13844–13851.

252. Xu, B.; Tao, N. 2003. Measurement of single-molecule resistance by repeated formation of molecular junctions. *Science* 301: 1221–1223.

253. Cui, X. D.; Primak, A.; Zarate, X.; Tomfohr, J.; Sankey, O. F.; Moore, A. L.; Moore, T. A.; Gust, D.; Harris, G.; Lindsay, S. M. 2001. Reproducible measurements of single-molecule conductivity. *Science* 249: 571–573.

254. Reed, M. A.; Zhou, C.; Muller, C. J.; Burgin, T. P.; Tour, J. M. 1997. Conductance of a molecular junction. *Science* 278: 252–253.

255. Jang, S.-Y.; Reddy, P.; Majumdar, A.; Segalman, R. A. 2006. Interpretation of stochastic events in single molecule conductance measurements. *Nano Lett.* 2006: 2362–2367.

256. Zhou, X.-S.; Wei, Y.-M.; Liu, L.; Chen, Z.-B.; Tang, J.; Mao, B. W. 2008. Extending the capability of STM break junction for conductance measurement of atomic-size nanowires: An electrochemical strategy. *J. Am. Chem. Soc.* 130: 13228–13230.

257. Haiss, W.; van Zalinge, H.; Higgins, S. J.; Bethell, D.; Höbenreich, H.; Schiffrin, D. J.; Nichols, R. J. 2003. Redox state dependence of single molecule conductivity. *J. Am. Chem. Soc.* 125: 15294–15295.

258. Xaio, X.; Xu, B.; Tao, N. J. 2004. Measurement of single molecule conductance: Benzenedithiol and benzenedimethanethiol. *Nano Lett.* 4: 267–271.

259. Xiao, X.; Nagahara, L. A.; Rawlett, A. M.; Tao, N. 2005. Electrochemical gate-controlled conductance of single oligo(phenylene ethynylene)s. *J. Am. Chem. Soc.* 127: 9235–9240.

260. Wierzbinksi, E.; Slowinski, K. 2006. In situ wiring of single molecules into an electrical circuit via electrochemical distance tunneling spectroscopy. *Langmuir* 22: 5205–5208.

261. Pobelov, I., V.; Li, Z.; Wandlowski, T. 2008. Electrolyte gating in redox-active tunneling junctions—An electrochemical STM approach. *J. Am. Chem. Soc.* 130: 16045–16054.

262. Leary, E.; Higgins, S. J.; van Zalinge, H.; Haiss, W.; Nichols, R. J.; Nygaard, S.; Jeppesen, J. O.; Ulstrup, J. 2008. Structure–property relationships in redox-gated single molecule junctions—A comparison of pyrrolo-tetrathiafulvalene and viologen redox groups. *J. Am. Chem. Soc.* 130: 12204–12205.

263. Hines, T.; Díez-Pérez, I.; Nakamura, H.; Shimazaki, T.; Asai, Y.; Tao, N. 2013. Controlling formation of single-molecule junctions by electrochemical reduction of diazonium terminal groups. *J. Am. Chem. Soc.* 135: 3319–3322.

264. Capozzi, B.; Chen, Q.; Darancet, P.; Kotiuga, M.; Buzzeo, M.; Neaton, J. B.; Nuckolls, C.; Venkataraman, L. 2014. Tunable charge transport in single-molecule junctions via electrolytic gating. *Nano Lett.* 14: 1400–1404.

265. Artés, J. M.; Díez-Pérez, I.; Gorostiza, P. 2012. Transistor-like behavior of single metalloprotein junctions. *Nano Lett.* 12: 2679–2684.

266. Alessandrini, A.; Corni, S.; Facci, P. 2006. Unravelling single metalloprotein electron transfer by scanning probe techniques. *Phys. Chem. Chem. Phys.* 8: 4383–4397.

267. Chi, Q.; Farver, O.; Ulstrup, J. 2005. Long-range protein electron transfer observed at the single molecule level: In situ mapping of redox-gated tunneling resonance. *Proc. Natl. Acad. Sci.* 102: 16203–16208.

21 Combined Atomic Force Microscopy–Scanning Electrochemical Microscopy

Christophe Demaille and Agnès Anne

CONTENTS

21.1 INTRODUCTION

Probing electrochemical reactions at surfaces with nanometer resolution is one of the major goals of nanoelectrochemistry. Scanning electrochemical microscopy (SECM)[1] is, as detailed in Chapter 18, a promising way to achieve this goal, provided experimental approaches are designed to endow SECM with nanometer resolution capabilities. Among these approaches, one is to couple SECM with some other local probe technique known for its robustness and inherent nanoscale resolution. A particularly attractive technique for this is atomic force microscopy (AFM),[2] introduced in 1986 by Binning et al., and which has now become one of the most popular local probe techniques for the nanoscale characterization of surfaces. Coupling SECM with AFM can seem a priori simple: it suffices to endow the nanoprobes commonly used in AFM with current measuring capabilities, that is, to design combined AFM tips which can also act as SECM-microelectrode probes. Of course things are not that simple and fabrication of such AFM-SECM probes has proved to be a formidable challenge.[3,4] Indeed to date, many of the publications related to AFM-SECM (or SECM-AFM, some authors preferring this later acronym) actually solely report on suitable fabrication methods for the combined probes. Here, we will primarily focus on the successful applications of AFM-SECM for the nanoscale characterization of electroactive systems while describing briefly the corresponding probe fabrication methods along the way. The newest techniques for fabricating AFM-SECM probes will then be examined separately in more detail. Most of all we intend to show that coupling SECM with AFM can bring much more than resolution enhancement: AFM-SECM makes possible the nanoscale exploration of electrochemical (or electrochemically transduced) processes impossible to reveal by mere AFM or SECM.

21.2 WORKING PRINCIPLE OF AFM-SECM: WHAT AFM BRINGS TO SECM?

To fully understand the benefits of coupling SECM with AFM, one of course first needs to understand the operating principle of SECM and be aware of its limitations for exploring nanosystems. The reader is referred to Chapter 18 and to the literature[1,5,6] for a detailed presentation of these topics that we only briefly recall in the following.

SECM relies on approaching a microelectrode, used as a local probe (tip), toward the surface of a substrate immersed in an electrolyte solution (i.e., *in situ*). Schematically, the electrochemical activity of the substrate is probed locally via the detection at the microelectrode of a flux of redox species generated at the surface. This detection is only effective provided the tip-to-substrate separation distance d is comparable to, or lower than, the microelectrode size a (see Figure 21.1).

In the most common mode of operation of SECM, depicted in Figure 21.1 and called the *feedback mode*, a redox species (P, the mediator) is introduced into the electrolyte solution and converted into its reduced or oxidized form (Q) at the suitably biased tip. Q diffuses toward the substrate and, provided the substrate displays some electrochemical activity, is converted back to P and fed back to the probe. This redox cycling of the mediator between the tip and substrate results in an increase of the tip current, i_{tip}, which is what is measured in SECM. The magnitude of this so-called positive feedback process depends intricately on *both* the tip–substrate separation d and the heterogeneous rate constant for the electron transfer at the substrate k_{ET}. This rate constant is the physical parameter of interest quantifying the electrochemical activity of the substrate. A particular case is met for

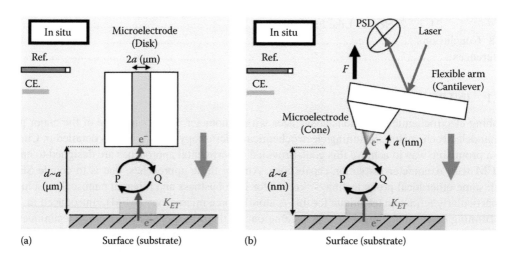

FIGURE 21.1 Working principle of (a) SECM and (b) AFM-SECM. The configuration depicted is the feedback SECM mode where a redox species (the mediator) is generated under its oxidized form (Q) at the microelectrode from its reduced form (P) initially introduced in solution. Upon reaching the reactive substrate, Q is converted back to P, at some rate characterized by a heterogeneous rate constant k_{ET}, and fed back to the tip. In (b), some interfacial force, F, is sensed as it bends the flexible arm of the combined probe.

electrochemically inactive substrates (i.e., $k_{ET}=0$) where i_{tip} decreases monotonously with decreasing d as a result of hindered diffusion of the mediator in a so-called negative feedback process.

Measurement of surface kinetics by SECM is typically carried out by recording approach curves, that is, by plotting the recorded tip current i_{tip} as a function of the tip–substrate distance d. Quantitative interpretation of the approach curves requires that it is adjusted to theoretical curves, which in turn imposes that d is accurately known. Because i_{tip} depends jointly on d and k_{ET}, d has to be determined *independently* from the current measurement. In conventional SECM, independent tip-to-substrate distance determination is not always possible and some uncertainty remains on the d value. The same limitation applies when SECM is used in modes other than feedback, such as the surface generator–tip collector mode (SG/TC), where the tip selectively detects species generated by surface reactions (e.g., enzymatic reactions).

As depicted in Figure 21.1a, microelectrode probes commonly used for SECM are of the disk-in-glass type and their size is in the micron range. Conventional SECM instrumentation easily allows these probes to be approached within less than a micron from a substrate. Much smaller nanometer-sized SECM microelectrodes are also available, but as detailed in Chapter 18, controlling the approach of such *nanoelectrodes* from a solid surface is much more challenging. One of the reasons for that is that physically aligning an electrode of this size perpendicularly to a surface is technically difficult. Hence, one of the first motivations for combining SECM with AFM is to benefit from the exquisite positioning capability offered by AFM for approaching nanometer-sized tips within a few nanometers (or less) from surfaces in a controlled way. This capability arises from the design of AFM probes, which integrate a flexible arm bearing the tip and whose position can be sensed with nanometric accuracy (see Figure 21.1b). Typically, this is achieved by bouncing a laser off the reflective cantilever and back to a position-sensitive detector (PSD). In the AFM-SECM configuration, the combined probe also integrates a nanoelectrode located at, or close to, the tip apex. Hence, upon approaching from a substrate, the combined probe can record an electrochemical current but can also physically sense and locate the substrate surface, simply because at contact, the cantilever bends upward. Importantly, because the cantilever acts as a very soft spring, the tip is not damaged when contacting the surface, unlike what typically happens in conventional SECM. Furthermore, this configuration allows the $d=0$ point to be unambiguously determined, permitting the accurate *a posteriori* calibration of the tip-to-substrate separation distance. Hence, AFM-SECM offers a straightforward *built-in* way of determining d *independently* from current measurement, which is a prerequisite for extracting quantitative information from SECM data. Typically, tip deflection *and* tip current AFM-SECM approach curves are recorded *simultaneously*, yielding the required i_{tip}/d data set.

Beyond its use for the tip-to-substrate distance calibration, the deflection approach curve can also yield valuable information regarding the interfacial forces sensed by the tip as it approaches the substrate. The deflection signal can be converted into actual force values provided the spring constant of the flexible cantilever arm is known. In such a case, the AFM-SECM configuration makes it uniquely possible to study the relation between local interfacial forces and electrochemical (or electrochemically transduced) processes occurring in the tip–substrate gap. This possibility, which can be regarded as the ultimate benefit of coupling AFM with SECM, has so far been underexploited.

SECM also offers an imaging mode which allows the electroactivity of composite surfaces to be mapped.[7] This is typically achieved by scanning (rastering) the SECM-microelectrode probe at some average height h above the surface, in search of current variations indicative of local heterogeneities in the electrochemical activity of the surface. Plotting the current as a function of the tip position ideally allows a 2D image of the surface activity to be reconstructed line by line, with a higher current contrast indicating a higher local activity. This ideal situation is met for an infinitely flat substrate; however, as depicted in Figure 21.2a, for a *real* substrate displaying corrugations, the actual tip-to-substrate distance d varies as a function of the local topography.

As a result, the variation of i_{tip} versus tip position is a convoluted function of the substrate topography and local reactivity (see Figure 21.2a'). Coupling SECM with AFM offers an elegant way to

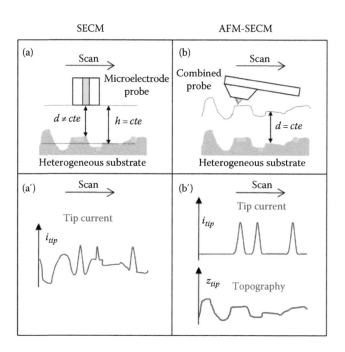

FIGURE 21.2 Schematic depiction of SECM (left) and AFM-SECM (right) feedback-mode imaging (scanning) of a rough heterogeneous substrate comprising electrochemically active sites (yellow) embedded into a nonactive matrix (gray). (a) In SECM, the microelectrode is scanned at a constant height h above the substrate, but the actual tip-to-substrate distance d varies along the line scan. (a') Consequently, the current scan line (and image) recorded is a convoluted function of topography and local activity of the substrate. (b) AFM-SECM allows the tip to be held at a constant distance from the substrate all along the scan line. (b') This constant-distance AFM-SECM imaging mode allows nonconvoluted current and topography scan lines (and images) to be simultaneously recorded.

solve this problem by enabling constant-distance SECM imaging. The principle of this much sought SECM imaging mode is that the tip is held at a predefined distance from the surface at *any point* of the line scan (see Figure 21.2b). This can be achieved by using the AFM-lift mode, which consists in scanning each line of the image twice.[8] In the first pass, the substrate topography is acquired and stored. In the second pass, the tip is first withdrawn from the surface by a preset distance d, and scanned again along the same line while being vertically moved to reproduce the recorded surface topography (Figure 21.2b'). Lift-mode AFM-SECM is also most useful if the surface to be interrogated is a conductor since it enables to avoid the occurrence of tip–substrate short-circuits during the electrochemical measurements. Constant-distance AFM-SECM imaging can also be implemented by using tapping mode AFM.[9] This AFM mode consists in using the surface-induced damping of a small oscillation imposed to the cantilever to control the tip–substrate distance. In this case, a constant tip–substrate distance is held via a feedback loop, which adjusts the tip vertical position in order to keep damping at a preset value. As compared to lift mode, tapping mode AFM-SECM displays the advantage of solely requiring a single pass; both topography and current data are acquired simultaneously. In any case, a valuable benefit of constant-distance AFM-SECM imaging is that two *independent* images are produced: a substrate topography image and a tip current image, *free from any artifacts due to topography* and reliably representing a map of the surface activity (see Figure 21.2b').

Overall, AFM-SECM imaging opens the unique possibility of locating topographically identifiable features, for example, nano-objects or nanodomains, on surfaces, to measure their lateral and/or vertical size, and to simultaneously probe their redox activity.

Finally, as indicated previously, a major expectation for coupling SECM with AFM is to endow SECM with the nanometer (or even subnanometer) resolution of AFM. This resolution comes from the sharpness of the AFM point probes. Hence, full benefit of coupling SECM with AFM is conditioned by the ability of fabricating combined AFM-SECM tips of nanometric dimensions. But even before the miniaturization issue comes the problem of integrating a microelectrode at the end of an AFM tip.

21.3 AVAILABILITY OF COMBINED AFM-SECM PROBES: A KEY ISSUE

In spite of the remarkable assets of AFM-SECM, its development has been so far hindered by the limited availability of dedicated probes, which are difficult to mass-produce complex microobjects. Conceptually, a combined AFM-SECM probe is fundamentally an AFM probe, capable of measuring local forces, bearing a microelectrode at or close to its tip. Since AFM-SECM experiments are conducted in situ, typically in a solution containing a redox mediator species, this requires that the whole body of the probe, and its microelectrode part, is insulated in order to avoid interfering electrochemical currents. However, since the microelectrode has to be connected to a potentiostat, a conducting path going from the microelectrode to some macroscopic connecting pad has also to be integrated in the probe. The probe material, if electrically conducting, can play this role; otherwise, a conducting layer can be deposited on the probe body. Hence, the whole problem is to perfectly insulate the conducting parts of the probe, which often amount to hundreds of square microns or more, while leaving the micrometer- or even nanometer-sized microelectrode uncoated and exposed to the solution. Several strategies have been proposed to achieve this, but until recently, no AFM-SECM probes were commercially available (see Section 21.7). Therefore, all the researchers who characterized nanosystems by AFM-SECM so far made use of AFM-SECM probes they designed and fabricated themselves. In the following, we review research works where AFM-SECM was actually used to investigate local electrochemical processes of interest. The techniques used to fabricate the probes which made these works possible are also described.

21.4 DEVELOPMENT AND APPLICATIONS OF AFM-SECM FOR NANOELECTROCHEMISTRY

21.4.1 IMAGING DISSOLUTION PROCESSES

In 1996, Macpherson, Unwin, Hillier, and Bard reported in a joint paper the first use of AFM-SECM.[10] The aim was to study the dissolution of KBr single crystals in situ, in a KBr-saturated acetonitrile solution. For this study, the authors fabricated a rudimentary, yet effective, AFM-SECM probe by coating the underside of a commercial silicon nitride AFM probe with a conductive evaporated layer of Cr/Pt. The probe was further insulated by a polystyrene coating deposited manually from a dichloroethane/polystyrene solution using a fine paintbrush. This process left a quite large electrochemically active area of 0.5–1 mm^2, which included the tip and the whole of the cantilever arm. This millimetric electrode was used to trigger the local dissolution of the crystal, by oxidizing Br$^-$ ions in solution and thus provoking local undersaturation, while the tip was simultaneously used in contact mode to image the resulting growth of spiral-shaped dissolution pits. The topographical resolution achieved was typical of that allowed by commercial AFM probes, a few tens of nanometers. Hence, even though no electrochemical current image was actually acquired, but solely a topographical image, this work showed that AFM-SECM could indeed be used to trigger and visualize surface dissolution reactions and to measure their kinetics. The same configuration was later used to study the proton-assisted dissolution of calcite.[11]

Macpherson and Unwin also reported a bit later the use of AFM-SECM for probing in situ the electrochemically induced dissolution of a single crystal of potassium ferrocyanide trihydrate

$(K_4Fe(CN)_6 \cdot 3H_2O)$.[12] In this work, a new design of AFM-SECM probe was first introduced. The probe described was hand-fabricated from a 50 μm Pt wire, electrochemically etched to form a sharp-point tip, bent at right angle, and finally flattened to form a flexible rectangular-shaped cantilever arm (spring constant ~0.06–0.5 N/m) (see Figure 21.3).

Insulation of the probe was insured by electrophoretic deposition of an anodic paint. The natural tendency of this kind of paint to retract upon curing resulted in the spontaneous exposure of the tip extremity, which formed an approximately hemispherical microelectrode characterized by a tip radius in the submicron range. In that sense, this hand-made probe was the first AFM-SECM probe integrating an actual (sub)-microelectrode. Again no electrochemical current images were recorded but contact mode AFM images allowed the electrochemically induced etch pits to be visualized (see Figure 21.4).

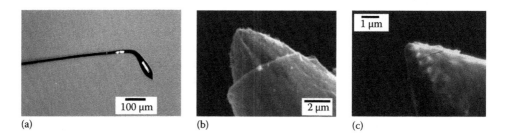

(a)   (b)   (c)

FIGURE 21.3 Combined AFM-SECM probe hand-fabricated from an etched, flattened, and bent platinum microwire. The probe is isolated by an electrophoretic anodic paint. (a) SEM micrograph of the probe, showing the cantilever and the tip. (b) and (c) Close-up SEM images of the tip extremity. Upon curing the insulating film shrunk, uncovering the tip apex, which behaved as a micrometer-sized (in b) or submicrometer-sized (in c) conical microelectrode. (From Macpherson, J.V. and Unwin, P.R., *Anal. Chem.*, 72, 276, 2000. With permission.)

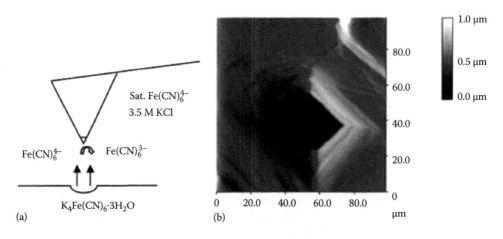

(a)   (b)

FIGURE 21.4 (a) Schematic of the principles involved in AFM-SECM-induced dissolution of a localized zone on a single crystal of potassium ferrocyanide trihydrate $K_4Fe(CN)_6 \cdot 3H_2O$. (b) AFM-SECM height image of the (010) surface of a potassium ferrocyanide trihydrate single crystal in a solution containing saturated potassium ferrocyanide, recorded immediately after the tip electrode had been used to electrochemically induce dissolution, with the tip positioned in the center of the scan. (From Macpherson, J.V. and Unwin, P.R., *Anal. Chem.*, 72, 276, 2000. With permission.)

21.4.2 Functional Imaging of Electroactive Sites on a Surface

Because of its combined nature, AFM-SECM makes uniquely possible the characterization of complex composite surfaces displaying *sites* (or objects) of differing electrochemical activity. AFM-SECM makes it notably possible to identify these sites (or objects) individually from the topography image while their electroactivity can be evidenced from the current image. Provided a suitable diffusional model is used, quantitative analysis of the current image can even allow the electroactivity of individual sites to be expressed in term of meaningful physical parameters, such as heterogeneous electron transfer rate constants.

In a pioneering paper, Macpherson et al. made use of the hand-made Pt AFM-SECM probes described previously (Figure 21.3), to characterize the electrochemical and structural properties of dimensionally stable Ti/TIO$_2$ anodes.[13] These so-called dimensionally stable anodes (DSAs) are widely used in industry because of their resistance to aggressive media. Their microscopic structure is quite complex since they consist of Pt microparticles (0.15–1 μm in size) embedded on, or within, a thin porous TiO$_2$ layer gown on a Ti substrate. Because of its ability to map electroactive sites on complex composite surfaces, AFM-SECM was a tool of choice for investigating locally the electrochemical behavior of DSA. The aim was to understand the macroscopic behavior of DSA, as observed by cyclic voltammetry, from the information gained at the microscale by AFM-SECM. Acquisition of combined topography and current AFM-SECM images, in contact mode AFM-fixed height SECM mode, and in an SECM SG/TC configuration, allowed individual Pt microparticles to be identified and their electrochemical activity to be probed (see Figure 21.5). Most interestingly, the current image showed that not all of the microparticles displayed the same level of activity: some were highly active, some only weakly, and some were inactive. *Being able to probe the dispersion in activity of otherwise topographically similar objects in such a way is a key benefit of AFM-SECM.*

The dispersion in particle activity was attributed to the variation in thickness of the TiO$_2$ layer separating the particles from the underlying Ti substrate, which allowed a more or less efficient electrical wiring of the particles.

Another early demonstration of the capability of AFM-SECM to identify extremely small active *sites* on a surface while probing their electrochemical activity was carried out by Macpherson and Unwin in 2005 and 2006.[14,15] The authors successfully imaged a model substrate consisting of an

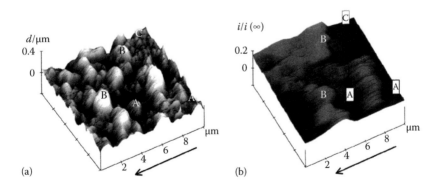

FIGURE 21.5 Combined contact mode AFM-fixed height SECM mode images of the surface of a Ti/TiO$_2$/Pt DSA. (a) Topography image. (b) Tip current image corresponding to the tip detection of the oxidized form of the mediator (IrCl$_6^{3-}$ complex) generated at the substrate. The tip current image was acquired with the tip positioned ~0.25 μm above the surface. Micrometer-sized platinum deposits are identified as large particles in the topography image. The current image shows that some of these particles (labeled A) are electrochemically inactive, whereas some others (labeled B) are active. The TiO$_2$ matrix (labeled C) is seen to be insulating, as expected. (From Macpherson, J.V. et al., *J. Electrochem. Soc.*, 149, B306, 2002. With permission.)

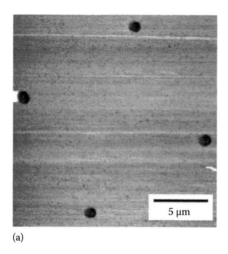
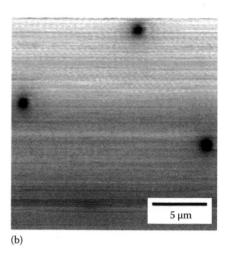

(a)   (b)

FIGURE 21.6   Lift-mode AFM-SG/TC mode SECM images of an array of 1 μm diameter disk microelectrodes. (a) Topography image. (b) Tip current image. The probe was biased so as to detect the reduced form of the mediator (ruthenium hexamine $Ru(NH_3)_6^{3+}$) produced at the disk microelectrodes from its oxidized form introduced in solution. The AFM-SECM configuration allowed individual microdisk electrodes to be resolved and their electrochemical activity to be probed. (From Dobson, P.S. et al., *Anal. Chem.*, 77, 424, 2005. With permission.)

array of ~1 μm diameter disk-shaped microelectrodes in the substrate generator–tip collector mode. Not only could individual microelectrodes be resolved in the topography images, but the electrochemical activity of the same microelectrodes could also be simultaneously visualized in the current image (see Figure 21.6). The presence of some electrochemically *inactive* microelectrodes in some of the microarray surfaces examined could be evidenced.

In order to achieve such a high imaging resolution, the authors developed an electron-beam-lithography–based (EBL) technique which allowed them to microfabricate sophisticated AFM-SECM probes.[15] These probes were actually *batch fabricated* from silicon wafers so that as many as ~60 probes could be produced in each run. This represented a clear improvement compared to other techniques reported so far, which allowed probes to be fabricated only one by one. The final probe has a triangular-shaped microelectrode at its extremity (base width 1 μm, height 0.65 μm), see Figure 21.7.

The spring constant of the probe cantilever was in the 1–1.5 N/m range; it could be operated in tapping or contact mode or coupled with lift mode, for acquiring current images. The authors demonstrated that, due to their triangular-shaped microelectrode apex, this kind of probes is preferably used in the substrate generator–tip collector mode. This means that such probes perform better when they are used to collect a redox species emitted by the electroactive sites to be probed, and more poorly when they are used to generate a diffusing species targeted at reacting with surface sites (feedback mode). Actually, as demonstrated very early on in the development of SECM microscopy, the same trend is to be expected with any conical-shaped SECM-microelectrode probe,[16] a configuration often encountered for AFM-SECM probes.

Davoodi et al. demonstrated that AFM-SECM is also particularly useful for characterizing localized corrosion processes in situ and in real time.[17,18] These authors were able to visualize the local electrochemical activity associated with the corrosion of Al alloys with micrometer resolution while following the resulting changes in topography. This enabled observation of local corrosion events such as the formation of pits or of intermetallic particles on the alloy surface. For these studies, the authors made use of an original combined AFM-SECM probe, then developed by Windsor Scientific Ltd.[19] This probe was formed by embedding a 1–5 μm diameter Pt wire in epoxy. Part of

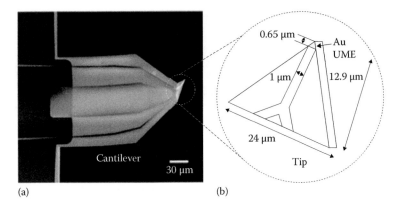

(a)   (b)

FIGURE 21.7   (a) SEM micrograph of a combined AFM-SECM probe microfabricated following Dobson's et al. process. (b) Schematic illustration of the tip, showing its dimensions and the location of the triangular gold microelectrode integrated at its apex (From Dobson, P.S. et al., *Anal. Chem.*, 77, 424, 2005. With permission.)

the probe was bent and coated with gold to act as a reflective and flexible cantilever. The ~10 µm diameter cylindrically shaped extremity of the Pt-in-epoxy probe was cut by focused ion beam (FIB), so as to form an inlaid disk-shaped microelectrode, while leaving a sharp epoxy cone in the immediate vicinity of the disk microelectrode (see Figure 21.8).

This cone, ~1–2 µm in height, was the effective AFM tip of the probe. As a result, the probe could be used in contact mode to characterize the conducting alloy surface while avoiding short-circuits. This probe design also allowed constant-distance SECM imaging since the disk microelectrode was de facto held away from the surface at a distance corresponding to height of the insulating conical tip (see Figure 21.8). Lift mode was used when a larger tip–substrate distance was required.

The heterogeneous electrochemical activity of another metallic alloy, a Ti-6Al-4V alloy commonly used for load-bearing medical implants, was investigated by Wittstock et al. using AFM-SECM.[20] The surface of this titanium-based alloy is naturally covered by a semiconducting passivating layer, a few nanometers in thickness. This layer is actually composed of two coexisting phases (α and β) of differing electronic properties. AFM-SECM was a priori a tool of choice for such a study: the α and β phases differ in height by some ~60 nm, and could thus potentially

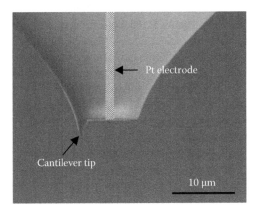

FIGURE 21.8   SEM micrograph of the extremity of a combined AFM-SECM probe fabricated by reshaping a Pt-in-epoxy disk microelectrode using FIB, to form a sharp epoxy cone to be used as an AFM tip. (From Davoodi, A. et al., *Electrochem. Solid State Lett.*, 8, B21, 2005. With permission.)

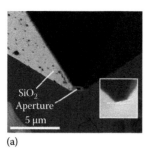
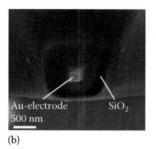
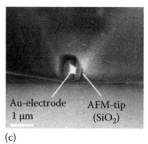

(a)   (b)   (c)

FIGURE 21.9   SEM micrographs of AFM-SECM probes microfabricated by filling up a hollow SiO_2 pyramid with evaporated gold. Depending on the way it is performed, FIB slicing of the pyramid apex can expose submicrometer-sized gold electrodes either at the extremity of the tip (b) or recessed from it as in (a), inset, and (c). (From Salomo, M. et al., *Microelectron. Eng.*, 87, 1537, 2010. With permission.)

be identified in the topography image, while their respective electrochemical properties could be simultaneously probed in situ by the combined tip. To achieve this goal, a new design of AFM-SECM probes was introduced.[21] Probes were fabricated by joining two silicon wafers: one carrying the cantilever membrane and tip and the other one playing the role of a rigid holder. The tip was a hollow fourfold silicon dioxide pyramid filled up by evaporating evaporated gold from its back side. The tip microelectrode was revealed by cutting the extremity of the pyramid using FIB milling. Depending on the cutting angle, square-shaped microelectrodes either protruding from the tip extremity, or recessed from it, could be produced. Recessed microelectrodes were preferably used; their widths were below 100 nm and in some instances as small as ~10–20 nm (see Figure 21.9).

Using these probes, AFM-SECM images of the Ti-6Al-4V alloy surface were acquired in contact mode AFM-feedback mode SECM using ruthenium hexa-amine $Ru(NH_3)_6^{3+}$ as a mediator. Most interestingly, even though the topographical image obtained did clearly show the two phases, the current image did not show any difference in contrast between the α and β phases (see Figure 21.10). Actually, the tip current varied slightly upon passing from one phase to the other, but this was solely an artefact due to the height of the phase boundary.

This result was unexpected since using a much larger ~2 μm sized microelectrode probe, the differing activity of these phases could be readily imaged using conventional SECM (see part (c) in Figure 21.10).[22] These seemingly conflicting AFM-SECM and SECM results actually exemplify a very important phenomenon, identified in the early days of SECM,[23,24] but often forgotten: *small* microelectrodes cannot probe *slow* electrochemical reactions, especially in SECM feedback mode. The reason is that slow local electrochemical reactions produce low fluxes of species, which cannot be efficiently collected by small microelectrodes.[6,25] Hence, in practical cases, a trade-off has to be found between the lateral SECM resolution achievable, which depends primarily on the microelectrode size, and the *kinetic* sensitivity of the probe (i.e., its ability to sense slow processes). This fundamental limitation is of particular importance in AFM-SECM where high spatial resolution is primarily sought. In the case of Wittstock's work, the electrochemical processes occurring at the α and β phases, albeit known to differ in rate, were both too slow to be sensed by the sub-100 nm sized probe, let alone be differentiated.

AFM-SECM has also been employed to study a most interesting, naturally nanostructured, conducting surface, which also displays a heterogeneous electrochemical activity: highly oriented pyrolytic graphite (HOPG).[26] This surface is known to be made of a series of stacked graphite (graphene) planes extending over a few microns wide or more, and separated by steps where the edges of the graphite planes are exposed. Electrochemists have long been interested in this carbon surface, and earlier studies had demonstrated that HOPG surfaces, used as macroscopic electrodes, displayed an electrochemical behavior, which was strongly dependent on the relative content in base planes versus step edges.[27] This behavior was attributed to enhanced electron transfer rate at

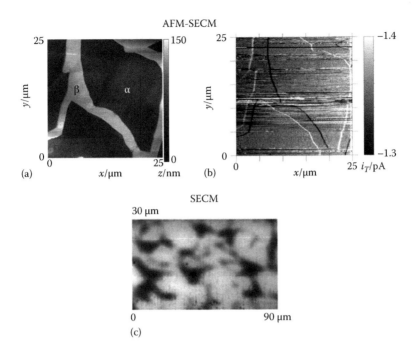

FIGURE 21.10 Contact mode AFM-feedback mode SECM images (a, topography) and (b, current); and constant-distance feedback mode SECM current image (c) of the same surface of Ti-6Al-4V alloy. The topography image (a) reveals the two phases, α and β, forming the oxide layer covering the surface. The AFM-SECM current image (b) shows no contrast difference between the two phases, the SECM current image (c) does. AFM-SECM images were acquired with a combined probe integrating a sub-100 nm sized microelectrode recessed from the tip apex. The SECM image (c) was acquired using a 1.7 μm diameter disk microelectrode. In both cases, ruthenium hexamine was used as a mediator in a 0.1M KCl solution. (From Pust, S.E. et al., *Nanotechnology*, 21, 105709, 2010; Pust, S.E. et al., *Adv. Mater.*, 19, 878, 2007. With permission.)

step edges, as compared to basal plane HOPG. However at that time, direct local scale probing of the electron transfer rates at step edges and at basal planes was not possible. The advent of AFM-SECM allowed this problem to be tackled. To achieve this, Frederix et al. developed extremely sophisticated combined AFM-SECM probes able to resolve HOPG steps, which are extremely steep and whose unit height, corresponding to a single graphene layer, is below 1 nm (~0.34 nm).[28] These probes were microfabricated starting from pyramidal SiO_2 mold formed into an Si wafer through an Si_xN_y layer. The mold was filled with a conducting layer of platinum or platinum silicide which was then fully encapsulated in another Si_xN_y layer. The cantilever was then defined by reactive-ion etching (RIE) of the Si_xN_y layers and bonded to a pyrex chip. The silicium matrix was dissolved chemically and the SiO_2 layer covering the tip was opened by timed etching in buffered hydrofluoric acid. Exposing the nanoelectrode integrated in the combined probe by such a simple etching step is a major advantage as compared to other AFM-SECM probe microfabrication processes which relied on RIE or FIB techniques.

The finally obtained combined probes (see Figure 21.11) had spring constants in the 0.02–0.1 N/m, making the measurement of pN forces possible. These probes were also remarkable because of their particularly small conducting conical nanoelectrode tip, characterized by a base radius of ~100 nm and a tip radius as low as 10 nm. It is also important to note that the robust fabrication process allowed wafer-scale production of these tips.

AFM-SECM images of a HOPG surface acquired in situ with this kind of probe in contact mode AFM-feedback mode SECM are shown in Figure 21.12.[29]

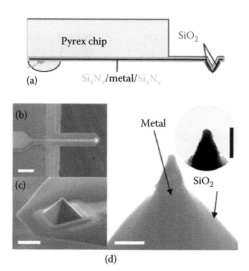

FIGURE 21.11 Combined AFM-SECM probe fabricated using state-of-the-art microtechnology. (a) Schematic view of the whole probe. (b) SEM micrograph showing the 130 μm long rectangular cantilever. The bright stripe in the middle is the conductive lane connecting tip and contact pad (scale bar: 50 μm). (c) SEM micrograph showing the pyramid shape of the tip entirely covered with the SiO_2 insulating layer. (d) TEM micrograph of the tip after the SiO_2 layer had been chemically etched over ~100 nm, exposing a conical nanoelectrode. (From Frederix, P.L.T.M. et al., *Nanotechnology*, 16, 997, 2005. With permission.)

One can see that the nanometer scale features of the HOPG surface are extremely well resolved in the topography image, in particular step edges are perfectly identified (Figure 21.12a). Most interestingly, the associated current image (Figure 21.12b) shows that a higher current contrast is recorded precisely at the level of many of the step edges. This finding was interpreted as resulting from a faster electron transfer rate at step edges as compared to basal planes. Modeling of the SECM feedback process suggested that the rate constant of heterogeneous electron transfer at step edges was ~100 times faster than at basal planes, at least for the mediator used, $Ru(NH_3)_6^{3+}$. Even though this result has to be put in perspective with recent works, which indicated that electron transfer at basal planes may not be much slower than at edges,[30] it was a clear demonstration of the capability of AFM-SECM to correlate the electrochemical activity of a surface with its topography at a *truly nanometric scale*. However, one intriguing aspect of this work was that it was actually conducted in contact mode, that is, the AFM-SECM probe was held in contact with the surface. In this configuration, intense short-circuit (or tunneling) currents should have been recorded, making the measurement of faradaic currents impossible, but none were reported. The absence of short-circuits was attributed to the presence of contamination on the tip apex. It would thus be most useful if this work could be reproduced with a non-contaminated probe and in lift-mode AFM-SECM. It is also worth indicating that, more than the characterization of HOPG, the final goal of Frederix et al. was to render possible the study of the electrical (and electrochemical) behavior of biological membranes at the nanometer scale by AFM-SECM. Even though no such work has been reported so far, the capability of the combined probes developed by this group for imaging nondestructively proteins embedded into membrane patches has been demonstrated.[28]

21.4.3 Imaging Enzyme Activity

Very early on in the development of SECM it was realized that this functional microscopy technique could be a unique tool for probing in situ the catalytic activity of enzymes on surfaces.[24] Probing of enzyme activity by SECM requires that one of the products of the enzymatic reaction is

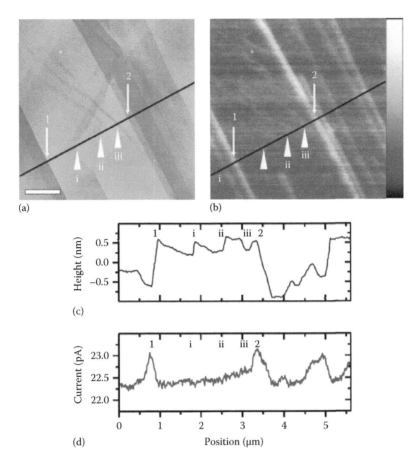

FIGURE 21.12 AFM-SECM imaging of a HOPG substrate. The images were acquired in contact mode AFM-feedback mode SECM with ruthenium hexa-amine as the redox mediator in aqueous solution. (a) Topography image. (b) Simultaneously acquired current image. Cross sections of the topography and current images taken along the blue line shown in (a) and (b) are, respectively, reproduced in (c) and (d). The tip and HOPG substrate were biased cathodically (−0.145 V) and anodically (+0.155 V) with respect to the standard potential of the mediator. Roman numbered arrow heads (i through iii) denote features only observed in topography. Arabic numbered arrows (1, 2) denote features observed both in topography and current. The asterisk denotes a feature only observed in the current image. (From Frederix, P.L.T.M. et al., *Nanotechnology*, 19, 384004, 2008. With permission.)

electrochemically detectable at the microelectrode probe. This condition is fulfilled in the case of redox enzymes but also for nonredox enzymes generating electroactive products, and many examples of SECM imaging of the activity of various enzymes immobilized on surfaces have indeed been reported.[31] Typically the resolution achieved is in the order of a few microns, due to the inherent limitations of SECM. This is frustratingly far from the ultimate resolution one can dream of for this kind of studies, which is the *single-enzyme molecule* resolution. Hence, it is not surprising that AFM-SECM, which potentially offers the nanometer resolution required for single-enzyme imaging, is very attractive for the investigation of enzyme activity on surfaces. Seminal work in that field came from Kranz et al. who imaged artificial enzyme patterns on surfaces using AFM-SECM.[32,33] For these studies, the authors made use of the very first AFM-SECM combined probes fabricated using micromachining technologies they had designed a few years before,[34] and have continuously improved ever since. These probes were formed by depositing a thin (~100–300 nm) gold layer onto a commercial silicon nitride (SI_3N_4) AFM probe, followed by complete insulation using SI_3N_4, SiO_2

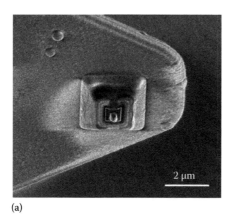 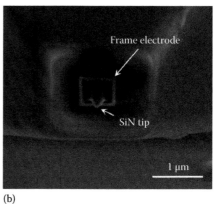

(a)   (b)

FIGURE 21.13 MEB micrographs of a combined AFM-SECM probe fabricated from a commercial AFM probe following Kranz's process. (a) Low magnification. (b) High magnification images showing the integrated frame microelectrode (edge length ~0.7 μm) centered on the FIB-reshaped sharp silicon nitride (Si_3N_4) AFM tip. This particular probe was insulated with alternating layers of silicon nitride/silicon oxide. (Unpublished data courtesy of Dr Christine Kranz, Institute of Analytical and Bioanalytical Chemistry, University of Ulm, Germany).

very early on in the development or parylene C deposits. The tip was then reshaped by FIB milling in order to form a sharp insulating SI_3N_4 cone intended to act as an AFM tip, and a submicron-sized gold frame microelectrode recessed from the tip apex (see Figure 21.13).

This design ensures a high topography resolution, and a built-in control of the substrate–microelectrode distance, set by the height of the SI_3N_4 cone, which is typically in the ~300–700 nm range. In their first work regarding enzyme imaging, these authors entrapped the redox enzyme glucose oxidase into roughly hemispherical polymer deposits (dots) electrophoretically grown from ~1 μm sized disk electrodes, and forming regular rectangular arrays of ~5 μm step-size. The array was imaged in tapping mode AFM-collection mode SECM with the probe biased at a potential suitable for detecting the enzymatic product (H_2O_2). The possibility of operating AFM-SECM in tapping mode, demonstrated earlier by the same authors,[9] was most useful for this study, because this mode is particularly gentle and as such recommended for imaging biological material. The topography images showed clearly the regularly arranged enzyme deposits, while the simultaneously acquired current images showed either no features in the absence of the enzymatic substrate (glucose), or clear current spots at the level of the enzyme deposits in the presence of glucose (see Figure 21.14).

These results demonstrated unambiguously the possibility of using AFM-SECM for imaging enzymatic activity. In a related paper, the same authors probed the local activity of another redox enzyme, horse radish peroxidase, by AFM-SECM, but this time operated in contact mode AFM-tip collection SECM mode.[33] The enzyme was immobilized on the surface in the form of isolated cross-linked protein gel spots, 1–2 μm in diameter. The enzyme activity was monitored via the collection by the tip of the oxidized form of the enzyme cosubstrate (ferrocene methanol) initially introduced in solution in its reduced from. There again the topography image showed the location of the spots, while the current image showed enzyme activity, only in the presence of the enzyme substrate (H_2O_2). It was emphasized that, due to the particular design of Kranz's probes, where the frame microelectrode is recessed from the tip extremity, the microelectrode–substrate separation could be kept very small (~400 nm in this paper). This is clearly an advantage brought by the AFM-SECM configuration since operating a conventional SECM probe within such a small distance of a substrate without tip crash would have been very difficult. Moreover, in the tip collection mode used here, a close imaging distance also translates into a higher lateral electrochemical resolution.

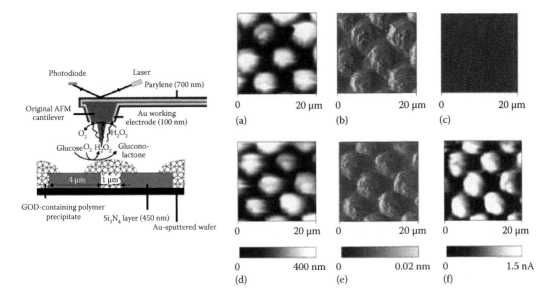

FIGURE 21.14   AFM-SECM imaging of enzyme activity. (Left): Schematic representation of AFM-SECM imaging of a sample micropatterned by a rectangular array of ~1 μm-sized glucose oxidase–containing polymer deposits (dots). The array was imaged in situ *in tapping mode* AFM-tip collection mode SECM with the probe biased at a potential suitable for detecting the enzymatic product (H_2O_2). (Right) Actual topography ((a) and (d)), deflection ((b) and (e)), and current ((c) and (f)) images recorded in the absence ((a through c)) or in the presence ((d through f)) of glucose in solution. (From Kueng, A. et al., *Angew. Chem. Int. Ed.*, 42, 3238, 2003. With permission.)

Another example of enzyme activity imaging by AFM-SECM has been reported by Hirata et al.[35] These authors imaged the surface of a polyelectrolyte thin film formed by deposition of successive layers of polystyrene sulfonate/glucose oxidase/poly-L-lysine onto a HOPG surface. The authors made use of a combined AFM-SECM probe they fabricated starting from a commercial AFM probe which was coated with gold and isolated with a photoresist layer. Focused illumination of the tip apex allowed its selective exposure, forming a microelectrode of an effective (electrochemical) size in the 50 nm–6 μm range. Interestingly, imaging was carried out in the AFM magnetic dynamic mode, which is similar to regular tapping mode but where the probe oscillation is induced magnetically. To this aim, a magnetic microbead was glued onto the back of the cantilever. The authors convincingly showed that the benefit of this mode is that magnetic excitation results in the probe displaying a single well-resolved vibration frequency, which can confidently be used for distance control in AFM-SECM imaging. On the contrary, mechanical excitation of AFM (or AFM-SECM) probes, as typically used in standard tapping mode, often gives rise to many spurious oscillations modes which, if selected for controlling the tip position, lead to tip crash. Moreover, in Hirata's work, the surface-induced shift of the probe oscillation frequency was used as the distance control mechanism (so-called FM detection mode), and not the damping of the probe oscillation damping as is normally the case (so-called AM detection mode). The probe was biased so as to detect the product of the enzymatic reaction H_2O_2, and the surface was sequentially imaged in the absence and in the presence of the enzyme substrate, glucose. Simultaneously acquired AFM-SECM images revealed that the film actually consisted of many aggregates, clearly visualized in the topography image, most of them associated with a tip current, but only when glucose was present in solution. Hence, the distribution of aggregates incorporating active enzyme molecules could be resolved with a submicron resolution.

Overall, the capability of AFM-SECM to image local enzyme activity at the micron and even submicron scale has been demonstrated, and further improvement in resolution is certainly possible. However, one should keep in mind that even the smallest enzyme *spots* analyzed in the works presented earlier actually contained a very large number of enzyme molecules forming multilayered structures. Hence, probing the activity of single-enzyme molecules by AFM-SECM is still not viable. Actually, as the pioneers of enzyme imaging by SECM initially realized, attainment of single enzyme resolution is hampered by kinetic rather than spatial limitations. This is a consequence of the principle evoked earlier and discussed in literature[25]: locating such a small thing as an enzyme requires a small microelectrode, but the smaller the microelectrode, the less efficient it is in collecting low fluxes of matter, and, alas, enzymes are slowly functioning machines. However, one can think of alternate strategies to make single-enzyme measurement possible in AFM-SECM, among them one strategy is confinement and redox recycling of the enzymatic products. An example of such a strategy based on tethering the redox substrate of the enzyme glucose oxidase to an AFM-SECM tip is reviewed in the following text (see Section 21.6).

21.4.4 IMAGING TRANSPORT PROCESSES THROUGH MICROPORES

What is typically measured in SECM is a local flux of species coming from surface sites. As described earlier, this flux can be generated by electroactive sites, which convert some redox species into its tip-detectable form. Yet local fluxes can also be due to the transport of species across pores, which may be artificial or biological pores. SECM has shown to be a valuable tool for studying such transport phenomena. Transport fluxes are probed either amperometrically, providing the transported species is redox active, or potentiometrically, providing the probe is capable of measuring local potentials. Yet, in order to evaluate the transport activities of submicron-sized and/or closely spaced *individual pores*, SECM has to be coupled with some technique, allowing pores to be located independently of their activity measurement. AFM-SECM, which allows small pores to be located from topography images, is therefore a priori particularly well suited to study transport across individual pores. Ideally, provided nanometer resolution is reached, AFM-SECM could even allow the action of individual nanopores in cell membranes to be probed, a major challenge in cell physiology. This goal is still distant but a solid body of work, reviewed in the following text, has now been published establishing the proof of concept of studying pore transport, so far only across artificial membranes, by AFM-SECM down to the ~100 nm scale.

In a pioneering work, Macpherson et al. made use of the hand-made AFM-SECM probes they designed previously (see Figure 21.3) and operated here in contact mode AFM-feedback mode SECM to acquire simultaneously the topographical and current images of a track-etched membrane.[12] Individual pores (0.6–1.2 μm in size) could be identified on the topography image, and their position was perfectly correlated with a decrease of the tip current, indicating hindered diffusion of the mediator at the tip. More interestingly, when the membrane was used as a separator between a compartment containing the redox mediator solution (the donor phase) and one containing solely the electrolyte (the receptor phase), AFM-SECM imaging of the membrane from the receptor side allowed the diffusional transport of the mediator through individual pores to be visualized, since this phenomenon obviously resulted in a marked tip current increase. *This was a clear demonstration of the unique capability of AFM-SECM to characterize both the topographical and transport properties of isolated micron size pores on a surface.* Using a simplified arrangement, where the receptor solution was removed, the membrane being solely *hydrated* by a thin film of the donor solution, the same authors were able to resolve the position and transport activity of individual 100 nm diameter pores in track-etched membranes (see Figure 21.15).[36] This represents the highest resolution achieved for the measurement of pore transport process by AFM-SECM.

Such an improvement in resolution compared to previous works partly comes from the fact that the configuration used was not strictly speaking in situ, since the sample was not immersed in an electrolyte but simply covered by a thin layer of solution. This configuration notably allowed

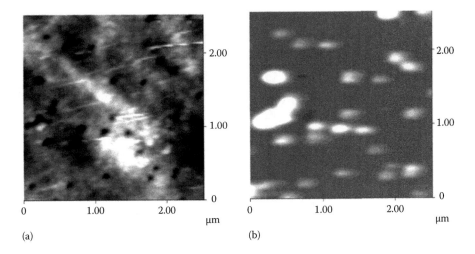

FIGURE 21.15 AFM-SECM imaging of diffusional transport of a redox mediator ($IrCl_6^{3-}$) through a porous membrane. (a) Topography image showing the ~100 nm diameter pores (b) Simultaneously acquired current image. The membrane separated a mediator containing solution from humid air and was simply wetted by a thin layer of solution. This simplified configuration allowed a noninsulated Pt-coated commercial AFM probe, biased to +1.0 V vs. an Ag/AgCl electrode inserted in the mediator solution, to be used to detect transport of the redox mediator through the pores. (From Macpherson, J.V. et al., *Anal. Chem.*, 74, 1841, 2002. With permission.)

unisolated sharp commercial Pt-coated AFM tips to be used as AFM-SECM probes. It was suggested that this same configuration could be used to investigate biological material in humid environments. Yet considering the fragility of such a material, it seems more recommendable to carry out this kind of investigation in actual in situ conditions, where both the sample and the isolated AFM-SECM probe are fully immersed in solution.

Transport of a biologically important metabolite, glucose, across artificial membranes was also investigated by Kranz et al. using AFM-SECM.[37] To this aim, these authors made use of their microfabricated AFM-SECM probes integrating a frame gold microelectrode recessed from the tip (see Figure 21.13). In order to make glucose electrochemically detectable by the probe, the frame gold microelectrode was chemically modified to bear the enzyme glucose oxidase embedded in a polymeric matrix. Glucose oxidase was selected because it selectively converts glucose into gluconolactone, while consuming oxygen and producing H_2O_2. The electrochemical detection of H_2O_2 at the frame electrode was used as a way of sensing the enzyme activity, which reflected the presence of glucose. Hence, the enzyme-modified AFM-SECM microelectrode can be seen as a miniaturized imaging biosensor. For imaging, a polycarbonate membrane with pores 200 nm in diameter was mounted so as to separate a donor compartment, containing 3 mM glucose, from an acceptor compartment, which contained solely buffer. AFM-SECM imaging of the acceptor side of the membrane was carried out in tapping mode AFM-tip collection mode SECM. The topography image allowed individual pores to be located while the current image displayed current spots around each of them (see Figure 21.16). Hence, diffusional transport of glucose across the membrane via the pores was evidenced and spatially resolved.

Moreover, the tip current could be converted into an actual (local) glucose concentration experienced by the probe, simply by precalibrating of the response of AFM probe/microsensor to glucose. Hence, the pore activity imaging by AFM-SECM was made quantitative, since it was possible to determine that the concentration of glucose in the center of the pore was ~0.66 mM, in agreement with theoretical models of diffusion through pores.

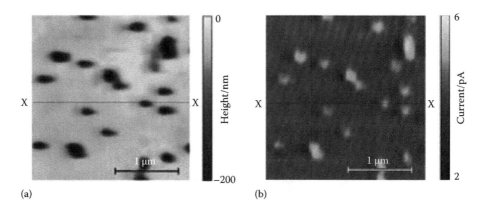

FIGURE 21.16 Tapping mode AFM-tip collector mode SECM imaging of glucose diffusion through a porous polycarbonate membrane (pore size: 200 nm). (a) Topography image. (b) Simultaneously acquired current image. The combined AFM-SECM probe was made selectively sensitive to glucose by immobilizing the enzyme glucose oxidase on the surface of the gold frame microelectrode integrated to the probe. The microelectrode was biased so as to detect H_2O_2, the product of the enzymatic oxidation of glucose. (From Kueng, A. et al., *Angew. Chem. Int. Ed.*, 44, 3419, 2005. With permission.)

21.4.5 SUBSTRATE PATTERNING USING AFM-SECM

In its basic principle, surface patterning by SECM relies on using the microelectrode probe to either trigger local substrate etching or localize deposition of material on the substrate surface. SECM patterning of surfaces by metal clusters, polymers, enzymes and other biomolecules, to form patterns typically in the ~10 μm range, has been described in literature and recently reviewed.[38] Combining AFM with SECM for patterning applications is obviously first expected to improve the patterning resolution, that is, to allow smaller patterns to be made. But the specific benefit one can expect from using AFM-SECM for patterning is that the combined probe can not only trigger local deposition but can also image in situ and in *real time* the deposited pattern. This is a crucial advantage over standard SECM patterning, which typically requires ex situ characterization of the deposits. So far, AFM-SECM has been used only twice for patterning applications.

Nishizawa et al. have described an AFM-SECM-based (bio)patterning technique, which consisted in using a combined AFM-SECM probe in contact mode to electrochemically locally etch away a protein-repellant layer covering a glass slide, to allow the subsequent adsorption of the cell adhesion promoting protein fibronectin in the etched area.[39] Etching was carried out by electrogenerating the HBrO species at the probe via the oxidation of Br^- ions present in the buffered etching medium. Successful etching of the protein-repellant layer, composed of adsorbed bovine serum albumin (BSA) or polyethylene imine/heparin, was evidenced, seemingly in situ, on the basis of topography AFM images, which showed a 2–3 nm deep pit in the layer corresponding to the etched area. Controlling the probe displacement allowed etched patterns of various geometries to be formed. Selective adsorption of fibronectin in the etched patterns was evidenced by fluorescence optical imaging. By modulating the etching time, that is, the duration of the anodic pulse applied to the probe, fribonectin patches, as small as 2 μm in diameter, could be patterned on the surface. For this work, the authors made use of an AFM-SECM probe fabricated by insulating a commercial Pt-coated AFM tip, further coated with a thin layer of Ti/TiO_2 and insulated by parylene C. Interestingly, the authors indicate that for a good insulation of the probe by parylene C, an adhesion promoting layer of 3-methacryloxypropultrimethoxysilane had to be self-assembled onto the Ti/TiO_2 surface. Exposure of the microelectrode tip was achieved by a gentle grinding of the tip. This resulted in a rather blunt tip bearing a ~0.5 mm size, most likely roughly ring-shaped microelectrode. In an ensuing paper, the same authors used the same experimental setup and strategy to fabricate fibronectin patterns *in the vicinity of cells* to direct their growth along the pattern.[40] Cell growth was monitored by optical microscopy.

Charlier et al. collaborated with Demaille's group in order to develop a lithographic mode for AFM-SECM, they labeled L-AFM-SECM, enabling to *draw* line patterns of polyacrylic acid onto a gold surface.[41] To this aim, a hand-made AFM-SECM probe, similar to the one originally described by Unwin et al. (see Figure 21.3), but made of gold,[42] was used to initiate locally the electrografting of vinylic monomers to the gold substrate. The probe was brought in contact with the substrate, subsequently retracted by ~5 µm, and scanned for patterning. Hence, deposition was carried out in constant-height mode (and not constant-distance mode). The deposition solution contained acrylic acid monomers and 4-nitrobenzene diazonium as an initiator. A two-electrode configuration was used: the gold substrate was biased negatively at −0.8 V with respect to the AFM-SECM probe. At such a negative potential, the diazonium was reduced at the substrate, generating nitrophenyl radicals which formed a chemically grafted layer on the surface, but also initiated polymerization of the acrylic acid monomers in close vicinity of the surface. Because of this proximity, grafting of the so-produced polyacrylic macroradicals to the nitrophenyl layer was favored. Overall a nitrophenyl/ acrylic acid copolymer-like thin film grew from the substrate surface. Yet, because of the two-electrode configuration used, film growth was confined to the area of the substrate located directly below the tip. Deposition of very thin lines of polymeric material, some ~200 nm in width and 35 nm in height, was then evidenced from ex situ AFM imaging of the surface (see Figure 21.17).

Importantly, it was shown that neither classical SECM nor AFM-SECM conducted with non-isolated probes could produce such small patterns. Hence, the unique ability of L-AFM-SECM for patterning submicron-sized features onto surfaces was demonstrated. One can only regret that the possibility offered by AFM-SECM of simultaneously depositing a pattern and imaging it was not exploited in this work or in Nishizawa's work.[39]

Obviously, further work is needed to demonstrate and to develop the full potentialities of AFM-SECM as a nanopatterning technique.

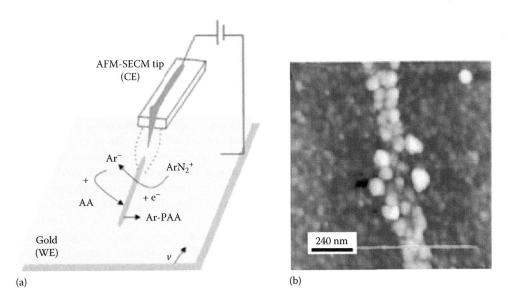

(a)   (b)

FIGURE 21.17   (a) Strategy for high-resolution patterning of a conducting surface (gold) by polyacrylic motifs using AFM-SECM as a lithography tool. A combined AFM-SECM probe is used as counter electrode for the local generation of nitrophenyl (aryl, Ar) radicals at the substrate by reduction of 4-nitrobenzene diazonium. These radicals initiate the polymerization and electrografting of vinylic monomers to the gold substrate. (b) Ex situ AFM taping mode images of the linear polymeric pattern thus formed. The pattern is ~200 nm wide and ~35 nm high. (From Ghorbal, A. et al., *ChemPhysChem*, 10, 1053, 2009. With permission.)

21.4.6 Quantitative Measurements with AFM-SECM

Beyond its unique imaging capabilities, SECM is also a quantitative technique, which allows the measurement of actual physicochemical parameters, such as the rate constants of (electro) chemical reactions occurring at interfaces. Such measurements typically rely on theoretical models, which quantitatively describe the SECM response in terms of approach curves. These models and corresponding data are available from literature for various modes of SECM operation, for numerous geometries of microelectrode probes, and for many differing chemical systems.[16] Using these models for a quantitative description of the response of a given AFM-SECM probe may be valid, provided the geometry of the microelectrode integrated in the probe is taken into account. In the best-case scenario, SECM approach curves for the geometry of the microelectrode are available. Yet, in many cases, the geometry of the AFM-SECM microelectrode is unusual and it is necessary to model, via numerical simulations, not only the theoretical SECM response but often even also the diffusional response of the microelectrode. Such is the case, for example, for the AFM-SECM probes shown in Figure 21.7, which comprises an unusually shaped *triangular* microelectrode. In this case, finite element method (FEM) simulations were required to calculate the diffusional current and the theoretical SECM feedback approach curves expected over a conducting and an insulating substrate.[14] It was shown that the corresponding positive and negative feedback effects were very low in magnitude (1.3 and 0.85 at most, respectively), a feature common to all *cone*-resembling SECM probes. This characteristic, together with their very low intrusiveness,[47] made these AFM-SECM probes better suited for the SECM tip-collecting mode. The complex architecture of Kranz' probes,[34] shown in Figure 21.13, also required that numerical simulations were ran for obtaining a quantitative model of their electrochemical response.[43] The boundary element method (BEM) was used to show that the SECM behavior of these probes, characterized by a square-shaped frame microelectrode centered around a sharp inert tip, was largely equivalent to the much simpler behavior of ring-shaped SECM microelectrodes of similar dimensions.[44] Establishing such equivalence, whenever possible, is highly desirable since it obviously greatly simplifies the quantitative interpretation of AFM-SECM data acquired with any given probe. Wittstock et al. also simulated the SECM response of an AFM-SECM probe bearing an inert tip and a recessed microelectrode,[20] but in a configuration where the microelectrode is offset from the tip axis, resembling the probe fabricated by Davodi et al.[17] and shown in Figure 21.8. As expected, these probes also displayed a poor feedback response. A better theoretical current feedback response was predicted for angled probes, such as the one reproduced in Figure 21.9.[21] Frederix et al. made use of 2D finite element modeling in order to calculate the diffusional current to the combined AFM-SECM probes they microfabricated and used for characterizing HOPG surfaces (see earlier text).[45] These probes bear at their apex a conically shaped nanoelectrode protruding from a conical insulating sheath (see Figure 21.11). They showed that the expression given in literature for calculating the diffusion current at conical SECM microelectrodes, which assumes a cylindrical insulation sheath, could be used, provided it was corrected by a factor 1.25. This factor reflects the better accessibility for diffusing species of probes with a conical rather than a cylindrical insulation. They also used simulations to calculate the theoretical approach curves over conducting and insulating substrates for this probe geometry. They showed that, upon approaching a substrate, their probes sensed the surface significantly sooner than expected for conical microelectrodes insulated by a cylindrical sheath. By resorting to much more complex 3D finite element simulations, the authors were also able to calculate the theoretical current response to be expected when such a probe was scanned over a sharp step on a conducting surface.[29] This modeling aimed at reproducing the complex experimental situation the authors encountered during the AFM-SECM characterization of the heterogeneous electrochemical activity of HOPG (see Figure 21.12). In a recent work, the diffusion current and SECM response of an AFM-SECM probe bearing a conical microelectrode sitting at the center of a flat tip apex was calculated by Denuault et al. using a finite element solver,[46] a situation corresponding to the probes fabricated by Pobelov et al. (see Figure 21.27 later in the chapter).[69] The effect of the cone aspect ratio and of the radius of the flat tip apex was evaluated. These simulations allowed the authors to derive useful empirical expressions for calculating the SECM response of these probes by simply knowing their dimensions (or conversely determining their dimensions from their SECM response).

Another often disregarded specificity of AFM-SECM probes is the fact that they not only integrate a microelectrode but also other bulkier parts such as the chip and the cantilever, which may influence their electrochemical response by shielding diffusion of species to/from the active sites on the surface. This latter problem has been specifically addressed both theoretically and experimentally by Unwin, Macpherson, and coworkers who recorded the diffusion controlled current at a disk-in-plane microelectrode, used as a model electroactive site, imaged by inert AFM probes of various geometries.[47] Plotting the microelectrode current as a function of the position of the AFM probe allowed visualizing the diffusion shielding effect. It was shown that for an AFM (and hence and AFM-SECM) probe to be minimally invasive (i.e., not to perturb diffusion), its cantilever should be beam-shaped long and narrow, and its tip conical, long, and characterized by a small cone angle. Fulfilling all these criteria when designing a particular AFM-SECM probe can thus allow alleviating diffusional shielding problems, which, for the probes considered by Unwin et al., was shown to decrease the feedback current by 15% at worse. In any case, the experimental strategy proposed by these authors may be used to quantify this effect for any given probe.

21.5 TOUCHING SURFACE–ATTACHED MACROMOLECULES WITH A MICROELECTRODE: Mᴛ/AFM-SECM

One of the remarkable benefits of the exquisite tip positioning capabilities of AFM-SECM is that extremely small, nanometer wide, microelectrode-to-substrate gaps can be created at will. Realizing that such nanogaps are of (macro)molecular dimension, Demaille and Anne introduced the concept of using a microelectrode, integrated at the apex of an AFM-SECM tip, to interrogate *directly,* by controlled physical contact, redox-labeled macromolecules immobilized onto surfaces.[48] Establishing such a contact between a solid tip and a fragile macromolecule necessitates the force sensing capability offered by AFM-SECM. The first step in developing this new type of AFM-SECM microscopy, now known as Molecule touching (Mt)/AFM-SECM, has been to fabricate appropriate combined AFM-SECM probes. This was carried out by adapting the microwire-based hand-fabrication technique previously reported by Unwin and Macpherson[12] (see Figure 21.3), modifying it in a few ways.[42] First, the probe material was changed from platinum to gold, in order to make the probe functionalizable by thiolated molecules when needed. Secondly, because the intensity of the current expected for such a new kind of AFM-SECM was at that time unknown, Demaille and coworkers proposed a technique to reshape the conical apex of the probes, with the aim of giving it a spherical geometry, maximizing the contact area between the microelectrode and the layer of surface-attached molecules. This reshaping technique was based on the controlled melting of the gold tip apex by generation of a spark using a dedicated apparatus. In order to further maximize the intensity of the current to be ultimately measured, the tip radius of the reshaped probe was kept in the 200–500 nm range. Thirdly, the probe was entirely insulated with a cathodic electrophoretic paint, which formed a robust pinhole-free coating. Last, the most useful innovation was the development of an extremely simple process to selectively uncover the extremity of the microelectrode, which is the most challenging step in the fabrication of AFM-SECM probes. The authors showed that by applying a high voltage pulse to the probe, the polymer coating covering the tip apex could be *selectively* blown away by the high electric field generated at the tip. The reliability of this process is ascertained by the fact that the authors have since been continuously using it not only to expose but also to clean the apex of their AFM-SECM probes (caution: the probes have to be kept away from metallic objects to avoid spark generation).

The first actual nanosystem probed by Mt/AFM-SECM consisted of a layer of home-synthesized polyethylene glycol (PEG)$_{3400}$ linear chains terminally attached to a gold surface and bearing a

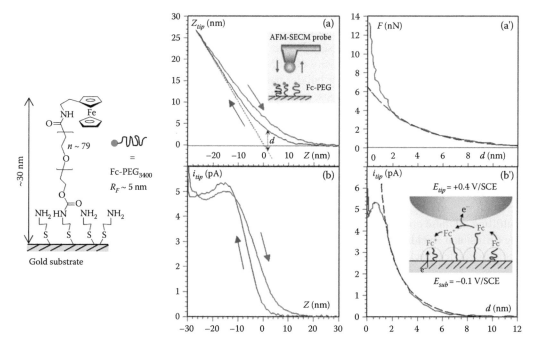

FIGURE 21.18 (Left): Chemical structure of a Fc-PEG$_{3400}$ layer grafted onto a gold substrate surface. R_F is the Flory radius of the chain, that is, its typical size when coiled. (Right) (a) Raw deflection and (b) tip current approach/retraction curves simultaneously recorded upon approaching an AFM-SECM probe from the Fc-PEG layer. The abscissa is the elongation/retraction length of the piezo supporting the probe. Analysis of the deflection approach curve allows d, the actual tip-to-substrate distance, to be derived as shown in (a). The raw approach curves can then be replotted as follows: (a') Force and (b') tip current versus d approach curves. The dotted lines in (a') and (b') are theoretical variations (see text). The inset in (b') shows the tip-to-surface cycling motion of Fc heads of the surface-attached PEG chains. (From Abbou, J. et al., *J. Am. Chem. Soc.*, 126, 10095, 2004. With permission.)

redox-ferrocene (Fc) group at their free end (see Figure 21.18, left panel).[48] These highly flexible chains, comprising ~80 ethylene oxide monomers, are nanometer-sized macromolecules; their fully extended length is ~30 nm, but in aqueous solution they form coils ~5 nm in diameter. Cyclic voltammetry at the modified surface allowed the presence of end-attached Fc-PEG chains on the gold surface to be confirmed and quantified. The gold surface and probe were biased at potentials, respectively, largely (>200 mV) anodic and cathodic, with respect to the standard potential of the Fc head of the chains (E^0 ~150 mV/SCE). Raw deflection and current approach/retraction curves such those shown in Figure 21.18 (right panel) were recorded upon approaching a hand-fabricated combined AFM-SECM probe toward the Fc-PEG modified surface. These curves are said to be raw, because their abscissa is the elongation/retraction length of the piezo tube supporting the probe.

The deflection approach curve was characterized by a rounded region, which in AFM is typical of the compression of an elastic layer, while the simultaneously recorded current approach curve displayed a broad, unexpectedly peak-shaped tip current variation (Figure 21.18a and b). As classical in AFM, the actual tip-to-substrate distance d could be derived from the raw deflection approach curve, as schematically depicted in Figure 21.18a. This allowed force and tip current versus d approach curves to be obtained (Figure 21.18a' and b'). These curves revealed that the tip started to sense the Fc-PEG layer, both in current and force, for a tip-to-substrate separation distance d ~10 nm, in agreement with the chain dimension in its coiled (so-called mushroom) state. The origin of the tip current was ascertained by stopping the tip approach a few nanometers (e.g., 5 nm) away from the surface and recording a tip cyclic voltammogram. A sigmoid-shaped voltammogram with a mid-height potential of ~150 mV/SCE was recorded, unambiguously demonstrating that the tip current was solely attributable to the electrochemical detection of the Fc heads by the AFM-SECM probe, and their ensuing tip-to-substrate redox cycling. This was the first demonstration of a positive feedback effect due to a redox species attached to a nanometer-sized macromolecule immobilized on a surface. Such results validated the proof of principle of Mt/AFM-SECM for which the redox label borne by the macromolecule (i.e., the Fc head) plays the role of SECM mediator (see inset in Figure 21.18b′). However, because the Fc heads are surface-tethered, rather than being dispersed in solution, conventional SECM theory could not be used to quantitatively account for the observed feedback. Hence, the authors developed a phenomenological model, termed the elastic bounded diffusion model, which described the tip-to-substrate cycling motion of the Fc head as a diffusional process limited by the PEG tether acting as a restoring spring.[49] This model introduced two parameters in addition to a classical effective diffusion coefficient (D_{eff}) for the Fc head: the chain spring constant and the Fc head equilibrium position in the layer. Realistic values of these two later chain-related parameters could be obtained by self-avoiding walk simulations, which sampled the statistical distribution of the conformations of the PEG_{3400} chains end-grafted to the surface and confined in the tip–substrate gap. A theoretical elastic bounded diffusion positive feedback approach curve could then be derived and fitted to the experimental current approach curve using only one adjustable parameter: the effective Fc head diffusion coefficient. As seen in Figure 21.18a′, a good quality fit was obtained, at least down to d ~2 nm, confirming the validity of the model and yielding a best fit value of $2 \cdot 10^{-8}$ cm²/s for D_{eff}. Such a low diffusion coefficient value, as compared to diffusion coefficients typical of small molecules such as ferrocene (~$7 \cdot 10^{-6}$ cm²/s), pointed to the large, additive, dragging effect of the chain monomers, which slowed the Fc head motion. Yet, the elastic bounded diffusion model did not account for the onset of a peak in the current approach curve in the d ~1–2 nm region. Insights into the origin of this peak were gained by analyzing the force approach curve. Whereas for $d > 2$ nm, the force curve could be nicely fitted by the exponential function typical of the compression of end-grafted polymers forming a mushroom layer (dotted line in Figure 21.18a′), an excess of force was sensed in the $d < 2$ nm region, which is precisely where the current peak was simultaneously detected. Moreover, self-avoiding walk simulations showed that in this same d region, the strongly confined PEG_{3400} chains were expected to abruptly elongate parallel to the surface, undergoing a so-called escape transition. Hence, it was concluded that the occurrence of a peak in the current approach curve revealed this elusive chain escape transition phenomenon, which had previously been predicted but never evidenced experimentally. Overall, this series of work is a unique example of how AFM-SECM can allow force and current data to be acquired and combined to address the behavior of a redox nano-object under confinement. These works also demonstrated that Mt/AFM-SECM allows both the conformation and *internal motional dynamics* of surface-attached macromolecules to be resolved simultaneously. Mt/AFM-SECM is thus ideally suited to study the complex interplay between confinement of nano-objects and their dynamics, with the only requirement that these objects are redox active (or labeled).

The same group, turning to a molecular system of higher biotechnological relevance, used Mt/AFM-SECM to study end-grafted molecular layers of short Fc-labeled DNA chains ((dT)$_{19}$ oligonucleotides), both in their single-stranded (ss) and double-stranded (ds) states, see Figure 21.19.[50]

They showed that the probe could physically contact the Fc-DNA chains, both in their single- and double-stranded state, generating an entropic force and a specific tip current due to oxidation of the Fc heads at the tip (and ensuing redox cycling). Most interestingly, it was observed that hybridization of surface-attached ss Fc-DNA resulted in a three- to fivefold *decrease* of the intensity of the tip current approach curve, whereas the force curve remained largely unaffected (see Figure 21.19). This decrease in tip current intensity was interpreted as indicating that the motional dynamics of Fc-dsDNA was slower than that of Fc-ssDNA. Two types of DNA chain motions allowing the tip-to-substrate cycling of the Fc head were envisioned: hinge motion of the chain around its relatively long (C_6) anchoring foot, and elastic deformation (bending) of the DNA chain. Considering the large difference in flexibility between single- and double-stranded DNA, the relatively modest

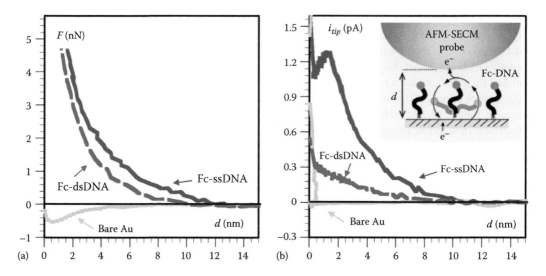

FIGURE 21.19 Mt/AFM-SECM force (a) and tip current (b) approach curves recorded upon approaching a combined probe from a layer of ferrocene-labeled DNA chains ((dT)$_{19}$ oligonucleotide) end-attached to a gold surface. The blue curves correspond to data acquired with the Fc-DNA in its single-stranded state; the red curves correspond to the same sample after hybridization of the DNA by its complementary strand. The inset shows the hinge motion of the Fc-DNA around its surface anchor, which is the motion permitting for the tip-to-substrate cycling of the Fc head. (From Wang, K. et al., *J. Phys. Chem. B.*, 111, 6051, 2007. With permission.)

magnitude of the current decrease upon hybridization suggested that, both for ss and Fc-dsDNA, hinge motion rather than chain bending was at play (see Figure 21.19b, inset). This work contributed to shine light on the mechanism of electron transport through redox DNA layers by demonstrating that it could simply be explained by hinge motion of the attached chain, with no need to invoke DNA conductivity effects. Mt/SECM-AFM was also used by the same authors for the in situ monitoring of enzymatic incorporation of a specialized Fc-labeled dideoxynucleotide[51] at the free end of oligonucleotides grafted on a gold surface. Force curves were used to demonstrate grafting of unlabeled (dT)$_{20}$—SH chains onto the gold surface. The onset of a current approach curve after the DNA layer had been exposed to the enzyme terminal transferase in the presence of the Fc-dideoxynucleotide, revealed the efficient single-base extension of the end-grafted DNA strands.[52]

Mt/AFM-SECM was later endowed with imaging capabilities.[53] As for any SECM-derived imaging technique, the key issue was to be able to hold a constant tip-to-surface distance during scanning, with the extra requirement that the tip had to be held in physical and electrochemical contact with the thin *molecular* layer of surface-immobilized redox-labeled macromolecules. Using a test-bed system consisting of gold band electrodes bearing a self-assembled layer of Fc-PEG$_{3400}$-disulfide, Demaille and coworkers showed that tapping mode allowed fulfilling these conditions, enabling constant-distance Mt/AFM-SECM imaging. Well-resolved topography and tip current images of the bands, such as those reproduced in Figure 21.20, were simultaneously obtained using ~50 nm tip radius hand-made conical gold probes.

The 500 nm wide gold bands were clearly seen in topography. The current image showed that the Fc-PEG chains were indeed covering the bands but also evidenced tiny coverage defects in the Fc-PEG layer, appearing as lower current regions and indicated by arrows in Figure 21.20b. Most of these defects could be associated with the presence of contaminating particles visible in the topography image, some of them being as small as ~100 nm. This proof-of-principle experiment demonstrated that Mt/AFM-SECM uniquely allows mapping *selectively* the surface distribution of

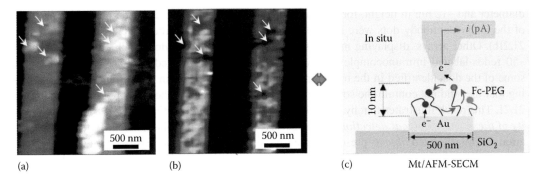

(a)   (b)   (c)   Mt/AFM-SECM

FIGURE 21.20 Tapping mode Mt/AFM-SECM imaging of ~500 nm wide gold bands bearing an end-attached layer of Fc-PEG$_{3400}$ disulfide. The topography (a) and current (b) images were acquired simultaneously. The arrows show submicrometer-sized contaminant particles visible both in topography and current (as low current regions). Z scale: topography 25 nm, current 2.5 pA. (c) Schematic description of the imaging process. (From Anne, A. et al., *Anal. Chem.*, 82, 6353, 2010. With permission.)

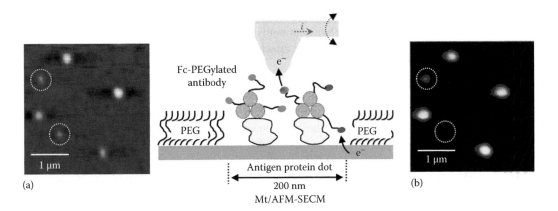

(a)   (b)
Mt/AFM-SECM

FIGURE 21.21 Tapping mode Mt/AFM-SECM topography (a) and tip current (b) images of an array of mouse IgG/antimouse IgG-PEG-Fc immunocomplexes dots on a gold surface. The array was fabricated by the bead lithography/PEG backfilling technique (see text). Among the two topographically similar dots circled in (a), only the lower one appears in (b), indicating that only this one contains the sought antigen (mouse IgG). Z scale: (a) 15 nm, (b) 0.4 pA. (From Anne, A. et al., *Anal. Chem.*, 83, 7924, 2010. With permission.)

nanometer-sized redox-labeled linear macromolecules onto electrode surfaces at the ~100 nm scale. This may be of interest for characterizing the surface of modern electrochemical biosensors which use end-attached DNA or peptide chains as sensing layers.

The imaging capabilities of Mt/AFM-SECM were later extended in order to enable mapping of the distribution of virtually *any type* of macromolecule immobilized onto an electrode, or simply a conducting surface.[54] The proposed universal imaging strategy relied on making surface-immobilized macromolecules (e.g., globular proteins) electrochemically visible via redox-immunomarking by specific antibodies conjugated to Fc-PEG chains (see cartoon in Figure 21.21).

In order to validate this imaging principle, a regular array of absorbed mouse IgG (acting as protein antigens to map) was formed on a gold surface, using the bead lithography technique combined with PEG backfilling, and subsequently recognized by Fc-PEG-labeled antimouse antibodies. The array was imaged in tapping mode Mt/AFM-SECM, and images such as those shown in Figure 21.21 were obtained. The topography image showed individual antigen/antibody dots, ~300 nm in diameter and ~12 nm in height, forming a regular pattern on the surface (see Figure 21.21a). Most of the antigen/antibody dots were also detected as current *spots* in the current image (see Figure 21.21b). Other arrays, displaying much smaller (~100 nm) diameter dots, each containing at most ~50 redox-labeled immunocomplexes, were also successfully imaged in this way. Interestingly, some of the dots identified in the topography images did not appear in the current images, indicating that they did not contain the sought antigen. Such is the case of the lower dot circled in Figure 21.21. This result illustrates that by *combining the nanometric resolution of AFM with the selectivity of the electrochemical detection*, Mt/AFM-SECM makes it possible to identify target antigens amidst similarly sized *objects* present on the surface, which could not be differentiated by conventional AFM. It was also emphasized that because redox cycling of the Fc label between the tip and the substrate surface is in itself a powerful amplification mechanism,[55] Mt/AFM-SECM is a priori amenable to the detection of single immunocomplex molecules.

Most recently, Mt/AFM-SECM was used to probe the electrochemical behavior and physical properties of individual nano-objects of great biotechnological importance: PEGylated gold nanoparticles.[56] The system studied was based on a random array of ~20 nm diameter citrate-stabilized gold nanoparticles immobilized on top of an amino-undecane thiol layer self-assembled onto a gold surface (see Figure 21.22, upper panel). The amino-undecane thiol layer acted as an insulating layer, which fully hampered electron transfer from the gold surface toward the solution. Yet because of their high density of electronic states, the nanoparticles could still rapidly exchange electrons with the gold surface *through* the aminothiol layer; they thus behaved as individual nanoelectrodes. Fc-PEG$_{3400}$ disulfide chains were attached to the surface-immobilized gold nanoparticles by selective thio-functionalization. Their presence on the surface (i.e., on the nanoparticles) could be ascertained and quantified by cyclic voltammetry at the substrate, which displayed the voltammetric signature of surface-attached Fc-PEG chains.

Tapping mode Mt/AFM-SECM imaging of the resulting Fc-PEGylated nanoparticle array yielded topography and current images such as those presented in Figure 21.22a and b. Individual isolated nanoparticles were clearly resolved in the topography image, and *some* of them were observed to be associated with well-defined *spots* in the current image. The origin of the current detected at the level of individual nanoparticles was ascertained by studying the dependence of the tip current as a function of the substrate potential: the obtained S-shaped voltammogram unambiguously showed that the tip selectively detected the Fc-PEG chains immobilized onto individual nanoparticles. As a result, the tip current was actually a measure of the Fc-PEG chain coverage of the nanoparticles. It was thus possible in a single-image scan to interrogate numerous well-resolved isolated nanoparticles and to measure both their *individual* size (from topography) and chain coverage (from the tip current). Because these two properties were simultaneously measured for the *same* nanoparticle, it was possible not only to construct nanoparticle size and chain coverage histograms, but also *to cross-correlate* these properties. This possibility was a unique benefit of the combined and high-resolution nature of Mt/AFM-SECM. Moreover, it was emphasized that Mt/AFM-SECM displays the particularity of allowing simultaneous single nanoparticle *and* ensemble measurements on the *same* sample: the current image yields information at the single-particle level, whereas the substrate voltammogram provides information about the chain coverage averaged over the particle ensemble. These multiscale information allowed the authors to show that the actual number of Fc-PEG chains by nanoparticle was broadly distributed in the 200–900 chain per particle range (average 530 chains/particle) and that as many as 20% of the particles were actually bare (i.e., did not bear any Fc-PEG chains). This broad distribution around the average was unexpected and of course could not have been evidenced from simple ensemble measurements.

The typical tip radius of the hand-made conical probes used in this work was in the order of ~50 nm. Yet, in order to achieve the high resolution required for sampling a large number of particles in a single scan of reasonable width, Demaille et al. had to improve their microwire-based hand-fabrication technique to produce even sharper AFM-SECM gold tips. By modifying the

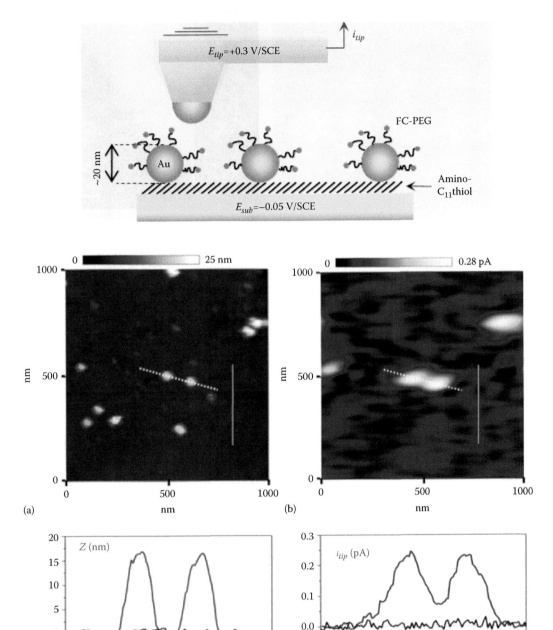

FIGURE 21.22 (Top): Schematic description of an array of Fc-PEGylated nanoparticles immobilized on an amino-undecanethiol layer assembled onto gold. (Bottom) Tapping mode Mt/AFM-SECM topography (a) and current (b) images of the array. Cross sections of the topography and current images taken along the short white line shown, passing through the center of two nanoparticles, are respectively plotted in red and blue. Corresponding cross sections taken in a zone free of nanoparticles, materialized by a vertical green line in the images, are plotted as black traces. The actual diameter of the nanoparticles is given by their apparent height, their width being overestimated due to tip convolution effects (tip radius ~50 nm). (From Huang, K. et al., *ACS Nano*, 7, 4151, 2013. With permission.)

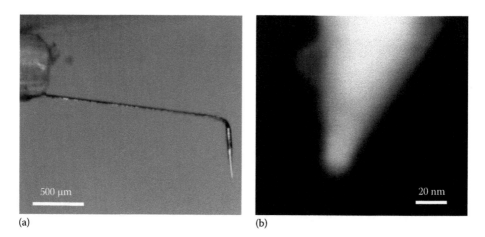

(a)   (b)

FIGURE 21.23 (a) Optical microscopy micrograph showing an AFM-SECM gold probe hand-fabricated following Demaille's process. (b) Scanning electron microscopy image of the apex of a last-generation probe displaying a ~20 nm tip radius.

etching step of the process, conical tips displaying a radius in the order ~20 nm were fabricated (see Figure 21.23).

These actual *nano* AFM-SECM probes were sufficiently sharp to allow visualizing simultaneously but independently the Fc-PEG corona and the gold core of individual PEGylated particle. The corona appeared as a broad current hallow surrounding the narrower nanoparticle core seen in the topography image.

This work was the first example of an SECM-derived technique allowing the size and electrochemical properties of individual nanometer-sized objects to be simultaneously probed. It illustrates the unique perspectives offered by high-resolution AFM-SECM for revealing the properties of individual nano-objects.

21.6 ATTACHING A REDOX MEDIATOR TO AN AFM-SECM TIP FOR THE FUNCTIONAL PROBING OF NANOSYSTEMS: Tarm/AFM-SECM

The spatial resolution of conventional SECM is primarily controlled by the size of the microelectrode probe. Hence, the fabrication of nanometer-sized SECM probes (nanoelectrodes) has been central to the efforts toward nanoscale SECM. As a result, SECM nanoelectrodes,[57] or AFM-SECM probes integrating nanoelectrodes, have been successfully fabricated using a variety of technologies (this chapter). Yet a paradox of this technological race is that if in SECM a smaller electrode means a higher spatial resolution, it also sets a *lower* bound to the kinetic window of surface processes explorable by feedback SECM. This comes from the fact that the kinetic window of SECM is set by the diffusional time of the mediator to/from the surface. Since the tip-to-substrate distance is necessarily in the order of the microelectrode size, a smaller microelectrode means a shorter diffusional time. However, only surface processes occurring within this time frame can be sensed, those occurring much slower giving rise to pure negative feedback responses, as if the surface was inactive.[25] Similar limitations also apply to the SG/TC mode. This fundamental limitation is of major importance for nanoelectrochemistry since it means that the activity of *nano* reactive sites on a surface *cannot* be sensed by SECM (nor any SECM-derived technique) if these sites function slower than some tip-size-dependent threshold. This notably hampers SECM probing of enzymatic reactions at the single-enzyme molecule level. Indeed, such an experiment would require using a nanoelectrode held a few nanometers away from the enzyme and diffusion across such a *nano* tip–substrate gap is much faster than enzymatic turnovers (microseconds versus milliseconds). As a way around this problem, which is rooted in the *diffusivity* of the SECM mediator, Demaille and Anne proposed to simply *tether* the redox mediator to the apex of the AFM-SECM tip.[58] In this configuration, diffusional dispersion of the mediator is avoided, the mediator being brought by the tip in the vicinity of the reactive site and confined there until it reacts with it. As a result, there is no lower bound to the kinetics of the sites, which can be probed by this so-called Tarm (Tip-attached redox mediator) AFM-SECM microscopy. Practical realization of this thought-provoking concept was carried out by functionalization of the apex of gold AFM-SECM tips by $Fc-PEG_{3400}$ chains. The presence of Fc-PEG chains at the tip apex was assessed by recording AFM-SECM deflection and current approach curves using a *bare* HOPG surface as the substrate. The deflection approach curve displayed the rounded aspect typical of PEG chains confined in the tip–substrate gap (e.g., Figure 21.5), while the current approach curve showed the typical i_{tip} vs. d variation expected for elastic bounded diffusion positive feedback of the Fc heads. Overall, it was shown that the Fc-PEG chains formed a saturated layer of hemispherical blobs (mushroom) on the surface of the tip apex, and that the Fc heads could indeed electrochemically probe the substrate surface when the tip was approached within ~10 nm of it. The imaging mode of Tarm/AFM-SECM was developed soon afterward.[59] It was shown that tapping mode was fully compatible with Tarm/AFM-SECM, allowing the Fc-PEGylated probe to be scanned over surfaces with the tip-attached Fc heads locally probing the electrochemical reactivity of the substrate. The actual possibility of using this innovative AFM-SECM technique for the functional imaging of nanosites on a surface was demonstrated by imaging the reputedly electrochemically heterogeneous surface of HOPG. The resolution achieved in topography was in the subnanometer range, making basal planes and step-edges sites of the surface perfectly identifiable (see Figure 21.24a).

The electrochemical resolution was in the sub-~100 nm range, limited more by the large radius of the spherical probe used than by the effective chain length (~10 nm), see Figure 21.24. Yet this resolution was more than sufficient to demonstrate that the basal planes of HOPG displayed a high electrochemical activity with respect to the Fc heads, but also to discern that the reactivity of plane edges was somewhat lower, see Figure 21.24.

Hence, this work showed that the kinetic restrictions for probing *slowly* functioning nanometer sites on surface by SECM can be alleviated by turning to Tarm/AFM-SECM. One can thus now reasonably envision, at least in principle, the possibility of probing single-enzyme kinetics by Tarm/AFM-SECM, even though this would require the measurement of sub-femto-ampere currents, which is barely possible today.

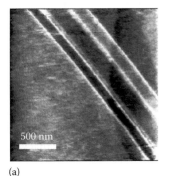
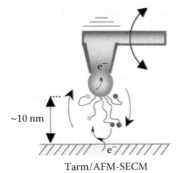
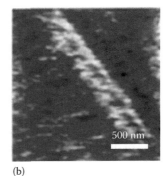

(a)

~10 nm

Tarm/AFM-SECM

(b)

FIGURE 21.24 Tapping mode Tarm/AFM-SECM imaging of a HOPG substrate. Simultaneously acquired (a) topography and (b) tip current images. The contrast in (b) shows that: (i) an electrochemical current due to the redox cycling of the tip-attached Fc heads is recorded all over the HOPG surface. (ii) Yet the tip current is more intense at the level of edge sites, suggesting an enhanced activity at these sites as compared to basal planes. Z scale: (a) 1.5 nm, (b) 0.4 pA. (From Anne, A. et al., *ACS Nano*, 3, 2927, 2009. With permission.)

21.7 ALTERNATIVE/IMPROVED STRATEGIES FOR FABRICATING AFM-SECM PROBES

As outlined earlier, the lack of availability of commercial AFM-SECM probes has so far forced the research groups willing to characterize nano-electrochemical systems by AFM-SECM to fabricate their own probes. The fabrication methods developed by these groups have been presented earlier, together with the results of their investigations. In such cases, the performances of the fabricated probes can be straightforwardly evaluated from the quality of the reported results. Yet, in order to help researchers choose the best probe design for their own needs, we find it also useful to review several alternative strategies that were proposed in literature to produce AFM-SECM probes. *Those probes were not used for the AFM-SECM characterization of actual nanosystems*, but at least validated to some degree. Two main probe designs can be delineated depending on whether the microelectrode is integrated at the tip apex, or recessed from it.

21.7.1 Probes with the Microelectrode Recessed from the (Insulating) Tip Extremity

This probe design allows constant-distance SECM imaging in AFM contact mode, since the insulating tip height keeps the microelectrode–surface separation constant, avoids short-circuits between the microelectrode and conducting surfaces, but also prevents fouling of the microelectrode by surface contaminants. Kranz's group pioneered the microfabrication of such combined AFM-SECM probes by integrating a gold frame microelectrode located at the base of a sharp silicon tip shaped by FIB milling, as described previously (see Figure 21.13). In recent years, this group has reported several improvements to the original design of their probes: The probe insulation was improved by making use of plasma-deposited fluorocarbon films instead of silicon nitride as the insulating material.[60] An ion beam–induced deposition process was also described for coating selectively the gold frame microelectrode by a platinum carbon (PtC) deposit. This resulted in an enhanced electrochemical sensitivity of the probe, due to the increased electroactive area of the microelectrode, but also to the catalytic properties of the PtC electrode material for the detection of the analyte of interest H_2O_2.[61] Another major recent evolution of this probe design consisted in replacing the gold electrode material by conductive boron-doped diamond (BDD) and the conical tip material by intrinsic (insulating) diamond.[62] To this aim, the undersize of a commercial silicon tip was coated by chemical vapor deposition (CVD) with heavily BDD, and subsequently by intrinsic diamond. Just as in the original process,[34] FIB milling was used to reshape the probe apex, forming a diamond-doped frame electrode recessed from a sharp diamond tip. The resulting AFM-SECM probe combines the exceptional electrochemical behavior of BDD microelectrodes (large accessible potential window, low fouling) with the extreme mechanical resistance of diamond AFM tips.

A major step toward the actual batch scale fabrication of AFM-SECM probes with recessed microelectrode has also been achieved by Krantz's group who were able to fabricate as many as 340 probes in a single batch with a success yield in the order of 60%.[63,64] Importantly, the microfabrication process developed to achieve this goal *does not* require the use of FIB nor electron beam lithography. This is a major improvement since these *writing* techniques, used in many of the AFM-SECM probe fabrication processes reported so far, notably for exposing the electroactive area of the probes, seriously limit the efficiency of batch-processing. The reason for that is that *writing* techniques require delicate and time-consuming alignment steps in order to focus the beam at precise positions near the tip. In Kranz's batch process, the AFM-SECM probes are fabricated in 10 conventional microfabrication steps, the most important of them being shown in Figure 21.25a and described in the following.

A cylindrical silicon cylinder (1.5 μm height, 3 μm diameter) is first etched in a 3 μm thick silicon on insulator (SOI) wafer, step (a) in Figure 21.25, and subsequently coated by an SiO_2 layer using plasma-enhanced chemical vapor deposition (PECVD), step (b). A conducting Pt/Ti layer is

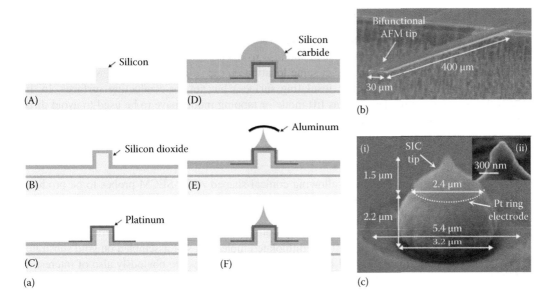

FIGURE 21.25 (a) Schematic steps for the fabrication of batch fabricated AFM cantilevers with integrated ring electrodes starting from a silicon on insulator (SOI) wafer: (A) cylindrical silicon tip etching, (B) PECVD silicon dioxide insulation at etched cylinder, (C) Pt electrode patterning, (D) PECVD silicon dioxide insulation and silicon carbide (SiC) deposition at electrode layer, (E) isotropic silicon carbide tip etching under aluminum mask, and (F) release of the AFM cantilever. (b) SEM micrograph of the actual cantilever bearing the bifunctional AFM-SECM tip. (c) (i) Zoomed-in SEM micrograph of the bifunctional tip showing the position of the Pt ring electrode and of the SiC cone, which plays the role of the AFM imaging tip. The inset (ii) shows the apex of the SiC cone, which is a few tens of nm in radius. (From Shin, H. et al., *Anal. Chem.*, 79, 4769, 2007; Shin, H. et al., *Sens. Actuat. B*, 134, 488, 2008. With permission).

then deposited to form the future electrode, conducting pad, and wires (step c). This layer is then insulated by PECVD deposition of a 400 nm thick SiO_2 layer, followed by a thick (1.4 µm) silicon carbide (SIC) layer, which results in giving the cylinder a dome-like shape (step d). A 3 µm aluminum disk is then patterned on the top of the dome as an etching mask. High pressure RIE is then used to etch away most of the SiC layer covering the surface, leaving only a thin SiC layer on the probe body while exposing a ring-shaped Pt electrode and forming a conical tip in the top part of the cylinder protected by the Al mask (step e). Further etching releases the cantilever and main chip from the wafer. The finally produced combined probe is shown in Figure 21.25b and c: it bears a Pt ring microelectrode, ~2–3 µm in diameter, located at the base of an SiC tip a few micrometers in height and whose radius can be as low as ~30 nm. Depending on the RIE etching conditions, the Pt-coated sidewalls of the silicon cylinder can also be exposed. The performance of the combined probes was evaluated by AFM-SECM imaging of test composite surfaces both in contact and tapping mode AFM-feedback mode SECM.

Most recently, fabrication of a *L-shaped* AFM-SECM probe, characterized by flat apex bearing a triangular frame Pt microelectrode in the center and an isolating AFM tip on the side, has been reported.[65] The final tip geometry resembles the one described previously by Davoodi et al.[17] (Figure 21.8), but the probe itself was fabricated from a commercial Pt/Cr-coated conductive AFM Si probe insulated by parylene C, and reshaped using FIB milling. The frame electrode was ~150 nm in size, while the tip height was 250 nm in height. Contact mode AFM-feedback mode SECM images of a carbon nanotube network was successfully acquired. The topography and current images were found to be laterally offset by ~400 nm, which corresponded to the distance separating the frame electrode from the off-centered AFM imaging conical tip.

21.7.2 PROBES WITH THE MICROELECTRODE LOCATED AT THE TIP EXTREMITY

This design is the most straightforward; it has the advantage of allowing real nanometer-sized microelectrode–surface gaps to be created and also to ensure that topographical and electrochemical information is collected at the exact same location on the surface. Simple AFM contact mode can be directly used to characterize insulating surfaces but not for conducting surfaces. In this case, other *noncontact* modes, such as lift mode or tapping mode, have to be used to avoid direct tip–substrate short-circuits. Three types of tip geometries are encountered: tips are often pyramid or conically shaped, but can also be needle shaped.

21.7.2.1 Conically Shaped Probes

The first microfabrication technique allowing conical-shaped AFM-SECM probes to be produced was reported by Frederix et al.[28] (see Figure 21.11). Since then, the process has been completed by a gold electrodeposition step, which allowed the PtSi microelectrode material to be fully gold-coated.[66] The aim is to render these probes suitable for probing in situ the electrical properties of molecular junctions formed between the tip itself and dithiolated molecules self-assembled on gold substrates. Although this is distinct from SECM measurements, these probes are obviously also of interest for AFM-SECM applications, requiring gold as the microelectrode material, for example, those involving chemical modification of the tip. The same group has very recently proposed a new process to fabricate AFM probes integrating a gold tip for AFM-SECM and in situ conducting AFM experiments.[67] The novelty of this process lies in the fact that the length and thickness of the cantilever are defined by photolithography and Si etching of the top surface of an *SOI* wafer (see Figure 21.26).

The integrated electrode, defined by electron beam lithography, is a gold nanowire, terminated by a ~20 nm in radius rounded apex, overhanging from the side of the cantilever (see Figure 21.26). It is isolated by a ~300 nm thick layer of silicon nitride. The quality of insulation was verified by metal electrodeposition experiments. No actual AFM-SECM experiments making use of these probes were reported so far. This new process allowed probes characterized by spring constants in the 0.05–14 N/m range to be batch fabricated.

Another microfabrication approach for producing AFM-SECM probes integrating sharp gold nanoelectrode tips has also been recently reported.[68] This approach, schematized in Figure 21.27, relies on the modification of standard commercial silicon AFM probes, which are coated with a conducting Ti/Au/Ti sandwich layer, and insulated with 600 nm of silicon nitride (Si_3N_4). The first and second Ti layers serve to promote the adhesion of gold to the silicon probe and to the silicon nitride

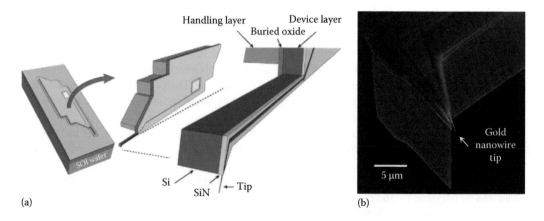

FIGURE 21.26 (a) Schematic view of the in-plane process for fabricating AFM-SECM probes bearing nanowire tips from silicon on insulator (SOI) wafers. (b) SEM micrograph of a fabricated probe showing the overhanging gold nanowire tip. (From Wu, Y. et al., *Sens. Actuat. A*, 215, 184, 2014. With permission).

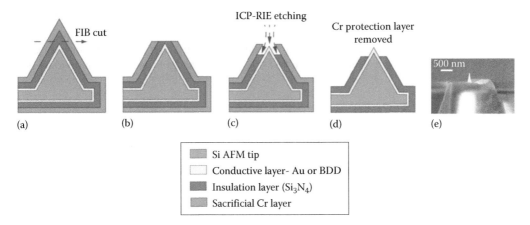

FIGURE 21.27 Process flow chart for AFM-SECM probe fabrication (a through d) and SEM image of the resulting tip (e). (a) A commercial AFM silicon probe coated with a conductive layer of Ti/Au/Ti or BDD, 600 nm silicon nitride (Si$_3$N$_4$) and 50 nm chromium (Cr), is FIB-cut. (b) Tip after FIB cutting. (c) Conductive gold or BDD tip exposed by selective ICP-RIE etching of the silicon nitride layer. (d) The Cr layer is completely removed by wet-etching. (e) MEB image of the resulting tip showing the protruding conical nanoelectrode. (From Avdic, A. et al., *Nanotechnology*, 22, 145306, 2011. With permission.)

layer, respectively. A protecting chromium layer is then evaporated onto the probe. The tip apex is then cut with FIB, and coupled plasma-RIE is used to selectively uncover the sharp conductive conical tip. As a last step, the chromium layer, which protected the rest of the probe from RIE etching, is chemically etched. The same process can also be carried out starting from conducting BDD AFM cantilevers; in such a case, the nanoelectrode tip material is BDD.

Voltammetric characterization of the resulting combined AFM-SECM probes showed that, in the case of gold, a large part of the nanoelectrode is covered with an interfering insulating layer, probably TiOx coming from the initial Ti adhesion layer. In the BDD case, where no Ti layer was present, no such effect was noted. Moreover, cyclic voltammetry experiments also indicated the presence of defects (pinholes) in the Si$_3$N$_4$ insulation later. In order to solve these problems, the fabrication process has been subsequently improved.[69] For a better tip insulation of the probe, the Si$_3$N$_4$ layer was further coated either with a ZrO$_2$ layer or by deposition of electrophoretic paint, and the gold tip was *cleaned* from the TiOx-contaminating layer. When ZrO$_2$ was used as a supplementary insulating layer, an extra FIB milling step had to be included in the process in order to remove this tough layer, immediately following RIE exposure of the tip. This extra step complicated the process but had the merit of removing the TiOx layer partly contaminating the gold tip. Alternatively, this layer could also be removed by wet-etching. Electrophoretic deposition of paint onto the whole body of the probe, including the connecting pad and wires, but excluding the cantilever, was also used as a way of achieving improved probe insulation. Contact mode AFM-feedback mode SECM approach curves, recorded over insulating and conducting surfaces, and combined imaging of a gold grating, were recorded to evaluate the performance of the probes. Successful formation and stretching of gold nanocontacts in an electrolyte solution demonstrated the usefulness of these probes for *in situ* current sensing (C-AFM) applications.

Another recent attempt aiming at producing combined AFM-SECM probes in a *simple* way from commercial AFM probes is also worth reporting[70]: It consisted in insulating a commercially available Pt-coated C-AFM probe with a CVD film of parylene C, and subsequently exposing its tip end by mechanical abrasion. Controlled abrasion of the insulating film covering the tip extremity was carried out by imaging a conducting Cu surface in contact/current sensing mode under a high force set-point in air. Exposure of the tip was monitored by the current passed between the tip and

Cu surface as the parylene coating was removed. Combined contact mode AFM-tip collection mode SECM images of diffusional transport of a redox mediator through a porous membrane validated the so-fabricated AFM-SECM probes. Even though such a fabrication process seems straightforward, its applicability seems limited by the fact that an AFM setup is needed for individual probe fabrication.

21.7.2.2 Needle-Like Probes

Giving an AFM tip a needle-like geometry finds its interest in the fact that it allows imaging high aspect ratio features such as deep trenches or holes. A particularly elegant way of fabricating needle-like AFM-SECM probes has been reported by Macpherson et al. who started by attaching a single-wall carbon nanotube (SWCNT) to the tip of a commercial, SiO_2 passivated, silicon probe.[71] A thin gold coating was then evaporated over the whole probe, which was subsequently insulated by a thin (~40–80 nm thick) conformal film of poly(oxyphenylene), formed by electropolymerization of phenol compounds. For a better insulation, an extra layer of silicon nitride was then CVD-deposited on the probe. FIB milling was then used to cut the nanotube overhanging from the tip to the desired length (~100 nm), thus exposing a disk-shaped Au nanoelectrode (~80 nm in diameter) at its extremity (see Figure 21.28).

The performance of this kind of combined probe was assessed by imaging a ~2 µm diameter Pt disk embedded in glass in tapping mode AFM-feedback mode SECM. The topographic resolution achieved, a few tens of nanometers, was surprisingly higher than what could have been expected from the *final* diameter of the nanotube which, because of the successive coating steps, had increased from ~10 nm for the bare SWCNT up to ~400 nm. The reason for such a high resolution was that, because the probe could not be perfectly positioned perpendicular to the surface, it was actually the edge of the insulated SWCNT, which played the role of the topography imaging tip. An extra benefit of this misalignment was that short-circuits between the disk-shaped microelectrode, located at the exact center of the cylindrical tip, and conducting surfaces were avoided, allowing contact mode AFM operation. These benefits are to be expected for all cylindrically shaped AFM-SECM probes displaying a flat tip apex.

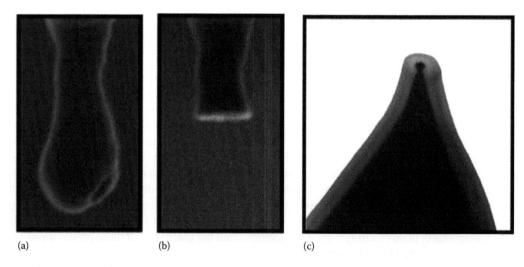

(a)   (b)   (c)

FIGURE 21.28 Focused ion beam (FIB) images of a needle-like AFM-SECM probe based on an SWCNT coated with gold and insulated by a conformal film of poly(oxyphenylene). (a) Prior and (b) after cutting by FIB. (c) TEM image of the final probe, the thin insulating coating, and disk-shaped gold nanoelectrode are clearly seen. Image widths: 1.3 µm in (a) and (b) and 3.7 µm in (c). (From Burt, D.P. et al., *Nano Lett.*, 5, 640, 2005. With permission.)

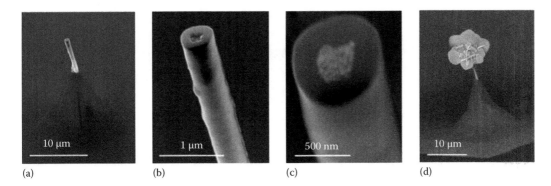

FIGURE 21.29 SEM images of a needle-like AFM-SECM probe derived from a commercial NaugaNeedle probe overcoated with a layer of electrophoretic paint. (a) and (b): Probe after FIB cutting, a slightly recessed Ag₂Ga disk is seen at the apex of the needle (c) Following platinum deposition, a slightly protruding Pt microelectrode is formed. (d) A silver *flower* electrodeposited at the tip apex to demonstrate the quality of the probe insulation. (From Wain, A.J. et al., *Electrochem. Commun.*, 13, 78, 2011. With permission.)

Producing high aspect ratio AFM-SECM probes starting from nanotubes attached to the extremity of a commercial AFM, as pioneered by Macpherson,[71] is a strategy that has been recently adopted by other groups.[72,73] Wain et al. proposed to start from AFM probes bearing Ag₂Ga alloy nanowires at their extremity.[72] These particular probes are fabricated by retracting a standard silver-coated AFM probe from liquid gallium drop. They are commercially available from NaugaNeedles, Louisville, KY, USA.[74] The Ag₂Ga wire is a few microns long and is electrically conducting and addressable. Furthermore, encapsulated probes, insulated by a ~200 nm thick layer of parylene C, are also available. In order to make these insulated probes suitable for AFM-SECM, Wain et al. had to modify them in the following ways. The authors first improved the quality of the insulation by depositing an electrophoretic paint, which filled out the defects in the parylene layer. FIB milling was then used to cut the extremity of the insulated nanowire in order to expose a disk-shaped Ag₂Ga electrode. Finally, since this material is not an ideal electrode material, it was replaced by platinum using an electroless galvanic exchange reaction. The result was a roughly disk-shaped Pt electrode ~300 nm in diameter, slightly protruding from the apex of the nanoneedle, whose overall diameter was around ~800 nm (see Figure 21.29).

Silver electrodeposition and cyclic voltammetry experiments confirmed the good electrochemical behavior of the probe. Finally, the application of these probes to AFM-SECM was demonstrated by imaging a gold grating, displaying ~2 μm large gold band electrodes, in lift-mode AFM-feedback SECM. *It is important to note that similar AFM-SECM probes are now available from the NaugaNeedle Company.* They differ from Wain's probes by the absence of electrophoretic paint overcoat, and are terminated by a protruding Pt microelectrode 20–100 nm in diameter.[74]

Another fabrication process based on growing nanoneedles at the tip of commercial AFM-SECM probes has been very recently described by Comstock et al.[73] These authors used FIB milling to cut the extremity of a silicon AFM probe, forming a flat platform at its apex. A Pt nanoneedle was then grown onto the platform by electron beam–induced deposition (EBID) using (trimethyl) methylcyclopentadienyl platinum as a precursor molecule. The diameter of the resulting nanoneedle was in the order of a few tens of nanometers, while its length ranged from a few hundreds of nanometers up to 1 μm. Because EBID-deposited platinum cannot be used as an electrode material, due to its too high resistivity resulting from residual carbon, the needle was coated by a 100–200 nm thick layer of gold. The entire probe was then insulated by atomic layer deposition (ALD) of a 20–50 nm thick Al₂O₃ film. The final step was to cut the nanoneedle extremity by FIB, exposing a disk-shaped gold nanoelectrode (~100 nm in diameter) at its apex. Silver electrodeposition was used to evidence

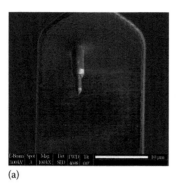
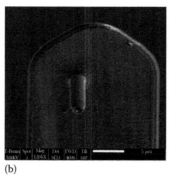
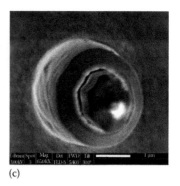

(a)   (b)   (c)

FIGURE 21.30 SEM pictures of microfabricated needle-like (a) and hollow needle (b) probes. (c) Zoomed-in SEM image of a needle-like probe showing the submicrometer-sized Pt electrode located at the needle apex. Scale bars in (a through c): 10, 5, 1 μm, respectively. (From Fasching, R.J. et al., *Sens. Actuat. B*, 108, 964, 2011. With permission.)

the proper insulation of the probe; contact mode AFM imaging of a calibration grating was also reported. No actual AFM-SECM experiments were performed.

Microfabrication techniques were used for *the batch scale* (as opposed to one by one) fabrication of needle-like AFM-SECM probes.[75] In this case, the entire probe, including the chip, cantilever, and tip, was microfabricated in a multistep process. Isotropic and anisotropic deep-reactive etch processes were used to form sharpened high aspect ratio silicon (HARS) tips on silicon wafers. Deposition of a silicon nitride layer body followed by a back etch step allowed embedding the silicon tips in an insulating layer through which they protruded. A platinum layer, a silicon nitride layer, and a gold layer were then successively deposited on the probe. The gold layer covering the tip apex was cut open by FIB to create a submicron-sized opening, exposing the underlying silicon nitride layer. Isotropic plasma etching was then used to remove the silicon nitride layer through the gold opening until the platinum layer was reached. Removing the gold layer by wet-etching then left the Pt-coated tip apex as the only electroactive area, forming a conically shaped microelectrode terminating a sharp insulated needle probe. The needle itself had a base radius of 600 nm, an aspect ratio larger than 20 (i.e., height > 1.2 μm), while the radius of the Pt tip was around 200 nm. Interestingly, hollow silicon nitride needle tips with a recessed platinum electrode at their apex were also obtained as a potentially useful by-product of the fabrication process (see Figure 21.30).

The good insulation of the probes was demonstrated by cyclic voltammetry experiments. Their AFM imaging capability was shown by imaging a metal stripe structure. However, no actual AFM-SECM characterization was carried out. Besides, the cantilevers of the probes were observed to be permanently bent due to the tensile stress of the multiple layers deposited on the probe; this may complicate the alignment of the laser used to track the tip position on many AFM systems.

21.8 CONCLUSION

It has been 17 years since the combination of SECM with AFM has been first reported.[10] During all these years, pioneering proof-of-principle experiments revealed the many possibilities offered by this combined AFM-SECM microscopy. AFM-SECM fulfilled its initial promise of allowing constant-distance SECM imaging combined with simultaneous topography imaging. The predicted improvement in SECM spatial resolution brought by this combination was also partly realized since submicron resolution is now typical for AFM-SECM, while nanometer resolution has been occasionally demonstrated. Other unexpected possibilities, such as gently touching and electrochemically interrogating surface-attached macromolecules using an AFM-SECM probe, have also emerged. Yet, the development of AFM-SECM has so far been slowed by the lack of availability of commercial probes due to the difficulty of batch-fabricating them. There is no doubt that it is only this lack of probes that has hampered many researchers from using AFM-SECM to explore nanosystems. However, as the first commercial AFM-SECM probes are now currently becoming available,[74] one can expect that new applications of AFM-SECM will emerge in the near future, making nanoscale SECM a reality to all. In particular, the exciting possibility offered by AFM-SECM of forming tip-to-substrate gaps of nanometer, of even subnanometer width, in an electrochemical configuration, will be of interest to many researchers. Forming such nanogaps, which can trap very few molecules between the tip and substrate, will likely generalize single-molecule experiments. Using even smaller gaps will permit to form molecular junctions, where molecules bridge the tip–substrate gap, allowing *intramolecular SECM* and/or molecular conductivity measurements to be carried out in situ.

REFERENCES

1. Bard, A. J. 2012. Introduction and principles. In *Scanning Electrochemical Microscopy. Second edition*, Bard, A. J. and Mirkin, M. V. (eds.), pp. 1–14. Boca Raton, FL: Taylor & Francis.
2. Binnig, G.; Quate, C. F., and Gerber, C. 1986. Atomic force microscope. *Phys. Rev. Lett.* 56: 930–933.
3. Macpherson, J. V. and Demaille, C. 2012. Hybrid scanning electrochemical techniques: Methods and applications. In *Scanning Electrochemical Microscopy. Second edition*, Bard, A. J. and Mirkin, M. V. (eds.), pp. 569–587. Boca Raton, FL: Taylor & Francis.
4. Kranz, C. 2014. Recent advancements in nanoelectrodes and nanopipettes used in combined scanning electrochemical microscopy techniques. *Analyst* 139: 336–352.
5. Sun, P.; Laforge, F. O., and Mirkin, M. V. 2007. Scanning electrochemical microscopy in the 21st century. *Phys. Chem. Chem. Phys.* 9: 802–823.
6. Wittstock, G.; Burchardt, M.; Pust, S. E.; Shen, Y., and Zhao, C. 2007. Scanning electrochemical microscopy for direct imaging of reaction rates. *Angew. Chem. Int. Ed.* 46: 1584–1617.
7. Fan, F.-R. F. 2012. Scanning electrochemical microscopic imaging. In *Scanning Electrochemical Microscopy. Second edition*, Bard, A. J. and Mirkin, M. V. (eds.), pp. 53–74. Boca Raton, FL: Taylor & Francis.
8. Macpherson, J. V. and Unwin, P. R. 2001. Noncontact electrochemical imaging with combined scanning electrochemical atomic force microscopy. *Anal. Chem.* 73: 550–557.
9. Kueng, A.; Kranz, C.; Mizaikoff, B.; Lugstein, A. and Bertagnolli, E. 2003. Combined scanning electrochemical atomic force microscopy for tapping mode imaging. *Appl. Phys. Lett.* 82: 1592–1594.
10. Macpherson, J. V.; Unwin, P. R.; Hillier, A. C., and Bard, A. J. 1996. In-situ imaging of ionic crystal dissolution using an integrated electrochemical/AFM probe. *J. Am. Chem. Soc.* 118: 6445–6452.
11. Jones, C. E.; Unwin, P. R., and Macpherson, J. V. 2003. In situ observation of the surface processes involved in dissolution from the cleavage surface of calcite in aqueous solution using combined scanning electrochemical-atomic force microscopy (SECM-AFM). *ChemPhysChem* 4: 139–146.
12. Macpherson, J. V. and Unwin, P. R. 2000. Combined scanning electrochemical-atomic force microscopy. *Anal. Chem.* 72: 276–285.
13. Macpherson, J. V.; Gueneau de Mussy, J.-P., and Delplancke, J.-L. 2002. High-resolution electrochemical, electrical, and structural characterization of a dimensionally stable Ti/TiO$_2$/Pt electrode. *J. Electrochem. Soc.* 149: B306–B313.
14. Dobson, P. S.; Weaver, J. M. R.; Holder, M. N.; Unwin, P. R., and Macpherson, J. V. 2005. Characterization of batch-microfabricated scanning electrochemical-atomic force microscopy probes. *Anal. Chem.* 77: 424–434.
15. Dobson, P. S.; Weaver, J. M. R.; Burt, D. P et al. 2006. Electron beam lithographically-defined scanning electrochemical-atomic force microscopy probes: Fabrication method and application to high resolution imaging on heterogeneously active surfaces. *Phys. Chem. Chem. Phys.* 8: 3909–3914.
16. Mirkin, M. V. and Wang, Y. 2012. Theory. In *Scanning Electrochemical Microscopy. Second edition*, Bard, A. J. and Mirkin, M. V. (eds.), pp. 92–96. Boca Raton, FL: Taylor & Francis.
17. Davoodi, A.; Pan, J.; Leygraf, C., and Norgren, S. 2005. In situ investigation of localized corrosion of aluminum alloys in chloride solution using integrated EC-AFM/SECM techniques. *Electrochem. Solid State Lett.* 8: B21–B24.

18. Davoodi, A.; Pan, J.; Leygraf, C., and Norgren, S. 2008. Multianalytical and in situ studies of localized corrosion of EN AW-3003 alloy—Influence of intermetallic particles. *J. Electrochem. Soc.* 155: C138–C146.

19. Davoodi, A.; Farzadi, A.; Pan, J.; Leygraf, C., and Zhu, Y. 2008. Developing an AFM-based SECM system; instrumental setup, SECM simulation, characterization, and calibration. *J. Electrochem. Soc.* 155: C474–C485.

20. Pust, S. E.; Salomo, M.; Oesterschulze, E., and Wittstock, G. 2010. Influence of electrode size and geometry on electrochemical experiments with combined SECM–SFM probes. *Nanotechnology* 21: 105709–[12pp].

21. Salomo, M.; Pust, S. E.; Wittstock, G., and Oesterschulze, E. 2010. Integrated cantilever probes for SECM/AFM characterization of surfaces. *Microelectron. Eng.* 87: 1537–1539.

22. Pust, S. E.; Scharnweber, D.; Kirchner, C. N., and Wittstock, G. 2007. Heterogeneous distribution of reactivity on metallic biomaterials: Scanning probe microscopy studies of the biphasic Ti alloy Ti6Al4V. *Adv. Mater.* 19: 878–882.

23. Bard, A. J.; Mirkin, M. V.; Unwin, P. R., and Wipf, D. O. 1992. Scanning electrochemical microscopy. 12. Theory and experiment of the feedback mode with finite heterogeneous electron-transfer kinetics and arbitrary substrate size. *J. Phys. Chem.* 96: 1861–1868.

24. Pierce, D. T. and Bard, A. J. 1993. Scanning electrochemical microscopy. 23. Retention localization of artificially patterned and tissue-bound enzymes. *Anal. Chem.* 65: 3598–3604.

25. Tefashe, U. M. and Wittstock, G. 2013. Quantitative characterization of shear force regulation for scanning electrochemical microscopy. *C. R. Chimie* 16: 7–14.

26. McCreery, R. L. 2008. Advanced carbon electrode materials for molecular electrochemistry. *Chem. Rev.* 108: 2646–2687.

27. Kneten, K. and McCreery, R. L. 1992. Effects of redox system structure on electron-transfer kinetics at ordered graphite and glassy carbon electrodes. *Anal. Chem.* 64: 2518–2524.

28. Frederix, P. L. T. M.; Gullo, M. R.; Akiyama, T. et al. 2005. Assessment of insulated conductive cantilevers for biology and electrochemistry. *Nanotechnology* 16: 997–[8pp].

29. Frederix, P. L. T. M.; Bosshart, P. D.; Akiyama, T. et al. 2008. Conductive supports for combined AFM–SECM on biological membranes. *Nanotechnology* 19: 384004–[10pp].

30. Patel, A. N.; Guille Collignon, M.; O'Connell, M. A. et al. 2012. A new view of electrochemistry at highly oriented pyrolytic graphite. *J. Am. Chem. Soc.* 134: 20117–20130.

31. Horrocks, B. R. and Wittstock, G. 2012. Biotechnological applications. In *Scanning Electrochemical Microscopy. Second edition*, Bard, A. J. and Mirkin, M. V. (eds.), pp. 317–378. Boca Raton, FL: Taylor & Francis.

32. Kueng, A.; Kranz, C.; Lugstein, A.; Bertagnolli, E., and Mizaikoff, B. 2003. Integrated AFM–SECM in tapping mode: Simultaneous topographical and electrochemical imaging of enzyme activity. *Angew. Chem. Int. Ed.* 42: 3238–3240.

33. Kranz, C.; Kueng, A.; Lugstein, A.; Bertagnolli, E., and Mizaikoff, B. 2004. Mapping of enzyme activity by detection of enzymatic products during AFM imaging with integrated SECM–AFM probes. *Ultramicroscopy* 100: 127–134.

34. Kranz, C.; Friedbacher, G.; Mizaikoff, B.; Lugstein, A.; Smoliner, J., and Bertagnolli, E. 2001. Integrating an ultramicroelectrode in an AFM cantilever: Combined technology for enhanced information. *Anal. Chem.* 73: 2491–2500.

35. Hirata, Y.; Yabuki, S., and Mizutani, F. 2004. Application of integrated SECM ultra-micro-electrode and AFM force probe to biosensor surfaces. *Bioelectrochemistry* 63: 217–224.

36. Macpherson, J. V.; Jones, C. E.; Barker, A. L., and Unwin, P. R. 2002. Electrochemical imaging of diffusion through single nanoscale pores. *Anal. Chem.* 74: 1841–1848.

37. Kueng, A.; Kranz, C.; Mizaikoff, B.; Lugstein, A., and Bertagnolli, E. 2005. AFM-tip-integrated amperometric microbiosensors: High-resolution imaging of membrane transport. *Angew. Chem. Int. Ed.* 44: 3419–3422.

38. Mandler, D. 2012. Micro- and nanopatterning using scanning electrochemical microscopy. In *Scanning Electrochemical Microscopy. Second edition*, Bard, A. J. and Mirkin, M. V. (eds.), pp. 489–524. Boca Raton, FL: Taylor & Francis.

39. Sekine, S.; Kaji, H., and Nishizawa, M. 2008. Integration of an electrochemical-based biolithography technique into an AFM system. *Anal. Bioanal. Chem.* 391: 2711–2716.

40. Sekine, S.; Kaji, H., and Nishizawa, M. 2009. Spatiotemporal sub-cellular biopatterning using an AFM-assisted electrochemical system. *Electrochem. Commun.* 11: 1781–1784.

41. Ghorbal, A.; Grisotto, F.; Charlier, J.; Palacin, S.; Goyer, C., and Demaille, C. 2009. Localized electrografting of vinylic monomers on a conducting substrate by means of an integrated electrochemical AFM probe. *ChemPhysChem* 10: 1053–1057.

42. Abbou, J.; Demaille, C.; Druet, M., and Moiroux, J. 2002. Fabrication of submicrometer-sized gold electrodes of controlled geometry for scanning electrochemical-atomic force microscopy. *Anal. Chem.* 74: 6355–6363.

43. Sklyar, O.; Kueng, A.; Kranz, C. et al. 2005. Numerical simulation of scanning electrochemical microscopy experiments with frame-shaped integrated atomic force microscopy-SECM probes using the boundary element method. *Anal. Chem.* 77: 764–771.

44. Lee, Y.; Amemiya, S., and Bard, A. J. 2001. Scanning electrochemical microscopy. 41. Theory and characterization of ring electrodes. *Anal. Chem.* 73: 2261–2267.

45. Gullo, M. R.; Frederix, P. L. T. M.; Akiyama, T.; Engel, A.; deRooij, N. F., and Staufer, U. 2006. Characterization of microfabricated probes for combined atomic force and high-resolution scanning electrochemical microscopy. *Anal. Chem.* 78: 5436–5442.

46. Leonhardt, K.; Avdic, A.; Lugstein, A. et al. 2013. Scanning electrochemical microscopy: Diffusion controlled approach curves for conical AFM-SECM tips. *Electrochem. Commun.* 27: 29–33.

47. Burt, D. P.; Wilson, N. R.; Janus, U.; Macpherson, J. V., and Unwin, P. R. 2008. In-situ atomic force microscopy (AFM) imaging: Influence of AFM probe geometry on diffusion to microscopic surfaces. *Langmuir* 24: 12867–12876.

48. Abbou, J.; Anne, A., and Demaille, C. 2004. Probing the structure and dynamics of end-grafted flexible polymer chain layers by combined atomic force-electrochemical microscopy. Cyclic voltametry within nanometer-thick macromolecular poly(ethylene glycol) layers. *J. Am. Chem. Soc.* 126: 10095–10108.

49. Abbou, J.; Anne, A., and Demaille, C. 2006. Accessing the dynamics of end-grafted flexible polymer chains by atomic force-electrochemical microscopy. Theoretical modeling of the approach curves by the elastic bounded diffusion model and Monte Carlo simulations. Evidence for compression-induced lateral chain escape. *J. Phys. Chem. B* 110: 22664–22675.

50. Wang, K.; Goyer, C.; Anne, A., and Demaille, C. 2007. Exploring the motional dynamics of end-grafted DNA oligonucleotides by in situ electrochemical atomic force microscopy. *J. Phys. Chem. B.* 111: 6051–6058.

51. Anne, A.; Blanc, B., and Moiroux, J. 2001. Synthesis of the first ferrocene-labeled dideoxynucleotide and its use for 3′-redox end-labeling of 5′-modified single-stranded oligonucleotides. *Bioconjugate Chem.* 12: 396–405.

52. Anne, A.; Bonnaudat, C.; Demaille, C., and Wang, K. 2007. Enzymatic redox 3′-end-labeling of DNA oligonucleotide monolayers on gold surfaces using terminal deoxynucleotidyl transferase (TdT)-mediated single base extension. *J. Am. Chem. Soc.* 129: 2734–2735.

53. Anne, A.; Cambril, E.; Chovin, A., and Demaille, C. 2010. Touching surface-attached molecules with a microelectrode: Mapping the distribution of redox-labeled macromolecules by electrochemical-atomic force microscopy. *Anal. Chem.* 82: 6353–6362.

54. Anne, A.; Chovin, A.; Demaille, C., and Lafouresse, M. 2011. High-resolution mapping of redox-immunomarked proteins using electrochemical-atomic force microscopy in molecule touching mode. *Anal. Chem.* 83: 7924–7932.

55. Fan, F.-R. F. and Bard, A. J. 1995. Electrochemical detection of single molecules. *Science* 267: 871–874.

56. Huang, K.; Anne, A.; Bahri, M. A., and Demaille, C. 2013. Probing individual redox PEGylated gold nanoparticles by electrochemical-atomic force microscopy. *ACS Nano* 7: 4151–4163.

57. Fan, F.-R. F. and Demaille, C. 2012. Preparation of tips for scanning electrochemical. In *Scanning Electrochemical Microscopy. Second edition*, Bard, A. J. and Mirkin, M. V. (eds.), pp. 25–51. Boca Raton, FL: Taylor & Francis.

58. Anne, A.; Demaille, C., and Goyer, C. 2009. Electrochemical atomic-force microscopy using a tip-attached redox mediator. Proof-of-concept and perspectives for functional probing of nanosystems. *ACS Nano* 3: 349–353.

59. Anne, A.; Cambril, E.; Chovin, A.; Demaille, C., and Goyer, C. 2009. Electrochemical atomic force microscopy using a tip-attached redox mediator for topographic and functional imaging of nanosystems. *ACS Nano* 3: 2927–2940.

60. Wiedemair, J.; Balu, B.; Moon, J.-S.; Hess, D. W.; Mizaikoff, B., and Kranz, C. 2008. Plasma-deposited fluorocarbon films: Insulation material for microelectrodes and combined atomic force microscopy-scanning electrochemical microscopy probes. *Anal. Chem.* 80: 5260–5265.

61. Wiedemair, J.; Moon, J.-S.; Reinauer, F.; Mizaikoff, B., and Kranz, C. 2010. Ion beam induced deposition of platinum carbon composite electrodes for combined atomic force microscopy-scanning electrochemical microscopy. *Electrochem. Commun.* 12: 989–991.

62. Smirnov, W.; Kriele, A.; Hoffmann, R. et al. 2011. Diamond-modified AFM probes: From diamond nanowires to atomic force microscopy-integrated boron-doped diamond electrodes. *Anal. Chem.* 83: 4936–4941.

63. Shin, H.; Hesketh, P.J.; Mizaikoff, B., and Kranz, C. 2007. Batch fabrication of atomic force microscopy probes with recessed integrated ring microelectrodes at a wafer level. *Anal. Chem.* 79: 4769–4777.

64. Shin, H.; Hesketh, P. J.; Mizaikoff, B., and Kranz, C. 2008. Development of wafer-level batch fabrication for combined atomic force-scanning electrochemical microscopy (AFM–SECM) probes. *Sens. Actuat. B* 134: 488–495.

65. Lee, E.; Kim, M.; Seong, J.; Shin, H. and Lim, G. 2013. An L-shaped nanoprobe for scanning electrochemical microscopy-atomic force microscopy. *Phys. Status Solidi RRL* 7: 406–409.

66. Wu, Y.; Akiyama, T.; van der Wal, P.; Gautsch, S., and deRooij, N. 2012. Development of insulated conductive AFM probes for experiments in electrochemical environment. *ECS Trans.* 50: 465–468.

67. Wu, Y.; Akiyama, T.; Gautsch, S.; van der Wal, P., and deRooij, N. 2014. In-plane fabrication of insulated gold-tip probes for electrochemical and force spectroscopy molecular experiments. *Sen. Actuat. A* 215: 184–188.

68. Avdic, A.; Lugstein, A.; Wu, M. et al. 2011. Fabrication of cone-shaped boron doped diamond and gold nanoelectrodes for AFM-SECM. *Nanotechnology* 22: 145306–[6pp].

69. Pobelov, I. V.; Mohos, M.; Yoshida, K. et al. 2013. Electrochemical current-sensing atomic force microscopy in conductive solutions. *Nanotechnology* 24: 115501–[10pp].

70. Derylo, M. A.; Morton, K. C., and Baker, L. A. 2011. Parylene insulated probes for scanning electrochemical-atomic force microscopy. *Langmuir* 27: 13925–13930.

71. Burt, D. P.; Wilson, R. R.; Weaver, J. M. R.; Dobson, P. S., and Macpherson, J. V. 2005. Nanowire probes for high resolution combined scanning electrochemical microscopy-atomic force microscopy. *Nano Lett.* 5: 640–643.

72. Wain, A. J.; Cox, D.; Zhou, S., and Turnbull, A. 2011. High-aspect ratio needle probes for combined scanning electrochemical microscopy-atomic force microscopy. *Electrochem. Commun.* 13: 78–81.

73. Comstock, D. J.; Elam, J. W.; Pellin, M. J., and Hersam, M. C. 2012. High aspect ratio nanoneedle probes with an integrated electrode at the tip apex. *Rev. Sci. Instrum.* 83: 113704–113708.

74. NaugaNeedles LCC, Louisville, Kentucky, USA. Website: http://www.nauganeedles.com.

75. Fasching, R. J.; Tao, Y., and Prinz, F. B. 2005. Cantilever tip probe arrays for simultaneous SECM and AFM analysis. *Sens. Actuat. B* 108: 964–972.

22 Nanoscale Potentiometry

Róbert E. Gyurcsányi and Ernö Pretsch

CONTENTS

22.1 INTRODUCTORY NOTES

Potentiometry is defined by the Merriam Webster dictionary as the measurement of electromotive force by means of a potentiometer as well as the use or application of such a measurement. For chemists, potentiometry is an electroanalytical method based on gathering chemical information by measuring electrical potentials at *zero current* with a high-input impedance voltmeter. Accordingly, this chapter will focus on the nanoscale aspects of potentiometry applied to chemical analysis. However, it is worth mentioning that potentiometry has also other uses than providing chemical information. In fact, the term *nanopotentiometry* is used for a special scanning probe technique applied to the determination of local electrical potentials and the distribution of electrical potentials, for example, for characterizing electronic and optoelectronic devices.[1]

Potentiometry looks back on a history of more than a century, being among the earliest instrumental methods of analysis. While its very beginning is marked by essential contributions toward understanding the electrode potential of electrodes of first and second type[2] as well as the potential difference arising at the interface of aqueous electrolytes of different concentrations and immiscible liquids,[3] the analytical aspects became dominant very early. This was due to the discovery of the potentiometric pH response of glasses by Cremer[4] in 1906, rapidly turned into a functional pH electrode by Haber and Klemensiewicz.[5] The ubiquitous importance of pH measurements gave a determinant impulse to potentiometry and, especially, to ion-selective potentiometry, which uses ion-selective electrodes (ISEs) to infer a selective potentiometric response. Potentiometry, by being ideally a *passive* technique in terms of measuring electrical potentials of spontaneous chemical processes with minimal electrical and chemical perturbations to the sample, lacks the wealth of measurement methodologies characteristic of other electroanalytical techniques, for example, voltammetry. Therefore, the essential progress in potentiometry went along the development of ion-selective membranes (ISMs) and electrodes and the implementation of new sensing schemes. This dominant direction profited from the progress in supramolecular chemistry, chemical synthesis of selective ligands, such as ionophores, as well as the use of advanced functional materials. In this latter aspect, the rational use of nanomaterials recently featured some spectacular solutions to inherent problems of ISEs, which will be the focus of this chapter.

22.2 CONVENTIONAL POTENTIOMETRIC ELECTRODES: STATE OF THE ART

The line of potentiometric electrodes evolved from electrodes based on electron exchange, such as simple electron conductors for redox potential measurements, to electrodes based on ion exchange. The latter are at the core of ISEs, featured with either solid-state or liquid sensing membranes that enable the determination of more than 60 ions[6] with major applications in the clinical analysis of body fluid electrolytes, environmental analysis, and process chemistry. The unique property of ISEs that the measured electrode potential is a function of the free ionic activity (a thermodynamic property) has distinctive advantages in studying biological systems, where the free ion activity, that is, the ionized species as referred to in biomedical literature, is of utmost importance. The same property enables the use of ISEs as indicator electrodes in complexometric titrations, during which, not considering dilution, only the free ion activity changes. While chemists still consider ISEs primarily as simple routine laboratory tools for direct potentiometry and potentiometric titrations, ISEs together with some other amperometric sensors are at the core of blood gas analyzers with an estimated market of 6 billion US $.[7] This clearly points out the major field of application of ISEs and their enormous impact on chemical analysis.

The solid-state ISMs are commonly nonporous solid membranes prepared from glasses, crystals, and compressed precipitate powders. They are generally low-capacity ion exchangers with mobile counterions that are exchanged at the membrane–aqueous solution interface and generate a phase boundary potential indicative of their activity in the aqueous solution (see Section 22.3.). The selectivity is governed either by the small size of the ions such as for H^+ (glass membrane), F^- (EuF_2-doped LaF_3 crystals), or by the solubility product in case of precipitate-type membranes (ions forming low-solubility precipitates are preferred). While excellent for some ions, the range of detected ions could be enlarged only by the implementation of liquid-membrane electrodes based on selective complexing agents.

The state-of-the-art ionophore-based ISEs are generally based on hydrophobic plasticized membranes with glass transition temperatures below room temperature. This ensures proper mechanical properties for handling and processing as well as reasonable ionic conductivities and diffusion of active membrane components (ca. 10^{-8} cm²/s). The poly(vinyl chloride) (PVC) membranes, most commonly used, are cast from a solution of plasticizer (66 wt%) and PVC (33 wt%) and as such are also termed liquid or solvent polymeric membranes. The first requirement with respect to these membranes is to exhibit permselectivity for ions of a given charge sign, which is induced by ion-exchanger sites in the membrane, that is, membranes with negative ionic sites are permselective to cations, while those with positive ionic sites are permselective to anions. The ionic sites may originate from the membrane matrix, but most commonly from lipophilic anion or cation exchangers added to the membrane. These externally added ionic sites are, with a few exception,[8] mobile sites, for example, tetra-alkyl ammonium compounds for establishing anionic and tetraphenylborate derivatives for cationic permselectivity. Such hydrophobic ion-exchanger membranes inherently prefer large ions of low charge density and, therefore, for some applications involving the determination of such ions (e.g., NO_3^-, ClO_4^-, Cs^+) are compounded only with an ion exchanger.[9] However, the selectivity of membranes can be tuned for a specific ion by formulating their composition with selective hydrophobic complexing agents, that is, electrically neutral or charged ionophores. The chemical structures of some of the most relevant membrane components for the fabrication of ion-selective nanoelectrodes are listed in Figure 22.1.

The general layout of liquid membrane–based ISE cell assemblies is shown in Figure 22.2. The sensing membrane of ionophore-based ISEs usually is plasticized PVC in which dioctyl sebacate (DOS) or 2-nitrophenyl octyl ether (o-NPOE) serves as plasticizer. In these highly plasticized polymeric films (liquid membranes), ion exchanger sites (R^-) and selective complexing agents (L) are trapped by their lipophilic properties. The membrane is interfaced with the electrical connection either through an inner solution or a solid contact.

22.3 THEORY

22.3.1 INTRODUCTION

The theory of potentiometric sensors is well established. For membranes with dimensions much larger than the Debye length, electroneutrality holds in the contacting phases so that the phase-boundary potential model[10–13] is capable of perfectly describing the experiments in all practically relevant cases. Among all analytical methods, the theory of potentiometry has unique predictive capabilities. Once the selectivity behavior is characterized, it is possible to predict the response for any mixture of the ions involved. This includes cases in which ion fluxes through the membrane (induced by current or concentration gradients) are not negligible as long as they are slow relative to the speed of the ion exchange at the phase boundary.

On the other hand, the Nernst–Planck theory is required for describing time- and space-dependent phenomena, for which the phase-boundary potential model is not suitable.[26,28–30] It is adequate for treating zero-current or dc experiments as well as influences of convection, flow, or stirring.

Finally, for very thin phases or nanopores, that is, for nanopotentiometry, as well as for high-frequency experiments, electroneutrality cannot be assumed any more so that the Nernst–Planck–Poisson model must be used.[14–17]

The Debye length, λ, depends on the permittivity (ε) of the solution, the concentration (c_i) and the ionic charge (z_i) of the species j, and the temperature (T):

$$\lambda = \sqrt{\frac{\varepsilon\varepsilon_0 RT}{F^2 \sum_i c_i z_i^2}}, \tag{22.1}$$

where
  ε_0 is the permittivity of vacuum
  R is the gas constant
  F is the Faraday constant

The Debye length is about 10 nm for a millimolar aqueous solution of a 1:1 electrolyte and is 2–4 times smaller for a typical PVC membrane depending on the polarity of the plasticizer.

For completeness, and in order to demonstrate the effects of the size reduction to nanoscale, first the model descriptions are reviewed that are valid for larger dimensions such as macroscopic electrodes and traditional microelectrodes.

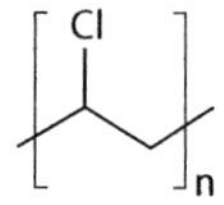

PVC, poly(vinyl chloride)

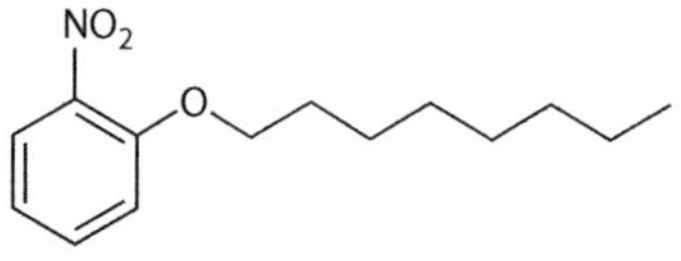

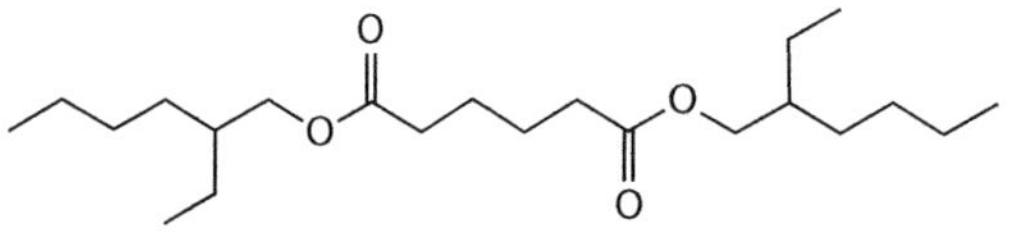

DOS, dioctyl sebacate, bis (2-ethylhexyl) sebacate

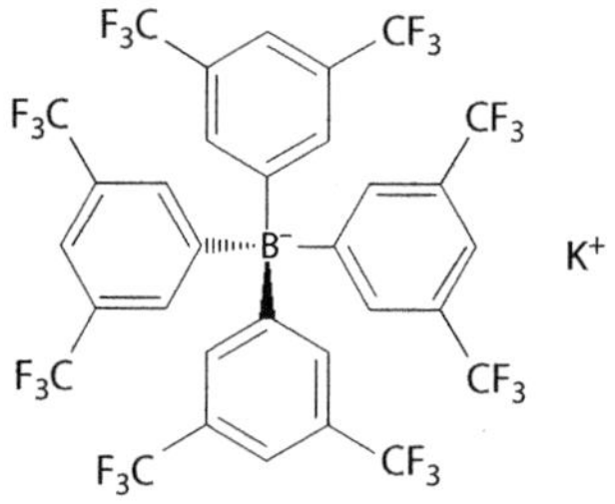

o-NPOE, o-nitrophenyl n-octyl ether

KTFPB, potassium tetrakis[3,5-bis(trifluoromethyl)
phenyl]borate

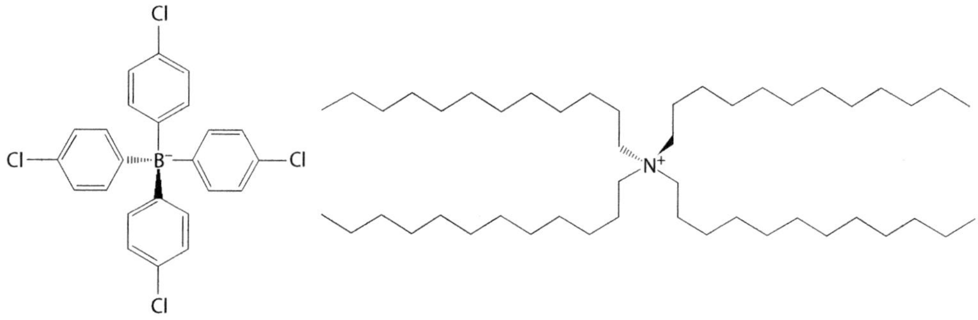

ETH 500, tetradodecyl ammonium tetrakis(4-chlorophenyl)borate

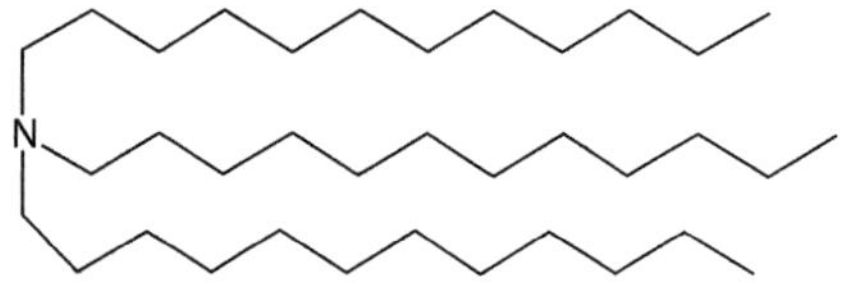

H⁺ ionophore, tridodecylamine

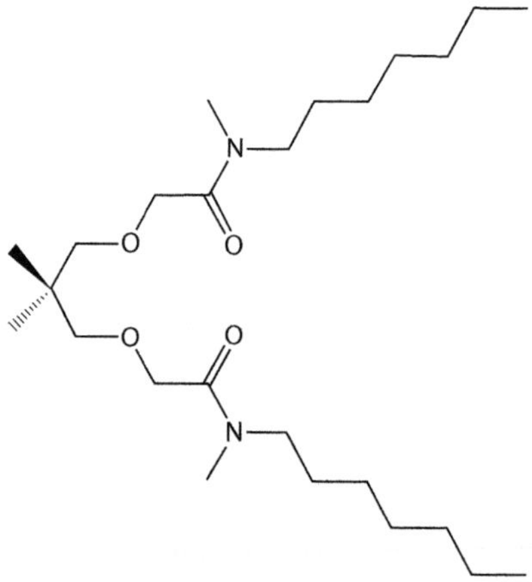

Li⁺ ionophore, ETH 149,
N,N'-diheptyl-N,N',5,5-tetramethyl-3,7-
dioxanonanediamide

FIGURE 22.1   Membrane components and ionophores used in ion-selective micro- and nanoelectrodes.
(Continued)

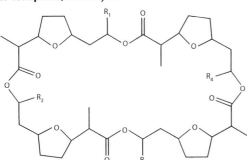

Na⁺ ionophore, monensin

Na⁺ ionophore, ETH 157, *N,N′*-dibenzyl-*N,N′*-diphenyl-1,2-phenylenedioxydiacetamide

K⁺ ionophore, BME 44
2-dodecyl-2-methyl-1,3-propanediyl bis[N-[5′-nitro(benzo-15-crown-5)-4′-yl]carbamate]

K⁺ ionophore, valinomycin

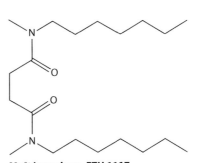

NH₄⁺ ionophore, nonactin (R₁, R₂, R₃, R₄: CH₃), monactin (R₁, R₂, R₃: CH₃, R₄: C₂H₅)

Mg²⁺ ionophore, ETH 1117,
N,N′-diheptyl-*N,N′*-dimethylsuccinamide

FIGURE 22.1 (CONTINUED) Membrane components and ionophores used in ion-selective micro- and nanoelectrodes.
(Continued)

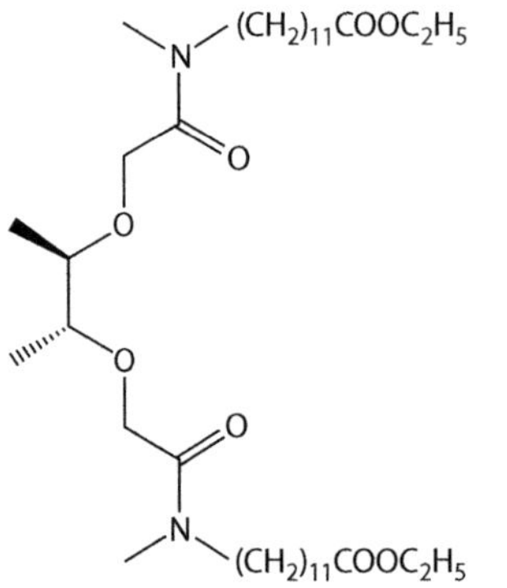

Ca^{2+} **ionophore, ETH 1001,** $(-)$-(R,R)-N,N'-bis-[11-(ethoxycarbonyl)undecyl]-$N,N',4,5$-tetramethyl -3, 6-dioxaoctanediamide

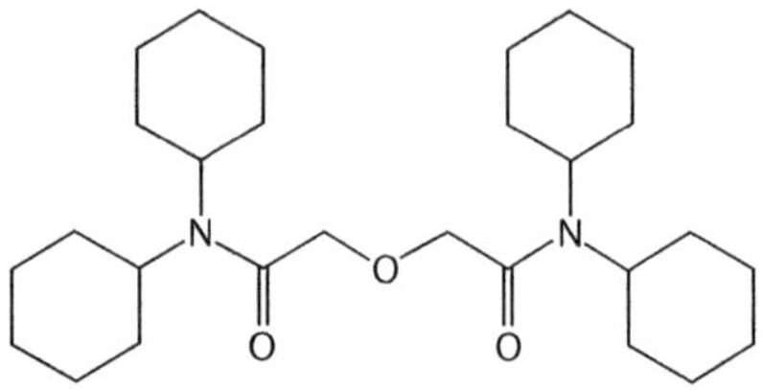

Ca^{2+} **ionophore, ETH 129,** N,N,N',N'-tetracyclohexyl-3-oxapentanediamide

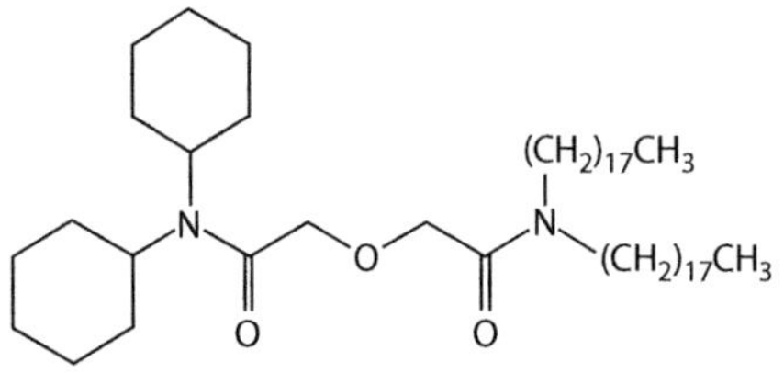

Ca^{2+} **ionophore, ETH 5234,** N,N-dicyclohexyl-N', N'-dioctadecyl-3-oxapentanediamide

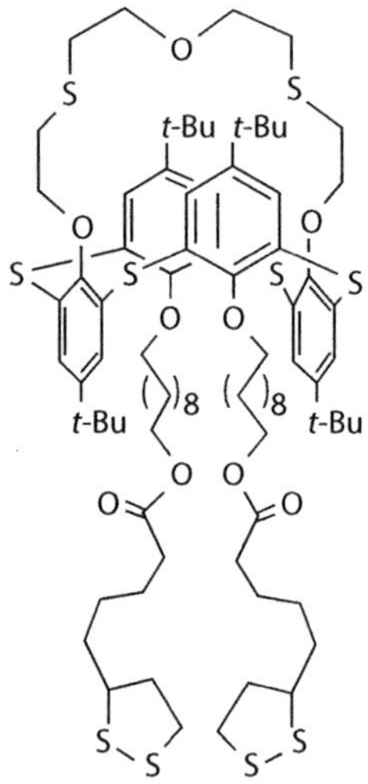

Ag$^+$ ionophore, SS-Ag-II

FIGURE 22.1 (CONTINUED) Membrane components and ionophores used in ion-selective micro- and nanoelectrodes.

22.3.2 PHASE-BOUNDARY POTENTIAL MODEL

It is a unique feature of ISEs that their potentiometric behavior can be described on the basis of a simple thermodynamic model. The basic assumptions are that there is a local equilibrium between the membrane (m) and the aqueous sample (aq) and that electroneutrality holds in both phases.[10,11] For charged species, this means that the electrochemical potential ($\tilde{\mu}_i$) of an ion i is equal in both phases:

$$\tilde{\mu}_i = \mu_i^0 + RT \ln a_i + z_i F \phi, \tag{22.2}$$

where
  μ_i^0 is the standard chemical potential
  a_i is the activity of the ion I; R, T, F, and ϕ are the gas constant, the absolute temperature, the Faraday constant, and the electrical potential, respectively

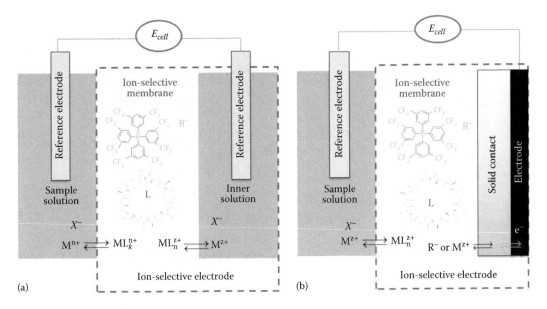

FIGURE 22.2 Schematic representations of cationic membrane–separated potentiometric cells with (a) a liquid electrolyte and (b) a solid (electron or mixed conductor-based) contact. M^{z+} is the primary ion, L is the ionophore (valinomycin), R^- is a mobile lipophilic anionic site (tetrakis[3,5-bis(trifluoromethyl)phenyl]borate), ML_n^{z+} is the ionophore–ion complex, and e^- indicates electrons.

The phase-boundary potential, $E_{PB} = \phi(m) - \phi(aq)$, then is

$$E_{PB} = \frac{\mu_i^0(aq) - \mu_i^0(m)}{z_i F} + \frac{RT}{z_i F} \ln \frac{a_i(aq)}{a_i(m)} = \varepsilon_i^0 + \frac{RT}{z_i F} \ln \frac{a_i(aq)}{a_i(m)}. \tag{22.3}$$

The standard potential, ε_i^0, defined in the earlier equation is constant for a given ion but varies from ion to ion.

To achieve a dependence of the phase-boundary potential on the activity of the sample ions, $a_i(aq)$, their activity in the membrane, $a_i(m)$, must be kept constant. This is accomplished by adding an ion exchanger to the membrane phase. If, for simplicity, the influence of ion-pair formation is neglected, then $a_i(m) = \gamma_i R_{tot}/z_i$ for a monovalent ion exchanger of the concentration R_{tot}, where γ_i is the activity coefficient of the ion i in the membrane. Thus, the phase-boundary potential is

$$E_{PB} = \varepsilon_i^0 - \frac{RT}{z_i F} \ln \frac{\gamma_i R_{tot}}{z_i} + \frac{RT}{z_i F} \ln a_i = E_i^0 + \frac{RT}{z_i F} \ln a_i = E_i^0 + \frac{s}{z_i} \log a_i. \tag{22.4}$$

The slope, s, incorporates the numerical values of R, T, and F as well as $1/\log e$ for changing the logarithmic base and is 59.2 mV at 25°C.

To achieve a selective response, usually, a complexing agent is incorporated in the ISM phase in addition to ion exchanger. The free ion activity in the membrane is reduced by the complex formation and thus, for cations, the term with R_{tot} decreases and, consequently, the phase-boundary potential increases (for quantitative descriptions, see, e.g., Refs. [12,18,19]).

The preceding equations are valid for any ion i. To describe the phase-boundary potential of an electrode for different ions, it is useful to apply one single constant term, E_I^0, which is defined for the so-called primary ion, I. For another ion, J, the E^0 value changes for two reasons: (1) the standard chemical potential, ε_J^0, is not equal to ε_I^0 and (2) the free ion activity in the membrane, $a_i(m)$, changes owing to the differences in the complex formation constants and the strength of ion pairs.

It is convenient to use the same E^0 term and to describe the difference in E^0 by a weighting factor of the activity term of Equation 22.5, the so-called potentiometric selectivity coefficient, K_{IJ}^{pot}:

$$E_{PB}(I) = E_I^0 + \frac{s}{z_I} \log a_I,$$ (22.5)

$$E_{PB}(J) = E_I^0 + \frac{s}{z_J} \log K_{IJ}^{pot} a_J^{z_J/z_I},$$ (22.6)

with

$$\log K_{IJ}^{pot} = \frac{z_I}{s} \left(E_J^0 - E_I^0 \right).$$ (22.7)

The selectivity coefficient can be quantitatively related to the complex and ion-pair formation constants.[12,18–20]

For a mixture of the two ions, I^z and J^z, having the same charge z, the phase-boundary potential is given by

$$E_{PB} = E_I^0 + \frac{s}{z} \log \left(a_I + K_{IJ}^{pot} a_J \right).$$ (22.8)

This equation was first derived by Nicolsky in 1937,[21] who thus introduced the potentiometric selectivity coefficient. Later some authors empirically extended it for ions of different charges using $K_{IJ}^{pot} a_J^{z_J/z_I}$ as the second term in Equation 22.8. The resulting empirical Nicolsky–Eisenman equation is, however, inconsistent since different results are obtained if the two ions are interchanged. This led to plenty of confusions, and different other selectivity measures were proposed since the simple fundamental meaning of K_{IJ}^{pot} was not recognized.

The correct use of the selectivity coefficient for describing the response to ions of different charges was first derived by Bakker in 1994.[19,22–24] Its validity has also been confirmed by corresponding experiments.[22]

The membrane potential, that is, the potential difference between the two phases separated by the membrane, is the sum of the two phase-boundary potentials and the transmembrane potential difference.[12,25,26] The latter is negligible in all practical cases since only one dominating exchangeable ion prevails in the membrane. The phase-boundary potential on the inner membrane side does not depend on the composition of the sample and is thus constant. The observed electromotive force is the sum of all contributions in a measuring cell. Besides the ISM, the only sample-dependent contribution comes from the reference electrode. In well-designed systems, its contribution is nearly constant and can be included together with the other contributions, such as the transmembrane potential and the phase-boundary potential on the inner membrane side, into one potential term of the cell, K_{cell}. Thus, the observed electromotive force, emf, is

$$emf \approx K_{cell} + E_{PB}.$$ (22.9)

22.3.3 NERNST–PLANCK THEORY

While the phase-boundary potential model perfectly describes the response of macroscopic electrodes even in the presence of any number of mono-, di-, and trivalent ions, more involved models are required if other membrane properties are of interest such as concentration profiles in the membrane or transient responses.

The Nernst–Planck equation describes the flux J_i of a charged species i in the presence of concentration and potential gradients. It is written here for a 1D system with mass transport in the x direction:

$$J_i(x,t) = -D_i \frac{\partial c_i}{\partial x} - z_i D_i c_i \frac{F}{RT} \frac{\partial \phi}{\partial x} + c_i v, \tag{22.10}$$

where

D_i is the diffusion coefficient
c_i and z_i are the concentration and the charge of the species i, respectively
x is the distance
t is the time
ϕ is the electrical potential
v is the flow velocity

Additionally, the continuity equation holds

$$\frac{\partial c_i}{\partial t} = -\frac{\partial J_i}{\partial x}, \tag{22.11}$$

and the total current density (j) is given by

$$j = F \sum z_i J_i. \tag{22.12}$$

Explicit solutions are available for selected cases including the time-dependent selectivity behavior[27] and memory effects[28] of polymeric membrane ISEs. Since, however, solutions are only available for a limited number of cases, the finite-difference approach is often used instead.[27,29–31] Here, the continuous liquid phase is replaced by N thin layers with individual but uniform properties. The differentials in Equations 22.10 and 22.11 are replaced by the differences between neighboring elements:

$$J_{i,v/v+1}(t) = \frac{D_i \left(c_{i,v} - c_{i,v+1} \right)}{\delta} + \frac{F}{RT} \frac{0.5 z_i D_i \left(c_{i,v} + c_{i,v+1} \right)\left(\phi_v - \phi_{v+1} \right)}{\delta} + 0.5 \left(c_{i,v} + c_{i,v+1} \right) v, \tag{22.13}$$

$$\frac{c_{i,v}(t+\Delta t) - c_{i,v}(t)}{\Delta t} = \frac{J_{i,v-1/v}(t) - J_{i,v/v+1}(t)}{\delta}, \tag{22.14}$$

where

$J_{i,v/v+1}(t)$ is the flux from the v-th to the $(v+1)$-th element
$c_{i,v}$ and $c_{i,v+1}$ are the concentrations in the respective elements
$\delta = d/N$ is the thickness of each element
Δt is the time interval

The time-dependent concentration changes are then

$$c_{i,v}(\tau + \Delta\tau) = c_{i,v}(\tau) + \frac{D_i}{D_o} \left(c_{i,v-1} - 2c_{i,v} + c_{i,v+1} \right)\Delta\tau$$

$$+ 0.5 z_i \frac{F}{RT} \frac{D_i}{D_o} \left(c_{i,v-1} + c_{i,v} \right)\left(\phi_{v-1} - \phi_v \right)\Delta\tau$$

$$- 0.5 z_i \frac{F}{RT} \frac{D_i}{D_o} \left(c_{i,v} + c_{i,v+1} \right)\left(\phi_v - \phi_{v+1} \right)\Delta\tau + 0.5 \left(c_{i,v-1} - c_{i,v+1} \right)\omega\Delta\tau, \tag{22.15}$$

with the dimensionless expression for the time

$$\tau = \frac{D_0}{\delta^2} t, \tag{22.16}$$

and the flow velocity

$$\omega = \frac{\delta}{D_0} v. \tag{22.17}$$

D_0 is an arbitrary diffusion coefficient.

Finally, the potential difference, $\phi_v - \phi_{v+1}$, is given by (cf. Equations 22.14 and 22.15):

$$\phi_v - \phi_{v+1} = -\frac{F}{RT} \frac{\sum z_i D_i \left(c_{i,v} - c_{i,v+1}\right) + 0.5 D_0 \sum z_i \left(c_{i,v} + c_{i,v+1}\right)\omega - j\delta/F}{0.5 \sum z_i^2 D_i \left(c_{i,v} + c_{i,v+1}\right)}. \tag{22.18}$$

Among others, this approach has been used to calculate concentration profiles and liquid junction potentials at controlled electrolyte flow,[27] current-induced concentration changes of mobile ions in liquid-membrane electrodes,[27] as well as for the computer simulation of membranes exposed to different sample ions.[27] It was also applied to calculate non-steady-state responses of membranes under current polarization.[32] Also, the response behavior of ISMs operated in a thin-layer coulometric detection mode has been simulated with this model.[24,33]

22.3.4 Nernst–Planck–Poisson Theory

The Poisson equation must be used if the electroneutrality condition does not hold, that is, when the relevant dimensions are not much larger than the Debye length. Examples include very thin membranes including bilayers[34–36] and nanochannels.[37] Multiple nanochannel structures may also occur in macroscopic ion-exchange membranes. For example, Nafion (tetrafluoroethylene-perfluoro-3,6-dioxa-4-methyl-7-octenesulfonic acid copolymer) membranes with 20 vol.% of water have water channels with diameters between 1.8 and 3.5 nm.[38] Also, the potentiometric response of nanoporous reference electrodes has been explained by the presence of nanopores.[39]

The Nernst–Planck–Poisson model relates the electric potential ϕ to the space-charge density ρ, which is defined by the local concentration c_i of all ions i at the distance x. For monovalent ions ($z_i = +1$ or -1),

$$\frac{d^2\phi}{dx^2} = -\frac{1}{\varepsilon\varepsilon_o} \rho(x) = -\frac{F}{\varepsilon\varepsilon_o} \sum_i z_i c_i(x), \tag{22.19}$$

where
  ε is the dielectric constant
  ε_0 is the permittivity of the vacuum

For describing the potential and concentration profiles at equilibrium, it is combined with the Boltzmann equation for the equilibrium distribution of cations (m) and anions (x) as a function of the potential ϕ:

$$c_m(x) = c_m(x_o)e^{-\frac{F}{RT}\left(\phi(x)-\phi(x_o)\right)}; \quad c_x(x) = c_x(x_o)e^{+\frac{F}{RT}\left(\phi(x)-\phi(x_o)\right)}, \tag{22.20}$$

where x_0 is an arbitrarily chosen reference position.[17] Explicit results have been obtained for a series of limiting cases including lipid bilayers[34,35] and membranes with or without ionic sites (space-charge membranes) in symmetric or asymmetric arrangements.[17] An important result of the model by Morf is that no Nernstian response is expected for a thin membrane with internal solid contact.[17] So far, no experimental results are available for such a system.

Numerical simulations on the basis of the Poisson equation combined with the Nernst–Planck model offer another possibility of describing such systems.[14–16,29,40] In the approaches by these groups, the finite-difference method was used to simultaneously solve the Nernst–Planck and the Poisson equations. Due to the complex mathematical procedure, the computing times are very large. So far, this approach has not yet been used for describing the response of thin membranes. Morf proposed a simpler finite-difference approach by combining the Poisson and Nernst–Planck equations in a stepwise way.[17] The Poisson model was used to numerically evaluate the potential profile (Equation 22.23) after each iteration step of the concentration profile (Equations 22.21 and 22.22). The updated potential was then used in the next step:

$$J_i(n, n+1) = D_i \frac{c_i(n) - c_i(n+1)}{\Delta x} + z_i D_i \frac{c_i(n) + c_i(n+1)}{2} \frac{\Psi(n) - \Psi(n+1)}{\Delta x}, \tag{22.21}$$

$$\frac{\Delta c_i(n)}{\Delta t} = \frac{J_i(n-1, n) - J_i(n, n+1)}{\Delta x}, \tag{22.22}$$

$$\frac{F}{RT} \frac{\phi(n-1) - 2\phi(n) + \phi(n+1)}{(\Delta x)^2} = -\frac{F^2}{RT\varepsilon\varepsilon_o} \sum_i z_i c_i(n). \tag{22.23}$$

The corresponding calculations can be simply performed by using a spreadsheet program. The validity of this approach is confirmed by the exact reproduction of the theoretical description based on explicit solutions.[17] It has also been used to evaluate a double-pulse protocol applied for thin-layer coulometric ISM electrodes in order to reduce the level of interference.[33]

22.4 MINIATURIZATION OF ION-SELECTIVE AND REFERENCE ELECTRODES TO THE NANOSCALE

Miniaturization in potentiometry was driven at first by the use of ISEs for ion measurements in physiology and medicine.[41,42] Practically all cellular, physiological, and pathological phenomena that occur in cells are accompanied by ionic changes (e.g., H^+, Ca^{2+}, Na^+, K^+, NH_4^+, Li^+, and Cl^-), and, therefore, the introduction of miniaturized ISEs gave a major boost to experimental physiology. While for measurements in the extracellular space electrodes with sensing areas of a few micrometers in diameter are already sufficient, intracellular measurements benefited from further reduction in size of both reference and indicator electrodes down to the nanometer scale to ensure minimal perturbation to the cellular environment. Generally, it is considered that all methods for intracellular ion measurements, either optical or electrochemical, have advantages and pitfalls;[43] however, the fact is that the use of ion-selective micro- and nanopipettes for extra- and intracellular ion measurement has fallen back with the implementation of fluorescent and bioluminescent ion indicators. One of the obvious advantages of these optical probes is that they offer spatial resolution and improved response times as compared to those of miniaturized ISEs. Still ISEs have essential benefits in terms of enabling ion activity measurements, providing excellent selectivity, straightforward calibration (quantification), and minimal perturbation of the chemical environment.

Other applications such as the measurement of local concentrations in environmental analysis[44,45] or using potentiometric probes as detectors in various types of scanning electrochemical microscopies[46–49] including probes with dual functionality are emerging,[50] but are still largely limited to micrometer-size electrodes due to inherent difficulties of using nanoelectrodes.

22.4.1 Fabrication Methods to Restrict the Active Sensing Area of Potentiometric Sensors

The dominant methodology for fabricating ion-selective nanoelectrodes is based on pulling glass or quartz capillaries in a controlled way to obtain micro- or nanopipettes. These can be filled with an electrolyte such as 1–3 M KCl and electrically contacted with a Ag/AgCl wire to obtain miniaturized reference electrodes or inserting, additionally, an ISM into the tip to obtain an ISE (Figure 22.3a).[51] The advantages of the micro- and nanopipette design, beside the simple and cost-effective fabrication, is their suitability for localized measurements and compatibility with positioning devices.

Nanometer-sized nanopipette-based reference electrodes have been reported as early as the 1960s, while micropipette-type electrodes have been used already in the 1940s.[52] At the start, these electrodes were made with the purpose of measuring cell membrane potentials rather than serving as a classical reference to an ISE. Chowdhury has fabricated short tapered nanopipette-type reference electrodes with only ca. 50 nm diameter tips by capillary pulling (Table 22.1a).[53] Electron microscopy was used to assess the tip diameter, but these measurements should be considered an estimate as only the outside tip diameter can be imaged and, additionally, the electron beam can melt fine tips as noted by the author. Therefore, in general, the actual diameter of the tip opening might be even smaller than the reported values. The first report was closely followed by studies aiming at the reproducible fabrication of such electrodes. Since the increased electrical resistance featured as one of the main problems of such nanopipette references, the pulling procedure was optimized to obtain short taper nanopipettes, a design that is beneficial both in terms of reducing the electrical resistance and alleviating fragility issues of the nanopipettes. To further reduce the electrical resistance of the nanopipettes, in some cases, the tips were beveled at an angle using an abrasive surface (Table 22.1b).[54] By this procedure, the area of the tip was increased while the tips remained sharp enough to penetrate the cell wall.

Clearly, the increased electrical resistance proved to be a more severe problem for miniaturized ISEs than for reference electrodes. This is due to the much higher resistivity of the ion-selectivity

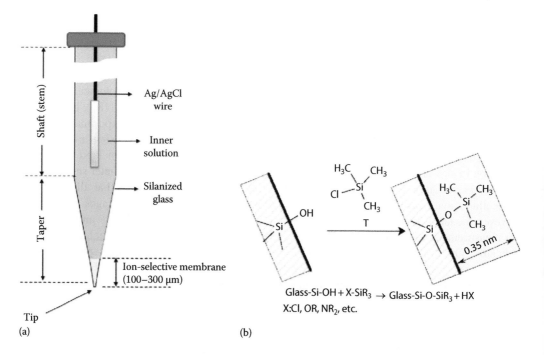

FIGURE 22.3   (a) Micro(nano)pipette-type ISE and (b) schematics of the silanization of the pipette surface.

TABLE 22.1

Representative Examples of Nanopipette-Type Reference Electrodes

Electrode	Size (nm)	Fabrication	Image	Reference
a Reference electrodes based on short tapered nanopipettes	~50	Obtained by pulling glass capillaries using a semiautomatic puller with air jet cooling.	$\theta \approx 11°$	[53]
b Beveled reference electrodes	~60	Fabricated from Pyrex glass capillaries by pulling. The tips are beveled to reduce the electrical resistance while conferring sharpness to enable penetration of cell walls.	250 nm	[54]
c Single-barrel reference nanoelectrodes	52	Nanopipettes fabricated by capillary pulling. The average diameter was 52 nm, with the largest not exceeding 70 nm and, occasionally, with tips as small as 20 nm.	Scale: 0.2 µm	[55]
d Double-barrel reference nanoelectrodes	70	Fabricated from theta capillaries by pulling.		[55]

membrane confined to the tip of the nanopipettes as compared with that of the highly concentrated electrolyte in the reference electrodes. Therefore, the reduction in size of the tip generally stopped at a few hundreds of nanometers as opposed to even 20 nm reported for nanopipette-type Ag/AgCl reference electrodes (Table 22.1c).[55] Since the electrical resistance of ion-selective nanopipettes, beside the tip geometry, also depends on the composition of the ISM, the resistances reported vary to a large extent. However, for guidance, a ca. 800 nm diameter neutral-carrier-based hydrogen ISE had a resistance of ca. 10^{11} Ω.[56] For comparison, a ca. 50 nm diameter nanopipette-based reference electrode filled with concentrated KCl has a resistance of roughly 10^8 Ω,[53] which is commensurable with that of common pH-sensitive glass electrodes. The use of various membrane additives or alternative membrane materials did not alleviate the problem of elevated resistances with ion-selective nanopipettes. Most notably, the use of lipophilic salts as organic electrolytes in ISMs, though proven to reduce the electrical resistance of uncharged ionophore-based macroelectrodes, failed to produce a significant resistance decrease when applied to micropipette-type ISEs.[51] As a consequence of their high electrical resistance, nanopipette-based ISEs require very high-input impedance voltmeters as well as, owing to their susceptibility to electrical noise, the use of Faraday shields and preamplifiers, the latter being generally integrated in the electrode holder. The closest possible integration of a preamplifier and an ISE was achieved by directly attaching a K^+-selective submicron diameter tip onto the gate of the amplifier to obtain a micro-ion-selective field-effect transistor.[57]

In order to stabilize the lipophilic ion-selective cocktail containing the selective ionophore and/or the ion exchanger, the hydrophilic glass surface must be turned into a hydrophobic one.[58] This proved to be an essential step to ensure proper lifetime and operation of the relevant ISEs; therefore, many procedures, most often employing various silane derivatives bearing hydrophobic groups to confer a hydrophobic coating upon silanization, were tested and applied.[59] During the silanization reaction, the silanol groups on the surface of the glass react with silane bearing hydrophobic groups and leaving groups (most often Cl or secondary amines [R_2N]), which results in the formation of Si–O–Si bonds (Figure 22.3b). Comparative studies suggest that the efficiency of silanization in terms of surface hydrophobicity increases in the order: trimethylchlorosilane, tributylchlorosilane, (dimethylamino)trimethylsilane, hexamethyldisilazane, dimethyldichlorosilane, and bis(dimethylamino)dimethylsilane.[60] Since these silanizing reagents react readily with water, it is essential to dry the glass surface; therefore, the silanizations are generally performed as a continuation of the drying step.

After proper hydrophobization, the ion-selective cocktail is generally loaded through the tip by capillary action and/or applying a slight suction to form a plug of ca. 100 μm at the very end of the pipette tip. The inner filling solution containing the primary ion is backfilled either before or after the ion-selective cocktail is loaded into the nanopipette and the construction is completed by inserting a Ag/AgCl wire and sealing the open end. Undoubtedly, filling the micropipette ISEs is the most difficult step in their preparation, which is rendered even more challenging if the pipette tip is reduced to submicron diameters.

The smallest ion-exchanger-based ISE was reported by Greger et al. (though no electron micrographs were provided in support) consisting of a 100 nm diameter tip filled with a commercially available ion exchanger to be used for the measurement of cellular K^+ activities (Table 22.2c).[63] ISEs based on ion exchangers were reduced to nanoscale dimensions also in double-barrel configuration, in which the joined reference and ISE had a total tip diameter of ca. 300 nm. The double-barrel configuration has the advantage that both the indicator and the reference electrodes are placed in the cell with a single puncturing. In contrast, if the reference electrode is outside the cell, the transmembrane potential is, additionally, part of the measured total potential and should be corrected for.

The size reduction of the nanopipette-type electrodes based on neutral ionophores stopped generally at a few hundreds of nanometers. The smallest tip diameter for a single-barrel nanopipette-based ISE of around 400 nm was reported for a Ca^{2+}-selective electrode featuring an ISM based on the ionophore ETH 1001 ((–)-(R,R)-N,N'-bis-[11-(ethoxycarbonyl)undecyl]-N,N',4,5-tetramethyl-3,6-dioxaoctanediamide) and was applied for intracellular Ca^{2+} measurements (Table 22.2a).[61] Using the same ionophore, the tip diameter of Ca^{2+}-selective electrodes was reduced further by using a double-barrel configuration made by pulling concentrically assembled round and triangular-shaped capillaries (Table 22.2d).[64] The average diameter of the double-barrel nanopipettes was 89 nm as determined by scanning electron microscopy (SEM), with the diameter of the inner ion-selective barrel within 50–70 nm and, as such, likely to be the smallest neutral-carrier-based ion-selective nanopipette.[66]

Submicron-size *solid-membrane ion-selective nanoelectrodes* were introduced already in the 1970s. Some of the earliest electrodes were fabricated by vacuum sputtering of a 250 nm thick Ag layer onto the tips of nanopipettes of ca. 300 nm tip diameter, followed by sealing the tips, except their very end (ca. 2–5 μm; Table 22.3b).[67] Thus, the smallest cross section of these electrodes used

TABLE 22.2
Representative Examples of Nanopipette-Type Liquid-Membrane Ion-Selective Electrodes

Electrode	Size	Fabrication	Image	Reference
a Ca^{2+}-selective nanoelectrode based on uncharged carrier	o.d. ~ 400 nm	Glass capillary filled with Ca^{2+}-selective membrane based on ETH 1001 ionophore.	Not available	[61]
b Ca^{2+}-selective nanoelectrode based on Ca^{2+}-selective resin	o.d. <180 nm	Made by pulling from omega-dot fiber-containing borosilicate capillary.	0.2 μm	[62]
c K$^+$-selective nanoelectrode based on ion exchanger	~100 nm	Glass capillary pulled to obtain nanopipettes with outer tip diameters of <100 nm and short tapers (~18 mm).	Not available	[63]
d Ca^{2+}-selective single- and double-barrel nanoelectrodes based on uncharged carrier (ETH 1001)	89 nm (double barrel), 50–70 nm (ISE)	Single- and double-barrel electrodes made by pulling concentrically assembled round and triangular-shaped aluminosilicate capillaries.	0.72 mm plane length Ø 1.2 mm	[64]
e Na$^+$-selective double-barrel ion-selective microelectrodes	~1 μm	Double-barrel ISEs fabricated from borosilicate theta glass capillaries with an improved method for better control.		[65]

TABLE 22.3

Representative Examples of Nanopipette-Type Solid-State Ion-Selective Electrodes

Electrode	Size (nm)	Fabrication	Image	Reference
a Submicrometer-size pH electrodes	~500	Micropipettes were pulled from quartz capillaries and beveled to submicron sharp tips. pH-sensitive glass was stretched to a very thin film pushed through by the beveled micropipette.		[69]
b Solid-state Ag$^+$ nanoelectrodes for Cl$^-$ measurements	~800	The tips of glass nanopipettes of ca. 0.3 μm diameter were coated under vacuum with a ~0.25 μm thick layer of spectroscopic grade silver. The tip was sealed in a tapered glass tube except the very end of the tip (~2–5 μm).		[67]
c Solid-state AgI nanoelectrodes for Ag$^+$ and I$^-$ measurements	~200 (Ag tip)	Glass capillaries with a 50 μm Ag wire are pulled. The Ag tip is exposed by removing a small part of the glass, which is then coated with a layer of AgCl (electrolytically) and Nafion.		[68]

for Cl⁻ measurements was likely to be of 800 nm in diameter. A more convenient procedure was proposed later based on directly pulling silver wires (Ø of 50 μm) into glass nanopipettes of 1.2 mm outer diameter and modifying the silver surface electrolytically (Table 22.3c).[68] After pulling, the metal wire was sealed in the glass at the pipette tip. To expose the tip of the wire, the glass at the tip is either etched away by using 40% HF or removed by micropolishing. The tip was further coated with a AgI layer by several potential cycles between +0.3 and −0.6 V versus Ag/AgCl in a solution containing 0.01 M KI and 0.1 M KNO₃. To stabilize the layer, that is, to avoid the fast dissolution of the minute amount of AgI precipitate, ultimately, a Nafion layer was deposited on top of the AgI coating. Such electrodes exhibited stable potentials for hours having close to Nernstian response for Ag⁺ (56 mV/decade) and slightly sub-Nernstian response for I⁻ (−47 mV/decade).

Another ingenious variation of obtaining an electrode embedded in the nanopipette barrel and subsequently modifying it with an ion-selective layer was reported for the preparation of nanometer-size pH electrodes. First, the barrel was filled with carbon by the pyrolytic decomposition of butane and subsequently, a thin layer of pH-sensitive hydrous iridium oxide was electrodeposited.[50] These electrodes were fabricated within double-barrel nanopipettes, with one of the barrels accommodating the pH electrode of ca. 100 nm in diameter, while the other barrel of the same size was filled with an electrolyte solution and electrically connected with a Ag/AgCl electrode. Such electrodes with dual functionality, that is, pH measurement and conductance cell, enabled high-resolution pH mapping by combining pH measurements with scanning ion conductance microscopy (SICM).[50]

Nanopipettes were also used to fabricate pH electrodes by first preparing and beveling the nanopipettes and then pushing them through an extremely thin layer of melted glass that closed the tip of the nanopipette.[69]

The fragility of the nanopipette-based potentiometric electrodes is an inherent consequence of their small size. Most notably, it is determined by the thickness of the glass/quartz at the tip. Generally, it is considered that the ratio of the inner and outer diameters of the original capillary remains fairly constant along the nanopipette with the glass wall generally becoming thinner at the very end of the tip.[70] This means that the thickness of the glass or quartz at the tip is comparable with the diameter of the tip and, accordingly, it is in the lower nanometer range, which explains the lack of robustness of such nanopipette-type electrodes. This problem can be addressed by using recessed type (micro[nano]cavity-based) ISEs. These were introduced first for fabricating solid-contact ISEs (SC ISEs) with diameters in the micrometer range by Sundfors et al.[71] and later applied by Shim et al.[72] to the construction of a wide range of nanometer-sized ISEs with solid-state and liquid membranes. The core element of the preparation is a sacrificial sharpened metal wire (Pt) sealed in a thick wall soda lime glass capillary with the glass polished away to expose a Pt nanodisk electrode (Table 22.4a). The controlled electrochemical etching of the metal wire in a 1.2 M CaCl₂ solution created a conical nanocavity of ca. 30 μm in the glass contacting the outer solution through an orifice of less than 500 nm while the Pt disk at the bottom of the cavity enabled the electrical contact with the subsequently deposited ISM. The concept was demonstrated through the fabrication of pH, Cl⁻, and K⁺-selective electrodes. In the former case, IrO$_x$ was electrodeposited on the surface of a Pt disk (Table 22.4a) while for Cl⁻ measurements, a Ag layer was electrolytically deposited onto the Pt disk and chlorinated with FeCl₃ to form a AgCl top layer. The preparation of K⁺-selective electrodes involved the subsequent deposition of a hydrogel-based liquid contact and, finally, of the valinomycin-based K⁺-selective liquid membrane on top of the Pt/Ag/AgCl layer (Table 22.4b). While the robustness of the nanocavity-type K⁺-selective electrodes was clearly improved and the electrodes were functional as potentiometric probes, it should be noted that the analytical performance parameters were considerably worse than those of the *conventional* nanopipette-type electrodes, that is, they exhibited sub-Nernstian slopes and a lifetime of only a few hours.

Wire-type ISEs are a somewhat less demanding way to prepare solid-state ion-selective nanoelectrodes. In this case, first an etched carbon fiber[73] or metal wire[74] is coated with an ion-selective layer and sealed in the tip of a micropipette such as that a part of the etched wire protrudes from the micropipette (Table 22.4c). The exposed lateral part of the wire of such electrodes, while having

TABLE 22.4

Representative Examples of Recessed (Nanocavity-) and Protruding Wire-Type (c) Solid-State Ion-Selective Nanoelectrodes

Electrode	Size (nm)	Fabrication	Image	Reference
a Solid-state IrO_x nanoelectrodes for pH measurements	Ø < 500	A sharp Pt tip is melted into a thick wall glass capillary. Etching away the exposed Pt tip gives way to electrolytic deposition of IrO_x.	Glass; Cavity; IrO_x; Pt	[72]
b Nanocavity-type liquid-membrane K^+-ISE based on valinomycin	Ø < 500	See previous text. K^+-ISE is formed by further deposition of an internal Ag/AgCl/KCl (hydrogel) reference and the ISM.	Glass; Pt; IS membrane; Hydrogel; AgCl; Ag	[72]
c Solid-state pH electrodes based on (1) polyaniline thin film electrodeposited onto ion-beam etched carbon fiber	~100–500 C fiber		Cu; Epoxy sealing; Glass micropipette; Etched W wire or carbon fiber coated with a pH sensitive layer	[73]
(2) W/WO_3 nanowire	~500–800 W wire			[74]

their tip in the nanometer range, sets them apart from true nano-ISEs. Sensors for pH were reported either by coating the ion beam–etched C fiber with a thin layer of polyaniline by electropolymerization or using an etched W wire with an electrolytically generated layer of WO_3. The analytical performance of these sensors approaches that of the conventional glass electrodes, though sub-Nernstian responses were typical for metal–metal oxide electrodes.

22.4.2 Characteristics of Nanoscale ISEs

In general, potentiometric electrodes upon downscaling to the micro- or nanodomain do not exhibit the same advantageous properties as voltammetric electrodes. While some of the negative consequences of nanoscaling potentiometric electrodes could be corrected by alternative fabrication methods, they are generally inferior to their conventional size counterparts in terms of mechanical robustness, lifetime, electrical resistance, and dynamic range.

Nano-ISEs fabricated by any of the methods described previously seem to suffer from very short lifetime, that is, ranging from a few hours[72] up to ca. 1 month (ca. 1 μm tip diameter Na+-ISE).[75] Thus, such electrodes are difficult to commercialize, and the task remains with the user to freshly prepare the electrodes, which constitutes a major limitation to their wider spreading. There are several reasons that might determine the shorter lifetime, that is, mechanical defects, electrical shunting, and membrane poisoning, but the most severe and ubiquitous factor seem to be the loss of membrane constituents.[66,76] This is a common problem with all types of nano-ISEs as the amounts of membrane components are minute and the mass transfer rate of these components diffusing away from micro- and nanoelectrodes is extremely high. Still, the situation is not completely clear, as in case of nanopipette ISEs, there is a larger reservoir of components, that is, the height of the membrane plug at the end of the nanopipette is ca. 100 μm as compared with an orifice of 100–500 nm. Finite-element simulation of the loss of a membrane component with a diffusion coefficient in the membrane and aqueous solution of 10^{-8} and 10^{-5} cm^2/s, respectively, from a nanopipette-type ISE (tip Ø: 100 nm; half opening angle: 5°, height of the IS membrane: 100 μm) is shown in Figure 22.4a. The boundary condition for the finite-element simulation was determined by the distribution/partition coefficient (K) between the aqueous phase and the membrane phase. The value of K was calculated from the logarithm of the partition coefficient between octanol and water (log P): log $K = (0.8 \pm 0.1)$ log $P + (0.4 \pm 0.4)$.[77]

The simulation shows that with a log P value above 8, the phase-boundary concentration of the lipophilic component in the ISM nanopipette stays above 90% of its initial concentration for at least 30 days. However, log P of some of the most extensively used uncharged ionophores, that is, ETH 1001, nonactin, and valinomycin, is 6.9–7.5, 5.8, and 7.8–8.6, respectively, and hence at or below this limit.[78] Additionally, the loss of plasticizer should also be considered as, for instance, o-nitrophenyl octyl ether (o-NPOE), which has also a relatively low lipophilicity (log $P = 5.9$). However, the estimation of its loss from a nanopipette needs a simulation with different input parameters as the self-diffusion of plasticizers and their concentration in the membrane is much higher than that of other membrane constituents, that is, of ionophore, ion exchanger, and lipophilic additives. A simple simulation not considering the change in viscosity upon the plasticizer release is shown in Figure 22.4b and is highly indicative of the severity of plasticizer loss, which negatively affects both the ion-selective response and the resistance of the electrode.

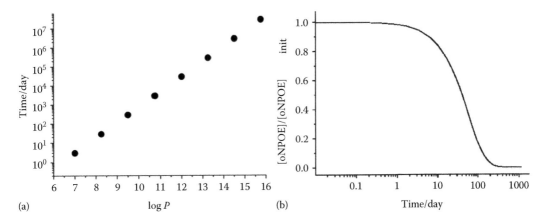

FIGURE 22.4 Loss of membrane components from a nanopipette-type ISE (tip Ø: 100 nm; half opening angle: 5°, height of the IS membrane: 100 μm, for other conditions, see main text): (a) Time until the phase-boundary concentration of the lipophilic component drops below 90% of the initial concentration. (b) Change in the normalized phase-boundary concentration of o-NPOE.

Thus, despite the combination of a large membrane reservoir and a small orifice typical of nano-pipette-type ISEs, the short lifetime of such electrodes seems to be well explained by a significant loss of membrane components.

22.4.3 NANOMETER-THICK SENSING MEMBRANES: BILAYER LIPID MEMBRANES

Lipid bilayers having a thickness of around 5 nm were probably the earliest nanoscale systems studied by potentiometry. Although initially the research was motivated by the understanding of ion transport through biological membranes, some potentiometric studies are also available. Mueller and coworkers pioneered the preparation of planar bilayer lipid membranes (BLMs) having a diameter of about 1 mm.[79–81] Such membranes, which are spontaneously formed from a solution of lipids, are called black lipid membranes because under reflected light the interference colors of the initially formed thicker membranes disappear and are replaced by nonreflecting (*secondary black*) when the bilayer is formed. After the discovery of the ion-selective behavior of some antibiotics by Stefanac and Simon,[82] Muller and Rudin demonstrated that their behavior is the same in BLMs.[83] Later, among others, the group of Eisenman extensively studied the carrier-mediated electrical properties of BLMs both theoretically[35] and experimentally.[84] An excellent review is available, which covers sensing devices, including potentiometric ones up to 2001.[85]

The group of Umezawa has started to use planar bilayer membranes, which are prepared by the so-called folding method.[86] Here, the solution is placed on each side of a Teflon film having a small aperture of about 200 μm diameter. Two monolayers are formed on each side of the Teflon film, which are unified to a bilayer at the aperture. This technique has the advantages: it is possible to use different lipids including synthetic ones and asymmetric bilayers can be also prepared if different solutions are used on the two sides of the Teflon sheet. Usually, in order to decrease the permeability of the bilayers to small molecules and to increase their stability, some additives are used, for example, 25 wt% of cholesterol. The membranes contained different concentrations of the K^+-selective antibiotic valinomycin. The bilayer membranes showed a cationic response even in the absence of charged sites.[36] This behavior is in accordance with the theory of Morf.[17] An increasing response range was found by increasing the concentration of valinomycin.[36] The addition of negative sites improved slightly the response, but, interestingly, a cationic response, though much suppressed, was also found in the presence of cationic sites (positively charged lipids) in the membrane. This behavior is in contrast to macroscopic membranes, for which no cationic response occurs in the absence of anionic sites. Based on this, it was postulated that such ultrathin membranes do not need ion-exchange sites beside the ionophore for a Nernstian potentiometric response as they do not confer permselectivity by themselves to the membrane, but their effect is due to simple charge attraction/repulsion with little relevance to permselectivity. Later, the same group investigated six other ionophores for Na^+, K^+, and Ca^{2+} and found only in one other case a Nernstian response in the absence of ionic sites.[87] More recently, Liu et al. prepared bilayer membranes in micron- to nanosized apertures and investigated their potentiometric responses using dibenzo-18-crown-6 as ionophore without adding ionic sites. Response slopes for K^+ were pronouncedly sub-Nernstian with 35, 20, and 16 mV/decade for apertures of 350 μm, 40 μm, and 660 nm, respectively.[88]

Other methods of preparing bilayer membranes, which lead to more robust systems and longer lifetimes include filter-supported, solid-supported, and gel-supported membranes.[85] Solid-supported membranes doped with ion-selective antibiotics were only used in conjunction with amperometric (porous Teflon membrane with nonactin for NH_3)[89] or conductivity (steel-supported membrane with valinomycin for K^+)[90] measurement modes. Agar gel–supported bilayer membranes with incorporated valinomycin showed Nernstian responses for different alkali cations.[91]

The potentiometric response of gel-supported bilayer membranes doped with different uncharged calixarenes was very disappointing.[92] Only in some cases, an irrelevant response slope of about 15 mV/decade could be observed, and this is only for a very restricted concentration range of about one order of magnitude. No response at all was obtained with a calix[4]aryl tetraester, similar to the one that has been so successful as Na^+-selective ionophores in bulk membranes.[93,94] In contrast, gel-supported bilayer membranes with 3-decyl-1,5,8-triazacyclodecane-2,4-dione, a cyclic amine, which is supposed to act as a charged ionophore,[6] showed satisfactory response to monohydrogen-phosphate[95] with the observed selectivity coefficients comparable to those of bulk membranes.[96]

The theoretical implications of using ultrathin sensing membranes is of utmost importance as these so-called space-charge membranes have a different behavior than conventional thick membranes with bulk electroneutrality and thin space charge regions only at their boundaries. In fact, the previously mentioned experimental studies could not benefit from the recent comprehensive theoretical description of the potentiometric behavior as well as potential and concentration profiles in ultrathin non-electroneutral membranes by Morf.[17] This study showed clearly that only space-charge membranes contacted with aqueous solutions on both sides will exhibit a Nernstian response. While unfortunately such membranes are rather unstable mechanically, their stabilized counterparts, that is, solid-contacted thin membranes, are theoretically predicted to have a sub-Nernstian slope.

Therefore, the use of BLMs for ion-selective potentiometry while an important chapter in the history of ISE development has nowadays little practical significance when compared with the robustness and analytical performance of conventional ISEs.

22.5 POTENTIOMETRIC NANOPORE SENSORS: NEW CONCEPT FOR THE MINIATURIZATION OF ION-SELECTIVE ELECTRODES

The miniaturization of traditional solvent polymeric membrane-based potentiometric sensors has obvious limitations in terms of reduced lifetimes and extremely high electrical resistances. Therefore, a completely different construction is needed to overcome these limitations. One approach that takes advantage of the unique opportunities provided by nanostructures and shown to be of perspective is based on the use of nanoporous membranes. The concept behind *nanopore potentiometry* is based on shrinking the nanopore restriction to approach molecular dimensions so that the physical-chemical properties of the inner wall of the nanopores determine the ion transport through the nanopore. Thus, in this approach, there is no solvent polymeric membrane, but solely the surface functionality of the nanopores is generating the ion-selective behavior.

The first step to achieve ion-selective response is to make the nanopores permselective, that is, to establish charge sign selectivity. This translates in having the nanopore surface either negatively or positively charged to induce cation or anion permselectivity, respectively. The surface of many materials is charged, either due to ionizable surface groups or due to adsorption of various ions from aqueous solution. However, the challenge to establish proper permselectivity is to control the surface charge density and the smallest restriction of the nanopores. To simplify the discussion, we will consider nanopores of cylindrical symmetry, that is, either cylindrical or conical, so the smallest constriction of circular geometry can be characterized by its diameter, d. In an aqueous electrolyte, the surface charge of the nanopore material will generate an electrical double layer, the thickness of which is characterized by the Debye length. As the Debye length is a rough measure of the range in which the surface charge of the wall is electrostatically compensated by an excess of counterions, a permselective behavior is expected if the radius of the nanopore restriction is smaller than the Debye length. On the contrary, for radii larger than the Debye length, the permselectivity is lost. As hydrophilic nanopores are accessible to aqueous solution, beside the surface charge density and the pore diameter, the ionic strength of the solution is also relevant for establishing a permselective potentiometric response. Thus, a charged nanoporous membrane will generate a membrane potential that is determined by the ion-exchange equilibrium in case of permselective membranes, while for membranes with pore radii exceeding considerably the Debye length, the measured potential is in fact the liquid–liquid junction potential at the interface of two miscible electrolyte solutions separated physically by the membrane.

A change from one type of response to the other may happen for the same nanoporous membrane in contact with solutions of different ionic strengths. For example, it has been shown recently that the potential of reference electrodes based on nanoporous glass plugs as salt bridges (e.g., Vycor with pore diameters of ca. 4–20 nm) is influenced even at moderately high ionic strengths (lower than 100 mM) by the composition of the sample.[39] These influences could not be explained by changes in the liquid junction potentials, and are attributed to the cationic response of the negatively charged pores that become cation-permselective at lower ionic strengths.

The feasibility of designing permselective nanopores was demonstrated by Martin's group[97] (cf. Figure 22.5[98]). They used electrolessly gold-plated track-etch membranes featuring straight-through cylindrical-shaped pores of a density of 6×10^8 pores/cm^2 within a ca. 6 µm thick polycarbonate membrane. By controlling the time of the gold plating, gold nanopores of various diameters were prepared. The gold spontaneously adsorbed anions from an aqueous solution, which imparted cation-exchanger properties to the gold nanoporous membrane.

Because the gold nanoporous membrane is electrically conductive, charge injection by applying an external potential was also shown to provide means for inducing either anionic or cationic permselectivity. The potentiometric response was derived using as starting point the expression of the liquid junction potential (E_j) between two miscible electrolytes composed of 1:1 salts:

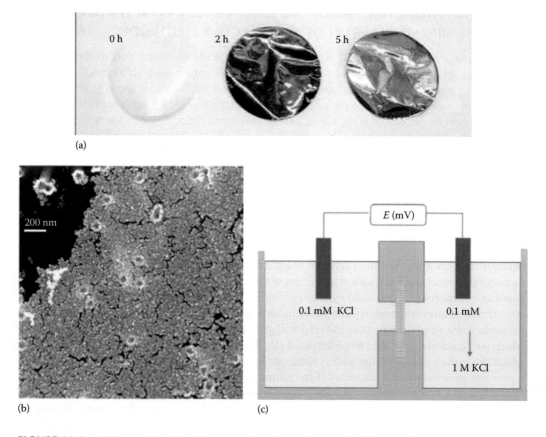

FIGURE 22.5 (a) Photos of a polycarbonate track-etch membrane prior (0 h) and after electroless gold plating for 2 and 5 h, respectively; (b) scanning electron micrograph of the gold-plated nanopores (50 nm diameter track-etch membranes coated for 190 min. Note the larger pores for their better visibility and that the outer gold coating is removed in the left corner to show the gold nanopores within the membrane); and (c) Schematic of the setup for measuring the potential of nanoporous membranes. The membranes are separating two electrolyte solutions each accommodating a Ag/AgCl reference electrode.

$$E_j = \left(t_+ - t_-\right)\frac{RT}{F}\ln\frac{a_{out}}{a_{in}}, \tag{22.24}$$

where

t_+ and t_- are the transference numbers for positive and negative ions, respectively

a_{out} and a_{in} are the activity of the electrolyte on the outer, sample side and inner side of the membrane, respectively

For classical microporous membranes and 1:1 salts ($z_+ = z_- = 1$ and $C_+ = C_-$), the relative magnitude of the transference numbers for the positive and negative ions is determined by the ratio of their mobility:

$$t_+ = \frac{|z_+|u_+C_+}{|z_+|u_+C_+ + |z_-|u_-C_-} = \frac{u_+}{u_+ + u_-}; \quad t_- = \frac{u_-}{u_+ + u_-}, \tag{22.25}$$

and accordingly, the magnitude of the liquid junction potential is also determined by the relative mobility of the ions. However, if a permselective nanoporous membrane separates the two electrolytes, this will change the transference numbers for the positive and negative ions beyond the values that would result from their relative mobility. Thus for an ideal cation-permselective membrane $t_+ = 1$ and $t_- = 0$, the membrane potential (E_m) will be

$$E_m = \frac{RT}{F}\ln\frac{a_{out}}{a_{in}} \tag{22.26}$$

as the transport of anions is fully restricted in the negatively charged nanopores. Similarly, for an ideally anion-permselective membrane ($t_- = 1$ and $t_+ = 0$),

$$E_m = -\frac{RT}{F}\ln\frac{a_{in}}{a_{out}}. \tag{22.27}$$

This simple intuitive model predicts Nernstian response for charged nanoporous membranes with ideal permselectivity, and sub-Nernstian potentiometric responses for membranes with *partial* permselectivities.

A more in-depth elucidation of the mechanism of permselective membranes can be done by using the Nernst–Planck/Poisson equations. As a model system, the gold nanoporous membranes introduced by Martin et al. were used; however, the surface charge was established by their chemical modification with self-assembled monolayers (SAMs) of chemisorbed electrically charged thiol derivatives. The average pore diameters of such membranes can be very precisely determined with gas permeation experiments based on the kinetic theory of gases[37,99,100]:

$$r = \sqrt[3]{\frac{3}{4}Q\left(\frac{MRT}{2\pi}\right)^{1/2}\frac{l}{n\,\Delta p}}, \tag{22.28}$$

where

r is the effective pore diameter (cm)

Q is the gas flux (mol/s)

Δp is the pressure drop (dyne/cm^2)

M is the molecular weight of the gas

R is the universal gas constant (erg/K/mol)

T is the absolute temperature (K)
l is the thickness of the membrane (cm)
n is the number of pores

Both anion- and cation-permselective membranes were synthesized by chemically modifying gold nanopores, having radii ranging from 3 to 25 nm, with thiol derivatives bearing either a quaternary ammonium (N,N,N-trimethyl-(11-mercaptoundecyl)ammonium, QT) or a sulfonic acid (10-mercaptodecane-1-sulfonic acid [MDSA]) functionality, respectively.[101] A simple experimental setup was used for potentiometric measurements in which the nanoporous membrane separates two electrolyte compartments each accommodating a Ag/AgCl reference electrode and a KCl solution. The EMF of the cell was measured as a function of the KCl concentration in the outer (sample) compartment at a constant 0.1 mM KCl concentration in the inner compartment. As shown in Figure 25.6, a potentiometric cationic response was observed for bare gold and MDSA-modified membranes and anionic responses for the QT-modified membranes. Very importantly, similar to conventional polymeric ion-exchanger electrodes, a co-ion interference region, that is, Donnan failure, was found at high concentrations. However, in contrary to polymeric membranes, the hydrophilic nanopores are wetted by aqueous electrolytes. Therefore, for nanopores, the deviation from the Nernstian response at high ionic strengths is due to an effective screening of the surface charge of the nanopores (decreased Debye length), leading to a gradual loss of permselectivity. The onset of the deviation is determined by the diameter of the nanopores and their surface charge density.

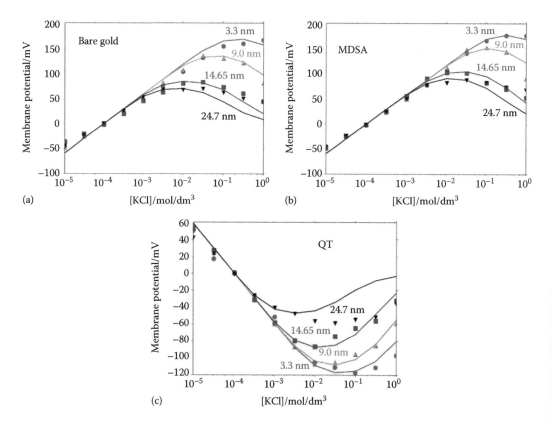

FIGURE 22.6 Potential response of (a) nonmodified, (b) MDSA-modified, and (c) QT-modified gold nanoporous membranes as a function of KCl concentration outside the electrodes (the concentration of the inner solution is fixed at $[KCl] = 10^{-4}$ M). Symbols represent measured potentials, while simulated responses are depicted as lines. The pore radii are indicated on the graphs.

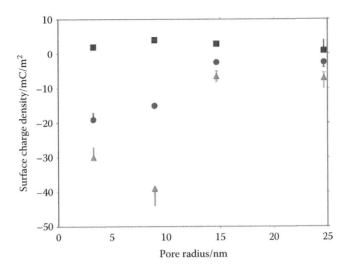

FIGURE 22.7 Effective surface charge densities on the inner surface of unmodified (•), MDSA-modified (▲), and QT-modified (■) gold nanopores as a function of the pore radius.

Thus, membranes with larger pore radii lose permselectivity and deviate from the Nernstian behavior at lower concentrations than those with smaller pore radii. The experimental responses were correlated with the numerical solutions of the Nernst–Planck/Poisson equation obtained by finite-element simulation. By inserting the experimentally determined pore radii, the surface charge density of the pore interiors could be determined. Figure 22.6 shows an excellent correlation between the experiments and the simulation. Significant deviations were found only at pores with large radius (25 nm) and for concentrations exceeding 0.1 M where due to significant transmembrane diffusion, the potential responses drifted considerably.

Thus, a simple potentiometric calibration in combination with the Nernst–Planck/Poisson equation allows the determination of the surface charge densities of nanoporous membranes as summarized in Figure 22.7 for the three types of nanopore modifications. The surface charge densities of MDSA-modified membranes, ranging from −6.5 up to −40 mC/m², were found to be significantly and consistently higher than that of the bare gold nanopores with the same pore diameter. The effect of MDSA is also visible in Figure 22.6b, as the MDSA modification of the nanopores extends the range of Nernstian potential response up to 0.1 M, while unmodified gold nanoporous membranes showed deviations from Nernstian response even at concentrations as low as 1 mM. Assuming that all negative charges originate from the singly charged MDSA molecules, the surface concentration of MDSA within the nanopores ranged from 7.4×10^{-12} mol/cm² ($r = 24.7$ nm) to 3.1×10^{-11} mol/cm² ($r = 3.27$ nm). These values correspond to less than 4% of a full monolayer coverage; thus, a densely packed monolayer is not formed within the nanopores. The surface charge densities of QT-modified nanopores (0.9–4.0 mC/m²) are much lower than those determined for negatively charged nanopores. However, in this case, the surface concentration of QT cannot be estimated given that most likely the effective positive surface charge is decreased by the charge neutralizing effect of the adsorbed anions on unmodified surfaces.

Simulations of the negatively charged nanopores based on the Nernst–Planck/Poisson equation show that as the KCl concentration in the outer solution increases, the outer phase-boundary Cl⁻ concentration in the nanopore increases gradually. However, in the Nernstian potential response range of the nanoporous membrane, the Cl⁻ concentration is orders of magnitude lower (note the logarithmic scale) than that of K⁺, which matches the surface charge equivalent concentration (c_{eq}) demarked with a horizontal dashed line in Figure 22.8a. The surface charge equivalent concentration is a conversion of the surface charge concentration on the inner pore wall to an equivalent volume

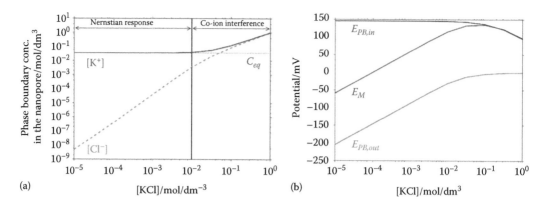

FIGURE 22.8 (a) Cross-section averaged phase-boundary concentrations of K^+ and Cl^- in the nanopore at the outer solution/pore interface as a function of KCl concentration in the outer solution (inner solution: 0.1 mM KCl). (b) Simulated membrane potential and phase-boundary potentials at the inner and outer interfaces of the nanopore. The simulations refer to a 6 µm long and 9 nm radius nanopore having a surface charge density of -15.0 mC/m^2.

concentration of monovalent ion; for example, a surface charge concentration of -15 mC/m^2 is converted to the amount of monovalent ion of equal charge in the volume of the pore, which corresponds in the particular case of a 6 µm long and 9 nm radius nanopore to a concentration of 0.035 M. The importance of this quantity is revealed by the fact that the onset of deviation from the Nernstian behavior occurs at outer KCl concentrations close to c_{eq}. At outer KCl concentrations higher than this value, both the K^+ and Cl^- concentrations increase in the nanopore, and ultimately, their phase-boundary concentrations in the membrane become equal, which indicates the total breakdown of the cation permselectivity. This corresponds to the negative slope region (Figure 22.8b) of the membrane potential, which ultimately should reach the liquid junction potential value (~ 0 mV in the case of KCl with close to equal mobilities of its counterions) at a complete breakdown of the permselectivity.[102]

Studies on solid-membrane electrodes stipulated rather clearly the basic requirements for membrane materials that, besides being practically insoluble in the sample solution, were expected to exhibit (1) rapid and reversible ion exchange of the primary ion at the membrane–solution interface, (2) ionic conduction, and (3) should be nonporous for the sample solution. This latter requirement is in obvious contradiction with the hydrophilic nanoporous construction. Thus, while the feasibility of generating cation-permselective nanopores has been demonstrated, further important prerequisites for an ion-selective response include the incorporation of a selective complexing agent, that is, an ionophore, and the establishment of a highly hydrophobic environment in the nanopores. As shown in Figure 22.9a, this concept has been demonstrated for an Ag^+-selective electrode. Thus, all functional components of a conventional polymeric ISM were matched with suitable thiol or disulfide derivate counterparts able to form a SAM on the surface of the nanopores. For a proof of principle, a synthetic Ag^+-selective thiacalixarene derivative bearing dithiolane moieties (SS-Ag-II) was used to induce Ag^+ selectivity.[103] Cation-exchanger sites were generated by using MDSA while, to take advantage of the latest results showing the superiority of fluorous ISMs,[104] a perfluorinated thiol (PFT) derivative was designed to confer hydrophobicity to the Au nanopores. The major challenge consisted in the modification of the interior of gold nanopores with three compounds of widely different polarities in the right a ratio to provide ion-selective response, for example, the ionophore to ion-exchanger ratio should be larger than 1, as the relative surface concentrations of the thiol derivatives do not necessarily reflect their composition in the solution phase used for the surface modification. However, impedance spectroscopy showed clear evidence for the formation of a mixed PFT–MDSA layer, the two derivatives with the largest polarity difference, in the nanopores.

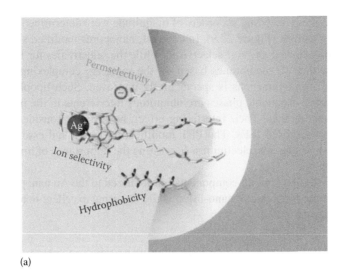

(a)

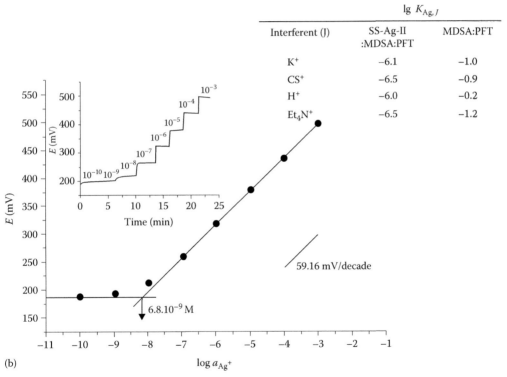

(b)

Interferent (J)	lg $K_{Ag,J}$	
	SS-Ag-II :MDSA:PFT	MDSA:PFT
K$^+$	−6.1	−1.0
CS$^+$	−6.5	−0.9
H$^+$	−6.0	−0.2
Et$_4$N$^+$	−6.5	−1.2

FIGURE 22.9 (a) Schematics of gold nanopore modification for selective ion-sensing and (b) the calibration curve with the respective potential-time traces and selectivity coefficients as insets.

Including also the Ag$^+$-selective ionophore SS-Ag-II in the modification trials revealed that the Au membranes could only be modified in one step with all components dissolved in methanol and the optimal composition of the modifying solution was SS-Ag-II/MDSA/PFT at a total concentration of 0.2 mM, and molar ratios of 11:10:1, respectively. The calibration curves for Ag$^+$ revealed detection limits in the lower nanomolar concentration range associated with fast and drift-free Nernstian potential responses (Figure 22.9b).

Excellent selectivities exceeding six orders of magnitude were determined for a range of representative interfering cations (Figure 22.9). In contrast, nanopores modified similarly, but without the ionophore, hardly showed any Ag^+ selectivity. While the selectivities for K^+ and H^+ lag behind those determined with PVC membranes based on the same Ag complexing unit[105] the selectivity over quaternary ammonium ions is spectacularly improved. Such lipophilic organic cations with high affinity for hydrophobic phases are ubiquitous interferents in the most commonly used solvent polymeric membranes. Their interfering effect, however, is considerably reduced in the PFT-modified nanopore-based ISEs. The PFT modification was found essential also for proper potentiometric response and ion selectivities, suggesting the importance of hindering the access of water into the nanopores.

Thus, the nanopore ISEs with all components immobilized to the Au nanoporous support membrane solve the lifetime problems of nano-ISEs, while featuring excellent potentiometric response and unique selectivity patterns.

22.6 NANOMATERIALS TO IMPROVE THE PERFORMANCE OF ION-SELECTIVE ELECTRODES

The motivations for implementing nanomaterials in polymeric membrane-based ISEs were to address some of their main limitations in terms of analyzing real samples. While this chapter focuses on improvements provided by nanomaterials in ion-selective potentiometry, it is worth mentioning that nanomaterials may help in other aspects of potentiometry as well. For instance, in case of redox electrodes, the requirement is to establish an electrode interface that is entirely nonselective and as such able to exchange electrons with all redox species in the solution. However, for many electrode materials, this may not be entirely the case and a differentiation based on the rates of electron transfer of the redox species and electrode material is likely to occur especially if the electrode reaction of some of the species is sluggish. This ultimately means that the electrode potential is most likely determined by the species with fast electron transfer rates and as such not automatically reflects the true concentration ratio of redox species in the solution. A solution for these problems was proposed by Mandler's group,[106] with the use of platinum and gold nanoparticles deposited on the surface of stainless steel electrodes. While no mechanistic treatment is available at this stage, the nanoparticle-modified electrodes exhibited significantly enhanced electron transfer rates and potentiometric responses for a variety of species. Thus, keeping in mind the wide range of possibilities offered by the use of nanomaterials, we will discuss in detail two of their most important application that provided significant progress beyond state of the art in ionophore-based ISEs, that is, nanostructured solid contacts and ionophore–nanomaterial conjugates.

22.6.1 ISEs with Nanomaterial-Based Solid Contact

The liquid contact separating the ISM from the inner reference electrode proved to be a limiting factor in a number of cases, such as in high-pressure conditions, for example, deep-water measurements or high-pressure sterilization, as well as in miniaturized ISEs. More recently, it was realized that the inner filling solution as a reservoir of primary ions, unless not properly optimized, can also have a detrimental effect on the lower detection limits of the respective electrodes. For all these reasons, the quest for proper solid contacts outlines as one of the major research direction in ISE development.

The coated-wire ISEs represented the first major approach to the fabrication of SC ISEs based on polymeric membranes.[107,108] These electrodes involved coating the reference element directly with the ISM without having a thermodynamically defined interface, which obviously impeded their long-term potential stability. The failure of this straightforward approach in a number of practical situations led to the formulation of the main criteria for stable-potential SC ISEs, which included the following[109]:

1. Reversible transition from ionic conduction in the ISM to the electronic conduction in the SC and supporting electrode
2. Ideally nonpolarizable interface with high exchange current density, which is not influenced by the input current of the measuring amplifier
3. Stable chemical composition

After several trials with little follow-up,[110–113] the field became dominated by electrically conducting polymers (ECPs) that, owing to their mixed ion and electron conduction, could interface the electron-conducting substrate electrode (reference element) and the ion-conductive ISM. While the ISE community considers the seminal paper on polypyrrole (PPy) solid contact by Cadogan et al.[114] as the onset of this research field, ECP solid contacts were, in fact, reported and used for pH ion-selective field effect transistors (ISFETs) 5 years earlier by Oyama et al.[115] The ECP-based SC ISEs seemed to be well suitable for solid-contact fabrication, and a plethora of publications appeared on introducing new ECP-doping ion combinations with PPy[114], poly(3,4-ethylenedioxythiophene) (PEDOT),[116] or poly(3-octylthiophene) (POT)[117] as the most promising materials. While the ECP-based solid contacts were performing reasonably well,[118–121] there were also signs of random erratic behavior, which were largely neglected. Some of these erratic behaviors could be traced back to the light[122] and temperature sensitivity[123] of the solid contacts and could be fairly easily corrected by protecting the SC ISEs from ambient light and temperature changes. However, the group of Pretsch, by introducing the so-called potentiometric aqueous layer test, found that a thin aqueous layer can also form beneath the polymeric ISM.[124] The composition of this layer can alter upon sample changes, resulting in characteristic potential drifts potentials of the relevant SC ISEs (Figure 22.10). The source of the drift is the transmembrane ion flux upon sample changes[125] and the consequent

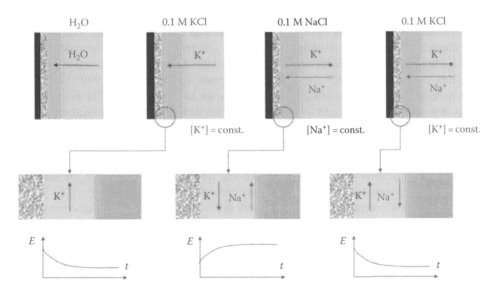

FIGURE 22.10 Aqueous layer test demonstrated schematically for a solid-contact (gray) K⁺-selective membrane (orange). The water uptake and transmembrane water transport result in the formation of an aqueous layer (thin blue phase) beneath the ISM. Upon contacting the membrane with high concentrations of K⁺, the transmembrane ion flux increases the K⁺ activity in that aqueous layer, generating a negative potential drift. When changing to high Na⁺ concentration samples, again owing to transmembrane ion fluxes, K⁺ in the aqueous solution gets exchanged with Na⁺ and the direction of the potential drift is reversed.

ion activity change in the aqueous inner layer. This is a general problem of SC ISEs and, in fact, many of the earlier claims become questionable in the light of this finding. Of note in membranes with low diffusivity such as silicon rubber and various polyacrylate membranes, the ion transport is extremely slow and the aqueous layer test as originally described cannot be used. Still in some cases, Malinowska's method[126] based on employing a pH-sensitive membrane coupled with fast transmembrane diffusion of CO_2 gas can be used (if the SC has no intrinsic potentiometric response to CO_2). In this case, the pH change in the aqueous inner layer due to transmembrane CO_2 flux is detected through the induced potential drift.

Thus, the main objective of developing a well-defined inner phase-boundary potential has got complemented with the requirement of hindering the formation of an aqueous layer beneath the ISM, for example, by using extremely hydrophobic solid-contact materials and/or ISM matrices. Also, since the origin of the aqueous layer is the water uptake and transport through the membrane toward its inner side, these two aspects became the subject of focused studies involving efforts to probe the layers beneath the ISM by De Marco[127,128] using synchrotron radiation as well as by Lindfors[129–133] employing Fourier transform infrared-attenuated total reflectance spectroscopy (FTIR-ATR). These measurements brought direct evidence on the existence of aqueous layers as well as to the dynamics of their formation due to the water uptake of ISMs and the transmembrane transport of water to the inner interface.

In this context, solid contacts based on nanomaterials resulted in a significant progress beyond state of art in SC ISEs. It was shown that carbon-based nanomaterials such as three dimensionally ordered macroporous (3DOM) carbon,[134,135] carbon nanotubes (CNTs)[136] as well as graphene[137–139] have unique advantages as solid contacts. These inherently hydrophobic materials offer good resistance against the formation of an inner aqueous layer, and they exhibit also low or no light sensitivity. It should be stressed, however, that the behavior of carbon nanomaterial–based solid contacts is very much dependent on their surface chemistry and oxidation of the surface may readily result in the formation of an aqueous layer beneath the ISM.[135] Therefore, purification of CNTs and proper pretreatment is required to ensure that the native hydrophobic form of the carbon nanomaterials is used, which might be not so straightforward if surfactant-stabilized aqueous suspensions of these materials are used.

The 3DOM carbon materials were prepared by colloidal crystal templating with monodisperse 420 nm diameter poly(methyl methacrylate) (PMMA) spheres. A resorcinol–formaldehyde precursor was infiltrated into the ordered array of PMMA deposited on a planar surface. After curing to cross-link the precursor at 85° for 3 days, the deposit was pyrolyzed at 900°C under flowing nitrogen to remove the template and convert the precursor to glassy carbon. The monolithic glassy carbon is of porous structure due to spherical shape voids generated by the removal of the spherical PMMA nanoparticles. Neither the synthesis of the 3DOM material nor its later integration in the electrode configuration seems simple, but the performance of the resulting electrodes is excellent and compensates for the cumbersome preparation. The CNT and graphene solid contacts benefit from an easier preparation method, that is, involving spraying a suspension on the surface of an electron conductor substrate, for example, a glassy carbon electrode.

The mechanism of potential stabilization is very different from the classical electron-to-ion conduction mediation. These nanomaterials are characterized by very large specific surface, and after deposition, they form a layer that is readily penetrated by the ISM membrane upon casting. Thus, the extremely large contact area between the interpenetrated layers results in a very large interfacial capacitance,[135] which renders difficult to polarize the interface. This is especially true given that the electrode potentials are commonly measured with high-input resistance voltmeters that allow only extremely tiny currents in the measuring circuits (i.e., pA or less). The large interfacial capacitance seems to be sufficient to ensure excellent potential stabilities of the relevant ISEs, and potential drifts as low as ~10–15 μV/h were reported for SC ISES based on 3DOM carbon and graphene solid contacts and in the range of 200–500 μV/h for those employing CNTs. As shown in Table 22.5, these nanomaterials exhibit excellent resistance to most common interfering effects arising from changes in the light intensity and partial pressures of O_2 and CO_2 in the measurement environment.

TABLE 22.5

Nanomaterial-Based Solid-Contact Ion-Selective Electrodes

Solid Contact	ISM	Detected Ion	Potential Drift, µV/h	Aqueous Inner Layer	Interfering Effects			Reference
					O_2	CO_2	Light	
Carbon-based nanomaterials								
3DOM carbon	PVC/*o*-NPOE	K$^+$	12	No	No	Yes	No	[134]
SWCNT	MMA/nBA	K$^+$, Choline, Ca^{2+}	224–500	No	No	No	No	[144–146]
Yarn/SWCNTs	PVC/DOS	K$^+$, NH$_4^+$, H$^+$	250	—	—	—	—	[147]
MWCNT	MMA/nBA	H$^+$, ClO$_4^-$	220	No	—	—	—	[136,148]
PEDOT/MWCNTs	PVC/DOS	K$^+$	—	No	No	No	—	[149]
Electrochemically reduced graphene oxide	PVC/*o*-NPOE	Ca^{2+}	15	No	No	No	No	[137]
Chemically reduced graphene oxide	PVC/*o*-NPOE	K$^+$	13	No	No	No	No	[138]
Graphene sheets	PVC/DOS	K$^+$	—	No	—	Yes	—	[139]
Graphene paper	PVC/*o*-NPOE	K$^+$, Ca^{2+}, H$^+$	2×10^3	No	No	No	No	[150]
Carbon black	PVC/*o*-NPOE	K$^+$	15	No	No	No	No	
Conducting polymer nanostructures								
Polyaniline nanoparticles	SR/DOS	Ca^{2+}, Ag$^+$	—	No	No	—	No	[140]
Gold nanoparticles								
Hydrophobic Au nanoparticles	PVC/DOS	K$^+$	380–660	No	—	—	—	[141]
Dithizone-modified Au nanoparticles	PVC/*o*-NPOE	Cu^{2+}	—	—	—	—	—	[151]
Tetrakis(4-chlorophenyl) borate anion-doped Au nanoclusters	PVC/DOS	K$^+$	10.1	No	No	No	No	[142]
Silver nanoparticles								
Ag nanoparticles by inkjet printing	PVC/*o*-NPOE	Ag$^+$	—	—	—	—	—	[152]

Source: Yin, T. and Qin, W., *Trends Anal. Chem.*, 51, 79, 2013.

The slight CO_2 sensitivity of 3DOM carbon–based solid contacts is similar with that of PEDOT SC ISEs (ca. 10 mV/h) and significantly less than of PPy-based ISEs (ca. 30 mV/h).

Solid contacts based on polyaniline (PANI) nanoparticles (∅ 8 nm), which consist of the electrically conducting emeraldine salt form of PANI, can be readily fabricated from a suspension of the nanoparticles in an organic solvent by drop-casting.[140] The subsequent drop-casting of the silicone rubber (SR)–based ISM results, in this case as well, in a slight intermixing of the two layers. While the PANI nanoparticle–based solid contacts seem to provide a similarly adequate interface as carbon nanomaterials, one of their main strengths proved to be the very good reproducibility of the E^0 values of the different electrodes. To assess the significance of this finding, it should be mentioned that whereas in case of liquid contact ISEs, the E^0 values of similarly constructed electrodes are easy to control, in case of solid-contact ISEs, these values in general show large variations among electrodes prepared in the exact same manner. Thus, even if the electrodes have a Nernstian response, the SC ISEs need to be calibrated in advance due to uncertainties in their E^0. This constitutes a major disadvantage if the respective SC ISEs are intended as single-use electrodes for point-of-care applications.

The Au nanoparticles with surface modifications conferring hydrophobic properties were also shown to have prospective as solid-contact materials.[141]Most notably, an arenethiolate monolayer-protected Au cluster (MPC) doped with tetrakis(4-chlorophenyl) borate (TB) was shown to provide excellent potential stabilities.[142] The surface-modifying material consisted of an equimolar mix of MPC^0 and MPC^+, in fact a redox couple with TB anions compensating the positive charge, which could be drop-casted onto the surface of an electrode. The redox couple stabilizes the electrode–SC interface inferring it with a well-defined redox potential while the TB anions that partition between K^+-selective membrane and the SC phase, provide a well-defined SC | ISM phase-boundary potential. These films are claimed to be unsusceptible to common gases (CO_2 and O_2), light, redox interferents, and formation of an aqueous layer beneath the ISM, though the notorious light sensitivity of tetraphenylborates may require protection from light, which however can be easily ensured.

22.6.2 Ion-Selective Membranes Based on Ionophore–Nanomaterial Conjugates

One major problem of polymeric ISMs is the gradual leaching of active components from the membrane into the sample solution.[77] This has detrimental effects on the lifetime and analytical performance of the respective ISEs, that is, selectivity and dynamic range. The leaching process is especially severe in the case of miniaturized electrodes (as discussed at nanopipette-based electrodes), flow-through systems, and samples with lipophilic character such as biological fluids favoring a more enhanced extraction of the membrane components.[78] In blood gas analyzers, these issues are overcome by frequent calibrations and discarding the electrodes that are outside specifications. The obvious solution to prevent the leaching of active components is their permanent confinement to the membrane material, for example, by covalent immobilization to the membrane matrix. Indeed, methods for immobilizing practically all membrane components (plasticizer,[153] lipophilic additive,[154] and ionophore) have been reported, with most of these studies focusing mainly on avoiding the loss of ionophore from the membrane. The covalent immobilization most often implied the use of functionalized polymers, for example, different functionalized PVCs,[155] polyurethanes,[156] silicon rubbers,[154,157] and polyacrylates,[158,159,160] bearing functional groups facilitating the covalent attachment of suitable ionophores. While rather convenient and simple, the polymeric reaction product often contained unreacted functional groups that have been reported to affect the selectivity and response time of the relevant ISEs.[156,161] The only attempt to directly copolymerize an ionophore with the *chemically inert* vinyl chloride monomer[162] owing to the low yield of ionophore attachment was also unable to provide a breakthrough.

However, even if successful, in all these immobilization approaches, the composition of the polymer and ionophore cannot be independently adjusted in the membrane, that is, the viscosity of the membrane and the ionophore loading are interdependent. This is disadvantageous in some circumstances such as case of mass transport–controlled ISEs for ultratrace analysis. The group of Gyurcsányi offered an elegant solution for this problem by immobilizing the ionophore to a nanoparticle carrier rather than attempting to attach the ionophore to any of the polymeric membrane components.[103] The concept was demonstrated through the subsequent modification of ca. 5.5 nm diameter Au nanoparticles with SS-Ag-II Ag^+-selective ionophore and, then, with 1-dodecanethiol (DDT) (Figure 22.11a). The role of DDT in the mixed SAM was to provide the nanoparticle conjugates with proper hydrophobicity to be soluble in hydrophobic ISMs without interfering with the ion complexation process. Due to their hydrophobicity, the ionophore–Au nanoparticle conjugates (IP–AuNPs) could be precipitated in ethanol and, after centrifugation, effectively purified by washing with ethanol. After resuspending in tetrahydrofuran, the IP–AuNPs were added to the membrane cocktail. They proved to be fully compatible with PVC membranes plasticized with *o*-NPOE and no phase segregation occurred if their content was kept <8 wt%. As determined by spectral imaging,[163] the IP–AuNPs were found to be practically immobile in the plasticized PVC membrane (Figure 22.11c).[103] The calculated upper limit for the diffusion coefficient of IP–AuNP was 5×10^{-12} cm^2/s, which is among the lowest reported till now,[164] and cannot be explained by the size increase of the

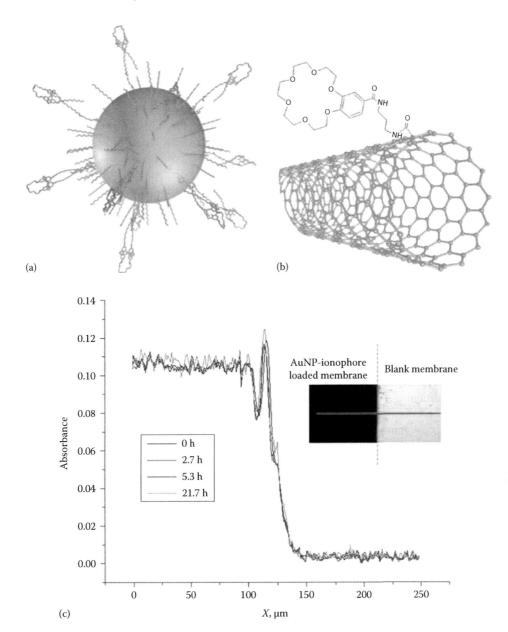

FIGURE 22.11 (a) Schematic representation of the ionophore–gold nanoparticle conjugate (the surface of a ca. 5.5 nm gold nanoparticle is modified with a mixed self-assembled layer of SS-Ag-II Ag$^+$ ionophore and DDT). (Adapted from Jágerszki, G., Grün, A., Bitter, I., Tóth, K., and Gyurcsányi, R.E., *Chem. Commun.*, 2010, 46, 607. With permission from The Royal Society of Chemistry.) (b) SWCNT modified with benzo-18-crown-6 Pb^{2+}-ionophore.[168] (c) Absorbance profiles recorded at 525 nm indicative of IP–AuNP diffusion from the loaded membrane (inset: left) into a blank one (right). The inset features the absorbance of the formulation of the membrane used for potentiometric measurements, which however was diluted 100-fold.

ionophore on the basis of the Einstein–Stokes equation. Formerly, diffusion coefficients in this range were only reported for nonplasticized methacrylic–acrylic ISMs.[164] However, in the case of IP–AuNP, it was achieved without increasing the viscosity of the ISM, which would affect the mobility of all membrane components.[164] Indeed, the bulk resistance of the IP–AuNP-based ISMs, determined by electrochemical impedance spectroscopy, was practically the same as that of an

ISM based on the free ionophore, having the identical amount of cation-exchanger sites. Another advantage of the IP–AuNP-compounded ISMs is the nontransparency of IP–AuNPs in the UV/visible range, which would inherently protect light-sensitive reference elements such as conducting polymer-based solid contacts.[122]

The control and assessment of the surface density of the ionophore on the AuNP, or any other carrier, is very critical for several reasons. First of all, the complexation of the ion by a surface-confined ionophore is expected to be hindered by the formation of other positively charged ion–ionophore complexes within the Debye length. Therefore, the synthesis was designed to provide ca. three times smaller surface concentration as the theoretically calculated 4.7×10^{-10} mol/cm^2. Second the surface concentration is required for the calculation of the ionophore loading to achieve optimal selectivity and dynamic range. Thus, a method was introduced to estimate the ionophore loading by titration of the IP–AuNPs in ISMs with a lipophilic anion. If the molar concentration of the latter exceeds that of the immobilized ionophore (the free ionophore concentration being zero), the membrane loses its selectivity. As with almost all covalently immobilized ligands, the selectivity was found to be lower than that of a membrane with free ionophore. This proved to be due to the smaller complex formation constant (β_{IL}) of the IP–AuNP with the Ag$^+$ ($\log\beta_{AgL} = 8.5$) as compared with the free ionophore (11.1). Of note, the selectivity coefficients are directly related to the ratio of the complex formation constants and the difference in the β_{IL} of free and conjugated ionophores with interfering ions was insignificant. At optimal composition after conventional conditioning of the ISM in 10 μM Ag$^+$ solution, the IP–AuNP-based ISMs were characterized by a lower detection limit of 5 nM, which closely matches the lowest value reported for a covalently immobilized ionophore-based membrane.[156] It should be emphasized that by using unconjugated Ag$^+$-selective ionophores, even lower detection could be obtained, however, after extensive conditioning of the ISMs (spanning over several days, and successively decreasing the primary ion concentration to 10 nM).[105]

The concept of immobilizing the ionophores on nanomaterials has been extended to other ligands and carriers. Most notably, the group of Rius synthesized a series of CNT–ionophore conjugates that were incorporated in polymeric membranes. The required synthetic effort was considerably larger than in case of AuNP modification with derivatives forming spontaneous Au–S bonds. The raw multi-walled CNTs were purified and oxidized to generate COOH surface functionalities of the CNT that after activation as acid chlorides were reacted with 1,3-diaminopropane to provide NH$_2$ groups.[165] These were reacted using standard coupling reagents to ultimately attach 4′-carboxy-benzo-18-crown-6, a Pb^{2+} ionophore (Figure 22.11b). To increase the ionophore loading, the synthetic method was further developed to attach several ionophores through first- and second-generation dendrimers to each COOH functional group on the CNT surface.[166] Finally, to reduce the complexity of the ionophore coupling to the CNTs, a noncovalent modification was also introduced that is based on π–π stacking interactions with the MWCNT walls of a pyrene-benzo-18-crown-6 ionophore. Apparently, this method solves some of the inherent problems of the earlier reported covalent modifications as well: for example, (1) it offers higher ionophore loadings, (2) avoids the use of the polymeric surfactant that is needed to disperse the CNTs in the polymeric membrane, and (3) reduces the likelihood of introducing nonreacted functional groups in the ISM. Improving on the latter two issues may lead to better selectivity of the relevant ISEs. The interesting feature of the synthesized hybrid materials is that after dispersing in a polymeric membrane, they can be deposited directly onto the surface of an electron conductor without the need for an additional solid contact to stabilize the potential. Thus, besides providing ion-selectivity, the new material acts also as a potential stabilizing inner contact. In fact, the use of dithizone-modified gold nanoparticles deposited as a self-standing film on an electron conductor, that is, without being incorporated in a polymeric film, was shown to provide Nernstian potentiometric responses for Cu^{2+}.[167] If coupled with an inherently more selective ligand, like 6-(bis(pyridin-2-ylmethyl)amino)hexane-1-thiol, then the same polymer-free modified AuNP layer-based ISEs were found to exhibit Nernstian potentiometric response and improved selectivities.[168]

22.7 POTENTIOMETRIC BIOASSAYS BASED ON NANOMATERIALS

The advent of ISEs with low detection limits had a major impact on potentiometric biosensing. The recognition that ISEs can be combined with reactive layers and membranes to provide new functionalities led to the introduction of the first potentiometric biosensor by Guilbault and Montalvo in 1969,[169] followed by the use of potentiometry for detecting bioaffinity interactions, initiated by the pioneering work of Rechnitz in the 1970s.[170–172] While potentiometric bioassays had a gradual liedown giving way to optical methodologies, the spectacular advances in mass transport–controlled polymer membrane ISEs with ultratrace detection capabilities led lately to a revisiting of potentiometry as a detection method in bioaffinity assays.

22.7.1 POTENTIOMETRIC BIOASSAYS BASED ON NANOPARTICLE-LABELED REAGENTS

Bioassays generally require a signal amplification mechanism to enable the detection of analytes at ultratrace levels. In practice, most often enzyme amplification is used that involves attaching an enzyme to a secondary bioreagent to form a sandwich complex together with the analyte and a primary capture probe immobilized onto a surface. From a suitable substrate each enzyme label catalyzes the formation of a large amount of chromophores or fluorophores detected optically, that is, for each analyte that bounds to the capture probe hundreds of thousands of optically detectable molecules are generated. It was recognized that enzyme amplification can be mimicked by using secondary reagents labeled with nanoparticles or nanocapsules as these nanostructures can be triggered to release a large amount of detectable components. Most often the release is made by dissolution of the nanomaterial.[173,174]

Replacing enzyme labels with nanoparticles has the advantage: (1) of reducing the susceptibility of the label to environmental conditions, and (2) may facilitate a faster release of detectable species than the enzyme-catalyzed reaction. Furthermore, potentiometric detection is a cost-effective method that can take advantage also of advances in the field of ISEs for ultratrace analysis[119,120,175,176] to provide sensitivities approaching that of fluorescence detection. Certainly, minimizing the volume in which the detected components are released is an essential requirement for ultrasensitive assays. Given the ease of electrode miniaturization potentiometric detection easily fulfills the requirements for low sample volumes. A very simple miniaturization of the measurement cells has already been shown to reduce the assessable sample volume to ca. 3 μL.[177] Thus, even rudimentary approaches to miniaturize SC ISEs and reference electrodes based on using conventional plastic micropipettes as electrode bodies made them fit in the wells of microtiter plates and enabled measurements in ca. 150 μL solution.[178,179]

One of the first relevant attempt toward potentiometric bioassays based on *nanomaterials* was the use of tetrapentylammonium ion (TPA$^+$)–loaded liposome labels, which released their content upon immunolysis.[180,181] The released TPA$^+$ could be detected by using a simple ion-exchanger-based electrode in a small volume solution (ca. 10 μL) entrapped between the electrode and a Ag/AgCl plate to maximize sensitivity.

The use of inorganic nanoparticles represented a step forward toward robust reagents able to generate large amounts of ions. Nanoparticles proposed for this purpose included gold nanoparticles[182] and quantum dots attached to a secondary bioreagent.[178,183,184] In case of Au nanoparticle labels, after the completion of the assay, the Au nanoparticles on the surface catalyzed the reduction of Ag from an acidic solution of Ag and hydroquinone with the formation of surface-confined silver clusters.[185] These were dissolved under mild conditions using dilute hydrogen peroxide and the released silver ions were ultimately measured with Ag$^+$-selective minielectrodes (Figure 22.12). To understand the magnitude of the amplification obtained with nanoparticles, one should bear in mind that a single Ag nanoparticle of 25 nm in diameter contains ca. 4.8×10^5 Ag atoms and the corresponding amount of Ag ions is released during oxidative dissolution. The same amplification concept was found to be applicable in more practical sensing setups. One of the first attempts in

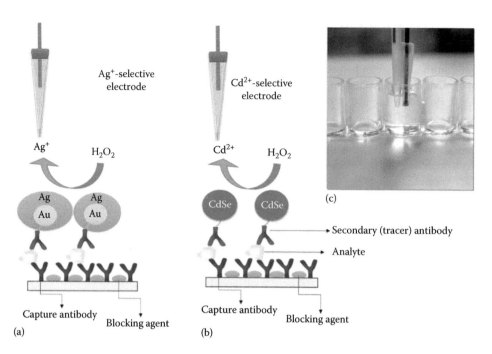

FIGURE 22.12 Schematic view of the final stage of a sandwich assay with potentiometric detection. After successive addition, incubation and washing away of the excess reagents, Ag^+ (a) or Cd^{2+} (b) released from the respective nanoparticles is measured in a small volume solution such as in the well of a microtiter plate using the relevant ion-selective and reference electrodes (c).

this direction has been a simple paper-based assay to detect IgE with a gold nanoparticle–labeled IgE-specific aptamer.[186] The assay used the concept of a blot-dot assay, that is, spotting the IgE-containing sample on nitrocellulose paper, blocking the paper with bovine serum albumin, and adding the gold labeled aptamer. However, after amplification by chemical plating of Ag, the released Ag^+ ions were measured directly in the wet paper sandwiched between a reference electrode and a solid-contact Ag^+-selective electrode. Interestingly, the detection of Ag^+ in paper versus measurements in aqueous solution was found to be ca. two orders of magnitude higher. The higher detection limit in paper as compared to solution phase was later confirmed also by a study that involved the potentiometric detection of K^+.[187] Even so, the analytical performance of potentiometric dot-blot assays matched that of reflectance-based assays, which suggests that potentiometric detection might represent a viable alternative to the conventionally used optical detection.

CdSe quantum dots were also explored as nanoparticle labels in potentiometric bioaffinity assays.[178] The quantum dots were attached to secondary bioreagents and at the end of the bioassay were dissolved in 150 µL of 3% H_2O_2 and 10^{-5} M $NaNO_3$. The constant Na^+ background allowed the use of a Na^+-selective minielectrode as reference while the released Cd^{2+} was detected by using a Cd^{2+}-selective minielectrode optimized for ultratrace analysis (Figure 22.12). The applicability of the concept was demonstrated for protein-,[178] nucleic acid-,[184] and aptamer-based[183] assays. The use of CdSe or CdS quantum dots has the advantage of direct ion release (without chemical plating), but the smaller size of the inorganic crystals was expected to result in lower signal amplifications. However, even so a lower detection limit of <10 fmol was achieved for IgG using cadmium-selenide quantum dots as labels, which represents an improvement of several orders of magnitude over the Au nanoparticle label-catalyzed Ag deposition approach. This seems to be due to the higher dilution volumes used in the latter approach and it is also likely that the amplification by chemical plating has not been optimized to the extent of fully avoiding nonspecific, concurrent Ag deposition at sites other than the nanoparticles of interest.

22.7.2 ION CHANNEL MIMETIC POTENTIOMETRIC BIOASSAYS

A completely different approach to design label-free potentiometric bioassays taking advantage of the unique amplification offered by nanostructures is based on the use of chemically modified nanopores. As early as 2003, Gyurcsányi and Pretsch[188] introduced the concept of using gold nanopores modified with bioreagents (biotin) that upon binding their cognate (avidin) caused the change in the permeability of the nanopore for marker ions (Ca^{2+}). Gold nanopores were obtained by electroless deposition of gold on polycarbonate track-etch membranes with 50 nm pore diameter. The approaches to the biotinylation of the gold nanopores included one- or two-step modifications using thiol, or disulfide-bearing reagents. Both methods involve the modification of the entire gold surface, that is, the pore walls as well as the external membrane surface. However, the biotinylated external membrane surface would bind a significant fraction of avidin from dilute samples that would not contribute to the signal change. Therefore, an original method was introduced for the differential modification of the external and internal surfaces of gold nanoporous membranes. The outer membrane surfaces were modified by soft lithography using a poly(dimethylsiloxane) stamp with (1-mercaptoundec-11-yl)hexa(ethylene glycol),[189] a thiol derivative that reduces nonspecific protein adsorption.[190] Meanwhile, the inner walls of the nanopores remained unaffected and susceptible to further modification by immersing the membrane in functionalized thiol derivatives that permit direct or indirect anchoring of bioprobes. The potentiometric measurement of the modulation of the Ca^{2+} flux through the nanopores induced by selective binding events made possible the indirect determination of avidin.

Later, this ion channel mimicking concept was used to detect the binding events directly on an ISM by using a covalently biotinylated membrane surface. The binding of avidin in this case directly hindered the electrochemically imposed ion transfer kinetics of sodium ion, and subsequently decreased the super-Nernstian step that is characteristic of a surface depletion of the marker ion. To enhance the sensitivity, which depends on the fraction of the surface blocked by avidin, the active surface area was reduced by applying a hydrophilic track-etch nanopore membrane on top of the biotinylated ISM surface.[191]

22.7.3 POTENTIOMETRIC BIOSENSING BASED ON SURFACE CHARGE MODIFICATIONS

An interesting approach for label-free direct potentiometric sensing of bacteria and proteins was introduced by the group of Rius.[192] The detection is based on the combined use of single-walled carbon nanotubes (SWCNTs) sprayed on the surface of an electrode and aptamers. Aptamers self-assemble on the wall of CNTs by π–π stacking interactions, which owing to the negatively charged aptamer backbone results in a negative surface charge excess on the SWCNTs. The selective binding of the target to the aptamer chains induces conformational changes in the aptamers, that is, most likely aptamers lying on the surface of CNTs are partially separated from the sidewalls of the nanotube. This separation is believed to reduce the negative surface charge of the SWCNTs with subsequent increase of the recorded potential. This sensing mechanism is quite unique and resembles that of field-effect transistor-based biosensors rather than the phase-boundary model of classical potentiometric sensors. The proof of concept was demonstrated first through the potentiometric detection of living *Salmonella typhi* (ST) using an ST-specific RNA aptamer covalently attached to the sidewalls of carboxylated CNTs. The aptamer-SWCNT-based electrodes responded from 0.2 CFU/mL to ca. 10^6 CFU/mL, which is remarkable. However, on the backside the linear range, the sensitivity was less than 2 mV/CFU decade in thermostated conditions even in a low ionic strength buffer (1.7 mM phosphate buffer saline), which is essential to maximize sensitivity. Nevertheless, the method could be applied with success for the determination of other pathogens from real samples, for example, *Escherichia coli*[193] and *Staphylococcus aureus* (*S. aureus*).[194] Interestingly, when comparing for the determination of *S. aureus*, the sensitivity of covalent versus noncovalent aptamer—SWCNT hybrids, the latter one offered higher sensitivity, but with an inferior detection limit.[194] The same result was obtained by using covalent and noncovalent immobilization of aptamers to graphene (graphene oxide and reduced graphene oxide), which could be used as a substitute for SWCNTs for the detection of *S. aureus*.[195]

The concept was also applied for the potentiometric detection of proteins, such as thrombin, which led to higher sensitivities as compared with pathogen targets,[196] that is, ca. 8 mV/decade thrombin concentration. The exact response mechanism still needs to be elucidated, but the enhanced sensitivity may be due to the positive charge of the thrombin molecule at physiological pH (charge neutralization effect) and the higher surface densities. Both effects can lead to a larger change in the surface charge. While the potential of SWCNTs is very sensitive to redox contaminants and ionic strength of the sample solution, an improved buffer composition was shown to enable extremely sensitive potentiometric determination of the variable surface glycoprotein (VSG) from African trypanosomes in diluted blood.[197] It has been reported that a dilute redox buffer (2 mM $[Fe(CN)_6]^{3-}/Fe(CN)_6]^{4-}$) can efficiently stabilize the signal without suppressing the potential changes due to the VSG binding to its specific aptamer receptor. Detection limits of 4 fM and 10 pM were obtained in buffered samples and in blood, respectively. The response curve exhibited two linear ranges, with 200 mV/decade of VSG concentration (low concentration range) and 34 mV/decade in the high concentration range.

ACKNOWLEDGMENTS

RGy gratefully acknowledges the support of the *Lendület* program of the Hungarian Academy of Sciences and TÁMOP-4.2.1/B-09/1/KMR-2010-0002. The authors thank L. Höfler for providing Figure 22.4 and I. Makra for Figures 22.6 through 22.8.

REFERENCES

1. Trenkler, T., P. De Wolf, W. Vandervorst, and L. Hellemans. 1998. Nanopotentiometry: Local potential measurements in complementary metal-oxide-semiconductor transistors using atomic force microscopy. *J. Vac. Sci. Technol., B* 16: 367–372.
2. Nernst, W.Z. 1889. Die elektromotorische Wirksamkeit der Ionen. *Z. Phys. Chem.* 4: 129–181.
3. Nernst, W. and E.H. Riesenfeld. 1902. Electrolytic appearances on the boundary of two solvents. *Ann. Phys.* 8: 600–608.
4. Cremer, M. 1906. The cause of the electromotor properties of tissue, and a contribution to the science of polyphasic electrolytes. *Z. Biol.* 47: 562–608.
5. Haber, F. and Z. Klemensiewicz. 1909. Concerning electrical phase boundary forces. *Z. Phys. Chem. Stoechiom. Verwandtschafts.* 67: 385–431.
6. Bühlmann, P., E. Pretsch, and E. Bakker. 1998. Carrier-based ion-selective electrodes and bulk optodes. 2. Ionophores for potentiometric and optical sensors. *Chem. Rev.* 98: 1593–1687.
7. Bakker, E. and E. Pretsch 2011. Advances in potentiometry. In *Electroanalytical Cchemistry: A Series of Advances*, eds.; Bard, A.J., Zoski, C., pp. 24, 1–74. Boca-Raton, FL: CRC Press, Taylor & Francis Group.
8. Qin, Y. and E. Bakker. 2003. A copolymerized dodecacarborane anion as covalently attached cation exchanger in ion-selective sensors. *Anal. Chem.* 75: 6002–6010.
9. Lindner, E., R.E. Gyurcsányi, and E. Pretsch 2012. Potentiometric ion sensors: Host–guest supramolecular chemistry in ionophore-based ion-selective membranes. In *Applications of Supramolecular Chemistry for 21st Century Technology*, ed.; Schneider, H.-J., xi, 441 p. Boca Raton, FL: Taylor & Francis.
10. Guggenheim, E.A. 1929. The conception of electrical potential difference between two phases and the individual activities of ions. *J. Phys. Chem.* 33: 842–849.
11. Guggenheim, E.A. 1930. On the conception of electrical potential difference between two phases. II. *J. Phys. Chem.* 34: 1540–1543.
12. Morf, W.E. 1981. *The Principles of Ion-Selective Electrodes and of Membrane Transport*. New York: Elsevier.
13. Bakker, E., P. Bühlmann, and E. Pretsch. 2004. The phase-boundary potential model. *Talanta* 63: 3–20.
14. Brumleve, T.R. and R.P. Buck. 1978. Numerical solution of the Nernst-Planck and Poisson equation system with applications to membrane electrochemistry and solid state physics. *J. Electroanal. Chem.* 90: 1–31.

15. Sokalski, T. and A. Lewenstam. 2001. Application of Nernst–Planck and poisson equations for interpretation of liquid-junction and membrane potentials in real-time and space domains. *Electrochem. Commun.* 3: 107–112.

16. Sokalski, T., P. Lingenfelter, and A. Lewenstam. 2003. Numerical solution of the coupled Nernst–Planck and Poisson equations for liquid junction and ion selective membrane potentials. *J. Phys. Chem. B* 107: 2443–2452.

17. Morf, W.E., E. Pretsch, and N.F.d. Rooij. 2010. Theoretical treatment and numerical simulation of potential and concentration profiles in extremely thin non-electroneutral membranes used for ion-selective electrodes. *J. Electroanal. Chem.* 641: 45–56.

18. Bakker, E., P. Bühlmann, and E. Pretsch. 1997. Carrier-based ion-selective electrodes and bulk optodes. 1. General characteristics. *Chem. Rev.* 97: 3083–3132.

19. Bakker, E. and E. Pretsch 2012. Advances in potentiometry. In *Electroanalytical Chemistry. A Series of Advances*, eds.; Bard, A.J., Zoski, C., pp. 24, 1–73. Boca Raton, FL: CRC Press.

20. Egorov, V.V., P.L. Lyaskovski, I.V. Il'inchik, V.V. Soroka, and V.A. Nazarov. 2009. Estimation of ion-pairing constants in plasticized poly(vinyl chloride) membranes using segmented sandwich membranes technique. *Electroanalysis* 21: 2061–2070.

21. Nicolsky, B.P. 1937. Theory of glass electrodes (in Russian). *Acta Physicochim. USSR* 7: 597–610.

22. Bakker, E., R.K. Meruva, E. Pretsch, and M.E. Meyerhoff. 1994. Selectivity of polymer membrane-based ion-selective electrodes: Self-consistent model describing the potentiometric response in mixed ion solutions of different charge. *Anal. Chem.* 66: 3021–3030.

23. Nägele, M., E. Bakker, and E. Pretsch. 1999. General description of the simultaneous response of potentiometric ionophore-based sensors to ions of different charge. *Anal. Chem.* 71: 1041–1048.

24. Bakker, E. 2010. Generalized selectivity description for polymeric ion-selective electrodes based on the phase boundary potential model. *J. Electroanal. Chem.* 639: 1–7.

25. Teorell, T. 1935. An attempt to formulate a qualitative theory of membrane permeability. *Proc. Soc. Exp. Biol. Med.* 33: 282–285.

26. Meyer, K.H. and J.-F. Sievers. 1936. La perméabiliteé des membranes I. Théorie de la perméabilité ionique. *Helv. Chim. Acta* 19: 649–664.

27. Morf, W.E., E. Pretsch, and N.F. de Rooij. 2008. Theory and computer simulation of the time-dependent selectivity behavior of polymeric membrane ion-selective electrodes. *J. Electroanal. Chem.* 614: 15–23.

28. Morf, W.E., E. Pretsch, and N.F. de Rooij. 2009. Memory effects of ion-selective electrodes: Theory and computer simulation of the time-dependent potential response to multiple sample changes. *J. Electroanal. Chem.* 633: 137–145.

29. Sandifer, J.R. and R.P. Buck. 1975. An algorithm for simulation of transient and alternating current electrical properties of conducting membranes, junctions, and one-dimensional, finite galvanic cells. *J. Phys. Chem.* 79: 384–392.

30. Stover, F.S. and R.P. Buck. 1976. Digital simulation of associated and nonassociated liquid membrane electrochemical properties. *Biophys. J.* 16: 753–770.

31. Nahir, T.M. and R.P. Buck. 1993. Transport processes in membranes containing neutral ion carriers, positive ion complexes, negative mobile sites, and ion pairs. *J. Phys. Chem.* 97: 12363–12372.

32. Höfler, L., I. Bedlechowicz, T.s. Vigassy, R.b.E. Gyurcsányi, E. Bakker, and E. Pretsch. 2009. Limitations of current polarization for lowering the detection limit of potentiometric polymeric membrane sensors. *Anal. Chem.* 81: 3592–3599.

33. Grygolowicz-Pawlak, E., A. Numnuam, P. Thavarungkul, P. Kanatharana, and E. Bakker. 2011. Interference compensation for thin layer coulometric ion-selective membrane electrodes by the double pulse technique. *Anal. Chem.* 84: 1327–1335.

34. Läuger, P. and B. Neumcke. 1973. Theoretical analysis of ion conductance in lipid bilayer membranes. In *Membranes*, ed.; Eisenman, G., pp. 2, 1–59. New York: Marcel Dekker.

35. Ciani, S.M., G. Eisenman, R. Laprade, and G. Szabo. 1973. Theoretical analysis of carrier-mediated electrical properties of bilayer membranes. In *Membranes*, ed.; Eisenman, G., pp. 2, 61–177. New York: Marcel Dekker.

36. Minami, H., N. Sato, M. Sugawara, and Y. Umezawa. 1991. Comparative study on the potentiometric responses between a valinomycin-based bilayer lipid membrane and a solvent polymeric membrane. *Anal. Sci.* 7: 853–862.

37. Jágerszki, G., A. Takács, I. Bitter, and R.E. Gyurcsányi. 2011. Solid-state ion channels for potentiometric sensing. *Angew. Chem. Int. Edit.* 50: 1656–1659.

38. Schmidt-Rohr, K. and Q. Chen. 2008. Parallel cylindrical water nanochannels in Nafion fuel-cell membranes. *Nat. Mater.* 7: 75–83.

39. Mousavi, M.P.S. and P. Bühlmann. 2013. Reference electrodes with salt bridges contained in nanoporous glass: An underappreciated source of error. *Anal. Chem.* 85: 8895–8901.

40. Sokalski, T., W. Kucza, M. Danielewski, and A. Lewenstam. 2009. Time-dependent phenomena in the potential response of ion-selective electrodes treated by the Nernst-Planck-Poisson model. Part 2: Transmembrane processes and detection limit. *Anal. Chem.* 81: 5016–5022.

41. Thomas, R.C. 1978. *Ion-Sensitive Intracellular Microelectrodes: How to Make and Use Them.* New York: Academic Press.

42. Kessler, M., J. Höper, and D.K. Harrison. 1985. *Ion Measurements in Physiology and Medicine.* New York: Springer-Verlag.

43. Takahashi, A., P. Camacho, J.D. Lechleiter, and B. Herman. 1999. Measurement of intracellular calcium. *Physiol. Rev.* 79: 1089–1125.

44. Revsbech, N.P., B.B. Jorgensen, T.H. Blackburn, and Y. Cohen. 1983. Microelectrode studies of the photosynthesis and O_2, H_2S, and pH profiles of a microbial mat. *Limnol. Oceanogr.* 28: 1062–1074.

45. Zhao, P. and W.-J. Cai. 1997. An improved potentiometric pCO_2 microelectrode. *Anal. Chem.* 69: 5052–5058.

46. Horrocks, B.R., M.V. Mirkin, D.T. Pierce, A.J. Bard, G. Nagy, and K. Tóth. 1993. Scanning electrochemical microscopy. 19. Ion-selective potentiometric microscopy. *Anal. Chem.* 65: 1213–1224.

47. Wei, C., A.J. Bard, G. Nagy, and K. Tóth. 1995. Scanning electrochemical microscopy. 28. Ion-selective neutral carrier-based microelectrode potentiometry. *Anal. Chem.* 67: 1346–1356.

48. Gray, N.J. and P.R. Unwin. 2000. Simple procedure for the fabrication of silver/silver chloride potentiometric electrodes with micrometre and smaller dimensions: Application to scanning electrochemical microscopy. *Analyst* 125: 889–893.

49. Etienne, M., A. Schulte, S. Mann, G. Jordan, I.D. Dietzel, and W. Schuhmann. 2004. Constant-distance mode scanning potentiometry. 1. Visualization of calcium carbonate dissolution in aqueous solution. *Anal. Chem.* 76: 3682–3688.

50. Nadappuram, B.P., K. McKelvey, R. Al Botros, A.W. Colburn, and P.R. Unwin. 2013. Fabrication and characterization of dual function nanoscale pH-scanning ion conductance microscopy (SICM) probes for high resolution pH mapping. *Anal. Chem.* 85: 8070–8074.

51. Ammann, D. 1986. *Ion-Selective Microelectrodes: Principles, Design, and Application.* New York: Springer-Verlag.

52. Ling, G. and R.W. Gerard. 1949. The normal membrane potential of frog sartorius fibers. *J. Cell. Comp. Physiol.* 34: 383–396.

53. Chowdhury, T.K. 1969. Fabrication of extremely fine glass micropipette electrodes. *J. Phys. E Sci. Instrum.* 2: 1087.

54. Brown, K.T. and D.G. Flaming. 1974. Beveling of fine micropipette electrodes by a rapid precision method. *Science* 185: 693–695.

55. Brown, K.T. and D.G. Flaming. 1977. New micro-electrode techniques for intracellular work in small cells. *Neuroscience* 2: 813–827.

56. Ammann, D., F. Lanter, R.A. Steiner, P. Schulthess, Y. Shijo, and W. Simon. 1981. Neutral carrier based hydrogen ion selective microelectrode for extra- and intracellular studies. *Anal. Chem.* 53: 2267–2269.

57. Haemmerli, A., J. Janata, and H.M. Brown. 1980. Ion-selective electrode for intracellular potassium measurements. *Anal. Chem.* 52: 1179–1182.

58. Walker, J.L. 1971. Ion specific liquid ion exchanger microelectrodes. *Anal. Chem.* 43: 89A–93A.

59. Munoz, J.L., F. Deyhimi, and J.A. Coles. 1983. Silanization of glass in the making of ion-sensitive microelectrodes. *J. Neurosci. Methods* 8: 231–247.

60. Deyhimi, F. and J.A. Coles. 1982. Rapid silylation of a glass surface: Choice of reagent and effect of experimental parameters on hydrophobicity. *Helv. Chim. Acta* 65: 1752–1759.

61. Tsien, R.Y. and T.J. Rink. 1980. Neutral carrier ion-selective microelectrodes for measurement of intracellular free calcium. *Biochim. Biophys. Acta Biomembr.* 599: 623–638.

62. Kelepouris, E., Z. Agus, and M. Civan. 1985. Intracellular calcium activity in split frog skin epithelium: Effect of cAMP. *J. Membr. Biol.* 88: 113–121.

63. Greger, R., C. Weidtke, E. Schlatter, M. Wittner, and B. Gebler. 1984. Potassium activity in cells of isolated perfused cortical thick ascending limbs of rabbit kidney. *Pflugers Arch.* 401: 52–57.

64. Yamaguchi, H. 1986. Recording of intracellular Ca^{2+} from smooth muscle cells by sub-micron tip, double-barrelled Ca^{2+}-selective microelectrodes. *Cell Calcium* 7: 203–219.

65. Semb, S.O., B. Amundsen, and O.M. Sejersted. 1997. A new improved way of making double-barrelled ion-selective micro-electrodes. *Acta Physiol. Scand.* 161: 1–5.

66. Buhrer, T., P. Gehrig, and W. Simon. 1988. Neutral-carrier-based ion-selective microelectrodes design and application—A review. *Anal. Sci.* 4: 547–557.

67. Armstrong, W.M., W. Wojtkowski, and W.R. Bixenman. 1977. A new solid-state microelectrode for measuring intra-cellular chloride activities. *Biochim. Biophys. Acta Biomembr.* 465: 165–170.

68. Shao, Y., M.V. Mirkin, G. Fish, S. Kokotov, D. Palanker, and A. Lewis. 1997. Nanometer-sized electrochemical sensors. *Anal. Chem.* 69: 1627–1634.

69. Pucacco, L.R. and N.W. Carter. 1978. A submicrometer glass-membrane pH microelectrode. *Anal. Biochem.* 89: 151–161.

70. Bils, R.F. and M. Lavallee. 1964. Measurement of glass microelectrodes. *Experientia* 20: 231–232.

71. Sundfors, F., R. Bereczki, J. Bobacka, K. Tóth, A. Ivaska, and R.E. Gyurcsányi. 2006. Microcavity based solid-contact ion-selective microelectrodes. *Electroanalysis* 18: 1372–1378.

72. Shim, J.H., J. Kim, G.S. Cha, H. Nam, R.J. White, H.S. White, and R.B. Brown. 2007. Glass nanopore-based ion-selective electrodes. *Anal. Chem.* 79: 3568–3574.

73. Zhang, X., B. Ogorevc, and J. Wang. 2002. Solid-state pH nanoelectrode based on polyaniline thin film electrodeposited onto ion-beam etched carbon fiber. *Anal. Chim. Acta* 452: 1–10.

74. Yamamoto, K., G. Shi, T. Zhou, F. Xu, M. Zhu, M. Liu, T. Kato, J.-Y. Jin, and L. Jin. 2003. Solid-state pH ultramicrosensor based on a tungstic oxide film fabricated on a tungsten nanoelectrode and its application to the study of endothelial cells. *Anal. Chim. Acta* 480: 109–117.

75. Steiner, R.A., M. Oehme, D. Ammann, and W. Simon. 1979. Neutral carrier sodium ion-selective microelectrode for intracellular studies. *Anal. Chem.* 51: 351–353.

76. Oesch, U., D. Ammann, and W. Simon. 1986. Ion-selective membrane electrodes for clinical use. *Clin. Chem.* 32: 1448–1459.

77. Oesch, U. and W. Simon. 1980. Lifetime of neutral carrier based ion-selective liquid-membrane electrodes. *Anal. Chem.* 52: 692–700.

78. Dinten, O., U.E. Spichiger, N. Chaniotakis, P. Gehrig, B. Rusterholz, W.E. Morf, and W. Simon. 1991. Lifetime of neutral-carrier-based liquid membranes in aqueous samples and blood and the lipophilicity of membrane components. *Anal. Chem.* 63: 596–603.

79. Mueller, P., D.O. Rudin, H.T. Tien, and W.C. Wescott. 1962. Reconstitution of excitable cell membrane structure in vitro. *Circulation* 26: 1167–1171.

80. Mueller, P. and D.O. Rudin. 1963. Induced excitability in reconstituted cell membrane structure. *J. Theor. Biol.* 4: 268–280.

81. Mueller, P., D.O. Rudin, H.T. Tien, and W.C. Wescott. 1963. Methods for the formation of single bimolecular lipid membranes in aqueous solution. *J. Phys. Chem.* 67: 534–535.

82. Stefanac, Z. and W. Simon. 1966. In-vitro-Verhalten von Makrotetroliden in Membranen als Grundlage für hochselektive kationenspezifische Elektrodensysteme. *Chimia* 20: 436–451.

83. Mueller, P. and D.O. Rudin. 1967. Development of K^+-Na^+ discrimination in experimental bimolecular lipid membranes by macrocyclic antibiotics. *Biochem. Biophys. Res. Commun.* 26: 398–404.

84. Szabo, G., G. Eisenman, R. Laprade, S. Ciani, and S. Krasne. 1973. ed.; Eisenman, G., Membranes, A series of advances, Vol. 2, New York: Marcel Dekker.

85. Trojanowicz, M. 2001. Miniaturized biochemical sensing devices based on planar bilayer lipid membranes. *Fresenius J. Anal. Chem.* 371: 246–260.

86. Montal, M. and P. Mueller. 1972. Formation of bimolecular membranes from lipid monolayers and a study of their electrical properties. *Proc. Natl. Acad. Sci. U.S.A.* 69: 3561–3566.

87. Sato, H., M. Wakabayashi, T. Ito, M. Sugawara, and Y. Umezawa. 1997. Potentiometric responses of ionophore-incorporated bilayer lipid membranes with and without added anionic sites. *Anal. Sci.* 13: 437–446.

88. Liu, B., D. Rieck, B.J. Van Wie, G.J. Cheng, D.F. Moffett, and D.A. Kidwell. 2009. Bilayer lipid membrane (BLM) based ion selective electrodes at the meso-, micro-, and nano-scales. *Biosens. Bioelectron.* 24: 1843–1849.

89. Thompson, M., U.J. Krull, and L.I. Bendell-Young. 1983. The bilayer lipid membrane as a basis for a selective sensor for ammonia. *Talanta* 30: 919–924.

90. Rehák, M., M. Šnejdárková, and M. Otto. 1993. Self-assembled lipid bilayers as a potassium sensor. *Electroanalysis* 5: 691–694.

91. Uto, M., M. Araki, T. Taniguhi, S. Hoshi, and S. Inoue. 1994. Stability of an agar-supported bilayer lipid membrane and its application to a chemical sensor. *Anal. Sci.* 10: 943–946.

92. Zhang, Y.-L., H.-X. Shen, Y. Liu, C.-X. Zhang, and L.-x. Chen. 2000. Salt-bridge supported bilayer lipid membrane modified with calix[n]arenes as alkali cation sensors. *Anal. Lett.* 33: 831–845.

93. Diamond, D., G. Svehla, E.M. Seward, and M.A. McKervey. 1988. A sodium ion-selective electrode based on p-t-butylcalix[4]aryl acetate as the ionophore. *Anal. Chim. Acta* 204: 223–231.

94. Cadogan, A.M., D. Diamond, M.R. Smyth, M. Deasy, M.A. McKervey, and S.J. Harris. 1989. Sodium-selective polymeric membrane electrodes based on calix[4]arene ionophores. *Analyst* 114: 1551–1554.

95. Zhang, Y.L., J. Dunlop, T. Phung, A. Ottova, and H.T. Tien. 2006. Supported bilayer lipid membranes modified with a phosphate ionophore. *Biosens. Bioelectron.* 21: 2311–2314.

96. Carey, C.M. and W.B. Riggan. 1994. Cyclic polyamine ionophore for use in a dibasic phosphate-selective electrode. *Anal. Chem.* 66: 3587–3591.

97. Nishizawa, M., V.P. Menon, and C.R. Martin. 1995. Metal nanotubule membranes with electrochemically switchable ion-transport selectivity. *Science* 268: 700–702.

98. Jágerszki, G., R.E. Gyurcsányi, L. Höfler, and E. Pretsch. 2007. Hybridization-modulated ion fluxes through peptide-nucleic-acid-functionalized gold nanotubes. A new approach to quantitative label-free DNA analysis. *Nano Lett.* 7: 1609–1612.

99. Petzny, W.J. and J.A. Quinn. 1969. Calibrated membranes with coated pore walls. *Science* 166: 751–753.

100. Martin, C.R., M. Nishizawa, K. Jirage, M.S. Kang, and S.B. Lee. 2001. Controlling ion-transport selectivity in gold nanotubule membranes. *Adv. Mater.* 13: 1351–1362.

101. Makra, I., G. Jágerszki, I. Bitter, and R.E. Gyurcsányi. 2012. Nernst-Planck/Poisson model for the potential response of permselective gold nanopores. *Electrochim. Acta* 73: 70–77.

102. Buck, R.P., K. Tóth, E. Gráf, G. Horvai, and E. Pungor. 1987. Donnan exclusion failure in low anion site density membranes containing valinomycin. *J. Electroanal. Chem.* 223: 51–66.

103. Jágerszki, G., A. Grün, I. Bitter, K. Tóth, and R.E. Gyurcsányi. 2010. Ionophore-gold nanoparticle conjugates for Ag+-selective sensors with nanomolar detection limit. *Chem. Commun.* 46: 607–609.

104. Boswell, P.G. and P. Bühlmann. 2005. Fluorous bulk membranes for potentiometric sensors with wide selectivity ranges: Observation of exceptionally strong ion pair formation. *J. Am. Chem. Soc.* 127: 8958–8959.

105. Szigeti, Z., A. Malon, T. Vigassy, V. Csokai, A. Grün, K. Wygladacz, N. Ye et al. 2006. Novel potentiometric and optical silver ion-selective sensors with subnanomolar detection limits. *Anal. Chim. Acta* 572: 1–10.

106. Noyhouzer, T., I. Valdinger, and D. Mandler. 2013. Enhanced potentiometry by metallic nanoparticles. *Anal. Chem.* 85: 8347–8353.

107. Cattrall, R.W. and I.C. Hamilton. 1984. Coated-wire ion-selective electrodes. *Ion-Sel. Electrode Rev.* 6: 125–172.

108. James, H., H. Freiser, and G. Carmack. 1972. Coated wire ion-selective electrodes. *Anal. Chem.* 44: 856–857.

109. Nikolskii, B.P. and E.A. Materova. 1985. Solid contact in membrane ion-selective electrodes. *Ion-Sel. Electrode Rev.* 7: 3–39.

110. Alegret, S. and A. Florido. 1991. Response characteristics of conductive polymer composite substrate all-solid-state poly(vinyl chloride) matrix membrane ion-selective electrodes in aerated and nitrogen-saturated solutions. *Analyst* 116: 473–476.

111. Khalil, S.A.H., G.J. Moody, J.D.R. Thomas, and J. Lima. 1986. Epoxy-based all-solid-state polyvinylchloride) matrix membrane calcium ion-selective microelectrodes. *Analyst* 111: 611–617.

112. Liu, D., R.K. Meruva, R.B. Brown, and M.E. Meyerhoff. 1996. Enhancing EMF stability of solid-state ion-selective sensors by incorporating lipophilic silver-ligand complexes within polymeric films. *Anal. Chim. Acta* 321: 173–183.

113. Simon, W., E. Pretsch, D. Ammann, W.E. Morf, M. Guggi, R. Bissig, and M. Kessler. 1975. Recent developments in field of ion-selective electrodes. *Pure Appl. Chem.* 44: 613–626.

114. Cadogan, A., Z.Q. Gao, A. Lewenstam, A. Ivaska, and D. Diamond. 1992. All-solid-state sodium-selective electrode based on a calixarene ionophore in a poly(vinyl chloride) membrane with a polypyrrole solid contact. *Anal. Chem.* 64: 2496–2501.

115. Oyama, N., T. Hirokawa, S. Yamaguchi, N. Ushizawa, and T. Shimomura. 1987. Hydrogen ion selective microelectrode prepared by modifying an electrode with polymers. *Anal. Chem.* 59: 258–262.

116. Bobacka, J. 1999. Potential stability of all-solid-state ion-selective electrodes using conducting polymers as ion-to-electron transducers. *Anal. Chem.* 71: 4932–4937.

117. Bobacka, J., M. McCarrick, A. Lewenstam, and A. Ivaska. 1994. All-solid-state poly(vinyl chloride) membrane ion-selective electrodes with poly(3-octylthiophene) solid internal contact. *Analyst* 119: 1985–1991.

118. Rubinova, N., K. Chumbimuni-Torres, and E. Bakker. 2007. Solid-contact potentiometric polymer membrane microelectrodes for the detection of silver ions at the femtomole level. *Sens. Actuators B* 121: 135–141.

119. Sutter, J., A. Radu, S. Peper, E. Bakker, and E. Pretsch. 2004. Solid-contact polymeric membrane electrodes with detection limits in the subnanomolar range. *Anal. Chim. Acta* 523: 53–59.

120. Sutter, J., E. Lindner, R.E. Gyurcsányi, and E. Pretsch. 2004. A polypyrrole-based solid-contact Pb^{2+}-selective PVC-membrane electrode with a nanomolar detection limit. *Anal. Bioanal. Chem.* 380: 7–14.

121. Gyurcsányi, R.E., N. Rangisetty, S. Clifton, B.D. Pendley, and E. Lindner. 2004. Microfabricated ISEs: Critical comparison of inherently conducting polymer and hydrogel based inner contacts. *Talanta* 63: 89–99.

122. Lindfors, T. 2009. Light sensitivity and potential stability of electrically conducting polymers commonly used in solid contact ion-selective electrodes. *J. Solid State Electrochem.* 13: 77–89.

123. Lindner, E. and R.E. Gyurcsányi. 2009. Quality control criteria for solid-contact, solvent polymeric membrane ion-selective electrodes. *J. Solid State Electrochem.* 13: 51–68.

124. Fibbioli, M., W.E. Morf, M. Badertscher, N.F. de Rooij, and E. Pretsch. 2000. Potential drifts of solid-contacted ion-selective electrodes due to zero-current ion fluxes through the sensor membrane. *Electroanalysis* 12: 1286–1292.

125. Gyurcsányi, R.E., E. Pergel, R. Nagy, I. Kapui, B.T.T. Lan, K. Tóth, I. Bitter, and E. Lindner. 2001. Direct evidence of ionic fluxes across ion selective membranes: A scanning electrochemical microscopic and potentiometric study. *Anal. Chem.* 73: 2104–2111.

126. Grygolowicz-Pawlak, E., K. Plachecka, Z. Brzozka, and E. Malinowska. 2007. Further studies on the role of redox-active monolayer as intermediate phase of solid-state sensors. *Sens. Actuators B* 123: 480–487.

127. De Marco, R., J.P. Veder, G. Clarke, A. Nelson, K. Prince, E. Pretsch, and E. Bakker. 2008. Evidence of a water layer in solid-contact polymeric ion sensors. *Phys. Chem. Chem. Phys.* 10: 73–76.

128. Veder, J.P., R. De Marco, G. Clarke, S.P. Jiang, K. Prince, E. Pretsch, and E. Bakker. 2011. Water uptake in the hydrophilic poly(3,4-ethylenedioxythiophene):poly(styrene sulfonate) solid-contact of all-solid-state polymeric ion-selective electrodes. *Analyst* 136: 3252–3258.

129. Lindfors, T., L. Höfler, G. Jágerszki, and R.E. Gyurcsányi. 2011. Hyphenated FT-IR-attenuated total reflection and electrochemical impedance spectroscopy technique to study the water uptake and potential stability of polymeric solid-contact ion-selective electrodes. *Anal. Chem.* 83: 4902–4908.

130. Lindfors, T., F. Sundfors, L. Höfler, and R.E. Gyurcsányi. 2009. FTIR-ATR study of water uptake and diffusion through ion-selective membranes based on plasticized poly(vinyl chloride). *Electroanalysis* 21: 1914–1922.

131. Lindfors, T., F. Sundfors, L. Höfler, and R.E. Gyurcsányi. 2011. The water uptake of plasticized poly(vinyl chloride) solid-contact calcium-selective electrodes. *Electroanalysis* 23: 2156–2163.

132. Sundfors, F., L. Höfler, R.E. Gyurcsányi, and T. Lindfors. 2011. Influence of poly(3-octylthiophene) on the water transport through methacrylic-acrylic based polymer membranes. *Electroanalysis* 23: 1769–1772.

133. Sundfors, F., T. Lindfors, L. Höfler, R. Bereczki, and R.E. Gyurcsányi. 2009. FTIR-ATR study of water uptake and diffusion through ion-selective membranes based on poly(acrylates) and silicone rubber. *Anal. Chem.* 81: 5925–5934.

134. Lai, C.Z., M.A. Fierke, A. Stein, and P. Bühlmann. 2007. Ion-selective electrodes with three-dimensionally ordered macroporous carbon as the solid contact. *Anal. Chem.* 79: 4621–4626.

135. Fierke, M.A., C.-Z. Lai, P. Buhlmann and A. Stein. 2010. Effects of architecture and surface chemistry of three-dimensionally ordered macroporous carbon solid contacts on performance of ion-selective electrodes. *Anal. Chem.* 82: 680–688.

136. Parra, E.J., G.A. Crespo, J. Riu, A. Ruiz, and F.X. Rius. 2009. Ion-selective electrodes using multi-walled carbon nanotubes as ion-to-electron transducers for the detection of perchlorate. *Analyst* 134: 1905–1910.

137. Ping, J., Y. Wang, Y. Ying, and J. Wu. 2012. Application of electrochemically reduced graphene oxide on screen-printed ion-selective electrode. *Anal. Chem.* 84: 3473–3479.

138. Ping, J., Y. Wang, J. Wu, and Y. Ying. 2011. Development of an all-solid-state potassium ion-selective electrode using graphene as the solid-contact transducer. *Electrochem. Commun.* 13: 1529–1532.

139. Li, F., J. Ye, M. Zhou, S. Gan, Q. Zhang, D. Han, and L. Niu. 2012. All-solid-state potassium-selective electrode using graphene as the solid contact. *Analyst* 137: 618–623.

140. Lindfors, T., J. Szűcs, F. Sundfors, and R.E. Gyurcsányi. 2010. Polyaniline nanoparticle-based solid-contact silicone rubber ion-selective electrodes for ultratrace measurements. *Anal. Chem.* 82: 9425–9432.

141. Jaworska, E., M. Wojcik, A. Kisiel, J. Mieczkowski, and A. Michalska. 2011. Gold nanoparticles solid contact for ion-selective electrodes of highly stable potential readings. *Talanta* 85: 1986–1989.
142. Zhou, M., S. Gan, B. Cai, F. Li, W. Ma, D. Han, and L. Niu. 2012. Effective solid contact for ion-selective electrodes: Tetrakis(4-chlorophenyl)borate (TB⁻) anions doped nanocluster films. *Anal. Chem.* 84: 3480–3483.
143. Yin, T. and W. Qin. 2013. Applications of nanomaterials in potentiometric sensors. *Trends Anal. Chem.* 51: 79–86.
144. Crespo, G.A., S. Macho, and F.X. Rius. 2008. Ion-selective electrodes using carbon nanotubes as ion-to-electron transducers. *Anal. Chem.* 80: 1316–1322.
145. Ampurdanés, J., G.A. Crespo, A. Maroto, M.A. Sarmentero, P. Ballester, and F.X. Rius. 2009. Determination of choline and derivatives with a solid-contact ion-selective electrode based on octaamide cavitand and carbon nanotubes. *Biosens. Bioelectron.* 25: 344–349.
146. Hernandez, R., J. Riu, and F.X. Rius. 2010. Determination of calcium ion in sap using carbon nanotube-based ion-selective electrodes. *Analyst* 135: 1979–1985.
147. Guinovart, T., M. Parrilla, G.A. Crespo, F.X. Rius, and F.J. Andrade. 2013. Potentiometric sensors using cotton yarns, carbon nanotubes and polymeric membranes. *Analyst* 138: 5208–5215.
148. Crespo, G., D. Gugsa, S. Macho, and F.X. Rius. 2009. Solid-contact pH-selective electrode using multi-walled carbon nanotubes. *Anal. Bioanal. Chem.* 395: 2371–2376.
149. Mousavi, Z., J. Bobacka, A. Lewenstam, and A. Ivaska. 2009. Poly(3,4-ethylenedioxythiophene) (PEDOT) doped with carbon nanotubes as ion-to-electron transducer in polymer membrane-based potassium ion-selective electrodes. *J. Electroanal. Chem.* 633: 246–252.
150. Ping, J., Y. Wang, K. Fan, W. Tang, J. Wu, and Y. Ying. 2013. High-performance flexible potentiometric sensing devices using free-standing graphene paper. *J. Mater. Chem. B* 1: 4781–4791.
151. Woźnica, E., M.M. Wójcik, J. Mieczkowski, K. Maksymiuk, and A. Michalska. 2013. Dithizone modified gold nanoparticles films as solid contact for Cu²⁺ ion-selective electrodes. *Electroanalysis* 25: 141–146.
152. Janrungroatsakul, W., C. Lertvachirapaiboon, W. Ngeontae, W. Aeungmaitrepirom, O. Chailapakul, S. Ekgasit, and T. Tuntulani. 2013. Development of coated-wire silver ion selective electrodes on paper using conductive films of silver nanoparticles. *Analyst* 138: 6786–6792.
153. Harrison, D.J., A. Teclemariam, and L.L. Cunningham. 1989. Photopolymerization of plasticizer in ion-sensitive membranes on solid-state sensors. *Anal. Chem.* 61: 246–251.
154. Brunink, J.A.J., R.J.W. Lugtenberg, Z. Brzozka, J.F.J. Engbersen, and D.N. Reinhoudt. 1994. The design of durable Na⁺-selective CHEMFETs based on polysiloxane membranes. *J. Electroanal. Chem.* 378: 185–200.
155. Daunert, S. and L.G. Bachas. 1990. Ion-selective electrodes using an ionophore covalently attached to carboxylated poly(vinyl chloride). *Anal. Chem.* 62: 1428–1431.
156. Püntener, M., T. Vigassy, E. Baier, A. Ceresa, and E. Pretsch. 2004. Improving the lower detection limit of potentiometric sensors by covalently binding the ionophore to a polymer backbone. *Anal. Chim. Acta* 503: 187–194.
157. Tsujimura, Y., T. Sunagawa, A. Yokoyama, and K. Kimura. 1996. Sodium ion-selective electrodes based on silicone-rubber membranes covalently incorporating neutral carriers. *Analyst* 121: 1705–1709.
158. Cross, G.G., T.M. Fyles, and V.V. Suresh. 1994. Coated-wire electrodes containing polymer immobilized ionophores blended with poly(vinyl chloride). *Talanta* 41: 1589–1595.
159. Heng, L.Y. and E.A.H. Hall. 2000. Producing "self-plasticizing" ion-selective membranes. *Anal. Chem.* 72: 42–51.
160. Qin, Y., S. Peper, A. Radu, A. Ceresa, and E. Bakker. 2003. Plasticizer-free polymer containing a covalently immobilized Ca²⁺-selective Ionophore for potentiometric and optical sensors. *Anal. Chem.* 75: 3038–3045.
161. Lindner, E., V.V. Cosofret, R.P. Kusy, R.P. Buck, T. Rosatzin, U. Schaller, W. Simon, J. Jeney, K. Tóth and E. Pungor. 1993. Responses of H⁺ selective solvent polymeric membrane electrodes fabricated from modified PVC membranes. *Talanta* 40: 957–967.
162. Bereczki, R., R.E. Gyurcsányi, B. Ágai, and K. Tóth. 2005. Synthesis and characterization of covalently immobilized bis-crown ether based potassium ionophore. *Analyst* 130: 63–70.
163. Gyurcsányi, R.E. and E. Lindner. 2006. Multispectral imaging of ion transport in neutral carrier-based cation-selective membranes. *Cytom. Part A* 69A: 792–804.
164. Heng, L.Y., K. Tóth, and E.A.H. Hall. 2004. Ion-transport and diffusion coefficients of non-plasticised methacrylic-acrylic ion-selective membranes. *Talanta* 63: 73–87.
165. Parra, E.J., P. Blondeau, G.A. Crespo, and F.X. Rius. 2011. An effective nanostructured assembly for ion-selective electrodes. An ionophore covalently linked to carbon nanotubes for Pb²⁺ determination. *Chem. Commun.* 47: 2438–2440.

166. Kerric, G., E.J. Parra, G.A. Crespo, F.X. Rius, and P. Blondeau. 2012. Nanostructured assemblies for ion-sensors: Functionalization of multi-wall carbon nanotubes with benzo-18-crown-6 for Pb^{2+} determination. *J. Mater. Chem.* 22: 16611–16617.

167. Woznica, E., M.M. Wojcik, M. Wojciechowski, J. Mieczkowski, E. Bulska, K. Maksymiuk, and A. Michalska. 2012. Dithizone modified gold nanoparticles films for potentiometric sensing. *Anal. Chem.* 84: 4437–4442.

168. Li, M., H. Zhou, L. Shi, D.W. Li, and Y.T. Long. 2014. Ion-selective gold-thiol film on integrated screen-printed electrodes for analysis of Cu(II) ions. *Analyst* 139: 643–648.

169. Guilbault, G.G. and J.G. Montalvo. 1969. A urea-specific enzyme electrode. *J. Am. Chem. Soc.* 91: 2164–2165.

170. Alexander, P.W. and G.A. Rechnitz. 1974. Ion-electrode based immunoassay and antibody-antigen precipitin reaction monitoring. *Anal. Chem.* 46: 1253–1257.

171. Dorazio, P. and G.A. Rechnitz. 1977. Ion electrode measurements of complement and antibody levels using marker-loaded sheep red blood-cell ghosts. *Anal. Chem.* 49: 2083–2086.

172. Meyerhoff, M. and G.A. Rechnitz. 1977. Antibody binding measurements with hapten-selective membrane electrodes. *Science* 195: 494–495.

173. Dequaire, M., C. Degrand, and B. Limoges. 2000. An electrochemical metalloimmunoassay based on a colloidal gold label. *Anal. Chem.* 72: 5521–5528.

174. Liu, G., J. Wang, J. Kim, M.R. Jan, and G.E. Collins. 2004. Electrochemical coding for multiplexed immunoassays of proteins. *Anal. Chem.* 76: 7126–7130.

175. Sokalski, T., A. Ceresa, T. Zwickl, and E. Pretsh. 1997. Large improvement of the lower detection limit of ion-selective polymer membrane electrode. *J. Am. Chem. Soc.* 119: 11347–11348.

176. Ceresa, A., A. Radu, S. Peper, E. Bakker, and E. Pretsch. 2002. Rational design of potentiometric trace level ion sensors. A Ag^+-selective electrode with a 100 ppt detection limit. *Anal. Chem.* 74: 4027–4036.

177. Malon, A., T. Vigassy, E. Bakker, and E. Pretsch. 2006. Potentiometry at trace levels in confined samples: Ion-selective electrodes with subfemtomole detection limits. *J. Am. Chem. Soc.* 128: 8154–8155.

178. Thürer, R., T. Vigassy, M. Hirayama, J. Wang, E. Bakker, and E. Pretsch. 2007. Potentiometric immunoassay with quantum dot labels. *Anal. Chem.* 79: 5107–5110.

179. Szűcs, J., E. Pretsch, and R.E. Gyurcsányi. 2009. Potentiometric enzyme immunoassay using miniaturized anion-selective electrodes for detection. *Analyst* 134: 1601–1607.

180. Shiba, K., Y. Umezawa, T. Watanabe, S. Ogawa, and S. Fujiwara. 1980. Thin-layer potentiometric analysis of lipid antigen-antibody reaction by tetrapentylammonium (TPA^+) ion loaded liposomes and TPA^+ ion-selective electrode. *Anal. Chem.* 52: 1610–1613.

181. Shiba, K., T. Watanabe, Y. Umezawa, S. Fujiwara, and H. Momoi. 1980. Liposome immunoelectrode. *Chem. Lett.*: 155–158.

182. Chumbimuni-Torres, K.Y., Z. Dai, N. Rubinova, Y. Xiang, E. Pretsch, J. Wang, and E. Bakker. 2006. Potentiometric biosensing of proteins with ultrasensitive ion-selective microelectrodes and nanoparticle labels. *J. Am. Chem. Soc.* 128: 13676–13677.

183. Numnuam, A., K.Y. Chumbimuni-Torres, Y. Xiang, R. Bash, P. Thavarungkul, P. Kanatharana, E. Pretsch, J. Wang, and E. Bakker. 2008. Aptamer-based potentiometric measurements of proteins using ion-selective microelectrodes. *Anal. Chem.* 80: 707–712.

184. Numnuam, A., K.Y. Chumbimuni-Torres, Y. Xiang, R. Bash, P. Thavarungkul, P. Kanatharana, E. Pretsch, J. Wang, and E. Bakker. 2008. Potentiometric detection of DNA hybridization. *J. Am. Chem. Soc.* 130: 410–411.

185. Taton, T.A., C.A. Mirkin, and R.L. Letsinger. 2000. Scanometric DNA array detection with nanoparticle probes. *Science* 289: 1757–1760.

186. Szűcs, J. and R.E. Gyurcsányi. 2012. Towards protein assays on paper platforms with potentiometric detection. *Electroanalysis* 24: 146–152.

187. Cui, J., G. Lisak, S. Strzalkowska, and J. Bobacka. 2014. Potentiometric sensing utilizing paper-based microfluidic sampling. *Analyst* 139: 2133–2136.

188. Gyurcsányi, R.E., T. Vigassy, and E. Pretsch. 2003. Biorecognition-modulated ion fluxes through functionalized gold nanotubules as a novel label-free biosensing approach. *Chem. Commun.*: 2560–2561.

189. Palegrosdemange, C., E.S. Simon, K.L. Prime, and G.M. Whitesides. 1991. Formation of self-assembled monolayers by chemisorption of derivatives of oligo(ethylene glycol) of structure $HS(CH_2)_{11}(OCH_2CH_2)$-OH on Gold. *J. Am. Chem. Soc.* 113: 12–20.

190. Xia, Y.N. and G.M. Whitesides. 1998. Soft lithography. *Annu. Rev. Mater. Sci.* 28: 153–184.

191. Xu, Y. and E. Bakker. 2008. Ion channel mimetic chronopotentiometric polymeric membrane ion sensor for surface-confined protein detection. *Langmuir* 25: 568–573.

192. Zelada-Guillen, G.A., J. Riu, A. Duzgun, and F.X. Rius. 2009. Immediate detection of living bacteria at ultralow concentrations using a carbon nanotube based potentiometric aptasensor. *Angew. Chem. Int. Edit.* 48: 7334–7337.

193. Zelada-Guillen, G.A., S.V. Bhosale, J. Riu, and F.X. Rius. 2010. Real-time potentiometric detection of bacteria in complex samples. *Anal. Chem.* 82: 9254–9260.

194. Zelada-Guillen, G.A., J.L. Sebastian-Avila, P. Blondeau, J. Riu, and F.X. Rius. 2012. Label-free detection of *Staphylococcus aureus* in skin using real-time potentiometric biosensors based on carbon nanotubes and aptamers. *Biosens. Bioelectron.* 31: 226–232.

195. Hernández, R., C. Vallés, A.M. Benito, W.K. Maser, F. Xavier Rius, and J. Riu. 2014. Graphene-based potentiometric biosensor for the immediate detection of living bacteria. *Biosens. Bioelectron.* 54: 553–557.

196. Duzgun, A., H. Imran, K. Levon, and F.X. Rius. 2013. Protein detection with potentiometric aptasensors: A comparative study between polyaniline and single-walled carbon nanotubes transducers. *Sci. World J.* 282756: 1–8.

197. Zelada-Guillen, G.A., A. Tweed-Kent, M. Niemann, H.U. Goringer, J. Riu, and F.X. Rius. 2013. Ultrasensitive and real-time detection of proteins in blood using a potentiometric carbon-nanotube aptasensor. *Biosens. Bioelectron.* 41: 366–371.

Index